Advances in Biochemistry in Health and Disease

Volume 33

Advances in Biochemistry in Health and Disease focus on the latest developments in biochemical research with implications for health and disease. This book series consists of original edited volumes and monographs, presented by leading experts in the field and provides an up to date and unique source of information for all those interested in the fundamental, biochemical processes of the latest and emerging topics and techniques.

Covering a wide variety of topics, this book series is a valuable source of information from those at the lab bench through to the Health Care workers.

Devendra K. Agrawal • Ranko Škrbić
Miloš P. Stojiljković • Dragan M. Djuric
Editors

Cardiovascular Toxicity

Incidence, Pathogenesis and Treatment Strategies

Springer

Editors
Devendra K. Agrawal
Department of Translational Research
Western University of Health Sciences
Pomona, CA, USA

Ranko Škrbić
Faculty of Medicine
University of Banja Luka
Banja Luka, Bosnia and Herzegovina

Miloš P. Stojiljković
Faculty of Medicine
University of Banja Luka
Banja Luka, Bosnia and Herzegovina

Dragan M. Djuric
Faculty of Medicine
University of Belgrade
Belgrade, Serbia

ISSN 2512-2142 ISSN 2512-2150 (electronic)
ISBN 978-3-031-90597-1 ISBN 978-3-031-90598-8 (eBook)
https://doi.org/10.1007/978-3-031-90598-8

Advances in Biochemistry in Health and Disease

This Springer imprint is published by the registered company Springer Nature Switzerland AG
The registered company address is: Gewerbestrasse 11, 6330 Cham, Switzerland

Preface

The physiological function of cardiovascular system in the body is the most critical pillar of human well-being. The heart and blood vessels play a major role in thermoregulation, transporting nutrients, gases, and waste products to body organs, maintain fluid balance within the body, and protect the body from infection and blood loss. Indeed, the cardiovascular diseases are the leading cause of morbidity and mortality. Pharmaceuticals, illicit drugs, toxins, poisons, and several environmental factors can induce toxic and adverse effects on the heart and blood vessels, leading to structural changes that impair the blood flow and circulation. Thus, the cardiovascular toxicity due to these factors is the culprit of the global surge in the rate and incidence of cardiovascular diseases and can significantly contribute to the overall burden of cardiovascular diseases. Co-morbidities, such as diabetes and obesity, epigenetic factors, and viral, bacterial, and parasite infections further enhance the cardiovascular toxicity. Thus, it is critical to recognize and understand the factors involved in the pathogenesis of cardiovascular toxicities and develop better therapeutic approaches.

The advent of new drugs and vaccines has been a boon in the treatment of various diseases. However, many drugs and vaccines can induce adverse effects, either directly or indirectly, by affecting the endothelium, coagulation factors, and platelet activation and thus inducing prothrombotic response, inducing oxidative stress and inflammation, interfering with the baroreceptors that regulate blood pressure and cardiovascular variability, altering the regulation of autonomic cardiovascular control, and disrupting electrophysiology and contractility. Unfortunately, this has been ignored for a long time and only lately realized with the attention and initiatives to expand and disseminate knowledge to general population and health practitioners, expand research activities to investigate the underlying cellular and molecular mechanisms, and develop better preventive and treatment strategies.

Cardiovascular adverse effects can be induced by many drug classes used in cancer treatment, such as kinase and proteasome inhibitors, as well as bacterial and viral infections, disorders of central nervous system and gastrointestinal system, and other inflammatory diseases. It is critical to understand the underlying mechanisms of the cardiovascular toxicity and how to prevent it. What are the effects of different genetics, epigenetics, and chemical, physical, and biological influences in cardiovascular toxicity?

In this very timely topic, a comprehensive and up-to-date information in a critical manner is presented on relevant topics related to incidence, pathogenesis, and treatment strategies of cardiovascular toxicity. The major complications due to chemotherapeutic agents and drugs used in the treatment of allergy, diabetes, inflammatory bowel disease, psychiatric problems, viral infections, and others may encompass a large spectrum of disorders, including the trigger or the development of arrhythmia/QTc Prolongation, atrial flutter, atrial and ventricular fibrillation, atrioventricular block, heart failure, hypertension, myocardial infarction, myocarditis, toxic myocardial injury, pericardial disease, peripheral and cerebrovascular ischemia, sinus tachycardia, thromboembolic events leading to limb ischemia, transient ischemic attack, stroke, valvular heart disease, vasculitis, and venous injury. Distinguished scientists, physicians, and surgeons from all over the world with expertise in the fields of cardiovascular diseases, cancer, and other diseases contributed their knowledge and expertise. The biological system in health and disease is complex and dynamic. It is therefore critical to better understand the biological complexity, interactions between genetic–epigenetic–environmental factors, interconnecting genes, interplay of proteins in the cellular and molecular responses in the underlying pathophysiologic mechanisms, and biologic pathways in cardiovascular toxicity. This information is clinically relevant in the mitigation and management of cardiovascular complications for prediction, prevention, and treatment. The identified gaps in our knowledge provide future directions for further investigation.

This book presents a total of 40 chapters that are divided into sub-sections titled "Disease- and Drug-Induced Cardiovascular Toxicity: Incidence and Underlying Pathophysiological Mechanisms," "Toxins and Poison-induced Cardiovascular Toxicity," "Genetics, Epigenetics, Environmental Pollution and Nutritional Effects in the Pathogenesis of Cardiovascular Toxicity," and "Prevention, Management and Treatment of Cardiovascular Toxicity."

The presented information will be a key resource for knowledge and guidance to researchers in the fields of cardiovascular, cancer, and other diseases, as well as healthcare professionals, including physicians, physician-assistants, nurses, nurse practitioners, pharmacists, and academics interested in public health. This compilation will be ideal as a textbook in various undergraduate, graduate, and postgraduate courses in the public health program, cardiovascular diseases, chemotherapeutic agents and cardiovascular toxicity, cardiovascular toxicity in infectious disease, the pathogenesis of drug-induced cardiovascular toxicity, seminars on cardiovascular toxicity, and epigenetics and genetics in the effect of drugs on cardiovascular toxicity.

Pomona, CA, USA — Devendra K. Agrawal
Banja Luka, Bosnia and Herzegovina — Ranko Škrbić
Banja Luka, Bosnia and Herzegovina — Miloš P. Stojiljković
Belgrade, Serbia — Dragan M. Djuric

Contents

Editors and Contributors

About the Editors

Devendra K. Agrawal holds multiple degrees, including MSc, PhD (Biochemistry), PhD (Medical Sciences), MBA, and MS (Information Technology and Management). He has served on several academic positions, including Clinical Biochemist, Professor, Founding Chairman of the Department of Clinical and Translational Science, Senior Associate Dean of Clinical and Translational Sciences, and Senior Vice President for Research and Biotechnology. Currently, Dr. Agrawal serves as Professor and Director, Department of Translational Research, Western University of Health Sciences, Pomona, CA, USA.

Dr. Agrawal is a member of the *Alpha Omega Alpha Honor Medical Society* and a Fellow of the AAAAI, AHA, APS, and IACS. He has been recognized with many awards, including the highest teaching award at Creighton University, Distinguished Research Career Award, Distinguished Professor Award, Distinguished Faculty Service Award, University Research Award, Distinguished Mentor Award, *Professor Bohuslav Ostadal Award* for Excellence in Cardiovascular Sciences, *Distinguished Leadership Award, Ricardo Gelpi Award, Andras Varro Award, Harpal Buttar Award for Excellence in Cardiovascular Sciences*, and *Cardiovascular Sciences Innovative Research Award.* Dr. Agrawal has been an invited speaker at many prestigious national and international scientific meetings. Dr. Agrawal is serving on the position of Secretary

General and Treasurer of the International Academy of Cardiovascular Sciences—North American section and the Director and Treasurer of the Academy of Cardiovascular Sciences Foundation USA, Inc.

Dr. Agrawal has served on many grant review panels, including the VA Merit Review, DoD, NIH-NIAID, NIH-NHLBI, MRC-UK, and US-Israel BSF Grant Review Committees, as well as the Science Foundation Ireland, Austrian Science Fund, British Lung Foundation, Swiss NSF, ViCi Netherlands, and the Asthma Research Foundation of Western Australia. Dr. Agrawal has served or is currently serving as an associate editor and on the editorial board of several prestigious journals. Dr. Agrawal contributed to science by publishing over 620 original research articles in peer-reviewed journals and book chapters. He has co-edited books on various topics, including allergy, asthma, and immunology, COVID-19 and stem cells, environmental factors in the pathogenesis of cardiovascular diseases, and flow cytometry. His major research areas include the cellular, molecular, and immunobiology of coronary and carotid artery disease, stem cell therapy, shoulder rotator cuff injuries, traumatic brain injury, diabetes and diabetic foot ulcers, renal diseases, and allergy and bronchial asthma. NIH funds his research educational programs and projects.

Dr. Agrawal has trained over 250 medical students, many of whom are now academic physician-scientists, and directly supervised and guided 73 graduate students to earn MD-PhD, MD-MS, MS, and PhD degrees, 74 postdoctoral fellows, and 15 assistant professors, many of whom are currently on high positions in academia and industries. Many clinical residents, fellows, and research investigators have received training under his direct supervision. He is passionate to train and mentor next generation of clinical and translational researchers.

Ranko Škrbić graduated in 1986 and received the MD degree from the Faculty of Medicine, University of Banja Luka, Bosnia and Herzegovina (ex-Yugoslavia), and MSc degree in 1991 from the Faculty of Natural Sciences, Zagreb University, Croatia. As a Japanese Government Research fellow, he spent 1 year at Shinshu University School of Medicine studying cardiovascular pharmacology. In 1994, he received his PhD degree from the Faculty of Medicine, Belgrade University, Serbia, and in the year 2000 the Clinical Pharmacology Specialist degree from the Faculty of Medicine, Novi Sad University, Serbia. He has also attended several short-term courses and trainings, including drug information training at Bristol Royal Infirmary, Bristol, UK in 1999; short-term course in pharmacoeconomics at the London School of Economics (LSE), London, UK; and a training in health management, the World Bank's Basic Health Project for Bosnia and Herzegovina, organized in collaboration with Imperial College London, UK, and Heidelberg University, Germany.

Dr. Škrbić is a Full Professor of Pharmacology, Toxicology, and Clinical Pharmacology since 2009. Notably, his professional career includes the following positions: Head of the Department of Pharmacology, Faculty of Medicine, University of Banja Luka (2002–2009); Head of the National Drug Information Centre, established in collaboration with WHO and EU-ECHO (1998–2006); Chairman of the Drug Registration Committee of the Republic of Srpska, Bosnia and Herzegovina (1999–2002); EU PHARE Regional Coordinator for Pharmaceutical Sector Development in Bosnia and Herzegovina (1998–1999); UNICEF Regional Coordinator for Essential Drugs and Rational Drug Use in the Republic of Srpska, Bosnia and Herzegovina (1999–2000); EU-CARDS Regional Coordinator for Health Care Reform in Bosnia and Herzegovina (2000–2002); Short-term consultant for the Development of Accreditation Standards and Clinical Guidelines, EPSILON Research, Development and Consulting, doo, Banja Luka (Basic Health Project of the World Bank; Accreditation and Quality Assurance) (2001–2003); Minister of Health and Social Welfare, Government of the Republic of Srpska, Bosnia and Herzegovina (2005–2013); Ambassador of Bosnia and

Herzegovina in Serbia (2013–2015); and Dean of the Faculty of Medicine, University of Banja Luka (2016–now).

Dr. Škrbić was the Founder and Chair of the PhD Program in Biomedical Sciences at the Faculty of Medicine, University of Banja Luka (2015); Visiting Professor, Faculty of Medicine, Belgrade University, Serbia (since 2019); and Visiting Professor at the First Moscow State Medical University, Moscow, Russia (2024). Dr. Škrbić has organized and chaired several international conferences, including *The 6th Xenobiotic Metabolism and Toxicity Workshop of Balkan Countries*, Banja Luka, Bosnia and Herzegovina, June 16–20, 2004; *The Third Health Ministers' Forum: Health in All Policies in South-eastern Europe—A Shared Goal and Responsibility*, Banja Luka, the Republic of Srpska, Bosnia and Herzegovina, October 13–14, 2011; *The International Conference on Medical and Biological Engineering 2019* (*CMBEBIH 2019*), Banja Luka, the Republic of Srpska, Bosnia and Herzegovina May 16–19, 2019; and *The 7th Meeting of the European Section and 8th Meeting of North American Section of the International Academy of Cardiovascular Sciences* (*IACS*), Banja Luka, the Republic of Srpska, Bosnia and Herzegovina, September 20–23, 2021

Dr. Škrbić is a member of several professional societies, including the Serbian Pharmacological Society, the Section for toxicology of the Serbian Medical Association, and the Association of Clinical Pharmacological Societies with Sections of Clinical Pharmacists and Pharmacoinformatics of Bosnia and Herzegovina. He is also an Honorary Member of the Cardiology Society of Serbia. Additionally, he serves as the President of the Association for Atherosclerosis and Cardiovascular Research, Banja Luka, the Republic of Srpska, Bosnia and Herzegovina. He is a Fellow of the International Academy of Cardiovascular Sciences (FIACS) and a Member of the Executive Council of the International Academy of Cardiovascular Sciences, European Section (IACS-ES). He is also a member of the Academy of Science and Arts of the Republic of Srpska. He has served as the associate editor or the member of editorial board of the several prestigious international journals.

Dr. Škrbić is the founder and head of the Centre for Biomedical Research, established in 2018 within the Faculty of Medicine, Banja Luka University. His research has been recognized mostly in the areas of cardiovascular pharmacology, pharmacoepidemiology, inflammation, ischemia-reperfusion injury, cardioprotection, and pharmacogenetics. Over the past 10 years, Dr. Škrbić's Centre has made significant contributions to the field by studying various well-established experimental methods and translating them into clinical practice. He has published more than 120 publications in peer-reviewed international journals, authored 4 chapters in international textbooks/monographs, and edited/co-edited 15 monographs/books.

Dr. Škrbić is a recipient of several awards and honors, including a Fellowship of the International Academy of Cardiovascular Sciences (IACS), the IACS Lifetime Achievement Award in Cardiovascular Science, Medicine and Surgery (2021), the Gold Medal for outstanding leadership in cardiovascular sciences (2022), a special award from the Ministry of Foreign Affairs of Japan (2021), and the Ricardo Gelpi Award for Excellence in Cardiovascular Sciences (2024).

Miloš P. Stojiljković, MD, PhD, FIACS holds MSc and PhD degrees. He is an experienced biomedical researcher who discharged numerous research, teaching, and managerial duties, including clinical pharmacologist, assistant professor, associate professor, and full professor of pharmacology, toxicology, and clinical pharmacology, as well as Vice-Dean for Scientific Research. Currently, Dr. Stojiljković is the Professor and Director, PhD Programme, Faculty of Medicine, University of Banja Luka.

Dr. Stojiljković was a Postdoctoral Fellow at the Medical University of South Carolina. He is a Fellow of the International Academy of Cardiovascular Sciences (IACS) and member of several distinguished scientific societies, such a German Society for Experimental and Clinical Pharmacology and Toxicology, British Pharmacological Society. He has been recognized with many awards, including several publication rewards from the Faculty of Medicine,

University of Banja Luka and *Distinguished Leadership Award* and *Vincenzo Panagia Distinguished Lecture Award* from the IACS. Dr. Stojiljković has been an invited speaker at many prestigious national and international scientific meetings. Dr. Stojiljković is currently a member of the Council of the IACS-European section.

Dr. Stojiljković has been serving as Editor-in-Chief of the biomedical journal *Scripta Medica* and as Associate Editor of the renowned international journals *Drug and Chemical Toxicology* and *Molecular and Cellular Biochemistry*. Dr. Stojiljković contributed to science by publishing over 200 original research articles in peer-reviewed journals and book chapters. His major research areas include the toxicology of carbamate and organophosphate acetylcholinesterase inhibitors, the influence of lipids on blood pressure, and the use of various chemical substances in cardioprotection.

Dr. Stojiljković has been an experienced teacher for 37 years and is dedicated to training and mentoring younger researchers.

Dragan M. Djuric earned his MD degree in 1987, an MS degree in 1991, a PhD degree in 1993, and a Clinical Physiologist degree in 2007, all from the Faculty of Medicine, University of Belgrade (Belgrade, Serbia). He finished his postdoctoral training in Germany at the Max Planck Institute for Physiological and Clinical Research and Kerckhoff Klinik GmbH, Bad Nauheim (1998 and 2001–2002) and in the USA at the Department of Physiology, College of Medicine, University of South Alabama, Mobile (2000). After serving in several academic positions, including Director and Chair, Institute of Medical Physiology, Faculty of Medicine, University of Belgrade (2006–2012), Founder and Chair, PhD Program in Physiological Sciences (2009–2024), leadership positions in Serbian Physiological Society (2003–2021), the FEPS Task Force Group on Education in Physiology (2007–2008), and national atherosclerosis societies (since 1998), he is currently a Professor of Medical Physiology, Faculty of Medicine, University of Belgrade (Belgrade, Serbia), and a Visiting Professor, Faculty of Medicine, University of Banja Luka (Bosnia and Herzegovina) (since 2019). Dr. Djuric is a Fellow of IACS and IUPS Academy of Physiology. He has been recognized with many awards, including the Naranjan Dhalla Award for Innovative Investigators in Cardiovascular Sciences (2024), the Oration Award from Prof. Harpal Buttar (2024), the Howard Morgan Award for Distinguished Achievements in Cardiovascular Research (2022), the James Willerson Award for Excellence in Cardiovascular Sciences (2021), the Distinguished Leadership Award in Cardiovascular Sciences (2019), the Andras Varro Award for Excellence in Cardiovascular Sciences (2018), and the Lifetime Achievement Award in Cardiovascular Science, Medicine and Surgery (2015), all from the International Academy of Cardiovascular Sciences (IACS), including European, North American, and Indian sections. Additionally, he has been honored with the Serbian Physiological Society Award for Lifetime Achievement in Physiological Sciences (2016); the Samuel Racz Medal and Honorary Membership for his contributions to physiology from the Hungarian Physiological Society (2010); Honorary Membership, Bulgarian Society for Cell Biology (2009); Honorary Membership, Romanian Society for Laboratory Medicine (2008); the Medal of the Yugoslav Society of Cardiology (2002); and the Belgrade City October Award (1987). Dr. Djuric has served on the editorial boards of several prestigious journals and in a few grants review panels. He was a reviewer in many journals, books, and meeting abstracts. Dr. Djuric contributed to science by publishing over 300 original research, review and editorial articles in peer-reviewed journals and book chapters. He has edited/co-edited 6 books on medical physiology, 7 monographs on endothelium, vascular biology, atherosclerosis, and nutrition, including Springer Nature book entitled *Environmental Factors in the Pathogenesis of Cardiovascular Diseases* (2024), 15 abstracts from scientific meetings which he organized/co-organized, and served as guest editor for 13 special or topic issues of prestigious international journals. He has been an invited speaker at many prestigious national and international scientific meetings and academic institutions. Dr. Djuric has trained over 30 medical students in research work and has directly supervised and guided 20 graduate students in earning their MD-MS and MD-PhD degrees. The National Ministry of Education, Science and Technological Development and EU COST funded his research projects. His major research areas include cardiovascular research, cardioprotection, experimental cardiometabolic models, endothelial dysfunction, experimental vascular models, gasotransmitters and oxidative stress, homocysteine and related vitamins, nutraceuticals, micronutrients and cardiovascular health, risk factors and cardiovascular health, vascular reactivity, mediators of inflammation, atherosclerosis, and environmental factors in the pathogenesis of cardiovascular diseases.

Contributors

Manas Aavula Department of Translational Research, College of Osteopathic Medicine of the Pacific, Western University of Health Sciences, Pomona, CA, USA

Faris Abusharkh Vanderbilt University, Nashville, TN, USA

Devendra K. Agrawal Department of Translational Research, College of Osteopathic Medicine of the Pacific, Western University of Health Sciences, Pomona, CA, USA

Vineet Agrawal Vanderbilt University, Nashville, TN, USA
Division of Cardiovascular Medicine, Department of Medicine, Vanderbilt University Medical Center, Nashville, TN, USA
Department of Veterans Affairs, Tennessee Valley Healthcare System, Nashville, TN, USA

Zubair Ahmed Department of Translational Research, College of Osteopathic Medicine of the Pacific, Western University of Health Sciences, Pomona, CA, USA

Dunja Aksentijevic William Harvey Research Institute, Barts Faculty of Medicine and Dentistry, Queen Mary University of London, London, UK

Biljana Antonijević University of Belgrade—Faculty of Pharmacy, Toxicological Risk Assessment Center, Belgrade, Serbia
Department of Toxicology "Akademik Danilo Soldatović", University of Belgrade-Faculty of Pharmacy, Belgrade, Serbia

Zorislava Bajic Department of Physiology, Faculty of Medicine, University of Banja Luka, Banja Luka, The Republic of Srpska, Bosnia and Herzegovina
Centre for Biomedical Research, Faculty of Medicine, University of Banja Luka, Banja Luka, The Republic of Srpska, Bosnia and Herzegovina

Surinder K. Batra Department of Biochemistry and Molecular Biology, University of Nebraska Medical Center, Omaha, NE, USA

Fred and Pamela Buffett Cancer Center, Eppley Institute for Research in Cancer and Allied Diseases, University of Nebraska Medical Center, Omaha, NE, USA

Sukhwinder K. Bhullar Department of Physiology and Pathophysiology, The Institute of Cardiovascular Sciences, St. Boniface Hospital Albrechtsen Research Centre, Winnipeg, MB, Canada
Department of Pharmacology and Therapeutics, Rady College of Medicine, Max Rady Faculty of Health Sciences, University of Manitoba, Winnipeg, MB, Canada

Antoinette Oliveira Blackman University Center of Central Planalto Apparecido dos Santos—UNICEPLAC, Gama, Federal District, Brazil
University Center of Brasilia—CEUB, Brasília, Federal District, Brazil
University Center Euro American—UNIEURO, Brasília, Brazil

Milovan Bojić Institute for Cardiovascular Diseases "Dedinje", Belgrade, Serbia

Roberto Bolli Institute of Molecular Cardiology, University of Louisville Physicians Outpatient Center, Louisville, KY, USA

Petar Bulat University of Belgrade—Faculty of Medicine, Belgrade, Serbia
Serbian Institute of Occupational Health, Belgrade, Serbia

Zorica Bulat Department of Toxicology "Akademik Danilo Soldatović", University of Belgrade—Faculty of Pharmacy, Belgrade, Serbia

Fihr Chaudhary Department of Translational Research, College of Osteopathic Medicine of the Pacific, Western University of Health Sciences, Pomona, CA, USA

Spencer Collins Department of Translational Research, College of Osteopathic Medicine of the Pacific, Western University of Health Sciences, Pomona, CA, USA

Marijana Curcic Department of Toxicology "Akademik Danilo Soldatović", University of Belgrade-Faculty of Pharmacy, Belgrade, Serbia

Anindita Das Pauley Heart Center, Department of Internal Medicine, School of Medicine, Virginia Commonwealth University, Richmond, VA, USA

Dragan M. Djuric Faculty of Medicine, Institute of Medical Physiology "Richard Burian", University of Belgrade, Belgrade, Serbia
Centre for Biomedical Research, Faculty of Medicine, University of Banja Luka, Banja Luka, The Republic of Srpska, Bosnia and Herzegovina

Marko Djuric Clinic for Anesthesiology and Intensive Care, University Clinical Hospital Center "Dr Dragisa Misovic-Dedinje", Belgrade, Serbia
Department of Anesthesiology, Reanimatology and Intensive Care, Faculty of Medicine, University of Belgrade, Belgrade, Serbia

Aleksandra Buha Đorđević Department of Toxicology "Akademik Danilo Soldatović", Faculty of Pharmacy, University of Belgrade, Belgrade, Republic of Serbia

Simona Dragan Department of Cardiology, University of Medicine and Pharmacy "Victor Babes", Timisoara, Romania

Danijela Đukić-Ćosić Department of Toxicology "Akademik Danilo Soldatović", Centre for Toxicological Risk Assessment, Faculty of Pharmacy, University of Belgrade, Belgrade, Serbia

Marcel P. Fraix Department of Translational Research, College of Osteopathic Medicine of the Pacific, Western University of Health Sciences, Pomona, CA, USA
Department of Physical Medicine and Rehabilitation, College of Osteopathic Medicine of the Pacific, Western University of Health Sciences, Pomona, CA, USA
Section of Health Policy and Management, Department of Population Health Science, College of Osteopathic Medicine of the Pacific, Western University of Health Sciences, Pomona, CA, USA

Adina V. Furdui-Linţa Department of Functional Sciences—Chair of Pathophysiology, Centre for Translational Research and Systems Medicine, "Victor Babeş" University of Medicine and Pharmacy, Timişoara, Romania

Ferenc Gallyas Jr Department of Biochemistry and Medicinal Chemistry, Medical School, University of Pecs, Pecs, Hungary

Sumner Gardner Vanderbilt University, Nashville, TN, USA

Silvia Giuchici Department of Functional Sciences—Chair of Pathophysiology, Centre for Translational Research and Systems Medicine, "Victor Babeş" University of Medicine and Pharmacy, Timişoara, Romania

Victor Gurevich Center of Atherosclerosis and Lipid Disorders, Saint-Petersburg State University, Medical Institution, St. Petersburg, Russia
Nord-Western State Medical University n.a. I.I. Mechnikov, Center of Atherosclerosis and Lipid Disorders, St. Petersburg, Russia

Robert Halmosi First Department of Internal Medicine, Clinical Centre, University of Pecs, Pecs, Hungary

Loredana N. Ionică Department of Internal Medicine—First University Clinic of Medical Semiotics, "Victor Babeş" University of Medicine and Pharmacy, Timişoara, Romania

Aleksandar Jovanović Department of Basic and Clinical Sciences, University of Nicosia Medical School, Nicosia, Cyprus
Center for Neuroscience and Integrative Brain Research (CENIBRE), University of Nicosia Medical School, Nicosia, Cyprus

Sugeeth Kandikattu Department of Translational Research, College of Osteopathic Medicine of the Pacific, Western University of Health Sciences, Pomona, CA, USA

Sina Khoshniyat Biomedicine, School of Sciences, La Trobe University, Melbourne, Australia

Lorrie A. Kirshenbaum Department of Physiology and Pathophysiology, The Institute of Cardiovascular Sciences, St. Boniface Hospital Albrechtsen Research Centre, Winnipeg, MB, Canada
Department of Pharmacology and Therapeutics, Rady College of Medicine, Max Rady Faculty of Health Sciences, University of Manitoba, Winnipeg, MB, Canada

Saisudha Koka Department of Pharmaceutical Sciences, Irma Lerma Rangel College of Pharmacy, Texas A&M University, Kingsville, TX, USA

Rakesh C. Kukreja Pauley Heart Center, Department of Internal Medicine, School of Medicine, Virginia Commonwealth University, Richmond, VA, USA
Division of Cardiology, Department of Internal Medicine, Pauley Heart Center, Virginia Commonwealth University Medical Center, Richmond, VA, USA

Nilima Rajpal Kundnani Department of Cardiology, University of Medicine and Pharmacy "Victor Babes", Timisoara, Romania

Branislav Kura Institute for Heart Research, Centre of Experimental Medicine, Slovak Academy of Sciences, Bratislava, Slovakia

Daniel F. Lighezan Department of Internal Medicine—First University Clinic of Medical Semiotics, "Victor Babeş" University of Medicine and Pharmacy, Timişoara, Romania

Melchior Luiz Lima Centrocor—Heart Clinic Ltd., Vitória, ES, Brazil

Suellen Keyze Almeida Lima University Center of Central Planalto Apparecido dos Santos—UNICEPLAC, Gama, Federal District, Brazil

Bogdan Lolescu Department of Cardiology, University of Medicine and Pharmacy "Victor Babes", Timisoara, Romania

Dragana Lončar-Stojiljković Institute for Cardiovascular Diseases "Dedinje", Belgrade, Serbia
Department of Anaesthesiology and Intensive Medicine, Faculty of Medicine, University of Banja Luka, Banja Luka, The Republic of Srpska, Bosnia and Herzegovina

Stefan Longobardi Internal Medicine, HCA Florida Blake Hospital, University of South Florida (USF) Morsani College of Medicine, Bradenton, FL, USA

Uglješa Maličević Department of Translational Research, Western University of Health Sciences, Pomona, CA, USA
Department of Pathophysiology and Centre for Biomedical Research, Faculty of Medicine, University of Banja Luka, Banja Luka, The Republic of Srpska, Bosnia and Herzegovina

Dana Man Department of Cardiology, University of Medicine and Pharmacy "Victor Babes", Timisoara, Romania

Sonja T. Marinković Centre for Biomedical Research, Faculty of Medicine, University of Banja Luka, Banja Luka, The Republic of Srpska, Bosnia and Herzegovina
Paediatric Clinic, University Clinical Centre of the Republic of Srpska, Banja Luka, The Republic of Srpska, Bosnia and Herzegovina

Shreyas Melanahalli Department of Translational Research, College of Osteopathic Medicine of the Pacific, Western University of Health Sciences, Pomona, CA, USA

Yssel Mendoza-Mari Department of Translational Research, College of Osteopathic Medicine of the Pacific, Western University of Health Sciences, Pomona, CA, USA

Evica Antonijević Miljaković University of Belgrade—Faculty of Pharmacy, Toxicological Risk Assessment Center, Belgrade, Serbia
Department of Toxicology "Akademik Danilo Soldatović", University of Belgrade-Faculty of Pharmacy, Belgrade, Serbia

Ali Moradi Internal Medicine, HCA Florida Blake Hospital, University of South Florida (USF) Morsani College of Medicine, Bradenton, FL, USA

Sakthivel Muniyan Department of Biochemistry and Molecular Biology, University of Nebraska Medical Center, Omaha, NE, USA

Danina M. Muntean Department of Functional Sciences—Chair of Pathophysiology, Centre for Translational Research and Systems Medicine, "Victor Babeş" University of Medicine and Pharmacy, Timişoara, Romania

Slavica S. Mutavdzin Krneta Institute of Medical Physiology "Richard Burian", Faculty of Medicine, University of Belgrade, Belgrade, Serbia

Irina Nenadic Clinic for Anesthesiology and Intensive Care, University Clinical Hospital Center "Dr Dragisa Misovic-Dedinje", Belgrade, Serbia
Department of Anesthesiology, Reanimatology and Intensive Care, Faculty of Medicine, University of Belgrade, Belgrade, Serbia

José Sobral Neto Centrocard, Brasília, Federal District, Brazil

Lana Nežić Department of Pharmacology, Toxicology and Clinical Pharmacology, Faculty of Medicine, University of Banja Luka, Banja Luka, The Republic of Srpska, Bosnia and Herzegovina

Andrey Obrezan Center of Atherosclerosis and Lipid Disorders, Saint-Petersburg State University, Medical Institution, St. Petersburg, Russia

Beatriz Montenegro Oliveira University Center of Central Planalto Apparecido dos Santos—UNICEPLAC, Gama, Federal District, Brazil

Marija Polovina Department of Cardiology, University Clinical Center of Serbia, Belgrade, Serbia
Faculty of Medicine, University of Belgrade, Belgrade, Serbia

Inna Rabinovich-Nikitin Department of Physiology and Pathophysiology, The Institute of Cardiovascular Sciences, St. Boniface Hospital Albrechtsen Research Centre, Winnipeg, MB, Canada
Department of Pharmacology and Therapeutics, Rady College of Medicine, Max Rady Faculty of Health Sciences, University of Manitoba, Winnipeg, MB, Canada

Nina M. Radisavljevic Institute of Medical Physiology "Richard Burian", Faculty of Medicine, University of Belgrade, Belgrade, Serbia

Mohamed M. Radwan Department of Translational Research, Western University of Health Sciences, Pomona, CA, USA

Vikrant Rai Department of Translational Research, Western University of Health Sciences, Pomona, CA, USA

Resmi Rajalekshmi Department of Translational Research, Western University of Health Sciences, Pomona, CA, USA

Kurt Ramey Cardiology, HCA Florida Blake Hospital, University of South Florida (USF) Morsani College of Medicine, Bradenton, FL, USA

Tatiana Ravingerova Institute for Heart Research, Centre of Experimental Medicine, Slovak Academy of Sciences, Bratislava, Slovakia

Jonatas M. Rolando Division of Metabolic and Cardiovascular Sciences, Burnett School of Biomedical Sciences, College of Medicine, University of Central Florida, Orlando, FL, USA

Naya Said Center for Translational Medicine, Faculty of Medicine, Semmelweis University, Budapest, Hungary

Arun Samidurai Pauley Heart Center, Department of Internal Medicine, School of Medicine, Virginia Commonwealth University, Richmond, VA, USA

David Jesse Sanchez Pharmaceutical Sciences Department, Western University of Health Sciences, Pomona, CA, USA

Inavolu Sriram Sandeep Department of Pharmaceutical Sciences, Irma Lerma Rangel College of Pharmacy, Texas A&M University, Kingsville, TX, USA

Daniel Marius Duda Seiman University of Medicine and Pharmacy, "Victor Babes", Timisoara, Romania

Dinender K. Singla Division of Metabolic and Cardiovascular Sciences, Burnett School of Biomedical Sciences, College of Medicine, University of Central Florida, Orlando, FL, USA

Ranko Škrbić Department of Pharmacology, Toxicology and Clinical Pharmacology, Faculty of Medicine, University of Banja Luka, Banja Luka, The Republic of Srpska, Bosnia and Herzegovina

Ján Slezák Institute for Heart Research, Centre of Experimental Medicine, Slovak Academy of Sciences, Bratislava, Slovakia

Raluca Şoşdean Department of Cardiology—2nd University Clinic of Cardiology, Research Centre of the Institute for Cardiovascular Diseases, "Victor Babeş" University of Medicine and Pharmacy, Timişoara, Romania

Sanja Stankovic Center for Medical Biochemistry, University Clinical Center of Serbia, Belgrade, Serbia
Faculty of Medical Sciences, University of Kragujevac, Kragujevac, Serbia

Marija Stojanovic Department of Translational Research, Western University of Health Sciences, Pomona, CA, USA
Faculty of Medicine, Institute of Medical Physiology "Richard Burian", University of Belgrade, Belgrade, Serbia

Miloš P. Stojiljković Department of Pharmacology, Toxicology and Clinical Pharmacology, Faculty of Medicine, University of Banja Luka, Banja Luka, The Republic of Srpska, Bosnia and Herzegovina
Institute for Cardiovascular Diseases "Dedinje", Belgrade, Serbia

Adrian Sturza Department of Functional Sciences—Chair of Pathophysiology, Centre for Translational Research and Systems Medicine, "Victor Babeş" University of Medicine and Pharmacy, Timişoara, Romania

Robert Subbiondo Cardiology, HCA Florida Blake Hospital, University of South Florida (USF) Morsani College of Medicine, Bradenton, FL, USA

Dongwei Sun Department of Bioengineering, University of California, Riverside, CA, USA

Wilson S. Tang Pharmaceutical Sciences Department, Western University of Health Sciences, Pomona, CA, USA

Finosh G. Thankam Department of Translational Research, College of Osteopathic Medicine of the Pacific, Western University of Health Sciences, Pomona, CA, USA

Raneeta Thingnam Department of Physiology and Pathophysiology, The Institute of Cardiovascular Sciences, St. Boniface Hospital Albrechtsen Research Centre, Winnipeg, MB, Canada
Department of Pharmacology and Therapeutics, Rady College of Medicine, Max Rady Faculty of Health Sciences, University of Manitoba, Winnipeg, MB, Canada

Kalman Toth First Department of Internal Medicine, Clinical Centre, University of Pecs, Pecs, Hungary

Belma Turan Department of Biophysics, Faculty of Medicine, Lokman Hekim University, Ankara, Turkey

Mihaela-Daniela Valcovici Department of Cardiology, University of Medicine and Pharmacy "Victor Babes", Timisoara, Romania

Lei Xi Pauley Heart Center, Department of Internal Medicine, School of Medicine, Virginia Commonwealth University, Richmond, VA, USA

Megan Young William Harvey Research Institute, Barts Faculty of Medicine and Dentistry, Queen Mary University of London, London, UK

Part I
Drug-Induced Cardiovascular Toxicity: Incidence and Underlying Pathophysiological Mechanisms

Chapter 1
Transient Ischemic Attack and Stroke in Cardiovascular Toxicity

Shreyas Melanahalli and Devendra K. Agrawal

Abstract Transient Ischemic Attack (TIA) and stroke are significant cerebrovascular events associated with cardiovascular toxicity. This paper examines the mechanisms and risks of TIA and stroke in relation to various medications, including combined oral contraceptives, tamoxifen, bevacizumab, antipsychotics, ponatinib, and nilotinib. These drugs, while effective in their primary indications, can increase the risk of thromboembolism and subsequent cerebrovascular events. Combined oral contraceptives alter the coagulation cascade, while tamoxifen and bevacizumab affect endothelial function and angiogenesis. Antipsychotics interact with fibrinogen and platelet function, potentially leading to a pro-thrombotic state. Tyrosine kinase inhibitors like ponatinib and nilotinib have been associated with vascular toxicity and atherosclerosis. The paper also discusses risk mitigation strategies, including dosage control, the use of anticoagulants and antiplatelets, and the role of ACE inhibitors and beta-blockers in managing cardiovascular toxicity. Understanding these mechanisms and implementing appropriate preventive measures are crucial for optimizing patient care and reducing the risk of TIA and stroke in individuals receiving these medications.

Keywords Angiogenesis · Angiotensin converting enzyme inhibitor · Anticoagulants · Antiplatelets · Antipsychotics · Bevacizumab · Cardiovascular toxicity · Combined oral contraceptives · Nilotinib · Oral contraceptives · Ponatinib · Stroke · Transient ischemic attack · Thromboembolism · Venous thromboembolism

---

S. Melanahalli · D. K. Agrawal (✉)
Department of Translational Research, College of Osteopathic Medicine of the Pacific, Western University of Health Sciences, Pomona, CA, USA
e-mail: shreyas.melanahalli@westernu.edu; DAgrawal@WesternU.edu

D. K. Agrawal et al. (eds.), *Cardiovascular Toxicity*, Advances in Biochemistry in Health and Disease 33, https://doi.org/10.1007/978-3-031-90598-8_1

Introduction

Nearly half (48%) of all U.S. adults over the age of 20 have some type of cardiovascular disease, based on 2016, and the risk increases with age [1]. Cardiovascular toxicity is a significant concern to modern medicine, encompassing a broad range of life-threatening complications, particularly in cancer treatment. Some of these life-threatening complications include heart failure, arrhythmia, myocarditis and other vascular events [2]. The toxicity often is an unintended consequence from drugs affecting normal cells within the cardiovascular system along with targeting those cancer cells.

Transient Ischemic Attack (TIA) is a medical emergency defined as a short-lasting blockage or inadequate blood supply to the focal brain, spinal cord, or retina causing neurological dysfunction without tissue injury and acute infarction. TIA's last less than an hour, more often last a few minutes, with a sudden onset. TIAs normally present as brief episodes of focal neurological dysfunction and/or speech impairment, which corresponds to specific vascular regions in the brain. These temporary symptoms indicate underlying cerebrovascular pathology, related to atherosclerosis or other vascular risk factors [3].

The significance of TIAs lies in their role as warning signs for future strokes. Recent studies have shown that patients who experience a TIA have a 6.1% incidence of ischemic stroke during a 5-year follow-up period, which is approximately fivefold higher than the general population [4]. This underscores the importance of prompt evaluation and management of TIAs to prevent more severe cerebrovascular events.

A stroke is defined as an acute cerebrovascular injury characterized by a sudden onset of focal neurological deficits persisting beyond 24 h, or leading to death, resulting from either ischemia or hemorrhage within the brain. It is caused by a disruption of blood supply to specific regions of the brain, leading to rapid loss of brain function and potential long-term neurological impairment. The diagnosis of stroke is confirmed through a combination of clinical presentation and neuroimaging evidence of brain tissue damage attributable to vascular pathology [5]. There is a substantial overlap between the risk factors for strokes and cardiovascular events. Hypertension and Atrial Fibrillation in addition to other risk factors—Diabetes, dyslipidemia, smoking, regular exercise, obesity, diet, alcohol, lower socioeconomic status—have been identified as major risk factors for strokes [6].

Understanding the relationship between TIA, stroke, and cardiovascular toxicity is crucial for improving patient outcomes. Cardiovascular toxicity can present as arterial and venous thrombosis, alterations in clotting factors, and embolism, all of which can contribute to the development of TIAs and stroke [2]. For instance, certain cancer therapies have been associated with an increased risk of venous thromboembolism and arterial toxicity, which can present as acute vasospasm, acute thrombosis, and accelerated atherosclerosis.

The interplay between cardiovascular toxicity and cerebrovascular events underscores the importance of a comprehensive approach to patient care, particularly in

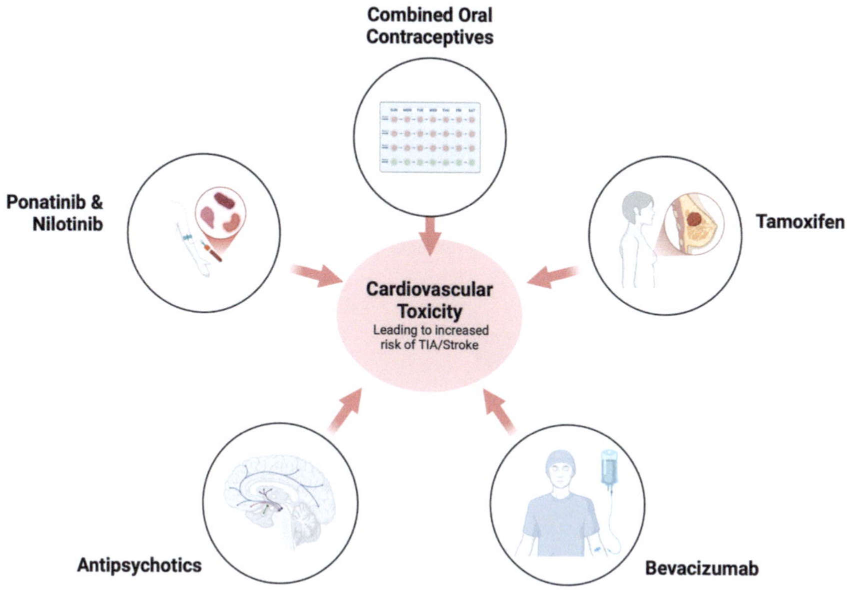

Fig. 1.1 Medications that are linked with increased risk of cardiovascular toxicity, which can increase the risk of TIA/stroke

the fields of cardio-oncology and neurology. Early recognition and management of TIAs and strokes, along with careful consideration of potential cardiovascular toxicities, are essential for reducing the risk of severe neurological outcomes and improving overall patient prognosis. Also, there are several medications, including oral contraceptivs, antipsychotics, and anti-cancer drugs, that are linked with increased risk of cardiovascular toxicity, which can increase the risk of TIA/stroke (Fig. 1.1). These are discussed in the following sections.

Combined Oral Contraceptives

Combined Oral Contraceptives (COCs) are widely used for contraception and management of menstrual disorders. These medications typically contain a combination of estrogen and progestin that effectively suppresses ovulation. However, their use has been associated with an increased risk of cardiovascular events, including transient ischemic attacks (TIAs) and strokes [7].

Combined oral contraceptives (COCs) primarily work by suppressing ovulation through the synergistic actions of estrogen and progestin components. The estrogen (typically ethinylestradiol) inhibits follicle-stimulating hormone (FSH) secretion from the pituitary gland, preventing follicular development and ovulation. The

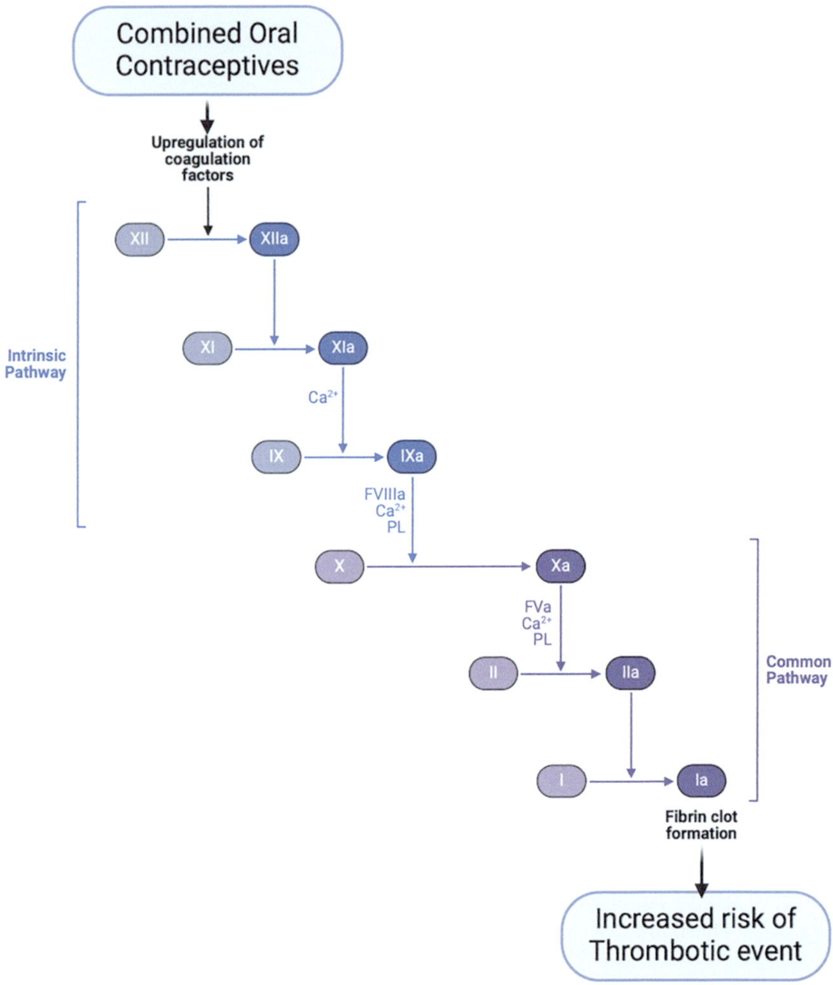

Fig. 1.2 Overview of the interactions between Combined Oral Contraceptives (COCs) and coagulation factors to increase risk of thrombotic event and cardiovascular toxicity

progestin component suppresses luteinizing hormone (LH) secretion, further preventing ovulation. Additionally, progestins thicken cervical mucus and alter the endometrium, making it less receptive to implantation [8].

Combined oral contraceptives (COCs) increase the risk of thrombosis through multiple mechanisms affecting the coagulation cascade. COC use leads to several procoagulant changes in blood proteins, including increased levels of coagulation factors II, VII, VIII, and X, as well as fibrinogen (Fig. 1.2). Important anticoagulant proteins such as antithrombin and protein S are simultaneously decreased. COC users also acquire resistance to activated protein C, a key anticoagulant that inactivates factors Va and VIIIa. The estrogen component of COCs, particularly ethinyl estradiol, is primarily responsible for these changes, with first-pass hepatic metabolism leading to increased hepatic synthesis of coagulation factors. Some progestins, such as desogestrel, may further contribute to the prothrombotic state by increasing procoagulant levels (factors VII, VIII, and X) and decreasing anticoagulant levels (protein S and antithrombin) compared to non-users. The risk of venous thrombosis varies depending on a multitude of factors, not limited to, age, family history, dose of estrogen and type of progestin. Risk factors such as factor V Leiden heterozygotes are much higher (up to 35 times) compared to the general population [9, 10]. The overall result is a shift in the hemostatic balance towards a more procoagulant state, increasing the risk of venous thromboembolism.

The increased risk of cardiovascular toxicity leading to thromboembolism has been linked with an increased risk of TIA/stroke. Recent meta-analyses have shown that COC use is associated with a 3.5-fold increased risk of venous thrombosis compared to non-use (relative risk 3.5, 95% confidence interval 2.9–4.3). Risk of venous thrombosis for combined oral contraceptives with 30–35 μg ethinylestradiol and gestodene, desogestrel, cyproterone acetate and drospirenone were similar, and about 50–80% higher than with levonorgestrel. Thirty micrograms ethinylestradiol with levonorgestrel should be the lowest recommended dose of combined oral contraceptive [11].

Tamoxifen

Tamoxifen is a breast cancer treatment after primary therapy in both males and females. It is used as adjuvant treatment after primary therapy, treatment of ductal carcinoma in situ, and for breast cancer risk reduction in high-risk patients [12]. Tamoxifen operates as a selective estrogen receptor modulator (SERM), primarily targeting breast cancer cells. Its main action involves competing with the body's natural estrogen to bind to estrogen receptors (ERs). When tamoxifen attaches to an ER, it causes a change in the receptor's shape, which alters how the receptor interacts with other cellular proteins. This modified ER-tamoxifen unit interferes with the normal processes of estrogen-driven gene expression and cell division, effectively slowing down the growth of ER-positive breast cancer cells. Tamoxifen's influence extends beyond just the estrogen signaling pathway, as it also affects other cellular communication networks involving growth factors. These include pathways like PI3K/AKT/mTOR and NFκB, which play roles in cell survival, growth, and inflammatory responses [13].

The PI3K/AKT/mTOR pathway plays a crucial role in various cellular processes and is frequently altered in breast cancer. This signaling cascade begins when PI3K phosphorylates PIP2 to produce PIP3 at cellular membranes. PIP3 then recruits and activates AKT and PDK-1, which are key mediators of the pathway. Activated AKT promotes protein synthesis, cell growth, survival, and motility by activating downstream kinases, including mTOR. The tumor suppressor PTEN counteracts this signaling by dephosphorylating PIP3. In breast cancer, genetic alterations often lead to increased PI3K activity or loss of inhibitory functions. Common alterations include mutations in PIK3CA (encoding the p110α catalytic subunit of PI3K), which occur in varying frequencies across different breast cancer subtypes. Other alterations can involve PIK3CB, PTEN, INPP4B, AKT1, and PDK1. These mutations or amplifications contribute to increased pathway activation, promoting tumor growth and potentially affecting treatment responses. The central role of the pathways in breast cancer has made it a target for therapeutic interventions with drugs such as tamoxifen [14]. The drug's effectiveness in treating and preventing ER-positive breast cancers is due to its ability to reduce both cancer recurrence and death rates. However, the intricate nature of tamoxifen's interactions within the cell can sometimes lead to drug resistance. This resistance may develop through various means, such as changes in how the ER functions, increases in certain growth factor receptors, or the activation of alternative cellular signaling routes [13].

In recent literature, two breast cancer cell lines, MCF7 and T47D, using two different models of tamoxifen exposure were examined. In Model 1, where cancer cells were treated with tamoxifen before exposure to blood components, both cell lines showed increased thrombin generation. MCF7 cells exposed to whole blood (WB) significantly increased thrombin—a clotting factor—levels ($p = 1.17E-06$) compared to untreated MCF7 cells. T47D cells initially produced higher thrombin levels (41.61 ng/mL ± 1.01E+03) than MCF7 cells (37.77 ng/mL ± 4.00E+01) but showed decreased levels when exposed to blood constituents ($p = 1.30E-06$). Both cell lines induced higher platelet activation marker expression (CD62P and CD63) in WB compared to non-tumorigenic cells. In platelet-rich plasma (PRP), MCF7 cells significantly increased both CD62P and CD63 expression ($p < 0.05$), indicating later stages of platelet activation [15]. The study demonstrated a clear association between tamoxifen treatment, breast cancer cell subtypes, and increased coagulation markers, suggesting an association between tamoxifen exposure and increased thrombosis. With the increased risk of TIA/stroke associated with increased risk of thrombosis, tamoxifen exposure has also been shown to be associated with an increased risk of ischemic stroke with women with breast cancer treated with tamoxifen at an 82% increased risk (Odds Ratio: 1.82 (95% CI, 1.41–2.36)) and 29% increased risk of any stroke (Odds Ratio: 1.40 (1.14–1.72)) [16].

Bevacizumab

Bevacizumab is a monoclonal antibody that is used to treat a variety of cancers, including but not limited to cervical, metastatic colorectal, non-squamous non-small cell lung, ovarian, and metastatic renal cancer. Binding of Vascular endothelial growth factor 1 (VEGF 1) and VEGF receptors (VEGFR) lead to increased endothelial cell proliferation by the formation of new blood vessels and angiogenesis. Bevacizumab acts as an inhibitor of VEGF 1 and VEGFR binding to restrict angiogenesis [17].

The vascular endothelial growth factor (VEGF) signaling pathway is a critical component in tumor angiogenesis and growth. In the hypoxic conditions typical of tumor microenvironments, hypoxia-inducible factor (HIF) becomes activated, triggering increased transcription of VEGF genes. Consequently, cancer cells produce and release VEGF-A into their surroundings. VEGF-A then binds to VEGF receptors, primarily VEGFR-1 and VEGFR-2, located on endothelial cells. This binding initiates a cascade of intracellular signaling events that promote endothelial cell proliferation, migration, and the formation of new blood vessels (angiogenesis). The vascular remodeling induced by bevacizumab not only reduces overall tumor blood supply but also has significant implications for the tumor microenvironment. The normalization of blood vessels can enhance the influx of immune cells into the tumor parenchyma, potentially boosting anti-tumor immune responses. Additionally, bevacizumab treatment may increase the expression of adhesion molecules like ICAM-1/2, VCAM-1, and CD34 on endothelial cells. This upregulation facilitates the trans-endothelial migration of leukocytes into the tumor parenchyma, further contributing to the anti-tumor immune response [18].

Bevacizumab, a monoclonal antibody targeting vascular endothelial growth factor (VEGF), has been associated with an elevated risk of thrombosis. This increased risk stems from the drug's mechanism of action, which interferes with VEGF-A signaling. By doing so, bevacizumab disrupts the normal functioning of endothelial cells and overall vascular health. One consequence is a decrease in the production of important anti-thrombotic factors, specifically nitric oxide and prostacyclin. Furthermore, the drug hinders the regeneration process of endothelial cells, leaving blood vessels more vulnerable to injury. These effects, combined with bevacizumab's influence on the coagulation system, create an environment that favors blood clot formation. As a result, patients receiving bevacizumab treatment face a higher likelihood of experiencing both venous and arterial thrombotic events. This underscores the importance of careful monitoring and risk assessment when administering this medication [19]. In previous meta-analysis, the prevalence of arterial and venous thromboembolic risk of bevacizumab in ovarian cancers were studied. In the study, it was found that patients treated with bevacizumab had an increased risk of arterial thromboembolic event (ATE) and venous thromboembolic event (VTE) with a relative risk ratio of 2.45 (95% CI, 1.27–4.27, $p = 0.008$) and 1.32 (95% CI, 1.02–1.79, $p = 0.04$), respectively, compared to the control [20]. These findings suggest that bevacizumab treatment in ovarian cancer patients is associated with a meaningful increase in both arterial and venous thromboembolic events, with the risk being more pronounced for arterial events.

A study conducted in 2014 looked at 12,917 patients from 17 randomized controlled trials and found there to be a significantly increased relative risk of 3.28 (95% CI, 1.97–5.48) of cerebrovascular events for patients treated with bevacizumab [21]. However, more recent studies have found no association between bevacizumab and increased risk of stroke/TIA. A recent meta-analysis showed that ovarian cancer patients receiving a combination and chemotherapy had a combined risk ratio of 2.14 (95% CI, 0.88–7.99). After further subgroup evaluation, when bevacizumab was used alongside carboplatin and paclitaxel, the occurrence of stroke-related complications was found to be 0.01% (95 CI, 0.00–0.01, $p < 0.01$). When distinguishing between types of stroke, ischemic events occurred at a rate of 0.01% (95% CI: 0.01–0.02, $p = 0.27$), while hemorrhagic strokes also had an incidence of 0.01% (95% CI: 0.00–0.01, $p < 0.01$) [22].

However, there have been more recent studies that provide another perspective. A recent retrospective cohort study done in 2020 evaluated 87,844 patients (69,007 bevacizumab, 10,895 ranibizumab, 7942 aflibercept) who had received an initial anti-VEGF injection for managing age related macular degeneration (nAMD), diabetic retinal disease (DRD) and retinal venous occlusive disease (RVO). Post-injection 180-day event rates for myocardial infarction, cerebrovascular disease, major bleeding and all cause hospitalization were measured for all 3 anti-VEGF agents. The study found no significant differences in the risk of acute cerebrovascular disease (which includes TIA and stroke) within 180 days of treatment initiation between bevacizumab and other anti-VEGF agents (ranibizumab and aflibercept) [23].

Antipsychotics

Antipsychotic medications are primarily indicated for the treatment of severe mental health disorders, with their use expanding beyond traditional applications in recent years. According to a systematic review and meta-analysis published in The Lancet in 2019, the primary indication for antipsychotics remains the acute treatment of schizophrenia, where they have shown significant efficacy in reducing overall symptoms compared to placebo [24]. However, their use has broadened to include other conditions. A 2023 study in The Lancet Psychiatry reported that in children and adolescents, antipsychotics are frequently prescribed for autism spectrum disorder, non-affective psychosis, anxiety disorders, attention-deficit/hyperactivity disorder (ADHD), depression, and conduct disorders [25]. Additionally, a 2021 study in Frontiers in Pharmacology found that antipsychotics are often prescribed off-label for controlling chronic diseases, sleeping disorders, benign paroxysmal positional vertigo, and irritable bowel syndrome [26]. These findings highlight the evolving landscape of antipsychotic use, emphasizing the need for careful consideration of their benefits and risks in various clinical contexts. Recent studies have highlighted the need for cautious use of antipsychotic medications due to their association with an increased risk of stroke and transient ischemic attack (TIA). Clinicians should consider alternative treatments where possible and closely monitor patients for signs of cerebrovascular events when antipsychotic use is deemed necessary.

Antipsychotic medications operate through intricate interactions with multiple neurotransmitter systems in the brain. The research by Amato et al. indicates that these drugs primarily function by blocking dopamine D2 receptors, which is considered the fundamental mechanism for their antipsychotic effectiveness. However, their influence extends beyond dopamine, encompassing several neurotransmitter systems. Notably, second-generation antipsychotics target both dopamine D2 and serotonin 5-HT2A receptors, a dual action that is believed to contribute to their enhanced efficacy and reduced motor side effects compared to their first-generation counterparts. The study also emphasizes the impact of antipsychotics on glutamatergic and GABAergic neurotransmission, affecting the equilibrium between brain excitation and inhibition. Moreover, these medications influence various intracellular signaling cascades, including those involved in neuroplasticity and neuroprotection. The researchers note that antipsychotics can modify gene expression and protein synthesis, potentially inducing long-term alterations in brain structure and function. This comprehensive approach to neurotransmitter modulation and cellular signaling underpins both the therapeutic benefits and adverse effects of antipsychotic drugs [27].

Antipsychotic medications have been associated with an increased risk of cardiovascular thrombosis through various mechanisms (Fig. 1.3). Recent research has found a specific biochemical pathway involving the interaction between certain antipsychotics, particularly clozapine, and the coagulation process. Clozapine has been found to directly interact with fibrinogen, influencing fibrin formation without compromising fibrinogen's structural integrity. This unintended side-effect results in the formation of thinner fibrin fibers and a slowed coagulation process, paradoxically conferring thrombogenic properties to fibrinogen. The effect appears to be dose-dependent, suggesting that higher doses of clozapine may increase thrombosis risk. Additionally, antipsychotics can elevate levels of hormones such as prolactin and leptin, which act as coactivators in adenosine diphosphate (ADP)-dependent platelet aggregation and increase P-selectin expression on platelets. This enhanced platelet activation and aggregation contribute to a pro-thrombotic state, potentially leading to both arterial and venous thrombosis. The complex interplay between antipsychotics, fibrin formation, and platelet function underscores the need for careful monitoring and individualized treatment approaches to mitigate thrombotic risks in patients receiving these medications [28].

The increased risk of cardiovascular toxicity leading to thromboembolism has been linked with an increased risk of TIA/stroke. A comprehensive study involving 7,267,673 individuals identified 25,532 cases of venous thromboembolism (VTE), comprising 15,975 deep vein thrombosis and 9557 pulmonary embolism instances, along with 89,491 matched controls. The research revealed that people who had been prescribed antipsychotic medications within the past 24 months exhibited a 32% higher likelihood of developing VTE compared to non-users, even after accounting for potential risk factors (odds ratio: 1.32, 95% CI: 1.23–1.42). Notably, patients who had begun a new antipsychotic regimen in the preceding 3 months faced approximately double the risk (odds ratio: 1.97, 95% CI: 1.66–2.33) [29].

The study also found that the risk was more pronounced for those prescribed atypical antipsychotics (adjusted odds ratio: 1.73, 95% CI: 1.37–2.17) compared to conventional drugs (1.28, 95% CI: 1.18–1.38). Furthermore, low-potency medications were associated with a higher risk (1.99, 95% CI: 1.52–2.62) than high-potency drugs (1.28, 95% CI: 1.18–1.38). The researchers estimated that for every 10,000 patients treated over a year, there would be 4 additional cases of VTE (95% CI: 3–5) across all age groups, and 10 extra cases (95% CI: 7–13) among patients aged 65 and older [29].

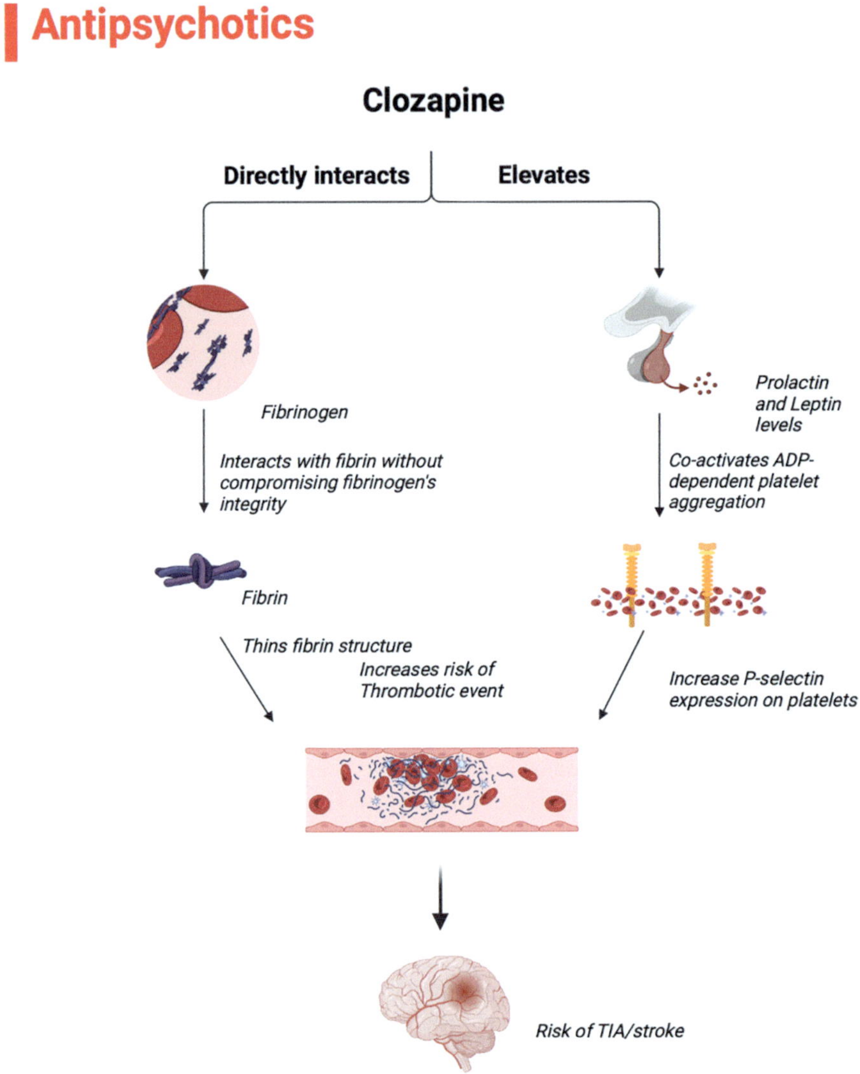

Fig. 1.3 Pathway of antipsychotics like clozapine leading to increased risk of cardiovascular toxicity and eventually increasing risk of stroke/TIA

Ponatinib and Nilotinib

Ponatinib, a third-generation BCR-ABL tyrosine kinase inhibitor (TKI), was intended for patients with chronic myeloid leukemia (CML) or Philadelphia chromosome-positive acute lymphoblastic leukemia (Ph+ ALL) who showed resistance or intolerance to previous TKI therapies. Following a brief market withdrawal due to safety concerns, ponatinib was reintroduced in December 2013 with a more restricted indication, focusing on patients with the T315I mutation or those without other suitable TKI options. A notable expansion of ponatinib's indication occurred on December 18, 2020, when the FDA approved its use in chronic-phase CML (CP-CML) patients who exhibited resistance or intolerance to at least two prior kinase inhibitors. Along with this expanded indication, a new dosing protocol of 45 mg daily until achieving $\leq$1% BCR-ABL IS (MR2), followed by a reduction to 15 mg daily, was established. This update provides an additional treatment avenue for CP-CML patients facing poor prognosis due to multidrug resistance or mutations that render other BCR-ABL targeting TKIs ineffective [30].

Nilotinib, a second-generation tyrosine kinase inhibitor, was also approved for patients with CP-CML. In a recent study, nilotinib showed rapid reduction in BCR-ABL1 levels, particularly in the first 12 months of treatment. The recommended initial dose is 300 mg twice daily, with potential dose escalation to 400 mg twice daily in cases of suboptimal response or treatment failure [31]. Comparing nilotinib to ponatinib, both are tyrosine kinase inhibitors used in CML treatment, but with distinct indications and safety profiles. While nilotinib is primarily used as a frontline therapy for newly diagnosed CML-CP, ponatinib is often reserved for patients with resistance or intolerance to other TKIs, particularly those with the T315I mutation. Nilotinib's dosing strategy allows for more flexibility, with options for dose escalation and reduction based on response and tolerability. In contrast, ponatinib typically follows a more rigid dosing regimen due to its potency and associated cardiovascular risks. However, both drugs aim to achieve deep molecular responses in CML patients.

Receptor Tyrosine Kinases (RTKs) activation leads to regulation of cell growth, tissue regeneration and repair. This pathway is highly regulated in normal cells, and if they are overexpressed can prompt tumor survival in several types of cancers. Activation in RTKs lead to activation of the STAT3, PI3K/AKT, and RAS/RAF/ERK pathways, which are known to drive cancers. Beyond its primary target, ponatinib, a multi-tyrosine kinase inhibitor, also demonstrates inhibitory effects on several other kinases, including FLT3, c-KIT, VEGFR, PDGFR, and members of the SRC family. This broad spectrum of kinase inhibition contributes to ponatinib's potency and its potential efficacy in treating various malignancies beyond chronic myeloid leukemia and Philadelphia chromosome-positive acute lymphoblastic leukemia. Its distinct structure of a carbon-carbon triple bond allows ponatinib to overcome the steric hindrance caused by mutations in the BCR-ABL kinase domain. Ponatinib binds to the ATP-binding site of the BCR-ABL kinase domain, inhibiting its kinase activity and reducing the signaling cascade [32]. Nilotinib, another TKI that targets BCR-ABL, binds to the inactive ABL kinase domain. Nilotinib's unique trifluoromethyl and imidazole-substituted phenyl group helps inhibit autophosphorylation of RTKs, reducing the downstream signaling pathway. It is primarily an antiproliferative rather than a proapoptotic [33]. These mechanisms allow both ponatinib and nilotinib to inhibit growth of various cancer cell lines, making it a valuable treatment option for chronic myeloid leukemia and potentially other malignancies driven by its target kinases.

Although the exact mechanism of action of how ponatinib and nilotinib causes cardiovascular toxicity is unclear, a recent study found both the drugs to affect the human umbilical vein endothelial cells (HUVECs) *in vitro*. Both drugs significantly reduced the HUVEC viability compared to the control, suggesting an impact on endothelial cell function. The endothelium integrity and angiogenesis are highly regulated for processes like apoptosis and proliferation with disruptions causing cardiovascular events and hypertension. Gene expression analysis revealed that ponatinib and nilotinib treatment alters pathways related to cardiovascular system development, blood vessel development, angiogenesis, and cell proliferation. These changes suggest a potential disruption in endothelial cell homeostasis and function. The impairment of endothelial cells' ability to form and maintain healthy blood vessels, combined with alterations in genes involved in vascular development and angiogenesis, may contribute to a pro-thrombotic environment [34]. In another study, the various TKIs and their risk ratio of causing cardiovascular toxicity was examined. Nilotinib and Ponatinib were both associated with an increased risk ratio for causing an embolic or thrombotic event at 1.3 (1.1–1.4) and 1.4 (1.2–1.6), respectively. Both drugs were also associated with cardiac failure with Nilotinib's risk ratio at 1.5 (1.3–1.7) and Ponatinib's risk ratio at 1.8 (1.4–2.4) [35]. Both drug's association with cardiovascular toxicity can lead to higher risk of cerebrovascular events such as TIA/strokes.

Research on nilotinib-related atherosclerosis has primarily focused on its association with progressive peripheral artery occlusive disease (PAOD), with fewer reported cases involving coronary artery disease (CAD) and cerebrovascular disease (CVD). The exact pathways through which nilotinib contributes to atherosclerosis development remain unclear and require further investigation [36]. However, in a series of three case studies, two of these individuals also exhibited peripheral artery disease and intracranial atherosclerosis, while the third developed a complication in the internal carotid artery. Notably, all three patients had elevated 10-year cardiovascular risk profiles when their strokes occurred. The researchers propose several potential mechanisms by which nilotinib might contribute to atherosclerosis, including its effects on blood vessel cells, surrounding tissues, mast cells, and metabolic factors like blood sugar and cholesterol levels. Given these findings, healthcare providers are advised to exercise caution when prescribing nilotinib, especially for patients with high cardiovascular risk. Close monitoring of metabolic changes and patient education about stroke symptoms are recommended to ensure timely medical intervention if necessary [37].

Although there are limited reports linking ponatinib and risk of TIA/strokes, a study examined TKIs and their effect on cerebrovascular events. Ponatinib had the highest incidence rate of 3.0 (0.8–12.1; $p = 0.04$), suggesting a statistically significant association between ponatinib and risk of cerebrovascular events [38]. In addition, a case study of a patient with no prior cardiovascular issues or history of ischemic events was put on the lowest maintenance dosage of ponatinib. The patient, after receiving 20 months of ponatinib treatment, experiences a middle cerebral artery infarction [37]. Given this case study in addition to the previous study examining ponatinib and risk of cerebrovascular events, it is crucial to implement regular monitoring and appropriate management of any existing cardiovascular conditions in patients receiving ponatinib. Furthermore, healthcare providers should ensure that patients undergoing ponatinib therapy are thoroughly informed about both the potential advantages and possible adverse effects associated with this medication. This approach allows for better-informed decision-making and heightened awareness of potential risks.

Risk Mitigation

Dosage Control

Dosage control is a critical strategy for reducing cardiovascular toxicity associated with various medications. By carefully managing drug doses, healthcare providers can help balance the therapeutic benefits of these medications against their potential cardiovascular risks. Specifically, for anti-cancer therapies, cardiac dose exposure and survival were negatively correlated [39]. Some studies have also shown that despite a longer exposure, lowering the dosages have suggested reduction in cardiotoxicity [40].

Ponatinib was often prescribed to combat the effects of Chronic Myeloid Leukemia (CML). As previously mentioned, ponatinib is a tyrosine kinase inhibitor known to have adverse cardiovascular effects. Balancing ponatinib use between efficacy and toxicity has been a challenge in clinical trials and in treating the disease. An article published in 2024 aims to compare outcomes between two groups of patients with chronic phase chronic myeloid leukemia (CP-CML) who initially received a daily 45 mg dose of ponatinib. The first group consists of patients from the Ponatinib Ph+ ALL and CML Evaluation (PACE) trial, which did not employ a response-based dosing strategy. The second group includes patients from the Optimizing Ponatinib Treatment In CP-CML (OPTIC) trial, which implemented a novel approach of planned dose reductions after patients reached a specific response threshold. This comparison found a 60% overall risk reduction in OPTIC trial patients compared to the PACE trial patients with a lower Treatment emergent adverse event (TEAE) in OPTIC trial patients (68%) compared to the PACE trial patients (84%). Although the median time on therapy was higher in the OPTIC group (19.5 months) compared to the PACE group (12.6 months), the reduction in risk of adverse occlusive events could provide insight on mitigating the risk of cardiovascular toxicity [41].

Anticoagulants and Antiplatelets

Anticoagulants and antiplatelets are two classes of antithrombotic drugs used to prevent and treat various cardiovascular disorders. Anticoagulants are primarily used for patients with venous thromboembolisms and atrial fibrillation. Antiplatelets are primarily used for patients with atherosclerosis. Recently, they have been used together to provide long-term anticoagulation in patients who need a more intensive anti-thrombotic therapy. Antiplatelet medications work by interfering with the processes of platelet activation and clumping, thus impeding the initial stages of blood clot development. In contrast, anticoagulant drugs primarily target the coagulation cascade, hindering the production of fibrin and consequently preventing clot formation [42–45].

In a recent study that explored the total arterial and venous thrombotic burden in patients with peripheral artery disease (PAD), patients were put into two groups to observe the rate of first and total thrombotic events. This study was conducted to show the effectiveness of a combination therapy. The first group was a placebo group on an aspirin background and the second group were given Rivaroxaban with aspirin. Aspirin is a powerful antiplatelet medication, and Rivaroxaban is an anticoagulant that directly inhibits Factor Xa coagulation factor [46]. This combination therapy resulted in a 23% reduction in total thrombotic events compared to aspirin alone (Hazard Ratio: 0.77, 95% CI: 0.67–0.89, $p = 0.0005$). The treatment prevented 6.1 total arterial and venous thrombotic events per 100 patients over 3 years [47].

It is important to note that these interactions can be complex and may vary depending on individual patient factors. Clinical decisions should be made on a case-by-case basis, considering the potential risks and benefits for each patient. Some of those risks may include increased risk of major bleeding and hemorrhage [48–53].

ACE Inhibitors and Beta Blocker

A common standard of care approach to reducing cardiovascular toxicity has been Angiotensin Converting Enzyme (ACE) inhibitors and beta blockers. ACE inhibitors and beta-blockers have emerged as the cornerstone therapies for managing chronic heart failure. These two drug classes have gained this status due to their proven ability to significantly decrease both the incidence of complications and the risk of death in patients with this condition. Their combined use has become fundamental in heart failure treatment protocols, reflecting their demonstrated effectiveness in improving patient outcomes and survival rates [54]. Starting from the 1980s and early 1990s, ACE inhibitors and beta blockers lead to notable improvements in left ventricular function and provide substantial benefits in terms of reducing adverse outcomes and extending life expectancy for patients with less severe forms of heart failure [55].

ACE inhibitors work by inhibiting the conversion of angiotensin I to angiotensin II, thereby reducing the vasoconstrictor effects of angiotensin II and decreasing aldosterone production. This leads to vasodilation and decreased blood pressure. Additionally, ACE inhibitors increase bradykinin levels, which can contribute to their cardiovascular benefits [56, 57]. Beta blockers act by antagonizing β-adrenergic receptors, primarily β1 receptors in the heart. This results in decreased heart rate, cardiac output, and blood pressure. Some beta blockers also have vasodilatory effects through α1-receptor blockade or β3-receptor activation [58]. Both drug classes have been shown to have beneficial effects on cardiovascular outcomes, though their specific mechanisms differ. ACE inhibitors tend to reverse the hemodynamic changes of hypertension, while beta blockers without intrinsic sympathomimetic activity primarily reduce heart rate and cardiac output.

Conclusion

The comprehensive review of TIA and stroke in the context of cardiovascular toxicity highlights the complex interplay between various medications and cerebrovascular risk. The mechanisms by which drugs like combined oral contraceptives, tamoxifen, bevacizumab, antipsychotics, ponatinib, and nilotinib contribute to TIA and stroke risk are multifaceted, involving alterations in coagulation cascades, endothelial function, and vascular integrity. The study of combined oral contraceptives reveals their impact on coagulation factors, potentially shifting the hemostatic balance towards a procoagulant state. Tamoxifen, while effective in breast cancer treatment, has shown associations with increased thrombin generation and platelet activation, contributing to thrombotic risk. Bevacizumab's effect on VEGF signaling, while beneficial in cancer treatment, can lead to endothelial dysfunction and increased thrombotic events. Antipsychotics present a unique challenge, with their cardiovascular risks stemming from interactions with fibrinogen and effects on platelet function. The tyrosine kinase inhibitors ponatinib and nilotinib, crucial in leukemia treatment, have demonstrated concerning associations with vascular toxicity and atherosclerosis, necessitating careful monitoring and management.

The exploration of risk mitigation strategies offers promising avenues for reducing cardiovascular toxicity. Dosage control, particularly evident in the case of ponatinib, shows potential in balancing efficacy and toxicity. The combined use of anticoagulants and antiplatelets presents a powerful approach to thrombosis prevention, albeit with careful consideration of bleeding risks. The established roles of ACE inhibitors and beta-blockers in managing heart failure and reducing cardiovascular toxicity provide a foundation for comprehensive care.

This review underscores the importance of individualized patient care and risk assessment. The balance between the therapeutic benefits of these medications and their potential cardiovascular risks requires careful consideration and ongoing monitoring. Future research should focus on refining risk prediction models, developing targeted prevention strategies, and exploring novel approaches to mitigate cardiovascular toxicity without compromising treatment efficacy.

In conclusion, the management of TIA and stroke risk in the context of cardiovascular toxicity demands a multidisciplinary approach. Clinicians must remain vigilant, considering patient-specific factors, medication interactions, and emerging evidence to optimize treatment outcomes. As our understanding of these complex interactions deepens, we can anticipate more tailored approaches to patient care, potentially leading to reduced incidence of TIA and stroke in vulnerable populations.

Key Points

- Transient Ischemic Attack (TIA) and stroke are significant cerebrovascular events associated with cardiovascular toxicity.
- Several medications, including combined oral contraceptives, tamoxifen, bevacizumab, antipsychotics, ponatinib, and nilotinib, can increase the risk of thromboembolism and subsequent cerebrovascular events..
- Combined oral contraceptives alter the coagulation cascade, while tamoxifen and bevacizumab affect endothelial function and angiogenesis.
- Antipsychotics interact with fibrinogen and platelet function, potentially leading to a pro-thrombotic state.
- Tyrosine kinase inhibitors like ponatinib and nilotinib have been associated with vascular toxicity and atherosclerosis.
- Dosage control, the use of anticoagulants and antiplatelets, and the role of ACE inhibitors and beta-blockers are potential mitigation strategies in managing cardiovascular toxicity.
- It is crucial to remain vigilant, considering patient-specific risk factors, medication interactions, and emerging evidence to optimize treatment outcomes.
- Additional careful studies are warranted to better understand the underlying cellular and molecular mechanism of the complex interplay between various medications and cerebrovascular risk and implement appropriate preventive measures.

Funding The research work of DKA is supported by R25AI179582 and R01 HL147662 grants from the National Institutes of Health, USA. The contents of this chapter are solely the responsibility of the authors and do not necessarily represent the official views of the National Institutes of Health.

Competing Interests Both authors have read the chapter and declare no conflict of interest. No writing assistance was utilized in the production of this manuscript.

Consent for Publication Both authors have read the chapter and consented for publication.

References

1. Benjamin EJ, Muntner P, Alonso A, Bittencourt MS, Callaway CW, Carson AP, Chamberlain AM, Chang AR, Cheng S, Das SR, Delling FN, Djousse L, Elkind MSV, Ferguson JF, Fornage M, Jordan LC, Khan SS, Kissela BM, Knutson KL et al (2019) Heart disease and stroke statistics—2019 update: a report from the American Heart Association. Circulation 139(10):e56–e528. https://doi.org/10.1161/CIR.0000000000000659
2. Barachini S, Ghelardoni S, Varga ZV, Mehanna RA, Montt-Guevara MM, Ferdinandy P, Madonna R (2023) Antineoplastic drugs inducing cardiac and vascular toxicity—an update. Vasc Pharmacol 153:107223. https://doi.org/10.1016/j.vph.2023.107223

3. Panuganti KK, Tadi P, Lui F (2024) Transient ischemic attack. In: StatPearls. StatPearls Publishing, Treasure Island, FL. https://www.ncbi.nlm.nih.gov/books/NBK459143/
4. Vinding NE, Butt JH, Lauridsen MD, Kristensen SL, Johnsen SP, Krøll J, Graversen PL, Kruuse C, Torp-Pedersen C, Køber L, Fosbøl EL (2023) Long-term incidence of ischemic stroke after transient ischemic attack: a nationwide study from 2014 to 2020. Circulation 148(13):1000–1010. https://doi.org/10.1161/CIRCULATIONAHA.123.065446
5. Sacco RL, Kasner SE, Broderick JP et al (2013) An updated definition of stroke for the 21st century: a statement for healthcare professionals from the American Heart Association/American Stroke Association. Stroke 44(7):2064–2089
6. Sabih A, Tadi P, Kumar A (2024) Stroke prevention. In: StatPearls. StatPearls Publishing, Treasure Island, FL. https://www.ncbi.nlm.nih.gov/books/NBK470234/
7. Champaloux SW, Tepper NK, Monsour M, Curtis KM, Whiteman MK, Marchbanks PA, Jamieson DJ (2017) Use of combined hormonal contraceptives among women with migraines and risk of ischemic stroke. Am J Obstet Gynecol 216(5):489.e1–489.e7. https://doi.org/10.1016/j.ajog.2016.12.019
8. Mode of action of oral contraceptives (1978). Fam Plan Inf Serv 1(1):11
9. Trenor CC III, Chung RJ, Michelson AD, Neufeld EJ, Gordon CM, Laufer MR, Emans SJ (2011) Hormonal contraception and thrombotic risk: a multidisciplinary approach. Pediatrics 127(2):347–357. https://doi.org/10.1542/peds.2010-2221
10. Stalas J, Morris R, Bu K, von Bargen K, Largmann R, Sanford K, Vandeventer J, Han W, Cheng F (2024) Comparing the risk of deep vein thrombosis of two combined oral contraceptives: norethindrone/ethinyl estradiol and drospirenone/ethinyl estradiol. Heliyon 10(5):e26462. https://doi.org/10.1016/j.heliyon.2024.e26462
11. de Bastos M, Stegeman BH, Rosendaal FR, Van Hylckama Vlieg A, Helmerhorst FM, Stijnen T, Dekkers OM (2014) Combined oral contraceptives: venous thrombosis. Cochrane Database Syst Rev 2014(3):CD010813. https://doi.org/10.1002/14651858.CD010813.pub2
12. Farrar MC, Jacobs TF (2023) Tamoxifen. In: StatPearls. StatPearls Publishing, Treasure Island, FL
13. Ali S, Rasool M, Chaoudhry H, Pushparaj PN, Jha P, Hafiz A, Mahfooz M, Abdus Sami G, Azhar Kamal M, Bashir S, Ali A, Sarwar Jamal M (2016) Molecular mechanisms and mode of tamoxifen resistance in breast cancer. Bioinformation 12(3):135–139. https://doi.org/10.6026/97320630012135
14. Li H, Prever L, Hirsch E, Gulluni F (2021) Targeting PI3K/AKT/mTOR signaling pathway in breast cancer. Cancers 13(14):3517. https://doi.org/10.3390/cancers13143517
15. Pather K, Augustine TN (2020) Tamoxifen induces hypercoagulation and alterations in ERα and ERβ dependent on breast cancer sub-phenotype ex vivo. Sci Rep 10(1):19256. https://doi.org/10.1038/s41598-020-75779-y
16. Bushnell CD, Goldstein LB (2004) Risk of ischemic stroke with tamoxifen treatment for breast cancer: a meta-analysis. Neurology 63(7):1230–1233. https://doi.org/10.1212/01.wnl.0000140491.54664.50
17. Gerriets V, Kasi A (2024) Bevacizumab. In: StatPearls. StatPearls Publishing, Treasure Island, FL. https://www.ncbi.nlm.nih.gov/books/NBK482126/
18. de Aguiar RB, de Moraes JZ (2019) Exploring the immunological mechanisms underlying the anti-vascular endothelial growth factor activity in tumors. Front Immunol 10:1023. https://doi.org/10.3389/fimmu.2019.01023
19. Mayenga M, Falvo N, Mahé I, Jannot AS, Gazeau B, Meyer G, Gendron N, Sanchez O, Djennaoui S, Planquette B (2023) Cancer-associated thrombosis on bevacizumab: risk of recurrence and bleeding when bevacizumab is stopped or continued. Cancers 15(15):3893. https://doi.org/10.3390/cancers15153893
20. Saerens M, De Jaeghere EA, Kanervo H, Vandemaele N, Denys H, Naert E (2021) Risk of thrombo-embolic events in ovarian cancer: does bevacizumab tilt the scale? A systematic review and meta-analysis. Cancers 13(18):4603. https://doi.org/10.3390/cancers13184603

21. Zuo PY, Chen XL, Liu YW, Xiao CL, Liu CY (2014) Increased risk of cerebrovascular events in patients with cancer treated with bevacizumab: a meta-analysis. PLoS One 9(7):e102484. https://doi.org/10.1371/journal.pone.0102484
22. Song L, Liu Y, Chen Z, Li Z, Zhu S, Zhao Y, Li H (2023) Association of bevacizumab and stroke in ovarian cancer: a systematic review and meta-analysis. Front Neurosci 17:1187957. https://doi.org/10.3389/fnins.2023.1187957
23. Maloney MH, Payne SR, Herrin J, Sangaralingham LR, Shah ND, Barkmeier AJ (2021) Risk of systemic adverse events after intravitreal bevacizumab, ranibizumab, and aflibercept in routine clinical practice. Ophthalmology 128(3):417–424. https://doi.org/10.1016/j.ophtha.2020.07.062
24. Huhn M, Nikolakopoulou A, Schneider-Thoma J, Krause M, Samara M, Peter N, Arndt T, Bäckers L, Rothe P, Cipriani A, Davis J, Salanti G, Leucht S (2019) Comparative efficacy and tolerability of 32 oral antipsychotics for the acute treatment of adults with multi-episode schizophrenia: a systematic review and network meta-analysis. Lancet 394(10202):939–951. https://doi.org/10.1016/S0140-6736(19)31135-3
25. Hales H, Gledhill J, Gajwani R, Callaghan A, Garralda ME, Downs J, Hayes RD (2022) Trends in antipsychotic prescribing to children and adolescents in England: cohort study using 2000–19 primary care data. Lancet Psychiatry 9(12):970–980. https://doi.org/10.1016/S2215-0366(22)00404-7
26. Bastaki D, Elhusein B, Jithesh PV, Abujaber A, Abdulmalik M, Syed N, Zakaria D, Elshafei M, Ullah E, Qoronfleh MW, Syed NI (2021) Prescription pattern of antipsychotics and off-label use in the Qatari population: a retrospective cross-sectional study. Front Pharmacol 12:753845. https://doi.org/10.3389/fphar.2021.753845
27. Brandão-Teles C, de Almeida V, Cassoli JS, Martins-de-Souza D (2019) Biochemical pathways triggered by antipsychotics in human [corrected] oligodendrocytes: potential of discovering new treatment targets. Front Pharmacol 10:186. https://doi.org/10.3389/fphar.2019.00186
28. Dho-Nagy EA, Brassai A, Lechsner P, Ureche C, Bán EG (2024) COVID-19 and antipsychotic therapy: unraveling the thrombosis risk. Int J Mol Sci 25(2):818. https://doi.org/10.3390/ijms25020818
29. Parker C, Coupland C, Hippisley-Cox J (2010) Antipsychotic drugs and risk of venous thromboembolism: nested case-control study. BMJ 341:c4245. https://doi.org/10.1136/bmj.c4245
30. Pulte ED, Chen H, Price LSL, Gudi R, Li H, Okusanya OO, Ma L, Rodriguez L, Vallejo J, Norsworthy KJ, de Claro RA, Theoret MR, Pazdur R (2022) FDA approval summary: revised indication and dosing regimen for ponatinib based on the results of the OPTIC trial. Oncologist 27(2):149–157. https://doi.org/10.1093/oncolo/oyab040
31. Hughes TP, Munhoz E, Aurelio Salvino M, Ong TC, Elhaddad A, Shortt J, Quach H, Pavlovsky C, Louw VJ, Shih LY, Turkina AG, Meillon L, Jin Y, Acharya S, Dalal D, Lipton JH (2017) Nilotinib dose-optimization in newly diagnosed chronic myeloid leukaemia in chronic phase: final results from ENESTxtnd. Br J Haematol 179(2):219–228. https://doi.org/10.1111/bjh.14829
32. Tan FH, Putoczki TL, Stylli SS, Luwor RB (2019) Ponatinib: a novel multi-tyrosine kinase inhibitor against human malignancies. Onco Targets Ther 12:635–645. https://doi.org/10.2147/OTT.S189391
33. Blay JY, von Mehren M (2011) Nilotinib: a novel, selective tyrosine kinase inhibitor. Semin Oncol 38 Suppl 1(0 1):S3–S9. https://doi.org/10.1053/j.seminoncol.2011.01.016
34. Gover-Proaktor A, Granot G, Pasmanik-Chor M, Pasvolsky O, Shapira S, Raz O et al (2018) Bosutinib, dasatinib, imatinib, nilotinib, and ponatinib differentially affect the vascular molecular pathways and functionality of human endothelial cells. Leuk Lymphoma 60(1):189–199. https://doi.org/10.1080/10428194.2018.1466294
35. Zeng P, Schmaier A (2020) Ponatinib and other CML tyrosine kinase inhibitors in thrombosis. Int J Mol Sci 21(18):6556. https://doi.org/10.3390/ijms21186556

36. Kakadia B, Thakkar R, Sanborn E, Suero-Abreu GA, Jovin TG, Then R (2021) Nilotinib-associated atherosclerosis presenting as multifocal intracranial stenosis and acute stroke. J Stroke Cerebrovasc Dis 30(8):105883
37. Azad F, Zhang J, Miranda CJ, Gravina M (2022) Ponatinib-induced cerebrovascular accident (CVA). Cureus 14(12):e32383. https://doi.org/10.7759/cureus.32383
38. Jain P, Kantarjian H, Boddu PC, Nogueras-González GM, Verstovsek S, Garcia-Manero G, Borthakur G, Sasaki K, Kadia TM, Sam P, Ahaneku H, O'Brien S, Estrov Z, Ravandi F, Jabbour E, Cortes JE (2019) Analysis of cardiovascular and arteriothrombotic adverse events in chronic-phase CML patients after frontline TKIs. Blood Adv 3(6):851–861. https://doi.org/10.1182/bloodadvances.2018025874
39. Luo J, Song J, Xiao L, Zhang J, Cao Y, Wang J, Wang P, Zhao L, Liu N (2023) Simultaneous integrated dose reduction intensity-modulated radiotherapy effectively reduces cardiac toxicity in limited-stage small cell lung cancer. Cancer Biol Med 20(6):452–464. https://doi.org/10.20892/j.issn.2095-3941.2022.0326
40. Shapira J, Gotfried M, Lishner M, Ravid M (1990) Reduced cardiotoxicity of doxorubicin by a 6-hour infusion regimen. A prospective randomized evaluation. Cancer 65(4):870–873. https://doi.org/10.1002/1097-0142(19900215)65:4<870::aid-cncr2820650407>3.0.co;2-d
41. Cortes JE, Apperley J, Lomaia E, Moiraghi B, Undurraga MS, Ossenkoppele GJ et al (2024) Dose modification dynamics of ponatinib in patients with chronic-phase chronic myeloid leukemia (CP-CML) from the PACE and OPTIC trials. Leukemia 38(3):639–647
42. Espinola-Klein C (2022) When and how to combine antiplatelet and anticoagulant drugs? Hamostaseologie 42(1):73–79. https://doi.org/10.1055/a-1724-4922
43. Mega JL, Simon T (2015) Pharmacology of antithrombotic drugs: an assessment of oral antiplatelet and anticoagulant treatments. Lancet 386(9990):281–291. https://doi.org/10.1016/S0140-6736(15)60243-4
44. Tsoumani ME, Tselepis AD (2017) Antiplatelet agents and anticoagulants: from pharmacology to clinical practice. Curr Pharm Des 23(9):1279–1293. https://doi.org/10.2174/1381612823666170124141806
45. Gauer JS, Riva N, Page EM, Philippou H, Makris M, Gatt A, Ariëns RAS (2020) Effect of anticoagulants on fibrin clot structure: a comparison between vitamin K antagonists and factor Xa inhibitors. Res Pract Thromb Haemost 4(8):1269–1281. https://doi.org/10.1002/rth2.12443
46. Samama MM (2011) The mechanism of action of rivaroxaban—an oral, direct Factor Xa inhibitor—compared with other anticoagulants. Thromb Res 127(6):497–504. https://doi.org/10.1016/j.thromres.2010.09.008
47. Berkowitz SD, Bauersachs RM, Szarek M, Nehler MR, Debus ES, Patel MR, Anand SS, Capell WH, Hess CN, Hsia J, Leeper NJ, Brasil D, Mátyás L, Diaz R, Brodmann M, Muehlhofer E, Haskell LP, Bonaca MP (2022) Prevention of arterial and venous thrombotic events in symptomatic peripheral arterial disease patients after lower extremity revascularization in the VOYAGER PAD trial: dual anticoagulant/antiplatelet regimen vs antiplatelet therapy alone. J Thromb Haemost 20(5):1193–1205. https://doi.org/10.1111/jth.15673
48. de Teresa E, Vera A, Ortigosa J, Pulpon LA, Arus AP, de Artaza M (1979) Interaction between anticoagulants and contraceptives: an unsuspected finding. Br Med J 2(6200):1260–1261. https://doi.org/10.1136/bmj.2.6200.1260
49. Panahi L, Saenz Ii JA, Nguyen N, Udeani G, Surani S (2020) Hypercoagulation from concomitant administration of tamoxifen and warfarin. Cureus 12(5):e8114. https://doi.org/10.7759/cureus.8114
50. Chen CM, Chang KH, Wang CL, Tu HT, Huang YT, Wu HC, Chang CH, Chang SH (2022) Major bleeding risk in atrial fibrillation patients co-medicated with non-vitamin K oral anticoagulants and antipsychotics. Front Pharmacol 13:819878. https://doi.org/10.3389/fphar.2022.819878
51. eviQ (n.d.) Chronic myeloid leukaemia—ponatinib. https://www.eviq.org.au/haematology-and-bmt/leukaemias/chronic-myeloid-leukaemia/1891-chronic-myeloid-leukaemia-ponatinib. Accessed 24 Dec 2024

52. Nalluri SR, Chu D, Keresztes R, Zhu X, Wu S (2014) Increased incidence of recurrent venous thromboembolism in patients with cancer receiving bevacizumab. J Hematol Oncol Pharm 4(3)
53. Noseworthy PA, Yao X, Gersh BJ, Hargraves I, Shah ND, Montori VM (2022) Drug interactions affecting oral anticoagulant use. Circ Arrhythm Electrophysiol 15(6):e007956. https://doi.org/10.1161/CIRCEP.121.007956
54. van der Horst IC, Voors AA, van Veldhuisen DJ (2007) Treatment of heart failure with ACE inhibitors and beta-blockers: what is next? Aldosterone receptor antagonists? Clin Res Cardiol 96(4):193–195. https://doi.org/10.1007/s00392-007-0487-y
55. Bristow MR (2000) Mechanistic and clinical rationales for using beta-blockers in heart failure. J Card Fail 6(2 Suppl 1):8–14
56. Peng H, Carretero OA, Vuljaj N, Liao TD, Motivala A, Peterson EL, Rhaleb NE (2005) Angiotensin-converting enzyme inhibitors: a new mechanism of action. Circulation 112(16): 2436–2445. https://doi.org/10.1161/CIRCULATIONAHA.104.528695
57. Henriksen EJ, Jacob S (2003) Modulation of metabolic control by angiotensin converting enzyme (ACE) inhibition. J Cell Physiol 196(1):171–179. https://doi.org/10.1002/jcp.10294
58. Oliver E, Mayor F Jr, D'Ocon P (2019) Beta-blockers: historical perspective and mechanisms of action. Revista espanola de cardiologia (English ed) 72(10):853–862. https://doi.org/10.1016/j.rec.2019.04.006

Chapter 2
Cardiac Dysfunction and Heart Failure Caused by Cancer Therapy-Induced Cardiotoxicity

Raluca Șoșdean, Silvia Giuchici, Loredana N. Ionică, Adina V. Furdui-Lința, Adrian Sturza, Danina M. Muntean, Daniel F. Lighezan, and Roberto Bolli

Abstract Cancer and heart failure rank among the leading causes of morbidity and mortality worldwide, posing significant public health challenges. Advances in cancer treatment, particularly for early-detected cases, have markedly improved survival rates. Over the past few decades, 5-year survival rates for many cancers have increased by nearly 10%. However, cancer therapies—alongside the disease itself—often result in cardiotoxicity, a critical complication that can lead to cardiac dysfunction and heart failure. The relationship between cancer and heart failure is multifaceted and interdependent, creating a vicious cycle where each condition worsens the progression and treatment outcomes of the other. The effects are cumulative, progressive and dose-dependent, with damage occurring during or years after therapy. Recognizing and managing cardiac dysfunction early is essential

R. Șoșdean
Department of Cardiology—2nd University Clinic of Cardiology, Research Centre of the Institute for Cardiovascular Diseases, "Victor Babeș" University of Medicine and Pharmacy, Timișoara, Romania
e-mail: sosdean.raluca@umft.ro

S. Giuchici · A. V. Furdui-Lința · A. Sturza · D. M. Muntean
Department of Functional Sciences—Chair of Pathophysiology, Centre for Translational Research and Systems Medicine, "Victor Babeș" University of Medicine and Pharmacy, Timișoara, Romania
e-mail: silvia.giuchici@umft.ro; adina.linta@umft.ro; sturza.adrian@umft.ro; daninamuntean@umft.ro

L. N. Ionică · D. F. Lighezan (✉)
Department of Internal Medicine—First University Clinic of Medical Semiotics, "Victor Babeș" University of Medicine and Pharmacy, Timișoara, Romania
e-mail: ionica.loredana@umft.ro; dlighezan@umft.ro

R. Bolli
Institute of Molecular Cardiology, University of Louisville Physicians Outpatient Center, Louisville, KY, USA
e-mail: rbolli@louisville.edu

D. K. Agrawal et al. (eds.), *Cardiovascular Toxicity*, Advances in Biochemistry in Health and Disease 33, https://doi.org/10.1007/978-3-031-90598-8_2

to preserve the survival benefits achieved with cancer care. Advances in heart failure treatments, coupled with the development of specialized cardio-oncology centres, have greatly improved the ability to address this complex interaction. This chapter summarizes the cardiac dysfunction and heart failure elicited by novel cancer therapeutics, emphasizing in the end the current management strategies aimed at optimizing patients' outcomes.

Keywords Cancer · Cardiac dysfunction · Heart failure · Cardiotoxicity · Cardio-oncology · Immune checkpoint inhibitors · CAR T-cell therapy · Human epidermal receptor 2-targeted therapy

Introduction

Cancer and heart failure are leading causes of morbidity and mortality worldwide. The treatment for both of these complex pathologies has evolved significantly during the last decades, improving patients' outcome. In some types of cancer, especially when detected early, the 5-year survival has increased almost 10%. For example, the 5-year survival for breast cancer diagnosed in the early stage has increased from 79% (1990) to 88% (2012), and this applies also to other solid and/or haematological cancers [1, 2].

Cancer is the major cause of cardiotoxicity, mainly due to current therapeutic approaches, but the disease itself may also be the cause of cardiac dysfunction and heart failure. A CDC WONDER database analysis from 1999 to 2020 showed increasing mortality rates in patients with cancer and heart failure in the United States [2]. Besides the direct effects on the myocardium, cancer therapy may also induce ischemia, hypertension, arrhythmias, and conduction abnormalities that may indirectly lead to heart failure. The incidence of these consequences depends largely on the patients' associated comorbidities and risk factors, type of cancer, type, combination and duration of therapy, and preexisting heart failure treatment. The association between cancer and heart failure is an important global public health problem, because it produces a vicious cycle in which these pathologies negatively affect each other's evolution and treatment [1, 3–7].

In some types of cancer, e.g. breast cancer, heart failure is an important contributor to mortality among cancer survivors, especially in the elderly, as recently reviewed [1, 8]. Besides myocardial dysfunction and heart failure, the cardiovascular side-effects of cancer/cancer therapy include promoting coronary artery disease, valvular disease, arrhythmias/conduction disorders, hypertension, pericardial disease, peripheral vascular disease, and thromboembolism [3].

Depending on the agent, the cardiac deleterious effects can be cumulative, progressive, and dose-dependent and were noted both during treatment as well as after the therapy was completed. Follow-up data in the cancer survivor paediatric population offer important information in this regard. In the Childhood Cancer Survivor Study [3, 4, 9, 10], cardiac death rate was about 8 times higher and congestive heart failure rate was about 15 times higher, as compared to the control group, during more than 15 years follow-up. This is the largest study in this regard, including more than 14,000 cancer survivors that were diagnosed between 1970 and 1986 in the United States and Canada, as well as about 10,000 diagnosed between 1987 and 1999.

Several agents may be responsible for cardiac dysfunction, with anthracycline occupying the leading positions. For breast cancer, the cumulative incidence of cardiac dysfunction was 38% with these agents and 32.5% with other agents in a 10-year follow-up, as demonstrated by the review of a Surveillance, Epidemiology, and End Results (SEER) database [1, 10]. On the other hand, Limat et al. demonstrated a heart failure incidence of 17% at 5 years follow-up in non-Hodgkin lymphoma survivors [11].

Early recognition of suggestive signs and symptoms along with comprehensive investigations for heart failure are of paramount importance for sustaining the survival benefit gained in cancer patients in the last decades. Although not always, cardiac dysfunction in cardiotoxicity may be reversible if detected and treated early, leading to a significant improvement in the patients' outcome, especially with the new advancements in heart failure treatment. The mainstay is to detect damage in the initial phase, when myocyte death and fibrosis are minimal. The development of cardio-oncological centres in the last decade is one of the most important achievements in this regard, facilitating better patient care from both sides [3, 4, 12].

The purpose of this chapter is to review the main factors that are involved in the cancer-heart failure crosstalk, along with several preventive strategies that are recommended by current guidelines in order to increase the awareness regarding this complex interaction.

Cancer-Related Cardiac Dysfunction and Heart Failure

Depending on its type and extension, cancer may induce besides the intense psychological burden, also an increase in the stress hormones, e.g., circulating catecholamines, along with systemic inflammatory responses, and thus neurohormonal changes. Some tumours, particularly those of the lung, breast, and pancreas, may release catecholamines or other stress mediators that precipitate myocardial injury [13, 14], often manifested as Takotsubo syndrome. Its pathophysiology is not completely elucidated but some key processes may be coronary artery spasm and severe microvascular dysfunction in the context of catecholamine excess, resulting in excessive reactive oxygen species production and inflammation. There are some hypotheses that adrenoreceptors in cardiac tissue may be modified by certain circulating paraneoplastic mediators, thus leading to contractile dysfunction. Giza et al. demonstrated that lymphoproliferative neoplasms were the most common type of cancer to induce Takotsubo cardiomyopathy (23.3%) followed by gastrointestinal (17%), ovarian (13.3%) and breast cancer (10%) [15]. High cardiac biomarker levels, repolarisation changes on the ECG suggesting acute myocardial infarction, typical apical ballooning aspect with a decreased ejection fraction in echography and/or cardiac magnetic resonance, lack of fibrosis and normal coronary arteries are key for diagnosis [4, 16]. Complete guideline directed heart failure treatment and transient interruption of the potential responsible and/or aggravating anticancer agent is needed when diagnosed [4]. Takotsubo syndrome appears to be more common than previously thought, highlighting the need for heightened awareness in cancer patients presenting with acute chest pain or unexplained heart failure symptoms [17]. Given its transient nature and the typically favourable prognosis, timely diagnosis and management are essential to optimize outcomes for these patients.

Over time, therapy of cancer has evolved from the administration of chemical compounds and radiation therapy to the use of targeted agents and immunotherapies.

Cardiac dysfunction is further augmented by different cancer therapies (chemo- and/or radiotherapy), as further detailed. The most studied cardiotoxic drugs are the anthracyclines whose deleterious effects are comprehensively detailed in another chapter.

Cancer Therapy-Related Cardiac Dysfunction and Heart Failure

Chimeric Antigen Receptor T Cell and Tumour Infiltrating Lymphocytes Therapies

Chimeric antigen receptor (CAR) T-cell therapy has demonstrated exceptional effectiveness in managing highly refractory and relapsing hematological malignancies in both pediatric and adult patients, offering a transformative option for cases that were previously challenging to treat. CAR T cells are patient-derived T lymphocytes that are genetically modified to recognize and target tumor-specific antigens, such as CD19 or B-cell maturation antigen (BCMA), thereby inducing apoptosis in tumor cells [18]. CAR T-cell therapy utilizes a genetically engineered CAR composed of three key components: an extracellular domain, a transmembrane domain, and an intracellular activation domain [19]. The extracellular domain, a single-chain variable fragment is derived from the variable heavy and light chains of an antibody and specifically binds to tumor antigens in a non-HLA-dependent manner. The transmembrane domain anchors the CAR in the T-cell membrane, while the intracellular domain integrates T-cell activation signaling domains, such as CD3, to enhance T-cell activation. The CAR construct is introduced into the patient's T cells using a lentiviral or retroviral vector, enabling the modified T cells to effectively recognize and attack tumor cells [20, 21].

Currently, CAR T-cell therapy is FDA-approved for treating patients with acute lymphoblastic leukemia (ALL), chronic lymphocytic leukemia, non-Hodgkin's lymphoma, and relapsed/refractory large B-cell lymphomas. However, interest in expanding its application has led to numerous studies and clinical trials exploring its use in other cancers, including multiple myeloma, mantle cell lymphoma, ovarian cancer, hepatocellular carcinoma, and other solid tumors. These investigations have shown promising results, suggesting a broader potential for CAR T-cell therapy in oncology [22].

Over the past decade, immunotherapy, particularly cell-based approaches, has revolutionized cancer treatment strategies. Recent clinical studies have shown promising efficacy in using tumor-infiltrating lymphocyte (TIL) therapy for advanced solid tumors. TIL therapy involves harnessing the patient's own immune cells, extracted from the tumor microenvironment, to target and destroy tumor cells, offering a personalized and effective approach to cancer immunotherapy [23].

One of the most frequent complications of CAR T-cell therapy is cytokine release syndrome (CRS), a systemic inflammatory response triggered by the excessive production of cytokines released by the proliferating and activated CAR T cells. CAR T-cell therapy and CRS are associated with several cardiovascular toxicities, driven by both direct and indirect mechanisms [24]. The direct mechanism involves the release of cytokines, which affects the cardiovascular system, commonly causing hypotension and tachycardia. In severe cases, these symptoms can progress to hemodynamic instability and shock. Hypotension has been reported in up to one-third of patients with CRS and generally responds to supportive care [25].The indirect mechanism arises from cross-reactivity against the cardiac protein titin, a key player in the contraction and relaxation of heart muscle. Affinity-enhanced T-cell receptors can mistakenly target an off-target peptide from titin, resulting in cardiac muscle toxicity, which is often severe and potentially fatal [26, 27]. Other cardiac toxicities associated with CAR T-cell therapy include elevated serum troponins and decreased left ventricular systolic function, potentially linked to tumor necrosis factor α release, which impairs myocyte shortening and induces mitochondrial dysfunction [28]. In a study by Kochenderfer et al., CAR T-cell therapy in patients with persistent B-cell malignancies post-allogenic hematopoietic cell transplantation showed a decrease in left ventricular ejection fraction (LVEF) within 12 days of infusion, lasting 4 months before returning to normal [29]. The JULIET trial reported no major cardiovascular events [30], but other trials demonstrated varying cardiovascular outcomes. In the ZUMA 1 trial, one patient died of cardiac arrest [31], while the ELIANA trial recorded three cardiac arrests (none fatal) and two cases of heart failure [32]. A study by Alvi et al. reported decreased LVEF in 28% of patients with echocardiograms pre- and post-CAR T-cell therapy, all of whom had CRS grade ≥ 2. Among these patients, there were 17 cardiovascular events, including 6 cardiovascular deaths (12%), 6 cases of decompensated heart failure (12%), and 5 cases of new-onset arrhythmias [33].

Karthikeyan et al. investigated and compared the cardiotoxicity profiles of CD19-targeted CAR T-cell therapies and blinatumomab, a CD19/CD3-targeted bispecific T-cell engager, using data from the World Health Organization pharmacovigilance database, VigiBase [34]. Both therapies have revolutionized the treatment of relapsed or refractory B-cell malignancies, but their distinct cardiac adverse event profiles remain underexplored. The study focused on adverse cardiac events reported in patients with diffuse large B-cell lymphoma (DLBCL) and acute lymphocytic leukemia (ALL). Tisagenlecleucel, a CAR T-cell therapy, was linked to high fatality rates associated with cardiac failure (85.7% in DLBCL and 80.0% in pediatric ALL) and severe hypotension, especially in DLBCL patients, where it demonstrated significantly higher fatality rates for hypotension (50.0%) compared to axicabtagene ciloleucel (5.6%). In pediatric ALL, hypotension fatality rates for tisagenlecleucel (34.7%) were similar to those for blinatumomab (20.0%). These findings highlight the substantial risk of life-threatening cardiac complications with CAR T-cell therapies, particularly tisagenlecleucel, and emphasize the need for interdisciplinary management strategies.

Zhai et al. analyzed the cardiotoxicity associated with CD19-directed CAR T-cell therapies using data from the US FDA's Adverse Event Reporting System database [35]. The findings highlight significant cardiac risks, with tisagenlecleucel demonstrating the highest rates of mortality (53.24%) and life-threatening events (13.39%). Both tisagenlecleucel and axicabtagene ciloleucel exhibited multiple cardiac event signals, although axicabtagene ciloleucel was associated with higher reporting rates for specific conditions such as atrial fibrillation, cardiomyopathy, cardiorenal syndrome, and sinus bradycardia.

In a prospective study, conducted by Hu et al., the changes in cardiac biomarkers were evaluated in 40 patients undergoing CAR T-cell therapy [36]. The authors focused on high-sensitivity troponin T (hsTropT) and N-terminal pro-B natriuretic peptide (NTproBNP) levels at baseline and on days 1,7, and 21 post-treatment. While hsTropT levels remained stable across all timepoints and were unaffected by grade 2 CRS, NTproBNP levels significantly increased on day 1 and day 7 and were further elevated in patients with grade 2 CRS. By day 21, NTproBNP levels returned to baseline. These findings suggest that NTproBNP may reflect acute cardiac stress, particularly in patients with more severe CRS, while hsTropT does not appear to indicate acute myocardial injury in this setting.

Goldman et al. performed a large retrospective pharmacovigilance study to investigate cardiovascular and pulmonary adverse events (CPAEs) associated with CD19-directed CAR-T therapy [37]. Among 2657 CAR-T recipients, 20.5% experienced CPAEs, with 68.3% overlapping CRS. Significant CPAEs included tachyarrhythmias (2.8%), cardiomyopathy (2.6%), pleural disorders (1.7%), and pericardial diseases (0.4%). Venous thromboembolic events (VTEs) were linked specifically to axicabtagene-ciloleucel. Atrial fibrillation was the most reported tachyarrhythmia, followed by ventricular arrhythmias. CPAEs occurred more frequently with axicabtagene-ciloleucel than tisagenlecleucel, with CPAEs carrying a 30.9% fatality rate.

In a novel study, Daryanani et al. explored the association between baseline echocardiographic parameters and the development of hemodynamically significant Cytokine Release Syndrome (CRS $\geq$2) in patients undergoing CD19 CAR T-cell therapy for relapsed or refractory B-cell hematological malignancies [38]. CRS is a common and potentially severe immune-related complication of CAR T-cell therapy, potentially leading to cardiac dysfunction. Given the critical role of cardiac reserve in tolerating CRS-related stress, this study aimed to identify echocardiographic predictors of severe CRS, assess cardiac function during follow-up, and evaluate the incidence of cancer therapy-related cardiac dysfunction (CTRCD). Among 291 patients analyzed, no significant baseline echocardiographic markers, including left ventricular ejection fraction and global longitudinal strain, predicted the development of CRS ≥2. Overall, systolic and diastolic function remained stable during follow-up in both CRS groups. CTRCD occurred in 4.5% of patients, predominantly in those with severe CRS, suggesting a multifactorial mechanism underlying hemodynamic instability rather than direct cardiac impairment.

Arrhythmias are among the most frequently reported complications of CAR T-cell therapy, often occurring in the context of CRS. The most commonly observed arrhythmias are asymptomatic prolonged QTc and atrial fibrillation [39, 40]. In the study by Alvi et al., five patients experienced new-onset arrhythmias, including supraventricular tachycardia in two cases and new-onset atrial fibrillation in three cases. Notably, three patients in the study died due to complications stemming from new-onset heart failure and/or tachyarrhythmias complicated by shock [33]. The mechanisms underlying these arrhythmias are not fully understood, but it is hypothesized that they may result from a direct cytokine effect on the cardiac electrical system. Similar phenomena have been observed in patients with inflammatory conditions such as rheumatoid arthritis [22], myocarditis, and those undergoing cardiac surgery [41]. A link between severe inflammatory states and arrhythmias has been observed, with elevated levels of inflammatory biomarkers such as C-reactive protein (CRP) [42], interleukin-6 (IL-6), leukocyte count and complement activation being associated with an increased incidence of atrial tachyarrhythmias.

Management of specific cardiovascular complications should follow the European Society of Cardiology (ESC) Guidelines, complemented by targeted treatment of CRS. This includes the use of therapies such as the anti-interleukin-6 receptor antibody tocilizumab and dexamethasone to mitigate inflammatory responses. Although cardiovascular complications are common with tumor-infiltrating lymphocyte (TIL) therapies, they do not appear to significantly impact survival. The most frequent CV events include hypotension, which may require treatment with intravenous fluids and pressors, atrial fibrillation, and, less commonly, elevations in cardiac troponin suggestive of myocardial damage. Further research is necessary to elucidate the mechanisms underlying these complications and to develop effective prevention strategies to aid clinicians in managing these CV events.

Human Epidermal Receptor 2-Targeted Therapy

Breast Cancer-Related Cardiotoxicity

Breast cancer, is divided according to the expression of oestrogen receptor, progesterone receptor and human epidermal growth factor receptor 2 (HER2). The resulting subtypes are Luminal A, Luminal B, HER2-positive and basal-like (triple negative), which are used to assess prognosis and guide oncologic therapy [43, 44]. Epidermal growth factor receptor 2 (ErbB2/HER2) together with ErbB1, ErbB3, and ErbB4, comprise the family of human epidermal growth factor receptor family. Their activation is closely linked with activation of the signalling pathways that stimulate tumour growth, being associated with a worse clinical outcome and prognosis [45].

In the past 20 years, the expression of HER2 in breast cancer has led to the use of targeted therapies that have had a tremendous impact on survival; however, side effects of these targeted therapies include cardiotoxicity, a clinical outcome that can still represent a challenge [46, 47]. At a cellular level, HER2 plays an important role in cardiomyocyte development. Studies involving HER2 blockade in murine models showed findings such as dilated cardiomyopathy, chamber dilation, wall thinning and decreased contractility, supporting the concept that ErbB2 signalling in cardiomyocytes is a requisite in the prevention of dilated cardiomyopathy [48]. Furthermore, HER2 plays a significant role in oxidative stress management: blockade of the anti-oxidative function mediated by neureregulin interaction with HER2 is considered responsible for the activation of the glutathione reductase system in the heart [49]. Perhaps the most important distinction in chemotherapy-related cardiac dysfunction (CRCD) is that anti-HER therapies are responsible for generating type II CRCD which is largely considered reversible and dose independent, compared to anthracycline regimens which generate type I CRCD. By inhibiting the ErbB2 pathway, growth and repair mechanisms required by cardiomyocytes for the maintenance of cardiac contractility, structure and function will be altered [50].

More than 25 years after it was approved, trastuzumab is still among the most used anti-HER therapies, either alone or in combination with other anti-HER therapies and/or chemotherapy with a broad spectrum in both early disease and locally advanced or metastatic settings [66]. A retrospective study that included 10,211 patients illustrated that among women younger than 65 treated with anthracyclines and/or trastuzumab, after a median of 5.79 years, 14.67% developed CVD and up to 16% developed CVD by 10 years [51]. Another study focused on cardiovascular events upon administering trastuzumab together with lapatinib (a reversible tyrosine kinase inhibitor) compared to trastuzumab monotherapy, and further showed that cardiac events are asymptomatic, occurring during anti-HER2 treatment and further increase from the 6th to the 12th month of treatment. Based on previous pre-clinical studies, it is thought that lapatinib provides a cardioprotective effect via activation of AMPK, protecting against apoptosis induced by TNFα, a known cytokine detected in cardiac failure [52]. Treatment with trastuzumab represents a viable option even in patients with pre-existing asymptomatic LVSD (LVEF = 40–53%), as suggested by a retrospective study showing that LVEF reduction of more than 10% occurred in 16.7% of the patients, and GLS (global longitudinal strain) reduction of more than 15% occurred in 11.1% [53]. The use of trastuzumab-deruxtecan, an antibody drug-conjugate has gained rapid approvals after promising results of DESTINY Breast01 Trial in women with metastatic breast cancer who were previously treated with other anti-HER2 regimens. In contrast with other anti-HER2 targeted therapies, trastuzumab deruxtecan was associated with fewer clinically significant cardiomyopathies, being infrequently observed, with only a total of four events of LVEF decrease occurring in the study [54]. Risk assessment of patients plays a crucial role in identifying risk factors, with previous anthracycline treatment being responsible for the majority of trastuzumab-induced cardiotoxicity. A well documented history together with proper evaluation by a cardio-oncology specialist including imaging techniques and use of biomarkers (troponins and pro-BNP) and an individualised treatment plan that incorporates effective prophylactic therapy such as beta blockers or angiotensin-converting enzyme inhibitors can provide mitigation of cardiac toxicity in selected cases [55].

Pertuzumab is another widely used anti-HER2 agent; however fewer studies were conducted regarding its potential cardiotoxic profile, compared to trastuzumab. A metaanalysis and systematic review noted that pertuzumab was associated with an approximately twofold increased risk of heart failure, with no detectable effect on asymptomatic or minimally symptomatic LVSD (left ventricular systolic dysfunction) [56]. The APHINITY trial showed that double HER2 blockade with trastuzumab and pertuzumab in HER2 early breast cancer patients together with chemotherapy resulted in an increased rate of complete pathological response. For cardiovascular monitoring, the study required an assessment of the LVEF every 3 months during treatment, every 6 months up to month 36, and yearly thereafter [57]. The approval of the combined regimen naturally raised the question of whether there will be an increase in cardiovascular related events. In the following study which included patients from the APHINITY trial in a 74-month median follow-up, there was no additional risk of the dual blockade with trastuzumab and pertuzumab, and most cardiovascular events were asymptomatic or mildly symptomatic decreases in LVEF occurring during treatment. However, the risk of cardiovascular events was higher in the group previously treated with anthracyclines [58].

In a very interesting study, researchers investigated the role of microRNA-130a (miR-130a) as a biomarker for cardiotoxicity in HER2-positive breast cancer patients undergoing epirubicin/cyclophosphamide followed by docetaxel plus trastuzumab (EC-D + T) adjuvant chemotherapy [59]. A total of 72 patients were monitored for cardiotoxic events, including declines in LVEF, heart failure, acute coronary syndromes, and arrhythmias, over a 15-month treatment period. The cumulative cardiotoxicity rate was 16.7%, primarily due to reductions in LVEF. Baseline miR-130a levels were negatively correlated with LVEF and positively correlated with cardiac troponin I, indicating myocardial stress. miR-130a expression increased throughout treatment, with more pronounced elevations in patients who developed cardiotoxicity. Further analysis identified miR-130a as an independent predictor of cardiotoxicity risk, suggesting its potential as a biomarker for cardiac complications in HER2-positive breast cancer patients receiving EC-D + T therapy.

Zhao et al. conducted a comprehensive analysis to assess the cardiotoxic effects of HER-2 inhibitors in breast cancer patients using real-world data from the FDA adverse event database [60]. Their study addressed prior limitations of smaller clinical trials and case reports by evaluating 167,639 patients diagnosed between 2004 and 2021. Patients were divided into three groups-HER-2 inhibitor, positive control, and control groups—and cardiovascular adverse events (CVAEs) were analyzed using a disproportionality method. Notably, 13.5% of patients treated with HER-2 inhibitors experienced CVAEs, with cardiomyopathy observed in all types of HER-2 inhibitors except trastuzumab deruxtecan. Trastuzumab was most frequently associated with CVAEs, occurring within a median of 80.5 days. These findings emphasize the importance of early cardiac monitoring, particularly within the first 6 months of anti-HER therapy, to mitigate associated cardiovascular risks.

Digestive Cancers-Related Toxicity

The overexpression of HER2 in gastric cancers in up to 20% of adenocarcinomas facilitates the use of Trastuzumab as targeted therapy [61, 62]. Currently, for gastric adenocarcinoma with HER2 overexpression the use of Trastuzumab together with a platinum and fluoropyrimidine based chemotherapy regimen should be regarded as first line treatment [63]. Moreover, the DESTINY Gastric 01 trial evaluated the use of an antibody drug-conjugate trastuzumab-deruxtecan in locally advanced or metastatic gastric or gastroesophageal junction cancer that had progressed after the patient had received at least two previous regimens, which included a fluoropyrimidine, a platinum agent, and trastuzumab [64]. The phase III ToGa study which evaluated the use of Trastuzumab together with chemotherapy has only reported a low occurrence of cardiotoxicity (5% in the trastuzumab plus chemotherapy group and 1% in the chemotherapy group) without the mentioning of LVEF profiles of these cases [63]. A Korean study conducted on 115 patients assessed trastuzumab-induced cardiotoxicity compared to chemotherapy treated patients. Findings showed that symptomatic heart failure was not observed in either group, although there was a significant asymptomatic drop in LVEF was noted in 5 (7.1%) of the trastuzumab combined-group patients (evaluated by echocardiography or by multigated acquisition scan). There is also a disparity between patient cardiotoxicity profiles, considering that factors like chest radiation, use of anthracyclines and obesity are not common findings in gastric cancer patients [65, 66]. Similar results were noted in another study where grade 1 cardiotoxicity was reported in 6% of the studied group treated with a combination of trastuzumab and capecitabine [67].

Immune Check Point Inhibitor Therapy

Immune check point inhibitors (ICIs) are humanised monoclonal antibodies that target immune checkpoint proteins, reactivating the immune response against tumour cells. The targeted proteins are co-inhibitory receptors that negatively regulate the immune response mediated by T cells; these receptors are implicated by various mechanisms in tumour tolerance and T cell exhaustion and are reactivated by ICIs [68–70].

ICIs are successfully used in different types of cancers like non-small lung cancer, melanoma, renal carcinoma, head and neck cancers, either alone or combined with chemotherapy and/or radiotherapy. There is an increasing number of patients in need for this type of therapy. In the United States alone there are almost 600,000 ICI-eligible patients [68, 71]. ICIs are classified in three categories: programmed cell death 1 (PD-1) inhibitors (i.e., Pembrolizumab), programmed death-ligand-1 (PD-L1) inhibitors (i.e. Atezolimumab), and cytotoxic T lymphocyte antigen 4 (CTLA-4) inhibitor (Ipilimumab) [68].

Unfortunately, by non-specifically activating the immune system, they also induce side effects various organs including, among others, cardiac toxicity and heart failure. Combination therapy is prone to a greater risk of cardiac adverse events that consist of myocarditis, but also of other non-inflammatory processes such as arrhythmias (ventricular and/or supraventricular), conduction defects including heart block, cardiomyopathy and pericardial effusion and/or pericarditis [72–75]. Approximately 45% of patients experience severe side effects, and cardiac toxicity, although rare, is often lethal [76], especially in the setting of myocarditis. ICI-induced myocarditis has a reported mortality rate of 40–50%, much higher than viral myocarditis [72, 74, 75]. Cardiotoxicity may be encountered as early as with the first cycle, but also late after the therapy has been interrupted, as the immune check point proteins inhibition may be prolonged and/or a chronic erroneous action of the immune system induced by ICIs may develop. The mean period till side effects manifest is after 2–3 cycles [74, 77, 78].

The mechanisms of cardiotoxicity are not fully elucidated, but several hypotheses exist regarding the erroneous immune responses, as T-cell infiltration was demonstrated in tissues with high expression of T cell receptors like tumour tissue, skeletal muscles and myocardium, frequently inducing myocyte degeneration and necrosis. These consist in targeting an antigen that is shared by these tissues (like desmin and/or troponin) or a different antigen (which may or may not be similar to the tumoral one) which is erroneously recognised by the T cells. T cell targets could also be generated by subclinical viral infections and/or certain HLA alleles, but these theories have not yet been demonstrated and further studies are warranted in this direction [71, 79].

Several early preclinical studies addressed the role of immune checkpoint proteins in the development of myocarditis. Tarrio et al. demonstrated that PD1 deficient mice are prone to a more severe disease, with increased inflammation, higher values of myocardial damage serum markers and a substantial infiltration of inflammatory cells including CD4 and CD8 positive T cells. The also demonstrated an in vitro increased response of CD4 T cells against myocardial cells [80]. Also, Okazaki et al. demonstrated that IgG autoantibodies against cardiac troponin I induces cardiomyopathy in PD-1 deficient mice, by chronically augmenting the calcium influx in normal cardiomyocytes [81]. Johnson et al. failed to demonstrate this mechanism in two patients with fulminant ICI-induced myocarditis, treated for melanoma, suggesting a significant variability of the incriminated mechanisms among patients. In their biopsy samples they demonstrated no autoantibody deposition in the injured myocardium, but an increased expression of PD-L1 which is sought to be a cardioprotective mechanism that may be reversed by ICIs. There was a notable infiltration of T-cell and macrophage in the myocardium, conduction system and skeletal muscle that indicated lymphocytic myocarditis and myositis [79]. This was consistent with the results of Grabie et al. that demonstrated that PD-L1 is very important for the control of immune mediated cardiac injury in PD-L1/L2 deficient mice. Deletion and/or blockade of PD- L1 induced a lethal myocarditis with polymorphonuclear leukocyte-rich abscesses [82]. Supplemental studies are warranted also in this direction.

ICI-induced myocarditis in itself proved to be highly arrhythmogenic and ECG abnormalities such as conduction blocks, repolarization anomalies, Q waves and decreased voltage seem to predispose to malignant arrhythmias and increased all-cause mortality. This may be explained by the fact they reflect a more advanced lymphocyte infiltration of the myocardium and conduction system, decreasing the electromotive force and acting on the inflammation is much more important than the use of antiarrhythmic drugs and/or devices in this situation [74, 83]. Other factors associated with a poor prognosis include preexisting cardiovascular diseases, a low GLS (modular values $<13\%$ when LVEF is $<50\%$, and $<16\%$ when LVEF is $>50\%$), increased T1 values and septal late gadolinium enhancement on CMR, increased troponin levels, absolute lymphocyte values decrease $\geq 35\%$ to admission and neutrophil/lymphocyte ratio increase $\geq 100\%$ to admission, increased CD3+ lymphocyte infiltration on byoptic fragments and treatment delay [84]. All of these investigations are recommended in this patients, according to their availability, although cardiac biopsy is less used as being an invasive method with a high rate of complications [4]. Generally, the treatment consists in high dose steroid anti inflammatory drugs and it was demonstrated that a higher initial dose is associated with a lower troponin level at discharge and a lower incidence, but not 100% prevention of major cardiac adverse events. Other recommended therapies consist in intravenous immunoglobulin, mycophenolate, antithymocyte globulin, infliximab [4, 71]. The ICI therapy is interrupted while investigations and treatment are being performed and the decision of reinitiation should be taken after multidisciplinary team discussion according to the risk/benefit balance [4].

Mascolo et al. investigated the occurrence of ICI-induced cardiac adverse drug reactions reported in the European pharmacovigilance database (Eudravigilance) until March 14, 2020. Among 2478 individual case safety reports involving ICIs, 3388 cardiac adverse effects were identified, with 99.4% classified as serious and 30.1% resulting in fatal outcomes. The most frequently reported events included myocarditis, cardiac failure, atrial fibrillation, pericardial effusion, and myocardial infarction. Nivolumab, pembrolizumab, and the combination of nivolumab/ ipilimumab were the most reported ICIs, with nivolumab showing a slightly higher reporting frequency compared to other ICIs. In the same year, Chen et al. published a study in which the authors investigated the association between ICIs and cardiotoxicity, analyzing data from the FDA Adverse Event Reporting System between 2014 and 2019 [85]. Among 7,443,137 cases and 36,326,611 drug-adverse event pairs, 9271 cases of ICI-induced cardiotoxicities were identified, with a higher prevalence and severity in males compared to females. Commonly reported cardiac events included dyspnea, myocarditis, atrial fibrillation, cardiac failure, and pericardial effusion, with myocarditis emerging as a high-risk complication associated with all ICIs.

In a recent study, Milutinovic et al. analyzed the adverse effects of ICIs using the Food and Drug Administration adverse event reporting database, with a focus on comparing pembrolizumab with other ICIs [86]. The investigation of 6719 ICI-related cardiac adverse events revealed that all cases were serious, with a mortality rate of 34.3%. Pembrolizumab was more frequently linked to myocarditis, pericardial disease, heart failure, and atrial fibrillation compared to other medications.

Heart failure without a direct inflammation of the myocardium may be induced by certain types of arrhythmias (i.e. atrial fibrillation secondary to ICI-induced thyrotoxicosis and/or severe systemic inflammatory syndromes), non-inflammatory left ventricular dysfunction, Tako-Tsubo syndrome, pericarditis with pericardial effusion, vasculitis, and even advanced coronary atherosclerosis by inducing an advanced plaque inflammation and modifying its composition, producing arterial and/or venous thrombosis, and inducing various types of coronary syndromes consequently affecting the myocardium [4, 84]. Inflammation of these other structures is treated with steroid therapy, whereas non-inflammatory heart failure syndromes should benefit from standard HF guideline directed therapy [4].

Haematopoietic Stem Cell Transplantation

Hematopoietic stem cell transplantation (HSCT) has become a cornerstone treatment for a range of hematologic malignancies, including leukaemias, lymphomas, and myelodysplastic syndromes. Chemotherapy and conditioning regimens include anthracyclines, alkylating agents, methotrexate and busulfan, and often total body irradiation. These serve to eradicate malignant cells, suppress immune function to prevent rejection, and prepare the body for engraftment of the newly transplanted hematopoietic stem cells. Despite its life-saving potential, HSCT may induce cardiotoxicity by various mechanisms oxidative stress, endothelial dysfunction, and direct damage to cardiac muscle cells, manifesting as congestive heart failure, arterial events (coronary syndromes), cardiac tamponade, arrhythmias (especially supraventricular- atrial fibrillation, flutter, tachycardias). The timing of cardiotoxicity in HSCT patients can vary, with both early and late manifestations observed. Acute cardiotoxicity often occurs during or shortly after the completion of the conditioning regimen, manifesting as asymptomatic left ventricular dysfunction and/or symptoms of heart failure, but chronic cardiotoxicity, on the other hand, may develop years after the transplant and often presents as a progressive form of heart failure [4, 87].

Risk factors for developing cardiotoxicity include cumulative drug doses, the type of chemotherapy used, pre-existing cardiac conditions, and the patient's age. Older patients are particularly vulnerable to chemotherapy-induced cardiac dysfunction due to age-related changes in myocardial resilience and reduced capacity for repair. Pre-existing cardiovascular comorbidities, such as hypertension or diabetes can exacerbate the negative effects of chemotherapeutic agents on the heart and may consequently increase the risk for developing cardiotoxicity [88].

Alkylating agents, such as cyclophosphamide, have been implicated in both acute and long-term cardiotoxicity, which is typically dose-dependent and it is usually encountered in patients receiving more than 140 mg/kg; however, even moderate doses can lead to long-term cardiac side effects, particularly in patients who have received regimens for severe malignancies [3, 89]. Incidence of cardiotoxicity ranges between 7% and 28% whereas mortality rates range between 11% and 43%. Cyclophosphamide is metabolised in the liver after administration, resulting in aldophosphamide production, which subsequently forms phosphoramide, an active antineoplastic agent and acrolein, a toxic metabolism byproduct which was demonstrated to induce myocardial and endothelial cell injury. The pathogenic mechanisms include alteration of the heart fatty acid binding proteins (H-FABP) with consequent alteration of cardio—energetics and depletion of antioxidants and ATP levels, inducing an increased production of reactive oxygen species (ROS), inflammation, mitochondrial and sarcoplasmic reticulum dysfunction and eventually cell apoptosis. These molecular disturbances will reflect in contractility alteration, cardiomyopathy, different types of coronary syndromes including myocardial infarction, and heart failure [90].

Other agents, including methotrexate and busulfan, have also been associated with cardiotoxicity, though their effects tend to be less prominent compared to anthracyclines and alkylating agents. Methotrexate was even demonstrated to generally have cardioprotective effects like improved endothelial cell function and reduced progression of atherosclerosis by its antiinflammatory, antiproliferative and immunosuppressive effects [91]. There are some case reports, though, incriminating methotrexate to induce cardiomyocyte apoptosis when used in high doses as part of a conditioning regimen [92], while busulfan, through its potent immunosuppressive properties, can predispose to vascular dysfunction and the development of secondary coronary artery disease [93].

The use of total body irradiation (TBI), often included in the conditioning regimen for certain malignancies, is difficult to evaluate as it is usually associated with different regimens of chemotherapy and the effects usually have a late onset after exposure. However, some studies found an increased relative risk of adverse cardiovascular events and absolute mortality. The risk of heart failure increased by 4.9-fold in survivors, due to interstitial myocardial fibrosis of variable size and distribution along with radiation induced endothelial damage with valvular heart disease and coronary artery disease. Left ventricular systolic dysfunction was usually encountered when associated with anthracycline therapy [3, 94].

Although rare, the cardiotoxic effects of chemotherapy and conditioning regimens used in HSCT are of particular concern, because they can lead to long-term cardiac morbidity and mortality and should be treated as per specific guideline-directed therapies. As for prevention, ACEi and beta-blocker may be useful, but data in the literature are scarce and further studies are needed in this direction [4]. In high-risk patients, the use of cardioprotective agents such as dexrazoxane has been shown to reduce the incidence of anthracycline-induced cardiotoxicity, although its use remains controversial due to potential concerns about its effect on stem cell engraftment [95].

Other Agents

Agents that inhibit vascular endothelial growth factor (VEGF) signalling and proteasome inhibitors also can induce cardiotoxicity, although the risk is lower compared with other agents previously described in this chapter. Anti VEGF agents were demonstrated to induce cardiac dysfunction in 3–15% of patients, whereas proteasome inhibitors in 4–25% depending on the agent being used [3].

Inhibition of Vascular Endothelial Growth Factor (VEGF) Signalling

VEGF inhibitors (anti-VEGF antibodies—i.e. bevacizumab and tyrosine kinase inhibitors—i.e. sunitinib, sorafenib, axitinib, pazopanib) are used for several types of solid cancers usually associated with metastases (glioblastoma, renal, colorectal, hepatocellular carcinoma, thyroid, sarcoma, etc.) and they act by inhibiting angiogenesis and neoformation blood vessels in the tumour. Sometimes they may induce reversible and/or irreversible cardiotoxicity, consisting in left ventricular dysfunction and heart failure especially when associated with other chemotherapeutic drugs and radiotherapy. The pathogenic mechanisms are not fully elucidated but the most likely is mitochondrial dysfunction along with inhibition of the AMP-kinase pathway. Since endomyocardial biopsies do not demonstrate fibrosis, there is a high probability of reversibility. Besides left ventricular dysfunction, these agents have been found to induce arterial hypertension and progressive atherosclerosis, resulting in myocardial ischemia, arrhythmias and thromboembolic diseases [3, 96].

Proteasome Inhibitors

These agents, represented by bortezomib and carfilzomib, are a new therapeutic line for multiple myeloma, Waldenström's macroglobulinemia and other malignancies. By inhibiting the proteasomes—the protein complexes responsible for cardiomyocyte maintenance—they may induce serious cardiac side effects secondary to cell cycle arrest and/or apoptosis, with a significantly higher risk associated to carfilzomib. Besides heart failure, cardiovascular toxicity also includes arrhythmias, arterial hypertension, and different types of coronary syndromes [3, 97].

In Table 2.1, the cardiotoxic effects of the novel chemotherapeutic agents are summarized.

Table 2.1 Overview of chemotherapy-related cardiotoxicity

Chimeric antigen receptor T cell and tumor infiltrating lymphocytes therapies		
Kochenderfer et al., 2013 [29]	10	CAR T-cell therapy linked to temporary decrease in LVEF, resolving in 4 months
ZUMA-1 Trial, 2017 [31]	101	1 cardiac arrest death reported in patients receiving CAR T-cell therapy
ELIANA Trial, 2018 [32]	75	3 cardiac arrests (non-fatal) and 2 cases of heart failure documented in pediatric CAR T-cell recipients
Alvi et al., 2020 [33]	137	17 cardiovascular events (12% mortality), with arrhythmias and decompensated heart failure in patients undergoing CAR T-cell therapy
Hu et al., 2021 [36]	40	No significant changes in hsTropT levels were observed across timepoints (baseline, days 1, 7, and 21) and no association was found between hsTropT levels and the presence of grade 2 cytokine release syndrome; significant increases in NTproBNP levels were noted (day 1 and 7) with NTproBNP elevation being higher in patients with grade 2 CRS; levels returned to baseline by day 21
Goldman et al. 2021 [37]	2657	20.5% of patients (n = 546) experienced cardiovascular and pulmonary adverse events (CPAEs); specific CPAEs with elevated risk: tachyarrhythmias (2.8%), with atrial fibrillation as the most common, cardiomyopathy (2.6%), pleural disorders (1.7%) and pericardial diseases (0.4%)
Karthikeyan et al., 2024 [34]	–	Tisagenlecleucel was associated with high fatality rates due to cardiac failure: 85.7% in DLBCL patients and 80.0% in pediatric ALL patients
Zhai et al., 2023 [35]	–	Tisagenlecleucel had the highest percentage of death (53.24%) and life-threatening events (13.39%)
Daryanani et al., 2024 [38]	297	The incidence of cancer therapy-related cardiac dysfunction was 4.5% and occurred mostly in the setting of cytokine release syndrome ≥2
Human epidermal receptor 2-targeted therapy		
Vo et al., 2024 [51]	2712	Higher risk of CVD (cardiomyopathy and/or HF) in patients receiving anthracyclines and/or trastuzumab compared with no chemotherapy, especially for those aged younger than 65 years, persisting at least 5 years postdiagnosis; increased ischemic heart disease risks after 5 or more years
Modi et al., 2020 [54]	184	4 events of LVEF decrease in patients receiving trastuzumab deruxtecan
Alhussein et al., 2021 [56]	–	An approximately twofold increased risk of HF, with no detectable effect on asymptomatic or minimally symptomatic LVSD in patients receiving pertuzumab

(continued)

Table 2.1 (continued)

de Azambuja et al., 2023 [58]	4769	Cardiac events observed in 159 patients: 83 in pertuzumab + trastuzumab and 76 in trastuzumab arms; Most cardiac events occurred during anti-HER2 therapy and were asymptomatic or mildly symptomatic decreases in LVEF; 2 cardiac deaths in each arm
Bang et al., 2010 [63]	584	Low occurrence of cardiotoxicity (5% in the trastuzumab plus chemotherapy group and 1% in the chemotherapy group)
Soo Park et al., 2017 [65]	115	Symptomatic heart failure was not observed in either group, but a significant asymptomatic drop in LVEF was noted in 5 (7.1%) of the trastuzumab combined-group patients and in 1 (2.2%) chemotherapy-only group patient Trastuzumab-induced cardiotoxicity was observed more frequently in elderly patients than in younger patients
Gürbüz et al., 2022 [67]	130	Grade 1 cardiotoxicity reported in 6% of the studied group treated with a combination of trastuzumab and capecitabine
Feng et al., 2021 [59]	72	Cardiotoxicity incidence was 16.7% (n = 12), primarily due to a ≥10% decline in LVEF and LVEF <53%; acute coronary syndrome occurred in 1.4% (n = 1), while no cases of heart failure or life-threatening arrhythmias were reported; baseline miR-130a expression was negatively correlated with LVEF and positively correlated with cardiac troponin I; miR-130a levels increased throughout treatment in all patients but showed a greater rise in cardiotoxicity cases
Zhao et al., 2023 [60]	167,639	13.5% of patients in the HER-2 inhibitor group (n = 2529) experienced CVAEs, with 81.5% of reports submitted by healthcare professionals; cardiomyopathy was linked to all HER-2 inhibitors except trastuzumab deruxtecan; trastuzumab had the highest number of CVAE reports (n = 2075)
Immune checkpoint inhibitor (ICI) therapy		
Li et al., 2022 [76]	5518 patients that received at least 1 ICI	12.5% developed cardiotoxicity in the Ist year; arrhythmias 9.3%, myocarditis 2.1% Monotherapy with ipilimumab or pembrolizumab had a higher risk of cardiotoxicity compared to nivolumab
Johnson et al., 2016 [79]	17,620	ICI-associated myocarditis 0.06% for nivolumab monotherapy (<0.01% fatal events); ICI-associated myocarditis 0.27% for a nivolumab and ipilimumab combination (0.17% fatal events)
Mahmood et al., 2018 [71]	140	ICI-associated myocarditis 1.14% across multiple ICIs and combinations
Power et al., 2021 [83]	125	ICI-associated myocarditis—highly arrhythmogenic; all-cause mortality associated with ECG markers (heart blocks, q waves, low voltage, ventricular arrhythmias)

(continued)

Table 2.1 (continued)

Salem et al., 2018 [98]	–	Pericardial diseases 0.36% of all reported adverse events, exact ICIs related incidence unknown; fatality rate of 21% Vasculitis 0.26% of all reported adverse events exact ICIs related incidence unknown; mortality rate of 6% in patients with ICI induced vasculitis
Drobni et al., 2022 [99]	2842	2-year incidence rate of atherosclerotic cardiovascular event after ICI treatment onset 6.55 per 100 person-years
Rubio-Infante et al., 2021 [100]	4751	ICI induced cardiovascular adverse events of 3.1% (monotherapy) and 5.8% (combination therapies)
Mascolo et al., 2021 [101]	2478	A total of 3388 cardiac ADRs were identified; the majority of cases (99.4%) were classified as serious; fatal outcomes were reported in 30.1% of cases; the most commonly reported cardiac events were myocarditis, cardiac failure, atrial fibrillation, pericardial effusion, and myocardial infarction; the most frequently reported ICIs were nivolumab (43.2%), pembrolizumab (32.5%), and the combination of nivolumab/ipilimumab (9.4%); nivolumab showed a slightly increased reporting frequency of cardiac ADRs compared to other ICIs (OR 1.09, 95% CI 1.01–1.18)
Chen et al., 2021 [85]	7,443,137 cases and 36,326,611 drug-adverse event pairs	Cardiotoxicities, identified in 9271 cases Male patients exhibited a higher frequency and severity of ICI-related cardiac adverse events compared to females; common cardiac adverse events: dyspnea (21.25%), myocarditis (5.16%), atrial fibrillation (4.84%), cardiac failure (4.00%) and pericardial effusion (3.56%)
Milutinovic et al., 2024 [86]	6719 ICI-related cardiac adverse events	Serious outcomes were reported in 100% of the cases, with 34.3% of the cases ending fatally; compared with all other medications in the database, pembrolizumab use was more frequently associated with myocarditis, pericardial disease, heart failure, and atrial fibrillation
Hematopoietic stem cell transplantation		
Baker et al., 2020 [87]	663	Cardiac events—9.8% of patients Atrial dilation—increased risk of arrhythmia Increased 1 year mortality in patients developing cardiac events
Armstrong et al., 2015 [94]	1820	Diastolic dysfunction in 22% of survivors exposed to radiotherapy alone Reduced exercise capacity in 27.4% exposed to radiotherapy alone

ADRs adverse drug reactions, *ALL* acute lymphocytic leukemia, *CVAEs* cardiovascular adverse events, *CRS* cytokine releasing syndrome, *DLBCL* diffuse large B-cell lymphoma, *ICI* immune checkpoint inhibitor

Long-Term Follow-Up and Chronic Cardiovascular Complications

Since certain chemotherapic agents may induce cardiac dysfunction and heart failure even long after therapy completion, current guidelines recommend long-term surveillance of these patients according to their risk of cancer therapy-induced cardiac dysfunction. This is critical for detecting recurrence, managing late effects of treatment, and improving survival outcomes. For example, in asyptomatic moderate risk patients it is recommended to perform a transtoracic echocardiography every 5 years, whereas in high and very high risk patients at 1, 3, and 5 years and every 5 years thereafter, along with cardiological reevaluation each time new symptoms are detected, cardiovascular toxicity reassessment every 5 years, and risk factor treatment/optimization [3, 4]. A proactive cardiotoxicity screening is pivotal in ensuring long-term cardiovascular health and improving the overall quality of life for cancer survivors.

Prevention and Monitoring of Cardiovascular Complications During Cancer Therapy

Besides the type of agent used, the combination with other cardiotoxic agents, dose, and duration of treatment, a very important aspect is the preexisting risk factors for cardiovascular diseases along with other comorbidities. Baseline evaluation of the patients is of paramount importance in order to establish the risk before starting the treatment. This will be a mainstay in the patients' follow-up and surveillance protocols. It is recommended to note the presence of preexisting cardiac disease including heart failure (independently of the ejection fraction), evidence of coronary artery disease, left ventricular hypertrophy either secondary to increased afterload or hypertrophic cardiomyopathy (including phenocopies), dilated cardiomyopathy, restrictive cardiomyopathy, moderate and/or severe valvular heart disease, arrhythmias and conduction problems, and to ensure that they are treated properly. At the same time, cardiovascular risk factors such as family history of early cardiovascular disease, age, prior chemotherapy (especially anthracyclines) and/or radiotherapy should be noted, diabetes, dyslipidemia, arterial hypertension should be detected and treated accordingly, and alcohol intake, smoking, obesity, sedentary lifestyle should be addressed and modified as possible. To do so, a very thorough clinical assessment combined with data including cardiac markers besides the standard blood tests, ECG and imaging techniques (echocardiography being the most widely available), are recommended so that asymptomatic changes may also be detected. According to the number and severity of risk factors and preexisting pathology, the patient may be at low, moderate or high risk to develop cancer therapy-related cardiac dysfunction [3, 4, 102]. Although limited, there are some studies that proposed risk score calculators in this regard [102]. A high risk patient should be evaluated by a cardiologist or even better, by a experienced cardiooncology team, whereas the other two groups need routine and/or closer oncology follow-up and cardiology specialist referral if cardiotoxicity is detected [3, 4].

Current guidelines recommend the use of angiotensin converting enzyme inhibitors (ACEi) or angiotensin II receptor 1 blockers (ARBs), mineralocorticoid receptor antagonists (MRAs) beta-blockers and statins in high risk patients for primary prevention of cardiotoxicity-induced heart failure, since several meta-analyses have proved a significant beneficial effect in preventing a LVEF decrease [4, 74, 103]. Li et al. proved this effect in a metaanalysis including 22 relevant studies and almost 2000 patients, but failed to prove a beneficial effect of ARBs in this regard [74]. On the other hand, other trial results showed a cardioprotective effect of saccubitril-valsartan, which reduced the risk of subclinical LV dysfunction [104]. The PRADA II trial results are awaited to confirm these results. Moreover, a cardioprotective effect of sodium-glucose transporter 2 inhibitors (SGLT2i) has been shown in patients with different types of cancer treated with potential cardiotoxic agents, and associated with diabetes mellitus type 2 [105]. For antracycline therapy, liposomal forms and dexrazoxane were proved to have a protective effect [4, 106]. Complete guideline recommended treatment of preexisting cardiac pathologies as secondary prevention [3, 4].

Baseline evaluation and subsequent periodic controls according to the known possible cardiotoxic effects of each anticancer agent are key to proper management of oncological patients aimed at achieving primary and/or secondary prevention of cardiovascular toxicity. According to both American and European guidelines, for patients receiving high-risk therapies, an ongoing surveillance every 3–6 months, particularly during treatment but also post-treatment, is recommended [4, 107]. Surveillance protocols consist of clinical evaluation, ECG, cardiac markers (natriuretic peptides, troponin) and imaging evaluation at selected time intervals. Transthoracic echocardiography, including advanced techniques, is the most widely used imaging technique, and the recommendation is to perform complete measurements according to the protocols. LVEF, assessed with tridimensional techniques if possible, and global longitudinal strain (GLS) are of paramount importance. A decrease in LVEF $>10\%$, with values below the normal limit, which should be confirmed at 2–3 weeks by repeated imaging, and of GLS $>15\%$ establish the diagnosis of cancer therapy-related cardiac dysfunction and should prompt initiation of guideline-directed heart failure treatment including ACEIs and betablockers. Moreover, GLS is important as it detects subclinical changes before the LVEF starts to decrease. More advanced imaging techniques such as cardiac magnetic resonance and/or nuclear imaging are recommended to complete the investigations, especially when in doubt or when more sensitive subclinical changes detection is warranted [108, 109].

Conclusions and Future Directions

Awareness of the cardiac dysfunction and heart failure caused by cancer therapy-induced cardiotoxicity, of the specific actions of each culprit agent including their mechanisms, and of the risk factors and timing of cardiotoxicity is key to a better management of the oncological patient, aimed at reducing morbidity and mortality as well as improving quality of life.

Specific subgroups of patients that require even higher attention are those with multiple cardiovascular risk factors, baseline cardiovascular comorbidities, and low normal and/or low baseline ejection fraction as well as those receiving anticancer regimens known to induce myocardial dysfunction, especially when in high doses and or for longer periods of time (high and very high-risk patients). The recent development of Cardio-Oncology Centres consisting of multidisciplinary teams facilitates better evaluation and monitoring and ensures better and more efficient decision making, so that the patient receives the best treatment from both an oncologic and a cardiologic point of view. Further refinement and increase in the availability of these centres are important future directions in this field. For now, treatment of cardiac dysfunction and heart failure caused by cancer therapy-induced cardiotoxicity consists of guideline-directed therapy for heart failure. For high and very high-risk patients receiving intensive cardiotoxic anticancer treatment, such as anthracyclines and or HER2 inhibitors, beta-blockers, RAASi, and statins along with proper management of cardiovascular risk factors are recommended for primary prevention as well. National and international scientific and research societies will play a very important role in establishing further evidence-based recommendations in this regard. There is emerging consensus that a cardio-oncology curriculum should be introduced in the near future that will include training in cardiology, oncology, hematology, and imaging techniques. Further randomised clinical trials and registries are needed that will focus on assessing the risk of cardiotoxicity and improving risk prediction through more sensitive imaging techniques, genetic testing, and use of artificial intelligence [3, 4].

References

1. Curigliano G, Cardinale D, Dent S, Criscitiello C, Aseyev O, Lenihan D et al (2016) Cardiotoxicity of anticancer treatments: epidemiology, detection, and management. CA Cancer J Clin 66(4):309–325
2. Qazi SU, Hamid A, Ansari HUH, Khouri MG, Anker MS, Hall ME et al (2024) Trends in cancer and heart failure related mortality in adult US population: a CDC WONDER database analysis from 1999 to 2020. Am Heart J 278:170–180
3. Zamorano JL, Lancellotti P, Rodriguez Muñoz D, Aboyans V, Asteggiano R, Galderisi M et al (2016) 2016 ESC Position Paper on cancer treatments and cardiovascular toxicity developed under the auspices of the ESC Committee for Practice Guidelines: The Task Force for cancer treatments and cardiovascular toxicity of the European Society of Cardiology (ESC). Eur Heart J 37(36):2768–2801

4. Lyon AR, López-Fernández T, Couch LS, Asteggiano R, Aznar MC, Bergler-Klein J et al (2022) 2022 ESC Guidelines on cardio-oncology developed in collaboration with the European Hematology Association (EHA), the European Society for Therapeutic Radiology and Oncology (ESTRO) and the International Cardio-Oncology Society (IC-OS). Eur Heart J 43(41):4229–4361
5. Siegel RL, Miller KD, Jemal A (2015) Cancer statistics, 2015. CA Cancer J Clin 65(1):5–29
6. Armstrong GT, Oeffinger KC, Chen Y, Kawashima T, Yasui Y, Leisenring W et al (2013) Modifiable risk factors and major cardiac events among adult survivors of childhood cancer. J Clin Oncol 31(29):3673–3680
7. Ewer MS, Ewer SM (2015) Cardiotoxicity of anticancer treatments. Nat Rev Cardiol 12(11): 620
8. Bodai BI, Tuso P (2015) Breast cancer survivorship: a comprehensive review of long-term medical issues and lifestyle recommendations. Perm J 19(2):48–79
9. Oeffinger KC, Mertens AC, Sklar CA, Kawashima T, Hudson MM, Meadows AT et al (2006) Chronic health conditions in adult survivors of childhood cancer. N Engl J Med 355(15): 1572–1582
10. Pinder MC, Duan Z, Goodwin JS, Hortobagyi GN, Giordano SH (2007) Congestive heart failure in older women treated with adjuvant anthracycline chemotherapy for breast cancer. J Clin Oncol 25(25):3808–3815
11. Limat S, Daguindau E, Cahn JY, Nerich V, Brion A, Perrin S et al (2014) Incidence and risk-factors of CHOP/R-CHOP-related cardiotoxicity in patients with aggressive non-Hodgkin's lymphoma. J Clin Pharm Ther 39(2):168–174
12. Cardinale D, Colombo A, Lamantia G, Colombo N, Civelli M, De Giacomi G et al (2010) Anthracycline-induced cardiomyopathy: clinical relevance and response to pharmacologic therapy. J Am Coll Cardiol 55(3):213–220
13. Ederhy S, Dolladille C, Thuny F, Alexandre J, Cohen A (2019) Takotsubo syndrome in patients with cancer treated with immune checkpoint inhibitors: a new adverse cardiac complication. Eur J Heart Fail 21(7):945–947
14. Ghadri JR, Wittstein IS, Prasad A, Sharkey S, Dote K, Akashi YJ et al (2018) International expert consensus document on takotsubo syndrome (part II): diagnostic workup, outcome, and management. Eur Heart J 39(22):2047–2062
15. Giza DE, Lopez-Mattei J, Vejpongsa P, Munoz E, Iliescu G, Kitkungvan D et al (2017) Stress-induced cardiomyopathy in cancer patients. Am J Cardiol 120(12):2284–2288
16. Desai A, Noor A, Joshi S, Kim AS (2019) Takotsubo cardiomyopathy in cancer patients. Cardiooncology 5:7
17. Barmore W, Patel H, Harrell S, Garcia D, Calkins JB Jr (2022) Takotsubo cardiomyopathy: a comprehensive review. World J Cardiol 14(6):355–362
18. June CH, Sadelain M (2018) Chimeric antigen receptor therapy. N Engl J Med 379(1):64–73
19. Ganatra S, Dani SS, Yang EH, Zaha VG, Nohria A (2022) Cardiotoxicity of T-cell antineoplastic therapies: JACC: CardioOncology Primer. JACC CardioOncol 4(5):616–623
20. Ye B, Stary CM, Li X, Gao Q, Kang C, Xiong X (2018) Engineering chimeric antigen receptor-T cells for cancer treatment. Mol Cancer 17(1):32
21. Sievers S, Watson G, Johncy S, Adkins S (2020) Recognizing and grading CAR T-cell toxicities: an advanced practitioner perspective. Front Oncol 10:885
22. Simbaqueba CC, Aponte MP, Kim P, Deswal A, Palaskas NL, Iliescu C et al (2020) Cardiovascular complications of chimeric antigen receptor T-cell therapy: the cytokine release syndrome and associated arrhythmias. J Immunother Precis Oncol 3(3):113–120
23. Zhao Y, Deng J, Rao S, Guo S, Shen J, Du F et al (2022) Tumor infiltrating lymphocyte (TIL) therapy for solid tumor treatment: progressions and challenges. Cancers 14(17):4160
24. Hay KA, Hanafi LA, Li D, Gust J, Liles WC, Wurfel MM et al (2017) Kinetics and biomarkers of severe cytokine release syndrome after CD19 chimeric antigen receptor-modified T-cell therapy. Blood 130(21):2295–2306

25. Lee DW, Kochenderfer JN, Stetler-Stevenson M, Cui YK, Delbrook C, Feldman SA et al (2015) T cells expressing CD19 chimeric antigen receptors for acute lymphoblastic leukaemia in children and young adults: a phase 1 dose-escalation trial. Lancet 385(9967):517–528
26. Cameron BJ, Gerry AB, Dukes J, Harper JV, Kannan V, Bianchi FC et al (2013) Identification of a Titin-derived HLA-A1-presented peptide as a cross-reactive target for engineered MAGE A3-directed T cells. Sci Transl Med 5(197):197ra03
27. Ganatra S, Redd R, Hayek SS, Parikh R, Azam T, Yanik GA et al (2020) Chimeric antigen receptor T-cell therapy-associated cardiomyopathy in patients with refractory or relapsed non-Hodgkin lymphoma. Circulation 142(17):1687–1690
28. Sato R, Nasu M (2015) A review of sepsis-induced cardiomyopathy. J Intensive Care 3:48
29. Kochenderfer JN, Dudley ME, Carpenter RO, Kassim SH, Rose JJ, Telford WG et al (2013) Donor-derived CD19-targeted T cells cause regression of malignancy persisting after allogeneic hematopoietic stem cell transplantation. Blood 122(25):4129–4139
30. Schuster SJ, Bishop MR, Tam CS, Waller EK, Borchmann P, McGuirk JP et al (2019) Tisagenlecleucel in adult relapsed or refractory diffuse large B-cell lymphoma. N Engl J Med 380(1):45–56
31. Neelapu SS, Locke FL, Bartlett NL, Lekakis LJ, Miklos DB, Jacobson CA et al (2017) Axicabtagene ciloleucel CAR T-cell therapy in refractory large B-cell lymphoma. N Engl J Med 377(26):2531–2544
32. Maude SL, Laetsch TW, Buechner J, Rives S, Boyer M, Bittencourt H et al (2018) Tisagenlecleucel in children and young adults with B-cell lymphoblastic leukemia. N Engl J Med 378(5):439–448
33. Alvi RM, Frigault MJ, Fradley MG, Jain MD, Mahmood SS, Awadalla M et al (2019) Cardiovascular events among adults treated with chimeric antigen receptor T-cells (CAR-T). J Am Coll Cardiol 74(25):3099–3108
34. Karthikeyan B, Sunder SS, Puzanov I, Olejniczak SH, Pokharel S, Sharma UC (2024) Cardiotoxic profiles of CAR-T therapy and bispecific T-cell engagers in hematological cancers. Commun Med (Lond) 4(1):116
35. Zhai Y, Hu F, Zhu B, Xu J, Guo X, Shi W et al (2023) Updated insights on cardiac risks of CD19-directed chimeric antigen receptor T-cell therapy: a pharmacovigilance study. Immunotherapy 15(6):443–456
36. Hu JR, Patel A, Huang S, Su YR, Dahlman KB, Tomasek K et al (2021) High sensitivity troponin T and NT-proBNP in patients receiving chimeric antigen receptor (CAR) T-cell therapy. Clin Hematol Int 3(3):96–102
37. Goldman A, Maor E, Bomze D, Liu JE, Herrmann J, Fein J et al (2021) Adverse cardiovascular and pulmonary events associated with chimeric antigen receptor T-cell therapy. J Am Coll Cardiol 78(18):1800–1813
38. Daryanani AE, Abbasi MA, Gomez Ardila MF, Tellez-Garcia E, Garzon-Dangond JM, Lin Y et al (2024) Baseline echocardiographic variables as predictors of hemodynamically significant cytokine release syndrome in adults treated with CD19 CAR T-cell therapy for hematological malignancies. Cardiooncology 10(1):91
39. Ganatra S, Parikh R, Neilan TG (2019) Cardiotoxicity of immune therapy. Cardiol Clin 37(4): 385–397
40. Brudno JN, Kochenderfer JN (2016) Toxicities of chimeric antigen receptor T cells: recognition and management. Blood 127(26):3321–3330
41. Hegemann N, Barth L, Döring Y, Voigt N, Grune J (2024) Implications for neutrophils in cardiac arrhythmias. Am J Physiol Heart Circ Physiol 326(3):H441–H458
42. Bruins P, te Velthuis H, Yazdanbakhsh AP, Jansen PG, van Hardevelt FW, de Beaumont EM et al (1997) Activation of the complement system during and after cardiopulmonary bypass surgery: postsurgery activation involves C-reactive protein and is associated with postoperative arrhythmia. Circulation 96(10):3542–3548
43. Perou CM, Sørlie T, Eisen MB, van de Rijn M, Jeffrey SS, Rees CA et al (2000) Molecular portraits of human breast tumours. Nature 406(6797):747–752

44. Wolff AC, Hammond ME, Hicks DG, Dowsett M, McShane LM, Allison KH et al (2014) Recommendations for human epidermal growth factor receptor 2 testing in breast cancer: American Society of Clinical Oncology/College of American Pathologists clinical practice guideline update. Arch Pathol Lab Med 138(2):241–256
45. Varricchi G, Ameri P, Cadeddu C, Ghigo A, Madonna R, Marone G et al (2018) Antineoplastic drug-induced cardiotoxicity: a redox perspective. Front Physiol 9:167
46. Pondé NF, Lambertini M, de Azambuja E (2016) Twenty years of anti-HER2 therapy-associated cardiotoxicity. ESMO Open 1(4):e000073
47. Cote GM, Sawyer DB, Chabner BA (2012) ERBB2 inhibition and heart failure. N Engl J Med 367(22):2150–2153
48. Crone SA, Zhao YY, Fan L, Gu Y, Minamisawa S, Liu Y et al (2002) ErbB2 is essential in the prevention of dilated cardiomyopathy. Nat Med 8(5):459–465
49. Sawyer DB, Zuppinger C, Miller TA, Eppenberger HM, Suter TM (2002) Modulation of anthracycline-induced myofibrillar disarray in rat ventricular myocytes by neuregulin-1beta and anti-erbB2: potential mechanism for trastuzumab-induced cardiotoxicity. Circulation 105(13):1551–1554
50. Ewer MS, Lippman SM (2005) Type II chemotherapy-related cardiac dysfunction: time to recognize a new entity. J Clin Oncol 23(13):2900–2902
51. Vo JB, Ramin C, Veiga LHS, Brandt C, Curtis RE, Bodelon C et al (2024) Long-term cardiovascular disease risk after anthracycline and trastuzumab treatments in US breast cancer survivors. J Natl Cancer Inst 116(8):1384–1394
52. Spector NL, Yarden Y, Smith B, Lyass L, Trusk P, Pry K et al (2007) Activation of AMP-activated protein kinase by human EGF receptor 2/EGF receptor tyrosine kinase inhibitor protects cardiac cells. Proc Natl Acad Sci U S A 104(25):10607–10612
53. Alizadehasl A, Mohseni M, Roudini K, Firoozbakhsh P (2024) Trastuzumab in patients with breast cancer and pre-existing left ventricular systolic dysfunction. Cardiooncology 10(1):16
54. Modi S, Saura C, Yamashita T, Park YH, Kim SB, Tamura K et al (2020) Trastuzumab deruxtecan in previously treated HER2-positive breast cancer. N Engl J Med 382(7):610–621
55. Dempsey N, Rosenthal A, Dabas N, Kropotova Y, Lippman M, Bishopric NH (2021) Trastuzumab-induced cardiotoxicity: a review of clinical risk factors, pharmacologic prevention, and cardiotoxicity of other HER2-directed therapies. Breast Cancer Res Treat 188(1): 21–36
56. Alhussein MM, Mokbel A, Cosman T, Aghel N, Yang EH, Mukherjee SD et al (2021) Pertuzumab cardiotoxicity in patients with HER2-positive cancer: a systematic review and meta-analysis. CJC Open 3(11):1372–1382
57. von Minckwitz G, Procter M, de Azambuja E, Zardavas D, Benyunes M, Viale G et al (2017) Adjuvant pertuzumab and trastuzumab in early HER2-positive breast cancer. N Engl J Med 377(2):122–131
58. de Azambuja E, Agostinetto E, Procter M, Eiger D, Pondé N, Guillaume S et al (2023) Cardiac safety of dual anti-HER2 blockade with pertuzumab plus trastuzumab in early HER2-positive breast cancer in the APHINITY trial. ESMO Open 8(1):100772
59. Feng Q, Ren Y, Hou A, Guo J, Mao Z, Liu S et al (2021) MicroRNA-130a increases and predicts cardiotoxicity during adjuvant chemotherapy in human epidermal growth factor receptor-2-positive breast cancer. J Breast Cancer 24(2):153–163
60. Zhao M, Chen C, Zhang C, Xu X, Tian F, Wu B et al (2023) Cardiotoxicity with human epidermal growth factor receptor-2 inhibitors in breast cancer: disproportionality analysis of the FDA adverse event reporting system. Int J Cardiol 375:87–93
61. Ariga S (2023) History and future of HER2-targeted therapy for advanced gastric cancer. J Clin Med 12(10):3391
62. Chua TC, Merrett ND (2012) Clinicopathologic factors associated with HER2-positive gastric cancer and its impact on survival outcomes—a systematic review. Int J Cancer 130(12): 2845–2856

63. Bang YJ, Van Cutsem E, Feyereislova A, Chung HC, Shen L, Sawaki A et al (2010) Trastuzumab in combination with chemotherapy versus chemotherapy alone for treatment of HER2-positive advanced gastric or gastro-oesophageal junction cancer (ToGA): a phase 3, open-label, randomised controlled trial. Lancet 376(9742):687–697
64. Shitara K, Bang YJ, Iwasa S, Sugimoto N, Ryu MH, Sakai D et al (2020) Trastuzumab deruxtecan in previously treated HER2-positive gastric cancer. N Engl J Med 382(25): 2419–2430
65. Soo Park J, Youn JC, Shim CY, Hong GR, Lee CK, Kim JH et al (2017) Cardiotoxicity of trastuzumab in patients with HER2-positive gastric cancer. Oncotarget 8(37):61837–61845
66. Pati S, Irfan W, Jameel A, Ahmed S, Shahid RK (2023) Obesity and cancer: a current overview of epidemiology, pathogenesis, outcomes, and management. Cancers 15(2):485
67. Gürbüz M, Akkuş E, Sakin A, Urvay S, Demiray AG, Şahin S et al (2022) Trastuzumab ± capecitabine maintenance after the first-line treatment of HER2-positive advanced gastric cancer: retrospective observational real-life data of Turkish oncology group. J Gastrointest Cancer 53(2):282–288
68. Shiravand Y, Khodadadi F, Kashani SMA, Hosseini-Fard SR, Hosseini S, Sadeghirad H et al (2022) Immune checkpoint inhibitors in cancer therapy. Curr Oncol 29(5):3044–3060
69. Sadeghi Rad H, Monkman J, Warkiani ME, Ladwa R, O'Byrne K, Rezaei N et al (2021) Understanding the tumor microenvironment for effective immunotherapy. Med Res Rev 41(3):1474–1498
70. Seidel JA, Otsuka A, Kabashima K (2018) Anti-PD-1 and Anti-CTLA-4 therapies in cancer: mechanisms of action, efficacy, and limitations. Front Oncol 8:86
71. Mahmood SS, Fradley MG, Cohen JV, Nohria A, Reynolds KL, Heinzerling LM et al (2018) Myocarditis in patients treated with immune checkpoint inhibitors. J Am Coll Cardiol 71(16): 1755–1764
72. Brahmer JR, Lacchetti C, Schneider BJ, Atkins MB, Brassil KJ, Caterino JM et al (2018) Management of immune-related adverse events in patients treated with immune checkpoint inhibitor therapy: American Society of Clinical Oncology clinical practice guideline. J Clin Oncol 36(17):1714–1768
73. Shalata W, Abu-Salman A, Steckbeck R, Mathew Jacob B, Massalha I, Yakobson A (2021) Cardiac toxicity associated with immune checkpoint inhibitors: a systematic review. Cancers 13(20):5218
74. Li X, Li Y, Zhang T, Xiong X, Liu N, Pang B et al (2020) Role of cardioprotective agents on chemotherapy-induced heart failure: a systematic review and network meta-analysis of randomized controlled trials. Pharmacol Res 151:104577
75. Escudier M, Cautela J, Malissen N, Ancedy Y, Orabona M, Pinto J et al (2017) Clinical features, management, and outcomes of immune checkpoint inhibitor-related cardiotoxicity. Circulation 136(21):2085–2087
76. Li C, Bhatti SA, Ying J (2022) Immune checkpoint inhibitors-associated cardiotoxicity. Cancers 14(5):1145
77. Yamaguchi S, Morimoto R, Okumura T, Yamashita Y, Haga T, Kuwayama T et al (2018) Late-onset fulminant myocarditis with immune checkpoint inhibitor nivolumab. Can J Cardiol 34(6):812.e1–812.e3
78. Zhang L, Reynolds KL, Lyon AR, Palaskas N, Neilan TG (2021) The evolving immunotherapy landscape and the epidemiology, diagnosis, and management of cardiotoxicity: JACC: CardioOncology Primer. JACC CardioOncol 3(1):35–47
79. Johnson DB, Balko JM, Compton ML, Chalkias S, Gorham J, Xu Y et al (2016) Fulminant myocarditis with combination immune checkpoint blockade. N Engl J Med 375(18): 1749–1755
80. Tarrio ML, Grabie N, Bu DX, Sharpe AH, Lichtman AH (2012) PD-1 protects against inflammation and myocyte damage in T cell-mediated myocarditis. J Immunol 188(10): 4876–4884

81. Okazaki T, Tanaka Y, Nishio R, Mitsuiye T, Mizoguchi A, Wang J et al (2003) Autoantibodies against cardiac troponin I are responsible for dilated cardiomyopathy in PD-1-deficient mice. Nat Med 9(12):1477–1483
82. Grabie N, Gotsman I, DaCosta R, Pang H, Stavrakis G, Butte MJ et al (2007) Endothelial programmed death-1 ligand 1 (PD-L1) regulates CD8+ T-cell mediated injury in the heart. Circulation 116(18):2062–2071
83. Power JR, Alexandre J, Choudhary A, Ozbay B, Hayek S, Asnani A et al (2021) Electrocardiographic manifestations of immune checkpoint inhibitor myocarditis. Circulation 144(18): 1521–1523
84. Thuny F, Naidoo J, Neilan TG (2022) Cardiovascular complications of immune checkpoint inhibitors for cancer. Eur Heart J 43(42):4458–4468
85. Chen C, Chen T, Liang J, Guo X, Xu J, Zheng Y et al (2021) Cardiotoxicity induced by immune checkpoint inhibitors: a pharmacovigilance study from 2014 to 2019 based on FAERS. Front Pharmacol 12:616505
86. Milutinovic S, Jancic P, Jokic V, Petrovic M, Dumic I, Rodriguez AM et al (2024) Pembrolizumab-associated cardiotoxicity: a retrospective analysis of the FDA adverse events reporting system. Pharmaceuticals (Basel) 17(10):1372
87. Baker JK, Shank-Coviello J, Zhou B, Dixon J, McCorkle R, Sarpong D et al (2020) Cardiotoxicity in hematopoietic stem cell transplant: keeping the beat. Clin Lymphoma Myeloma Leuk 20(4):244–51.e4
88. Bocchi EA, Avila MS, Ayub-Ferreira SM (2019) Aging, cardiotoxicity, and chemotherapy. Aging 11(2):295–296
89. Cardinale D, Iacopo F, Cipolla CM (2020) Cardiotoxicity of anthracyclines. Front Cardiovasc Med 7:26
90. Iqubal A, Iqubal MK, Sharma S, Ansari MA, Najmi AK, Ali SM et al (2019) Molecular mechanism involved in cyclophosphamide-induced cardiotoxicity: old drug with a new vision. Life Sci 218:112–131
91. Bălănescu AR, Bojincă VC, Bojincă M, Donisan T, Bălănescu SM (2019) Cardiovascular effects of methotrexate in immune-mediated inflammatory diseases. Exp Ther Med 17(2): 1024–1029
92. Shah S, Haeger-Overstreet K, Flynn B (2022) Methotrexate-induced acute cardiotoxicity requiring veno-arterial extracorporeal membrane oxygenation support: a case report. J Med Case Rep 16(1):447
93. Ciurea SO, Andersson BS (2009) Busulfan in hematopoietic stem cell transplantation. Biol Blood Marrow Transplant 15(5):523–536
94. Armstrong GT, Joshi VM, Ness KK, Marwick TH, Zhang N, Srivastava D et al (2015) Comprehensive echocardiographic detection of treatment-related cardiac dysfunction in adult survivors of childhood cancer: results from the St. Jude Lifetime Cohort Study. J Am Coll Cardiol 65(23):2511–2522
95. Zheng H, Zhan H (2024) Preventing anthracycline-associated heart failure: what is the role of dexrazoxane?: JACC: CardioOncology Controversies in Cardio-Oncology. JACC CardioOncol 6(2):318–321
96. Mihalcea D, Memis H, Mihaila S, Vinereanu D (2023) Cardiovascular toxicity induced by vascular endothelial growth factor inhibitors. Life (Basel) 13(2):366
97. Georgiopoulos G, Makris N, Laina A, Theodorakakou F, Briasoulis A, Trougakos IP et al (2023) Cardiovascular toxicity of proteasome inhibitors: underlying mechanisms and management strategies: JACC: CardioOncology State-of-the-Art Review. JACC CardioOncol 5(1):1–21
98. Salem JE, Manouchehri A, Moey M, Lebrun-Vignes B, Bastarache L, Pariente A et al (2018) Cardiovascular toxicities associated with immune checkpoint inhibitors: an observational, retrospective, pharmacovigilance study. Lancet Oncol 19(12):1579–1589

99. Drobni ZD, Alvi RM, Taron J, Zafar A, Murphy SP, Rambarat PK et al (2020) Association between immune checkpoint inhibitors with cardiovascular events and atherosclerotic plaque. Circulation 142(24):2299–2311
100. Rubio-Infante N, Ramírez-Flores YA, Castillo EC, Lozano O, García-Rivas G, Torre-Amione G (2021) Cardiotoxicity associated with immune checkpoint inhibitor therapy: a meta-analysis. Eur J Heart Fail 23(10):1739–1747
101. Mascolo A, Scavone C, Ferrajolo C, Rafaniello C, Danesi R, Del Re M et al (2021) Immune checkpoint inhibitors and cardiotoxicity: an analysis of spontaneous reports in eudravigilance. Drug Saf 44(9):957–971
102. Ezaz G, Long JB, Gross CP, Chen J (2014) Risk prediction model for heart failure and cardiomyopathy after adjuvant trastuzumab therapy for breast cancer. J Am Heart Assoc 3(1):e000472
103. Fang K, Zhang Y, Liu W, He C (2021) Effects of angiotensin-converting enzyme inhibitor/angiotensin receptor blocker use on cancer therapy-related cardiac dysfunction: a meta-analysis of randomized controlled trials. Heart Fail Rev 26(1):101–109
104. De Vecchis R, Paccone A (2020) A case series about the favorable effects of sacubitril/valsartan on anthracycline cardiomyopathy. SAGE Open Med Case Rep 8:2050313x20952189
105. Avula V, Sharma G, Kosiborod MN, Vaduganathan M, Neilan TG, Lopez T et al (2024) SGLT2 inhibitor use and risk of clinical events in patients with cancer therapy-related cardiac dysfunction. JACC Heart Fail 12(1):67–78
106. Macedo AVS, Hajjar LA, Lyon AR, Nascimento BR, Putzu A, Rossi L et al (2019) Efficacy of dexrazoxane in preventing anthracycline cardiotoxicity in breast cancer. JACC CardioOncol 1(1):68–79
107. Armenian SH, Lacchetti C, Barac A, Carver J, Constine LS, Denduluri N et al (2017) Prevention and monitoring of cardiac dysfunction in survivors of adult cancers: American Society of Clinical Oncology Clinical Practice Guideline. J Clin Oncol 35(8):893–911
108. Oikonomou EK, Kokkinidis DG, Kampaktsis PN, Amir EA, Marwick TH, Gupta D et al (2019) Assessment of prognostic value of left ventricular global longitudinal strain for early prediction of chemotherapy-induced cardiotoxicity: a systematic review and meta-analysis. JAMA Cardiol 4(10):1007–1018
109. Lang RM, Badano LP, Mor-Avi V, Afilalo J, Armstrong A, Ernande L et al (2015) Recommendations for cardiac chamber quantification by echocardiography in adults: an update from the American Society of Echocardiography and the European Association of Cardiovascular Imaging. J Am Soc Echocardiogr 28(1):1–39.e14

Chapter 3
Echocardiographic Aspects in Cardio-oncology

Mihaela-Daniela Valcovici, Bogdan Lolescu, and Daniel Marius Duda Seiman

Abstract Echocardiography is an essential non-invasive, accessible and reproducible technique, contributing to the correct assessment of cardiac structure and function in cardio-oncologic patients, both in the pre-oncologic treatment stage, during the oncological therapy course and after if necessary. Echocardiographic examinations of neoplastic patients are effective in early detection of "iatrogenic" cardiovascular dysfunctions, or, in case of pre-existing cardiovascular conditions, echocardiography is one of the most accessible and reliable tool to monitor these patients.

Specific oncologic therapies may determine a wide range of cardiovascular effects, identified as "*cancer-therapeutics-related cardiac dysfunction*" (CTRCD): left ventricular dysfunction with all its underlying mechanisms, QT interval prolongation, hypertension, pulmonary hypertension, thromboembolic events, valvular dysfunctions, and pericardial disease (Moran and Plana, ASE's comprehensive echocardiography, Elsevier, 2022). The risk-benefit assessment of cancer treatment led to the term "*accepted cardiotoxicity*", defining those situations when cancer treatment may and should be given because its benefits are superior in comparison to a further possible cardiovascular impact (Lennep and Mitchell, Manualul Washington de Cardio-Oncologie. Ghid practic pentru diagnosticul si tratamentul complicatiilor cardiovasculare la pacientii cu cancer, Prior, 2024). Early cardioprotective therapy initiation (inhibitors of the renin-angiotensin-aldosterone system, beta-blockers) may reduce the risk of left ventricular dysfunction in patients receiving active oncologic therapy regimens (Fine and Mitchell, Manualul Washington de Cardio-Oncologie. Ghid practic pentru diagnosticul si tratamentul complicatiilor cardiovasculare la pacientii cu cancer, Prior, 2024).

M.-D. Valcovici · B. Lolescu
Department of Cardiology, University of Medicine and Pharmacy "Victor Babes", Timisoara, Romania
e-mail: Mihaela.Valcovici@umft.ro; bogdan.lolescu@umft.ro

D. M. D. Seiman (✉)
University of Medicine and Pharmacy, "Victor Babes", Timisoara, Romania
e-mail: daniel.duda-seiman@umft.ro

D. K. Agrawal et al. (eds.), *Cardiovascular Toxicity*, Advances in Biochemistry in Health and Disease 33, https://doi.org/10.1007/978-3-031-90598-8_3

Keywords Cancer-related cardiac dysfunction · Dosatinib · Echocardiography · Global circumferential strain · Global longitudinal strain · Global radial strain · Left ventricular diastolic dysfunction · Left ventricular ejection fraction · Myocarditis · Ponatinib · Radiotherapy · Speckle tracking echocardiography

Left Ventricular Ejection Fraction

In absolute terms, The American Society of Echocardiography (ASE) and the European Association of Cardiovascular Imaging (EACVI) agreed in their joint consensus to define CTRCD, the situation when the left ventricular ejection fraction (LVEF) decreases more than 10 points to the threshold of less than 53% while receiving oncologic therapy regimens [1]. The European Society of Cardiology (ESC) considers CTRCD if LVEF is reduced and/or relative changes in global longitudinal strain (GLS) is documented [2]. Baseline (before therapy) LVEF is considered a risk factor for CTRCD development: thus, a quite normal or borderline LVEF (50–54%) and, obviously, a value <50% predicts a future CTRCD in case of most cancer therapy regimens, especially when using anthracyclines or trastuzumab [2]. As a practical aspect, LVEF is mandatory to be determined in all oncological patients before any specific therapy regimen, as well as within therapy course (specific time ranges are defined), also when cardiotoxicity is suspected, and as a follow-up method after completing oncological treatment [3]. It is well-known that two-dimensional (2D) determination of LVEF is subjected to the law of inter-observer variability (about 10%), but there is also a range of values obtained by the same assessor. And this is due to an imperfect delimitation of the endocardium and because mathematic formulas to calculate 2D-LVEF do not take into account the systolic longitudinal contraction of the left ventricle. Several echocardiographic techniques are available to improve the LVEF calculation: contrast echocardiography, three-dimensional (3D) echocardiography, stress echocardiography [3, 4].

Current guidelines [2] create a practical and objective hierarchy of imaging techniques to assess LVEF and ventricular volumes: 3D echocardiography is preferred; if this technique is unavailable, 2D echocardiography is recommended (Simpson's biplane method). Contrast echocardiography should be used if standard transthoracic images are flawed because of low-quality windows, considering that at least two left ventricular walls cannot be properly described. An alternative to this method would be cardiac magnetic resonance (CMR) if available. Multigated acquisition nuclear imaging (MUGA) might be a third-line imaging technique if echocardiography and CMR are both limited or unavailable, but it should be very carefully indicated because high radiation exposure and obtained data is limited (e.g., impossibility to assess the pulmonary arterial pressure or the ventricular longitudinal movement).

The hemodynamic state is determinant for the LVEF assessment. Hence, it can be stated that the hydration state influences LVEF [2]. An interesting relation was observed in patients with atrial fibrillation: LVEF depends on the RR interval length variability, and this is more obvious in cases with higher average heart rate and lower LVEF [5].

In order to estimate an echocardiographic outcome after oncological therapy in terms of reversibility of myocardial dysfunction, it is worth to remind the fundamental types of CTRCD [6]:

- Type I: induced by anthracyclines, leading to permanent and irreversible cardiomyocyte destruction;
- Type II: induced by anti-HER2 antibodies and tyrosine kinase inhibitors; usually, cardiomyocyte dysfunction is reversible, with possible *restitutio ad integrum* 2–4 months after treatment ending; there is a 20% chance to develop irreversible lesions due to mechanism aspects that are similar to type I drugs.

Resting LVEF may not be affected in early stages of using anthracyclines regimens, so complementary imaging techniques are necessary to identify systolic myocardial dysfunction as toxic effect [7].

3D-echocardiography can acquire high-quality images, which are also reproducible, but unfortunately, there is not a good availability of the technique within the examination centers. So, there is not an established cut-off value of LVEF calculated using 3D-echocardiography [6]. Recommendations are formulated to use 3D-LVEF when possible and feasible (good quality acquisitions), the interpretation should focus also on clinical aspects, ventricular volumes and myocardial strain [2, 8, 9]. To obtain reliable data using 3D-LVEF, it is important to have an as much as lower variability between observers. Medical personnel and technicians have to be trained for real-time 3D images acquisitions. A recent Denish study concluded that it is important to identify those situations of +/−10 percentage points change in LVEF measurements, to optimize therapy decisions in oncologic patients [10]. Data is conflicting: in some studies, 2D contrast enhanced echocardiography may be superior to 3D techniques [11, 12].

Global Longitudinal Strain (GLS)

Speckle tracking echocardiography (STE) uses transthoracic windows and analyses the deformation capacity of myocardial segments (known as myocardial strain). *Speckle* considers a distribution point of grey values within the obtained ultrasound image, while *speckle tracking* determines the deformation between such two points (identified as *strain*), calculating a deformation score. *Strain* is defined as the percentual difference concerning a positive or negative deformation of an object corresponding to its initial state [13]. So, *myocardial strain* is calculated as a negative value, representing the change of the myocardium length within its contraction, corresponding to its original length [11]. Considering the three contractility movements of the myocardium, there are three possible global strain scores to be calculated: *Global Longitudinal Strain—GLS* (validated and used in clinical practice); *Global Radial Strain—GRS; Global Circumferential Strain—GCS* [13].

It is advised that the same machine or software be used for each patient's serial GLS measurement because strain measurements may vary throughout vendors [2].

GLS has the capacity to detect subclinical myocardial damage (extremely important in oncologic patients), whereas it changes before a significant decrease of the LVEF might be documented. Some researchers have observed that GLS falls by 9–19% from its initial value in case of anthracycline-based therapy regimens [7]. It is recommended to determine baseline GLS (before starting oncologic therapy regimens) [2], because normal or decreased baseline GLS values are strong predictors to develop myocardial cardiotoxic effects in case of anthracycline-class based therapy [14]. As higher baseline GLS is, it may be a strong predictor for cardiac death or decompensated heart failure (GLS $>-17.5\%$) [15], or it may be suggestive for increased all-cause mortality (GLS $>-15\%$) [16]. This aspect was also reinforced in another study, concluding that baseline GLS values that are closer to the origin on the numeric axis might be related with higher risk of mortality in patients attending anthracycline-based therapy regimens [17].

With a 95% upper limit of GLS reduction of 15%, a future decline in LVEF was predicted by a median GLS change of 13.6%. The 15% cut-off is the threshold advised for GLS monitoring during cancer treatment since it increases specificity [2].

Another issue concerns the comparison to 3D-LVEF determination: GLS is more suitable to detect early cardiotoxic effects of anthracyclines and trastuzumab than 3D-LVEF significant change is seen, but this early decrease of GLS is strongly correlated to further 3D-LVEF impairment [18]. In this study [18], "traditional" cardiovascular risk factors (age, hypertension) were not associated with impairment of echocardiographic parameters.

Obesity influences negatively GLS, proven by its improvement after performing obesity surgical techniques. Other conditions that lead to a lower GLS are type 2 diabetes mellitus, dyslipidemia, and, possibly in those patients using beta-blockers, while pre-treatment with ACE inhibitors might have an initial over-estimated GLS [13, 19].

It has been demonstrated that a GLS drop of more than 15% has a sensitivity of 73.3–92.5% and a specificity of 78.5–83% for identifying future cardiotoxicity, which is defined as a decline in LVEF of more than 10% [20].

Myocarditis is not predicted by GLS estimations. At least 10% change in terms of impairment of basal segment longitudinal strain was a common finding in patients with myocarditis (patients receiving immune checkpoint inhibitors) [8].

Radiotherapy is also recognized to determine cardio toxic effects and GLS is able to detect subclinical myocardial changes. In addition, it is important to determine the so-called "cardiac doses". For instance, a target of 5 Gy would be sufficient in breast cancer with left-sided radiotherapy intention [14].

In Fig. 3.1 we show the case of a 72 years old female patient diagnosed with left breast cancer, at baseline and after 4 cycles of epirubicin treatment: we used the same software to estimate GLS, showing that there was a slight impairment of GLS, while 2D-LVEF has not consistently diminished.

A promising tool in the early detection of cardiotoxic effects is the 3D-GLS determination, with strain changes being observed after low doses of anthracycline derivatives. This promotes in next period of time a further LVEF decrease which is

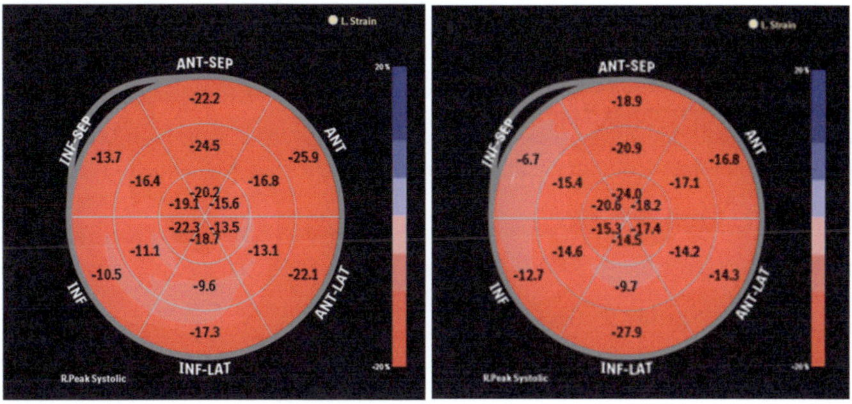

Fig. 3.1 GLS behavior after 4 cycles of epirubicin in a 72 years female patient with left breast cancer

determined using available techniques [21]. By removing apical foreshortening and simultaneously obtaining data from all myocardial segments in a shorter amount of time, 3D speckle tracking echocardiography shows potential [7].

For assessing LV function and identifying subclinical LV disease early on, the strain area (*area strain*) is a potential deformation metric. Due to the incompressibility of the myocardium, the radial dimension must grow during systole as the longitudinal dimension decreases with the circumferential dimension. That is to say, we can determine the deformation of the subendocardial surface, which is inversely proportional to the radial deformation and typically has negative values, by combining the longitudinal and circumferential deformation in a specific segment of the myocardium [22].

Left Ventricular Diastolic Dysfunction

It is possible to evaluate LV diastolic function in addition to LV systolic function, which includes grading diastolic function and estimating LV filling pressures [20]. When breast cancer patients with normal baseline diastolic function were treated with different chemotherapy regimens, diastolic dysfunction occurred an average of 73 days before systolic failure [20, 23].

Tissue Doppler Imaging in left ventricular diastolic dysfunction assessment has several advantages, allowing to describe the diastolic dysfunction type (degree), as well as to observe the increased ventricular filling pressure. It also is feasible in differentiating constrictive vs. restrictive diastolic dysfunction type. In a former study [24], a decreased isovolumic relaxation time measured at the mitral annulus (<80 ms) shortly after the initiation of an anthracycline-based therapy regimen was a powerful predictor for later important alteration of LVEF (<50%).

A measure of diastolic function that has attracted a lot of attention is left atrial (LA) strain, which is linked to invasive filling pressure measurements and poorer clinical results. Age affects left atrial strain values, which are not yet consistent among suppliers. Lastly, when atrial fibrillation is present, left atrial strain cannot be used to estimate LV filling pressures precisely. Endocardial definition that permits LA strain measurement may be challenging due to the thin-walled nature of the left atrial wall [8].

Right Ventricular Function

Apart from the modifications in left ventricular (LV) function, changes in right ventricular (RV) diastolic and systolic function may be early indicators of anthracycline-induced cardiotoxicity [20]. Direct neoplastic involvement or the cardiotoxic effects of chemotherapy can both result in right ventricular (RV) abnormalities. It has been seen that dasatinib can result in pulmonary hypertension, and that anthracyclines, trastuzumab, and cyclophosphamide can impair right ventricular function [7].

The broad and consistent impact of anti-HER2 medications on heart function is confirmed by the deformation mechanics of the LV and RV, which exhibit comparable temporal patterns and degrees of dysfunction following trastuzumab therapy. For the RV GLS percent change, a cutoff value of -14.8% was suggested, which excluded patients with cardiotoxicity from trastuzumab [14].

Pulmonary tumor thrombotic microangiopathy, a cancer-related condition marked by an abrupt increase in pulmonary pressure, can also be diagnosed by echocardiography [6].

Echocardiographic Monitoring Protocols [2]

In order to have an integrated oncology—cardiology approach, it is mandatory to estimate the cardiovascular risk and the cardiovascular toxicity risk, at the time of cancer is being established and before the initiation of specific therapy. A comprehensive tool was elaborated by the Heart Failure Association (HFA) of the ESC in collaboration with the International Cardio-Oncology Society (ICOS) called HFA-ICOS risk assessment defining four risk categories, according to which monitoring strategies while and after treatment are formulated:

- Low risk = no risk factors OR one moderate risk factor;
- Moderate risk (M) = moderate risk factors with a total of 2–4 points (Moderate 1 [M1] = 1 point; Moderate [M2] = 2 points);
- High risk (H) = moderate risk factors with a total of ≥5 points OR any high-risk factor;
- Very-high risk (VH) = any very-high risk factor.

Anthracycline Therapy Monitoring Protocol

Anthracycline therapy monitoring protocol

Risk category	Baseline	C2	C4	C6	3 months post therapy	12 months post therapy
Low risk	🟢		🟠			🟢
Moderate risk	🟢		🟡			🟢
High and very high risk	🟢	🟢	🟢	🟢	🟢	🟢

Ci = anthracycline cycle; echocardiography should be performed before;
🟢 = class I;
🟡 = class IIa;
🟠 = class IIb.

HER-2-Targeted Therapy Monitoring Protocol

HER-2-targeted therapy monitoring protocol

Risk category	Baseline	3 months	6 months	9 months	12 months	3 months post therapy	12 months post therapy
Low and moderate risk	🟢	🟢	🟢	🟢	🟢		🟢
High and very high risk	🟢	🟢	🟢	🟢	🟢	🟢	🟢

🟢 = class I.

Vascular Endothelial Growth Factor Inhibitors Therapy Monitoring Protocol

Vascular endothelial growth factor inhibitors therapy monitoring protocol

Risk category	Baseline	3 months	4 months	6 months	8 months	9 months	12 months	Every 6-12 months post therapy
Low risk	🟡							
Moderate risk	🟡		🟠		🟠		🟠	🟡
High and very high risk	🟢	🟡		🟡		🟡	🟡	🟡

🟢 = class I;
🟡 = class IIa;
🟠 = class IIb.

Fluoropyrimidines—baseline echocardiographic examination in all patients with positive history for cardiovascular disease, before treatment start (class I C).

Multitargeted Kinase Inhibitors Targeting BCR-ABL Monitoring Protocol

Multitargeted kinase inhibitors targeting BCR-ABL monitoring protocol

Risk category	Baseline	3 months	6 months	9 months	12 months	Every 6-12 months post therapy
Low risk	Dasatinib					Dasatinib; Ponatinib
Moderate risk	Dasatinib					Dasatinib; Ponatinib
High and very high risk	Dasatinib	Dasatinib; Ponatinib	Dasatinib; Ponatinib	Dasatinib; Ponatinib	Dasatinib; Ponatinib	Dasatinib; Ponatinib

Green color – class I C;
Yellow color – class IIa;
Orange color – class IIb.

References

1. Moran TB, Plana JC (2022) Cardiotoxic effects of cancer therapy. In: Lang RM, Goldstein SA, Kronzon I, Khandheria BK, Saric M, Mor-Avi V (eds) ASE's comprehensive echocardiography, 3rd edn. Elsevier, Philadelphia, pp 849–857
2. Lyon AR et al (2022) 2022 ESC Guidelines on cardio-oncology developed in collaboration with the European Hematology Association (EHA), the European Society for Therapeutic Radiology and Oncology (ESTRO) and the International Cardio-Oncology Society (IC-OS): developed by the task force on cardio-oncology of the European Society of Cardiology (ESC). Eur Heart J 43(41):4229–4361
3. Rădulescu D, Pârv A, Buzdugan E, Rădulescu ML, Grosu AI, Crișan S, Crișan D, Stoicescu L (2020) Rolul ecocardiografiei în cardio-oncologie. In: Vlad C, Rădulescu D, Drăgan S, Burz C (eds) Manual de Cardio-Oncologie. Editura Medicală Universitară "Iuliu Hațieganu" Cluj-Napoca, Cluj-Napoca, pp 144–155
4. Lennep BW, Mitchell JD (2024) Cardiotoxicitate acceptata. In: Lenihan DJ (ed) Manualul Washington de Cardio-Oncologie. Ghid practic pentru diagnosticul si tratamentul complicatiilor cardiovasculare la pacientii cu cancer. Prior, Bucuresti, pp 268–279
5. Lin CP et al (2012) Left ventricular systolic function is sensitive to cycle-length irregularity in patients with atrial fibrillation and systolic dysfunction. Acta Cardiol Sin 28(2):103–110
6. Onishi T et al (2021) Practical guidance for echocardiography for cancer therapeutics-related cardiac dysfunction. J Echocardiogr 19(1):1–20
7. Venneri L, Zoppellaro G, Khattar RS (2018) Cardio-oncology: the role of advanced echocardiography in cancer patients. Expert Rev Cardiovasc Ther 16(4):249–258
8. El-Rayes M et al (2023) Recent advances in the use of echocardiography in cardio-oncology. Curr Treat Options Cardiovasc Med 25(11):655–674
9. Esmaeilzadeh M et al (2022) A combined echocardiography approach for the diagnosis of cancer therapy-related cardiac dysfunction in women with early-stage breast cancer. JAMA Cardiol 7(3):330–340

10. Nielsen MO et al (2024) Reproducibility and repeatability in focus: evaluating LVEF measurements with 3D echocardiography by medical technologists. Diagnostics 14(16):15
11. Fine C, Mitchell JD (2024) Tehnici ecografice pentru evaluarea sigurantei cardiologice a tratamentului oncologic. In: Lenihan DJ (ed) Manualul Washington de Cardio-Oncologie. Ghid practic pentru diagnosticul si tratamentul complicatiilor cardiovasculare la pacientii cu cancer. Prior, Bucuresti, pp 192–207
12. Hoffmann R et al (2014) Analysis of left ventricular volumes and function: a multicenter comparison of cardiac magnetic resonance imaging, cine ventriculography, and unenhanced and contrast-enhanced two-dimensional and three-dimensional echocardiography. J Am Soc Echocardiogr 27(3):292–301
13. Grosu AI, Buzdugan E, Rădulescu ML, Stoicescu L, Crişan S, Rădulescu D (2020) Ecocardiografia speckle tracking. In: Vlad C, Rădulescu D, Drăgan S, Burz C (eds) Manual de Cardio-Oncologie. Editura Medicală Universitară "Iuliu Haţieganu" Cluj-Napoca, Cluj-Napoca, pp 162–165
14. Slawinski G et al (2023) Global longitudinal strain in cardio-oncology: a review. Cancers 15(3): 12
15. Ali MT et al (2016) Myocardial strain is associated with adverse clinical cardiac events in patients treated with anthracyclines. J Am Soc Echocardiogr 29(6):522
16. Kang Y et al (2019) Symptomatic heart failure in acute leukemia patients treated with anthracyclines. JACC CardioOncol 1(2):208–217
17. Araujo-Gutierrez R et al (2021) Baseline global longitudinal strain predictive of anthracycline-induced cardiotoxicity. Cardiooncology 7(1):8
18. van der Linde D et al (2023) Global longitudinal strain: an early marker for cardiotoxicity in patients treated for breast cancer. Neth Hear J 31(3):103–108
19. Collier P, Phelan D, Klein A (2017) A test in context: myocardial strain measured by speckle-tracking echocardiography. J Am Coll Cardiol 69(8):1043–1056
20. Bhat S et al (2019) The role of echocardiography in cardio-oncology patients: contemporary indications and future directions. Curr Cardiovasc Imaging Rep 12(12):42
21. Sławiński G et al (2023) Global longitudinal strain in cardio-oncology: a review. Cancers 15(3): 986
22. Tun HN, Vlad C (2020) Tehnici ecocardiografice noi în cardio-oncologie. In: Vlad C, Rădulescu D, Drăgan S, Burz C (eds) Manual de Cardio-Oncologie. Editura Medicală Universitară "Iuliu Haţieganu" Cluj-Napoca, Cluj-Napoca, pp 166–171
23. Klein R et al (2019) Diastolic dysfunction can precede systolic dysfunction on MUGA in cancer patients receiving trastuzumab-based therapy. Nucl Med Commun 40(1):22–29
24. Tassan-Mangina S et al (2006) Tissue Doppler imaging and conventional echocardiography after anthracycline treatment in adults: early and late alterations of left ventricular function during a prospective study. Eur J Echocardiogr 7(2):141–146

Chapter 4
Arrhythmia/QTc Prolongation in Cardiotoxicity

Antoinette Oliveira Blackman, Beatriz Montenegro Oliveira, Suellen Keyze Almeida Lima, and José Sobral Neto

Abstract QT prolongation and QT dispersion (QTd) are key electrocardiographic markers for assessing cardiotoxicity, particularly in drug-induced cardiac complications. QT prolongation, characterized by an extended QT interval, indicates delayed ventricular repolarization, increasing the risk of severe arrhythmias such as Torsades de Pointes. QT dispersion, defined as the difference between the longest and shortest QT intervals recorded across ECG leads, reflects regional variations in repolarization and indicates electrical instability and arrhythmogenic potential. Drug-induced cardiotoxicity is a significant concern, particularly with chemotherapeutic agents, antipsychotics, and certain antibiotics. The primary mechanism involves inhibiting potassium ion channels, leading to prolonged repolarization and increased susceptibility to arrhythmias. Given these risks, continuous ECG monitoring and effective risk stratification are essential for patients receiving QT-prolonging therapies. Modern cardiologists must recognize the cardiotoxic effects of conventional and emerging therapeutic agents, including chemotherapy, targeted therapies, and immunotherapies. A multidisciplinary, collaborative approach is essential for optimizing cardiovascular health in patients undergoing treatment for various conditions. This strategy ensures that potentially life-saving therapies are administered safely while minimizing the risk of cardiotoxicity.

In memoriam of José Sobral Neto.

A. O. Blackman (✉)
University Center of Central Planalto Apparecido dos Santos—UNICEPLAC, Gama, Federal District, Brazil

University Center of Brasilia—CEUB, Brasília, Federal District, Brazil

University Center Euro American—UNIEURO, Brasília, Brazil
e-mail: antoinette.blackman@uniceplac.edu.br

B. M. Oliveira · S. K. A. Lima
University Center of Central Planalto Apparecido dos Santos—UNICEPLAC, Gama, Federal District, Brazil

J. S. Neto
Centrocard, Brasília, Federal District, Brazil

D. K. Agrawal et al. (eds.), *Cardiovascular Toxicity*, Advances in Biochemistry in Health and Disease 33, https://doi.org/10.1007/978-3-031-90598-8_4

Keywords Electrocardiography · Long-QT syndrome · Rate corrected QT · Cardiotoxicity · Arrhythmia · Torsade de Pointes · Prolonged QT interval · QTc-prolongation

Introduction

The World Health Organization reported in 2020 data that cardiovascular and oncological diseases are the leading causes of global mortality [1]. Sudden cardiac death (SCD) accounts for millions of fatalities annually and continues to be a significant public health concern. To mitigate its impact, it is essential to identify individuals at high risk and apply disease-specific risk strategies.

QT prolongation is associated with increased mortality risk in the general population [2]. It is recognized as a surrogate marker for cardiotoxicity, particularly in the context of drug safety evaluation [3]. To achieve the best possible patient outcomes, it is essential to thoroughly assess the risk of QT prolongation in relation to the benefits of treatment.

The heart has an electrical system that allows it to contract. A vital aspect of this system consists of depolarization and repolarization processes. Electrical activity is conducted through the sinoatrial (SA) and atrioventricular (AV) nodes to the ventricles. Following AV nodal conduction, impulses travel through the His-Purkinje system—comprising the bundle of His, right and left bundle branches, and Purkinje fibers—ensuring rapid and coordinated depolarization of the ventricular myocardium [4, 5]. Cardiac electrical activity is delineated on the electrocardiogram (ECG), where the P wave, QRS complex, and T wave are observed. The QRS complex represents ventricular depolarization. Finally, the T wave illustrates ventricular repolarization (Fig. 4.1).

This chapter focuses on the QT interval, an electrophysiological marker that reflects the duration of the cardiac action potential, more specifically, ventricular repolarization. It represents the time between the start of the QRS complex and the end of the T wave and is measured in milliseconds (ms).

When measured on the ECG, the QT interval lengthens if the heart rate is slower (bradycardia) and shortens when the heart rate is faster (tachycardia) (Fig. 4.1). The QT interval is considered normal whether it is equal to or less than 440 ms in men and equal to or less than 460 ms in women. Patients with a QTc >500 ms or an increase of >60 ms from baseline are considered at significantly elevated risk for TdP [6]. It is a relevant parameter because alterations suggest a risk of heart outcomes. Prolongation of the QT interval on the electrocardiogram (ECG) is a significant clinical concern because it is associated with an increased risk of potentially fatal ventricular arrhythmias, such as Torsade de Pointes (TdP) [7].

Torsade de Pointes is a myocardial arrhythmia commonly associated with Long QT Syndrome. It is characterized by periodic twists in the amplitude and direction of the QRS complex around its axis, i.e. the wave pattern of the QRS complex rotates

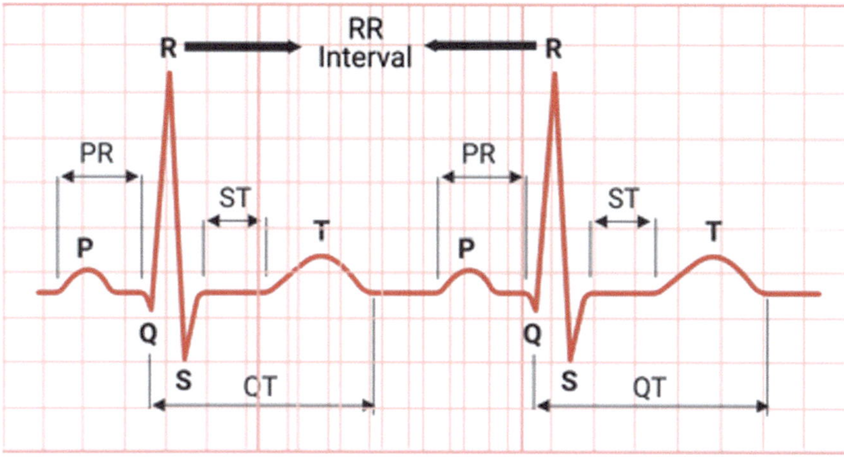

Fig. 4.1 Schematic diagram showing cardiac electrical activity on the electrocardiogram (ECG), where the P wave, QRS complex, and T wave are observed. The QRS complex represents ventricular depolarization, and the T wave illustrates ventricular repolarization. (Created in Biorender.com)

around the baseline. As previously explained, prolonging the QT interval delays ventricular repolarization, which increases the chance of early afterdepolarizations, allowing a new action potential before the cardiac cell is completely repolarized. It consists of a polymorphic ventricular tachycardia of paramount importance since it can evolve into ventricular fibrillation and, ultimately, cardiac arrest [8]. Premature activation of the myocardium, represented by after-depolarizations, can be a trigger for Torsades de Pointes, especially in the case of risk factors such as the use of drugs that predispose to prolongation of the QT interval. It is diagnosed by analyzing the ECG. The patient may show signs of hypotension, syncope, change in heart rate, and signs of heart failure usually sudden onset [9, 10].

Because it varies according to heart rate, it is necessary to use correction formulas to establish an average QT interval, known as QTc. This allows for a more accurate QT interval, especially at higher heart rates [11, 12].

The Danish physician, Louis Siquard Fridericia proposed in a normal persona at rest that the duration of electrical systole on an ECG is proportional to the cube root of the duration of the pulse period, as shown in the following equation, known as Fridericia equation:

$$\mathrm{QTc} = QT\sqrt[3]{\mathrm{RR}}$$

Where, RR refers to an interval between consecutive R waves in an ECG, measured in seconds.

Evaluation of the QTc is essential for determining the real risk of QT prolongation. Among the most widely used formulas for this correction are Bazett's Formula, which adjusts the QT according to the square root of the RR interval, and the Fridericia Formula, which proposes a correction based on the cube root of the RR interval [13, 14]. Both are widely used in clinical practice and research, although they have limitations, especially in patients with extreme heart rates, both for bradycardia and tachycardia. The Framingham and Hodges formulas are also used to determine QTc duration. For better understanding, the formulas mentioned below are:

Bazett formula: $QTc = QT/\sqrt{RR}$
Fridericia formula: $QTc = QT/RR\ 1/3$
Framingham formula: $QTc = QT + 0.154\ (1 - RR)$
Hodges formula: $QTc = QT + 1.75\ (\text{heart rate} - 60)$

QT dispersion (QTd), defined as the difference between the maximum and minimum QT intervals recorded in different ECG le msads, reflects hetergeneity in ventricular repolarization regional variations in repolarization and serves as a predictor of electrical instability and arrhythmogenic potential. Elevated QTd has been identified as an independent predictor of sudden cardiac death (SCD) in large-scale cohort studies, with pre-event QTd > 50 ms associated with a sixfold increase in SCD risk compared to individuals with QTd < 30 ms [15]. In patients receiving chemotherapy particularly anthracyclines - QTd and QTc dispersion have been demonstrated to increase significantly post-treatment, even when changes in mean QTc are modest, suggesting early electrophysiological effects preceding structural cardiomyophaty [16]. Despite its methodological limitations QT dispersion remains a valuable non-invasive marker for identifying patients at elevated risk of arrhythmia and SCD in the context of cardiotoxic therapy.

Several factors can contribute to the prolongation of the corrected QT interval (QTc), including demographic characteristics (age and gender), genetic predisposition, hormonal imbalances, electrolyte disturbances (such as hypokalemia and hypocalcemia), and drug-induced cardiotoxicity [17, 18].

In addition, various classes of drugs, such as antibiotics, chemotherapy, antidepressants, and antipsychotics, have been implicated in inducing cardiotoxicity by prolonging the QT interval. Understanding these phenomena is crucial for pharmacological and clinical safety, given that these drugs are widely used in clinical practice and can induce cardiotoxicity, prolonging the QT interval [19, 20].

Given the growing concern about the cardiovascular safety of widely prescribed drugs, it is essential to apply rigorous methods to assess the extent of cardiotoxicity and the impact of QT prolongation on patient morbidity and mortality.

A detailed analysis of these aspects is essential to guide safer therapeutic strategies and support updated clinical guidelines. This chapter aims to comprehensively analyze the latest scientific evidence on the relationship between the drugs mentioned and QT prolongation, highlighting the pathophysiological mechanisms involved, associated risk factors, and monitoring and clinical management strategies to prevent serious cardiac complications.

Pathophysiological Mechanisms

Patients with predisposing factors, such as hereditary long QT syndrome, hypokalemia, and concomitant use of multiple QT-prolonging drugs, require close monitoring. Serial ECGs and risk stratification based on predictive models are fundamental for the early identification of vulnerable patients [21]. Drug-induced QT interval prolongation involves diverse pathophysiological mechanism.

Ventricular repolarization is a complex process mediated by various ion channels, including sodium and potassium channels, especially hERG (human Ether-à-go-go-Related Gene) potassium channels. Blockage of these channels by the drugs aforementioned and further explored below can reduce electrical conduction through the potassium ion channels resulting in prolongation of the QT interval and increased risk of arrhythmias with potencial for sudden cardiac death [22].

The impact of drug-induced cardiotoxicity has been the subject of numerous meta-analyses and highly relevant studies in the scientific literature due to its influence on cardiac pharmacokinetics.

Drug Classes Involved in QT Interval Prolongation

Drugs from different classes modulate the activity of the hERG channel, including antipsychotics, chemotherapeutic agents, antimycobacterial, antiarrhythmics, antibiotics and antidepressants.

Antipsychotics: Pimozide, Haloperidol, Chorpromazine

Primozide is a typical neuroleptic or antipsychotic drug used in the management of Tourette syndrome and other psychiatric disorders. However, its use is associated with QT interval prolongation and an increased risk of severe ventricular arrhythmias, such as *torsades de pointes*. Studies indicate that the coadministration of clarithromycin may impair pimozide metabolism by inhibiting CYP3A enzyme activity, leading to increased plasma drug levels and enhanced cardiotoxicity [23].

Research has shown that pimozide significantly prolongs the QTc interval, an effect that becomes more pronounced when its metabolism is inhibited by clarithromycin. This mechanism raises the plasma concentration of the drug, increasing the risk of adverse cardiac events. Although pimozide can be metabolized through other pathways, CYP3A inhibition plays a central role in its elimination and is the primary factor involved in this process [23].

A single 6 mg dose of pimozide is sufficient to induce measurable changes in the QTc interval. Concomitant use with clarithromycin reduces its clearance, prolonging its pharmacological effects and increasing the potential for cardiac toxicity. Given this risk, it is recommended to avoid the combination of pimozide with potent CYP3A inhibitors, such as clarithromycin, to minimize the likelihood of fatal arrhythmias and ensure greater therapeutic safety [23]. Haloperidol , is a potent hERG channel blocker. Both oral and intravenous formulations can lead to significant QT prolongation and TdP, with the intravenous route posing higher risk due to peak plasma levels. The Food Drug Administration recommends cardiac monitoring for all patients receiving it. Chlorpromazine, another first-generation antipsychotic, exhibits moderate hERG blockade. Its QT-prolonging effect is dose-dependent and may be exacerbated by hypokalemia or concurrent medications. It can also interfere with calcium homeostasis and autonomic control, increasing TdP susceptibility in predisposed patients.

Chemotherapeutic Agents

Although cancer survival rates have increased due to advances in treatment, cardiotoxicity resulting from antineoplastic therapy represents a significant challenge [24, 25].

In chemotherapy, for example, anthracyclines, such as Doxorubicin and Epirubicin, are among the main agents implicated in QT prolongation, as evidenced by studies analyzing their association with cardiotoxicity in patients with Non-Hodgkin's Lymphoma (NHL) [21].

Careful assessment of the need to use drugs with QT prolongation potential, regular electrocardiographic monitoring, and correction of electrolyte disturbances are the main preventive strategies.

Adequate cardiac monitoring is essential for the early detection of cardiovascular complications and for improving the clinical outcomes of patients undergoing pharmacological treatments in oncology [26, 27].

Anthracyclines

Anthracyclines, including Doxorubicin, are widely used in treating various types of cancer, including soft tissue sarcomas. However, their use is associated with a significant risk of cardiotoxicity, which can lead to ventricular dysfunction and heart failure [28].

Anthracyclines are effective against various tumors, but their cardiotoxicity limits clinical use, potentially causing heart failure and cardiomyopathy. One of the main adverse effects is increased QT interval dispersion, which is associated with a higher risk of ventricular arrhythmias and cardiac mortality. This effect results from oxidative stress induced by free radicals, which affects ion channels and alters ventricular repolarization. Dexrazoxane, by chelating iron and reducing free radical formation, mitigates these alterations, contributing to electrophysiological stability and suggesting a potential cardioprotective effect [29].

Doxorubicin

A phase III study aimed to assess the incidence and risk factors for cardiotoxicity in patients with advanced soft tissue sarcoma treated with doxorubicin, who had not previously received chemotherapy treatment. Participants were randomized to receive D in doxorubicin at doses of 75 or 90 mg/m^2, administered every 3 weeks for up to 6 cycles [28].

Cardiac function was monitored using serial echocardiograms, assessing left ventricular ejection fraction (LVEF) and the incidence of congestive heart failure (CHF). Of the 326 patients included in the study, 319 were evaluable for cardiotoxicity. The incidence of CHF was 5% in the group treated with 75 mg/m^2 and 9% in the 90 mg/m^2 group. A reduction in LVEF was observed in 10% of patients in the 75 mg/m^2 group and 15% in the 90 mg/m^2 group [28].

Risk factors for cardiotoxicity included advanced age, pre-existing hypertension and concomitant use of other cardiotoxic agents. The results indicate that Doxorubicin is associated with a significant risk of cardiotoxicity in patients with advanced soft tissue sarcoma, especially at higher doses [28].

Early identification of risk factors, such as advanced age and hypertension, is crucial for implementing monitoring and prevention strategies. Regular monitoring of cardiac function during treatment with Doxorubicin is recommended to detect early signs of ventricular dysfunction and enable timely therapeutic interventions [28].

Doxorubicin is effective in the treatment of advanced soft tissue sarcomas, but its use is associated with a significant risk of cardiotoxicity. Therefore, regular monitoring of cardiac function and identification of risk factors are essential to minimize cardiovascular adverse effects and optimize therapeutic results [28].

Epirubicin

Epirubicin is an anthracycline widely used in the treatment of NHL, but its use is also associated with cardiotoxicity, including alterations in the QT interval, which can predispose to ventricular arrhythmias. QT dispersion is a measure of the heterogeneity of ventricular repolarization and a risk marker for arrhythmias. Dexrazoxane is a cardioprotective agent that can mitigate the cardiotoxic effects of anthracyclines [30].

Patients with NHL scheduled to receive chemotherapy containing Epirubicin were divided into two groups: one received only standard chemotherapy and the other received Dexrazoxane before each cycle of chemotherapy. Twelve-lead electrocardiograms were performed before and after each cycle to measure QTd. Cardiac function was also monitored by echocardiography. In the group receiving chemotherapy alone, a significant increase in QTd was observed after the initial cycles, indicating greater heterogeneity in ventricular repolarization. However, in the group that received concomitant Dexrazoxane, QTd remained stable throughout treatment [30].

Furthermore, left ventricular function was preserved in the Dexrazoxane group, while the ejection fraction was significantly reduced in the group without Dexrazoxane [30].

The findings suggest that Epirubicin-based chemotherapy increases QTd, potentially increasing the risk of ventricular arrhythmias in patients with NHL. Dexrazoxane administration prevents these electrocardiographic changes and preserves cardiac function, possibly due to its iron-chelating properties, which reduce the formation of free radicals and myocardial damage [30].

Dexrazoxane supplementation in NHL patients undergoing Epirubicin-based chemotherapy can prevent increased QT dispersion and deterioration in cardiac function, thus reducing the risk of treatment-associated ventricular arrhythmias [30].

Given the growing concern about the cardiovascular safety of widely prescribed drugs, it is essential to apply rigorous methods to assess the extent of cardiotoxicity and the impact of QT prolongation on patient morbidity and mortality.

A detailed analysis of these aspects is essential to guide safer therapeutic strategies and support up-to-date clinical guidelines. This chapter aims to comprehensively analyze the latest scientific evidence on the relationship between the drugs mentioned and QT prolongation, highlighting the pathophysiological mechanisms involved, the associated risk factors, and the monitoring and clinical management strategies to avoid serious cardiac complications.

Trametinib

A selective MEK inhibitor, is widely used in the treatment of solid tumors, especially in melanomas with BRAF V600 mutation. Despite its effectiveness, cardiotoxicity is a rare but severe side effect that can lead to heart failure, left ventricular dysfunction, and hypertension. QT interval prolongation, a risk marker for potentially fatal arrhythmias like torsades de pointes, underscores the need to investigate the drug's impact on cardiac repolarization [31].

Oncological patients, frequently exposed to multiple therapies and comorbidities, are at increased risk for electrophysiological changes. However, studies have shown that trametinib does not cause significant QT interval prolongation, even at supratherapeutic doses. Pharmacokinetic analysis revealed no relevant correlation between the drug's plasma concentrations and ventricular repolarization, reinforcing the safety of the treatment [31].

These findings are crucial for clinical management, allowing for the continued use of trametinib with appropriate monitoring. The confirmation that the drug does not significantly interfere with the heart's electrical stability provides additional safety for patients already facing the challenges of a complex illness, enabling better therapeutic outcomes with reduced cardiovascular risk [31].

Immune Checkpoint Inhibitors (ICIs)

Immune Checkpoint Inhibitors (ICIs) have been introduced as therapeutic options for a variety of cancers. This class of medication enhances antitumor activity through the activation of the immune system. ICIs can induce QT interval prolongation due to drug-induced cardiotoxicity and also myocarditis, which, although uncommon, is associated with an increased risk of Major Adverse Cardiac Events (MACE). In this regard, the ECG should be used to assess the QRS interval to stratify the risk of patients with myocarditis secondary to ICI use, as ECG is a widely available tool [32, 33].

Antimycobacterial: Bedaquiline

Bedaquiline is a groundbreaking antibiotic used to treat multidrug-resistant tuberculosis (MDR-TB). While it demonstrates significant efficacy, its use is associated with potential cardiovascular adverse effects, particularly QT interval prolongation. This electrocardiographic change serves as a marker for severe ventricular arrhythmias, including torsades de pointes [34].

Clinical studies have shown that bedaquiline administration can cause an average increase in the corrected QT interval (QTc) of up to 15.7 ms over 24 weeks of treatment. The risk of arrhythmias may be further heightened when bedaquiline is used in combination with other medications that also prolong the QT interval [34].

Bedaquiline works by inhibiting bacterial ATP synthase, a key enzyme for Mycobacterium tuberculosis replication. However, this mechanism can also affect human proteins, potentially causing cardiac toxicity. Additionally, bedaquiline may impair mitochondrial function, contributing to its cardiac adverse effects, and interfere with liver metabolism, leading to hepatotoxicity [34].

Due to these risks, patient monitoring during bedaquiline therapy is critical. Regular electrocardiograms should be performed to assess the QT interval, along with liver function tests to detect potential transaminase elevations. It is also recommended to avoid concomitant use of bedaquiline with other drugs that prolong the QT interval or are hepatotoxic, when possible [34].

While bedaquiline offers a significant advancement in the treatment of resistant tuberculosis, its use demands careful clinical monitoring to manage potential cardiac and hepatic adverse effects. A comprehensive understanding of these risks is essential to optimize treatment and ensure patient safety [34].

Antiarrhythmics

Amiodarone

Amiodarone is widely used in the treatment of cardiac arrhythmias, such as atrial fibrillation and ventricular tachycardia, due to its effectiveness. However, its use is associated with adverse effects, with QT interval prolongation being one of the most significant. This prolongation can predispose the patient to severe ventricular arrhythmias, such as ventricular fibrillation [35].

QT interval prolongation induced by amiodarone is a well-recognized side effect, which can be exacerbated by factors such as the administered dose, the patient's pre-existing health conditions, and interactions with other medications. These factors increase the risk of fatal complications, such as ventricular fibrillation. Therefore, rigorous monitoring of the QT interval is essential when managing patients undergoing amiodarone treatment [35].

It is crucial for healthcare professionals to remain vigilant, particularly in patients with a history of arrhythmias or those taking medications that affect the QT interval. Early recognition of QT interval changes allows for immediate intervention, reducing the incidence of severe adverse events and optimizing clinical management during amiodarone therapy [35].

Pathophysiological Mechanisms

Patients with predisposing factors, such as hereditary long QT syndromes, hypokalemia, and concomitant use of several QT-prolonging drugs, require close monitoring. Using serial ECGs and risk stratification based on predictive models are pivotal for the early identification of vulnerable patients [21].

Ventricular repolarization is a complex process mediated by various ion channels, including sodium and potassium channels, especially hERG (human Ether-à-go-go-Related Gene) potassium channels. Blockage of these channels by the drugs mentioned above and explored in more detail below can prolong the QT interval and increase the risk of arrhythmias [15].

The impact of drug-induced cardiotoxicity has been the subject of numerous meta-analyses and highly relevant studies in the scientific literature due to its influence on cardiac pharmacokinetics [21].

Classes of Drugs Involved in QT Interval Prolongation

Drugs from different classes modulate the activity of the hERG channel, including chemotherapy, antibiotics, antidepressants, and antipsychotics.

Although cancer survival rates have increased due to advances in treatment, cardiotoxicity resulting from antineoplastic therapy represents a significant challenge [24, 25].

In chemotherapy, for example, anthracyclines such as Doxorubicin and Epirubicin are among the main agents implicated in QT interval prolongation, as evidenced by studies that analyzed their association with cardiotoxicity in patients with Non-Hodgkin Lymphoma (NHL) [21].

The main preventive strategies are careful assessment of the need to use drugs with the potential to prolong the QT interval, regular electrocardiographic monitoring, and correction of electrolyte disturbances. Recent studies point to the role of Dexrazoxane in minimizing anthracycline-induced cardiotoxicity, making it a promising approach to preserving cardiac safety [19, 20].

Adequate cardiac monitoring is essential for the early detection of cardiovascular complications and for improving the clinical outcomes of patients undergoing pharmacological treatments in oncology [26, 27].

Anthracyclines

Anthracyclines, including doxorubicin, are widely used in treating various types of cancer, including soft tissue sarcomas. However, their use is associated with a significant risk of cardiotoxicity, which can lead to ventricular dysfunction and heart failure [28].

Doxorubicin

A phase III study aimed to assess the incidence and risk factors for cardiotoxicity in patients with advanced soft tissue sarcoma treated with doxorubicin who had not previously received chemotherapy treatment. Participants were randomized to receive doxorubicin at doses of 75 or 90 mg/m^2, administered every 3 weeks for up to 6 cycles [28].

Cardiac function was monitored using serial echocardiograms, assessing left ventricular ejection fraction (LVEF) and congestive heart failure (CHF) incidence. Of the 326 patients included in the study, 319 were assessed for cardiotoxicity. The incidence of CHF was 5% in the group treated with 75 mg/m^2 and 9% in the 90 mg/m^2 group. A reduction in LVEF was observed in 10% of patients in the 75 mg/m^2 group and 15% in the 90 mg/m^2 group [28].

Risk factors for cardiotoxicity included advanced age, pre-existing hypertension, and concomitant use of other cardiotoxic agents. The results indicate that doxorubicin is associated with a significant risk of cardiotoxicity in patients with advanced soft tissue sarcoma, especially at higher doses [28].

Early identification of risk factors, such as advanced age and hypertension, is crucial for implementing monitoring and prevention strategies. Regular monitoring of cardiac function is recommended during treatment with Doxorubicin to detect early signs of ventricular dysfunction and enable timely therapeutic interventions [28].

Doxorubicin is effective in the treatment of advanced soft tissue sarcomas, but its use is associated with a significant risk of cardiotoxicity. Therefore, regular monitoring of cardiac function and identification of risk factors are essential to minimize cardiovascular adverse effects and optimize therapeutic results [28].

Epirubicin

Epirubicin is an anthracycline widely used in the treatment of NHL. Still, its use is also associated with cardiotoxicity, including alterations in the QT interval, which can predispose to ventricular arrhythmias. QT dispersion measures the heterogeneity of ventricular repolarization and is a risk marker for arrhythmias. Dexrazoxane is a cardioprotective agent that can attenuate the cardiotoxic effects of anthracyclines [30].

Patients with NHL scheduled to receive chemotherapy containing epirubicin were divided into two groups: one received only standard chemotherapy, and the other received dexrazoxane before each cycle of chemotherapy. Twelve-lead electrocardiograms were performed before and after each cycle to measure QTd. Cardiac function was also monitored by echocardiography. A significant increase in QTd was observed in the group receiving chemotherapy alone after the initial cycles, indicating greater heterogeneity in ventricular repolarization. However, in the group that received Dexrazoxane concomitantly, QTd remained stable throughout treatment [30].

Furthermore, left ventricular function was preserved in the Dexrazoxane group, while the ejection fraction was significantly reduced in the group without Dexrazoxane [30].

The findings suggest that epirubicin-based chemotherapy increases QTd, potentially increasing the risk of ventricular arrhythmias in patients with NHL. The administration of dexrazoxane seems to prevent these electrocardiographic alterations and preserve cardiac function, possibly due to its iron-chelating properties, which reduce the formation of free radicals and myocardial damage [30].

Supplementation with dexrazoxane in NHL patients undergoing epirubicin-based chemotherapy can prevent an increase in QT dispersion and deterioration in cardiac function, thus reducing the risk of ventricular arrhythmias associated with treatment [30].

Drugs Withdrawn from the Market Due to QT Interval Prolongation

Terodiline is a drug that has been withdrawn from the market due to severe ventricular arrhythmias associated with QT interval prolongation [36].

In a double-blind, placebo-controlled, crossover study involving nine healthy volunteers, the cardiovascular and electrocardiographic effects of the R(+)- and S(−)-terodiline enantiomers were investigated. Each participant received single oral doses of racemic Terodiline (200 mg), R(+)-Terodiline (100 mg), S(−)-Terodiline (100 mg) or placebo. Plasma concentrations of each enantiomer and cardiovascular and electrocardiographic effects, including QT intervals and QT dispersion, were measured over 14 days after each treatment [36].

Both racemic terodiline and the R(+) enantiomer significantly increased the QT interval, corrected QT interval (QTc) and QRS duration (all $p < 0.05$), without affecting QT dispersion. The enantiomer S(−)-Terodiline (100 mg) had no effect on QTc. The maximum effects occurred 8 h after administration, with increases in QTc in relation to the baseline value of 23 ms for Racemic Terodiline, 19 ms for the R(+) enantiomer and 0 ms for the S(−) enantiomer. Although differences were observed in the pharmacokinetics of the two enantiomers, these were not sufficient to explain the differences in electrocardiographic effects, and the elimination half-lives were similar. The elimination of the Terodiline enantiomers was not significantly delayed in two poor metabolizers of debrisoquine (CYP2D6) [36].

The QT prolongation associated with racemic Terodiline is caused exclusively by the R(+) enantiomer, which therefore appears to be responsible for the ventricular arrhythmias induced by the drug [36].

The cardiotoxic effects of Terodiline are stereoselective, with the R(+) enantiomer being mainly responsible for the prolongation of the QT interval and the ventricular arrhythmias associated with the drug [36].

Contraindications

Patients diagnosed with congenital Long QT Syndrome or other genetic etiologies associated with QT interval prolongation should be extremely cautious when using drugs known to prolong this interval. In addition, individuals with electrolyte disorders, such as hypokalemia, hypomagnesemia, or hypocalcemia, should be closely monitored, as such imbalances can exacerbate QT interval prolongation, increasing the risk of malignant ventricular arrhythmias [37].

It is imperative to consider that patients taking diuretics or those with symptoms such as diarrhea can develop significant electrolyte alterations. Many of these individuals may not have their electrolyte levels assessed before taking drugs that prolong the QT interval, which reinforces the need for caution when prescribing these drugs [7].

Drug interactions are another important concern. Certain drugs that prolong the QT interval are metabolized by the cytochrome P450 (CYP450) enzyme system. Coadministration of CYP450 inhibitors can result in elevated plasma concentrations of these drugs, significantly increasing the risk of QT interval prolongation and, consequently, potentially fatal ventricular arrhythmias [7].

Clinical Management and Prevention Strategies

A detailed understanding of these drugs' mechanisms of action, potential adverse effects, specific contraindications, and the need for close monitoring is imperative for safe clinical practice. Healthcare professionals should be alert to drug interactions that may exacerbate QT prolongation and implement monitoring strategies, such as periodic ECGs and assessment of serum electrolyte levels, to mitigate the risk of cardiac toxicity [7].

The therapeutic approach to patients taking multiple agents with the potential to prolong the QT interval requires careful interprofessional coordination. This includes careful drug selection, dose adjustment as necessary, and patient education about the signs and symptoms of arrhythmia. Collaboration between cardiologists, pharmacologists, and other healthcare professionals is key to optimizing clinical outcomes and ensuring patient safety [7].

References

1. Naghavi M, Abajobir AA, Abbafati C et al (2017) Global, regional, and national age-sex specific mortality for 264 causes of death, 1980–2016: a systematic analysis for the Global Burden of Disease Study 2016. Lancet 390(10100):1151–1210. https://doi.org/10.1016/S0140-6736(17)32152-9
2. Chan CH, Liu CM, Chen PF et al (2024) Association between QT prolongation and cardiovascular mortality in cancer patients. Cardiooncology 10:69. https://doi.org/10.1186/s40959-024-00271-9
3. Zareba W, Moss AJ, Schwartz PJ (2003) Prolonged QT interval and risk of sudden cardiac death. Circulation 107(14):187–194
4. Draghici AE, Taylor JA (2016) A base fisiológica e a medição da variabilidade da frequência cardíaca em humanos. J Physiol Anthropol 35(1):22
5. Gianni C, Burkhardt JD, Trivedi C, Mohanty S, Natale A (2018) O papel da rede de Purkinje na fibrilação ventricular prematura desencadeada pelo complexo ventricular. J Interv Card Electrophysiol 52(3):375–383
6. Chien HT, Lin FJ, Juang JJ, Lin SW (2025) Clin Pharmacol Ther 117(2):495–505. https://doi.org/10.1002/cpt.3469. Accession Number: 39380486. https://www.ncbi.nlm.nih.gov/pubmed/39380486
7. Farzam K, Tivakaran VS (2025) QT prolonging drugs. In: StatPearls. StatPearls Publishing, Treasure Island, FL. Disponível em: https://www.ncbi.nlm.nih.gov/books/NBK534864/
8. Dessertenne F (1966) La tachycardie ventriculaire à deux foyers opposés variables. Arch Mal Coeur Vaiss 59(2):263–272

9. Berul C (2023) Acquired long QT syndrome: definitions, causes, and pathophysiology. UpToDate, Inc. Acesso em: 16 de fevereiro de 2025
10. Berul C (2022) Acquired long QT syndrome: clinical manifestations, diagnosis, and management. UpToDate, Inc. Acesso em: 16 de fevereiro de 2025
11. Postema PG, Wilde AA (2014) A medição do intervalo QT. Curr Cardiol Rev 10(3):287–294
12. Al-Akchar M, Siddique MS (2022) Síndrome do QT longo. In: StatPearls. StatPearls Publishing, Treasure Island, FL. [citado em 15 de fevereiro de 2025]. Disponível em: https://www.ncbi.nlm.nih.gov/books/NBK526060/
13. Bazett HC (1920) An analysis of the time-relations of electrocardiograms. Heart 7:353–370
14. Fridericia LS (1920) The duration of systole in the electrocardiogram of normal subjects and of patients with heart disease. Acta Med Scand 53(1):469–486
15. Malik M, Batchvarov VN (2000) Measurement, interpretation and clinical potential of QT dispersion. J Am Coll Cardiol 36(6):1749–1766
16. Zareba W, Lin DA. (2003) QT dispersion: is it a valid predictor of mortality? Pacing Clin Electrophysiol. (26):1048-1051. https://doi.org/10.1046/j.1460-9592.26.s5.6x
17. Heemskerk CPM, Pereboom M, van Stralen K et al (2018) Risk factors for QTc interval prolongation. Eur J Clin Pharmacol 74(2):183–191. https://doi.org/10.1007/s00228-017-2381-5
18. Mangoni AA, Kinirons MT, Swift CG, Jackson SH (2003) Impact of age on QT interval and QT dispersion in healthy subjects: a regression analysis. Age Ageing 32(3):326–331. https://doi.org/10.1093/ageing/32.3.326
19. Zareba W, Lin DA (2020) Effect of clarithromycin on pharmacokinetics & pharmacodynamics... Br J Clin Pharmacol 86(2):190–198
20. Zareba W,Lin D Antipsychotic drugs & QT interval prolongation. Cardiol Clin (20)38:(02) 201_213
21. Mesitskaya DF, Fashafsha ZZ, Poltavskaya MG et al (2021) A single-lead ECG-based cardiotoxicity detection in patients on polychemotherapy. CardioOncol 9(1):45–52
22. Recanatini M, Poluzzi E, Masetti M, Cavalli A, De Ponti F (2005) Prolongamento do QT através do bloqueio do canal hERG K(+): conhecimento atual e estratégias para a previsão precoce durante o desenvolvimento de medicamentos. Med Res Rev 25(2):133–166. https://doi.org/10.1002/med.20019
23. Desta Z, Kerbusch T, Flockhart DA (1999) Efeito da claritromicina na farmacocinética e farmacodinâmica do pimozida em indivíduos saudáveis com metabolizadores pobres e extensivos do citocromo P450 2D6 (CYP2D6). Clin Pharmacol Ther 65(1):10–20. https://doi.org/10.1016/S0009-9236(99)70117-7
24. Hashim D, Boffetta P, La Vecchia C et al (2016) The global decrease in cancer mortality: trends and disparities. Ann Oncol 27(5):926–933. https://doi.org/10.1093/annonc/mdw027
25. World Health Organization (WHO). Cancer. Published online February 3, 2022. https://www.who.int/news-room/fact-sheets/detail/cancer
26. Lyon AR, López-Fernández T, Couch LS et al (2022) 2022 ESC Guidelines on cardio-oncology developed in collaboration with the European Hematology Association (EHA), the European Society for Therapeutic Radiology and Oncology (ESTRO) and the International Cardio-Oncology Society (IC-OS). Eur Heart J 43(41):4229–4361. https://doi.org/10.1093/eurheartj/ehac244
27. Kuznetsova N, Gubina A, Sagirova Z, Dhif D, Gognieva D, Melnichuk O et al (2022) Left ventricular diastolic dysfunction screening by a smartphone-case based on single lead ECG. Clin Med Insights Cardiol 16:11795468221120088. https://doi.org/10.1177/11795468221120088
28. Jones RL, Wagner A, Kawai A, Shahir A, Soldatenkova V, Wright J, Tap WD (2016) Avaliação da cardiotoxicidade associada à doxorrubicina em pacientes com sarcoma de tecidos moles avançado em um estudo randomizado de fase III. Lancet Oncol 17(1):181–189. https://doi.org/10.1016/S1470-2045(15)00492-0

29. Jaillon P, Morganroth J, Brumpf I, Talbot G, Sparfloxacin Safety Group (1999) Visão geral dos dados de segurança eletrocardiográfica e cardiovascular da sparfloxacina. J Antimicrob Chemother 44(6):759–767. https://doi.org/10.1093/jac/44.6.759
30. Galetta F, Franzoni F, Cervetti G, Cecconi N, Carpi A, Petrini M, Santoro G (2005) Efeito da quimioterapia baseada em epirrubicina e suplementação com dexrazoxano na dispersão do QT em pacientes com linfoma não-Hodgkin. Biomed Pharmacother 59(10):541–545. https://doi.org/10.1016/j.biopha.2005.07.006
31. Patnaik A, Tolcher A, Papadopoulos KP, Beeram M, Rasco D, Werner TL, Bauman JW, Scheuber A, Cox DS, Patel BR, Zhou Y, Hamid M, Schramek D, Sharma S (2015) Estudo de Fase 1 para avaliar o efeito do inibidor de MEK trametinibe na repolarização cardíaca em pacientes com tumores sólidos. Cancer Chemother Pharmacol 75(5):999–1007. https://doi.org/10.1007/s00280-015-2741-0
32. Zlotoff DA et al (2021) Electrocardiographic features of immune checkpoint inhibitor associated myocarditis. J Immunother Cancer 9(3):e002007
33. Duan J et al (2018) Anticancer drugs-related QTc prolongation, torsade de pointes and sudden death: current evidence and future research perspectives. Oncotarget 9(39):25738
34. Traoré AN, Rikhotso MC, Banda NT, Mashilo MS, Ngandu JPK, Mavumengwana V, Loxton AG, Kinnear C, Potgieter N, Heysell S, Warren R (2022) Effectiveness of the novel anti-TB bedaquiline against drug-resistant TB in Africa: a systematic review of the literature. Pathogens 11(6):636. https://doi.org/10.3390/pathogens11060636. PMID: 35745490; PMCID: PMC9229213
35. Stojanovic M, Nikolic M, Nedeljkovic I, Gojkovic D (2023) QT prolongation preceding ventricular fibrillation after amiodarone administration: a case report. Cureus 15(12):e63763. https://doi.org/10.7759/cureus.63763. PMID: 39099957; PMCID: PMC11296664
36. Sicouri S, Antzelevitch C(2004). Mechanism underlying the actions of antidepressant and antipsychotic drugs that cause cardiac sudden arrest. J Clin Psychopharmacol. 24(5):S5–S14
37. Golzari H, Dawson NV, Speroff T, Thomas C (2007) Intervalos QTc prolongados em eletrocardiogramas de admissão: prevalência e correspondência com anormalidades eletrolíticas de admissão. Conn Med 71(7):389–397
38. Mark PB, Wang L, Arnold K et al (2006) Corrected QT interval assessment in clinical practice: the Peter Mark formula. Am J Cardiol 98(9):1177–1183
39. Malik M, Batchvarov VN (2000) Measurement, interpretation and clinical potential of QT dispersion. J Am Coll Cardiol 36(6):1749–1766.

Chapter 5
Impact of Cardiotoxicity on Cardiac Bioenergetics

Megan Young and Dunja Aksentijevic

Abstract Cardiotoxicity represents a significant adverse effect associated with various therapeutic interventions and the misuse of substances, profoundly disrupting cardiac bioenergetics. This chapter examines the underlying mechanisms of cardiotoxicity, emphasizing the roles of mitochondrial dysfunction and oxidative stress, which are instrumental in mediating effects on cardiac metabolism and ATP synthesis. We investigate how various pharmacological agents, including anthracyclines and HER2-targeted therapies, in addition to radiation exposure and the consumption of recreational drugs, induce oxidative stress and compromise mitochondrial integrity. Notably, anthracyclines, such as doxorubicin, cause mitochondrial dysfunction by promoting the generation of excessive reactive oxygen species (ROS). In parallel, HER2-targeted therapies disrupt the AKT signaling pathway, resulting in significant metabolic dysregulation. Moreover, radiation-induced cardiotoxicity is characterized by alterations in DNA integrity, mitochondrial impairment, and modifications in protein expression, all leading to compromised cardiac function. Additional classes of drugs, including antipsychotic medications, antibiotics, and anti-retroviral agents, have been implicated in cardiotoxicity through mechanisms such as the inhibition of mitochondrial protein synthesis and disturbances in energetic homeostasis. Furthermore, chronic alcohol consumption exacerbates oxidative stress and mitochondrial impairment, contributing to the development of alcoholic cardiomyopathy. Overall, these various cardiotoxic agents converge on mitochondrial dysfunction, highlighting the urgent need for targeted strategies aimed at mitigating their detrimental effects on cardiac energetics, optimizing therapeutic effectiveness, and ultimately preventing heart failure.

M. Young · D. Aksentijevic (✉)
William Harvey Research Institute, Barts Faculty of Medicine and Dentistry, Queen Mary University of London, London, UK
e-mail: d.aksentijevic@qmul.ac.uk

D. K. Agrawal et al. (eds.), *Cardiovascular Toxicity*, Advances in Biochemistry in Health and Disease 33, https://doi.org/10.1007/978-3-031-90598-8_5

Keywords Anthracyclines · Antibiotics · Anti-cancer drugs · Antipsychotics · Antiretroviral therapy · Cardiotoxicity · Energetics · HER2 · Metabolism · Mitochondria · Mitochondrial dysfunction · Oxidative stress · Radiation therapy · Substance abuse

Introduction

Cardiotoxicity can arise from therapeutic drug treatments and radiotherapy, as well as substance abuse including alcohol and recreational drugs. In the case of therapeutic drugs, cardiotoxicity is a leading cause of drug discontinuation thus limiting clinical use as well as effectiveness of some therapies. Mitochondrial dysfunction and oxidative stress are key mechanism responsible for cardiotoxicity as they ultimately govern cardiac bioenergetics and can trigger cellular death [1]. Therefore, there is an unmet need for therapeutic treatments to prevent cardiotoxicity and targeting cardiac metabolism and mitochondria may be a potential solution.

Mitochondrial Dysfunction and Oxidative Stress.

Mitochondria are abundant in cardiomyocytes and play a vital role in supporting cardiac function. They are central to cardiac metabolism and ATP production, as well as important mediators of stress response and cellular death. To maintain continuous myocardial contractility and sufficient cardiac output, the energetic demand of the heart is vast, requiring around 6 kg of ATP a day [2]. The breakdown of metabolic substrates, primarily fatty acids by beta-oxidation and glucose by glycolysis, fuels the tricarboxylic acid (TCA) cycle located in the mitochondrial matrix (Fig. 5.1). The reducing equivalents produced by the TCA cycle, NADH and $FADH_2$, then drive aerobic ATP production through the electron transport chain (ETC). Electrons are donated between complexes of the ETC located within the inner mitochondrial membrane, creating a proton gradient that is harnessed by ATP synthase to produce ATP in a process termed oxidative phosphorylation [3, 4] (Fig. 5.1). Cardiac metabolism is intrinsically linked with cardiac function such that disruption to mitochondria or any component of cardiac metabolism, be it substrate utilisation, oxidative phosphorylation, or ATP utilisation, will have profound effects on cardiac bioenergetics and energetic failure can ensue.

Aside from energy production, mitochondria play other physiological and pathophysiological roles, including metabolic signalling, regulation of redox balance and apoptosis. Mitochondria are a source of reactive oxygen species (ROS), in part as a by-product of the ETC, which in physiological conditions are kept under control by intracellular antioxidants. However, in pathological states excessive ROS production overruns the antioxidant defence mechanisms leading to oxidative stress [5]. ROS are capable of damaging mitochondrial DNA as well as compromising mitochondrial structure through lipid peroxidation. Furthermore, excessive ROS production

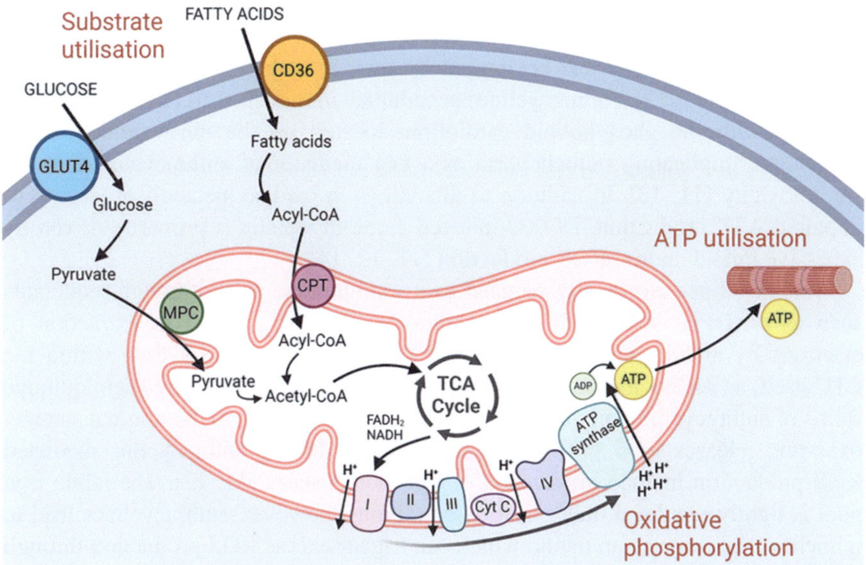

Fig. 5.1 Overview of cardiac metabolism. Fatty acids and glucose, primary substrates utilised in energy production, are transported into the mitochondria matrix where acetyl-CoA is produced to feed into the TCA cycle. The reducing equivalents $FADH_2$ and NADH produced by the TCA cycle donate electrons to the complexes of the ETC, generating a proton gradient across the mitochondrial membrane. This proton gradient is harnessed by ATP synthase to produce ATP which can then be utilised by the myofibrils to produce cardiac contraction. *ADP* adenosine diphosphate, *ATP* adenosine triphosphate, *CPT* carnitine palmitoyltransferase, *FADH* flavin adenine nucleotide, *GLUT4* glucose transporter 4, *MPC* mitochondrial pyruvate carrier, *NADH* nicotinamide adenine nucleotide, *TCA* tricarboxylic acid

can trigger opening of the mitochondrial permeability transition pore (mPTP), leading to a collapse of the mitochondrial membrane potential and ETC function, release of caspase enzymes into the cytosol ultimately triggering cell death.

Anti-cancer Therapies

Cardiotoxicity is a limiting factor of several anti-cancer therapies hampering the clinical use of such treatments [6, 7]. As cardiovascular disease is a leading cause of non-cancer deaths among survivors [8–10], understanding the mechanisms underlining cardiotoxicity is imperative to limit the mortality. Anti-cancer therapies often target metabolic pathways of tumor cells, which can impact substrate utilisation, as well as cause mitochondrial dysfunction in cardiomyocytes.

Anthracyclines

Anthracyclines, such as doxorubicin (DOXO), are potent anti-cancer drugs however their clinical application can be limited by their cumulative and dose-dependent cardiotoxicity [11, 12]. Anthracyclines accumulate in mitochondria as they bind in a complex with the phospholipid cardiolipin located on the inner mitochondrial membrane, implicating mitochondria as a key mediator of anthracycline-induced cardiotoxicity [11, 13]. In addition to alterations in cardiac metabolism leading to impaired ATP production, DOXO-induced cardiomyopathy is primarily driven by excessive mitochondrial ROS production [11, 12, 14].

Anthracyclines are readily reduced to a semiquinone form, through reductants such as NADPH or the ETC in the mitochondria [11, 14]. This extraction of electrons by anthracyclines from the ETC may disrupt electron flow within the ETC itself, as such compromising ATP production [14]. The produced semiquinone forms of anthracyclines are unstable and transformation back to the original state by oxidation releases ROS [14]. Other mechanisms driving anthracycline mediated ROS production include impairment of iron homeostasis [11, 12]. The labile iron pool is tightly regulated under healthy conditions; however, anthracyclines lead to mitochondrial iron accumulation which can trigger excess ROS production through the Fenton reaction and lead to cell death by ferroptosis [11, 14]. Targeting iron ferroptosis to mitigate DOXO-cardiotoxicity has been limited by inability to reach mitochondrial iron storage, reduced DOXO therapeutic efficiency as well as off-target pro-cancer effects [12]. In addition to iron mediated ROS production, the AKT signalling pathway is also implicated in the mechanism behind DOXO-cardiotoxicity leading to impaired ETC function and ROS production [11, 15]. However, despite the overwhelming evidence of increased ROS production and oxidative stress in DOXO-induced cardiotoxicity, antioxidant therapies have failed to show beneficial effects in mitigating cardiac injury [14].

DOXO has been reported to impact substrate utilisation and uptake [14]. DOXO was shown to trigger a switch in primary substrate utilisation from fatty acids to glucose [12, 16]. The healthy heart is metabolically flexible when it comes to substrate utilisation, however prolonged glucose utilisation is known to be maladaptive due to inefficient energy production, leading to a decline in function [2, 3]. A longitudinal study investigating DOXO-induced cardiac dysfunction reported a down-regulation of genes involved in fatty acid and glucose transport, as early as 1-week post-DOXO treatment [17]. Therefore, in addition to a switch in substrate preference, DOXO treatment causes reduced availability of both cardiac glucose and fatty acids resulting in impaired ATP production and oxygen consumption preceding a significant drop in ejection fraction [18, 19].

Inhibitors of the sodium-glucose linked-transporter (SGLT) 2 which are used to combat cardiovascular disease in patients with type 2 diabetes and chronic kidney disease, have shown promise in pre-clinical models of DOXO-induced cardiomyopathy [20–22]. Empagliflozin inhibition of SGLT2 was associated with increased levels of circulating ketone body, beta hydroxybutyrate (βOHB). βOHB treatment was shown to result in beneficial bioenergetic impact in DOXO-treated cells by reducing ROS generation [20].

HER2 Targeted Treatments

Human epidermal growth factor receptor 2 (HER2) protein is over-expressed in up to 30% of breast cancers and has also been identified in several other forms of cancer. HER2 drives tumour proliferation and overexpression is associated with lower survival rates [23, 24]. Therapeutically targeting HER2 is therefore an effective anti-cancer therapy [25]. HER2, part of the ErbB family of transmembrane growth factor receptor tyrosine kinases, plays a physiological role in the heart and cardiotoxicity is a side-effect of HER2 targeted treatments [26]. HER2 signalling is activated by formation of heterodimers with other HER proteins, the most potent combination being with HER3 following ligand binding [27]. Downstream pathways activated include PI3K/AKT and MAPK/ERK which are known to play a cardioprotective role in the heart [28]. Reduction of AKT signalling leads to mitochondrial impairment and alterations in metabolism, as well as increasing ROS production [28]. HER2 targeted therapies such as monoclonal antibodies, which interact with extracellular domains of the HER2 protein, suppress these downstream signalling pathways triggering cardiotoxic effects. HER2 associated cardiotoxicity in some instances can be alleviated when treatment ceases. However, during treatment periods cardiac function has been shown to decline rapidly [29]. For instance, Trastuzumab was shown to lead to a reduction in mitochondrial density, increased apoptosis, as well as the morphological changes in cardiac myofibrils limiting their ability to utilise ATP efficiently and contract [29]. As a result, a drop in cardiac function was observed within 7 days in Trastuzumab-treated mice [29]. Furthermore, HER2 targeted treatment given in combination with anthracyclines worsens cardiotoxicity [23].

Radiation Therapy

Radiation therapy utilised in cancer treatment can cause toxic effects to surrounding tissues, which when used in the thoracic region such as in treatment for breast and lung cancer can lead to cardiotoxicity [30]. Radiated-induced heart disease describes a spectrum of cardiac abnormalities, the prevalence of which can be hard to determine given different latency times however incidence has been reported as high as 40–50% in breast cancer and Hodgkin's lymphoma patients [31]. Radiation therapy directly modifies DNA, leading to mitochondrial damage and ROS production, subsequently having implications for mitochondrial function [30, 31]. It is known that gross morphological changes occur in the myocardium following radiation therapy, including alterations in mitochondrial structure [32–35]. Proteomic analysis of cardiac tissue revealed susceptibility of mitochondrial related proteins to irradiation, with changes identified as early as 24 hours, with upregulation of proteins associated with oxidative stress [36]. Furthermore, long-term changes in the cardiac proteome, with alterations in proteins involved in lipid metabolism, particularly PPARα, and oxidative phosphorylation have been reported

[35, 37]. Mitochondria isolated from rodent hearts exposed to radiation show extensive dysregulation including decreased respiration, changes in mitochondrial membrane potential, enhanced susceptibility to mPTP opening and apoptosis [37, 38]. Currently there are no therapeutic treatments to mitigate radiation induced cardiotoxicity, other than to minimise and localise radiation exposure [31].

Antipsychotics

Antipsychotic-induced cardiotoxicity has a range of presentations including arrythmia, myocarditis, ischemic injury as well as heart failure, and can be life-threatening with an increased risk of sudden cardiac death [39]. Mitochondrial dysfunction has been implicated in the mechanism underlying this cardiotoxicity. Clozapine, an effective antipsychotic that is associated with cardiotoxicity, undergoes bioactivation to a reactive nitrenium ion by cytochrome P450s which is readily detoxified by conjugation with reduced glutathione [40]. Cardiac mitochondria have been shown to catalyse the transformation from clozapine to reactive nitrenium ion, which may result in toxicity leading to mitochondrial damage, including complexes of the ETC [41].

Furthermore, 4-weeks of antipsychotic treatment in mice revealed changes in the cardiac proteome, including proteins comprising ETC complex 1, mitochondrial function and oxidative phosphorylation [42]. These changes in mitochondria will lead to a decline in mitochondrial respiration and energy production. In non-cardiac cell lines, as well as isolated cardiac mitochondria, antipsychotics have been reported to inhibit mitochondrial bioenergetics with reduced intracellular ATP levels and a decline in oxygen consumption rate [41–43].

Aside from direct cardiac effects, antipsychotic medications have been associated with worsening of metabolic syndromes in patients including insulin resistance and type 2 diabetes [44]. These metabolic phenotypes are associated with poor cardiovascular outcomes, with known changes in substrate utilisation and cardiac metabolism. However, the metabolic side-effects of antipsychotic medications are associated with clinical anti-psychotic improvements making the use of them determined on an individual basis [45].

Antibiotics

Tetracycline antibiotics, such as doxycycline, are known to inhibit mitochondrial protein synthesis through inhibition of mitochondrial DNA translation [46]. Doxycycline treatment reduced oxidative capacity, and ^{13}C labelling of glucose revealed a shift towards glucose use in anaerobic glycolysis away from incorporation into the TCA cycle in a dose-dependent manner [46]. These changes in metabolism were associated with higher diastolic calcium levels, higher susceptibility to arrhythmia and cardiac dysfunction, highlighting the link between bioenergetic capacity and cardiac function [46].

Anti-retroviral Therapy

Highly active anti-retrovirus therapy (HAART) is used in the treatment of human immunodeficiency virus (HIV) infection to block disease progression and alleviate symptoms. HAART therapies widely use inhibitors of nucleoside reverse transcriptase such as AZT (3′-azido 3′-deoxythymidine) which are associated with cardiotoxicity [47]. AZT is linked with mitochondrial dysfunction in part due to inhibiting mitochondrial DNA replication, which can impact the complexes of the ETC affecting energy production as well as ROS production [47–49]. This is reflected in the decline in cytochrome oxidase activity which comprises complex IV of the ETC, as well as the observed reduction in mitochondrial oxygen consumption rate following treatment with AZT [47, 50]. Confirming a role for ROS generation in AZT-induced cardiotoxicity, reduction in oxygen consumption rate is alleviated with pharmacological antioxidant treatment, alongside improvement in mitochondrial structure [48]. This is supported by findings from transgenic mouse models in which mice with enhance antioxidant capacity were protected protection against AZT-induced oxidative stress and cardiomyopathy involving cardiac dilation and hypertrophy [51]. Furthermore, AZT is reported to reduce mPTP opening which in part occurs due to excessive ROS generation [47]. A case-study reporting a patient treated with nucleoside-reverse transcriptase inhibitor reflects these pre-clinical findings, with the patient suffering from dilated cardiomyopathy and hypertrophied myocardial tissue with abnormal mitochondria structured observed in biopsied cardiac tissue [52].

Substance Abuse

Recreational Drugs

Cardiovascular complications are a common side effect of chronic use of all classes of recreational drugs [53, 54]. Research into the mechanisms behind the cardiovascular effects of these drugs is not extensive, but there are some implications on mitochondria and cardiac metabolism.

3,4-methylenedioxymethamphetamine (MDMA), commonly known as ecstasy, alters levels of serotonin, and can directly interact with serotonin receptors which are present on cardiomyocytes [55]. Cardiovascular side effects of MDMA use include hypertension and tachycardia which will demand an increased energy supply [56]. Acute MDMA exposure in rodents led to alterations in fatty acid metabolism including increased carnitine [55]. Chronic MDMA exposure initiates an oxidative stress response with identified nitration of mitochondrial proteins and myofilament proteins, associated with LV dilation and diastolic dysfunction [57]. Complexes of the ETC were among proteins identified to undergo nitration following MDMA treatment, which may affect ATP production [57]. Furthermore, antioxidant defence has been shown to be impaired following only an acute MDMA exposure, which will further exacerbate oxidative stress [58].

Impaired mitochondrial function, oxidative stress and reduced energetic production has also been associated with the abuse of cocaine [59].

Alcohol

Chronic alcohol consumption is associated with alcoholic cardiomyopathy (ACM), involving structural changes progressing to functional decline in the absence of confounding cardiovascular disease such as coronary artery disease [60]. Excessive ROS production leading to oxidative stress in ACM is reported in several studies, associated with changes in cardiac structure and function. Ethanol metabolism leads to the formation of acetaldehyde which has been shown to increase ROS production through a NOX2 mediated pathway [61]. Furthermore, sarcoplasmic reticulum calcium leak and contractile dysfunction induced by ethanol could be abolished by NOX2 deletion [62]. Compensatory increases in antioxidants, particularly catalase, are also seen in ACM to counteract ROS production, however these defence mechanisms are not sufficient to prevent oxidative stress but could be harnessed as a therapeutic strategy [63]. Excessive alcohol intake has also been linked to induction of mPTP opening as well as cellular death by ferroptosis and apoptosis [59, 64–66].

Chronic alcohol consumption has also been found to increase myocardial uptake of long-chain fatty acids via transporters CD36 and Slc27a1, concurrent with a decline in ATP and reduced cardiac function [67]. Furthermore, accumulation of mitochondrial free fatty acids with chronic alcohol consumption has been associated with reduced carnitine [68]. Capacity of the TCA cycle and complexes of the ETC decline with excess chronic alcohol consumption [69]. Reduction in myocardial oxygen consumption and myocardial external efficiency, which links myocardial metabolism to generated mechanical output, has been shown in patients with alcoholic cardiomyopathy [70, 71].

Summary

Cardiotoxicity is a consequence of pharmacological treatments, as well as abused substances. However, the mechanisms triggering cardiotoxicity converge on mitochondria with alterations in substrate utilisation, mitochondrial dysfunction, enhanced production of ROS and cell death. Collectively, these lead to reduced cardiac bioenergetic availability, ultimately leading to cardiac dysfunction (Fig. 5.2).

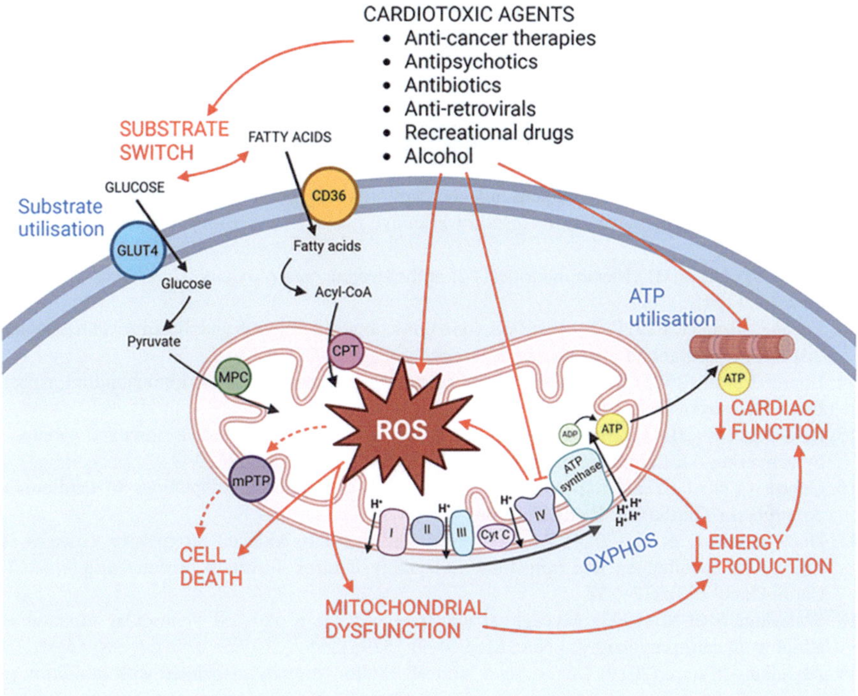

Fig. 5.2 Cardiac metabolism and mitochondria are key mediators of cardiotoxicity**.** ROS production, impairment of oxidative metabolism via the ETC, alterations in substrate utilisation and changes in cardiac structure are all implicated in cardiotoxicity. These changes can lead to mitochondrial dysfunction leading to reduced energy production impairing cardiac function, as well as triggering cell death. *ADP* adenosine diphosphate, *ATP* adenosine triphosphate, *CPT* carnitine palmitoyltransferase, *FADH* flavin adenine nucleotide, *GLUT4* glucose transporter 4, *MPC* mitochondrial pyruvate carrier, *mPTP* mitochondrial permeability transition pore, *NADH* nicotinamide adenine nucleotide, *OXPHOS* oxidative phosphorylation, *ROS*, reactive oxygen species

References

1. Varga ZV et al (2015) Drug-induced mitochondrial dysfunction and cardiotoxicity. Am J Physiol Heart Circ Physiol 309(9):H1453–H1467
2. Neubauer S (2007) The failing heart–an engine out of fuel. N Engl J Med 356(11):1140–1151
3. Lopaschuk GD et al (2021) Cardiac energy metabolism in heart failure. Circ Res 128(10): 1487–1513
4. Doenst T, Nguyen TD, Abel ED (2013) Cardiac metabolism in heart failure: implications beyond ATP production. Circ Res 113(6):709–724
5. D'Oria R et al (2020) The role of oxidative stress in cardiac disease: from physiological response to injury factor. Oxidative Med Cell Longev 2020:5732956
6. Abdul-Rahman T et al (2023) Chemotherapy induced cardiotoxicity: a state of the art review on general mechanisms, prevention, treatment and recent advances in novel therapeutics. Curr Probl Cardiol 48(4):101591

7. Morelli MB et al (2022) Cardiotoxicity of anticancer drugs: molecular mechanisms and strategies for cardioprotection. Front Cardiovasc Med 9:847012
8. Gon Y et al (2023) Heart disease mortality in cancer survivors: a population-based study in Japan. J Am Heart Assoc 12(23):e029967
9. Zaorsky NG et al (2017) Causes of death among cancer patients. Ann Oncol 28(2):400–407
10. Armanious MA et al (2018) Cardiovascular effects of radiation therapy. Curr Probl Cancer 42(4):433–442
11. Rawat PS et al (2021) Doxorubicin-induced cardiotoxicity: an update on the molecular mechanism and novel therapeutic strategies for effective management. Biomed Pharmacother 139: 111708
12. Russo M et al (2021) Metabolic aspects of anthracycline cardiotoxicity. Curr Treat Options in Oncol 22(2):18
13. Paradies G et al (2019) Role of cardiolipin in mitochondrial function and dynamics in health and disease: molecular and pharmacological aspects. Cells 8(7):728
14. Huang J et al (2022) Understanding anthracycline cardiotoxicity from mitochondrial aspect. Front Pharmacol 13:811406
15. Liao ZQ et al (2021) Rutaecarpine inhibits doxorubicin-induced oxidative stress and apoptosis by activating AKT signaling pathway. Front Cardiovasc Med 8:809689
16. Guerra G et al (2024) Exploring the mechanisms of metabolic adaptations to cardiotoxic doxorubicin. Cardiovasc Res 120(1):44
17. Diaz-Guerra A et al (2024) Anthracycline cardiotoxicity induces progressive changes in myocardial metabolism and mitochondrial quality control: novel therapeutic target. JACC CardioOncol 6(2):217–232
18. Wakasugi S et al (1993) Myocardial substrate utilization and left ventricular function in adriamycin cardiomyopathy. J Nucl Med 34(9):1529–1535
19. Abdullah CS et al (2019) Doxorubicin-induced cardiomyopathy associated with inhibition of autophagic degradation process and defects in mitochondrial respiration. Sci Rep 9(1):2002
20. Oh CM et al (2019) Cardioprotective potential of an SGLT2 inhibitor against doxorubicin-induced heart failure. Korean Circ J 49(12):1183–1195
21. Dabour MS et al (2024) The cardioprotective and anticancer effects of SGLT2 inhibitors: JACC: cardiooncology state-of-the-art review. JACC CardioOncol 6(2):159–182
22. Sabatino J et al (2020) Empagliflozin prevents doxorubicin-induced myocardial dysfunction. Cardiovasc Diabetol 19(1):66
23. Slamon DJ et al (2001) Use of chemotherapy plus a monoclonal antibody against HER2 for metastatic breast cancer that overexpresses HER2. N Engl J Med 344(11):783–792
24. English DP, Roque DM, Santin AD (2013) HER2 expression beyond breast cancer: therapeutic implications for gynecologic malignancies. Mol Diagn Ther 17(2):85–99
25. Slavcheva SE, Angelov A (2023) HER2-targeted therapy-from pathophysiology to clinical manifestation: a narrative review. J Cardiovasc Dev Dis 10(12):489
26. Ponde NF, Lambertini M, de Azambuja E (2016) Twenty years of anti-HER2 therapy-associated cardiotoxicity. ESMO Open 1(4):e000073
27. Rubin I, Yarden Y (2001) The basic biology of HER2. Ann Oncol 12(1):3–8
28. Eaton H, Timm KN (2023) Mechanisms of trastuzumab induced cardiotoxicity – is exercise a potential treatment? Cardio-Oncology 9(1):22
29. ElZarrad MK et al (2013) Trastuzumab alters the expression of genes essential for cardiac function and induces ultrastructural changes of cardiomyocytes in mice. PLoS One 8(11): e79543
30. Livingston K et al (2020) The role of mitochondrial dysfunction in radiation-induced heart disease: from bench to bedside. Front Cardiovasc Med 7:20
31. Wang H et al (2019) Radiation-induced heart disease: a review of classification, mechanism and prevention. Int J Biol Sci 15(10):2128–2138
32. Khan MY (1973) Radiation-induced cardiomyopathy. I. An electron microscopic study of cardiac muscle cells. Am J Pathol 73(1):131–146

33. Burch GE et al (1968) Effects of radiation on the human heart. An electron microscopic study. Arch Intern Med 121(3):230–234
34. Ferreira-Machado SC et al (2013) Caspase-3 activation and increased procollagen type I in irradiated hearts. An Acad Bras Cienc 85(1):215–222
35. Azimzadeh O et al (2013) PPAR alpha: a novel radiation target in locally exposed mus musculus heart revealed by quantitative proteomics. J Proteome Res 12(6):2700–2714
36. Azimzadeh O et al (2011) Rapid proteomic remodeling of cardiac tissue caused by total body ionizing radiation. Proteomics 11(16):3299–3311
37. Barjaktarovic Z et al (2011) Radiation-induced signaling results in mitochondrial impairment in mouse heart at 4 weeks after exposure to X-rays. PLoS One 6(12):e27811
38. Sridharan V et al (2014) Radiation-induced alterations in mitochondria of the rat heart. Radiat Res 181(3):324–334
39. Li XQ, Tang XR, Li LL (2021) Antipsychotics cardiotoxicity: what's known and what's next. World J Psychiatry 11(10):736–753
40. Dragovic S et al (2013) Characterization of human cytochrome P450s involved in the bioactivation of clozapine. Drug Metab Dispos 41(3):651–658
41. Arzuk E, Karakus F, Orhan H (2021) Bioactivation of clozapine by mitochondria of the murine heart: possible cause of cardiotoxicity. Toxicology 447:152628
42. Beauchemin M et al (2020) Exploring mechanisms of increased cardiovascular disease risk with antipsychotic medications: risperidone alters the cardiac proteomic signature in mice. Pharmacol Res 152:104589
43. Elmorsy E et al (2017) Effect of antipsychotics on mitochondrial bioenergetics of rat ovarian theca cells. Toxicol Lett 272:94–100
44. Scigliano G, Ronchetti G (2013) Antipsychotic-induced metabolic and cardiovascular side effects in schizophrenia: a novel mechanistic hypothesis. CNS Drugs 27(4):249–257
45. Pillinger T et al (2020) Comparative effects of 18 antipsychotics on metabolic function in patients with schizophrenia, predictors of metabolic dysregulation, and association with psychopathology: a systematic review and network meta-analysis. Lancet Psychiatry 7(1):64–77
46. Wust RCI et al (2021) The antibiotic doxycycline impairs cardiac mitochondrial and contractile function. Int J Mol Sci 22(8):4100
47. Nomura R et al (2017) Azidothymidine-triphosphate impairs mitochondrial dynamics by disrupting the quality control system. Redox Biol 13:407–417
48. Liu Y et al (2014) Tempol protects cardiomyocytes from nucleoside reverse transcriptase inhibitor-induced mitochondrial toxicity. Toxicol Sci 139(1):133–141
49. de la Asuncion JG et al (2004) AZT induces oxidative damage to cardiac mitochondria: protective effect of vitamins C and E. Life Sci 76(1):47–56
50. McCurdy DT, Kennedy JM (1998) AZT decreases rat myocardial cytochrome oxidase activity and increases beta-myosin heavy chain content. J Mol Cell Cardiol 30(10):1979–1989
51. Kohler JJ et al (2009) Transgenic mitochondrial superoxide dismutase and mitochondrially targeted catalase prevent antiretroviral-induced oxidative stress and cardiomyopathy. Lab Investig 89(7):782–790
52. Frerichs FC, Dingemans KP, Brinkman K (2002) Cardiomyopathy with mitochondrial damage associated with nucleoside reverse-transcriptase inhibitors. N Engl J Med 347(23):1895–1896
53. Roset-Altadill A, Wat D, Radike M (2024) Cardiovascular and pulmonary complications of recreational drugs: a pictorial review. Eur J Radiol 178:111648
54. Ghuran A, van Der Wieken LR, Nolan J (2001) Cardiovascular complications of recreational drugs. BMJ 323(7311):464–466
55. Perrine SA et al (2009) Cardiac effects of MDMA on the metabolic profile determined with 1H-magnetic resonance spectroscopy in the rat. NMR Biomed 22(4):419–425
56. Fonseca DA et al (2021) Ecstasy (3,4-methylenedioxymethamphetamine): cardiovascular effects and mechanisms. Eur J Pharmacol 903:174156
57. Shenouda SK et al (2008) Ecstasy produces left ventricular dysfunction and oxidative stress in rats. Cardiovasc Res 79(4):662–670

58. Cerretani D et al (2008) Cardiac oxidative stress determination and myocardial morphology after a single ecstasy (MDMA) administration in a rat model. Int J Legal Med 122(6):461–469
59. Martins MJ et al (2018) Ethanol addictively enhances the in vitro cardiotoxicity of cocaine through oxidative damage, energetic deregulation, and apoptosis. Arch Toxicol 92(7): 2311–2325
60. Piano MR, Phillips SA (2014) Alcoholic cardiomyopathy: pathophysiologic insights. Cardiovasc Toxicol 14(4):291–308
61. Brandt M et al (2016) NOX2 amplifies acetaldehyde-mediated cardiomyocyte mitochondrial dysfunction in alcoholic cardiomyopathy. Sci Rep 6:32554
62. Mustroph J et al (2018) SR Ca(2+)-leak and disordered excitation-contraction coupling as the basis for arrhythmogenic and negative inotropic effects of acute ethanol exposure. J Mol Cell Cardiol 116:81–90
63. Mustroph J et al (2019) Mechanisms of cardiac ethanol toxicity and novel treatment options. Pharmacol Ther 197:1–10
64. Song C et al (2024) Role of ferroptosis regulation by Nrf2/NQO1 pathway in alcohol-induced cardiotoxicity in vitro and in vivo. Chem Res Toxicol 37(6):1044–1052
65. Hajnoczky G et al (2005) Alcohol and mitochondria in cardiac apoptosis: mechanisms and visualization. Alcohol Clin Exp Res 29(5):693–701
66. Steiner JL, Lang CH (2017) Etiology of alcoholic cardiomyopathy: mitochondria, oxidative stress and apoptosis. Int J Biochem Cell Biol 89:125–135
67. Hu C et al (2013) Chronic ethanol consumption increases cardiomyocyte fatty acid uptake and decreases ventricular contractile function in C57BL/6J mice. J Mol Cell Cardiol 59:30–40
68. Jing L et al (2011) Carnitine regulates myocardial metabolism by peroxisome proliferator-activated receptor-alpha (PPARalpha) in alcoholic cardiomyopathy. Med Sci Monit 17(1): BR1–BR9
69. Matyas C et al (2016) Chronic plus binge ethanol feeding induces myocardial oxidative stress, mitochondrial and cardiovascular dysfunction, and steatosis. Am J Physiol Heart Circ Physiol 310(11):H1658–H1670
70. Shi X et al (2022) Characterization of myocardial oxidative metabolism and myocardial external efficiency in high-risk alcohol cardiotoxicity and alcoholic cardiomyopathy via dynamic 11C-Acetate positron emission tomography. J Nucl Cardiol 29(1):278–288
71. Liu S et al (2020) Myocardial tissue and metabolism characterization in men with alcohol consumption by cardiovascular magnetic resonance and 11C-acetate PET/CT. J Cardiovasc Magn Reson 22(1):23

Chapter 6
Shared Pathophysiologic Mechanisms Between Cancer and Pulmonary Hypertension

Sumner Gardner, Faris Abusharkh, and Vineet Agrawal

Abstract Pulmonary hypertension comprises a heterogeneous collection of disorders that all collectively result in the development of elevated mean resting pressures in the pulmonary artery. Key pathobiological manifestations of pulmonary hypertension include pathologic vascular remodeling of the arteries and veins of the pulmonary vasculature and eventual load stress-induced failure of the right ventricle. Regardless of the upstream cause of pulmonary hypertension, there has been a growing recognition of both shared risk factors and similar pathophysiologic mechanisms that drive both cancer and pulmonary hypertension This book chapter discusses the existing literature that demonstrates similar multi-hit genetic mutations that can lead to both pulmonary hypertension and cancer, similar pathways that drive metabolic reprogramming of affected cells in both conditions, and abnormal responses to oxygen tension changes in the environment. Finally, this book chapter also discusses the existing literature that supports how cancer can itself drive pathologic pulmonary hypertension through increased venous thromboembolic risk, tumor metastasis, and because of novel cancer therapeutics.

Keywords Pulmonary circulation · Pulmonary hypertension · Cancer · Glycolysis · Lipid metabolism · Mitochondrial dysfunction · Thrombosis · Chronic thromboembolic pulmonary hypertension · Metastasis · Therapy complications

---

Sumner Gardner and Faris Abusharkh contributed equally with all other contributors.

S. Gardner · F. Abusharkh
Vanderbilt University, Nashville, TN, USA

V. Agrawal (✉)
Vanderbilt University, Nashville, TN, USA

Division of Cardiovascular Medicine, Department of Medicine, Vanderbilt University Medical Center, Nashville, TN, USA

Department of Veterans Affairs, Tennessee Valley Healthcare System, Nashville, TN, USA
e-mail: vineet.agrawal@va.gov; vineet.agrawal@vumc.org

D. K. Agrawal et al. (eds.), *Cardiovascular Toxicity*, Advances in Biochemistry in Health and Disease 33, https://doi.org/10.1007/978-3-031-90598-8_6

Introduction

Pulmonary hypertension is defined by a heterogenous collection of disorders that all result in the development of an elevated resting mean arterial pressure above 20 mmHg in the pulmonary arteries [1]. This diagnosis can be suggested by non-invasive modalities such as echocardiography [2–4], but ultimately requires direct catheter-based measurement of pulmonary pressure through procedures such as right heart catheterizations [1]. Increases in pulmonary pressure can occur from heterogeneous causes that variably lead to remodeling of both the arterial (or pre-capillary) vascular bed or venous (post-capillary) vascular bed within the pulmonary vasculature [5].

The World Health Organization has classified patients into 5 clinical subgroups of pulmonary hypertension: [1] Group 1—Pulmonary arterial hypertension (PAH), [2] Group 2—Pulmonary hypertension due to left heart disease, [3] Group 3—Pulmonary hypertension due to intrinsic lung disease, [4] Group 4—Pulmonary hypertension due to chronic thromboembolic disease or obstruction, and [5] Group 5—pulmonary hypertension due to multifactorial mechanisms (Fig. 6.1). Epidemiologically, the true incidence and prevalence of pulmonary hypertension worldwide is poorly understood. In part due to incomplete data outside of the United States of America and Europe for diagnosis, and in part due to recent changes in the definition of pulmonary hypertension lowering the threshold for diagnosis to 20 mmHg [1], pulmonary hypertension incidence and prevalence worldwide are likely are likely underestimated. More contemporary estimates suggest that the incidence of WHO Group 1 pulmonary arterial hypertension is anywhere between 1 and eight million adults per year in developed countries, and prevalence from 6–30 million adults per year [6]. The prevalences of Groups 2 and 3 pulmonary hypertension, due to heart disease and lung disease respectively, are considered to be far higher both in developed countries and worldwide, with some estimates extrapolating that up to 60 million adults worldwide may suffer from pulmonary hypertension due to heart disease (e.g. Group 2) worldwide alone [6].

There have been many aspects of pulmonary hypertension that have been shown to be intricately connected with malignancies and cancers. The goal of this book chapter is to synthesize and review these various aspects. The book chapter will cover overlapping and similar pathophysiologic mechanisms that have been to contribute to growth and metastasis in the context of cancer as well as pathologic vascular remodeling in the context of pulmonary hypertension. The book chapter will also cover how the presence of cancer, either through metastatic occlusion of pulmonary vasculature or through promotion of a pro-thrombotic state, can give rise to pulmonary hypertension. Finally, the book chapter will discuss how emerging precision medicine-based cancer therapeutics may contribute to the development of pulmonary hypertension in cancer survivors as well as patients actively undergoing treatment.

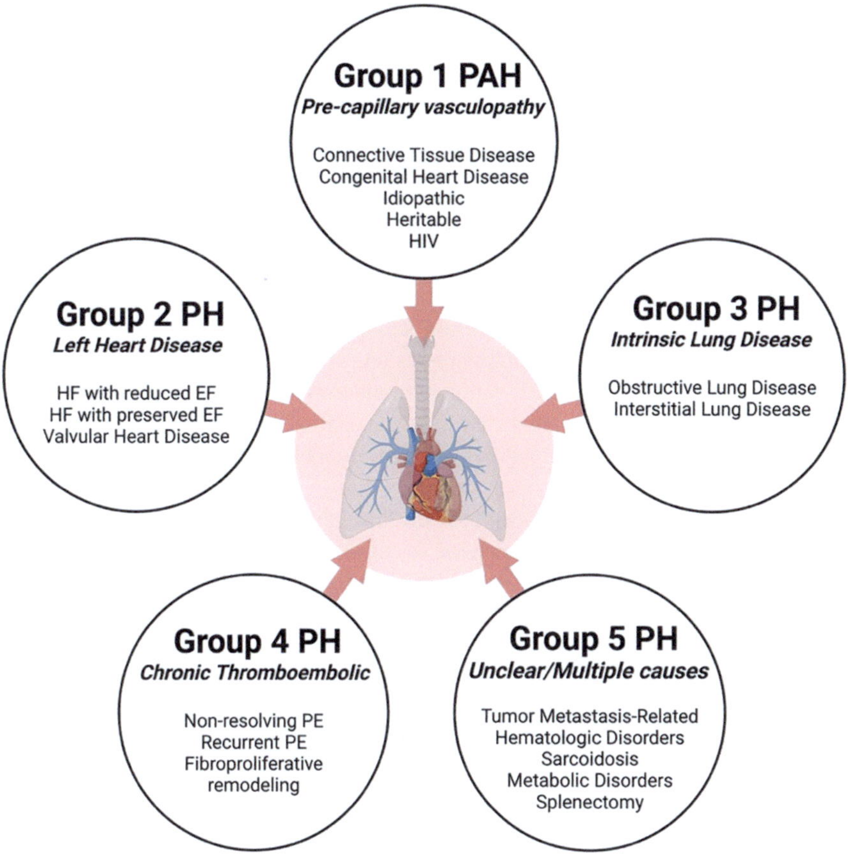

Fig. 6.1 Clinical classification system for pulmonary hypertension based on the World Health Organization (WHO) classification that includes group 1 (pulmonary arterial hypertension), group 2 (pulmonary hypertension due to left heart disease), group 3 (pulmonary hypertension due to intrinsic lung disease), group 4 (pulmonary hypertension due to chronic thromboembolic pulmonary hypertension), and group 5 (pulmonary hypertension due to miscellaneous, unclear, or multifactorial causes). *HIV* human immunodeficiency virus, *HF* heart failure, *EF* ejection fraction, *PE* pulmonary embolus

Pathophysiologic Mechanisms that Contribute to Pulmonary Hypertension and Cancer

Much of our understanding of mechanisms governing pathologic pulmonary vascular remodeling are derived from experimental, preclinical models that approximate features of Group 1 pulmonary arterial hypertension. Though distinct in the organs affected and in primary means by which molecular changes occur, studies have found many molecular similarities between mechanisms that drive cancer progression and pulmonary vascular remodeling.

PAH is thought to result in a hyperproliferative state that involves both endothelial cells and smooth muscle cells in the pulmonary vasculature after repetitive and sustained insults such as vascular injury, inflammation, shear stress, and pseudohypoxia in the setting of underlying predispositions such as genetic mutations, termed the "multiple hit hypothesis" [7]. Ultimately, these various factors lead to adaptive changes in cells through metabolic reprogramming and activation of pro-proliferative pathways, leading to a hyperproliferative state in certain cell populations. These various factors and their similarities to cancer pathogenesis are discussed below.

The concept that PAH may share pathophysiologic mechanisms with cancer was first considered after studies identified clonal expansions of endothelial cell populations in the occlusive pulmonary lesions of patients with PAH, known as plexiform lesions [8]. Using methylation patterns from samples derived patients with PAH vs pulmonary hypertension due to other causes, they identified that 77% of lesions in PAH patients displayed evidence of monoclonal expansion, compared to all lesions in patients with PH from secondary causes all being polyclonal in origin. Supporting this finding, more recent studies have shown through direct isolation of endothelial cells from the catheters used for clinically conducted right heart catheterizations that endothelial cells derived from patients with PAH were more resistant to apoptosis [9]. Similar in concept to the multi-hit hypothesis of cancer development [10], a similar multi-hit hypothesis has been proposed in the development of PAH to include predisposing genetic factors [11], inflammation, repetitive cell injury, and metabolic reprogramming.

Genetic Predispositions to PAH Development

With respect to predisposing factors, it is well established through genetic studies of patients with pulmonary arterial hypertension that a subset of patients harbor pathogenic mutations in genes that contribute to the development of PAH [12, 13]. The first such gene was identified in 2000 to be the bone morphogenic protein receptor 2, or BMPR2 gene. Mutations in BMPR2 are thought to be responsible for nearly 70% of cases of familial or heritable PAH [14]. Introduction of a dominant negative patient-derived mutation in BMPR2 in the smooth muscle of mice alone was sufficient to promote influx of inflammatory cells in the lungs as well as an imbalance between hyperproliferation and apoptosis in the smooth muscles of pulmonary arteries [15]. Furthermore, studies have demonstrated that loss of BMPR2 drives a paradoxical hyperproliferative response as well as an increased susceptibility to apoptosis in pulmonary arterial endothelial cells when stimulated by BMP9, a ligand known to bind BMPR2 [11]. Additionally, loss of BMPR2 impairs monolayer integrity in vitro and endothelial cell barrier function in vivo in pulmonary endothelium, consistent with a phenotypic "dedifferentiation" or partial endothelial-to-mesenchymal transition that is thought to occur in malignancies as well [16]. Consistent with the popular multi-hit theory of cancer development [10], there is also growing recognition that the presence of PAH-associated mutations in BMPR2 represent an initial hit that then requires additional hits prior to the development of pathologic pulmonary vascular remodeling. Such stimuli have been proposed to include endothelial injury, inflammation, prolonged exposure to hypoxia, and exposure to serotonin (findings reviewed in [17]). Finally, since the identification of BMPR2, other genetic mutations have been identified and recognized by the World Symposium on Pulmonary Hypertension as pathogenic contributors to PAH. These include mutations in other genes known to the affect the BMP/TGF pathways (ALK1, ENG, SMAD9, SMAD1, CAV1) as well as novel genes such as KCNK3, TBX4, EIF2KA4, and SOX18 [12, 13, 18]. More work remains to understand the role that these genetic mutations play in contributing to pulmonary vascular remodeling in PAH.

Metabolic Reprogramming in PAH

Metabolic reprogramming is thought to be central to tumor initiation and progression. Due to cancer driving mutations, genomic instability (including mitochondrial genomic instability), and changes in environmental nutrient availability, cancer cells adopt wide scale metabolic changes in order to maintain cell proliferation, cell survival, and counteract oxidative stress [19]. The most famous example of altered metabolism within cancers comes from Otto Warburg, who first described the Warburg effect in cancer cells. The Warburg effect refers to an observation that cancer cells preferentially utilize glucose to undergo glycolysis despite the presence of oxygen, generating large amounts of lactic acid in the process [20]. The molecular basis of this observation, which was confirmed both in vitro and in vivo [21], was later found to be due, in part, to alterations in key pathways activated in cancer cells. These pathways include hypoxia inducible factor (HIF), AKT, and mTOR pathways, respectively [22–24].

A similar phenomenon has been noted in both endothelial cells and smooth muscle cells from the pulmonary vasculature of patients with pulmonary arterial hypertension [25, 26]. Studies have demonstrated that the smooth muscles in pulmonary arteries suppress glucose oxidation and increase cytoplasmic glycolysis in part through inhibition of the pyruvate dehydrogenase enzyme (PDH). As PDH is a key enzyme in converting pyruvate to Acetyl-CoA to activate the TCA cycle, the result of inhibition of PDH is decreased Krebs cycle activity and downstream oxidative phosphorylation through mitochondrial complexes [27–29]. Ultimately, this results in over-reliance on glycolysis in the cytoplasm for generation of ATP, which occurs even in the absence of hypoxia (Fig. 6.2).

Concomitant with abnormal use of glucose is the dysregulation of utilization of other substrates for metabolism in PAH as well. These include over-reliance on amino acid substrates to supply the TCA cycle as well as dysregulated metabolism. While much of this work has been focused on the right ventricle of preclinical models of pulmonary hypertension, in part due to the high metabolic demands of the

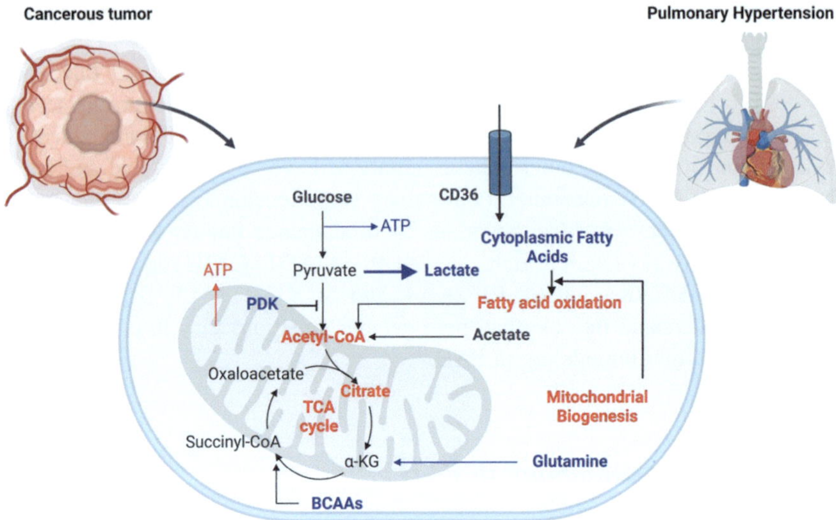

Fig. 6.2 Shared metabolic pathways that are dysregulated in both cancers and pulmonary hypertension include over-reliance on ATP generation through glycolysis, diminished production of Acetyl-CoA through glucose to fuel the Tricarboxylic Acid (TCA) cycle, increased fatty acid uptake, diminished fatty acid oxidation due to transcriptional downregulation of fatty acid oxidation-related genes, and increased utilization of branched-chain amino acids (BCAA) and glutamine as fuel sources for the TCA cycle. *ATP* adenosine triphosphate, *BCAA* branched chain amino acids, *TCA* tricarboxylic acid

heart, many of these same findings have been applied to endothelial cell populations in PAH models as well. In the presence of dysregulated glucose metabolism, studies have identified an increased reliance on amino acid substrates for metabolic function in various cell types in PAH. Glutamine is actively transported into the cell via the SLC1A5 transporter [30]. An important first step in the utilization of glutamine as an energy substrate is the breakdown of glutamine into glutamate and ammonia by glutaminase. Studies have shown increased glutaminase expression in the lung tissue of patients with PAH [31–33]. This has been shown through both transcript level expression and through post-translational modifications through processes such as serotonylation [33]. In the setting of pulmonary hypertension, either due to mechanical stimuli such as increased stiffness from extracellular matrix or through hypoxia-mediated signaling through HIF1a, increased uptake of glutamine is modulated via HIF1a or c-myc signaling to continue to supply the TCA cycle for ATP production. This occurs at the cost of utilization of glutamate to generate glutathione or NADPH to regulate oxidative stress-mediated generation of oxygen free radicals in the cell, further disrupting the redox balance of cells [30] (Fig. 6.2).

Evidence of dysregulation lipid metabolism in both cancer and pulmonary hypertension has recently garnered greater attention. Lipid metabolites include fatty acids, phospholipids, and cholesterol comprise an important class of metabolites that serve as important nutrients for cellular and organ-level homeostasis. In addition to representing the building blocks of cellular components such as the plasma membrane, a number of lipid metabolites undergo fatty acid oxidation to generate Acetyl coenzyme A (Acetyl-CoA), an important intermediate for oxidative phosphorylation pathways [34]. A key feature of cancer cells is dysregulated and aggressive proliferation, a process that requires not only adequate energy for growth but also adequate supply of building blocks of cells [19]. Thus, it is not surprising that there is a widespread reprogramming of fatty acid metabolic programs that accommodates this growth.

While typically cells regulate the uptake of lipid species as a primary source of fatty acids in cells, there is evidence that cancer cells exhibit many changes to enhance lipid species through both increased synthesis of lipid species as well as increased uptake. Supporting the prognostic importance of lipid metabolic pathways in cancers, studies have shown that the metastatic potential of cells is dependent on the ratio of saturated to unsaturated fatty acids, such that inhibitors of stearoyl-CoA desaturase, a key membrane-associated enzymes in the ER that catalyzes formation of monounsaturated fatty acids, can inhibit metastatic potential of cancers and decrease cancer cell survival [35]. Untargeted omics-based studies have also identified the an association between lipid metabolites and progression of cancers. Metabolomic studies of hepatocellular carcinoma have additionally demonstrated that metabolites specifically associated with the stearoyl-CoA desaturase pathway were significantly associated with progression of HCC, potentially corroborating the role of this pathway in regulation of cancer metastasis [36].

Mechanistically, more work is required to understand precisely how alterations in lipid pathways either may be an upstream cause or consequence of cancer progression. However, preclinical studies have pointed to both alterations in synthesis of lipid intermediates as well as uptake of fatty acids as important regulators of cancer progression. Increased synthesis occurs through stimulation of pathways such as mTOR signaling and AMPK pathways [37, 38]. A consequence of the activation of these pathways includes transcriptomic changes that upregulate a number of fatty acid synthase-related genes, which in turn promote proliferation and angiogenesis in cancer-associated endothelial cells [39]. Increased uptake is thought to be driven by upregulation of fatty acid transport-related receptors such as CD36. Studies have shown that palmitic acid or western (high lipid content) diets increase metastatic potential of cancers through increased lipid uptake mediated by CD36 [38].

Recent studies in the last 10 years have similarly noted abnormalities in lipid handling in the setting of pulmonary hypertension [40–42]. These findings were first identified based on a clinical study that demonstrated an unusually high prevalence of glucose intolerance in patients with pulmonary arterial hypertension (56%) in the absence of typical risk factor for diabetes [43]. This was later followed up with further studies that demonstrated abnormal increases in lipid deposition in the setting of pulmonary arterial hypertension, most notably in the right ventricle of patients with pulmonary hypertension [44–46]. Further studies have demonstrated that an imbalance between lipid uptake into cells and mitochondrial utilization of lipid species leads to the cytoplasmic accumulation of toxic lipid intermediates such as ceramides, long-chain fatty acids, and some medium chain fatty acids, leading to cellular lipotoxicity [42, 47, 48] (Fig. 6.2).

Similar to what has been studied in cancer, increased CD36 expression have been identified in the setting of pulmonary hypertension, which leads to an increase uptake and of transport of fatty acids into the cytoplasm of cells [44]. Beyond models of heritable PAH due to BMPR2 mutation, similar findings of dysregulated fatty acid uptake have been associated with worse outcomes in CTEPH patients as well [49]. Beyond increased uptake, studies have also demonstrated that a number of transcriptional changes occur intracellularly in the setting of pulmonary hypertension that lead to decreases in fatty acid utilization. An important mechanism that has been found to contribute to decreased utilization is the downregulation of genes known to regulate fatty acid beta oxidation such as PPARα, PPARγ, and PGC1α [50]. Additional studies have shown that counter-regulation of metabolism through the Randle cycle, a metabolic process in which glucose and fatty acids counter-regulate each other by directly competing as substrates for ATP generation within a given cell, also leads to decreased utilization of fatty acids with a greater preference for glucose [41, 51]. Finally, dysregulation of mechanisms that drive transport of fatty acids from the cytoplasm into mitochondria, and particularly through genes such as Cpt1, have been implicated in the impaired ability of cells to utilize fatty acids in the setting of pulmonary hypertension [42].

Pulmonary hypertension is also now known to be a condition associated with an unexpectedly high incidence of insulin resistance across multiple studies [43, 52–54]. A similar phenomenon is now suggested to occur in patients with cancer as well [55]. While less understood in the setting of cancers, mechanistically insulin resistance is thought to be secondary to fatty acid metabolic defects in the setting of pulmonary hypertension [52]. Direct studies of insulin administration or insulin resistance that shown in pre-clinical models that insulin dysregulation can, in and of itself, contribute to both pathologic pulmonary vascular remodeling as well as downstream right heart failure [54, 56, 57].

Clinical Causes of Pulmonary Hypertension Associated with Cancer

Of the various causes of pulmonary hypertension, certain groups of pulmonary hypertension have been associated with cancer. These groups include WHO Group 4 pulmonary hypertension due to chronic thromboembolic pulmonary hypertension (CTEPH) and WHO Group 5 pulmonary hypertension, a heterogeneous group of causes of pulmonary hypertension due to unclear or multifactorial causes including pulmonary tumor thrombotic microangiopathy. Finally, it is well recognized that therapies for pulmonary hypertension can have off-target effects on other organs such as the heart, which can predispose patients to the development of WHO Group

2 pulmonary hypertension due to left heart disease. These conditions are discussed below.

Chronic thromboembolic pulmonary hypertension, or CTEPH, is a rare complication that can occur after patients develop pulmonary embolism. Varying studies have suggested that the rates of CTEPH development range anywhere from 0.1 to 11.8%, highlighting the limited epidemiologic data in this space [58]. The clinical hallmark of this disease is the persistence and/or re-development of symptoms such as dyspnea on exertion, shortness of breath, difficulty with breathing, and edema. Less common symptoms such as syncope and hemoptysis have also been reported. Pathologically, the disease is characterized by major vessel fibro-thrombotic obstructions that can affect one or more branches of pulmonary artery trees. Epidemiologically, patients with CTEPH have been noted to be younger than other groups of PH (mean age 56 vs 64 for Group 3 PH and 69 for Group 2 PH). Unlike other causes of PH, CTEPH patients were 51% female and 49% male, compared to Group 1 PH which showed a female predilection with 73% of Group 1 PH patients being female [59].

Diagnosis of this condition requires not only invasive confirmation of elevated pulmonary pressures consistent with pulmonary hypertension but also imaging evidence of distal obstruction in the lung. This is typically achieved with a V/Q scan non-invasive, which is characterized by the loss of distal perfusion in lung territories despite normal ventilation. In consideration of therapeutic intervention, invasive pulmonary angiography is also often performed, which will typically demonstrate abrupt loss of vascular architecture with selective angiography of different segments of the left and right pulmonary artery trunks [1]. Management of CTEPH often involves conservative treatments such as diuretics, but definitive treatment typically requires pulmonary endarterectomy, a difficult surgery only performed at certain experienced centers around the country where the fibro-thrombotic lesions are peeled away from the pulmonary arteries to restore flow through the obstructed pulmonary vasculature [1]. In patients not considered surgical candidates, balloon pulmonary angioplasty can be performed. Recent data suggests that concomitant treatment with soluble guanylyl cyclase activator therapy can further improve outcomes in this patient population. This was initially studied in the CHEST-1 and PATENT-1 trials, which showed improvements in 6 min walk distance, pulmonary vascular resistance, functional class, and NT-pro-BNP (a biomarker of stress on the heart) in patients treated with riociguat compared to placebo [60, 61].

Studies have noted a connection between pulmonary hypertension and cancer, particularly as it relates to venous thromboembolism (Fig. 6.3). It is well established that cancer is a highly prothrombotic state, and patients with active malignancy are thought to be at especially risk of venous thromboembolism development, with rates as high as 15% annual risk [62, 63]. Based on risk scores such as the Ottawa risk score for recurrent venous thromboembolism, solid tumor malignancies appeared to be more common among patients who presented with recurrent venous thromboembolism, with lung cancer ultimately identified as an independent risk factor [64]. Other retrospective studies have identified increased risk of recurrnet

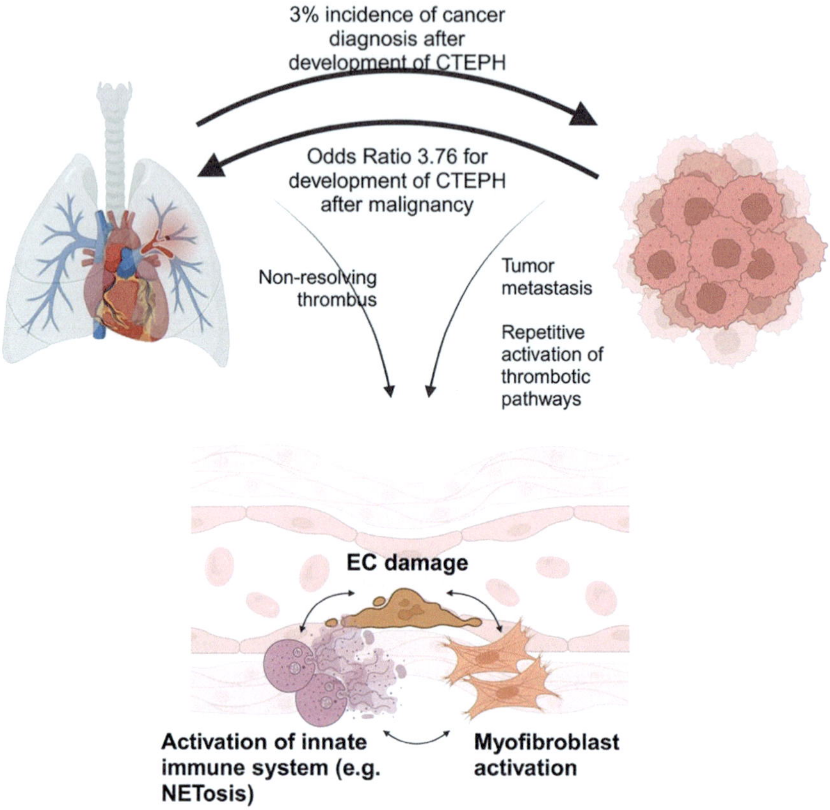

Fig. 6.3 Shared risk for development of pulmonary hypertension due to chronic thromboembolic pulmonary hypertension (CTEPH) and cancer-associated venous thromboembolism. Both conditions increase the risk for future development of the other condition. Mechanistically, both conditions ultimately lead to a vicious cycle of endothelial cell damage, activation of local inflammatory cells such as neutrophils to promote neutrophil extracellular trap (NET) formation, local dedifferentiation of smooth muscle cells into myofibroblast-like phenotypes. *CTEPH* chronic thromboembolic pulmonary hypertension, *EC* endothelial cell, *NET* neutrophil extracellular trap

thromboembolism with other solid tumor cancers such as pancreatic cancer, brain cancer, and ovarian cancer, as well [65]. Finally, this strong association between maligancy and thromboembolic risk has led to studies investigating the question of whether undiagonsed malignancy should be sought after in patients who present with venous thromboembolisms such as pulmonary emboli. Despite identifying a nearly 10% occult malignancy detection rate [66], subsequent studies have failed to show a clear benefit beyond age appropriate screening, leading to current guidelines suggesting no more than limited screening in patients who present with an unprovoked venous thromboembolism [67].

Studies have also identified a connection between CTEPH development and cancer. In a retrospective study of 687 patients in three European centers over

11 years, risk factors for the development of CTEPH included the presence of known risk factors such as ventriculo-arterial shunts, infected pacemakers, splenectomy, recurrent venous thromboembolism, and pro-thrombotic conditions such as lupus anticoagulant presence [68]. However, the study also found novel risk factors which included thyroid replacement therapy (OR 6.10) and history of malignancy (OR 3.76), which represented one of the first studies to connect cancer to CTEPH development. Subsequent studies have suggested that, among patients with CTEPH, the frequency of a history of malignancy ranges from 5–15% [68–71]. Not unexpectedly, the presence of cancer is associated with an increased risk of all-cause death in patients with CTEPH, even if undergoing the appropriate treatment [72]. Finally, studies have noted that the diagnosis of CTEPH in patients is associated with an increased likelihood of receiving a cancer diagnosis in subsequent years, with a nearly 3.3% incidence of new cancer diagnoses in the first year of pulmonary hypertension diagnosis (standardized incidence ratio of 1.96) [73]. The standardized incidence ratio dropped to 1.15 in subsequent years, but remained higher at 1.41 for patients with CTEPH, again highlighting the inter-connection between cancer risk and CTEPH (Fig. 6.3).

Mechanistically, while cancer is thought to promote thrombosis through a variety of mechanisms such as tumor metastasis and release of tumor-associated cytokines and growth factors that promote endothelial injury and inflammation, downstream pathways that result in chronic remodeling of thrombus and fibroproliferative changes in the thrombosed vessel are thought to be similar between CTEPH and cancer [74]. Likewise, in CTEPH, while the initial insult may be a deep vein thrombus that embolizes to the lung to cause an acute pulmonary embolism, non-resolution of the thrombus and/or inhibition of pro-resolving pathways in inflammatory cells, platelets, and endothelial cells leads to a feed forward cycle that results in similar fibroproliferative pathways that lead to CTEPH [75]. Ultimately, these pathways lead to a chronic cycle of activated endothelial cells, myofibroblast proliferation, and further activation of local inflammatory cells such as neutrophils to promote the generation of neutrophil extracellular traps (NETs) in the local vasculature [74–76] (Fig. 6.3).

Tumor Metastasis as a Cause of Pulmonary Hypertension

The World Health Organization Group 5 of pulmonary hypertension includes the subgroup of patients who develop pulmonary hypertension due to pulmonary tumor embolic disease or pulmonary tumor thrombotic microangiopathy. These occlusions are often found evidence of pulmonary tumor embolic on clinically performed autopsies [77], but also have the potential to present with very similar symptoms of shortness of breath, dyspnea on exertion, fatigue, and evidence of increased right sided pressures in the heart with end-organ manifestations of liver and/or kidney dysfunction. While the true incidence of these tumors is not well known, studies have suggested that this diagnosis is likely underestimated with estimates of 3–26%

of patients with solid tumors having evidence of neoplastic emboli [77, 78]. The most commonly reported cancers associated with pulmonary tumor emboli include liver, renal, breast, gastric, and cholangiocarcinoma tumors, with proposed mechanism of embolus through hematogenous spread [77]. In contrast to pulmonary tumor emboli, which are thought to be largely asymptomatic in most patients, pulmonary tumor thrombotic microangiopathy is associated with fibrocellular occlusions in small vessels with associated with intimal proliferation [79]. While associated with the same tumors, studies have suggested the presence of evidence of more disease metastasis with lymphangitic spread noted on histopathologic examination [80], although such studies have been limited in numbers of samples studies and often case reports. Diagnostically, while some imaging studies such as MRIs, high resolution CTs, or PET imaging may point to a possible pulmonary tumor embolism, diagnostically patients undergo a very similar set of tests to all comers who present for evaluation of pulmonary hypertension due to other causes. Treatment-wise, there is limited evidence that chemotherapy improves outcomes in late presentations of PTE or PTTM, with some studies also proposed anti-inflammatory therapies, with limited success thus far. The supporting studies for therapeutic trials in PTE and PTTM are reviewed in [77].

Cancer Therapy-Related Pulmonary Hypertension

With the large growth in more personalized therapies available to patients who suffer from cancer, there has been a concomitant growth in recognition of unique side effect profiles to each cancer therapeutic available. Many of these therapies have been shown to be associated with an increased risk of pulmonary hypertension due to a variety of reasons. While small studies have implicated other agents as proteasome inhibitors and interferon therapies [81, 82], the greatest body of literature supports off-target effects of alkylating agents, tyrosine kinase inhibitors, and immunomodulatory therapies on the development of pulmonary hypertension.

Alkylating agents have long been used as chemotherapeutic agents for a variety of malignancies. Agents such as mitomycin-C and cyclophosphamide have been associated with an increased risk for WHO Group 1 PH due to pulmonary veno-occlusive disease (PVOD) [83], with further studies recapitulating pathologic changes of PVOD in pre-clinical animal models with administration of agents such as mitomycin-C [84]. Mechanistically, while poorly understood, alkylating agents are thought to directly promote endothelial cell injury. However, whether there is a selective effect of alkylating agents on certain vascular beds (e.g. pulmonary venous vs arterial beds), remain poorly understood and require further study.

Small molecule inhibitors of tyrosine kinases have revolutionized the personalized treatment of cancer [85]. Receptor tyrosine kinases are a broad and diverse family of receptors that characterized by a single transmembrane helix domain that harbor a cytoplasmic kinase domain that is capable of phosphorylating tyrosine residues on intracellular proteins [85]. This internal phosphorylation process can activate or repress a number of signal transduction mediators and/or transcription factors to exert control over a broad collection of intermediate pathways such as mitogen-activated protein kinase (MAPK), phosphoinositide 3-kinase (PI3K), mechanistic target of rapamycin (MTOR), calcium signaling, Janus kinase/signal transducer and activator of transcription protein (JAK/STAT), and/or protein kinase C (PKC) dependent pathways [86]. These pathways can regulate broad cellular functions such as metabolism, proliferation, survival, differentiation, migration/motility, response to cytokines and growth factors. Based on known mechanisms that underlie pulmonary hypertension as well, it is therefore not surprising that small molecular inhibitors of receptor tyrosine kinases can have off-target effects that result in abnormal pulmonary vascular remodeling.

The tyrosine kinase inhibitor, dasatinib, is used in the treatment of hematologic malignancies such as chronic myeloid leukemia (CML) or acute lymphoblastic leukemia (ALL). The incidence of pulmonary hypertension, primarily defined by non-invasive echocardiography, has been estimated to be as high as 12% when compared to imatinib, another tyrosine kinase inhibitor that is known to target the Philadelphia chromosome Bcr-Abl fusion protein [87–89]. Physiologically, dasatinib is known to be associated with polyserositis, resulting in development of both pericardial and pleural effusions when compared to other receptor tyrosine kinase inhibitor therapies [90]. Notably, studies have also demonstrated that, in the vast majority of patients, withdrawal of dasatinib therapy will result in normalization of estimated pulmonary pressure by echocardiography [87–89]. In experimental models of pulmonary hypertension, dasatinib was found to worsen pulmonary vascular remodeling in two models of experimental pulmonary hypertension induced by monocrotaline and chronic hypoxia, respectively. In vitro, dasatinib, but not imatinib, was found to promote endothelial cell dysfunction through induction of endoplasmic reticulum stress, and ultimately apoptosis through increased production of reaction oxygen species, and ultimately apoptosis [91]. While less data exists for pulmonary hypertension related to other receptor tyrosine kinase inhibitors, case reports have suggested that treatment with ponatinib and bosutinib have been associated with development of pulmonary hypertension as well, each associated with serositis as well [92–94].

Contrary to side effects of dasatinib, ponatinib, and bosutinib, other studies have also noted that treatment with receptor tyrosine kinase inhibitors may also contribute to improvement in pulmonary hypertension in certain settings. Specifically, there are pre-clinical studies have demonstrated that inhibition of epidermal growth factor receptors (EGFR) can lead to improvement in pulmonary vascular remodeling through inhibition of smooth muscle cell survival and proliferation. However, this effect was noted to differ by the specific inhibitor used, suggesting potential off-target effects of the specific drug affect pulmonary vascular remodeling [95]. Finally, imatinib has been studied as a therapeutic target in the treatment of pulmonary hypertension. While initially designed as a drug to target the Bcr-Abl fusion product that characterizes chronic myeloid leukemia due to the Philadelphia chromosome, its off target effects on platelet derived growth factor have been exploited to study the role of PDGF inhibition in experimental pulmonary hypertension [96]. Mechanistically, imatinib was found to be an acute pulmonary vasodilator in a rat model [97] and promote apoptosis of smooth muscle cells through a PDGF-mediated mechanism [98]. In the randomized, double-blind, placebo-controlled IMPRES trial of patients with prevalent pulmonary arterial hypertension, imatinib treatment led to improved exercise capacity and hemodynamics in patients with pulmonary arterial hypertension, but also led to many side effects including coagulopathies and life-threatening hematomas [99]. Findings from these studies and others, however, have led to development and testing of novel receptor tyrosine kinase inhibitors for treatment of pulmonary hypertension, including seralutinib (a PDGF and c-kit receptor inhibitor) [100].

Finally, immune checkpoint inhibitor (ICI) therapies have revolutionized the care of cancer patients [101]. In particular, many of these therapies function by suppressing aspects of the immune system, which ultimately leads to greater anti-tumor effects through increased immune surveillance identifying and destroying malignant cells [101]. However, these therapies are known to also be associated with a wide variety of side effects that include gastrointestinal toxicity, pulmonary pneumonitis, and cardiotoxicity through autoimmune mechanisms [102]. While less studied, case reports and small studies have also suggested there may be evidence of adverse pulmonary vascular remodeling in response to ICI therapy. In a pharmacovigilance study, Palassin et al. identified 42 patients treatment with ICI therapy between 2015 and 2021 who were found to have pulmonary hypertension after treatment [103]. Of the 42, 30 (71.4%) were found to have pulmonary hypertension, 11 (26.2%) were diagnosed with pulmonary arterial hypertension, and 1 (2.4%) patient was diagnosed with pulmonary veno-occlusive disease. Of note, nearly half of these patients received nivolumab for their ICI therapy and nearly 43% of these patients suffered from lung cancer. The median time of diagnosis was 77 days after initiation of therapy. Of note, the diagnoses reported in this retrospective study were based on reported terms in the VigiBase pharmacovigilance database, and did not represent the gold standard methods for diagnosis of pulmonary hypertension and further characterization [103]. Other case reports have also reported ICI-therapy-related pulmonary hypertension with imaging and catheterization-based diagnosis of pulmonary hypertension [104, 105], but further studies are required to more completely understand how ICI therapy may lead to both subclinical and clinical evidence of pulmonary vascular remodeling in patients.

In summary, studies dating as far back as 30 years ago have identified important similarities between pathophysiologic mechanisms that drive progression of cancers and those that contribute to dysregulated pulmonary vascular remodeling. In part through activation of hypoxia-mediated transcription factors as well as through separate insults that cause repetitive endothelial cell injury, many similarities have been identified at the level of cellular metabolism that regulate pro-proliferative and de-differentiation phenotypes in both conditions. Clinical studies have identified conditions related to pulmonary hypertension that are intimately connected with cancers and have been shown to be associated with a greater risk of cancer. Conversely, studies have also demonstrated that both incident and prevalent cancer increase the risk of development of conditions that are associated with downstream development of pulmonary hypertension. Finally, through studies of novel precision medicine-based therapeutics for the treatment of cancer, intimate links between cancer and pulmonary hypertension pathophysiology have been further identified with respect to regulation of receptor tyrosine kinase mediated signaling and regulation of the local inflammatory milieu through immune checkpoint inhibitor therapies. With the growth of precision medicine therapeutics in the field of oncology, molecular understanding of pulmonary hypertension in patients is expected to accelerate in the coming decades.

Acknowledgments Figures were created using BioRender.com.

References

1. Humbert M, Kovacs G, Hoeper MM, Badagliacca R, Berger RM, Brida M, Carlsen J, Coats AJS, Escribano-Subias P, Ferrari P et al (2022) 2022 ESC/ERS guidelines for the diagnosis and treatment of pulmonary hypertension. Eur Heart J 43:3618–3731. https://doi.org/10.1093/eurheartj/ehac237
2. Huston JH, Maron BA, French J, Huang S, Thayer T, Farber-Eger EH, Wells QS, Choudhary G, Hemnes AR, Brittain EL (2019) Association of mild echocardiographic pulmonary hypertension with mortality and right ventricular function. JAMA Cardiol 4:1112–1121. https://doi.org/10.1001/jamacardio.2019.3345
3. Agrawal V, D'Alto M, Naeije R, Romeo E, Xu M, Assad TR, Robbins IM, Newman JH, Pugh ME, Hemnes AR et al (2019) Echocardiographic detection of occult diastolic dysfunction in pulmonary hypertension after fluid challenge. J Am Heart Assoc 8:e012504. https://doi.org/10.1161/jaha.119.012504
4. Agrawal V, Byrd BF, Brittain EL (2019) Echocardiographic evaluation of diastolic function in the setting of pulmonary hypertension. Pulm Circ 9:2045894019826043. https://doi.org/10.1177/2045894019826043
5. Ameri P, Mercurio V, Pollesello P, Anker MS, Backs J, Bayes-Genis A, Borlaug BA, Burkhoff D, Caravita S, Chan SY et al (2024) A roadmap for therapeutic discovery in pulmonary hypertension associated with left heart failure. A scientific statement of the Heart Failure Association (HFA) of the ESC and the ESC Working Group on Pulmonary Circulation & Right Ventricular Function. Eur J Heart Fail 26:707–729. https://doi.org/10.1002/ejhf.3236
6. Suzuki T, Carrier EJ, Talati MH, Rathinasabapathy A, Chen X, Nishimura R, Tada Y, Tatsumi K, West J (2018) Isolation and characterization of endothelial-to-mesenchymal transition cells in pulmonary arterial hypertension. Am J Physiol Lung Cell Mol Physiol 314:L118–L126. https://doi.org/10.1152/ajplung.00296.2017
7. Suzuki T, Tada Y, Nishimura R, Kawasaki T, Sekine A, Urushibara T, Kato F, Kinoshita T, Ikari J, West J et al (2016) Endothelial-to-mesenchymal transition in lipopolysaccharide-induced acute lung injury drives a progenitor cell-like phenotype. Am J Physiol Lung Cell Mol Physiol 310:L1185–L1198. https://doi.org/10.1152/ajplung.00074.2016
8. Lee SD, Shroyer KR, Markham NE, Cool CD, Voelkel NF, Tuder RM (1998) Monoclonal endothelial cell proliferation is present in primary but not secondary pulmonary hypertension. J Clin Invest 101:927–934. https://doi.org/10.1172/JCI1910
9. Ventetuolo CE, Aliotta JM, Braza J, Chichger H, Dooner M, McGuirl D, Mullin CJ, Newton J, Pereira M, Princiotto A et al (2020) Culture of pulmonary artery endothelial cells from

pulmonary artery catheter balloon tips: considerations for use in pulmonary vascular disease. Eur Respir J 55:1901313. https://doi.org/10.1183/13993003.01313-2019
10. Vogelstein B, Kinzler KW (1993) The multistep nature of cancer. Trends Genet 9:138–141. https://doi.org/10.1016/0168-9525(93)90209-z
11. Theilmann AL, Hawke LG, Hilton LR, Whitford MKM, Cole DV, Mackeil JL, Dunham-Snary KJ, Mewburn J, James PD, Maurice DH et al (2020) Endothelial BMPR2 loss drives a proliferative response to BMP (bone morphogenetic protein) 9 via prolonged canonical signaling. Arterioscler Thromb Vasc Biol 40:2605–2618. https://doi.org/10.1161/ATVBAHA.119.313357
12. Chew JD, Loyd JE, Austin ED (2017) Genetics of pulmonary arterial hypertension. Semin Respir Crit Care Med 38:585–595. https://doi.org/10.1055/s-0037-1606201
13. Morrell NW, Aldred MA, Chung WK, Elliott CG, Nichols WC, Soubrier F, Trembath RC, Loyd JE (2019) Genetics and genomics of pulmonary arterial hypertension. Eur Respir J 53: 1801899. https://doi.org/10.1183/13993003.01899-2018
14. Evans JD, Girerd B, Montani D, Wang XJ, Galie N, Austin ED, Elliott G, Asano K, Grunig E, Yan Y et al (2016) BMPR2 mutations and survival in pulmonary arterial hypertension: an individual participant data meta-analysis. Lancet Respir Med 4:129–137. https://doi.org/10.1016/S2213-2600(15)00544-5
15. Majka S, Hagen M, Blackwell T, Harral J, Johnson JA, Gendron R, Paradis H, Crona D, Loyd JE, Nozik-Grayck E et al (2011) Physiologic and molecular consequences of endothelial Bmpr2 mutation. Respir Res 12:84. https://doi.org/10.1186/1465-9921-12-84
16. Clere N, Renault S, Corre I (2020) Endothelial-to-mesenchymal transition in cancer. Front Cell Dev Biol 8:747. https://doi.org/10.3389/fcell.2020.00747
17. Boucherat O, Agrawal V, Lawrie A, Bonnet S (2022) The latest in animal models of pulmonary hypertension and right ventricular failure. Circ Res 130:1466–1486. https://doi.org/10.1161/circresaha.121.319971
18. West JD, Austin ED, Rizzi EM, Yan L, Tanjore H, Crabtree AL, Moore CS, Muthian G, Carrier EJ, Jacobson DA et al (2021) KCNK3 mutation causes altered immune function in pulmonary arterial hypertension patients and mouse models. Int J Mol Sci 22:5014. https://doi.org/10.3390/ijms22095014
19. Martinez-Reyes I, Chandel NS (2021) Cancer metabolism: looking forward. Nat Rev Cancer 21:669–680. https://doi.org/10.1038/s41568-021-00378-6
20. Warburg O (1956) On the origin of cancer cells. Science 123:309–314. https://doi.org/10.1126/science.123.3191.309
21. Gatenby RA, Gillies RJ (2004) Why do cancers have high aerobic glycolysis? Nat Rev Cancer 4:891–899. https://doi.org/10.1038/nrc1478
22. Hoxhaj G, Manning BD (2020) The PI3K-AKT network at the interface of oncogenic signalling and cancer metabolism. Nat Rev Cancer 20:74–88. https://doi.org/10.1038/s41568-019-0216-7
23. Hurst JH (2016) William Kaelin, Peter Ratcliffe, and Gregg Semenza receive the 2016 Albert Lasker basic medical research award. J Clin Invest 126:3628–3638. https://doi.org/10.1172/JCI90055
24. Sabatini DM (2017) Twenty-five years of mTOR: uncovering the link from nutrients to growth. Proc Natl Acad Sci USA 114:11818–11825. https://doi.org/10.1073/pnas.1716173114
25. Caruso P, Dunmore BJ, Schlosser K, Schoors S, Dos Santos C, Perez-Iratxeta C, Lavoie JR, Zhang H, Long L, Flockton AR et al (2017) Identification of MicroRNA-124 as a major regulator of enhanced endothelial cell glycolysis in pulmonary arterial hypertension via PTBP1 (polypyrimidine tract binding protein) and pyruvate kinase M2. Circulation 136: 2451–2467. https://doi.org/10.1161/CIRCULATIONAHA.117.028034
26. Xu W, Koeck T, Lara AR, Neumann D, DiFilippo FP, Koo M, Janocha AJ, Masri FA, Arroliga AC, Jennings C et al (2007) Alterations of cellular bioenergetics in pulmonary artery

endothelial cells. Proc Natl Acad Sci USA 104:1342–1347. https://doi.org/10.1073/pnas.0605080104
27. Piao L, Sidhu VK, Fang YH, Ryan JJ, Parikh KS, Hong Z, Toth PT, Morrow E, Kutty S, Lopaschuk GD et al (2013) FOXO1-mediated upregulation of pyruvate dehydrogenase kinase-4 (PDK4) decreases glucose oxidation and impairs right ventricular function in pulmonary hypertension: therapeutic benefits of dichloroacetate. J Mol Med 91:333–346. https://doi.org/10.1007/s00109-012-0982-0
28. Piao L, Fang YH, Cadete VJ, Wietholt C, Urboniene D, Toth PT, Marsboom G, Zhang HJ, Haber I, Rehman J et al (2010) The inhibition of pyruvate dehydrogenase kinase improves impaired cardiac function and electrical remodeling in two models of right ventricular hypertrophy: resuscitating the hibernating right ventricle. J Mol Med 88:47–60. https://doi.org/10.1007/s00109-009-0524-6
29. Tian L, Wu D, Dasgupta A, Chen KH, Mewburn J, Potus F, Lima PDA, Hong Z, Zhao YY, Hindmarch CCT et al (2020) Epigenetic metabolic reprogramming of right ventricular fibroblasts in pulmonary arterial hypertension: a pyruvate dehydrogenase kinase-dependent shift in mitochondrial metabolism promotes right ventricular fibrosis. Circ Res 126:1723–1745. https://doi.org/10.1161/CIRCRESAHA.120.316443
30. Bertero T, Perk D, Chan SY (2019) The molecular rationale for therapeutic targeting of glutamine metabolism in pulmonary hypertension. Expert Opin Ther Targets 23:511–524. https://doi.org/10.1080/14728222.2019.1615438
31. Rachedi NS, Tang Y, Tai YY, Zhao J, Chauvet C, Grynblat J, Akoumia KKF, Estephan L, Torrino S, Sbai C et al (2024) Dietary intake and glutamine-serine metabolism control pathologic vascular stiffness. Cell Metab 36:1335–1350. https://doi.org/10.1016/j.cmet.2024.04.010
32. Egnatchik RA, Brittain EL, Shah AT, Fares WH, Ford HJ, Monahan K, Kang CJ, Kocurek EG, Zhu S, Luong T et al (2017) Dysfunctional BMPR2 signaling drives an abnormal endothelial requirement for glutamine in pulmonary arterial hypertension. Pulm Circ 7:186–199. https://doi.org/10.1086/690236
33. Penumatsa KC, Fanburg BL (2014) Transglutaminase 2-mediated serotonylation in pulmonary hypertension. Am J Physiol Lung Cell Mol Physiol 306:L309–L315. https://doi.org/10.1152/ajplung.00321.2013
34. Chandel NS (2021) Lipid metabolism. Cold Spring Harb Perspect Biol 13:a040576. https://doi.org/10.1101/cshperspect.a040576
35. Williams KJ, Argus JP, Zhu Y, Wilks MQ, Marbois BN, York AG, Kidani Y, Pourzia AL, Akhavan D, Lisiero DN et al (2013) An essential requirement for the SCAP/SREBP signaling axis to protect cancer cells from lipotoxicity. Cancer Res 73:2850–2862. https://doi.org/10.1158/0008-5472.CAN-13-0382-T
36. Budhu A, Roessler S, Zhao X, Yu Z, Forgues M, Ji J, Karoly E, Qin LX, Ye QH, Jia HL et al (2013) Integrated metabolite and gene expression profiles identify lipid biomarkers associated with progression of hepatocellular carcinoma and patient outcomes. Gastroenterology 144:1066–1075. https://doi.org/10.1053/j.gastro.2013.01.054
37. Chen WL, Jin X, Wang M, Liu D, Luo Q, Tian H, Cai L, Meng L, Bi R, Wang L et al (2020) GLUT5-mediated fructose utilization drives lung cancer growth by stimulating fatty acid synthesis and AMPK/mTORC1 signaling. JCI Insight 5:131596. https://doi.org/10.1172/jci.insight.131596
38. Pascual G, Avgustinova A, Mejetta S, Martin M, Castellanos A, Attolini CS, Berenguer A, Prats N, Toll A, Hueto JA et al (2017) Targeting metastasis-initiating cells through the fatty acid receptor CD36. Nature 541:41–45. https://doi.org/10.1038/nature20791
39. Zaytseva YY, Elliott VA, Rychahou P, Mustain WC, Kim JT, Valentino J, Gao T, O'Connor KL, Neltner JM, Lee EY et al (2014) Cancer cell-associated fatty acid synthase activates endothelial cells and promotes angiogenesis in colorectal cancer. Carcinogenesis 35:1341–1351. https://doi.org/10.1093/carcin/bgu042

40. Ryan JJ, Archer SL (2015) Emerging concepts in the molecular basis of pulmonary arterial hypertension: part I: metabolic plasticity and mitochondrial dynamics in the pulmonary circulation and right ventricle in pulmonary arterial hypertension. Circulation 131:1691–1702. https://doi.org/10.1161/CIRCULATIONAHA.114.006979
41. Archer SL, Fang YH, Ryan JJ, Piao L (2013) Metabolism and bioenergetics in the right ventricle and pulmonary vasculature in pulmonary hypertension. Pulm Circ 3:144–152. https://doi.org/10.4103/2045-8932.109960
42. Agrawal V, Hemnes AR, Shelburne NJ, Fortune N, Fuentes JL, Colvin D, Calcutt MW, Talati M, Poovey E, West JD et al (2022) L-carnitine therapy improves right heart dysfunction through Cpt1-dependent fatty acid oxidation. Pulm Circ 12:e12107. https://doi.org/10.1002/pul2.12107
43. Pugh ME, Robbins IM, Rice TW, West J, Newman JH, Hemnes AR (2011) Unrecognized glucose intolerance is common in pulmonary arterial hypertension. J Heart Lung Transplant 30:904–911. https://doi.org/10.1016/j.healun.2011.02.016
44. Talati MH, Brittain EL, Fessel JP, Penner N, Atkinson J, Funke M, Grueter C, Jerome WG, Freeman M, Newman JH et al (2016) Mechanisms of lipid accumulation in the bone morphogenetic protein receptor type 2 mutant right ventricle. Am J Respir Crit Care Med 194:719–728. https://doi.org/10.1164/rccm.201507-1444OC
45. Brittain EL, Talati M, Fessel JP, Zhu H, Penner N, Calcutt MW, West JD, Funke M, Lewis GD, Gerszten RE et al (2016) Fatty acid metabolic defects and right ventricular lipotoxicity in human pulmonary arterial hypertension. Circulation 133:1936–1944. https://doi.org/10.1161/circulationaha.115.019351
46. Hemnes AR, Brittain EL, Trammell AW, Fessel JP, Austin ED, Penner N, Maynard KB, Gleaves L, Talati M, Absi T et al (2014) Evidence for right ventricular lipotoxicity in heritable pulmonary arterial hypertension. Am J Respir Crit Care Med 189:325–334. https://doi.org/10.1164/rccm.201306-1086OC
47. Tuder RM, Robinson JC, Graham BB (2014) Fat and cardiotoxicity in hereditary pulmonary hypertension. Am J Respir Crit Care Med 189:247–249. https://doi.org/10.1164/rccm.201312-2240ED
48. Brittain EL, Talati M, Fortune N, Agrawal V, Meoli DF, West J, Hemnes AR (2019) Adverse physiologic effects of western diet on right ventricular structure and function: role of lipid accumulation and metabolic therapy. Pulm Circ 9:2045894018817741. https://doi.org/10.1177/2045894018817741
49. Lankeit M, Dellas C, Panzenbock A, Skoro-Sajer N, Bonderman D, Olschewski M, Schafer K, Puls M, Konstantinides S, Lang IM (2008) Heart-type fatty acid-binding protein for risk assessment of chronic thromboembolic pulmonary hypertension. Eur Respir J 31:1024–1029. https://doi.org/10.1183/09031936.00100407
50. Gomez-Arroyo J, Mizuno S, Szczepanek K, Van Tassell B, Natarajan R, dos Remedios CG, Drake JI, Farkas L, Kraskauskas D, Wijesinghe DS et al (2013) Metabolic gene remodeling and mitochondrial dysfunction in failing right ventricular hypertrophy secondary to pulmonary arterial hypertension. Circ Heart Fail 6:136–144. https://doi.org/10.1161/CIRCHEARTFAILURE.111.966127
51. Randle PJ, Garland PB, Hales CN, Newsholme EA (1963) The glucose fatty-acid cycle. Its role in insulin sensitivity and the metabolic disturbances of diabetes mellitus. Lancet 1:785–789. https://doi.org/10.1016/s0140-6736(63)91500-9
52. Hemnes AR, Luther JM, Rhodes CJ, Burgess JP, Carlson J, Fan R, Fessel JP, Fortune N, Gerszten RE, Halliday SJ et al (2019) Human PAH is characterized by a pattern of lipid-related insulin resistance. JCI Insight 4:123611. https://doi.org/10.1172/jci.insight.123611
53. Zamanian RT, Hansmann G, Snook S, Lilienfeld D, Rappaport KM, Reaven GM, Rabinovitch M, Doyle RL (2009) Insulin resistance in pulmonary arterial hypertension. Eur Respir J 33:318–324. https://doi.org/10.1183/09031936.00000508

54. West J, Niswender KD, Johnson JA, Pugh ME, Gleaves L, Fessel JP, Hemnes AR (2013) A potential role for insulin resistance in experimental pulmonary hypertension. Eur Respir J 41: 861–871. https://doi.org/10.1183/09031936.00030312
55. Berlier C, Saxer S, Lichtblau M, Schneider SR, Schwarz EI, Furian M, Bloch KE, Carta AF, Ulrich S (2022) Influence of upright versus supine position on resting and exercise hemodynamics in patients assessed for pulmonary hypertension. J Am Heart Assoc 11:e023839. https://doi.org/10.1161/jaha.121.023839
56. Hemnes AR, Fessel JP, Chen X, Zhu S, Fortune NL, Jetter C, Freeman M, Newman JH, West JD, Talati MH (2020) BMPR2 dysfunction impairs insulin signaling and glucose homeostasis in cardiomyocytes. Am J Physiol Lung Cell Mol Physiol 318:L429–L441. https://doi.org/10.1152/ajplung.00555.2018
57. Trammell AW, Talati M, Blackwell TR, Fortune NL, Niswender KD, Fessel JP, Newman JH, West JD, Hemnes AR (2017) Pulmonary vascular effect of insulin in a rodent model of pulmonary arterial hypertension. Pulm Circ 7:624–634. https://doi.org/10.1086/689908
58. Ende-Verhaar YM, Cannegieter SC, Vonk Noordegraaf A, Delcroix M, Pruszczyk P, Mairuhu AT, Huisman MV, Klok FA (2017) Incidence of chronic thromboembolic pulmonary hypertension after acute pulmonary embolism: a contemporary view of the published literature. Eur Respir J 49:1601792. https://doi.org/10.1183/13993003.01792-2016
59. Hemnes AR, Leopold JA, Radeva MK, Beck GJ, Abidov A, Aldred MA, Barnard J, Rosenzweig EB, Borlaug BA, Chung WK et al (2022) Clinical characteristics and transplant-free survival across the spectrum of pulmonary vascular disease. J Am Coll Cardiol 80:697–718. https://doi.org/10.1016/j.jacc.2022.05.038
60. Ghofrani HA, D'Armini AM, Grimminger F, Hoeper MM, Jansa P, Kim NH, Mayer E, Simonneau G, Wilkins MR, Fritsch A et al (2013) Riociguat for the treatment of chronic thromboembolic pulmonary hypertension. N Engl J Med 369:319–329. https://doi.org/10.1056/NEJMoa1209657
61. Ghofrani HA, Galie N, Grimminger F, Grunig E, Humbert M, Jing ZC, Keogh AM, Langleben D, Kilama MO, Fritsch A et al (2013) Riociguat for the treatment of pulmonary arterial hypertension. N Engl J Med 369:330–340. https://doi.org/10.1056/NEJMoa1209655
62. Kearon C, Akl EA, Ornelas J, Blaivas A, Jimenez D, Bounameaux H, Huisman M, King CS, Morris TA, Sood N et al (2016) Antithrombotic therapy for VTE disease: CHEST guideline and expert panel report. Chest 149:315–352. https://doi.org/10.1016/j.chest.2015.11.026
63. Stevens SM, Woller SC, Baumann Kreuziger L, Doerschug K, Geersing GJ, Klok FA, King CS, Murin S, Vintch JRE, Wells PS et al (2024) Antithrombotic therapy for VTE disease: compendium and review of CHEST guidelines 2012-2021. Chest 166:388–404. https://doi.org/10.1016/j.chest.2024.03.003
64. Louzada ML, Carrier M, Lazo-Langner A, Dao V, Kovacs MJ, Ramsay TO, Rodger MA, Zhang J, Lee AY, Meyer G et al (2012) Development of a clinical prediction rule for risk stratification of recurrent venous thromboembolism in patients with cancer-associated venous thromboembolism. Circulation 126:448–454. https://doi.org/10.1161/CIRCULATIONAHA.111.051920
65. Chee CE, Ashrani AA, Marks RS, Petterson TM, Bailey KR, Melton LJ, Heit JA (2014) Predictors of venous thromboembolism recurrence and bleeding among active cancer patients: a population-based cohort study. Blood 123:3972–3978. https://doi.org/10.1182/blood-2014-01-549733
66. Carrier M, Le Gal G, Wells PS, Fergusson D, Ramsay T, Rodger MA (2008) Systematic review: the Trousseau syndrome revisited: should we screen extensively for cancer in patients with venous thromboembolism? Ann Intern Med 149:323–333. https://doi.org/10.7326/0003-4819-149-5-200809020-00007
67. Delluc A, Antic D, Lecumberri R, Ay C, Meyer G, Carrier M (2017) Occult cancer screening in patients with venous thromboembolism: guidance from the SSC of the ISTH. J Thromb Haemost 15:2076–2079. https://doi.org/10.1111/jth.13791

68. Bonderman D, Wilkens H, Wakounig S, Schafers HJ, Jansa P, Lindner J, Simkova I, Martischnig AM, Dudczak J, Sadushi R et al (2009) Risk factors for chronic thromboembolic pulmonary hypertension. Eur Respir J 33:325–331. https://doi.org/10.1183/09031936.00087608
69. Delcroix M, Lang I, Pepke-Zaba J, Jansa P, D'Armini AM, Snijder R, Bresser P, Torbicki A, Mellemkjaer S, Lewczuk J et al (2016) Long-term outcome of patients with chronic thromboembolic pulmonary hypertension: results from an International Prospective Registry. Circulation 133:859–871. https://doi.org/10.1161/CIRCULATIONAHA.115.016522
70. Pepke-Zaba J, Delcroix M, Lang I, Mayer E, Jansa P, Ambroz D, Treacy C, D'Armini AM, Morsolini M, Snijder R et al (2011) Chronic thromboembolic pulmonary hypertension (CTEPH): results from an international prospective registry. Circulation 124:1973–1981. https://doi.org/10.1161/CIRCULATIONAHA.110.015008
71. Kerr KM, Elliott CG, Chin K, Benza RL, Channick RN, Davis RD, He F, LaCroix A, Madani MM, McLaughlin VV et al (2021) Results from the United States chronic thromboembolic pulmonary hypertension registry: enrollment characteristics and 1-year follow-up. Chest 160:1822–1831. https://doi.org/10.1016/j.chest.2021.05.052
72. Nakamura J, Tsujino I, Masaki K, Hosokawa K, Funakoshi K, Taniguchi Y, Adachi S, Inami T, Yamashita J, Ogino H et al (2024) Cancer as an independent mortality risk in chronic thromboembolic pulmonary hypertension. J Heart Lung Transplant 10:22. https://doi.org/10.1016/j.healun.2024.10.022
73. Sorensen HT, Skajaa N, Klok FA, Laugesen K, Farkas DK (2022) Cancer risk in pulmonary hypertension patients. Clin Epidemiol 14:173–177. https://doi.org/10.2147/CLEP.S345054
74. Pantazi D, Alivertis D, Tselepis AD (2024) Underlying mechanisms of thrombosis associated with cancer and anticancer therapies. Curr Treat Options in Oncol 25:897–913. https://doi.org/10.1007/s11864-024-01210-7
75. Simonneau G, Torbicki A, Dorfmuller P, Kim N (2017) The pathophysiology of chronic thromboembolic pulmonary hypertension. Eur Respir Rev 26:160112. https://doi.org/10.1183/16000617.0112-2016
76. Sharma S, Hofbauer TM, Ondracek AS, Chausheva S, Alimohammadi A, Artner T, Panzenboeck A, Rinderer J, Shafran I, Mangold A et al (2021) Neutrophil extracellular traps promote fibrous vascular occlusions in chronic thrombosis. Blood 137:1104–1116. https://doi.org/10.1182/blood.2020005861
77. Price LC, Seckl MJ, Dorfmuller P, Wort SJ (2019) Tumoral pulmonary hypertension. Eur Respir Rev 28:180065. https://doi.org/10.1183/16000617.0065-2018
78. Kane RD, Hawkins HK, Miller JA, Noce PS (1975) Microscopic pulmonary tumor emboli associated with dyspnea. Cancer 36:1473–1482. https://doi.org/10.1002/1097-0142(197510)36:4<1473::aid-cncr2820360440>3.0.co;2-d
79. von Herbay A, Illes A, Waldherr R, Otto HF (1990) Pulmonary tumor thrombotic microangiopathy with pulmonary hypertension. Cancer 66:587–592. https://doi.org/10.1002/1097-0142(19900801)66:3<587::aid-cncr2820660330>3.0.co;2-j
80. Gru AA, Pai RK, Roma AA (2012) Pulmonary tumor thrombotic microangiopathy in patients with low-grade ovarian serous neoplasm: a clinicopathologic review of 2 cases of a previously unknown association. Int J Gynecol Pathol 31:438–442. https://doi.org/10.1097/PGP.0b013e318249287d
81. Al-Qazazi R, Lima PDA, Prisco SZ, Potus F, Dasgupta A, Chen KH, Tian L, Bentley RET, Mewburn J, Martin AY et al (2022) Macrophage-NLRP3 activation promotes right ventricle failure in pulmonary arterial hypertension. Am J Respir Crit Care Med 206:608–624. https://doi.org/10.1164/rccm.202110-2274OC
82. Savale L, Sattler C, Gunther S, Montani D, Chaumais MC, Perrin S, Jais X, Seferian A, Jovan R, Bulifon S et al (2014) Pulmonary arterial hypertension in patients treated with interferon. Eur Respir J 44:1627–1634. https://doi.org/10.1183/09031936.00057914
83. Ranchoux B, Gunther S, Quarck R, Chaumais MC, Dorfmuller P, Antigny F, Dumas SJ, Raymond N, Lau E, Savale L et al (2015) Chemotherapy-induced pulmonary hypertension:

role of alkylating agents. Am J Pathol 185:356–371. https://doi.org/10.1016/j.ajpath.2014.10.021

84. Perros F, Gunther S, Ranchoux B, Godinas L, Antigny F, Chaumais MC, Dorfmuller P, Hautefort A, Raymond N, Savale L et al (2015) Mitomycin-induced pulmonary veno-occlusive disease: evidence from human disease and animal models. Circulation 132:834–847. https://doi.org/10.1161/CIRCULATIONAHA.115.014207
85. Pottier C, Fresnais M, Gilon M, Jerusalem G, Longuespee R, Sounni NE (2020) Tyrosine kinase inhibitors in cancer: breakthrough and challenges of targeted therapy. Cancer 12:731. https://doi.org/10.3390/cancers12030731
86. Schlessinger J (2000) Cell signaling by receptor tyrosine kinases. Cell 103:211–225. https://doi.org/10.1016/s0092-8674(00)00114-8
87. Jin W, Yang S, Yu C, Zhu T, Jiang Q (2022) Variables associated with pulmonary hypertension screened by echocardiography in chronic myeloid leukemia patients on dasatinib therapy. Front Cardiovasc Med 9:960531. https://doi.org/10.3389/fcvm.2022.960531
88. Kantarjian H, Shah NP, Hochhaus A, Cortes J, Shah S, Ayala M, Moiraghi B, Shen Z, Mayer J, Pasquini R et al (2010) Dasatinib versus imatinib in newly diagnosed chronic-phase chronic myeloid leukemia. N Engl J Med 362:2260–2270. https://doi.org/10.1056/NEJMoa1002315
89. Montani D, Bergot E, Gunther S, Savale L, Bergeron A, Bourdin A, Bouvaist H, Canuet M, Pison C, Macro M et al (2012) Pulmonary arterial hypertension in patients treated by dasatinib. Circulation 125:2128–2137. https://doi.org/10.1161/CIRCULATIONAHA.111.079921
90. Agrawal V, Christenson ES, Showel MM (2015) Tyrosine kinase inhibitor induced isolated pericardial effusion. Case Rep Oncol 8:88–93. https://doi.org/10.1159/000375484
91. Guignabert C, Phan C, Seferian A, Huertas A, Tu L, Thuillet R, Sattler C, Le Hiress M, Tamura Y, Jutant EM et al (2016) Dasatinib induces lung vascular toxicity and predisposes to pulmonary hypertension. J Clin Invest 126:3207–3218. https://doi.org/10.1172/JCI86249
92. Quilot FM, Georges M, Favrolt N, Beltramo G, Foignot C, Grandvuillemin A, Montani D, Bonniaud P, Camus P (2016) Pulmonary hypertension associated with ponatinib therapy. Eur Respir J 47:676–679. https://doi.org/10.1183/13993003.01110-2015
93. Seegobin K, Babbar A, Ferreira J, Lyons B, Cury J, Seeram V (2017) A case of worsening pulmonary arterial hypertension and pleural effusions by bosutinib after prior treatment with dasatinib. Pulm Circ 7:808–812. https://doi.org/10.1177/2045893217733444
94. Riou M, Seferian A, Savale L, Chaumais MC, Guignabert C, Canuet M, Magro P, Rea D, Sitbon O, Jais X et al (2016) Deterioration of pulmonary hypertension and pleural effusion with bosutinib following dasatinib lung toxicity. Eur Respir J 48:1517–1519. https://doi.org/10.1183/13993003.01410-2016
95. Dahal BK, Cornitescu T, Tretyn A, Pullamsetti SS, Kosanovic D, Dumitrascu R, Ghofrani HA, Weissmann N, Voswinckel R, Banat GA et al (2010) Role of epidermal growth factor inhibition in experimental pulmonary hypertension. Am J Respir Crit Care Med 181:158–167. https://doi.org/10.1164/rccm.200811-1682OC
96. Schermuly RT, Dony E, Ghofrani HA, Pullamsetti S, Savai R, Roth M, Sydykov A, Lai YJ, Weissmann N, Seeger W et al (2005) Reversal of experimental pulmonary hypertension by PDGF inhibition. J Clin Invest 115:2811–2821. https://doi.org/10.1172/JCI24838
97. Abe K, Toba M, Alzoubi A, Koubsky K, Ito M, Ota H, Gairhe S, Gerthoffer WT, Fagan KA, McMurtry IF et al (2011) Tyrosine kinase inhibitors are potent acute pulmonary vasodilators in rats. Am J Respir Cell Mol Biol 45:804–808. https://doi.org/10.1165/rcmb.2010-0371OC
98. Nakamura K, Akagi S, Ogawa A, Kusano KF, Matsubara H, Miura D, Fuke S, Nishii N, Nagase S, Kohno K et al (2012) Pro-apoptotic effects of imatinib on PDGF-stimulated pulmonary artery smooth muscle cells from patients with idiopathic pulmonary arterial hypertension. Int J Cardiol 159:100–106. https://doi.org/10.1016/j.ijcard.2011.02.024
99. Hoeper MM, Barst RJ, Bourge RC, Feldman J, Frost AE, Galie N, Gomez-Sanchez MA, Grimminger F, Grunig E, Hassoun PM et al (2013) Imatinib mesylate as add-on therapy for pulmonary arterial hypertension: results of the randomized IMPRES study. Circulation 127:1128–1138. https://doi.org/10.1161/CIRCULATIONAHA.112.000765

100. Frantz RP, McLaughlin VV, Sahay S, Escribano Subias P, Zolty RL, Benza RL, Channick RN, Chin KM, Hemnes AR, Howard LS et al (2024) Seralutinib in adults with pulmonary arterial hypertension (TORREY): a randomised, double-blind, placebo-controlled phase 2 trial. Lancet Respir Med 12:523–534. https://doi.org/10.1016/S2213-2600(24)00072-9
101. Naimi A, Mohammed RN, Raji A, Chupradit S, Yumashev AV, Suksatan W, Shalaby MN, Thangavelu L, Kamrava S, Shomali N et al (2022) Tumor immunotherapies by immune checkpoint inhibitors (ICIs); the pros and cons. Cell Commun Signal 20:44. https://doi.org/10.1186/s12964-022-00854-y
102. Conroy M, Naidoo J (2022) Immune-related adverse events and the balancing act of immunotherapy. Nat Commun 13:392. https://doi.org/10.1038/s41467-022-27960-2
103. Palassin P, Maria ATJ, Ladhari C, Quantin X, Montani D, Bourdin A, Boissin C, Fesler P, Hillaire-Buys D, Guilpain P et al (2022) Pulmonary hypertension reported with immune checkpoint inhibitors: a pharmacovigilance study. Cancer Immunol Immunother 71:3093–3097. https://doi.org/10.1007/s00262-022-03208-2
104. Acar C, Sahin G, Yuksel HC, Karaca B (2024) Rare immune-related adverse effect of pembrolizumab: pulmonary hypertension. J Chemother 37(2):193–198. https://doi.org/10.1080/1120009X.2024.2349858
105. Glick M, Baxter C, Lopez D, Mufti K, Sawada S, Lahm T (2020) Releasing the brakes: a case report of pulmonary arterial hypertension induced by immune checkpoint inhibitor therapy. Pulm Circ 10:2045894020960967. https://doi.org/10.1177/2045894020960967

Chapter 7
Cardiovascular Toxicity of Catecholamines

Miloš P. Stojiljković, Dragana Lončar-Stojiljković, Sonja T. Marinković, Milovan Bojić, and Ranko Škrbić

Abstract Endogenous catecholamines noradrenaline, adrenaline and dopamine have a significant role in the functioning of several organ systems including the cardiovascular one. Noradrenaline is the principal neurotransmitter in the post-ganglionic neurons of the sympathetic nerve system, while adrenaline and, to a lesser extent, noradrenaline are secreted from the adrenal medulla. All these mechanisms are involved in the stressful reaction to various stimuli as part of the fight-or-flight reaction. There are a number of pathological conditions where excessive production of catecholamines results in overstimulation of the adrenergic receptors, which are responsible for hypertension, tachycardia, coronary vasospasm and cardiac ischaemia, resulting in myocardial injury. Such conditions include Pheochromocytoma and takotsubo syndrome. Xenobiotics can also damage the

M. P. Stojiljković (✉)
Department of Pharmacology, Toxicology and Clinical Pharmacology, Faculty of Medicine, University of Banja Luka, Banja Luka, The Republic of Srpska, Bosnia and Herzegovina

Institute for Cardiovascular Diseases "Dedinje", Belgrade, Serbia
e-mail: milos.stojiljkovic@med.unibl.org

D. Lončar-Stojiljković
Institute for Cardiovascular Diseases "Dedinje", Belgrade, Serbia

Department of Anaesthesiology and Intensive Medicine, Faculty of Medicine, University of Banja Luka, Banja Luka, The Republic of Srpska, Bosnia and Herzegovina

S. T. Marinković
Centre for Biomedical Research, Faculty of Medicine, University of Banja Luka, Banja Luka, The Republic of Srpska, Bosnia and Herzegovina

Paediatric Clinic, University Clinical Centre of the Republic of Srpska, Banja Luka, The Republic of Srpska, Bosnia and Herzegovina

M. Bojić
Institute for Cardiovascular Diseases "Dedinje", Belgrade, Serbia

R. Škrbić
Department of Pharmacology, Toxicology and Clinical Pharmacology, Faculty of Medicine, University of Banja Luka, Banja Luka, The Republic of Srpska, Bosnia and Herzegovina

D. K. Agrawal et al. (eds.), *Cardiovascular Toxicity*, Advances in Biochemistry in Health and Disease 33, https://doi.org/10.1007/978-3-031-90598-8_7

cardiovascular system via the adrenergic receptors and they can be divided into direct adrenoceptor agonists (eg, isoprenaline), indirect sympathomimetics, acting via the excessively liberated noradrenaline (eg, amphetamines, cocaine), and mixed-action sympathomimetics that act in both manners (eg, ephedrine). The main recognised mechanisms of catecholamine-induced myocardial injury are myocardial ischemia, coronary insufficiency, intracellular calcium (Ca^{2+}) overload, mitochondrial dysfunction, depletion of high-energy phosphate stores, increased membrane permeability, electrolyte content changes, alterations in metabolism and oxidative stress. Oxidation products of catecholamines—adrenochromes—can also be the cause of myocardial necrosis.

Keywords Catecholamines · Cardiotoxic agents · Oxidative stress · Adrenochrome · Takotsubo cardiomyopathy · Adrenergic receptor · MDMA · Phaeochromocytoma

Introduction

Catecholamines in the Cardiovascular System

As neurotransmitters of the sympathetic nervous system (SNS), catecholamines, including adrenaline (epinephrine), noradrenaline (norepinephrine), and dopamine, are indispensable molecules in the human body. Coupled with the second branch of the autonomic nervous system (ANS)—the parasympathetic nervous system (PNS) and its transmitter the acetylcholine, catecholamines are responsible for the unconscious control of the body's homeostasis, both in physiologic and pathologic conditions. They regulate vital functions such as heart rate, cardiac contractility, and vascular tone. Catecholamines additionally influence metabolic processes, such as glycogenolysis and lipolysis, ensuring an adequate energy supply during stress.

Catecholamines were first identified at the beginning of the XX century, with the isolation of adrenaline from the adrenal medulla and its characterisation as a pressor substance. Gradually, during the next half a century, the investigation of the catecholamine functional role led to the identification of catecholamines as sympathetic neurotransmitters by von Euler, in the 1940s. Technological advances, such as the measurement of catecholamines in tissues by fluorescent spectroscopy and the visualisation of catecholamines in the nerves by histofluorometric procedure, led to further advances and discoveries of the catecholamine metabolism and the process of the noradrenaline reuptake at the sympathetic nerve endings [1]. In the 1970s, the first catecholamine agonists and antagonists were synthesised, which enabled the investigation of the catecholamine receptors [2]. At that point two adrenergic (β1 and β2) and two dopamine (D1, D2) receptors were identified. This boosted the development of antihypertensive drugs, leading to further advances in the field [1, 3].

Centrally regulated by the hypothalamus, both SNS and PNS have acetylcholine-mediated nicotinic receptors between the preganglionic and postganglionic neurons [4]. Postganglionic neurons, however, are different. PNS achieves its effects by activating muscarinic receptors on the target organs. Sympathetic effects, on the other hand, are mediated by adrenergic receptors, including alpha receptors and beta receptors [5]. Regarding the cardiovascular system, alpha receptors are predominantly found in the blood vessels, whilst beta receptors are primarily located in the myocardium. Both α1 and α2 receptor activation in the smooth muscle cells of the peripheral arteries, promote smooth muscle contraction, and subsequent vasoconstriction. However, presynaptic α2 receptor activation, exerts sympatholytic effects, inhibiting the release of noradrenaline. In the myocardium, activation of β1 and β2 receptors induces positive chronotropic, dromotropic and inotropic effects. As for the β3 receptors, they have been shown to induce both negative and positive inotropic effects, depending on the different pathway activations [6].

Elucidation of the structures of the G-protein coupled receptors with which the catecholamines interact as first messengers was marked as a most important discovery of the field at the Tenth International Catecholamine Symposium [7], which was confirmed later that year with a Nobel prize award to Robert J. Lefkowitz and Brian K. Kobilka (*Catecholamine research in the twenty-first century*, 2013).

All 3 subtypes of β-adrenergic receptors (AR) belong to the G protein-coupled receptor (GPCR) superfamily of membrane receptors and share several structural and functional features, however, the 3 subtypes display different affinities for given ligands, specific cellular expression and subcellular localisation patterns, differential coupling to signalling cascades, and distinct regulatory mechanisms [8]. Both β1-AR and β2-AR can couple to Gs protein. The activation of the Gαs subunit leads to the activation of adenylyl cyclase (AC), which in turn catalyses the formation of cyclic adenosine monophosphate (cAMP) from adenosine triphosphate (ATP). Local increases in cAMP trigger PKA activation, thus releasing the functional catalytic subunit, which coordinately phosphorylates the plasma membrane L-type calcium channel (LTCC). This increases Ca^{2+} influx, which in turn activates the ryanodine receptor 2 (RyR2) in the sarcoplasmic reticulum (SR) membrane, resulting in increased cytosolic Ca^{2+} concentration available for contraction. In addition to these inotropic and lusitropic effects, adrenergic stimulation also promotes direct cAMP modulation of hyperpolarization-activated cyclic nucleotide-gated channels that carry the pacemaker current, raising heart rate (chronotropic effect) [9]. Beta 3 adrenergic receptor function in the myocardium is however not completely elucidated. It has been shown to have negative inotropic effects via the Gi protein/eNOS/cGMP/PKG pathway. However, β3 receptor downstream signalling in the heart might also involve L-type Ca^{2+} channel stimulation through the cAMP/PKA pathway which leads to positive inotropic effects in the human atria [6].

Additionally, agonist stimulation of β-AR sequentially promotes GPCR kinase-mediated phosphorylation of β-ARs and recruitment of β-arrestins leading to termination of G protein signalling, a process termed GPCR desensitisation, receptor internalisation, and downregulation. Therefore, the overall cellular effects of GPCR stimulation would result from the balance between the G protein-dependent and GRKs/β-arrestin-dependent branches of GPCR signalling [10] (Fig. 7.1).

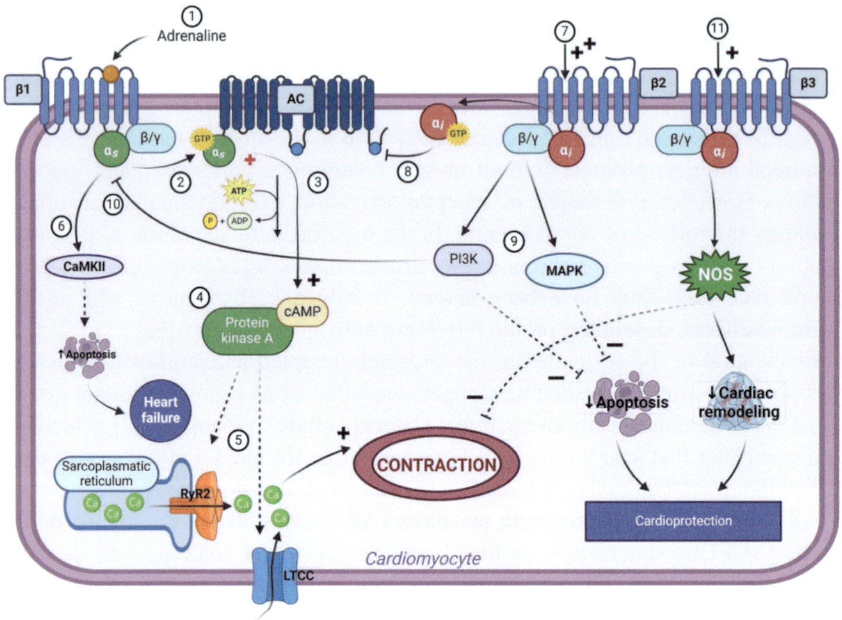

Fig. 7.1 β-adrenergic receptors mediated intracellular pathways. (1) Main pathway: catecholamine binding to β1 receptor induces coupling of the receptor with the Gs protein; the Gαs-GTP subunit dissociates from the complex and activates adenylyl cyclase (AC) (2); AC catalyses cyclic adenosine monophosphate (cAMP) synthesis (3), which leads to protein kinase A (PKA) activation (4). PKA then phosphorylates plasma membrane L-type calcium channel (LTCC) or the RyR2 calcium channel in the sarcoplasmic reticulum (5), resulting in increased cytosolic Ca^{2+} concentration available for contraction of cardiac muscle. (6) In chronic heart failure, continuous stimulation of β1 receptors induces apoptosis via Ca^{2+}/calmodulin-dependent protein kinase II (CaMKII), leading to heart damage. (7) If β2 receptors are continuously stimulated, i.e. when using selective β1-blockers, they couple with the Gi protein. The Gαi-GTP subunit inhibits AC (8), while the 5β subunit of Gi induces both the inhibition of apoptosis via stimulation of mitogen-activated protein kinases (MAPK) and the phosphatidylinositol 3-kinase (PI3K)-protein kinase B (AKT) pathway (9), and of Gs-mediated deleterious effects (10), leading to cardioprotection. Additionally, in heart failure, stimulation of the β3 receptors activates nitric oxide synthase (NOS) and leads to a decrease in apoptosis and cardiac remodelling (11). Created in https://BioRender.com

Early Animal Studies and Initial Investigations of Catecholamine Cardiotoxicity

In the years following the identification of catecholamines, a variety of catecholamine analogues were synthesised. In the late 1930s, a group of scientists working at the Department of Pharmacology at Vienna University was engaged in pharmacological analysis of these newly synthesised sympathomimetics. The investigations were mainly focused on the bronchodilatory and pressor activity of the substances, since adrenaline was still the most commonly used bronchodilator, despite having undesirable or even dangerous side effects on the cardiovascular system. An isopropyl analogue of adrenaline was recognised for its strong bronchodilator effects, which it produced in tenfold lower doses than adrenaline. Further analysis regarding its cardiovascular effects showed that the isopropyl-noradrenaline had pronounced chronotropic and positive inotropic effects on the heart. The importance of the substance was recognised immediately, and only two months into experiments, an application for the grant of a patent was made [2].

Even in the early years following the discovery of catecholamines, Josué described myocardial hypertrophy following adrenaline administration [11]. During the following years, several publications discussed the cardiotoxic effects of catecholamines. This culminated with the introduction of an animal model in the 1960s. Rona et al. demonstrated the potential of isoprenaline overdose to induce myocardial necrosis in animal models. The seminal work provided a foundation for subsequent investigations into the pathophysiological mechanisms of catecholamine-induced cardiac injury [12]. The potential mechanisms that were investigated during the next quarter of the century were summarised in a review by Rona [13]. He described the major mechanisms such as hypoxia, altered membrane permeability, coronary microcirculation disruption, and Ca^{2+} overload. He also mentioned a new potential pathomechanism of catecholamine-induced myocardial injury, developed later by Dhalla and associates. In those studies, it was observed that spontaneously oxidised isoprenaline produces extensive myocardial damage and functional cardiac impairment. They showed, both in vivo and in vitro, that an oxidation product of adrenaline, named adrenochrome, depresses the contractile activity of the myocardium and affects the cardiac muscle cell membranes with a dose- and time-dependent phenomenon. This led to the hypothesis that oxidation products of the catecholamine, and not catecholamines themselves, are the cause of myocardial necrosis.

Pathophysiological Mechanisms of Catecholamine Cardiotoxicity, Acute and Chronic Consequences

Nowadays, several mechanisms of the cardiotoxic effect of excessive catecholamine stimulation are explained, and even though all of them can be described as an individual pathological state, they also influence one another and are therefore understood as complex entities. The main recognised mechanisms are myocardial ischemia, coronary insufficiency, intracellular calcium (Ca^{2+}) overload, mitochondrial dysfunction, depletion of high-energy phosphate stores, increased membrane permeability, electrolyte content changes, alterations in metabolism and oxidative stress [11]. Among them, reactive oxygen species (ROS) and oxidative stress are considered to have the central role in subsequent DNA damage, apoptosis, inflammation, fibrosis and hypertrophy of the cardiomyocytes that follow the initial stages of myocardial damage, and ultimately lead to chronic heart failure [14, 15]. Because of that special consideration of oxidative stress associated with catecholamine

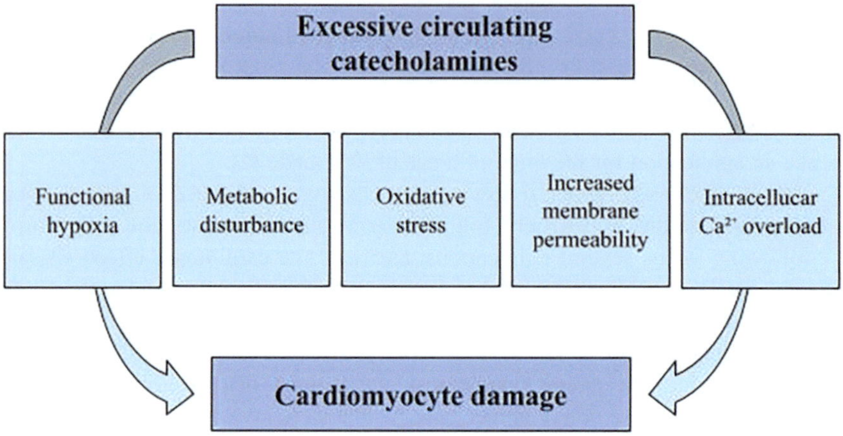

Fig. 7.2 Graphical representation of the main mechanism of catecholamine-induced myocardial damage

overstimulation will be made after a more succinct description of the other mechanisms (Fig. 7.2).

Mechanism of Acute Catecholamine Cardiotoxicity

As mentioned before, one of the earliest described mechanisms of cardiotoxic effects of catecholamines is hypoxia and ischemia [13, 16]. The myocardial hypoxia that occurs during the catecholamine overstimulation, is referred to as functional hypoxia, as there is no actual interruption of the blood flow like in the case of myocardial infarction. This is considered to be a consequence of two concomitant events, one being the increase in oxygen demand and the other being coronary artery spasm. Both events are direct effects of adrenoreceptor activation in the cardiomyocytes and coronary arteries' smooth muscle cells, respectively [17]. The rise in cardiac oxygen demand is explained via the catecholamine-induced increase of contractility (positive inotropic effect), heart rate (positive chronotropic effect), the threshold of excitation (positive bathmotropic effect) and conductivity (positive dromotropic effect) of catecholamines. These effects lead to an increase in cardiac work, and therefore to greater oxygen demand. On the other hand, increased dilatation of AV shunts, and decrease of the aortic blood pressure, alongside coronary artery constriction compromise blood flow to the myocardium and consequent decrease in oxygen supply [17].

Another important mechanism of catecholamine-induced cardiac damage is the alteration of cell metabolism. Described in the beginnings of the catecholamine research as the 'oxygen-wasting effect' [18] it is explained by the mobilisation of free fatty acids by the excessive circulating catecholamines. Mobilised free fatty acids disrupt the process of oxidative phosphorylation, thus increasing oxygen consumption and depleting the high-energy phosphate stores in the myocardial cells [11]. This process overlaps with the previously described coronary insufficiency, but it has been described that it can occur in the absence of coronary vasospasm. Investigations showed that there is a maximal inotropic effect of the catecholamines, and therefore further increase in circulating catecholamines will have little or no effect on the oxygen consumption via the increased workload. However, a further increase in oxygen consumption was described [18, 19], thus confirming the oxygen-wasting theory. Later studies considered the role of the catecholamine oxidation products, rather than catecholamines themselves, in the uncoupling of oxidative phosphorylation in the mitochondria, which will be referred to later on.

Further on, catecholamine-induced loss of cellular Mg^{2+} supports the above-mentioned reduction of the high-energy phosphate stores. Magnesium is vital for activating enzymes that transfer phosphate into mitochondria for ATP production. A decrease in Mg^{2+} disrupts this transfer, leading to impaired oxidative phosphorylation and mitochondrial dysfunction which leads to irreversible damage of the myocardiocytes [11]. Mg^{2+} depletion is also linked to increased Ca^{2+} uptake by the mitochondria, which indirectly influences ATP production. Early studies showed a double increase in net myocardial Ca^{2+} concentration [20, 21]. This overload triggers the excessive activation of Ca^{2+}-dependent ATPases, impairing mitochondrial function and causing mechanical and electrical disturbances in cardiomyocytes [22]. Recent studies clearly connect excessive amounts of circulating catecholamines with mitochondrial Ca^{2+} overload, depression of mitochondrial function, generation of oxidative stress, activation of proteases, fragmentation of DNA and impaired cardiovascular function [23].

A key factor in catecholamine-induced cardiotoxicity is considered to be oxidative stress. This process is a result of the auto-oxidation of catecholamines under excessive stimulation of the sympathetic nervous system. Catecholamine catabolism is normally regulated via two enzymes—monoamine oxidase (MAO) and catechol-O-methyltransferase (COMT). When these enzymes are saturated, excessive catecholamines undergo auto-oxidation, forming ortho-semiquinones that convert to ortho-quinones. These undergo intracellular cyclization, creating unstable leucoaminochromes, which are rapidly oxidized into aminochromes and subsequently polymerized into aminolutins [24]. Nowadays, MAO-A is known as a key factor in the development of acute and chronic heart diseases. Exposure to MAO-A leads to the accumulation of LC311, p62 and ubiquitylated proteins, blocking the apoptosis, and causing mitochondrial fission and cellular necrosis. Besides, MAO-A, through the accumulation of cathepsin D and Lamp 1 proteins, reduces lysosomal acidification and blocks an autophagy regulator — transcription factor-EB (TFEB) [25].

The toxic effects of oxidized catecholamines manifest through several mechanisms. They interact with sulfhydryl groups in proteins, altering membrane permeability, mitochondrial calcium uptake, oxygen consumption, and ATP production. Interestingly, the previously mentioned increased magnesium concentrations can mitigate these effects by stabilizing energy production [26]. Additionally, quinonoid compounds raise the levels of superoxide anions and hydrogen peroxide, leading to hydroxyl radical formation, which damages proteins, lipids, and DNA [27]. This was confirmed in studies that used antioxidants, that have shown efficacy in preventing oxyradical-induced cellular changes. Prolonged abnormalities in Ca^{2+} handling, lead to the loss of myofibrillar sensitivity to calcium and further derangements and degeneration of the myofilaments, explaining contractile dysfunction of the myocardium. A modification of myosin gene expression is also connected to the decrease in myofibrillar Ca^{2+} stimulated ATPase activity in the infarcted myocardium [23].

The complexity of the Ca^{2+}-influx and Ca^{2+}-efflux mechanisms has not yet been fully clarified. Although oxidative stress certainly increases the permeability of the sarcolemmal membrane via lipid peroxidation, thus allowing the Ca^{2+} entry, additional mechanisms are to be considered. Among them, a decrease in phospholipid N-methylation, induced by different oxyradical generating systems can influence the membrane fluidity. Sarcolemmal phospholipase C and D activity also can be modified during oxidative stress, supporting the abnormalities in Ca^{2+} handling [23, 25].

Free radicals formed during the catecholamine oxidation, also induce tumor necrosis factor (TNF)-α production, initiating apoptosis independent of β-adrenoceptor pathways. Both catecholamine overstimulation and TNFα cause an increase in nonphagocytic NOS activity, leading to cardiac hypertrophy, collagen deposition, metalloprotease activation, fibrosis and heart failure [28, 29].

Chronic Consequences of Catecholamine-Induced Cardiotoxicity, Experimental Implications

The vast majority of the experiments regarding catecholamines published until today, investigate the cardiotoxicity of isoprenaline and the potential of various compounds to diminish it. As mentioned before, the isoprenaline model was established in the 1960s, and over time several different approaches to this acute myocardial injury model were developed, as well as a delayed model of isoprenaline-induced heart failure, which will be considered in the following text.

Isoprenaline is chemically known as c4-[1-hydroxy-2-[(1-methylethyl)amino] ethyl]-1,2-benzenediol monohydrochloride is classified as a nonselective β-adrenergic agonist. Clinically it is mostly used to treat arrhythmias such as bradycardia, thioridazine-induced torsade de pointes, and heart block [30]. When administered in consecutive fractionated doses, it causes irreversible damage to the myocardium, which ultimately leads to chronic heart failure [31]. Although there is a significant amount of different dosing regimens, 5–10 mg/kg/day for 7–28 consecutive days is used in most studies. Rodents are the preferred experimental animal used, mostly Wistar albino rats or Swiss mice [30]. Studies suggest that male rats are the preferred sex, as there is information that they are more susceptible to fibrosis due to multiple factors i.e. increased activation of fibroblasts, increased expression of β1 and β2 receptors and increased expression of adrenergic cascade mediators [31]. Finally, the isoprenaline model of chronic heart failure is the simplest available model, as it does not require additional expenses as genetic-based models, or complex equipment as surgical models.

The previously described mechanism of acute cardiotoxicity is, as mentioned before, followed by subsequent DNA damage, apoptosis, inflammation, fibrosis and hypertrophy of the cardiomyocytes that ultimately lead to chronic heart failure [14, 15]. The central role in this process is attributed to oxidative stress and reactive oxygen species. Exposure to superoxide anions and other ROS stimulates the differentiation of fibroblasts and the release of TGF-β, which initiates the fibrogenic response. Isoprenaline administration additionally decreases the expression of sirtuin 1 (SIRT1), a histone deacetylase that acts as a transcription factor. Reduced SIRT1 expression leads to the loss of competing signalling pathways and the predomination of TGF-β signalling. This leads to the activation of Smad-2 and Smad-3 intermediates, and further stimulates cardiac hypertrophy and fibrosis [32]. Moreover, studies have demonstrated the role of PPAR-γ downregulation after isoprenaline administration in the development of cardiac fibrosis [33].

Modulation of immune-inflammatory cascades and proinflammatory activity of isoprenaline is also one of the mechanisms that leads towards cardiac fibrosis and chronic heart failure. Firstly, overstimulation of adrenergic receptors by isoprenaline leads to the release of proinflammatory mediators, and activation of the NF-κB signalling pathway. A decrease of anti-inflammatory cells such as CD206+, combined with infiltration with proinflammatory cells like CD86+ was also noted after isoprenaline administration. Additionally, a rise in CCR2 receptors was also noted after isoprenaline administration. Macrophage migration into the damaged myocardium is stimulated via the CCR2 signalling, which stimulates cardiac remodelling and fibrosis [31, 33].

An additional mechanism of isoprenaline-induced cardiac fibrosis is the activation of the renin-angiotensin system. Studies confirm that isoprenaline administration increases the levels of angiotensin-converting enzyme (ACE), which converts angiotensin I into angiotensin II. A rise in angiotensin II levels promotes fibroblast differentiation and hyperplasia in the myocardium [34].

Finally, isoprenaline administration induces the expression of fibrotic proteins and subsequent accumulation of cardiac collagen and extracellular matrix proteins in the cardiac interstitium, which is a primary characteristic of cardiac fibrosis. As mentioned, isoprenaline activates the NF-κB signalling pathway, activating p38, ERK and JNK kinases that aggravate fibrosis [35]. Additionally, a direct phosphorylation of the p38, ERK and JNK kinases by isoprenalin, leads to activation and proliferation of fibroblasts and to imbalance of cytokines tissue inhibitor of metalloproteinase-1 (TIMP-1) and matrix metalloproteinase-9 (MMP9) in favor of the aftermentioned, further promoting the fibrotic process. There are numerous additional mechanisms that are discussed or investigated in the studies using the isoprenaline model of heart failure, however further insight surpasses the scope of the present chapter.

Catecholamine Induced Cardiotoxicity, Clinical Implications

In various conditions, overactivation of the sympatho-adrenergic system is an essential mechanism that provides short-term adaptation to stressful conditions. This is also known as the *fight-or-flight response* [36]. In the same way, administration of exogenous catecholamines is a life-saving measure in conditions with reduced cardiac output and/or hypotension. Although beneficial and necessary in the short term, prolonged elevation of circulating catecholamines initiates significant cardiotoxic effects. This can be observed in several conditions such as critical illness, stress-induced or takotsubo cardiomyopathy, chronic heart failure, phaeochromocytoma, and also during prolonged therapy with exogenous catecholamines.

Catecholamines Cardiotoxicity in the Critically Ill

As mentioned, in several critical conditions such as septic shock, brain injury, cardiogenic shock, and cardiac arrest, a reflex sympathetic overstimulation provides a life-saving response. The response of the sympathetic nervous system can be classified into short-term and long-term reflexes. Short-term reflexes, which maintain blood flow to vital organs, are mediated by baroreceptors in the large systemic arteries (aorta, carotid sinus, pulmonary artery) and veins, as well as in cardiac atria and ventricles, whilst long-term reflexes that maintain vascular tone and circulating blood act through the renin-angiotensin-aldosterone system. In contrast to these short-term benefits, prolonged elevated catecholamine levels in these settings significantly increase myocardial oxygen demand, trigger arrhythmias, and cause direct damage to myocytes through mechanisms such as oxidative stress and calcium overload.

In patients with sepsis, high catecholamine levels coupled with the systemic inflammatory response, further strain the myocardium and contribute to the development of septic cardiomyopathy. Likewise, in patients with polytrauma, the protective adrenergic storm exacerbates multi-organ failure. Also, in the state of acute myocardial infarction, a surge of catecholamines, intended to maintain blood supply to vital organs, on the contrary, increases the risk of ventricular arrhythmias and even expansion of infarct size [3]. The mechanisms behind these are partially elucidated. Catecholamine overstimulation damages cardiomyocytes by inducing mitochondrial dysfunction, which is explained through two main mechanisms. Firstly, the overactivation of beta receptors induces intracellular calcium overload, via the previously explained cAMP/PKA/RyR2 pathway. Secondly, catecholamines increase oxidative stress in the myocardium, via the formation of "aminochromes", that further undergo redox cycling in the mitochondria, and result in the production of excessive amounts of free radicals. Both processes increase the permeability of the mitochondrial membrane, which promotes cardiomyocyte cell death via the apoptotic and necrotic pathways.

Additionally, excess catecholamines lead to desensitisation and downregulation of adrenergic receptors. The two processes are considered to be two mechanisms of so-called catecholamine-induced autonomic dysfunction. Receptor desensitisation happens via phosphorylation of beta receptors by PKA, leading to Gαs and Gβγ uncoupling, which allows G protein-coupled kinase-2 (GPK2) to bind to Gβγ. Beta receptor phosphorylation results in conformational changes in beta receptors that allow a scaffolding protein called B-arrestin-1 to bind to the beta receptors, forming a protein-receptor complex that leads to conformational changes in beta receptors. This results in three effects. First, the conformational changes impair beta receptor coupling to Gs protein and enhance its binding to the Gi proteins, consequently suppressing the activity of adenyl cyclase (AC) and preventing cAMP production. Second, cyclic nucleotide phosphodiesterase (cnPDE) attaches to the beta receptor plus B-arrestin complex, leading to the hydrolyzation of and degradation of the pre-existing cAMP. Finally, beta receptor internalization and degradation are induced via the B-arrestin as previously described [3, 36].

The described mechanisms of beta receptor desensitisation and downregulation, weaken the compensatory effects produced by SNS activation, leading to the deterioration of blood pressure, despite the high levels of circulating catecholamines. This is why some studies suggest that the administration of beta-blockers in the critically ill may reverse these effects via the upregulation of adrenergic receptors and enhancement of receptor sensitivity [37, 38].

Catecholamines and Takotsubo Cardiomyopathy

Firstly described in 1990 and later on defined as a clinical condition in the early 2000s, takotsubo syndrome (TTS) represents a quintessential example of catecholamine-induced cardiotoxicity. Also known as „broken heart syndrome, TTS predominantly affects postmenopausal women and is usually triggered by intense physical and/or emotional stress [39, 40]. Patients with TTS present with acute chest pain, electrocardiographic changes, and elevated cardiac biomarkers, mimicking an acute myocardial infarction. However, coronary angiography typically reveals unobstructed coronary arteries. This is why takotsubo syndrome is recognised as a separate entity. Typical findings in TTS are transient regional left ventricular dysfunction, with akinesia of apical segments, that is associated with a metabolic derangement of the affected myocardium. This results in typical ballooning of the left ventricle, which can be seen in the imaging studies. It was first recognised by Dr. Hikaru Sato, from the Hiroshima City Hospital, who then proposed the term "takotsubo-like left ventricular dysfunction" because of the similarities in the shapes of the left ventricle in TTS and the Japanese fishing pot for catching octopuses named "takotsubo" (Fig. 7.3). Later on, due to its clinical heterogeneity (clinical profile, different triggers, different anatomic forms, left

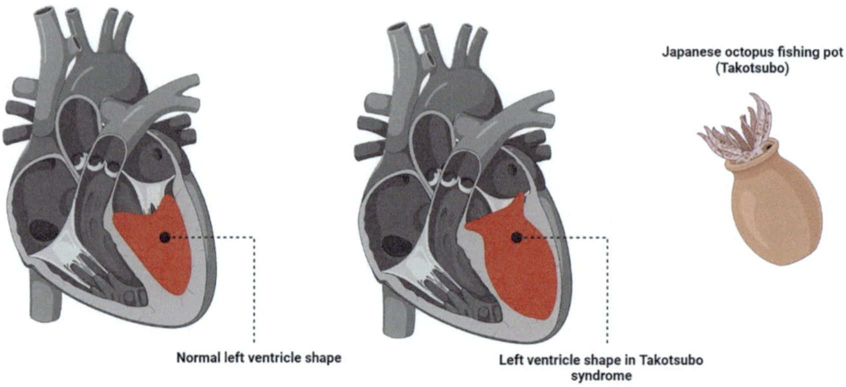

Fig. 7.3 Graphical representation of (left) normal left ventricle shape, (central) typical apical ballooning of the left ventricle in the takotsubo syndrome (TTS) that bears resemblance with the Japanese octopus catching pot or *Tako-Tsubo* (right). Created in https://BioRender.com

ventricle or right ventricle or RV variable involvement, different pathophysiology) takotsubo is thought to be a cluster of different diseases with similar expression (phenocopies), fulfilling the definition of "syndrome" [41, 42].

Pathophysiologically, TTS development is attributed to excessive catecholamine release during emotional and/or physical stress, which produces a catecholamine injury of the myocardium, that most researchers describe as myocardial stunning [43]. This correlates with reports of conditions such as phaeochromocytoma and thyrotoxicosis causing TTS, as well as with findings that both adrenaline and noradrenaline levels are notably more elevated in TTS than in AMI. Additionally, neuroimaging studies have revealed heightened sympathetic nervous system activity in TTS patients, further linking stress responses with cardiac dysfunction [41].

The initial stress-induced increase in noradrenaline secretion stimulates the hypothalamic-pituitary-adrenal axis, and release of primarily adrenaline from the cells of the adrenal medulla. Studies have also demonstrated an increased noradrenaline spillover in the coronary sinus, which suggests increased release of catecholamines in the myocardium, from the cardiac nerve endings. This induces calcium overload, resulting in contraction band necrosis, one of the pathological hallmarks of TTS. The typical pattern of left ventricular dysfunction in TTS, can on the other hand be explained by the regional differences in the adrenergic receptor density. The apical segment of the left ventricle has a denser adrenaline β-AR expression compared to the basal segments, whereas a reverse distribution is present for noradrenaline β1-AR. As previously mentioned, the overstimulation of beta2-adrenoceptors with adrenaline triggers a switch from Gs to Gi coupling, which is associated with a negative inotropic response of the left ventricular apex, thus protecting the basal myocardium from the catecholamine storm, and limiting the degree of myocardial injury in the acute phase of TTS.

Additionally, coronary artery spasm and/or diffuse coronary vasoconstriction have commonly been reported in patients with TTS, suggesting transient myocardial ischaemia as an additional pathogenetic mechanism of TTS. Studies have also indicated that microvascular dysfunction and endothelial injury contribute to the pathogenesis of TTS, further implicating catecholamines in the cascade of events leading to cardiac dysfunction [41, 44].

Phaeochromocytoma, Cardiotoxic Manifestations

A rare neuroendocrine tumour of the adrenal medulla, phaeochromocytoma, is another example of excessive endogenous catecholamine secretion. The typical clinical presentation of phaeochromocytoma is paroxysmal episodes of palpitations, headaches, hypertension and diaphoresis, all resulting from surges in catecholamine secretion. The clinical features of phaeochromocytoma-related cardiomyopathy include hypertension, dilated or hypertrophic cardiomyopathy, pulmonary oedema due to cardiogenic and noncardiogenic factors, cardiac arrhythmias, and even cardiac arrest. Whether the cardiotoxic effects will manifest as a reversible myocardial dysfunction or will progress to irreversible damage, depends only on the timely diagnosis and treatment. Similarly as previously described, catecholamines cause a direct toxic effect on the myocardium, via their oxidation products. Besides, through receptor-mediated calcium dysregulation, catecholamines promote mitochondrial dysfunction and cell death. Studies have shown that prolonged catecholamine surges lead to structural and functional alterations in the myocardium, including interstitial fibrosis, myocyte hypertrophy, and impaired contractility [45].

Indirect Catecholamine Agonists and Cardiotoxicity

Sympathomimetic drugs are divided into three groups: direct-acting, indirect-acting and mixed-action. While the first one directly stimulates adrenergic receptors, the second one does the same, but via increased availability of the endogenous neurotransmitter noradrenaline, with the third one combining the effects of the former two [46]. Cardiotoxic effects of indirect catecholamines also have significant clinical implications, and will therefore be explained further. Indirect catecholamines are a group of drugs that exert their toxicity either completely or partially via the increased availability of endogenous catecholamines. Their mechanisms of action include increased liberation of noradrenaline from the presynaptic sympathetic nerve endings and decreased removal of noradrenaline from synaptic clefts [47]. The resulting build-up of noradrenaline results in the overstimulation of postsynaptic adrenergic receptors wherever they are, including the ones localised in the myocardium and blood vessels [14]. Some other molecules act as mixed-action adrenergic agonists, including both increased intrasynaptic availability of noradrenaline and direct stimulation of adrenergic receptors.

Amphetamine

Amphetamine acts mainly via the accumulated increased concentrations of catecholamines, causing an array of cardiovascular adverse effects, including hypertension, tachycardia, ventricular fibrillation, myocardial infarction and cardiac arrest [48], but also accelerated atherosclerosis, acute coronary syndrome and sudden cardiac death [49]. Longer abuse of amphetamine results in damage of cardiomyocytes, myocardial inflammation and fibrosis, leading to dilated cardiomyopathy and extremely decreased left-ventricular ejection fraction (LVEF) [50]. It seems that the major mechanism of cardiotoxic action of amphetamine is the catecholamine-induced increased production of myocardial reactive oxygen species (ROS) [51].

Methamphetamine (N-methyl-alpha-methylphenethylamine)

Abuse of crystalline methamphetamine hydrochloride ("ice", "crystal meth") can result in numerous harmful effects on the cardiovascular system, including hypertension, accelerated atherosclerosis, vasospasm-induced acute coronary syndromes, sudden cardiac death, and dilated cardiomyopathy with congestive heart failure and left ventricular dysfunction [52]. All these effects are the result of the excessive release of catecholamines that cause vasoconstriction and vasospasm, which ultimately lead to hypertension, tachycardia, endothelial dysfunction, and cardiotoxicity [53]. It seems that the Janus kinase–signal transducer and activator of transcription 3 (JAK–STAT3) pathway that is otherwise involved in ischemia/reperfusion injury in the myocardium, may also be activated by the vasospasm-inducing action of the drug [54].

Methamphetamine increases cardiac microvascular permeability via the VEGF-PI3K-Akt-eNOS signalling pathway. This effect can be blocked with bevacizumab, a humanised monoclonal antibody that prevents the binding of vascular endothelial growth factor (VEGF) to its receptors [55].

In a series of 100 patients who died because of a methamphetamine overdose, 68% had clear myocardial histopathological changes, including myocardial fibre hypertrophy, perivascular fibrosis and acute myocardial infarction. The proposed mechanism is an excess of catecholamines, with consequential coronary vasospasm and ischaemia, ROS production, mitochondrial dysfunction, accelerated apoptosis, cardiomyocyte necrosis, impaired intracellular calcium hemostasis and abnormality in cardiac protein production [56].

It was also shown in the isolated rat cardiac mitochondria that methamphetamine disrupted mitochondrial function and induced ROS formation, lipid peroxidation, GSH depletion, MMP collapse and mitochondrial swelling [57].

In a similar manner, it was shown in mice that methamphetamine induces myocardial injury, oxidative stress and apoptosis via the nuclear factor-erythroid factor 2-related factor 2/heme oxygenase-1 (Nrf2/HO-1) pathway and that these phenomena can be alleviated by Nrf2-specific agonist sulforaphane. These findings also support the catecholaminergic involvement in the development of methamphetamine cardiac toxicity [58].

3,4-Methylenedioxymethamphetamine (MDMA)

MDMA is an empathogenic-entactogenic drug and a popular substance of abuse that comes in the form of tablets (*ecstasy*) or crystalline powder (*mandy*, *molly*). It releases all and blocks the reuptake of all three major monoamines in the central nervous system representing a strong serotonin-noradrenaline-dopamine releasing agent (SNDRA) and serotonin-noradrenaline-dopamine reuptake inhibitor (SNDRI) [59]. As a result, various untoward cardiovascular effects may occur, including hypertension, tachycardia, cardiac arrhythmias, myocardial infarction or, after prolonged use, myocardial hypertrophy, fibrosis and dilatatory heart failure [60]. Although the proposed mechanism includes the catecholaminergic mechanism of cardiotoxicity, it seems that other factors also contribute to the deleterious effect of MDMA (Fig. 7.4).

It seems, based on the in vitro studies on isolated rat cardiomyocytes, that neither MDMA nor its metabolite 3,4-methylenedioxyamphetamine (MDA) could damage the cardiomyocytes. On the contrary, two other metabolites of MDMA, N-methyl-R-methyldopamine (N-Me-R-MeDA), and R-methyl-dopamine (R-MeDA) induced

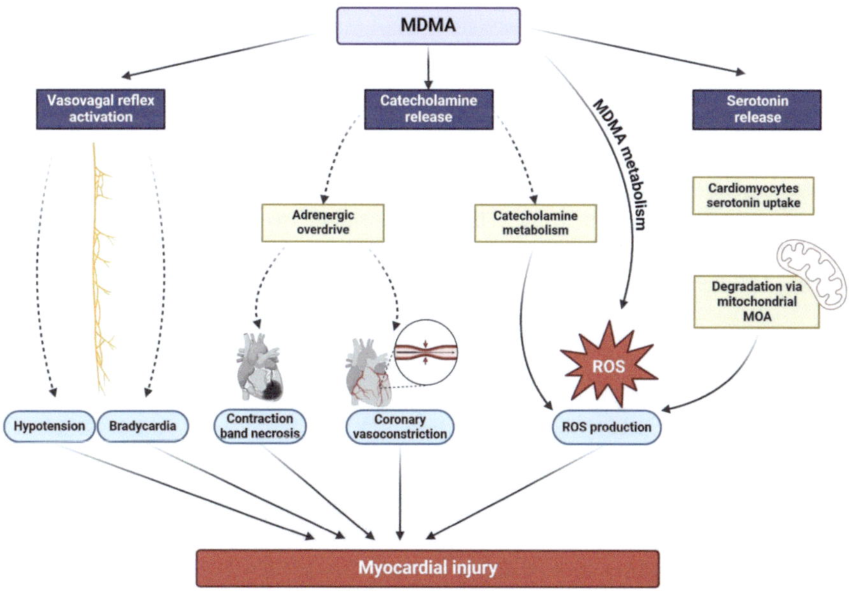

Fig. 7.4 Mechanisms of MDMA-induced myocardial injury. Created in https://BioRender.com

the loss in cell morphology, accompanied by the decrease in levels of the reduced glutathione (GSH), increase in the calcium ion and decrease in the ATP intracellular levels and antioxidative enzyme activities. These findings indicated a more complex mechanism of MDMA-induced cardiac injury, involving the metabolites of a parental molecule [61].

Moreover, it seems that MDMA has a more deleterious effect on the brains, hearts and kidneys of senescent rats, in comparison with young adults [62].

Mixed-Action Adrenergic Agonists and Cardiotoxicity

Some molecules act as mixed-action adrenergic agonists, including both increased intrasynaptic availability of noradrenaline and direct stimulation of adrenergic receptors. One of the oldest known members of this group is ephedrine [63].

Ephedrine

Ephedrine alkaloids (eg, ephedrine, pseudoephedrine, methyl ephedrine, norephedrine, and norpseudoephedrine) are present in many species of the plant Ephedra (Ma Huang). They act as sympathomimetics stimulating the release of endogenous catecholamines (eg, noradrenaline) that act as agonists of ⟨- and β-adrenergic receptors. In addition, ephedrine can act directly upon the β1-adrenergic receptor subtypes leading to increased heart rate and cardiac output that together result in increased blood pressure. Such direct actions may also promote cardiac arrhythmias via stimulation of the sinoatrial node [64].

It was shown in experiments in rats that both the Chinese plant Ma Huang and its main alkaloid ephedrine, when combined with caffeine, induce haemorrhage, necrosis and myocardial degeneration in the interventricular septum even as early as 2–4 h after their administration [65]. This effect seems to be age-dependent since cardiotoxicity of ephedrine was more pronounced in 14-week-old rats than in 7-week-old ones [66].

Conclusion

Catecholamines, including adrenaline, noradrenaline, and dopamine, play a pivotal role in the regulation of the cardiovascular function through their actions as neurotransmitters of the sympathetic nervous system. While they are essential for maintaining homeostasis and adapting to acute stress, their excessive or prolonged activation can lead to significant cardiovascular toxicity. The pathophysiological mechanisms underlying catecholamine-induced cardiotoxicity are complex and interconnected. Experimental models, particularly the isoprenaline-induced acute myocardial injury and chronic heart failure models, have provided valuable insights into these mechanisms. A key factor in catecholamine-induced cardiotoxicity is considered to be oxidative stress, while increased myocardial oxygen demand, coronary vasospasm, and metabolic disturbances are additional important mechanisms of the acute cardiotoxicity. Oxidative stress is a result of the auto-oxidation of catecholamines under excessive stimulation of the sympathetic nervous system, with consequent production of adrenochromes. These mechanisms are followed by subsequent DNA damage, apoptosis, inflammation, fibrosis and hypertrophy of the cardiomyocytes which are the mechanisms of chronic cardiotoxicity. Cellular damage, structural remodelling, and functional decline of the myocardium, ultimately lead to chronic heart failure. Clinically, the balance between the short-term benefits of catecholamines in maintaining cardiovascular function and their long-term detrimental effects is critical. In conditions such as sepsis, myocardial infarction, and stress-induced cardiomyopathy or takotsubo cardiomyopathy, excessive catecholamine release exacerbates myocardial injury, arrhythmias, and heart failure. This underscores the need for careful management of catecholamine levels in clinical settings, particularly in critically ill patients. Indirect and mixed-action catecholamine agonists, such as cocaine, amphetamines and ephedrine, exacerbate cardiovascular damage through similar mechanisms. These substances, often abused or misused, pose significant public health challenges due to their widespread availability and potent adrenergic effects. Their impact on the cardiovascular system further highlights the broader implications of catecholamine toxicity beyond endogenous overstimulation.

References

1. Axelrod J (1985) Foreword. In: Elsevier eBooks. Elsevier, Cham, pp xv–xviii. https://doi.org/10.1016/b978-0-409-95184-4.50004-7
2. Konzett H (1981) On the discovery of isoprenaline. Trends Pharmacol Sci 2:47–49. https://doi.org/10.1016/0165-6147(81)90259-5
3. Eraky AM et al (2024) Beta-blockers as an immunologic and autonomic manipulator in critically ill patients: a review of the recent literature. Int J Mol Sci 25(15):8058. https://doi.org/10.3390/ijms25158058
4. Gibbons CH (2019) Basics of autonomic nervous system function. In: Handbook of clinical neurology. Elsevier, Amsterdam, pp 407–418. https://doi.org/10.1016/b978-0-444-64032-1.00027-8
5. Gatica S et al (2023) Effects of adrenergic receptor stimulation on human hemostasis: a systematic review. Adv Exp Med Biol 1408:49–63. https://doi.org/10.1007/978-3-031-26163-3_3
6. Schena G, Caplan MJ (2019) Everything you always wanted to know about β3-AR * (* but were afraid to ask). Cells 8(4):357. https://doi.org/10.3390/cells8040357
7. Eiden LE (2013) Catecholamine research in the 21st century. Elsevier, Amsterdam. https://doi.org/10.1016/c2012-0-07405-4

8. Oliver E, Mayor F, D'Ocon P (2019) Beta-blockers: historical perspective and mechanisms of action. Rev Esp Cardiol 72(10):853–862. https://doi.org/10.1016/j.rec.2019.04.006
9. Rockman HA, Koch WJ, Lefkowitz RJ (2002) Seven-transmembrane-spanning receptors and heart function. Nature 415(6868):206–212. https://doi.org/10.1038/415206a
10. De Lucia C, Eguchi A, Koch WJ (2018) New insights in cardiac Β-adrenergic signaling during heart failure and aging. Front Pharmacol 9:904. https://doi.org/10.3389/fphar.2018.00904
11. Adameova A, Abdellatif Y, Dhalla NS (2009) Role of the excessive amounts of circulating catecholamines and glucocorticoids in stress-induced heart disease. Can J Physiol Pharmacol 87(7):493–514. https://doi.org/10.1139/y09-042
12. Rona G et al (1959) An infarct-like myocardial lesion and other toxic manifestations produced by isoproterenol in the rat. Arch Pathol 67(4):443–455
13. Rona G (1985) Catecholamine cardiotoxicity. J Mol Cell Cardiol 17(4):291–306. https://doi.org/10.1016/s0022-2828(85)80130-9
14. Du Y, Demillard LJ, Ren J (2021) Catecholamine-induced cardiotoxicity: a critical element in the pathophysiology of stroke-induced heart injury. Life Sci 287:120106. https://doi.org/10.1016/j.lfs.2021.120106
15. Ljubunčić P et al (1991) Importance of oxygen free radical for physiology, pathology and therapy. Srp Arh Celok Lek 119(11–12):314–321
16. Ljubunčić P et al (1991) The effect of preliminary administration of dimethylsulfoxide on the development of experimental myocardial necrosis in rats treated with isoprenaline. Srp Arh Celok Lek 119(9–10):238–242
17. Beamish RE, Dhalla NS (1985) Involvement of catecholamines in coronary spasm under stressful conditions. In: Beamish RE, Singal PK, Dhalla NS (eds) Stress and heart disease. Developments in cardiovascular medicine, vol 45. Springer, Boston. https://doi.org/10.1007/978-1-4613-2587-1_10
18. Raab W (1943) Pathogenetic significance of adrenalin and related substances in heart muscles. Exp Med Surg 1:188–225
19. Clancy RL et al (1967) Inotropic augmentation of myocardial oxygen consumption. Am J Phys 212(5):1055–1061. https://doi.org/10.1152/ajplegacy.1967.212.5.1055
20. Fleckenstein A et al (1974) Myocardial fiber necrosis due to intracellular Ca overload—a new principle in cardiac pathophysiology. Recent Adv Stud Cardiac Struct Metab 4:563–580
21. Ljubunčić P et al (1992) The effect of calcium antagonists on the cardiotoxic effects of isoprenaline in rats. Vojnosanit Pregl 49(4):305–311
22. Vassalle M, Lin CI (2004) Calcium overload and cardiac function. J Biomed Sci 11(5):542–565. https://doi.org/10.1007/bf02256119
23. Shah AK et al (2021) Oxidative stress as a mechanism for functional alterations in cardiac hypertrophy and heart failure. Antioxidants 10(6):931. https://doi.org/10.3390/antiox10060931
24. Behonick GS et al (2001) Toxicology update: the cardiotoxicity of the oxidative stress metabolites of catecholamines (aminochromes). J Appl Toxicol 21(S1):S15–S22. https://doi.org/10.1002/jat.793
25. Dhalla NS et al (2022) Involvement of oxidative stress in the development of subcellular defects and heart disease. Biomedicines 10(2):393. https://doi.org/10.3390/biomedicines10020393
26. Yates JC et al (1980) Protection against adrenochrome-induced myocardial damage by various pharmacological interventions. Br J Exp Pathol 61(3):242–255
27. Dhalla NS, Temsah RM, Netticadan T (2000) Role of oxidative stress in cardiovascular diseases. J Hypertens 18(6):655–673. https://doi.org/10.1097/00004872-200018060-00002
28. Bajic Z et al (2022) Homocysteine, vitamins B6 and folic acid in experimental models of myocardial infarction and heart failure—how strong is that link? Biomol Ther 12(4):536. https://doi.org/10.3390/biom12040536
29. Sobot T et al (2024) Effect of folic acid on isoprenaline-induced myocardial injury in rats. Physiol Int 111(1):80–96. https://doi.org/10.1556/2060.2023.00291
30. Allawadhi P et al (2018) Isoproterenol-induced cardiac ischemia and fibrosis: plant-based approaches for intervention. Phytother Res 32(10):1908–1932. https://doi.org/10.1002/ptr.6152

31. Eddin LB et al (2024) Isoproterenol mechanisms in inducing myocardial fibrosis and its application as an experimental model for the evaluation of therapeutic potential of phytochemicals and pharmaceuticals. Anim Models Exp Med. https://doi.org/10.1002/ame2.12496
32. Bugyei-Twum A et al (2018) Sirtuin 1 activation attenuates cardiac fibrosis in a rodent pressure overload model by modifying Smad2/3 transactivation. Cardiovasc Res 114(12):1629–1641. https://doi.org/10.1093/cvr/cvy131
33. Ma Z-G et al (2017) Piperine attenuates pathological cardiac fibrosis via PPAR-Γ/AKT pathways. EBioMedicine 18:179–187. https://doi.org/10.1016/j.ebiom.2017.03.021
34. Garvin AM et al (2021) Transient ACE (angiotensin-converting enzyme) inhibition suppresses future fibrogenic capacity and heterogeneity of cardiac fibroblast subpopulations. Hypertension 77(3):904–918. https://doi.org/10.1161/hypertensionaha.120.16352
35. Dai H et al (2019) Phosphocreatine attenuates isoproterenol-induced cardiac fibrosis and cardiomyocyte apoptosis. Biomed Res Int 2019:5408289. https://doi.org/10.1155/2019/5408289
36. Liaudet L, Calderari B, Pacher P (2014) Pathophysiological mechanisms of catecholamine and cocaine-mediated cardiotoxicity. Heart Fail Rev 19(6):815–824. https://doi.org/10.1007/s10741-014-9418-y
37. Kastratović DA et al (2007) Carvedilol increases copper-zinc superoxide dismutase activity in patients with acute myocardial infarction. Basic Clin Pharmacol Toxicol 101(2):138–142. https://doi.org/10.1111/j.1742-7843.2007.00094.x
38. Ljubunčić P et al (1992) Efficacy of pretreatment with adrenergic beta-receptor blockers in the prevention of cardiotoxic effects of isoprenaline in rats. Vojnosanit Pregl 49(4):297–304
39. Bajic Z et al (2023) Cardioprotective effects of liraglutide pretreatment on isoprenaline-induced myocardial injury in rats. Can J Physiol Pharmacol 101(5):258–267. https://doi.org/10.1139/cjpp-2022-0534
40. Marinković ST et al (2023) Pomegranate peel extract attenuates isoprenaline-induced takotsubo-like myocardial injury in rats. Pharmaceutics 15(6):1697. https://doi.org/10.3390/pharmaceutics15061697
41. Dias A et al (2018) Takotsubo syndrome: state-of-the-art review by an expert panel—part 1. Cardiovasc Revasc Med 20(1):70–79. https://doi.org/10.1016/j.carrev.2018.11.015
42. Stănescu C, Branidou K (2006) Takotsubo cardiomyopathy. Rom J Intern Med 44(2):97–116
43. Bajic Z et al (2024) Liraglutide protects cardiomyocytes against isoprenaline-induced apoptosis in experimental takotsubo syndrome. Biomedicines 12(6):1207. https://doi.org/10.3390/biomedicines12061207
44. Dias A et al (2018) Takotsubo syndrome: state-of-the-art review by an expert panel—part 2. Cardiovasc Revasc Med 20(2):153–166. https://doi.org/10.1016/j.carrev.2018.11.016
45. Kassim TA et al (2008) Catecholamine-induced cardiomyopathy. Endocr Pract 14(9): 1137–1149. https://doi.org/10.4158/ep.14.9.1137
46. Rahman R, Saleem S, Khan R (2023) Mechanism of action of adrenergic drugs and recent updates. In: Kazmi I, Karmakar SA, Shaharyar MA, Afzal M, Al-Abbasi FA (eds) How synthetic drugs work. Academic, New York, pp 47–60
47. Mladěnka P, Applová L, Patočka J, Costa VM, Remiao F, Pourová J et al (2018) Comprehensive review of cardiovascular toxicity of drugs and related agents. Med Res Rev 38(4): 1332–1403. https://doi.org/10.1002/med.21476
48. Alzeer AA, Suliman I, Altamimi M, Alshudukhi AM, Alzeer AA, Alwasidi EO (2023) Acute myocardial infarction associated with amphetamine use and smoking in a young healthy individual. Cureus 15(12):e50323. https://doi.org/10.7759/cureus.50323
49. Böttner J, Fischer-Schaepmann T, Werner S, Knauth S, Jahnke HG, Thiele H et al (2024) Amphetamine increases vascular permeability by modulating endothelial actin cytoskeleton and NO synthase via PAR-1 and VEGF-R. Sci Rep 14(1):3596. https://doi.org/10.1038/s41598-024-53470-w

50. Berlot B, Williams MGL, Bucciarelli-Ducci C (2020) The colours of amphetamine-induced cardiomyopathy. J Cardiovasc Med 21(5):396–397. https://doi.org/10.2459/JCM.0000000000000922
51. Costa VM, Carvalho F, Duarte JA, Bastos ML, Remião F (2013) The heart as a target for xenobiotic toxicity: the cardiac susceptibility to oxidative stress. Chem Res Toxicol 26(9):1285–1311. https://doi.org/10.1021/tx400130v
52. Tobolski J, Sawyer DB, Song SJ, Afari ME (2022) Cardiovascular disease associated with methamphetamine use: a review. Heart Fail Rev 27(6):2059–2065. https://doi.org/10.1007/s10741-022-10261-7
53. Kevil CG, Goeders NE, Woolard MD, Bhuiyan MS, Dominic P, Kolluru GK et al (2019) Methamphetamine use and cardiovascular disease. Arterioscler Thromb Vasc Biol 39(9):1739–1746. https://doi.org/10.1161/ATVBAHA.119.312461
54. Annawald K, Streckfuss-Bömeke K, Meyer T (2024) Methamphetamine-induced cardiotoxicity: in search of protective transcriptional mechanisms. Herz 49(6):434–440. https://doi.org/10.1007/s00059-024-05279-6
55. Chen R, Huang P, Wei S, Zhang C, Lai X, Wang H et al (2022) Methamphetamine exposure increases cardiac microvascular permeability by activating the VEGF-PI3K-Akt-eNOS signaling pathway, reversed by bevacizumab. Hum Exp Toxicol 41:9603271221121795. https://doi.org/10.1177/09603271221121795.3-632
56. Akhgari M, Mobaraki H, Etemadi-Aleagha A (2017) Histopathological study of cardiac lesions in methamphetamine poisoning-related deaths. Daru 25(1):5. https://doi.org/10.1186/s40199-017-0170-4
57. Shabani M, Jamali Z, Bayrami D, Salimi A (2023) Vanillic acid alleviates methamphetamine-induced mitochondrial toxicity in cardiac mitochondria via antioxidant activity and inhibition of MPT pore opening: an in-vitro study. BMC Pharmacol Toxicol 24(1):33. https://doi.org/10.1186/s40360-023-00676-9
58. Yu H, Peng Y, Dong W, Shen B, Yang G, Nie Q et al (2023) Nrf2 attenuates methamphetamine-induced myocardial injury by regulating oxidative stress and apoptosis in mice. Hum Exp Toxicol 42:9603271231219488. https://doi.org/10.1177/09603271231219488
59. Docherty JR, Alsufyani HA (2021) Pharmacology of drugs used as stimulants. J Clin Pharmacol 61(Suppl 2):S53–S69. https://doi.org/10.1002/jcph.1918
60. Bonsignore A, Barranco R, Morando A, Fraternali Orcioni G, Ventura F (2019) MDMA induced cardio-toxicity and pathological myocardial effects: a systematic review of experimental data and autopsy findings. Cardiovasc Toxicol 19(6):493–499. https://doi.org/10.1007/s12012-019-09526-9
61. Carvalho M, Remião F, Milhazes N, Borges F, Fernandes E, Monteiro MC et al (2004) Metabolism is required for the expression of ecstasy-induced cardiotoxicity in vitro. Chem Res Toxicol 17(5):62
62. Feio-Azevedo R, Costa VM, Barbosa DJ, Teixeira-Gomes A, Pita I, Gomes S et al (2018) Aged rats are more vulnerable than adolescents to "ecstasy"-induced toxicity. Arch Toxicol 92(7):2275–2295. https://doi.org/10.1007/s00204-018-2226-8
63. Ibrahim R, Nyska A, Dunnick J, Ramot Y (2021) The toxicologic pathology aspects of selected natural herbal products and related compounds. J Toxicol Pathol 34(3):181–211. https://doi.org/10.1293/tox.2021-0016
64. Brown CE, Trauth SE, Grippo RS, Gurley BJ, Grippo AA (2012) Combined effects of ephedrine-containing dietary supplements, caffeine, and nicotine on morphology and ultrastructure of rat hearts. J Caffeine Res 2(3):123–132. https://doi.org/10.1089/jcr.2012.0021
65. Dunnick JK, Kissling G, Gerken DK, Vallant MA, Nyska A (2007) Cardiotoxicity of Ma Huang/caffeine or ephedrine/caffeine in a rodent model system. Toxicol Pathol 35(5):657–664. https://doi.org/10.1080/01926230701459978
66. Howden R, Hanlon PR, Petranka JG, Kleeberger S, Bucher J, Dunnick J et al (2005) Ephedrine plus caffeine causes age-dependent cardiovascular responses in Fischer 344 rats. Am J Physiol Heart Circ Physiol 288(5):H2219–H2224. https://doi.org/10.1152/ajpheart.01164.2004

Chapter 8
Effect of Antipsychotic and Antidepressant Drugs in Cardiovascular Toxicity

Marija Stojanovic and Devendra K. Agrawal

Abstract Antipsychotics and antidepressants are widely used in clinical practice for treatment of mental disorders. In comparison to general population patients diagnosed with schizophrenia and depression are at higher risk for developing some of cardiovascular adverse effects related to prescribed therapy. Adverse effects after antipsychotics and antidepressants treatment may vary in clinical presentation, starting from asymptomatic ECG changes to serious arrhythmia and in some cases sudden cardiac death. One of the most common reported adverse effects regarding both antipsychotics and antidepressants is orthostatic hypotension. In addition, QT prolongation has been reported as a ECG finding associated with ventricular arrhythmia such as torsade de pointes. Even though, not that frequent, torsade de pointes has potential to evolve in lethal arrhythmia leading to sudden cardiac death. Fatal outcome of antipsychotic and antidepressant medication has been reported in some cases, presented with myocarditis and cardiomyopathy at pathological basis. Underlying pathophysiological mechanisms od antipsychotics and antidepressants induced cardiotoxicity are mediated by different ion channel blockade, adrenergic and muscarinic receptor blockade. Prolonged QT interval is associated with HERG potassium channel blockade. This inhibition affect physiological repolarization of cardiomyocytes leading to action potential prolongation, manifested on ECG as QT prolongation. Aimed to prevent cardiovascular adverse effects during treatment of mental disorders special attention should be given on preventive cardiac function monitoring, especially in subjects with cardiovascular comorbidities.

M. Stojanovic
Department of Translational Research, Western University of Health Sciences, Pomona, CA, USA

Faculty of Medicine, Institute of Medical Physiology "Richard Burian", University of Belgrade, Belgrade, Serbia

D. K. Agrawal (✉)
Department of Translational Research, Western University of Health Sciences, Pomona, CA, USA
e-mail: DAgrawal@WesternU.edu

D. K. Agrawal et al. (eds.), *Cardiovascular Toxicity*, Advances in Biochemistry in Health and Disease 33, https://doi.org/10.1007/978-3-031-90598-8_8

Keywords Antipsychotics · Antidepressants · Orthostatic hypotension · QT prolongation · ECG monitoring · Co-existing heart disorders · Cardiovascular risk factors

Introduction

Cardiovascular toxicity (CVT) refers to adverse effects of numerous drugs that affect normal heart physiology leading to arrhythmias, myocardial dysfunction and even sudden cardiac death. It is recognized as one of the most common adverse effects in clinical practice [1]. Cardiac adverse effects of drugs had been shown to be the reason for both withdrawal of the drug from clinical practice and also exclusion of it from clinical trials [2]. That is one of the reasons why the pharmacovigilance is of great importance to ensure drugs with high safety profile on the market, as well as to give better insight into the mechanisms underlying cardiotoxicity. Nevertheless, precise molecular pathways underlying drugs induced cardiotoxicity are still to be elucidated.

Numerous drugs for the treatment of central nervous system (CNS) diseases have been shown to exhibit a certain degree of cardiotoxicity, such as antiepileptic drugs, drugs used in the treatment of neurodegenerative diseases (Alzheimer's disease, Parkinson's disease), multiple sclerosis or autism spectrum disorder [3–8]. Nevertheless, large amount of scientific evidence indicates the importance of cardiac adverse effects of drugs used in the treatment of mental disorders, for both schizophrenia and depression. Additionally, it has been reported that dual therapy including new antidepressants and atypical antipsychotics is proved to be effective in patients suffer from schizoaffective disorder or psychotic depression [9, 10]. Undoubtedly, this leads to drug-drug interactions which represents one of the major concerns subjected to the treatment of different clinical entities of schizophrenia and depressive disorders.

Taken all the above into account and having in mind increasing number of newly diagnosed patients with schizophrenia and depression, this chapter aims to review and analyze the most common cardiac adverse effects followed by antipsychotics and antidepressants treatment.

Antipsychotics and Cardiovascular Toxicity (CVT)

It is worth mentioning that not all patients are at the same risk of development of cardiovascular toxicity following antipsychotic treatment. Factors that influence reactions to a drug are affected by different determinants, such as (1) personal characteristics of the patients, (2) potential drug interactions as well as the (3) drug properties itself. (1) Factors related to the patients usually include older age, associated comorbidities and polypharmacy, such as history of heart disease or potassium and magnesium disbalance [11]. (2) Drug interactions of antipsychotics are reported with antiseizure medications, anticholinergic drugs, tricyclic antidepressants, antiarrhythmics and antihypertensives [12–15]. (3) From the perspective of a drug specific profile, for instance, increased clinical application of atypical antipsychotic drugs is preferable, due to their better treatment of antipsychotic symptoms, even though these agents are associated with weight gain and adverse cardiovascular events [16].

Antipsychotics are diverse group of drugs used to treat schizophrenia, bipolar disorder, and major depressive disorder [17]. They are traditionally divided into first-generation antipsychotics (FGAs) and second-generation antipsychotics (SGAs) also known as atypical antipsychotics [18]. The mechanism of action of the latter (clozapine, olanzapine, quetiapine, risperidone) includes blockade of D2 dopaminergic receptors, as well as blockade of 5-HT_{2A} serotonin receptors [19]. Blockade of serotonin or dopamine receptors is responsible for their antipsychotic effects. On the other side, mechanism of action of the first-generation antipsychotics (chlorpromazine, fluphenazine, haloperidol, loxapine, molindone, pimozide) includes almost exclusively modulation of dopaminergic neurotransmission by D2 receptors antagonism in the brain [20, 21]. Some of the new generation antipsychotics (aripiprazole) have unique properties acting as partial agonist of D2 and 5-HT_{1A} receptors, in combination with antagonism at 5-HT_{2A} receptors [22–24]. Even though SGAs have been shown to exert better safety in comparison to FGAs, their cardiovascular and metabolic effects raise several concerns [25].

Cardiovascular adverse effects of antipsychotics are mediated by receptors, mostly presented in the heart (muscarinic receptors-M2) and blood vessels (α1-adrenoreceptors) [26]. Some others adverse effects in forms of life threatening arrythmias are not that common and are usually mediated by blockade of cardiac potassium channels, such as HERG channels, encoded by a human ether-a-go-go related gene-HERG [27, 28]. Additional mechanisms of cardiovascular adverse effects have been proposed, such as blockade of calmodulin or calcium and sodium channel blockade [26]. Recent data reported by Le et al. highlighted NLRP3 inflammasome-mediated pyroptosis as an underlying mechanism of antipsychotics induced cardiotoxicity [29]. Oxidative stress, mitochondrial disfunction or DNA damage are subcellular findings related to antipsychotic induced CVT [30, 31].

Tachycardia, bradycardia, hypertension or electrocardiographic abnormalities are among most common adverse effects followed by antipsychotic treatment of mental disorders [32–34] (Tables 8.1 and 8.2). In addition, it is worth mentioning that also sudden cardiac death has been reported as the most severe adverse effect, even though the precise mechanism underlying this phenomenon is missing, due to lack of sufficient post-mortem analysis (Fig. 8.1). In addition, some conditions are associated with increased cardiovascular risk in schizophrenia patients, due to their sedentary lifestyle and metabolic dysfunctions, such as insulin resistance, obesity and diabetes mellitus type 2 (DM2) [53]. In the following text the mechanisms underlying the most important cardiac adverse effects of antipsychotics, representing their cardiotoxicity, will be discussed.

Table 8.1 Cardiovascular (CV) adverse effects of typical antipsychotics presented by the mechanism of action

Drug	CV adverse effect	Underlying mechanism	Study type	Reference
Typical antipsychotics				
Chlorpromazine	Torsade de pointes; QT prolongation	Not reported	Case report, woman received a moderate dose	[35]
	Prolonged APD	Inhibition of HERG K^+ channels	Xenopus oocyte	[36]
Haloperidol	Prolonged APD	Inhibition of HERG K^+ channels	Xenopus oocytes	[37]
	Torsade de pointes and QT prolongaton	Inhibition of HERG K^+ channels	Literature review on normal and schizophrenic human subjects	[38]
	Sudden death			
Droperidol	Torsade de pointes and QT prolongaton	Inhibition of HERG K^+ channels	Literature review on normal and schizophrenic human subjects	[38]
	Sudden death			
Thioridazine	Prolonged APD	Inhibition of HERG K^+ channels	Guinea-pig ventricular myocytes	[39]
	Torsade de pointes and QT prolongaton		Literature review on normal and schizophrenic human subjects	[38]
	Sudden death			
Pimozide	Prolonged APD	Inhibition of Ca^{2+} channels; inhibition of HERG K^+ channels	Rat ventricular myocytes	[40, 41]
	Torsade de pointes and QT prolongaton	Inhibition of HERG K^+ channels	Literature review on normal and schizophrenic human subjects	[38]
	Sudden death			

APD action potential duration

Mechanisms Underlying CVT of Antipsychotic Drugs

Orthostatic Hypotension

Orthostatic hypotension (OH) is recognized as the most common adverse effect of antipsychotics [54]. It refers to the inadequate maintenance of blood pressure (BP) in terms of sudden drop of BP when moving from laying or sitting to standing position. It can be asymptomatic, but it can also be presented with various clinical

Table 8.2 Cardiovascular (CV) adverse effects of atypical antipsychotics presented by the mechanism of action

Drug	CV adverse effect	Underlying mechanism	Study type	References
Atypical antipsychotics				
Risperidone	Prolonged APD	Inhibition of HERG K^+ channels	Rabbit ventricular myocytes	[42]
	Multifocal atrial tachycardia	Not reported	Case report, 13 years old male patient with autism	[43]
	Torsade de pointes			
	Bradycardia	Associated CAD	Case report, 82 years old woman	[44]
Clozapine	Reduced HRV	Blockade of M_2 muscarinic receptors in the heart	Clinical study with chronic schizophrenia patients	[45]
		ANS dysregulation	Study with 56 schizophrenia patients	[46]
	Myocarditis (eosinophilic infiltrate)	IgE-mediated hypersensitivity	Necropsy and other records about patients receiving the drug	[47]
	Cardiomyopathy	Accumulated effects of other antipsychotic agents		
	Atrial fibrilation	Not reported	Case report, woman with schizophrenia	[48]
	Tachycardia	Vagolytic effects of clozapine	Review of side effects of clozapine	[49]
	Orthostatic hypotension	α1-adrenergic blockade	Review of clinical studies	[50]
Sertindole	Prolonged APD	Inhibition of HERG K^+ channels	Human cardiomyocytes	[40]
	Torsade de pointes and QT prolongaton		Literature review on normal and schizophrenic human subjects	[38]
	Sudden death			
Olanzapine	Orthostatic hypotension	α1-adrenergic blockade	Case report, woman with paranoid schizophrenia	[51]
	Atrial fibrilation	Not reported	Case report, woman with schizophrenia	[48]
Aripiprazole	Orthostatic hypotension	Competitive inhibition of aripiprazole metabolizing enzyme -CYP3A4	Case report, woman diagnosed with bipolar I manic episode	[52]

APD action potential duration, *HRV* heart rate variability, *CYP3A4* cytochrome P450 3A4, *ANS* autonomic nervous system, *CVS* cardiovascular, *CAD* coronary artery disease

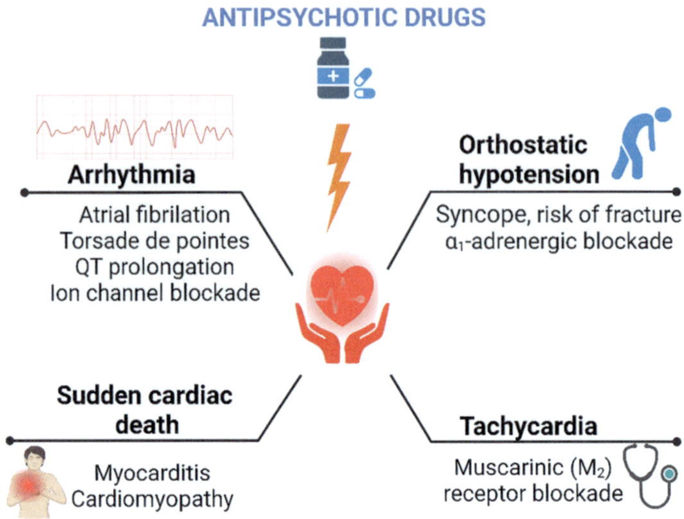

Fig. 8.1 Antipsychotic treatment of mental disorders causes cardiovascular adverse effects presented in different clinical entities: arrythmia, hypotension, tachycardia or even sudden cardiac death. Eosinophilic myocarditis and cardiomyopathy represent the pathological substrate for sudden cardiac death. Underlaying pathophysiological mechanisms include ion channel blockade, adrenergic and cholinergic receptors blockade

manifestations from mild cognitive impairment to presyncope and syncope [55]. When clinically presented in a form of syncope, OH represents the risk of injury due to falling. Additionally, around one third of all falls in the elderly homes are associated with use of antipsychotics [56]. Antipsychotics have also been associated with increased risk of hip fracture [57]. It has been shown that during the antipsychotic's treatment a worsening of associated cardiac diseases occurs, such as angina pectoris. This is due to increased oxygen consumption followed by reflex tachycardia, which occurs as a reactive response to vasodilation. Vasodilation occurs in a response to a blockade of α1-adrenergic receptors by antipsychotics. This is exactly the mechanism responsible for antipsychotics induced orthostatic hypotension. Namely, in physiological conditions activation of α1-adrenergic receptors in blood vessels is responsible for vasoconstriction in response to circulating catecholamines, and by that, for maintaining sympathetic tone. Logically, by blockade of these receptors, vasodilation occurs leading to hypotension. Hypotensive effect has been found to be dose dependent and to decrease with prolonged therapy [58]. Not all antipsychotic drugs have the same affinity to α1-adrenergic receptors. Namely, higher incidence of OH occurrence have been reported in association with low-potency FGAs (chlorpromazine) in comparison to high-potency FGAs (haloperidol) or most of SGAs except clozapine [50, 59].

Heart Rate Variability (HRV)

Variation in time between two heartbeats is known as HRV. This parameter serves as an indicator of cardiovascular pathologies, as well as determinant of autonomic nervous system function. HRV analysis is recognized as a non-invasive method for estimation of sympathetic and parasympathetic system interactions [60]. HRV is a marker of poor prognosis, since it is a predictor of sudden cardiac death in population of patients with myocardial infarction and heart failure [61]. In psychiatry patients suffering from mental disorders, reduced HRV has been found after treatment with antipsychotics [62, 63]. This effect is mediated by cholinergic receptors, which when blocked are responsible for the anticholinergic effects. Anticholinergic properties are reflected in decreased parasympathetic regulation of the cardiac function. One study, observing acute changes of HRV after single dose antipsychotic treatment in healthy man volunteers, showed that olanzapine increased and thioridazine decrease HRV, while risperidone didn't affect is. According to these results, authors suggest HRV as a potential index of cardiotoxicity, since it is sensitive to acute one dose of antipsychotic drugs [64]. In order to estimate HRV, ECG recordings in this study were conducted for 10 h. Another literature data also showed decreased HRV in patients on monotherapy with haloperidol, clozapine and olanzapine compared to controls. It is worth mentioning that in comparison to the previous one, in this study ECG recording was performed for 10 min which is more comfortable from the perspective of clinical setting [46]. Nevertheless, to investigate the drug effects on HRV, a methodological approach is still to be improved for the optimum method.

Tachycardia

As previously mentioned, anticholinergic effects of antipsychotics came from blockade of M_2 muscarinic receptors in the heart. In this case, when vagal tone is decreased, predominance of the sympathetic tone occurs, and therefore consequent sinus tachycardia. Sometimes in addition to tachycardia other anticholinergic side effects, such as constipation or dry mouth, take place [65]. So far, reported tachycardia following antipsychotic treatment was mild in intensity, usually with tolerance developed over time. In addition, according to the literature data, antipsychotics induced tachycardia is not related to the serious cardiovascular complications. It is worth mentioning that in the case of clozapine treatment persistent tachycardia may occur, with heart rate values higher than 110–120 bpm, when β-blockers are used for managing the patient with cardiovascular symptoms. Aside from clozapine, as the only one SGA related to tachycardia, usually are low-potency typical antipsychotics such as chlorpromazine and thioridazine associated with this condition [66]. These effects are dose-dependent, and sometimes potentiated by α1-receptors blockers, which cause vasodilation and reflex tachycardia. Since tachycardia increases cardiac tissue oxygen demand, it could contribute to the worsening of pre-existed cardiovascular diseases. Therefore, patients suffering from schizophrenia or other mental disorders treated with antipsychotics, should undergo ECG monitoring in case of cardiovascular comorbidities [67].

Lengthening of QT Interval

One of the ECG manifestations of antipsychotic treatment is reflected in ECG recording as a QT prolongation [68]. This refers to a longer duration of QT interval, measuring from the beginning of Q wave until the end of the T wave. Different ion currents, such as potassium or sodium, responsible for the cardiac action potential, has been found to participate in the mechanism of QT prolongation. Nevertheless, in most cases, QT prolongation induced by antipsychotics is mediated by potassium channel blockade, specifically potassium rapid delayed rectifier channel-Ikr (also known as HERG channel) leading to prolonged cardiac repolarization as a result of blockade of outward movement of potassium [69]. Specifically, some FGAs such as haloperidol and chlorpromazine in addition to potassium channel blockade, also inhibit sodium and calcium channels [70, 71]. It is worth noting that haloperidol and sertindole have been described as a potent blockers of HERG channels [37, 40]. In humans, sertindole was shown to have a high affinity antagonism towards cardiac HERG potassium channels [40, 72]. Experiments performed on guinea-pig ventricular myocytes showed prolongation of the action potential following haloperidol and droperidol treatment by inhibiting HERG. Additionally, the inhibition of potassium current by haloperidol were five or twenty times more potent in comparison to its effects on sodium and calcium channels [73, 74].

QT prolongation as an ECG substrate is found to be associated with ventricular arrythmias such as torsade de pointes in patients treated with antipsychotic drugs (TdP) [75]. TdP is defined as a dangerous polymorphic ventricular arrhythmia which may be spontaneously recovered into the sinus-rhythm or, in rare cases, can lead to syncope of even sudden cardiac death [76]. Moreover, sudden cardiac deaths reported following antipsychotic treatment are usually associated with complicated torsade de pointes [77]. This is the reason why antipsychotics, in addition to other types of drugs (antidepressants, antimicotics, antiarrhythmics, antihypertensives, antihistamines) are considered torsadogenic [78]. Nevertheless, the correlation between QT prolongation and TdP turns out to be very complex, since it is affected by various factors including age, female sex, electrolyte (dis)balance such as hypokaliemia, calciemia or magnesiemia, co-existing cardiac pathologies, diabetes mellitus or polypharmacy [79, 80]. One of the most dangerous risk factors is congenital prolongation of QT interval, known as long QT (LQT) [81]. In addition, it has been reported that torsadogenic potential of antipsychotics is directly related to the high dose of the drug, even though it has been shown that psychiatry patients itself already have higher risk of sudden cardiac death due to their sedentary lifestyle and associated metabolic and cardiovascular disorders [82]. Literature data proposed a list of antipsychotics with very strong torsdogenic signal such as: olanzapine, chlorpromazine, clozapine, haloperidol, quetiapine, risperidone and ziprasidone [83].

What is shown so far is that all the drugs that lead to the torsade de pointes act by inhibiting the potassium channels, in particular potassium rectifier channel (IKr). It is worth mentioning that experimental studies performed in an in vivo animal (rabbit) model reported anti-arrhythmic properties of antipsychotics, by reducing the incidence of drug induced TdP. These effects were mainly mediated by adrenergic-α1-receptor blockade [84]. Literature data suggests some other receptors likewise muscarinic or β-adrenergic receptors or ion channels as a mediator of anti-arrhythmic properties of antipsychotic drugs [85]. On the other side, one study including human volunteer paricipants receiving one of six antipsychotics compared QT intervals durations between FGAs and SGAs and reported that thioridazine had the greatest lengthening in QT interval in comparison to other antipsychotics [86]. It is noteworthy that according to the results of this study FDA (Food and Drug Administration) assigned a black box warning for thioridazine. Some other FGAs (mesoridazine, droperidol, pimozide) also have black box warning, since they have been found to prolong QT interval for more than 30 ms and therefore contribute to the risk of sudden cardiac death.

Myocarditis and Cardiomyopathy

Myocarditis is one of the uncommon adverse effects following antipsychotic treatment, clinically presented with a variety of symptoms starting from subclinical manifestations to cardiac decompensation such in a form of cardiomyopathy. Both cardiomyopathy and myocarditis were documented after treatment of mental disorders with therapeutical dosages of antipsychotics. One study demonstrated eosinophilic or lymphocytic myocarditis following clozapine therapy [47]. It has been reported that the first symptoms were presented two weeks from the beginning of the therapy [53]. Furthermore, eosinophilic myocarditis has found to be pathological substrate for the clozapine induced sudden cardiac death. Cardiomyopathy following therapy with other atypical antipsychotics (quetinapine) has also been described, presented on ECG as an acute STEMI (ST-segment elevation myocardial infarction) [87]. Even though, myocarditis and cardiomyopathy are rare adverse effects of antipsychotic treatment, they could be misdiagnosed since their unspecific clinical manifestations, leading to potential heart failure of sudden cardiac death.

Sudden Cardiac Death

Recent literature findings indicate that prolonged use of FGAs and SGAs is associated with higher rate of sudden cardiac death (SCD) in comparison to non-users of neuroleptics [88]. Similarly, one cohort study reported increased incidence of sudden cardiac death in subjects taking antipsychotics in comparison to non-treated patients with antipsychotics in a dose-dependent manner [89]. The incidence of SCD in schizophrenic patients is estimated to be four times higher in comparison to general population. It has been found that sudden cardiac death in schizophrenic patients has multifactorial nature, considering both environmental factors (sedentary lifestyle, poor diet, alcohol abuse) and associated cardiac diseases [90]. Furthermore, co-existing cardiovascular pathologies counts for 60% of SCD in patients diagnosed with schizophrenia including ischemic heart disease, arrhythmia, myocarditis and cardiomyopathy [91]. Hattori et al. reported decreased parasympathetic tone according to the medical records of deceased individuals suffering from schizophrenia, indicating association between autonomic nervous system disfunction and mortality in schizophrenia [92]. Among all antipsychotics, clozapine is found to be the most prominent trigger of sudden cardiac death [93]. Identifying and managing some of the above-mentioned risk factors could be one of the future strategies to prevent SCD in patient suffering from schizophrenia treated with antipsychotics.

Since, depression episodes are in some cases presented in individuals diagnosed with schizophrenia, antidepressant drugs have been subscribed to schizophrenic patients, and therefore found to potentiate the worsening of the cardiac adverse effects of antipsychotics. These effects have been reported after therapy with the most common prescribed antidepressants, selective serotonin reuptake inhibitors (SSRIs) [94].

Considering all the above, the question arises about the antidepressant's effects on the cardiovascular functions, aside from their co-administration with antipsychotics. In the following text, we aimed to summarize the previously published knowledge about the inter-connection between antidepressants and cardiotoxicity.

Antidepressants and Cardiovascular Toxicity (CVT)

Antidepressants are medications of choice for the medical treatment of depression. Even though psychotherapy is sufficient to treat some mild forms of depression, combination of pharmacotherapy and psychotherapy turns out to be the most effective treatment option [95]. Depression could also be present in patients suffering from schizophrenia or anxiety disorders, which could mask depressive symptoms, making depression misdiagnosed and therefore mistreated [96]. Nevertheless, 65% of the patients that are properly treated respond effectively to antidepressants. Depression refractory to any kind of therapy has been found to be presented in 15% of all diagnosed cases [97].

Antidepressants are typically classified into several classes [95]:

- Selective serotonin reuptake inhibitors (SSRIs): citalopram, escitalopram, fluoxetine, paroxetine, sertraline

- Serotonin and norepinephrine reuptake inhibitors (SNRIs): desvenlafaxine, duloxetine, levomilnacipran, venlafaxine
- Atypical antidepressants: bupropion, mirtazapine, nefazodone, trazodone, vilazodone, vortioxetine
- Tricyclic antidepressants (TCAs): amitriptyline, amoxapine, desipramine, doxepin, imipramine, trimipramine
- Monoamine oxidase inhibitors (MAOIs): isocarboxazid, phenelzine, selegiline.

Their therapeutical effects antidepressants exert mainly through modulation of neurotransmission in the brain on the pre/postsynaptic level. If consider their cardiovascular (adverse) effects some other mechanisms are of importance. Even though new generation of antidepressants have satisfactory safety profile (SSRIs) in light of their cardiac adverse effects, literature findings demonstrate some of their cardiotoxic properties, as well. Most of the previously published data reported cardiovascular adverse effects of old antidepressants such as TCAs [98, 99].

Cardiovascular adverse effects of antidepressants include spectrum of clinical manifestations starting from mild forms of hypotension and different forms of arrhythmia to sudden cardiac death [100]. Blockade of different ion channels (K^+, Na^+ or Ca^{2+}) responsible for physiological propagation of action potential in the heart, are underlying mechanism for some of the cardiac adverse effects. Additionally, some other mechanisms, likewise blockade of α-adrenergic receptors has also been proposed. In the following text the most important cardiovascular complications related to antidepressants medication will be described in more detail.

Mechanisms Underlying CVT of Antidepressant Drugs

Orthostatic Hypotension

As already, mentioned, orthostatic hypotension (OH) represents huge decrease in blood pressure in the standing position, which is frequently presented in the elderly. Aside from cardiovascular medications, pharmacological treatment of depression could frequently lead to OH, making it one of the most common side effects of antidepressants. TCAs have been found to induce OH by the blockade of α-adrenergic receptors [101] (Table 8.3). By blockade of these receptors responsible for vasoconstriction, vasodilation will appear, leading to hypotension. Therefore, antidepressants share the same mechanism of OH as previously described for antipsychotics. Literature finding, also proposed hypotensive effects of SSRIs, without evidence regarding responsible mechanism [104]. One case-control study reported significant association between old individuals taking fluoxetine (SSRI) and OH [105]. Another study performed by Briggs et al. also investigated occurrence of OH in the older population. This cross-sectional study showed twofold higher prevalence of OH in patient taking antidepressants in comparison to non-treated subjects. It is noteworthy that method of choice for blood pressure

Table 8.3 Cardiovascular (CV) adverse effects of antidepressants presented by the mechanism of action

Drug	CV adverse effect	Underlying mechanism	Study type	References
TCAs	Orthostatic hypotension	Blockade of α-adrenergic receptors	Hyman study in elderly patients	[101]
	QT prolongation	Na^+ channel blockade, Ca^{2+} channel blockade	Case reports, review	[102]
TCAs-imipramine	Sudden cardiac death	Impaired metabolism due to cytochrome CYP2D6, and/or concurrent therapy with phenothiazines	Case reports, 7 years old boy and 21 years old woman	[103]
SSRI-fluoxetine	Orthostatic hypotension	Not reported	Human study in elderly subjects	[104]
			Case-control study in elderly	[105]
	Hypotension	Not reported	Experimental model of hypertensive rats	[106]
	Vasodilation by reduction of vascular myogenic tone	Blockade of Ca^{2+} channels	Experimental model of isolated rat skeletal muscle arterioles	[107]
	Bradycardia	A direct effect of serotonin on medullary cardiovascular center	Case reports on humans	[108]
	QT prolongation	Combined HERG channel inhibition and disruption of trafficking	Case study of a patient on fluoxetine	[109]
SSRI-citalopram	Bradycardia	Not reported	Retrospective clinical study	[110]
Atypical AD-Trazodone	QT prolongation Hypotension	Not reported	Case report, 45 years old woman with MDD and anxiety	[111]

MDD major depressive disorder, *AD* antidepressant

(BP) determination in this study was continuous beat-to-beat finometry, which was found to be more sensitive in comparison to traditional BP measurement using sphygmomanometer [112]. Results from animal models, also speaks in favor of hypotensive effects of fluoxetine [106]. Authors proposed interaction of fluoxetine with vasomotor center as an explanation of its hypotensive effect. Additional mechanisms of fluoxetine action include blockade of Ca^{2+} influx into the vascular smooth muscle cells, leading to reduction of arterial myogenic tone and subsequent vasodilation [107]. Considering all above mentioned, special attention should be given to older patients treated with antidepressants, since OH as an adverse effect is considerable risk factor for falling and fractures in this population.

Arrhythmogenic Properties of Antidepressants

ECG tracings following antidepressants intake, both in therapeutical and overdose, showed changes indicating different types of hearth rhythmicity abnormalities. This has been found in patients without cardiovascular comorbidities, as well as in individuals with coexisting heart disease [113]. Namely, QRS complex and QT interval prolongation are described in patients treated with TCAs [114]. These manifestations are of special importance in those individuals already on treatment with antiarrhythmics (Na^+ channel blockers), since the combination of drugs could deteriorate the symptoms, leading in some cases to life threatening arrythmias [115]. Population specifically vulnerable to cardiotoxic effects of TCAs are shown to be children [116]. In comparison to TCAs, therapy with SSRIs is considered more favorable from the aspect of fewer cardiotoxic side effects. Nevertheless, authors reported some ECG changes such as nonsignificant QT prolongation, as well as lowering of the T-wave amplitude after citalopram (SSRIs) treatment [110]. According to Rotterdam study, citalopram showed the highest QT prolongation in comparison to other antidepressants [94]. Therefore, it is advisable to avoid citalopram prescription to patients already diagnosed with cardiac arrhythmia or presented with risk factors for QT prolongation and torsade de pontes (TdP) (Fig. 8.2). Additionally, SSRIs have been showed to induce rhythm abnormalities and syncope [117]. Inhibition of HERG potassium channels is underlying mechanism for QT prolongation induced by SSRIs. The same mechanism is reported responsible for arrhythmogenic properties of trazodone, typical antidepressant [111]. It has been shown that desipramine (TCAs) inhibit HERG channels by reducing their membrane expression, with long QT consequently [118]. According to published guidelines the special approach to patient prior to antidepressant

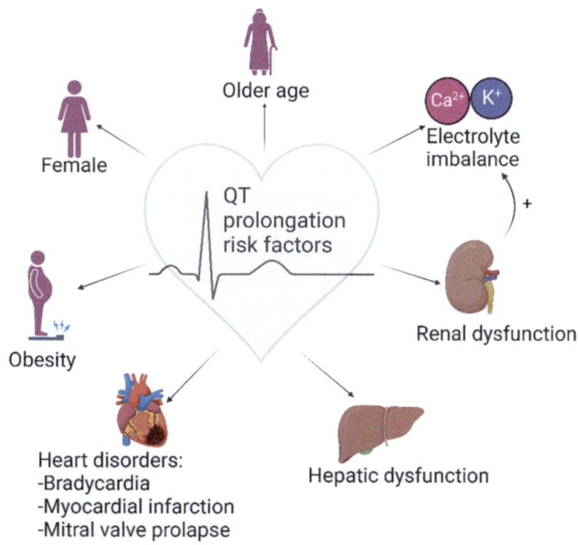

Fig. 8.2 Risk factors for developing a prolonged QT interval. Patients at risk for QT prolongation on antidepressants therapy should undergo screening for ECG markers of cardiac damage in order to prevent serious cardiac adverse effects. ECG monitoring prior introduction of antidepressants and during therapy is advisable for patients with associated risk factors and co-existing cardiac diseases

prescription is needed, especially in cases of increased risk of cardiac arrhythmias. These recommendations advise drug monitoring and assessment of QT interval [119]. It is not surprising that authors described QT interval as a surrogate marker of arrhythmia risk [120]. It is noteworthy that also inhibition of sodium channels has been found to participate arrhythmogenic properties of venlafaxine, another SSRI [121]. As previously stated, inhibition of ion channels in the heart disrupts the physiological process of heart tissue depolarization and repolarization, leading to ECG abnormalities and finally clinical manifestations. Torsade de pointes is shown to be one of the most important since it is potentially life-threatening arrhythmia and has been identified as a contributing factor to longer hospitalization and higher mortality.

Sudden Cardiac Death

Sudden cardiac death (SCD) has been reported after treatment with higher doses of antidepressants, usually in individuals already diagnosed with some of the cardiovascular diseases or presented with CVS risk factors. In most of the cases cause of death was fatal ventricular arrhythmia such as Tdp. A network meta-analysis, aimed to estimate the risk of sudden cardiac death in patient receiving antidepressants, were performed and reported by Prasitlumkum et al. According to their findings low risk for SCD has been identified among patients on antidepressants therapy with TCAs, SSRI and SNRI [122]. Additionally, one retrospective cohort study reported no risk of SCD associated with citalopram (SSRI) treatment [123]. In contrast to this, some other literature data reported lethal side effects of SSRIs such as fatal arrhythmia [124]. It is worth mentioning that SSRIs are considered safe for clinical usage when prescribed in therapeutic dosage. Fatal clinical manifestations after antidepressants medication are usually described in patient with electrolyte disbalance, cardiovascular comorbidities and therefore associated polypharmacy.

Conclusion and Future Perspective

According to our knowledge, patients with mental disorders have found to have higher mortality rates from CV disease in comparison to general population. Cardiac adverse effects of antipsychotics and antidepressants are presented in wide array of clinical manifestations from asymptomatic ECG abnormalities to serious arrhythmia and even sudden cardiac death. Of special importance for both antipsychotics and antidepressants treatment is that cardiac adverse effects are deteriorated if patients are already diagnosed with CV disease, such as arrhythmia or coronary artery disease. Therefore, clinicians should be aware of these facts and act proactively in order to prevent adverse cardiac outcomes following the treatment of mental disorders. One of suggested methods is ECG monitoring prior and during the treatment, mostly for the patients with co-existing CV disorders, or with CV risk factors. This will help to detect asymptomatic changes in the ECG tracing, especially QT prolongation, which has been found to be associated to severe and sometimes fatal cardiac arrhythmias. To implement aforementioned in clinical setting, multidisciplinary approach is recommended, indicating cooperation of different specialist for the universal goal—to provide the best possible patient treatment and care. Additional careful studies are warranted to better understand the underlying cellular and molecular mechanism of the complex interplay between various medications and cerebrovascular risk and implement appropriate preventive measures.

Funding The research work of DKA is supported by R25AI179582 and R01 HL147662 grants from the National Institutes of Health, USA. The contents of this chapter are solely the responsibility of the authors and do not necessarily represent the official views of the National Institutes of Health.

Competing Interests Both authors have read the chapter and declare no conflict of interest. No writing assistance was utilized in the production of this manuscript.

Consent for Publication Both authors have read the chapter and consented for publication.

References

1. Mamoshina P, Rodriguez B, Bueno-Orovio A (2021) Toward a broader view of mechanisms of drug cardiotoxicity. Cell Rep Med 2(3):100216
2. Ferri N, Siegl P, Corsini A, Herrmann J, Lerman A, Benghozi R (2013) Drug attrition during pre-clinical and clinical development: understanding and managing drug-induced cardiotoxicity. Pharmacol Therap 138(3):470–484
3. Heranval A, Lefaucheur R, Fetter D, Rouillé A, Le Goff F, Maltête D (2016) Drugs with potential cardiac adverse effects: retrospective study in a large cohort of parkinsonian patients. Rev Neurol 172(4–5):318–323
4. Guldiken B, Rémi J, Noachtar S (2016) Cardiovascular adverse effects of phenytoin. J Neurol 263:861–870
5. Zavarella M, Villatore A, Rocca MA, Peretto G, Filippi M (2023) The heart–brain interplay in multiple sclerosis from pathophysiology to clinical practice: a narrative review. J Cardiovasc Dev Dis 10(4):153
6. Ahir BK, Pratten MK (2014) Developmental cardiotoxicity effects of four commonly used antiepileptic drugs in embryonic chick heart micromass culture and embryonic stem cell culture systems. Toxicol In Vitro 28(5):948–960
7. Su C-M, Kung C-T, Wang Y-C, Lu C-H (2009) Life-threatening cardiotoxicity due to chronic oral phenytoin overdose. Neurol India 57(2):200–202
8. Waller EA, Kaplan J (2006) Pergolide-associated valvular heart disease. Compr Ther 32(2):94–101
9. Kantrowitz JT, Citrome L (2011) Schizoaffective disorder: a review of current research themes and pharmacological management. CNS Drugs 25:317–331
10. Farahani A, Correll CU (2012) Are antipsychotics or antidepressants needed for psychotic depression? A systematic review and meta-analysis of trials comparing antidepressant or antipsychotic monotherapy with combination treatment. J Clin Psychiatry 73(4):1269

11. Reilly J, Ayis S, Ferrier I, Jones S, Thomas S (2000) QTc-interval abnormalities and psychotropic drug therapy in psychiatric patients. Lancet 355(9209):1048–1052
12. Besag FM, Berry D, Vasey MJ, Patsalos PN (2023) Drug-drug interactions between antiseizure medications and antipsychotic medications: a narrative review and expert opinion. Expert Opin Drug Metab Toxicol 19(11):829–847
13. Spina E, de Leon J (2014) Clinically relevant interactions between newer antidepressants and second-generation antipsychotics. Expert Opin Drug Metab Toxicol 10(5):721–746
14. Buzea CA, Dima L, Correll CU, Manu P (2022) Drug–drug interactions involving antipsychotics and antihypertensives. Expert Opin Drug Metab Toxicol 18(4):285–298
15. Conley RR, Kelly DL (2007) Drug-drug interactions associated with second-generation antipsychotics: considerations for clinicians and patients. Psychopharmacol Bull 40(1):77
16. Kameg B, Champion C (2022) Atypical antipsychotics: managing adverse effects. Perspect Psychiatr Care 58(2):691–695
17. Ertugrul A, Meltzer HY (2003) Antipsychotic drugs in bipolar disorder. Int J Neuropsychopharmacol 6(3):277–284
18. Kaar SJ, Natesan S, Mccutcheon R, Howes OD (2020) Antipsychotics: mechanisms underlying clinical response and side-effects and novel treatment approaches based on pathophysiology. Neuropharmacology 172:107704
19. Siafis S, Tzachanis D, Samara M, Papazisis G (2018) Antipsychotic drugs: from receptor-binding profiles to metabolic side effects. Curr Neuropharmacol 16(8):1210–1223
20. Grinchii D, Dremencov E (2020) Mechanism of action of atypical antipsychotic drugs in mood disorders. Int J Mol Sci 21(24):9532
21. Swathy B, Polakkattil BK, Banerjee M (2019) Pharmacoepigenetics of antipsychotic drugs. In: Pharmacoepigenetics. Elsevier, Amsterdam, pp 733–739
22. Davies MA, Sheffler DJ, Roth BL (2004) Aripiprazole: a novel atypical antipsychotic drug with a uniquely robust pharmacology. CNS Drug Rev 10(4):317–336
23. Burris KD, Molski TF, Xu C, Ryan E, Tottori K, Kikuchi T et al (2002) Aripiprazole, a novel antipsychotic, is a high-affinity partial agonist at human dopamine D2 receptors. J Pharmacol Exp Ther 302(1):381–389
24. Taylor D (2003) Aripiprazole: a review of its pharmacology and clinical use. Int J Clin Pract 57(1):49–54
25. Orsolini L, De Berardis D, Volpe U (2020) Up-to-date expert opinion on the safety of recently developed antipsychotics. Expert Opin Drug Saf 19(8):981–998
26. Buckley NA, Sanders P (2000) Cardiovascular adverse effects of antipsychotic drugs. Drug Saf 23:215–228
27. Taglialatela M, Castaldo P, Pannaccione A, Giorgio G, Annunziato L (1998) Human ether-a-gogo related gene (HERG) K channels as pharmacological targets: present and future implications. Biochem Pharmacol 55(11):1741–1746
28. Silvestre J, Prous J (2007) Comparative evaluation of hERG potassium channel blockade by antipsychotics. Methods Find Exp Clin Pharmacol 29(7):457–466
29. Li L, Gao P, Tang X, Liu Z, Cao M, Luo R et al (2022) CB1R-stabilized NLRP3 inflammasome drives antipsychotics cardiotoxicity. Signal Transduct Target Ther 7(1):190
30. Leucht S, Cipriani A, Spineli L, Mavridis D, Örey D, Richter F et al (2013) Comparative efficacy and tolerability of 15 antipsychotic drugs in schizophrenia: a multiple-treatments meta-analysis. Lancet 382(9896):951–962
31. Pacher P, Kecskemeti V (2004) Cardiovascular side effects of new antidepressants and antipsychotics: new drugs, old concerns? Curr Pharm Des 10(20):2463–2475
32. Michelsen JW, Meyer JM (2007) Cardiovascular effects of antipsychotics. Expert Rev Neurother 7(7):829–839
33. Muench J, Hamer AM (2010) Adverse effects of antipsychotic medications. Am Fam Physician 81(5):617–622

34. De Berardis D, Rapini G, Olivieri L, Di Nicola D, Tomasetti C, Valchera A et al (2018) Safety of antipsychotics for the treatment of schizophrenia: a focus on the adverse effects of clozapine. Therap Adv Drug Saf 9(5):237–256
35. Ochiai H, Kashiwagi M, Usui T, Oyama Y, Tokita Y, Ishikawa T (1990) Torsade de pointes with T wave alternans in a patient receiving moderate dose of chlorpromazine: report of a case. Kokyu Junkan Resp Circ 38(8):819–822
36. Thomas D, Wu K, Kathöfer S, Katus HA, Schoels W, Kiehn J et al (2003) The antipsychotic drug chlorpromazine inhibits HERG potassium channels. Br J Pharmacol 139(3):567–574
37. Suessbrich H, Schönherr R, Heinemann S, Attali B, Lang F, Busch A (1997) The inhibitory effect of the antipsychotic drug haloperidol on HERG potassium channels expressed in Xenopus oocytes. Br J Pharmacol 120(5):968–974
38. Glassman AH, Bigger JT Jr (2001) Antipsychotic drugs: prolonged QTc interval, torsade de pointes, and sudden death. Am J Psychiatry 158(11):1774–1782
39. Drolet B, Vincent F, Rail J, Chahine M, Deschênes D, Nadeau S et al (1999) Thioridazine lengthens repolarization of cardiac ventricular myocytes by blocking the delayed rectifier potassium current. J Pharmacol Exp Ther 288(3):1261–1268
40. Rampe D, Murawsky MK, Grau J, Lewis EW (1998) The antipsychotic agent sertindole is a high affinity antagonist of the human cardiac potassium channel HERG. J Pharmacol Exp Ther 286(2):788–793
41. Enyeart J, Dirksen R, Sharma V, Williford D, Sheu S (1990) Antipsychotic pimozide is a potent Ca^{2+} channel blocker in heart. Mol Pharmacol 37(5):752–757
42. Gluais P, Bastide M, Caron J, Adamantidis M (2002) Risperidone prolongs cardiac action potential through reduction of K^{+} currents in rabbit myocytes. Eur J Pharmacol 444(3): 123–132
43. Oner T, Akdeniz C, Adaletli H (2016) Multifocal atrial tachycardia caused by risperidone. Elsevier, Amsterdam, pp 855–857
44. Sharma N, Bhat S, Ravi D, Ochieng P (2020) Severe hypothermia, bradycardia and cardiac arrest in association with risperidone. BMJ Case Rep 13(5):e234999
45. Zahn TP, Pickar D (1993) Autonomic effects of clozapine in schizophrenia: comparison with placebo and fluphenazine. Biol Psychiatry 34(1–2):3–12
46. Cohen H, Loewenthal U, Matar M, Kotler M (2001) Association of autonomic dysfunction and clozapine: heart rate variability and risk for sudden death in patients with schizophrenia on long-term psychotropic medication. Br J Psychiatry 179(2):167–171
47. Kilian JG, Kerr K, Lawrence C, Celermajer DS (1999) Myocarditis and cardiomyopathy associated with clozapine. Lancet 354(9193):1841–1845
48. Cam B, Gülseren L, Mete L, Gülseren Ş (2015) Clozapine and olanzapine associated atrial fibrillation: a case report. Turk J Psychiatry 26(3):221–226
49. Safferman A, Lieberman JA, Kane JM, Szymanski S, Kinon B (1991) Update on the clinical efficacy and side effects of clozapine. Schizophr Bull 17(2):247–261
50. Mackin P (2008) Cardiac side effects of psychiatric drugs. Hum Psychopharmacol Clin Exp 23 (S1):S3–S14
51. Jana AK, Praharaj SK, Roy N (2015) Olanzapine-induced orthostatic hypotension. Clin Psychopharmacol Neurosci 13(1):113
52. Takahashi K, Nakamura T, Sasayama D, Washizuka S (2021) Orthostatic hypotension: uncommon side effect of aripiprazole. Psychiatr Danub 33(1):63–64
53. Li KJ, Greenstein AP, Delisi LE (2018) Sudden death in schizophrenia. Curr Opin Psychiatry 31(3):169–175
54. Ferinauli F, Narulita S, Hijriyati Y (2021) Effect of antipsychotic drugs and orthostatic hypotension on the risk of falling in schizophrenic patients. J Public Health Res 10(2):184
55. Wieling W, Kaufmann H, Claydon VE, van Wijnen VK, Harms MP, Juraschek SP et al (2022) Diagnosis and treatment of orthostatic hypotension. Lancet Neurol 21(8):735–746
56. Westbury JL (2017) Antipsychotic drug prescribing in nursing homes. JAMA 318(18): 1829–1830

57. Ray WA, Griffin MR, Schaffner W, Baugh DK, Melton LJ III (1987) Psychotropic drug use and the risk of hip fracture. N Engl J Med 316(7):363–369
58. Hansen TE, Casey DE, Hoffman WF (1997) Neuroleptic intolerance. Schizophr Bull 23(4): 567–582
59. Tandon R (2002) Safety and tolerability: how do newer generation "atypical" antipsychotics compare? Psychiatry Q 73:297–311
60. Tiwari R, Kumar R, Malik S, Raj T, Kumar P (2021) Analysis of heart rate variability and implication of different factors on heart rate variability. Curr Cardiol Rev 17(5): e160721189770
61. Tuininga Y, Van Veldhuisen D, Brouwer J, Haaksma J, Crijns H, Lie K (1994) Heart rate variability in left ventricular dysfunction and heart failure: effects and implications of drug treatment. Heart 72(6):509–513
62. Aguirre RR, Mustafa MZ, Dumenigo A, Schulz S, Voss A, Goubran B et al (2018) Influence of acute antipsychotic treatment on cardiorespiratory coupling and heart rate variability. Cureus 10(1):e2066
63. Montaquila JM, Trachik BJ, Bedwell JS (2015) Heart rate variability and vagal tone in schizophrenia: a review. J Psychiatr Res 69:57–66
64. Silke B, Campbell C, King D (2002) The potential cardiotoxicity of antipsychotic drugs as assessed by heart rate variability. J Psychopharmacol 16(4):355–360
65. Leung JY, Barr AM, Procyshyn RM, Honer WG, Pang CC (2012) Cardiovascular side-effects of antipsychotic drugs: the role of the autonomic nervous system. Pharmacol Ther 135(2): 113–122
66. Huhn M, Arndt T, Schneider-Thoma J, Leucht S (2022) Effects of antipsychotics on heart rate in treatment of schizophrenia: a systematic review and meta-analysis. Therap Adv Psychopharmacol 12:20451253221097261
67. De Hert M, Vancampfort D, Correll C, Mercken V, Peuskens J, Sweers K et al (2011) Guidelines for screening and monitoring of cardiometabolic risk in schizophrenia: systematic evaluation. Br J Psychiatry 199(2):99–105
68. Bordet C, Garcia P, Salvo F, Touafchia A, Galinier M, Sommet A et al (2023) Antipsychotics and risk of QT prolongation: a pharmacovigilance study. Psychopharmacology 240(1): 199–202
69. Nemati M, Hosseinzadeh Z, Nemati F, Ebrahimi B (2023) Impact of antipsychotics and antidepressants drugs on long QT syndrome induction related to hERG channel dysfunction: a systematic review. Biochem Biophys Res Commun 681:90–96
70. Tarabová B, Nováková M, Lacinová L (2009) Haloperidol moderately inhibits cardiovascular L-type calcium current. Gen Physiol Biophys 28(3):249
71. Chen W, Zhu F, Guo J, Sheng J, Li W, Zhao X et al (2014) Chronic haloperidol increases voltage-gated Na^+ currents in mouse cortical neurons. Biochem Biophys Res Commun 450(1): 55–60
72. Kongsamut S, Kang J, Chen X-L, Roehr J, Rampe D (2002) A comparison of the receptor binding and HERG channel affinities for a series of antipsychotic drugs. Eur J Pharmacol 450(1):37–41
73. Ogata N, Narahashi T (1989) Block of sodium channels by psychotropic drugs in single guinea-pig cardiac myocytes. Br J Pharmacol 97(3):905
74. Drolet B, Zhang S, Deschênes D, Rail J, Nadeau S, Zhou Z et al (1999) Droperidol lengthens cardiac repolarization due to block of the rapid component of the delayed rectifier potassium current. J Cardiovasc Electrophysiol 10(12):1597–1604
75. Beach SR, Celano CM, Noseworthy PA, Januzzi JL, Huffman JC (2013) QTc prolongation, torsades de pointes, and psychotropic medications. Psychosomatics 54(1):1–13
76. Hassaballa HA, Balk RA (2003) Torsade de pointes associated with the administration of intravenous haloperidol. Am J Ther 10(1):58–60

77. Ruiz Diaz JC, Frenkel D, Aronow WS (2020) The relationship between atypical antipsychotics drugs, QT interval prolongation, and torsades de pointes: implications for clinical use. Expert Opin Drug Saf 19(5):559–564
78. Testai L, Bianucci A, Massarelli I, Breschi MC, Martinotti E, Calderone V (2004) Torsadogenic cardiotoxicity of antipsychotic drugs: a structural feature, potentially involved in the interaction with cardiac HERG potassium channels. Curr Med Chem 11(20):2691–2706
79. Crumb WJ, Ekins S, Sarazan RD, Wikel JH, Wrighton SA, Carlson C et al (2006) Effects of antipsychotic drugs on I to, I Na, I sus, I K1, and hERG: QT prolongation, structure activity relationship, and network analysis. Pharm Res 23:1133–1143
80. Nielsen J, Graff C, Kanters JK, Toft E, Taylor D, Meyer JM (2011) Assessing QT interval prolongation and its associated risks with antipsychotics. CNS Drugs 25:473–490
81. Tamargo J (2000) Drug-induced torsade de pointes: from molecular biology to bedside. Jpn J Pharmacol 83(1):1–19
82. Warner J, Barnes T, Henry J (1996) Electrocardiographic changes in patients receiving neuroleptic medication. Acta Psychiatr Scand 93(4):311–313
83. Raschi E, Poluzzi E, Godman B, Koci A, Moretti U, Kalaba M et al (2013) Torsadogenic risk of antipsychotics: combining adverse event reports with drug utilization data across Europe. PLoS One 8(11):e81208
84. Mow T, Frederiksen K, Thomsen MB (2015) Assessment of anti-arrhythmic activity of antipsychotic drugs in an animal model: influence of non-cardiac α1-adrenergic receptors. Eur J Pharmacol 748:10–17
85. Antzelevitch C, Burashnikov A, Sicouri S, Belardinelli L (2011) Electrophysiologic basis for the antiarrhythmic actions of ranolazine. Heart Rhythm 8(8):1281–1290
86. Harrigan EP, Miceli JJ, Anziano R, Watsky E, Reeves KR, Cutler NR et al (2004) A randomized evaluation of the effects of six antipsychotic agents on QTc, in the absence and presence of metabolic inhibition. J Clin Psychopharmacol 24(1):62–69
87. Wassef N, Khan N, Munir S (2015) Quetiapine-induced myocarditis presenting as acute STEMI. Case Rep 2015:bcr2014207151
88. Ray WA, Chung CP, Murray KT, Hall K, Stein CM (2009) Atypical antipsychotic drugs and the risk of sudden cardiac death. N Engl J Med 360(3):225–235
89. Hennessy S, Bilker WB, Knauss JS, Margolis DJ, Kimmel SE, Reynolds RF et al (2002) Cardiac arrest and ventricular arrhythmia in patients taking antipsychotic drugs: cohort study using administrative data. BMJ 325(7372):1070
90. Vohra J (2020) Sudden cardiac death in schizophrenia: a review. Heart Lung Circ 29(10): 1427–1432
91. Straus SM, Bleumink GS, Dieleman JP, van der Lei J, Jong G, Kingma JH et al (2004) Antipsychotics and the risk of sudden cardiac death. Arch Intern Med 164(12):1293–1297
92. Hattori S, Suda A, Kishida I, Miyauchi M, Shiraishi Y, Fujibayashi M et al (2018) Association between dysfunction of autonomic nervous system activity and mortality in schizophrenia. Compr Psychiatry 86:119–122
93. Li L, Ye X, Zhao Z, Gao P, Jiang Y (2018) Overlooked fatal infectious diseases after long-term antipsychotic use in patients with psychiatric illness. Schizophr Res 195:258–259
94. Maljuric NM, Noordam R, Aarts N, Niemeijer MN, van den Berg ME, Hofman A et al (2015) Use of selective serotonin re-uptake inhibitors and the heart rate corrected QT interval in a real-life setting: the population-based Rotterdam Study. Br J Clin Pharmacol 80(4):698–705
95. Richelson E (2001) Pharmacology of antidepressants. Mayo Clin Proc 76:511–527
96. Keller MB, Lavori PW, Mueller TI, Endicott J, Coryell W, Hirschfeld RM et al (1992) Time to recovery, chronicity, and levels of psychopathology in major depression: a 5-year prospective follow-up of 431 subjects. Arch Gen Psychiatry 49(10):809–816
97. Anderson IM (2000) Selective serotonin reuptake inhibitors versus tricyclic antidepressants: a meta-analysis of efficacy and tolerability. J Affect Disord 58(1):19–36
98. Goldberg RJ, Capone RJ, Hunt JD (1985) Cardiac complications following tricyclic antidepressant overdose: issues for monitoring policy. JAMA 254(13):1772–1775

99. Pacher P, Ungvari Z, Nanasi PP, Furst S, Kecskemeti V (1999) Speculations on difference between tricyclic and selective serotonin reuptake inhibitor antidepressants on their cardiac effects. Is there any? Curr Med Chem 6(6):469–480
100. Varley CK (2001) Sudden death related to selected tricyclic antidepressants in children: epidemiology, mechanisms and clinical implications. Paediatr Drugs 3:613–627
101. Gaxatte C, Faraj E, Lathuillerie O, Salleron J, Deramecourt V, Pardessus V et al (2017) Alcohol and psychotropic drugs: risk factors for orthostatic hypotension in elderly fallers. J Hum Hypertens 31(4):299–304
102. Vieweg WVR, Wood MA (2004) Tricyclic antidepressants, QT interval prolongation, and torsade de pointes. Psychosomatics 45(5):371–377
103. Robert Swanson J, Jones GR, Krasselt W, Denmark LN, Ratti F (1997) Death of two subjects due to imipramine and desipramine metabolite accumulation during chronic therapy: a review of the literature and possible mechanisms. J Forensic Sci 42(2):335–339
104. Poon I, Braun U (2005) High prevalence of orthostatic hypotension and its correlation with potentially causative medications among elderly veterans. J Clin Pharm Ther 30(2):173–178
105. Cherin P, Colvez A, de Periere GD, Sereni D (1997) Risk of syncope in the elderly and consumption of drugs: a case-control study. J Clin Epidemiol 50(3):313–320
106. Fuller RW, Holland DR, Yen TT, Bemis KG, Stamm NB (1979) Antihypertensive effects of fluoxetine and L-5-hydroxytryptophan in rats. Life Sci 25(14):1237–1242
107. Ungvari Z, Pacher P, Koller A (2000) Serotonin reuptake inhibitor fluoxetine decreases arteriolar myogenic tone by reducing smooth muscle $[Ca^{2+}]_i$. J Cardiovasc Pharmacol 35(6): 849–854
108. Ellison JM, Milofsky JE, Ely E (1990) Fluoxetine-induced bradycardia and syncope in two patients. J Clin Psychiatry 51(9):385–386
109. Hancox J, Mitcheson J (2006) Combined hERG channel inhibition and disruption of trafficking in drug-induced long QT syndrome by fluoxetine: a case-study in cardiac safety pharmacology. Br J Pharmacol 149(5):457–459
110. Enemark B (1993) The importance of ECG monitoring in antidepressant treatment. Nord J Psychiatry 47(30):57–65
111. Mohan G, Ajitkumar A, Bhide P, Ravilla J, Kramer V (2023) Trazodone overdose manifesting as hypotension and QT prolongation. Cureus 15(3):e36871
112. Briggs R, Carey D, McNicholas T, Claffey P, Nolan H, Kennelly S et al (2018) The association between antidepressant use and orthostatic hypotension in older people: a matched cohort study. J Am Soc Hypertens 12(8):597–604. e1
113. Glassman AH (1984) Cardiovascular effects of tricyclic antidepressants. Annu Rev Med 35: 503–511
114. Giardina E, Bigger JT Jr, Glassman AH, Perel JM, Kantor SJ (1979) The electrocardiographic and antiarrhythmic effects of imipramine hydrochloride at therapeutic plasma concentrations. Circulation 60(5):1045–1052
115. Roose SP, Glassman AH, Giardina EG, Walsh BT, Woodring S, Bigger JT (1987) Tricyclic antidepressants in depressed patients with cardiac conduction disease. Arch Gen Psychiatry 44(3):273–275
116. Biederman J (1991) Sudden death in children treated with a tricyclic antidepressant. J Am Acad Child Adolesc Psychiatry 30(3):495–498
117. Funk KA, Bostwick JR (2013) A comparison of the risk of QT prolongation among SSRIs. Ann Pharmacother 47(10):1330–1341
118. Dennis AT, Nassal D, Deschenes I, Thomas D, Ficker E (2011) Antidepressant-induced ubiquitination and degradation of the cardiac potassium channel hERG. J Biol Chem 286(39):34413–34425
119. Goldberger JJ, Cain ME, Hohnloser SH, Kadish AH, Knight BP, Lauer MS et al (2008) American Heart Association/American College of Cardiology Foundation/Heart Rhythm Society scientific statement on noninvasive risk stratification techniques for identifying patients at risk for sudden cardiac death: a scientific statement from the American Heart

Association Council on Clinical Cardiology Committee on Electrocardiography and Arrhythmias and Council on Epidemiology and Prevention. Circulation 118(14):1497–1518
120. Waring WS (2012) Clinical use of antidepressant therapy and associated cardiovascular risk. Drug Healthc Patient Saf 4:93–101
121. Lenkey N, Karoly R, Lukacs P, Vizi ES, Sunesen M, Fodor L et al (2010) Classification of drugs based on properties of sodium channel inhibition: a comparative automated patch-clamp study. PLoS One 5(12):e15568
122. Prasitlumkum N, Cheungpasitporn W, Tokavanich N, Ding KR, Kewcharoen J, Thongprayoon C et al (2021) Antidepressants and risk of sudden cardiac death: a network meta-analysis and systematic review. Med Sci 9(2):26
123. Ray WA, Chung CP, Murray KT, Hall K, Stein CM (2017) High-dose citalopram and escitalopram and the risk of out-of-hospital death. J Clin Psychiatry 78(2):19944
124. Grace SL, Medina-Inojosa JR, Thomas RJ, Krause H, Vickers-Douglas KS, Palmer BA et al (2018) Antidepressant use by class: association with major adverse cardiac events in patients with coronary artery disease. Psychother Psychosom 87(2):85–94

Chapter 9
Cardiovascular Toxicity in Diabetes Mellitus: Focus on Diabetogenic and Anti-diabetogenic Agents

Nina M. Radisavljevic, Slavica S. Mutavdzin Krneta, and Dragan M. Djuric

Abstract Cardiovascular toxicity refers to a broad spectrum of harmful effects from various agents on the heart and blood vessels, including structural damage, functional disturbances, and electrophysiological abnormalities. In diabetes mellitus (DM), these effects are further exacerbated by the complex interaction of chronic hyperglycemia and associated metabolic disorders. Key mechanisms underlying cardiovascular toxicity in DM include disruption of cardiac energy metabolism, chronic inflammation, oxidative stress, accumulation of advanced glycation end products (AGEs), mitochondrial dysfunction, endoplasmic reticulum stress, apoptosis of cardiomyocytes, and endothelial dysfunction. Novel biomarkers, such as high-sensitivity troponin (hs-cTn), homocysteine, natriuretic peptides (B-type natriuretic peptide—BNP and inactive component of its precursor—NT-proBNP), and heart-type fatty acid-binding protein (h-FABP), enable early detection and intervention in cardiovascular toxicity. Cardiovascular diseases remain the leading cause of morbidity and mortality in DM, despite intensive glycemic control. Novel antihyperglycemic agents, such as sodium-glucose cotransporter 2 (SGLT2) inhibitors, glucagon-like peptide-1 receptor agonists (GLP-1 RAs), and dipeptidyl peptidase-4 (DPP-4) inhibitors, along with well-known drugs like metformin, demonstrate promising cardioprotective effects that extend beyond their primary antihyperglycemic action. However, cardiovascular impact varies among different drug classes, even within the same class. Statins, in addition to their basic hypolipidemic effect, exhibit numerous pleiotropic effects that significantly contribute to reducing cardiovascular risk in both DM patients and the general population.

N. M. Radisavljevic · S. S. Mutavdzin Krneta
Institute of Medical Physiology "Richard Burian", Faculty of Medicine, University of Belgrade, Belgrade, Serbia

D. M. Djuric (✉)
Institute of Medical Physiology "Richard Burian", Faculty of Medicine, University of Belgrade, Belgrade, Serbia

Centre for Biomedical Research, Faculty of Medicine, University of Banja Luka, Banja Luka, Bosnia and Herzegovina
e-mail: dragan.djuric@med.bg.ac.rs

D. K. Agrawal et al. (eds.), *Cardiovascular Toxicity*, Advances in Biochemistry in Health and Disease 33, https://doi.org/10.1007/978-3-031-90598-8_9

When combined with statins, agents like ezetimibe, fibrates, and proprotein convertase subtilisin/kexin type 9 (PCSK9) inhibitors further enhance cardiovascular risk reduction. Additionally, research has indicated potential cardioprotective effects of certain vitamins and trace elements in the context of DM. Understanding the mechanisms of cardiovascular toxicity in DM and the cardioprotective effects of various therapeutic modalities opens new perspectives in the prevention and treatment of cardiovascular complications associated with this condition.

Keywords Anti-diabetic agents · Biomarkers · Cardiovascular toxicity · Diabetes mellitus · Diabetogenic agents

Introduction

Cardiovascular toxicity refers to the direct and indirect harmful effects of various compounds such as drugs, chemicals, toxins, environmental pollutants, metabolic products, and others on the heart and blood vessels [1]. The concept of cardiovascular toxicity is broad and is primarily recognized in the literature as a consequence of the harmful effects of exogenous chemical substances, particularly drugs. The term "cardiovascular toxicity"was first used in 1946 to describe the toxic effects of local anesthetics, mercurial diuretics, and digitalis [2]. Today, the list of potentially cardiotoxic drugs has expanded, with cytostatic drugs at the forefront. Thanks to advancements in chemotherapy for malignant diseases, the survival rate of cancer patients has increased. Simultaneously, there has been an increase in the incidence of cardiovascular diseases among survivors and toxic effects on the cardiovascular system have been documented with the use of nearly all cytostatic agents [3]. Cells within the cardiovascular system are continuously exposed to various potentially harmful agents. These agents can be of exogenous origin (such as the aforementioned pharmaceutical drugs) and endogenous origin, including signaling molecules that maintain homeostasis but act as toxic stimuli at high concentrations [4]. The toxic effects of these agents include structural damage to the heart, functional disorders, electrophysiological abnormalities, induction of oxidative stress, inflammation, and other alterations in the normal functioning of the cardiovascular system, resulting in reduced blood flow, circulatory disorders, and the development of cardiovascular diseases [5]. Metabolic changes associated with diabetes mellitus (DM), including hyperglycemia and hyperlipidemia, as well as reduced insulin production or secretion, are critical factors in developing cardiovascular diseases in individuals with DM. These metabolic disorders contribute to cardiovascular toxicity by producing oxidative stress, inflammation, and endothelial dysfunction [6].

DM represents a global health problem. Current data indicate that 537 million adults suffer from DM, which is expected to rise to 643 million by 2030 and to 783 million by 2045 [7]. Although the incidence of both type 1 diabetes (T1D) and type 2 diabetes (T2D) is increasing, T2D disproportionately contributes to the global burden on healthcare systems, accounting for up to 90% of all cases

[8, 9]. Cardiovascular diseases are the leading cause of morbidity and mortality in individuals with DM [10, 11]. People with DM have a significantly higher risk of developing atherosclerotic coronary heart disease, peripheral arterial disease, cerebrovascular diseases, cardiovascular autonomic neuropathy, diabetic cardiomyopathy, and heart failure [11].

Hyperglycemia, Dyslipidemia, and Cardiovascular Toxicity

Prolonged exposure to elevated blood glucose levels harms various tissues and organs. The pathophysiological process by which chronic hyperglycemia exerts harmful effects on the cardiovascular system involves increased glucose metabolism, leading to the production of free radicals, decreased activity of antioxidant enzymes, activation of protein kinase C (PKC), and the formation of advanced glycation end products (AGEs). Oxidative stress is central to all these processes (Fig. 9.1) [12–14].

There is a strong link between chronic hyperglycemia, endothelial dysfunction, and the progression of atherosclerosis. Under conditions of chronic hyperglycemia, the accumulation of AGEs leads to endothelial activation and surface expression of adhesion molecules, promoting the adherence and entry of monocytes/macrophages into the subendothelial space during the initial stages of plaque formation. AGEs also stimulate the release of cytokines from macrophages, potentiating inflammation within the plaque [15]. However, a study by Mastrogiacamo et al. showed that hyperglycemia alone was not sufficient for the progression of atherosclerosis in the absence of dyslipidemia [16].

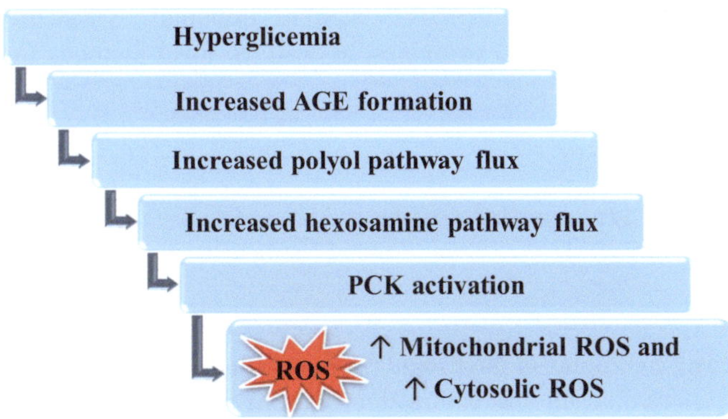

Fig. 9.1 Abnormal metabolic pathways caused by hyperglycemia. Hyperglycemia triggers a cascade of events starting with the formation of advanced glycation end-products (AGE), leading to increased polyol pathway flux, increased hexosamine biosynthesis pathway, and activation of protein kinase C (PKC). These events generate reactive oxygen species (ROS) in mitochondrial and cytosolic compartments

Prolonged myocardial exposure to elevated levels of circulating metabolic substrates, such as glucose, free fatty acids, and triglycerides, as well as hormones like insulin and leptin, accelerates pathological changes in the myocardium that lead to the development of hypertrophy and heart failure [17]. The accumulation of intracellular fatty acids and triglycerides promotes endoplasmic reticulum stress, cell apoptosis, and inflammatory response [18]. Using rat cardiomyocyte cultures, Hickson-Bick et al. showed that cardiomyocyte exposure to palmitic acid leads to mitochondrial dysfunction and cellular apoptosis [19]. Furthermore, it has been demonstrated that hyperglycemia and high levels of palmitic acid induce reactive oxygen species (ROS) formation in H9c2 myoblasts [17]. The synergy of these pathological conditions—hyperglycemia and dyslipidemia—is present in DM, obesity, and heart failure.

Cardiovascular Toxicity Biomarkers

Given the global burden of cardiovascular diseases, there is an increasing need for their early detection and monitoring. Biomarkers are significant in identifying cardiovascular risk and toxicity, enabling timely and effective interventions. Traditional biomarkers, such as lipid profile, glycemia, and hormones, remain essential diagnostic tools. However, current approaches involve advanced methods that provide deeper insight into the pathophysiological processes of cardiovascular damage. Biochemical markers of myocardial injury, cardiac function biomarkers, endothelial dysfunction biomarkers, and oxidative stress and inflammation biomarkers, individually or in combination, provide valuable information about the type and degree of cardiovascular damage [1, 20, 21].

In the following sections, special attention will be given to specific well-established cardiovascular biomarkers, such as troponins, natriuretic peptides, C-reactive protein (CRP), and cytokines, as well as novel methods for assessing cardiovascular toxicity.

Biomarkers of Myocardial Injury

Troponin

Troponin (cTn), a complex of three regulatory proteins—troponin I, C, and T-plays a crucial role in cardiac muscle contraction. Due to its sensitivity and specificity, cardiac troponin is considered the gold standard in diagnosing cardiac damage [22]. Under physiological conditions, the blood level of troponin is almost negligible, but in conditions of heart damage, such as myocardial infarction, the troponin level rises [23].

Troponin tests have gone through several developmental phases. The latest high-sensitivity troponin tests (hs-cTn) detect minimal myocardial damage, even in asymptomatic patients [24]. Numerous studies have indicated that hs-cTn is an independent risk factor for cardiovascular diseases [25]. This has led to investigations of its role in high-risk populations for cardiovascular disease development, such as those with T2D [26]. One meta-analysis showed that elevation of hs-cTn is associated with a significantly higher risk of adverse cardiovascular events in the diabetic population [27].

Homocysteine

Homocysteine is a sulfur-containing, non-protein, semi-essential amino acid produced as an intermediate product in the metabolism of the amino acids methionine and cysteine [28–30]. Homocysteine has been identified as a risk factor and marker for cardiovascular diseases [31]. Elevated serum concentrations of homocysteine produce atherothrombotic effects, causing endothelial dysfunction, smooth muscle cell proliferation, inflammation, and increased oxidative stress [32]. Hyperhomocysteinemia is associated with the development of other pathological conditions, including DM and insulin resistance. Additionally, high serum concentrations of homocysteine are considered a significant risk factor for the development of chronic complications of this disease [33, 34].

B-Type Natriuretic Peptide and Incative Component of Its Precursor

B-type natriuretic peptide (BNP) is a neurohormone synthesized and released from the cardiac ventricles. Its precursor (proBNP) is enzymatically cleaved into an active component (BNP) and an inactive component (NT-proBNP). Concentrations of both BNP and NT-proBNP are measured in blood [35]. BNP is synthesized and released in response to excessive pressure or volume-induced stretching of the ventricle. Given its longer half-life compared to BNP, NT-proBNP is often preferred in research [31]. BNP and NT-proBNP are highly sensitive and specific markers of myocardial damage and represent the gold standard in diagnosing heart failure [36]. One study showed that NT-proBNP levels were inversely proportional to the risk of developing T2D in healthy individuals, especially women, but also that elevated concentrations of this neurohormone were associated with a higher risk of microvascular and macrovascular complications of DM [37].

Heart-Type Fatty Acid-Binding Protein

Heart-type fatty acid-binding protein (h-FABP), a small cytoplasmic protein, transports free fatty acids from the plasma membrane to the mitochondria of cardiomyocytes [1]. Because of its low molecular weight, water solubility, and high concentration in cardiomyocytes, the potential role of h-FABP as a marker of early myocardial damage has been investigated in numerous studies [38, 39]. Compared to cTn, h-FABP is considered a superior marker in the early diagnosis of acute myocardial infarction due to its faster release into circulation following myocardial damage [40]. H-FABP has also been identified as an essential factor in monitoring therapy and prognosis of acute myocardial infarction [41]. One experimental study showed that h-FABP is a sensitive marker of myocardial damage, and its serum values correlate with the development of myocardial lesions [42]. Additionally, one study indicated that h-FABP is a promising marker for assessing cardiovascular risk in prediabetes [43].

Oxidative Stress Biomarkers

Oxidative stress, defined as an imbalance in the production of free radicals and antioxidants, favoring free radicals, plays a significant role in the development of cardiovascular diseases. Excess ROS can lead to damage to proteins, lipids, and DNA. Pathological increase in ROS is associated with arterial hypertension, acute myocardial infarction, and heart failure [44, 45]. Most biomarkers of oxidative stress can be divided into three major groups depending on the substrate or targets of oxidation (lipids, proteins, nucleic acids). The most commonly used markers of lipid peroxidation are thiobarbituric acid reactive substances (TBARS) and malondialdehyde (MDA). Nitrotyrosine and carbonyl groups are often used as markers of oxidative protein damage, although their occurrence is not specific to oxidative protein modification. 8-hydroxy-2'deoxyguanosine (8-OHdG) is the most commonly used marker of oxidative DNA damage [45, 46].

Biomarkers of Inflammation

Local and systemic inflammation plays a central role in the development of cardiovascular diseases, from initial endothelial dysfunction to clinical manifestation [47]. Inflammatory markers such as CRP and interleukins are commonly used for cardiovascular disease screening.

Studies show that CRP predicts the development of adverse cardiovascular outcomes, including myocardial infarction, ischemic stroke, and sudden cardiac death in individuals with or without previously diagnosed coronary heart disease. CRP is involved in atherogenesis through activation of the complement pathway, release of proinflammatory cytokines, induction of endothelial dysfunction, and inhibition of nitric oxide (NO) synthesis [48]. CRP values directly correlate with cardiovascular risk factors, such as body mass index, smoking, systolic blood pressure, total cholesterol levels, triglycerides, and glycemia [49].

With the development of high-sensitivity tests for detecting lower CRP values (hs-CRP), studies have shown that even a mild increase in CRP values, undetected by conventional tests, predicts the development of cardiovascular events [50]. hs-CRP has been recognized as a significant indicator of subclinical low-grade inflammation. One study indicated that low-grade inflammation, measured by hs-CRP, is a significant predictor of cardiovascular incidents and overall mortality in patients with T2D [51]. However, a study by Han et al. showed that conventional CRP tests highly correlate with hs-CRP tests and can replace the more expensive hs-CRP measurement for cardiovascular risk assessment [52]. These results suggest that, although hs-CRP is superior in detecting low-grade inflammation, conventional CRP tests represent a significant alternative due to their availability and lower costs.

Cytokines

Cytokines are mediators of inflammatory and immune reactions and are involved in the development of cardiovascular pathology. Given their short-term action and relatively simple measurement, cytokines are considered useful biomarkers of cardiovascular toxicity [53].

Interleukin-6

Interleukin-6 (IL-6) is a significant contributor to the development of vascular inflammation. Its actions, including potentiating smooth muscle cell proliferation, endothelial dysfunction, and the release of other proinflammatory cytokines, procoagulant factors, and matrix metalloproteinases (MMPs), lead to the progression and destabilization of atherosclerotic plaques [54]. The maintenance of elevated IL-6 levels for 24 h after acute ST-elevation myocardial infarction (STEMI) is associated with a larger infarct lesion size and impairment of cardiac function, underscoring its detrimental effects [55].

Interleukin-1

The interleukin-1 (IL-1) family comprises 11 cytokines and 10 receptors and is particularly notable for its role in inflammatory processes. Among them, IL-1β and IL-18 are the most researched ones. In experimental models of atherosclerosis, it has been shown that inhibition of IL-1 (both genetic and pharmacological inhibition) slows the development of atherosclerosis, while enhanced release of IL-1 accelerates the progression of this disease [56]. Furthermore, IL-1β stimulates nitric oxide synthase expression in cardiomyocytes, leading to increased nitric oxide activity, energy deficit, and cardiac contraction weakening [57]. The critical role of IL-1β in

T2D is also highlighted. Specifically, chronically elevated levels of IL-1β can lead to pancreatic β-cell dysfunction and apoptosis and ultimately to the development of DM [58].

Interleukin-10

Interleukin-10 (IL-10) is a pleiotropic cytokine synthesized by diverse immune cells, including Th2 T lymphocytes, T-regulatory lymphocytes, B lymphocytes, and macrophages [59]. IL-10 significantly influences the pathogenesis of atherosclerosis, cardiac fibrosis, cardiac hypertrophy, and hypertension [60]. Despite its recognized anti-inflammatory and anti-atherogenic properties, elevated IL-10 levels were initially hypothesized to confer protection against cardiovascular diseases. However, research has not substantiated this hypothesis [61]. Conversely, IL-10 is postulated to enhance pancreatic β-cell function and decelerate the progression of DM in animal models [62, 63].

Tumor Necrosis Factor-Alpha

Tumor necrosis factor-alpha (TNF-α) is a polypeptide hormone produced by activated monocytes and macrophages [64]. TNF-α plays a significant role in the progression of heart failure, acting as a mediator of cardiac muscle dysfunction and cardiac remodeling [65]. The serum concentration of TNF-α positively correlates with the size of the infarct zone and left ventricular dysfunction [1].

Transforming Growth Factor-Beta

Although known for its anti-inflammatory role, transforming growth factor-beta (TGF-β) is an important mediator of cellular damage in various cardiac diseases [66]. Experimental studies using the model of myocardial ischemia-reperfusion injury have indicated protective effects of TGF-β through reduction of oxidative stress and decreased release of proinflammatory cytokines, such as TNF-α [67]. Similarly, released TGF-β shows a protective role under cardiac pressure overload. However, prolonged and excessive activation of this cytokine can lead to cardiac fibrosis and subsequent dysfunction [68].

The Role of Other Cytokines as Markers of Inflammation

In a recent study, Pan et al. showed that high serum levels of soluble interleukin-2 receptor and interleukin-8 significantly predict the development of adverse cardiovascular events [69]. Moreover, research has indicated that high concentrations of interleukin-2 can induce the onset of acute myocardial infarction [70]. Similar results come from studies examining the role of interleukin-17 in cardiovascular diseases. Interleukin-17 induces the development of cardiac fibrosis by activating MMPs and programmed cell death [70, 71].

Fibrinogen

Fibrinogen plays a crucial role in blood clotting and is an important marker of inflammation [1, 72]. One meta-analysis showed that elevated fibrinogen levels positively correlate with poor prognosis of coronary artery disease. In this context, routine measurement of fibrinogen can identify patients at high risk for developing coronary artery disease [73]. Hyperfibrinogenemia has also been observed in the diabetic population [74]. Under conditions of chronic hyperglycemia, fibrinogen glycation occurs, disrupting its structural and functional characteristics and contributing to an increased risk of thrombotic complications in DM [75]. Overall, research results suggest a potential role of fibrinogen in predicting cardiovascular risk [70].

Cardiac Remodeling Biomarkers

Galectin-3

Galectin-3 (Gal-3), a β-galactosidase-binding lectin, modulates cell adhesion through interaction with extracellular matrix proteins [76]. In healthy myocardium, Gal-3 expression is very low. However, in case of cardiac damage of any kind, there is a rapid expression of Gal-3 [77]. Gal-3 plays a key role in cardiac remodeling, particularly in cardiac hypertrophy and fibrosis [76]. In experimental models, pharmacological inhibition of Gal-3 significantly preserved systolic and diastolic function by inhibiting fibrosis, collagen synthesis, and deposition [78].

Matrix Metalloproteinases

MMPs are a family of 25 proteolytic enzymes that regulate the process of cardiac tissue remodeling and extracellular matrix degradation [79]. MMP-2 and MMP-9 are the most studied MMPs, given their simple measurement by gelatin zymography [80]. In animal models of myocardial infarction, MMP-9 expression significantly increases and is associated with inflammation, diabetic microvascular complications, and cardiac dysfunction [81]. MMPs are critical factors in all stages of atherosclerosis, mediating vascular inflammation, endothelial dysfunction, smooth muscle cell migration, vascular calcification, extracellular matrix degradation, and consequent plaque destabilization [82].

Diabetes Mellitus and Cardiovascular Toxicity

DM, the most common endocrine disease nowadays, is recognized as an independent risk factor in the development of cardiovascular diseases. DM affects all cardiovascular system components—the microvasculature, large arteries, heart, and kidneys [83]. The increased risk of cardiovascular complications in DM is explained by a complex interaction of hyperglycemia-induced endothelial dysfunction, chronic inflammation, and oxidative stress [8].

The heart of diabetic patients undergoes structural and functional changes that can evolve into heart failure. Although coronary artery disease is the primary cause of heart failure in DM patients, heart failure can also develop in the absence of coronary artery disease or hypertension. This has led to the emergence of the concept of a primary heart muscle disease specific to DM—diabetic cardiomyopathy. Data from the Framingham study indicate that the incidence of heart failure is increased in both men and women with DM compared to healthy individuals of the same age, and this association was independent of obesity, hypertension, dyslipidemia, and coronary heart disease [84]. The pathophysiological mechanisms underlying diabetic cardiomyopathy are the formation and accumulation of AGEs, mitochondrial dysfunction, endoplasmic reticulum stress, oxidative stress, chronic inflammation, and cellular apoptosis. These factors are crucial for understanding cardiovascular toxicity in DM [85].

Cardiovascular Toxicity in Experimental Diabetes Mellitus

Numerous experimental models of DM have been developed to study the structural and functional changes in the cardiovascular system. Despite limitations, animal models of DM provide significant insight into the pathophysiological mechanisms of cardiovascular toxicity in this disease [86]. According to the methodological approach, animal models of DM are classified as chemically induced, spontaneous autoimmune, genetically induced, diet-induced, virus-induced, and surgically induced animal models [87].

Chemically Induced Models of Diabetes Mellitus: The Role of Diabetogenic Agents

Streptozotocin

Among several available chemical substances for DM induction, streptozotocin (STZ) is the most commonly used diabetogenic agent [88]. STZ enters pancreatic islet β cells via GLUT2 transporters as a glucose analog. In intracellular conditions, STZ causes DNA alkylation and fragmentation, thus exhibiting its diabetogenic properties. Additionally, STZ stimulates the production of free radicals, accumulation of AGEs, and secretion of cytokines. Through these mechanisms, STZ leads to β cell necrosis and the development of DM [89].

STZ was initially used to induce T1D, but it has been shown that the application of low doses of STZ, or combinations of STZ and nicotinamide, or STZ and a high-fat diet, as well as the administration of STZ injection in the neonatal period, can also induce T2D [86, 89, 90]. This flexibility in STZ use has allowed researchers to study various aspects of the pathophysiological mechanisms of DM and associated cardiovascular complications. The STZ-induced DM model has been widely used in studying diabetic cardiomyopathy.

Alloxan

Alloxan induces DM through two distinct mechanisms. The first mechanism involves the inhibition of glucose-induced insulin secretion, mediated by specific inhibition of the glucokinase enzyme. The second mechanism relates to its ability to induce the formation of ROS, resulting in selective β cell necrosis [86, 91].

Although alloxan, like STZ, is most commonly used to induce T1D, under certain conditions, it can also be used to induce T2D. Alloxan leads to impaired glucose tolerance at lower doses, mimicking the initial phase of T2D. Combined with a high-fat diet or nicotinamide, alloxan leads to progressive β cell dysfunction and insulin resistance, corresponding to T2D [86].

Pathophysiological Mechanisms of Cardiovascular Toxicity in Diabetes Mellitus: Evidence from Experimental Studies

Hyperglycemia Induces Cardiac Apoptosis

Results from preclinical studies unequivocally indicate that prolonged exposure to elevated blood glucose levels, independent of other factors, causes cardiomyocyte damage. In one study, 8 weeks after induction of DM by STZ, flow cytometry measurements showed a high degree of cardiomyocyte apoptosis as a consequence of hyperglycemia. Their results are interpreted as hyperglycemia-induced disturbance of energy metabolism and increased metabolism of branched-chain amino acids [92]. Apoptosis is considered a key driver of hypertrophy in residual cardiomyocytes and the proliferation of fibroblasts, leading to cardiac remodeling and dysfunction [93].

Hyperglycemia Induces Cardiomyocyte Fibrosis

The presence of myocardial fibrosis is documented in animal models of DM. The accumulation and cross-linking of collagen in the cardiac interstitium increase myocardial rigidity and are closely associated with impairment of myocardial contractile function [94]. It has been shown that in conditions of STZ-induced diabetes, there is reduced degradation of the extracellular matrix and development of a high degree of interstitial fibrosis compared to controls [95]. Additionally, hyperglycemia and accumulation of AGEs promote profibrotic reactions in fibroblasts and myofibroblasts by stimulating TGF-β, TNF, and interleukins [93]. Described processes are central mechanisms of left ventricular dysfunction and the development of heart failure with preserved ejection fraction, which is the basis of diabetic cardiomyopathy [13, 96].

Hyperglycemia Induces Disturbance of Cardiac Energy Metabolism

Abundant evidence indicates that cardiac dysfunction in DM patients is associated with metabolic abnormalities. DM is characterized by high levels of free fatty acids and decreased glucose utilization [97–99]. Insulin resistance in T2D and insulin deficiency in T1D result in reduced glucose uptake by cells, while increased levels of free fatty acids suppress glucose oxidation through the Randle cycle [100, 101]. This leads to the accumulation of free fatty acids and triglycerides and increased fatty acid oxidation capacity in cardiomyocytes. The increase in free fatty acid oxidation capacity requires more oxygen and results in high-intensity generation of ROS [102]. These mechanisms collectively lead to the development of a cardiac phenotype corresponding to diabetic cardiomyopathy [103]. On the other hand, normalization of cardiac metabolism in animal models of DM can reverse the development of diabetic cardiomyopathy, emphasizing the role of disturbed metabolism in the onset of this disease (Fig. 9.2) [102, 104, 105].

Hyperglycemia Induces Endothelial Dysfunction

Under physiological conditions, endothelial cells balance vasodilatory and vasoconstrictive factors, enabling tissue perfusion proportional to metabolic demands. Additionally, endothelial cells regulate hemostasis, angiogenesis, and vascular permeability [106, 107].

Endothelial dysfunction is considered an early indicator of vascular pathology and is defined as a reduction or absence of vasodilatory response to endothelium-dependent vasodilators, including acetylcholine [108]. Chronic exposure of the endothelium to hyperglycemia leads to adverse effects that compromise vascular function [109]. From a pathophysiological perspective, the development of endothelial dysfunction in DM is a complex process involving various mechanisms.

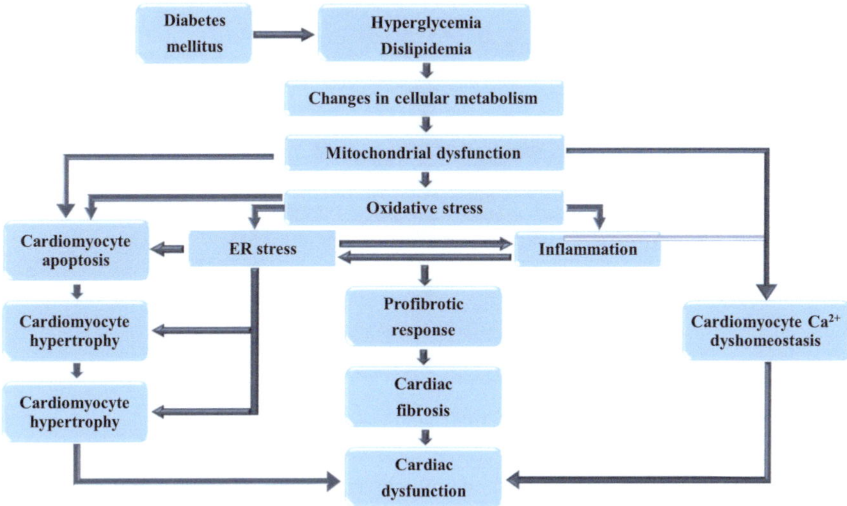

Fig. 9.2 Pathophysiological mechanisms leading to cardiac dysfunction in diabetes mellitus (DM). The complex mechanisms of cardiac dysfunction in diabetes begin with hyperglycemia and dyslipidemia, leading to altered cellular metabolism, mitochondrial dysfunction, and oxidative stress generation. This triggers endoplasmic reticulum (ER) stress and inflammation, producing a profibrotic response and cardiac fibrosis. Additionally, cardiomyocyte calcium dyshomeostasis contributes to cardiomyocyte apoptosis and hypertrophy, culminating in cardiac dysfunction

Mechanisms of Endothelial Dysfunction

The Role of Gasotransmitters: Nitric Oxide

NO participates as a mediator in several physiological processes, including vasodilation, inhibition of platelet aggregation, and regulation of apoptosis, inflammation, and angiogenesis (Fig. 9.3) [110]. Endothelial dysfunction in DM is closely associated with reduced available NO [108, 111]. In an STZ-induced DM model, Li et al. showed that endothelial dysfunction in carotid arteries results from reduced NO-dependent vasodilation. The same study also indicated decreased basal NO concentration in the diabetic environment [112].

The availability of NO under DM conditions varies depending on the duration of the disease. In the initial stages, compensatory mechanisms may lead to a relative normalization of NO levels. However, as the disease progresses, NO production decreases, resulting in NO deficiency in the later stages [110]. In an animal model of DM, Stadler et al. detected significantly elevated NO levels in the aorta 3 weeks after disease induction. This elevation persisted even after 7 weeks of the disease [113]. Reduced NO production in advanced DM can be explained in several ways, including decreased activity and expression of endothelial nitric oxide synthase (eNOS), reduced sensitivity of eNOS to acetylcholine, increased arginase activity, and L-arginine deficiency [110]. NO is considered a vasculoprotective molecule.

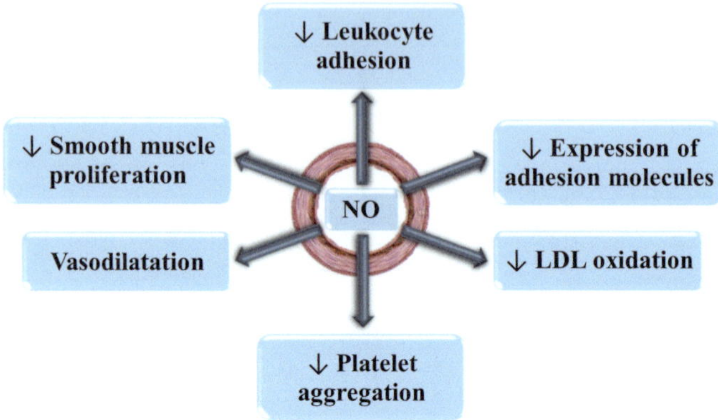

Fig. 9.3 Atheroprotective effects of NO; LDL—low-density lipoprotein

However, given that NO readily reacts with superoxide, one of its multiple effects is the nitrosylation of proteins at thiol groups and the formation of reactive nitrogen species (RNS), such as peroxynitrite [114]. Peroxynitrite exerts its harmful effects on cells through multiple mechanisms. One of the fundamental mechanisms is the depletion of tetrahydrobiopterin (BH4), an essential cofactor for eNOS necessary for NO synthesis. Additionally, peroxynitrite causes degradation of various biomolecules, resulting in damage to vascular endothelium, vascular smooth muscles, and cardiac muscle [115]. Peroxynitrite is believed to play a role in the development of microvascular complications of DM, particularly diabetic polyneuropathy [116, 117].

The Role of Gasotransmitters: Hydrogen Sulfide (H_2S)

H_2S is recognized as an essential signaling molecule and mediator of biological processes. Only recently has the significance of physiological H_2S metabolites, including persulfides, polysulfides, and other reactive sulfur species (RSS), been acknowledged [118]. H_2S is a product of enzymes involved in cysteine metabolism—cystathionine β-synthase (CBS), cystathionine γ-lyase (CSE), and cystathionine aminotransferase (CAT) [119]. In addition to its vasodilatory effect, this physiologically active molecule also possesses antioxidant properties through the neutralization of ROS [118]. In an experimental study using an animal model of DM, a significantly increased synthesis of H_2S was recorded in the pancreas and liver. It is believed that this increased synthesis of H_2S in conditions of DM-induced endothelial dysfunction is a compensatory mechanism aimed at mitigating the deficit of other vasodilatory molecules, such as NO [120]. However, the expression of CSE and CBS enzymes and the production of H_2S were significantly reduced in a genetically induced model of DM [121]. Additionally, another study showed that genetic and pharmacological inhibition of CSE delayed the onset of DM induced by STZ. The findings presented in this study suggest that inhibition of the CSE/H_2S system may have a favorable effect in preventing or slowing the progression of DM

[122]. Research indicates a complex role of H_2S in regulating glucose homeostasis. In an animal model of DM, Wu et al. documented increased levels of CSE in the pancreas along with enhanced production of H_2S. The intraperitoneal administration of DL-propargylglycine (a selective inhibitor of CSE activity—PPG) over 4 weeks resulted in improved glycemic control, increased serum insulin levels, and reduced levels of glycosylated hemoglobin (HbA1c) [123]. The results of this study suggest a potential link between elevated H_2S production in the pancreas and impaired insulin secretion in DM [124, 125].

The Role of Gasotransmitters: Carbon Monoxide (CO)

Over the past few decades, CO has been recognized as a significant molecule in biological systems. CO is produced as a product of heme degradation, catalyzed by heme oxygenases (HO). Three isoforms of this enzyme have been identified: HO-1, HO-2, and HO-3. While HO-1 and HO-2 are physiologically active, the physiological significance of the HO-3 isoform has yet to be confirmed [126]. Initially considered solely as a toxic product, CO exerts critical vasculoprotective effects through vasodilation, angiogenesis stimulation, immune response modulation, platelet aggregation inhibition, and anti-inflammatory effects [127].

Research has indicated a complex relationship between CO and metabolic homeostasis in a diabetic environment. Specifically, a reduced concentration of CO is associated with the development of insulin resistance and endothelial dysfunction. In contrast, higher levels of CO exhibit protective effects in DM [126]. In an experimental model of T2D, treatment with a CO-releasing agent (CORM-401) led to significant improvements in metabolic parameters, including weight reduction in diabetic mice, improved glucose tolerance, and insulin sensitivity, along with reduced inflammation in adipose tissue [128]. Similar results were obtained in a study by Nickolai et al. Applying cobalt protoporphyrin (an inducer of HO-1 activity) increased CO levels, reduced glycemia, and enhanced insulin sensitivity [129]. These findings suggest that improved insulin sensitivity is a crucial mechanism of CO's vasculoprotective effect in DM [126]. Moreover, CO is believed to have a significant protective effect in the context of microvascular complications of DM. In animal models of DM, the antioxidant property of HO-1 is crucial for renoprotection. Conversely, a deficit of the HO-2 isoform leads to an increase in superoxide anions and deterioration of kidney function, indicating the complex role of HO isoforms in the development of diabetic nephropathy [117, 126].

Hyperglycemia Induces Thrombosis

As previously mentioned, a structurally and functionally intact endothelium is crucial for maintaining vascular homeostasis. Damage to endothelial integrity disrupts the delicate balance between procoagulant and anticoagulant mediators, favoring the development of thrombosis [130, 131]. DM is considered a prothrombotic state characterized by alterations in the levels and activities of prothrombotic and fibrinolytic factors, increasing the risk of thrombogenesis [132]. Increased platelet reactivity is one of the most important causes of vascular toxicity in DM [133]. Under hyperglycemic conditions, the non-enzymatic glycation of proteins on the platelet membrane surface reduces membrane fluidity and promotes platelet adhesion. Additionally, exposure to hyperosmolarity activates the expression of platelet GP IIb/IIIa receptors and P-selectin, indicating the direct osmotic effect of hyperglycemia [134]. Research by Fidler et al. in an animal model of T1D indicated increased platelet glucose uptake and metabolism and increased expression of the GLUT3 transporter. This metabolic change led to platelet hyperactivation while blocking glucose utilization negated hyperglycemia-induced platelet hyperreactivity [135]. In a study by Hadas et al., the impact of oxidative stress on platelet activity in a diabetic environment was examined. It was shown that methylglyoxal (MGO) directly increases platelet aggregation and reduces thrombus stability [136]. MGO is the most significant compound among reactive carbonyl species (RCS) responsible for the non-enzymatic glycation process and AGE formation [137]. Moreover, in T2D, there is an increased expression of adhesion molecules such as CD63, CD62p, CD49b, and CD36 on the platelet surface. Improving glycemic control decreases the expression of these adhesion molecules and reduces platelet hyperreactivity [138]. In addition to platelet hyperreactivity, physiological anti-aggregation mechanisms are significantly weakened in DM [139]. The combined effect of endothelial dysfunction and vascular inflammation increases the release of von Willebrand factor (vWF) and plasminogen activator inhibitor-1 (PAI-1), which in turn promote platelet adhesion and interfere with blood clot breakdown [133]. Elevated systemic levels of vWF are associated with vascular complications in DM, and vWF is considered a valuable marker of cardiovascular risk in the diabetic population [133, 140, 141]. An imbalance in the concentration of procoagulant and anticoagulant factors contributes to hypercoagulability in DM. In the diabetic environment, the concentration of procoagulant factors, including fibrinogen, prothrombin, factors V, VII, VIII, X, XI, and prekallikrein, increases. Simultaneously, levels of anticoagulant factors, including protein C, protein S, and antithrombin, are reduced [133, 142, 143]. The reduced fibrinolytic capacity in DM further exacerbates the prothrombotic state and increases the risk of vascular toxicity. Research by Bryk et al. indicated that increased protein carbonylation in T2D significantly reduces the activity of the fibrinolytic system. These findings suggest a close link between oxidative stress and prothrombotic changes, particularly impaired fibrinolysis in this disease (Fig. 9.4) [144].

Hyperglycemia and the Vascular Wall

Research has shown that hyperglycemia and associated insulin resistance cause structural and functional changes in the vascular wall, likely contributing to accelerated vascular aging and increased cardiovascular risk in DM [145]. Increased vascular stiffness results from different pathophysiological abnormalities in endothelial cells, the extracellular matrix, and other structural elements of the vascular

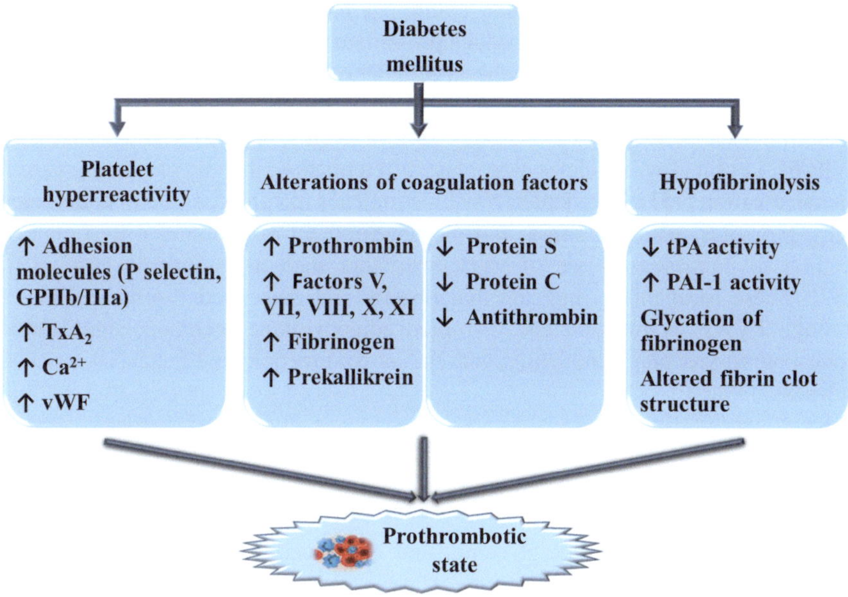

Fig. 9.4 Hemostatic abnormalities in DM. *GPIIb/IIIa* glycoprotein IIb/IIIa, *TxA2* tromboxan A2, *vWF* von Willebrand factor, *tPA* tissue plasminogen activator, *PAI-1* plasminogen activator inhibitor-1

wall. Additionally, increased vascular stiffness is considered an early prognostic marker of cardiovascular diseases, particularly in populations with DM and patients with insulin resistance and associated obesity [146]. Hyperglycemia and inadequate insulin secretion synergistically contribute to arterial thickening and are involved in the pathophysiological process of developing hypertension in T1D and T2D [147]. Chronic hyperglycemia and hyperinsulinemia increase local renin-angiotensin-aldosterone system (RAAS) activity and the expression of type 1 angiotensin receptors in blood vessels, promoting the development of vascular wall hypertrophy and fibrosis. Chronic hyperglycemia also promotes non-enzymatic glycation of proteins, leading to the covalent cross-linking of collagen and the formation of AGEs. This process affects the mechanical properties of the interstitial tissue of the arterial wall, contributing to increased vascular stiffness [148]. In an experimental study by Gao et al., histological analysis of the abdominal aorta in diabetic rats indicated significant tunica media thickening and damage, with partial detachment of endothelial cells, hypertrophy, and disorganized arrangement of smooth muscle cells in the tunica media. These changes indicate significant disruption of the typical structure of the blood vessel wall in diabetic conditions [149]. Furthermore, there is a clear connection between obesity, insulin resistance, and increased vascular stiffness. Under physiological conditions, perivascular adipose tissue (PVAT; tunica adiposa) releases numerous bioactive substances that maintain vascular homeostasis, including adipocytokines, NO, H_2S, and others. However, in obesity, PVAT dysfunction occurs, exhibiting harmful effects on blood vessels [150]. Hypertrophic adipocytes produce pro-inflammatory cytokines. IL-6 promotes increased collagen synthesis and the accumulation of AGEs in the vascular wall. Elevated circulating levels of leptin stimulate the proliferation and migration of smooth muscle cells, increase the expression of MMPs, and promote the release of TGF-β1. Leptin also stimulates osteoblast differentiation, thereby promoting vascular calcification [151]. Like leptin, resistin stimulates the proliferation of endothelial cells, promotes oxidative stress, and the development of endothelial dysfunction, leading to increased release of vasoconstrictors such as endothelin [146, 151, 152]. These mechanisms indicate that obesity and associated insulin resistance, through PVAT dysfunction and disrupted adipocytokine secretion, significantly contribute to vascular remodeling, endothelial dysfunction, and increased vascular stiffness.

Hyperglycemia Induces Atherosclerosis

The chronic effects of hyperglycemia and associated dyslipidemia lead to the formation of ROS, increased creation and deposition of AGEs, and the expression of receptors for AGEs. This creates a vicious cycle of chronic inflammation within the vascular wall, contributing to the development and progression of atherosclerosis [16, 153, 154]. Initial insights into the pathophysiological mechanisms of atherosclerosis development and progression in a diabetic environment come from experimental studies. In the $ApoE^{-/-}$ knockout mouse model, under conditions of hypercholesterolemia and hypertriglyceridemia, the development of atherosclerosis is observed as early as 8 weeks of age. When these mice are additionally exposed to STZ-induced diabetes, there is a significant acceleration in the progression of atherosclerosis [155].

Cardiovascular Toxicity in Clinical Diabetes Mellitus

Coronary Artery Disease

Numerous studies, starting with the Framingham study, have linked coronary artery disease (CAD) with DM [84]. Moreover, DM is considered an equivalent of CAD. CAD is responsible for up to 75% of total mortality in the population with DM [156].

Various pathophysiological mechanisms underlie the connection between DM and CAD. As previously mentioned, the diabetic environment contributes to the development and progression of atherosclerosis, causing endothelial dysfunction, oxidative stress, and chronic inflammation within the vascular wall. Additionally, due to associated insulin resistance, hyperinsulinemia, and vascular calcification in diabetics, there is destabilization of atherosclerotic plaque and consequent thrombosis, factors precipitating the development of acute coronary syndrome (ACS)

[157, 158]. A particular challenge in clinical practice is the atypical presentation of ACS in individuals with DM, which complicates diagnosis and timely treatment in this population [157, 159].

Peripheral Arterial Disease

Peripheral arterial disease (PAD) is an atherosclerotic occlusive disease of the lower extremities. It represents a manifestation of systemic atherosclerosis and is one of the most common macrovascular complications of DM. Most diabetics with PAD also have concomitant coronary artery disease. Therefore, a significant proportion of morbidity and mortality in this population is closely associated with myocardial infarction, ischemic stroke, and sudden cardiac death [160]. In most patients, the disease is asymptomatic, but nearly a quarter of those affected experience rapid disease progression within five years, with amputations occurring in 4% of cases. It is important to emphasize that DM affects not only the large peripheral blood vessels but also the distal arterioles (peripheral microangiopathy) [9, 161].

Diabetic Cardiomyopathy and Heart Failure

Diabetic cardiomyopathy is now recognized as a specific form of cardiomyopathy that develops independently of other cardiovascular risk factors. Given that most patients with DM already have some associated cardiovascular comorbidities, defining this condition is particularly challenging [162]. There are two main accepted models for the development of diabetic cardiomyopathy. According to the traditional model, diabetic cardiomyopathy is characterized by early asymptomatic diastolic dysfunction, accompanied by left ventricular hypertrophy, which gradually progresses to systolic dysfunction. However, novel development models suggest the existence of two independent phenotypes of this disease: diastolic dysfunction that evolves into heart failure with preserved ejection fraction and systolic dysfunction that evolves into heart failure with reduced ejection fraction [163, 164]. Both T1D and T2D are independent risk factors for heart failure. Data indicate that the incidence of heart failure in people with DM is increased by 2–5 times compared to those without diabetes [165]. Additionally, 30–40% of those with heart failure show signs of impaired glucose tolerance or overt T2D [166]. These data suggest a bidirectional relationship between these two conditions. Chronic hyperglycemia, systemic and cardiac insulin resistance, and associated dyslipidemia, through their complex interactions, initiate a series of pathophysiological processes underlying diabetic cardiomyopathy and subsequent heart failure [90]. Disturbances in cardiac energy metabolism, chronic inflammation, oxidative stress, accumulation of AGE, mitochondrial dysfunction, endoplasmic reticulum stress, and cardiomyocyte apoptosis play central roles in developing cardiac dysfunction in DM [6]. These processes promote fibrosis and hypertrophy of cardiomyocytes, leading to myocardial rigidity, LV dysfunction, and the development of heart failure [167]. Despite the global increase in experimental and clinical studies on diabetic cardiomyopathy over the past few decades, there is still no consensus on a specific therapeutic approach to treating diabetic cardiomyopathy and heart failure associated with DM [85, 168]. Consequently, the incidence and mortality rate from heart failure in this population remain high [168].

Cardiovascular Toxicity and New Antihyperglycemic Agents

Previous studies have shown that the extent of cardiovascular complications in DM depends on the duration and level of hyperglycemia [169]. Consequently, research has focused on demonstrating that intensive glycemic control can reduce the number of vascular events. While intensive glycemic control has decreased the incidence of microvascular complications of DM, it has not significantly impacted macrovascular complications of the disease, such as peripheral artery disease, ischemic heart disease, and stroke [161]. Prompted by the adverse cardiovascular effects of rosiglitazone, an oral antihyperglycemic drug, the US Food and Drug Administration (FDA) and the European Medicines Agency (EMA) have established stricter guidelines for assessing the cardiovascular safety of antihyperglycemic agents before approving their use [170, 171]. As a result, numerous cardiovascular outcome trials (CVOTs) of new antidiabetic drugs have been conducted, providing insight into the cardiometabolic spectrum of these drugs. This has ushered in a new era in glucose-lowering drugs and diabetes treatment, focusing on cardiovascular benefits [161, 172].

Sodium-Glucose Cotransporter 2 Inhibitors

Sodium-glucose cotransporter 2 (SGLT2) inhibitors represent a novel class of oral antihyperglycemic drugs used to regulate glycemia in the T2D population. Unlike well-established, traditional antihyperglycemic therapeutic approaches that increase insulin sensitivity, preserve pancreatic β-cell function, or enhance glucose utilization, SGLT2 inhibitors have a unique mechanism of action [173].

Sodium-Glucose Cotransporter 2 Inhibitors Mechanism of Action

SGLT2 inhibitors exert their pharmacodynamic effects through the selective inhibition of the SGLT2 protein, the primary glucose transporter on the luminal membrane of proximal renal tubules [174]. The benefits of SGLT2 inhibitors are driven by several mechanisms. These include promoting diuresis and natriuresis, which reduce blood pressure and alleviate cardiac load, potentially through the inhibition of sodium-hydrogen exchanger 3 and sympathetic activity. Additionally, SGLT2 inhibitors decrease proximal sodium and glucose reabsorption, normalize tubuloglomerular feedback, and reduce hyperfiltration, while also mimicking systemic hypoxia to stimulate erythropoiesis and improve organ oxygenation. SGLT2 inhibitors demonstrate diverse beneficial effects on the cardiovascular system, effects that extend beyond their glucose-lowering properties. Several mechanisms likely contribute to these multifaceted cardiovascular benefits [175].

Cardiovascular Benefits of Sodium-Glucose Cotransporter 2 Inhibitors

Sodium-Glucose Cotransporter 2 Inhibitors Reduce Oxidative Stress

Oxidative stress is a central mechanism of myocardial damage associated with DM and is present in all stages of diabetic cardiomyopathy development. Increased production of ROS induces the accumulation of AGEs, expression and activation of receptors for AGEs (RAGE), activates PKC, and increases hexosamine pathway flux, contributing to the development of cardiomyopathy. SGLT2 inhibitors reduce oxidative stress by increasing the activity of NOS, thereby increasing NO production in cardiomyocytes [176]. Dapagliflozin decreases free radical formation by reducing the expression of NADPH oxidase 4 (NOX4) enzyme [177]. Additionally, empagliflozin can reduce oxidative stress by decreasing levels of cyclic guanosine monophosphate (cGMP)-dependent protein kinase (PKG) [178]. Defects in the antioxidant defense system further increase oxidative stress in diabetic cardiomyopathy. In experimental DM models, dapagliflozin restores the expression of superoxide dismutase (SOD) and catalase in kidney tissue, reducing oxidative damage [177, 179]. However, the beneficial effect of SGLT2 inhibitors on reducing oxidative damage may be lost when combined with other medications [180]. Mitochondria are the primary source of ROS. Under hyperglycemic conditions, mitochondria increase ROS production, leading to tissue damage [181]. In an experimental DM model, 20 weeks of empagliflozin therapy prevented the development of vascular complications by suppressing oxidative stress in mitochondria [177, 182]. In human endothelial cells, empagliflozin significantly reduced mitochondrial calcium load and superoxide production under hyperglycemic conditions [183]. In conditions of hyperglycemia, insulin resistance, and hyperlipidemia, there is an increased flux of nicotinamide adenine dinucleotide (NAD+) and flavin adenine dinucleotide (FAD+) into the mitochondrial electron transport chain. As electron donors, NAD and FAD increase electron flow through the transport chain, leading to hyperpolarization of the inner mitochondrial membrane and excess ROS production [184]. Through their antihyperglycemic effect, SGLT2 inhibitors reduce the level of electron donors and overall oxidative damage [177]. Increased expression of the Na^+/H^+ exchanger (NHE1) contributes to the development of vascular abnormalities in DM, most likely through insulin resistance, endothelial dysfunction, and accelerated atherosclerosis [185]. Additionally, increased NHE1 activity is associated with the development of heart failure. Specifically, activation of NHE1, followed by an increase in intracellular Na^+ concentration, activationn of the Na^+/Ca^{2+} exchanger, and subsequent Ca^{2+} influx, leads to cardiomyocyte damage and cardiomyopathy [173, 174, 185]. By inhibiting NHE1, empagliflozin reduces cytosolic Na^+ and Ca^{2+} concentrations while increasing mitochondrial Ca^{2+} concentration. SGLT2 inhibitors stabilize mitochondrial function through this mechanism and prevent further oxidative damage [176, 186]. This effect of empagliflozin appears to be responsible for reducing hypertrophy, fibrosis, and myocardial remodeling in heart failure [176, 185]. Hyperglycemia promotes oxidative stress by forming AGEs, leading to an increase in intracellular oxidants and a decrease in available NO [187]. By reducing glucose levels in peripheral blood, SGLT2 inhibitors suppress hyperglycemia-induced toxic effects. This effect manifests through a reduction in AGE levels and decreased expression of their receptors [174]. In studies using animal models of DM, SGLT2 inhibitors have demonstrated suppression of AGEs in kidney tissue and the aorta. However, this effect was not observed in the myocardium of treated animals [173].

Sodium-Glucose Cotransporter 2 Inhibitors Reduce Inflammation

Recent research indicates that SGLT2 inhibitors possess anti-inflammatory effects in the T2D population, independent of their antihyperglycemic effect. These anti-inflammatory effects represent one of the cardioprotective mechanisms of SGLT2 inhibitors, particularly in the context of atherosclerosis prevention [188]. The anti-inflammatory effect of SGLT2 inhibitors is partly achieved by activating AMP-dependent protein kinase (AMPK). By activating AMPK, an enzyme that plays a central role in energy balance and inflammation at the cellular level, these drugs lower circulating levels of pro-inflammatory cytokines such as IL-6 and TNF-α [189].

Additionally, SGLT2 inhibitors reduce the activation of nuclear factor kappa B (NF-κB), a key transcription factor in regulating the inflammatory response, leading to a concomitant reduction in IL-6, TNF-α, and hs-CRP [178, 190]. The use of SGLT2 inhibitors is also associated with decreased activation of the NLRP3 inflammasome, which promotes atherosclerosis in DM [178, 191]. Hyperuricemia promotes oxidative stress, pro-inflammatory response, and endothelial dysfunction, thereby contributing to the development of diabetic nephropathy and heart failure [174]. Therefore, the anti-inflammatory properties of SGLT2 inhibitors can be partly explained by their effect of increased uric acid urinary excretion [192].

Effects of Sodium-Glucose Cotransporter 2 Inhibitors on Cardiac Metabolism

Under physiological conditions, the heart uses fatty acids (40–60%) and glucose (20–40%) to produce energy in the form of ATP, while lactate, ketone bodies, and amino acids serve as alternative energy substrates [193]. However, in cardiovascular diseases and DM, significant changes occur in the utilization of various energy substrates in cardiomyocytes. Under diabetic conditions, there is a significant 'metabolic shift'. This shift is characterized by an increased rate of lipid oxidation and decreased glucose oxidation in cardiomyocytes. The resulting accumulation and increased oxidation capacity of free fatty acids lead to cardiac steatosis and cytotoxicity [194]. SGLT2 inhibitors increase ketogenesis and the concentration of ketone bodies in circulation. Ketone bodies are considered a more efficient energy substrate, as their oxidation requires less oxygen to generate the same amount of ATP than free fatty acids [195]. Moderate increases in ketone body metabolism in the heart benefit overall cardiovascular system function in various pathological conditions, including DM, obesity, and heart failure [192]. Furthermore, the oxidation of ketone bodies is linked to reduced production of reactive oxygen species (ROS) compared to the oxidation of free fatty acids. This difference exerts a protective effect on mitochondrial function [196]. In animal studies of diabetic cardiomyopathy, empagliflozin has been shown to improve cardiac function through increased oxidation of ketone bodies [197]. This 'metabolic shift' enhances the heart's energy balance and left ventricle systolic function [173].

Clinical Studies Results

Many randomized, double-blind, placebo-controlled CVOTs have provided insight into the broad spectrum of cardioprotective effects of SGLT2 inhibitors. These studies have shown that the use of SGLT2 inhibitors significantly reduces the risk of cardiovascular events and cardiovascular death in populations with T2D and without this disease [173, 174]. These effects are believed to extend beyond their primary antihyperglycemic action [171]. The EMPA-REG OUTCOME study investigated the effect of the SGLT2 inhibitor empagliflozin on cardiovascular mortality and morbidity in a T2D population with established cardiovascular risk. The study demonstrated a reduction in the risk of major adverse cardiovascular events (MACE), including cardiovascular death, non-fatal myocardial infarction, and non-fatal stroke, in the group of subjects receiving empagliflozin compared to placebo [198]. The risk of hospitalization due to heart failure and all-cause mortality was significantly reduced with empagliflozin use [173, 199]. Similar results come from the CANVAS study, where canagliflozin showed superiority in reducing MACE compared to placebo. However, no reduction in the risk of death from cardiovascular diseases and all-cause mortality was observed [200]. A consistent finding across CVOTs is a significant reduction in the risk of hospitalization due to heart failure in the T2D population, as well as in populations without this disease [173]. In the DAPA-HF study, dapagliflozin significantly reduced the risk of worsening heart failure and death from cardiovascular causes in a population of patients with heart failure with reduced ejection fraction, both with and without associated DM [201]. Recent research has shown that SGLT2 inhibitors significantly reduced the risk of hospitalization due to heart failure in patients with varying LV ejection fraction values. One meta-analysis suggests that the potential cardiovascular and renal benefits of SGLT2 inhibitors in the T2D population with high cardiovascular risk significantly outweigh the risk of adverse effects of these drugs [199].

Glucagon-Like Peptide-1 Receptor Agonists

Glucagon-like peptide-1 receptor agonists (GLP-1 RAs) are crucial in managing DM and obesity. Originally developed to control blood glucose levels in DM, GLP-1 RAs have also shown cardioprotective benefits that extend beyond their primary antihyperglycemic function [176].

Mechanism of Action of Glucagon-Like Peptide-1 Receptor Agonists

The development of GLP-1 RAs was prompted by the discovery of the GLP-1 peptide, an incretin secreted from L cells in the small intestine in response to food intake. Due to rapid degradation by dipeptidyl peptidase-4 (DPP-4) and renal clearance, native GLP-1 has a very short half-life, which made its application in T2D ineffective [202, 203]. Mimicking the physiological role of GLP-1, GLP-1 RAs stimulate glucose-dependent insulin secretion from pancreatic β-cells and inhibit glucagon release from pancreatic α-cells while increasing satiety and reducing body weight [204]. The first synthetic GLP-1 RA developed was exenatide, which showed 53% sequence homology with human GLP-1 and resistance to DPP-4 enzyme degradation. Analogous to exenatide, other GLP-1 RAs were developed, such as semaglutide, liraglutide, and dulaglutide [205]. These drugs have shown high efficacy in reducing glycemia and body weight, with a high safety profile and milder side effects [206, 207]. Given the wide expression of GLP-1 receptors, including cardiomyocytes and endothelial cells, the direct effects of GLP-1 RAs on the cardiovascular system have attracted particular attention [204, 208, 209]. These direct impacts are reflected in their antioxidant, anti-inflammatory, and anti-atherosclerotic actions.

Cardiovascular Benefits of Glucagon-Like Peptide-1 Receptor Agonists: Focus on Oxidative Stress, Inflammation, and Atherosclerosis

GLP-1 RAs reduce the generation of ROS in cardiomyocytes and endothelial cells and reduce circulating markers of oxidative stress, such as 8-OHdG [210]. GLP-1 RAs exert their antioxidant effects by suppressing the enzyme NOX4 while simultaneously increasing the levels of superoxide dismutase 1 (SOD1) and glutathione peroxidase (GSH-Px), both critical enzymes in antioxidant defense [211]. In an experimental study by Li et al., liraglutide reduced serum levels of AGEs and subsequently decreased the expression of RAGE. This effect is believed to be due to the direct influence of GLP-1 RAs on AGE metabolism rather than their antihyperglycemic action [212]. Furthermore, Romero et al. demonstrated that liraglutide therapy can alleviate endoplasmic reticulum stress, thereby improving pre-existing vascular insulin resistance in DM [213]. These findings highlight the multifaceted cardioprotective mechanisms of GLP1-RAs, emphasizing their potential to mitigate cardiovascular complications in diabetic patients beyond glycemic control. GLP-1 RAs decrease the expression of endothelial and leukocyte adhesion molecules such as vascular cell adhesion molecule 1 (VCAM-1), intercellular adhesion molecule 1 (ICAM-1), monocyte chemoattractant protein 1 (MCP-1), and E-selectin. They also inhibit the transformation of monocytes/macrophages into foam cells. Additionally, GLP-1 RAs reduce the secretion of pro-inflammatory cytokines like TNF-α, IL-1, and IL-6. Through these mechanisms, GLP-1 RAs achieve their anti-inflammatory and anti-atherosclerotic effects [208, 210]. The antiatherosclerotic effects of GLP1-RAs can be partially ascribed to their impact on specific cardiovascular biomarkers. A study by Ahmad et al. demonstrated that liraglutide therapy enhances circulating levels of vascular endothelial growth factor (VEGF-1) and stromal factor 1 compared to sitagliptin. These biomarkers are vital for angiogenesis, vascular repair, and slowing the progression of atherosclerosis [214]. Furthermore, GLP-1 RAs exhibit vasodilatory properties by promoting NO production and modulating the RAAS activation. This reduces systemic vascular resistance and arterial blood pressure [204, 215]. Studies have shown that GLP-1 RAs can lower concentrations of low-density lipoprotein (LDL) cholesterol and triglycerides while increasing high-density lipoprotein (HDL) cholesterol levels. Thus, GLP-1 RAs directly impact the lipid profile, reducing cardiovascular risk [176, 215, 216].

Clinical Study Results

It is important to note that a significant number of clinical studies have unequivocally confirmed the cardiovascular benefits of GLP-1 RAs in populations with DM, especially those with high cardiovascular risk or pre-existing cardiovascular disease. Additionally, it is believed that these medications may benefit the cardiovascular system in individuals without developed DM. In studies focused on cardiovascular safety, the use of GLP-1 RAs has shown a significant reduction in the incidence of cardiovascular events, including non-fatal myocardial infarction, non-fatal stroke, and cardiovascular disease-related mortality [204]. Data from clinical trials such as LEADER, REWIND, and SUSTAIN-6 confirm the cardiovascular benefits of GLP-1 RAs. These benefits were observed in patients with established atherosclerotic cardiovascular disease and those at high risk for its development [216–219]. A meta-analysis demonstrated that GLP-1 RAs significantly reduced the risk of non-fatal myocardial infarction compared to placebo [220]. A seven-day GLP-1 RA therapy preserved cardiac muscle viability and reduced infarct size after STEMI in patients treated with primary percutaneous coronary intervention [221–223]. In a recent study, Chowdhary et al. demonstrated that four months of liraglutide therapy results in significant improvement in myocardial perfusion in patients with T2D, without previously established cardiovascular pathology. Additionally, the use of liraglutide correlated with a substantial increase in the distance covered during a six-minute walk test, compared to pioglitazone [224]. Additionally, GLP-1 RAs showed greater efficacy in reducing the risk of cerebrovascular stroke than myocardial infarction [220, 225].

Evidence regarding the potential benefits of GLP-1 RAs in treating heart failure has been inconsistent. In a meta-analysis by Ferreira et al., these medications reduced hospitalization due to newly developed heart failure. However, they did not improve outcomes in patients with existing heart failure at the start of the study [226]. Moreover, it was shown that GLP-1 RAs might have harmful effects in patients with heart failure with reduced ejection fraction [227, 228]. Further research is necessary to define the potential role and safety profile of GLP-1 RAs in the therapeutic approach to different heart failure phenotypes [209].

Metformin

Metformin is one of the oldest oral antihyperglycemic drugs and represents the gold standard in T2D treatment. Its efficient glycemic control, favorable impact on body weight, high safety profile, absence of hypoglycemia as a side effect, and low cost make metformin the drug of choice for the treatment of T2D, whether used as monotherapy or in combination with other antihyperglycemics [229]. Besides its antidiabetic properties, metformin has shown potential in treating other conditions, including polycystic ovary syndrome, neurodegenerative diseases, cancer, and cardiovascular diseases [230].

Mechanism of Action of Metformin

Metformin lowers blood glucose levels through three primary mechanisms. Primarily, it inhibits gluconeogenesis and glycogenolysis, reducing glucose production in the liver. Additionally, it increases glucose utilization in peripheral tissues and moderately delays glucose absorption in the gastrointestinal tract [231, 232]. Despite its long-term use, there are still controversies regarding the exact mechanism of action of metformin. On the molecular level, metformin inhibits complex I of the mitochondrial respiratory chain in the liver, reducing cellular ATP levels and leading to the activation of AMPK. The activation of AMPK is considered the critical mechanism for inhibiting hepatic gluconeogenesis and the increase in insulin sensitivity of peripheral tissues. Moreover, metformin also exhibits AMPK-independent effects [231–234]. Additionally, metformin exerts antihyperglycemic effects by modulating GLP-1 activity in the duodenum [235].

Cardiovascular Benefits of Metformin: Focus on Antioxidant and Anti-inflammatory Properties

The cardiovascular protective effects of metformin were first documented in the United Kingdom Prospective Diabetes Study (UKPDS). In the studied population, metformin reduced the risk of developing macrovascular complications of DM compared to standard therapy [236]. Other clinical studies have confirmed this effect of metformin [237–239]. Metformin achieves its cardioprotective properties by reducing oxidative stress in mitochondria. In an experimental study by Vilela et al., metformin reduced the production of ROS by inhibiting complex I of the mitochondrial respiratory chain [240]. In another study, using an alloxan-induced DM model, metformin improved the antioxidant defense system by increasing serum concentrations of glutathione (GSH), SOD, catalase, and decreasing serum concentrations of MDA [241]. In a study by Liu et al., metformin demonstrated a protective effect against hyperglycemia-induced oxidative stress through an AMPK-dependent mechanism [242]. Clinical studies have shown that metformin exhibits superior antioxidant properties compared to lifestyle and dietary interventions in populations with newly diagnosed T2D [243, 244]. Metformin's antioxidant properties can also be partly explained by reducing non-enzymatic protein glycation under hyperglycemic conditions [245, 246]. Additionally, recent research suggests that metformin's antioxidant properties may be attributed to the inhibition of mitochondrial monoamine oxidase A and B (MAO-A, MAO B), which are significant sources of oxidative stress in diabetic cardiomyopathy [247]. Metformin's anti-inflammatory properties are indirectly achieved by affecting cardiovascular risk factors such as hyperglycemia, dyslipidemia, and insulin resistance. However, recent studies indicate a direct anti-inflammatory effect of this drug [248]. AMPK is central in modulating various cellular functions, including proliferation and inflammation. The activation of AMPK partly explains the direct anti-inflammatory effect of metformin. NF-kB regulates the activation of numerous pro-inflammatory factors, including TNF-α, IL-1β, and IL-6. Recent studies suggest that metformin directly binds to NF-kB, reducing inflammation [232, 248, 249]. Additionally, metformin inhibits the expression of pro-inflammatory cytokines such as TNF-α, TGF-β, and IL-6. The mechanism of these effects is not fully elucidated [250]. These effects are important for understanding metformin's anti-atherogenic actions.

Other Cardioprotective Effects of Metformin

In an experimental model of ischemic heart disease, the administration of low doses of metformin improved the structure and contractile function of the left ventriculum [251]. Additionally, metformin prevented and mitigated the development of myocardial fibrosis in animal models of heart failure [252]. This effect of metformin can be partly explained by the inhibition of collagen synthesis. A similar effect was documented in an experimental study on ischemia-reperfusion injury, where metformin reduced fibrosis and cardiac remodeling following acute myocardial infarction [253].

Cardiovascular Benefits of Other Drugs in the Treatment of DM

Dipeptidyl Peptidase 4 Inhibitors

DPP-4 inhibitors exert their antihyperglycemic effect by inhibiting the DPP-4 enzyme, thereby reducing the breakdown of GLP-1 incretins. It is believed that DPP-4 inhibitors exhibit cardioprotective effects by reducing oxidative stress and inflammation, improving endothelial function, and enhancing myocardial perfusion [254]. Additionally, experimental studies have shown that DPP-4 inhibitors slow down myocardial remodeling and heart dysfunction [255]. Furthermore, DPP-4 inhibitors have slowed the progression of atherosclerosis and improved endothelial function in apoE-deficient mice [256]. On the other hand, in three large randomized, placebo-controlled studies, DPP-4 inhibitors did not significantly reduce the risk of myocardial infarction and ischemic stroke [257–259]. Moreover, saxagliptin increased the risk of hospitalization due to heart failure [258]. Additionally, one meta-analysis suggests that DPP-4 inhibitors have a neutral effect on the risk of myocardial infarction, cardiac arrhythmias, heart failure, cerebrovascular stroke, and cardiovascular death. In contrast, the risk of developing atrial flutter increases compared to placebo [260]. DPP-4 inhibitor therapy has reduced the risk of heart failure with preserved ejection fraction and cardiovascular death in the population with DM. However, such effect was absent in patients with heart failure with moderately or significantly reduced ejection fraction [261]. These clinical study results do not correlate with findings from preclinical research and require further explanation.

Sulfonylureas

Sulfonylurea compounds exert their hypoglycemic effect by stimulating insulin secretion in pancreatic β-cells through the inhibition of potassium flux via ATP-dependent potassium channels [262]. There is evidence suggesting that sulfonylurea therapy is associated with an increased risk of ventricular arrhythmias and sudden cardiac death in patients with T2D [263]. Additionally, sulfonylurea monotherapy is linked to higher rates of ischemic stroke, cardiovascular mortality, and all-cause mortality compared to metformin [264]. However, Wang et al. found that sulfonylurea therapy was not associated with a higher risk of adverse cardiovascular events [265]. Overall, the existing evidence on the cardiovascular safety of sulfonylureas remains controversial.

Thiazolidinediones

Thiazolidinediones are hypoglycemic agents that increase tissue sensitivity to insulin. These drugs are highly selective agonists of peroxisome proliferator-activated (PPAR) γ receptors, which are involved in the metabolism of carbohydrates, fats, and proteins. Thiazolidinediones have shown significant antiatherogenic properties by reducing circulating levels of pro-inflammatory markers and slowing the thickening of carotid artery intima [266, 267]. In an animal model of DM, pioglitazone reduced ROS production in mitochondria, demonstrating antioxidant and anti-inflammatory effects [268]. In a meta-analysis, Zhu et al. reported that pioglitazone use significantly reduced the risk of adverse cardiovascular events, specifically myocardial infarction and ischemic stroke. However, the risk of congestive heart failure was increased [269]. This is explained by dose-dependent fluid retention, a side effect of thiazolidinediones [270]. Due to the high risk of developing heart failure or worsening pre-existing heart failure, these drugs are contraindicated in this population [267].

Synthetic Insulins

In addition to its vital role in metabolism, insulin has significant protective effects on the cardiovascular system, acting as an antioxidant and anti-inflammatory agent [271, 272]. Despite these beneficial effects, insulin use is associated with a risk of fluid retention and hypoglycemia, potentially exacerbating heart failure [273]. However, a recent meta-analysis suggests that insulin use did not impact the incidence of cardiovascular events, including hospitalization due to heart failure [274].

Cardiovascular Toxicity and Other Agents

Lipid-Lowering Agents

Statins

Statins, primarily known as inhibitors of hydroxymethylglutaryl-coenzyme A reductase (HMG-CoA reductase), represent the gold standard in dyslipidemia pharmacotherapy. Beyond their primary lipid-lowering effect, statins exhibit numerous pleiotropic effects that significantly reduce cardiovascular risk in populations with and without DM [275–277].

Research has shown that statins benefit vascular homeostasis through increased NO bioavailability [278]. In an experimental model of myocardial infarction, atorvastatin therapy significantly improved LV and endothelial function. This effect of atorvastatin results from increased NO bioavailability [279]. Independent of their lipid-lowering effect, statins also exhibit antithrombotic effects by reducing platelet activation, aggregation, and adhesion [280]. A profibrinolytic effect of statins has also been established by increasing tissue plasminogen activator (tPA) activity as a result of increased tPA expression and decreased PAI-1 expression [281]. Additionally, statin therapy is associated with lower blood levels of vWF and D-dimer [282]. It is hypothesized that the anticoagulant effects of statins contribute to reducing the frequency of venous thromboembolism recurrence [280]. Cardiovascular risk reduction is also attributed to the anti-inflammatory properties of these drugs. Statins reduce circulating CRP, IL-1, and TNF-α levels while decreasing nuclear factor kappa B (NF-κB) expression [275, 283]. The anti-inflammatory effect of statins can be partially explained by increased adiponectin levels, an adipocytokine with predominantly anti-inflammatory properties. A recent meta-analysis indicated the significant potential of statins in reducing epicardial adipose tissue. These results are significant in cardiovascular health, as excessive accumulation of epicardial adipose tissue is associated with an increased risk of cardiovascular complications [284]. On the other hand, statin use may increase levels of leptin and resistin, hormones associated with increased body weight and development of insulin resistance. Therefore, the impact of statins on adipocytokines is complex and varies depending on patient characteristics and drug dosage [285]. Despite undoubted cardiovascular benefits, long-term use of statins may be associated with a slight increase in the risk of developing DM, especially in individuals with pre-existing risk factors for this disease [286]. Results from the JUPITER study indicated that rosuvastatin usage was associated with a 25% increase in the incidence of newly diagnosed diabetes compared to placebo [287, 288]. These findings were further confirmed by a meta-analysis by Thakker et al., which indicated an increased risk of developing DM with statin use. This risk was more pronounced with the use of more potent drugs in this group and at higher doses [289]. Despite these findings, it is essential to emphasize that the consensus in the professional community is that the benefit of reducing cardiovascular risk significantly outweighs the potential risk of developing DM, especially in high-risk populations [286]. However, these results highlight the need for careful monitoring of glycemia in patients on long-term statin therapy, especially in those predisposed to developing DM.

Ezetimibe

The combined therapy of ezetimibe and statins allows additional reduction of serum LDL cholesterol. This synergistic effect is achieved through ezetimibe-induced inhibition of the Niemann-Pick C1-Like 1 (NPC1L1) receptor in the gastrointestinal tract, resulting in decreased intestinal cholesterol absorption [290]. In the IMPROVE-IT study, the combination of ezetimibe and simvastatin, besides significantly reducing LDL cholesterol, decreased the risk of adverse cardiovascular events. These effects were particularly pronounced in patients with DM [290, 291]. Ezetimibe's cardiovascular benefits are based on its pleiotropic effects, including anti-inflammatory and antioxidant properties and inhibition of platelet aggregation and activation. These effects are considered particularly significant for slowing the progression of atherosclerosis and stabilizing atherosclerotic plaques [292].

Fibrates

Activation of PPAR-α receptors and subsequent reduction in serum triglyceride levels represents the critical mechanism of action for fibrates in treating dyslipidemias [276]. Evidence for the cardiovascular benefits of fibrates is inconsistent. In the PROMINENT study, fibrate therapy improved the lipid profile of patients with T2D, while no impact on reducing the risk of adverse cardiovascular events was observed [293]. However, a recent meta-analysis showed that fibrate therapy significantly reduced the risk of major adverse cardiovascular events (MACE). This cardioprotective effect of fibrates is primarily the result of LDL cholesterol level reduction, while the decrease in triglyceride levels has a less significant impact [294].

Proprotein Convertase Subtilisin/Kexin Type 9 Inhibitors

Proprotein convertase subtilisin/kexin type 9 (PCSK9) inhibitors represent a novel class of lipid-lowering drugs. This class consists of monoclonal antibodies that, by blocking the activity of PCSK9 convertase, reduce the degradation of LDL receptors and increase the removal of LDL cholesterol [290, 295]. Research has shown the benefits of PCSK9 inhibitor therapy in improving lipid profiles, stabilizing atherosclerotic plaques, and preventing adverse cardiovascular events [296].

Inclisiran

An alternative path to inhibiting PCSK9 convertase is the new drug inclisiran, a small interfering RNA molecule that prevents the expression of the PCSK9 gene [290]. A pooled analysis of randomized, double-blind clinical studies examining the efficacy of inclisiran in treating dyslipidemias provided insight into the potential cardiovascular benefits of this drug through the reduction of adverse cardiovascular events [297]. Further research is needed for more detailed examinations of the cardiovascular benefits of inclisiran.

The Role of Vitamins in Cardiovascular Toxicity

Due to their regulatory role in numerous metabolic processes, hydro-soluble and lipo-soluble vitamins are essential in maintaining cardiovascular health. Deficiency of certain vitamins, such as B-group vitamins, vitamins C, D, A, and E, is associated with various cardiovascular diseases, including atherosclerosis, hypertension, myocardial infarction, and heart failure [298]. Additionally, patients with DM often have reduced levels of so-called antioxidant vitamins, most likely due to increased oxidative stress in this disease [299].

Vitamin B6

Vitamin B6 is an essential hydro-soluble vitamin with diverse roles in the body. As a coenzyme, vitamin B6 participates in over 100 metabolic reactions [300]. Vitamin B6 encompasses three chemically similar compounds—pyridoxine, pyridoxal, and pyridoxamine, along with their phosphorylated metabolites [299]. The active form of vitamin B6 is pyridoxal 5′-phosphate.

Research indicates a potential role of vitamin B6 in blood pressure regulation and cardiovascular disease prevention. Moreover, individuals with reduced blood levels of the active form of vitamin B6 may be at increased risk for developing atherosclerosis, stroke, and thrombotic complications [301]. Huang et al. demonstrated that increased dietary intake of vitamin B6 is associated with a reduced incidence of cardiovascular diseases [302]. A similar association was observed in a cohort study by Minović et al. However, the correlation between higher vitamin B6 levels and reduced cardiovascular risk was more pronounced in women than men [303]. Furthermore, there is a strong correlation between vitamin B6 and DM. Studies have shown that vitamin B6 reduces blood glucose levels in STZ-induced DM model and HbA1c levels in patients with T2D [304]. Recent research has demonstrated that pyridoxamine, a vitamin B6 analog, significantly alleviates cardiovascular complications associated with T2D. D'Haese et al. showed that pyridoxamine can prevent adverse cardiac remodeling in prediabetic rats by reducing fasting plasma glucose levels and interstitial fibrosis, as well as limiting oxidative stress biomarkers, such as 3-nytrotirosine and 4-hydroxynonenal [305]. The cardioprotective effects of vitamin B6 are primarily attributed to its antioxidant properties, which inhibit the formation of free radicals, AGEs, and lipid peroxidation [306]. Beyond its direct role in neutralizing free radicals, vitamin B6 indirectly enhances the body's antioxidant defense system. Vitamin B6 is involved in the cysteine synthesis process, an amino acid essential for producing GSH, a significant antioxidant [307]. Consequently, maintaining adequate levels of vitamin B6 contributes significantly to bolstering the organism's overall antioxidant capacity. Vitamin B6 also exhibits anti-inflammatory effects. Specifically, vitamin B6 reduces the expression of genes that encode certain pro-inflammatory cytokines [308].

Vitamin B12 and Folate

Vitamin B12 and folate play indispensable roles in the proper function of the central nervous system. These B-group vitamins are crucial in the one-carbon metabolic pathway, which involves transferring methyl groups for DNA synthesis and methylation [309]. Vitamin B12 deficiency is associated with an increased risk of cardiovascular diseases, including atherosclerosis, myocardial infarction, and stroke. Similarly, folic acid deficiency has been observed in vascular pathology and congenital heart diseases [298]. The primary mechanism by which vitamin B12 and folate deficiency contribute to increased cardiovascular risk is through their essential role in homocysteine metabolism [310]. Hyperhomocysteinemia accelerates atherosclerosis progression and is an independent risk factor for cardiovascular pathology and adverse cardiovascular events. Given their effectiveness in lowering serum homocysteine concentrations, research has focused on the potential role of vitamin B12 and folate in reducing cardiovascular disease incidence [311]. However, different studies have not consistently supported this hypothesis. One study showed that B-vitamin supplementation did not impact cardiovascular outcomes, including stroke, coronary heart disease, and cardiovascular mortality. One possible explanation for these results is the relatively short duration of B-vitamin therapy [312]. Conversely, in a study by Schnyder et al., folic acid, vitamin B6, and B12 supplementation and the subsequent reduction in serum homocysteine concentrations improved cardiovascular outcomes in patients following percutaneous coronary intervention [313]. Vitamin B12 deficiency has also been linked to T2D, with metformin therapy being a primary contributing factor [314]. The exact mechanism by which chronic metformin use leads to vitamin B12 deficiency remains unclear. However, factors such as reduced small intestine motility, alterations in the gastrointestinal microbiota, and inhibition of the absorption of the vitamin B12-intrinsic factor complex are thought to play significant roles [315]. Moreover, vitamin B12 deficiency has also been observed in individuals with DM who do not use metformin, particularly those with diabetic neuropathy [299].

Vitamin D

Vitamin D, a liposoluble vitamin, is crucial for bone metabolism regulation. Given the widespread presence of vitamin D receptors (VDR) in nearly all tissues, recent research suggests a potential role for this vitamin in various physiological processes. Increasing evidence highlights the importance of vitamin D in autoimmune diseases, infectious diseases, cardiovascular diseases, DM, and various malignancies [316, 317]. In the cardiovascular system, vitamin D exerts pleiotropic effects by activating its receptors, which are expressed on the surface of cardiomyocytes and endothelial cells of blood vessels. Additionally, vitamin D regulates blood pressure through the RAAS [318]. Studies in VDR knockout mouse models and vitamin D-deficient rats have demonstrated cardiomyocyte hypertrophy and collagen accumulation in the heart muscle, resulting in impaired contractility and cardiac remodeling [319]. In humans, vitamin D deficiency is associated with vascular dysfunction, increased arterial stiffness, and left ventricular hypertrophy [318]. Vitamin D deficiency is also linked to insulin resistance, while supplementation with this vitamin improves glucose metabolism and insulin sensitivity [299]. Furthermore, a recent study by Isa et al. demonstrated a correlation between microvascular and macrovascular complications of DM and vitamin D levels [320].

The Role of Trace Elements in Cardiovascular Toxicity

Although present in living tissues in small amounts, trace elements play a crucial role in different metabolic processes at the cellular and subcellular levels. They are essential for the proper functioning of the immune system, nerve conduction, muscle contractions, regulation of membrane potential, and numerous enzymatic reactions, thereby significantly contributing to maintaining homeostasis and optimal organism functioning [321, 322]. Trace elements have a dual function. Their optimal concentration is necessary for maintaining physiological processes. However, disturbances in their serum concentrations can lead to the activation of alternative metabolic pathways and the development of diseases [322].

Zinc

Zinc (Zn) plays multiple roles in the body's antioxidant defense system. As a cofactor, Zn activates vital enzymes in the antioxidant system while protecting the cell membrane from oxidative damage [323]. Zn deficiency can negatively affect cardiac structure and function, contributing to the development of cardiovascular diseases, including diabetic cardiomyopathy [324]. In an animal model of T2D, Wang et al. investigated the impact of Zn deficiency on cardiac function. Their results indicated that Zn deficiency significantly exacerbates cardiac damage caused by oxidative stress, inflammation, and fibrosis [325]. Additionally, research has suggested a potential link between Zn deficiency and an increased risk of developing DM [326, 327].

Magnesium

Magnesium (Mg) is an essential mineral with multiple roles in the human body. As a cofactor in more than 300 enzymatic reactions, Mg participates in numerous physiological processes, including muscle contraction, neuromuscular conduction, and cardiac function. It is necessary for all energy-dependent transport systems in cells, including glycolysis and oxidative phosphorylation [328, 329]. Mg regulates vascular smooth muscle tone, endothelial function, and cardiac muscle excitability.

Therefore, Mg is of vital importance in the etiopathology of various cardiovascular disorders, including hypertension, atherosclerosis, coronary artery disease, congestive heart failure, and cardiac arrhythmias [330]. In addition to its essential role in the cardiovascular system, Mg is of particular significance in the DM population. Mg deficiency may further increase the risk of developing and progressing cardiovascular complications of DM [331]. Studies have found a correlation between low Mg concentrations and increased frequency of arrhythmias in obese patients with T2D. Additionally, a link has been established between Mg deficiency and accelerated decline in renal function [332].

Chromium

Chromium (Cr) is important in metabolizing carbohydrates, proteins, and lipids [333]. It is also an essential cofactor in insulin action and a structural component of an organic complex known as the glucose tolerance factor [321]. The hypothesis regarding Cr′s potential to enhance insulin action has directed research toward evaluating the impact of increased intake of this trace element on preventing glucose tolerance disorders. However, studies have yielded conflicting results. One study demonstrated that Cr supplementation did not affect lipid and glucose metabolism in patients without DM, while improved glycemic control was observed in patients with T2D [334]. However, in another study, Cr supplementation did not improve insulin sensitivity in normoglycemic subjects. Moreover, in subjects with high serum Cr concentrations, a paradoxical decrease in insulin sensitivity was observed [335].

Conclusion

DM significantly increases the risk of cardiovascular toxicity and, consequently, cardiovascular diseases through complex pathophysiological mechanisms, with chronic hyperglycemia-induced oxidative stress, endothelial dysfunction, and inflammation playing key roles.

While traditional antihyperglycemic drugs have provided varying cardiovascular safety profiles, novel agents such as SGLT2 inhibitors and GLP-1 RAs have shown promising cardioprotective effects beyond their primary antihyperglycemic action. These medications have reduced the incidence of cardiovascular events and mortality not only in the diabetic population but also in individuals without the disease. However, the cardiovascular benefits differ not only between drug classes but also within the same class.

Additionally, research has indicated potential cardioprotective effects of specific vitamins and trace elements in the context of DM. A deeper understanding of the mechanisms driving cardiovascular toxicity and the protective effects of various therapeutic strategies opens new perspectives in the prevention and treatment of cardiovascular complications of DM.

References

1. Zoltani CK (2019) Cardiovascular toxicity biomarkers. In: Gupta RC (ed) Biomarkers in toxicology, 2nd edn. Academic, London, pp 209–228
2. Chung R, Ghosh AK, Banerjee A (2018) Cardiotoxicity: precision medicine with imprecise definitions. Open Heart 5(2):e000774
3. Kim H, Chung WB, Cho KI et al (2018) Diagnosis, treatment, and prevention of cardiovascular toxicity related to anti-cancer treatment in clinical practice: an opinion paper from the working group on Cardio-Oncology of the Korean Society of Echocardiography. J Cardiovasc Ultrasound 26(1):1–25
4. Roberts JA, Rainbow RD, Sharma P (2023) Mitigation of cardiovascular disease and toxicity through NRF2 signalling. Int J Mol Sci 24(7):6723
5. Singh AK (2016) Mechanisms of nanoparticle tocixity. In: Singh AK (ed) Engineered nanoparticles. Academic, London, pp 295–341
6. Salvatore T, Pafundi PC, Galiero R et al (2021) The diabetic cardiomyopathy: the contributing pathophysiological mechanisms. Front Med 8:695792
7. Magliano DJ, Boyko EJ (2021) IDF diabetes atlas 10th edition scientific committee IDF diabetes atlas, 10th edn. International Diabetes Federation, Brussels. https://diabetesatlas.org/atlas/tenth-edition/. Accessed 10 June 2024
8. Leon BM, Maddox TM (2015) Diabetes and cardiovascular disease: epidemiology, biological mechanisms, treatment recommendations and future research. World J Diabetes 6(13): 1246–1258
9. Dal Canto E, Ceriello A, Rydén L et al (2019) Diabetes as a cardiovascular risk factor:an overview of global trends of macro and micro vascular complications. Eur J Prev Cardiol 26(2):25–32
10. Schmidt AM (2019) Diabetes mellitus and cardiovascular disease. Arterioscler Thromb Vasc Biol 39(4):558–568
11. Swamy S, Noor SM, Mathew RO (2023) Cardiovascular disease in diabetes and chronic kidney disease. J Clin Med 12(22):6984
12. Campos C (2012) Chronic hyperglycemia and glucose toxicity: pathology and clinical sequelae. Postgrad Med 124(6):90–97
13. Battault S, Renguet E, Van Steenbergen A et al (2020) Myocardial glucotoxicity: mechanisms and potential therapeutic targets. Arch Cardiovasc Dis 113(11):736–748
14. Kawahito S, Kitahata H, Oshita S (2009) Problems associated with glucose toxicity: role of hyperglycemia-induced oxidative stress. World J Gastroenterol 15(33):4137–4142
15. Poznyak A, Grechko AV, Poggio P et al (2020) The diabetes mellitus-atherosclerosis connection: the role of lipid and glucose metabolism and chronic inflammation. Int J Mol Sci 21(5): 1835
16. Mastrogiacomo L, Ballagh R, Venegas-Pino DE et al (2023) The effects of hyperglycemia on early endothelial activation and the initiation of atherosclerosis. Am J Pathol 193(1):121–133
17. Cerf ME (2018) Cardiac glucolipotoxicity and cardiovascular outcomes. Medicina 54(5):70
18. Chuanbao L, Huiruo L, Feng X et al (2022) The role of lipotoxicity in cardiovascular disease. Emerg Crit Care Med 2:214–218
19. Hickson-Bick DL, Buja LM, McMillin JB (2000) Palmitate-mediated alterations in the fatty acid metabolism of rat neonatal cardiac myocytes. J Mol Cell Cardiol 32(3):511–519
20. Kim SY, Lee JP, Shin WR et al (2022) Cardiac biomarkers and detection methods for myocardial infarction. Mol Cell Toxicol 18(4):443–455
21. Upadhyay RK (2015) Emerging risk biomarkers in cardiovascular diseases and disorders. J Lipids 2015:971453
22. Park KC, Gaze DC, Collinson PO et al (2017) Cardiac troponins: from myocardial infarction to chronic disease. Cardiovasc Res 113(14):1708–1718
23. Zhang X, Sun Y, Zhang Y et al (2022) Cardiac biomarkers for the detection and management of cancer therapy-related cardiovascular toxicity. J Cardiovasc Dev Dis 9(11):372

24. Kontos MC, Turlington JS (2020) High-sensitivity troponins in cardiovascular disease. Curr Cardiol Rep 22:30
25. Lazar DR, Lazar FL, Homorodean C et al (2022) High-sensitivity troponin: a review on characteristics, assessment, and clinical implications. Dis Markers 2022:9713326
26. Bellia C, Lombardo M, Della-Morte D (2020) Use of troponin as a predictor for cardiovascular diseases in patients with type 2 diabetes mellitus. Clin Chim Acta 507:54–61
27. Song T, Lan Y, Li K et al (2023) Prognostic value of high-sensitivity cardiac troponin for major adverse cardiovascular events in patients with diabetes: a systematic review and meta-analysis. PeerJ 11:e16376
28. Hermann A, Sitdikova G (2021) Homocysteine: biochemistry, molecular biology and role in disease. Biomol Ther 11(5):737
29. Djuric D, Jakovljevic V, Zivkovic V et al (2018) Homocysteine and homocysteine-related compounds: an overview of the roles in the pathology of the cardiovascular and nervous systems. Can J Physiol Pharmacol 96(10):991–1003
30. Hasan T, Arora R, Bansal AK et al (2019) Disturbed homocysteine metabolism is associated with cancer. Exp Mol Med 51(2):1–13
31. Bajic Z, Sobot T, Skrbic R et al (2022) Homocysteine, vitamins B6 and folic acid in experimental models of myocardial infarction and heart failure-how strong is that link? Biomol Ther 12(4):536
32. Liu B, Ma S, Wang T et al (2016) A novel rat model of heart failure induced by high methionine diet showing evidence of association between hyperhomocysteinemia and activation of NF-kappaB. Am J Transl Res 8(1):117–124
33. Muzurović E, Kraljević I, Solak M et al (2021) Homocysteine and diabetes: role in macrovascular and microvascular complications. J Diabetes Complicat 35(3):107834
34. Mursleen MT, Riaz S (2017) Implication of homocysteine in diabetes and impact of folate and vitamin B12 in diabetic population. Diabetes Metab Syndr 11(Suppl 1):141–146
35. Wettersten N, Maisel AS (2016) Biomarkers for heart failure: an update for practitioners of internal medicine. Am J Med 129(6):560–567
36. Goryacheva OA, Ponomaryova TD, Drozd DD et al (2022) Heart failure biomarkers BNP and NT-proBNP detection using optical labels. TrAC Trends Anal Chem 146:116477
37. Birukov A, Eichelmann F, Kuxhaus O et al (2020) Opposing associations of NT-proBNP with risks of diabetes and diabetes-related complications. Diabetes Care 43(12):2930–2937
38. Rezar R, Jirak P, Gschwandtner M et al (2020) Heart-type fatty acid-binding protein (H-FABP) and its role as a biomarker in heart failure: what do we know so far? J Clin Med 9(1):164
39. Das UN (2015) Heart-type fatty acid-binding protein (H-FABP) and coronary heart disease. Indian Heart J 68(1):16–18
40. Salman Habib SM, Khan AR, Habib S et al (2021) Role of heart fatty acid binding protein in early detection of non ST-elevation myocardial infarct and its comparison with other cardiac markers. J Pak Med Assoc 71(1):233–238
41. Huang X, Bai S, Luo Y (2024) Advances in research on biomarkers associated with acute myocardial infarction: a review. Medicine 103(15):e37793
42. Hasić S, Jadrić R, Cosović E et al (2011) Heart-type fatty acid-binding protein and its relation with morphological changes in rat myocardial damage model induced by isoproterenol. Bosn J Basic Med Sci 11(4):240–244
43. Ramesh P, Chauhan A, Goyal P et al (2020) H-FABP: a beacon of hope for prediabetic heart disease. J Family Med Prim Care 9(7):3421–3428
44. Abdelazim AM, Abomughaid MM (2024) Oxidative stress: an overview of past research and future insights. All Life 17:2316092
45. Panda P, Verma HK, Lakkakula S et al (2022) Biomarkers of oxidative stress tethered to cardiovascular diseases. Oxidative Med Cell Longev 2022:9154295
46. Ng ML, Ang X, Yap KY et al (2023) Novel oxidative stress biomarkers with risk prognosis values in heart failure. Biomedicines 11(3):917

47. Henein MY, Vancheri S, Longo G et al (2022) The role of inflammation in cardiovascular disease. Int J Mol Sci 23(21):12906
48. Shrivastava AK, Singh HV, Raizada A et al (2015) C-reactive protein, inflammation and coronary heart disease. Egypt Heart J 67(2):89–97
49. Silva D, Pais de Lacerda A (2012) High-sensitivity C-reactive protein as a biomarker of risk in coronary artery disease. Rev Port Cardiol 31(11):733–745
50. Sellmayer A, Limmert T, Hoffmann U (2003) High sensitivity C-reactive protein in cardiovascular risk assessment. CRP mania or useful screening? Int Angiol 22(1):15–23
51. Sharif S, Van der Graaf Y, Cramer MJ et al (2021) Low-grade inflammation as a risk factor for cardiovascular events and all-cause mortality in patients with type 2 diabetes. Cardiovasc Diabetol 20:220
52. Han E, Fritzer-Szekeres M, Szekeres T et al (2022) Comparison of high-sensitivity C-reactive protein vs C-reactive protein for cardiovascular risk prediction in chronic cardiac disease. J Appl Lab Med 7:1259–1271
53. Haybar H, Bandar B, Torfi E et al (2023) Cytokines and their role in cardiovascular diseases. Cytokine 169:156261
54. Schuett H, Luchtefeld M, Grothusen C et al (2009) How much is too much? Interleukin-6 and its signalling in atherosclerosis. Thromb Haemost 102:215–222
55. Feng Y, Ye D, Wang Z et al (2022) The role of interleukin-6 family members in cardiovascular diseases. Front Cardiovasc Med 9:818890
56. Bhaskar V, Yin J, Mirza AM et al (2011) Monoclonal antibodies targeting IL-1 beta reduce biomarkers of atherosclerosis in vitro and inhibit atherosclerotic plaque formation in apolipoprotein E-deficient mice. Atherosclerosis 216:313–320
57. Szekely Y, Arbel Y (2018) A review of interleukin-1 in heart disease: where do we stand today? Cardiol Ther 7:25–44
58. Alfadul H, Sabico S, Al-Daghri NM (2022) The role of interleukin-1β in type 2 diabetes mellitus: a systematic review and meta-analysis. Front Endocrinol 13:901616
59. Welsh P, Murray MP, Ford I et al (2011) Circulating interleukin-10 and risk of cardiovascular events: a prospective study in the elderly at risk. ATVB 31:2338–2344
60. Xu S, Zhang J, Liu J et al (2021) The role of interleukin-10 family members in cardiovascular diseases. Int Immunol 94:107475
61. Goldwater D, Karlamangla A, Merkin SS et al (2019) Interleukin-10 as a predictor of major adverse cardiovascular events in a racially and ethnically diverse population: multi-ethnic study of atherosclerosis. Ann Epidemiol 30:9–14
62. Lu J, Liu J, Li L et al (2020) Cytokines in type 1 diabetes: mechanisms of action and immunotherapeutic targets. Clin Transl Immunol 9:e1122
63. dos Santos Haber JF, Barbalho SM, Sgarbi JA et al (2023) The relationship between type 1 diabetes mellitus, TNF-α, and IL-10 gene expression. Biomedicines 11:1120
64. Garza HHE, Garza HLJ, Gonzalez RH et al (2002) Importance of tumor necrosis factor-alpha in the pathogenesis of heart failure. Rev Esp Cardiol 55:61–66
65. Dunlay SM, Weston SA, Redfield MM et al (2008) Tumor necrosis factor-alpha and mortality in heart failure: a community study. Circulation 118:625–631
66. Zhang H, Dhalla NS (2024) The role of pro-inflammatory cytokines in the pathogenesis of cardiovascular disease. Int J Mol Sci 25:1082
67. Hanna A, Frangogiannis NG (2019) The role of the TGF-β superfamily in myocardial infarction. Front Cardiovasc Med 18:140
68. Frangogiannis NG (2022) Transforming growth factor-β in myocardial disease. Nat Rev Cardiol 19:435–455
69. Pan K, Xu C, Chen C et al (2023) Soluble interleukin-2 receptor combined with interleukin-8 is a powerful predictor of future adverse cardiovascular events in patients with acute myocardial infarction. Front Cardiovasc Med 10:1110742
70. Liu Y, Zhang D, Yin D (2023) Pathophysiological effects of various interleukins on primary cell types in common heart disease. Int J Mol Sci 24:6497

71. Robert M, Miossec P (2017) Effects of interleukin 17 on the cardiovascular system. Autoimmun Rev 16:984–991
72. Liu Y, Wang M, Su JB et al (2024) Potential clinical value of fibrinogen-like protein 1 as a serum biomarker for the identification of diabetic cardiomyopathy. Sci Rep 14:10311
73. Cui Z, Zhao G, Liu X (2022) Blood fibrinogen level as a biomarker of adverse outcomes in patients with coronary artery disease: a systematic review and meta-analysis. Medicine 101: e30117
74. Bembde AS (2012) A study of plasma fibrinogen level in type-2 diabetes mellitus and its relation to glycemic control. Indian J Hematol Blood Transfus 28:105–108
75. Abdul Razak MK, Sultan AA (2019) The importance of measurement of plasma fibrinogen level among patients with type-2 diabetes mellitus. Diabetes Metab Syndr 13:1151–1158
76. Zaborska B, Sikora-Frąc M, Smarż K et al (2023) The role of galectin-3 in heart failure-the diagnostic, prognostic and therapeutic potential-where do we stand? Int J Mol Sci 24:13111
77. Mitic VT, Stojanovic DR, Deljanin Ilic MZ et al (2020) Cardiac remodeling biomarkers as potential circulating markers of left ventricular hypertrophy in heart failure with preserved ejection fraction. Tohoku J Exp Med 250:233–242
78. Yu L, Ruifrok WP, Meissner M et al (2013) Genetic and pharmacological inhibition of galectin-3 prevents cardiac remodeling by interfering with myocardial fibrogenesis. Circ Heart Fail 6:107–117
79. Bostan MM, Stătescu C, Anghel L et al (2020) Post-myocardial infarction ventricular remodeling biomarkers-the key link between pathophysiology and clinic. Biomol Ther 10: 1587
80. Nielsen SH, Mouton AJ, DeLeon-Pennell KY et al (2019) Understanding cardiac extracellular matrix remodeling to develop biomarkers of myocardial infarction outcomes. Matrix Biol 75-76:43–57
81. Halade GV, Jin YF, Lindsey ML (2013) Matrix metalloproteinase (MMP)-9: a proximal biomarker for cardiac remodeling and a distal biomarker for inflammation. Pharmacol Ther 139:32–40
82. Olejarz W, Łacheta D, Kubiak-Tomaszewska G (2020) Matrix metalloproteinases as biomarkers of atherosclerotic plaque instability. Int J Mol Sci 21:3946
83. American Diabetes Association, National Heart, Lung, and Blood Institute, Juvenile Diabetes Foundation International (1999) Diabetes mellitus: a major risk factor for cardiovascular disease. A joint editorial statement. Circulation 100:1132–1133
84. Kannel WB, McGee DL (1979) Diabetes and cardiovascular risk factors: the Framingham study. Circulation 59(1):8–13
85. Tan Y, Zhang Z, Zheng C et al (2020) Mechanisms of diabetic cardiomyopathy and potential therapeutic strategies: preclinical and clinical evidence. Nat Rev Cardiol 17:585–607
86. Singh R, Gholipourmalekabadi M, Shafikhani SH (2024) Animal models for type 1 and type 2 diabetes: advantages and limitations. Front Endocrinol 15:1359685
87. Kleinert M, Clemmensen C, Hofmann SM et al (2018) Animal models of obesity and diabetes mellitus. Nat Rev Endocrinol 14:140–162
88. Ghasemi A, Jeddi S (2023) Streptozotocin as a tool for induction of rat models of diabetes: a practical guide. EXCLI J 22:274–294
89. Goyal SN, Reddy NM, Patil KR et al (2016) Challenges and issues with streptozotocin-induced diabetes - a clinically relevant animal model to understand the diabetes pathogenesis and evaluate therapeutics. Chem Biol Interact 244:49–63
90. Mutavdzin S, Gopcevic K, Stankovic S et al (2019) The effects of folic acid administration on cardiac oxidative stress and cardiovascular biomarkers in diabetic rats. Oxidative Med Cell Longev 2019:1342549
91. Ighodaro OM, Adeosun AM, Akinloye OA (2017) Alloxan-induced diabetes, a common model for evaluating the glycemic-control potential of therapeutic compounds and plants extracts in experimental studies. Medicina 53:365–374

92. Zhang X, Lin Q, Chen J et al (2018) High glucose-induced cardiomyocyte death may be linked to unbalanced branched-chain amino acids and energy metabolism. Molecules 23:807
93. Huo JL, Feng Q, Pan S et al (2023) Diabetic cardiomyopathy: early diagnostic biomarkers, pathogenetic mechanisms, and therapeutic interventions. Cell Death Discov 9:256
94. Marino F, Salerno N, Scalise M et al (2023) Streptozotocin-induced type 1 and 2 diabetes mellitus mouse models show different functional, cellular and molecular patterns of diabetic cardiomyopathy. Int J Mol Sci 24:1132
95. Adeghate E, Singh J (2014) Structural changes in the myocardium during diabetes-induced cardiomyopathy. Heart Fail Rev 19:15–23
96. Pan KL, Hsu YC, Chang ST et al (2023) The role of cardiac fibrosis in diabetic cardiomyopathy: from pathophysiology to clinical diagnostic tools. Int J Mol Sci 24:8604
97. Ménard SL, Croteau E, Sarrhini O et al (2010) Abnormal in vivo myocardial energy substrate uptake in diet-induced type 2 diabetic cardiomyopathy in rats. Am J Physiol Endocrinol Metab 298:E1049–E1057
98. Ashrafi R, Davis G (2011) Cardiomyopathy in diabetics: a review of current opinion on the underlying pathological mechanisms. Adv Diabetol 27:175–181
99. Mereweather LJ, Montes Aparicio CN, Heather LC (2020) Positioning metabolism as a central player in the diabetic heart. J Lipid Atheroscler 9:92–109
100. Randle PJ, Garland PB, Hales CN, Newsholme EA (1963) The glucose fatty-acid cycle. Its role in insulin sensitivity and the metabolic disturbances of diabetes mellitus. Lancet 1:785–789
101. Sowton AP, Griffin JL, Murray AJ (2019) Metabolic profiling of the diabetic heart: toward a richer picture. Front Physiol 10:639
102. An D, Rodrigues B (2006) Role of changes in cardiac metabolism in development of diabetic cardiomyopathy. Am J Physiol Heart Circ Physiol 291:H1489–HH506
103. Heather LC, Gopal K, Srnic N et al (2024) Redefining diabetic cardiomyopathy: perturbations in substrate metabolism at the heart of its pathology. Diabetes 73:659–670
104. Belke DD, Larsen TS, Gibbs EM et al (2000) Altered metabolism causes cardiac dysfunction in perfused hearts from diabetic (db/db) mice. Am J Physiol Endocrinol Metab 279:E1104–E1113
105. Mazumder PK, O'Neill BT, Roberts MW et al (2004) Impaired cardiac efficiency and increased fatty acid oxidation in insulin-resistant ob/ob mouse hearts. Diabetes 53:2366–2374
106. Sena CM, Pereira AM, Seiça R (2013) Endothelial dysfunction - a major mediator of diabetic vascular disease. Biochim Biophys Acta 1832:2216–2231
107. Clyne AM (2021) Endothelial response to glucose: dysfunction, metabolism, and transport. Biochem Soc Trans 49:313–325
108. Triggle CR, Ding H, Marei I, Anderson TJ, Hollenberg MD (2020) Why the endothelium? The endothelium as a target to reduce diabetes-associated vascular disease. Can J Physiol Pharmacol 98:415–430
109. Meza CA, La Favor JD, Kim DH et al (2019) Endothelial dysfunction: is there a hyperglycemia-induced imbalance of NOX and NOS? Int J Mol Sci 20:3775
110. Bahadoran Z, Mirmiran P, Kashfi K et al (2023) Vascular nitric oxide resistance in type 2 diabetes. Cell Death Dis 14:410
111. Ozturk Z (2019) Diabetes, oxidative stress and endothelial dysfunction. Bezmialem Sci 7:52–55
112. Li JC, Velagic A, Qin CX, Li M et al (2021) Diabetes attenuates the contribution of endogenous nitric oxide but not nitroxyl to endothelium dependent relaxation of rat carotid arteries. Front Pharmacol 11:585740
113. Stadler K, Jenei V, von Bölcsházy G, Somogyi A, Jakus J (2003) Increased nitric oxide levels as an early sign of premature aging in diabetes. Free Radic Biol Med 35:1240–1251
114. Mutavdžin S (2022) Effects of vitamin B6 and folic acid on the cardiovascular system of rats with diabetes mellitus induced by streptozotocin. Doctoral dissertation, University of Belegrade, Faculty of Medicine

115. Pitocco D, Tesauro M, Alessandro R et al (2013) Oxidative stress in diabetes: implications for vascular and other complications. Int J Mol Sci 14:21525–21550
116. Vareniuk I, Pavlov IA, Drel VR et al (2007) Nitrosative stress and peripheral diabetic neuropathy in leptin-deficient (ob/ob) mice. Exp Neurol 205:425–436
117. van den Born JC, Hammes HP, Greffrath W et al (2016) Diabetic microvascular complications (DIAMICOM). Gasotransmitters in vascular complications of diabetes. Diabetes 65:331–345
118. Kolluru GK, Shen X, Kevil CG (2020) Reactive sulfur species: a new redox player in cardiovascular pathophysiology. Arterioscler Thromb Vasc Biol 40:874–884
119. Hendriks KD, Maassen H, van Dijk PR et al (2019) Gasotransmitters in health and disease: a mitochondria-centered view. Curr Opin Pharmacol 45:87–93
120. Yusuf M, Kwong Huat BT, Hsu A et al (2005) Streptozotocin-induced diabetes in the rat is associated with enhanced tissue hydrogen sulfide biosynthesis. Biochem Biophys Res Commun 333:1146–1152
121. Kundu S, Pushpakumar SB, Tyagi A et al (2013) Hydrogen sulfide deficiency and diabetic renal remodeling: role of matrix metalloproteinase-9. Am J Physiol Endocrinol Metab 304: E1365–E1378
122. Yang G, Tang G, Zhang L et al (2011) The pathogenic role of cystathionine γ-lyase/hydrogen sulfide in streptozotocin-induced diabetes in mice. Am J Pathol 179:869–879
123. Wu L, Yang W, Jia X et al (2009) Pancreatic islet overproduction of H2S and suppressed insulin release in Zucker diabetic rats. Lab Investig 89:59–67
124. Zhang H, Huang Y, Chen S et al (2020) Hydrogen sulfide regulates insulin secretion and insulin resistance in diabetes mellitus, a new promising target for diabetes mellitus treatment? A review. J Adv Res 27:19–30
125. Munteanu C, Rotariu M, Turnea M et al (2022) Recent advances in molecular research on hydrogen sulfide (H_2S) role in diabetes mellitus (DM)-a systematic review. Int J Mol Sci 23: 6720
126. Juin SK, Ouseph R, Gondim DD et al (2023) Diabetic nephropathy and gaseous modulators. Antioxidants 12(5):1088
127. Choi YK, Kim YM (2021) Regulation of endothelial and vascular functions by carbon monoxide *via* crosstalk with nitric oxide. Front Cardiovasc Med 8:649630
128. Braud L, Pini M, Muchova L, Manin S et al (2018) Carbon monoxide-induced metabolic switch in adipocytes improves insulin resistance in obese mice. JCI Insight 3:e123485
129. Nicolai A, Li M, Kim DH et al (2009) Heme oxygenase-1 induction remodels adipose tissue and improves insulin sensitivity in obesity-induced diabetic rats. Hypertension 53:508–515
130. Dubsky M, Veleba J, Sojakova D et al (2023) Endothelial dysfunction in diabetes mellitus: new insights. Int J Mol Sci 24:10705
131. Pilard M, Ollivier EL, Gourdou-Latyszenok V et al (2022) Endothelial cell phenotype, a major determinant of venous thrombo-inflammation. Front Cardiovasc Med 9:864735
132. Vazzana N, Ranalli P, Cuccurullo C et al (2012) Diabetes mellitus and thrombosis. Thromb Res 129:371–377
133. Vaidya AR, Wolska N, Vara D et al (2021) Diabetes and thrombosis: a central role for vascular oxidative stress. Antioxidants 10:706
134. Kaur R, Kaur M, Singh J (2018) Endothelial dysfunction and platelet hyperactivity in type 2 diabetes mellitus: molecular insights and therapeutic strategies. Cardiovasc Diabetol 17:121
135. Fidler TP, Marti A, Gerth K et al (2019) Glucose metabolism is required for platelet hyperactivation in a murine model of type 1 diabetes. Diabetes 68:932–938
136. Hadas K, Randriamboavonjy V, Elgheznawy A et al (2013) Methylglyoxal induces platelet hyperaggregation and reduces thrombus stability by activating PKC and inhibiting PI3K/Akt pathway. PLoS One 8:e74401
137. Stratmann B (2022) Dicarbonyl stress in diabetic vascular disease. Int J Mol Sci 11:6186
138. Batten L, Sathyapalan T, Palmer TM (2023) Molecular mechanisms linking diabetes with increased risk of thrombosis. Int J Mol Sci 24:7465
139. Yngen M (2005) Platelet hyperactivity in diabetes mellitus. Eur Cardiol 1:1–6

140. Peng X, Wang X, Fan M et al (2020) Plasma levels of von Willebrand factor in type 2 diabetes patients with and without cardiovascular diseases: a meta-analysis. Diabetes Metab Res Rev 36:e3193
141. Umadevi B, Roopakala MS, Wilma Silvia CR, Kumar P (2016) Role of von Willebrand factor in type 2 diabetes mellitus patients. J Evol Med Dent Sci 5:6075–6079
142. Ilyas S, Waheed U, Saboor M (2013) Coagulation abnormalities in diabetes mellitus. J Public Health Biol Sci 2:308–312
143. Kearney K, Tomlinson D, Smith K et al (2017) Hypofibrinolysis in diabetes: a therapeutic target for the reduction of cardiovascular risk. Cardiovasc Diabetol 16:34
144. Bryk AH, Konieczynska M, Rostoff P et al (2019) Plasma Protein Oxidation as a determinant of impaired fibrinolysis in type 2 diabetes. Thromb Haemost 119:213–222
145. Kozakova M, Palombo C (2016) Diabetes mellitus, arterial wall, and cardiovascular risk assessment. Int J Environ Res Public Health 13:201
146. Jia G, Aroor AR, DeMarco VG et al (2015) Vascular stiffness in insulin resistance and obesity. Front Physiol 14:231
147. Jia G, Bai H, Mather B et al (2024) Diabetic vasculopathy: molecular mechanisms and clinical insights. Int J Mol Sci 25:804
148. Zieman SJ, Melenovsky V, Kass DA (2005) Mechanisms, pathophysiology, and therapy of arterial stiffness. Arterioscler Thromb Vasc Biol 25:932–943
149. Gao JR, Qin XJ, Fang ZH et al (2019) To explore the pathogenesis of vascular lesion of type 2 diabetes mellitus based on the PI3K/Akt signaling pathway. J Diabetes Res 2019:4650906
150. Man AWC, Zhou Y, Xia N et al (2023) Perivascular adipose tissue oxidative stress in obesity. Antioxidants 12:1595
151. Para I, Albu A, Porojan MD (2021) Adipokines and arterial stiffness in obesity. Medicina 57: 653
152. Jaganathan R, Ravindran R, Dhanasekaran S (2018) Emerging role of adipocytokines in type 2 diabetes as mediators of insulin resistance and cardiovascular disease. Can J Diabetes 42: 446–456
153. Venegas-Pino DE, Lagrotteria A, Wang PW et al (2018) Evidence of extensive atherosclerosis, coronary artery disease and myocardial infarction in the ApoE$^{-/-}$:Ins2$^{+/Akita}$ mouse fed a western diet. Atherosclerosis 275:88–96
154. Venegas-Pino DE, Wang PW, Stoute HK et al (2016) Sex-specific differences in an ApoE$^{-/-}$: Ins2$^{+/Akita}$ mouse model of accelerated atherosclerosis. Am J Pathol 186:67–77
155. Schleicher E, Friess U (2007) Oxidative stress, AGE, and atherosclerosis. Kidney Int Suppl 106:S17–S26
156. Babes EE, Bustea C, Behl T et al (2022) Acute coronary syndromes in diabetic patients, outcome, revascularization, and antithrombotic therapy. Biomed Pharmacother 148:112772
157. Gupta N, Elnour AA, Sadeq A et al (2022) Diabetes and the heart: coronary artery disease. Eur Soc Cardiol 22:10
158. Al-Nozha MM, Ismail HM, Al Nozha OM et al (2016) Coronary artery disease and diabetes mellitus. J Taibah Univ Med Sci 11:330–338
159. Khafaji HA, Suwaidi JM (2014) Atypical presentation of acute and chronic coronary artery disease in diabetics. World J Cardiol 6:802–813
160. Mwebaze RM, Kibirige D (2014) Peripheral arterial disease among adult diabetic patients attending a large outpatient diabetic clinic at a national referral hospital in Uganda: a descriptive cross sectional study. PLoS One 9:e105211
161. Maranta F, Cianfanelli L, Cianflone D (2021) Glycaemic control and vascular complications in diabetes mellitus type 2. Adv Exp Med Biol 1307:129–152
162. Radzioch E, Dąbek B, Balcerczyk-Lis M et al (2024) Diabetic cardiomyopathy-from basics through diagnosis to treatment. Biomedicines 12:765
163. Zhao X, Liu S, Wang X et al (2022) Diabetic cardiomyopathy: clinical phenotype and practice. Front Endocrinol 13:1032268

164. Seferović PM, Paulus WJ (2015) Clinical diabetic cardiomyopathy: a two-faced disease with restrictive and dilated phenotypes. Eur Heart J 36:1718–1727
165. Elendu C, Amaechi DC, Elendu TC et al (2023) Heart failure and diabetes: understanding the bidirectional relationship. Medicine 102:e34906
166. Hoek AG, Dal Canto E, Wenker E et al (2024) Epidemiology of heart failure in diabetes: a disease in disguise. Diabetologia 67:574–601
167. Oktay AA, Paul TK, Koch CA et al (2023) Diabetes, cardiomyopathy, and heart failure. In: Feingold KR, Anawalt B, Blackman MR et al (eds) Endotext. MDText.com, Inc., South Dartmouth. https://www.ncbi.nlm.nih.gov/books/NBK560257/. Accessed 14 June 2024
168. Shou Y, Li X, Fang Q et al (2024) Progress in the treatment of diabetic cardiomyopathy, a systematic review. Pharmacol Res Perspect 12:e1177
169. Turnbull FM, Abraira C, Anderson RJ et al (2009) Intensive glucose control and macrovascular outcomes in type 2 diabetes. Diabetologia 52:2470
170. Kaul S, Diamond GA (2008) Rosiglitazone and cardiovascular risk. Curr Atheroscler Rep 10: 398–404
171. Sharma A, Pagidipati NJ, Califf RM et al (2020) Impact of regulatory guidance on evaluating cardiovascular risk of new glucose-lowering therapies to treat type 2 diabetes mellitus: lessons learned and future directions. Circulation 141:843–862
172. Schnell O, Barnard-Kelly K, Battelino T et al (2024) CVOT summit report 2023: new cardiovascular, kidney, and metabolic outcomes. Cardiovasc Diabetol 23:104
173. Xiang B, Zhao X, Zhou X (2021) Cardiovascular benefits of sodium-glucose cotransporter 2 inhibitors in diabetic and nondiabetic patients. Cardiovasc Diabetol 20:78
174. Cowie MR, Fisher M (2020) SGLT2 inhibitors: mechanisms of cardiovascular benefit beyond glycaemic control. Nat Rev Cardiol 17:761–772
175. Gao YM, Feng ST, Wen Y et al (2022) Cardiorenal protection of SGLT2 inhibitors-perspectives from metabolic reprogramming. EBioMedicine 83:104215
176. Myasoedova VA, Bozzi M, Valerio V et al (2023) Anti-inflammation and anti-oxidation: the key to unlocking the cardiovascular potential of SGLT2 inhibitors and GLP1 receptor agonists. Antioxidants 13:16
177. Yaribeygi H, Atkin SL, Butler AE et al (2019) Sodium-glucose cotransporter inhibitors and oxidative stress: an update. J Cell Physiol 234:3231–3237
178. Preda A, Montecucco F, Carbone F et al (2024) SGLT2 inhibitors: from glucose-lowering to cardiovascular benefits. Cardiovasc Res 120:443–460
179. Shin SJ, Chung S, Kim SJ et al (2016) Effect of sodium–glucose co-transporter 2 inhibitor, dapagliflozin, on renal renin–angiotensin system in an animal model of type 2 diabetes. PLoS One 11:e0165703
180. Heyman SN, Khamaisi M, Rosen S et al (2017) Potential hypoxic renal injury in patients with diabetes on SGLT2 inhibitors: caution regarding concomitant use of NSAIDs and iodinated contrast media. Diabetes Care 40:e40–e41
181. Rovira-Llopis S, Bañuls C, Diaz-Morales N et al (2017) Mitochondrial dynamics in type 2 diabetes: pathophysiological implications. Redox Biol 11:637–645
182. Zhou B, Tian R (2018) Mitochondrial dysfunction in pathophysiology of heart failure. J Clin Invest 128:3716–3726
183. Mone P, Varzideh F, Jankauskas SS et al (2022) SGLT2 inhibition via empagliflozin improves endothelial function and reduces mitochondrial oxidative stress: insights from frail hypertensive and diabetic patients. Hypertension 79:1633–1643
184. Teshima Y, Takahashi N, Nishio S et al (2014) Production of reactive oxygen species in the diabetic heart. Roles of mitochondria and NADPH oxidase. Circ J 78:300–306
185. Packer M (2017) Activation and inhibition of sodium-hydrogen exchanger is a mechanism that links the pathophysiology and treatment of diabetes mellitus with that of heart failure. Circulation 136:1548–1559

186. Baartscheer A, Schumacher CA, Wüst RC et al (2017) Empagliflozin decreases myocardial cytoplasmic Na^+ through inhibition of the cardiac Na^+/H^+ exchanger in rats and rabbits. Diabetologia 60:568–573
187. Salim HM, Fukuda D, Yagi S et al (2016) Glycemic control with ipragliflozin, a novel selective SGLT2 inhibitor, ameliorated endothelial dysfunction in streptozotocin-induced diabetic mouse. Front Cardiovasc Med 3:43
188. Scisciola L, Cataldo V, Taktaz F et al (2022) Anti-inflammatory role of SGLT2 inhibitors as part of their anti-atherosclerotic activity: data from basic science and clinical trials. Front Cardiovasc Med 9:1008922
189. Alsereidi FR, Khashim Z, Marzook H et al (2024) Targeting inflammatory signaling pathways with SGLT2 inhibitors: Insights into cardiovascular health and cardiac cell improvement. Curr Probl Cardiol 49:102524
190. Chen B, Guo J, Ye H et al (2024) Role and molecular mechanisms of SGLT2 inhibitors in pathological cardiac remodeling (review). Mol Med Rep 29:73
191. Ye J, Li L, Wang M et al (2022) Diabetes mellitus promotes the development of atherosclerosis: the role of NLRP3. Front Immunol 13:900254
192. La Grotta R, de Candia P, Olivieri F et al (2022) Anti-inflammatory effect of SGLT-2 inhibitors via uric acid and insulin. Cell Mol Life Sci 79:273
193. Su S, Ji X, Li T et al (2023) The changes of cardiac energy metabolism with sodium-glucose transporter 2 inhibitor therapy. Front Cardiovasc Med 10:1291450
194. Zeng Y, Li Y, Jiang W, Hou N (2024) Molecular mechanisms of metabolic dysregulation in diabetic cardiomyopathy. Front Cardiovasc Med 11:1375400
195. Lauritsen KM, Nielsen BRR, Tolbod LP et al (2021) SGLT2 inhibition does not affect myocardial fatty acid oxidation or uptake, but reduces myocardial glucose uptake and blood flow in individuals with type 2 diabetes: a randomized double-blind, placebo-controlled crossover trial. Diabetes 70:800–808
196. Kaplan A, Abidi E, El-Yazbi A et al (2018) Direct cardiovascular impact of SGLT2 inhibitors: mechanisms and effects. Heart Fail Rev 23:419–437
197. Lopaschuk GD, Verma S (2020) Mechanisms of cardiovascular benefits of sodium glucose co-transporter 2 (SGLT2) inhibitors: a state-of-the-art review. JACC Basic Transl Sci 5:632–644
198. Zinman B, Wanner C, Lachin JM et al (2015) Empagliflozin, cardiovascular outcomes, and mortality in type 2 diabetes. N Engl J Med 373:2117–2128
199. Marilly E, Cottin J, Cabrera N et al (2022) SGLT2 inhibitors in type 2 diabetes: a systematic review and meta-analysis of cardiovascular outcome trials balancing their risks and benefits. Diabetologia 65:2000–2010
200. Neal B, Perkovic V, de Zeeuw D et al (2013) Rationale, design, and baseline characteristics of the Canagliflozin Cardiovascular Assessment Study (CANVAS)–a randomized placebo-controlled trial. Am Heart J 166:217–223
201. McMurray JJV, Solomon SD, Inzucchi SE et al (2019) Dapagliflozin in patients with heart failure and reduced ejection fraction. N Engl J Med 381:1995–2008
202. Brunton SA, Wysham CH (2020) GLP-1 receptor agonists in the treatment of type 2 diabetes: role and clinical experience to date. Postgrad Med 132:3–14
203. Andreasen CR, Andersen A, Knop FK et al (2021) How glucagon-like peptide 1 receptor agonists work. Endocr Connect 10:R200–R212
204. Ferhatbegović L, Mršić D, Macić-Džanković A (2023) The benefits of GLP1 receptors in cardiovascular diseases. Front Clin Diabetes Health 4:1293926
205. Marx N, Husain M, Lehrke M et al (2022) GLP-1 receptor agonists for the reduction of atherosclerotic cardiovascular risk in patients with type 2 diabetes. Circulation 146:1882–1894
206. Antza C, Nirantharakumar K, Doundoulakis I et al (2019) The development of an oral GLP-1 receptor agonist for the management of type 2 diabetes: evidence to date. Drug Des Devel Ther 13:2985–2996

207. Jensterle M, Rizzo M, Haluzík M et al (2022) Efficacy of GLP-1 RA approved for weight management in patients with or without diabetes: a narrative review. Adv Ther 39:2452–2467
208. Wu Q, Li D, Huang C et al (2022) Glucose control independent mechanisms involved in the cardiovascular benefits of glucagon-like peptide-1 receptor agonists. Biomed Pharmacother 153:113517
209. Nesti L, Trico D (2024) Cardioprotective effects of glucagon-like peptide 1 receptor agonists in heart failure: myth or truth? World J Diabetes 15:818–822
210. Gallo G, Volpe M (2024) Potential mechanisms of the protective effects of the cardiometabolic drugs type-2 sodium–glucose transporter inhibitors and glucagon-like peptide-1 receptor agonists in heart failure. Int J Mol Sci 25:2484
211. Petersen KE, Rakipovski G, Raun K et al (2016) Does glucagon-like peptide-1 ameliorate oxidative stress in diabetes? Evidence based on experimental and clinical studies. Curr Diabetes Rev 12:331–358
212. Li P, Tang Z, Wang L et al (2017) Glucagon-like peptide-1 analogue liraglutide ameliorates atherogenesis via inhibiting advanced glycation end product-induced receptor for advanced glycosylation end product expression in apolipoprotein-E deficient mice. Mol Med Rep 16: 3421–3426
213. Bretón-Romero R, Weisbrod RM, Feng B et al (2018) Liraglutide treatment reduces endothelial endoplasmic reticulum stress and insulin resistance in patients with diabetes mellitus. J Am Heart Assoc 7:e009379
214. Ahmad E, Waller HL, Sargeant JA et al (2021) Effects of liraglutide versus sitagliptin on circulating cardiovascular biomarkers, including circulating progenitor cells, in individuals with type 2 diabetes and obesity: analyses from the LYDIA trial. Diabetes Obes Metab 23: 1409–1414
215. Solini A, Tricò D, Del Prato S et al (2023) Incretins and cardiovascular disease: to the heart of type 2 diabetes? Diabetologia 66:1820–1831
216. Gerstein HC, Colhoun HM, Dagenais GR et al (2019) Dulaglutide and cardiovascular outcomes in type 2 diabetes (REWIND): a double-blind, randomised placebo-controlled trial. Lancet 394(10193):121–130
217. Zinman B, Nauck MA, Bosch-Traberg H et al (2018) Liraglutide and glycaemic outcomes in the LEADER trial. Diabetes Ther 9:2383–2392
218. Marso SP, Bain SC, Consoli A et al (2016) Semaglutide and cardiovascular outcomes in patients with type 2 diabetes. N Engl J Med 375:1834–1844
219. Pillai P, Modarressi T (2024) GLP1RAs in clinical practice: therapeutic advances and safety perspectives. American College of Cardiology, Washington
220. Pahud de Mortanges A, Sinaci E, Salvador D Jr et al (2022) GLP-1 receptor agonists and coronary arteries: from mechanisms to events. Front Pharmacol 13:856111
221. Trombara F, Cosentino N, Bonomi A et al (2023) Impact of chronic GLP-1 RA and SGLT-2I therapy on in-hospital outcome of diabetic patients with acute myocardial infarction. Cardiovasc Diabetol 22:26
222. Lønborg J, Vejlstrup N, Kelbæk H et al (2012) Exenatide reduces reperfusion injury in patients with ST-segment elevation myocardial infarction. Eur Heart J 33:1491–1499
223. Ogbu IR, Ngwudike C, Lal K et al (2021) Role of glucagon-like peptide-1 agonist in patients undergoing percutaneous coronary intervention or coronary artery bypass grafting: a meta-analysis. Am Heart J Plus 11:100063
224. Chowdhary A, Thirunavukarasu S, Joseph T et al (2024) Liraglutide improves myocardial perfusion and energetics and exercise tolerance in patients with type 2 diabetes. JACC 84:540–557
225. Sattar N, Lee MMY, Kristensen SL et al (2021) Cardiovascular, mortality, and kidney outcomes with GLP-1 receptor agonists in patients with type 2 diabetes: a systematic review and meta-analysis of randomised trials. Lancet Diabetes Endocrinol 9:653–662

226. Ferreira JP, Saraiva F, Sharma A et al (2023) Glucagon-like peptide 1 receptor agonists in patients with type 2 diabetes with and without chronic heart failure: a meta-analysis of randomized placebo-controlled outcome trials. Diabetes Obes Metab 25:1495–1502
227. Neves JS, Packer M, Ferreira JP (2023) Increased risk of heart failure hospitalization with GLP-1 receptor agonists in patients with reduced ejection fraction: a meta-analysis of the EXSCEL and FIGHT trials. J Card Fail 29:1107–1109
228. Jorsal A, Kistorp C, Holmager P et al (2017) Effect of liraglutide, a glucagon-like peptide-1 analogue, on left ventricular function in stable chronic heart failure patients with and without diabetes (LIVE)-a multicentre, double-blind, randomised, placebo-controlled trial. Eur J Heart Fail 19:69–77
229. Sanchez-Rangel E, Inzucchi SE (2017) Metformin: clinical use in type 2 diabetes. Diabetologia 60:1586–1593
230. Feng J, Wang X, Ye X et al (2022) Mitochondria as an important target of metformin: the mechanism of action, toxic and side effects, and new therapeutic applications. Pharmacol Res 177:106114
231. Pernicova I, Korbonits M (2014) Metformin–mode of action and clinical implications for diabetes and cancer. Nat Rev Endocrinol 10:143–156
232. Rena G, Hardie DG, Pearson ER (2017) The mechanisms of action of metformin. Diabetologia 60:1577–1585
233. Vial G, Detaille D, Guigas B (2019) Role of mitochondria in the mechanism(s) of action of metformin. Front Endocrinol 10:294
234. Apostolova N, Iannantuoni F, Gruevska A et al (2020) Mechanisms of action of metformin in type 2 diabetes: effects on mitochondria and leukocyte-endothelium interactions. Redox Biol 34:101517
235. Mohammed I, Hollenberg MD, Ding H et al (2021) A critical review of the evidence that metformin is a putative anti-aging drug that enhances healthspan and extends lifespan. Front Endocrinol 12:718942
236. King P, Peacock I, Donnelly R (1999) The UK prospective diabetes study (UKPDS): clinical and therapeutic implications for type 2 diabetes. Br J Clin Pharmacol 48:643–648
237. Yu SF, Hong CT, Chen WT et al (2023) Metformin adherence and the risk of cardiovascular disease: a population-based cohort study. Ther Adv Chronic Dis 14:20406223231163115
238. Han Y, Xie H, Liu Y et al (2019) Effect of metformin on all-cause and cardiovascular mortality in patients with coronary artery diseases: a systematic review and an updated meta-analysis. Cardiovasc Diabetol 18:96
239. Dihoum A, Rena G, Pearson ER et al (2023) Metformin: evidence from preclinical and clinical studies for potential novel applications in cardiovascular disease. Expert Opin Investig Drugs 32:291–299
240. Diniz Vilela D, Gomes Peixoto L, Teixeira RR et al (2016) The role of metformin in controlling oxidative stress in muscle of diabetic rats. Oxidative Med Cell Longev 2016: 6978625
241. Chukwunonso Obi B, Chinwuba Okoye T, Okpashi VE et al (2016) Comparative study of the antioxidant effects of metformin, glibenclamide, and repaglinide in alloxan-induced diabetic rats. J Diabetes Res 2016:1635361
242. Liu XD, Li YG, Wang GY et al (2020) Metformin protects high glucose-cultured cardiomyocytes from oxidative stress by promoting NDUFA13 expression and mitochondrial biogenesis via the AMPK signaling pathway. Mol Med Rep 22:5262–5270
243. Esteghamati A, Eskandari D, Mirmiranpour H et al (2013) Effects of metformin on markers of oxidative stress and antioxidant reserve in patients with newly diagnosed type 2 diabetes: a randomized clinical trial. Clin Nutr 32:179–185
244. Bulatova N, Kasabri V, Qotineh A et al (2018) Effect of metformin combined with lifestyle modification versus lifestyle modification alone on proinflammatory-oxidative status in drug-naïve pre-diabetic and diabetic patients: a randomized controlled study. Diabetes Metab Syndr 12:257–267

245. Schernthaner G, Brand K, Bailey CJ (2022) Metformin and the heart: update on mechanisms of cardiovascular protection with special reference to comorbid type 2 diabetes and heart failure. Metabolism 130:155160
246. Adeshara KA, Bangar NS, Doshi PR et al (2020) Action of metformin therapy against advanced glycation, oxidative stress and inflammation in type 2 diabetes patients: 3 months follow-up study. Diabetes Metab Syndr 14:1449–1458
247. Ionică LN, Gaiță L, Bînă AM et al (2021) Metformin alleviates monoamine oxidase-related vascular oxidative stress and endothelial dysfunction in rats with diet-induced obesity. Mol Cell Biochem 476:4019–4029
248. Saisho Y (2015) Metformin and inflammation: its potential beyond glucose-lowering effect. Endocr Metab Immune Disord Drug Targets 15:196–205
249. Chiang CF, Chao TT, Su YF et al (2017) Metformin-treated cancer cells modulate macrophage polarization through AMPK-NF-κB signaling. Oncotarget 8:20706–20718
250. Sakata N (2024) The anti-inflammatory effect of metformin: the molecular targets. Genes Cells 29:183–191
251. Vettor R, Valerio A, Ragni M et al (2014) Exercise training boosts eNOS-dependent mitochondrial biogenesis in mouse heart: role in adaptation of glucose metabolism. Am J Physiol Endocrinol Metab 306:E519–E528
252. Nesti L, Natali A (2017) Metformin effects on the heart and the cardiovascular system: a review of experimental and clinical data. Nutr Metab Cardiovasc Dis 27:657–669
253. Loi H, Kramar S, Laborde C et al (2021) Metformin attenuates postinfarction myocardial fibrosis and inflammation in mice. Int J Mol Sci 22:9393
254. Zakaria EM, Tawfeek WM, Hassanin MH et al (2022) Cardiovascular protection by DPP-4 inhibitors in preclinical studies: an updated review of molecular mechanisms. Naunyn Schmiedeberg's Arch Pharmacol 395:1357–1372
255. Wu L, Wang K, Wang W et al (2018) Glucagon-like peptide-1 ameliorates cardiac lipotoxicity in diabetic cardiomyopathy via the PPARα pathway. Aging Cell 17:e12763
256. Matsubara J, Sugiyama S, Akiyama E et al (2013) Dipeptidyl peptidase-4 inhibitor, sitagliptin, improves endothelial dysfunction in association with its anti-inflammatory effects in patients with coronary artery disease and uncontrolled diabetes. Circ J 77:1337–1344
257. White WB, Cannon CP, Heller SR et al (2013) Alogliptin after acute coronary syndrome in patients with type 2 diabetes. N Engl J Med 369:1327–1335
258. Scirica BM, Bhatt DL, Braunwald E et al (2013) Saxagliptin and cardiovascular outcomes in patients with type 2 diabetes mellitus. N Engl J Med 369:1317–1326
259. Green JB, Bethel MA, Armstrong PW et al (2015) Effect of sitagliptin on cardiovascular outcomes in type 2 diabetes. N Engl J Med 373:232–242
260. Patoulias DI, Boulmpou A, Teperikidis E et al (2021) Cardiovascular efficacy and safety of dipeptidyl peptidase-4 inhibitors: a meta-analysis of cardiovascular outcome trials. World J Cardiol 13:585–592
261. Enzan N, Matsushima S, Kaku H et al (2023) Beneficial effects of dipeptidyl peptidase-4 inhibitors on heart failure with preserved ejection fraction and diabetes. JACC Asia 3:93–104
262. Thulé PM, Umpierrez G (2014) Sulfonylureas: a new look at old therapy. Curr Diab Rep 14:473
263. Lee TTL, Hui JMH, Lee YHA et al (2022) Sulfonylurea is associated with higher risks of ventricular arrhythmia or sudden cardiac death compared with metformin: a population-based cohort study. J Am Heart Assoc 11:e026289
264. Filion KB, Douros A, Azoulay L et al (2019) Sulfonylureas as initial treatment for type 2 diabetes and the risk of adverse cardiovascular events: a population-based cohort study. Br J Clin Pharmacol 85:2378–2389
265. Wang H, Cordiner RLM, Huang Y et al (2023) Cardiovascular safety in type 2 diabetes with sulfonylureas as second-line drugs: a nationwide population-based comparative safety study. Diabetes Care 46:967–977

266. Xue CY, Zhou MQ, Zheng QY et al (2022) Thiazolidinediones play a positive role in the vascular endothelium and inhibit plaque progression in diabetic patients with coronary atherosclerosis: a systematic review and meta-analysis. Front Cardiovasc Med 9:1043406
267. Ahmad BA, Sanghani IM, Sayabugari R et al (2023) Beyond blood sugar: investigating the cardiovascular effects of antidiabetic drugs. Cureus 15:e46373
268. Zhang Z, Zhang X, Meng L et al (2021) Pioglitazone inhibits diabetes-induced atrial mitochondrial oxidative stress and improves mitochondrial biogenesis, dynamics, and function through the PPAR-γ/PGC-1α signaling pathway. Front Pharmacol 12:658362
269. Zhu J, Yu X, Zheng Y et al (2020) Association of glucose-lowering medications with cardiovascular outcomes: an umbrella review and evidence map. Lancet Diabetes Endocrinol 8:192–205
270. Eggleton JS, Jialal I (2023) Thiazolidinediones. In: StatPearls. StatPearls Publishing, Treasure Island. https://www.ncbi.nlm.nih.gov/books/NBK551656/
271. Ng KW, Allen ML, Desai A et al (2012) Cardioprotective effects of insulin: how intensive insulin therapy may benefit cardiac surgery patients. Circulation 125:721–728
272. Fu J, Yu MG, Li Q (2021) Insulin's actions on vascular tissues: physiological effects and pathophysiological contributions to vascular complications of diabetes. Mol Metab 52:101236
273. Liu J, Hu X (2022) Impact of insulin therapy on outcomes of diabetic patients with heart failure: a systematic review and meta-analysis. Diab Vasc Dis Res 19:14791641221093175
274. Mannucci E, Targher G, Nreu B et al (2022) Effects of insulin on cardiovascular events and all-cause mortality in patients with type 2 diabetes: a meta-analysis of randomized controlled trials. Nutr Metab Cardiovasc Dis 32:1353–1360
275. Choudhary A, Rawat U, Kumar P, Mittal P (2023) Pleotropic effects of statins: the dilemma of wider utilization of statin. Egypt Heart J 75:1
276. Kosmas CE, Silverio D, Sourlas A et al (2018) Impact of lipid-lowering therapy on glycemic control and the risk for new-onset diabetes mellitus. Drugs Context 7:212562
277. Ward NC, Watts GF, Eckel RH (2019) Statin toxicity. Circ Res 124:328–350
278. Chen WH, Chen CH, Hsu MC et al (2024) Advances in the molecular mechanisms of statins in regulating endothelial nitric oxide bioavailability: interlocking biology between eNOS activity and L-arginine metabolism. Biomed Pharmacother 171:116192
279. Landmesser U, Engberding N, Bahlmann FH et al (2004) Statin-induced improvement of endothelial progenitor cell mobilization, myocardial neovascularization, left ventricular function, and survival after experimental myocardial infarction requires endothelial nitric oxide synthase. Circulation 110:1933–1939
280. Siniscalchi C, Basaglia M, Riva M et al (2023) Statins effects on blood clotting: a review. Cells 12:2719
281. Marchi Cappelletti R (2011) Statins therapy: effects on plasma fibrinogen levels and fibrinolysis. J Nutr Disorders Ther S6:001
282. Sahebkar A, Serban C, Ursoniu S et al (2016) The impact of statin therapy on plasma levels of von Willebrand factor antigen. Systematic review and meta-analysis of randomised placebo-controlled trials. Thromb Haemost 115:520–532
283. German CA, Liao JK et al (2023) Understanding the molecular mechanisms of statin pleiotropic effects. Arch Toxicol 97:1529–1545
284. Jamialahmadi T, Simental-Mendia LE, Eid AH et al (2024) Efficacy of statin therapy in reducing epicardial adipose tissue: a systematic review and meta-analysis. Arch Med Sci 20: 997–1001
285. Al-hussaniy HA, Al-Shammari AH, Naji Sameer H et al (2023) The relationship between statin therapy and adipocytokine/inflammatory mediators in dyslipidemic nondiabetic patients: a comparative study. Pharmacia 70(3):581–585
286. Low Wang CC, Hess CN, Hiatt WR et al (2016) Clinical update: cardiovascular disease in diabetes mellitus atherosclerotic cardiovascular disease and heart failure in type 2 diabetes mellitus—mechanisms, management, and clinical considerations. Circulation 133:2459–2502

287. Ridker PM, Danielson E, Fonseca FAH et al (2008) Rosuvastatin to prevent vascular events in men and women with elevated C-reactive protein. N Engl J Med 359:2195–2207
288. Laakso M, Fernandes Silva L (2023) Statins and risk of type 2 diabetes: mechanism and clinical implications. Front Endocrinol 14:1239335
289. Thakker D, Nair S, Pagada A et al (2016) Statin use and the risk of developing diabetes: a network meta-analysis. Pharmacoepidemiol Drug Saf 25:1131–1149
290. Velidakis N, Stachteas P, Gkougkoudi E et al (2024) Classical and novel lipid-lowering therapies for diabetic patients with established coronary artery disease or high risk of coronary artery disease-a narrative clinical review. Pharmaceuticals 17:568
291. Cannon CP, Blazing MA, Giugliano RP et al (2015) Ezetimibe added to statin therapy after acute coronary syndromes. N Engl J Med 372:2387–2397
292. Crea F, Niccoli G (2015) Ezetimibe and plaque regression: cholesterol lowering or pleiotropic effects? JACC 66:508–510
293. Pradhan AD, Glynn RJ, Fruchart JC et al (2022) Triglyceride lowering with pemafibrate to reduce cardiovascular risk. N Engl J Med 387:1923–1934
294. Kim KA, Kim NJ, Choo EH (2024) The effect of fibrates on lowering low-density lipoprotein cholesterol and cardiovascular risk reduction: a systemic review and meta-analysis. Eur J Prev Cardiol 31:291–301
295. Page MM, Watts GF (2016) PCSK9 inhibitors - mechanisms of action. Aust Prescr 39:164–167
296. Chen H, Chen X (2023) PCSK9 inhibitors for acute coronary syndrome: the era of early implementation. Front Cardiovasc Med 10:1138787
297. Ray KK, Raal FJ, Kallend DG et al (2023) Inclisiran and cardiovascular events: a patient-level analysis of phase III trials. Eur Heart J 44:129–138
298. Shah AK, Dhalla NS (2021) Effectiveness of some vitamins in the prevention of cardiovascular disease: a narrative review. Front Physiol 12:729255
299. Raghuvanshi DS, Chakole S, Kumar M (2023) Relationship between vitamins and diabetes. Cureus 15:e36815
300. Institute of Medicine (US) Standing Committee on the Scientific Evaluation of Dietary Reference Intakes and its Panel on Folate, Other B Vitamins, and Choline (1998) Dietary reference intakes for thiamin, riboflavin, niacin, vitamin B6, Folate, vitamin B12, pantothenic acid, biotin, and choline. National Academies Press, Washington. Available from https://www.ncbi.nlm.nih.gov/books/NBK114313/
301. Stach K, Stach W, Augoff K (2021) Vitamin B6 in health and disease. Nutrients 13:3229
302. Huang J, Khatun P, Xiong Y et al (2023) Intakes of folate, vitamin B6, and vitamin B12 and cardiovascular disease risk: a national population-based cross-sectional study. Front Cardiovasc Med 10:1237103
303. Minović I, Kieneker LM, Gansevoort RT et al (2020) Vitamin B6, inflammation, and cardiovascular outcome in a population-based cohort: the prevention of renal and vascular end-stage disease (PREVEND) study. Nutrients 12:2711
304. Mascolo E, Vernì F (2020) Vitamin B6 and diabetes: relationship and molecular mechanisms. Int J Mol Sci 21:3669
305. D'Haese S, Claes L, Jaeken E et al (2024) Pyridoxamine alleviates cardiac fibrosis and oxidative stress in western diet-induced prediabetic rats. Int J Mol Sci 25:8508
306. Dhalla NS, Takeda S, Elimban V (2013) Mechanisms of the beneficial effects of vitamin B6 and pyridoxal 5-phosphate on cardiac performance in ischemic heart disease. Clin Chem Lab Med 51:535–543
307. Hsu CC, Cheng CH, Hsu CL et al (2015) Role of vitamin B6 status on antioxidant defenses, glutathione, and related enzyme activities in mice with homocysteine-induced oxidative stress. Food Nutr Res 59:25702
308. Mikkelsen K, Dargahi N, Fraser S et al (2023) High-dose vitamin B6 (pyridoxine) displays strong anti-inflammatory properties in lipopolysaccharide-stimulated monocytes. Biomedicines 11:2578

309. Troen AM (2012) Folate and vitamin B12: function and importance in cognitive development. Nestle Nutr Inst Workshop Ser 70:161–171
310. Pawlak R (2015) Is vitamin B12 deficiency a risk factor for cardiovascular disease in vegetarians? Am J Prev Med 48:e11–e26
311. Herrmann W, Herrmann M (2022) The controversial role of HCY and vitamin B deficiency in cardiovascular diseases. Nutrients 14:1412
312. Toole JF, Malinow MR, Chambless LE et al (2004) Lowering homocysteine in patients with ischemic stroke to prevent recurrent stroke, myocardial infarction, and death: the Vitamin Intervention for Stroke Prevention (VISP) randomized controlled trial. JAMA 291:565–575
313. Schnyder G, Roffi M, Flammer Y, Pin R, Hess OM (2002) Effect of homocysteine-lowering therapy with folic acid, vitamin B12, and vitamin B6 on clinical outcome after percutaneous coronary intervention: the Swiss Heart study: a randomized controlled trial. JAMA 288:973–979
314. Kibirige D, Mwebaze R (2013) Vitamin B12 deficiency among patients with diabetes mellitus: is routine screening and supplementation justified? J Diabetes Metab Disord 12:17
315. Alvarez M, Sierra OR, Saavedra G et al (2019) Vitamin B12 deficiency and diabetic neuropathy in patients taking metformin: a cross-sectional study. Endocr Connect 8:1324–1329
316. Pilz S, Tomaschitz A, März W et al (2011) Vitamin D, cardiovascular disease and mortality. Clin Endocrinol 75:575–584
317. Hung M, Birmingham WC, Ocampo M et al (2023) The role of vitamin D in cardiovascular diseases. Nutrients 15:3547
318. de la Guía-Galipienso F, Martínez-Ferran M, Vallecillo N et al (2021) Vitamin D and cardiovascular health. Clin Nutr 40:2946–2957
319. Mancuso P, Rahman A, Hershey SD et al (2008) 1,25-dihydroxyvitamin-D3 treatment reduces cardiac hypertrophy and left ventricular diameter in spontaneously hypertensive heart failure-prone (cp/+) rats independent of changes in serum leptin. J Cardiovasc Pharmacol 51:559–564
320. Isa Z, Amsah N, Ahmad N (2023) The impact of vitamin D deficiency and insufficiency on the outcome of type 2 diabetes mellitus patients: a systematic review. Nutrients 15:2310
321. Rusnak IT (2015) Diabetes mellitus and some microelements. Int J Endocrinol 11:36–38
322. Wechselberger C, Messner B, Bernhard D (2023) The role of trace elements in cardiovascular diseases. Toxics 11:956
323. Marreiro DDN, Cruz KJC, Morais JBS et al (2017) Zinc and oxidative stress: current mechanisms. Antioxidants 6:24
324. Cai L, Tan Y, Watson S et al (2023) Diabetic cardiomyopathy - zinc preventive and therapeutic potentials by its anti-oxidative stress and sensitizing insulin signaling pathways. Toxicol Appl Pharmacol 477:116694
325. Wang S, Wang B, Wang Y et al (2017) Zinc prevents the development of diabetic cardiomyopathy in db/db mice. Int J Mol Sci 13:580
326. Ranasinghe P, Wathurapatha WS, Galappatthy P et al (2018) Zinc supplementation in prediabetes: a randomized double-blind placebo-controlled clinical trial. J Diabetes 10:386–397
327. MacKenzie S, Bergdahl A (2022) Zinc homeostasis in diabetes mellitus and vascular complications. Biomedicines 10:139
328. Bo S, Pisu E (2008) Role of dietary magnesium in cardiovascular disease prevention, insulin sensitivity and diabetes. Curr Opin Lipidol 19:50–56
329. Al Alawi AM, Majoni SW, Falhammar H et al (2018) Magnesium and human health: perspectives and research directions. Int J Endocrinol 2018:9041694
330. Kolte D, Vijayaraghavan K, Khera S et al (2014) Role of magnesium in cardiovascular diseases. Cardiol Rev 22:182–192
331. Oost LJ, Tack CJ, de Baaij JHF (2023) Hypomagnesemia and cardiovascular risk in type 2 diabetes. Endocr Rev 44:357–378
332. Barbagallo M, Dominguez LJ et al (2015) Magnesium and type 2 diabetes. World J Diabetes 6:1152–1157

333. Alkhalidi F (2023) A comparative study to assess the use of chromium in type 2 diabetes mellitus. J Med Life 16:1178–1182
334. Balk EM, Tatsioni A, Lichtenstein AH et al (2007) Effect of chromium supplementation on glucose metabolism and lipids: a systematic review of randomized controlled trials. Diabetes Care 30:2154–2163
335. Masharani U, Gjerde C, McCoy S et al (2012) Chromium supplementation in non-obese non-diabetic subjects is associated with a decline in insulin sensitivity. BMC Endocr Disord 12:31

Chapter 10
Diabetic Foot Ulcer and Cardiac Autonomic Neuropathy

Vikrant Rai

Abstract Diabetes, a chronic inflammatory disease and a constituent of metabolic disorder has multiple complications including retinopathy, nephropathy, angiopathy, gastroparesis, diabetic foot ulcers, atherosclerosis, increased propensity to infections, hypogonadism, psychological issues, and cardiac autonomic neuropathy. Diabetic foot ulcers (DFU) are a debilitating socioeconomic clinical issue resulting in increased amputation rate, morbidity, and mortality. Cardiac autonomic neuropathy (CAN), a chronic condition, occurs in people with type 1 and type 2 diabetes. CAN damages the nerves controlling heart and blood vessels leading to abnormal heart rate and blood circulation. Studies have supported the notion that CAN increases the incidence of DFU and patients with CAN and DFU have increased mortality. Both DFU and CAN are associated with diabetes and patients with DFUs have a higher prevalence of CAN. However, the literature discussing the association between DFUs and CAN and how each other affects the outcome is scarce. This chapter discusses various aspects of DFU and CAN, the importance of screening in the presence of CAN or diabetes to prevent DFU development, followed by the strategies to manage when they coexist.

Keywords Diabetes · Diabetic foot ulcer · Cardiac autonomic neuropathy · Screening

Introduction

Diabetic foot ulcer (DFU) and cardiac autonomic neuropathy (CAN), both are complications of diabetes mellitus [1]. The global prevalence of diabetes is estimated to rise to 10.2% and 10.9% of total world population by 2030 and 2045, respectively [2]. For type II diabetes, the global prevalence is projected to increase to 7079

V. Rai (✉)
Department of Translational Research, Western University of Health Sciences, Pomona, CA, USA
e-mail: vrai@westernu.edu

D. K. Agrawal et al. (eds.), *Cardiovascular Toxicity*, Advances in Biochemistry in Health and Disease 33, https://doi.org/10.1007/978-3-031-90598-8_10

individuals per 100,000 by 2030 [3]. Not only in adults, but childhood diabetes is also rising, and the global incidence rate increased from 9.31 to 11.61 between 1990 to 2019 [4]. In the United States, as per the CDC, 11.6% of the population was suffering from diabetes and 14.7% of all US adults had diabetes in 2021. Clinically, these figures are worrisome because the increasing prevalence of diabetes will increase the costs of treatment not only for diabetes but also for associated complications including microangiopathy (hypertension, atherosclerosis, coronary artery disease, strokes, and erectile dysfunction), retinopathy, nephropathy, neuropathy, diabetic foot ulcers (DFUs), myopathy, increased susceptibility to infections, osteoporosis, arthropathies, and liver damage [5]. Of these complications, diabetic neuropathy is important because it is not only associated with DFUs but also with other cardiac complications. Diabetic foot ulcers are chronic inflammatory diseases in the presence of prolonged diabetes resulting in decreased sensation. Persistent inflammation and prolonged inflammatory phase of the wound healing in diabetic foot ulcers renders them nonhealing resulting in increased incidences of amputation. Nonhealing of DFUs is mainly due to chronic inflammation, decreased angiogenesis, and attenuated extracellular matrix formation. Despite the presence of recent treatment strategies for DFUs including wound dressing, oxygen therapy, wound debridement, and use of antibiotics, healing of diabetic ulcers remains a concern and various emerging therapies have been postulated but none have improved clinical outcomes significantly [6–10].

Cardiac autonomic neuropathy is due to impaired autonomic function and is followed by the cardiovascular system's nervous system imbalance resulting in symptoms like palpitations, dizziness, presyncope, and syncopal attacks. It is a serious complication of type 1 and type 2 diabetes and is associated with major cardiovascular events, morbidity, and mortality. In patients suffering from type I diabetes, duration of hyperglycemia and diabetes are the strongest risk factors while in type II diabetes, insulin resistance, obesity, hypertension, and hyperlipidemia are common risk factors [11]. DFUs patient has a high prevalence of CAN ranging between 43% to 66% and patients with CAN could also increase the incidence of developing DFUs. CAN may significantly reduce DFU healing time; however, it is not clear whether CAN increases the incidence of amputation or not and the results are inconsistent. CAN may be associated with increased mortality in patients with DFUs. Both DFUs and CAN increase the risk of cardiovascular events [1, 12].

Both DFUs and CAN are associated with diabetes and patients with DFUs have a higher prevalence of CAN. However, the interplay between DFUs and CAN is not well understood yet. This chapter discusses various aspects of DFU and CAN followed by the interaction between them, the approach to managing them when they coexist, and measures to attenuate DFU incidence in the presence of CAN.

Diabetic Neuropathy

Diabetic neuropathy is segmental demyelination and axonal degeneration of the peripheral neurons in diabetics [13]. Numbness, tingling, pain, or weakness are common clinical manifestations and the only available treatment for diabetic neuropathy is glucose control and pain management [14]. Distal symmetric polyneuropathy is very common in distal limbs and sensory loss begins distally in lower limbs and then progresses often accompanied by pain increasing morbidity. Approximately 50% of diabetics develop diabetic neuropathy [15, 16]. Diabetic neuropathy can cause both orthostatic hypotension and diabetic foot ulcers, together increasing cardiovascular mortality [17]. Diabetic neuropathy is a broad term referring to the signs and symptoms of neuropathy in patients with diabetes without any other causes of neuropathy. Diabetic neuropathy may be symmetric such as diabetic polyneuropathy, painful autonomic neuropathy, painful distal neuropathy with weight loss "diabetic cachexia", insulin neuritis, polyneuropathy after ketoacidosis, polyneuropathy with glucose impairment, chronic inflammatory demyelinating polyneuropathy with diabetes mellitus, and asymmetric neuropathy such as radiculoplexoneuropathies (lumbosacral, thoracic, cervical), mononeuropathies, median neuropathy at the wrist, ulnar neuropathy at the elbow, peroneal neuropathy at the fibular head, and cranial neuropathy [18]. High blood sugar levels, hyperlipidemia, duration of diabetes, smoking, alcohol abuse, obesity, microcirculatory dysfunction, and impaired insulin signaling are common risk factors for diabetic neuropathy. Pain, numbness, tingling, and loss of sensation in the feet, legs, and hands are common symptoms while problems with the digestive system, urinary tract, blood vessels, and heart are common in diabetic neuropathy [15]. Improved glycemic control, diet, and lifestyle intervention with exercise disease-modifying therapies using α-lipoic acid, benfotiamine, seal oil v-3 polyunsaturated fatty acid, aldose reductase inhibitors, and managing pain are the strategies to manage diabetic neuropathy [15].

Orthostatic Hypotension

Orthostatic hypotension in diabetes occurs due to efferent sympathetic vasomotor denervation contributing to attenuated splanchnic and other peripheral vascular bed vasoconstriction [19]. Orthostatic hypotension is characterized by a decrease in systolic blood pressure $\geq$20 mmHg or diastolic blood pressure $\geq$10 mmHg when someone changes his position from supine to standing. Orthostatic hypotension manifests with debilitating symptoms including falls, syncope, and cognitive impairment, and the patient may develop myocardial infarction, heart failure, stroke, and fatal coronary heart disease, all increasing cardiovascular mortality [20, 21]. Diabetic neuropathies contribute to DFUs by sensory, motor, or autonomic impairment increasing the risk of foot trauma and ulcers. Impaired pain and temperature sensation also contribute to DFUs [22]. DFUs not only increase morbidity by itself but also contribute to increased mortality by increasing the risk of cardiovascular diseases (CVDs) including myocardial infarction and stroke. DFU also makes it difficult to manage peripheral artery disease associated with DFUs and other comorbid conditions [23, 24].

Diabetic Foot Ulcers

DFUs, the open wounds on the foot of a diabetic patient, are the most debilitating clinical concern accounting for increased morbidity and amputation. The patient with diabetes has a 15% lifetime risk of developing DFUs. The presence of hyperglycemia, obesity, diabetes-related renal disease, foot on weight-wearing surfaces, history of foot ulcers, visual loss, renal transplant, plantar callus, elevated foot pressure, foot deformities, and poor social background are common risk factors for DFUs. Peripheral arterial disease, peripheral neuropathy, and trauma to the foot are common causes of DFUs. Chronic inflammation, decreased angiogenesis, and attenuated extracellular matrix formation due to the presence of persistent inflammation with a prolonged inflammatory phase of wound healing are common causes for a nonhealing pattern of DFUs [6–10, 25, 26] (Fig. 10.1). Offloading, removing pressure from the wound, and managing infection using antibiotics are existing treatment strategies for DFUs, however, the prevalence of amputation varies between 5–33% worldwide and is associated with race, ethnicity, and rurality [27, 28] and 50% of those undergoing amputation die within 5 years [29]. Immune cells interacting with fibroblasts, endothelial cells, epithelial cells, and endothelial cells in the presence of cytokines, growth factors, and chemokines play a critical role in regulating DFU healing [30]. Additionally, levels of various factors including growth factors, transcription factors, and microRNAs in the ulcer regulate DFU healing by regulating angiogenesis and extracellular matrix remodeling. The role of exosomes, stem cells negative pressure oxygen therapy, hyperbaric oxygen therapy, and platelet-rich plasma therapy have been suggested to improve DFU healing by various studies [7, 9, 10, 31, 32] (Fig. 10.1).

Cardiac Autonomic Neuropathy

Heart rate and blood pressure are regulated by the autonomic nervous system. Stimulation of the sympathetic nervous system results in tachycardia and vasoconstriction while smooth muscle relaxation resulting in attenuated heart rate and vasodilation are due to parasympathetic nervous system stimulation [33]. Parasympathetic stimulation is mediated by the vagus nerve and impaired vagal activity results in CAN impairing cardiac sympathetic tone in the early stage resulting in reduced heart rate variability, resting tachycardia, and reduced exercise tolerance

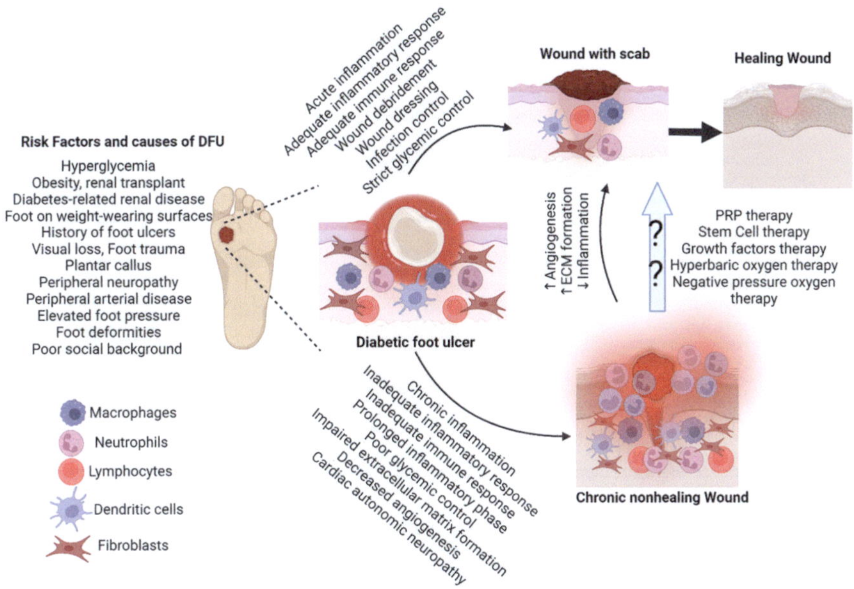

Fig. 10.1 Risk factors and pathophysiology of diabetic foot ulcer (DFU). Glycemic control, attenuating inflammation, and promoting angiogenesis and extracellular matrix (ECM) secretion promote wound healing while chronic inflammation, decreased angiogenesis, and ECM lead to nonhealing DFU due to the prolonged inflammatory phase of wound healing. The role of platelet-rich plasma therapy, oxygen therapy, stem cell therapy, and exosome therapy are being considered and developing but still, the results are not convincing and more studies are needed to improve clinical outcomes and the underlying molecular mechanisms

while in the later phase sympathetic denervation results in baroreceptor reflex impairment, orthostatic hypotension, and ultimately nonresponsive heart rate [1] (Fig. 10.2). Increasing age, duration of diabetes, poor glycemic control, obesity, smoking, diabetic microvascular complications (neuropathy, nephropathy, microalbuminuria, and retinopathy), female sex, African-American race, impaired renal function, coronary artery calcification, autoimmunity, obstructive sleep apnea, inflammation, and a history of cardiovascular event are common risk factors and are involved in pathogenesis of CAN [34–36] (Fig. 10.2). CAN is associated with various cardiovascular system and peripheral vascular functions, resting tachycardia, abnormal blood pressure regulation, orthostatic hypotension, orthostatic tachycardia or bradycardia, and exercise intolerance [34] (Fig. 10.2).

CAN may also be due to gut dysbiosis. Impaired hormone and neurotransmitter secretion due to gut dysbiosis may dysregulate or modulate the feedback mechanism between gut-brain's cardioregulatory regions leading to autonomic dysregulation and cardiac autonomic neuropathy [37]. Additionally, gut dysbiosis leading to overactive inflammation, impaired mucosal barrier function, and immune dysfunction contribute to cardiovascular diseases [38] and it is noteworthy that these factors are also the risk factors for CAN. Metabolites like trimethylamine N-oxide (TMAO)

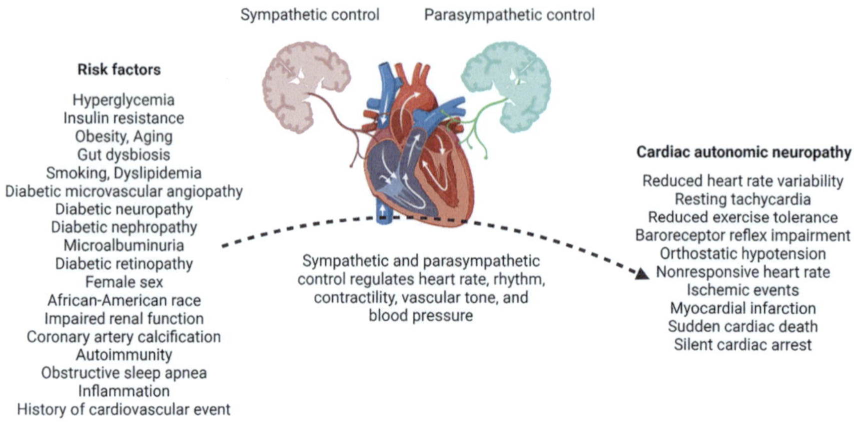

Fig. 10.2 Risk factors and manifestations of cardiac autonomic neuropathy

and phenylacetylglutamine (PAG) are produced by gut microbiota. TMAO is associated with atherosclerosis, hypertension, diabetes, and platelet aggregation while PAG promotes adverse cardiovascular phenotypes by interacting with adrenergic receptors regulating cardiovascular homeostasis. Both TAMO and PAG are associated with an increased risk of CVD [39, 40]. It is interesting to note that TAMO activates the cardiac autonomous nervous system aggravating ischemia-induced ventricular arrhythmia [41].

CAN screening should be performed in all patients with uncontrolled diabetes type II and after 5 years of onset of type I diabetes and is done by assessing the patients for the presence of symptoms associated with CAN including resting tachycardia, exercise intolerance, orthostatic hypotension, orthostatic tachycardia syndrome, QTi prolongation, reverse dipping and non-dipping pattern, impaired heart rate variability, and silent myocardial ischemia [34, 42]. Dhumad et al. [43] investigated the correlation between risk factors for diabetic CAN such as age, gender, duration of diabetes, body mass index (BMI), waist-hip ratio (WHR), and glycemic control (HbA1c) and reported a strong association between CAN severity and age, duration of disease, HbA1c severity, and WHR but not with BMI. The study also reported a significantly longer duration of disease in patients with diabetes with CAN compared to those without CAN. Further, the authors also highlighted a higher sensitivity of parasympathetic impairment to the detection of autonomic dysfunctions compared to sympathetic impairment. Similar findings on the importance of parasympathetic function were also reported by Ahmed et al. [44] and the study concluded that autonomic neuropathy contributes to the development of diabetic foot ulcers. Treatment of CAN involves glucose control, lifestyle modification with exercise, management of hypercholesterolemia using statins, decreasing intestinal absorption of low-density lipoproteins with ezetimibe, PCSK9 inhibition using alirocumab and evolocumab, using fibrates, nicotinic acid, and bile acid sequestrants among other medications [34, 45].

Diabetic Foot Ulcers and Cardiac Autonomic Neuropathy

CAN in diabetes has a prevalence of 1–90% and this wide variation is due to different parameters studies in various studies [1]. The prevalence of CAN increases with age and duration of diabetes while strict glycemic control decreases the prevalence and studies suggested that the presence of CAN is a predictor of DFU development. Microangiopathy, increased secretion of inflammatory cytokines and receptors for advanced glycation endproducts, and increased expression of reactive oxygen contribute to the pathophysiology of CAN in diabetic patients though the exact mechanism is not clearly understood. CAN increases the risk of mortality in diabetic patients [1]. This increase in mortality may be due to increased cardiovascular events and often precipitates "silent myocardial infarction" in patients with diabetes with CAN. Increased ischemic events in diabetic patients with CAN may be due to increased development of atherosclerosis. A prospective observational single-center cohort study evaluated if CAN can predict DFU healing and found a similar proportion of DFU healing while significantly different duration of healing among individuals with and without CAN [46]. The same study found no significant difference in amputation in DFU between patients with or without CAN. However, another study found a significant trend between CAN severity and frequency of amputations of lower limbs with increasing incidence of amputation with CAN severity [47]. The findings suggest that the presence of CAN decreases the healing time of DFU. This may be due to the avoidance of strenuous exercise and the presence of hyperemia due to sympathetic impairment in the lower limb vasculature in DFU patients with severe CAN [42, 48]. Additionally, the DFU outcome depends on CAN severity.

Menon et al. reported a higher prevalence of CAN in patients with diabetes with an increased risk of foot ulcers [12]. Yun et al. [47] studied the factors influencing the development of DFU in type II diabetes patients without diabetic polyneuropathy with a total of 595 patients and found a significant independent association between DFU development and the presence of CAN. The study also concluded that CAN is a predictor of DFU even in the absence of diabetic polyneuropathy. Fisher et al. reported that CAN prevalence is approximately 17–66% in type I diabetes patients and 31–73% in patients with type II diabetes [36]. It should also be noted that CAN may affect the feet by causing loss of sweat leading to dry skin resulting in cracks and splits making it prone to infection contributing to DFUs [49].

Overall, the association of CAN with DFU development and CAN as a predictor of DFU emphasize the need to screen patients with definite CAN and type 2 diabetes or CAN alone for foot injuries, and foot status, and educate patients on strict glucose control to prevent development of DFUs, amputations, and mortality due to cardiovascular events. Clinicians should be more vigilant with patients with type 2 diabetes and definite CAN, especially those with uncontrolled glycemic status. Early screening is important because CAN may remain silent in its sub-clinical phase and heart rate variability can help in early diagnosis of CAN in diabetic patients [48]. Early detection is important because of the lack of specific treatment to slow the progression or prevent CAN development. Maser et al. [50] emphasized that after the identification of CAN in a patient with diabetes, it is important for the healthcare provider to alert the patient, motivate him to exercise, increase surveillance for cardiac ischemia, reexamine the list of medication the patient is taking, and aggressively treating the risk factors such as hypertension and other which are associated with CAN.

Conclusion

CAN is a predictor of DFU and diabetes in the common denominator for both these entities. The presence of CAN promotes the development of DFU and both CAN and DFU increase cardiovascular mortality. The studies suggest that patients with CAN should be thoroughly screened and investigated for cardiovascular complications and the development of DFUs. A lack of specific treatment to slow the progression or prevent CAN development and the significant association of CAN with DFU development underscores the importance of screening in such patients.

Acknowledgments None to declare.

Funding No extramural funding was received for this work.

Conflicts of Interest "The author declares no conflicts of interest."

References

1. Basra R, Papanas N, Farrow F, Karalliedde J, Vas P (2022) Diabetic foot ulcers and cardiac autonomic neuropathy. Clin Ther 44:323–330. https://doi.org/10.1016/j.clinthera.2021.12.002
2. Saeedi P, Petersohn I, Salpea P, Malanda B, Karuranga S, Unwin N, Colagiuri S, Guariguata L, Motala AA, Ogurtsova K, Shaw JE, Bright D, Williams R (2019) Global and regional diabetes prevalence estimates for 2019 and projections for 2030 and 2045: results from the International Diabetes Federation Diabetes Atlas, 9(th) edition. Diabetes Res Clin Pract 157:107843. https://doi.org/10.1016/j.diabres.2019.107843
3. Khan MAB, Hashim MJ, King JK, Govender RD, Mustafa H, Al Kaabi J (2020) Epidemiology of type 2 diabetes - global burden of disease and forecasted trends. J Epidemiol Glob Health 10:107–111. https://doi.org/10.2991/jegh.k.191028.001
4. Zhang K, Kan C, Han F, Zhang J, Ding C, Guo Z, Huang N, Zhang Y, Hou N, Sun X (2023) Global, regional, and national epidemiology of diabetes in children from 1990 to 2019. JAMA Pediatr 177:837–846. https://doi.org/10.1001/jamapediatrics.2023.2029
5. Farmaki P, Damaskos C, Garmpis N, Garmpi A, Savvanis S, Diamantis E (2020) Complications of the type 2 diabetes mellitus. Curr Cardiol Rev 16:249–251. https://doi.org/10.2174/1573403X1604201229115531

6. Rai V, Moellmer R, Agrawal DK (2022) The role of CXCL8 in chronic nonhealing diabetic foot ulcers and phenotypic changes in fibroblasts: a molecular perspective. Mol Biol Rep 49:1565–1572. https://doi.org/10.1007/s11033-022-07144-3
7. Rai V, Moellmer R, Agrawal DK (2022) Stem cells and angiogenesis: implications and limitations in enhancing chronic diabetic foot ulcer healing. Cells 11:152287. https://doi.org/10.3390/cells11152287
8. Littig JPB, Moellmer R, Estes AM, Agrawal DK, Rai V (2022) Increased population of CD40+ fibroblasts is associated with impaired wound healing and chronic inflammation in diabetic foot ulcers. J Clin Med 11:216335. https://doi.org/10.3390/jcm11216335
9. Rai V, Moellmer R, Agrawal DK (2023) Role of fibroblast plasticity and heterogeneity in modulating angiogenesis and healing in the diabetic foot ulcer. Mol Biol Rep 50:1913–1929. https://doi.org/10.1007/s11033-022-08107-4
10. Littig JPB, Moellmer R, Agrawal DK, Rai V (2023) Future applications of exosomes delivering resolvins and cytokines in facilitating diabetic foot ulcer healing. World J Diabetes 14:35–47. https://doi.org/10.4239/wjd.v14.i1.35
11. Williams S, Raheim SA, Khan MI, Rubab U, Kanagala P, Zhao SS, Marshall A, Brown E, Alam U (2022) Cardiac autonomic neuropathy in type 1 and 2 diabetes: epidemiology, pathophysiology, and management. Clin Ther 44:1394–1416. https://doi.org/10.1016/j.clinthera.2022.09.002
12. Menon AS, Dixit A, Garg MK, Girish R (2017) Cardiac autonomic neuropathy in patients with type 2 diabetes mellitus at high risk for foot ulcers. Indian J Endocrinol Metab 21:282–285. https://doi.org/10.4103/ijem.IJEM_542_16
13. Ahmed N (2005) Advanced glycation endproducts–role in pathology of diabetic complications. Diabetes Res Clin Pract 67:3–21. https://doi.org/10.1016/j.diabres.2004.09.004
14. Callaghan BC, Cheng HT, Stables CL, Smith AL, Feldman EL (2012) Diabetic neuropathy: clinical manifestations and current treatments. Lancet Neurol 11:521–534. https://doi.org/10.1016/S1474-4422(12)70065-0
15. Feldman EL, Callaghan BC, Pop-Busui R, Zochodne DW, Wright DE, Bennett DL, Bril V, Russell JW, Viswanathan V (2019) Diabetic neuropathy. Nat Rev Dis Primers 5:42. https://doi.org/10.1038/s41572-019-0097-9
16. Feldman EL, Callaghan BC, Pop-Busui R, Zochodne DW, Wright DE, Bennett DL, Bril V, Russell JW, Viswanathan V (2019) Diabetic neuropathy. Nat Rev Dis Primers 5:41. https://doi.org/10.1038/s41572-019-0092-1
17. Sambataro M, Sambado L, Spinetti G, Seganfreddo E, Trevisiol E, Marcon M, Stefani P, Furlan A, Cacciatore M, Paccagnella A (2019) Cardiovascular mortality and orthostatic hypotension in type 2 diabetic patients (T2D) with lower limb lesions—Italian diabetic patients' Association-Section of Treviso Project. Diabetes 68:621
18. Bansal V, Kalita J, Misra UK (2006) Diabetic neuropathy. Postgrad Med J 82:95–100. https://doi.org/10.1136/pgmj.2005.036137
19. Pop-Busui R (2010) Cardiac autonomic neuropathy in diabetes: a clinical perspective. Diabetes Care 33:434–441. https://doi.org/10.2337/dc09-1294
20. Juraschek SP, Daya N, Appel LJ, Miller ER, McEvoy JW, Matsushita K, Ballantyne CM, Selvin E (2018) Orthostatic hypotension and risk of clinical and subclinical cardiovascular disease in middle-aged adults. J Am Heart Assoc 7:8884. https://doi.org/10.1161/JAHA.118.008884
21. Fedorowski A, Ricci F, Hamrefors V, Sandau KE, Hwan Chung T, Muldowney JAS, Gopinathannair R, Olshansky B (2022) Orthostatic hypotension: management of a complex, but common, medical problem. Circ Arrhythm Electrophysiol 15:e010573. https://doi.org/10.1161/CIRCEP.121.010573
22. Tecilazich F, Veves A (2018) Role of peripheral neuropathy in the development of foot ulceration and impaired wound healing in diabetes mellitus. In: Nutritional and therapeutic interventions for diabetes and metabolic syndrome. Elsevier, Amsterdam, pp 95–104

23. Brownrigg JR, Davey J, Holt PJ, Davis WA, Thompson MM, Ray KK, Hinchliffe RJ (2012) The association of ulceration of the foot with cardiovascular and all-cause mortality in patients with diabetes: a meta-analysis. Diabetologia 55:2906–2912. https://doi.org/10.1007/s00125-012-2673-3
24. Gallagher KA, Mills JL, Armstrong DG, Conte MS, Kirsner RS, Minc SD, Plutzky J, Southerland KW, Tomic-Canic M (2024) Current status and principles for the treatment and prevention of diabetic foot ulcers in the cardiovascular patient population: a scientific statement from the American Heart Association. Circulation 149:e232–e253. https://doi.org/10.1161/CIR.0000000000001192
25. Wang X, Yuan CX, Xu B, Yu Z (2022) Diabetic foot ulcers: classification, risk factors and management. World J Diabetes 13:1049–1065. https://doi.org/10.4239/wjd.v13.i12.1049
26. Boulton AJ, Whitehouse RW (2017) The diabetic foot. In: Endotext. MDText.com, South Dartmouth
27. Brennan MB, Powell WR, Kaiksow F, Kramer J, Liu Y, Kind AJH, Bartels CM (2022) Association of race, ethnicity, and rurality with major leg amputation or death among medicare beneficiaries hospitalized with diabetic foot ulcers. JAMA Netw Open 5:e228399. https://doi.org/10.1001/jamanetworkopen.2022.8399
28. Aalaa M, Vahdani AM, Mohajeri Tehrani M, Mehrdad N, Zohdirad M, Sadati M, Amini M, Mehrpour S, Ebrahimi M, Larijani B, Amini MR, Sanjari M (2024) Epidemiological insights into diabetic foot amputation and its correlates: a provincial study. Clin Med Insights Endocrinol Diabetes 17:11795514241227618. https://doi.org/10.1177/11795514241227618
29. Weledji EP, Fokam P (2014) Treatment of the diabetic foot - to amputate or not? BMC Surg 14:83. https://doi.org/10.1186/1471-2482-14-83
30. Smith J, Rai V (2024) Novel factors regulating proliferation, migration, and differentiation of fibroblasts, keratinocytes, and vascular smooth muscle cells during wound healing. Biomedicines 12:91939. https://doi.org/10.3390/biomedicines12091939
31. Rai V, Le H, Agrawal DK (2023) Novel mediators regulating angiogenesis in diabetic foot ulcer healing. Can J Physiol Pharmacol 101:488–501. https://doi.org/10.1139/cjpp-2023-0193
32. Kim J, Nomkhondorj O, An CY, Choi YC, Cho J (2023) Management of diabetic foot ulcers: a narrative review. J Yeungnam Med Sci 40:335–342. https://doi.org/10.12701/jyms.2023.00682
33. Vaseghi M, Shivkumar K (2008) The role of the autonomic nervous system in sudden cardiac death. Prog Cardiovasc Dis 50:404–419. https://doi.org/10.1016/j.pcad.2008.01.003
34. Serhiyenko VA, Serhiyenko AA (2018) Cardiac autonomic neuropathy: risk factors, diagnosis and treatment. World J Diabetes 9:1–24. https://doi.org/10.4239/wjd.v9.i1.1
35. Haas AV, McDonnell ME (2018) Pathogenesis of cardiovascular disease in diabetes. Endocrinol Metab Clin N Am 47:51–63. https://doi.org/10.1016/j.ecl.2017.10.010
36. Fisher VL, Tahrani AA (2017) Cardiac autonomic neuropathy in patients with diabetes mellitus: current perspectives. Diabetes Metab Syndr Obes 10:419–434. https://doi.org/10.2147/DMSO.S129797
37. Zubcevic J, Richards EM, Yang T, Kim S, Sumners C, Pepine CJ, Raizada MK (2019) Impaired autonomic nervous system-microbiome circuit in hypertension. Circ Res 125:104–116. https://doi.org/10.1161/CIRCRESAHA.119.313965
38. Rashid S, Sado AI, Afzal MS, Ahmed A, Almaalouli B, Waheed T, Abid R, Majumder K, Kumar V, Tejwaney U, Kumar S (2024) Role of gut microbiota in cardiovascular diseases - a comprehensive review. Ann Med Surg 86:1483–1489. https://doi.org/10.1097/MS9.0000000000001419
39. Witkowski M, Weeks TL, Hazen SL (2020) Gut microbiota and cardiovascular disease. Circ Res 127:553–570. https://doi.org/10.1161/CIRCRESAHA.120.316242
40. Ahmadmehrabi S, Tang WHW (2017) Gut microbiome and its role in cardiovascular diseases. Curr Opin Cardiol 32:761–766. https://doi.org/10.1097/HCO.0000000000000445

41. Meng G, Zhou X, Wang M, Zhou L, Wang Z, Wang M, Deng J, Wang Y, Zhou Z, Zhang Y, Lai Y, Zhang Q, Yang X, Yu L, Jiang H (2019) Gut microbe-derived metabolite trimethylamine N-oxide activates the cardiac autonomic nervous system and facilitates ischemia-induced ventricular arrhythmia via two different pathways. EBioMedicine 44:656–664. https://doi.org/10.1016/j.ebiom.2019.03.066
42. Vinik AI, Ziegler D (2007) Diabetic cardiovascular autonomic neuropathy. Circulation 115: 387–397
43. Dhumad MM, Hamdan FB, Khudhair MS, Al-Matubsi HY (2021) Correlation of staging and risk factors with cardiovascular autonomic neuropathy in patients with type II diabetes mellitus. Sci Rep 11:3576. https://doi.org/10.1038/s41598-021-80962-w
44. Ahmed ME, Delbridge L, Le Quesne LP (1986) The role of autonomic neuropathy in diabetic foot ulceration. J Neurol Neurosurg Psychiatry 49:1002–1006. https://doi.org/10.1136/jnnp.49.9.1002
45. Vinik AI, Casellini C, Parson HK, Colberg SR, Nevoret ML (2018) Cardiac autonomic neuropathy in diabetes: a predictor of cardiometabolic events. Front Neurosci 12:591. https://doi.org/10.3389/fnins.2018.00591
46. Farrow F, Fountoulakis N, Cummins K, Sharma A, Mahmood R, Vas P, Thomas S, Karalliedde J (2021) Cardiac autonomic neuropathy is associated with improved diabetic foot ulcer healing: a single centre prospective cohort study. Exp Clin Endocrinol Diabetes 129:895–898. https://doi.org/10.1055/a-1158-9130
47. Yun J-S, Cha S-A, Lim T-S, Lee E-Y, Song K-H, Ahn Y-B, Yoo K-D, Kim J-S, Park Y-M, Ko S-H (2016) Cardiovascular autonomic dysfunction predicts diabetic foot ulcers in patients with type 2 diabetes without diabetic polyneuropathy. Medicine 95:e3128
48. Dimitropoulos G, Tahrani AA, Stevens MJ (2014) Cardiac autonomic neuropathy in patients with diabetes mellitus. World J Diabetes 5:17
49. Sudo SZ, Montagnoli TL, Rocha BS, Santos AD, de Sa MPL, Zapata-Sudo G (2022) Diabetes-induced cardiac autonomic neuropathy: impact on heart function and prognosis. Biomedicines 10:123258. https://doi.org/10.3390/biomedicines10123258
50. Maser RE, Lenhard MJ (2005) Cardiovascular autonomic neuropathy due to diabetes mellitus: clinical manifestations, consequences, and treatment. J Clin Endocrinol Metab 90:5896–5903. https://doi.org/10.1210/jc.2005-0754

Chapter 11
Statins and Cardiotoxicity

Victor Gurevich and Andrey Obrezan

Abstract Clinical trials have demonstrated the efficacy of statins in reducing cardiovascular mortality and morbidity. Key studies showed a significant reduction in myocardial infarction and stroke when patients were treated with different statins. While statins are generally well-tolerated, a small percentage of patients experience side effects. The most serious risk is rhabdomyolysis, but this is very rare. Statin-associated muscle symptoms (SAMS), including myalgia and myopathy, are more commonly reported. Monitoring creatine phosphokinase levels before and during statin therapy is crucial for patient safety. Statins have also been implicated in possible cardiac toxicity, with some studies suggesting a link to a condition referred to as statin-associated cardiomyopathy. Reports indicate that long-term use of statins could negatively impact heart function, especially in susceptible individuals. However, other studies have demonstrated that statins can have protective effects in patients undergoing chemotherapy for cancer, mitigating cardiotoxic effects of certain antitumor drugs. In summary, statins play a significant role in cardiovascular health through their lipid-lowering and pleiotropic effects. Although they can cause side effects, their benefits in preventing heart-related issues outweigh the risks for many patients. Regular monitoring and individualized approaches to therapy can help mitigate adverse effects while maximizing cardiovascular protection.

Keywords Antimalignant therapy · Cancer · Cardiotoxicity · Side effects of statins · Statins

---

V. Gurevich (✉)
Center of Atherosclerosis and Lipid Disorders, Saint-Petersburg State University, Medical Institution, St. Petersburg, Russia

Nord-Western State Medical University n.a. I.I. Mechnikov, Center of Atherosclerosis and Lipid Disorders, St. Petersburg, Russia

A. Obrezan
Center of Atherosclerosis and Lipid Disorders, Saint-Petersburg State University, Medical Institution, St. Petersburg, Russia

D. K. Agrawal et al. (eds.), *Cardiovascular Toxicity*, Advances in Biochemistry in Health and Disease 33, https://doi.org/10.1007/978-3-031-90598-8_11

Introduction

Cardiotoxicity refers to heart damage or temporary dysfunction resulting from treatments for various diseases, including cardiac conditions. While cardiotoxicity is typically linked to chemotherapy, other cancer therapies, such as chest irradiation, can also adversely affect the heart. Patients receiving antitumor treatments should have regular cardiac evaluations to identify cardiotoxicity early. The risk of cardiac injury may be mitigated through pharmacological interventions.

The ways in which drugs affect the heart can vary: deterioration of myocardial contractility, compromised microcirculation, thromboembolic issues, mitochondrial dysfunction, non-infectious inflammation, and more.

The most concerning outcome of drug-induced cardiotoxicity is the onset and advancement of heart failure. With prolonged use of certain medications, the myocardium continually faces negative impacts and cannot recover, eventually causing detrimental remodeling that mainly involves replacing functioning cardiomyocytes with fibrous tissue. Reversible cardiac dysfunction can also occur. This is caused by disturbances in the metabolism and functionality of contractile proteins, often influenced by HER2 inhibitors, which are commonly used in cancer treatment.

Consequently, when using cardiotoxic medications, especially anthracyclines, individuals require routine assessments to quickly identify any heart issues. ACE inhibitors, beta blockers, and statins are administered for cardiac protection. The choice of medication depends on the anticipated type of cardiotoxicity. Regarding statins, despite clear positive outcomes related to the primary and secondary prevention of atherosclerosis complications, the concern over statins' cardiotoxicity remains a topic of discussion. This is largely due to the relatively high occurrence of skeletal-muscular side effects associated with statins, which often leads to their discontinuation, even when there is a need for intense lipid-lowering treatment. Therefore, assessing all factors for and against statin use, considering the issue of drug-induced cardiotoxicity, continues to be important today.

Hypolipidemic and Pleiotropic Actions of Statins

The molecular mechanism of statins action is based on inhibition of the cholesterol synthesis key enzyme—HMG-CoA reductase. To compensate for short-term intracellular cholesterol deficiency caused by a decrease in the enzymatic activity of HMG-CoA reductase, hepatocytes, due to the activation of the corresponding genes according to the principle of negative feedback, express an increased number of low density lipoprotein (LDL) receptors, increasing the capture of LDL from plasma. As a result, a dose-dependent decrease in the level of total cholesterol and LDL cholesterol in blood plasma is observed. At the same time, due to a change in the ratio of intermediate forms of lipoproteins, the level of triglycerides (TG) is moderately reduced, which in turn contributes to an insignificant but reliable increase in the

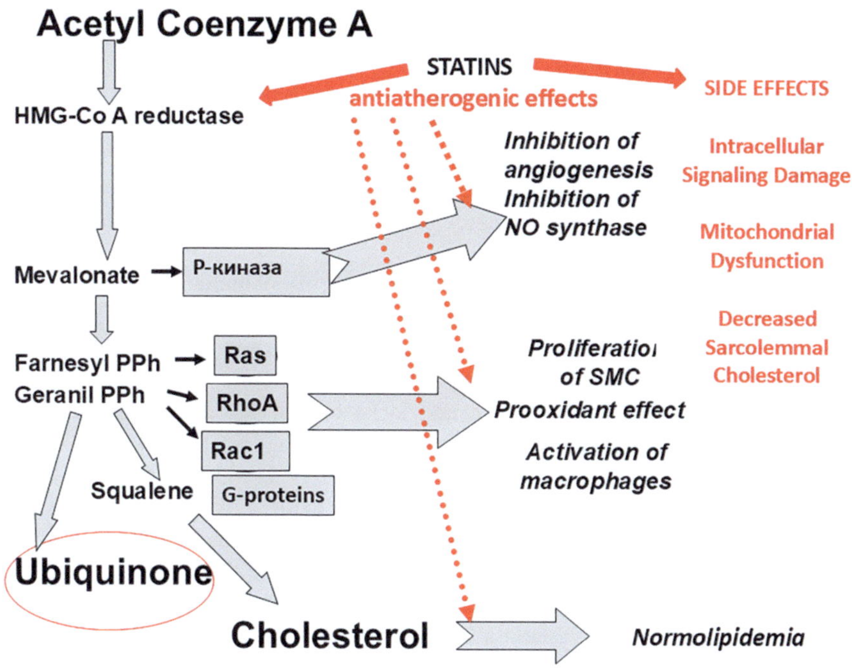

Fig. 11.1 Hypolipidemic and pleiotropic effect of statins. Original to the author

content of high density lipoproteins (HDL) cholesterol. It should be noted that other cholesterol precursors: mevalonate, geranyl-pyrophosphate and farnesyl-pyrophosphate are also effector molecules that regulate cellular and intercellular signaling through kinase reactions and post-translational modification of G proteins, leading to the activation of proatherogenic reactions: oxidative stress, monocyte migration, smooth muscle cell proliferation, inhibition of angiogenesis and suppression of NO synthase activity. Inhibition of HMG-CoA reductase by statins leads to a decrease in the synthesis of not only cholesterol, but also its precursors. This causes suppression of a number of proatherogenic reactions and is related to the so-called pleiotropic effects of statins (Fig. 11.1).

The entire spectrum of additional effects of statins, not directly related to the hypolipidemic effect, are usually called their pleiotropic effects. These properties, along with a significant decrease in the level of atherogenic lipoproteins, determine the key role of statins in the primary and secondary prevention of coronary heart disease and other clinical manifestations of atherosclerosis.

The leading pleiotropic effects of statins are:

- improvement of impaired endothelial function
- suppression of the synthesis of inflammation mediators
- prevention of oxidative modification of LDL
- the ability to stabilize the fibrous capsule of an atherosclerotic plaque

- antiplatelet action
- immunomodulatory action

The pleiotropic effects of statins develop quite quickly and their effect lasts for a long time. For this reason, for example, the use of statins in the maximum tolerated dose has long been practiced, regardless of the cholesterol level in the acute period of myocardial infarction, since it has been proven that at this stage of coronary heart disease, the leading therapeutic effect of statins is both anti-inflammatory and antithrombotic action.

Statins differ in their pharmacodynamic and pharmacokinetic properties.

Lovastatin is the first drug in this group, currently almost never used in clinical practice due to its low efficiency and the need for double administration.

Simvastatin is similar to lovastatin in its molecular structure, but has an additional methyl group, is much more active and lacks one of the major drawbacks of lovastatin: a short-term increase in spontaneous platelet aggregation observed during the first two weeks of administration in approximately 25–45% of patients. Lovastatin and simvastatin are prodrugs, their conversion to the active form occurs in vivo in the liver. Prescribing simvastatin at a dose of 10–40 mg reduces LDL-C by 25–45%, TG—by 20–25%.

Pravastatin is a derivative of the very first statin—mevastatin, which is a product of fungal metabolism. Pravastatin's activity in inhibiting HMG-CoA reductase is slightly lower than that of lovastatin and simvastatin. At the same time, it is one of the most well-studied statins. A large amount of data has been obtained on the effectiveness of pravastatin in preventing recurrent coronary episodes, as well as evidence of the pleiotropic properties of the drug. Fluvastatin is the first synthetic statin that does not require conversion to an active form in vivo and, therefore, has no active metabolites in the blood. This causes a lower frequency of side effects of fluvastatin compared to other statins. The drug is hydrophilic, almost does not penetrate the blood-brain barrier, its action is almost completely localized in the liver, and it has the shortest half-life. Due to its lower lipid-lowering activity compared to other statins, the drug has not been widely used. Another synthetic drug, atorvastatin, is much more effective and is one of the most widely used hypolipidemic agents at present. The drug causes a significant decrease in the level of total cholesterol (by 30–46%) and LDL cholesterol (by 41–61%) when prescribed in doses from 10 to 80 mg. In addition, when it is used, there is a significant decrease in the level of TG—by 23–45% and an increase in the content of HDL cholesterol by 7–10%. The high comparative effectiveness of atorvastatin was demonstrated in the CURVES study, in which patients with elevated LDL cholesterol were treated with different doses of atorvastatin, pravastatin, fluvastatin, lovastatin and simvastatin. As a result of 8 weeks of therapy, the LDL-C level decreased by 51% when prescribing 40 mg of atorvastatin, taking the same dose of pravastatin, lovastatin and fluvastatin led to a decrease in this indicator by 34, 31 and 23%, respectively. and only simvastatin demonstrated a similar effect. As shown in the multicenter open comparative study TREAT-TO-TARGET, most patients achieve the target LDL-C level using the initial dose of atorvastatin 10 mg.

Table 11.1 Outcomes of most important statin trials

Trial study medication	Hypolipidemic effect	Relative risk reduction
4 S Simvastatin	25% ↓ total C 35% ↓ LDL-C	30% total mortality 34% coronary complications 42% coronary mortality
LIPID Pravastatin	19% ↓ total C 27% ↓ LDL-C	24% coronary mortality 23% total mortality 29% fatal and nonfatal MI 24% revascularisation
ASCOT-LLA Atorvastatin	51% ↓ LDL-C 19% ↓ TG	36% nonfatal MI rate 33% ischemic stroke 42% revascularisation
JUPITER Rosuvastatin	50% ↓ LDL-C 17% ↓ TG	44% primary endpoint 47% fatal and nonfatal MI,Stroke, CV death 47% revascularization or hospital admission with unstable angina 20% total mortality

Original to the author

Rosuvastatin, according to the results of clinical trials, has high lipid-lowering activity and can be used in lower doses. The initial dose of rosuvastatin 5 mg and 10 mg/day leads after 12 weeks to a decrease in LDL-C by 46% and 50%, respectively. Prescribing the maximum dose of rosuvastatin 40 mg made it possible to achieve a decrease in LDL-C by more than 55%. According to the results of completed clinical trials, the appointment of rosuvastatin at a dose of 10–20 mg allows achieving target values of LDL-C in 80–95% of patients. At the same time, it should be noted that about 30% of rosuvastatin is excreted by the kidneys, unlike other statins, which may explain the development of dose-dependent moderate proteinuria during treatment with maximum doses of this statin. Rosuvastatin at a maximum dose of 40 mg is not used if the SCF decreases below 60 ml/min, and in severe renal failure (SCF <30 ml/min), the appointment of any dose of rosuvastatin is contraindicated.

Pitavastatin showed very encouraging results during clinical trials: the lipid-lowering activity of the drug allows it to be classified as a statin with moderate activity. Its effectiveness in reducing the level of cholesterol and LDL-C at a dose of 4 mg is not inferior to atorvastatin at a dose of 20 mg. In the LIVES study, pitavastatin demonstrated a statistically significant reduction in the incidence of recurrent coronary events, the leading criterion for the clinical efficacy of statins. Numerous randomized clinical trials have convincingly demonstrated the effect of statins on reducing overall and cardiovascular (CV) mortality; they significantly reduce the risk of myocardial infarction and stroke, as well as the need for myocardial revascularization surgery [1]. Table 11.1 presents the results of the largest studies of statin efficacy, confirming their positive impact on survival and quality of life.

The Heart Protection Study (HPS) in the UK involved over 20,000 patients aged 40–80 years with a high risk of vascular complications of atherosclerosis. Half of the patients received 40 mg of simvastatin for 5.5 years, the rest took a placebo. Statin treatment resulted in a third reduction in the risk of myocardial infarction and stroke and a decrease in the need for revascularization operations, including in patients with low and low baseline total cholesterol levels (below 5.0 mmol/l), as well as in patients with type 1 and type 2 diabetes. The HPS study was the first clinical trial to involve over 5000 women, and the beneficial effect of statin therapy did not depend on the patient's gender. A significant reduction in the risk of major vascular complications was registered in all age groups, including patients over 75 years of age. The aim of the AVERT study was to compare the effectiveness of treating patients with coronary artery disease with hyperlipidemia using angioplasty and aggressive therapy with atorvastatin at a dose of 80 mg. As a result of 18 months of treatment with atorvastatin, the level of LDL-C decreased by 46%, while the overall frequency of ischemic episodes decreased by 36%. For the first time, it was demonstrated that the effectiveness of treating patients with stable coronary artery disease by intensively lowering the level of LDL-C is not inferior in prognosis to angioplasty. However, it is fair to note that although patients who underwent coronary angioplasty also received lipid-lowering therapy, they, unlike patients in the conservative treatment group who took 80 mg of atorvastatin daily, did not achieve the target values of LDL-C.

The use of statins for secondary prevention of coronary artery disease has become generally accepted. The most effective is the early administration of statins to patients with acute coronary syndrome. In this case, the pleiotropic effects of statins are fully realized.

The largest study devoted to the primary prevention of cardiovascular disease (CVD) is the JUPITER study—a randomized, double-blind, placebo-controlled clinical trial, the purpose of which is to establish whether treatment with rosuvastatin in middle-aged and elderly individuals without previous CVD, with low (below generally accepted normal values) LDL-C levels—<130 mg/dL (3.4 mmol/L), but with an elevated level of highly sensitive CRP (>2 mg/L), will lead to a decrease in CV events. The study was conducted in 26 countries worldwide, in 1315 centers 17,802 men over 50 years old and women over 60 years old were randomized in a 1: 1 ratio to rosuvastatin 20 mg/day and placebo. The study showed a cumulative reduction in the combined primary endpoint by 44% ($p < 0.00001$), as well as the risk of MI, stroke, death from CV causes—by 47% ($p < 0.00001$), revascularization or hospitalization with a diagnosis of "unstable angina"—by 47% ($p < 0.00001$) and death from any cause—by 20% in the rosuvastatin group ($p = 0.02$). Since statins interfere with the synthesis of cholesterol at very early stages, the formation of a number of intermediate metabolites is reduced, thereby realizing the antiatherogenic effect of this class of drugs. The antiatherogenic effect of statins is confirmed by the results of studies with angiographic or ultrasound control, performed using various drugs: Fluvastatin, Simvastatin, Atorvastatin, Rosuvastatin and Pitavastatin. A necessary condition for slowing the progression and even regression of atherosclerosis is a sufficient duration of therapy and an effective decrease in the level of atherogenic lipoproteins. METEOR study is the only placebo-controlled trial that has shown the possibility of therapeutic effects on atherosclerosis in individuals with initial signs of this disease and a low risk of developing coronary heart disease (10-year risk according to the Framingham scale <10%). This study examined rosuvastatin 40 mg for the prevention of atherosclerosis in asymptomatic patients at low risk of coronary heart disease. In the rosuvastatin group, the rate of atherosclerosis progression in all segments of the carotid arteries was significantly lower than in the placebo group. A significant slowdown in the progression of atherosclerosis was noted against the background of a significant decrease in the level of LDL-C (by 49%) and an increase in the level of HDL-C by 8%.

All the above data convincingly confirm the high efficiency of the class of statins in reducing cardiovascular risks. This class of drugs not only confirmed the validity of the proatherogenic effect of LDL-C, but also allowed to significantly reduce population indicators of cardiovascular morbidity and mortality. At the same time, like any highly effective drugs, statins cause a number of side effects that should not escape the attention of the doctor.

In general, drugs of this group are well tolerated and adverse reactions during treatment with statins are rare. According to multicenter studies, the number of patients who stopped taking statins due to the development of side effects does not exceed 2–4%. The most common within these limits are: abdominal pain, constipation, diarrhea, flatulence, asthenia, headache, dizziness, increased transaminase activity, skin rash, itching, very rarely muscle symptoms are observed—myalgia, myopathy and extremely rarely—rhabdomyolysis [2].

Musculoskeletal Side Effects of Statins

The most serious complication of statin therapy is rhabdomyolysis, which can lead to the development of acute renal failure. This complication is extremely rare, for most statins the incidence does not exceed 0.15 cases per one million prescriptions issued. A higher incidence of fatal rhabdomyolysis with cerivastatin (3.6 cases per one million prescriptions issued) was the basis for recalling the drug from clinical practice. It should be noted that the absolute incidence of rhabdomyolysis in patients treated with cerivastatin was low and, perhaps, if there were no other statins with a better safety profile, this would only serve as a basis for greater caution when taking the drug. In addition, in most cases, rhabdomyolysis with this drug developed in patients receiving combination therapy with cerivastatin and gemfibrozil.

In clinical practice, we mainly encounter myalgia and myopathies, which are referred to as the so-called "statin-associated muscle symptoms" (SAMS). In 2015, the European Atherosclerosis Society developed a special document devoted to the tactics of managing patients who develop SAMS. It is believed that in general clinical practice they can be encountered quite often, therefore, before prescribing statin therapy, it is necessary to monitor not only the activity of transaminases, but also the activity of creatine phosphokinase (CPK). Statins are not prescribed for an

Table 11.2 Statin-associated muscle adverse effects

Diagnosis	Signs and symptoms
Myalgia	Unexplained muscle pain with normal CPK levels
Myopathy	Muscle weakness with or without elevated CPK levels
Myositis	Muscle inflammation
Muscle necrosis	Mild: CPK 3–9 times ULN Moderate: CPK 10–49 times ULN Severe: CPK >50 times ULN
Clinical rhabdomyolysis	Muscle necrosis with myoglobinuria or acute renal failure

Original to the author

unexplained four-fold excess of the upper limit of normal (ULN) of CPK activity. During treatment, special attention should be paid to the development of muscle weakness, fatigue, and myalgia. Table 11.2 presents the classification of statin-associated myopathies. Naturally, when symptoms such as muscle pain or weakness appear, monitoring of CPK activity is mandatory.

Clinical and laboratory signs of impending rhabdomyolysis include:

- Muscle pain, cramps
- Weakness of unknown origin
- Dark urine
- Elevated CPK levels (>10 times normal)

Diagnosis of SAMS, despite its apparent simplicity, is quite a complex task, since a number of patients tend to associate almost all their muscle and joint symptoms with the intake of these drugs. In order to conclude that muscle pain or muscle weakness are caused by the therapy, the doctor needs to trace a clear cause-and-effect relationship between the symptoms and the intake of a statin drug. When SAMS develops, it is very important to choose the right tactics for selecting statin therapy. It involves prescribing smaller doses of the same drug or a more effective one with gradual titration of the dose from the minimum. An intermittent (never a course) administration regimen is also possible, that is, every other day or 2 times a week. Naturally, statins with a long half-life are used for this: rosuvastatin, atorvastatin or pitavastatin. It is interesting to note that in double-blind studies of other classes of lipid-lowering agents, which included patients with declared statin intolerance, and the study involved titration of statins, approximately 70% of patients were able to take the therapy.

In case of muscle symptoms, as well as every 3 months in some patients: elderly, with hypothyroidism, type 2 diabetes, frail body type, chronic renal failure, receiving combination therapy of statins with fibrates, it is necessary to monitor the activity of CPK. In other cases, when a preliminary study does not show a significant increase in CPK activity, preventive monitoring of this indicator during treatment is not considered necessary.

An indication for discontinuing the drug or reducing the dose is an increase in CPK activity by 4 times the ULN in the presence of muscle symptoms or by 10 times the ULN in the absence of muscle symptoms.

Specific Effect of Statins on the Myocardium

There are several studies that have demonstrated the negative effect of statins on myocardial damage markers. Thus, Unlu et al. found a significant increase in highly sensitive troponin I after physical exercise in patients taking statins, which allowed them to conclude that statins have a cardiotoxic effect [3]. Other researchers did not confirm these data, but agreed that statins can have an adverse effect on the myocardium if other factors are present [4].

For example, Langsøen et al. suggest the existence of statin-associated cardiomyopathy (SACM) as a common and yet unrecognized cause of iatrogenic heart failure (HF). In their own study, the authors prospectively identified patients receiving long-term statin therapy who developed HF in the absence of any identifiable cause. Of the 142 patients with HF identified, 94% had preserved ejection fraction (EF). In these patients, statins were discontinued and CoQ10 was added at a dose of 300 mg/day.

After a mean follow-up of 2.8 years, New York Heart Association class 1 increased from 8% to 79% ($p < 0.0001$). In patients with preserved EF, 34% showed normalization of diastolic function, and 25% showed an improvement ($p < 0.0001$). In patients with reduced EF at baseline, EF improved from a mean of 35–47% ($p = 0.02$). Mortality within 1 year was 0%, and mortality within 3 years was 3%, which is significantly lower compared to the generally known indicators for this nosology. Based on the data obtained, the authors suggested that patients receiving long-term statin therapy may develop statin-associated cardiomyopathy, which can be safely treated by stopping statin therapy and supplementing with CoQ10. Moreover, it was suggested that statin-associated cardiomyopathy may be a factor explaining the increasing prevalence of HF with preserved EF at present. The authors suggest that the supraphysiological levels of CoQ10 achieved in this study (mean = 6.6 μg/mL) not only correct possible tissue CoQ10 deficiency but also support the function of statin-damaged mitochondria [5].

Other authors also link the mechanisms of the negative effect of statins on cardiomyocytes with mitochondrial dysfunction and believe that risk groups include elderly patients and individuals with diseases associated with changes in the quantity and quality of function of these cellular organelles [6].

On the other hand, according to a number of other researchers, the above mentioned pleiotropic mechanisms, may explain the positive effects of statins on the hypertrophic cardiomyopathy and chronic heart failure improvement due to stimulation of angiogenesis, antioxidant and anti-apoptotic action [7].

Due to the well-known and well-studied influence of statins on the development of new cases of diabetes, there is an opinion that the cause of diabetic

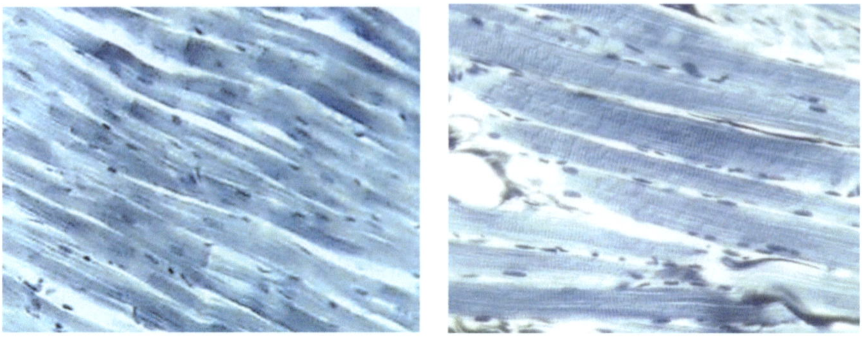

Fig. 11.2 Microscopic preparations of human myocardium (left) and striated skeletal muscle (right), stained with iron hematoxylin, magnification ×100. Prepared by Yu.Andreev, PhD and E. Kazantseva, PhD. Original to the author

cardiomyopathy may be the long-term use of high doses of statins [8]. At the same time, other authors, on the contrary, have shown the therapeutic effect of statins at the initial stage of the development of diabetic cardiomyopathy. This is supporting by recent findings that atorvastatin alleviated myocardial fibrosis under experimental conditions by regulating the antioxidant stress and anti-inflammatory effects of macrophage polarization on diabetic myocardium, and thereby improving DCM [9].

It is interesting that clinically manifested side effects of statins are characteristic mainly in relation to skeletal muscles. As for cardiomyocytes, the negative effect of statins on them, despite a significant amount of experimental data, has practically not been confirmed in clinical practice, with the exception of a number of observations described above.

This can be explained by the existence of a number of differences in the morphology and metabolism of skeletal muscles and myocardium. The myocardium is represented by cardiomyocytes that have all the characteristics of individual cells with one nucleus and their own cellular organelles. In contrast, skeletal muscles are morphologically multinucleated symplasts (Fig. 11.2).

Therefore, any negative effects on them can more easily generalize and take on an avalanche-like character. In addition, contractions of skeletal muscles and myocardium, despite common mechanisms, differ significantly. Skeletal muscles controlled by somatic nervous system, their contraction is fast and needed more energy consumption while cardiomyocytes are under autonomic nervous control by sinus node, contract involuntary and have more reliable safety from injury likely due to primary aerobic metabolism, more mitochondria quantity and higher redox potential [10].

Cardioprotective Effects of Statins in Antitumor Therapy

Among drugs with cardiovascular toxicity, antitumor agents occupy a leading position. As it turned out, statins are able to reliably reduce the cardiotoxic effect of a number of antitumor drugs regardless of their mechanism of action. It is well known for example that anthracyclines can cause serious therapy-related cardiac dysfunction by mitochondrial DNA damage and some other mechanisms. At the same time, a number of single-center and multicenter studies have demonstrated a mitigation of the cardiotoxic effect of anthrocycline in patients receiving medium and high doses of statins.

The current meta-analysis, encompassing 808 cancer patients from five randomized clinical trials (RCT), shows a beneficial effect of statins in preventing left ventricular dysfunction after anthracyclines treatment, evidenced by both lower incidence of cardiotoxicity and smaller mean change in left ventricular ejection fraction (LVEF) in statin-treated patients compared to placebo [11].

In other multicenter double-blinded, placebo-controlled trial, patients at increased risk of anthracycline-related cancer therapy-related cardiac dysfunction were randomly assigned to atorvastatin 40 mg or placebo once-daily. It was shown that pre- and post-anthracycline LVEF or ventricular remodeling did not differ significantly between the statin and placebo groups when adjusted for pre-anthracycline values. In addition the investigators observed no significant difference in the increase in hsTnI and other tissue biomarkers between the groups during treatment suggesting that statins do not prevent myocardial injury edema and/or fibrosis [12]. The results of one more meta-analysis demonstrate that anthracycline-based chemotherapy for lymphoma and breast cancer in patients concomitantly prescribed with statins showed a lower risk of cardiotoxicity than without statins. However, evidence supporting the benefit of statins to prevent a decline in LVEF relative to baseline in anthracycline-treated cancer patients was nonsignificant [13]. The authors of recent review summarized that although the available data do not conclusively support the universal use of cardioprotective statin therapy, the current guideline recommendation to consider statins for patients at high or very high risk, such as those receiving high-dose anthracyclines, seems appropriate [14].

Trastuzumab therapy for HER2-positive cancers also is associated with cardiotoxicity. It concluded from umbrella review synthesized evidence from systematic reviews and meta-analyses that statins significantly reduce the risk of cardiac complications during trastuzumab therapy [15].

In several trials was convincingly shown that statin therapy is associated with a reduced risk of major adverse cardiovascular events (MACE) in patients with breast cancer undergoing breast-conserving surgery and adjuvant whole breast radiotherapy. Hydrophilic statins rosuvastatin and pravastatin in these patients also exhibit the most pronounced cardioprotective effects [16, 17].

Conclusions

It is obvious that statins are effective agents for lowering cholesterol through the inhibition of HMG-CoA reductase, which leads to increased LDL receptor expression in hepatocytes and a subsequent decrease in total and LDL cholesterol levels. Their pleiotropic effects, including improved endothelial function, reduced inflammation, stabilized atherosclerotic plaques, antiplatelet effects, and immunomodulation, contribute significantly to their role in preventing coronary heart disease and atherosclerosis. Statins vary in pharmacodynamic and pharmacokinetic properties, with newer agents like atorvastatin, rosuvastatin, and pitavastatin demonstrating greater efficacy in lipid reduction compared to older statins like lovastatin and pravastatin. Clinical trials indicate that these modern statins significantly reduce the incidence of cardiovascular events and improve survival rates. While statins may present varying profiles of side effects and metabolization, their overall impact on cardiovascular health underscores their importance in both primary and secondary prevention strategies for patients at risk of atherosclerotic disease.

Nevertheless, statin therapy can pose musculoskeletal side effects, with rhabdomyolysis being the most severe yet rare complication, particularly associated with cerivastatin, which led to its market recall due to higher incidences of fatal cases. Generally, myalgia and myopathies, collectively known as statin-associated muscle symptoms (SAMS), are more commonly observed. Management of SAMS has been outlined by the European Atherosclerosis Society, emphasizing the need for monitoring both transaminase and creatine phosphokinase (CPK) levels before and during therapy. A significant increase in CPK levels, particularly a fourfold elevation, warrants reevaluation of statin therapy. SAMS can manifest as myalgia, myopathy, myositis, or varying degrees of muscle necrosis, and clinical signs indicating impending rhabdomyolysis include muscle pain, weakness, dark urine, and elevated CPK levels. Diagnosing SAMS is complex due to patients often attributing all muscle-related symptoms to statins, necessitating a clear cause-and-effect identification. Appropriate management strategies involve dose adjustment, switching to different statins, or employing intermittent dosing schedules. Regular CPK monitoring is advised for high-risk groups and following the emergence of muscle symptoms, whereas routine monitoring may not be necessary in other cases unless significant CPK elevations are noted. Overall, judicious monitoring and management can help mitigate the risks associated with statins while allowing for continued therapy.

The effects of statins on the myocardium reveal both negative and positive aspects, with a substantial number of studies indicating potential cardiotoxicities. It was identified significant elevations in troponin I levels in patients on statins, suggesting cardiotoxic effects, while other researchers acknowledged that these drugs may adversely affect the myocardium in the presence of additional risk factors.

Some researchers highlighted statin-associated cardiomyopathy (SACM) as a largely unrecognized cause of iatrogenic heart failure, particularly HFpEF, evidenced by a significant proportion of HFpEF patients undergoing lipid-lowering therapy. The mechanisms behind these adverse effects are theorized to stem from statins' inhibition of HMG-CoA reductase, leading to decreased levels of essential mitochondrial compounds such as CoQ10. While some studies indicate a decline in mitochondrial function due to statins, others have proposed potential therapeutic benefits, including mitigation of diabetic cardiomyopathy through antioxidant effects. Interestingly, reported side effects of statins are predominantly associated with skeletal muscle rather than the myocardium, possibly attributed to differences in cellular morphology and metabolism. While the evidence for adverse effects on cardiomyocytes remains less pronounced clinically, the potential mechanisms linking statins to cardiac impairment and their multifaceted role in various cardiac conditions necessitate further investigation.

And at last statins demonstrate significant cardioprotective effects in patients undergoing antitumor therapy, particularly concerning the cardiotoxicity associated with anthracyclines and trastuzumab. Despite the potential for anthracyclines to cause serious cardiac dysfunction through mitochondrial DNA damage, various studies, including a meta-analysis cancer patients, indicate that patients receiving statins experience a lower incidence of cardiotoxicity and less change in left ventricular ejection fraction (LVEF). However, results from additional placebo-controlled trials suggest that while there is no distinguishable impact on ventricular remodeling or myocardial injury markers between statin and placebo groups, the overall risk of cardiotoxicity appears reduced when statins are co-administered with anthracycline-based chemotherapy. Evidence supporting the decline of LVEF prevention is minor, leading to an ambiguous consensus on the universal recommendation of statin therapy for all anthracycline-treated patients. Nevertheless, guidelines favor considering statins for high-risk patients, especially those receiving high-dose anthracyclines. Furthermore, statins significantly lower the risk of cardiac complications in HER2-positive cancers during trastuzumab therapy and have been shown to reduce major adverse cardiac events (MACE) in breast cancer patients receiving specific surgical and radiotherapy treatments. Hydrophilic statins, such as rosuvastatin and pravastatin, are particularly noted for their pronounced cardioprotective effects in these patients.

References

1. Chou R, Cantor A, Dana T et al (2022) Statin use for the primary prevention of cardiovascular disease in adults: updated evidence report and systematic review for the US Preventive Services Task Force. JAMA 328(8):754–771. https://doi.org/10.1001/jama.2022.12138
2. Zeng W, Deng H, Luo Y, Zhong S, Huang M, Tomlinson B (2024) Advances in statin adverse reactions and the potential mechanisms: a systematic review. J Adv Res. https://doi.org/10.1016/j.jare.2024.12.020
3. Ünlü S, Nurkoç SG, Sezenöz B, Cingirt M, Gülbahar Ö, Abacı A (2019) Impact of statin use on high sensitive troponin T levels with moderate exercise. Acta Cardiol 74:380–385

4. Mital S, Liao JK (2004) Statins and the myocardium. Semin Vasc Med 4(4):377–384. https://doi.org/10.1055/s-2004-869594
5. Langsjoen PH, Langsjoen JO, Langsjoen AM, Rosenfeldt F (2019) Statin-associated cardiomyopathy responds to statin withdrawal and administration of coenzyme Q10. Perm J 23:257. https://doi.org/10.7812/TPP/18.257
6. Chaulin A (2023) Cardiotoxicity as a possible side effect of statins. Rev Cardiovasc Med 24(1): 22. https://doi.org/10.31083/j.rcm2401022
7. Somers T, Siddiqi S, Morshuis WJ, Russel FGM, Schirris TJJ (2023) Statins and cardiomyocyte metabolism, friend or foe? J Cardiovasc Dev Dis 10(10):417. https://doi.org/10.3390/jcdd10100417
8. Huang TS, Wu T, Wu YD et al (2023) Long-term statins administration exacerbates diabetic nephropathy via ectopic fat deposition in diabetic mice. Nat Commun 14(1):390. https://doi.org/10.1038/s41467-023-35944-z
9. Lei XT, Pu DL, Shan G, Wu QN (2024) Atorvastatin ameliorated myocardial fibrosis by inhibiting oxidative stress and modulating macrophage polarization in diabetic cardiomyopathy. World J Diabetes 15(6):1070–1073. https://doi.org/10.4239/wjd.v15.i6.1070
10. Guo Z, Tian Y, Gao J et al (2024) Enhancement of mitochondrial homeostasis: a novel approach to attenuate hypoxic myocardial injury. Int J Med Sci 21(15):2897–2911. https://doi.org/10.7150/ijms.103986
11. D'Amario D, Laborante R, Bianchini E et al (2023) Statins as preventive therapy for anthracycline cardiotoxicity: a meta-analysis of randomized controlled trials. Int J Cardiol 391:131219. https://doi.org/10.1016/j.ijcard.2023.131219
12. Thavendiranathan P, Houbois C, Marwick TH et al (2023) Statins to prevent early cardiac dysfunction in cancer patients at increased cardiotoxicity risk receiving anthracyclines. Eur Heart J Cardiovasc Pharmacother 9(6):515–525. https://doi.org/10.1093/ehjcvp/pvad031
13. Titus A, Cheema HA, Shafiee A, Seighali N, Shahid A, Bhanushali KB, Ganatra S (2023) Statins for attenuating cardiotoxicity in patients receiving anthracyclines: a systematic review and meta-analysis. Curr Probl Cardiol 48(10):101885. https://doi.org/10.1016/j.cpcardiol.2023.101885
14. Zadok OIB, Nohria A. Statins for cardioprotection fromanthracycline cardiotoxicity. https://www.acc.org/Latest-in-Cardiology/Articles/2023/09/07/12/43/
15. Wang Y, Xu J, Xie Y et al (2024) Interventions for prevention and treatment of trastuzumab-induced cardiotoxicity: an umbrella review of systematic reviews and meta-analyses. Front Pharmacol 15:1479983. https://doi.org/10.3389/fphar.2024.1479983
16. Huang YJ, Lin JA, Chen WM, Shia BC, Wu SY (2024) Statin therapy reduces radiation-induced cardiotoxicity in patients with breast cancer receiving adjuvant radiotherapy. J Am Heart Assoc 13(20):e036411. https://doi.org/10.1161/JAHA.124.036411
17. Walls GM, O'Connor J, Harbinson M et al (2023) Association between statin therapy dose intensity and radiation cardiotoxicity in non-small cell lung cancer: results from the NI-HEART study. Radiother Oncol 186:109762. https://doi.org/10.1016/j.radonc.2023.109762

Chapter 12
Anesthetics and Cardiovascular Toxicity

Marko Djuric and Irina Nenadic

Abstract Anesthetics are widely used in clinical practice to ensure safe anesthesia and enable surgical procedures, as well as for sedating patients in intensive care units. This chapter examines the cardiotoxic effects of intravenous, inhalational, and local anesthetics, focusing on their mechanisms of action and associated risks. While their cardioprotective actions are often highlighted, certain clinical conditions reveal their potential for cardiotoxicity. Intravenous anesthetics like propofol may cause hypotension, myocardial depression, and, in rare cases, propofol infusion syndrome, while ketamine's sympathomimetic effects increase blood pressure and cardiac output, posing risks for patients with cardiovascular disease. Both can affect mitochondrial function and the cardiovascular system. Inhalational anesthetics, such as sevoflurane, isoflurane, and desflurane, have varied effects: sevoflurane reduces cardiac contractility, whereas isoflurane and desflurane are linked to oxidative stress and DNA damage. Local anesthetics, including lidocaine, bupivacaine, and ropivacaine, block sodium channels but at high doses may cause severe cardiovascular reactions, such as arrhythmias and reduced contractility. Monitoring hemodynamic parameters is crucial when using these anesthetics, particularly in patients with compromised cardiovascular functions. This chapter underscores the importance of dose adjustment, risk identification, and further research to minimize toxic effects, enhancing patient safety in anesthesiology practice.

Keywords Cardiovascular toxicity · Hemodynamic stability · Inhalational anesthetics · Intravenous anesthetics · Local anesthetics

M. Djuric (✉) · I. Nenadic
Clinic for Anesthesiology and Intensive Care, University Clinical Hospital Center "Dr Dragisa Misovic-Dedinje", Belgrade, Serbia

Department of Anesthesiology, Reanimatology and Intensive Care, Faculty of Medicine, University of Belgrade, Belgrade, Serbia

D. K. Agrawal et al. (eds.), *Cardiovascular Toxicity*, Advances in Biochemistry in Health and Disease 33, https://doi.org/10.1007/978-3-031-90598-8_12

Introduction

Anesthetics are drugs that enable safe surgical intervention by causing reversible loss of sensation and consciousness. Although these agents have improved significantly over time, their use remains associated with various adverse effects on the cardiovascular system. Modern medicine cannot be imagined without anesthetics, which are used not only for the safe performance of surgical procedures but also for the sedation of critically ill patients in intensive care units, as well as for invasive outpatient procedures. Anesthetics used during surgery or sedation prevent patients from feeling pain and remaining conscious. However, in addition to their anesthetic properties, these drugs also exhibit numerous non-anesthetic properties. They can affect the cardiovascular system, affecting heart rate, arterial blood pressure, and cardiac output. One of the roles of the anesthesiologist is to balance the delivery of anesthesia and minimize the adverse cardiovascular effects that these drugs can cause. Different groups of anesthetics can have different effects on the cardiovascular system. Some of them can cause vasodilatation of blood vessels, thus reducing blood flow to the right heart and consequently leading to a drop in arterial blood pressure [1]. Some older studies performed on experimental animals also report myocardial depression and vasodilation that anesthetics can induce [2]. A problem in clinical practice can be elderly patients with a compromised cardiovascular system. These are usually high-risk patients in whom strict attention must be paid to the method, speed and dose of anesthetic administration, all in order to avoid cardiovascular depression. It is necessary to take care of patients with heart diseases and manage their medications. Anesthetics can often interact with these drugs, which can alter the effectiveness and toxicity of the anesthetic drugs used during surgery [3]. Because of such problems during clinical practice, cooperation between cardiologists and anesthesiologists is often necessary. Doctors of these specialties should participate together in preparing high-risk patients, all in order for the patient to undergo surgery safely. There are several reasons why the effects of anesthetics on the patient's cardiovascular system should be well known. The primary goal is to keep the patient safe. As is known, cardiovascular complications are among the most severe risks associated with anesthesia [4]. Hypotension, arrhythmias and cardiac arrest are most commonly referred to here. That is why knowing the adverse effects of anesthetics allows the anesthesiologist to make a better anesthesia management plan in order to reduce these adverse effects of anesthetics. Also, creating a treatment plan that should be adapted to each patient individually is necessary. For example, etomidate can be used for induction of anesthesia in patients with compromised cardiovascular function because of its minimal effects on the heart and blood vessels. On the other hand, propofol should be avoided in hypotensive patients. During the surgical intervention, the patient must be hemodynamically stable. Fluctuations in blood pressure or heart rate can lead to a drop in tissue perfusion and oxygenation, compromising organ function and wound healing. Oxygen is essential for various processes such as cell growth, the formation of new blood vessels, and the production of extracellular matrix, all of which are necessary to restore function and tissue integrity at the wound site [5]. It should be remembered that each patient has a unique physiological response to anesthetic administration. Factors that can affect this are race, comorbidities, age, medications and others. Elderly patients are known to have reduced cardiovascular reserve [6]. They may be more susceptible to hypotensive episodes due to anesthetic administration. Unwanted cardiovascular problems that can occur before or during surgery often occur in the postoperative period as well. That is why the anesthesiologist must start treatment of cardiovascular disorders that occur as a consequence not only of anesthetic administration but also of surgical intervention. This chapter will discuss different groups of anesthetics and their mechanisms of adverse effects on the cardiovascular system.

Intravenous Anesthetics and Cardiovascular Toxicity

Propofol is one of the most commonly used anesthetics in clinical practice. Due to its pharmacokinetic profile, this drug is known for its rapid onset and short duration of action. Propofol may be associated with cardiovascular toxicity, particularly in susceptible populations such as elderly patients with multiple comorbidities and children. The damaging property of propofol is well known, which, when administered continuously in the form of infusions for an extended period and in high doses, can develop propofol infusion syndrome. This condition is extremely rare but potentially fatal. The onset of metabolic acidosis, rhabdomyolysis, or ECG changes characterizes the syndrome. It may also include the onset of acute kidney injury, hyperkalemia, lipidemia, heart failure, fever, elevated liver enzymes, and elevated lactate levels [7]. The exact mechanism by which this syndrome occurs has not been fully elucidated. Some older studies suggest that the accumulation of inactive propofol metabolites, lipid microembolization, and impaired lactate metabolism in the liver lead to the appearance of this syndrome [8–10]. Some recent studies in rats have shown that propofol inhibits coenzyme Q, which transfers electrons from complex II to complex III in the mitochondrial electron transport chain [11]. This is explained by the similarity of the structure of propofol and coenzyme Q. A recent study by Vollmer et al., using electron microscopy found electron-dense bodies associated with mitochondria in muscle and liver cells corresponding to free fatty acid accumulation, which provide direct evidence for mitochondrial damage in propofol infusion syndrome [12]. Unfortunately, the exact pathophysiological path of the origin of this syndrome has not been thoroughly investigated, and researchers mainly focus on mitochondria and the respiratory chain. Recent research is moving towards developing an ex-vivo model using isolated perfused neonatal mouse hearts to study propofol-induced cardiotoxicity, aiming to replicate the cardiac features of this syndrome [13]. The study showed that a toxic dose of propofol rapidly leads to heart rhythm problems, reduces the heart's ability to contract, and impairs mitochondrial function without causing cell death [13]. Unfortunately, children and newborns are a sensitive part of the population, in which the onset of propofol infusion syndrome has already been seen, and the mitochondria of the heart, which are parts of the cells that produce energy, are most often damaged. In newborn mice, exposure to propofol led to increased proton leakage in the heart's mitochondria, meaning that their work and energy production were impaired [14]. This is one of the reasons that leads to impaired heart function. This study also showed that propofol interferes with a part of the mitochondria called coenzyme Q, which is crucial for energy production. This study on newborn mice may explain why children and infants are more likely to develop this syndrome. That is why it is necessary to approach alternative treatments in pediatric practice. Studies conducted on cardiomyocytes derived from human pluripotent stem cells have shown that high concentrations of propofol (10 μg/mL or more) impair mitochondrial function and reduce cell viability [15]. In the same study, propofol-induced mitochondrial dysfunction was shown to be associated with downregulation of genes such as PGC-1 alpha. This gene is known to be crucial for mitochondrial energy metabolism. What was particularly interesting was that coenzyme Q10 was found to protect cardiomyocytes from the toxic effects of propofol by restoring the expression of these essential genes and improving mitochondrial function. This study supports coenzyme Q10 as a protective agent against propofol-induced cytotoxicity that may have important clinical implications. Propofol infusion syndrome occurs most often when propofol is used in high doses. Research by Chisato Sumi et al. has shown, similar to some other research, that propofol disrupts mitochondrial function, leading to a shift from oxidative phosphorylation to glycolysis, which can subsequently lead to cell death [16]. In this way, propofol acts on mitochondrial complexes I, II and III, creating reactive oxygen species and leading to cell apoptosis. In this same study, propofol at concentrations of 50 μM or higher was shown to be cytotoxic within 6 h, while 25 μM propofol increased cell death and caspase activity after 12 h. Lactic acidosis that occurs in propofol infusion syndrome can be explained by the fact that propofol increases glycolysis and increases lactate production. This mechanism can explain the lactic acidosis that occurs in this syndrome. As mentioned, propofol suppresses oxygen consumption in mitochondria by inhibiting complexes I, II and III of the electron transport chain, which leads to decreased ATP production and increased reliance on glycolysis for energy. Certain studies show that cells with active hypoxia-inducible factor I (HIF-1) exhibit higher glycolysis and ATP levels, which may indicate a shift from aerobic to anaerobic metabolism. This may help cells resist the toxicity of propofol [17]. From this, it can be seen that the activation of the HIF-1 factor leads to a more excellent resistance of cells to the toxic action of propofol, showing that HIF-1 has a protective role and can potentially be used in a new therapeutic approach. Ketamine and propofol belong to the group of intravenous anesthetics. Unlike propofol, which has a depressing effect on the cardiovascular system, ketamine increases cardiac output and arterial blood pressure. In addition to its hypnotic effect, this drug has an analgesic effect. In recent years, its effects on the cardiovascular system have caused concern in patients who already have cardiovascular disease. It is known for its sympathomimetic properties, which can lead to significant cardiovascular changes. Consequently, it can lead to increased levels of circulating catecholamines. It has been shown in experimental animals that ketamine can alter the expression of genes related to hypoxia and oxidative stress pathways, which may play a key role in cardiovascular function. Research results showed that ketamine alters the activity of hypoxia-related genes p53, Wnt, Notch, TGFβ, PPAR and oxidative stress pathways, suggesting that these pathways may be involved in the toxic effects of ketamine [18]. It is interesting that the 2-week administration of ketamine in experimental animals caused a significant increase in the weight of the heart and kidneys, as well as a higher level of apoptosis, which indicates a toxic action in these organs [19]. In the same study, it was shown that cessation of ketamine for 2–4 weeks only partially reduces heart and kidney damage. Experimental animal studies show that chronic abuse of ketamine can lead to significant cardiovascular toxicity, including ventricular structural changes and sympathetic and electrophysiological changes, increasing the risk of sudden cardiac death [20]. Ketamine-induced effects include apoptosis of the ventricular myocardium, fibrosis, and altered conduction properties, which favor the development of malignant arrhythmias. Ketamine also causes abnormal growth of nerve fibers in the heart. It was exciting that the administration of metoprolol (a β blocker) helped maintain normal heart function and structure, suggesting that it may have a protective role against the toxic effects of ketamine. Unfortunately, ketamine is a drug that is often taken recreationally and abused on the streets. Cardiovascular effects in people who take this drug recreationally include the onset of sinus tachycardia, which is usually self-regulating, and chest pain and palpitations, which may occur less often [21]. Ketamine can also be toxic to human neurons, as shown by some experiments that use neurons derived from human-induced pluripotent stem cells. When neurons were exposed to high concentrations of ketamine (500 μM) for 24 h, there was a significant increase in cell death markers and production of reactive oxygen species, indicating cell damage [22]. The same study showed that lower concentrations of ketamine lead to a decrease in ATP levels and changes in mitochondrial function without directly causing cell death. The study found that ketamine directly inhibits mitochondrial complexes I and V, crucial for cell energy production. From this, it can be concluded that there is a similarity between propofol and ketamine and their negative effect on mitochondria. Older studies show that ketamine leads to a decrease in left ventricular systolic and diastolic function in patients with ischemic heart disease [23]. This indicates potential cardiovascular toxicity. When it comes to ketamine, it is necessary to conduct further research in the direction of toxic effects on the cardiovascular system in patients who have associated cardiovascular diseases. Another possible mechanism by which ketamine leads to cardiovascular toxicity is to inhibit ATP-sensitive potassium channels in the heart, which are crucial for the protective effects of ischemic preconditioning in the rabbit heart [24]. By blocking KATP channels, ketamine prevents the heart from using its natural defense mechanism against ischemia. Therefore, the interaction between ketamine and KATP channels may lead to potential cardiovascular toxicity by impairing the ability of the heart to undergo preconditioning, which is a protective mechanism against ischemic damage. Developing new anesthesiology protocols that do not endanger the heart's natural protective mechanisms during operations is necessary. Etomidate is another drug that can be used intravenously to induce anesthesia. This drug is a favorite in anesthesiology practice due to its rapid onset of action, short duration of action and hemodynamic stability. Etomidate is known for its minimal cardiovascular effects. It can maintain normal arterial blood pressure and cardiac output in patients with compromised cardiovascular function. This makes it an ideal intravenous anesthetic in hemodynamically unstable patients. It is known that patients with diabetes, especially those with heart problems, have an increased risk of adverse cardiac events during surgery and anesthesia. Anesthetics can worsen pre-existing cardiovascular problems. In patients with diabetes, advanced glycation end products (AGEs) accumulate and contribute to diabetic complications in small and large blood vessels. Studies have shown that etomidate and a specific AGE (Glu-BSA) synergistically reduce the viability of Eahy926 human endothelial cells and induce apoptosis [25]. Also, the same study showed that the two agents synergistically promoted oxidative stress by reducing mitochondrial respiratory chain complex IV and mitochondrial membrane potential while regulating reactive oxygen species and mitochondrial superoxide. Here, we show that oxidative stress plays a vital role in the onset of apoptosis when etomidate and AGEs act together in endothelial Eahy926 cells. In anesthesiology practice, it is known that etomidate can affect adrenal function and alter cortisol levels. Studies have shown that the use of etomidate for induction of anesthesia is associated with higher 30-day mortality and cardiovascular complications compared with propofol, possibly due to effects on adrenal function and cortisol release during surgery [26]. There is an association between the use of etomidate and cardiovascular morbidity, which may indicate a potential cardiovascular toxicity with the use of etomidate during induction of anesthesia. An interesting study on experimental animals showed that etomidate can have cytotoxic effects and induce apoptosis. The study found that etomidate caused toxic effects on mouse leukemia cells, leading to changes in cell shape and decreased viability, and also contributed to apoptosis [27]. In this study, its harmful effect on the cardiovascular system was not shown, but it was clearly shown that this drug can potentially lead to cellular toxicity and apoptosis. It is necessary to conduct further research on the effect of this drug on the human cardiovascular system. Benzodiazepines are a special group of drugs used in anesthesia. These drugs are used for their anxiolytic, hypnotic, anticonvulsant, sedative effects, and muscle relaxant properties. In clinical practice, they are most often used to sedate patients in intensive care units, but they can also be used to introduce patients to anesthesia. They are most often used in combination with other intravenous anesthetics. Moreover, while they are generally considered safe drugs, there is growing evidence to suggest potential cardiovascular toxicity, especially with long-term use or overdose. In one interesting study on benzodiazepine poisoning, it was shown that the majority of poisoned patients, when admitted to the emergency room, had average heart rate, ECG and blood pressure. Only a small number of patients had hypotension, hypertension, bradycardia and tachycardia [28]. The toxicity of benzodiazepines to the cardiovascular system can sometimes be assessed using electrocardiographic manifestations. Research has shown that diazepam is the only benzodiazepine that does not cause a widening of the QRS complex, and oxazepam is the only one that does not cause a prolongation of the PR interval [29]. It can be concluded from this that if the patient presents with a reduced level of consciousness and accompanying signs of benzodiazepine poisoning, ECG changes may speak in favor of poisoning with certain drugs from the benzodiazepine group. There are cases in clinical practice where benzodiazepine poisoning has led to cardiovascular toxicity, which can be manifested by atrioventricular block [30]. Benzodiazepines are thought to act as calcium channel blockers, potentially causing AV blocks. Diazepam has been shown to suppress AV nodal conduction directly. However, in clinical practice, there is an antidote for benzodiazepine overdose, namely flumazenil. This medicine can reverse the unwanted cardiovascular effects caused by benzodiazepines. When it comes to hospital conditions, the most commonly used drug from this group is midazolam. This drug, although generally considered safe, may increase the risks of an adverse cardiovascular event in patients who already have heart problems. In one case study, a 5-year-old boy with diphtheria experienced airway obstruction and myocarditis, requiring mechanical ventilation and continuous intravenous midazolam sedation. Six hours after the infusion, the child developed pulsus bigeminus and bradycardia, which led to death [31]. This may indicate caution when using midazolam for myocarditis and possibly other heart problems due to the potential cardiotoxic effect of the drug. Although there is insufficient work to indicate a direct toxic effect of midazolam on the cardiovascular system, there is evidence obtained in experimental animals showing that midazolam reduces cell viability in mouse Leydig progenitor cells, indicating its potential cytotoxic effects [32]. The same study also mentions its effects on inducing cell apoptosis. Midazolam has been shown to activate caspase 8 and 3 proteins and phosphorylate JNK, EPK ½ and p38, suggesting that it initiates apoptosis through these pathways and can also induce the expression of p-EIF2α, ATF4, ATF3 and CHOP, indicating that it can cause apoptosis through endoplasmic reticulum stress in mouse TM3 cells [32]. From this, it can also be concluded that midazolam can injure the cells, which may give an idea for further research regarding the effects on the cardiovascular system.

Inhalational Anesthetics and Cardiovascular Toxicity

Nitrous oxide, or "laughing gas", is a colorless, non-flammable gas with a slightly sweet smell. It is used in anesthesia, but more as an addition to a more potent inhalational anesthetic such as sevoflurane. It is used in medical and dental practice due to its anesthetic and analgesic properties. Unfortunately, the recreational use of laughing gas has increased in recent years. Chronic exposure to nitrous oxide (N_2O) can lead to various health problems, the most common being neurological problems such as sensory and motor disorders, and cardiovascular problems such as thrombosis can also occur [33]. Exposure to N_2O can lead to oxidative stress by increasing the generation of reactive oxygen species. Unfortunately, occupational exposure to N_2O can lead to oxidative DNA damage. In a study involving nurses exposed to N_2O, oxidative stress markers such as 8-iso-PGF2α and thiobarbituric acid-reactive substances (TBARS) were significantly higher compared to unexposed individuals

[34]. This study also showed that glutathione peroxidase (an antioxidant enzyme) activity was reduced in those exposed to N_2O. It was concluded that oxidative stress is a crucial factor linking N_2O to genotoxicity. When it comes to the cardiovascular system, it is known that oxidative stress can lead to endothelial dysfunction, inflammation, and atherosclerosis and thus contribute to the development of cardiovascular toxicity. It is assumed that the toxic effect of N_2O on the cardiovascular system may be indirect through the inactivation of vitamin B12. Vitamin B12 is an essential cofactor in metabolic pathways involved in DNA and myelin synthesis [35]. Vitamin B12 deficiency can increase homocysteine levels, an amino acid associated with cardiovascular disorders. Elevated homocysteine can damage blood vessels and contribute to the development of atherosclerosis and thrombosis. N_2O can cause cardiovascular toxicity through the development of hypoxia, which leads to acute anoxemic damage to the myocardium and pulmonary and cerebral edema [36]. On the other hand, sevoflurane is an inhalational anesthetic that anesthesiologists favor because of its rapid onset of action and rapid recovery from anesthesia. It is the most commonly used inhalational anesthetic due to its favorable pharmacokinetic profile. There is evidence to support its adverse effects on the cardiovascular system, such as cardiovascular toxicity. Studies are mainly based on the effects of sevoflurane on the cardiovascular system during prolonged anesthesia and operations. Certain studies show no significant changes in troponin I, creatine kinase MB, and myoglobin levels in people receiving sevoflurane when comparing their values before induction with those measured 4 and 24 h after induction [37]. From the same study, it could be concluded that long-term anesthesia with sevoflurane in concentrations ranging from 1.6% to 3% does not lead to changes in markers of myocardial damage or toxicity, which makes it a safe anesthetic for extended operations. The literature mainly mentions the cardioprotective properties of sevoflurane. However, there are mechanisms by which sevoflurane can indirectly hurt the cardiovascular system. Research in rats has shown that postconditioning with sevoflurane can inhibit L-type calcium channels in cardiac myocytes [38]. Because of this, it can occur with a decrease in calcium influx during the action potential. This decrease in intracellular calcium can disrupt myocardial contractility, causing adverse inotropic effects. Sevoflurane is known to activate sarcolemmal and mitochondrial KATP channels in the heart [39, 40]. During low ATP conditions, such as ischemia, these channels open and allow K to leave the cell, which can shorten the duration of the action potential and slow the electrical conduction of the heart, potentially leading to arrhythmias [41]. Alterations in the K channel function may prolong the QT interval [42]. Although there is no direct evidence that sevoflurane administration can lead to cardiac disorders in this way, this may be an idea for further research regarding the negative effect of sevoflurane on the cardiovascular system. What was observed in the experimental model was that NO bioavailability and increased matrix metalloproteinase-2 (MMP-2) activity were reduced in the sevoflurane group compared to the isoflurane group, and reduced NO levels and increased MMP-2 activity may be involved in vascular dysfunction after anesthesia with sevoflurane [43]. This result may suggest that sevoflurane anesthesia may lead to increased production of reactive oxygen species (ROS) in the cardiovascular system. Excess ROS can

damage cellular components, including lipids, proteins and DNA, and potentially lead to myocardial damage. Oxidative stress can disrupt normal cellular signaling pathways and contribute to the development of myocardial arrhythmias and dysfunction. Sevoflurane-guided anesthesia may activate inflammatory pathways. There is evidence that sevoflurane can increase levels of interleukin 6 [44]. The inflammatory response itself can contribute to myocardial damage and contribute to postoperative complications, especially in patients who already have certain cardiovascular diseases. While its cardioprotective properties are mainly discussed, some evidence suggests that sevoflurane can reduce mitochondrial activity in heart cells by affecting the electron transport chain [45]. This decrease in mitochondrial activity caused by sevoflurane is dose-dependent. A higher dose administered leads to a more significant decrease in activity. The same paper discusses that ROS plays a significant role in reducing mitochondrial activity, especially in complex I of the electron transport chain. This study shows that it is essential to understand how anesthetics like sevoflurane affect mitochondrial function, as this may have clinical implications. In addition to sevoflurane, volatile inhalational anesthetics also include isoflurane and desflurane. They are used less often in clinical practice than sevoflurane. Although there is no direct evidence that isoflurane can have a cardiotoxic effect, some studies talk about its cytotoxic potential. Isoflurane is thought to cause cell death by causing excessive calcium release from the endoplasmic reticulum by activating the inositol 1,4,5 triphosphate receptor [46]. Although the study was performed on rat pheochromocytoma cells, it was found that exposure to 1 MAC (Minimum Alveolar Concentration) of isoflurane for 12 h led to significant changes in cellular structure, including swelling of the mitochondria and endoplasmic reticulum, as well as changes in the shape of the nucleus. Research like this suggests that inhaled anesthetics like isoflurane can cause cell damage, and the inositol 1,4,5 triphosphate receptor may be a key site for preventing anesthesia-induced cytotoxicity. Some of the mechanisms by which isoflurane can lead to cardiovascular toxicity can be explained by different mechanisms of how it acts on the cardiovascular system. This anesthetic can affect vascular tone, myocardial function and heart electrical conduction. Isoflurane is known to cause peripheral vasodilatation, leading to a decrease in systemic vascular resistance and hypotension. One of the mechanisms by which it can act as a vasodilator is by reducing arginine vasopressin. Arginine vasopressin is otherwise a potent vasoconstrictor. It works by increasing the amount of calcium inside the cells of the blood vessel walls and making the muscle fibers more sensitive to calcium. The researchers showed that isoflurane and sevoflurane could reduce the contraction caused by arginine vasopressin in a dose-dependent manner, meaning that the more anesthetic was used, the fewer blood vessels contracted [47]. One possible mechanism addressed here is the inhibition of arginine vasopressin-induced vasoconstriction by lowering intracellular calcium in rat aortic smooth muscle, as demonstrated in this study. Some older studies have also suggested that isoflurane reduces baroreflex control of heart rate and may inhibit the sympathetic nervous system [48]. This anesthetic can also have a negative inotropic effect, which reduces the force of myocardial contraction. This side effect can disrupt cardiac output, especially in patients with pre-existing heart disease. One of the mechanisms is the already mentioned decrease in the influx and release of calcium from the endoplasmic reticulum, which may be crucial for myocardial contraction. A study in experimental animals showed that sevoflurane can reduce the ability of the heart muscle to respond to calcium, which may be crucial for muscle contraction [49]. From this experimental research, it could be concluded that isoflurane reduces the sensitivity of myofilaments to calcium in damaged myocardium, which may have clinical implications if this anesthetic is used in patients with damaged myocardium. It is not uncommon in clinical practice that isoflurane can cause a dose-dependent increase in heart rate. In the past, the phenomenon of coronary steal that could be caused by isoflurane was often discussed. Isoflurane was reported to be a potent coronary vasodilator that could induce coronary steal and myocardial ischemia in patients with coronary artery disease. Further research rejected this theory and found that isoflurane has a preconditioning, cardioprotective effect [50]. Isoflurane can also act on voltage-gated sodium channels, which are crucial for the excitability of nerve and muscle cells. The isoform-selective effects of isoflurane on the main Na channel isoforms expressed in excitable tissues were investigated. It is concluded that the effects of isoflurane on Na channels involve two mechanisms: enhanced inactivation due to a hyperpolarizing change in the voltage dependence of fast steady-state inactivation and tonic block [51]. The isoflurane's isoform-selective effects on Na channels suggest a potential pharmacological basis for isoform discrimination to improve anesthetic specificity in cardiac cells. There are indications that the inhaled anesthetic isoflurane may increase myocardial oxygen demand. A study done on diabetic rats showed that isoflurane can reduce the force of cardiac muscle contraction in both healthy and diabetic rats. However, it has a more severe effect on diabetic rats by lowering intracellular calcium levels. Also, diabetic rats had reduced myocardial oxygen consumption, which may indicate heart muscle damage in diabetic conditions [52]. Therefore, it is necessary to be especially careful if this anesthetic is used in diabetic patients with impaired heart function. Desflurane also belongs to the group of volatile halogenated anesthetics. This drug is favored for its pharmacokinetic properties, but there are still concerns about its potential cardiovascular toxicity. Moreover, if its direct toxic action on the cardiovascular system is rarely mentioned in the literature, some findings speak of its pro-inflammatory and genotoxic action. The research results showed a significant genotoxic effect and an increase in the levels of IL-6 and IL-8, and the gene expression of these cytokines remained unchanged, suggesting that desflurane can cause inflammation and DNA damage without affecting gene expression [53]. Some findings suggest that desflurane anesthesia in healthy individuals undergoing minor surgery may cause DNA damage [54]. These findings suggest that further research is needed regarding the potential genotoxic effects of desflurane on the cardiovascular system. This is not the only mechanism by which desflurane can potentially impair the function of the cardiovascular system; exposure to this anesthetic can cause oxidative stress in patients undergoing surgery. In one clinical study, it was proven that desflurane can increase the levels of lipid peroxidation. On the other hand, it can decrease the levels of α-tocopherol (a form of vitamin E). The study also showed that the activities of erythrocyte superoxide dismutase and serum glutathione peroxidase did not show significant differences between the preoperative and postoperative periods, which only suggests that the oxidative stress caused by desflurane affected the levels of vitamin E but not these antioxidative enzymes [55]. In some cases, its hepatotoxic effect is also described, especially for elderly patients [56]. If the literature mainly talks about the direct cardioprotective effects of inhalational anesthetics, it can be concluded from these studies that they can also cause potential toxicity to the cardiovascular system. A thorough understanding of all their effects, along with the careful use of these anesthetics in clinical practice, is essential for optimizing patient outcomes. As our knowledge of molecular mechanisms continues to evolve, future research into the cardiovascular toxicity of these drugs may reveal new strategies to mitigate the risks of inhalational anesthetics in clinical practice. The goal is to ensure safer and more efficient anesthetic and surgical care, especially in patients with associated cardiovascular diseases.

Local Anesthetics and Cardiovascular Toxicity

Local anesthetics are drugs that are mainly used for the treatment of acute intra- or postoperative pain, but also chronic pain, as well as in the treatment of cardiac arrhythmias. They can be applied topically by tissue infiltration or injection around the nerves or neuraxial (within spinal and peridural analgesia). At the same time, certain local anesthetics can also be administered intravenously for local, intravenous analgesia. According to their chemical structure, they are weak bases (pKa 7.5–8.5) consisting of a lipophilic aromatic ring, a hydrophilic tertiary amine and an ester or amide bond between them, based on which the classification into aminoesters and aminoamides was made [57]. They are formulated as the hydrochloride salt to make local anesthetics stable in solution. As such, the molecules exist in a water-soluble quaternary state upon injection and cannot penetrate the neuron. The time of onset of action of local anesthetics directly depends on the percentage of molecules that are transformed into a tertiary liposoluble structure at physiological pH values. Since the pKa of most local anesthetics is between 7.5 and 8.5, many molecules are in the quaternary cationic form. This slows down the onset of action of the drug. However, after entering the intracellular space, the molecules dissociate to achieve a new balance of ionized and non-ionized parts, depending on the intracellular pH and pKa of the local anesthetic. The ionized form binds reversibly to the intracellular domain of voltage-dependent NaV+ channels, inhibiting neuronal ion transfer and cell membrane depolarization. These channels consist of one α unit that is the binding site for local anesthetics and one or more β subunits that modulate the cell membrane's kinetics and voltage-dependent activation/inactivation [58]. Depending on the type of α subunit, these channels have at least ten different types. Mutation of the gene for the α subunit leads to the development of various pathological conditions, such as congenital insensitivity to pain, paroxysmal pain syndrome, cardiac arrhythmias or epilepsy [59]. In addition, local anesthetics bind to and block potassium (K^+), calcium (Ca^{2+}), and sodium-potassium ATPase channels.

They can also interfere with intracellular and transmembrane cell signaling, affecting cyclic adenosine monophosphate (cAMP), protein kinase B (PK-B), and 5-adenosine monophosphate-activated protein kinase. They also interfere with mitochondrial metabolism, affect the production of adenosine triphosphate (ATP), inhibit the ryanodine receptor on the sarcoplasmic reticulum and reduce the sensitivity of myofilaments to Ca^{2+} [60]. Acidosis, i.e., low arterial blood pH values, can cause intracellular trapping of ions, as a result of which the toxic effect of the drug can be enhanced [61]. Voltage-gated sodium channels (NaV+) also initiate and conduct action potentials in the cardiac conduction system. Some local anesthetics have been shown to interact with cardiac L-type calcium channels and inhibit cell membrane repolarization, leading to β-adrenergic-mediated blockade of cAMP production and mitochondrial dysfunction [62]. In a clinical study performed on cultured human cardiomyocytes, calcium supplementation reduced bupivacaine-induced cardiotoxicity but enhanced ropivacaine-induced cardiotoxicity [63]. By suppressing the activity of protein kinase B and protein kinase activated by 5-adenosine monophosphate, the intracellular glucose metabolism facilitated by insulin is interrupted, and the intracellular ATP reserve is reduced. Local anesthetics directly affect heart tissue (sinus node, Purkinje fibers, myocardium). The direct suppressive effect of local anesthetics is exerted on the neural control mechanisms of baroreceptors, which affects vascular tone [64]. With the application of high doses of local anesthetics or with their continuous long-term application, intoxication (LAST—local anesthetics system toxicity) can occur, which is most often manifested by symptoms from the CNS and CVS. Cardiovascular toxicity caused by local anesthetics depends on several different parameters. From the type of local anesthetic, the amount of drug used and the route of administration, but also from the characteristics of the patient himself, such as age, conditions such as pregnancy or associated diseases. It has long been known that during pregnancy, the level of protein decreases, and thus, the level of the protein-bound fraction of local anesthetics, which is why the possibility of developing systemic toxicity is increased [65]. Conditions of extreme malnutrition also lead to the same. Although the most common consequences of an overdose of local anesthetics are central nervous system effects such as epileptic seizures, isolated cardiovascular effects occur in as many as 20% of overdose cases [66]. Most often, the first cardiovascular effects of an overdose of local anesthetics occur due to neurointoxication and are manifested in the form of hypertension and tachycardia. In the future, local anesthetics can affect the heart's conducting system (the bundle of His fibers is most often affected), leading to myocardial dysfunction and disturbance of vascular tone. Lidocaine and mepivacaine most often affect myocardial contractility, while ropivacaine, levobupivacaine and bupivacaine show a negative inotropic effect and lead to arrhythmias [67]. The first clinical symptom of cardiovascular toxicity is usually the appearance of arrhythmias. By driving the resting membrane potential to a more negative level, action potential propagation is impaired, resulting in prolonged PR, QRS, and ST intervals. Re-entrance tachyarrhythmias and bradyarrhythmias appear, which can be aggravated by further blockade of potassium channels, prolonging the QT interval.

Lidocaine

Lidocaine belongs to the group of amide local anesthetics. It shows analgesic, anti-inflammatory and antiarrhythmic effects. It was discovered in 1943 and patented as an analgesic, and already in the mid-50s of the twentieth century, it began to be used as an antiarrhythmic. Until the mid-1980s, lidocaine was the drug of choice in the treatment of ventricular arrhythmias [68]. Like other local anesthetics, lidocaine exerts its effect by blocking voltage- and pH-dependent sodium channels in a dose-dependent manner. In this way, it leads to a decrease in the speed of signal conduction in the heart. Lidocaine stabilizes the myocardial cell membrane and exhibits an antiarrhythmic effect by reducing myocardial automaticity: Lidocaine exhibits a dose-dependent effect on the automaticity of myocardial pacemaker cells by slowing down the rate of phase 4 depolarization, thus prolonging the refractory period, i.e. the resting phase [69]. This drug binds predominantly, but not exclusively, to inactive NaV+ channels. Also, it should be noted that lidocaine shows a small but negligible effect on voltage-gated potassium channels [70]. **Prevention of reentry arrhythmias**: Lidocaine helps prevent reentry arrhythmias by suppressing premature ventricular contractions, thus restoring a normal rhythm. A decrease in the duration of the action potential and an increase in the effective refractory period can terminate the reentry tachycardia. **Shortening action potential duration**: Lidocaine shortens action potential duration and refractory period, which is instrumental in ischemic myocardium where action potentials tend to be prolonged. The threshold for the occurrence of ventricular fibrillation is increased during the administration of lidocaine in therapeutic concentrations and in patients with ischemic heart disease, even at lower doses. The most pronounced effect was recorded on Purkinje fibers, followed by ventricular cells. The effect on the atria is less pronounced and is present only at higher drug concentrations in the blood [69]. In several studies carried out in the 70s of the twentieth century, it was proven that the effect of lidocaine on the degree of depolarization depends not only on the dose of the drug but also on the concentration of extracellular potassium [71, 72]. Therefore, in ischemic myocardium, in which hypoxia and subsequent acidosis lead to the release of potassium from the cell and the emergence of extracellular hyperkalemia, there is a progressive loss of the resting membrane potential, a decrease in the speed and refractoriness of the action potential, and a transient increase, and then reducing excitability. Hyperkalemia increases the affinity of lidocaine for sodium channels and potentiates its effect at lower concentrations. Lidocaine also causes hypotension in the following ways:

Direct myocardial depression Lidocaine reduces myocardial contractility at high doses, leading to a decrease in cardiac output.

Vasodilation It may cause mild vasodilation by blocking sodium channels in efferent vasoconstrictor sympathetic nerves or directly acting on the smooth vascular muscle. This contributes to the reduction of systemic vascular resistance and blood pressure [73]. A study conducted in a population of patients undergoing vascular surgery showed that intraoperative use of lidocaine infusion as an adjunctive analgesic did not affect intraoperative blood pressure variability [74]. High doses of lidocaine, especially administered as a continuous infusion, can lead to bradycardia. Lidocaine shows minimal effect on the SA node but can prolong conduction time through the AV node and, consequently, in toxic doses, can lead to AV block [75]. While lidocaine is often used to treat arrhythmias, paradoxically, at toxic levels, it can cause ventricular arrhythmias, including ventricular fibrillation and ventricular tachycardia. This proarrhythmic effect is rare but is more likely in cases of overdose or patients with pre-existing severe heart disease. Lidocaine has been proven not to affect the onset of endothelial dysfunction [76]. In cases of lidocaine overdose, severe cardiovascular toxicity can occur, including:

1. Pronounced hypotension
2. Bradycardia
3. Asystole
4. Cardiac arrest

Toxic levels can also lead to neurological symptoms such as epileptic seizures, which can exacerbate cardiovascular effects through increased sympathetic activity.

Bupivacaine

Bupivacaine is a long-acting local anesthetic with an amide structure that exhibits stereoisomerism due to a chiral center attached to groups in different spatial orientations. A pair of stereoisomers are optically active and differ in their effect on polarized light's clockwise (R+) or counterclockwise (S−) rotation. Bupivacaine is a racemic mixture of both enantiomers in a 1:1 ratio, while levobupivacaine is the S and dextrobupivacaine is the R isomer. Bupivacaine can have significant cardiovascular effects, especially when administered in high doses or accidentally injected into a blood vessel as part of local anesthetic toxicity (LAST). Levobupivacaine shows the best safety profile and causes the least CNS and cardiotoxicity [77]. This is caused by the lower affinity of the drug for sodium voltage-gated channels compared to bupivacaine and dextrobupivacaine. Cardiovascular effects of toxicity generally begin with tachycardia, hypertension, and arrhythmias but continue with bradycardia, hypotension, conduction block, and myocardial depression. In the most severe cases, they can cause cardiac arrest [78]. Bupivacaine binds rapidly to Na voltage-dependent channels during action potential generation in the heart. However, it is slowly released from the receptor during diastole, in contrast to lidocaine, which has the property of rapid binding and release. Bupivacaine, therefore, exhibits a greater degree of cardiotoxicity than lidocaine [79]. In the treatment of these complications, symptomatic therapy with vasoactive drugs is used, along with intralipid infusion. However, recent research has shown that, although there is a positive effect on survival in animal models, it is limited to a favorable pharmacokinetic effect in the human population [80]. Cardiac toxicity for bupivacaine enantiomers does not fully correlate with the effect caused by the drug at the level of sodium channels, which could suggest that there are alternative mechanisms for this effect. High concentrations of bupivacaine cause abnormal mitochondrial autophagy and inhibition of ATP synthase, as well as a negative effect on oxidative phosphorylation and intracellular chemical energy exchange [81]. Bupivacaine blocks the enzyme carnitine acylcarnitine transferase, which transports acylcarnitine across mitochondrial membranes into fatty acids during aerobic metabolism. Oxidation of fatty acid derivatives in the mitochondrial matrix typically provides more than 70% of the energy needs of the myocardium. By inhibiting this enzyme, it is assumed that bupivacaine causes cardiotoxic effects, such as heart rhythm disturbances or even cardiac arrest [82]. Hypoxia has been shown to exacerbate bupivacaine-induced cardiotoxicity because oxygen limitation could further impair oxidative phosphorylation that is already inhibited by reduced transport of essential substrates [83]. Levobupivacaine and bupivacaine can block activation of the nucleus tractus solitarus, leading to hypotension, bradycardia, and arrhythmia. Compared with bupivacaine and dextrobupivacaine, levobupivacaine shows a lower frequency of these complications, even at higher doses [84].

Ropivacaine

Ropivacaine belongs to the group of amide local anesthetics. Unlike most others, ropivacaine is synthesized as an isolated S enantiomer, which has a less toxic effect than the R isomeric form. Ropivacaine has been shown to cause motor blockade to a lesser degree than bupivacaine, while the sensory effect of these two drugs is similar. Ropivacaine exhibits a biphasic vascular effect. Lower concentrations (0.063–0.5%) lead to vasoconstriction, while higher concentrations (1%) cause vasodilation [85]. Compared to bupivacaine, ropivacaine causes fewer cardiovascular side effects [86]. Reduces maximal depolarization frequency in the isolated heart in animal model studies to a lesser extent than bupivacaine but more than lidocaine. If the mentioned drugs are applied directly to the coronary arteries of anesthetized pigs, the decrease in the frequency of depolarization of the cell membrane and the prolongation of the QRS interval is caused mainly by bupivacaine, then ropivacaine, then lidocaine [87]. In addition to the cardiac effects of local anesthetics, which are manifested by adverse inotropic effects, bradycardia and AV blocks of varying degrees, cardiac function is controlled by the central nervous system. Numerous studies on animal models, but also the human population, have shown that the toxic doses for the CNS are significantly higher for ropivacaine compared to racemic bupivacaine [88]. In the case of ropivacaine overdose and cardiotoxicity, sodium bicarbonate reverses the electrophysiological manifestations of the block better than intralipid intravenous infusion [89].

Chloroprocaine

Chloroprocaine is a short-acting local anesthetic of ester structure. Because of its low potential for systemic toxicity, chloroprocaine is used to identify accidental intravascular epidural placement [90]. It is most commonly used in obstetric anesthesia because of its low potential for toxicity to mother and child [91]. Doses of chloroprocaine higher than 11 mg/kg can cause intoxication. Due to its rapid metabolism, toxic cardiovascular and CNS effects are low. Self-limiting bradycardia of an expansive QRS complex lasting 30 s has been described after the administration of chloroprocaine to infants [92]. Narrow QRS complex bradycardia has been described in adult patients undergoing intravenous regional anesthesia with chloroprocaine [93]. The safety profile of chloroprocaine is good, and no long-term morbidity or mortality has been described as a consequence of the use of this drug.

Conclusion

The interplay between anesthetic agents and cardiovascular functions remains a critical area of investigation, with far-reaching implications for clinical practice. Modern anesthetics, while highly effective, exhibit a range of cardiotoxic and vasoactive effects mediated through diverse mechanisms, including ion channel modulation, oxidative stress, and mitochondrial dysfunction. Isoflurane, sevoflurane, and desflurane, among others, demonstrate significant impacts on cardiac contractility, vascular reactivity, and endothelial function, which can be exacerbated in patients with preexisting cardiovascular or systemic conditions. Local anesthetics, though indispensable in regional anesthesia, pose risks of systemic toxicity, including arrhythmias and cardiac depression, emphasizing the need for vigilance and the implementation of mitigation strategies such as lipid emulsion therapy. The findings underscore the importance of tailoring anesthetic choice and dose to the individual patient's physiological status and the surgical context to minimize adverse outcomes. Advances in understanding the molecular and cellular mechanisms underlying these effects provide opportunities for safer anesthetic practices and targeted therapies. Future research should prioritize elucidating patient-specific risk factors and developing strategies to mitigate the cardiovascular side effects of both general and local anesthetics, ensuring optimal outcomes in perioperative care.

References

1. Sinha S, Sinharoy P, Bratz IN, Damron DS (2015) Propofol causes vasodilation in vivo via TRPA1 ion channels: role of nitric oxide and BKCa channels. PLoS One 10(4):e0122189. https://doi.org/10.1371/journal.pone.0122189
2. Royse CF, Liew DF, Wright CE, Royse AG, Angus JA (2008) Persistent depression of contractility and vasodilation with propofol but not with sevoflurane or desflurane in rabbits. Anesthesiology 108(1):87–93. https://doi.org/10.1097/01.anes.0000296077.32685.26
3. Elvir-Lazo OL, White PF, Cruz Eng H, Yumul F, Chua R, Yumul R (2021) Impact of chronic medications in the perioperative period—anesthetic implications (part II). Postgrad Med 133(8): 920–938. https://doi.org/10.1080/00325481.2021.1982298
4. Jeffrey P, Cardinale N, Gillespie L, Germond L (2019) Complications of general anesthesia. In: Catastrophic perioperative complications and management. Springer, Berlin, pp 95–103. https://doi.org/10.1007/978-3-319-96125-5_7
5. Garima, Agarwal T, Costantini M, Pal S, Kumar A (2022) Oxygenation therapies for improved wound healing: current trends and technologies. J Mater Chem B 10(39):7905–7923. https://doi.org/10.1039/d2tb01498j
6. Rich MW (2014) Cardiovascular problems in the elderly. In: Evidence-based cardiology consult. Springer, London, pp 267–281. https://doi.org/10.1007/978-1-4471-4441-0_19
7. Hemphill S, McMenamin L, Bellamy MC, Hopkins PM (2019) Propofol infusion syndrome: a structured literature review and analysis of published case reports. Br J Anaesth 122(4): 448–459. https://doi.org/10.1016/j.bja.2018.12.025
8. Bray RJ (1999) Propofol-infusion syndrome in children. Lancet 353(9169):2074–2075. https://doi.org/10.1016/S0140-6736(05)77896-X
9. Marinella MA (1996) Lactic acidosis associated with propofol. Chest 109(1):292. https://doi.org/10.1378/chest.109.1.292
10. Parke TJ, Stevens JE, Rice AS, Greenaway CL, Bray RJ, Smith PJ, Waldmann CS, Verghese C (1992) Metabolic acidosis and fatal myocardial failure after propofol infusion in children: five case reports. BMJ 305(6854):613–616. https://doi.org/10.1136/bmj.305.6854.613
11. Vanlander AV, Okun JG, de Jaeger A, Smet J, De Latter E, De Paepe B, Dacremont G, Wuyts B, Vanheel B, De Paepe P, Jorens PG, Van Regenmortel N, Van Coster R (2015) Possible pathogenic mechanism of propofol infusion syndrome involves coenzyme Q. Anesthesiology 122(2):343–352. https://doi.org/10.1097/ALN.0000000000000484
12. Vollmer JP, Haen S, Wolburg H, Lehmann R, Steiner J, Reddersen S, Fend F, Fallier-Becker P (2018) Propofol-related infusion syndrome: ultrastructural evidence for a mitochondrial disorder. Crit Care Med 46(1):e91–e94. https://doi.org/10.1097/CCM.0000000000002802
13. Barajas MB, Wang A, Griffiths KK, Sun L, Yang G, Levy RJ (2022) Modeling propofol-induced cardiotoxicity in the isolated-perfused newborn mouse heart. Physiol Rep 10(15): e15402. https://doi.org/10.14814/phy2.15402
14. Barajas MB, Brunner SD, Wang A, Griffiths KK, Levy RJ (2022) Propofol toxicity in the developing mouse heart mitochondria. Pediatr Res 92(5):1341–1349. https://doi.org/10.1038/s41390-022-01985-1
15. Kido K, Ito H, Yamamoto Y, Makita K, Uchida T (2018) Cytotoxicity of propofol in human induced pluripotent stem cell-derived cardiomyocytes. J Anesth 32(1):120–131. https://doi.org/10.1007/s00540-017-2441-0
16. Sumi C, Okamoto A, Tanaka H, Nishi K, Kusunoki M, Shoji T, Uba T, Matsuo Y, Adachi T, Hayashi JI, Takenaga K (2018) Propofol induces a metabolic switch to glycolysis and cell death in a mitochondrial electron transport chain-dependent manner. PLoS One 13(2):e0192796. https://doi.org/10.1371/journal.pone.0192796
17. Sumi C, Okamoto A, Tanaka H, Kusunoki M, Shoji T, Uba T, Adachi T, Iwai T, Nishi K, Harada H, Bono H, Matsuo Y, Hirota K (2018) Suppression of mitochondrial oxygen metabolism mediated by the transcription factor HIF-1 alleviates propofol-induced cell toxicity. Sci Rep 8(1):8987. https://doi.org/10.1038/s41598-018-27220-8

18. Gu Q, Kanungo J (2021) Effect of ketamine on gene expression in zebrafish embryos. J Appl Toxicol 41(12):2083–2089. https://doi.org/10.1002/jat.4199
19. Li X, Li S, Zheng W, Pan J, Huang K, Chen R, Pan T, Liao G, Chen Z, Zhou D, Shen W, Zhou W, Liu Y (2015) Environmental enrichment and abstinence attenuate ketamine-induced cardiac and renal toxicity. Sci Rep 5:11611. https://doi.org/10.1038/srep11611
20. Li Y, Shi J, Yang BF, Liu L, Han CL, Li WM, Dong DL, Pan ZW, Liu GZ, Geng JQ, Sheng L, Tan XY, Sun DH, Gong ZH, Gong YT (2012) Ketamine-induced ventricular structural, sympathetic and electrophysiological remodeling: pathological consequences and protective effects of metoprolol. Br J Pharmacol 165(6):1748–1756. https://doi.org/10.1111/j.1476-5381.2011.01635.x
21. Kalsi SS, Wood DM, Dargan PI (2011) The epidemiology and patterns of acute and chronic toxicity associated with recreational ketamine use. Emerg Health Threats J 4:7107. https://doi.org/10.3402/ehtj.v4i0.7107
22. Ito H, Uchida T, Makita K (2015) Ketamine causes mitochondrial dysfunction in human induced pluripotent stem cell-derived neurons. PLoS One 10(5):e0128445. https://doi.org/10.1371/journal.pone.0128445
23. Jakobsen CJ, Torp P, Vester AE, Folkersen L, Thougaard A, Sloth E (2010) Ketamine reduce left ventricular systolic and diastolic function in patients with ischaemic heart disease. Acta Anaesthesiol Scand 54(9):1137–1144. https://doi.org/10.1111/j.1399-6576.2010.02283.x
24. Han J, Kim N, Joo H, Kim E (2002) Ketamine abolishes ischemic preconditioning through inhibition of K(ATP) channels in rabbit hearts. Am J Phys Heart Circ Phys 283(1):H13–H21. https://doi.org/10.1152/ajpheart.01064.2001
25. Wang X, Jin A, An M, Ding Y, Tuo Y, Qiu Y (2014) Etomidate deteriorates the toxicity of advanced glycation end products to human endothelial Eahy926 cells. J Toxicol Sci 39(6):887–896. https://doi.org/10.2131/jts.39.887
26. Komatsu R, You J, Mascha EJ, Sessler DI, Kasuya Y, Turan A (2013) Anesthetic induction with etomidate, rather than propofol, is associated with increased 30-day mortality and cardiovascular morbidity after noncardiac surgery. Anesth Analg 117(6):1329–1337. https://doi.org/10.1213/ANE.0b013e318299a516
27. Wu RS, Wu KC, Yang JS, Chiou SM, Yu CS, Chang SJ, Chueh FS, Chung JG (2011) Etomidate induces cytotoxic effects and gene expression in a murine leukemia macrophage cell line (RAW264.7). Anticancer Res 31(6):2203–2208
28. Eizadi-Mood N, Montazeri K, Dehghani Dastjerdi M (2011) Evaluation of cardiovascular manifestations in benzodiazepine poisoning. Iran J Toxicol 5(2):373–376
29. Kazemzadeh N, Mohammadi S, Emamhadi M, Amirfarhangi A, Sanaei-Zadeh H (2014) Electrocardiographic manifestations of benzodiazepine toxicity. Iran J Toxicol 8(24):952–955
30. Anand K, Kumar M (2013) Benzodiazepine overdose associated atrioventricular block. Anesth Essays Res 7(3):419–420. https://doi.org/10.4103/0259-1162.123280
31. Gohil JR, Jindal S (2019) Cardiotoxicity associated with midazolam in a patient with diphtheria. Curr Drug Saf 14(1):57–59. https://doi.org/10.2174/1574886313666181105122154
32. Kang FC, Wang SC, Chang MM, Pan BS, Wong KL, Cheng KS, Huang BM (2018) Midazolam activates caspase, MAPKs and endoplasmic reticulum stress pathways, and inhibits cell cycle and Akt pathway, to induce apoptosis in TM3 mouse Leydig progenitor cells. Onco Targets Ther 11:1475–1490. https://doi.org/10.2147/OTT.S154442
33. Gernez E, Lee GR, Niguet JP, Zerimech F, Bennis A, Grzych G (2023) Nitrous oxide abuse: clinical outcomes, pharmacology, pharmacokinetics, toxicity, and impact on metabolism. Toxics 11(12):962. https://doi.org/10.3390/toxics11120962
34. Wrońska-Nofer T, Nofer JR, Jajte J, Dziubałtowska E, Szymczak W, Krajewski W, Wąsowicz W, Rydzyński K (2012) Oxidative DNA damage and oxidative stress in subjects occupationally exposed to nitrous oxide (N2O). Mutat Res 731(1–2):58–63. https://doi.org/10.1016/j.mrfmmm.2011.10.010
35. Thayabaran D, Burrage D (2021) Nitrous oxide-induced neurotoxicity: a case report and literature review. Br J Clin Pharmacol 87(9):3622–3626. https://doi.org/10.1111/bcp.14779

36. Potocka-Banaś B, Dembińska T, Borowiak K (2016) Chapter 91: Acute and long-term toxicity caused by addictive inhalation of nitrous oxide and impact on neuropathology. In: Neuropathology of drug addictions and substance misuse. Elsevier, Amsterdam, pp 998–1003. https://doi.org/10.1016/B978-0-12-800212-4.00091-1
37. Zhang YL, Guan C.Y., Hu SF, Zhou DC (2009) [Effect of long-term sevoflurane anesthesia on markers of myocardial damage]. Zhonghua Yi Xue Za Zhi 89(27):1916–1918 (in Chinese)
38. Gong JS, Yao YT, Fang NX, Huang J, Li LH (2012) Sevoflurane postconditioning alleviates action potential duration shortening and L-type calcium current suppression induced by ischemia/reperfusion injury in rat epicardial myocytes. Chin Med J (English) 125(19):3485–3491
39. Pagel PS (2008) Additive cardioprotection by ethanol and sevoflurane: are sarcolemmal KATP channels also involved? Anesth Analg 106(6):1926–1927. https://doi.org/10.1213/01.ane.0000312680.16252.56
40. Kaneda K, Miyamae M, Sugioka S, Okusa C, Inamura Y, Domae N, Kotani J, Figueredo VM (2008) Sevoflurane enhances ethanol-induced cardiac preconditioning through modulation of protein kinase C, mitochondrial KATP channels, and nitric oxide synthase, in Guinea pig hearts. Anesth Analg 106(1):9–16. https://doi.org/10.1213/01.ane.0000297298.93627.36
41. Farid TA, Bhaskaran A, Nanthakumar K (2018) KATP channel blockade as a novel antiarrhythmic strategy: evolving from tachy to Brady therapy. Endocrinology 159(8):3081–3082. https://doi.org/10.1210/en.2018-00525
42. Steffensen AB, Refaat MM, David JP, Mujezinovic A, Calloe K, Wojciak J, Nussbaum RL, Scheinman MM, Schmitt N (2015) High incidence of functional ion-channel abnormalities in a consecutive Long QT cohort with novel missense genetic variants of unknown significance. Sci Rep 5:10009. https://doi.org/10.1038/srep10009
43. Rocha TLA, Borges TF, Rodrigues SD, Martins LZ, da Silva MLS, Bonacio GF, Rizzi E, Dias-Junior CA (2023) Sevoflurane and isoflurane anesthesia induce redox imbalance, but only sevoflurane impairs vascular contraction. Fundam Clin Pharmacol 37(5):937–946. https://doi.org/10.1111/fcp.12901
44. Silva MAP, Braz LG, Braz JRC, Braz MG (2023) Modulation of gene expression and inflammation but not DNA damage after sevoflurane anesthesia. Environ Mol Mutagen 64(5):315–320. https://doi.org/10.1002/em.22539
45. Riess ML, Eells JT, Kevin LG, Camara AK, Henry MM, Stowe DF (2004) Attenuation of mitochondrial respiration by sevoflurane in isolated cardiac mitochondria is mediated in part by reactive oxygen species. Anesthesiology 100(3):498–505. https://doi.org/10.1097/00000542-200403000-00007
46. Zhai WH, Zhao J, Huo SP, Chen XG, Li YD, Zhang ZL, Yu LL, Song S, Wang QJ (2015) Mechanisms of cytotoxicity induced by the anesthetic isoflurane: the role of inositol 1,4,5-trisphosphate receptors. Genet Mol Res 14(2):6929–6942. https://doi.org/10.4238/2015.June.26.1
47. Shimogai M, Ogawa K, Tokinaga Y, Yamazaki A, Hatano Y (2010) The cellular mechanisms underlying the inhibitory effects of isoflurane and sevoflurane on arginine vasopressin-induced vasoconstriction. J Anesth 24(6):893–900. https://doi.org/10.1007/s00540-010-1033-z
48. Lee JS, Morrow D, Andresen MC, Chang KS (2002) Isoflurane depresses baroreflex control of heart rate in decerebrate rats. Anesthesiology 96(5):1214–1222. https://doi.org/10.1097/00000542-200205000-00026
49. Meng T, Ren X, Chen X, Yu J, Agrimi J, Paolocci N, Gao WD (2019) Anesthetic agents isoflurane and propofol decrease maximal Ca2+-activated force and thus contractility in the failing myocardium. J Pharmacol Exp Ther 371(3):615–623. https://doi.org/10.1124/jpet.119.259556
50. Crystal GJ (2017) Isoflurane and the coronary steal controversy of the 1980s: origin, resolution, and legacy. J Anesth Hist 3(2):56–62. https://doi.org/10.1016/j.janh.2017.04.002
51. OuYang W, Hemmings HC Jr (2007) Isoform-selective effects of isoflurane on voltage-gated Na+ channels. Anesthesiology 107(1):91–98. https://doi.org/10.1097/01.anes.0000268390.28362.4a

52. Shen X, Bhatt N, Xu J, Meng T, Aon MA, O'Rourke B, Berkowitz DE, Cortassa S, Gao WD (2014) Effect of isoflurane on myocardial energetic and oxidative stress in cardiac muscle from Zucker diabetic fatty rat. J Pharmacol Exp Ther 349(1):21–28. https://doi.org/10.1124/jpet.113.211144
53. Arruda NM, Braz LG, Nogueira FR, Souza KM, Aun AG, Figueiredo DBS et al (2019) Inflammation and DNA damage induction in surgical patients maintained with desflurane anesthesia. Mutat Res Genet Toxicol Environ Mutagen 846:403073. https://doi.org/10.1016/j.mrgentox.2019.07.003
54. Nogueira FR, Braz LG, Andrade LR, Carvalho AL, Vane LA, Módolo NS et al (2016) Evaluation of genotoxicity of general anesthesia maintained with desflurane in patients under minor surgery. Environ Mol Mutagen 57(4):312–316. https://doi.org/10.1002/em.22012
55. Eroglu F, Yavuz L, Ceylan BG, Yilmaz F, Eroglu E, Delibas N, Naziroğlu M (2010) New volatile anesthetic, desflurane, reduces vitamin E level in blood of operative patients via oxidative stress. Cell Biochem Funct 28(3):211–216. https://doi.org/10.1002/cbf.1641
56. Tung D, Yoshida EM, Wang CS, Steinbrecher UP (2005) Severe desflurane hepatotoxicity after colon surgery in an elderly patient. Can J Anaesth 52(2):133–136. https://doi.org/10.1007/BF03027717
57. Becker DE, Reed KL (2012) Local anesthetics: review of pharmacological considerations. Anesth Prog 59(2):90–101; quiz 102–103. https://doi.org/10.2344/0003-3006-59.2.90
58. Arumugam S, Contino V, Kolli S (2020) Local anesthetics systemic toxicity (LAST): a review and update. Curr Anesthesiol Rep 10:218–226. https://doi.org/10.1007/s40140-020-00381-x
59. Lirk P, Picardi S, Hollmann MW (2014) Local anaesthetics: 10 essentials. Eur J Anaesthesiol 31(11):575–585. https://doi.org/10.1097/EJA.0000000000000137
60. El-Boghdadly K, Pawa A, Chin KJ (2018) Local anesthetic systemic toxicity: current perspectives. Local Reg Anesth 11:35–44. https://doi.org/10.2147/LRA.S154512
61. Gitman M, Fettiplace MR, Weinberg GL, Neal JM, Barrington MJ (2019) Local anesthetic systemic toxicity: a narrative literature review and clinical update on prevention, diagnosis, and management. Plast Reconstr Surg 144(3):783–795. https://doi.org/10.1097/PRS.0000000000005989
62. Butterworth JFIV (2010) Models and mechanisms of local anesthetic cardiac toxicity: a review. Reg Anesth Pain Med 35(2):167–176. https://doi.org/10.1097/AAP.0b013e3181d231b9
63. Plakhotnik J, Zhang L, Estrada M, Coles JG, Lonnqvist PA, Maynes JT (2022) Local anesthetic cardiac toxicity is mediated by cardiomyocyte calcium dynamics. Anesthesiology 137(6): 687–703. https://doi.org/10.1097/ALN.0000000000004389
64. Poon YY, Liu YW, Huang YH, Chan SHH, Tsai CY (2021) Postoperative stroke after spinal anesthesia and responses of carotid or cerebral blood flow and baroreflex functionality to spinal bupivacaine in rats. Biology (Basel) 10(7):617. https://doi.org/10.3390/biology10070617
65. Santos AC, Pedersen H, Harmon TW, Morishima HO, Finster M, Arthur GR, Covino BG (1989) Does pregnancy alter the systemic toxicity of local anesthetics? Anesthesiology 70(6): 991–995. https://doi.org/10.1097/00000542-198906000-00018
66. Gitman M, Barrington MJ (2018) Local anesthetic systemic toxicity: a review of recent case reports and registries. Reg Anesth Pain Med 43(2):124–130. https://doi.org/10.1097/AAP.0000000000000721
67. Wolfe JW, Butterworth JF (2011) Local anesthetic systemic toxicity: update on mechanisms and treatment. Curr Opin Anaesthesiol 24(5):561–566. https://doi.org/10.1097/ACO.0b013e32834a9394
68. Gordh T (2010) Lidocaine: the origin of a modern local anesthetic. Anesthesiology 113(6): 1433–1437. https://doi.org/10.1097/ALN.0b013e3181fcef48
69. Güler S, Könemann H, Wolfes J, Güner F, Ellermann C, Rath B, Frommeyer G, Lange PS, Köbe J, Reinke F, Eckardt L (2023) Lidocaine as an anti-arrhythmic drug: are there any indications left? Clin Transl Sci 16(12):2429–2437. https://doi.org/10.1111/cts.13650
70. Rolf S, Haverkamp W, Borggrefe M, Musshoff U, Eckardt L, Mergenthaler J, Snyders DJ, Pongs O, Speckmann EJ, Breithardt G, Madeja M (2000) Effects of antiarrhythmic drugs on

cloned cardiac voltage-gated potassium channels expressed in Xenopus oocytes. Naunyn Schmiedeberg's Arch Pharmacol 362(1):22–31. https://doi.org/10.1007/s002100000257
71. Singh BN, Williams EM (1971) Effect of altering potassium concentration on the action of lidocaine and diphenylhydantoin on rabbit atrial and ventricular muscle. Circ Res 29(3): 286–295. https://doi.org/10.1161/01.RES.29.3.286
72. Wittig J, Harrison LA, Wallace AG (1973) Electrophysiological effects of lidocaine on distal Purkinje fibers of canine heart. Am Heart J 86(1):69–78. https://doi.org/10.1016/0002-8703(73)90011-2
73. Newton DJ, McLeod GA, Khan F, Belch JJ (2007) Mechanisms influencing the vasoactive effects of lidocaine in human skin. Anaesthesia 62(2):146–150. https://doi.org/10.1111/j.1365-2044.2006.04901.x
74. Gajniak D, Mendrala K, König-Widuch G, Parzonka S, Gierek D, Krzych ŁJ (2024) Effect of lidocaine on intraoperative blood pressure variability in patients undergoing major vascular surgery. BMC Anesthesiol 24(1):170. https://doi.org/10.1186/s12871-024-02550-5
75. Pustetto M, Goldsztejn N, Touihri K, Engelman E, Ickx B, Van Obbergh L (2020) Intravenous lidocaine to prevent endothelial dysfunction after major abdominal surgery: a randomized controlled pilot trial. BMC Anesthesiol 20(1):155. https://doi.org/10.1186/s12871-020-01075-x
76. Odor PM, Cavalier AG, Reynolds ND, Ang KS, Parrington SJ, Xu H, Johnston A, Sage FJ (2019) Safety and pharmacokinetics of levobupivacaine following fascia iliaca compartment block in elderly patients. Drugs Aging 36(6):541–548. https://doi.org/10.1007/s40266-019-00652-1
77. Heppolette CAA, Brunnen D, Bampoe S, Odor PM (2020) Clinical pharmacokinetics and pharmacodynamics of levobupivacaine. Clin Pharmacokinet 59(6):715–745. https://doi.org/10.1007/s40262-020-00868-0
78. Clarkson CW, Hondeghem LM (1985) Mechanism for bupivacaine depression of cardiac conduction: fast block of sodium channels during the action potential with slow recovery from block during diastole. Anesthesiology 62(4):396–405. https://doi.org/10.1097/00000542-198504000-00006
79. Harvey M, Cave G (2017) Lipid emulsion in local anesthetic toxicity. Curr Opin Anaesthesiol 30(5):632–638. https://doi.org/10.1097/ACO.0000000000000498
80. Sanford M, Keating GM (2010) Levobupivacaine: a review of its use in regional anaesthesia and pain management. Drugs 70(6):761–791. https://doi.org/10.2165/11203250-000000000-00000
81. Nouette-Gaulain K, Jose C, Capdevila X, Rossignol R (2011) From analgesia to myopathy: when local anesthetics impair the mitochondrion. Int J Biochem Cell Biol 43(1):14–19. https://doi.org/10.1016/j.biocel.2010.10.005
82. Weinberg GL, Palmer JW, VadeBoncouer TR, Zuechner MB, Edelman G, Hoppel CL (2000) Bupivacaine inhibits acylcarnitine exchange in cardiac mitochondria. Anesthesiology 92(2): 523–528. https://doi.org/10.1097/00000542-200002000-00036
83. Heavner JE, Dryden CF Jr, Sanghani V, Huemer G, Bessire A, Badgwell JM (1992) Severe hypoxia enhances central nervous system and cardiovascular toxicity of bupivacaine in lightly anesthetized pigs. Anesthesiology 77(1):142–147. https://doi.org/10.1097/00000542-199207000-00020
84. Denson DD, Behbehani MM, Gregg RV (1992) Enantiomer-specific effects of an intravenously administered arrhythmogenic dose of bupivacaine on neurons of the nucleus tractus solitarius and the cardiovascular system in the anesthetized rat. Reg Anesth 17(6):311–316
85. Hansen TG (2004) Ropivacaine: a pharmacological review. Expert Rev Neurother 4(5): 781–791. https://doi.org/10.1586/14737175.4.5.781
86. Moussaoui A, Boudjahfa S, Benbernou S et al (2023) Cardiovascular toxicity: comparison between ropivacaine and bupivacaine in spinal anesthesia. Reg Anesthes Pain Med 48:A184

87. Arlock P (1988) Actions of three local anesthetics: lidocaine, bupivacaine and ropivacaine on Guinea pig papillary muscle sodium channels (Vmax). Pharmacol Toxicol 63(2):96–104. https://doi.org/10.1111/j.1600-0773.1988.tb00918.x
88. Huang YF, Pryor ME, Mather LE, Veering BT (1998) Cardiovascular and central nervous system effects of intravenous levobupivacaine and bupivacaine in sheep. Anesth Analg 86(4): 797–804. https://doi.org/10.1097/00000539-199804000-00023
89. Zaballos M, Fernández I, Melone A, Rodríguez L, Varela O, García S et al (2023) Sodium bicarbonate reverts electrophysiologic cardiotoxicity of ropivacaine faster than lipid emulsions in a porcine model. Basic Clin Pharmacol Toxicol 132(2):211–222. https://doi.org/10.1111/bcpt.13822
90. Guay J (2006) The epidural test dose: a review. Anesth Analg 102(3):921–929. https://doi.org/10.1213/01.ane.0000196687.88590.6b
91. Kuhnert BR, Kuhnert PM, Philipson EH, Syracuse CD, Kaine CJ, Yun CH (1986) The half-life of 2-chloroprocaine. Anesth Analg 65(3):273–278. https://doi.org/10.1097/00132582-198609000-00037
92. Cladis FP, Litman RS (2004) Transient cardiovascular toxicity with unintentional intravascular injection of 3% 2-chloroprocaine in a 2-month-old infant. Anesthesiology 100(1):181–183. https://doi.org/10.1097/00000542-200401000-00030
93. Marsch SC, Sluga M, Studer W, Barandun J, Scharplatz D, Ummenhofer W (2004) 0.5% versus 1.0% 2-chloroprocaine for intravenous regional anesthesia: a prospective, randomized, double-blind trial. Anesth Analg 98(6):1789–1793. https://doi.org/10.1213/01.ANE.0000116929.45557.CE

Chapter 13
Anti-inflammatory Bowel Diseases Drugs and Cardiovascular Toxicity

Uglješa Maličević and Devendra K. Agrawal

Abstract Drug-induced cardiovascular toxicity represents a significant clinical challenge, especially in the context of Inflammatory Bowel Disease (IBD) therapies, where medications essential for controlling inflammation can, paradoxically, compromise cardiovascular health. Despite rigorous and extensive preclinical and clinical trials, cardiotoxic effects may still manifest in larger, diverse patient populations, leading to severe complications such as myocarditis, cardiomyopathy, arrhythmias, myocardial ischemia, heart failure, and other conditions. This chapter provides a comprehensive analysis of the cardiovascular risks associated with commonly used anti-IBD therapies, including aminosalicylates, corticosteroids, immunomodulators, and biologics. These drugs, essential for managing chronic inflammation in IBD, may not only exacerbate pre-existing cardiovascular conditions but also induce new cardiac complications. Mechanisms of drug-induced cardiotoxicity often involve oxidative stress, free radical generation, and inflammatory pathways, which compromise myocardial contractility and tissue integrity, increasing both morbidity and mortality. A deeper understanding of such adverse effects allows clinicians to personalize treatment approaches, balancing the therapeutic benefits of IBD management with ensuring cardiovascular safety. Furthermore, this review emphasizes the critical importance of ongoing monitoring and vigilance in clinical practice, as early detection and intervention can significantly mitigate cardiotoxic risks, thereby enhancing the overall quality of care for patients with IBD.

U. Maličević
Department of Translational Research, Western University of Health Sciences, Pomona, CA, USA

Department of Pathophysiology and Centre for Biomedical Research, Faculty of Medicine, University of Banja Luka, Banja Luka, The Republic of Srpska, Bosnia and Herzegovina
e-mail: ugljesa.malicevic@med.unibl.org

D. K. Agrawal (✉)
Department of Translational Research, Western University of Health Sciences, Pomona, CA, USA
e-mail: DAgrawal@WesternU.edu

D. K. Agrawal et al. (eds.), *Cardiovascular Toxicity*, Advances in Biochemistry in Health and Disease 33, https://doi.org/10.1007/978-3-031-90598-8_13

Keywords Anti-IBD drugs · Cardiotoxicity · Cardiovascular safety · Drug-induced cardiovascular toxicity · Heart failure · Inflammation · Oxidative stress

Introduction

Drug-induced cardiovascular toxicity (DICT) refers to the harmful impact of certain medications on the heart, potentially leading to severe cardiovascular complications [1]. DICT is a significant concern for both prescription and illicit drugs, often emerging only after routine clinical use, which can pose serious clinical challenges [2]. Despite comprehensive preclinical testing and extensive clinical trials, some cardiotoxic effects may only manifest when these drugs are administered to a broader, more diverse patient population. Such delayed recognition of cardiovascular toxicity can result in serious cardiovascular complications, ultimately limiting the therapeutic application of otherwise effective treatments and necessitating closer monitoring or even market withdrawal [3]. The mechanisms driving cardiovascular toxicity typically involve oxidative stress, free radical generation, and hypoxia, which can lead to apoptosis and dysregulation of myocardial contractility following prolonged exposure to cardiotoxic drugs [4]. Clinically, DICT manifests a spectrum of cardiovascular conditions, including myocarditis, cardiomyopathy, arrhythmias, myocardial ischemia, and heart failure [5] (Fig. 13.1). Additionally, it may induce structural and functional changes in the heart, such as ventricular dysfunction, reduced ejection fraction, and compromised myocardial tissue integrity. These complications not only impair cardiac function but also significantly increase the risk of morbidity and mortality [6].

Inflammatory Bowel Disease (IBD), encompassing Crohn's disease (CD) and ulcerative colitis (UC), is characterized by chronic inflammation of the gastrointestinal tract [7]. CD can affect any part of the GI tract, leading to transmural inflammation, whereas UC is confined to the colon and rectum, with inflammation typically restricted to the mucosal layer. The chronic and relapsing nature of IBD, combined with its complex management challenges, significantly impacts patients' quality of life. Frequent flare-ups, serious complications, and the necessity for long-term, multidisciplinary care make IBD a clinical concern of great importance [8, 9].

Managing IBD is complicated due to the disease's unpredictable course and individual variability in response to treatment. A personalized approach is crucial to achieve therapeutic goals such as reducing inflammation, controlling symptoms, inducing remission, and preventing complications [10]. Treatment strategies typically include anti-inflammatory drugs such as aminosalicylates and corticosteroids, alongside immunomodulators and biologic therapies to alter the immune response and target specific inflammatory pathways [11] (Fig. 13.1). While these therapies are crucial in managing inflammation and controlling symptoms, they also carry potential risks, including cardiovascular toxicity. The mechanisms of cardiotoxicity associated with IBD treatments often involve inflammatory pathways and metabolic disturbances that can exacerbate preexisting cardiovascular conditions or create new heart-related issues. This chapter aims to explore and describe the most

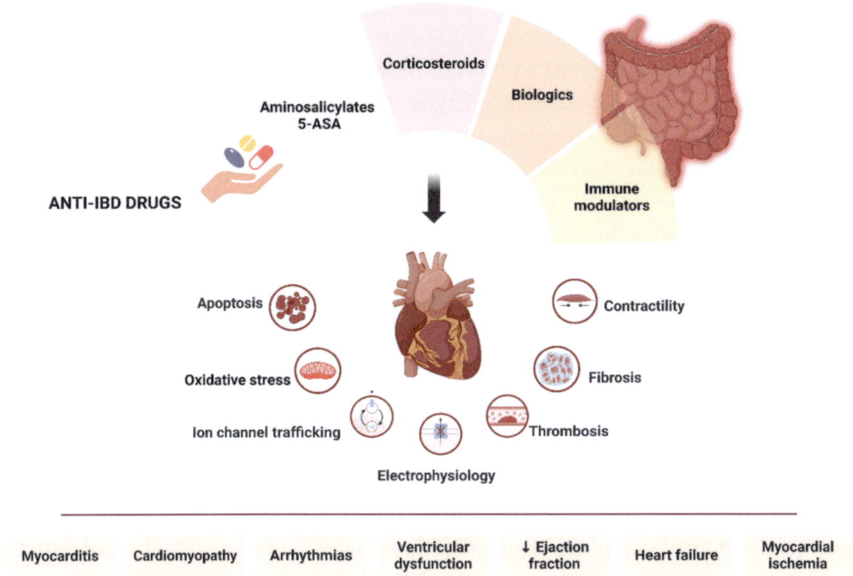

Fig. 13.1 This figure illustrates various drugs commonly used in the treatment of Inflammatory Bowel Disease (IBD) and highlights their potential cardiovascular toxicities. The drugs are categorized into four classes: aminosalicylates, corticosteroids, immunomodulators, and biologics. Each class is associated with specific cardiovascular risks, such as arrhythmias, cardiomyopathy, ventricular dysfunction, and reduced ejection fraction. In more severe cases, these medications may lead to serious conditions, including myocarditis, myocardial ischemia, and heart failure

common adverse cardiovascular effects associated with each group of drugs used in the treatment of IBD. Understanding these risks helps clinicians tailor treatments for individual patients, ensuring an optimal balance between therapeutic effectiveness and safety.

Aminosalicylate-Induced Cardiovascular Toxicity in IBD

Sulfasalazine and mesalazine (mesalamine or 5-aminosalicylic acid) are two commonly used anti-inflammatory drugs in the 5-ASA class of aminosalicylates, particularly effective in managing IBD. These medications play a key role in controlling active disease, maintaining remission, and preventing flare-ups by reducing inflammation in the intestines [12]. Although their precise mechanisms of action are not fully understood, both drugs are believed to inhibit the cyclooxygenase (COX) enzyme, which reduces the production of prostaglandins, substances that contribute to inflammation [13]. This reduction is thought to decrease activation of the peroxisome proliferator-activated receptor gamma (PPAR-γ) pathway, ultimately lowering the activity of nuclear factor kappa B (NF-κB), a key driver of intestinal inflammation [14]. Despite their overall good tolerance, sulfasalazine and mesalazine are still associated with a variety of potential adverse effects. Common side effects include headaches and gastrointestinal disturbances such as abdominal pain, nausea, vomiting, and diarrhea [15]. In some cases, patients may develop more serious reactions, including toxic hepatitis, pancreatitis, peripheral neuropathy, leukopenia, hemolytic anemia, and agranulocytosis [16]. However, research by Sehgal et al. indicates that higher doses of mesalazine are just as safe as lower doses, with no significant increase in side effects [17]. This is believed to be due to the hypersensitivity nature of most reactions rather than dose-dependent toxicity.

A growing concern with these medications, particularly mesalazine, is their potential for cardiovascular toxicity (Fig. 13.2). While cardiovascular side effects were not traditionally recognized as a major risk, recent evidence suggests they may indeed occur in rare cases. Reported cardiac complications include myocarditis, pericarditis, cardiomyopathy, myocardial infarction (MI), atrioventricular (AV) block, and ventricular dysfunction, with an estimated incidence ranging from 0% to 0.3% [17]. Among these, pericarditis and myocarditis are the most frequently reported adverse effects [18]. Although rarely performed, myocardial biopsy remains the gold standard for diagnosing myocarditis, often revealing eosinophilic infiltration in the inflamed myocardium [19]. Concerningly, up to 30% of biopsy-confirmed myocarditis cases may progress to dilated cardiomyopathy, a condition associated with poor outcomes [20]. Diagnosing myocarditis typically involves several tools in addition to a biopsy. Blood tests can detect elevated cardiac biomarkers and signs of inflammation, while electrocardiograms (ECGs), although often nonspecific, may reveal ST-segment changes, T-wave abnormalities, or arrhythmias [21]. Imaging techniques like cardiac MRI and echocardiography provide valuable insights into the extent of inflammation, structural changes, myocardial edema, and overall heart function [22] (Fig. 13.2). Identifying whether cardiac symptoms are drug-related can be challenging. The Naranjo algorithm provides a valuable tool for evaluating the likelihood of a drug-induced reaction. A score above nine strongly indicates a drug-induced adverse reaction [23]. However, it is also important to consider that myocarditis can be a rare extra-intestinal manifestation of IBD, which complicates differentiating between IBD-related myocarditis and 5-ASA-induced myocarditis [24]. Mesalazine-induced cardiotoxicity usually manifests within 2–4 weeks of starting treatment, although the timeline can vary [25]. In some cases, particularly when patients are also receiving steroids, symptoms may not appear until months or even years after therapy begins [26]. This variability underscores the need for ongoing cardiovascular monitoring in patients during mesalazine therapy, as delayed reactions still pose significant clinical risks.

Several reports have highlighted that mesalazine-induced cardiotoxicity can affect individuals of any age or gender. Common symptoms include chest pain, fever, shortness of breath, palpitations, and fatigue. Importantly, in all documented cases, symptoms rapidly resolve after discontinuation of mesalazine, strongly suggesting a direct link between the drug and cardiac inflammation [14, 23, 24, 27–30].

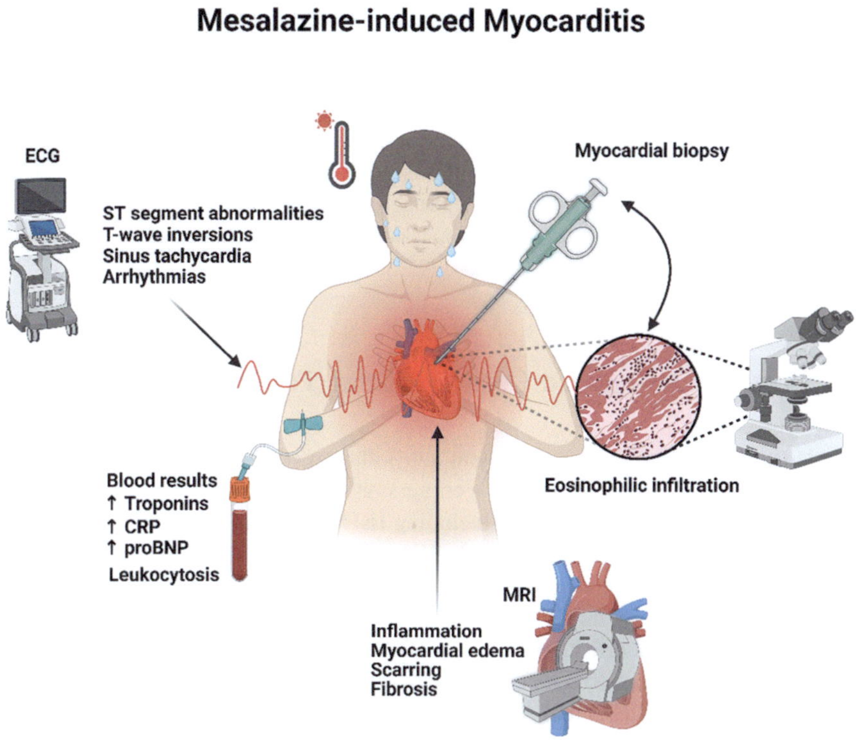

Fig. 13.2 Clinical presentation and diagnostic approach for mesalazine-induced myocarditis in IBD. Patients with mesalazine-induced myocarditis often present with nonspecific symptoms such as chest pain, shortness of breath, fever, palpitations, and fatigue, raising suspicion for myocarditis. A myocardial biopsy is crucial for diagnosis, commonly revealing eosinophilic infiltration, which is a hallmark of drug-induced myocarditis. Additionally, blood tests can detect elevated cardiac biomarkers, including troponins and the N-terminal prohormone of brain natriuretic peptide (NT-proBNP), along with signs of systemic inflammation, such as elevated CRP and leukocytosis. While ECG findings may be subtle, they can show ST-segment changes, T-wave abnormalities, or arrhythmias. Imaging techniques like cardiac MRI and echocardiography offer detailed insights into myocardial inflammation and structural changes, further aiding in diagnosing and evaluating heart function. These diagnostic tools collectively offer a comprehensive approach to identifying and managing this rare complication in IBD patients treated with mesalazine

In a recent study, Andrei et al. described two IBD patients with diverse presentations of mesalazine-induced cardiotoxicity. The first case involved a 28-year-old woman with UC who developed fever and chest pain 18 days after starting mesalazine. An ECG showed tachycardia and an incomplete right bundle branch block, while blood tests revealed leukocytosis and elevated levels of C-reactive protein (CRP), cardiac troponin T, and NT-proBNP, suggesting myocarditis (Fig. 13.2). After discontinuing the medication, her symptoms resolved, but they reappeared upon drug reintroduction, confirming a direct connection between mesalazine and the cardiac symptoms. The second case was a 79-year-old man newly diagnosed with IBD who developed symptoms of heart failure 4 weeks after starting 5-ASA therapy. Following drug discontinuation and the initiation of heart failure treatment, his condition significantly improved [31].

The consistency of clinical presentations across case reports suggests a common underlying mechanism, most likely immune-mediated. These findings underscore the importance for clinicians to be cautious when prescribing mesalazine, even to low-risk patients, and highlight the importance of closely monitoring cardiac symptoms, particularly in IBD patients who may already be susceptible to inflammatory complications.

Corticosteroids-Induced Cardiovascular Toxicity in IBD

Corticosteroids (CSs) have been potent anti-inflammatory and immunomodulatory agents for over 70 years and remain a cornerstone in the management of both acute and chronic inflammatory diseases, including IBD [32]. They are particularly effective in managing moderate to severe IBD, effectively controlling inflammation during acute exacerbations, and inducing remission when other treatments are either insufficient or slower-acting [33]. Despite their longstanding use, some aspects of their mechanisms and effects remain incompletely understood. CSs are believed to exert their anti-inflammatory effects by inhibiting the synthesis and transcription of pro-inflammatory proteins and suppressing some key inflammatory pathways. This results in a downregulation of nuclear factor kappa-B (NF-κB) activity and a consequent decrease in the production of pro-inflammatory cytokines, including interleukin (IL)-1, IL-6, and tumor necrosis factor-alpha (TNF-α), crucial cytokines involved in IBD pathophysiology [34, 35]. Additionally, CSs promote the upregulation of anti-inflammatory mediators, helping to restore immune balance and reduce inflammation. This multifaceted action contributes to their effectiveness in managing acute inflammatory responses in IBD.

First-generation synthetic CSs, such as prednisone, methylprednisolone, and hydrocortisone, have traditionally been the most commonly used treatments for IBD. However, these drugs exert systemic effects, affecting the entire body rather than specifically targeting inflamed areas [36]. In response to these limitations, second-generation CSs, such as budesonide and beclometasone dipropionate (BDP), were developed. These newer agents act locally, providing targeted anti-inflammatory effects while minimizing systemic exposure and reducing the risks associated with long-term use of first-generation CSs [37–39]. Katz famously described CSs as the "Dr. Jekyll and Mr. Hyde" of IBD therapy, reflecting their dual nature. While CSs provide rapid and effective symptom relief, they are also associated with a broad range of potential adverse effects, underscoring the challenge of balancing their therapeutic benefits against their risks [40]. Long-term use and high doses of CSs are well-known to increase the risk of serious complications, including immune and adrenal suppression, osteoporosis, fractures, diabetes,

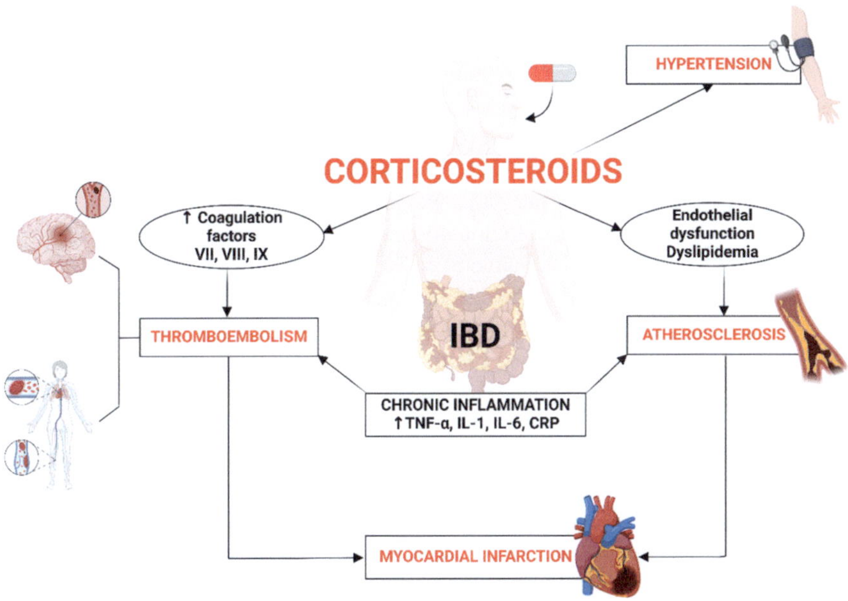

Fig. 13.3 Long-term use of corticosteroids (CS) in managing IBD significantly increases the risk of cardiovascular complications, especially at high doses. These complications include hypertension, thromboembolism, atherosclerosis, and myocardial infarction (MI). CSs elevate levels of coagulation factors VII, VIII, IX, and when combined with increased pro-inflammatory cytokines such as TNF-α, IL-1, and IL-6, along with elevated C-reactive protein (CRP) during chronic IBD inflammation, they promote a hypercoagulable state. This elevated coagulability further increases the risk of thromboembolic events, including venous and arterial thrombosis. Additionally, CSs impair endothelial function by inducing oxidative stress and reducing nitric oxide (NO) bioavailability, leading to endothelial dysfunction and increased vascular tone. The resulting dyslipidemia and altered cholesterol levels further exacerbate the development of atherosclerosis. Collectively, these mechanisms, exacerbated by the chronic inflammatory state characteristic of IBD, initiate a cascade of severe cardiovascular complications, including venous thromboembolism (VTE), pulmonary embolism, stroke, and MI

cardiovascular disease (CVD), dyslipidemia, and various dermatological and gastrointestinal issues [41].

One of the most significant cardiovascular side effects associated with CS use, particularly glucocorticoids, is hypertension (Fig. 13.3). Since both generations of CSs used in IBD treatment are primarily glucocorticoids [36], these drugs are believed to affect kidney function and vascular health through multiple mechanisms. Glucocorticoids increase the filtration fraction, contributing to glomerular hypertension, and stimulate the production of angiotensinogen and atrial natriuretic peptide, both of which disrupt fluid balance and raise blood pressure [42]. Additionally, glucocorticoids inhibit the synthesis and release of vasodilators such as nitric oxide (NO) and prostacyclin (PGI2), leading to further vasoconstriction and resulting in hypertension [43]. They also influence vascular tone by downregulating calcium-activated potassium channels, promoting sustained vasoconstriction. In a large cohort study that examined glucocorticoid use in various inflammatory diseases, Mebrahtu et al. reported a high prevalence of hypertension among patients undergoing glucocorticoid therapy. The study found a dose-dependent increase in hypertension risk, with higher cumulative doses of oral glucocorticoids significantly impacting blood pressure regulation [44]. Similarly, a retrospective cohort study in patients with rheumatoid arthritis (RA) showed that glucocorticoid use, particularly at doses exceeding 7.5 mg per day, was associated with an increased risk of hypertension [45]. Given that common doses of oral prednisone and methylprednisolone for mild to moderate IBD range from 20 to 40 mg per day and up to 60 mg for severe cases [46], prolonged CS use in IBD patients significantly increases the risk of hypertension. These findings underscore the cardiovascular risks associated with long-term glucocorticoid therapy in IBD populations. While CS use in IBD is known to cause hypertension, its association with other severe cardiovascular outcomes, such as thromboembolism and MI, raises significant concern. The reported incidence of thromboembolism in IBD patients varies between 1% and 8%, with some autopsy studies suggesting an incidence as high as 39% [47]. A meta-analysis by Sarlos et al. confirmed that CS use in IBD patients doubles the risk of venous thromboembolism (VTE) [48]. Furthermore, Higgins et al. found that patients receiving CS therapy have nearly a fivefold higher risk of VTE compared to those treated with a biologic agent alone. Additionally, they noted that when CSs are used alongside biologic agents, the risk of VTE remains similar to that of CS therapy alone [49]. As a chronic condition, IBD is characterized by elevated levels of pro-inflammatory cytokines, including TNF-α, IL-1, and IL-6. These cytokines, along with increased C-reactive protein (CRP), activate the coagulation cascade and inhibit fibrinolysis [50]. The pro-coagulation state induced by inflammation in IBD may be further exacerbated by CS use, which increases the levels of coagulation factors VII, VIII, and IX, thereby creating a vicious cycle of heightened coagulation risk [51] (Fig. 13.3). In a study by Bollen et al., a significant prevalence of thromboembolic events was observed in IBD patients, with 83% experiencing venous thrombosis and 17% suffering from arterial thromboembolism. Importantly, a substantial portion of this cohort (71%) had active disease at the time of the thromboembolic event, suggesting a potential correlation between disease activity and the risk of thromboembolism [52].

The previously mentioned cytokines, along with elevated CRP levels, are also pivotal in driving atherosclerosis in IBD patients [53]. This process unfolds in several critical stages, including persistent vascular inflammation, lipid accumulation within the arterial walls, intimal thickening, and fibrosis [54]. Over time, these changes can lead to plaque formation, which may rupture or erode, ultimately increasing the risk of MI (Fig. 13.3). A recent study by D'Ascenzo et al. revealed that individuals with IBD face a significantly higher risk of arterial ischemic events, particularly MI, compared to the general population. Notably, this elevated risk persists despite a lower prevalence of conventional MI risk factors such as hypertension, diabetes, and dyslipidemia. This suggests that non-traditional factors, especially chronic inflammation, play a key role in driving cardiovascular risk in IBD patients [55]. The American College of Cardiology outlines several potential mechanisms linking IBD to an increased risk of atherosclerotic cardiovascular disease

(ASCVD). These factors include local and systemic inflammation, abnormalities in the gut microbiome, endothelial dysfunction, thrombosis, and lipid dysfunction. Among these, CSs, commonly used in the treatment of IBD, are particularly noteworthy as they are known to elevate the risk of ASCVD, especially when used long-term or at high doses [56]. CSs can impair endothelial function, disrupting the balance between vasodilation and vasoconstriction. They specifically enhance the production and activity of vasoconstrictors such as endothelin-1 (ET-1) and angiotensin II (AngII). Additionally, these drugs contribute to increased reactive oxygen species (ROS) generation, promoting oxidative stress that decreases the bioavailability of NO [43]. Reduced NO levels impair the endothelium's ability to induce vasodilation, exacerbating vascular tone, and increasing the risk of conditions such as hypertension and atherosclerosis.

Another critical aspect of CS therapy is its effect on lipid metabolism, which is particularly relevant in patients with IBD, where lipid metabolism is already altered due to chronic inflammation [57]. CSs often exacerbate this condition, resulting in elevated levels of total cholesterol and low-density lipoprotein (LDL) cholesterol [58]. This dyslipidemia significantly contributes to the pathogenesis of atherosclerosis, as the increased LDL cholesterol accumulates in arterial walls, promoting plaque formation. Consequently, these lipid abnormalities, coupled with the pre-existing dyslipidemia associated with IBD and endothelial dysfunction, heighten the risk of cardiovascular events, such as MI, particularly in patients undergoing long-term CS treatment (Fig. 13.3).

Immunomodulators and Cardiovascular Toxicity in IBD

Immunomodulators, such as methotrexate and thiopurines (azathioprine and 6-mercaptopurine), are commonly used in the management of IBD to maintain remission and reduce reliance on steroid therapy [59]. These agents modulate the immune response by suppressing inflammation associated with IBD. Generally, immunomodulators are not typically associated with significant cardiovascular adverse effects compared to other classes of IBD medications. However, it is important to note that while they have a favorable safety profile, the use of immunomodulators may still carry some risks, including hepatotoxicity and an increased risk of infections, which could indirectly affect cardiovascular health [60, 61].

The Janus kinase (JAK) pathway plays a critical role in the signal transduction of several pro-inflammatory cytokines involved in gastrointestinal inflammation, making the JAK-STAT pathway a promising target for IBD treatment [62]. JAK inhibitors are one of the most famous immunomodulators used to treat UC, with tofacitinib being the most widely prescribed. Newer JAK inhibitors, such as upadacitinib and filgotinib, have primarily been studied for UC but have also shown promising results in CD [63]. Common safety concerns with JAK inhibitors include the risks of infections, thromboembolism, and malignancies [64]. Although cardiovascular events are rare, they can still occur and are more frequently associated with tofacitinib compared to other JAK inhibitors. Recent studies do not suggest a heightened risk of cardiovascular events with tofacitinib and other JAK inhibitors in the IBD population [65–67]. However, a U.S. Food and Drug Administration (FDA) review of a large safety trial found an increased risk of serious cardiovascular events, such as heart attack, stroke, cancer, blood clots, and death, associated with Xeljanz and Xeljanz XR (tofacitinib) [68]. Although this trial was conducted in RA patients, given that IBD is also an immune-mediated inflammatory disease (IMID), there is potential for similar adverse effects in IBD patients.

Biologics-Induced Cardiovascular Toxicity in IBD

Biologics are an advanced class of medications that have revolutionized the treatment of IBD, especially in patients unresponsive to traditional therapies like aminosalicylates, corticosteroids, or immunomodulators [7]. First introduced in the 1990s, biologics are derived from living cells and are designed to precisely target specific molecules that drive inflammation in IBD [69]. By focusing on key components of the immune system, such as cytokines and immune cell pathways, this group of medications helps reduce inflammation, promote healing of the intestinal lining, and support long-term remission. Currently, the approved biologic therapies for IBD encompass a variety of targeted agents that modulate different immune response pathways. These include anti-tumor necrosis factor (TNF) therapies, anti-adhesion molecules, and anti-interleukin (IL)-12/23 agents [70]. Each therapy class targets specific immune mechanisms to reduce inflammation and control the disease, offering a more precise and tailored approach to managing IBD.

Anti-TNF therapies, including infliximab, adalimumab, and certolizumab pegol, are widely used to neutralize TNF-α, a key pro-inflammatory cytokine involved in the pathogenesis of IBD [71] (Fig. 13.4). TNF-α plays a pivotal role in driving the chronic inflammation that characterizes IBD, contributing to tissue damage and disease progression [72]. By inhibiting TNF-α, these drugs aim to reduce inflammation, promote mucosal healing, and help patients achieve and maintain clinical remission. They are particularly beneficial for those with moderate to severe IBD who have not responded to conventional treatments. While anti-TNF therapies have undeniably transformed the therapeutic approach to IBD, their administration is accompanied by notable risks. Common adverse effects include infections, autoimmune reactions, injection site reactions, and immunogenicity, which can attenuate the long-term efficacy of treatment [73]. Moreover, these agents have been implicated in severe complications, including malignancies, liver function abnormalities, demyelinating disorders, and various cardiovascular events. Dermatologic manifestations, such as skin eruptions and psoriasis-like lesions, have also been frequently reported [73, 74]. Of particular concern are the cardiovascular risks associated with anti-TNF therapy. Although these agents generally exhibit favorable safety profiles over prolonged use, they possess the potential to exacerbate heart failure and other cardiovascular-related conditions. This concern was notably raised by Anti-TNF

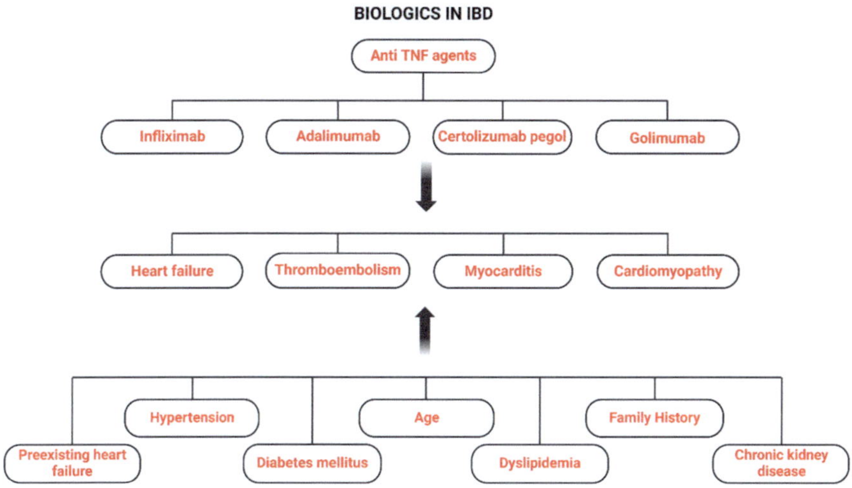

Fig. 13.4 Cardiovascular risks associated with anti-TNF agents in IBD patients. Anti-TNF agents, including infliximab, adalimumab, certolizumab pegol, and golimumab, play a crucial role in managing IBD by inhibiting tumor necrosis factor-alpha (TNF-α), thereby reducing inflammation and promoting healing. The figure outlines potential cardiovascular risks in IBD patients associated with these therapies, including exacerbation of heart failure, increased thromboembolic events, and rare cases of myocarditis or cardiomyopathy. Additionally, it emphasizes the importance of identifying patient risk factors that may elevate cardiovascular risks, such as pre-existing heart conditions, hypertension, diabetes, dyslipidemia, chronic kidney disease, age, and family history

Alpha Therapy Against Congestive Heart Failure (ATTACH) trial. In this Phase 2 trial involving 150 patients with stable heart failure, infliximab not only demonstrated no significant therapeutic advantage over placebo but, at a higher dose (10 mg/kg), was associated with increased all-cause mortality and a higher rate of heart failure-related hospitalizations [75]. Several case reports highlight the cardiovascular risks associated with anti-TNF therapy in IBD patients. For example, Grilo et al. described a 50-year-old woman with CD who likely developed heart failure due to infliximab use. After undergoing an ileocolectomy, this patient, who had a history of hypertension and diabetes, began treatment with infliximab and azathioprine, during which she exhibited signs of heart failure [76]. Similarly, Keating reported a 32-year-old woman with Turner syndrome and CD who developed drug-induced cardiomyopathy after initiating adalimumab, following the failure of corticosteroid therapy. An echocardiogram confirmed severe cardiac dysfunction, revealing a markedly reduced ejection fraction of less than 20% and a grossly dilated left ventricle [77]. Another report by Slattery et al. involved a 21-year-old male with acute severe UC who developed hypersensitivity myocarditis following infliximab administration [78]. Although very rare, infliximab-induced myocarditis is a serious complication that requires prompt recognition and management, highlighting the need for clinicians to be aware of this potential risk to reduce the likelihood of significant morbidity (Fig. 13.4).

Both anti-adhesion molecules and anti-IL-12/23 agents play key roles in the treatment of IBD. Natalizumab, an anti-adhesion agent, targets α4-integrin, effectively blocking lymphocyte migration into inflamed tissues, making it particularly beneficial for patients with moderate to severe CD [79, 80]. However, despite its efficacy, natalizumab has raised concerns, primarily due to its association with serious infections [81]. Additionally, there are worries about its potential link to thromboembolic events, which could increase the risk of cardiovascular complications such as heart attacks or strokes, particularly in patients with preexisting cardiovascular risk factors. While some case reports have suggested a possible connection between natalizumab and thromboembolic events [82], the evidence remains limited and is insufficient to establish a clear or common association. In contrast, vedolizumab, another anti-adhesion agent, specifically targets α4β7 integrin, focusing its anti-inflammatory effects on the gut in both CD and UC [83, 84]. Vedolizumab has demonstrated a favorable safety profile, with no significant cardiovascular toxicity, making it a well-tolerated option for managing IBD without the added concerns of cardiovascular risk.

By targeting the p40 subunit of both IL-12 and IL-23, ustekinumab effectively reduces inflammation and promotes healing of the intestinal lining in patients with IBD [85]. Like the anti-adhesion agents, ustekinumab has a favorable safety profile, demonstrating minimal cardiovascular risk compared to other biologic therapies. However, some reports suggest a potential link between ustekinumab and cardiovascular events [86, 87], indicating a need for caution in individuals with a personal or family history of heart disease. Further investigation is needed to clarify these associations and understand the potential underlying mechanisms.

Conclusions and Future Directions

Managing Inflammatory Bowel Disease (IBD) presents a distinct challenge, especially when balancing the therapeutic benefits of anti-IBD drugs with the risk of potential cardiovascular toxicity. While medications such as aminosalicylates, corticosteroids, immunomodulators, and biologics effectively control inflammation and induce remission, their use carries potential risks. Cardiovascular complications, including myocarditis, arrhythmias, and heart failure, can arise due to mechanisms like oxidative stress, free radical production, and the activation of inflammatory pathways, all of which compromise myocardial function and tissue integrity. Although these cardiotoxic effects are uncommon in clinical studies, they may become more apparent in diverse patient populations.

Due to the significant cardiovascular risks associated with IBD medications, a personalized treatment strategy is essential. Clinicians must carefully evaluate the benefits of anti-IBD drugs against their potential to exacerbate pre-existing cardiovascular conditions or trigger new cardiac issues. This requires continuous monitoring and vigilance in clinical practice, as early detection of cardiotoxicity can enable timely intervention, reducing the risk of severe outcomes. A comprehensive understanding of drug-induced cardiovascular toxicity (DICT) allows healthcare practitioners to design individualized treatments that not only manage IBD effectively but also protect cardiovascular health, ultimately improving overall patient care.

Key Points

- **Cardiovascular risks in IBD therapies**: Anti-IBD drugs, including aminosalicylates, corticosteroids, immunomodulators, and biologics, are essential for managing chronic inflammation, but their use carries cardiovascular risks such as myocarditis, cardiomyopathy, arrhythmias, myocardial ischemia, and heart failure.
- **Mechanisms of cardiotoxicity**: Drug-induced cardiovascular toxicity often arises from oxidative stress, free radical generation, and the activation of inflammatory pathways. These processes can result in myocardial dysfunction, reduced contractility, and tissue damage, ultimately increasing morbidity and mortality.
- **Importance of personalized treatment**: A personalized approach is essential to balance the benefits of IBD therapies with the risk of exacerbating pre-existing cardiovascular conditions or triggering new cardiac complications.
- **Continuous monitoring and early intervention**: Ongoing clinical monitoring is critical for detecting early signs of cardiotoxicity. Prompt identification and intervention can mitigate severe cardiovascular outcomes, enhancing patient safety and overall care quality.
- **Clinical implications and future directions:** A comprehensive understanding of the cardiovascular effects of anti-IBD drugs would allow healthcare professionals to improve treatment strategies, ensuring both effective management of IBD and the protection of cardiovascular health. Additional research is needed to create safer, more tailored medicines that minimize cardiotoxic risks.

Funding The research work of DKA is supported by the R25 AI179582 and R01 HL147662 grants from the National Institutes of Health, USA. The contents of this article are solely the responsibility of the authors and do not necessarily represent the official views of the National Institutes of Health.

Competing Interests All authors have read the manuscript and declare no conflict of interest. No writing assistance was utilized in the production of this manuscript.

Consent for Publication All authors have read the manuscript and consented for publication.

References

1. Ma W, Wei S, Zhang B, Li W (2020) Molecular mechanisms of cardiomyocyte death in drug-induced cardiotoxicity. Front Cell Dev Biol 8:434. https://pubmed.ncbi.nlm.nih.gov/32582710/
2. Ferdinandy P, Baczkó I, Bencsik P, Giricz Z, Görbe A, Pacher P et al (2019) Definition of hidden drug cardiotoxicity: paradigm change in cardiac safety testing and its clinical implications. Eur Heart J 40(22):1771–1777. https://pubmed.ncbi.nlm.nih.gov/29982507/

3. Brenner GB, Makkos A, Nagy CT, Onódi Z, Sayour NV, Gergely TG et al (2020) Hidden cardiotoxicity of Rofecoxib can be revealed in experimental models of ischemia/reperfusion. Cells 9(3):551. /pmc/articles/PMC7140447/
4. D'Oria R, Schipani R, Leonardini A, Natalicchio A, Perrini S, Cignarelli A et al (2020) The role of oxidative stress in cardiac disease: from physiological response to injury factor. Oxidative Med Cell Longev 2020:5732956. /pmc/articles/PMC7244977/
5. Li MY, Peng LM, Chen XP (2022) Pharmacogenomics in drug-induced cardiotoxicity: current status and the future. Front Cardiovasc Med 9:966261. /pmc/articles/PMC9606405/
6. Morelli MB, Bongiovanni C, Da Pra S, Miano C, Sacchi F, Lauriola M et al (2022) Cardiotoxicity of anticancer drugs: molecular mechanisms and strategies for cardioprotection. Front Cardiovasc Med 9:847012. /pmc/articles/PMC9051244/
7. Muzammil MA, Fariha F, Patel T, Sohail R, Kumar M, Khan E et al (2023) Advancements in inflammatory bowel disease: a narrative review of diagnostics, management, epidemiology, prevalence, patient outcomes, quality of life, and clinical presentation. Cureus. 15(6):e41120. /pmc/articles/PMC10382792/
8. Mirza RM, MacKean G, Shaffer SR, Sewitch MJ, Heisler C, McLeod J et al (2022) Patient experiences in the management of inflammatory bowel disease: a qualitative study. J Can Assoc Gastroenterol 5(6):261. /pmc/articles/PMC9713627/
9. Schoenfeld R, Nguyen GC, Bernstein CN (2020) Integrated care models: optimizing adult ambulatory care in inflammatory bowel disease. J Can Assoc Gastroenterol. 3(1):44. /pmc/articles/PMC8218536/
10. Cai Z, Wang S, Li J (2021) Treatment of inflammatory bowel disease: a comprehensive review. Front Med 8:765474. /pmc/articles/PMC8720971/
11. Liu J, Di B, Xu L-L (2023) Recent advances in the treatment of IBD: targets, mechanisms and related therapies. Cytokine Growth Factor Rev 71–72:1–12. https://pubmed.ncbi.nlm.nih.gov/37455149/
12. Nielsen OH, Munck LK (2007) Drug insight: aminosalicylates for the treatment of IBD. Nat Clin Pract Gastroenterol Hepatol 4(3):160–170. https://pubmed.ncbi.nlm.nih.gov/17339853/
13. Miyoshi J, Yajima T, Shimamura K, Matsuoka K, Okamoto S, Higuchi H et al (2012) 5-Aminosalicylic acid mediates expression of cyclooxygenase-2 and 15-hydroxyprostaglandin dehydrogenase to suppress colorectal tumorigenesis. Anticancer Res 32(4):1193–1202. https://pubmed.ncbi.nlm.nih.gov/22493349/
14. Perez-Colon E, Dadlani GH, Wilmot I, Miller M (2011) Mesalamine-induced myocarditis and coronary vasculitis in a pediatric ulcerative colitis patient: a case report. Case Rep Pediatr 2011: 1–5. https://pubmed.ncbi.nlm.nih.gov/22611512/
15. Loftus EV, Kane SV, Bjorkman D (2004) Systematic review: short-term adverse effects of 5-aminosalicylic acid agents in the treatment of ulcerative colitis. Aliment Pharmacol Ther 19(2):179–189. https://pubmed.ncbi.nlm.nih.gov/14723609/
16. Bergman R, Parkes M (2006) Systematic review: the use of mesalazine in inflammatory bowel disease. Aliment Pharmacol Ther 23(7):841–855. https://pubmed.ncbi.nlm.nih.gov/16573787/
17. Sehgal P, Colombel JF, Aboubakr A, Narula N (2018) Systematic review: safety of mesalazine in ulcerative colitis. Aliment Pharmacol Ther 47(12):1597–1609. https://pubmed.ncbi.nlm.nih.gov/29722441/
18. Brown G (2016) 5-Aminosalicylic acid–associated myocarditis and pericarditis: a narrative review. Can J Hosp Pharm 69(6):466. /pmc/articles/PMC5242279/
19. Dominguez F, Kühl U, Pieske B, Garcia-Pavia P, Tschöpe C (2016) Update on myocarditis and inflammatory cardiomyopathy: reemergence of endomyocardial biopsy. Rev Esp Cardiol (Engl Ed) 69(2):178–187. https://pubmed.ncbi.nlm.nih.gov/26795929/
20. Caforio ALP, Pankuweit S, Arbustini E, Basso C, Gimeno-Blanes J, Felix SB et al (2013) Current state of knowledge on aetiology, diagnosis, management, and therapy of myocarditis: a position statement of the European Society of Cardiology Working Group on myocardial and pericardial diseases. Eur Heart J 34(33):2636–2648. https://pubmed.ncbi.nlm.nih.gov/23824828/

21. Buttà C, Zappia L, Laterra G, Roberto M (2020) Diagnostic and prognostic role of electrocardiogram in acute myocarditis: a comprehensive review. Ann Noninvasive Electrocardiol 25(3): e12726. /pmc/articles/PMC7958927/
22. Haq MI, Ahmed S, Pasha W, Zaidi SAA (2019) Mesalazine-induced cardiotoxicity. J Rare Disord Diagnosis Ther 4(6):24. https://raredisorders.imedpub.com/articles/mesalazineinduced-cardiotoxicity.php?aid=23957
23. Chaudhry A, Noor J, Fatima G, Batool S, Noor R (2023) Mesalamine—a revered drug for inflammatory bowel disease with detrimental effects on the lung and heart. Cureus 15(8): e43567. /pmc/articles/PMC10503225/
24. Shergill S (2020) Mesalazine-induced myopericarditis: a case report. Eur Hear J Case Rep 5(2): ytaa508. https://pubmed.ncbi.nlm.nih.gov/33738396/
25. Moss AC, Peppercorn MA (2007) The risks and the benefits of mesalazine as a treatment for ulcerative colitis. Expert Opin Drug Saf 6(2):99–107. https://pubmed.ncbi.nlm.nih.gov/17367256/
26. Chen J, Duan T, Fang W, Liu S, Wang C (2022) Analysis of clinical characteristics of mesalazine-induced cardiotoxicity. Front Pharmacol 13:970597. /pmc/articles/PMC9520406/
27. Kyriakou M, Sakellaropoulos S, Constantinides T, Chatzantonis G, Avraamides P, Mitsis A (2023) Myocarditis related to the use of Mesalazine. J Med Cases 14(7):237–243. /pmc/articles/PMC10409542/
28. Sorleto M, Dürrwald S, Wiemer M (2015) Mesalazine-induced myopericarditis in a patient with a recent diagnosis of Crohn's disease: apropos of a case. Case Rep Cardiol 2015:1–2. https://pubmed.ncbi.nlm.nih.gov/26266055/
29. Ali A, Kanagavelu ASK, Rahimi A, Mehmood Z, Tariq AB, Das T et al (2021) A rare case of mesalazine-induced acute myocarditis in a 19-year-old female with ulcerative colitis. Cureus. 13(11):e20036. /pmc/articles/PMC8717744/
30. Radhakrishnan ST, Mohanaruban A, Hoque S (2018) Mesalazine-induced myocarditis: a case report. J Med Case Rep 12(1):44. /pmc/articles/PMC5822631/
31. Andrei V, D'ettore N, Scheggi V, Di Mario C (2023) Mesalazine-induced myopericarditis: a case series. Eur Hear J Case Rep. 7(9):ytad424. https://pubmed.ncbi.nlm.nih.gov/37719002/
32. Dorrington AM, Selinger CP, Parkes GC, Smith M, Pollok RC, Raine T (2020) The historical role and contemporary use of corticosteroids in inflammatory bowel disease. J Crohns Colitis 14(9):1316–1329. https://pubmed.ncbi.nlm.nih.gov/32170314/
33. Dubois-Camacho K, Ottum PA, Franco-Muñoz D, De La Fuente M, Torres-Riquelme A, Díaz-Jiménez D et al (2017) Glucocorticosteroid therapy in inflammatory bowel diseases: from clinical practice to molecular biology. World J Gastroenterol 23(36):6628–6638. /pmc/articles/PMC5643284/
34. Hussenbux A, De Silva A (2021) Steroids in inflammatory bowel disease: a clinical review. J Prescrib Pract 3(3):107–111. https://doi.org/10.12968/jprp202133107. https://www.magonlinelibrary.com/doi/10.12968/jprp.2021.3.3.107
35. Coutinho AE, Chapman KE (2011) The anti-inflammatory and immunosuppressive effects of glucocorticoids, recent developments and mechanistic insights. Mol Cell Endocrinol 335(1):2. /pmc/articles/PMC3047790/
36. Bruscoli S, Febo M, Riccardi C, Migliorati G (2021) Glucocorticoid therapy in inflammatory bowel disease: mechanisms and clinical practice. Front Immunol 12:691480. /pmc/articles/PMC8209469/
37. Ferretti F, Cannatelli R, Monico MC, Maconi G, Ardizzone S (2022) An update on current pharmacotherapeutic options for the treatment of ulcerative colitis. J Clin Med 11(9):2302. /pmc/articles/PMC9104748/
38. Silverman J, Otley A (2011) Budesonide in the treatment of inflammatory bowel disease. Expert Rev Clin Immunol 7(4):419–428. https://pubmed.ncbi.nlm.nih.gov/21790284/
39. Rizzello F, Mazza M, Salice M, Calabrese C, Calafiore A, Campieri M et al (2018) The safety of beclomethasone dipropionate in the treatment of ulcerative colitis. Expert Opin Drug Saf 17(9): 963–969. https://pubmed.ncbi.nlm.nih.gov/30101623/

40. Katz JA (2004) Treatment of inflammatory bowel disease with corticosteroids. Gastroenterol Clin N Am 33(2):171–189. https://pubmed.ncbi.nlm.nih.gov/15177533/
41. Liu D, Ahmet A, Ward L, Krishnamoorthy P, Mandelcorn ED, Leigh R et al (2013) A practical guide to the monitoring and management of the complications of systemic corticosteroid therapy. Allergy Asthma Clin Immunol 9(1):30. /pmc/articles/PMC3765115/
42. Hunter RW, Ivy JR, Bailey MA (2014) Glucocorticoids and renal Na+ transport: implications for hypertension and salt sensitivity. J Physiol 592(Pt 8):1731–1744. /pmc/articles/PMC400174
43. Macleod C, Hadoke PWF, Nixon M (2021) Glucocorticoids: fuelling the fire of atherosclerosis or therapeutic extinguishers? Int J Mol Sci 22(14):7622. /pmc/articles/PMC8303333/
44. Mebrahtu TF, Morgan AW, West RM, Stewart PM, Pujades-Rodriguez M (2020) Oral glucocorticoids and incidence of hypertension in people with chronic inflammatory diseases: a population-based cohort study. CMAJ 192(12):E295–E301. https://pubmed.ncbi.nlm.nih.gov/32392512/
45. Costello RE, Yimer BB, Roads P, Jani M, Dixon WG (2021) Glucocorticoid use is associated with an increased risk of hypertension. Rheumatology (Oxford) 60(1):132. /pmc/articles/PMC7785301/
46. Prednisolone (oral route) proper use. Mayo Clinic. https://www.mayoclinic.org/drugs-supplements/prednisolone-oral-route/proper-use/drg-20075189
47. Tezel A, Demir M (2012) Inflammatory bowel disease and thrombosis. Turk J Hematol 29(2):111. /pmc/articles/PMC3986948/
48. Sarlos P, Szemes K, Hegyi P, Garami A, Szabo I, Illes A et al (2018) Steroid but not biological therapy elevates the risk of venous thromboembolic events in inflammatory bowel disease: a meta-analysis. J Crohns Colitis 12(4):489–498. https://pubmed.ncbi.nlm.nih.gov/29220427/
49. Higgins PDR, Skup M, Mulani PM, Lin J, Chao J (2015) Increased risk of venous thromboembolic events with corticosteroid vs biologic therapy for inflammatory bowel disease. Clin Gastroenterol Hepatol 13(2):316–321. https://pubmed.ncbi.nlm.nih.gov/25038374/
50. Magro F, Soares JB, Fernandes D (2014) Venous thrombosis and prothrombotic factors in inflammatory bowel disease. World J Gastroenterol 20(17):4857. /pmc/articles/PMC4009517/
51. van Zaane B, Nur E, Squizzato A, Gerdes VEA, Büller HR, Dekkers OM et al (2010) Systematic review on the effect of glucocorticoid use on procoagulant, anti-coagulant and fibrinolytic factors. J Thromb Haemost 8(11):2483–2493. https://pubmed.ncbi.nlm.nih.gov/20735729/
52. Bollen L, Vande Casteele N, Ballet V, Van Assche G, Ferrante M, Vermeire S et al (2016) Thromboembolism as an important complication of inflammatory bowel disease. Eur J Gastroenterol Hepatol 28(1):1–7. https://pubmed.ncbi.nlm.nih.gov/26513610/
53. Weissman S, Sinh P, Mehta TI, Thaker RK, Derman A, Heiberger C et al (2020) Atherosclerotic cardiovascular disease in inflammatory bowel disease: the role of chronic inflammation. World J Gastrointest Pathophysiol 11(5):104. /pmc/articles/PMC7403753/
54. Jebari-Benslaiman S, Galicia-García U, Larrea-Sebal A, Olaetxea JR, Alloza I, Vandenbroeck K et al (2022) Pathophysiology of atherosclerosis. Int J Mol Sci 23(6):3346. /pmc/articles/PMC8954705/
55. D'Ascenzo F, Bruno F, Iannaccone M, Testa G, De Filippo O, Giannino G et al (2023) Patients with inflammatory bowel disease are at increased risk of atherothrombotic disease: a systematic review with meta-analysis. Int J Cardiol 378:96–104. https://pubmed.ncbi.nlm.nih.gov/36863421/
56. Goldfine AB, Fonseca V, Jablonski KA, Chen YDI, Tipton L, Staten MA et al (2013) Salicylate (salsalate) in patients with type 2 diabetes: a randomized trial. Ann Intern Med 159(1):1–12. https://pubmed.ncbi.nlm.nih.gov/23817699/
57. Scoville EA, Allaman MM, Brown CT, Motley AK, Horst SN, Williams CS et al (2018) Alterations in lipid, amino acid, and energy metabolism distinguish Crohn's disease from ulcerative colitis and control subjects by serum metabolomic profiling. Metabolomics 14(1):17. /pmc/articles/PMC5907923/

58. Ross IL, Marais AD (2014) The influence of glucocorticoids on lipid and lipoprotein metabolism and atherosclerosis. S Afr Med J 104(10):671–674. https://pubmed.ncbi.nlm.nih.gov/25363050/
59. Ooi CJ, Hilmi I, Banerjee R, Chuah SW, Ng SC, Wei SC et al (2019) Best practices on immunomodulators and biologic agents for ulcerative colitis and Crohn's disease in Asia. Intest Res 17(3):285. /pmc/articles/PMC6667368/
60. Cohn HM, Dave M, Loftus EV (2017) Understanding the cautions and contraindications of immunomodulator and biologic therapies for use in inflammatory bowel disease. Inflamm Bowel Dis 23(8):1301. /pmc/articles/PMC5731468/
61. Sattler L, Hanauer SB, Malter L (2021) Immunomodulatory agents for treatment of patients with inflammatory bowel disease (review safety of anti-TNF, anti-integrin, anti IL-12/23, JAK inhibition, sphingosine 1-phosphate receptor modulator, azathioprine/6-MP and methotrexate). Curr Gastroenterol Rep 23(12):30. https://pubmed.ncbi.nlm.nih.gov/34913108/
62. Cohen BL (2021) Janus kinase signaling in inflammatory bowel disease. Gastroenterol Hepatol (N Y) 17(6):278–281. https://pubmed.ncbi.nlm.nih.gov/34776802/
63. Herrera-deGuise C, Serra-Ruiz X, Lastiri E, Borruel N (2023) JAK inhibitors: a new dawn for oral therapies in inflammatory bowel diseases. Front Med 10:1089099. /pmc/articles/PMC10017532/
64. Spiewak TA, Patel A (2022) User's guide to JAK inhibitors in inflammatory bowel disease. Curr Res Pharmacol drug Discov 3:100096. https://pubmed.ncbi.nlm.nih.gov/35300073/
65. Shehab M, Alrashed F, Alkazemi A, Lakatos PL, Bessissow T (2023) Impact of biologic therapies and small molecules on the risk of major adverse cardiovascular events in patients with inflammatory bowel diseases: systematic review and meta-analysis of randomized controlled trials. Expert Rev Gastroenterol Hepatol 17(5):469–477. https://pubmed.ncbi.nlm.nih.gov/36961082/
66. Olivera PA, Lasa JS, Peretto G, Zuily S, Danese S, Peyrin-Biroulet L (2023) Review article: risk of cardiovascular events in patients with inflammatory bowel disease receiving small molecule drugs. Aliment Pharmacol Ther 57(11):1231–1248. https://pubmed.ncbi.nlm.nih.gov/37038269/
67. Xie W, Xiao S, Huang Y, Sun X, Zhang Z (2019) Effect of tofacitinib on cardiovascular events and all-cause mortality in patients with immune-mediated inflammatory diseases: a systematic review and meta-analysis of randomized controlled trials. Ther Adv Musculoskelet Dis 11: 1759720X19895492. https://pubmed.ncbi.nlm.nih.gov/31897092/
68. FDA requires warnings about increased risk of serious heart-related events, cancer, blood clots, and death for JAK inhibitors that treat certain chronic inflammatory conditions | FDA. https://www.fda.gov/drugs/fda-drug-safety-podcasts/fda-requires-warnings-about-increased-risk-serious-heart-related-events-cancer-blood-clots-and-death
69. Nag A, Singh M, Thomas J, Ravichandran R, Gupta L, Panjiyar BK (2023) Role of biologic therapies in the rheumatic manifestations of inflammatory bowel disease: a systematic analysis. Cureus. 15(9):e45195. /pmc/articles/PMC10576538/
70. Torres J, Bonovas S, Doherty G, Kucharzik T, Gisbert JP, Raine T et al (2020) ECCO guidelines on therapeutics in Crohn's disease: medical treatment. J Crohns Colitis 14(1): 4–22. https://pubmed.ncbi.nlm.nih.gov/31711158/
71. Lee JU, Shin W, Son JY, Yoo KY, Heo YS (2017) Molecular basis for the neutralization of tumor necrosis factor α by Certolizumab Pegol in the treatment of inflammatory autoimmune diseases. Int J Mol Sci 18(1):228. /pmc/articles/PMC5297857/
72. Souza RF, Caetano MAF, Magalhães HIR, Castelucci P (2023) Study of tumor necrosis factor receptor in the inflammatory bowel disease. World J Gastroenterol 29(18):2733–2746. https://pubmed.ncbi.nlm.nih.gov/37274062/
73. Gerriets V, Goyal A, Khaddour K (2024) Tumor necrosis factor inhibitors. https://pubmed.ncbi.nlm.nih.gov/29494032/
74. D'Haens G (2007) Risks and benefits of biologic therapy for inflammatory bowel diseases. Gut 56(5):725–732. /pmc/articles/PMC1942157/

75. Chung ES, Packer M, Lo KH, Fasanmade AA, Willerson JT (2003) Randomized, double-blind, placebo-controlled, pilot trial of infliximab, a chimeric monoclonal antibody to tumor necrosis factor-alpha, in patients with moderate-to-severe heart failure: results of the anti-TNF therapy against congestive heart failure (ATTACH) trial. Circulation 107(25):3133–3140. https://pubmed.ncbi.nlm.nih.gov/12796126/
76. Grillo TG, Almeida LR, Beraldo RF, Marcondes MB, Queiróz DAR, da Silva DL et al (2021) Heart failure as an adverse effect of infliximab for Crohn's disease: a case report and review of the literature. World J Clin Cases 9(33):10382–10391. /pmc/articles/PMC8638044/
77. Keating E, Barry Kelleher T, Lahiff C (2020) De novo anti-TNF-α-induced congestive heart failure in a patient with turner syndrome and Crohn's disease. Inflamm Bowel Dis 26(12): E161–E162. https://pubmed.ncbi.nlm.nih.gov/32619005/
78. Slattery E, Ismail N, Sheridan J, Eustace K, Harewood G, Patchett S (2011) Myocarditis associated with infliximab: a case report and review of the literature. Inflamm Bowel Dis 17(7):1633–1634. https://pubmed.ncbi.nlm.nih.gov/21674722/
79. Selewski DT, Shah GV, Segal BM, Rajdev PA, Mukherji SK (2010) Natalizumab (Tysabri). AJNR Am J Neuroradiol 31(9):1588–1590. /pmc/articles/PMC7964985/
80. Ghosh S, Goldin E, Gordon FH, Malchow HA, Rask-Madsen J, Rutgeerts P et al (2003) Natalizumab for active Crohn's disease. N Engl J Med 348(1):24–32. https://pubmed.ncbi.nlm.nih.gov/12510039/
81. Luna G, Alping P, Burman J, Fink K, Fogdell-Hahn A, Gunnarsson M et al (2020) Infection risks among patients with multiple sclerosis treated with Fingolimod, Natalizumab, rituximab, and injectable therapies. JAMA Neurol 77(2):184–191. https://pubmed.ncbi.nlm.nih.gov/31589278/
82. Soltys J, Padarti A, Martin B, Butler T, Kilgo W (2022) Cerebral venous thrombosis and pulmonary embolism following natalizumab infusion: case series. Neuroimmunol Rep 2: 100145
83. Sandborn WJ, Feagan BG, Rutgeerts P, Hanauer S, Colombel JF, Sands BE et al (2013) Vedolizumab as induction and maintenance therapy for Crohn's disease. N Engl J Med 369(8):711–721. https://pubmed.ncbi.nlm.nih.gov/23964933/
84. Feagan BG, Rutgeerts P, Sands BE, Hanauer S, Colombel JF, Sandborn WJ et al (2013) Vedolizumab as induction and maintenance therapy for ulcerative colitis. N Engl J Med 369(8):699–710. https://pubmed.ncbi.nlm.nih.gov/23964932/
85. Pugliese D, Privitera G, Fiorani M, Parisio L, Calvez V, Papa A et al (2022) Targeting IL12/23 in ulcerative colitis: update on the role ofustekinumab. Ther Adv Gastroenterol 15: 17562848221102283. /pmc/articles/PMC9201364/
86. Beroukhim K, Danesh M, Nguyen C, Koo J, Leon A (2015) A case report of heart failure after therapy with ustekinumab. J Dermatol Dermatologic Surg 19(2):117–119
87. Abdelnabi M, ElNawaa S, Benjanuwattra J, Elmassry M, Nair N (2022) Ustekinumab-induced fatal acute heart failure in a young female: a case report. Methodist Debakey Cardiovasc J 18(1): 54–58. https://pubmed.ncbi.nlm.nih.gov/35891701/

Chapter 14
Antiviral Therapy and Cardiovascular Toxicity

Wilson S. Tang and David Jesse Sanchez

Abstract Antiviral therapies play an important role in managing both acute and chronic viral infections. Concerns about the impact of antivirals on cardiovascular health have grown over the years, particularly due to their widespread use during the COVID-19 pandemic and in long-term HIV treatment, which has highlighted potential cardiotoxic risks. This chapter explores the mechanisms underlying antiviral-induced cardiovascular toxicity, including mitochondrial dysfunction, metabolic dysregulation, endothelial dysfunction, and QT prolongation. A key example of antiviral-induced cardiotoxicity is the chronic administration of antiretroviral therapy (ART), which has been linked to dyslipidemia, insulin resistance, and an increased risk of myocardial infarction. This effect is particularly evident in ART regimens containing protease inhibitors such as ritonavir. Acute antiviral therapies, such as remdesivir that has been used for SARS-CoV-2 treatment, can lead to the development of bradycardia and arrhythmias. These examples show that careful patient monitoring, especially for individuals with pre-existing cardiovascular risk factors, is required when balancing antiviral efficacy and cardiovascular safety necessitates. Large clinical studies such as the D:A:D cohort have shown incidence, reinforcing the need for cardiovascular risk stratification in HIV treatment. Similarly, concerns about cardiovascular events related to COVID-19 antiviral treatments highlight the need to understand drug-induced endothelial activation and arrhythmias, particularly in acute infections like COVID-19. As antivirals remain integral to treatment of viral disease, a stronger understanding of the cardiovascular risks inherent in the use of antivirals will be critical for optimizing patient outcomes while minimizing cardiovascular risks.

Keywords Antiviral · Antiretroviral · ART · HIV · AIDS · SARS-CoV-2 · COVID-19 · Cardiotoxicity

---

W. S. Tang · D. J. Sanchez (✉)
Pharmaceutical Sciences Department, Western University of Health Sciences, Pomona, CA, USA
e-mail: sanchezd@westernu.edu

D. K. Agrawal et al. (eds.), *Cardiovascular Toxicity*, Advances in Biochemistry in Health and Disease 33, https://doi.org/10.1007/978-3-031-90598-8_14

Introduction

Current approaches to the management of both acute and chronic viral infections have increasingly shifted towards using antiviral drugs as part of the standard of care for patients. The past decade has seen increased introduction of new antivirals especially during the COVID-19 pandemic and the continuing expansion of access to pre-exposure prophylaxis to prevent HIV infection. With indications that span both acute and chronic viral infections, the use of antivirals is becoming more common in individuals who could be at risk of developing cardiovascular complications. This makes understanding the possible mechanisms and connections between antivirals and cardiotoxicity increasingly important.

The interplay between antiviral therapy and cardiovascular health represents an important aspect of clinical management for patients with viral infections. While new antiviral drugs have advanced effective treatment of viral diseases, the impact of antivirals on cardiovascular systems has emerged as a significant concern. This chapter explores the commons mechanisms that induce cardiotoxicity as well as examples of antiviral therapies that could impact cardiovascular health, with a particular emphasis on antiretroviral therapy (ART) for HIV, compared to acute antiviral therapy of SARS-CoV-2.

Overview of Acute Versus Chronic Antiviral Administration on Cardiotoxicity

As with any drug, it is essential to identify any comorbidities or potential risk factors that could impact cardiovascular health prior and during treatment. In addition, monitoring for adverse events that occur during antiviral administration is important even as the focus is often on the viral infection itself. As reviewed in Table 14.1, differences in how antivirals are administered will lead to different clinical manifestations of cardiovascular toxicity. Consequently, one must weigh the benefits of antiviral use versus the impact of the drug on cardiovascular health.

Mechanisms of Antiviral-Induced Cardiovascular Toxicity

Mitochondrial Toxicity

Mitochondrial toxicity occurs when disruption in mitochondrial function or integrity leads to cellular damage. This is especially important to consider when evaluating the impact of antiviral therapies on organs such as the heart that require large amounts of mitochondria function [8]. The significance of proper mitochondria function can be attributed to their role in ATP production, calcium homeostasis,

Table 14.1 Comparison of acute and chronic use of antivirals on cardiovascular health

	Acute antiviral administration	Chronic antiviral administration
Time course	Days to weeks	Months to years
Risk	Low frequency but potential life threatening	High frequency and built on progressive cardiac dysfunction
Reversibility	Often reversible upon discontinuation of antiviral therapy	Often irreversible due to long-term cardiovascular change
Common symptoms	Sudden arrhythmias, bradycardia, hypotension, myocarditis	Gradual heart failure, cardiac fibrosis, metabolic syndrome
Potential mechanisms	Ion channel blockade, direct myocarditis, electrolyte imbalance	Mitochondrial dysfunction, chronic inflammation, metabolic dysregulation
Examples	***Remdesivir***: bradycardia, hypotension [1] ***Lopinavir/ritonavir***: QT prolongation [2] ***Sofosbuvir***: bradycardia [3] ***Interferon-α***: myocarditis [4]	***Zidovudine***: cardiomyopathy [5] ***Protease inhibitors***: atherosclerosis [6] and aortic stiffness [7]

and balancing of cellular oxidation reduction reactions. Several classes of drugs, including anthracyclines (e.g., doxorubicin), nucleoside reverse transcriptase inhibitors (NRTIs), and tyrosine kinase inhibitors (TKIs), have been shown to induce mitochondrial toxicity by interfering with these processes. These drugs can impair electron transport chain (ETC) function, reducing ATP production, or promoting mitochondrial permeability transition pore (mPTP) opening [9, 10]. These examples of drug-induced disruption of normal mitochondria function can result in energy depletion, oxidative stress, and apoptosis. In cardiomyocytes the consequences of mitochondrial function disruption are especially significant to consider as they may impair oxidative phosphorylation, leading to excessive reactive oxygen species (ROS) production, mitochondrial DNA (mtDNA) damage, and activation of cell death pathways. Any or all of these pathways may promote cardiotoxicity.

As mitochondrial dysfunction continues, cardiomyocytes will shift away from fatty acid oxidation (FAO), the primary energy source of the heart, toward a glycolysis-focused environment, which is less efficient for ATP generation [11]. This change in the energy production of mitochondria may weaken the heart and increase oxidative stress, further damaging mitochondrial components and worsening the myocardial dysfunction. In addition, disruption in mitochondrial metabolism promotes cardiac hypertrophy, fibrosis, and arrhythmias, all of which contribute to progressive heart failure.

Nucleoside Reverse Transcriptase Inhibitors (NRTIs), such as zidovudine and stavudine, impair mitochondrial DNA replication, resulting in oxidative stress and cardiomyopathy [12]. Clinical studies have confirmed these findings, showing an increased incidence of dilated cardiomyopathy in patients on long-term zidovudine therapy [13]. Oxidative stress and chronic inflammation in HIV-infected individuals receiving antiretroviral therapy (ART) have been implicated in the increased risk of cardiovascular disease, including atherosclerosis [14].

Dyslipidemia and Insulin Resistance

Metabolic syndrome is a set of often interconnected conditions that significantly increase the risk of cardiovascular disease [15]. Conditions that are encompassed within metabolic syndrome include elevated triglycerides and low high-density lipoprotein (HDL) cholesterol, hypertension, and impaired glucose metabolism all of which contribute to cardiovascular disease by endothelial cell dysfunction and atherosclerosis [16]. During long-term antiviral drug treatment, it is important to determine if there is development of dyslipidemia, including elevated triglycerides and low-density lipoprotein (LDL) cholesterol along with lessened HDL, as well as insulin resistance. These two indications are key parts of metabolic syndrome development and can enhance each other such as with chronic insulin resistance leading to enhanced lipogenesis and dyslipidemia, synergistically compounding cardiovascular risk [17]. Also, dyslipidemia can induce oxidative stress and systemic inflammation that will enhance myocardial damage and vascular remodeling, increasing cardiotoxicity during antiviral administration [18].

This mechanism of cardiotoxicity is especially stressed for individuals on antiretroviral therapy, particularly protease inhibitors. Protease inhibitors are well-known to induce metabolic derangements, including increased triglycerides and LDL cholesterol levels, alongside insulin resistance [19]. These metabolic changes accelerate atherosclerosis and increase cardiovascular disease risk. Another example is ritonavir, which has been shown to inhibit the activity of GLUT4, a glucose transporter, thereby increasing hyperglycemia and insulin resistance [20]. The addition of new antiretroviral drugs, such as integrase inhibitors, has led to changes in HIV treatment that may help reduce its impact on cardiovascular disease [21–23].

Endothelial Dysfunction

Endothelial dysfunction is part of the pathologic process that leads to cardiotoxicity. Endothelial cells promote vascular homeostasis, regulate nitric oxide signaling, and help control thrombosis [24]. Inflammatory or oxidative stress induced activation of endothelial cells will trigger increased vascular permeability, platelet aggregation, and reduced coronary blood flow, all of which increase the risk of cardiovascular complications. An important biomarker for endothelial cell activation is von Willebrand factor (vWF). Increases in the levels of vWF may indicate a heightened state of endothelial cell activations that may be a contributing factor to thrombotic and ischemic events [25]. Another endothelial cell protein, thrombomodulin, plays an important role in endothelial function by regulating coagulation and has been found to promote nitric oxide synthesis [26]. Disruptions in thrombomodulin activity and nitric oxide production have been implicated in endothelial dysfunction, leading to vascular complications. During COVID-19, endothelial activation and dysfunction were linked to severe inflammatory responses and cardiovascular complications [27]. Endothelial cell activation, mediated by drug-induced inflammation, may be a key driver of these cardiovascular complications [28]. These findings show the importance of endothelial cells in cardiac health, noting the need for close monitoring of markers of endothelial function to limit cardiotoxic effects.

QT Prolongation and Arrhythmias

QT prolongation is a lengthening of the QT interval on an electrocardiogram (ECG), which represents the time it takes for the heart ventricles to depolarize and repolarize during a single heartbeat. This is a key indicator that the heart is maintaining normal cardiac rhythm; QT prolongation can be a marker of instability in cardiac electrophysiology. Many drugs, including antiviral agents, can interfere with cardiac ion channels, particularly those regulating potassium efflux, leading to repolarization delays and an extended QT interval [29, 30]. As a result, QT prolongation is widely recognized as an important early indicator of drug-induced cardiotoxicity and worth monitoring. However, while this may represent a clinically relevant concern, the number of patients that have been reported to present with QT prolongation is low in most clinical settings [31, 32].

Examples of Cardiotoxicity in HIV Antiretroviral Therapy (ART)

Antiretroviral drugs have become common in the therapeutic regimes for people living with HIV infection (PLWH) or in prophylaxis for individuals at risk of being exposed to HIV through pre-exposure prophylaxis (PrEP). According to UNAIDS, at the end of 2023 approximately 30.7 million people were utilizing ART while the WHO estimated that more than 3.5 million people were receiving PrEP. This amount of access to antiretrovirals for extended time necessitates determining if there is an impact on cardiovascular health.

Numerous studies have demonstrated an increased risk of cardiovascular events in individuals on long-term ART. Early studies determined that antiretroviral treatment induced changes that could be associated with increased cardiovascular disease. Use of protease inhibitors had shown association with hyperlipidemia and insulin resistance in PLWH as well as individuals with no HIV infection [13, 33]. Further studies showed that there was an increased risk of myocardial infarction overtime, especially past 18 months, in individuals taking protease inhibitors [34].

The Data Collection on Adverse Events of Anti-HIV Drugs (D:A:D) study, was one of the largest cohort studies to assess the impact of antiretrovirals on the health of PLWH. Initial findings of this study used a cohort of 23,468 patients showing that

Table 14.2 Antiretrovirals used in HIV therapy that may increase cardiovascular risk

Antiretroviral	Antiretroviral drug class	Cardiotoxicity effect
Abacavir (ABC)	Nucleoside reverse transcriptase inhibitor (NRTI)	Increased risk of myocardial infarction; findings are mixed [37]
Ritonavir (RTV)	Protease inhibitor	Pro-atherosclerosis effects and worsened lipid profiles [38, 39]
Dolutegravir (DTG)	Integrase inhibitor	Weight gain and metabolic syndrome, increasing cardiovascular risk [40, 41]
Bictegravir (BIC)	Integrase inhibitor	Possible links to weight gain and metabolic changes [42]
Cobicistat (COBI)	Pharmacokinetic booster used in ART	Can cause renal impairment potentially leading to hypertension and increased cardiovascular risk [43, 44]

the incidence of myocardial infarction has an adjusted relative rate per year of antiretroviral exposure of 1.26 (95% CI, 1.12–1.41; P <0.001) [35]. However, the overall number of myocardial infarction events in this report was small compared to the total number of people taking ART (3.5 cases per 1000 person-years among PLWH receiving antiretroviral therapy) suggesting that cardiovascular Based on these findings ART may contribute to dyslipidemia and insulin resistance, thereby contributing to increased MI risk. Further analysis showed that protease inhibitors and some nucleoside reverse transcriptase inhibitors (NRTIs) may be the cause of higher rates of myocardial infarction [36].

As shown in Table 14.2, several currently used antiretrovirals may increase the risk for cardiotoxicity. For example, patients on ritonavir-boosted regimens showed a twofold increase in the risk of myocardial infarction compared to those not receiving ART. Clinical guidelines now recommend careful cardiovascular risk stratification before initiating ART containing abacavir. These findings support the need for monitoring cardiovascular health during ART.

Examples of Cardiotoxicity in SARS-COV-2 Antiviral Therapies

The rapid onset of the COVID-19 pandemic led to use of therapies like remdesivir, lopinavir/ritonavir, and even the antimalarial drug hydroxychloroquine to find therapies that could effectively stop SARS-CoV-2 infection. These drugs have varying efficacy against SARS-CoV-2 but sometimes demonstrated potential for arrhythmias, myocarditis, and endothelial dysfunction. For instance, remdesivir has been associated with bradycardia in a significant subset of patients, with bradycardia rates ranging from 10% to 15% in clinical trials [45]. However, when considering antivirals, the risk of cardiovascular events must be weighed against the viral disease itself. With the use of remdesivir, while bradycardia was seen in some patients it has been seen to associate with better clinical outcomes from the perspective of the

SARS-CoV-2 infection, possibly due to the activity of the antiviral itself [46, 47]. On the other hand, myocarditis has been observed in COVID-19 patients, and its potential association with antiviral treatments like remdesivir remains important to understand [48].

The proarrhythmic effects of hydroxychloroquine have been well-documented, with observational studies linking its use to QT prolongation in approximately 8% of treated patients [49]. Systematic review has found that chloroquine and hydroxychloroquine treatment is significantly linked to QT prolongation and an elevated risk cardiac events such as MI [50, 51]

Conclusion

The intersection of cardiovascular health and antiviral therapies represents a critical area of concern in the management of both acute and chronic viral infections. Although the introduction of novel antiviral drugs has substantially improved the outcomes for patients with infections like COVID-19 and HIV, the potential for these treatments to induce cardiotoxicity cannot be neglected. The mechanisms underlying antiviral-induced cardiovascular toxicity are multifold, ranging from mitochondrial dysfunction and dyslipidemia to endothelial dysfunction and arrhythmias. As the use of antiviral medications becomes more common, particularly among individuals with an increased risk of cardiovascular diseases, it is crucial that clinicians monitor for signs of cardiovascular damage throughout the treatment. A fuller understanding of these mechanisms, as well as the long-term effects of antiviral drugs on heart health, will be imperative in developing strategies to mitigate risks, ensuring that the benefits of antiviral therapies outweigh their potential cardiovascular damage. Careful cardiovascular risk stratification and monitoring are essential for patients receiving long-term antiviral regimes, especially in the context of HIV therapy and COVID-19, where the balance between treating the underlying infection and managing the potential cardiovascular side effects must be carefully considered.

References

1. Singla K et al (2022) Remdesivir induced bradycardia and QT prolongation: a rare side effect of a ubiquitous drug of the COVID-19 era. J Anaesthesiol Clin Pharmacol 38(Suppl 1):S148–S149
2. Zhu S et al (2020) QTc prolongation during antiviral therapy in two COVID-19 patients. J Clin Pharm Ther 45(5):1190–1193
3. Monoe C et al (2020) Severe bradycardia induced by sofosbuvir and amiodarone which resolved after the discontinuation of both drugs. Intern Med 59(20):2619–2622
4. Khakoo AY et al (2005) Reversible cardiomyopathy caused by administration of interferon alpha. Nat Clin Pract Cardiovasc Med 2(1):53–57

5. Frerichs FC, Dingemans KP, Brinkman K (2002) Cardiomyopathy with mitochondrial damage associated with nucleoside reverse-transcriptase inhibitors. N Engl J Med 347(23):1895–1896
6. Rhew DC et al (2003) Association between protease inhibitor use and increased cardiovascular risk in patients infected with human immunodeficiency virus: a systematic review. Clin Infect Dis 37(7):959–972
7. Schillaci G et al (2005) Impact of treatment with protease inhibitors on aortic stiffness in adult patients with human immunodeficiency virus infection. Arterioscler Thromb Vasc Biol 25(11): 2381–2385
8. Zong Y et al (2024) Mitochondrial dysfunction: mechanisms and advances in therapy. Signal Transduct Target Ther 9(1):124
9. Tokarska-Schlattner M et al (2006) New insights into doxorubicin-induced cardiotoxicity: the critical role of cellular energetics. J Mol Cell Cardiol 41(3):389–405
10. Varga ZV et al (2015) Drug-induced mitochondrial dysfunction and cardiotoxicity. Am J Physiol Heart Circ Physiol 309(9):H1453–H1467
11. Bertero E, Maack C (2018) Metabolic remodelling in heart failure. Nat Rev Cardiol 15(8): 457–470
12. Lewis W, Dalakas MC (1995) Mitochondrial toxicity of antiviral drugs. Nat Med 1(5):417–422
13. Carr A et al (1998) A syndrome of peripheral lipodystrophy, hyperlipidaemia and insulin resistance in patients receiving HIV protease inhibitors. AIDS 12(7):F51–F58
14. Dirajlal-Fargo S, Funderburg N (2022) HIV and cardiovascular disease: the role of inflammation. Curr Opin HIV AIDS 17(5):286–292
15. Grundy SM et al (2004) Definition of metabolic syndrome: report of the National Heart, Lung, and Blood Institute/American Heart Association conference on scientific issues related to definition. Circulation 109(3):433–438
16. Alberti KG et al (2009) Harmonizing the metabolic syndrome: a joint interim statement of the International Diabetes Federation Task Force on Epidemiology and Prevention; National Heart, Lung, and Blood Institute; American Heart Association; World Heart Federation; International Atherosclerosis Society; and International Association for the Study of Obesity. Circulation 120(16):1640–1645
17. Ormazabal V et al (2018) Association between insulin resistance and the development of cardiovascular disease. Cardiovasc Diabetol 17(1):122
18. Vekic J et al (2023) Oxidative stress, atherogenic dyslipidemia, and cardiovascular risk. Biomedicines 11(11):2897
19. Friis-Moller N et al (2003) Cardiovascular disease risk factors in HIV patients—association with antiretroviral therapy.. Results from the DAD study. AIDS 17(8):1179–1193
20. Hresko RC, Hruz PW (2011) HIV protease inhibitors act as competitive inhibitors of the cytoplasmic glucose binding site of GLUTs with differing affinities for GLUT1 and GLUT4. PLoS One 6(9):e25237
21. Alvi RM et al (2018) Protease inhibitors and cardiovascular outcomes in patients with HIV and heart failure. J Am Coll Cardiol 72(5):518–530
22. Neesgaard B et al (2022) Associations between integrase strand-transfer inhibitors and cardiovascular disease in people living with HIV: a multicentre prospective study from the RESPOND cohort consortium. Lancet HIV 9(7):e474–e485
23. O'Halloran JA et al (2020) Brief report: integrase strand transfer inhibitors are associated with lower risk of incident cardiovascular disease in people living with HIV. J Acquir Immune Defic Syndr 84(4):396–399
24. Cyr AR et al (2020) Nitric oxide and endothelial dysfunction. Crit Care Clin 36(2):307–321
25. Vischer UM (2006) von Willebrand factor, endothelial dysfunction, and cardiovascular disease. J Thromb Haemost 4(6):1186–1193
26. David-Dufilho M et al (2005) Endothelial thrombomodulin induces Ca2+ signals and nitric oxide synthesis through epidermal growth factor receptor kinase and calmodulin kinase II. J Biol Chem 280(43):35999–36006

27. Gu SX et al (2021) Thrombocytopathy and endotheliopathy: crucial contributors to COVID-19 thromboinflammation. Nat Rev Cardiol 18(3):194–209
28. Hanser S et al (2022) Increased endothelial biomarkers are associated with HIV antiretroviral therapy and C-reactive protein among a African rural population in Limpopo Province, South Africa. Front Public Health 10:980754
29. Castillo R et al (2002) Efavirenz-associated QT prolongation and torsade de pointes arrhythmia. Ann Pharmacother 36(6):1006–1008
30. Ly T, Ruiz ME (2007) Prolonged QT interval and torsades de pointes associated with atazanavir therapy. Clin Infect Dis 44(6):e67–e68
31. Esmel-Vilomara R et al (2022) QTc interval prolongation in patients infected with SARS-CoV-2 and treated with antiviral drugs. An Pediatr (Engl Ed) 96(3):213–220
32. Schiavone M et al (2022) Arrhythmogenic risk and mechanisms of QT-prolonging drugs to treat COVID-19. Card Electrophysiol Clin 14(1):95–104
33. Mulligan K et al (2000) Hyperlipidemia and insulin resistance are induced by protease inhibitors independent of changes in body composition in patients with HIV infection. J Acquir Immune Defic Syndr 23(1):35–43
34. Mary-Krause M et al (2003) Increased risk of myocardial infarction with duration of protease inhibitor therapy in HIV-infected men. AIDS 17(17):2479–2486
35. Friis-Moller N et al (2003) Combination antiretroviral therapy and the risk of myocardial infarction. N Engl J Med 349(21):1993–2003
36. DAD Study Group et al (2007) Class of antiretroviral drugs and the risk of myocardial infarction. N Engl J Med 356(17):1723–1735
37. Sabin CA et al (2016) Is there continued evidence for an association between abacavir usage and myocardial infarction risk in individuals with HIV? A cohort collaboration. BMC Med 14: 61
38. Purnell JQ et al (2000) Effect of ritonavir on lipids and post-heparin lipase activities in normal subjects. AIDS 14(1):51–57
39. Dressman J et al (2003) HIV protease inhibitors promote atherosclerotic lesion formation independent of dyslipidemia by increasing CD36-dependent cholesteryl ester accumulation in macrophages. J Clin Invest 111(3):389–397
40. Sax PE et al (2020) Weight gain following initiation of antiretroviral therapy: risk factors in randomized comparative clinical trials. Clin Infect Dis 71(6):1379–1389
41. Jemal M (2024) A review of dolutegravir-associated weight gain and secondary metabolic comorbidities. SAGE Open Med 12:20503121241260613
42. Bai R et al (2022) Effects of different integrase strand transfer inhibitors on body weight in patients with HIV/AIDS: a network meta-analysis. BMC Infect Dis 22(1):118
43. Greig SL, Deeks ED (2016) Elvitegravir/cobicistat/emtricitabine/tenofovir alafenamide: a review in HIV-1 infection. Drugs 76(9):957–968
44. Casado JL et al (2019) Evaluation of kidney function in HIV-infected patients receiving an antiretroviral regimen containing one or two inhibitors of the tubular secretion of creatinine. HIV Med 20(10):648–656
45. Gubitosa JC et al (2020) Marked sinus bradycardia associated with remdesivir in COVID-19: a case and literature review. JACC Case Rep 2(14):2260–2264
46. Umeh CA et al (2023) Bradycardia and outcomes in COVID-19 patients on remdesivir: a multicenter retrospective study. Cardiol Res 14(3):192–200
47. Ai MY, Chang WL, Yang CJ (2023) Remdesivir-induced bradycardia and mortality in SARS-CoV-2 infection, potential risk factors assessment: a systematic review and meta-analysis. J Clin Med 12(24):7518
48. Urban S et al (2022) COVID-19 related myocarditis in adults: a systematic review of case reports. J Clin Med 11(19):5519

49. Mercuro NJ et al (2020) Risk of QT interval prolongation associated with use of hydroxychloroquine with or without concomitant azithromycin among hospitalized patients testing positive for coronavirus disease 2019 (COVID-19). JAMA Cardiol 5(9):1036–1041
50. Tleyjeh IM et al (2021) Cardiac toxicity of chloroquine or hydroxychloroquine in patients with COVID-19: a systematic review and meta-regression analysis. Mayo Clin Proc Innov Qual Outcomes 5(1):137–150
51. Oren O et al (2020) Use of chloroquine and hydroxychloroquine in COVID-19 and cardiovascular implications: understanding safety discrepancies to improve interpretation and design of clinical trials. Circ Arrhythm Electrophysiol 13(6):e008688.

Chapter 15
Immune Checkpoint Inhibitors and Myocarditis: Clinical Impact and Underlying Mechanisms

Ali Moradi, Naya Said, Sina Khoshniyat, Kurt Ramey, Stefan Longobardi, and Robert Subbiondo

Abstract Immune checkpoint inhibitors (ICIs) represent a novel class of cancer therapy that has significantly extended patient survival compared to traditional treatments like chemotherapy. In addition to improving mortality, ICIs also enhance quality of life by avoiding many of the common side effects associated with chemotherapy, which indiscriminately targets both healthy and cancerous cells. ICIs work by enhancing the immune system's ability to target cancer cells, primarily through three pathways: cytotoxic T-lymphocyte-associated protein 4 (CTLA-4), programmed cell death protein 1 (PD-1), and programmed death-ligand 1 (PD-L1). These agents can be administered as monotherapy or in combination, with each approach presenting a unique side effect profile linked to immune-related adverse events (irAEs) due to immune system modulation. These side effects can affect multiple organs, including the heart, where myocarditis is the most common cardiovascular manifestation. While the exact pathophysiology of ICI-induced myocarditis is not well understood, molecular mimicry may play a role. Timely diagnosis of ICI-induced myocarditis is crucial, as it has significant implications for patient mortality. The diagnosis of ICI-induced myocarditis is based on clinical presentation and cardiac evaluation, including measurements of troponin and BNP, as well as

---

A. Moradi (✉) · S. Longobardi
Internal Medicine, HCA Florida Blake Hospital, University of South Florida (USF) Morsani College of Medicine, Bradenton, FL, USA
e-mail: ali.moradi@hcahealthcare.com; Stefan.longobardi@hcahealthcare.com

N. Said
Center for Translational Medicine, Faculty of Medicine, Semmelweis University, Budapest, Hungary

S. Khoshniyat
Biomedicine, School of Sciences, La Trobe University, Melbourne, Australia

K. Ramey · R. Subbiondo
Cardiology, HCA Florida Blake Hospital, University of South Florida (USF) Morsani College of Medicine, Bradenton, FL, USA
e-mail: kurt.ramey@hcahealthcare.com; Robert.subbiondo@hcahealthcare.com

D. K. Agrawal et al. (eds.), *Cardiovascular Toxicity*, Advances in Biochemistry in Health and Disease 33, https://doi.org/10.1007/978-3-031-90598-8_15

EKG, echocardiography, cardiac MRI (CMR), and, in severe cases where the cause remains unclear, endomyocardial biopsy. Management typically involves discontinuing the ICI therapy and initiating immunosuppressive treatment. A wide range of immunosuppressive therapies can be considered, starting with corticosteroids as the first-line treatment. Other options include infliximab, tocilizumab, mycophenolate mofetil, etc. The potential side effects of ICIs can be life-threatening and must be well understood by physicians. It is essential for healthcare providers to distinguish cardiac irAEs from other common conditions, such as myocardial infarction.

Keywords Immune checkpoint inhibitors · Cardiotoxicity · Myocarditis · Immune-related adverse events · PD-1 inhibitor · PD-L1 inhibitors · CTLA-4 inhibitors · ICIs · irAEs · Nivolumab

Introduction

Immune checkpoint inhibitors (ICIs) have transformed cancer treatment by targeting key regulatory proteins in the immune system, allowing T-cells to recognize and attack tumor cells more effectively [1]. ICIs block immune checkpoints like cytotoxic T-lymphocyte antigen-4 (CTLA-4), programmed cell death protein-1 (PD-1), and programmed death-ligand 1 (PD-L1), which tumors exploit to suppress the immune system and evade destruction [2]. By inhibiting these checkpoints, ICIs re-engage the immune system's ability to detect and destroy cancer cells. These therapies have demonstrated significant improvements in survival and clinical outcomes [3]. The increased immune activation induced by ICIs can lead to a range of immune-related adverse events (irAEs), affecting multiple organ systems. These irAEs result from immune dysregulation and can involve various organs, including the heart [4]. While many irAEs are manageable with immunosuppressive therapies such as corticosteroids, ICIs can be associated with life-threatening complications, particularly cardiotoxicity. Among these, myocarditis, a rare but highly fatal side effect, has garnered significant attention due to its association with ICI therapy [5]. Myocarditis is an inflammatory condition of the myocardium, which can be caused by viral infections, autoimmune diseases, or drug reactions. It is characterized by inflammation of the heart muscle, which can lead to arrhythmias, heart failure, or even sudden cardiac death [5, 6]. In the context of ICI therapy, myocarditis represents a rare but particularly severe form of irAEs. Though the incidence of ICI-associated myocarditis is relatively low—ranging from 0.06% to 1% but can be life-threatening [7]. In this chapter, we explore the importance of novel cancer therapies and the life-threatening adverse effects of myocarditis induced by ICIs. It is crucial for physicians to be aware of this complication to ensure timely diagnosis and management.

Immune Checkpoint Inhibitors

ICIs fall under the category of targeted drug therapy and work as monoclonal antibodies to target different pathways commonly utilized by tumor cells as a method of immune evasion. Most commonly, pathways like PD-1, PDL-1, or CTLA-4 are targeted by this novel drug class [8]. Concisely, these drugs work by preventing the binding of checkpoint proteins with their respective protein partners in the interaction between tumor cells and T cells. Normally, their binding prevents T cells from killing harmful tumor cells, so when these drugs inhibit this process, T cells regain the ability to destroy tumor cells [9]. The United States Food and Drugs Administration (FDA) has approved 11 checkpoint inhibitors as of January 2024, with Ipilimumab being the first of these drugs to be approved in 2011 [10]. In contrast to traditional therapeutic modalities, which attack the cancer itself, immune targeting by ICIs tries to reprogram the host immune system to allow it to fight tumor cells better [9]. This enhances and reinstates the immune T cell's ability to inhibit tumor survival and proliferation. For such reasons, ICIs are often associated with milder side effects, including rash, fatigue, colitis, and fever. Rarely do patients experience irAEs involving various systems, including the lungs, heart, and gastro-intestinal tract [11]. Pembrolizumab is one of the most used ICIs. This drug shows strong efficacy against malignancies like melanoma and non-small cell lung cancer and greatly improves patient outcomes such as overall survival and progression-free survival. Not only that, but ICIs targeting this PD-1/PD-L1 interaction have the additional benefit of reinstating the function of tumor-associated macrophages (TAMS) by re-establishing their phagocytic abilities against tumor cells, allowing them to phagocytose cancerous cells [12, 13]. Compared to CTLA-4 ICIs, PD-1 inhibitors appear to be more selective but advantageously associated with fewer irAEs. PD-1 ICI-associated side effects tend to appear in the first few months of the treatment [11]. Other examples of PD-1 antagonists are Nivolumab for tumors like breast neoplasms and melanoma and Atezolizumab for non-small cell lung cancer and bladder cancer [13]. The first ICI to gain marketing authorization in 2011 as an immune-oncological agent was CTLA-4 targeting, known as Ipilimumab. Its main indication was its use for metastatic melanoma. While side effects from PD-1 inhibitors typically arise early in treatment, CTLA-4 antagonists tend to produce side effects only after several months. Additionally, severe side effects are observed in about 10% of patients treated with PD-1 blockers, but this rate increases to a concerning 30% with CTLA-4 antagonists, which limits their use [11]. Table 15.1 lists various types of immune checkpoint inhibitors (ICIs) and their FDA-approved indications.

In 2022, Upadhaya et al. reported approximately 5000 clinical trials investigating the use of different ICIs, either alone, in combination, or concomitantly with other treatment modalities for the treatment of different malignancies. This stems from the unparalleled success of this novel drug group, which has even managed to achieve long-term remission in particular patient subgroups [19]. A study by Hodi et al. reported an average survival of 10.1 months in patients taking either Ipilimumab

Table 15.1 Common immune checkpoint inhibitors and their FDA-approved indication. ICI: immune checkpoint inhibitor, FDA: food and drug administration, PD-1: program cell death protein-1, PD-L1: programmed death-ligand 1, CTLA-4: cytotoxic T-lymphocyte antigen-4, non-squamous cell lung cancer

ICI	Target	FDA approved indication
Pembrolizumab	PD-1	Stage IV non-squamous and squamous NSCLC [14]
Nivolumab	PD-1	Stage III-B or IV Squamous NSCLC [15]
Cemiplimab	PD-1	Metastatic cutaneous squamous cell carcinoma [16]
Atezolizumab	PD-L1	Stage III-B or IV nonsquamous and squamous NSCLC [17]
Durvalumab	PD-L1	Stage III NSCLC [18]
Ipilimumab	CTLA-4	Stage 3 or 4 malignant melanoma [16]

monotherapy or a combination of Ipilimumab and glycoprotein 100 peptide, compared to a 6.4-month overall average survival in the patient cohort who received solely the glycoprotein 100 peptide without the ICI addition. More recent research suggests that a combination of a PD-1 antagonist and a CTLA-4 antagonist appears to amplify the immune response against melanoma at different phases. Such a combination has retained FDA approval for use against BRAF-negative melanoma as first-line therapeutic agents, again highlighting their success [20].

Mechanism of Action of Immune Checkpoint Inhibitors

It is believed that the CTLA-4-based response occurs at a preliminary level of the immune checkpoint pathway, regulating T cell proliferation in the early stages of the immune response, particularly in lymph nodes. In opposition, PD-1-based response is believed to mainly act on peripheral tissue, particularly in the later stages of the immune response [21]. Research concludes that PD-1 receptors have been found in T cells, B cells, dendritic cells, natural killer cells, and activated monocytes. On the other hand, CTLA-4 receptors are found solely on the T cell surface and, precisely, are essential in the regulation of T cell proliferation in T cells that have been priorly activated, suggesting that its absence results in unregulated T cell proliferation. At the core of the immune recognition of tumor cells is the ability of T cells to detect these potentially harmful tumor cells- what is known as immune tolerance; however, over time, tumor cells have developed mechanisms to evade the immune response. Tumor cells now induce what is known as tumor tolerance and attempt to exhaust T cells to overcome the host's immune system. Taking the PD-1/PD-L1 receptor-ligand interaction as an example, some tumors overexpress the PD-L1 ligand to avoid being recognized and subsequently destructed by the host's immune system, hence "turning off" the host's immune response. This is where the role of ICIs comes into play by blocking these checkpoints and turning on the switch to reactivate the immune system. In contrast to conventional chemotherapy, which was once deemed the gold standard of cancer treatment, which works by targeting rapidly dividing cells, this type of immune therapy works by essentially rebooting the immune response to reintroduce its own tumor defense [9].

Cardiovascular Side Effects of Immune-Checkpoint Inhibitors

As advantageous as ICIs are in oncological treatment, these therapies can present with different toxicity profiles, with the rate of adverse events rising to 70–80%. Despite the infrequency of their occurrence, cardiovascular irAEs of ICI have a mortality rate of up to 60% [22]. Not only that, but a combined use of different ICIs concomitantly increases the risk and mortality of these adverse reactions, particularly with myocarditis, at almost twice the rate. The most common manifestations on the cardiovascular spectrum are myocarditis, arrhythmia, cardiomyopathies, pericarditis, pericardial effusion, and vasculitis [8]. IrAEs appear to be agent-specific, meaning that the rate of these adverse events and their prognosis are related to the type of ICI used [23].

Myocarditis Induced by Immune-Checkpoint Inhibitors

In a study involving over 20,000 cancer patients treated with ICIs, either as monotherapy or combination therapy, the incidence of myocarditis was approximately 0.9%, with the incidence of fatal cases reported at 0.3% [24]. In the other study, the prevalence of myocarditis induced by ICIs among 964 patients who were started on ICI therapy was 1.14%, with a median time of onset of 34 days from the first dose of ICI given to the patient [25]. While the occurrence of this condition remains rare, the number of patients eligible for ICIs has significantly increased [26]. For instance, in 2015, approximately 600,000 patients in the US were eligible to receive ICIs for cancer treatment [27]. Given the expanding use of ICIs in oncology, diagnosis and managing ICI-associated myocarditis has become increasingly important. The timing of myocarditis onset is also critical, as it often develops within the first few weeks or months after initiating therapy [26].

Pathophysiology of ICI-Associated Myocarditis

Pathogenesis of ICI-mediated cardiovascular adverse events remains blurred. With a specific focus on ICI-associated myocarditis, current research suggests that molecular mimicry could be the possible culprit behind this new syndrome [28]. T-cells activated against tumor antigens may cross-react with structurally similar cardiac antigens, triggering an autoimmune response against the myocardium. Evidence from PD-1 knockout mice shows that, in the absence of PD-1, mice spontaneously develop autoimmune myocarditis, underscoring the protective role of PD-1 signaling in the heart [29, 30]. It has also been reported that T-cell infiltration of the myocardium is present in ICI myocarditis [8]. ICIs can disrupt the normal balance of immune tolerance, leading to uncontrolled T-cell activation. Under normal circumstances, PD-1 interacts with its ligands to downregulate immune responses during inflammation. Without PD-1 or CTLA-4 regulation, T-cells become hyperactive and may target not only tumor cells but also healthy tissues, including the myocardium [31, 32]. Others have proposed specific human leukocyte antigens (HLA) subtype association that increases the possibility of developing irAEs, although, for precise determination, genome-wide studies would be required to solidify a link [23].

Evaluation and Diagnosis of ICI-Associated Myocarditis

According to prior studies, myocarditis is more common after receiving the first dose of immune therapy and more common in combination therapy [33]. In terms of the clinical presentation of ICI-induced myocarditis, the degree of symptoms can be categorized according to severity by the clinical practice guidelines presented by the American Society of Clinical Oncology. The severity of myocarditis can range from subclinical presentations, detected only by elevated troponin levels or abnormal ECG findings, to fulminant myocarditis with severe left ventricular dysfunction and life-threatening arrhythmias. In general, accompanying symptoms of angina, dyspnea, shortness of breath, palpitations, fatigue, and, in severe form, cardiogenic shock with concomitant arrhythmias are not uncommon [8, 26]. The heterogeneity in presentation often complicates diagnosis, as myocarditis can mimic more prevalent cardiovascular conditions. Timely recognition of ICI-associated myocarditis is crucial for effective intervention and management, as delays in diagnosis are associated with increased morbidity and mortality [34]. ICI-associated myocarditis can be categorized into three clinical groups. First, asymptomatic patients present with isolated elevations in cardiac biomarkers, such as troponin I or troponin T, indicating myocardial injury without clinical symptoms. These elevations are often detected incidentally during routine monitoring of cancer patients on ICI therapy. Despite the absence of symptoms, these patients require close surveillance due to the potential for progression to symptomatic myocarditis. Subclinical myocarditis may evolve into more severe forms if left untreated [35]. The second group includes patients with mild to moderate symptomatic myocarditis who present with nonspecific symptoms such as fatigue, dyspnea, chest pain, and palpitations. These symptoms may be misattributed to other conditions, such as pulmonary embolism, acute coronary syndrome (ACS), or anxiety, all of which are common in cancer patients undergoing multiple therapies. Myocarditis manifests in this mild to moderate form in approximately 30–40% of cases, with elevated cardiac biomarkers and electrocardiographic (ECG) abnormalities, including ST-segment changes or T-wave inversions. Further diagnostic workup is often required to differentiate myocarditis from myocardial ischemia [36]. Severe or fulminant myocarditis represents the most dangerous form of ICI-associated myocarditis, occurring in approximately 10–20% of cases. Fulminant myocarditis can progress rapidly, leading to cardiogenic shock, life-threatening arrhythmias, or sudden cardiac arrest. Patients may present with acute heart failure, severe dyspnea, hypotension, and syncope. Ventricular arrhythmias or complete heart block can develop in some cases, necessitating urgent medical intervention. Mortality rates for fulminant myocarditis are alarmingly high, with estimates ranging from 50 to 65%, even with aggressive treatments, such as mechanical circulatory support using extracorporeal membrane oxygenation or ventricular assist devices [37]. The frequency of symptoms in patients with ICI-associated myocarditis varies. Chest pain is reported in 50–60% of cases, dyspnea in 30–50%, and palpitations in 10–20%. Syncope is less common but often indicates severe arrhythmias or conduction disturbances, such as atrioventricular block or ventricular tachycardia. In some instances, myocarditis co-presents with other irAEs, such as myositis or myasthenia gravis, forming an "overlap syndrome." This syndrome, which occurs in 20–30% of cases, is characterized by a combination of myocarditis, muscle weakness, ptosis, and dysphagia, further complicating diagnosis and management [38]. The diagnosis of myocarditis typically involves elevated cardiac biomarkers (such as troponin), ECG, echocardiography, and advanced imaging techniques like cardiac magnetic resonance imaging. However, the gold standard for diagnosing ICI-induced myocarditis is an endometrial biopsy, which is invasive and not commonly performed [24]. ECG abnormalities, such as ST-segment elevation, T-wave inversions, or sinus tachycardia, are frequently observed in ICI-associated myocarditis. However, these findings are nonspecific and may not always correlate with the extent of myocardial involvement. Conduction disturbances, including AV block and more severe arrhythmias, are also common. In severe cases, patients may experience ventricular fibrillation or complete heart block, which may require the insertion of temporary or permanent pacemakers [39]. Cardiac imaging is a crucial tool in diagnosing ICI-associated myocarditis. Echocardiography is often the first-line imaging modality and may reveal left ventricular systolic dysfunction, regional wall motion abnormalities, or pericardial effusion. However, echocardiographic findings can be normal in early or mild cases, limiting its sensitivity. Cardiac magnetic resonance imaging (CMR) is more sensitive for detecting myocardial inflammation and fibrosis. CMR findings, such as late gadolinium enhancement and T2-weighted hyperintensity, are highly suggestive of active myocarditis and are often used to guide therapeutic decisions [39]. Guidelines from the American Society of Clinical Oncology and the European Society for Medical Oncology recommend baseline cardiovascular assessments for patients receiving ICI therapy, including ECG and troponin level measurement. In patients with pre-existing cardiovascular disease or risk factors, echocardiography and close monitoring during treatment are advised [40]. In 94% of patients with myocarditis induced by ICIs who presented with clinical symptoms, troponin levels were elevated. Natriuretic peptides were abnormal in 66%. Echocardiography revealed a normal ejection fraction in 51% of cases, while ECG abnormalities were observed in 89% of patients

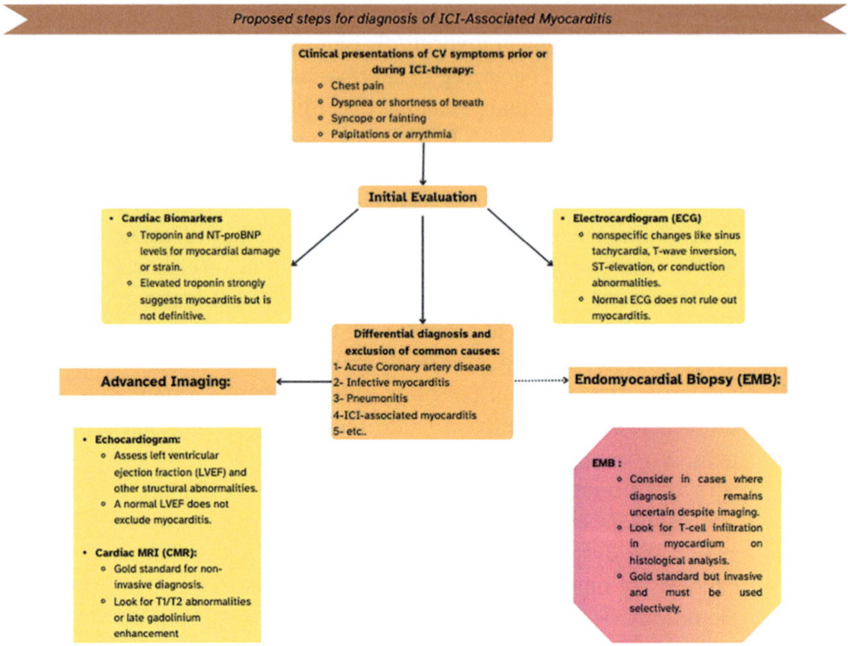

Fig. 15.1 Differential diagnosis of myocarditis induced by ICIs based on the information presented in [42]. ICI: immune checkpoint inhibitor, CMR: Cardiac magnetic resonance imaging, EMB: endometrial biopsy

[24]. Although these tests are important in helping physicians diagnose myocarditis, they are not specific or sensitive to myocarditis [25, 41]. Early detection is crucial, as ICI-associated myocarditis can rapidly progress to heart failure or arrhythmias without timely intervention. Figure 15.1 gives valuable insight into the differential diagnosis of myocarditis induced by myocarditis, and the information presented in this figures is based on the published report [42].

Management of ICI-Associated Myocarditis

Most irAEs associated with ICIs are self-limiting. According to previous studies, most patients who develop myocarditis present with no or mild symptoms and are treated with a short course of corticosteroids and/or mycophenolate. However, severe myocarditis may require extensive interdisciplinary collaboration involving both oncologists and cardiologists [43, 44]. Treatment for ICI-related myocarditis typically involves discontinuing ICI therapy upon diagnosis. The first-line treatment is corticosteroid therapy, as it suppresses the immune response altered by ICIs. Depending on the severity of the patient's condition, various forms of corticosteroids can be used—from intravenous methylprednisolone to oral prednisone. The dosage should be adjusted based on the patient's clinical condition and severity of symptoms. If biomarkers do not improve with first-line therapy, second-line options, such as infliximab, tocilizumab, immunoglobulins, mycophenolate mofetil, alemtuzumab, abatacept, and plasmapheresis can also be considered [45, 46]. In cases of severe or fulminant myocarditis, hospitalization in an intensive care unit is required upon immediate discontinuation of the ICI. High-dose intravenous corticosteroids (methylprednisolone 1 g/day for 3–5 days) are the cornerstone of treatment. Refractory cases may necessitate additional immunosuppressive therapies, such as mycophenolate mofetil, anti-thymocyte globulin, or infliximab. Patients in cardiogenic shock or with severe hemodynamic instability may require mechanical circulatory support, including ECMO or VADs, to maintain cardiac output while addressing the underlying inflammation. Long-term follow-up is crucial, as patients may develop late-onset heart failure or arrhythmias, even after acute myocarditis has resolved [47]. According to a prior study by Mahmood et al., patients who received high-dose glucocorticoid for myocarditis induced by ICIs had lower major cardiac events when compared with patients who were treated with low doses [25]. Infliximab suppresses inflammatory responses, and some case reports have documented improvement in patients with ICI-induced myocarditis. It has also been shown to resolve arrhythmias and restore cardiac function in certain cases [46, 48]. Immunoglobulins are mostly used in the case of fulminant myocarditis, which can suppress proinflammatory cytokines at the mRNA level. Immunoglobulin is usually added as a second agent to glucocorticoid therapy for more aggressive therapy [46, 49]. Mycophenolate mofetil is primarily used to prevent acute allograft rejection by inhibiting the proliferation of lymphocytes. They also have been used for treating myocarditis induced by ICIs with promising improvement. It has been used both as monotherapy and in combination therapy with glucocorticoids [5, 46]. IL-6 plays a significant role in myocarditis induced by ICIs, and tocilizumab is an IL-6 receptor inhibitor; a case report showed improvement upon using them [50]. The third-line therapy for treating ICI-induced myocarditis is plasmapheresis, reserved for patients who do not respond to other treatments. Its mechanism of action involves removing circulating immune complexes from the patient's blood [51, 52]. Although the guidelines for managing ICI-induced myocarditis are not very clear, Fig. 15.2 provides valuable insights into the treatment of patients suspected of having myocarditis induced by ICIs.

Conclusion

ICIs are novel cancer therapies that have significantly improved life expectancy in patients with various deadly cancers, such as melanoma. However, these therapies are associated with rare but potentially life-threatening side effects, including myocarditis. ICI-induced myocarditis may result from mechanisms such as T-cell cross-reactivity and molecular mimicry, leading to inappropriate T-cell activation. Early recognition and management are critical to prevent fatal outcomes. Physicians must

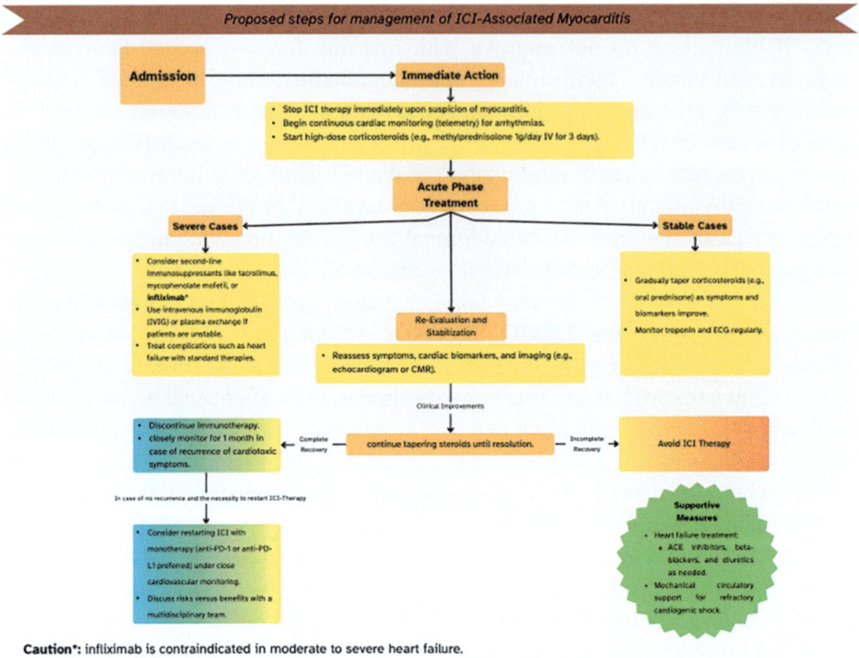

Fig. 15.2 A schematic representing a stepwise approach for the management of ICI-associated myocarditis [42]. ICI: immune checkpoint inhibitor, ECG: echocardiography, CMR: cardiac magnetic resonance imaging, EMB: endometrial biopsy, IVIG: intravenous immunoglobulin

be aware of the diagnostic approach and treatment strategies, which include promptly discontinuing the therapy, employing a multidisciplinary team (involving oncologists and cardiologists), and initiating immunosuppressive treatment such as high-dose corticosteroids and other immunosuppressive agents.

References

1. Galluzzi L, Humeau J, Buqué A, Zitvogel L, Kroemer G (2020) Immunostimulation with chemotherapy in the era of immune checkpoint inhibitors. Nat Rev Clin Oncol 17(12):725–741
2. Wei SC, Duffy CR, Allison JP (2018) Fundamental mechanisms of immune checkpoint blockade therapy. Cancer Discov 8(9):1069–1086
3. Wieder T, Eigentler T, Brenner E, Röcken M (2018) Immune checkpoint blockade therapy. J Allergy Clin Immunol 142(5):1403–1414
4. Postow MA, Sidlow R, Hellmann MD (2018) Immune-related adverse events associated with immune checkpoint blockade. N Engl J Med 378(2):158–168
5. Moradi A, Kodali A, Okoye C, Klein DH, Mohamoud I, Olanisa OO et al (2023) A systematic review of myocarditis induced by immune checkpoint inhibitors: how concerning is the most common cardiotoxicity of immune checkpoint inhibitors? Cureus 15(7):e42071

6. Caforio ALP, Malipiero G, Marcolongo R, Iliceto S (2017) Myocarditis: a clinical overview. Curr Cardiol Rep 19(7):63
7. Tajiri K, Aonuma K, Sekine I (2018) Immune checkpoint inhibitor-related myocarditis. Jpn J Clin Oncol 48(1):7–12
8. Palaskas N, Lopez-Mattei J, Durand JB, Iliescu C, Deswal A (2020) Immune checkpoint inhibitor myocarditis: pathophysiological characteristics, diagnosis, and treatment. J Am Heart Assoc 9(2):e013757
9. Shiravand Y, Khodadadi F, Kashani SMA, Hosseini-Fard SR, Hosseini S, Sadeghirad H et al (2022) Immune checkpoint inhibitors in cancer therapy. Curr Oncol 29(5):3044–3060
10. Zotova L (2023) Immune checkpoint inhibitors-related myocarditis: a review of reported clinical cases. Diagnostics (Basel) 13(7):1243
11. Alturki NA (2023) Review of the immune checkpoint inhibitors in the context of cancer treatment. J Clin Med 12(13):4301
12. Hargadon KM, Johnson CE, Williams CJ (2018) Immune checkpoint blockade therapy for cancer: an overview of FDA-approved immune checkpoint inhibitors. Int Immunopharmacol 62:29–39
13. Tan S, Li D, Zhu X (2020) Cancer immunotherapy: pros, cons and beyond. Biomed Pharmacother 124:109821
14. Rolfo C, Caglevic C, Santarpia M, Araujo A, Giovannetti E, Gallardo CD et al (2017) Immunotherapy in NSCLC: a promising and revolutionary weapon. Immunotherapy 995:97–125
15. Wu G (2021) Therapeutic effects of pembrolizumab combined with paclitaxel and cisplatin chemotherapy on advanced non-squamous non-small cell lung cancer and influencing factors. Indian J Pharm Sci 83(1):120–126
16. Migden MR, Rischin D, Schmults CD, Guminski A, Hauschild A, Lewis KD et al (2018) PD-1 blockade with cemiplimab in advanced cutaneous squamous-cell carcinoma. N Engl J Med 379(4):341–351
17. Mansh M (2011) Ipilimumab and cancer immunotherapy: a new hope for advanced stage melanoma. Yale J Biol Med 84(4):381–389
18. Paz-Ares L, Spira A, Raben D, Planchard D, Cho B, Özgüroğlu M et al (2020) Outcomes with durvalumab by tumour PD-L1 expression in unresectable, stage III non-small-cell lung cancer in the PACIFIC trial. Ann Oncol 31(6):798–806
19. Tan S, Day D, Nicholls SJ, Segelov E (2022) Immune checkpoint inhibitor therapy in oncology: current uses and future directions: JACC: CardioOncology state-of-the-art review. JACC CardioOncol 4(5):579–597
20. Singh S, Numan A, Agrawal N, Tambuwala MM, Singh V, Kesharwani P (2020) Role of immune checkpoint inhibitors in the revolutionization of advanced melanoma care. Int Immunopharmacol 83:106417
21. Buchbinder EI, Desai A (2016) CTLA-4 and PD-1 pathways: similarities, differences, and implications of their inhibition. Am J Clin Oncol 39(1):98–106
22. Zotova L (2023) Immune checkpoint inhibitors-related myocarditis: a review of reported clinical cases. Diagnostics 13(7):1243
23. Ball S, Ghosh RK, Wongsaengsak S, Bandyopadhyay D, Ghosh GC, Aronow WS et al (2019) Cardiovascular toxicities of immune checkpoint inhibitors. J Am Coll Cardiol 74(13): 1714–1727
24. Johnson DB, Balko JM, Compton ML, Chalkias S, Gorham J, Xu Y et al (2016) Fulminant myocarditis with combination immune checkpoint blockade. N Engl J Med 375(18):1749–1755
25. Mahmood SS, Fradley MG, Cohen JV, Nohria A, Reynolds KL, Heinzerling LM et al (2018) Myocarditis in patients treated with immune checkpoint inhibitors. J Am Coll Cardiol 71(16): 1755–1764
26. Neilan TG, Rothenberg ML, Amiri-Kordestani L, Sullivan RJ, Steingart RM, Gregory W et al (2018) Myocarditis associated with immune checkpoint inhibitors: an expert consensus on data gaps and a call to action. Oncologist 23(8):874–878

27. Andrews A (2015) Treating with checkpoint inhibitors-figure $1 million per patient. Am Health Drug Benefits 8(Spec Issue):9
28. Moslehi J, Salem JE (2022) Immune checkpoint inhibitor myocarditis treatment strategies and future directions. JACC CardioOncol 4(5):704–707
29. Baik AH, Tsai KK, Oh DY, Aras MA (2021) Mechanisms and clinical manifestations of cardiovascular toxicities associated with immune checkpoint inhibitors. Clin Sci (Lond) 135(5):703–724
30. Tarrio ML, Grabie N, Bu DX, Sharpe AH, Lichtman AH (2012) PD-1 protects against inflammation and myocyte damage in T cell-mediated myocarditis. J Immunol 188(10): 4876–4884
31. Asano M, Toda M, Sakaguchi N, Sakaguchi S (1996) Autoimmune disease as a consequence of developmental abnormality of a T cell subpopulation. J Exp Med 184(2):387–396
32. Latchman Y, Wood CR, Chernova T, Chaudhary D, Borde M, Chernova I et al (2001) PD-L2 is a second ligand for PD-1 and inhibits T cell activation. Nat Immunol 2(3):261–268
33. Kindermann I, Kindermann M, Kandolf R, Klingel K, Bültmann B, Müller T et al (2008) Predictors of outcome in patients with suspected myocarditis. Circulation 118(6):639–648
34. Wang Y, Sun Z, Wang X, Liu F, Wu Y, Wei Q et al (2024) Myasthenia gravis and myocarditis induced by chemoimmunotherapy in locally advanced gastric cancer: a case report. Exp Ther Med 28(5):426
35. Zhou B, Qin Q, Fang Y, Liu X, Zhang M, Wang S et al (2024) Exosomes from human bone marrow MSCs alleviate PD-1/PD-L1 inhibitor-induced myocardial injury in melanoma mice by regulating macrophage polarization and pyroptosis. Life Sci 358:123108
36. Kalinoski H, Daoud A, Rusinkevich V, Jurčová I, Talor MV, Welsh RA et al (2024) Injury-induced myosin-specific tissue-resident memory T cells drive immune checkpoint inhibitor myocarditis. Proc Natl Acad Sci 121(42):e2323052121
37. Frascaro F, Bianchi N, Sanguettoli F, Marchini F, Meossi S, Zanarelli L et al (2023) Immune checkpoint inhibitors-associated myocarditis: diagnosis, treatment and current status on rechallenge. J Clin Med 12(24):7737
38. Balanescu DV, Donisan T, Palaskas N, Lopez-Mattei J, Kim PY, Buja LM et al (2020) Immunomodulatory treatment of immune checkpoint inhibitor-induced myocarditis: pathway toward precision-based therapy. Cardiovasc Pathol 47:107211
39. Lehmann LH, Cautela J, Palaskas N, Baik AH, Meijers WC, Allenbach Y et al (2021) Clinical strategy for the diagnosis and treatment of immune checkpoint inhibitor-associated myocarditis: a narrative review. JAMA Cardiol 6(11):1329–1337
40. Brahmer JR, Abu-Sbeih H, Ascierto PA, Brufsky J, Cappelli LC, Cortazar FB et al (2021) Society for Immunotherapy of Cancer (SITC) clinical practice guideline on immune checkpoint inhibitor-related adverse events. J Immunother Cancer 9(6):e002435
41. Friedrich Matthias G, Sechtem U, Schulz-Menger J, Holmvang G, Alakija P, Cooper Leslie T et al (2009) Cardiovascular magnetic resonance in myocarditis: a JACC white paper. J Am Coll Cardiol 53(17):1475–1487
42. Ganatra S, Neilan TG (2018) Immune checkpoint inhibitor-associated myocarditis. Oncologist 23(8):879–886
43. Esfahani K, Elkrief A, Calabrese C, Lapointe R, Hudson M, Routy B et al (2020) Moving towards personalized treatments of immune-related adverse events. Nat Rev Clin Oncol 17(8): 504–515
44. Furukawa A, Tamura Y, Taniguchi H, Kawamura A, Nagase S, Hayashi A et al (2023) Prospective screening for myocarditis in cancer patients treated with immune checkpoint inhibitors. J Cardiol 81(1):63–67
45. Vicino A, Hottinger AF, Latifyan S, Boughdad S, Becce F, Prior JO et al (2024) Immune checkpoint inhibitor-related myositis and myocarditis: diagnostic pitfalls and imaging contribution in a real-world, institutional case series. J Neurol 271(4):1947–1958
46. Wu Y, Xu Y, Xu L (2023) Drug therapy for myocarditis induced by immune checkpoint inhibitors. Front Pharmacol 14:1161243

47. Coustal C, Vanoverschelde J, Quantin X, Lesage C, Michot J-M, Lappara A et al (2023) Prognosis of immune checkpoint inhibitors-induced myocarditis: a case series. J Immunother Cancer 11(5):e004792
48. Martinez-Calle N, Rodriguez-Otero P, Villar S, Mejías L, Melero I, Prosper F et al (2018) Anti-PD1 associated fulminant myocarditis after a single pembrolizumab dose: the role of occult pre-existing autoimmunity. Haematologica 103(7):e318–e321
49. Shioji K, Kishimoto C, Sasayama S (2001) Fc receptor-mediated inhibitory effect of immunoglobulin therapy on autoimmune giant cell myocarditis: concomitant suppression of the expression of dendritic cells. Circ Res 89(6):540–546
50. Doms J, Prior JO, Peters S, Obeid M (2020) Tocilizumab for refractory severe immune checkpoint inhibitor-associated myocarditis. Ann Oncol 31(9):1273–1275
51. Patel RP, Parikh R, Gunturu KS, Tariq RZ, Dani SS, Ganatra S et al (2021) Cardiotoxicity of immune checkpoint inhibitors. Curr Oncol Rep 23(7):79
52. Frigeri M, Meyer P, Banfi C, Giraud R, Hachulla A-L, Spoerl D et al (2018) Immune checkpoint inhibitor-associated myocarditis: a new challenge for cardiologists. Can J Cardiol 34(1):92.e1–92.e3

Chapter 16
Hormonal Therapy and Cardiotoxicity in Prostate Cancer Patients

Sakthivel Muniyan, Lei Xi, Arun Samidurai, Anindita Das, Rakesh C. Kukreja, and Surinder K. Batra

Abstract Prostate cancer (PCa) is a common cancer among men in the US and globally, and the incidence is gradually increasing due to an increase in the aged population. With the help of early diagnosis by prostate-specific antigen (PSA) and the advent of novel therapeutics, PCa is managed well. Among them, androgen deprivation therapy (ADT) is the cornerstone and has transformed the management of advanced PCa. However, ADT also comes with the risk of cardiotoxicity, which has become a major cause of death among PCa patients. This chapter highlights the benefits of ADT in treating PCa while examining the clinical and epidemiological evidence linking to cardiovascular risk. We have also reviewed the cardiotoxicity of other emerging PCa treatments such as abiraterone, and second-generation antiandrogens, highlighting their potential impact on the cardiovascular health of advanced PCa patients.

Keywords Prostate cancer · Androgen deprivation therapy · Cardiotoxicity · Gonadotropin-releasing hormone analogues · Abiraterone acetate

---

S. Muniyan
Department of Biochemistry and Molecular Biology, University of Nebraska Medical Center, Omaha, NE, USA

L. Xi · A. Samidurai · A. Das · R. C. Kukreja (✉)
Pauley Heart Center, Department of Internal Medicine, School of Medicine, Virginia Commonwealth University, Richmond, VA, USA
e-mail: rakesh.kukreja@vcuhealth.org

S. K. Batra (✉)
Department of Biochemistry and Molecular Biology, University of Nebraska Medical Center, Omaha, NE, USA

Fred and Pamela Buffett Cancer Center, Eppley Institute for Research in Cancer and Allied Diseases, University of Nebraska Medical Center, Omaha, NE, USA
e-mail: sbatra@unmc.edu

D. K. Agrawal et al. (eds.), *Cardiovascular Toxicity*, Advances in Biochemistry in Health and Disease 33, https://doi.org/10.1007/978-3-031-90598-8_16

Introduction

Targeted treatments and precision medicine have significantly improved survival in patients with cancer. However, therapy-induced cardiotoxicity is a growing concern, adversely impacting long-term quality of life and overall survival. This issue is particularly relevant in prostate cancer (PCa), which has high incidence and survival rates among men primarily due to androgen deprivation therapy (ADT). While ADT remains the cornerstone in managing PCa and extending survival, it is also associated with notable cardiovascular risks. ADT has been linked to hypertension, myocardial infarction, stroke, heart failure, and arrhythmia, which may contribute substantially to mortality in PCa patients. Although growing literature suggests ADT-induced cardiotoxic events, the evidence is far from clear due to multiple ADT agents and evolving treatment modalities in advanced PCa. This chapter examines not only the efficacy of current ADT agents but also their cardiotoxic effects. A comprehensive understanding of the therapy-specific cardiovascular risks is critical for early detection, effective management, and the development of novel strategies to protect the heart during cancer treatment.

Prostate Cancer, Its Current Prevalence and Clinical Management

Prostate cancer (PCa) remains the leading cause of cancer in men in the US and globally [1, 2]. In 2024, it is estimated that 299,010 new cases of PCa will be diagnosed in the US alone [1], and of those, 35,250 patients are expected to succumb to this disease [1]. Fortunately, with early screening and novel therapeutic options, the number of PCa survivors in the US was expected to increase from 3.6 million patients in 2019 to 5 million in 2030 [3]. Also worldwide, this increase in PCa incidence between 974,000 in 2005 to 1.5 million in 2022 suggests the number of PCa survivors will go up gradually [2, 4]. PCa is diagnosed by means of a serum biomarker—prostate-specific antigen (PSA) levels and/or a digital rectal exam. When the patients have confined tumors, there are 8 major treatment options with 22 FDA-approved therapeutic (including 11 GnRH analogues) agents for the clinical management of PCa (https://www.cancer.gov/about-cancer/treatment/drugs/prostate; Fig. 16.1). In most cases, the tumors progressed from indolent to aggressive and the treatment also varies according to the stage of the disease. In brief, when the tumor develops and grows in the prostate, these organ-confined cases of PCa are managed well either by surgery or radiation, or by both. Once the tumor recurs and/or develops metastatic (advanced) PCa, these tumors are managed by androgen deprivation therapy (ADT). Advanced PCa often relapses as castration-resistant (CR) tumors, leading to aggressive progression.

Most cancer therapies focus on the androgen receptor (AR) signaling targeting agents. These therapies include ADT which is achieved either surgically (bilateral

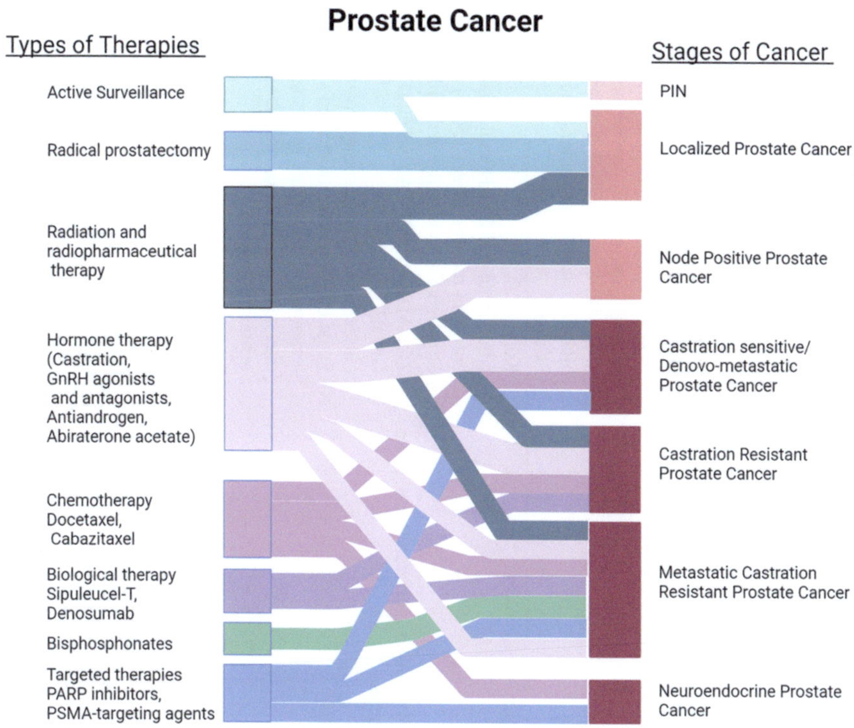

Fig. 16.1 Prostate cancer (PCa) management in clinic. This figure illustrates the therapeutic landscape for PCa across various disease stages. The left side stacked bar depicts the different therapeutic options for PCa, the thick line connects to the right stacked column indicating different stages of cancer. Thick colored arrows connect each stage of cancer to the most commonly used and clinically relevant therapeutic options

orchiectomy) or chemically with gonadotropin-releasing hormone (GnRH) analogues), antiandrogens, androgen synthesis inhibitors (*e.g.* abiraterone acetate), etc. Traditionally, ADT is given to relapsed PCa patients as monotherapy. The goal of ADT is to decrease the circulating testosterone to the level of 20–50 ng/dL (0.7–1.7 nmol/L) [5], thereby reducing the availability of ligands, which is essential for the growth of the PCa cells. Since the 1990s, ADT has been the cornerstone of locally advanced and metastatic PCa. Numerous clinical studies have consistently demonstrated the benefit of ADT in the effective management of advanced PCa (Table 16.1). Recently, ADT has also been given as a neoadjuvant, either concurrent, or adjuvant therapy with radiation for localized or locally advanced PCa Traditional chemotherapies such as docetaxel manage castration-resistant prostate cancer (CRPC), but newer therapeutic agents introduced since 2010 such as second-generation antiandrogens, PARP inhibitors, bone-seeking agents and prostate specific membrane antigen (PSMA) targeting agents have increased treatment options [27, 28]. Despite this, most PCa patients die within 3 years of CRPC diagnosis.

Table 16.1 Androgen receptor pathway-targeted therapeutic approaches for different stages of prostate cancer

Stage of prostate cancer	Drug FDA approved	Class/target	Mechanism of action	Phase III study and outcome (Reference)
Advanced PCa to achieve CAB	Bicalutamide (Casodex) 1995	Non-steroidal antiandrogen	Inhibits AR signaling	CASODEX Combination Study Group [6]. Bicalutamide + LHRH agonist (goserelin) increased median time to progression and death
Advanced PCa	Flutamide (Eulexin) 1996	Non-steroidal antiandrogen	Inhibits AR signaling	EORTC study [7]. Flutamide and LHRH agonist (goserelin) improved the time to progression and survival compared to bilateral orchiectomy among metastatic PCa
Metastatic PCa, who already undergone surgical castration	Nilutamide (Nilandron) 1996	Non-steroidal antiandrogen	Inhibits AR signaling	Kuhn et al. [8] Nilutamide and GnRH analogue buserelin reduced PSA (>50%) in 64% of patients who were treated with MAB
Advanced PCa	Goserelin acetate (Zoladex) 1996	GnRH agonists	Inhibits GnRH mediated testosterone synthesis	ZOLADEX PROSTATE STUDY GROUP [9] Goserelin and orchiectomy treatment in Stage D2 PCa patients showed similar responses in decreasing serum testosterone, acid phosphatase, and time to treatment failure
Advanced PCa	Leuprolide acetate (Lupron) 1998	GnRH agonists	Inhibits GnRH mediated testosterone synthesis	Leuprolide Study Group [10] Comparable efficacy and survival at 1 year between leuprolide and DES
Advanced PCa	Triptorelin pamoate (Trelstar) 2001	GnRH agonists	Inhibits GnRH mediated testosterone synthesis	Kuhn et al. [11] Triptorelin demonstrated higher testosterone reduction with similar PSA reduction than leuprorelin at 1 year
Metastatic PCa		GnRH agonists		

	Histrelin acetate 2004		Inhibits GnRH mediated testosterone synthesis	Schlegel et al. [12] Histrelin maintains serum testosterone levels at castrate level as long as the patients were with the therapy
Advanced PCa	Degarelix (Firmozon) 2008	GnRH antagonist	GnRH receptor-mediated inhibition of testosterone synthesis	Ferring pharma study [13] Degarelix and leuprolide showed similar efficacy in maintaining low testosterone levels
Advanced PCa	Abiraterone Acetate (Zytiga) Post chemotherapy 2011	17 a-hydroxylase/C17,20-lyase (CYP17) inhibitor	Inhibits adrenal and intra-tumoral cholesterol to androgen biosynthesis	COU-AA-301 [14] Abiraterone acetate increased OS (14.8 versus 10.9 months) compared to placebo in metastatic CRPC patients
Metastatic CRPC	Enzalutamide (Xtandi) Post chemotherapy 2012	AR antagonist	Blocks androgen binding, inhibits AR nuclear translocation, and DNA binding and transcription	AFFIRM [15] Enzalutamide prolongs OS in men with CRPC after chemotherapy (18.4 versus 13.6 months) compared with placebo
Metastatic CRPC chemotherapy-naïve	Abiraterone	17 a-hydroxylase/C17,20-lyase (CYP17) inhibitor	Inhibits adrenal and intra-tumoral cholesterol to androgen biosynthesis	COU-AA-302 Median OS in the abiraterone acetate of 34.7 months vs. 30.3 months in the placebo group ($p = 0.0033$) [16]
Metastatic CRPC, chemotherapy-naïve	Enzalutamide (Xtandi) 2014	AR antagonist	Blocks androgen binding, inhibits AR nuclear translocation, and DNA binding and transcription	Significant reduction of radiographic PFS at 12 months among the patients treated with enzalutamide, as compared with 2.8 months who received placebo [17]
Metastatic CRPC, chemotherapy-naïve	Enzalutamide (Xtandi) 2016	AR antagonist	Blocks androgen binding, inhibits AR nuclear translocation, and DNA binding and transcription	Significant improvement in radiographic PFS at 15.7 months among the patients treated with enzalutamide compared with the bicalutamide group of

(continued)

Table 16.1 (continued)

Stage of prostate cancer	Drug FDA approved	Class/target	Mechanism of action	Phase III study and outcome (Reference)
				5.8 months (HR 0·44 [95% CI 0.34–0.57]; p < 0.0001) [18]
Non-metastatic CRPC	Enzalutamide (Xtandi) 2018	AR antagonist	Blocks androgen binding, inhibits AR nuclear translocation, and DNA binding and transcription	The median MFS was 36.6 months in the enzalutamide group versus 14.7 months in the placebo group (HR for metastasis or death, 0.29 (95% CI, 0.24–0.35; P < 0.001)) [19]
Non-metastatic CRPC	Apalutamide (Erleada) 2018	AR antagonist	Blocks androgen binding, inhibits AR nuclear translocation, and DNA binding and transcription	Metastasis-free survival: 24.3 months (40.5 vs. 16.2 months) [20]
Non-metastatic CRPC	Darolutamide 2019	AR antagonist	Blocks androgen binding, inhibits AR nuclear translocation, and DNA binding and transcription	The median MFS was 40.4 vs. 18.4 months for patients treated with darolutamide with GnRH analog compared to placebo with GnRH analog (p < 0.0001) [21]
Metastatic CSPC	Apalutamide (Erleada) 2019	AR antagonist	Blocks androgen binding, inhibits AR nuclear translocation, and DNA binding and transcription	Radiographic PFS at 24 months was 68.2% in the apalutamide group vs. 47.5% in placebo group. HR = 0.48; (95% CI, 0.39–0.60; P < 0.001) [22]
Advanced PCa	Relugolix (Orgovyx) 2020	Oral GnRH antagonist	GnRH receptor-mediated inhibition of testosterone synthesis	Relugolix achieved rapid and sustained suppression of testosterone compared to leuprolide (P < 0.001) [23]
Metastatic HSPC	Darolutamide in combination with docetaxel 2022	AR antagonist	Blocks androgen binding, inhibits AR nuclear translocation, and DNA binding and transcription	Darolutamide treatment significantly lowered the risk of death among metastatic, HSPC (HR 0.68; 95% CI, 0.57–0.80; P < 0.001) [24]

Non-metastatic CSPC	Enzalutamide (Xtandi) 2023	AR antagonist	Blocks androgen binding, inhibits AR nuclear translocation, and DNA binding and transcription	Significant improvement in metastasis-free survival for enzalutamide plus leuprolide vs. placebo plus leuprolide (HR 0.42; 95% CI: 0.30, 0.61; p-value <0.0001) [25]
BRCA-mutated metastatic CRPC	Niraparib and abiraterone acetate 2024	17 a-hydroxylase/ C17,20-lyase (CYP17) inhibitor	Inhibits adrenal and intra-tumoral cholesterol to androgen biosynthesis	Significant improvement in rPFS of 25 vs. 17 months (HR = 0.66 (0.54–0.81)) for olaparib + abiraterone versus placebo + abiraterone [26]

AR—androgen receptor; BRCA—breast cancer susceptibility gene; CAB—combined androgen blockade; CRPC—castration-resistant prostate cancer; CSPC—castration sensitive prostate cancer; CYP17—cytochrome P450 17α-hydroxylase/C-17,20-lyase; DES—diethylstilbestrol; DNA—deoxyribonucleic acid; GnRH—gonodotrophin releasing hormone; HR—hazard ratio; HSPC—hormone sensitive prostate cancer; LHRH—luteinising hormone-releasing hormone; MAB—maximum androgen blockade; MFS—metastasis-free survival; PCa—prostate cancer; PFS—progression-free survival; PSA—prostate-specific antigen; OS—overall survival

Hormonal Therapy and Prostate Cancer

Androgen Biosynthesis

Approximately 90% of circulatory androgens, including testosterone are synthesized and regulated by the hypothalamus/pituitary/gonadal axis in a pulsatile mode. The elevated levels of testosterone provide negative feedback to the hypothalamus signaling it to temporarily pause the production. In testicular Leydig cells, testosterone is derived from the steroidogenic precursor, cholesterol through the classical Δ5 pathway [29]. The first step is the conversion of cholesterol to pregnenolone by the P450 family protein CYP11AI [29]. The pregnenolone is then converted into testosterone mainly through intermediates, 17-OH pregnenolone and DHEA with a smaller contribution through progesterone [30, 31]. Additionally, a limited portion (5–10%) of androgens are synthesized from adrenal glands. Similar to cholesterol to pregnenolone conversion, adrenal biosynthesis of androgens and their precursors also occurs through the Δ5 pathway in zona reticularis [32, 33]. However, the abundance of cytochrome P450 17α-hydroxylase/17,20-lyase (CYP17A1) and relatively low HSD3B2 expression in the zona reticularis favors the synthesis of androgen precursors rather than active androgens [34]. These androgen precursors are secreted into the bloodstream and subsequently converted into their active forms in peripheral tissues. Circulating testosterone can be further converted into 5-α-dihydrotestosterone (5α-DHT) by 5 α-reductase in peripheral tissues such as the prostate (Fig. 16.2).

Androgen Receptor Signaling

Androgens primarily exert their effects by more potent DHT and to a lesser extent by testosterone through classical androgen/AR signaling pathways (Fig. 16.3a) [35]. The steroid hormone receptor family member AR generally resides in the cytoplasm, bound with heat shock proteins. Upon binding with its ligands, AR homodimerizes, undergoes phosphorylation and translocates into the nucleus, where it binds with androgen response elements (ARE) and induces transcription of certain genes [36–38]. The ubiquitous expression of ARs in PCa cells and their vital role in driving prostate tumor growth forms the biological basis of targeting the AR signaling axis as a unanimous choice of physicians to control tumor growth [36, 38, 39]. The purpose of ADT is to reduce the physiological concentration of testosterone (300–1000 ng/dL) into the castrated levels (20–50 ng/dL) [40–42].

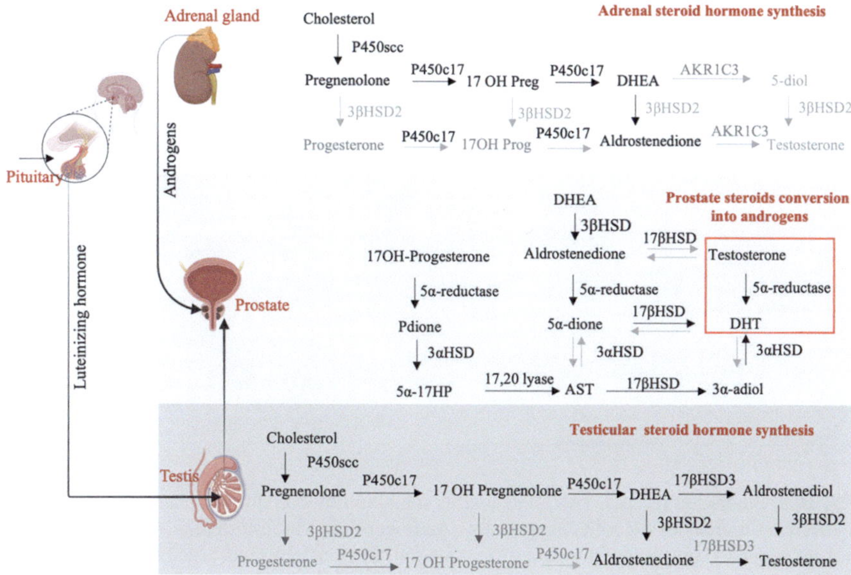

Fig. 16.2 Hypothalamus-pituitary-testis (GnRH signaling) axis and androgen biosynthesis. This figure depicts the hypothalamus-pituitary-testis axis and the key steps involved in androgen biosynthesis. The anterior pituitary releases GnRH, which stimulates the Leydig cells in the testis to synthesize testosterone which is then released in the circulation to reach the target organ like the prostate in men. In testis, cholesterol acts as a precursor and converted into a series of byproducts by cytochrome P450 and dehydrogenase before ending as testosterone. In the prostate, testosterone is mainly derived from 17-OH progesterone and DHEA mainly with the action of reductase and dehydrogenase. In addition a minor portion is also contributed from the cholesterol. In the adrenal gland which contributes to 5–10% of androgens mainly derived by cholesterol through cytochrome P450 and dehydrogenases. Cholesterol forms as the precursor molecule for all steroid hormones, including androgens. Thick black arrows: Indicate the major pathway of testosterone synthesis. Masked intermediary products and light arrows represent other components and pathways involved in androgen biosynthesis. The conversion of testosterone to potent DHT within prostate is driven by 5α-reductase. GnRH—gonadotrophin releasing hormone, DHEA—dehydroepiandrosterone, DHT—dihydrotestosterone

ADT and Prostate Cancer

The cells of prostate origin depend on androgen for their differentiation and survival, which formed the rationale for ADT over the last eight decades. ADT is achieved through various mechanisms, including bilateral orchiectomy (surgical castration), and medical castration, the utilization of hormonal therapies such as luteinizing hormone-releasing hormone (LHRH) or gonadotrophin-releasing hormone (GnRH) agonists or antagonists [43]. ADT is a cornerstone in PCa patient management, particularly in conditions where there is metastatic spread or when patients present with high-risk localized disease. ADT is given in combination with radiation

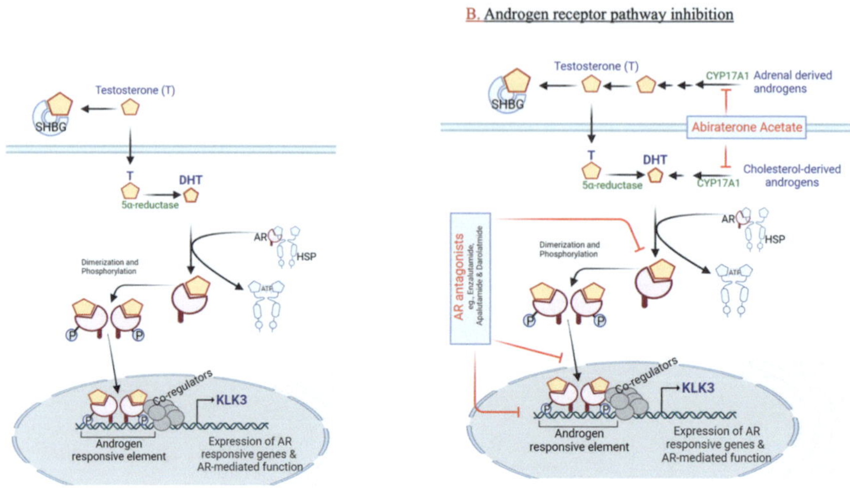

Fig. 16.3 Androgen receptor (AR) axis and its inhibition by current therapeutic agents. (**a**) The illustrative summary of AR signaling pathway. Androgens, such as testosterone, enter the target cell and are converted into potent dihydrotestosterone (DHT) by 5alpha reductase. DHT then displaces heat shock proteins (HSPs) and bind to the AR. The androgen-bound AR translocates to the nucleus and binds to specific DNA sequences called androgen response elements (AREs) on target genes. This AR binding initiates the transcription of genes involved in prostate cell growth, proliferation, and survival. (**b**) Antiandrogens (e.g., enzalutamide, apalutamide etc.,) compete with androgens for binding to the AR, blocking androgen-mediated activation. Additionally the AR signaling also blocked by AR pathway inhibitors (e.g., abiraterone acetate). Abiraterone acetate inhibit enzymes involved in androgen biosynthesis, such as CYP17A1, thereby reducing the levels of androgens available to bind to the AR. CYP17A1—cytochrome P450 17A1, SHBG—sex hormone-binding globulin, KLK3—kallikrein-related peptidase 3, a gene that encodes prostate-specific antigen (PSA) protein

or surgery (radical prostatectomy) for localized PCa, for the tumor biochemically recurred, and for nodal or metastatic PCa. Addition of ADT with EBRT improved disease-specific survival (DSS) and overall survival (OS) compared with individual treatments with either ADT or EBRT [44–46]. In addition, GnRH agonist leads to a transient surge in circulating testosterone, which is largely controlled by first-generation antiandrogens. The first-generation antiandrogens directly inhibit androgen receptor (AR) translocation and AR-mediated activity. A combined analysis of the phase III clinical trial revealed that even after controlling for prognostic factors, 7-month treatment with ADT (Goserelin and Bicalutamide) increased the median survival for 13, 44, and 75 months among the newly diagnosed metastatic PCa patients with the PSA level of >4 ng/mL, between 0.2 and 4 ng/mL or less, and 0.2 ng/mL or less, respectively [47]. However, the efficacy of ADT generally reduces after 18 months of treatment, leading to the development of castration-resistant prostate cancer (CRPC) [27].

The extra-gonadal androgen synthesis and prostate AR signaling are regulated by non-steroidal antiandrogens including bicalutamide, flutamide, and nilutamide, as well as steroidal antiandrogen cyproterone acetate [48]. However, ADT-induced AR mutations and amplification results in the conversion of antiandrogens flutamide and nilutamide into agonistic ligands rendering them ineffective over time. Additionally, the rare occurrence of hepatic toxicities has limited the clinical use of cyproterone acetate, leaving bicalutamide as the most commonly used antiandrogen for achieving complete androgen blockade (CAB) [49]. The extra hypothalamus/pituitary/testis axis is also responsible for 5–10% of circulating androgens primarily derived from the adrenal gland.

Recent advancements have also introduced novel agents like AR pathway inhibitors (ARPI), which have shown promise in improving outcomes for patients who develop resistance to standard therapies [24–26, 50]. These second-generation antiandrogens have three different mechanisms, which overcome the limitations observed with first-generation antiandrogens (Fig. 16.3b). First the second-generation antiandrogens such as enzalutamide function by binding to the AR and displacing androgen binding thereby blocking AR-mediated transcriptional activity. Additionally, these antiandrogens prevent the AR translocalization and binding with the androgen-responsive element (ARE), thereby disrupting AR variant-mediated cellular signaling. Further, these second-generation antiandrogens work at the nanomolar range with higher potency. Enzalutamide, in particular has shown a significant survival advantage in both post-chemotherapy and chemotherapy naïve metastatic CRPC [15, 17]. Further, the recent approval of apalutamide and darolutamide also adds to the list of AR-targeting agents to combat CRPC [20, 21]. Based on the superior activity of these second-generation antiandrogens, darolutamide was first tested in non-metastatic CRPC. In the ARAMIS study, darolutamide showed significant overall survival benefit in nonmetastatic CRPC (HR = 0.71 (0.50–0.99], P = .045) [21]. Recently, enzalutamide [19] and apalutamide [20] showed similar significant survival benefits of reduced hazard ratios for death (HR = 0.73 (0.0.61–0.89), P = .0011 and HR = 0.75 (0.50–0.99), P = 0.0045, respectively).

The interest in agents targeting the conversion of steroids to androgen derivatives has grown in recent years because of the success of abiraterone acetate in improving the survival of CRPCa patients [51]. Abiraterone acetate belongs to the CYP17 lyase inhibitor family which provides a survival advantage in both pre and post-chemotherapeutic conditions. Abiraterone along with GnRH analogues have shown significant survival improvement (OS) in castration-naïve (HR = 0.63 [0.52–0.76], $P < 0.001$) and CSPC (HR = 0.62 [95% CI 0.51–0.76], $P < 0.001$) [52, 53]. Similarly, enzalutamide has been shown to improve the survival of PCa patients with low-volume disease (HR = 0.43 (0.26–0.72), P = 0.0012) [54]. Apalutamide showed significant overall survival improvement over placebo in metastatic CSPC (HR = 0.67, (0.51–0.89), P = 0.005) [22]. The overall survival benefits of AR pathway inhibitors in advanced PCa patients are summarized in Table 16.1. Despite the development of a novel and refined repertoire of drugs targeting androgen deprivation at various stages, failure of ADT leads to approximately 28,000 PCa deaths; and ever increasingly decreased quality of life including cardiovascular toxicity among PCa survivors.

ADT and Cardiovascular Dysfunction

Despite the survival benefit to ADT, unfortunately, in the majority of the cases, there is also increasing evidence that ADT is associated with significant adverse effects, including sexual dysfunction, vasomotor flushing, loss of libido, fatigue, gynecomastia, anemia, osteoporosis [55], insulin insensitivity, diabetes, cardiovascular risk [55–58] and acute kidney injury [59]. Many epidemiological and population studies has shown that ADT is associated with an increased incidence of cardiovascular complications [55–60]. Epidemiological studies estimate that ~50% of patients will undergo ADT at some point in their treatment course [14, 15], and at least 20–30% of these patients will develop the risk for cardiovascular events, particularly when receiving GnRH agonists [6–11, 16]. A study using the SEER database and Swedish cancer registry estimated that cardiovascular disease (CVD) is the single most common cause of death in PCa patients who have greater than 10-year survival after diagnosis [17]. Recent results from two Phase 3 studies further show that CVD is the second-leading cause of death among PCa patients who underwent ADT [18, 19].

In recent years, one downside effect increasingly shown in many epidemiological and population studies is that 'the use of GnRH agonists and or abiraterone acetate, is associated with increased incidence of cardiovascular complications' [55–60].

GnRH Analogues and Cardiovascular Toxicity

The study led by Tayek et al. [61] is among the first one which shows the use of GnRH agonists is associated with altered lipid metabolism in a small group of men with PCa. Later studies have revealed a link between the use of ADT, particularly GnRH agonists, and heightened risk of CVD events. In 2006 using SEER database, Keating et al. [56] showed that GnRH agonist use is associated with increased risk of developing diabetes (adjusted hazard ratio [HR], 1.44; $P = 0.001$), coronary heart disease (16%, HR 1.16; $P = 0.001$), myocardial infarction (MI) (HR 1.11; $P = 0.03$), and sudden cardiac death (HR 1.16; $p = 0.004$) among men diagnosed with localized/regional PCa. The association was significant even after adjusting for patient and tumor characteristics when compared to men without ADT. In 2007, Tsai et al. [55] analyzed Cancer of the Prostate Strategic Urologic Research Endeavor (CAPSURE) database and found that GnRH agonists and/or antiandrogens use is associated with increased risk of cardiovascular mortality ($HR = 2.6$; 95% CI, 1.4–4.7; $P = 0.002$). Baseline heart disease ($HR = 1.1$; 95% $CI = 0.5–2.4$; $P = 0.8$) or diabetes ($HR = 1.3$; 95% $CI = 0.5–3.2$; $P = 0.6$) showed no significant association with cardiovascular-related mortality. Study compared patients treated with radical prostatectomy ($n = 3262$) and radiotherapy ($n = 1630$), including 1015 patients who received neoadjuvant or adjuvant ADT [55]. The findings further revealed a significant association among patients who underwent radical prostatectomy as primary therapy, but not among those treated with external beam radiation therapy (EBRT), brachytherapy, or cryotherapy. Similarly, Saigal et al. [60] analyzed SEER national cancer registry and reported that the patients who received ADT for 1 year had 20% higher cardiovascular-related morbidity when compared to patients who did not receive ADT. On multivariate analysis, after adjusting various factors, patients who received ADT had 1.2 times higher cardiovascular morbidity risk than those who did not receive ADT. The cardiovascular morbidity risk was also significantly higher during the first 12 months of the period than those who received more than 12 months (HR = 1.37; 95% CI, 1.29–1.46). This time, Keating et al., using Veterans Healthcare Administration showed that GnRH agonist use is significantly associated with increased risk of incidental coronary heart disease (CHD) (HR = 1.19, 95% CI = 1.10– 1.28), MI (HR = 1.28, 95% CI = 1.08–1.52), sudden cardiac death (HR = 1.35, 95% CI = 1.18–1.54), and stroke (HR = 1.22, 95% CI = 1.10–1.36) and diabetes (HR = 1.28, 95% CI = 1.19–1.38). Orchiectomy and CAB are also associated with an increased incidence of CHD and MI, respectively.

The National Prostate Cancer Register of Sweden, PCBaSe, which contains more than 96% of patients data was used to investigate the possible associations between PCa treatment (any form) and CV risk among 76,600 men [62]. Data from this observational study indicated that ADT increases the risk of CVD, with the highest risk associated with GnRH agonists and the lowest risk observed with antiandrogens. Moreover, the study found that even after adjusting for baseline CVD, the standardized incidence ratio was higher for ADT as was mortality ratio. Among individual treatments, GnRH agonists were associated with increased risk for MI (1.32, 95% CI = 1.23–1.42), IHD (1.37, 95% CI = 1.29–1.45), Heart failure (1.35, 95% CI = 1.26–1.45) and stroke (1.25, 95% CI = 1.16–1.33), but not arrhythmia. With the updated number of patients, Keating and colleagues [63] analyzed the SEER database and reported that ADT (both GnRH agonists and orchiectomy) is associated with an increased risk of incidental peripheral artery disease (PAD) (HR: 1.16; 95% CI, 1.12–1.21) and venous thromboembolism (VTE) (HR: 1.10; 95% CI, 1.04–1.15). Similarly, orchiectomy was significantly associated with an increased risk of PAD (HR: 1.13; 95% CI, 1.02–1.26) and VTE (HR: 1.27; 95% CI, 1.11–1.45) [63]. In contrast, a study by Jaspersen et al. from the Danish cancer registry, [64] found that PCa patients treated with GnRH agonists or antiandrogens is associated with increased risk of MI (HR = 1.31, 95% CI, 1.16–1.49) and stroke (HRs 1.19, 95% CI, 1.06–1.35) compared to those not exposed to ADT, but not with orchiectomy. Similarly, multivariate analyses of SEER data revealed that the GnRH use as part of ADT regimen was associated with 1.11-fold higher risk of CAD ($P < 0.001$), a higher risk of acute MI (HR 1.09; $P < 0.001$), SCD (HR: 1.18; $P < 0.001$) whereas bilateral orchiectomy did not show a significant association with these risks [65].

A study by Nanda et al. [66], showed that 4-month neoadjuvant ADT use was associated with an increased risk of all-cause mortality among men with a history of CAD-induced CHF and MI, but not among those without the comorbidity. Additionally, using multi-institutional data, the same authors [11] reported that 4-month of neoadjuvant ADT (GnRH) was significantly associated with an increased risk of all-cause mortality among men with low-risk PCa (HR:1.27; 95% CI, 1.07–1.51; $p < 0.01$), but not among those with intermediate- or high-risk PCa. In contrast, Alibhai et al. [67] showed that continuous ADT of 6 months or more increased the likelihood of developing diabetes and fragility fractures, but was not associated with any cardiac events. This study, which matched subjects 1:1, analyzed the incidence of acute MI, sudden cardiac death, and diabetes without differentiating the type of ADT.

Not only in the US population, but in other countries where PCa is a major cause of cancers similar observations are made with GnRH agonist treatment. In a retrospective study using 13,343 Lithuanian PCa patients, GnRH agonists were associated with a higher risk of CVD death than those not treated with ADT (HR 2.14, 95% CI [1.86–2.45], $p < 0.001$). Moreover, these patient populations showed increased risk of IHD and stroke among GnRH agonists-treated patients [68]. A retrospective study of PCa patients who received GnRH agonists between 1999 and 2021 revealed a direct association between GnRH agonist use and treatment duration. Patients with ≥2 years of GnRH agonist use had an estimated 23% increase in the sub-hazard of the endpoint (sub-hazard ratio 1.23 [1.04–1.46], $p = 0.017$) [69]. A nationwide study in Norway found that ADT was associated with a higher risk of composite CV events, heart-related issues, and all-cause mortality. This risk was higher in patients with moderate CV risk and those on ADT for longer than 7 months [70]. An Asian cohort of ~2000 patients revealed GnRH antagonists were associated with a lower risk of composite CV events (HR, 0.48; 95% CI, 0.25–0.90) and CV-related death (HR 0.21; 95% CI, 0.06–0.70) than GnRH agonists therapy [71]. In the Swedish cohort, patients with greater than 2 baseline CV events were twice as likely to experience CV-related events within 6 months of GnRH agonist treatment compared to an age-matched cohort of men without PCa [72]. The increased CV risk events persist among the men with PCa without a prior history of CVD (HR 1.19; 95% CI, 1.14–1.24) [72].

Besides observational studies, most randomized controlled trials (RCT) meta-analyses show mixed observation on the association between ADT and CV risk. In a prospective controlled randomized trial, 6-month use of GnRH agonists with radiation therapy significantly increased the risk of cardiac death, but only among the patients with preexisting comorbidities ($HR = 4.1; p\ 0.01$) [73]. Several studies have reported conflicting associations between ADT and adverse CVD outcomes. Short-term GnRH agonists treatment did not show any association with cardiovascular mortality after 10 years of followup compared with radiation therapy (RT) alone [74]. The European Organization for Research and Treatment of Cancer [EORTC] randomized controlled trial did not show any cardiovascular risk associated with ADT [75]. Similarly, the three phase III randomized controlled trial with a follow-up of at least 8 years did not find any significant association between GnRH agonist use and increased risk of CV mortality [45, 76–80]. From a meta-analysis of 8 randomized trials comprising 4141 patients, Nguyen et al. [81], did not find an increased risk of cardiovascular death among GnRH agonists therapy. Similarly, 6-month neo-adjuvant use of GnRH agonist did not show any association with cardiac events when compared with RT alone [82]. Whereas, pooled analysis of six phase

3 prospective, randomized clinical trials that included 2328 patients showed the patients who received GnRH antagonists had significantly lower cardiac events (HR = 0.44, 95% CI, 0.26–0.74; $p = 0.002$) when compared to patients who received GnRH agonists [83]. Posthoc analysis further revealed GnRH agonists contribute to as much as 50% increase in cardiac events in patients with preexisting CVD [83]. A meta-analysis showed that GnRH agonists were associated with 38% increased risk of CVD [84], including nonfatal or fatal MI (57%) and stroke (51%). Unlike the prior studies based on either US-based or Europe-based databases, Bosco et al. [84] represents most of the international population database. In a recent retrospective real-world observational study, GnRH agonist use is associated with increased CV risk events within 3 years of ADT initiation among patients who had a baseline history of CVD [85].

Overall, most studies suggest that there is a clinical association between ADT and CV risk events. Whereas, the Prostate, Lung, Colorectal, and Ovarian (PLCO) Randomized Controlled Trial prospective data collection from 1993 to 2001 did not find an association between ADT + RT and cardiovascular mortality compared to the RT group [86]. Similarly, the combined registry of the National Cancer Center Singapore and the National Heart Center Singapore from 2000 to 2019 has 1446 patients who undergone radiation therapy with more than 7 years of follow-up did not reveal any association between ADT and major cardiovascular events after curative radiotherapy [87].

Cardiotoxicity of GnRH Agonists Versus Antagonists

With the widespread use of GnRH antagonists, studies have sought to compare the cardiovascular risks associated with GnRH agonists and antagonists. A study by Iacovelli et al. [88] revealed that second-generation antiandrogen Abiraterone acetate significantly increases the risk of cardiac events when used in conjunction with ADT. This finding is further supported by a phase II study involving men with PCa and preexisting CVD who received GnRH agonists or antagonists for 1 year. The primary goal was to measure endothelial function, while the secondary outcome focused on new cardiovascular events. Although no difference in endothelial function (primary outcome) was observed between the two groups, patients treated with GnRH agonists experienced a higher incidence of major CV events compared to those on GnRH antagonists [89]. This study suggests that the choice between GnRH agonists and antagonists may impact cardiac outcomes in PCa patients with existing cardiovascular issues. Further, with the higher number of patient data and longer follow-up, recent meta-analysis and population-based studies revealed similar observations of increased CV risk events among the PCa patients who received GnRH agonists. Pooled data from six prospective clinical trials showed that GnRH antagonists had a significantly lower risk for any CV event or CV death among patients with preexisting CV disease within 1 year of initiating ADT [83]. Meta-analysis of clinical trial data revealed that GnRH antagonists show reduced CV risk events (Risk ratio (RR) 0.57 (0.39–0.81) and CV deaths RR 0.49 (0.25–0.96)) compared with GnRH agonists [90]. Similarly, other meta-analyses of clinical trial participants revealed reduced cardiovascular events ((RR: 0.52, 95% CI: 0.34–0.80) among the GnRH antagonists treated group [91]. In the Asian population, GnRH antagonist-treated patients show reduced CV risk events compared to GnRH agonist-treated patients [92]. The CV risk is further reduced with GnRH antagonists in patients with pre-existing CVD (aHR 0.16; 95% CI, 0.05–0.50; $p = 0.0017$) compared to GnRHagonists-treated patients [92]. A recent study using Quebec administrative databases of 10,785 patients revealed most patients were treated with GnRH agonists as ADT. This study found that patients on the GnRH antagonist had a decreased risk of heart failure (HF) compared to those on GnRH agonists, particularly among patients with prior cardiovascular disease (HR 0.46; 95% CI, 0.26–0.79). Moreover, the risk of IHD in patients without prior CVD is decreased in patients receiving GnRH antagonist (HR 0.26; 95% CI, 0.11–0.65) [93]. The PRONOUNCE trial, originally designed to determine the CVD as one of the major outcomes was stopped due to reduced related patient enrollment. Whereas, a recent meta-analysis of real-world evidence studies indicates that GnRH agonists are associated with an increase in major adverse cardiovascular events (MACEs) compared to GnRH antagonists [94]. Based on all this information it is understood that ADT may be associated with some form of CV risk events. However, the detailed understanding of cardiac events associated with individual ADT is incomplete, which impedes the design of preventive or curative interventions for CV risks/toxicity.

Second-Generation Antiandrogens and Cardiovascular Toxicities

Recent advancements in PCa treatment have introduced novel agents, including AR pathway inhibitors, which show promise in improving outcomes for patients resistant to standard therapies [15–19, 21, 23–26, 52, 95]. For example, agents such as abiraterone and enzalutamide have proven effective in further suppressing androgen signaling pathways. These treatments provide hope for patients with CRPC and enhance the efficacy of conventional ADT in castration-sensitive conditions. However, these agents are associated with adverse effects, and the CV risks vary across different ADT modalities.

ADT for PCa is associated with a higher risk of acute MI and CVD but not SCD. A study by Iacovelli et al. [88], revealed that the use of second-generation antiandrogen Abiraterone acetate significantly increases the risk for hypertension. A meta-analysis examined the risk of CV events and hypertension in nonmetastatic CRPC patients treated with enzalutamide, apalutamide, and darolutamide. The study found a significantly increased risk of CV events and grade 3–4 hypertension with these hormonal agents. The findings suggest the need for caution and underscore the importance of conducting further large-cohort studies to validate these results [96].

Patients receiving abiraterone had longer progression-free survival (PFS) and overall survival (OS) compared to those who did not, albeit with a modest increase in adverse effects, particularly hypertension [97]. Another study compared the risks of major adverse cardiovascular events (MACE) between enzalutamide and abiraterone among PCa patients in Hong Kong. A total of 1015 patients were analyzed. Enzalutamide users showed significantly lower risks of MACE and adverse CV events compared to abiraterone users [98]. The risk of acute MI and CVD was higher for those on new hormonal agents like abiraterone and enzalutamide compared to traditional ADT (AMI: RR = 1.31; 95% CI, 0.66–2.63, P = 0.44, for >5 year group; CVD: RR = 1.12, 95% CI, 0.97–1.30, P = 0.12, for >5 year group) [99]. Another retrospective study of the Asian population revealed that enzalutamide users had a lower risk of MI (wHR 0.57 [95% CI: 0.33–0.97], $P = 0.040$) and all-cause mortality (wHR 0.71 [95% CI: 0.59–0.85], $p < 0.001$) compared to abiraterone acetate users. A follow-up study indicated that the higher risk is associated with higher prednisone intake [100].

A recent study compared the incidence of hospitalization among metastatic patients who received abiraterone acetate and enzalutamide. Administrative claims data indicate that hospitalization for heart failure increased among the patients who received Abiraterone acetate (HR = 2.56), 95% CI (1.32–4.94). Corresponding HRs for AMI and ischemic stroke were 1.94 (95% CI: 0.90, 4.18) and 1.25 (95% CI: 0.54, 2.85), respectively [101]. In metastatic HSPC, ADT + docetaxel + Abiraterone acetate with prednisone (P) showed a higher risk for hypertension (RR = 2.85, 95% CI, 1.67–4.89) and arrhythmias (RR 2.01, 95% CI 1.17–3.44) compared to ADT + DTX [102]. This association suggests that abiraterone users had higher risks of CV risk events like myocardial infarction.

Summary and Perspective

From the above in depth review, it is clear that both GnRH agonists and antagonists are effective in reducing testosterone levels and managing PCa. However, their impacts on cardiovascular health are markedly different. Clinical evidence suggests that GnRH agonists are associated with a higher risk of cardiotoxicity compared to GnRH antagonists, particularly in patients with preexisting cardiovascular conditions. This disparity may be attributed to several factors, including the distinct mechanisms of action of these drugs.

GnRH receptors, expressed in extragonadal tissues such as the heart, regulate intracellular calcium ion concentrations and cardiac contractility via GnRH receptor/PKA signaling [90]. GnRH agonists have been shown to increase resting intracellular Ca^{2+} levels while reducing electrostimulation-induced Ca^{2+} changes, potentially disrupting cardiac function Additionally, the initial testosterone surge caused by GnRH agonists can destabilize atherosclerotic plaques, increasing the risk of cardiovascular events. Elevated follicle-stimulating hormone (FSH) levels induced by GnRH agonists may further promote vascular inflammation and plaque formation.

In contrast, GnRH antagonists do not trigger an initial testosterone surge, contributing to their lower cardiovascular risk profile. However, both GnRH agonists and antagonists are indirectly linked to cardiotoxicity through metabolic complications such as weight gain, insulin resistance, dyslipidemia, and fat accumulation [3, 59, 83], which collectively exacerbate cardiovascular risk.

Compounding these risks, newer AR pathway inhibitors, particularly abiraterone acetate, have been associated with a higher incidence of cardiovascular events in patients with advanced PCa [103]. This underscores the urgent need for comprehensive investigations into the cardiotoxic mechanisms of AR pathway inhibitors to optimize treatment strategies and minimize adverse effects. Future research should focus on elucidating these mechanisms and identifying novel drug targets to improve the safety and efficacy of AR pathway inhibitors. Such advancements could transform PCa therapy, enhancing patient outcomes while reducing cardiovascular risks, particularly in those with advanced disease.

Acknowledgments The authors were partially supported by grants from the Department of Defense-Idea Award (W81XWH-18-1-0308 to SKB and RCK). Biorender.com was used to create part of the figures.

References

1. Siegel RL, Giaquinto AN, Jemal A (2024) Cancer statistics, 2024. CA Cancer J Clin 74:12–49
2. Bray F, Laversanne M, Sung H, Ferlay J, Siegel RL, Soerjomataram I, Jemal A (2024) Global cancer statistics 2022: GLOBOCAN estimates of incidence and mortality worldwide for 36 cancers in 185 countries. CA Cancer J Clin 74:229–263
3. Miller KD, Nogueira L, Mariotto AB, Rowland JH, Yabroff KR, Alfano CM, Jemal A, Kramer JL, Siegel RL (2019) Cancer treatment and survivorship statistics, 2019. CA Cancer J Clin 69: 363–385
4. Global Burden of Disease Cancer Collaboration, Fitzmaurice C, Allen C, Barber RM, Barregard L, Bhutta ZA, Brenner H, Dicker DJ, Chimed-Orchir O, Dandona R, Dandona L, Fleming T, Forouzanfar MH, Hancock J, Hay RJ, Hunter-Merrill R, Huynh C, Hosgood HD, Johnson CO, Jonas JB, Khubchandani J, Kumar GA, Kutz M, Lan Q, Larson HJ, Liang X, Lim SS, Lopez AD, MacIntyre MF, Marczak L, Marquez N, Mokdad AH, Pinho C, Pourmalek F, Salomon JA, Sanabria JR, Sandar L, Sartorius B, Schwartz SM, Shackelford KA, Shibuya K, Stanaway J, Steiner C, Sun J, Takahashi K, Vollset SE, Vos T, Wagner JA, Wang H, Westerman R, Zeeb H, Zoeckler L, Abd-Allah F, Ahmed MB, Alabed S, Alam NK, Aldhahri SF, Alem G, Alemayohu MA, Ali R, Al-Raddadi R, Amare A, Amoako Y, Artaman A, Asayesh H, Atnafu N, Awasthi A, Saleem HB, Barac A, Bedi N, Bensenor I, Berhane A, Bernabe E, Betsu B, Binagwaho A, Boneya D, Campos-Nonato I, Castaneda-Orjuela C, Catala-Lopez F, Chiang P, Chibueze C, Chitheer A, Choi JY, Cowie B, Damtew S, das Neves J, Dey S, Dharmaratne S, Dhillon P, Ding E, Driscoll T, Ekwueme D, Endries AY, Farvid M, Farzadfar F, Fernandes J, Fischer F, TT GH, Gebru A, Gopalani S, Hailu A, Horino M, Horita N, Husseini A, Huybrechts I, Inoue M, Islami F, Jakovljevic M, James S, Javanbakht M, Jee SH, Kasaeian A, Kedir MS, Khader YS, Khang YH, Kim D, Leigh J, Linn S, Lunevicius R, El Razek HMA, Malekzadeh R, Malta DC, Marcenes W, Markos D,

Melaku YA, Meles KG, Mendoza W, Mengiste DT, Meretoja TJ, Miller TR, Mohammad KA, Mohammadi A, Mohammed S, Moradi-Lakeh M, Nagel G, Nand D, Le Nguyen Q, Nolte S, Ogbo FA, Oladimeji KE, Oren E, Pa M, Park EK, Pereira DM, Plass D, Qorbani M, Radfar A, Rafay A, Rahman M, Rana SM, Soreide K, Satpathy M, Sawhney M, Sepanlou SG, Shaikh MA, She J, Shiue I, Shore HR, Shrime MG, So S, Soneji S, Stathopoulou V, Stroumpoulis K, Sufiyan MB, Sykes BL, Tabares-Seisdedos R, Tadese F, Tedla BA, Tessema GA, Thakur JS, Tran BX, Ukwaja KN, Uzochukwu BSC, Vlassov VV, Weiderpass E, Terefe MW, Yebyo HG, Yimam HH, Yonemoto N, Younis MZ, Yu C, Zaidi Z, Zaki MES, Zenebe ZM, Murray CJL, Naghavi M (2017) Global, regional, and national cancer incidence, mortality, years of life lost, years lived with disability, and disability-adjusted life-years for 32 cancer groups, 1990 to 2015: a systematic analysis for the global burden of disease study. JAMA Oncol 3: 524–548

5. Nishiyama T (2014) Serum testosterone levels after medical or surgical androgen deprivation: a comprehensive review of the literature. Urol Oncol 32(38):e17–e28
6. Schellhammer PF, Sharifi R, Block NL, Soloway MS, Venner PM, Patterson AL, Sarosdy MF, Vogelzang NJ, Schellenger JJ, Kolvenbag GJ (1997) Clinical benefits of bicalutamide compared with flutamide in combined androgen blockade for patients with advanced prostatic carcinoma: final report of a double-blind, randomized, multicenter trial. Casodex Combination Study Group. Urology 50:330–336
7. Denis LJ, Carnelro de Moura JL, Bono A, Sylvester R, Whelan P, Newling D, Depauw M (1993) Goserelin acetate and flutamide versus bilateral orchiectomy: a phase III EORTC trial (30853). EORTC GU Group and EORTC Data Center. Urology 42:119–129; discussion 129–130.
8. Kuhn JM, Billebaud T, Navratil H, Moulonguet A, Fiet J, Grise P, Louis JF, Costa P, Husson JM, Dahan R et al (1989) Prevention of the transient adverse effects of a gonadotropin-releasing hormone analogue (buserelin) in metastatic prostatic carcinoma by administration of an antiandrogen (nilutamide). N Engl J Med 321:413–418
9. Soloway MS, Chodak G, Vogelzang NJ, Block NL, Schellhammer PF, Smith JA Jr, Scott M, Kennealey G, Gau TC (1991) Zoladex versus orchiectomy in treatment of advanced prostate cancer: a randomized trial. Zoladex Prostate Study Group. Urology 37:46–51
10. Leuprolide Study Group (1984) Leuprolide versus diethylstilbestrol for metastatic prostate cancer. N Engl J Med 311:1281–1286
11. Kuhn JM, Abourachid H, Brucher P, Doutres JC, Fretin J, Jaupitre A, Jorest R, Lambert D, Petit J, Pin J, Blumberg J, Dufour-Esquerre F (1997) A randomized comparison of the clinical and hormonal effects of two GnRH agonists in patients with prostate cancer. Eur Urol 32:397–403
12. Schlegel PN, Kuzma P, Frick J, Farkas A, Gomahr A, Spitz I, Chertin B, Mack D, Jungwirth A, King P, Nash H, Bardin CW, Moo-Young A (2001) Effective long-term androgen suppression in men with prostate cancer using a hydrogel implant with the GnRH agonist histrelin. Urology 58:578–582
13. Klotz L, Boccon-Gibod L, Shore ND, Andreou C, Persson BE, Cantor P, Jensen JK, Olesen TK, Schroder FH (2008) The efficacy and safety of degarelix: a 12-month, comparative, randomized, open-label, parallel-group phase III study in patients with prostate cancer. BJU Int 102:1531–1538
14. de Bono JS, Logothetis CJ, Molina A, Fizazi K, North S, Chu L, Chi KN, Jones RJ, Goodman OB Jr, Saad F, Staffurth JN, Mainwaring P, Harland S, Flaig TW, Hutson TE, Cheng T, Patterson H, Hainsworth JD, Ryan CJ, Sternberg CN, Ellard SL, Flechon A, Saleh M, Scholz M, Efstathiou E, Zivi A, Bianchini D, Loriot Y, Chieffo N, Kheoh T, Haqq CM, Scher HI, C.-A.-. Investigators (2011) Abiraterone and increased survival in metastatic prostate cancer. N Engl J Med 364:1995–2005
15. Scher HI, Fizazi K, Saad F, Taplin ME, Sternberg CN, Miller K, de Wit R, Mulders P, Chi KN, Shore ND, Armstrong AJ, Flaig TW, Flechon A, Mainwaring P, Fleming M, Hainsworth JD,

Hirmand M, Selby B, Seely L, de Bono JS, AFFIRM Investigators (2012) Increased survival with enzalutamide in prostate cancer after chemotherapy. N Engl J Med 367:1187–1197

16. Ryan CJ, Smith MR, Fizazi K, Saad F, Mulders PF, Sternberg CN, Miller K, Logothetis CJ, Shore ND, Small EJ, Carles J, Flaig TW, Taplin ME, Higano CS, de Souza P, de Bono JS, Griffin TW, De Porre P, Yu MK, Park YC, Li J, Kheoh T, Naini V, Molina A, Rathkopf DE, C.-A.-. Investigators (2015) Abiraterone acetate plus prednisone versus placebo plus prednisone in chemotherapy-naive men with metastatic castration-resistant prostate cancer (COU-AA-302): final overall survival analysis of a randomised, double-blind, placebo-controlled phase 3 study. Lancet Oncol 16:152–160
17. Beer TM, Armstrong AJ, Rathkopf DE, Loriot Y, Sternberg CN, Higano CS, Iversen P, Bhattacharya S, Carles J, Chowdhury S, Davis ID, de Bono JS, Evans CP, Fizazi K, Joshua AM, Kim CS, Kimura G, Mainwaring P, Mansbach H, Miller K, Noonberg SB, Perabo F, Phung D, Saad F, Scher HI, Taplin ME, Venner PM, Tombal B, PREVAIL Investigators (2014) Enzalutamide in metastatic prostate cancer before chemotherapy. N Engl J Med 371: 424–433
18. Shore ND, Chowdhury S, Villers A, Klotz L, Siemens DR, Phung D, van Os S, Hasabou N, Wang F, Bhattacharya S, Heidenreich A (2016) Efficacy and safety of enzalutamide versus bicalutamide for patients with metastatic prostate cancer (TERRAIN): a randomised, double-blind, phase 2 study. Lancet Oncol 17:153–163
19. Hussain M, Fizazi K, Saad F, Rathenborg P, Shore N, Ferreira U, Ivashchenko P, Demirhan E, Modelska K, Phung D, Krivoshik A, Sternberg CN (2018) Enzalutamide in men with nonmetastatic, castration-resistant prostate cancer. N Engl J Med 378:2465–2474
20. Smith MR, Saad F, Chowdhury S, Oudard S, Hadaschik BA, Graff JN, Olmos D, Mainwaring PN, Lee JY, Uemura H, Lopez-Gitlitz A, Trudel GC, Espina BM, Shu Y, Park YC, Rackoff WR, Yu MK, Small EJ, S. Investigators (2018) Apalutamide treatment and metastasis-free survival in prostate cancer. N Engl J Med 378:1408–1418
21. Fizazi K, Shore N, Tammela TL, Ulys A, Vjaters E, Polyakov S, Jievaltas M, Luz M, Alekseev B, Kuss I, Kappeler C, Snapir A, Sarapohja T, Smith MR, A. Investigators (2019) Darolutamide in nonmetastatic, castration-resistant prostate cancer. N Engl J Med 380:1235–1246
22. Chi KN, Agarwal N, Bjartell A, Chung BH, de Santana Gomes AJP, Given R, Soto AJ, Merseburger AS, Ozguroglu M, Uemura H, Ye D, Deprince K, Naini V, Li J, Cheng S, Yu MK, Zhang K, Larsen JS, McCarthy S, Chowdhury S, T. Investigators (2019) Apalutamide for metastatic, castration-sensitive prostate cancer. N Engl J Med 381:13–24
23. Shore ND, Saad F, Cookson MS, George DJ, Saltzstein DR, Tutrone R, Akaza H, Bossi A, van Veenhuyzen DF, Selby B, Fan X, Kang V, Walling J, Tombal B, H.S. Investigators (2020) Oral relugolix for androgen-deprivation therapy in advanced prostate cancer. N Engl J Med 382:2187–2196
24. Smith MR, Hussain M, Saad F, Fizazi K, Sternberg CN, Crawford ED, Kopyltsov E, Park CH, Alekseev B, Montesa-Pino A, Ye D, Parnis F, Cruz F, Tammela TLJ, Suzuki H, Utriainen T, Fu C, Uemura M, Mendez-Vidal MJ, Maughan BL, Joensuu H, Thiele S, Li R, Kuss I, Tombal B, A.T. Investigators (2022) Darolutamide and survival in metastatic, hormone-sensitive prostate cancer. N Engl J Med 386:1132–1142
25. Freedland SJ, de Almeida Luz M, De Giorgi U, Gleave M, Gotto GT, Pieczonka CM, Haas GP, Kim CS, Ramirez-Backhaus M, Rannikko A, Tarazi J, Sridharan S, Sugg J, Tang Y, Tutrone RF Jr, Venugopal B, Villers A, Woo HH, Zohren F, Shore ND (2023) Improved outcomes with enzalutamide in biochemically recurrent prostate cancer. N Engl J Med 389: 1453–1465
26. Fallah J, Xu J, Weinstock C, Brave MH, Bloomquist E, Fiero MH, Schaefer T, Pathak A, Abukhdeir A, Bhatnagar V, Chiu HJ, Ricks T, John C, Hamed S, Lee C, Pierce WF, Kalavar S, Philip R, Tang S, Amiri-Kordestani L, Pazdur R, Kluetz PG, Suzman D (2024) FDA approval summary: olaparib in combination with Abiraterone for treatment of patients with BRCA-mutated metastatic castration-resistant prostate cancer. J Clin Oncol 42:605–613

27. Muniyan S, Li B, Batra SK (2022) Editorial: metastatic castration resistant prostate cancer: prognosis and treatment. Front Oncol 12:913630
28. Muniyan S, Siddiqui JA, Batra SK (2021) Therapeutic options for prostate cancer: a contemporary update. In: Surinder MPP, Batra K (eds) Gene regulation and therapeutics for cancer. CRC Press, New York, p 328
29. Miller WL (1988) Molecular biology of steroid hormone synthesis. Endocr Rev 9:295–318
30. Fluck CE, Miller WL, Auchus RJ (2003) The 17, 20-lyase activity of cytochrome p450c17 from human fetal testis favors the delta5 steroidogenic pathway. J Clin Endocrinol Metab 88: 3762–3766
31. Sherbet DP, Tiosano D, Kwist KM, Hochberg Z, Auchus RJ (2003) CYP17 mutation E305G causes isolated 17,20-lyase deficiency by selectively altering substrate binding. J Biol Chem 278:48563–48569
32. Miller WL, Auchus RJ (2011) The molecular biology, biochemistry, and physiology of human steroidogenesis and its disorders. Endocr Rev 32:81–151
33. Turcu AF, Auchus RJ (2015) Adrenal steroidogenesis and congenital adrenal hyperplasia. Endocrinol Metab Clin N Am 44:275–296
34. Schiffer L, Arlt W, Storbeck KH (2018) Intracrine androgen biosynthesis, metabolism and action revisited. Mol Cell Endocrinol 465:4–26
35. Askew EB, Gampe RT Jr, Stanley TB, Faggart JL, Wilson EM (2007) Modulation of androgen receptor activation function 2 by testosterone and dihydrotestosterone. J Biol Chem 282: 25801–25816
36. Taplin ME (2007) Drug insight: role of the androgen receptor in the development and progression of prostate cancer. Nat Clin Pract Oncol 4:236–244
37. Mostaghel EA, Page ST, Lin DW, Fazli L, Coleman IM, True LD, Knudsen B, Hess DL, Nelson CC, Matsumoto AM, Bremner WJ, Gleave ME, Nelson PS (2007) Intraprostatic androgens and androgen-regulated gene expression persist after testosterone suppression: therapeutic implications for castration-resistant prostate cancer. Cancer Res 67:5033–5041
38. Feldman BJ, Feldman D (2001) The development of androgen-independent prostate cancer. Nat Rev Cancer 1:34–45
39. Harris WP, Mostaghel EA, Nelson PS, Montgomery B (2009) Androgen deprivation therapy: progress in understanding mechanisms of resistance and optimizing androgen depletion. Nat Clin Pract Urol 6:76–85
40. Conteduca V, Di Lorenzo G, Tartarone A, Aieta M (2013) The cardiovascular risk of gonadotropin releasing hormone agonists in men with prostate cancer: an unresolved controversy. Crit Rev Oncol Hematol 86:42–51
41. Bubley GJ, Carducci M, Dahut W, Dawson N, Daliani D, Eisenberger M, Figg WD, Freidlin B, Halabi S, Hudes G, Hussain M, Kaplan R, Myers C, Oh W, Petrylak DP, Reed E, Roth B, Sartor O, Scher H, Simons J, Sinibaldi V, Small EJ, Smith MR, Trump DL, Wilding G et al (1999) Eligibility and response guidelines for phase II clinical trials in androgen-independent prostate cancer: recommendations from the Prostate-Specific Antigen Working Group. J Clin Oncol 17:3461–3467
42. Oefelein MG, Feng A, Scolieri MJ, Ricchiutti D, Resnick MI (2000) Reassessment of the definition of castrate levels of testosterone: implications for clinical decision making. Urology 56:1021–1024
43. Muniyan S, Xi L, Datta K, Das A, Teply BA, Batra SK, Kukreja RC (2020) Cardiovascular risks and toxicity—the Achilles heel of androgen deprivation therapy in prostate cancer patients. Biochim Biophys Acta Rev Cancer 1874:188383
44. Bolla M, Van Tienhoven G, Warde P, Dubois JB, Mirimanoff RO, Storme G, Bernier J, Kuten A, Sternberg C, Billiet I, Torecilla JL, Pfeffer R, Cutajar CL, Van der Kwast T, Collette L (2010) External irradiation with or without long-term androgen suppression for prostate cancer with high metastatic risk: 10-year results of an EORTC randomised study. Lancet Oncol 11:1066–1073

45. Pilepich MV, Winter K, Lawton CA, Krisch RE, Wolkov HB, Movsas B, Hug EB, Asbell SO, Grignon D (2005) Androgen suppression adjuvant to definitive radiotherapy in prostate carcinoma—long-term results of phase III RTOG 85-31. Int J Radiat Oncol Biol Phys 61: 1285–1290
46. Warde P, Mason M, Ding K, Kirkbride P, Brundage M, Cowan R, Gospodarowicz M, Sanders K, Kostashuk E, Swanson G, Barber J, Hiltz A, Parmar MK, Sathya J, Anderson J, Hayter C, Hetherington J, Sydes MR, Parulekar W, N.C.P.M.U.P. Investigators (2011) Combined androgen deprivation therapy and radiation therapy for locally advanced prostate cancer: a randomised, phase 3 trial. Lancet 378:2104–2111
47. Hussain M, Tangen CM, Higano C, Schelhammer PF, Faulkner J, Crawford ED, Wilding G, Akdas A, Small EJ, Donnelly B, MacVicar G, Raghavan D, Southwest Oncology Group Trials (2006) Absolute prostate-specific antigen value after androgen deprivation is a strong independent predictor of survival in new metastatic prostate cancer: data from Southwest Oncology Group Trial 9346 (INT-0162). J Clin Oncol 24:3984–3990
48. Lam JS, Leppert JT, Vemulapalli SN, Shvarts O, Belldegrun AS (2006) Secondary hormonal therapy for advanced prostate cancer. J Urol 175:27–34
49. Ricci F, Buzzatti G, Rubagotti A, Boccardo F (2014) Safety of antiandrogen therapy for treating prostate cancer. Expert Opin Drug Saf 13:1483–1499
50. Orme JJ, Pagliaro LC, Quevedo JF, Park SS, Costello BA (2022) Rational second-generation antiandrogen use in prostate cancer. Oncologist 27:110–124
51. Crawford ED, Shore ND, Petrylak DP, Higano CS, Ryan CJ (2017) Abiraterone acetate and prednisone in chemotherapy-naive prostate cancer patients: rationale, evidence and clinical utility. Ther Adv Med Oncol 9:319–333
52. Fizazi K, Tran N, Fein L, Matsubara N, Rodriguez-Antolin A, Alekseev BY, Ozguroglu M, Ye D, Feyerabend S, Protheroe A, De Porre P, Kheoh T, Park YC, Todd MB, Chi KN, L. Investigators (2017) Abiraterone plus prednisone in metastatic, castration-sensitive prostate cancer. N Engl J Med 377:352–360
53. James ND, de Bono JS, Spears MR, Clarke NW, Mason MD, Dearnaley DP, Ritchie AWS, Amos CL, Gilson C, Jones RJ, Matheson D, Millman R, Attard G, Chowdhury S, Cross WR, Gillessen S, Parker CC, Russell JM, Berthold DR, Brawley C, Adab F, Aung S, Birtle AJ, Bowen J, Brock S, Chakraborti P, Ferguson C, Gale J, Gray E, Hingorani M, Hoskin PJ, Lester JF, Malik ZI, McKinna F, McPhail N, Money-Kyrle J, O'Sullivan J, Parikh O, Protheroe A, Robinson A, Srihari NN, Thomas C, Wagstaff J, Wylie J, Zarkar A, Parmar MKB, Sydes MR, Investigators S (2017) Abiraterone for prostate cancer not previously treated with hormone therapy. N Engl J Med 377:338–351
54. Davis ID, Martin AJ, Stockler MR, Begbie S, Chi KN, Chowdhury S, Coskinas X, Frydenberg M, Hague WE, Horvath LG, Joshua AM, Lawrence NJ, Marx G, McCaffrey J, McDermott R, McJannett M, North SA, Parnis F, Parulekar W, Pook DW, Reaume MN, Sandhu SK, Tan A, Tan TH, Thomson A, Tu E, Vera-Badillo F, Williams SG, Yip S, Zhang AY, Zielinski RR, Sweeney CJ, ENZAMET Trial Investigators and the Australian and New Zealand Urogenital and Prostate Cancer Trials Group (2019) Enzalutamide with standard first-line therapy in metastatic prostate cancer. N Engl J Med 381:121–131
55. Tsai HK, D'Amico AV, Sadetsky N, Chen MH, Carroll PR (2007) Androgen deprivation therapy for localized prostate cancer and the risk of cardiovascular mortality. J Natl Cancer Inst 99:1516–1524
56. Keating NL, O'Malley AJ, Smith MR (2006) Diabetes and cardiovascular disease during androgen deprivation therapy for prostate cancer. J Clin Oncol 24:4448–4456
57. Keating NL, O'Malley AJ, Freedland SJ, Smith MR (2010) Diabetes and cardiovascular disease during androgen deprivation therapy: observational study of veterans with prostate cancer. J Natl Cancer Inst 102:39–46
58. Saylor PJ, Smith MR (2013) Metabolic complications of androgen deprivation therapy for prostate cancer. J Urol 189:S34–S42; discussion S43-34.

59. Gandaglia G, Sun M, Hu JC, Novara G, Choueiri TK, Nguyen PL, Schiffmann J, Graefen M, Shariat SF, Abdollah F, Briganti A, Montorsi F, Trinh QD, Karakiewicz PI (2014) Gonadotropin-releasing hormone agonists and acute kidney injury in patients with prostate cancer. Eur Urol 66:1125–1132
60. Saigal CS, Gore JL, Krupski TL, Hanley J, Schonlau M, Litwin MS, Urologic Diseases in America Project (2007) Androgen deprivation therapy increases cardiovascular morbidity in men with prostate cancer. Cancer 110:1493–1500
61. Tayek JA, Heber D, Byerley LO, Steiner B, Rajfer J, Swerdloff RS (1990) Nutritional and metabolic effects of gonadotropin-releasing hormone agonist treatment for prostate cancer. Metabolism 39:1314–1319
62. Van Hemelrijck M, Garmo H, Holmberg L, Ingelsson E, Bratt O, Bill-Axelson A, Lambe M, Stattin P, Adolfsson J (2010) Absolute and relative risk of cardiovascular disease in men with prostate cancer: results from the population-based PCBaSe Sweden. J Clin Oncol 28:3448–3456
63. Hu JC, Williams SB, O'Malley AJ, Smith MR, Nguyen PL, Keating NL (2012) Androgen-deprivation therapy for nonmetastatic prostate cancer is associated with an increased risk of peripheral arterial disease and venous thromboembolism. Eur Urol 61:1119–1128
64. Jespersen CG, Norgaard M, Borre M (2014) Androgen-deprivation therapy in treatment of prostate cancer and risk of myocardial infarction and stroke: a nationwide Danish population-based cohort study. Eur Urol 65:704–709
65. Gandaglia G, Sun M, Popa I, Schiffmann J, Abdollah F, Trinh QD, Saad F, Graefen M, Briganti A, Montorsi F, Karakiewicz PI (2014) The impact of androgen-deprivation therapy (ADT) on the risk of cardiovascular (CV) events in patients with non-metastatic prostate cancer: a population-based study. BJU Int 114:E82–E89
66. Nanda A, Chen MH, Braccioforte MH, Moran BJ, D'Amico AV (2009) Hormonal therapy use for prostate cancer and mortality in men with coronary artery disease-induced congestive heart failure or myocardial infarction. JAMA 302:866–873
67. Alibhai SM, Duong-Hua M, Sutradhar R, Fleshner NE, Warde P, Cheung AM, Paszat LF (2009) Impact of androgen deprivation therapy on cardiovascular disease and diabetes. J Clin Oncol 27:3452–3458
68. Jonusas J, Drevinskaite M, Patasius A, Kincius M, Janulionis E, Smailyte G (2022) Androgen-deprivation therapy and risk of death from cardio-vascular disease in prostate cancer patients: a nationwide Lithuanian population-based cohort study. Aging Male 25:173–179
69. Chan JSK, Tang P, Hui JMH, Lee YHA, Dee EC, Ng K, Liu K, Tse G, Ng CF (2022) Association between duration of gonadotrophin-releasing hormone agonist use and cardio-vascular risks: a population-based competing-risk analysis. Prostate 82:1477–1480
70. Forster RB, Engeland A, Kvale R, Hjellvik V, Bjorge T (2022) Association between medical androgen deprivation therapy and long-term cardiovascular disease and all-cause mortality in nonmetastatic prostate cancer. Int J Cancer 151:1109–1119
71. Chen DY, Su PJ, See LC, Liu JR, Chuang CK, Pang ST, Tseng CN, Chen SW, Hsieh IC, Chu PH, Lin YC, Hsu CL, Chang JW, Lin MS, Pang JS, Hsieh MJ, Huang WK (2021) Gonadotropin-releasing hormone antagonist associated with lower cardiovascular risk compared with gonadotropin-releasing hormone agonist in prostate cancer: a nationwide cohort and in vitro study. Prostate 81:902–912
72. O'Farrell S, Garmo H, Holmberg L, Adolfsson J, Stattin P, Van Hemelrijck M (2015) Risk and timing of cardiovascular disease after androgen-deprivation therapy in men with prostate cancer. J Clin Oncol 33:1243–1251
73. D'Amico AV, Chen MH, Renshaw AA, Loffredo M, Kantoff PW (2008) Androgen suppression and radiation vs radiation alone for prostate cancer: a randomized trial. JAMA 299:289–295
74. Voog JC, Paulus R, Shipley WU, Smith MR, McGowan DG, Jones CU, Bahary JP, Zeitzer KL, Souhami L, Leibenhaut MH, Rotman M, Husain SM, Gore E, Raben A, Chafe S, Sandler

HM, Efstathiou JA (2016) Cardiovascular mortality following short-term androgen deprivation in clinically localized prostate cancer: an analysis of RTOG 94-08. Eur Urol 69:204–210
75. Studer UE, Whelan P, Albrecht W, Casselman J, de Reijke T, Hauri D, Loidl W, Isorna S, Sundaram SK, Debois M, Collette L (2006) Immediate or deferred androgen deprivation for patients with prostate cancer not suitable for local treatment with curative intent: European Organisation for Research and Treatment of Cancer (EORTC) Trial 30891. J Clin Oncol 24: 1868–1876
76. Efstathiou JA, Bae K, Shipley WU, Hanks GE, Pilepich MV, Sandler HM, Smith MR (2008) Cardiovascular mortality and duration of androgen deprivation for locally advanced prostate cancer: analysis of RTOG 92-02. Eur Urol 54:816–823
77. Efstathiou JA, Bae K, Shipley WU, Hanks GE, Pilepich MV, Sandler HM, Smith MR (2009) Cardiovascular mortality after androgen deprivation therapy for locally advanced prostate cancer: RTOG 85-31. J Clin Oncol 27:92–99
78. Roach M III, Bae K, Speight J, Wolkov HB, Rubin P, Lee RJ, Lawton C, Valicenti R, Grignon D, Pilepich MV (2008) Short-term neoadjuvant androgen deprivation therapy and external-beam radiotherapy for locally advanced prostate cancer: long-term results of RTOG 8610. J Clin Oncol 26:585–591
79. Pilepich MV, Winter K, John MJ, Mesic JB, Sause W, Rubin P, Lawton C, Machtay M, Grignon D (2001) Phase III radiation therapy oncology group (RTOG) trial 86-10 of androgen deprivation adjuvant to definitive radiotherapy in locally advanced carcinoma of the prostate. Int J Radiat Oncol Biol Phys 50:1243–1252
80. Horwitz EM, Bae K, Hanks GE, Porter A, Grignon DJ, Brereton HD, Venkatesan V, Lawton CA, Rosenthal SA, Sandler HM, Shipley WU (2008) Ten-year follow-up of radiation therapy oncology group protocol 92-02: a phase III trial of the duration of elective androgen deprivation in locally advanced prostate cancer. J Clin Oncol 26:2497–2504
81. Nguyen PL, Je Y, Schutz FA, Hoffman KE, Hu JC, Parekh A, Beckman JA, Choueiri TK (2011) Association of androgen deprivation therapy with cardiovascular death in patients with prostate cancer: a meta-analysis of randomized trials. JAMA 306:2359–2366
82. Wilcox C, Kautto A, Steigler A, Denham JW (2012) Androgen deprivation therapy for prostate cancer does not increase cardiovascular mortality in the long term. Oncology 82: 56–58
83. Albertsen PC, Klotz L, Tombal B, Grady J, Olesen TK, Nilsson J (2014) Cardiovascular morbidity associated with gonadotropin releasing hormone agonists and an antagonist. Eur Urol 65:565–573
84. Bosco C, Bosnyak Z, Malmberg A, Adolfsson J, Keating NL, Van Hemelrijck M (2015) Quantifying observational evidence for risk of fatal and nonfatal cardiovascular disease following androgen deprivation therapy for prostate cancer: a meta-analysis. Eur Urol 68: 386–396
85. Lowentritt B, Fallick M, Pruett J, Jiang T, Li E, Brown B, Dufour R (2024) Incidence of cardiovascular events in patients with prostate cancer and treated with androgen deprivation therapy. Urol Pract 11:154–161
86. Butler SS, Mahal BA, Moslehi JJ, Nohria A, Dee EC, Makkar R, Whitbeck A, Wangoe J, Mouw KW, Nguyen PL, Muralidhar V (2021) Risk of cardiovascular mortality with androgen deprivation therapy in prostate cancer: a secondary analysis of the prostate, lung, colorectal, and ovarian (PLCO) randomized controlled trial. Cancer 127:2213–2221
87. Li Y, Ong WS, Shwe MTT, Yit NLF, Quek SZH, Pang EPP, Looi WS, Nei WL, Wang MLC, Chua MLK, Tan TWK, Chua ET, Ng CT, Tuan JKL (2022) Cardiovascular toxicities of androgen deprivation therapy in Asian men with localized prostate cancer after curative radiotherapy: a registry-based observational study. Cardiooncology 8:4
88. Iacovelli R, Ciccarese C, Bria E, Romano M, Fantinel E, Bimbatti D, Muraglia A, Porcaro AB, Siracusano S, Brunelli M, Mazzarotto R, Artibani W, Tortora G (2018) The cardiovascular toxicity of abiraterone and enzalutamide in prostate cancer. Clin Genitourin Cancer 16:e645–e653

89. Margel D, Peer A, Ber Y, Shavit-Grievink L, Tabachnik T, Sela S, Witberg G, Baniel J, Kedar D, Duivenvoorden WCM, Rosenbaum E, Pinthus JH (2019) Cardiovascular morbidity in a randomized trial comparing GnRH agonist and GnRH antagonist among patients with advanced prostate cancer and preexisting cardiovascular disease. J Urol 202:1199–1208
90. Nelson AJ, Lopes RD, Hong H, Hua K, Slovin S, Tan S, Nilsson J, Bhatt DL, Goodman SG, Evans CP, Clarke NW, Shore ND, Margel D, Klotz LH, Tombal B, Leong DP, Alexander JH, Higano CS (2023) Cardiovascular effects of GnRH antagonists compared with agonists in prostate cancer: a systematic review. JACC CardioOncol 5:613–624
91. Abufaraj M, Iwata T, Kimura S, Haddad A, Al-Ani H, Abusubaih L, Moschini M, Briganti A, Karakiewicz PI, Shariat SF (2021) Differential impact of gonadotropin-releasing hormone antagonist versus agonist on clinical safety and oncologic outcomes on patients with metastatic prostate cancer: a meta-analysis of randomized controlled trials. Eur Urol 79:44–53
92. Shao YJ, Hong JH, Chen CK, Huang CY (2023) Cardiovascular risk of gonadotropin-releasing hormone antagonist versus agonist in men with prostate cancer: an observational study in Taiwan. Prostate Cancer Prostatic Dis 26:722–729
93. Dragomir A, Touma N, Hu J, Perreault S, Aprikian AG (2023) Androgen deprivation therapy and risk of cardiovascular disease in patients with prostate cancer based on existence of cardiovascular risk. J Natl Compr Cancer Netw 21:163–171
94. Patel S, Zhu K, Dave CV, Ghajar M, Zhang Y, Saraiya B, Bandera EV, Khosrow-Khavar F (2024) Comparative cardiovascular safety of gonadotropin-releasing hormone antagonists and agonists among patients diagnosed with prostate cancer: a systematic review and meta-analysis of real-world evidence studies. Eur Urol Oncol 8:510–519
95. de Bono JS, Scher HI, Montgomery RB, Parker C, Miller MC, Tissing H, Doyle GV, Terstappen LW, Pienta KJ, Raghavan D (2008) Circulating tumor cells predict survival benefit from treatment in metastatic castration-resistant prostate cancer. Clin Cancer Res 14:6302–6309
96. Rizzo A, Merler S, Sorgentoni G, Oderda M, Mollica V, Gadaleta-Caldarola G, Santoni M, Massari F (2021) Risk of cardiovascular toxicities and hypertension in nonmetastatic castration-resistant prostate cancer patients treated with novel hormonal agents: a systematic review and meta-analysis. Expert Opin Drug Metab Toxicol 17:1237–1243
97. Fizazi K, Foulon S, Carles J, Roubaud G, McDermott R, Flechon A, Tombal B, Supiot S, Berthold D, Ronchin P, Kacso G, Gravis G, Calabro F, Berdah JF, Hasbini A, Silva M, Thiery-Vuillemin A, Latorzeff I, Mourey L, Laguerre B, Abadie-Lacourtoisie S, Martin E, El Kouri C, Escande A, Rosello A, Magne N, Schlurmann F, Priou F, Chand-Fouche ME, Freixa SV, Jamaluddin M, Rieger I, Bossi A, P.-. investigators (2022) Abiraterone plus prednisone added to androgen deprivation therapy and docetaxel in de novo metastatic castration-sensitive prostate cancer (PEACE-1): a multicentre, open-label, randomised, phase 3 study with a 2 x 2 factorial design. Lancet 399:1695–1707
98. Lee YHA, Hui JMH, Leung CH, Tsang CTW, Hui K, Tang P, Chan JSK, Dee EC, Ng K, McBride S, Nguyen PL, Tse G, Ng CF (2024) Major adverse cardiovascular events of enzalutamide versus abiraterone in prostate cancer: a retrospective cohort study. Prostate Cancer Prostatic Dis 27:776–782
99. Liang Z, Zhu J, Chen L, Xu Y, Yang Y, Hu R, Zhang W, Song Y, Lu Y, Ou N, Liu X (2020) Is androgen deprivation therapy for prostate cancer associated with cardiovascular disease? A meta-analysis and systematic review. Andrology 8:559–574
100. Chan JSK, Lee YHA, Leung CH, Leung DKW, Dee EC, Ng K, Tse G, Ng CF (2024) Associations between glucocorticoid use and major adverse cardiovascular events in patients with prostate cancer receiving antiandrogen: a retrospective cohort study. Prostate Cancer Prostatic Dis
101. Conover MM, Weaver J, Fan B, Leitz G, Richarz U, Li Q, Gifkins D (2023) Cardiovascular outcomes among patients with castration-resistant prostate cancer: a comparative safety study using US administrative claims data. Prostate 83:729–739

102. Aziz MK, Molony D, Monlezun D, Holder T, Brunckhorst O, Higgason N, Roland J, Magill R, Fatakdawala M, Iacobucci A, Mody-Bailey N, Owen C, Zarker A, Thames E, Swaby J, Xiao D, Choi L, Desai S, Galan J, Deng B, Hartshorne T, Nichols A, Zhang A, Imber J, Song J, Jones W, Rivas A, Sanchez D, Guhan M, Gandaglia G, Ranganath S, Jacob J, Howell S, Plana J, van den Bergh R, Roberts M, Sommer SG, Oldenburg J, Ploussard G, Tilki D, Schoots I, Briers E, Stranne J, Rouviere O, van Oort I, Oprea-Lager D, De Santis M, Cornford P, P. European Association of Urology Prostate Cancer Guidelines, Koutroumpakis E, Ziaolhagh A, Ali A, Yusuf SW, Iliescu C, Canfield S (2025) Prostate cancer therapy cardiotoxicity map (PROXMAP) for advanced disease states: a systematic review and network meta-analysis with Bayesian modeling of treatment histories. Eur Urol 87:15–26
103. Chan JSK, Satti DI, Lee YHA, Hui JMH, Dee EC, Ng K, Liu K, Tse G, Ng CF (2023) Temporal trends in cardiovascular burden among patients with prostate cancer receiving androgen deprivation therapy: a population-based cohort study. Br J Cancer 128:2253–2260

Chapter 17
Doxorubicin Induced Mitochondrial Dysfunction and Cardiotoxicity

Sukhwinder K. Bhullar, Raneeta Thingnam, Inna Rabinovich-Nikitin, and Lorrie A. Kirshenbaum

Abstract Although anthracyclines have markedly improved treatment outcomes for cancer patients, their use is associated with a wide variety of cardiovascular complications. These include, arrhythmias, ventricular enlargement, hypertension, dilated cardiomyopathy, and congestive heart failure. Such adverse conditions may be further exacerbated by inflammatory processes, specifically those initiated by mitochondrial dysfunction, which leads to cardiac cell death and ventricular dysfunction. Understanding the mechanisms that underlie the cardiotoxicity of anthracyclines is of paramount importance in reducing the cardiovascular morbidity associated with cancer therapy. In this article we provide an up to date overview of the role played by mitochondrial dysfunction in the pathogenesis of cancer therapies-induced cardiotoxicity. Given the significant clinical implications and wide usage of Doxorubicin therapy, this article primarily focuses on the influence of this chemotherapeutic agent and its cytotoxic effects on mitochondrial dysfunction and cardiac injury. The updated insights compiled in this article may support developing preventive approaches against cardiotoxicity by strategically targeting mitochondrial signaling pathways and inhibiting the inflammatory responses as potential therapeutic interventions. Ultimately, such initiatives may support and enhance patient outcomes while preserving the therapeutic efficacy of cancer treatments.

Keywords Doxorubicin · Mitochondrial dysfunction · Inflammation · Cardiotoxicity

---

S. K. Bhullar · R. Thingnam · I. Rabinovich-Nikitin · L. A. Kirshenbaum (✉)
Department of Physiology and Pathophysiology, The Institute of Cardiovascular Sciences, St. Boniface Hospital Albrechtsen Research Centre, Winnipeg, MB, Canada

Department of Pharmacology and Therapeutics, Rady College of Medicine, Max Rady Faculty of Health Sciences, University of Manitoba, Winnipeg, MB, Canada
e-mail: irabinovich-nikitin@sbrc.ca; Lkirshenbaum@sbrc.ca

D. K. Agrawal et al. (eds.), *Cardiovascular Toxicity*, Advances in Biochemistry in Health and Disease 33, https://doi.org/10.1007/978-3-031-90598-8_17

Introduction

While cardiovascular disease and cancer etiologies are fundamentally distinct, a bidirectional relationship exists in which the risk associated with one condition can markedly influence the other, which is now well recognized [1]. The global incidence of these diseases is increasing, significantly affecting overall morbidity and mortality statistics, and projected healthcare costs associated with these conditions will exceed $1 trillion for cardiovascular diseases and $8.3 trillion for cancer by the year 2030 [2, 3]. Notably, cancer therapies, including traditional cytotoxic chemotherapies, targeted molecular therapies, and immunotherapies, have significantly enhanced treatment effectiveness for patients diagnosed with cancer [4, 5]. Indeed, survival rates for cancer patients continue to improve in the long term; however, life-threatening side effects of these therapies, such as cardiotoxicity, have emerged as a significant contributor to morbidity and mortality among survivors [6–10]. It is worth noting that most (98%) of cardiotoxicity possibilities, including arrhythmias, ventricular enlargement, congestive heart failure, and dilated cardiomyopathy, occur in approximately 9% of cancer patients, thus limiting the drug's clinical use [11–21]. It is essential to note that cardiotoxicity observed in cancer patients arises from the effectiveness of cancer drugs on healthy cells, which results from their inherent lack of specificity. Certainly, several factors, dosage, administration route, duration of anthracycline therapy, and individual patient risk factors, including sex, age, ethnicity, diabetes, dyslipidaemia, obesity, hypertension, cardiovascular disease, and co-medications are considered to be a key in influence onset of cardiotoxicity in cancer patients [16, 22–29]. Since cardiomyocytes possess limited regenerative capacity, they are particularly susceptible to the long-term adverse effects of these therapies, primarily resulting in damage to cardiac muscle. Although an early indicator is a decrease in left ventricular ejection fraction from baseline, acute heart failure or subclinical left ventricular dysfunction develops into heart failure over several years following cancer treatment progressively [22, 30–34]. Notably, although discontinuation of the cancer therapy has been shown to reverse some of the cardiotoxic effects, the ensuing cardiac injury maybe become irreversible if the intrinsic cell death mechanisms within cardiomyocytes are impaired [35–38].

Though complex and not fully elucidated, the underlying mechanisms in the pathogenesis of anthracycline-induced cardiotoxicity, the significant role of mitochondria has been evidenced. Anthracyclines inhibit mitochondrial fusion, promote fission, and disrupt function, leading to oxidative stress, metabolic dysfunction, and inflammation, collectively contributing to the progression of heart failure [39–43]. In view of these findings, targeting mitochondrial dynamics and functionality, may signify a capable approach to mitigate the cardiotoxic effects associated with anthracycline treatment. Given that the underlying mechanisms are not yet fully understood [44], this review article aims to consolidate knowledge, specifically regarding the critical roles of Doxorubicin (Dox)-induced cardiotoxicity responses attributed to mitochondrial dysfunction and cardiac cell death.

Doxorubicin-Induced Cardiotoxicity

Dox is one of the most widely utilized agents within the anthracycline class for treating a variety of human malignancies [45]. Multiple mechanisms have been attributed to its anticancer activity [46, 47], however, it primarily, exerts its effects on neoplastic cells by intercalating with DNA, resulting in double-nucleotide stranded breaks, and inhibition of topoisomerase II enzymes, which ostensibly inhibits DNA replication. In fact, morphological defects induced by Dox are consistent with apoptotic cell death [48–52]. Increased apoptosis associated with Dox in cardiac myocytes has been associated with acute and/or chronic defects that manifest as supraventricular arrhythmias, changes in QT-interval and left ventricular dysfunction. While these symptoms often improve over time, they may progress to early cardiotoxicity (within the first year) or late cardiotoxicity (over decades), both of which can lead to congestive heart failure [21, 30, 53–55]. Studies have shown that Dox-induced cardiac complications, primarily result from elevated oxidative stress and increased generation of reactive oxygen species (ROS) which damage mitochondria and other organelles. Moreover, Dox can amplify oxidative stress by inhibiting iron regulatory protein-1 (IRP-1) resulting in elevated intracellular iron [38, 56]. Since the heart has inherently limited antioxidative capacity, it is particularly susceptible to this oxidative damage from increased ROS. Moreover, Dox was shown to inhibit antioxidative enzymes glutathione and catalase, as well as levels of sulfhydryl groups further exacerbating cellular damage [41, 57]. In addition, Dox-cardiotoxicity is also associated with altered calcium ions (Ca2+) homeostasis, which can reportedly induce apoptosis through caspase activation in cardiomyocytes and endothelial cells [8, 41, 58]. Additionally, Dox has been shown to target key signaling molecules involved in apoptosis, including the cellular proteins p53, GATA-4 and others reviewed in [59–61]. Dox also inhibits the cell cycle regulator p300 by enhancing its degradation through the ubiquitin-proteasome system [62, 63]. Also, the cardiotoxic mechanisms of Dox overlap with its anticancer effects, as the drug causes more significant DNA damage via topoisomerase IIα compared to topoisomerase IIβ, which predominates in cancer cells and immature cardiomyocytes [64, 65]. Thus, while Dox treatment has been proven effective for treating cancers, it also causes significant cardiotoxic side effects.

Dox-Induced Mitochondrial Dysfunction and Cardiotoxicity

The mitochondrion is a vital cellular organelle that comprises 36–40% of the total volume of cardiomyocytes, and major source of adenosine triphosphate (ATP) source through oxidative metabolism. Mitochondrial damage resulting from modifications in cellular metabolism, epigenetic signaling, and systemic effects, have been reported in cancer patients and mice treated with Dox [66–70]. Notably, since Dox is highly lipophilic and readily permeates cell membrane, it can trigger

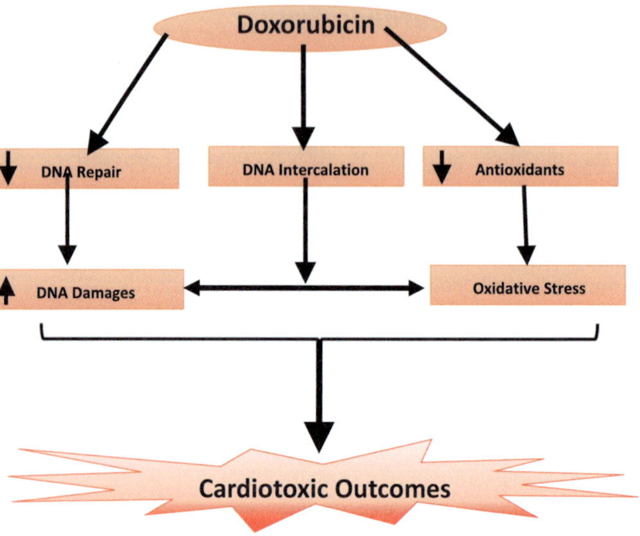

Fig. 17.1 A schematic representation illustrating the mitochondrial dysfunction induced by doxorubicin cancer treatment which contributes to cardiotoxic outcomes

mitochondrial damage within a few hours of cardiomyocytes treated with Dox [50] (Fig. 17.1).

Cardiomyocytes depend heavily on mitochondrial oxidative phosphorylation as the primary source of ATP to maintain ion homeostasis, metabolism, and contractile function. However, mitochondria are especially vulnerable to oxidative damage, which adversely impacts mitochondrial ion transport, mtDNA integrity, and overall mitochondrial function [71, 72]. Following the administration, Dox, rapidly binds to the phospholipid cardiolipin which is highly abundant in mitochondria, which disrupts the outer membrane interface resulting in the generation of superoxide radicals. Moreover, other mitochondrial electron complex proteins are adversely affected by Dox treatment, which impairs mitochondrial bioenergetics [73–76]. Furthermore, Dox decreases the oxidation of long-chain fatty acids in cardiac mitochondria while elevating glucose metabolism (Fig. 17.1). This metabolic shift indicates a transition towards anaerobic metabolism in concurrence with aerobic metabolism, which may promote heart failure though metabolic mis-matching in mitochondrial energy production and expenditure [73, 77–79]. Furthermore, Dox directly interferes with mitochondrial function by promoting the opening of the mitochondrial permeability transition pore (mPTP), which culminates in cellular apoptosis [80]. In Dox-treated cardiac myocytes, mitochondrial respiration (state four) was increased while active (state three) respiration was decreased, indicating mitochondrial bioenergetic failure which is crucial for ATP production [81]. Moreover, the ability of Dox to bind to cardiolipin and trigger cytochrome c release is believed to be a critical step in the processing of caspase-3 and cell death. Further, Dox-induced mitochondrial ROS production can also contribute to endothelial dysfunction, impaired vascularization [82, 83], enhanced microvascular permeability [84], and disrupted endothelial paracrine signaling [85, 86]. Since Dox significantly impacts the cardiac antioxidant defense system, it increases the production of ROS while concurrently suppressing the cardiac antioxidant mechanisms, which involve modifications in sirtuin proteins, particularly SIRT1 and SIRT3, that inhibit the function of endogenous antioxidant enzymes, have been reported. The significance of the mPTP opening induced by BNIP3 (BCL2/adenovirus E1B 19 kD protein-interacting protein3), is notable. Studies involving mice with BNIP3 knockdown treated with Dox have shown that the inhibition of mPTP opening effectively restored mitochondrial function [87, 88]. Moreover, impaired NF-κB signaling in cardiac myocytes exposed to Dox activated the Bnip3 gene, leading to the mPTP opening, resulting in necrotic cell death dependent on cyclophilin D has been reported Restoring NF-κB signaling has been proven to inhibit the opening of the mPTP and mitigate necrotic cell death induced by Dox [89]. Decreased activity of critical enzymes, including heme oxygenase 1, superoxide dismutase, glutathione peroxidase, and nuclear respiratory factors (NRF1 and NRF2) have been reported to be reduced following Dox treatment. In addition, Dox can induce endoplasmic reticulum (ER) stress by disrupted equilibrium between ER protein load and its folding capacity resulting in the accumulation of misfolded or unfolded proteins within the ER lumen. Alterations in autophagy and mitophagy have been reported with Dox-cardiotoxicity. Studies involving human adult ventricular cardiomyocyte cells (AC16) have revealed suppressed mitochondrial biogenesis proteins, including PGC-1α, NRF-1, and TFAM [90, 91]. Concordantly, Dox impairs autophagy by altering expression levels of mitochondrial clearance proteins PINK and Parkin levels as well as decreasing LC3-II levels, resulting in the accumulation of autophagosomes and autolysosomes resulting in cell death. Specifically, mitochondrial fusion and fission disrupt structural and functional integrity, which has been shown to lead to significant mitochondrial dysfunction [79, 92]. The stabilizing protein mitofusin2 has evidence of suppressing dox-induced cardiotoxicity [93].

Thus, strategic targeting of the mitochondria dysfunction-associated pathway may enhance patient outcomes by alleviating the cardiotoxic effects associated with Dox administration. The subsequent section will address the specific attributes relevant to the existing approaches to prevent Dox-induced cardiotoxicity.

Pharmacological Inhibition of Dox-Induced Cardiotoxicity

Co-medications

Drugs such as dexrazoxane, renin-angiotensin-aldosterone system (RAAS) inhibitors, and β-blockers have shown efficacy in supporting high-risk patients undergoing cancer therapy, with substantial evidence indicating their ability to prevent and improve cardiomyopathy [94–97]. Specific pharmacological agents, such as candesartan, carvedilol, lisinopril and bisoprolol have demonstrated effectiveness in preserving left ventricular ejection fraction and in reducing cardiac biomarkers, suggesting they may provide cardioprotective effects when administered concurrently with Dox treatment [98, 99]. Statins, such as pitavastatin, confer protection against Dox-induced cardiotoxicity by modulating the reactive nitrogen species (RNS) and Ras-relatedC3 botulinum toxin substrate 1 (Rac1)-dependent AMPK pathway [100]. Metformin has also demonstrated a significant capacity to activate adenosine monophosphate-activated protein kinase (AMPK), influencing apoptotic pathways, mitigating oxidative stress, and preserving cardiac function in Dox-treated rats [101]. Likewise, resveratrol has cardioprotective activity through activating AMPK, attenuating oxidative stress apoptosis, ameliorating fibrosis, and showing anti-inflammatory activity in Dox-induced cardiotoxicity [102]. This drug has also shown prevention of Dox-induced cardiotoxicity, activating PGC-1α and reducing oxidative stress via enhancing AMPK and SIRT1 cascades [103]. Furthermore, phosphodiesterase-5 (PDE-5) inhibitors, including sildenafil, preserve mitochondrial integrity and reduce ventricular dysfunction, demonstrating a protective effect in Dox-induced cardiotoxicity [104, 105]. Additionally, Alpha-1 adrenergic agonists reduced apoptosis and fibrosis in cardiac muscle, improving mitochondrial function and protecting against Dox-induced adverse effects have been demonstrated by increasing the anti-apoptotic factor Bcl-2, which is crucial for maintaining mitochondrial health [106]. Dexrazoxane, a free radical scavenger used as an adjunctive agent, has shown chelator action on iron-mediated oxygen-free radical activation and lipid peroxidation, reducing iron-free radical levels; thus, it shows cardioprotective action against Dox treatment [107]. The preservation of heart function in individuals treated with Dox has been attributed to the effects of carvedilol [108]. The drug probucol has demonstrated effective antioxidant properties, contributing to the reduction of cardiac muscle injury and heart failure while preserving the anticancer efficacy of Dox [109]. Erythropoietin, when used prophylactically, demonstrates protective effects against cardiotoxicity induced by Dox by diminishing apoptosis and cardiomyopathy [110]. Moreover, the adiponectin agonist can potentially alleviate dox-induced cardiotoxicity by reducing cardiomyocyte apoptosis and mitigating oxidative stress, which is considerable [111]. Additionally, numerous supplements and natural compounds, including vitamin E, allicin, curcumin, visnagin, and schisandrin B, have demonstrated significant potential in mitigating the cardiotoxic effects induced by Dox to reduce oxidative stress, inhibit apoptosis, and decrease levels of ROS, thus diminishing injury to cardiomyocytes. Additionally, these compounds inhibit the activity of mitochondrial malate dehydrogenase, which aids in preserving mitochondrial function and reversing the detrimental effects of Dox on cardiac tissue [29, 112–114]. Moreover, modifying Dox administration from bolus injection to continuous infusion can significantly reduce peak plasma concentrations, minimizing cardiac tissue exposure to the drug [29, 46, 115]. Applying a liposomal formulation of Dox represents a promising strategy to mitigate the toxic effects associated with conventional Dox while preserving its therapeutic efficacy [116]. Notably, integrating antioxidant therapies like alpha-tocopherol, N-acetylcysteine, and targeted mitochondrial division inhibitors can improve mitochondrial function and reduce cardiomyocyte apoptosis caused by

Dox therapy. Existing heart failure therapies, especially SGLT2 inhibitors, have exhibited promise in protecting cardiac tissue from the harmful effects of Dox [117].

Non-pharmacological Approaches

In addition to pharmacological interventions, such as exercise have proven effective in reducing oxidative stress. This reduction occurs through the enhancement of endogenous antioxidants, the preservation of mitochondrial function, and the prevention of mitochondrial-mediated apoptosis [36]. Exercise-promoted upregulation of heat shock proteins has been reported to maintain the ultrastructure of cardiomyocytes, alleviate fibrosis, preserve cardiac size and function, and enhance vascular health, thereby providing cardioprotective effects against the detrimental impacts of Dox [118–120]. Recent studies have explored the influence of greenspaces and nature-based interventions on cardiovascular health outcomes among individuals who have survived cancer [121].

Although, various pharmacological and non-pharmacological strategies have effectively reduced the risk of cardiotoxicity associated with Dox treatment. However, it is essential to develop comprehensive preventive measures that enhance patient outcomes while preserving the efficacy of cancer therapies.

Conclusion

Although cancer treatments have greatly improved efficacy and survival rates, cardiotoxicity has become a significant factor affecting morbidity and mortality in cancer survivors. It is now well-documented that treatment with Dox can induce mitochondrial dysfunction, which is a primary driver of delayed cardiac inflammation and progression of cardiotoxicity. This article highlighted the importance in understanding Dox-cardiotoxicity which is instrumental in developing therapeutic interventions to mitigate heart failure. The development of such future strategies may prove beneficial in reducing morbidity with better outcomes in cancer patients undergoing Dox treatment.

Acknowledgements The authors thank Floribeth Aguilar and Victoria Margulets for technical assistance. This research was supported by project grants from the Canadian Institutes of Health Research (CIHR) to L.A.K., and to I.R.N. I.R.N. is the Evelyn Wyrzykowski Family Professor in Cardiovascular Sciences. L.A.K. is the Naranjan Dhalla Chair in Cardiovascular Sciences at the University of Manitoba and holds a Canada Research Chair in Molecular Cardiology.

References

1. Bertero E, Ameri P, Maack C (2019) Bidirectional relationship between cancer and heart failure: old and new issues in cardio-oncology. Card Fail Rev 5:106–111
2. National Academies of Sciences, Medicine Division, Board on Global Health, et al (2017) Global health and the future role of the United States. National Academies Press, Washington, DC
3. Dunbar SB, Khavjou OA, Bakas T et al (2018) Projected costs of informal caregiving for cardiovascular disease: 2015 to 2035: a policy statement from the American Heart Association. Circulation 137:e558–e577
4. Dallavalle S, Dobričić V, Lazzarato L et al (2020) Improvement of conventional anti-cancer drugs as new tools against multidrug resistant tumors. Drug Resist Updat 50:100682
5. Bukowski K, Kciuk M, Kontek R (2020) Mechanisms of multidrug resistance in cancer chemotherapy. Int J Mol Sci 21:3233
6. Menna P, Salvatorelli E (2017) Primary prevention strategies for anthracycline cardiotoxicity: a brief overview. Chemotherapy 62:159–168
7. Felker GM, Thompson RE, Hare JM et al (2000) Underlying causes and long-term survival in patients with initially unexplained cardiomyopathy. N Engl J Med 342:1077–1084
8. Tocchetti CG, Cadeddu C, Di Lisi D et al (2019) From molecular mechanisms to clinical management of antineoplastic drug-induced cardiovascular toxicity: a translational overview. Antioxid Redox Signal 30:2110–2153
9. Vejpongsa P, Yeh ETH (2014) Prevention of anthracycline-induced cardiotoxicity: challenges and opportunities. J Am Coll Cardiol 64:938–945
10. Klee NS, McCarthy CG, Martinez-Quinones P, Webb RC (2017) Out of the frying pan and into the fire: damage-associated molecular patterns and cardiovascular toxicity following cancer therapy. Ther Adv Cardiovasc Dis 11:297–317
11. Qin A, Thompson CL, Silverman P (2015) Predictors of late-onset heart failure in breast cancer patients treated with doxorubicin. J Cancer Surviv 9:252–259
12. Mattioli R, Ilari A, Colotti B et al (2023) Doxorubicin and other anthracyclines in cancers: activity, chemoresistance and its overcoming. Mol Asp Med 93:101205
13. Menna P, Gonzalez Paz O, Chello M et al (2012) Anthracycline cardiotoxicity. Expert Opin Drug Saf 11:S21–S36
14. Cardinale D, Iacopo F, Cipolla CM (2020) Cardiotoxicity of anthracyclines. Front Cardiovasc Med 7:26
15. Minotti G, Menna P, Salvatorelli E et al (2004) Anthracyclines: molecular advances and pharmacologic developments in antitumor activity and cardiotoxicity. Pharmacol Rev 56:185–229
16. Dempsey N, Rosenthal A, Dabas N et al (2021) Trastuzumab-induced cardiotoxicity: a review of clinical risk factors, pharmacologic prevention, and cardiotoxicity of other HER2-directed therapies. Breast Cancer Res Treat 188:21–36
17. Swain SM, Whaley FS, Ewer MS (2003) Congestive heart failure in patients treated with doxorubicin: a retrospective analysis of three trials. Cancer 97:2869–2879
18. Chatterjee K, Zhang J, Honbo N, Karliner JS (2010) Doxorubicin cardiomyopathy. Cardiology 115:155–162
19. Nebigil CG, Désaubry L (2018) Updates in anthracycline-mediated cardiotoxicity. Front Pharmacol 9:1262
20. van der Zanden SY, Qiao X, Neefjes J (2021) New insights into the activities and toxicities of the old anticancer drug doxorubicin. FEBS J 288:6095–6111
21. Herrmann J (2020) Adverse cardiac effects of cancer therapies: cardiotoxicity and arrhythmia. Nat Rev Cardiol 17:474–502
22. Pardo Sanz A, Zamorano JL (2020) 'Cardiotoxicity': time to define new targets? Eur Heart J 41:1730–1732

23. Belger C, Abrahams C, Imamdin A, Lecour S (2024) Doxorubicin-induced cardiotoxicity and risk factors. Int J Cardiol Heart Vasc 50:101332
24. Qiu S, Zhou T, Qiu B et al (2021) Risk factors for anthracycline-induced cardiotoxicity. Front Cardiovasc Med 8:736854
25. Saleh Y, Abdelkarim O, Herzallah K, Abela GS (2021) Anthracycline-induced cardiotoxicity: mechanisms of action, incidence, risk factors, prevention, and treatment. Heart Fail Rev 26: 1159–1173
26. Von Hoff DD, Layard MW, Basa P et al (1979) Risk factors for doxorubicin-induced congestive heart failure. Ann Intern Med 91:710–717
27. Philip LJ, Findlay SG, Gill JH (2022) Baseline blood pressure and development of cardiotoxicity in patients treated with anthracyclines: a systematic review. Int J Cardiol Cardiovasc Risk Prev 15:200153
28. Sheppard RJ, Berger J, Sebag IA (2013) Cardiotoxicity of cancer therapeutics: current issues in screening, prevention, and therapy. Front Pharmacol 4:19
29. Haug KG, Weber B, Hochhaus G, Butterweck V (2012) Nonlinear pharmacokinetics of visnagin in rats after intravenous bolus administration. Eur J Pharm Sci 45:79–89
30. Thorn CF, Oshiro C, Marsh S et al (2011) Doxorubicin pathways: pharmacodynamics and adverse effects. Pharmacogenet Genomics 21:440–446
31. Sawyer DB (2013) Anthracyclines and heart failure. N Engl J Med 368:1154–1156
32. Sheibani M, Azizi Y, Shayan M et al (2022) Doxorubicin-induced cardiotoxicity: an overview on pre-clinical therapeutic approaches. Cardiovasc Toxicol 22:292–310
33. Carvalho C, Santos RX, Cardoso S et al (2009) Doxorubicin: the good, the bad and the ugly effect. Curr Med Chem 16:3267–3285
34. Mercurio V, Pirozzi F, Lazzarini E et al (2016) Models of heart failure based on the cardiotoxicity of anticancer drugs. J Card Fail 22:449–458
35. Cardinale D, Colombo A, Bacchiani G et al (2015) Early detection of anthracycline cardiotoxicity and improvement with heart failure therapy. Circulation 131:1981–1988
36. Hahn VS, Lenihan DJ, Ky B (2014) Cancer therapy-induced cardiotoxicity: basic mechanisms and potential cardioprotective therapies. J Am Heart Assoc 3:e000665
37. An J, Li P, Li J et al (2009) ARC is a critical cardiomyocyte survival switch in doxorubicin cardiotoxicity. J Mol Med 87:401–410
38. Minotti G, Ronchi R, Salvatorelli E et al (2001) Doxorubicin irreversibly inactivates iron regulatory proteins 1 and 2 in cardiomyocytes: evidence for distinct metabolic pathways and implications for iron-mediated cardiotoxicity of antitumor therapy. Cancer Res 61:8422–8428
39. Osataphan N, Phrommintikul A, Chattipakorn SC, Chattipakorn N (2020) Effects of doxorubicin-induced cardiotoxicity on cardiac mitochondrial dynamics and mitochondrial function: insights for future interventions. J Cell Mol Med 24:6534–6557
40. Bikomeye JC, Terwoord JD, Santos JH, Beyer AM (2022) Emerging mitochondrial signaling mechanisms in cardio-oncology: beyond oxidative stress. Am J Physiol Circ Physiol 323: H702–H720
41. Cappetta D, De Angelis A, Sapio L et al (2017) Oxidative stress and cellular response to doxorubicin: a common factor in the complex milieu of anthracycline cardiotoxicity. Oxidative Med Cell Longev 2017:1521020
42. Zorov DB, Juhaszova M, Sollott SJ (2014) Mitochondrial reactive oxygen species (ROS) and ROS-induced ROS release. Physiol Rev 94:909–950
43. de Barros Wanderley MR Jr, Ávila MS, Fernandes-Silva MM et al (2022) Plasma biomarkers reflecting high oxidative stress in the prediction of myocardial injury due to anthracycline chemotherapy and the effect of carvedilol: insights from the CECCY Trial. Oncotarget 13: 214–223
44. Robinson EL, Azodi M, Heymans S, Heggermont W (2020) Anthracycline-related heart failure: certain knowledge and open questions: where do we stand with chemotherapyinduced cardiotoxicity? Curr Heart Fail Rep 17:357–364

45. Wenningmann N, Knapp M, Ande A et al (2019) Insights into doxorubicin-induced cardiotoxicity: molecular mechanisms, preventive strategies, and early monitoring. Mol Pharmacol 96:219–232
46. Rawat PS, Jaiswal A, Khurana A et al (2021) Doxorubicin-induced cardiotoxicity: an update on the molecular mechanism and novel therapeutic strategies for effective management. Biomed Pharmacother 139:111708
47. Gabani M, Castañeda D, Nguyen QM et al (2021) Association of cardiotoxicity with doxorubicin and trastuzumab: a double-edged sword in chemotherapy. Cureus 13:e18194
48. Nukala SB, Jousma J, Cho Y et al (2022) Long non-coding RNAs and microRNAs as crucial regulators in cardio-oncology. Cell Biosci 12:24
49. Speth PAJ, Van Hoesel Q, Haanen C (1988) Clinical pharmacokinetics of doxorubicin. Clin Pharmacokinet 15:15–31
50. Renu K, Abilash VG, Tirupathi Pichiah PB, Arunachalam S (2018) Molecular mechanism of doxorubicin-induced cardiomyopathy—an update. Eur J Pharmacol 818:241–253
51. Unverferth DV, Magorien RD, Unverferth BP et al (1981) Human myocardial morphologic and functional changes in the first 24 hours after doxorubicin administration. Cancer Treat Rep 65:1093–1097
52. Moscoso I, Rodríguez-Mañero M, Cebro-Márquez M et al (2024) Transforming cardiotoxicity detection in cancer therapies: the promise of MicroRNAs as precision biomarkers. Int J Mol Sci 25:11910
53. Zhang S, Liu X, Bawa-Khalfe T et al (2012) Identification of the molecular basis of doxorubicin-induced cardiotoxicity. Nat Med 18:1639–1642
54. Curigliano G, Lenihan D, Fradley M et al (2020) Management of cardiac disease in cancer patients throughout oncological treatment: ESMO consensus recommendations. Ann Oncol 31:171–190
55. Ky B, Vejpongsa P, Yeh ETH et al (2013) Emerging paradigms in cardiomyopathies associated with cancer therapies. Circ Res 113:754–764
56. Minotti G, Cairo G, Monti E (1999) Role of iron in anthracycline cardiotoxicity: new tunes for an old song? FASEB J 13:199–212
57. Šimůnek T, Štěrba M, Popelová O et al (2009) Anthracycline-induced cardiotoxicity: overview of studies examining the roles of oxidative stress and free cellular iron. Pharmacol Rep 61:154–171
58. Octavia Y, Tocchetti CG, Gabrielson KL et al (2012) Doxorubicin-induced cardiomyopathy: from molecular mechanisms to therapeutic strategies. J Mol Cell Cardiol 52:1213–1225
59. Liu J, Mao W, Ding B, Liang C (2008) ERKs/p53 signal transduction pathway is involved in doxorubicin-induced apoptosis in H9c2 cells and cardiomyocytes. Am J Physiol Circ Physiol 295:H1956–H1965
60. Aries A, Paradis P, Lefebvre C et al (2004) Essential role of GATA-4 in cell survival and drug-induced cardiotoxicity. Proc Natl Acad Sci 101:6975–6980
61. Park A-M, Nagase H, Liu L et al (2011) Mechanism of anthracycline-mediated down-regulation of GATA4 in the heart. Cardiovasc Res 90:97–104
62. Poizat C, Puri PL, Bai Y, Kedes L (2005) Phosphorylation-dependent degradation of p300 by doxorubicin-activated p38 mitogen-activated protein kinase in cardiac cells. Mol Cell Biol 25: 2673–2687
63. Ranek MJ, Wang X (2009) Activation of the ubiquitin-proteasome system in doxorubicin cardiomyopathy. Curr Hypertens Rep 11:389–395
64. Cui N, Wu F, Lu W et al (2019) Doxorubicin-induced cardiotoxicity is maturation dependent due to the shift from topoisomerase IIα to IIβ in human stem cell derived cardiomyocytes. J Cell Mol Med 23:4627–4639
65. Durbecq V, Paesmans M, Cardoso F et al (2004) Topoisomerase-IIα expression as a predictive marker in a population of advanced breast cancer patients randomly treated either with single-agent doxorubicin or single-agent docetaxel. Mol Cancer Ther 3:1207–1214

66. Tokarska-Schlattner M, Zaugg M, Zuppinger C et al (2006) New insights into doxorubicin-induced cardiotoxicity: the critical role of cellular energetics. J Mol Cell Cardiol 41:389–405
67. Papazoglou P, Peng L, Sachinidis A (2021) Epigenetic mechanisms involved in the cardiovascular toxicity of anticancer drugs. Front Cardiovasc Med 8:658900
68. Dhingra R, Rabinovich-Nikitin I, Rothman S et al (2022) Proteasomal degradation of TRAF2 mediates mitochondrial dysfunction in doxorubicin-cardiomyopathy. Circulation 146:934–954
69. Zhong F, Liang S, Zhong Z (2019) Emerging role of mitochondrial DNA as a major driver of inflammation and disease progression. Trends Immunol 40:1120–1133
70. Ghigo A, Li M, Hirsch E (2016) New signal transduction paradigms in anthracycline-induced cardiotoxicity. Biochim Biophys Acta 1863:1916–1925
71. Wu L, Wang L, Du Y et al (2023) Mitochondrial quality control mechanisms as therapeutic targets in doxorubicin-induced cardiotoxicity. Trends Pharmacol Sci 44:34–49
72. Bartlett JJ, Trivedi PC, Yeung P et al (2016) Doxorubicin impairs cardiomyocyte viability by suppressing transcription factor EB expression and disrupting autophagy. Biochem J 473: 3769–3789
73. Chen Y, Huang T, Shi W et al (2020) Potential targets for intervention against doxorubicin-induced cardiotoxicity based on genetic studies: a systematic review of the literature. J Mol Cell Cardiol 138:88–98
74. Goormaghtigh E, Chatelain P, Caspers J, Ruysschaert JM (1980) Evidence of a specific complex between adriamycin and negatively-charged phospholipids. Biochim Biophys Acta 597:1–14
75. Schlame M, Rua D, Greenberg ML (2000) The biosynthesis and functional role of cardiolipin. Prog Lipid Res 39:257–288
76. Kashfi K, Israel M, Sweatman TW et al (1990) Inhibition of mitochondrial carnitine palmitoyltransferases by adriamycin and adriamycin analogues. Biochem Pharmacol 40: 1441–1448
77. Carvalho RA, Sousa RPB, Cadete VJJ et al (2010) Metabolic remodeling associated with subchronic doxorubicin cardiomyopathy. Toxicology 270:92–98
78. Raj S, Franco VI, Lipshultz SE (2014) Anthracycline-induced cardiotoxicity: a review of pathophysiology, diagnosis, and treatment. Curr Treat Options Cardiovasc Med 16:1–14
79. Ding M, Shi R, Cheng S et al (2022) Mfn2-mediated mitochondrial fusion alleviates doxorubicin-induced cardiotoxicity with enhancing its anticancer activity through metabolic switch. Redox Biol 52:102311
80. Nicolay K, de Kruijff B (1987) Effects of adriamycin on respiratory chain activities in mitochondria from rat liver, rat heart and bovine heart. Evidence for a preferential inhibition of complex III and IV. Biochim Biophys Acta 892:320–330
81. Goormaghtigh E, Huart P, Brasseur R, Ruysschaert J-M (1986) Mechanism of inhibition of mitochondrial enzymatic complex I–III by adriamycin derivatives. Biochim Biophys Acta 861:83–94
82. Huang C, Zhang X, Ramil JM et al (2010) Juvenile exposure to anthracyclines impairs cardiac progenitor cell function and vascularization resulting in greater susceptibility to stress-induced myocardial injury in adult mice. Circulation 121:675–683
83. Galan-Arriola C, Vílchez-Tschischke JP, Lobo M et al (2022) Coronary microcirculation damage in anthracycline cardiotoxicity. Cardiovasc Res 118:531–541
84. Wilkinson EL, Sidaway JE, Cross MJ (2016) Cardiotoxic drugs Herceptin and doxorubicin inhibit cardiac microvascular endothelial cell barrier formation resulting in increased drug permeability. Biol Open 5:1362–1370
85. Räsänen M, Degerman J, Nissinen TA et al (2016) VEGF-B gene therapy inhibits doxorubicin-induced cardiotoxicity by endothelial protection. Proc Natl Acad Sci 113: 13144–13149
86. Vásquez-Vivar J, Martasek P, Hogg N et al (1997) Endothelial nitric oxide synthase-dependent superoxide generation from adriamycin. Biochemistry 36:11293–11297

87. Dhingra A, Jayas R, Afshar P et al (2017) Ellagic acid antagonizes Bnip3-mediated mitochondrial injury and necrotic cell death of cardiac myocytes. Free Radic Biol Med 112:411–422
88. Dhingra R, Margulets V, Chowdhury SR et al (2014) Bnip3 mediates doxorubicin-induced cardiac myocyte necrosis and mortality through changes in mitochondrial signaling. Proc Natl Acad Sci 111:E5537–E5544
89. Dhingra R, Guberman M, Rabinovich-Nikitin I et al (2020) Impaired NF-κB signalling underlies cyclophilin D-mediated mitochondrial permeability transition pore opening in doxorubicin cardiomyopathy. Cardiovasc Res 116:1161–1174
90. Yin J, Guo J, Zhang Q et al (2018) Doxorubicin-induced mitophagy and mitochondrial damage is associated with dysregulation of the PINK1/parkin pathway. Toxicol In Vitro 51: 1–10
91. Sishi BJN, Loos B, van Rooyen J, Engelbrecht A-M (2013) Autophagy upregulation promotes survival and attenuates doxorubicin-induced cardiotoxicity. Biochem Pharmacol 85:124–134
92. Xia Y, Chen Z, Chen A et al (2017) LCZ696 improves cardiac function via alleviating Drp1-mediated mitochondrial dysfunction in mice with doxorubicin-induced dilated cardiomyopathy. J Mol Cell Cardiol 108:138–148
93. Guberman M, Dhingra R, Cross J et al (2024) IKKβ stabilizes Mitofusin 2 and suppresses doxorubicin cardiomyopathy. Cardiovasc Res 120:164–173
94. Bansal N, Adams MJ, Ganatra S et al (2019) Strategies to prevent anthracycline-induced cardiotoxicity in cancer survivors. Cardiooncology 5:1–22
95. Liesse K, Harris J, Chan M et al (2018) Dexrazoxane significantly reduces anthracycline-induced cardiotoxicity in pediatric solid tumor patients: a systematic review. J Pediatr Hematol Oncol 40:417–425
96. Sobczuk P, Czerwińska M, Kleibert M, Cudnoch-Jędrzejewska A (2022) Anthracycline-induced cardiotoxicity and renin-angiotensin-aldosterone system—from molecular mechanisms to therapeutic applications. Heart Fail Rev 27:295–319
97. Palvia AR, Damera AR, Nandi AR et al (2024) Cardio-oncology's modern approaches to prevent doxorubicin-induced cardiotoxicity: a systematic review. Cureus 16:e66215
98. Lee M, Chung W, Lee J et al (2021) Candesartan and carvedilol for primary prevention of subclinical cardiotoxicity in breast cancer patients without a cardiovascular risk treated with doxorubicin. Cancer Med 10:3964–3973
99. Wihandono A, Azhar Y, Abdurahman M, Hidayat S (2021) The role of lisinopril and bisoprolol to prevent anthracycline induced cardiotoxicity in locally advanced breast cancer patients. Asian Pac J Cancer Prev 22:2847–2853
100. Choi HC, Song P, Xie Z et al (2008) Reactive nitrogen species is required for the activation of the AMP-activated protein kinase by statin in vivo. J Biol Chem 283:20186–20197
101. Kobashigawa LC, Xu YC, Padbury JF et al (2014) Metformin protects cardiomyocyte from doxorubicin induced cytotoxicity through an AMP-activated protein kinase dependent signaling pathway: an in vitro study. PLoS One 9:e104888
102. Timm KN, Tyler DJ (2020) The role of AMPK activation for cardioprotection in doxorubicin-induced cardiotoxicity. Cardiovasc drugs Ther 34:255–269
103. Liu D, Ma Z, Xu L et al (2019) PGC1α activation by pterostilbene ameliorates acute doxorubicin cardiotoxicity by reducing oxidative stress via enhancing AMPK and SIRT1 cascades. Aging (Albany, NY) 11:10061
104. Fisher PW, Salloum F, Das A et al (2005) Phosphodiesterase-5 inhibition with sildenafil attenuates cardiomyocyte apoptosis and left ventricular dysfunction in a chronic model of doxorubicin cardiotoxicity. Circulation 111:1601–1610
105. Das A, Durrant D, Mitchell C et al (2010) Sildenafil increases chemotherapeutic efficacy of doxorubicin in prostate cancer and ameliorates cardiac dysfunction. Proc Natl Acad Sci 107: 18202–18207
106. Beak JY, Huang W, Parker JS et al (2017) An oral selective alpha-1A adrenergic receptor agonist prevents doxorubicin cardiotoxicity. Basic Transl Sci 2:39–53

107. Schroeder PE, Hasinoff BB (2005) Metabolism of the one-ring open metabolites of the cardioprotective drug dexrazoxane to its active metal-chelating form in the rat. Drug Metab Dispos 33:1367–1372
108. Kalay N, Basar E, Ozdogru I et al (2006) Protective effects of carvedilol against anthracycline-induced cardiomyopathy. J Am Coll Cardiol 48:2258–2262
109. Siveski-Iliskovic N, Hill M, Chow DA, Singal PK (1995) Probucol protects against adriamycin cardiomyopathy without interfering with its antitumor effect. Circulation 91:10–15
110. Ramond A, Sartorius E, Mousseau M et al (2008) Erythropoietin pretreatment protects against acute chemotherapy toxicity in isolated rat hearts. Exp Biol Med 233:76–83
111. Zhao D, Xue C, Li J et al (2020) Adiponectin agonist ADP355 ameliorates doxorubicin-induced cardiotoxicity by decreasing cardiomyocyte apoptosis and oxidative stress. Biochem Biophys Res Commun 533:304–312
112. Abdel-Daim MM, Kilany OE, Khalifa HA, Ahmed AAM (2017) Allicin ameliorates doxorubicin-induced cardiotoxicity in rats via suppression of oxidative stress, inflammation and apoptosis. Cancer Chemother Pharmacol 80:745–753
113. dos Santos DS, dos Santos Goldenberg RC (2018) Doxorubicin-induced cardiotoxicity: from mechanisms to development of efficient therapy. In: Cardiotoxicity. IntechOpen, London, pp 3–24
114. Li L, Pan Q, Han W et al (2007) Schisandrin B prevents doxorubicin-induced cardiotoxicity via enhancing glutathione redox cycling. Clin Cancer Res 13:6753–6760
115. Bin WB, Leung KT, Poon EN-Y (2022) Mitochondrial-targeted therapy for doxorubicin-induced cardiotoxicity. Int J Mol Sci 23:1912
116. Franco YL, Vaidya TR, Ait-Oudhia S (2018) Anticancer and cardio-protective effects of liposomal doxorubicin in the treatment of breast cancer. Breast Cancer Targets Ther 10: 131–141
117. Oh C-M, Cho S, Jang J-Y et al (2019) Cardioprotective potential of an SGLT2 inhibitor against doxorubicin-induced heart failure. Korean Circ J 49:1183–1195
118. Dozic S, Howden EJ, Bell JR et al (2023) Cellular mechanisms mediating exercise-induced protection against cardiotoxic anthracycline cancer therapy. Cells 12:1312
119. Marques-Aleixo I, Santos-Alves E, Torrella JR et al (2018) Exercise and doxorubicin treatment modulate cardiac mitochondrial quality control signaling. Cardiovasc Toxicol 18:43–55
120. Ventura-Clapier R (2009) Exercise training, energy metabolism, and heart failure. Appl Physiol Nutr Metab 34:336–339
121. Bikomeye JC, Beyer AM, Kwarteng JL, Beyer KMM (2022) Greenspace, inflammation, cardiovascular health, and cancer: a review and conceptual framework for greenspace in cardio-oncology research. Int J Environ Res Public Health 19:2426

Chapter 18
Ponatinib in Chronic Myeloid Leukemia and Philadelphia Chromosome-Positive Acute Lymphoblastic Leukemia: Efficacy, Mechanisms, and Cardiotoxic Challenges

Jonatas M. Rolando and Dinender K. Singla

Abstract Ponatinib, a third-generation tyrosine kinase inhibitor (TKI), is pivotal in treating chronic myeloid leukemia (CML) and Philadelphia chromosome-positive acute lymphoblastic leukemia (Ph+ ALL), particularly in cases involving the T315I mutation, which promotes resistance to earlier TKI generations. Ponatinib effectively inhibits the mutated breakpoint cluster region (BCR)-Abelson murine leukemia viral oncogene homolog (ABL) kinase, by circumventing the steric hindrance from this mutation. However, its extensive off-target effects often lead to cardiotoxicity, presenting as arterial occlusive events, heart failure, and hypertension, which have raised significant clinical concerns. Studies indicate that ponatinib-induced cardiotoxicity may result from its impact on cell survival pathways, inflammation, and the integrated stress response. Despite Food and Drug Administration (FDA) approval, adverse cardiovascular outcomes have led to dose adjustments and an exploration of alternative treatment strategies. Current research focuses on mitigating ponatinib's cardiotoxic effects through co-therapies and structural analogs, aiming to balance efficacy against the T315I mutation with a safer cardiovascular profile. This review explores ponatinib's development as a TKI, clinical trials, structure, mechanisms of action, and toxic side effects, especially cardiotoxic effects, highlighting potential therapeutic strategies for reducing cardiovascular risks while maintaining anticancer efficacy.

Keywords Arterial occlusion · BCR-ABL kinase · Blood cancer · Cardiac failure · Cardiotoxicity · Hypertension · Iclusig · Tyrosine kinases · Tyrosine kinase inhibitor · T315I mutation

---

J. M. Rolando · D. K. Singla (✉)
Division of Metabolic and Cardiovascular Sciences, Burnett School of Biomedical Sciences, College of Medicine, University of Central Florida, Orlando, FL, USA
e-mail: dinender.singla@ucf.edu

D. K. Agrawal et al. (eds.), *Cardiovascular Toxicity*, Advances in Biochemistry in Health and Disease 33, https://doi.org/10.1007/978-3-031-90598-8_18

Introduction

Tyrosine kinases (TKs) are a group of enzymes that utilize adenosine triphosphate (ATP) as a substrate to phosphorylate specific tyrosine residues on target proteins [1]. They play a crucial role in signal transduction, which is related to cellular mechanisms such as cell growth, survival, differentiation, proliferation, migration, adhesion, and metabolism [2–4]. There are 90 different TKs identified, divided into 58 receptor tyrosine kinases (RTKs) and 32 non-receptor tyrosine kinases (NRTKs) [3]. RTKs are divided into 20 subfamilies and share a related structure comprising an extracellular binding domain followed by a single transmembrane helix and an intracellular tyrosine kinase domain [5–7]. RTKs are activated by ligand binding to the extracellular domain, leading to activation of the intracellular tyrosine kinase domain, which catalyzes the transport of the γ-phosphate from ATP into tyrosine hydroxyl groups of different proteins [1, 8, 9]. NRTKs are divided into nine subfamilies and, unlike RTKs don't have an extracellular ligand-binding domain and a transmembrane-spanning region, being either anchored to the cell membrane or a cytoplasmic kinase [10, 11]. NRTKs have a larger C-terminal compared to the N-terminal, with the N-terminal region displaying different kinase domains that confer structural variability to the subfamilies [10, 12, 13]. NRTKs are activated through a multifaceted process involving protein-protein interactions that allow tyrosine kinase phosphorylation [8, 14]. Therefore, TKs are fundamental for regulating cellular functioning and metabolism; however, since these kinases are involved in crucial cellular roles, mutations that lead to disruption of TKs, such as gain of function, deletions, and insertions, are deeply associated with cancer development [15–17].

Since the early 90s, many researchers have been studying these kinases to develop cancer therapies. Eventually, tyrosine kinase inhibitors (TKIs) were developed as a targeted therapy to inhibit tyrosine kinases. TKIs are classified as irreversible or reversible, with irreversible ones forming a covalent bond with the kinase domain, blocking the ATP site. Additionally, TKIs are divided into five types according to their binding patterns to the kinase domain [18–20]. TKIs are one of the most studied subjects in the twenty-first century and one of the most effective treatments developed to combat cancer, raising the life expectancy of many cancer patients in the last two decades [21].

TKIs are effective in ceasing the action of aberrant TKs that cause many different cancers [22]. Currently, there are 80 TKIs approved by the Food and Drug Administration (FDA) to treat different types of cancers such as human epidermal growth factor receptor 2 (HER2)-positive breast cancer, leukemias, kidney cancer, melanoma, and non-small cell lung cancer [23]. In 2001, imatinib was the first tyrosine kinase inhibitor (TKI) to get FDA approval for the treatment of chronic myeloid leukemia (CML) [24].

CML is generated by chromosomal translocation, which is responsible for the fusion of the Abelson murine leukemia viral oncogene homolog (ABL) gene on chromosome 9 with the breakpoint cluster region (BCR) gene on chromosome

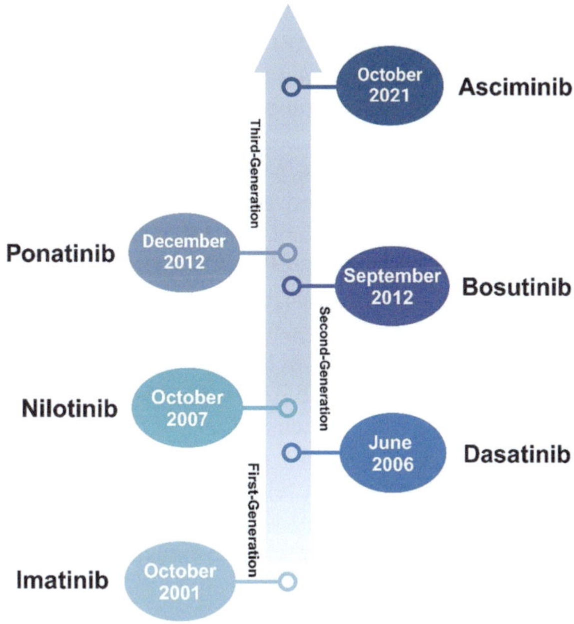

Fig. 18.1 Approval timeline of different generations TKIs for the treatment of CML and Ph+ ALL. The vertical arrow represents the chronological progression of events. (Created with BioRender.com)

22, resulting in the formation of the tyrosine kinase oncoprotein called BCR-ABL in the translocated Philadelphia chromosome [25, 26]. This oncoprotein is a constitutive active tyrosine kinase, which promotes cell proliferation *via* different signaling pathways such as Janus kinase (JAK)/signal transducer and activator of transcription (STAT) and phosphatase and tensin homolog deleted on chromosome ten (PTEN)/phosphoinositide 3-kinase (PI3K)/protein kinase B (AKT) [27, 28].

Nowadays, with the advent of TKIs, studies are showing an increase in the survival rate of CML patients, which is very close to the general population [29]. Frequent kinase domain mutations lead to drug resistance, thus, since 2001, there have been three different generations of TKIs developed for the treatment of CML and Philadelphia chromosome-positive acute lymphoblastic leukemia (Ph+ ALL) to overcome drug resistance [30], which are shown in Fig. 18.1.

Imatinib, a first-generation drug, was approved by the FDA in 2001, being the first-line therapy for patients with all phases of CML and Ph+ ALL [31, 32]. Second-generation drugs dasatinib and nilotinib were approved in 2006 and 2007, respectively. Both were prescribed for patients with CML and Ph+ ALL who were intolerant to imatinib [30, 33, 34]. Another second-generation drug, bosutinib, was approved in 2012 and was prescribed for patients non-responsive to previous TKIs [35, 36]. Ponatinib was the first third-generation TKI approved by the FDA, being mainly prescribed for patients who are carriers of the T315I mutation (mutation of threonine to isoleucine at codon 315) [37, 38]. In 2021, the latest third-generation TKI, asciminib, was approved by the FDA for CML patients in the chronic phase (CP) who were previously unresponsive to two or more TKIs and for carriers of the

T315I mutation [39, 40]. Asciminib was designed to specifically target the ABL myristoyl pocket (STAMP) inhibitor, being a pioneering allosteric inhibitor [41, 42]. Except for asciminib and ponatinib, all other TKIs are approved to treat pediatric CML, while imatinib and dasatinib are approved to treat pediatric Ph+ ALL [43–50].

Among all generations, ponatinib is defined by many researchers as the most cardiotoxic TKI ever approved by FDA. Approved in 2012, ponatinib was, until 2021 the only practical option for CML patients with the gatekeeper mutation T315I, being still the only option for Ph+ ALL patients with the mutation [30, 51]. However, asciminib showed a decreased potency against T315I mutation, leaving ponatinib as the main therapy for those patients [52, 53]. This mutation changes the conformation of the hydrophobic pocket in BCR-ABL protein, making first and second-generation drugs ineffective [30, 54, 55].

Although TKIs have been effective in treating CML and Ph+ ALL patients, their cardiotoxic side effects affect many patients after cancer treatment [56]. This review will discuss ponatinib's development as a TKI, its structure, and its mechanism of action. We will also emphasize ponatinib's side effects during CML treatment, especially those associated with the cardiovascular system.

Ponatinib Development as a Tyrosine Kinase Inhibitor

The kinase domain of the BCR-ABL oncoprotein is susceptible to point mutations. One specific mutation results in the substitution of cytosine with thymine, which changes threonine to isoleucine at codon 315. The threonine side chain is crucial for the binding of first- and second-generation tyrosine kinase inhibitors (TKIs) within the ATP-binding pocket. In contrast, the isoleucine substitution introduces steric hindrance, which prevents TKIs from entering the hydrophobic pocket and inhibits their effectiveness. To address this issue, Ariad Pharmaceuticals specifically designed the structure of ponatinib to overcome drug resistance caused by this gatekeeper mutation, effectively avoiding the steric hindrance associated with the isoleucine side chain [57–59].

The preclinical data showed that ponatinib was able to block wild-type (WT) and many mutant forms of BCR-ABL kinase domains, including T315I mutation, in Ba/F3 cell lines with an inhibitory concentration 50 (IC50) ranging from 0.37 to 2 nM [60]. In addition, it was observed that ponatinib was able to reduce cell viability in different CML cell lines with an IC50 approximately 500-fold lower than normal cells [60]. Furthermore, they used a mice xenograft model with Ba/F3 cells expressing native BCR-ABL or BCR-ABLT315I and administered them with ponatinib through oral gavage for 19 days in different dosages. Ponatinib was able to significantly increase the median survival rate of mice in a dose-dependent manner for both native BCR-ABL or BCR-ABLT315I when compared to vehicle-treated mice, with 50 mg/kg dosage showing a 96% reduction in tumor volume [60]. Additionally, they observed that BCR-ABLT315I mice administered with high doses of dasatinib had no increase in survival rate compared to vehicle-treated mice, showing the potency of ponatinib in inhibiting mutated BCR-ABL tyrosine kinase as well as the inefficacy of dasatinib towards T315I mutation [60].

In 2008, after completing preclinical studies, ponatinib began its first phase of clinical trials. This phase involved 81 patients with refractory hematologic malignancies, and the aim was to determine the maximum tolerated dose. The study found that 45 mg was the maximum dose tolerated, with pancreatitis identified as the dose-limiting toxicity. The most common side effects reported were constipation and rash. Overall, ponatinib demonstrated effectiveness in patients who had previously resisted other tyrosine kinase inhibitors (TKIs), with 72% of the cohort achieving a major cytogenetic response [61, 62].

Following the completion of the phase I trial, the phase II trial entitled Ponatinib Ph+ ALL and CML Evaluation (PACE) commenced in 2010. This trial was conducted across multiple countries, involving 449 patients with CML at all stages, as well as those with Ph+ ALL. These participants were either carriers of the T315I mutation or were resistant or intolerant to dasatinib or nilotinib. Throughout the trial, it was found that ponatinib demonstrated antileukemia activity in both CML and Ph+ ALL across different disease stages, with 83% of the cohort achieving a major cytogenetic response. However, 12% of patients discontinued their participation in the study, and 26% experienced arterial occlusive events (AOEs), which included fatal myocardial infarctions and strokes. Notably, no new mutations emerged during ponatinib treatment [63, 64].

Following the initial 2-year follow-up with promising results, FDA gave accelerated approval for ponatinib in December 2012 as a third-line treatment option for all phases of CML and Ph+ ALL patients resistant to prior TKIs or diagnosed with T315I mutation. However, due to the increased occurrence of thrombosis and vessel narrowing in patients, the FDA decided to suspend sales in October 2013. This suspension was removed later due to the lack of options for mutation carriers [65, 66]. The 5-year report showed that 82% of patients in the chronic phase of CML (CP-CML) maintained major cytogenetic response after 5 years and 40% of all patients achieved major molecular response at some point of treatment during the timeframe of the trial. In all groups, a 73% overall survival rate was observed. However, these favorable results came with increased toxicity, with patients experiencing an increased risk of developing AOEs, venous thromboembolic events (VTEs), heart failure, hepatotoxicity, and hypertension, among other problems [67].

A phase III clinical trial entitled Evaluation of Ponatinib versus Imatinib in Chronic Myeloid Leukemia (EPIC) started in August 2012, following the FDA's approval. This study had participants from all around the world and was designed to compare the efficacy of 45 mg of ponatinib against 400 mg of imatinib in CP-CML [68]. Due to the FDA market suspension of ponatinib, the study was suspended in October 2013. Only 23 patients reached major molecular response after 12 months of treatment, moreover, ponatinib showed increased efficacy in reducing BCR-ABL transcripts compared to imatinib but showed more AOEs compared to imatinib [68, 69].

Following the EPIC trial, another phase II trial was started by Ariad Pharmaceuticals and completed by Takeda Pharmaceuticals after Takeda acquired Ariad. This trial entitled Optimizing Ponatinib Treatment in CP-CML (OPTIC) aimed to define the response and safety of ponatinib in 3 different dosages (45, 30, and 15 mg daily) in patients in CP-CML resistant to prior TKIs or carriers of T315I mutation [70]. In this trial, 283 patients were divided into 3 groups with 60.2% in the 45 mg group, 40.9% in the 30 mg group, and 39.6% in the 15 mg group reaching ≤1% BCR-ABL transcripts after 48 months. Forty-five milligrams group had the highest progression-free survival, and the 15 mg had lower rates of AOEs. The most severe adverse events observed were thrombocytopenia, hypertension, neutropenia, heart failure, and hepatoxicity. Most of the patients that reached ≤1% BCR-ABL in the 45 mg group reduced their dose to 15 mg, displaying a 14% incidence of AOEs. Hence, a starting dose of 45 mg per day with a posterior reduction to 15 mg per day after reaching ≤1% BCR-ABL was observed to bring the best benefit/risk ratio [70–72].

Finally, another phase III clinical trial was initiated by Takeda Pharmaceuticals in 2018 entitled Philadelphia Chromosome-Positive Acute Lymphoblastic Leukemia with Concomitant Chemotherapy (PhALLCON). In this study, the authors aimed to check the efficacy of ponatinib in combination with vincristine, a reduced-intensity chemotherapy, compared to imatinib in newly diagnosed Ph+ ALL patients [73]. The first reports showed that in a group of 245 patients, ponatinib (164) and imatinib (81), the ponatinib group had significantly higher minimal residual disease (MRD) negativity (BCR-ABL ≤0.01%) complete remission (CR) compared to imatinib. Treatment-emergent adverse events (TEAEs) rates were similar between the two drugs, including AOEs. Thus, ponatinib in combination with vincristine was more effective than imatinib for the treatment of Ph+ ALL patients [73, 74].

This summarizes the development and clinical trials performed by Ariad and Takeda Pharmaceuticals for ponatinib (Table 18.1), other clinical trials concerning ponatinib use for cancers other than CML and Ph+ ALL were performed over the years, which will be discussed later in this review.

Ponatinib Structure and Mechanism of Actions

As previously mentioned ponatinib was designed to overcome resistance from the gatekeeper mutation T315I, which results in steric hindrance in the ATP binding region, inhibiting the entrance of other TKIs into the hydrophobic pocket [57, 58].

Ponatinib structure is comprised basically of five critical parts that promote better inhibitory potency against mutant BCR-ABLT315I. The five parts are the hinge region, ethynyl linker, A ring, linker 2, and B ring (Fig. 18.2). The hinge region is formed by a fusion of an imidazole ring with a pyridazine ring which binds to the ATP binding region in the kinase domain. The imidazole ring is responsible for the hydrogen bond formation with the hinge of the pocket, while the pyridazine ring forms hydrophobic interactions with different regions of the ATP binding site [82, 83].

Table 18.1 Clinical trials of ponatinib in CML and Ph+ ALL patients. A summary of ponatinib's clinical trials assessing study phases, patient cohorts, response rates, and reported adverse events

Trial name and phase	Trial status	Type of patients	Outcome
Safety phase I study (NCT00660920)	Completed	Adults ≥18 years old diagnosed with hematologic malignancy (other than lymphoma)	The study identified 45 mg as the dose-limiting toxicity and 72% of the patients had a major cytogenic response [61, 62, 75]
PACE phase II study (NCT01207440)	Completed	Adults ≥18 years old diagnosed with CML in any phase or Ph+ ALL, resistant to prior TKI treatment	83% of the cohort displayed a major cytogenic response. However, patients were discontinued from the study due to severe adverse events and 26% presented AOEs, including fatal myocardial infarction [63, 64, 67, 76]
EPIC phase III study (NCT01650805)	Terminated	Adults ≥18 years old with newly diagnosed CP-CML	EPIC trial was terminated due to the accumulation of severe adverse events, especially related to AOEs, lacking data on the comparison between ponatinib and imatinib for patients with newly diagnosed CP-CML [65, 66, 68, 69, 77, 78]
OPTIC phase II study (NCT02467270)	Active, not recruiting	Adults ≥18 years old in CP-CML resistant to prior TKI treatment	Patients in the 45 mg group showed the highest progression-free survival, while the 15 mg group had lower rates of AOEs. In this way, a starting dose of 45 mg per day with a posterior reduction to 15 mg per day after major molecular response promotes a better benefit/risk ratio [70–72, 79]
PhALLCON phase III study (NCT03589326)	Active, not recruiting	Adults ≥18 years old newly diagnosed with Ph+ ALL	Ponatinib group had a significantly higher minimal residual disease-negative complete remission compared to imatinib, being more effective than imatinib for the treatment of newly diagnosed Ph+ ALL patients in combination with vincristine [73, 74, 80, 81]

AOEs: arterial occlusive events; CML: chronic myeloid leukemia; CP-CML: chronic phase of chronic myeloid leukemia; EPIC: evaluation of ponatinib versus imatinib in chronic myeloid leukemia; OPTIC: optimizing ponatinib treatment in CP-CML; PACE: ponatinib in Ph+ ALL and CML evaluation; Ph+ ALL: Philadelphia chromosome-positive acute lymphoblastic leukemia; PhALLCON: Philadelphia chromosome-positive acute lymphoblastic leukemia with concomitant chemotherapy; TKI tyrosine kinase inhibitor

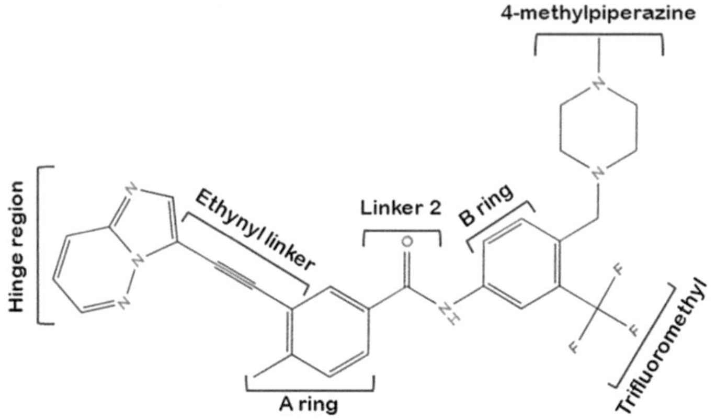

Fig. 18.2 Structure of ponatinib. This diagram illustrates the structural elements of ponatinib, designed to circumvent the T315I mutation in BCR-ABL kinase. Key features, including the hinge region, ethynyl linker, A and B rings, and linker 2 regions, contribute to its potency and binding affinity, enabling effective inhibition despite steric hindrance caused by T315I mutation

The ethynyl linker is a very important part of ponatinib's structure, this triple bond ethynyl linker is what differs its structure compared to other previous TKIs. This linker is responsible for connecting the hinge region and A ring allowing ponatinib to avoid the steric hindrance created by the isoleucine side chain in mutant BCR-ABLT315I [84, 85]. The isoleucine side chain extends towards the hydrophobic pocket active site, blocking the entrance of other TKIs, however, the triple bond ethynyl linker can compass the side chain and bind into the active site of the hydrophobic pocket in a tenfold increased strength compared to linkers with single or double bonds [82, 86].

Following, the A ring is formed of a methylphenyl moiety and is essential for binding with the active site of the hydrophobic pocket in the kinase domain of the mutant BCR-ABLT315I [87]. The linker 2, an amide linkage, is important for the connection between A ring and B ring, spatially arranging ponatinib structure to the BCR-ABL kinase domain [88]. At last, B ring is important to enhance the inhibitory potency of ponatinib against BCR-ABLT315I, consisting of a trifluoromethyl ring bound to a 4-methylpiperazine [87]. The piperazine group is important for hydrogen bond formation with some residues of BCR-ABL protein, allowing B ring to bind in aspartate-phenylalanine-glycine (DFG)-out pocket of the BCR-ABL active site [86, 88]. In this way, the ponatinib structure retain its particularities while sharing similarities with previous TKIs, which allows it to be effective against other BCR-ABL mutations as well as native BCR-ABL oncoprotein.

Ponatinib's daily intake dosage ranges from 15 to 45 mg through the oral route, with peak concentration levels being reached after 6 h of administration. It is known that ponatinib does not interact with food, however, higher stomach pH levels are associated with decreased bioavailability of the drug [89]. Furthermore, ponatinib showed a dose-dependent half-life clearance with doses higher than 30 mg having a half-life of 24 h, with most of its excretion happening through feces and some through urine [90].

The metabolization of the drug happens in the liver through multiple human cytochrome P450 (CYP) enzymes, such as CYP3A4, CYP3A5, CYP2C8, and CYP2D6, mainly by CYP3A4. Therefore, medications such as azoles (CYP3A4 inhibitors) or rifampicin (CYP3A4 inducers) can interfere with ponatinib metabolization, disrupting its efficacy when used together, and is not recommended for coadministration with ponatinib [37, 89, 91]. In an experiment with Sprague Dawley rats administered with ponatinib, it was observed that thirteen different phase I metabolites and one phase II metabolite were excreted in urine with at least 64% of the dosage going through phase I and II metabolism [92]. Metabolized ponatinib has a broad distribution in the body, with higher concentrations in the thyroid and lungs and lower concentrations in the liver, brain, and bones [37]. Patients with previous hepatic impairment showed a higher occurrence of adverse reactions. Additionally, animal studies showed that ponatinib has fetal toxicity, lacks safety during lactation, and might promote infertility in females [90].

As mentioned earlier, ponatinib can inhibit not only the mutated BCR-ABLT315I kinase but the native type as well as other single-point mutations, making ponatinib a pan-mutational TKI for BCR-ABL kinase. Furthermore, ponatinib can inhibit other types of TKs from different families, including vascular endothelial growth factor receptor (VEGFR), platelet-derived growth factor receptor (PDGFR), erythropoietin-producing human hepatocellular (EPH) receptors, steroid receptor coactivator (SRC) families of kinases, mast/stem cell growth factor receptor (KIT), rearranged during transfection (RET), fibroblast growth factor receptor (FGFR), tunica interna endothelial cell kinase 2 (TIE2), and Feline McDonough Sarcoma-like tyrosine kinase 3 (FLT3) [60, 90, 93–100].

Ponatinib is capable of binding to FGFR in a similar way as it binds to BCR-ABL kinase in the DFG motif of the kinase domain through hydrogen bonds, leading to inhibition of the kinase [101]. Ponatinib can also bind to PDGFRα, RET, and KIT through multiple hydrogen bonds at different residues of their kinase domain, resulting in inhibition of triple-negative breast cancer cell line Hs578T proliferation, reduction of mitotic activity and RET phosphorylation in medullary thyroid carcinoma cells, and inhibition of KIT kinase in mast cell leukemia cells, respectively [102–104].

Therefore, the multiple TKs inhibition performed by ponatinib led to many studies and clinical trials of ponatinib against different types of hematological and solid tumors, other than CML and Ph+ ALL [37], which are listed in Table 18.2.

Ponatinib-Induced Cardiovascular Toxicity

Since ponatinib is a TKI that can inhibit multiple tyrosine kinases it can affect different organs and systems driving many side effects that vary from skin problems, leukopenia, and thrombocytopenia to heart failure [90, 118–120]. The most serious

Table 18.2 Clinical trials of ponatinib in other types of cancer. An overview of clinical trials investigating ponatinib's off-label uses in various hematologic and solid tumors beyond CML and Ph+ ALL

Type of cancers	Type of patients	Targeted kinases	Clinical trials
Acute myeloid leukemia	Adults ≥18 years old	Mutated FLT3	NCT02428543 and NCT02829840 [105, 106]
Advanced solid tumors	Adults ≥18 years old	Multiple kinases	NCT02272998 [107]
Cholangiocarcinoma	Adults ≥18 years old	Mutated FGFR2	NCT02265341 [108]
Endometrial carcinoma	Adults ≥18 years old	Mutated FGFR2	NCT01888562 [107]
Gastrointestinal stromal tumor	Adults ≥18 years old	Mutated KIT	NCT03171389 and NCT01874665 [109–111]
Glioblastoma	Adults ≥18 years old	Mutated KIT	NCT02478164 [112]
Lung cancers	Adults ≥18 years old	Mutated FGFR1	NCT01935336 [113]
Medullary thyroid cancers	Adults ≥18 years old	Mutated RET	NCT01838642 [107]
Non-small cell lung cancer	Adults ≥18 years old	Mutated FGFR1 and RET	NCT01813734, NCT01761747, and NCT03704688 [114–116]
Refractory leukemias, lymphomas, or solid tumors	Children ≥1 to <18 years old	Multiple kinases	NCT01281735 and NCT03934372 [49, 117]

FGFR1: fibroblast growth factor receptor 1; FGFR2: fibroblast growth factor receptor 2; FLT3: feline McDonough sarcoma-like tyrosine kinase 3; KIT: mast/stem cell growth factor receptor; RET: rearranged during transfection

side effects related to ponatinib usage were liver failure, vascular occlusive events (VOE), hypertension, and heart failure [90], which are summarized in Table 18.3.

Venous thromboembolic events have been reported in different clinical trials such as PhALLCON and PACE, in which serious or severe VTEs were observed, including venous thrombosis, embolism, jugular vein thrombosis, and retinal vein thrombosis with vision loss experienced [90, 123, 131].

Arterial occlusive events were the most prominent fatal side effects related to ponatinib during the clinical trials. An *in vitro* study showed that ponatinib induced arterial apoptosis, thrombosis, and activation of platelets [132]. An *in vivo* study showed that WT mice treated with ponatinib had five to sixfold higher von Willebrand factor (VWF) expression and platelet adhesion. While, ApoE−/− mice treated with ponatinib showed VWF and platelet signals two to four times higher compared to control, showing that ponatinib displays vascular toxicity through VWF-mediated platelet adhesion leading to a secondary microvascular angiopathy [133]. Moreover, in PACE clinical trial 26% of all patients experienced AOEs with 14% of all patients having grade 3 or 4 events and 2% experiencing fatal AOEs.

Table 18.3 Ponatinib's severe side effects. A table displaying cardiovascular and hepatic side effects observed in ponatinib clinical trials with CML and Ph+ ALL patients, detailing the severity and frequency of adverse effects in each clinical trial. OPTIC includes patients who had a starting dose of 45 mg with a posterior reduction to 15 mg

Adverse side effect	Clinical trial	Incidence of events
Liver failure [90, 121, 122]	PACE	32% of 449 patients (13% of 449 patients displayed serious or severe events)
	OPTIC	28% of 94 patients (6% of 94 patients displayed serious or severe events)
	PhALLCON	66% of 163 patients (30% of 163 patients displayed serious or severe events)
VTEs [90, 123]	PACE	6% of 449 patients (5.8% of 449 patients displayed serious or severe events)
	OPTIC	1.1% of 94 patients (no serious or severe events reported)
	PhALLCON	12% of 163 patients (3.1% of 163 patients displayed serious or severe events)
AOEs [76, 90, 124]	PACE	26% of 449 patients (14% of 449 patients displayed serious or severe events)
	OPTIC	14% of 94 patients (6% of 94 patients displayed serious or severe events)
	PhALLCON	6% of 163 patients (3.7% of 163 patients displayed serious or severe events)
Hypertension [90, 125–127]	PACE	32% of 449 patients (13% of 449 patients displayed serious or severe events)
	OPTIC	32% of 94 patients (12% of 94 patients displayed serious or severe events)
	PhALLCON	34% of 163 patients (14% of 163 patients displayed serious or severe events)
Heart failure [90, 128–130]	PACE	9% of 449 patients (7% of 449 patients displayed serious or severe events)
	OPTIC	13% of 94 patients (1.1% of 94 patients displayed serious or severe events)
	PhALLCON	6% of 163 patients (1.2% of 163 patients displayed serious or severe events)

AOEs: arterial occlusive events; OPTIC: optimizing ponatinib treatment in CP-CML; PACE: ponatinib in Ph+ ALL and CML evaluation; PhALLCON: Philadelphia chromosome-positive acute lymphoblastic leukemia with concomitant chemotherapy; VTEs: venous thromboembolic events

Furthermore, it was observed that during the PACE trial, some patients had recurrent or multisite vascular occlusion. In the OPTIC trial, 14% of all patients had AOEs, 6% experienced grade 3 or 4 events with 2.1% having fatal AOEs. Finally, in PhALLCON, 6% of all patients developed AOEs with 3.7% of all patients having grade 3 or 4 events. 0.6% of the cohort in the PhALLCON trial experienced fatal AOEs. Events reported included peripheral vascular, cardiovascular, and cerebrovascular events, being more frequent with advanced age. Common risk factors included hypercholesterolemia, non-ischemic cardiac disease, and hypertension.

However, younger patients who haven't reported previous risk factors also developed AOEs [76, 90, 124, 134, 135].

Hypertension was documented in various clinical trials, including instances of hypertensive crises. Tyrosine kinase inhibitors (TKIs) are linked to the development of hypertension, particularly those that interact with the VEGF signaling pathway, with ponatinib being a notable example. A meta-analysis showed that ponatinib carries the highest risk of developing hypertension compared to other TKIs, with a pooled incidence rate of 17%. In clinical trials, patients with chronic myeloid leukemia (CML) reported hypertensive events at an incidence ranging from 20% to 30% [125–127, 136].

Another prominent side effect of ponatinib use was the development of heart failure, including severe and fatal heart failure events. In the PACE trial, 9% of 449 patients developed some type of heart failure event, with 7% having serious or severe (grade 3 or 4) events. In addition, decreased ejection fraction and congestive cardiac failure, were the most frequent events recorded during the PACE trial. Following, in OPTIC, 13% of 94 patients developed heart failure events with 1.1% experiencing grade 3 or 4 events. OPTIC patients had left ventricular hypertrophy and increased levels of brain natriuretic peptide (BNP) as the most common heart failure event. In PhALLCON clinical trial 6% of 163 patients experienced heart failure events, with 1.2% of the cohort having serious or severe events. Like the OPTIC trial, the most common heart failure event in PhALLCON was increased levels of BNP [90, 128–130].

Therefore, ponatinib usage can lead to many side effects related to the cardiovascular system, being potentially fatal in some cases of VOEs and heart failure.

Mechanisms of Ponatinib-Induced Cardiotoxicity

Recent studies using various models have examined the molecular mechanisms of ponatinib-induced cardiotoxicity, with their findings summarized in Fig. 18.3.

In a recent study conducted by investigators at the University of Chicago Medical Center, it was found that ponatinib can induce endothelial permeability and dysfunction in vitro through the activation of Rho-associated protein kinase (ROCK). The researchers collected peripheral blood mononuclear cells from 53 adult patients diagnosed with either chronic myeloid leukemia (CML) or Philadelphia chromosome-positive acute lymphoblastic leukemia (Ph+ ALL), all of whom were treated with various generations of tyrosine kinase inhibitors (TKIs), including ponatinib. Notably, patients treated with ponatinib exhibited increased ROCK activity compared to the control group, which was linked to a higher incidence of cardiovascular adverse events (CVAEs) [137]. Other than this study there is a lack of clinical data on the mechanism of ponatinib-induced cardiotoxicity. Thus, in the following paragraphs, the review will discuss major findings from recent *in vitro* and *in vivo* studies.

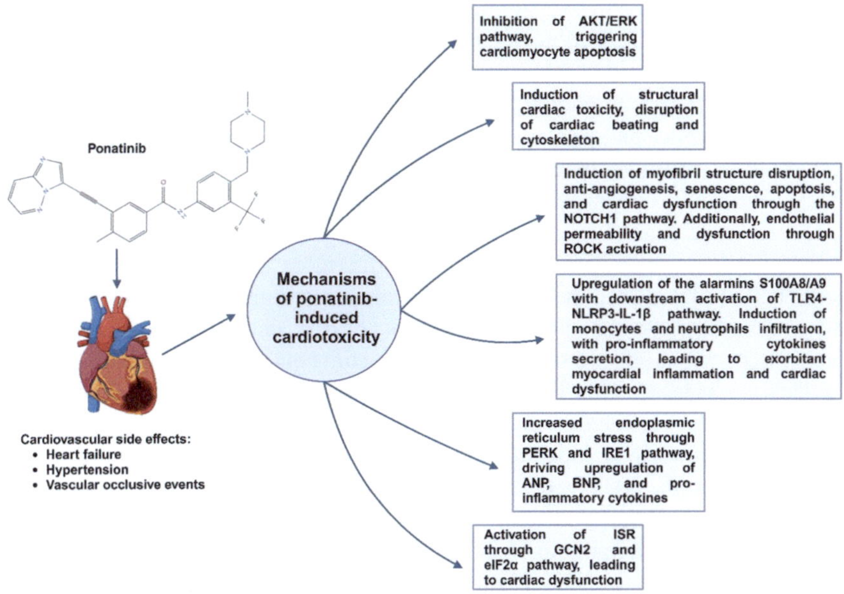

Fig. 18.3 Mechanisms of ponatinib-induced cardiotoxicity. This figure illustrates the molecular pathways implicated in ponatinib-induced cardiotoxicity. (Created with BioRender.com). Akt: protein kinase B; ANP: atrial natriuretic peptide; BNP: brain natriuretic peptide; eIF2α: eukaryotic translation initiation factor 2 alpha subunit; ERK: extracellular signal-regulated kinase; GCN2: general control nonderepressible 2; IL-1β: interleukin-1 beta; IRE1: serine/threonine-protein kinase/endoribonuclease; ISR: integrated stress response; NLRP3: nucleotide-binding domain, leucine-rich–containing family, pyrin domain–containing-3; NOTCH1: neurogenic locus notch homolog protein 1; PERK: protein kinase R-like endoplasmic reticulum Kinase; ROCK: rho-associated protein kinase; S100A8: S100 calcium-binding protein A8; S100A9: S100 calcium-binding protein A9; TLR4: Toll-like receptor 4

A study has identified that ponatinib inhibits AKT/extracellular signal-regulated kinase (ERK) cell survival signaling pathway, triggering apoptosis in cardiomyocytes, resulting in reduced fractional shortening in zebrafish model [138], while another zebrafish model study revealed that ponatinib showed a dose-dependent heart toxicity with increased expression of cardiac failure markers atrial natriuretic peptide (ANP) and BNP [139]. Another study with human-induced pluripotent stem cell-derived cardiomyocytes (hiPSC-CM) showed that ponatinib can inhibit cell survival signaling pathways as well as induce structural cardiac toxicity, disrupting cardiac beating and cytoskeleton [140]. Additionally, it was observed that ponatinib can exert its cardiotoxic effects through the neurogenic locus notch homolog protein 1 (NOTCH1) pathway, leading to cardiac dysfunction in females and males. Male mice were more susceptible to ponatinib-induced cardiotoxicity due to a different modulation of cardiac connexins [141]. Furthermore, in another study from the same group, ponatinib induced apoptosis, myofibril structure disruption, and anti-angiogenesis in the heart, through blockage of the

NOTCH1 signaling pathway. Ultimately, they observed that ejection fraction was reduced in ponatinib-treated mice [142].

A recent study revealed that wild-type (WT) mice treated with ponatinib did not exhibit any reduction in cardiac function. However, in a model featuring high-fat diet ApoE−/− mice and pressure overload WT mice, which mimicked some of the comorbidities observed in clinical trial patients experiencing heart failure side effects, cardiac dysfunction was evident. It was also found that high-fat diet ApoE−/− mice treated with ponatinib showed an upregulation of the alarmins calcium-binding protein A8 (A8) and A9. Furthermore, the pressure overload WT model demonstrated an upregulation of alarmins that triggered a downstream activation cascade involving toll-like receptor 4 (TLR4), nucleotide-binding domain, leucine-rich–containing family, pyrin domain–containing-3 (NLRP3), and interleukin-1 beta (IL-1β). This cascade led to the secretion of pro-inflammatory cytokines and significant myocardial inflammation, characterized by the upregulation of various inflammatory cells such as monocytes and neutrophils [143]. Wang et al., 2023, recently showed that ponatinib can drive endoplasmic reticulum stress through both protein kinase R-like endoplasmic reticulum kinase (PERK) and serine/threonine-protein kinase/endoribonuclease (IRE1), driving ANP and BNP expression as well pro-inflammatory cytokines, ultimately leading to cardiotoxic effects [144]. Finally, a recent study showed that ponatinib can activate the integrated stress response (ISR), through activation of general control nonderepressible 2 (GCN2) and eukaryotic translation initiation factor 2 alpha subunit (eIF2α), being a major contributor to cardiac dysfunction both *in vitro* and *in vivo* [145].

Thus, recent studies have shown that ponatinib-induced cardiotoxicity is related to cardiac inflammation and stress response, driving cardiac cell death and structural disruption through different signaling pathways. However, more studies are necessary to understand the full picture of ponatinib-induced cardiotoxicity and possible ways to circumvent ponatinib side effects on the heart.

Conclusions and Future Perspectives

Ponatinib is a third-generation TKI developed specifically to treat CML and Ph+ ALL patients who are carriers of the T315I mutation [146, 147]. Although ponatinib has proven to be effective against T315I mutation, being able to avoid the steric hindrance in the ATP binding region created by the mutation, its structure can bind and inhibit many tyrosine kinases, leading to off-target side effects [83, 87, 148]. These side effects are more observed in the heart where ponatinib can target different pathways related to cardiac cell survival, inflammation, and stress response, leading to cardiotoxic effects [138, 143, 145]. Nowadays, different studies are searching for alternatives to counteract ponatinib toxic side effects, especially focusing on cardiac toxicity, to provide better options for patients whose only treatment option remains ponatinib.

A recent study showed that empagliflozin a sodium/glucose cotransporter 2 (SGLT2) inhibitor was able to restore cardiac autophagy disrupted by ponatinib, increase cell viability, and ameliorate cardiac function disrupted in ponatinib-treated mice [149]. Moreover, the use of siRNA-NOTCH1 was able to prevent ponatinib-induced cardiac apoptosis, senescence, and vascular toxicity [142]. Neuregulin-1β an enhancer of AKT/ERK expression was able to rescue cardiomyocytes from apoptosis under ponatinib treatment [138]. Another recent study showed that dexamethasone, an anti-inflammatory corticosteroid, inhibited NLRP3 inflammasome signaling, leading to decreased cardiac inflammation and pro-inflammatory M1 macrophage proliferation, as well as cardiac function preservation. They also observed that paquinimod, an S100A8/S100A9 inhibitor, and CY-09, an NLRP3 inflammasome inhibitor, could reduce the proliferation of inflammatory cells in the heart and spleen, reducing cardiac and systemic inflammation promoted by ponatinib, as well as preserve cardiac function under ponatinib treatment. Yan et al., 2024, showed that the administration of an integrated stress response inhibitor (ISRIB) with ponatinib preserved cardiac function, decreased activation of ISR, and reduced cardiomyocyte apoptosis [145]. The authors of a recent study aimed to develop ponatinib analogs with lower cardiovascular toxicity but preserved effectiveness against the mutated BCR-ABLT315I kinase [150]. The researchers noted that two analogs, designated as 33a and 36a, exhibited reduced cardiotoxicity in both in vitro and in vivo studies, leading to decreased levels of troponin-T and troponin-I compared to ponatinib. Both analogs demonstrated the ability to inhibit the proliferation of wild-type (WT) chronic myeloid leukemia (CML) cells and T315I cells, while also preserving the contractility of hiPSC-derived cardiomyocytes. Additionally, they contributed to increased survival rates in mice and decreased tumor growth, similar to ponatinib [150].

Future studies and clinical trials aiming at reducing ponatinib cardiotoxicity while maintaining its effectiveness against the mutated BCR-ABL T315I oncoprotein could significantly improve the quality of life for patients with CML and Ph+ ALL undergoing ponatinib treatment.

Acknowledgments We thank Caitlin Beech, Chandrakala Narasimhulu, and Omonzejie Imaralu for assistance in proofreading the manuscript.

Grants This study was partly supported by Internal Endowed Chair Advent Health support (to D. K.S.).

Disclosures The authors declare no conflicts of interest.

Author Contributions Conceptualization: DKS; Data curation: JMR; Formal analysis: JMR; Methodology: DKS; Resources: DKS; Supervision: DKS; Visualization: DKS; Manuscript original draft: JMR; Manuscript review & editing: DKS; Approved final version of manuscript: DKS.

References

1. Paul MK, Mukhopadhyay AK (2004) Tyrosine kinase—role and significance in cancer. Int J Med Sci 1(2):101–115. https://doi.org/10.7150/ijms.1.101
2. Mongre RK, Mishra CB, Shukla AK, Prakash A, Jung S, Ashraf-Uz-Zaman M, Lee MS (2021) Emerging importance of tyrosine kinase inhibitors against cancer: quo vadis to cure? Int J Mol Sci 22(21):11659. https://doi.org/10.3390/ijms222111659
3. Robinson DR, Wu YM, Lin SF (2000) The protein tyrosine kinase family of the human genome. Oncogene 19(49):5548–5557. https://doi.org/10.1038/sj.onc.1203957
4. Schlessinger J (2000) Cell signaling by receptor tyrosine kinases. Cell 103(2):211–225. https://doi.org/10.1016/S0092-8674(00)00114-8
5. Hubbard SR (1999) Structural analysis of receptor tyrosine kinases. Prog Biophys Mol Biol 71(3–4):343–358. https://doi.org/10.1016/s0079-6107(98)00047-9
6. Manning G, Plowman GD, Hunter T, Sudarsanam S (2002) Evolution of protein kinase signaling from yeast to man. Trends Biochem Sci 27(10):514–520. https://doi.org/10.1016/s0968-0004(02)02179-5
7. Pottier C, Fresnais M, Gilon M, Jérusalem G, Longuespée R, Sounni NE (2020) Tyrosine kinase inhibitors in cancer: breakthrough and challenges of targeted therapy. Cancers (Basel) 12(3):731. https://doi.org/10.3390/cancers12030731
8. Du Z, Lovly CM (2018) Mechanisms of receptor tyrosine kinase activation in cancer. Mol Cancer 17(1):58. https://doi.org/10.1186/s12943-018-0782-4
9. Pawson T, Gish GD, Nash P (2001) SH2 domains, interaction modules and cellular wiring. Trends Cell Biol 11(12):504–511. https://doi.org/10.1016/s0962-8924(01)02154-7
10. Siveen KS, Prabhu KS, Achkar IW, Kuttikrishnan S, Shyam S, Khan AQ et al (2018) Role of non receptor tyrosine kinases in hematological malignances and its targeting by natural products. Mol Cancer 17(1):31. https://doi.org/10.1186/s12943-018-0788-y
11. Li E, Hristova K (2006) Role of receptor tyrosine kinase transmembrane domains in cell signaling and human pathologies. Biochemistry 45(20):6241–6251. https://doi.org/10.1021/bi060609y
12. Schenk PW, Snaar-Jagalska BE (1999) Signal perception and transduction: the role of protein kinases. Biochim Biophys Acta 1449(1):1–24. https://doi.org/10.1016/s0167-4889(98)00178-5
13. Eshaq AM, Flanagan TW, Hassan SY, Al Asheikh SA, Al-Amoudi WA, Santourlidis S et al (2024) Non-receptor tyrosine kinases: their structure and mechanistic role in tumor progression and resistance. Cancers (Basel) 16(15):2754. https://doi.org/10.3390/cancers16152754
14. Ana A, Susana S, José R (2019) Non-receptor tyrosine kinases role and significance in hematological malignancies. In: Huan R (ed) Tyrosine kinases as druggable targets in cancer. IntechOpen, Rijeka, p 2
15. Fouad YA, Aanei C (2017) Revisiting the hallmarks of cancer. Am J Cancer Res 7(5):1016–1036

16. McDonell LM, Kernohan KD, Boycott KM, Sawyer SL (2015) Receptor tyrosine kinase mutations in developmental syndromes and cancer: two sides of the same coin. Hum Mol Genet 24(R1):R60–R66. https://doi.org/10.1093/hmg/ddv254
17. Yamaoka T, Kusumoto S, Ando K, Ohba M, Ohmori T (2018) Receptor tyrosine kinase-targeted cancer therapy. Int J Mol Sci 19(11):3491. https://doi.org/10.3390/ijms19113491
18. Fabbro D, Cowan-Jacob SW, Moebitz H (2015) Ten things you should know about protein kinases: IUPHAR Review 14. Br J Pharmacol 172(11):2675–2700. https://doi.org/10.1111/bph.13096
19. Norman RA, Toader D, Ferguson AD (2012) Structural approaches to obtain kinase selectivity. Trends Pharmacol Sci 33(5):273–278. https://doi.org/10.1016/j.tips.2012.03.005
20. Zhao Z, Liu Q, Bliven S, Xie L, Bourne PE (2017) Determining cysteines available for covalent inhibition across the human kinome. J Med Chem 60(7):2879–2889. https://doi.org/10.1021/acs.jmedchem.6b01815
21. Amir M, Javed S (2021) A review on the therapeutic role of TKIs in case of CML in combination with epigenetic drugs. Front Genet 12:742802. https://doi.org/10.3389/fgene.2021.742802
22. Metibemu DS, Akinloye OA, Akamo AJ, Ojo DA, Okeowo OT, Omotuyi IO (2019) Exploring receptor tyrosine kinases-inhibitors in cancer treatments. Egypt J Med Hum Genet 20(1):35. https://doi.org/10.1186/s43042-019-0035-0
23. Roskoski R Jr (2024) Properties of FDA-approved small molecule protein kinase inhibitors: a 2024 update. Pharmacol Res 200:107059. https://doi.org/10.1016/j.phrs.2024.107059
24. Wu P, Nielsen TE, Clausen MH (2015) FDA-approved small-molecule kinase inhibitors. Trends Pharmacol Sci 36(7):422–439. https://doi.org/10.1016/j.tips.2015.04.005
25. Deininger MWN, Goldman JM, Melo JV (2000) The molecular biology of chronic myeloid leukemia. Blood 96(10):3343–3356. https://doi.org/10.1182/blood.V96.10.3343
26. Walker LC, Ganesan TS, Dhut S, Gibbons B, Lister TA, Rothbard J, Young BD (1987) Novel chimaeric protein expressed in Philadelphia positive acute lymphoblastic leukaemia. Nature 329(6142):851–853. https://doi.org/10.1038/329851a0
27. Amarante-Mendes GP, Rana A, Datoguia TS, Hamerschlak N, Brumatti G (2022) BCR-ABL1 tyrosine kinase complex signaling transduction: challenges to overcome resistance in chronic myeloid leukemia. Pharmaceutics 14(1):215. https://doi.org/10.3390/pharmaceutics14010215
28. Helgason GV, Karvela M, Holyoake TL (2011) Kill one bird with two stones: potential efficacy of BCR-ABL and autophagy inhibition in CML. Blood 118(8):2035–2043. https://doi.org/10.1182/blood-2011-01-330621
29. Hehlmann R (2020) Chronic myeloid leukemia in 2020. Hemasphere 4(5):e468. https://doi.org/10.1097/hs9.0000000000000468
30. Singh AP, Umbarkar P, Tousif S, Lal H (2020) Cardiotoxicity of the BCR-ABL1 tyrosine kinase inhibitors: emphasis on ponatinib. Int J Cardiol 316:214–221. https://doi.org/10.1016/j.ijcard.2020.05.077
31. Cohen MH, Johnson JR, Pazdur R (2005) U.S. Food and Drug Administration Drug Approval Summary: conversion of imatinib mesylate (STI571; Gleevec) tablets from accelerated approval to full approval. Clin Cancer Res 11(1):12–19
32. Dagher R, Johnson J, Williams G, Keegan P, Pazdur R (2004) Accelerated approval of oncology products: a decade of experience. JNCI J Natl Cancer Inst 96(20):1500–1509. https://doi.org/10.1093/jnci/djh279
33. Conchon M, Freitas CM, Rego MA, Braga Junior JW (2011) Dasatinib - clinical trials and management of adverse events in imatinib resistant/intolerant chronic myeloid leukemia. Rev Bras Hematol Hemoter 33(2):131–139. https://doi.org/10.5581/1516-8484.20110034
34. Kantarjian HM, Giles F, Gattermann N, Bhalla K, Alimena G, Palandri F et al (2007) Nilotinib (formerly AMN107), a highly selective BCR-ABL tyrosine kinase inhibitor, is effective in patients with Philadelphia chromosome-positive chronic myelogenous leukemia in chronic phase following imatinib resistance and intolerance. Blood 110(10):3540–3546. https://doi.org/10.1182/blood-2007-03-080689

35. Jabbour E, Kantarjian H, Cortes J (2015) Use of second- and third-generation tyrosine kinase inhibitors in the treatment of chronic myeloid leukemia: an evolving treatment paradigm. Clin Lymphoma Myeloma Leuk 15(6):323–334. https://doi.org/10.1016/j.clml.2015.03.006
36. Doan V, Wang A, Prescott H (2015) Bosutinib for the treatment of chronic myeloid leukemia. Am J Health Syst Pharm 72(6):439–447. https://doi.org/10.2146/ajhp140221
37. Gao Y, Ding Y, Tai XR, Zhang C, Wang D (2023) Ponatinib: an update on its drug targets, therapeutic potential and safety. Biochim Biophys Acta Rev Cancer 1878(5):188949. https://doi.org/10.1016/j.bbcan.2023.188949
38. Pulte ED, Chen H, Price LSL, Gudi R, Li H, Okusanya OO et al (2022) FDA approval summary: revised indication and dosing regimen for ponatinib based on the results of the OPTIC Trial. Oncologist 27(2):149–157. https://doi.org/10.1093/oncolo/oyab040
39. Pamuk GE, Chow ECY, Ionan AC, Chen H, Lee SL, Hsu V et al (2024) FDA approval summary: asciminib for Ph+ CML in chronic phase treated with two or more tyrosine kinase inhibitors and for the T315I mutation. Clin Cancer Res 30(19):4266–4271. https://doi.org/10.1158/1078-0432.Ccr-24-1086
40. Cortes JE, Sasaki K, Kim D-W, Hughes TP, Etienne G, Mauro MJ et al (2024) Asciminib monotherapy in patients with chronic-phase chronic myeloid leukemia with the T315I mutation after ≥1 prior tyrosine kinase inhibitor: 2-year follow-up results. Leukemia 38(7):1522–1533. https://doi.org/10.1038/s41375-024-02278-8
41. Choi EJ (2023) Asciminib: the first-in-class allosteric inhibitor of BCR::ABL1 kinase. Blood Res 58(S1):S29–S36. https://doi.org/10.5045/br.2023.2023017
42. Réa D, Mauro MJ, Boquimpani C, Minami Y, Lomaia E, Voloshin S et al (2021) A phase 3, open-label, randomized study of asciminib, a STAMP inhibitor, vs bosutinib in CML after 2 or more prior TKIs. Blood 138(21):2031–2041. https://doi.org/10.1182/blood.2020009984
43. Lee JW, Chung NG (2011) The treatment of pediatric chronic myelogenous leukemia in the imatinib era. Korean J Pediatr 54(3):111–116. https://doi.org/10.3345/kjp.2011.54.3.111
44. Millot F, Baruchel A, Guilhot J, Petit A, Leblanc T, Bertrand Y et al (2011) Imatinib is effective in children with previously untreated chronic myelogenous leukemia in early chronic phase: results of the french national phase IV trial. J Clin Oncol 29(20):2827–2832. https://doi.org/10.1200/jco.2010.32.7114
45. Suttorp M, Schulze P, Glauche I, Göhring G, von Neuhoff N, Metzler M et al (2018) Front-line imatinib treatment in children and adolescents with chronic myeloid leukemia: results from a phase III trial. Leukemia 32(7):1657–1669. https://doi.org/10.1038/s41375-018-0179-9
46. Gore L, Kearns PR, de Martino ML, Lee, Souza CAD, Bertrand Y et al (2018) Dasatinib in pediatric patients with chronic myeloid leukemia in chronic phase: results from a phase II trial. J Clin Oncol 36(13):1330–1338. https://doi.org/10.1200/jco.2017.75.9597
47. Hijiya N, Maschan A, Rizzari C, Shimada H, Dufour C, Goto H et al (2023) The long-term efficacy and safety of nilotinib in pediatric patients with CML: a 5-year update of the DIALOG study. Blood Adv 7(23):7279–7289. https://doi.org/10.1182/bloodadvances.2023010122
48. Hijiya N, Maschan A, Rizzari C, Shimada H, Dufour C, Goto H et al (2021) A phase 2 study of nilotinib in pediatric patients with CML: long-term update on growth retardation and safety. Blood Adv 5(14):2925–2934. https://doi.org/10.1182/bloodadvances.2020003759
49. Brivio E, Pennesi E, Willemse ME, Huitema ADR, Jiang Y, van Tinteren HDR et al (2024) Bosutinib in resistant and intolerant pediatric patients with chronic phase chronic myeloid leukemia: results from the phase I part of study ITCC054/COG AAML1921. J Clin Oncol 42(7):821–831. https://doi.org/10.1200/jco.23.00897
50. Hoy SM (2024) Bosutinib: pediatric first approval. Paediatr Drugs 26(2):209–214. https://doi.org/10.1007/s40272-023-00608-4
51. Cortes J, Lang F (2021) Third-line therapy for chronic myeloid leukemia: current status and future directions. J Hematol Oncol 14(1):44. https://doi.org/10.1186/s13045-021-01055-9
52. Senapati J, Sasaki K, Issa GC, Lipton JH, Radich JP, Jabbour E, Kantarjian HM (2023) Management of chronic myeloid leukemia in 2023 – common ground and common sense. Blood Cancer J 13(1):58. https://doi.org/10.1038/s41408-023-00823-9

53. Padala S, Cortes J (2023) Asciminib in chronic myeloid leukemia: a STAMP for expedited delivery? Haematologica 108(11):2913–2918. https://doi.org/10.3324/haematol.2022.282361
54. Nicolini FE, Ibrahim AR, Soverini S, Martinelli G, Müller MC, Hochhaus A et al (2013) The BCR-ABLT315I mutation compromises survival in chronic phase chronic myelogenous leukemia patients resistant to tyrosine kinase inhibitors, in a matched pair analysis. Haematologica 98(10):1510–1516. https://doi.org/10.3324/haematol.2012.080234
55. Mian AA, Schüll M, Zhao Z, Oancea C, Hundertmark A, Beissert T et al (2009) The gatekeeper mutation T315I confers resistance against small molecules by increasing or restoring the ABL-kinase activity accompanied by aberrant transphosphorylation of endogenous BCR, even in loss-of-function mutants of BCR/ABL. Leukemia 23(9):1614–1621. https://doi.org/10.1038/leu.2009.69
56. Shyam Sunder S, Sharma UC, Pokharel S (2023) Adverse effects of tyrosine kinase inhibitors in cancer therapy: pathophysiology, mechanisms and clinical management. Signal Transduct Target Ther 8(1):262. https://doi.org/10.1038/s41392-023-01469-6
57. Bose P, Park H, Al-Khafaji J, Grant S (2013) Strategies to circumvent the T315I gatekeeper mutation in the Bcr-Abl tyrosine kinase. Leuk Res Rep 2(1):18–20. https://doi.org/10.1016/j.lrr.2013.02.001
58. Hoy SM (2014) Ponatinib: a review of its use in adults with chronic myeloid leukaemia or Philadelphia chromosome-positive acute lymphoblastic leukaemia. Drugs 74(7):793–806. https://doi.org/10.1007/s40265-014-0216-6
59. Tan FH, Putoczki TL, Stylli SS, Luwor RB (2019) Ponatinib: a novel multi-tyrosine kinase inhibitor against human malignancies. Onco Targets Ther 12:635–645. https://doi.org/10.2147/ott.S189391
60. O'Hare T, Shakespeare WC, Zhu X, Eide CA, Rivera VM, Wang F et al (2009) AP24534, a pan-BCR-ABL inhibitor for chronic myeloid leukemia, potently inhibits the T315I mutant and overcomes mutation-based resistance. Cancer Cell 16(5):401–412. https://doi.org/10.1016/j.ccr.2009.09.028
61. Cortes JE, Kantarjian H, Shah NP, Bixby D, Mauro MJ, Flinn I et al (2012) Ponatinib in refractory Philadelphia chromosome-positive leukemias. N Engl J Med 367(22):2075–2088. https://doi.org/10.1056/NEJMoa1205127
62. Mauro MJ, Cortes JE, Kantarjian HM, Shah NP, Bixby D, Flinn I et al (2013) Safety and durability of ponatinib in patients with Philadelphia chromosome-positive (Ph+) leukemia: long-term follow-up of an ongoing phase I study. J Clin Oncol 31(15 Suppl):7063. https://doi.org/10.1200/jco.2013.31.15_suppl.7063
63. Cortes JE, Kim DW, Pinilla-Ibarz J, le Coutre P, Paquette R, Chuah C et al (2013) A phase 2 trial of ponatinib in Philadelphia chromosome-positive leukemias. N Engl J Med 369(19): 1783–1796. https://doi.org/10.1056/NEJMoa1306494
64. Cortes JE, Kim D-W, Pinilla-Ibarz J, Paquette R, le Coutre PD, Chuah C et al (2012) PACE: a pivotal phase II trial of ponatinib in patients with CML and Ph+ALL resistant or intolerant to dasatinib or nilotinib, or with the T315I mutation. J Clin Oncol 30(15 Suppl):6503. https://doi.org/10.1200/jco.2012.30.15_suppl.6503
65. Prasad V, Mailankody S (2014) The accelerated approval of oncologic drugs: lessons from ponatinib. JAMA 311(4):353–354. https://doi.org/10.1001/jama.2013.284531
66. Senior M (2014) FDA halts then allows sales of Ariad's leukemia medication. Nat Biotechnol 32(1):9–11. https://doi.org/10.1038/nbt0114-9a
67. Cortes JE, Kim DW, Pinilla-Ibarz J, le Coutre PD, Paquette R, Chuah C et al (2018) Ponatinib efficacy and safety in Philadelphia chromosome-positive leukemia: final 5-year results of the phase 2 PACE trial. Blood 132(4):393–404. https://doi.org/10.1182/blood-2016-09-739086
68. Lipton JH, Chuah C, Guerci-Bresler A, Rosti G, Simpson D, Assouline S et al (2016) Ponatinib versus imatinib for newly diagnosed chronic myeloid leukaemia: an international, randomised, open-label, phase 3 trial. Lancet Oncol 17(5):612–621. https://doi.org/10.1016/s1470-2045(16)00080-2

69. Poch Martell M, Sibai H, Deotare U, Lipton JH (2016) Ponatinib in the therapy of chronic myeloid leukemia. Expert Rev Hematol 9(10):923–932. https://doi.org/10.1080/17474086.2016.1232163
70. Cortes JE, Apperley J, Lomaia E, Moiraghi B, Sutton MU, Pavlovsky C et al (2021) OPTIC primary analysis: a dose-optimization study of 3 starting doses of ponatinib (PON). J Clin Oncol 39(15 Suppl):7000. https://doi.org/10.1200/JCO.2021.39.15_suppl.7000
71. Cortes J, Deininger M, Lomaia E, Moiraghi B, Undurraga S, Pavlovsky C et al (2023) Long-term results from the optic trial: a dose-optimization study of 3 starting doses of ponatinib. Blood 142:3164. https://doi.org/10.1182/blood-2023-178790
72. Jabbour E, Apperley J, Cortes J, Rea D, Deininger M, Abruzzese E et al (2024) Dose modification dynamics of ponatinib in patients with chronic-phase chronic myeloid leukemia (CP-CML) from the PACE and OPTIC trials. Leukemia 38(3):475–481. https://doi.org/10.1038/s41375-024-02159-0
73. Aldoss I, Ribera J-M, Kantarjian H, Montesinos P, Leonard JT, Gomez-Almaguer D et al (2023) Ponatinib versus imatinib in patients with newly diagnosed Ph+ ALL: subgroup analysis of the phase 3 Phallcon Study. Blood 142(Suppl 1):2871. https://doi.org/10.1182/blood-2023-179537
74. Jabbour E, Kantarjian HM, Aldoss I, Montesinos P, Leonard JT, Gomez-Almaguer D et al (2023) First report of PhALLCON: a phase 3 study comparing ponatinib (pon) vs imatinib (im) in newly diagnosed patients with Philadelphia chromosome-positive acute lymphoblastic leukemia (Ph+ALL). J Clin Oncol 41(36 Suppl):398868. https://doi.org/10.1200/JCO.2023.41.36_suppl.398868
75. Cortes J, Talpaz M, Bixby D, Deininger M, Shah N, Flinn IW et al (2010) A phase 1 trial of oral ponatinib (AP24534) in patients with refractory chronic myelogenous leukemia (CML) and other hematologic malignancies: emerging safety and clinical response findings. Blood 116(21):210. https://doi.org/10.1182/blood.V116.21.210.210
76. Januzzi JL, Garasic JM, Kasner SE, McDonald V, Petrie MC, Seltzer J et al (2022) Retrospective analysis of arterial occlusive events in the PACE trial by an independent adjudication committee. J Hematol Oncol 15(1):1. https://doi.org/10.1186/s13045-021-01221-z
77. Lipton JH, Deininger MWN, Lustgarten S, Turner CD, Rivera VM, Clackson T et al (2013) EPIC: a phase III randomized, open-label study of ponatinib versus imatinib in adult patients with newly diagnosed chronic myeloid leukemia in chronic phase. J Clin Oncol 31(15 Suppl):TPS7129. https://doi.org/10.1200/jco.2013.31.15_suppl.tps7129
78. Lipton JH, Chuah C, Guerci-Bresler A, Rosti G, Simpson D, Assouline S et al (2014) Epic: a phase 3 trial of ponatinib compared with imatinib in patients with newly diagnosed chronic myeloid leukemia in chronic phase (CP-CML). Blood 124(21):519. https://doi.org/10.1182/blood.V124.21.519.519
79. Cortes J, Apperley J, Lomaia E, Moiraghi B, Undurraga Sutton M, Pavlovsky C et al (2021) Ponatinib dose-ranging study in chronic-phase chronic myeloid leukemia: a randomized, open-label phase 2 clinical trial. Blood 138(21):2042–2050. https://doi.org/10.1182/blood.2021012082
80. Jabbour E, Kantarjian HM, Aldoss I, Montesinos P, Leonard JT, Gomez-Almaguer D et al (2024) In-depth analysis of responders in the phase 3 PhALLCON trial of ponatinib vs imatinib in newly diagnosed Ph+ ALL. J Clin Oncol 42(16 Suppl):6533. https://doi.org/10.1200/JCO.2024.42.16_suppl.6533
81. Jabbour E, Kantarjian HM, Aldoss I, Montesinos P, Leonard JT, Gómez-Almaguer D et al (2024) Ponatinib vs imatinib in frontline philadelphia chromosome-positive acute lymphoblastic leukemia: a randomized clinical trial. JAMA 331(21):1814–1823. https://doi.org/10.1001/jama.2024.4783
82. Huang W-S, Metcalf CA, Sundaramoorthi R, Wang Y, Zou D, Thomas RM et al (2010) Discovery of 3-[2-(imidazo[1,2-b]pyridazin-3-yl)ethynyl]-4-methyl-N-{4-[(4-methylpiperazin-1-yl)methyl]-3-(trifluoromethyl)phenyl}benzamide (AP24534), a potent, orally active pan-inhibitor of breakpoint cluster region-abelson (BCR-ABL) kinase including

the T315I gatekeeper mutant. J Med Chem 53(12):4701–4719. https://doi.org/10.1021/jm100395q
83. Molica M, Scalzulli E, Colafigli G, Foà R, Breccia M (2019) Insights into the optimal use of ponatinib in patients with chronic phase chronic myeloid leukaemia. Ther Adv Hematol 10: 2040620719826444. https://doi.org/10.1177/2040620719826444
84. Pavlovsky C, Chan O, Talati C, Pinilla-Ibarz J (2019) Ponatinib in the treatment of chronic myeloid leukemia and philadelphia chromosome positive acute lymphoblastic leukemia. Future Oncol 15(3):257–269. https://doi.org/10.2217/fon-2018-0371
85. Guo M, Duan Y, Dai S, Li J, Chen X, Qu L et al (2022) Structural study of ponatinib in inhibiting SRC kinase. Biochem Biophys Res Commun 598:15–19. https://doi.org/10.1016/j.bbrc.2022.02.001
86. Zhou T, Commodore L, Huang W-S, Wang Y, Thomas M, Keats J et al (2011) Structural mechanism of the pan-BCR-ABL inhibitor ponatinib (AP24534): lessons for overcoming kinase inhibitor resistance. Chem Biol Drug Des 77(1):1–11. https://doi.org/10.1111/j.1747-0285.2010.01054.x
87. Miller GD, Bruno BJ, Lim CS (2014) Resistant mutations in CML and Ph+ALL—role of ponatinib. Biologics 8:243–254. https://doi.org/10.2147/BTT.S50734
88. Buffa P, Romano C, Pandini A, Massimino M, Tirrò E, Di Raimondo F et al (2014) BCR-ABL residues interacting with ponatinib are critical to preserve the tumorigenic potential of the oncoprotein. FASEB J 28(3):1221–1236. https://doi.org/10.1096/fj.13-236992
89. Jain H, Thorat J, Sengar M, Dubey A (2019) Ponatinib: a drug review. Cancer Res Stat Treat 2(2):190–196. https://doi.org/10.4103/crst.Crst_98_19
90. Takeda Pharmaceuticals America (2022). ICLUSIG® (ponatinib) tablets for oral use. https://www.iclusig.com/sites/default/files/2023-02/iclusig-prescribing-information.pdf. Accessed 3 May 2024
91. Hanley MJ, Diderichsen PM, Narasimhan N, Srivastava S, Gupta N, Venkatakrishnan K (2022) Population pharmacokinetics of ponatinib in healthy adult volunteers and patients with hematologic malignancies and model-informed dose selection for pediatric development. J Clin Pharmacol 62(4):555–567. https://doi.org/10.1002/jcph.1990
92. Attwa MW, Kadi AA, Darwish HW, Amer SM, AlRabiah H (2018) LC-ESI-MS/MS identification and characterization of ponatinib in vivo phase I and phase II metabolites. Clin Chim Acta 485:144–151. https://doi.org/10.1016/j.cca.2018.06.035
93. Li H, Wang Y, Chen Z, Lu J, Pan J, Yu Y et al (2017) Novel multiple tyrosine kinase inhibitor ponatinib inhibits bFGF-activated signaling in neuroblastoma cells and suppresses neuroblastoma growth in vivo. Oncotarget 8(4):5874–5884. https://doi.org/10.18632/oncotarget.11580
94. Whittle SB, Patel K, Zhang L, Woodfield SE, Du M, Smith V, Zage PE (2016) The novel kinase inhibitor ponatinib is an effective anti-angiogenic agent against neuroblastoma. Investig New Drugs 34(6):685–692. https://doi.org/10.1007/s10637-016-0387-y
95. Kanojia D, Garg M, Martinez J, Anand MT, Luty SB, Doan NB et al (2017) Kinase profiling of liposarcomas using RNAi and drug screening assays identified druggable targets. J Hematol Oncol 10(1):173. https://doi.org/10.1186/s13045-017-0540-x
96. Gozgit JM, Wong MJ, Wardwell S, Tyner JW, Loriaux MM, Mohemmad QK et al (2011) Potent activity of ponatinib (AP24534) in models of FLT3-driven acute myeloid leukemia and other hematologic malignancies. Mol Cancer Ther 10(6):1028–1035. https://doi.org/10.1158/1535-7163.Mct-10-1044
97. Ren M, Qin H, Ren R, Cowell JK (2013) Ponatinib suppresses the development of myeloid and lymphoid malignancies associated with FGFR1 abnormalities. Leukemia 27(1):32–40. https://doi.org/10.1038/leu.2012.188
98. Gozgit JM, Squillace RM, Wongchenko MJ, Miller D, Wardwell S, Mohemmad Q et al (2013) Combined targeting of FGFR2 and mTOR by ponatinib and ridaforolimus results in synergistic antitumor activity in FGFR2 mutant endometrial cancer models. Cancer Chemother Pharmacol 71(5):1315–1323. https://doi.org/10.1007/s00280-013-2131-z

99. Li SQ, Cheuk AT, Shern JF, Song YK, Hurd L, Liao H et al (2013) Targeting wild-type and mutationally activated FGFR4 in rhabdomyosarcoma with the inhibitor ponatinib (AP24534). PLoS One 8(10):e76551. https://doi.org/10.1371/journal.pone.0076551
100. Han Y, Gu Z, Wu J, Huang X, Zhou R, Shi C et al (2019) Repurposing ponatinib as a potent agent against KIT mutant melanomas. Theranostics 9(7):1952–1964. https://doi.org/10.7150/thno.30890
101. Fauster A, Rebsamen M, Huber KVM, Bigenzahn JW, Stukalov A, Lardeau CH et al (2015) A cellular screen identifies ponatinib and pazopanib as inhibitors of necroptosis. Cell Death Dis 6(5):e1767. https://doi.org/10.1038/cddis.2015.130
102. Kim S, You D, Jeong Y, Yoon SY, Kim SA, Lee JE (2021) Inhibition of platelet-derived growth factor C and their receptors additionally increases doxorubicin effects in triple-negative breast cancer cells. Eur J Pharmacol 895:173868. https://doi.org/10.1016/j.ejphar.2021.173868
103. De Falco V, Buonocore P, Muthu M, Torregrossa L, Basolo F, Billaud M et al (2013) Ponatinib (AP24534) is a novel potent inhibitor of oncogenic RET mutants associated with thyroid cancer. J Clin Endocrinol Metab 98(5):E811–E819. https://doi.org/10.1210/jc.2012-2672
104. Jin B, Ding K, Pan J (2014) Ponatinib induces apoptosis in imatinib-resistant human mast cells by dephosphorylating mutant D816V KIT and silencing β-catenin signaling. Mol Cancer Ther 13(5):1217–1230. https://doi.org/10.1158/1535-7163.Mct-13-0397
105. Larrosa-Garcia M, Baer MR (2017) FLT3 inhibitors in acute myeloid leukemia: current status and future directions. Mol Cancer Ther 16(6):991–1001. https://doi.org/10.1158/1535-7163.Mct-16-0876
106. Dumas PY, Bertoli S, Bérard E, Largeaud L, Bidet A, Delabesse E et al (2020) Real-world outcomes of patients with refractory or relapsed FLT3-ITD acute myeloid leukemia: a Toulouse-Bordeaux DATAML Registry Study. Cancers (Basel) 12(8):2044. https://doi.org/10.3390/cancers12082044
107. Musumeci F, Greco C, Grossi G, Molinari A, Schenone S (2018) Recent studies on ponatinib in cancers other than chronic myeloid leukemia. Cancers (Basel) 10(11):430. https://doi.org/10.3390/cancers10110430
108. DeLeon T, Alberts SR, McWilliams RR, Hubbard JM, Ahn DH, Bekaii-Saab TS et al (2018) A pilot study of ponatinib in cholangiocarcinoma patients with FGFR2 fusions. J Clin Oncol 36(4 Suppl):TPS532. https://doi.org/10.1200/JCO.2018.36.4_suppl.TPS532
109. George S, von Mehren M, Fletcher JA, Sun J, Zhang S, Pritchard JR et al (2022) Phase II study of ponatinib in advanced gastrointestinal stromal tumors: efficacy, safety, and impact of liquid biopsy and other biomarkers. Clin Cancer Res 28(7):1268–1276. https://doi.org/10.1158/1078-0432.Ccr-21-2037
110. Heinrich MC, Mehren M, von Demetri GD, Fletcher JA, Sun JG, Kerstein D et al (2015) Ponatinib efficacy and safety in patients (pts) with advanced gastrointestinal stromal tumors (GIST) after tyrosine kinase inhibitor (TKI) failure: results from a phase 2 study. J Clin Oncol 33(15 Suppl):10535. https://doi.org/10.1200/jco.2015.33.15_suppl.10535
111. Falkenhorst J, Hamacher R, Reichardt P, Ivanyi P, Kasper B, Hohenberger P et al (2020) Lower-dosing ponatinib in pre-treated GIST: results of the POETIG phase II trial. J Clin Oncol 38(15 Suppl):11536–11536. https://doi.org/10.1200/JCO.2020.38.15_suppl.11536
112. Lee EQ, Muzikansky A, Reardon DA, Dietrich J, Nayak L, Duda DG et al (2018) Phase II trial of ponatinib in patients with bevacizumab-refractory glioblastoma. J Clin Oncol 36(15 Suppl):2032. https://doi.org/10.1200/JCO.2018.36.15_suppl.2032
113. Ng TL, Yu H, Smith D, Boyle TA, York ER, Leedy S et al (2018) Preselection of lung cancer cases using FGFR1 mRNA and gene copy number for treatment with ponatinib. J Clin Oncol 36(15 Suppl):12095. https://doi.org/10.1200/JCO.2018.36.15_suppl.12095
114. Gainor JF, Gadgeel S, Ou SI, Yeap B, Otterson GA, Shaw AT (2020) A phase II study of the multikinase inhibitor ponatinib in patients with advanced, RET-rearranged NSCLC. JTO Clin Res Rep 1(3):100045. https://doi.org/10.1016/j.jtocrr.2020.100045

115. Ren M, Hong M, Liu G, Wang H, Patel V, Biddinger P et al (2013) Novel FGFR inhibitor ponatinib suppresses the growth of non-small cell lung cancer cells overexpressing FGFR1. Oncol Rep 29(6):2181–2190. https://doi.org/10.3892/or.2013.2386
116. Arbour KC, Manchado E, Bott MJ, Ahn L, Tobi Y, Ni AA et al (2022) Phase 1 clinical trial of trametinib and ponatinib in patients with NSCLC harboring KRAS mutations. JTO Clin Res Rep 3(1):100256. https://doi.org/10.1016/j.jtocrr.2021.100256
117. Millot F, Suttorp M, Versluys AB, Kalwak K, Nelken B, Ducassou S et al (2020) Ponatinib in childhood Philadelphia chromosome–positive leukaemias: an international registry of childhood chronic myeloid leukaemia study. Eur J Cancer 136:107–112. https://doi.org/10.1016/j.ejca.2020.05.020
118. Jain P, Kantarjian H, Jabbour E, Gonzalez GN, Borthakur G, Pemmaraju N et al (2015) Ponatinib as first-line treatment for patients with chronic myeloid leukaemia in chronic phase: a phase 2 study. Lancet Haematol 2(9):e376–e383. https://doi.org/10.1016/s2352-3026(15)00127-1
119. Chan O, Talati C, Isenalumhe L, Shams S, Nodzon L, Fradley M et al (2020) Side-effects profile and outcomes of ponatinib in the treatment of chronic myeloid leukemia. Blood Adv 4(3):530–538. https://doi.org/10.1182/bloodadvances.2019000268
120. Viganò M, La Milia M, Grassini MV, Pugliese N, De Giorgio M, Fagiuoli S (2023) Hepatotoxicity of small molecule protein kinase inhibitors for cancer. Cancers (Basel) 15(6):1766. https://doi.org/10.3390/cancers15061766
121. Wang Z, Wang X, Wang Z, Feng Y, Jia Y, Jiang L et al (2021) Comparison of hepatotoxicity associated with new BCR-ABL tyrosine kinase inhibitors vs imatinib among patients with chronic myeloid leukemia: a systematic review and meta-analysis. JAMA Netw Open 4(7): e2120165. https://doi.org/10.1001/jamanetworkopen.2021.20165
122. Boo YL, Liam CCK, Toh SG, Lim SM (2019) Rechallenge of ponatinib in chronic myeloid leukaemia after hepatotoxicity. Hong Kong Med J 25(2):162–163. https://doi.org/10.12809/hkmj187420
123. Haguet H, Douxfils J, Mullier F, Chatelain C, Graux C, Dogné JM (2017) Risk of arterial and venous occlusive events in chronic myeloid leukemia patients treated with new generation BCR-ABL tyrosine kinase inhibitors: a systematic review and meta-analysis. Expert Opin Drug Saf 16(1):5–12. https://doi.org/10.1080/14740338.2017.1261824
124. Caocci G, Mulas O, Abruzzese E, Luciano L, Iurlo A, Attolico I et al (2019) Arterial occlusive events in chronic myeloid leukemia patients treated with ponatinib in the real-life practice are predicted by the Systematic Coronary Risk Evaluation (SCORE) chart. Hematol Oncol 37(3): 296–302. https://doi.org/10.1002/hon.2606
125. Roa-Chamorro R, Jaén-Águila F, Puerta-Puerta JM, Torres-Quintero L, González-Bustos P, Mediavilla-García JD (2021) Arterial hypertension assessment in a population with chronic myeloid leukemia. Sci Rep 11(1):14637. https://doi.org/10.1038/s41598-021-94127-2
126. Mulas O, Caocci G, Mola B, La Nasa G (2021) Arterial hypertension and tyrosine kinase inhibitors in chronic myeloid leukemia: a systematic review and meta-analysis. Front Pharmacol 12:674748. https://doi.org/10.3389/fphar.2021.674748
127. Agarwal M, Thareja N, Benjamin M, Akhondi A, Mitchell GD (2018) Tyrosine kinase inhibitor-induced hypertension. Curr Oncol Rep 20(8):65. https://doi.org/10.1007/s11912-018-0708-8
128. Casavecchia G, Galderisi M, Novo G, Gravina M, Santoro C, Agricola E et al (2020) Early diagnosis, clinical management, and follow-up of cardiovascular events with ponatinib. Heart Fail Rev 25(3):447–456. https://doi.org/10.1007/s10741-020-09926-y
129. Sacha T, Krawczyk K (2025) Ponatinib in the treatment of patients with chronic myeloid leukemia and increased cardiovascular risk: a review of management strategies. Hematol Transfus Cell Ther 47(1):103675. https://doi.org/10.1016/j.htct.2024.04.124
130. Kanbayashi Y, Uchida M, Nakano K, Wakabayashi H, Shimizu T (2023) Evaluation of cardiac adverse events with ponatinib using a spontaneous reporting database. Oncology 101(6):397–405. https://doi.org/10.1159/000529768

131. Groarke JD, Cheng S, Moslehi J (2013) Cancer-drug discovery and cardiovascular surveillance. N Engl J Med 369(19):1779–1781. https://doi.org/10.1056/NEJMp1313140
132. Merkulova A, Mitchell SC, Stavrou EX, Forbes GL, Schmaier AH (2019) Ponatinib treatment promotes arterial thrombosis and hyperactive platelets. Blood Adv 3(15):2312–2316. https://doi.org/10.1182/bloodadvances.2019000034
133. Latifi Y, Moccetti F, Wu M, Xie A, Packwood W, Qi Y et al (2019) Thrombotic microangiopathy as a cause of cardiovascular toxicity from the BCR-ABL1 tyrosine kinase inhibitor ponatinib. Blood 133(14):1597–1606. https://doi.org/10.1182/blood-2018-10-881557
134. Takahashi N, Kondo T, Ikari Y, Fukumoto Y, Hatake K, Masunari A et al (2024) Real-world outcomes of ponatinib treatment in 724 patients with CML and Ph+ ALL: a post-marketing surveillance study with a special interest in arterial occlusive events in Japan. Jpn J Clin Oncol 54(8):930–938. https://doi.org/10.1093/jjco/hyae061
135. Januzzi JL, Garasic J, Kasner S, McDonald V, Petrie MC, Seltzer J et al (2020) An independent review of arterial occlusive events (AOEs) in the ponatinib (PON) phase II PACE trial (NCT01207440) in patients (pts) with Ph+ leukemia. J Clin Oncol 38(15 Suppl):7550. https://doi.org/10.1200/JCO.2020.38.15_suppl.7550
136. Ahmed K, Kaddoura R, Yassin MA (2022) A practical guide to managing hypertension, hyperlipidemia, and hyperglycemia in patients with chronic myeloid leukemia. Front Med (Lausanne) 9:1025392. https://doi.org/10.3389/fmed.2022.1025392
137. Yu B, Osman AEG, Sladojevic N, Prabhu N, Tai HC, Chen D et al (2022) Involvement of rho-associated coiled-coil containing kinase (ROCK) in BCR-ABL1 tyrosine kinase inhibitor cardiovascular toxicity. JACC CardioOncol 4(3):371–383. https://doi.org/10.1016/j.jaccao.2022.06.004
138. Singh AP, Glennon MS, Umbarkar P, Gupte M, Galindo CL, Zhang Q et al (2019) Ponatinib-induced cardiotoxicity: delineating the signalling mechanisms and potential rescue strategies. Cardiovasc Res 115(5):966–977. https://doi.org/10.1093/cvr/cvz006
139. Zakaria ZZ, Suleiman M, Benslimane FM, Al-Badr M, Sivaraman S, Korashy HM et al (2024) Imatinib- and ponatinib-mediated cardiotoxicity in zebrafish embryos and H9c2 cardiomyoblasts. Mol Med Rep 30(4):187. https://doi.org/10.3892/mmr.2024.13311
140. Talbert DR, Doherty KR, Trusk PB, Moran DM, Shell SA, Bacus S (2015) A multi-parameter in vitro screen in human stem cell-derived cardiomyocytes identifies ponatinib-induced structural and functional cardiac toxicity. Toxicol Sci 143(1):147–155. https://doi.org/10.1093/toxsci/kfu215
141. Madonna R, Moscato S, Polizzi E, Pieragostino D, Cufaro MC, Del Boccio P et al (2021) Connexin 43 and Connexin 26 involvement in the ponatinib-induced cardiomyopathy: sex-related differences in a murine model. Int J Mol Sci 22(11):5815. https://doi.org/10.3390/ijms22115815
142. Madonna R, Pieragostino D, Cufaro MC, Del Boccio P, Pucci A, Mattii L et al (2022) Sex-related differential susceptibility to ponatinib cardiotoxicity and differential modulation of the Notch1 signalling pathway in a murine model. J Cell Mol Med 26(5):1380–1391. https://doi.org/10.1111/jcmm.17008
143. Tousif S, Singh AP, Umbarkar P, Galindo C, Wheeler N, Toro Cora A et al (2023) Ponatinib drives cardiotoxicity by S100A8/A9-NLRP3-IL-1β mediated inflammation. Circ Res 132(3):267–289. https://doi.org/10.1161/circresaha.122.321504
144. Wang H, Wang Y, Li J, He Z, Boswell SA, Chung M et al (2023) Three tyrosine kinase inhibitors cause cardiotoxicity by inducing endoplasmic reticulum stress and inflammation in cardiomyocytes. BMC Med 21(1):147. https://doi.org/10.1186/s12916-023-02838-2
145. Yan G, Han Z, Kwon Y, Jousma J, Nukala SB, Prosser BL et al (2024) Integrated stress response potentiates ponatinib-induced cardiotoxicity. Circ Res 134(5):482–501. https://doi.org/10.1161/circresaha.123.323683
146. Müller MC, Cervantes F, Hjorth-Hansen H, Janssen J, Milojkovic D, Rea D, Rosti G (2017) Ponatinib in chronic myeloid leukemia (CML): consensus on patient treatment and

management from a European expert panel. Crit Rev Oncol Hematol 120:52–59. https://doi.org/10.1016/j.critrevonc.2017.10.002
147. Kantarjian HM, Jabbour E, Deininger M, Abruzzese E, Apperley J, Cortes J et al (2022) Ponatinib after failure of second-generation tyrosine kinase inhibitor in resistant chronic-phase chronic myeloid leukemia. Am J Hematol 97(11):1419–1426. https://doi.org/10.1002/ajh.26686
148. Rossari F, Minutolo F, Orciuolo E (2018) Past, present, and future of Bcr-Abl inhibitors: from chemical development to clinical efficacy. J Hematol Oncol 11(1):84. https://doi.org/10.1186/s13045-018-0624-2
149. Mattii L, Moscato S, Ippolito C, Polizzi E, Novo G, Zucchi R et al (2024) Empagliflozin mitigates ponatinib-induced cardiotoxicity by restoring the connexin 43-autophagy pathway. Biomed Pharmacother 178:117278. https://doi.org/10.1016/j.biopha.2024.117278
150. Hnatiuk AP, Bruyneel AAN, Tailor D, Pandrala M, Dheeraj A, Li W et al (2022) Reengineering ponatinib to minimize cardiovascular toxicity. Cancer Res 82(15):2777–2791. https://doi.org/10.1158/0008-5472.Can-21-3652

Chapter 19
Nutrient-Drug Interactions and Cardiotoxicity in Oncologic Patients

Simona Dragan, Dana Man, and Nilima Rajpal Kundnani

Abstract The field of cardio-oncology aims to manage the complex interplay between cancer therapies and cardiovascular health. A critical yet underexplored aspect of this discipline is the impact of food-drug interactions (FDIs) on therapeutic efficacy and safety. This chapter highlights the mechanisms, clinical implications, and management strategies of FDIs in cardio-oncology. Understanding these interactions is essential for optimizing patient outcomes, minimizing adverse effects, and ensuring the effectiveness of both oncologic and cardiovascular treatments.

Keywords Cardio-oncology · Nutrient-drug interaction · Food-drug interaction · Cancer therapies

Introduction

Cancer represents a significant global health challenge, with its prevalence and mortality rates increasing rapidly. In recent years, cancer has emerged as the leading cause of death worldwide. Cancer patients often require a variety of medications throughout their treatment journey, including anticancer drugs, supportive therapies, and other prescribed medicines, as well as herbal remedies and specific products. This extensive use of substances increases the risk of significant drug interactions (DIs). This chapter explores the prevalence, characteristics, and predictors of DIs in cancer patients.

A drug interaction (DI) occurs when the clinical effect of a targeted drug is altered by another agent, impacting its pharmacokinetics, pharmacodynamics, or pharmaceutical properties. Such interactions can reduce drug efficacy or exacerbate adverse effects, ultimately diminishing the patient's quality of life. Cancer patients are especially susceptible to DIs for several reasons. They often receive multiple

S. Dragan · D. Man · N. R. Kundnani (✉)
Department of Cardiology, University of Medicine and Pharmacy "Victor Babes", Timisoara, Romania
e-mail: simona.dragan@umft.ro; man.dana@umft.ro; knilima@umft.ro

D. K. Agrawal et al. (eds.), *Cardiovascular Toxicity*, Advances in Biochemistry in Health and Disease 33, https://doi.org/10.1007/978-3-031-90598-8_19

anticancer agents, are prescribed additional medications for chronic conditions, and use supportive therapies to manage treatment-related side effects or disease symptoms. Additionally, many cancer patients turn to alternative medicines, such as herbal remedies, or over-the-counter (OTC) products. Notably, severe drug-related events, including DIs, are responsible for approximately 4% of cancer-related deaths.

Numerous studies from both developed and developing countries have explored the prevalence, predictors, and characteristics of drug-drug interactions (DDIs) among cancer patients, revealing significant variations in DDI prevalence. These differences are likely attributable to variations in prescribing patterns and the tools used to assess interactions.

Drug-drug interactions refer to events or conditions during pharmacotherapy that can compromise the anticipated therapeutic outcomes for patients.

This chapter examines the prevalence and risk classification of drug-herb and drug-food interactions, while also identifying the predictors of DDIs in oncology patients. Food-drug interactions represent a critical frontier in cardio-oncology, influencing therapeutic efficacy and safety. A deeper understanding of these interactions, coupled with proactive management strategies, is vital for improving outcomes in this vulnerable patient population. Interdisciplinary approaches integrating dietary science and pharmacology are essential to advance care in this field.

The use of herbs, food as medicine, and dietary supplements (HFDSs) has been steadily increasing in Western countries. In regions such as Asia, Africa, and South America, HFDSs form a central part of medical treatments, especially where access to pharmaceutical drugs is limited. Between 1997 and 2015, the prevalence of herbal medicine use in the United States rose from 12% to nearly one-third of the population. Furthermore, an estimated 40–60% of cancer patients are reported to use HFDSs. Additionally, HFDSs are frequently utilized in phase I clinical trials, where up to 93% of participants report using pharmacologically active substances, predominantly vitamins and mineral supplements.

Drug-food or drug-nutrient interactions refer to the physical, chemical, biochemical, physiological, and/or pathophysiological reactions between the active substances in a medication and those present in certain foods. Most often, compounds in food interfere with the proper functioning of medications, leading to outcomes such as diminished therapeutic effects, enhanced therapeutic effects with a risk of toxicity, or the emergence of a new side effect that is typically not caused by either of the substances involved.

These interactions can alter the pharmacokinetics and/or pharmacodynamics of a medication. Pharmacodynamic interactions may produce additive, synergistic, or antagonistic effects on the receptors involved in the therapeutic action. In contrast, pharmacokinetic interactions affect the absorption, distribution, metabolism, and excretion of the medication, as well as the nutrients entering the body simultaneously, shortly before, or shortly after the drug is administered. Unlike pharmacodynamic interactions, which involve specific receptor actions, pharmacokinetic interactions can often be more easily identified by measuring drug levels in plasma.

Patients most susceptible to drug interactions are the elderly, primarily due to polypharmacy and reduced clearance and elimination of certain substances resulting from age-related organ function decline. Additionally, other at-risk groups for drug-food interactions include post-transplant patients, cancer patients, individuals with HIV, malnourished patients, and those receiving enteral nutrition.

Given the increasing number of patients with chronic illnesses requiring pharmacological treatment, combined with the prevalence of unhealthy eating habits, drug-food interactions have become a growing concern. In response, recent years have seen the development of numerous software tools and websites designed to identify potential interactions between prescribed medications and specific dietary components for individual patients.

Intensive research in the field of drug-food and drug-nutrient interactions holds the potential to identify the most effective methods for safely and precisely administering pharmacological therapies. These efforts can also facilitate the creation of individualized dietary plans that ensure optimal intake of macro- and micronutrients without causing adverse effects.

However, until an ideal method for personalized planning in the context of polypharmacy is achieved, it remains crucial to understand the foods and bioactive substances that can alter the effects of medications.

Mechanisms of Nutrient-Drug Interactions

FDIs can alter drug absorption, metabolism, distribution, and excretion through various mechanisms:

Absorption Modulation

Certain foods delay gastric emptying or alter gastric pH, affecting drug solubility.

High-fat meals can enhance the bioavailability of lipophilic drugs, while dietary fiber may reduce drug absorption by binding to active compounds.

Enzyme Inhibition or Induction

Grapefruit juice is a potent inhibitor of CYP3A4, leading to increased plasma levels of drugs metabolized by this enzyme, such as certain tyrosine kinase inhibitors (TKIs).

Cruciferous vegetables induce phase II enzymes, potentially reducing the efficacy of drugs requiring CYP-mediated activation.

Altered Drug Transport

P-glycoprotein (P-gp) transporters, essential for drug efflux, are influenced by dietary components like flavonoids, affecting the intracellular concentration of drugs.

Microbiome Modulation

The gut microbiome metabolizes dietary and pharmaceutical compounds, impacting drug bioavailability and toxicity. Antibiotics or dietary changes can disrupt microbiome-drug interactions.

Chemotherapeutic Agents

Anthracyclines (e.g., doxorubicin): Alcohol consumption exacerbates cardiotoxicity. Antioxidant-rich diets may reduce oxidative stress but could interfere with the pro-oxidative mechanisms essential for anthracycline efficacy.

Platinum-based agents (e.g., cisplatin): Magnesium and potassium deficiencies, exacerbated by dietary restrictions, increase the risk of arrhythmias.

Targeted Therapies

Tyrosine Kinase Inhibitors (e.g., imatinib, sunitinib): Grapefruit and Seville oranges inhibit CYP3A4, increasing toxicity risks. Conversely, St. John's wort induces CYP3A4, reducing drug efficacy.

Immune Checkpoint Inhibitors (e.g., nivolumab): Probiotic-rich diets may influence the gut microbiota, enhancing immune responses but potentially increasing autoimmune-related cardiotoxicity.

Cardiovascular Drugs

Anticoagulants (e.g., warfarin): Vitamin K-rich foods (e.g., leafy greens) antagonize warfarin's anticoagulant effect, requiring consistent dietary intake.

Statins: Grapefruit juice inhibits CYP3A4, raising statin levels and increasing myopathy risk.

Clinical Implications

Therapeutic Failure or Toxicity

FDIs can lead to subtherapeutic drug levels, reducing cancer control or increasing cardiovascular risks.

Overexposure to drugs can exacerbate toxicities such as myelosuppression or cardiotoxicity.

Adverse Events and Hospitalizations

Unmanaged FDIs contribute to treatment discontinuations and hospital admissions, undermining patient quality of life.

Personalized Nutrition

Nutrigenomics may provide insights into patient-specific FDIs, guiding dietary recommendations tailored to genetic profiles.

Management Strategies

Patient Education

Patients should receive clear instructions regarding food restrictions and consistent dietary habits during therapy.

Pharmacological Monitoring

Regular assessment of drug plasma levels and biomarkers can help identify FDIs early.

Interdisciplinary Collaboration

Dietitians, oncologists, and cardiologists must collaborate to develop integrated care plans addressing FDIs.

Dietary Interventions

Structured meal plans and supplementation can mitigate nutrient-drug imbalances.

Beverage-Medication Interactions

Fruit Juices

Grapefruit Juice

While grapefruit (*Citrus paradisi*) is not typically consumed as a medicinal herbal supplement, its impact on drug pharmacokinetics when consumed as a beverage is well-documented. The phytochemical constituents of grapefruit juice (GFJ) are known to be potent inhibitors of cytochrome P450 enzymes (CYPs) and P-glycoprotein (P-gp). These inhibitory effects can significantly alter the metabolism and absorption of various drugs. Numerous comprehensive reviews have explored the interactions between GFJ and prescription medications, highlighting its potential to influence drug efficacy and safety.

Among all beverages, grapefruit juice is responsible for the majority of drug-food interactions. Its active compounds, known as furanocoumarins (naringin, naringenin, bergamottin, -dihydroxybergamottin, and psoralens), have the ability to interfere with the primary metabolic pathway of most xenobiotics—hepatic metabolism—by inhibiting cytochrome P450 enzymes, particularly CYP3A4.

CYP3A4 is located in enterocytes lining the small intestine and colon, as well as in hepatocytes. This means that an orally administered drug is metabolized twice before entering systemic circulation. For example, felodipine has an oral bioavailability of only 15%, despite being 100% absorbed in the digestive tract, as it is extensively metabolized first in the enterocytes of the small intestine and then in hepatocytes. When felodipine is taken with grapefruit juice, its plasma concentration increases significantly, posing a high risk of overdose due to the inhibition of CYP3A4 activity, primarily in the small intestine.

Even micromolar concentrations of naringin have been implicated in reducing the bioavailability of fexofenadine.

In a study involving 21 healthy volunteers, concurrent consumption of grapefruit juice (GFJ) with nilotinib resulted in a 60% increase in the peak concentration of nilotinib and a 29% increase in its area under the curve (AUC). Participants received 400 mg of nilotinib along with either 250 mL of double-strength GFJ or water in a two-period crossover study, separated by a 10-day washout period. The observed interaction was attributed to the inhibitory effects of GFJ phytochemicals on cytochrome P450 enzymes (CYPs).

A case report by Valenzuela et al. highlighted a similar interaction in a patient with esophageal squamous cell carcinoma who consumed 250 mL of GFJ daily for over 3 months while undergoing treatment with docetaxel. The patient exhibited unusual pharmacokinetics of docetaxel, with an abnormally slow elimination rate and a plasma clearance of 13.2 L/h—significantly lower than the typical clearance rate of 36.7 L/h.

After GFJ discontinuation, there was a 60% reduction in the AUC (to infinity), a 36% increase in plasma clearance, and a 10% decrease in the terminal half-life of docetaxel. These findings confirmed that GFJ suppressed docetaxel clearance, likely due to its inhibitory effects on CYP enzymes responsible for the drug's metabolism. This case underscores the potential for GFJ to significantly alter the pharmacokinetics of chemotherapeutic agents, necessitating caution in its consumption during treatment.

According to the literature, grapefruit juice affects the therapeutic effects of multiple classes of drugs, leading to various adverse effects. Below are the main drug classes impacted by grapefruit juice consumption, along with associated adverse effects:

1. Antineoplastic Agents

   Drugs: Crizotinib, dasatinib, erlotinib, everolimus, lapatinib, nilotinib, pazopanib, sunitinib, vandetanib, vemurafenib.

   Adverse Effects: Myelotoxicity, torsades de pointes (a specific type of polymorphic ventricular tachycardia).
2. Anti-Infective Agents

   Drugs: Erythromycin, halofantrine, primaquine, quinine, rilpivirine.

   Adverse Effects: Torsades de pointes, orthostatic hypotension, syncope, myelotoxicity.
3. Lipid-Lowering Agents

   Drugs: Atorvastatin, lovastatin, simvastatin.

   Adverse Effects: Myalgia, rhabdomyolysis.
4. Cardiovascular Medications

   Drugs: Amiodarone, apixaban, clopidogrel, dronedarone, eplerenone, felodipine, manidipine, nifedipine, quinidine, rivaroxaban, ticagrelor.

   Adverse Effects: Torsades de pointes, gastrointestinal and renal bleeding, hyperkalemia, severe arrhythmias, peripheral edema, hypotension.
5. Central Nervous System Medications

   Drugs: Alfentanil, buspirone, dextromethorphan, fentanyl, ketamine, lurasidone, oxycodone, pimozide, quetiapine, triazolam, ziprasidone.

   Adverse Effects: Dizziness, sedation, drowsiness, hallucinations, hypoventilation, torsades de pointes, orthostatic hypotension, syncope.
6. Gastrointestinal Medications

   Drug: Domperidone.

   Adverse Effect: Torsades de pointes.
7. Immunosuppressants

   Drugs: Cyclosporine, everolimus, sirolimus, tacrolimus.

   Adverse Effects: Nephrotoxicity, myelotoxicity.
8. Urinary Medications

   Drugs: Darifenacin, fesoterodine, solifenacin, silodosin, tamsulosin.

Adverse Effects: Urinary retention, constipation, orthostatic hypotension, dizziness.

These interactions highlight the importance of monitoring grapefruit juice consumption in patients taking these medications to avoid potentially severe side effects.

While grapefruit juice is the most well-known for its interactions with pharmacological therapy, other citrus juices (such as orange, lemon, or pomelo juice) should not be overlooked.

1. **Pomelo Juice**
   (a) Alters the pharmacokinetics of cyclosporine (increasing AUC and Cmax while decreasing t1/2) and type 5 phosphodiesterase inhibitors (PDE5).
2. **Orange Juice**
   (a) Due to its hesperidin content, it reduces alendronate absorption by approximately 60% and interacts with antihistamines (e.g., fexofenadine), beta-blockers (e.g., atenolol, celiprolol), and anti-asthmatics (e.g., montelukast).
3. **Apple Juice**
   (a) Interacts with antihistamines (e.g., fexofenadine), beta-blockers (e.g., atenolol), anti-asthmatics (e.g., montelukast), and antihypertensives (e.g., aliskiren).
   (b) **Example:** A study examining the pharmacokinetics of atenolol (50 mg) taken with apple juice (600 mL and 1200 mL) found an 82% reduction in atenolol's AUC at 1200 mL compared to water. The interaction is thought to result from inhibition of the intestinal OATP2B1 transporter, with gastric acidity from high juice volumes also playing a role.
4. **Calcium-Fortified Juices**
   (a) Certain calcium-enriched fruit juices can chelate medications, reducing their therapeutic effect. This is most studied with antibiotics (e.g., fluoroquinolones) but can also occur with aluminum and iron supplements.
5. **Potassium-Rich Juices and Fruits**
   (a) Juices or fruits high in potassium (e.g., bananas, oranges, plums, apricots) can interact with medications affecting electrolyte balance, such as ACE inhibitors (e.g., ramipril), diuretics (e.g., spironolactone, amiloride, triamterene), and angiotensin receptor blockers (ARBs). This is particularly concerning for patients with renal impairments undergoing hemodialysis or those with hypoaldosteronism, as hyperkalemia induced by such combinations can lead to fatal cardiac arrest.
   (b) **Prune Juice:** Commonly used as a laxative, it may alter drug absorption when taken shortly before or after medications, requiring patient caution.

6. **Cranberry Juice**
   (a) Highly beneficial for urinary tract infections, but studies investigating its interactions with warfarin (a low-therapeutic-index anticoagulant) have yielded conflicting results, with some failing to confirm significant interactions.
7. **Tropical Fruit Juices**
   (a) Juices from fruits like pineapple, papaya, kiwi, and passion fruit have been studied in vitro for potential drug interactions due to their inhibitory effects on CYP2C9 and CYP3A4 enzymes.
   (b) **Pineapple Juice:** Contains high bromelain concentrations, demonstrating the strongest CYP2C9 inhibition among tropical fruit juices, with the inhibitory effect increasing with consumption volume.
8. **Blueberry Juice**
   (a) When combined with etanercept (used for juvenile idiopathic arthritis), it has shown improvement in symptoms and a reduction in the drug's adverse effects.

These examples emphasize the significance of monitoring juice and fruit consumption during pharmacotherapy, as even seemingly harmless dietary choices can significantly impact drug efficacy and safety.

Green Tea

In vitro studies have demonstrated that certain green tea components, specifically epigallocatechin gallate (EGCG) and epicatechin gallate (ECG), inhibit estrone-3-sulfate uptake mediated by OATP2B1 and OATP1A2 transporters by approximately 70% and 75%, respectively. Consuming 240–300 mL (1–2 cups) of green tea increases intestinal concentrations of ECG and EGCG to levels sufficient to inhibit OATP activity. In vivo studies have further shown that green tea can inhibit hepatic OATP2, leading to reduced hepatic uptake and elevated plasma concentrations of atorvastatin.

Coffee

Coffee is the second most consumed beverage, typically enjoyed during breakfast and sometimes at lunch. While moderate consumption provides cardioprotective and cognitive benefits and exhibits antioxidant properties due to its abundance of bioactive compounds, certain restrictions apply when combined with specific pharmacological therapies. Coffee influences the pharmacokinetics of several medications as follows:

Reduced absorption: escitalopram oxalate, iron and iron-containing drugs, midazolam, phenothiazine, butyrophenone, thyroxine, calcium, vitamin D, glucose.

Increased absorption: aspirin, felodipine, halofantrine, ergotamine, levodopa.

Reduced distribution: memantine, donepezil.

Increased distribution: levodopa.

Inhibited/reduced metabolism: clozapine, amitriptyline, lithium, theophylline, warfarin, clomipramine, fluvoxamine, imipramine, haloperidol, olanzapine, lidocaine, mexiletine, propranolol, triamterene, verapamil, ropivacaine, melatonin, dextromethorphan, methotrexate, pefloxacin, propafenone.

Increased metabolism: paracetamol.

Increased excretion: minerals (calcium, magnesium, sodium, chloride), water-soluble vitamins (B-complex), oxandrolone, and epi-oxandrolone.

These interactions highlight the need to monitor coffee intake, especially in individuals undergoing complex pharmacological treatments, to ensure both safety and efficacy.

Alcoholic Beverages

Alcohol consumption is known to interact with a wide range of medications. While moderate alcohol intake, such as one or two glasses of wine or beer during social events, generally poses minimal risk, chronic excessive alcohol consumption can significantly increase the likelihood and severity of such interactions, often leading to harmful health effects.

Wines are rich in polyphenols, with red wine containing the highest concentrations, including resveratrol, anthocyanins, tannins, and catechins. These polyphenols are known to modulate certain metabolic processes; however, their effects are minimal with moderate consumption compared to excessive alcohol intake.

Chronic excessive alcohol consumption leads to both pharmacokinetic and pharmacodynamic interactions:

- **Pharmacokinetic interactions:** Chronic alcohol intake induces enzymes such as CYP2E1, CYP3A3, and CYP2A1, altering the metabolism of various medications.
- **Pharmacodynamic interactions:** These occur when alcohol is combined with specific medications, such as:
  - **CNS-active drugs:** Sedatives, antidepressants, antiepileptics, and antipsychotics, increasing sedation or other CNS effects.
  - **Aspirin and anticoagulants:** Elevated risk of bleeding, particularly in the digestive tract.
  - **Analgesics like acetaminophen:** Increased risk of hepatotoxicity.
  - **Antidiabetic drugs:** Potential to induce hypoglycemia.

These interactions highlight the importance of limiting alcohol consumption, especially for individuals undergoing pharmacological treatment, to avoid adverse effects and ensure medication efficacy.

Disulfiram-like reactions (Antabuse reactions) may occur with the use of certain medications, such as metronidazole, ketoconazole, griseofulvin, and cephalosporins (cefuroxime, cefotetan, ceftriaxone, cefoperazone), leading to symptoms such as throbbing headache, nausea, sweating, abdominal pain, palpitations, and tachycardia.

Understanding the interactions between medications and daily beverages is critically important. Many people consume fruit juice, tea, or coffee with breakfast, which, due to their abundance of bioactive compounds, can significantly interact with certain medications. This is especially relevant for drugs commonly taken in the morning, as these interactions can alter their absorption, metabolism, or efficacy.

Coca Cola

- Single or multiple doses of Coca-Cola have been found to increase the absorption of ibuprofen.

Diet and Food Groups in Interaction with Medications

Similar to beverages, foods—whether of plant or animal origin—exhibit specific interactions with certain classes of medications. Their abundance in nutrients and bioactive compounds, combined with the complexity of biochemical and physiological processes in the body, often interferes with the pharmacological mechanisms of certain therapies.

Drug-Herbal Interaction

One of the most significant risks associated with the simultaneous use of herbal products and chemotherapeutic agents is herb-drug interactions. This risk is heightened in patients with chronic conditions who are on multiple medications. In cancer management, herb-drug interactions are particularly concerning due to the steep dose-response relationship and the potential toxicity of chemotherapeutic agents. The primary mechanism involves herbal-induced inhibition or activation of drug-metabolizing enzymes (DMEs) and/or transport proteins, which can alter the pharmacokinetics of the affected drug. While much of the research on this topic is laboratory-based—investigating the impact of herbal products on DMEs and predicting their clinical relevance—not all instances of DME or transporter protein modulation lead to significant clinical interactions.

Similar to synthetic and natural drugs, phytochemicals can influence physiological processes and potentially cause toxicity. Despite limited data on the safety of herbal preparations, the global sales and use of medicinal herbs and complementary medicines have risen significantly. In the United States, the passage of the Dietary Supplement Health and Education Act 25 years ago is thought to have increased the popularity of herbal products and bolstered public confidence in the quality of commercial supplements. A key concern with herbal supplementation is its concurrent use with prescription medications.

The benefits of concurrent herb-drug use remain uncertain, but one significant clinical consequence is the potential for herb-drug interactions. Initially brought to light by the accidental discovery of the grapefruit juice-felodipine interaction, which resulted in a 2.8-fold increase in the oral bioavailability of felodipine, herb-drug interactions have become a critical consideration in pharmacotherapy. This phenomenon has since emerged as a distinct area of research within the field of pharmacological studies.

Herb-drug interactions occur when the pharmacokinetics, pharmacodynamics, or overall effect of a drug are altered by the concurrent use of an herbal product. While many herb-drug interactions are mild and may have minimal clinical significance, some cases necessitate therapeutic intervention. Examples of such interactions include bleeding caused by the combination of garlic (*Allium sativum*) with warfarin, extrapyramidal symptoms triggered by betel nuts (*Areca catechu*) in patients on neuroleptic medications, and episodes of mania in individuals taking antidepressants alongside ginseng (*Panax ginseng*). These and other clinically significant interactions underscore the importance of recognizing potential herb-drug interactions in medical practice.

Chemotherapeutic agents are inherently toxic and often associated with a wide range of side effects. Many cancer patients turn to herbal products alongside their anti-cancer medications to alleviate chemotherapy-related side effects and improve overall well-being. However, the potential risks of herb-drug interactions often outweigh the perceived benefits. Herb-drug interactions are particularly problematic in chemotherapy due to several factors. Chief among these is the narrow therapeutic window of most chemotherapeutic agents; even slight alterations in their delicate dose-response balance can result in significant toxicity.

Secondly, the plasma concentrations of some chemotherapeutic agents are often unreliable predictors of safety and efficacy. Variability in individual pharmacokinetic (PK) profiles has shown significant differences in responses to chemotherapy, further complicated by discrepancies between measurable plasma drug concentrations and those at target sites of action. As a result, even minor alterations in the disposition of a chemotherapeutic agent, following a carefully calibrated dosing regimen, can lead to treatment failure or toxicity.

Thirdly, the efficacy of certain chemotherapeutic agents, such as ifosfamide and cyclophosphamide, relies on their biotransformation into active metabolites by cytochrome P450 (CYP) enzymes. Since many herb-drug interactions stem from the inhibitory or inductive effects of phytochemicals on these metabolic enzymes, such interactions can render these prodrugs either ineffective or toxic.

Finally, cancer patients frequently have comorbid conditions that require multiple medications, in addition to antiemetics and other chemotherapy-related drugs, significantly increasing the risk of herb-drug interactions.

Understanding the mechanisms underlying herb-drug interactions is crucial for predicting and mitigating potentially harmful effects. These mechanisms can be broadly categorized into pharmaceutical, pharmacodynamic, and pharmacokinetic interactions.

Pharmaceutical interactions typically occur due to physicochemical incompatibilities among drugs and formulations when they are in close physical proximity, such as within an IV bag. However, information about such incompatibilities between herbal formulations and prescription drugs is largely lacking. Furthermore, these interactions are less likely with herbal products, as they generally do not come into direct contact with prescription drugs before simultaneous administration.

Pharmacodynamic interactions involve the potentiation, additive, or antagonistic effects of a drug when an herbal product is present. Predicting these interactions requires a thorough understanding of the biomolecular and pharmacological effects of individual herbs and their phytoconstituents. Unfortunately, little is known about the identities and biological activities of the active phytochemicals in many of the herbs commonly used by patients. Despite this, the potential for pharmacodynamic herb-drug interactions remains significant, as phytochemicals can interact with biological receptors. For example, the antidepressant effect of St. John's Wort may exhibit an additive effect in patients already taking prescription antidepressants.

The most significant category of herb-drug interactions is pharmacokinetic interactions, which primarily arise from the inhibition or induction of drug metabolism and clearance by enzymes or transport proteins. These interactions are often mediated at the molecular level by drug-metabolizing enzymes, particularly cytochrome P450 enzymes (CYPs), and transport proteins. Since most anti-cancer drugs are substrates of CYPs and transport proteins, they are particularly susceptible to such interactions.

Phytochemicals in herbal products can inhibit or induce CYP enzymes, leading to substantial changes in drug metabolism. CYP inhibition slows the clearance of CYP substrates, causing drug accumulation—a significant concern in cancer chemotherapy, where many drugs have a narrow therapeutic window. This inhibition is especially problematic for CYP-dependent prodrugs like ifosfamide and cyclophosphamide, as stalled biotransformation can result in therapy failure. Conversely, CYP induction increases metabolic activity, reducing drug exposure, which can lead to sub-therapeutic drug levels, immediate treatment failure, and long-term drug resistance.

Both enzyme inhibition and induction impact the bioavailability and clearance of cancer drugs. Herbal products such as St. John's wort, ginkgo, ginseng, licorice, kava, garlic, cranberry, grape seed, germander, goldenseal, valerian, and black cohosh have been shown to influence CYP activity through either inhibitory or inductive effects. Similar effects on phase II enzymes have also been reported. Additionally, these interactions can affect renal excretion and tissue distribution, often by altering protein binding and displacement dynamics.

Pharmacokinetic herb-drug interactions can also occur through the modulation of transport proteins, with P-glycoprotein (P-gp) being a key player. Also known as multidrug resistance protein 1 (MDR1) or ATP-binding cassette sub-family B member 1 (ABCB1), P-gp is a 160-kDa ATP-dependent efflux surface glycoprotein originally identified in Chinese hamster ovary cells.

P-gp is frequently expressed in tumors exhibiting the multidrug resistance (MDR) phenotype. In normal cells, it is primarily localized on the apical or luminal membranes of tissues involved in excretory and barrier functions, including the liver, kidneys, intestines, and adrenal glands. It is also a critical component of physiological barriers such as the blood-brain barrier, blood-testes barrier, and blood-ovary barrier.

This strategic localization enhances P-gp's protective and detoxification roles by limiting the cellular penetration and tissue distribution of drugs and other xenobiotics through its efflux activity.

As a high-capacity transport protein, P-glycoprotein (P-gp) influences a wide variety of structurally and pharmacologically diverse drugs, including chemotherapeutic agents, antiretrovirals, immunosuppressants, cardioactive drugs, and centrally acting medications. Numerous drugs are known to inhibit P-gp activity, with verapamil and cyclosporine often used as standard controls in P-gp studies. Other inhibitors, such as ketoconazole, quinidine, and ritonavir, have been implicated in adverse drug interactions due to their P-gp inhibitory effects.

Herbal products and phytochemicals also modulate P-gp activity. Inhibitors include silymarin and milk thistle extracts, ginseng-derived ginsenosides, piperine, capsaicin, and various other compounds. Similar to cytochrome P450 enzymes (CYPs), the expression and activity of P-gp can be induced. St. John's Wort is a well-known herbal inducer of P-gp.

Herbal Products with Documented Clinical Interactions with Chemotherapeutic Drugs

Echinacea

Echinacea formulations are widely used as complementary treatments for respiratory infections and the common cold. Among cancer patients, echinacea is particularly popular as an immunomodulatory supplement. Animal studies suggest that echinacea may have potential benefits in mitigating certain forms of cancer, such as leukemia. However, the active constituents and pharmacological mechanisms underlying these effects remain poorly understood.

In the United States, echinacea ranks among the most widely sold herbal products, with most preparations derived from *Echinacea purpurea*, one of the nine commonly recognized species. Several preclinical studies have highlighted potential herb-drug interactions between echinacea and anticancer drugs. Notably, echinacea extracts have been shown to induce P-glycoprotein (P-gp) and CYP3A4, a critical enzyme-transporter combination involved in the biotransformation and pharmacokinetics of many anticancer agents. These interactions may influence the efficacy and safety of chemotherapeutic treatments.

Echinacea has also been shown to inhibit CYP3A4. This dual capability to both inhibit and induce drug-metabolizing enzymes complicates the prediction of clinically significant herb-drug interactions with various CYP and P-glycoprotein (P-gp) substrates. In human studies, echinacea significantly increased the systemic clearance of midazolam, a CYP3A4 substrate, by 34%, indicating the potential for interactions with anticancer drugs.

While preclinical studies provide strong evidence that echinacea interacts with CYP enzymes and transport proteins, clinical data on its interactions with anticancer drugs are limited. For instance, in a study involving 10 cancer patients, echinacea did not significantly alter the pharmacokinetics of docetaxel, a drug metabolized by CYP3A4 and transported by P-gp. In this study, patients received an intravenous dose of docetaxel on day 1, followed by echinacea supplementation (20 oral drops three times daily of a commercially available product) from days 7 to 21, and a second dose of docetaxel on day 22. No notable changes were observed in docetaxel pharmacokinetics with or without echinacea supplementation.

In contrast, echinacea has been shown to decrease the plasma concentration of darunavir, an antiretroviral drug, in HIV/AIDS patients, highlighting the variability of its effects across different drugs. This underscores the need for more extensive research to fully understand echinacea's clinical interactions, particularly with anticancer therapies.

A case report documented a significant interaction between echinacea and etoposide in a cancer patient. The patient, an adult newly diagnosed with squamous cell carcinoma of the lung, received cisplatin and etoposide on the first day of treatment, with normal bloodwork initially recorded. However, by day 8 of the first chemotherapy cycle, the patient experienced a severe drop in platelet count—over two-thirds—necessitating a platelet transfusion.

In the second cycle of chemotherapy, echinacea was discontinued, and the patient did not require further platelet transfusions. No additional complications were reported during the patient's 20-day hospital stay, and they were discharged with instructions to avoid herbal supplements.

Etoposide, a cytotoxic agent and CYP substrate, has myelosuppression as its dose-limiting toxicity. The interaction is believed to have occurred due to echinacea-induced CYP inhibition, resulting in etoposide accumulation and subsequent thrombocytopenia. This case highlights the potential risks of combining herbal supplements with chemotherapy drugs.

Garlic

Garlic (*Allium sativum*) is a widely used herbal supplement, commonly taken to support the treatment of infections, diabetes, and heart diseases. Its use is particularly prevalent among individuals with chronic conditions, including cancer. The primary bioactive compound in garlic is allicin (diallyl thiosulfinate).

Studies have demonstrated that whole garlic extracts can inhibit the CYP3A4-mediated conversion of testosterone to 6β-hydroxytestosterone in vitro, using liver microsomal incubations. This suggests that garlic may influence the metabolism of drugs that are substrates of CYP3A4, raising the potential for herb-drug interactions.

In a study investigating the impact of garlic supplementation on the pharmacokinetics of docetaxel, Cox and colleagues administered docetaxel weekly to women with metastatic breast cancer over 3–4 weeks. Beginning 3 days after the initial dose of docetaxel, participants received a 12-day regimen of garlic supplementation, consisting of 600 mg taken twice daily.

By Day 15, garlic supplementation had reduced the clearance of docetaxel by 36%, from 30.8 to 20.0 L/h/m^2. While other pharmacokinetic parameters were reported to show no significant changes, the reduction in docetaxel clearance raises concerns about the risk of toxicity due to drug accumulation. This interaction is likely attributable to the inhibition of CYP enzymes by garlic phytochemicals, as these enzymes are crucial for the metabolism of docetaxel.

Ginseng

Ginseng is one of the most widely consumed herbal products worldwide, particularly in the United States. Commercial ginseng products are primarily derived from three species: *Panax ginseng* (Asian ginseng), *Panax quinquefolius* (American ginseng), and *Panax japonicus* (Japanese ginseng). While ginseng is often promoted for benefits such as energy enhancement, immunomodulation, increased sexual desire, and pain relief, most of these claims are anecdotal.

The pharmacological activity of ginseng is largely attributed to ginsenosides, a group of steroidal saponins that are its primary phytochemical components. Studies have reported antioxidant and cardiovascular protective effects of ginseng. Additional pharmacological activities include immunomodulatory and anticarcinogenic effects, neurotransmitter modulation, and antimitogenic activity, further suggesting its potential therapeutic benefits.

The effects of ginseng on drug-metabolizing enzymes (DMEs) and P-glycoprotein (P-gp) have been inconsistent across studies. Some in vitro research has reported no inhibitory activity on cytochrome P450 enzymes (CYPs), while others have demonstrated inhibitory effects on DMEs.

In a study involving eight healthy volunteers, researchers examined the impact of *Panax ginseng* extracts on the pharmacokinetics of midazolam (a CYP3A4 substrate) and fexofenadine (a P-gp substrate). The results revealed a significant reduction in the area under the curve (AUC) and maximum plasma concentration (Cmax) of midazolam. The authors attributed these findings to the inductive effects of ginseng on CYP3A4/5, suggesting a potential interaction that may alter drug metabolism.

Ginseng, a widely used herbal supplement among cancer patients, has the potential to mediate clinically significant interactions with chemotherapeutic agents. In one case report, a patient undergoing treatment with imatinib for chronic myelogenous leukemia (CML) developed hepatotoxicity after concurrent use of an energy drink containing *Panax ginseng*. The patient, who had been on imatinib for 7 years without issues, exhibited liver dysfunction confirmed by abnormal liver function test results (elevated alanine aminotransferase, aspartate aminotransferase, alkaline phosphatase, total bilirubin, and albumin) and liver biopsy findings.

Hepatotoxicity symptoms resolved after discontinuing the energy drink. While imatinib is known to cause hepatotoxicity in some patients, typically within the first two years of therapy, this patient's long-term tolerance to the drug suggests that the energy drink may have contributed to the liver dysfunction. However, given the multicomponent nature of the energy drink, it remains unclear whether ginseng alone was responsible for the adverse reaction. This case underscores the complexity of herb-drug interactions and highlights the need for caution when combining herbal supplements with chemotherapeutic agents.

Milk Thistle

Milk thistle (*Silybum marianum*) is a widely used herbal product, often employed as a complementary therapy by cancer patients and to support immunity in individuals with HIV/AIDS. It is also commonly used for the prevention and treatment of liver diseases. The active component of milk thistle is silymarin, a complex mixture of biologically active flavonolignans found primarily in the leaves, seeds, and fruit of the plant.

Commercial milk thistle products are typically standardized to contain silymarin, which includes flavonolignans and a flavonoid. A specific fraction of silymarin, known as silibinin (comprising silybin A and silybin B), is also available commercially. A recent publication provides a detailed review of the phytochemical components of milk thistle and their nomenclature.

Silymarin has shown promising anticancer activity in clinical investigations. However, in vitro studies have demonstrated that silymarin can inhibit the activity of cytochrome P450 (CYP) enzymes and phase II metabolizing enzymes. This inhibitory potential has been observed in clinical studies as well, such as one in which silymarin significantly reduced the CYP2C9-mediated metabolism of losartan, highlighting its potential for herb-drug interactions.

The widespread use of milk thistle among cancer patients raises concerns about potential herb-drug interactions; however, clinical data on this subject remains limited. One study investigated whether the inhibitory effects of milk thistle extract on CYP3A4 would alter the pharmacokinetics of irinotecan, a CYP3A4 substrate, when taken concurrently.

The study involved six cancer patients undergoing weekly irinotecan treatment, during which they were also administered milk thistle three times daily for 12 days. The pharmacokinetics of irinotecan and its metabolites were assessed. The findings revealed that neither short-term (4 days) nor prolonged use (12 days) of milk thistle significantly altered the pharmacokinetics of irinotecan. Only a slight, statistically insignificant reduction in clearance was observed, with values of 31.2, 25.4, and 25.6 L/h in the first, second, and third weeks, respectively.

Nevertheless, the potential risk of clinically significant herb-drug interactions between milk thistle and chemotherapeutic agents cannot be entirely ruled out. This is largely due to variations in silymarin concentrations across different commercially available milk thistle formulations, which may influence their interaction profiles. Further research is needed to provide conclusive evidence.

St. John's Wort

St. John's wort (*Hypericum perforatum*) is a widely used herbal supplement commonly employed in the treatment of depression, anxiety, sleep disorders, and nervousness. Official guidelines in several countries recommend St. John's wort for managing depression, contributing to its growing popularity among diverse patient populations. It is also frequently used for conditions such as premenstrual syndrome, alcohol withdrawal, and somatoform disorders.

St. John's wort contains several active phytochemical constituents, including naphthodianthrones (e.g., hypericin), phloroglucinols (e.g., hyperforin), and flavonol glycosides (e.g., hyperosides), which have been isolated and characterized. The antidepressant activity of St. John's wort is primarily attributed to hyperforin, a compound with potent inhibitory effects on the synaptic reuptake of central neurotransmitters such as dopamine, norepinephrine, and serotonin.

Several in vitro studies have demonstrated that extracts of St. John's wort can modulate the activity of cytochrome P450 (CYP) enzymes and major drug transporters. St. John's wort is a potent inducer of CYP2B6, CYP2C19, CYP2E1, and CYP3A4. Hyperforin, one of its primary active constituents, not only induces several CYP isoforms but also acts as a potent inhibitor of CYP2C9 and CYP2D6. Other constituents of St. John's wort also exhibit inhibitory effects on CYP enzymes. For instance, biapigenin, a flavonoid found in the herb, strongly inhibits CYP1A2, CYP2C9, and CYP3A4, while hypericin acts as a competitive inhibitor of CYP2C9, CYP2D6, and CYP3A4.

Mechanistic studies in cell lines and animal models have highlighted the potential for herb-drug interactions involving St. John's wort. Human studies have investigated the effects of its concurrent administration with clinically significant substrates of these CYP enzymes and transporters. Additionally, clinical case reports have documented significant herb-drug interactions between St. John's wort and various prescription medications, underscoring the need for careful consideration of its use in patients undergoing pharmacological treatments.

Human studies and clinical case reports have demonstrated that St. John's wort can significantly alter the pharmacokinetics of various CYP3A4 and P-glycoprotein (P-gp) substrates. These include drugs such as omeprazole, simvastatin, cyclosporine, indinavir, verapamil, and tacrolimus.

Four clinically relevant studies from the literature highlight the effects of St. John's wort on the pharmacokinetics of anticancer drugs. One study evaluated its impact on docetaxel, a CYP3A4 substrate, in 10 cancer patients. Participants received an intravenous dose of 135 mg of docetaxel on day 1, followed by blood sampling for pharmacokinetic analysis. From days 7 to 21, they were administered 300 mg tablets of St. John's wort extract (*Hyperiplant*®), three times daily.

The results showed a 12% reduction in the mean area under the curve (AUC∞) for docetaxel and a 14% increase in total clearance due to St. John's wort supplementation. Additionally, the maximum plasma concentration (Cmax) and half-life (T1/2) of docetaxel were slightly decreased, though the changes were not statistically significant. Interestingly, the study also reported a lower incidence of docetaxel-related adverse effects in patients taking St. John's wort.

These findings align with the known ability of St. John's wort to induce CYP3A4, thereby accelerating the metabolism of its substrates and potentially reducing their therapeutic effects.

High-Fiber Foods

Simvastatin, ezetimibe, pravastatin, and fluvastatin can generally be administered without specific considerations regarding food intake. However, high-fiber foods and diets may reduce the efficacy of these medications. Concurrent administration of statins with food can alter the pharmacokinetics or pharmacodynamics of these drugs, increasing the risk of adverse reactions such as myopathy or rhabdomyolysis or reducing their pharmacological effect.

For instance, consuming pectin or oat bran alongside lovastatin may decrease its absorption. Nevertheless, lovastatin should be taken with food to enhance gastrointestinal absorption and bioavailability. In contrast, the absorption of rosuvastatin, another lipid-lowering agent, is significantly reduced when taken in a fed state compared to fasting conditions, highlighting the need to administer rosuvastatin prior to meals.

Conclusions

Epidemiological studies suggest that approximately 10% of general hospital admissions may be attributed to the effects of multiple drug use, leading to adverse drug interactions and reactions. Drug interactions can significantly impact drug concentrations in the body, posing particular challenges for chemotherapeutic agents, which are typically administered at doses near their maximum tolerable levels. On one hand, drug interactions that increase the clearance of cytotoxic drugs may result in subtherapeutic exposure, potentially fostering drug resistance and treatment failure. On the other hand, interactions causing drug accumulation can lead to dangerously high drug concentrations, triggering life-threatening toxicities.

Cancer patients are particularly vulnerable to such interactions due to the frequent use of multiple medications to manage co-morbidities and cancer-related conditions. This risk is further compounded by the use of herbal products, which can introduce additional herb-drug interactions. The pharmacodynamic effects of herbal products, including organ-specific actions and long-term receptor interactions, may pose additional risks and complexities for cancer treatment, potentially undermining therapeutic outcomes. Despite limited evidence supporting the therapeutic benefits of herbal supplements in cancer treatment, their use is widespread among cancer patients. Studies report that 50–66% of patients use one or more forms of complementary or alternative medicine, with herbal preparations being the most common, often alongside conventional cancer therapies.

As previously noted, combining herbal supplements with conventional cancer therapies poses significant risks due to potential herb-drug interactions. These interactions are primarily pharmacokinetic in nature, arising from the induction or inhibition of drug-metabolizing enzymes and transport proteins.

Echinacea, for example, can both inhibit and induce cytochrome P450 (CYP) enzymes, making its effects on cancer therapy unpredictable. Current evidence is limited and conflicting, with some studies suggesting it may either enhance or diminish the efficacy of chemotherapy. Similarly, ginseng, as an inducer of CYP enzymes, could increase the risk of adverse effects when used alongside chemotherapy. Although it remains unclear whether ginseng was the definitive cause of hepatotoxicity in the reported case, its enzyme-inducing properties suggest a potential role in the observed liver dysfunction. Further research is needed to clarify ginseng's interactions with chemotherapeutic agents.

Garlic and milk thistle, on the other hand, inhibit CYP enzymes, which can slow the metabolism of certain chemotherapeutic drugs. This inhibition may alter the pharmacokinetics of these agents, potentially leading to drug accumulation and heightened toxicity. Evidence from the literature supports their potential to clinically influence chemotherapy outcomes.

Grapefruit juice presents a unique interaction risk, as it is commonly consumed for non-medicinal purposes. Studies indicate that grapefruit juice inhibits CYP enzymes and P-glycoprotein (P-gp), leading to the accumulation of their substrates. This can increase the risk of adverse effects from chemotherapy, emphasizing the critical role of dietary considerations during cancer treatment.

Overall, these interactions underscore the need for careful evaluation of herbal supplement and dietary intake in cancer patients undergoing chemotherapy to minimize risks and optimize therapeutic outcomes.

To provide effective care, healthcare providers should engage in open conversations with patients to document their use of herbal products and offer appropriate counseling. Patients should be educated about both the potential benefits and risks of using herbal products in cancer treatment. This education should emphasize the lack of robust scientific evidence supporting many claims, as well as the aggressive marketing strategies often used to promote these products. Most importantly, patients need to understand the potential for herb-drug interactions, which can lead to toxicity or treatment failure, and how these risks may impact their overall therapy outcomes.

These examples emphasize the significance of monitoring juice and fruit consumption during pharmacotherapy, as even seemingly harmless dietary choices can significantly impact drug efficacy and safety.

Food-drug interactions are a crucial consideration in the care of cardio-oncology patients. These interactions can significantly influence treatment outcomes by affecting drug absorption, metabolism, and efficacy. Clinicians should maintain a high level of awareness regarding common food-drug interactions and provide patients with education and guidance on managing these interactions. By meticulously evaluating both drug-drug and food-drug interactions, healthcare providers can optimize treatment strategies and enhance outcomes for cardio-oncology patients.

Furthermore, interdisciplinary collaboration among oncologists, cardiologists, pharmacists, and registered dietitians is essential to ensure comprehensive patient care. Future research is needed to further elucidate food-drug interactions in this population and develop more specific guidelines.

Future Directions

Emerging technologies such as metabolomics and microbiome profiling offer promising avenues for elucidating FDIs in cardio-oncology. Research into dietary interventions and precision medicine approaches will further enhance patient outcomes. Randomized controlled trials exploring dietary modulation in cancer patients undergoing cardiotoxic treatments are urgently needed..

As cardio-oncology continues to evolve, the impact of nutrient-drug interactions (NDIs) on cardiovascular health and cancer treatment outcomes is becoming increasingly relevant. Given the complex interplay between chemotherapy, targeted therapies, and cardiovascular medications, optimizing nutritional strategies to mitigate adverse effects and enhance therapeutic efficacy is a growing area of interest.

Developing individualized dietary recommendations based on pharmacogenomics and metabolic profiling to minimize NDIs. Precision nutrition strategies to optimize drug absorption, metabolism, and elimination in cancer patients at risk for cardiovascular complications.

Investigating how specific nutrients (e.g., antioxidants, omega-3 fatty acids, polyphenols) influence the pharmacokinetics and pharmacodynamics of cardio-oncology drugs. Understanding the role of gut microbiota in modulating NDIs, particularly in patients receiving immune checkpoint inhibitors and chemotherapy.

Exploring the protective effects of dietary interventions, such as the Mediterranean diet or plant-based diets, in reducing chemotherapy-induced cardiotoxicity.

Evaluating the role of micronutrients like Coenzyme Q10, magnesium, and vitamin D in preserving cardiac function during cancer treatment.

Assessing the impact of commonly prescribed cardiovascular drugs (e.g., beta-blockers, statins, ACE inhibitors) on the metabolism of chemotherapy and targeted therapies.

Identifying dietary modifications that may improve drug tolerance and effectiveness in patients with concurrent cardiovascular disease and cancer.

Conducting large-scale studies to establish evidence-based dietary guidelines for cardio-oncology patients.Investigating the long-term cardiovascular effects of specific dietary patterns in cancer survivors.

Utilizing artificial intelligence (AI) and digital health tools to predict and monitor NDIs in real time. Developing mobile applications and wearable devices to track dietary intake and medication responses in cancer patients at risk for cardiovascular events.

Enhancing collaboration between oncologists, cardiologists, dietitians, and pharmacologists to create comprehensive treatment plans.

Establishing standardized protocols for managing NDIs in cardio-oncology clinics.

By advancing research in these areas, the field of cardio-oncology can improve patient outcomes, minimize treatment-related cardiovascular risks, and optimize the therapeutic potential of both oncologic and cardiovascular interventions.

References

1. Beavers CJ et al (2022) Cardio-oncology drug interactions: a scientific statement from the American Heart Association. Circulation 145(15):e811–e838
2. Buajordet I et al (2001) Fatal adverse drug events: the paradox of drug treatment. J Intern Med 250(4):327–341
3. Cancer, W.H.O. 23.12.2024. https://www.who.int/news-room/fact-sheets/detail/cancer
4. He SM et al (2010) Effects of herbal products on the metabolism and transport of anticancer agents. Expert Opin Drug Metab Toxicol 6(10):1195–1213
5. Kamaraju S et al (2021) Interactions between cardiology and oncology drugs in precision cardio-oncology. Clin Sci 135(11):1333–1351
6. Lapidari P et al (2021) Determinants of use of oral complementary-alternative medicine among women with early breast cancer: a focus on cancer-related fatigue. Breast Cancer Res Treat 190(3):517–529
7. Lee RT et al (2021) Prevalence of potential interactions of medications, including herbs and supplements, before, during, and after chemotherapy in patients with breast and prostate cancer. Cancer 127(11):1827–1835
8. Meijerman I, Beijnen JH, Schellens JH (2006) Herb-drug interactions in oncology: focus on mechanisms of induction. Oncologist 11(7):742–752
9. Petric Z et al (2020) Food-drug interactions with fruit juices. Foods 10:1
10. Riechelmann R, Girardi D (2016) Drug interactions in cancer patients: a hidden risk? J Res Pharm Pract 5(2):77–78
11. Spanakis M, Patelarou E, Patelarou A (2022) Drug-food interactions with a focus on Mediterranean diet. Appl Sci 12(20):10207
12. Kenakin T (2017) Pharmacokinetics I: permeation and metabolism. In: Pharmacology in drug discovery and development
13. Valenzuela B et al (2011) Effect of grapefruit juice on the pharmacokinetics of docetaxel in cancer patients: a case report. Br J Clin Pharmacol 72(6):978–981

14. Wolf C et al (2022) Interactions in cancer treatment considering cancer therapy, concomitant medications, food, herbal medicine and other supplements. J Cancer Res Clin Oncol 148(2): 461–473
15. Kamaraju S et al (2021) Interactions between cardiology and oncology drugs in precision cardio-oncology. Clin Sci (Lond) 135(11):1333–1351
16. D'Andre SD et al (2022) Dietary supplement use and recommendations for discontinuation in an integrative oncology clinic. Support Care Cancer 31(1):40
17. Nana Poku SR, Andres MS, Gevaert S, Alexander R (2022) Lyon monitoring and treatment of cardiovascular complications during cancer therapies. In: Part II: Tyrosine kinase inhibitors. European Society of Cardiology
18. Collado-Borrell R et al (2016) Oral antineoplastic agent interactions with medicinal plants and food: an issue to take into account. J Cancer Res Clin Oncol 142(11):2319–2330
19. Food and drink to avoid during cancer treatment. Cancer Research UK. (2023)
20. Anon (2023) Cancer therapy interactions with foods and dietary supplements (PDQ®). National Cancer Institute
21. Cancer Drugs, Supplements, Interactions & Food. (2023) Breastcanceroptions.org
22. Cancer Drug Therapy and Nutrition (2023) Leukemia & Lymphoma Society.
23. Typical Drug Interactions in Oncology—U.S. Pharmacist. Accessed 4 Jan 2025
24. Cardio-Oncology Drug Interactions—Professional Heart Daily [cited 2025 05.01]. https://professional.heart.org/en/science-news/cardio-oncology-drug-interactions
25. Raisi-Estabragh Z et al (2024) Cardiovascular considerations before cancer therapy. JACC: CardioOncology 6(5):631–654
26. Rashrash M, Schommer JC, Brown LM (2017) Prevalence and predictors of herbal medicine use among adults in the United States. J Patient Exp 4(3):108–113
27. Luo Q, Asher GN (2018) Use of dietary supplements at a Comprehensive Cancer Center. J Altern Complement Med 24(9–10):981–987
28. Bossaer JB, Odle BL (2012) Probable etoposide interaction with Echinacea. J Diet Suppl 9(2): 90–95
29. Goey AK et al (2013) The effect of Echinacea purpurea on the pharmacokinetics of docetaxel. Br J Clin Pharmacol 76(3):467–474
30. Goey AK et al (2014) The effect of St John's wort on the pharmacokinetics of docetaxel. Clin Pharmacokinet 53(1):103–110
31. Binns CW, Lee MK, Lee AH (2018) Problems and prospects: public health regulation of dietary supplements. Ann Rev Public Health 39:403–420
32. Awortwe C et al (2015) Echinacea purpurea up-regulates CYP1A2, CYP3A4 and MDR1 gene expression by activation of pregnane X receptor pathway. Xenobiotica 45(3):218–229
33. Zheng YF et al (2014) Evaluation of the in vitro/in vivo drug interaction potential of BST204, a purified dry extract of ginseng, and its four bioactive ginsenosides through cytochrome P450 inhibition/induction and UDP-glucuronosyltransferase inhibition. Food Chem Toxicol 68:117–127
34. Alsanad SM, Williamson EM, Howard RL (2014) Cancer patients at risk of herb/food supplement-drug interactions: a systematic review. Phytother Res 28(12):1749–1755
35. Ali-Shtayeh MS et al (2016) Complementary and alternative medicine use among cancer patients in Palestine with special reference to safety-related concerns. J Ethnopharmacol 187: 104–122
36. Fang Y et al (2020) Key considerations in the design of real-world studies. Contemp Clin Trials 96:106091
37. Fugh-Berman A (2000) Herb-drug interactions. Lancet 355(9198):134–138

Part II
Toxins and Poison-Induced Cardiovascular Toxicity

Chapter 20
Cardiotoxicity of Pesticides

Biljana Antonijević and Miloš P. Stojiljković

Abstract Pesticides are used to protect crops from various pests and against the vectors of life threathening communicable diseases. They are classified into various groups based on their chemical structures, such as organochlorine compounds, organophosphosphates, carbamates, pyrethroids, coumarin derivatives, dipyridyles, triazoles, triazines, etc. These chemicals have intrinsic toxicity, so the impact of pesticide exposure on human health and the environment remains a significant concern. Pesticides' toxicological profiles include a broad spectrum of possible adverse outcomes such as neurotoxicity, hepatotoxicity, cardiotoxicity, immunotoxicity, endocrine disruption, carcinogenicity, etc. Cardiotoxicity is of particular importance since cardiovascular diseases (CVDs) are among the most significant causes of mortality in humans, e.g. among the top 10 global causes of death in 2021, ischaemic heart disease held the first place. Concerning environmental risk factors, almost all pesticides can induce cardiotoxicity linked to a variety of molecular pathways including an imbalance in Ca2+ homeostasis, mitochondrial dysfunction, and mechanisms such as oxidative stress and apoptosis. Pesticide cardiotoxicity covers different clinical features such as electrocardiogram abnormalities, higher incidences of myocardial infarction, impaired systolic and diastolic performance, increased levels of coronary heart disease markers, and blood pressure disorders. Concerning the cardiovascular risk associated with pesticide exposure, the utmost goal is to build capacity in sound pesticide management across both agriculture and public health sectors following the principles of One Health.

B. Antonijević (✉)
Faculty of Pharmacy, Department of Toxicology, University of Belgrade, Belgrade, Serbia
e-mail: biljana.antonijevic@pharmacy.bg.ac.rs

M. P. Stojiljković
Faculty of Medicine, University of Banja Luka, Banja Luka, The Republic of Srpska, Bosnia and Herzegovina
e-mail: milos.stojiljkovic@med.unibl.org

D. K. Agrawal et al. (eds.), *Cardiovascular Toxicity*, Advances in Biochemistry in Health and Disease 33, https://doi.org/10.1007/978-3-031-90598-8_20

Keywords Carbamates · Carcinogenicity · Cardiotoxicity · Dipyridyles · Endocrine disruption · Hepatotoxicity · Immunotoxicity · Neurotoxicity · Organochlorine insecticides · Organophosphates · Pyrethoroids · Triazines · Triazoles

Introduction

A pesticide is defined as a substance or a mixture of chemical or biological components designed to repel, destroy, or control pests, or to regulate plant growth. According to the application area and regulatory procedures, those that are used in agriculture are known as plant protection products, whereas the biocides are commonly used in medicine, industry, and public sector. They are classified into various groups based on their chemical structures, such as organochlorine compounds, organophosphosphates, carbamates, pyrethroids, coumarin derivatives, dipyridyles, triazoles, triazines, etc. On the basis of the target organism, they can be grouped as insecticides, fungicides, herbicides, acaricides, algaecides, rodenticides, larvicides, repellants, and desiccants. Pesticides are chemicals massively used to control pests that harm plants, animals, and humans. Globally, over 1000 pesticides are used to protect food not to be damaged or destroyed by pests and to protect against the vectors of various life threathening communicable diseases.

The United Nations Population Division estimates that by the year 2050 there will be 9.7 billion people on Earth. Nearly all of this population growth will occur in developing countries. The Food and Agriculture Organization of the United Nations [1] estimates that in developing countries, 80% of the increase in food production needed to keep pace with population growth, is projected to come from either increases in yields and/or the number of times each year crops can be grown on the same land. Only 20% of extra food production is expected to result from an expansion of farming land. Pesticides can prevent large crop losses and will therefore continue to play a role in agriculture.

Between 2000 and 2021, global pesticide use rose by 62%, reaching 3.5 million tonnes in 2021. Most of this increase occurred between 2000 and 2016, followed by a slight decline until 2018 and renewed growth thereafter. The Americas contributed the most to this rise, followed by Asia, Europe, Africa and Oceania. The share of the Americas, the largest contributor, increased from 41% to 50% of global pesticides consumption while that of Asia and Europe decreased by 5–6% points to 28% and 14%, respectively. Africa and Oceania applied small amounts of pesticides over time, but Africa nonetheless had the highest growth in pesticides applications (+176%). Brazil was the largest pesticide user in 2021 with 0.72 million tonnes, or 20% of the world total, far ahead of the United States (0.46 million tonnes) and Indonesia (0.28 million tonnes) [1].

Pesticide Toxicity

The impact of pesticide exposure on human health and the environment remains a significant concern. Their global use has been linked to environmental contamination, affecting food, soil, and ground or surface water systems. Monitoring of chemicals in European rivers has identified over 400 distinct substances. Among these, pesticides accounted for the largest proportion (42%), followed by pharmaceuticals and illicit drugs (34%), and industrial chemicals (17%) [2]. Some older, less expensive pesticides, such as dichlorodiphenyltrichloroethane (DDT) and lindane, can persist in soil and water for many years. These substances have been banned or restricted by nations adhering to the Stockholm Convention (2001), an international treaty aimed to eliminate or restrict the production and use of persistent organic pollutants (POPs). POPs are characterized by their toxic properties, resistance to degradation, bioaccumulation, and their ability to transport across international borders through air, water, and migratory species. These pollutants can be deposited far from their original sources, accumulating in terrestrial and aquatic ecosystems.

Since pesticides have intrinsic toxicity, they are potentially toxic to humans and can have both acute and chronic health effects, depending on the quantity, exposure duration and ways in which a person is exposed. Pesticides have different properties and toxicological effects. The same chemical can have different effects at different doses, that is, the amount of chemical to which a person is exposed. Toxicity can also depend on the route by which the exposure occurs, such as by swallowing, inhaling or direct contact with the skin.

Based on the globally harmonized system (GHS) of classification and labeling of chemicals, 28%, 15%, 14% and 11% of pesticides have the following hazardous properties: acute toxicity, reproductive toxicity or skin sensitisation, specific target organ toxicity after repeated exposure and carcinogenicity, respectively [3].

People who face the greatest health risks from pesticide exposure are those who are directly exposed to them at work, i.e. agricultural workers during mixing, loading, and applying pesticides and re-entry workers or anyone else in close proximity during and shortly after pesticide tratment (Fig. 20.1). Using pesticides, people can be exposed in their homes or gardens. The general population, those not in the area where pesticides are used, is exposed to significantly lower levels of pesticide residues through food and water.

However, pesticides are among the leading causes of death by self-poisoning, particularly in low- and middle-income countries [4]. Approximately 3,000,000 cases of pesticide poisoning occur every year worldwide, with about 220,000 deaths per year [5]. Comprehensive evaluations of the pesticides' toxicological profiles have revealed a broad spectrum of possible adverse outcomes such as cardiotoxicity linked to the variety of molecular pathways including imbalance in Ca^{2+} homeostasis, mitochondrial dysfunction, and mechanisms such as oxidative stress and apoptosis.

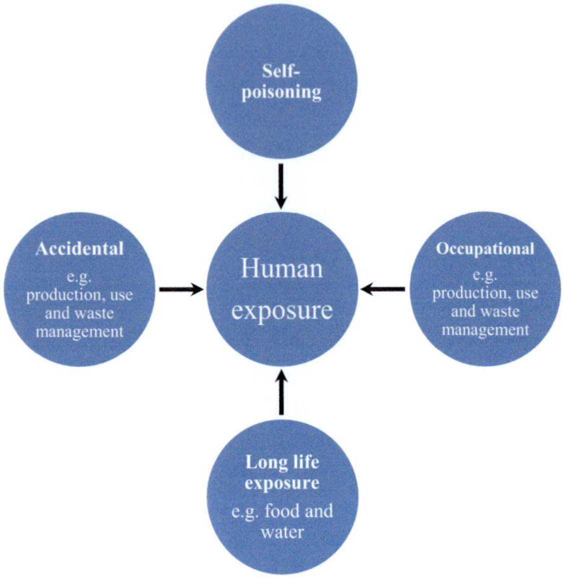

Fig. 20.1 Human exposure to pesticides throughout their life cycle

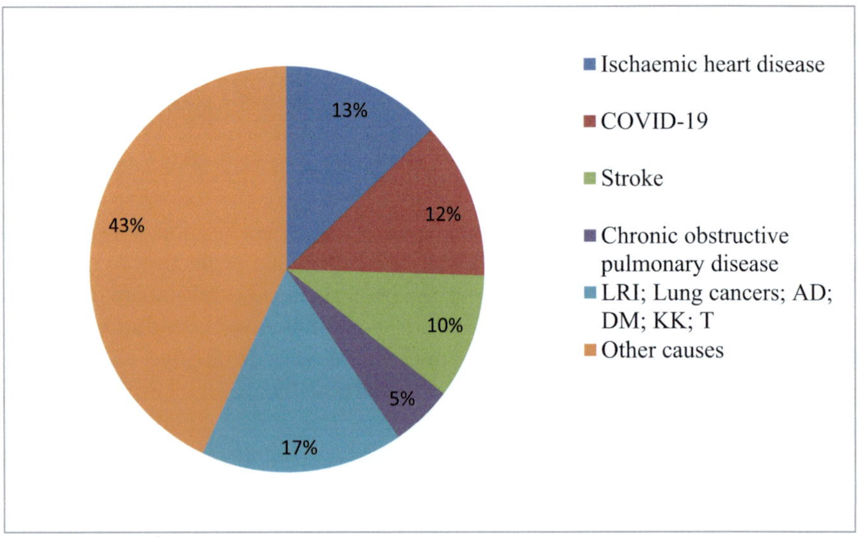

Fig. 20.2 Leading causes of death in 2021 globally

Cardiotoxicity

Cardiovascular diseases (CVDs) are among the most significant causes of mortality in humans. According to the WHO data (Fig. 20.2), among the top 10 global causes of death in 2021, ischaemic heart disease was on the first place followed by

COVID-19, stroke, chronic obstructive pulmonary disease, lower respiratory infections (LRI), lung cancers (LC), Alzheimer disease and other dementias (AD), diabetes mellitus (DM), kidney diseases (KD) and tuberculosis (T). The top 10 causes of death accounted for 39 million deaths, or 57% of the total 68 million deaths worldwide. The ischaemic heart disease is responsible for 13% of the world's total deaths [6]. Since 2000, the largest increase in deaths has been for this disease, rising by 2.7 million to 9.1 million deaths in 2021. CVDs are the leading cause of death globally, taking an estimated 17.9 million lives each year. CVDs are a group of disorders of the heart and blood vessels and include coronary heart disease, cerebrovascular disease, rheumatic heart disease and other conditions. More than four out of five CVD deaths are due to heart attacks and strokes, and one third of these deaths occur prematurely in people under 70 years of age. Amongst environmental risk factors, air pollution along with long life exposure to food and water contaminants are important factors. It has been shown that approximately 60 pesticides were involved in cardiotoxicity: 30, 13, and 17 were insecticides, herbicides, and fungicides, respectively [3, 7].

Organochlorine Insecticides

Some of the most widely used insecticides in history are members of the organochlorine class and include DDT, aldrin, dieldrin, lindane, heptachlor, chlordane, and toxaphene. They are GABA-gated chlorine channel blockers, delay the closing of the sodium ion channel, prevent the full opening of the potassium gates, inhibit a specific neuronal adenosine triphosphatase (ATPase), inhibit the ability to transport calcium ions in nerves possibly by binding with a hydrophobic site on calmodulin, induce a deficiency of brain serotonin, cause the changes in the levels of biogenic amines and amino acids [8]. After a relatively high doses, the hyperexcitability of the nerve system results in trains of repetitive discharges, leading to tremors and convulsions. Due to a number of molecular pathways, they lead to the impairment of nerve impulse conduction, hormonal imbalance, decrease in fertility, embryotoxicity, fetotoxicity, suppress the humoral immune response across multiple species, increase susceptibility to infection [9]. Organochlorines are endocrine dusruptors *i.e.* interact with steroid hormone receptors for estrogens and androgens and disturb the homeostasis of thyroid hormones. For some of them (DDT), positive associations between exposure and certain types of cancers have been shown [10].

The cardiotoxicity of organochlorines, observed in both animal models and human cases, includes: electrocardiogram abnormalities (hypotension and ECG abnormalities such as non-ST segment elevation myocardial infarction), higher incidences of myocardial infarction, impaired systolic and diastolic performance, changes in heart structure and function, histopathological findings (haemorrhage, cytoplasmic edema, swelling and vacuolization of mitochondria in myocardial cells, and degeneration of myocardial fibers), oxidative stress (increased oxidative stress markers like lipid peroxidation, altered antioxidant enzyme activities, and increased levels of malondialdehyde (MDA)), atherogenic and inflammatory molecular mechanism involving fatty acid and glucose metabolism [11], increased levels of coronary heart disease markers, blood pressure disorders (hypotension and changes in blood pressure regulation) [3, 7, 12].

Organophosphates and Carbamates

Organophosphorus (OP) insecticides, recognised by their high acute toxicity, are a large family of compounds, sharing the same mode of action (e.g. parathion, malathion, dichlorvos, monocrotophos, chlorpyriphos, dimethoate, diazinon, etc.). The primary molecular mechanism of action of the OP pesticides is inhibition of acetylcholinesterase (AChE) producing excessive acetylcholine (ACh) accumulation and overstimulation of cholinergic neurons. The organophosphates (OPs) or their active metabolites are electrophilic compounds with moderate to high potency for phosphorylating the serine hydroxyl group located in the active site of AChE. The reactivity of OP compounds varies depending upon the chemical structure. The clinical manifestations of acute poisoning reflect hypercholinergic stimulation, with consequent muscarinic, nicotinic and central nervous system effects [13]. OPs may also inhibit some other serine esterases: neuropathy target esterase (NTE), butyrylcholinesterase (BuChE,), carboxylesterases (CarbE), phosphorylphosphatases (A-esterases), fatty acid amide hydrolase (FAAH), arylformamidase (AFMID), acylpeptide hydrolase (APH), lysyl oxidase (LyO) [14]. Carbamates are commonly known as reversible inhibitors of AChE [15]. Depending on the level of exposure, like OPs, they may produce similar clinical picture resulted from muscarinic and nicotinic overstimulation.

The reported cardiotoxicity of organophosphates includes: electrocardiogram abnormalities (QT prolongation, ST-T abnormalities, and sinus tachycardia), myocardial infarction, impaired systolic and diastolic performance (reduced systolic and diastolic performances, decreased left ventricular fractional shortening, and ejection fraction), functional remodeling (changes in heart structure and function, such as left ventricular dysfunction), histopathological findings (haemorrhage, vacuolization, signs of apoptosis, degeneration, and cytoplasmic edema in myocardial cells), oxidative stress (increased oxidative stress markers like lipid peroxidation, protein carbonyls, and altered antioxidant enzyme activities), biochemical complications (distorted lipid metabolism, increased cytotoxicity enzyme activities, and altered lipidemic profiles), coronary artery disease (increased levels of coronary heart disease markers), blood pressure disorders (hypotension and changes in blood pressure regulation). Tachycardia and hypertension as signs of excessive stimulation of nicotinic receptors (vs. signs of muscarinic overstimulation - brachycardia and hypertension) are most often of central origin and signs of severe poisoning [16, 17]. These effects were observed in both animal models and human cases, indicating a significant risk of cardiovascular complications due to organophosphate exposure [12, 18, 19].

Pyrethoroids

The primary site of action for pyrethroids (e.g., cypermethrin, permethrin, allethrin, resmethrin, cyfluthrin or esfenvalerate) has generally been considered to be interference with the sodium gate in the nerve membrane. Pyrethroide insecticides slow the closing of sodium channel gates following an initial influx of sodium during the depolarizing phase of an action potential, which results in the excitatory phenomena similar to those that appear in case of organochlorine poisonings [8]. Two different types of pyrethroids are recognized, based on differences in basic structure (the presence or absence of a cyano group in the alpha position), and the symptoms of poisoning. The term T-syndrome (from tremor) has been applied to type I responses, while the term CS-syndrome (from choreoathetosis and salivation) has been applied to type II pyrethroids which have a cyano group in their chemical structure. The direct action of pyrethroids on sensory nerve endings frequently leads to paresthesia, usually seen in occupational exposure to pyrethroids [20]. Pyrethroids may prolong the ventricular action potentials and evoke after-depolarizations, indicating an arrythmogenic activity. The cardiotoxicity of pyrethroids in humans and animals also includes risk of coronary heart disease, cardiac muscle hypotrophy along with increased heart cells' calcium, reactive oxygen species (ROS) generation in heart tissues, increased lipid peroxidation, altered the antioxidant capacity in heart cells' plasma membrane and oxidative DNA base modifications [19].

Dipyridyles

In this group of pesticides used for the control of a wide range of grasses and weeds, paraquat is one of the most commonly used broad spectrum herbicides worldwide. It is highly toxic to both humans and animals. It is capable to form a superoxide anion, and directly or indirectly induces NO synthase-mediated nitric oxide production. Its toxicity results from the ROS generation and nitrite species via the redox cycle process and causes oxidative stress-related damage to cellular organelles, proteins, nucleic acids, and lipids [21]. It potentially causes acute respiratory distress syndrome. Severe paraquat toxicity causes multi-organ failure with histological and functional changes in the lungs, kidneys, liver, myocardium, and adrenal glands. Paraquat cardiotoxicity is associated with oxidative stress, myocardial inflammation, apoptosis and endoplasmic reticulum (ER) stress.

Along with myocardial contractility suppression, which seems to be the main myocardial effect, paraquat intoxication may also induce a geometric transformation of the left ventricle including enlarged left ventricular end systolic diameter [19].

Triazines

Atrazine ia the most studied triazine herbicide, not only because its intrinsic toxicity but also because it is stable in the environment and can thus particularly contaminate surface and ground waters. It can interfere with the binding of some ligands to the $GABA_A$ receptor in a noncompetitive manner. Atrazine has the properties of an endocrine disruptor as it may interfere with female and male hormone regulation and activity [22]. It induced cardiotoxicity via the modulation of cardiac ATPase including Na^+ -K^+ -ATPase, Ca^{2+} -ATPase, Mg^{2+} -ATPase and changes in the transcription of pump subunits, leading to ionic disorder [23]. Additionally, it significantly decreases glutathione GSH and total thiol content both in serum and in cardiac tissue promoting thus oxidative damage and apoptosis [3].

Triazoles

Triazole fungicides (e.g. difenoconazole, epoxiconazole, metconazole, paclobutrazol, tebuconazole, tetraconazole, triadimenol, triticonazole) are widely used in agriculture and are stable in water and soil. These fungicides have been connected to induce carcinogenicity, endocrine disruption and impairment of neurological, immunological, and reproductive functions [24]. They trigger the oxidative stress and the inhibition of antioxidant-defense enzyme. By these mechanisms, they may provoke acute systolic heart failure [3].

Evidence on Cardiovascular Toxicity of Pesticides in Humans

The majority of human data on cardiotoxicity of pesticides come from acute poisoning cases and epidemiological data on worker exposure, showing a close association between exposure to pesticides and cardiovascular diseases (CVDs) such as congestive heart failure, arrhythmia, direct cardiac electrophysiological changes, myocardial infarction, and sudden death in humans [3, 5, 11, 12, 25, 26]. In occupational settings, significant exposure mainly comes from inadequat protection measures or the level of compliance with them. However, long-term low-level exposure is very often underestimated in assessing the risk from pesticide exposure.

Agricultural studies have also shown that exposure to pesticides is associated with increased blood pressure, elevated serum lipoproteins, and atherosclerosis which are known risk factors for CVD as well as a significant correlation between pesticide exposure and some symptoms of CVDs like chest pain, palpitation, and leg swelling. In a cohort study on occupational exposure that involved more than 7500 participants, a significant relationship has been found between exposure to pesticides and an increase in the incidence of CVDs [27]. Mohammadkhani et al. (2023) have shown that high levels of circulating p,p'-DDE were associated with hypertension [11]. In a retrospective study, the clinical data on acute exposure to organophosphorus pesticide have shown a significantly higher incidence of cardiac arrhythmia and heart failure compared to the control group [28]. In the acute poisonings, glyphosate induced heart conduction problems leading to the development of life-threatening arrhythmias [12]. An intentional cypermethrin poisoning caused prolonged bradycardia explained by its complex interactions with voltage gated sodium and calcium channels and gamma amino butyric acid mediated chloride channels [29].

Cardiovascular Toxicity of Pesticides in Animals

Results from animal studies also support pesticide induced cardiovascular adverse effects. Research on animal models, such as zebrafish and rodents, underpin the findings of pesticide-induced cardiovascular dysfunctions, including cardiac malformations and histopathological alterations. In embryonic zebrafish exposed to methidathion, cardiotoxicity, and cell apoptosis have been ascribed to the development of OS by increasing SOD activity, ROS, and MDA, as well as the expression of pro-inflammatory factors [30]. In zebrafish glyphosate caused cardiotoxicity was associated with DNA damage and mitochondrial damage [31]. Oral administration of atrazine in mice led to cardiac fibrosis, evidenced by histopathological scores and elevated heart index, interstitial collagen deposition, and extensive myofiber damage [32]. Like atrazine, tebuconazole an effective systemic fungicide of the triazoles family, given orally for 28 consecutive days in rats caused development of the fibrosis in the myocardium. This finding was accompanied with a decrease in cardiac AChE, an increase in serum marker enzymes such as creatinine phosphokinase (CPK) and lactate dehydrogenase (LDH), and serum levels of total cholesterol (T-CHOL), triglyceride (TG), low-density lipoprotein cholesterol (LDL-C) and reduced high-density lipoprotein cholesterol (HDL-C) levels [33]. Tebuconazole also caused significant electrocardiogram (ECG) remodeling with prolonged PR and QTc interval duration due to inteferencies with the transport of the calcium, sodium and potassium ions in the cardiomyocytes [24]. Subacute oral exposure to chlodecone in rats was associated with significantly increased atrial fibrillation (AF) incidence, with fibrosis and inflammation being important contributors to AF vulnerability [34].

Mechanistic Insight into the Pesticide Cardiotoxicity

Oxidative stress is identified as a major mediator of pesticide-induced CVDs. Pesticides generate reactive oxygen species (ROS) [35], leading to endothelial dysfunction, increased vascular reactivity, and atherosclerosis. The activity of the antioxidant enzymes including superoxide dismutase (SOD), catalase (CAT), and glutathione peroxidase (GPx), and downregulated the mRNA expression levels of Nrf2 have been typically used to reveal the mechanistic insight into OS generated by the pesticide [36]. Besides, cell migration, apoptotic process, protein phosphorylation, and signal transduction might also be linked to pesticide-related cardiotoxicity [37].

It was demonstrated that cell exposure to DDVP, imidacloprid, and thiamethoxam induced cell injury/death by necroptosis in cardiomyoblast H9c2 cell line [38, 39]. The toxicity of acetamide on human cardiomyocyte (AC16) cell line and zebrafish embryonic heart has been related to the induction of ROS, and subsequent activation of the caspase-3 and caspase-9 signaling pathways, causing the apoptosis [40]. A pyrethroid insecticide, fenpropathrin, caused dose-dependent cardiotoxicity in zebrafish associated to its high affinity for various voltage-gated sodium channels [41]. Similarly, chlorantraniliprole can cause decrease in voltage-activated Ca^{2+} currents densities [42]. Calcium ion release and reabsorption in the cardiomyocyte sarcoplasmic reticulum was mediated by DDT, *i.e.* its interaction with Ryanodine receptor type 2 (RyR2) and SR/ER Ca2 + ATPase (SERCA2a) [11]. Biomarkers of toxicity in rats given a pyrethroid permethrin were increased Nrf2 gene expression levels, calcium, and cardiac hypotrophy [43]. Atrazine promoted cardiomyocyte pyroptosis via the activation of NF-κB pathway and induced excessive production of inflammatory factors [32]. Morphological enlargement and senescence of AC16 human cardiomyocytes exposed to glyphosate was caused by ROS-mediated DNA damage. Glyphosate also altered endoplasmic reticulum protein processing signals, causing endoplasmic reticulum stress [31]. Metabolomic analyses of the atrial fibrillation in rats induced by several commonly used pesticides showed that impared mitochondrial function and energy metabolism were involved in their cardiotoxicity [39].

Conclusion

Pesticide poisoning results from occupational, accidental and intentional exposure. On the other hand, their environmental presence is directly associated with long-life low-level exposure. Generally, pesticide exposure represents one of the risk factors accompanied by the development of CVDs, a prominent global threat to human health. In this respect, the utmost challenge is to build capacity in sound pesticide management across both agriculture and public health sectors following the principles of One Health as an integrated, unifying approach that aims to sustainably balance and optimize the health of people, animals and ecosystems.

References

1. FAO (2023) World food and agriculture—statistical yearbook 2023. In: World food and agriculture—statistical yearbook 2023. FAO. https://doi.org/10.4060/cc8166en
2. Busch W, Schmidt S, Kühne R, Schulze T, Krauss M, Altenburger R (2016) Micropollutants in European rivers: A mode of action survey to support the development of effect-based tools for water monitoring. Environ Toxicol Chem 35:1887–1899. https://doi.org/10.1002/etc.3460
3. Georgiadis N, Tsarouhas K, Dorne JLCM, Kass GEN, Laspa P, Toutouzas K, Koulaouzidou EA, Kouretas D, Tsitsimpikou C (2022) Cardiotoxicity of chemical substances: an emerging Hazard class. J Cardiovasc Dev Dis 9:226. https://doi.org/10.3390/jcdd9070226
4. Maksimović ŽM, Jović-Stošić J, Vučinić S, Perković-Vukčević N, Vuković-Ercegović G, Škrbić R, Stojiljković MP (2023) Acute organophosphate and carbamate pesticide poisonings—a five-year survey from the National Poison Control Center of Serbia. Drug Chem Toxicol 46:113–121. https://doi.org/10.1080/01480545.2021.2012481
5. Adeyemi JA, Ukwenya VO, Arowolo OK, Olise CC (2021) Pesticides-induced cardiovascular dysfunctions: prevalence and associated mechanisms. Curr Hypertens Rev 17:27–34. https://doi.org/10.2174/1573402117666210111102508
6. WHO (2024) WHO methods and data sources for global burden of disease estimates 2000-2011. WHO, p 86
7. El-Nahhal Y, El-Nahhal I (2021a) Cardiotoxicity of some pesticides and their amelioration. Environ Sci Pollut Res 28:44726–44754. https://doi.org/10.1007/s11356-021-14999-9
8. Coats JR (1990) Mechanisms of toxic action and structure-activity relationships for organochlorine and synthetic pyrethroid insecticides. Environ Health Perspect 87:255–262. https://doi.org/10.1289/ehp.9087255
9. Faroon O, Ruiz P, Jones D, Keith S, Scinicariello F, Zaccaria K, Citra M, Herber D, Honey P, Ingerman L, Llados F, McClure P, Stevens S (2022) Health effects. In: Introduction. Agency toxic Subst. Dis. Regist, pp 15–255
10. IARC (2018) IARC monographs on the evaluation of carcinogenic risks to humans: DDT, LINDANE, and 2,4-D. IARC
11. Ali Mohammadkhani M, Shahrzad S, Haghighi M, Ghanbari R, Mohamadkhani A (2023) Insights into organochlorine pesticides exposure in the development of cardiovascular diseases: a systematic review. Arch Iran Med 26:592–599. https://doi.org/10.34172/aim.2023.86
12. Le Quilliec E, Fundere A, Al-U'datt DGF, Hiram R (2023) Pollutants, including organophosphorus and organochloride pesticides, may increase the risk of cardiac remodeling and atrial fibrillation: a narrative review. Biomedicines 11:2427. https://doi.org/10.3390/biomedicines11092427
13. Antonijevic B, Stojiljkovic MP (2007) Unequal efficacy of pyridinium oximes in acute organophosphate poisoning. Clin Med Res 5:71–82. https://doi.org/10.3121/cmr.2007.701
14. Casida JE, Quistad GB (2005) Serine hydrolase targets of organophosphorus toxicants. Chem Biol Interact 157–158:277–283. https://doi.org/10.1016/j.cbi.2005.10.036
15. Stojiljković MP, Škrbić R, Jokanović M, Kilibarda V, Bokonjić D, Vulović M (2018) Efficacy of antidotes and their combinations in the treatment of acute carbamate poisoning in rats. Toxicology 408:113–124. https://doi.org/10.1016/j.tox.2018.08.017
16. Maksimović ŽM, Duka D, Bendarčuk N, Škrbić R, Stojiljković MP (2021) Onset rate and intensity of signs of organophosphate poisoning related to paraoxon dose and survival in rats. Scr Med 52:49–58. https://doi.org/10.5937/scriptamed52-31191
17. Stojiljković MP, Škrbić R, Maksimović ŽM (2024) In: Djuric DM, Agrawal DK (eds) Cardiovascular effects of acetylcholinesterase inhibitors. Springer, Cham, pp 233–257
18. El-Nahhal Y, El-Nahhal I (2021b) Cardiotoxicity among children accidentally exposed to organophosphate insecticides coupled with scorpion bites in Gaza. Health (Irvine) 13:1045–1063. https://doi.org/10.4236/health.2021.139078

19. Georgiadis N, Tsarouhas K, Tsitsimpikou C, Vardavas A, Rezaee R, Germanakis I, Tsatsakis A, Stagos D, Kouretas D (2018) Pesticides and cardiotoxicity. Where do we stand? Toxicol Appl Pharmacol 353:1–14. https://doi.org/10.1016/j.taap.2018.06.004
20. Agency for Toxic Substances and Disease Registry, (ATSDR) (2014) Toxicological profile for pyrethrins and pyrethroids. S. Dep. Heal. Hum. Serv. Public Heal. Serv, Atlanta, GA U
21. Onur B, Çavuşoğlu K, Yalçin E, Acar A (2022) Paraquat toxicity in different cell types of Swiss albino mice. Sci Rep 12:1–11. https://doi.org/10.1038/s41598-022-08961-z
22. Agency for Toxic Substances and Disease Registry (ATSDR) (2003) Toxicological Profile for Atrazine. In: ATSDR's Toxicol Profiles. ATSDR. https://doi.org/10.1201/9781420061888_ch141
23. Yang Y, Wei S, Zhang B, Li W (2021) Recent Progress in environmental toxins-induced cardiotoxicity and protective potential of natural products. Front Pharmacol 12:1–17. https://doi.org/10.3389/fphar.2021.699193
24. Santos-Miranda A, Joviano-Santos JV, Cruz-Nascimento T, Neri EA, Souza DS, Marques LP, Krieger JE, Roman-Campos D (2022) The fungicide Tebuconazole induces electromechanical cardiotoxicity in murine heart and human cardiomyocytes derived from induced pluripotent stem cells. Toxicol Lett 359:96–105. https://doi.org/10.1016/j.toxlet.2022.02.005
25. Bulka CM, Daviglus ML, Persky VW, Durazo-Arvizu RA, Lash JP, Elfassy T, Lee DJ, Ramos AR, Tarraf W, Argos M (2019) Association of occupational exposures with cardiovascular disease among US Hispanics/Latinos. Heart 105:439–448. https://doi.org/10.1136/heartjnl-2018-313463
26. Mills KT, Blair A, Freeman LEB, Sandler DP, Hoppin JA (2009) Pesticides and myocardial infarction incidence and mortality among male pesticide applicators in the agricultural health study. Am J Epidemiol 170:892–900. https://doi.org/10.1093/aje/kwp214
27. Berg ZK, Rodriguez B, Davis J, Katz AR, Cooney RV, Masaki K (2019) Association between occupational exposure to pesticides and cardiovascular disease incidence: the Kuakini Honolulu heart program. J Am Heart Assoc 8:1–9. https://doi.org/10.1161/JAHA.119.012569
28. Chen KX, Zhou XH, Sun CA, Yan PX, Bil J (2019) Manifestations of and risk factors for acute myocardial injury after acute organophosphorus pesticide poisoning. Medicine (United States) 98:1–5. https://doi.org/10.1097/MD.0000000000014371
29. Shilpakar O, Karki B (2021) Cypermethrin poisoning manifesting with prolonged bradycardia: a case report. Tox Rep 8:10–12. https://doi.org/10.1016/j.toxrep.2020.12.005
30. Wu Y, Chen T, Xia Y, Wang J, Wang A, Wang B, Wang J, Yao W (2023) Developmental toxicity, immunotoxicity and cardiotoxicity induced by methidathion in early life stages of zebrafish. Pestic Biochem Physiol 194:105526. https://doi.org/10.1016/j.pestbp.2023.105526
31. Lu J, Zhang C, Wang W, Xu W, Chen W, Tao L, Li Z, Zhang Y, Cheng J (2023) Exposure to environmental concentrations of glyphosate induces cardiotoxicity through cellular senescence and reduced cell proliferation capacity. Ecotoxicol Environ Saf 261:115112. https://doi.org/10.1016/j.ecoenv.2023.115112
32. Zhao X, Luo H, Yao S, Yang T, Fu F, Yue M, Ruan H (2024) Atrazine exposure promotes cardiomyocyte pyroptosis to exacerbate cardiotoxicity by activating NF-κB pathway. Sci Total Environ 915:170028. https://doi.org/10.1016/j.scitotenv.2024.170028
33. Ben Othmène Y, Hamdi H, Annabi E, Amara I, Ben Salem I, Neffati F, Najjar MF, Abid-Essefi S (2020) Tebuconazole induced cardiotoxicity in male adult rat. Food Chem Toxicol 137: 111134. https://doi.org/10.1016/j.fct.2020.111134
34. Fundere A, Rose A, Xiong F, Muthukumarasamy KM, Altuntas Y, Dasari H, Villeneuve L, Sirois MG, Tanguay JF, Tardif JC, Hiram R (2024) Daily exposure to chlordecone, an organochlorine pesticide, increases cardiac fibrosis and atrial fibrillation vulnerability. J Hazard Mater 478:135533. https://doi.org/10.1016/j.jhazmat.2024.135533
35. Chen T, Chen H, Wang A, Yao W, Xu Z, Wang B, Wang J, Wu Y (2023) Methyl parathion exposure induces development toxicity and cardiotoxicity in zebrafish embryos. Toxics 11:1–18. https://doi.org/10.3390/toxics11010084

36. Peritore AF, Franco GA, Molinari F, Arangia A, Interdonato L, Marino Y, Cuzzocrea S, Gugliandolo E, Britti D, Crupi R (2023) Effect of pesticide vinclozolin toxicity exposure on cardiac oxidative stress and myocardial damage. Toxics 11:1–13
37. Guo Y, Li N (2024) Network toxicology and molecular docking to investigative the non-acetylcholinesterase mechanisms and targets of cardiotoxicity injury induced by organophosphorus pesticides. Medicine (United States) 103:e39963. https://doi.org/10.1097/MD.0000000000039963
38. Ben Salem I, Boussabbeh M, Pires Da Silva J, Saidi NE, Abid-Essefi S, Lemaire C (2023) Effects of Dichlorvos on cardiac cells: toxicity and molecular mechanism of action. Chemosphere 330:138714. https://doi.org/10.1016/j.chemosphere.2023.138714
39. Ma Y, Tao Q, Huang L, Wang Q (2024) Comparative cytotoxicity and mitochondrial disruption in H9c2 cardiomyocytes induced by common pesticides. Chemosphere 367:143627. https://doi.org/10.1016/j.chemosphere.2024.143627
40. Wang X, Geng L, Wu M, Xu W, Cheng J, Li Z, Tao L, Zhang Y (2023) Molecular mechanisms of cardiotoxicity induced by acetamide and its chiral isomers. Sci Total Environ 900:166349. https://doi.org/10.1016/j.scitotenv.2023.166349
41. Saputra F, Lai YH, Roldan MJM, Alos HC, Aventurado CA, Vasquez RD, Der Hsiao C (2023) The effect of the Pyrethroid pesticide Fenpropathrin on the cardiac performance of zebrafish and the potential mechanism of toxicity. Biology (Basel) 12:1214. https://doi.org/10.3390/biology12091214
42. Kaabeche M, Charreton M, Kadala A, Mutterer J, Charnet P, Collet C (2024) Cardiotoxicity of the diamide insecticide chlorantraniliprole in the intact heart and in isolated cardiomyocytes from the honey bee. Sci Rep 14:1–13. https://doi.org/10.1038/s41598-024-65007-2
43. Dhivya Vadhana MS, Siva Arumugam S, Carloni M, Nasuti C, Gabbianelli R (2013) Early life permethrin treatment leads to long-term cardiotoxicity. Chemosphere 93:1029–1034. https://doi.org/10.1016/j.chemosphere.2013.05.073

Chapter 21
Cardiotoxicity of Gaseous Poisons

Marijana Curcic, Evica Antonijević Miljaković, and Biljana Antonijević

Abstract Cardiotoxicity resulting from gaseous pollutants can present diverse symptoms, contingent upon the gas kind, concentration, and exposure length. The issue is a major concern in medicine, especially in nations where air quality fails to meet legal standards and levels of gaseous pollutants exceed permissible limits. Common symptoms of cardiotoxicity due to gaseous poisons include chest pain, a sensation of pressure or tightness in the chest, palpitations, shortness of breath potentially accompanied by a feeling of suffocation, dizziness or lightheadedness resulting from diminished oxygen delivery to the brain, syncope, fatigue or weakness often attributable to reduced oxygen supply and cardiac function, nausea or vomiting, cyanosis, hypotension, and arrhythmias. Some well-known gaseous substances that are cardiotoxic include carbon monoxide (CO), sulfur dioxide (SO_2), nitrogen dioxide (NO_2), chlorine (Cl), cyanide gas (HCN) etc. Interference with the oxigen transport could affect almost all tissues, while other four mechanisms, direct myocardial damage, electrophysiological alterations, inflammatory responses, and autonomic nervous system disruption could be hazardous, as well, and cause severe carditoxicity.

Keywords Carbon monoxide · Sulfur dioxide · Nitrogen dioxide · Chlorine · Cyanide gas

---

M. Curcic (✉) · E. A. Miljaković · B. Antonijević
University of Belgrade—Faculty of Pharmacy, Belgrade, Serbia

University of Belgrade—Faculty of Pharmacy; Toxicological Risk Assessment Center, Belgrade, Serbia
e-mail: Marijana.curcic@pharmacy.bg.ac.rs

D. K. Agrawal et al. (eds.), *Cardiovascular Toxicity*, Advances in Biochemistry in Health and Disease 33, https://doi.org/10.1007/978-3-031-90598-8_21

Introduction

Cardiotoxicity denotes the detrimental effects on the heart, as well as other parts of cardiovascular system or brain centers, resulting from exposure to gaseous poisons. This disorder may result in many cardiovascular issues, such as cardiomyopathy, arrhythmias, hypertension, and perhaps heart failure. The issue is a major concern in medicine, especially in nations where air quality fails to meet legal standards and levels of gaseous pollutants exceed permissible limits. Cardiotoxicity resulting from gaseous pollutants can present diverse symptoms, contingent upon the gas kind, concentration, and exposure length. Common symptoms of cardiotoxicity due to gaseous poisons include chest pain, a sensation of pressure or tightness in the chest, palpitations, shortness of breath potentially accompanied by a feeling of suffocation, dizziness or lightheadedness resulting from diminished oxygen delivery to the brain, syncope, fatigue or weakness often attributable to reduced oxygen supply and cardiac function, nausea or vomiting, cyanosis, hypotension, and arrhythmias. The symptoms may differ significantly based on the individual's health status, the particular gas involved, and the degree of exposure. In the event of suspected exposure to a hazardous gas, it is imperative to obtain prompt medical assistance.

Cardiotoxicity Mechanisms of Gaseous Poisons

Gaseous poisons may induce cardiotoxicity through several cellular and physiological mechanisms (Fig. 21.1). These include:

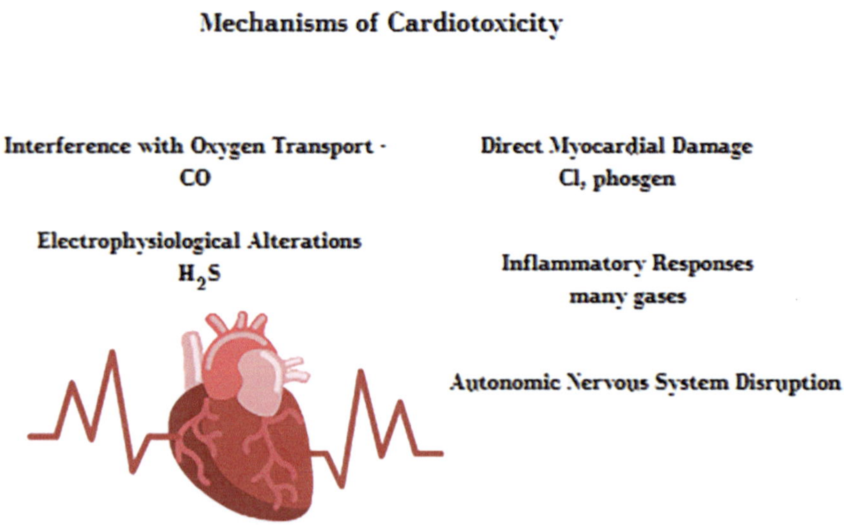

Fig. 21.1 Major cellular and physiological changes in the underlying mechanisms of poisons-induced cardiotoxicity

1. Interference with Oxygen Transport—Numerous gaseous poisons, including CO, hinder the blood's capacity to convey oxygen. Carbon monoxide binds to hemoglobin with a higher affinity than oxygen, resulting in the formation of carboxyhemoglobin, which diminishes oxygen transport to the cardiac muscle, causing hypoxia and consequent myocardial damage.
2. Direct Myocardial Damage—Certain gases, including chlorine and phosgene, have the potential to directly harm myocardial cells. These gases can induce oxidative stress, resulting in lipid peroxidation of cell membranes, enzyme inactivation, and apoptosis of cardiac cells.
3. Electrophysiological Alterations—Interference with cardiac ion channels at the electrophysiological level can result in arrhythmias. Hydrogen sulfide has the capacity to modulate cardiac ion channels, resulting in alterations to cardiac action potential and arrhythmogenic activity.
4. Inflammatory Responses—Exposure to certain gases may provoke an inflammatory reaction in the heart. Cytokine release can facilitate this process, resulting in myocarditis and negatively affecting heart function over time.
5. Autonomic Nervous System Disruption—Certain gases influence the autonomic nervous system, potentially resulting in abnormalities in cardiac sympathetic and parasympathetic activity. This disturbance may induce changes in heart rate and increase the risk of arrhythmias.

As mentioned, cardiotoxic gaseous poisons can affect the heart's function and may lead to serious cardiac issues. Some well-known gaseous substances that are cardiotoxic include carbon monoxide (CO), sulfur dioxide (SO_2), nitrogen dioxide (NO_2), chlorine (Cl), cyanide gas (HCN), and phosgene (ClCOCl).

Carbon Monoxide (CO)

Carbon monoxide is a colorless, non-irritating, odorless, and tasteless gas. It is present in both outdoor and indoor air [1].

Exposure to carbon monoxide in the general population can arise from multiple sources. Motor vehicles are a major source of carbon monoxide emissions, particularly in urban regions characterized by severe traffic congestion. Inadequately ventilated garages or tunnels might result in increased carbon monoxide exposure. Appliances include gas stoves, ovens, barbecues, and water heaters that utilize natural gas, oil, coal, or wood can generate carbon monoxide if they malfunction or lack sufficient ventilation. Indoor heating systems, especially those utilizing fossil fuels, may produce carbon monoxide if inadequately maintained. This encompasses furnaces, fireplaces, and wood-burning stoves. Tobacco smoke contains carbon monoxide (CO), resulting in higher CO levels in the atmosphere and bloodstream of active smokers and individuals exposed to secondhand smoke indoors. Utilizing charcoal grills indoors or in enclosed areas without enough ventilation can result in indoor carbon monoxide accumulation, presenting a poisoning hazard. Residential fires generate carbon monoxide as a consequence of incomplete combustion. Individuals in proximity to a burning structure may be exposed to carbon monoxide, presenting a health hazard. Industries including steel making, petroleum refining, and chemical production can emit CO into the surrounding environment. Utilizing gasoline or diesel-powered generators, lawnmowers, and other small engines in confined areas, such as basements or garages, may result in dangerous carbon monoxide concentrations. Engines on vessels and RV generators lacking adequate ventilation can emit carbon monoxide, resulting in potential exposure for those aboard [1–6].

Upon entering the human body by breathing, carbon monoxide induces cardiotoxicity through the following processes. Carbon monoxide attaches to hemoglobin in red blood cells with an affinity 250 times greater than that of oxygen. This forms carboxyhemoglobin, which markedly diminishes the capacity of red blood cells to convey oxygen throughout the body. Consequently, tissues, especially the heart, endure significant hypoxia, undermining cellular metabolism, function, and energy production. Oxygen is essential for oxidative phosphorylation in mitochondria, where ATP, the cell's energy currency, is generated. Without sufficient oxygen, the cardiac muscle fails to produce adequate ATP, reducing its capacity to contract and effectively circulate blood [1].

CO exposure can produce hypoxic conditions that result in heightened production of reactive oxygen species (ROS) in cardiac cells. Oxidative stress can harm cellular components such as lipids, proteins, and DNA, resulting in heart cell injury and inflammation. Inflammatory reactions worsen myocardial injury and may facilitate the onset of heart disorders. Carbon monoxide directly interacts with cellular mitochondria, impairing energy generation and the regulation of apoptosis. Carbon monoxide‘s interference with mitochondrial enzyme activity disrupts electron transport and ATP generation, critical processes for the viability and function of cardiac muscle cells.The heart depends on a constant and adequate supply of oxygen to sustain its contractile function; hence, carbon monoxide impairs myocardial contractility, diminishing the heart's pumping efficiency. Prolonged and severe exposure might result in myocardial depression, potentially leading to cardiogenic shock or heart failure. The interruption of oxygen supply and mitochondrial function, along with the accumulation of acidic metabolites from anaerobic respiration, disrupts normal cardiac electrical activity, potentially leading to arrhythmias. Carbon monoxide can impact the endothelial cells of blood arteries, impairing the control of vascular tone and inducing vasoconstriction, which may elevate blood pressure and diminish coronary blood flow, so exacerbating cardiac stress. The aforementioned mechanisms demonstrate the intricate ways in which carbon monoxide exposure can adversely affect the cardiovascular system, especially at elevated or prolonged exposure levels, underscoring the necessity of preventing such exposure and providing adequate treatment and care for affected individuals.

Acute carbon monoxide poisoning can impact nearly all tissues by affecting the cardiovascular system and neurological system, resulting in a wide range of symptoms. The symptoms of mild carbon monoxide poisoning include headache, nausea, vomiting, dizziness, blurred vision, and occasionally cherry red lips and skin.

Headache and dizziness are the most prevalent symptoms. Mild carbon monoxide poisoning can be misdiagnosed due to its similarity to flu-like viral diseases. Symptoms of moderate carbon monoxide poisoning include confusion, syncope, chest discomfort, dyspnea, weakness, tachycardia, tachypnea, and rhabdomyolysis. Severe poisoning can cause life-threatening symptoms such as cardiac arrhythmias, myocardial ischemia, cardiac arrest, hypotension, respiratory arrest, noncardiogenic pulmonary edema, convulsions, and coma. Exposure can cause neuropsychiatric impairment, with symptoms such as inappropriate euphoria, impaired judgment, poor concentration, memory loss, cognitive and personality changes, psychosis, and Parkinsonism, occurring within days to 3–4 weeks. Acute carbon monoxide poisoning symptoms in children are similar to those in adults [1, 7].

The non-hypoxic mechanisms of action of carbon monoxide are primarily attributed to its binding to heme in proteins other than hemoglobin. Carbon monoxide significantly affects components of various critical physiological regulatory systems, particularly those related to brain and muscle oxygenation. Alongside inhalation exposure to carbon monoxide in the atmosphere, internal exposures arise from the endogenous generation of carbon monoxide through several precursors, such as heme breakdown, auto-oxidation of phenols, and photo-oxidation of organic molecules. and lipid peroxidation of cellular membrane lipids, as well as from the oxidative metabolism of exogenous precursors (e.g., carbon tetrachloride, dichloromethane, and other dihalomethanes). The two metabolic sources of carbon monoxide lead to a body burden of carbon monoxide without external exposure to carbon monoxide in the air. The endogenous generation rate of carbon monoxide is predicted to be roughly 0.42 mL at standard temperature and pressure [1].

Storage and utilization mechanisms include myoglobin and neuroglobin. The nitric oxide cell signaling pathway involves nitric oxide synthase and guanylyl cyclase. The prostaglandin cell signaling pathway encompasses cyclooxygenase and prostaglandin H synthase. Energy metabolism and mitochondrial respiration are facilitated by cytochrome c oxidase, cytochrome c, and NADPH oxidase. Steroid and drug metabolism is mediated by cytochrome P450. Cellular redox balance and reactive oxygen species (ROS) are regulated by catalase and peroxidases. Various transcription factors, such as neuronal PAS domain protein and $NPAS_2$, are implicated in the regulation of circadian rhythm [1].

Recent toxicological and epidemiological studies have concentrated on investigating the health impacts of low-level carbon monoxide exposure that does not lead to evident carbon monoxide poisoning, while also seeking to elucidate the relationships between carbon monoxide toxicity and its in vivo production and metabolism. This research has demonstrated that the heart, circulatory system, brain, and developing nervous system exhibit heightened sensitivity to carbon monoxide. These investigations have additionally demonstrated that individuals with chronic cardiovascular and/or respiratory conditions may be especially susceptible to carbon monoxide exposure [1].

In conclusion, the cardiovascular system exhibits sensitivity to changes in oxygen supply. Inhaled carbon monoxide restricts oxygen availability, potentially impacting the cardiovascular system. The severity of hypoxia and the level of tissue damage are contingent upon the carbon monoxide dosage. Numerous reviews have examined the impact of carbon monoxide on the cardiovascular system particularly assessing the impact of carbon monoxide on ventricular fibrillation, hemodynamics, cardiomegaly, hematology, and atherosclerosis ([1]; INCHEM, 1999).

Sulfur Dioxide (SO_2)

Sulfur dioxide (SO_2) is a colorless gas possessing a distinctive, unpleasant, acrid odor. Exposure to SO_2 within the general populace may arise from various causes. SO_2 is a prevalent byproduct of industrial activities, especially those that involve the combustion of low-grade fossil fuels containing sulfur impurities in power plants and oil refineries. Industries that extract or refine metal ores, including copper and zinc, also emit SO_2. The combustion of coal and oil in power plants, industrial facilities, and domestic heating systems can emit sulfur dioxide into the atmosphere. In certain localities, the combustion of sulfur-containing fuels in automobiles, trucks, and buses contributes to SO_2 concentrations, although the use of cleaner fuels and emission controls has mitigated this source in numerous areas. In certain areas, particularly in underdeveloped nations, the combustion of biomass or other sulfur-laden substances for cooking can emit SO_2. Natural sources of SO_2, such as volcanic eruptions, can emit substantial quantities of sulfur dioxide into the atmosphere, impacting air quality throughout extensive regions. In addition to volcanic activity, other natural processes, such as the decomposition of organic materials, can generate sulfur compounds that may transform into SO_2 in the atmosphere [8].

Sulfur dioxide is employed in numerous industries. It is used to produce sulfuric acid, paper, and food preservatives. Some examples of workers at danger of being exposed to SO_2 are as follows: factory workers in industries where it occurs as a byproduct, such as copper smelting or power plants, industry workers who produce sulfuric acid, workers in paper manufacturing and food processing, including dry fruit preservation,workers who create fertilizer etc. [8].

Exposure to SO_2 can lead to irritation of the eyes, nose, and throat. Symptoms encompass nasal mucus, choking, cough, reflex bronchoconstriction, and in the presence of fluids, frostbite. Workers may suffer adverse effects from exposure to SO_2. The extent of exposure is contingent upon the dosage, duration, and nature of the task performed [8].

The majority of effects are localized in the upper respiratory tract, as 90% of inhaled SO_2 is quickly deposited there. However, in cases of significant exposure, enough gas can reach the lower airways, leading to chemical pneumonitis and pulmonary edema [8–12].

Sulfur dioxide can have cardiotoxic effects via multiple routes, affecting cardiovascular health. Inhalation of SO_2 can provoke oxidative stress by the production of reactive oxygen species (ROS). Reactive oxygen species can harm cardiac cells by oxidizing lipids, proteins, and DNA, resulting in cellular injury and malfunction in cardiac tissues. Exposure to SO_2 can initiate inflammatory pathways, resulting in the secretion of pro-inflammatory cytokines [11]. The inflammatory reaction may harm endothelial cells and facilitate the progression of atherosclerosis, hence elevating the risk of cardiovascular illnesses. Sulfur dioxide can compromise endothelial function, essential for sustaining vascular homeostasis. This malfunction may cause diminished nitric oxide generation, a vasodilator, leading to heightened vascular resistance and hypertension. Exposure to SO_2 can disrupt the autonomic nervous system, resulting in imbalances that influence heart rate and blood pressure regulation, potentially leading to arrhythmias and other cardiac complications [12]. SO_2 may exert direct toxic effects on myocardial cells, diminishing their contractility and resulting in impaired heart function. This may impair the heart's capacity to pump blood efficiently, leading to heart failure. Sulfur dioxide has a negative impact on the human, cardiovascular, respiratory and nervous systems, resulting in type 2 diabetes and non-accidental mortality. Prolonged exposure to SO_2 can result in detrimental cardiac remodeling, marked by alterations in the heart's size, shape, and function. This renovation may elevate the chance of heart disease development.Although some data suggests that sulfur dioxide at certain concentrations has no negative health consequences, its synergistic impact with other air pollutants may be considerable [8–10].

These processes underscore the possible cardiovascular hazards linked to SO_2 exposure, stressing the necessity of air quality surveillance and reducing exposure to this pollutant, especially among at-risk groups such as individuals with pre-existing cardiovascular ailments [8].

Nitrogen Dioxide (NO_2)

The general population is exposed to NO_2 from a variety of sources. Vehicle emissions are the most common source of NO_2 in metropolitan areas, with cars, buses, and trucks accounting for the majority. NO_2 is a major byproduct of combustion engines that use fossil fuels. The concentration of NO_2 is especially high during peak traffic hours and in locations with regular congestion. Efforts to minimize NO_2 emissions from automobiles include promoting electric vehicles, creating low-emission zones in cities, and improving public transit infrastructure. The use of fossil fuels in industries such as power generating, oil refining, and manufacturing produces NO_2. Power stations, particularly those run on coal or oil, contribute significantly to NO_2 emissions. Indoor air quality can suffer greatly as a result of combustion-related activities such as smoking, burning candles, or utilizing wood-burning stoves. Increased NO_2 levels indoors can be harmful to one's health, especially for people who already have respiratory issues. To improve indoor air quality, avoid smoking indoors and use air filters or purifiers. The use of gas stoves, ovens, and heaters can increase indoor NO_2 levels, particularly in locations without adequate ventilation. Regular maintenance, coupled with the installation of carbon monoxide detectors, is recommended to maintain safety and reduce exposure to dangerous emissions [4].

The breakdown of nitrogen molecules in soils and the application of nitrogen-based fertilizers are two agricultural techniques that can raise NO_2 levels, albeit they are not as important as urban sources. The creation of controlled-release fertilizers and precision farming are two methods that contribute to a decrease in atmospheric nitrogen emissions.Although not as much as anthropogenic sources, natural occurrences do contribute to NO_2. For example, the high temperatures created following a lightning strike cause the production of nitrogen oxides, including NO_2. Nitrogen oxides are also released by several soil microorganisms during their metabolic activities. Significant increases in NO_2 and other pollutants can result from large-scale wildfires, like those that have recently occurred in several parts of the world. Emissions from wildfires are complicated and include a variety of pollutants that can have an impact on air quality across long distances, occasionally even affecting areas that are distant from the fire's source [13].

Low amounts of NO_2 in the air can irritate eyes, nose, throat, and lungs, which could cause cough and make person feel tired, sick, and short of breath. The lungs can also fill with fluid one or 2 days following low-level contact. Inhaling plenty of NO_2 might burn, spasm, and enlarge throat and upper respiratory tract tissues. Different tissues will get less oxygen, lungs will fill with fluid, and they may die. When it touches skin or eyes, heavy NO_2 or liquid nitrogen dioxide can burn [13].

Nitrogen dioxide can have several cardiotoxic effects, primarily mediated by diverse biological pathways impacting the cardiovascular system. Exposure to NO_2 induces the generation of reactive oxygen species (ROS), which can result in oxidative damage to cells and tissues, including those within the cardiovascular system. Oxidative stress can lead to lipid peroxidation, DNA damage, and changes in cellular function, all of which contribute to cardiovascular disease. Nitrogen dioxide (NO_2) can induce inflammatory reactions in the body, elevating the concentrations of inflammatory cytokines and adhesion molecules. This may result in endothelial dysfunction, hindering the capacity of blood vessels to dilate and elevating the risk of atherosclerosis, a significant precursor to cardiovascular disease. The endothelium is essential for vascular health, and NO_2 can compromise endothelial function by diminishing nitric oxide bioavailability. Nitric oxide is crucial for vasodilation and the regulation of vascular tone. Malfunction in this pathway can lead to elevated blood pressure and further cardiovascular complications [14]. Exposure to NO_2 has been associated with disturbances in the autonomic nervous system, which governs heart rate and blood pressure. Alterations in autonomic equilibrium may result in arrhythmias and hypertension, hence exacerbating cardiovascular risk. NO_2 can influence lipid metabolism, potentially resulting in changes in cholesterol levels that promote the onset of atherosclerosis. The buildup of low-density lipoprotein (LDL) cholesterol in the arteries can result in plaque development, so constricting arteries and hindering blood flow. Prolonged exposure to elevated concentrations of NO_2 can impair cardiomyocyte function, diminishing their contractility and resulting in decreased cardiac output. This may aggravate heart failure and other chronic cardiac problems. Nitrogen dioxide may augment blood coagulation and platelet aggregation, hence elevating the risk of thrombosis. Coronary artery thrombosis can result in myocardial infarctions, rendering this process more perilous for persons with pre-existing cardiac problems. Extended exposure to NO_2 may cause genetic and epigenetic alterations that influence gene expression related to cardiovascular health. These alterations may have enduring impacts on the structure and function of the cardiovascular system, thereby elevating the risk of heart disease development. Together, these mechanisms illustrate an interplay between NO_2 exposure and cardiovascular health, highlighting the importance of minimizing exposure to this pollutant through environmental policies and personal preventive measures [15, 16].

Children are likely to be impacted by nitrogen oxides exposure similarly to adults. However, it remains uncertain if youngsters exhibit different vulnerability to nitrogen oxides compared to adults. Exposure of gestating animals to nitrogen oxides has caused detrimental consequences in developing embryos. Nitrogen oxides have induced alterations in the genetic material of animal cells. However, it remains uncertain whether exposure to nitrogen oxides may induce developmental consequences in people [13].

Given the possible health consequences of NO_2 exposure, such as respiratory problems, worsened asthma, and other cardiovascular disorders, minimizing exposure via collaborative efforts in policy, technology, and human actions is critical. Public awareness and education may play critical roles in propelling community-based activities to address air pollution and enhance environmental health.

Chlorine (Cl_2/$Cl^{\cdot}$) Gas

Chlorine gas is a highly reactive and possibly dangerous chemical, with exposure largely occurring in industrial and domestic contexts. Chlorine is widely utilized in businesses for the production of items such as paper, textiles, and insecticides. Unintentional discharges or leaks from industrial facilities can result in exposure in adjacent regions, particularly if safety measures are disregarded. Chlorine is frequently utilized in water treatment procedures to purify potable water and swimming pools. Personnel managing chlorine at these sites may face exposure risks if adequate protective measures are not implemented. Numerous domestic cleaning agents, particularly bleaches and disinfectants, comprise chlorine compounds. Combining these chemicals with other cleansers, such as ammonia or hydrochloric acid, may generate hazardous chlorine gas, posing a risk of exposure in residential environments. Chlorine disinfection activities, such as those in public swimming pools or hot tubs, may occasionally result in exposure, particularly due to faulty management or inadvertent spilling [17].

Pressurized tanks and cylinders deliver chlorine. Leaks from shipping or storage accidents can expose workers and the public. Chemical labs employ chlorine for reactions and studies. If equipment fails or spills, lab workers may be exposed to chlorine gas. In WWI, chlorine gas was employed in chemical warfare, and exposure can occur in military situations where chemical weapons are kept or handled. Hurricanes and earthquakes can destroy chlorine storage facilities, causing unintended leaks and exposure. Chlorine-related industrial mishaps are also dangerous [17].

Unexpected exposure to high chlorine concentrations can cause tachycardia and raised blood pressure in emergency care patients. However, the extent of this effect in relation to stress is unclear. In some cases, EKGs conducted quickly after exposure showed no significant changes [17].

Most individuals with the disorders experienced persistent tachypnea and tachycardia throughout evaluation. Studies on the long-term consequences of chlorine exposure did not reveal any specific cardiovascular problems. Research indicates that acute Cl_2 exposure leads to systemic vascular damage and subsequent myocardial contractile impairment. This review examines lung and cardiac pathology linked to acute Cl_2 inhalation and discusses new findings indicating that mitochondrial dysfunction is fundamental to the pathophysiology of Cl_2-induced toxicity. Finally, presented data indicating that the activation of autophagy mitigates Cl_2-induced lung inflammation and may serve as a viable therapeutic target for alleviating the deleterious consequences of Cl_2 exposure [17].

Cl_2-Toxicity; Cardiovascular Effects

Some individuals have suffered systemic harm from brain, liver, and heart gas exposure. White matter hemorrhages, hepatic transaminitis, and cardiomegaly have been reported with high-dose Cl_2 exposure. Systemic vasculature and vascular function are also damaged by Cl_2 exposure in animal studies. After 30 min of Cl_2 (250–400 ppm), rats showed considerably lower eNOS protein expression. These mice also showed substantial reduction in eNOS-dependent vasodilation (driven by acetylcholine) 24–48 h after exposure, showing that NO dysregulation may cause Cl_2 inhalation–induced systemic endothelial dysfunction. In the early phases of Cl_2-induced damage, inflammatory, cell-derived, inducible NOS (iNOS) may compensate for eNOS loss and avoid mean blood pressure rises. Inhibiting iNOS increased systemic blood pressures considerably [17].

Cardiotoxicity is another risk factor for worse Cl_2 damage outcomes immediately. In rats, inhaling 500 ppm Cl_2 for 30 min raises coronary sinus lactate, suggesting heart anaerobic metabolism. Systolic and diastolic blood pressures drop, myocardial contractile power decreases, and biventricular failure and death result with Cl_2 inhalation. These findings may be replicated by exposure to chloramine, a possible circulating Cl_2 reactant product, showing that Cl_2 (and its reactants) cause cardiac damage independently and uniquely [17].

Eight of nine immediate victims of the January 2005 Graniteville, South Carolina train derailment's acute Cl_2 emission had cardiomegaly on postmortem. A hospital chest X-ray and autopsy revealed cardiomegaly in the one victim who died. Chest X-rays revealed cardiomegaly in four of the hospitalized individuals. Additionally, 64% of hospitalized patients had tachycardia and 52% had hypertension in the ED. There was minimal acknowledgment of Cl_2's cardiovascular effects in humans until the Graniteville accident. High Cl_2 gas levels may cause vascular damage and cardiac depression in humans and rats, which might worsen stroke symptoms. However, no human cohort research has found such long-term cardiovascular consequences [17].

The Graniteville cohort, with 305 patients, is the only systematic long-term longitudinal cohort study of a population exposed to a single high-concentration Cl_2 gas event with pre–post individual exposure and health comparisons and prospective longitudinal lung function assessment. Some cohort members' acute lung function loss between 2004 and 2005 had not fully recovered at 7–9 years postevent. Interestingly, 12% of participants were medically ineligible for spirometry testing due to high blood pressure, despite attempts to control it. The modeled Cl_2 dose was significantly associated with increased blood pressure 7–9 years later ($P < 0.05$). Hospitalizations for high blood pressure jumped from 110 per 10,000 Graniteville inhabitants in 2005 to 220 in 2012. These findings suggest that Cl_2 exposure may have long-term cardiovascular consequences [17].

In a Swedish chlorine factory research, 44 workers exposed to chlorine and 33 white-collar workers matched for age and smoking status were examined for changes in vital capacity (VC) and FEV1 during a 10-year period [18]. Throughout the years, chlorine levels was consistently below 0.5 ppm, a specified guideline. Test results indicate that white-collar workers have greater reduction in VC and FEV1 over time (significantly for FEV1) compared to workers exposed to chlorine. Hyback [18] suggested that low chlorine gas concentrations may prevent workers from developing respiratory infections, which can lead to respiratory function deterioration [17, 18].

Cyanide Gas (HCN)

Cyanides, a varied group of chemicals including the extremely reactive cyanide anion (CN^-), are generated from both human activities and natural sources. Cyanogenic compounds are referred to as chemicals that emit cyanide. The prevalent cyanide chemicals in the environment include sodium cyanide, potassium cyanide, and gaseous hydrogen cyanide, with the latter being the primary form found in the atmosphere The word 'cyanide' in this article denotes the cyanide ion or the CN^- within a chemical. Cyanides may be emitted into the environment during industrial processes or from smoke or vehicle exhaust containing the incomplete combustion byproducts of nitrogen-containing organic polymers [19].

The general population is primarily exposed to cyanides through inhalation of cigarette smoke and ingestion of specific foods. Outdoor air cyanide levels have been reported to range from 0.33 to 0.76 ppb [20]. Cyanide concentrations in smoke produced by U.S. commercial cigarettes vary from 10 to 400 μg per cigarette. Mainstream (inhaled) smoke contains, while sidestream smoke ranges from 0.006 to 0.27 μg per cigarette [19].

Despite not being a direct cardiotoxin, it prevents cells from breathing, which results in cellular hypoxia. This can have a serious impact on the heart's capacity to work.

Cyanide primarily exerts its toxicological effects by attaching to the metallic cofactor in metalloenzymes, hence disrupting enzyme and cellular activity. Cytochrome c oxidase, an enzyme in the mitochondrial respiratory chain, is the primary target of cyanide exposure. Its inhibition obstructs tissues from using oxygen. The outcome is a decrease in oxygen enough to induce tissue damage (histotoxic hypoxia) across the body, particularly affecting tissues with elevated oxygen requirements and/or a lack of detoxifying enzymes like rhodanese. The suppression of cellular oxygen use leads to elevated oxygen tensions in peripheral tissues, resulting in a diminished unloading gradient for oxyhemoglobin. Consequently, oxyhemoglobin is transported in venous blood, serving as a biomarker for cyanide exposure. Cyanide not only binds to cytochrome c oxidase but also inhibits the activities of catalase, peroxidase, hydroxocobalamin, phosphatase, tyrosinase, ascorbic acid oxidase, xanthine oxidase, and succinic dehydrogenase, potentially contributing to the manifestations of cyanide toxicity [19].

A comprehensive review of these endpoints led to the following conclusions regarding hazard identification for oral exposure to cyanide: the effects of the thyroid are considered a potential health concern for humans, neurological effects are recognized as a health impact for humans; and Male reproductive effects are a potential health concern for humans.

Evidence of cardiovascular effects, namely bradycardia and arrhythmia, following exposure to cyanide chemicals in people and animals is minimal. Certain observations are ascribed to the overall harmful effects of cyanide, including decreased cellular oxygen use and/or cyanide-induced CNS depression instead of direct harmful effects on the cardiovascular system [19].

It is documented that four persons killed by inhalation of hydrogen cyanide gas (concentration unspecified) had a significant deceleration of heart rate after 1–3 min of exposure, accompanied by further alterations in heart rate, sinus abnormalities, and audio-visual dissociation. Palpitations and hypotension were the most commonly documented cardiovascular consequences in patients following accidental cyanide inhalation poisoning; nevertheless, precise exposure amounts remained unknown [19].

Workers exposed to cyanide have reported subjective chest discomfort; nevertheless, it remains uncertain if this generic symptom is directly associated with possible cardiovascular repercussions [19].

Thirteen out of fifty-six workers exposed to an unspecified quantity of hydrogen cyanide vapor during heat treatment (case hardening) and electroplating reported subjective symptoms of chest discomfort. In a separate occupational investigation, 7 out of 36 workers exposed to an undefined type of cyanide at mean concentrations ranging from 6.416 to 10.375 ppm for 5 to 15 years in air.

The origin of cyanide exposure was an electroplating bath of sodium cyanide and copper cyanide. Among 36 workers previously exposed to 15 ppm hydrogen cyanide for an average of 11 months at a silver-reclaiming plant, 31% reported experiencing chest discomfort and 14% reported palpitations throughout their employment [19].

Clinical Considerations of Cardiotoxic Effects Due to Gaseous Poison Exposure

Exposure to gaseous poisons can have cardiotoxic consequences that range in severity from modest symptoms like palpitations and chest pain to severe ones like arrhythmias, heart failure, and cardiac arrest. A complete medical history is frequently necessary for the diagnosis in order to determine possible exposure, and imaging, electrocardiograms, and blood tests may be used to evaluate heart damage. Prompt cessation of exposure and the provision of 100% oxygen are essential for patients impacted by gasses such as carbon monoxide. Subsequent treatment is contingent upon the particular gas implicated, with management approaches possibly encompassing antidotes, symptomatic intervention, and supportive care, including oxygen therapy and cardiovascular assistance.

Conclusion

In conclusion, the cardiotoxicity of gaseous poisons is a significant medical concern that necessitates prompt recognition and intervention. Understanding the underlying mechanisms of toxicity aids in the development of effective treatment strategies and prevention measures. Based on toxicity mechanisms CO, CN, H_2S are the most affective for the cardiovascular system disorders, while indirect impact of SO_2, NO_2 or chlorine could not be diminihed. Interference with the oxigen transport could affect almost all tissues, while other four mechanisms, direct myocardial damage, electrophysiological alterations, inflammatory responses, and autonomic nervous system disruption could be hazardous, as well, and cause severe carditoxicity.

References

1. ATSDR (2012) Toxicological profile for carbon-monoxide. U.S. Department of Health and Human services Public Health Service Agency for Toxic Substances and Disease Registry
2. Chu LM, Shaefi S, Byrne JD, de Souza RW, Otterbein LE (2021) Carbon monoxide and a change of heart. Redox Biol 48:102183
3. Gandini C, Castoldi AF, Candura SM, Locatelli C, Butera R, Priori S, Manzo L (2001) Carbon monoxide cardiotoxicity. J Toxicol Clin Toxicol 39(1):35–44
4. Ismail MM, El-Ghamry H, Shaker OG, Fawzi MM, Ibrahim SF (2013) Some biomarkers in carbon monoxide-induced cardiotoxicity. J Environ Anal Toxicol 3(176):2161–0525

5. Saraçoğlu E, Vuruşkan E, Kılıç S, Çekici Y, Onur B, Arslan Y, Kılıç E, Aykut Ö (2018) Predicting cardiotoxic effects of carbon monoxide poisoning using speckle tracking echocardiography. Cardiovasc Toxicol 18(2):175–183
6. Temrel TA, Bilge S (2020) Myocardial repolarization parameters and neutrophil-to-lymphocyte ratio are associated with cardiotoxicity in carbon monoxide poisoning. Cardiovasc Toxicol 20(2):190–196
7. Chen TM, Kuschner WG, Gokhale J, Shofer S (2007) Outdoor air pollution: nitrogen dioxide, sulfur dioxide, and carbon monoxide health effects. Am J Med Sci 333(4):249–256
8. ATSDR (2007) Toxicological profile for sulfur-dioxide. U.S. Department of Health and Human Services Public Health Service Agency for Toxic Substances and Disease Registry
9. Qin G, Wu M, Wang J, Xu Z, Xia J, Sang N (2016) Sulfur dioxide contributes to the cardiac and mitochondrial dysfunction in rats. Toxicol Sci 151(2):334–346
10. Song Y, Xu Z, Zhong Q, Zhang R, Sun X, Chen G (2023) Sulfur signaling pathway in cardiovascular disease. Front Pharmacol 14:1303465
11. Yang L, Zhang H, Chen P (2018) Sulfur dioxide attenuates sepsis-induced cardiac dysfunction via inhibition of NLRP3 inflammasome activation in rats. Nitric Oxide 81:11–20
12. Zhang S, Lu W, Li Z, Wang S, Jiang M, Yuan Q, Wei Z, Zhang H (2020) Mechanisms underlying sulfur dioxide pollution induced ventricular arrhythmia: a simulation study. In: 2020 IEEE international conference on bioinformatics and biomedicine (BIBM). IEEE, pp 381–386
13. Weinberger B, Laskin DL, Heck DE, Laskin JD (2001) The toxicology of inhaled nitric oxide. Toxicol Sci 59(1):5–16
14. Bahadır A, Kurucu N, Kadıoğlu M, Yenilme E (2014) The role of nitric oxide in doxorubicin-induced cardiotoxicity: experimental study. Turk J Hematol 31(1):68–74
15. Mukhopadhyay P, Rajesh M, Bátkai S, Kashiwaya Y, Hasko G, Liaudet L, Szabó C, Pacher P (2009) Role of superoxide, nitric oxide, and peroxynitrite in doxorubicin-induced cell death in vivo and in vitro. Am J Phys Heart Circ Phys 296(5):H1466–H1483
16. Vásquez-Vivar J, Martasek P, Hogg N, Masters BS, Pritchard KA, Kalyanaraman B (1997) Endothelial nitric oxide synthase-dependent superoxide generation from adriamycin. Biochemistry 36(38):11293–11297
17. ATSDR (2010) Toxicological profile for chlorine. U.S. Department of Health and Human Services Public Health Service Agency for Toxic Substances and Disease Registry
18. Hyback B (1999) A long-term study of pulmonary function at low exposures to chlorine. Int Arch Occup Environ Health 72:M024–M028
19. ATSDR (2024) Toxicological profile for cyanides—draft. U.S. Department of Health and Human Services Public Health Service Agency for Toxic Substances and Disease Registry
20. Jaszczak E, Narkowicz S, Namieśnik J, Polkowska Ż (2017) Determination of cyanide in urine and saliva samples by ion chromatography with pulsed amperometric detection. Monatsh für Chem 148:1645–1649

Chapter 22
Cardiovascular Toxicity of Volatile Organic Compounds

Evica Antonijević Miljaković, Marijana Curcic, and Biljana Antonijević

Abstract . Volatile organic compounds (VOCs) represent a large and diverse group of chemicals, typically liquids characterized by varying degrees of lipophilicity and volatility. VOCs are most commonly used as industrial solvents, making occupational exposure a predominant route of human contact. In addition, intentional inhalation of these substances—known as volatile substance abuse (VSA)—poses a significant health concern. VOCs also enter the environment primarily through evaporation, contributing—along with other sources such as tobacco smoke, hazardous waste sites, vehicle emissions, industrial discharges, and household products—to ambient air pollution. This environmental presence has raised concerns about the exposure of the general population. Exposure to VOCs has been associated with a range of adverse health effects in both humans and experimental animals. This chapter summarizes current evidence on the cardiovascular toxicity of solvents/VOCs, emphasizing the need to investigate VOC exposure as a potential risk factor in the development of cardiovascular disease (CVD).

Keywords Cardiovascular disease · Mortality · Volatile organic compounds · Solvents · Urinary metabolites

Introduction

Volatile organic compounds (VOCs) represent a large group of chemicals characterized by high vapor pressure and low water solubility. They are most commonly used as industrial solvents—substances capable of dissolving, diluting, dispersing, or extracting other materials without chemically altering them. Owing to this versatility, solvents are essential across virtually all industrial sectors, making

E. A. Miljaković (✉) · M. Curcic · B. Antonijević
Department of Toxicology "Akademik Danilo Soldatović", University of Belgrade-Faculty of Pharmacy, Belgrade, Serbia
e-mail: evica.antonijevic@pharmacy.bg.ac.rs; marijana.curcic@pharmacy.bg.ac.rs; biljana.antonijevic@pharmacy.bg.ac.rs

D. K. Agrawal et al. (eds.), *Cardiovascular Toxicity*, Advances in Biochemistry in Health and Disease 33, https://doi.org/10.1007/978-3-031-90598-8_22

occupational exposure the most prevalent route of human contact.In the latest global review, the International Labour Organization confirmed solvents as a priority group of chemical hazards in the workplace based on human health impacts, along side asbestos, silica, heavy metals, dyes, nanomaterials, perfluorinated chemicals, endocrine-disrupting chemicals, pesticides and workplace air-pollution [1]. Industries with high solvent use include engineering, construction, chemicals, printing, rubber and plastics production, footwear and textiles, food processing, woodworking, dry cleaning, ink and pharmaceutical manufacturing and paint production. Beyond occupational settings, intentional inhalation of solvents—known as **volatile substance abuse** (VSA)—is another significant route of exposure. Additionally, many solvents are released into the environment through evaporation. Combined with emissions from sources such as tobacco smoke, hazardous waste sites, vehicle exhaust, industrial processes, and household products, these contribute significantly to ambient VOC pollution, raising concerns about **general population exposure**. Growing awareness of the toxicological and ecotoxicological risks associated with solvent exposure has led to regulatory actions and technological innovations, particularly in Europe and the United States. For example, the 1987 Montreal Protocol prompted the restriction or phase-out of many ozone-depleting solvents, while water-based paints have increasingly replaced traditional solvent-based coatings [2]. In some sectors, such as dry cleaning, advancements in equipment and processes have reduced solvent use [3]. Since 1990s , significant efforts have been made to develop "green solvents" as safer and environmentally friendly alternatives. However, only a limited number have reached industrial-scale adoption [4]. Moreover, challenges remain in demonstrating the sustainability of these alternatives, particularly in sourcing non-fossil organic carbon for solvent production [4, 5]. In contrast, low-and middle-income countries (LMICs) often face critical gaps in regulation, availability of personal protective equipment (PPE), and health standards in industries, making occupational solvent exposure in these regions disproportionately high.

Solvents are broadly classified based on their molecular structure or functional groups. Common classes include aliphatic hydrocarbons, many of which are chlorinated (i.e., halocarbons); aromatic hydrocarbons; alcohols; ethers; esters/acetates; amides/amines; aldehydes and ketones; as well as complex mixtures. Solvents are usually organic liquids with varying degrees of lipophilicity and volatility, depending on the numbers of carbon and/or halogen atoms. These properties, coupled with small molecular size and lack of charge, make **inhalation** the primary route of exposure, enabling rapid absorption across the pulmonary epithelium. In addition to inhalation, **dermal** exposure and percutaneous uptake contribute significantly to total body burden especially in certain industries, such as painting and industrial degreasing [6]. Exposure to solvents has been associated a wide range of adverse health effects in humans. **Acute,** high-dose exposures can casue cause respiratory irritation, skin and eye irritation, central nervous system (CNS) effects (dizziness, headaches, confusion, unconsciousness, respiratory depression) and potentially death. **Chronic**, low-dose exposures to solvents were associated with cancer, neurotoxicity, reproductive and developmental toxicity, liver and kidney

Table 22.1 Overview of VOCs and associated cardiovascular effects

VOC/group	Study type	Key outcomes
Carbon disulfide	Cohort/meta-analysis, human and animal studies	↑ Cardiovascular mortality, ↑ CHD, ↑ metabolic syndrome
Benzene	Cross-sectional and animal studies	↓ Circulating angiogenic cells, ↑ LDL: HDL ratio, endothelial dysfunction
Toluene	Case-control and case reports	↑ QTc interval, ↑ risk of cardiomyopathy, ↑ P-wave duration
BTEX (occupational exposure)	Cross-sectional, prospective cohort	↑ Diastolic BP, ↑ myocardial infarction/ ischemic heart disease
BTEX (environmental exposure)	Time-series, cross-sectional, cohort studies	↑ Heart failure hospitalizations, ↑ circulatory mortality
Other VOCs (e.g., acrolein, styrene)	Human biomonitoring, cross-sectional, cohort studies	↑ Systolic BP, ↓ endothelial function, ↓ angiogenic cells, ↑ CVD mortality

BTEX benzene, toluene, ethylbenzene, xylene; *CHD* coronary heart disease; *CVD* cardiovascular disease; *BP* blood pressure

damage, and sensitization or allergic reactions. The International Agency for Research on Cancer (IARC) has classified benzene and trichloroethylene (TCE) as carcinogenic to humans (group 1) based on sufficient evidences in epidemiological and experimental studies [7, 8]. Additionally, methylene chloride and tetrachloroethylene were classified as probably carcinogenic (group 2A), according to limited evidences in humans, but sufficient evidences in experimental studies [8, 9]. Finally, the existing evidences regarding cardiovascular toxicity of solvents/ volatile organic compounds were described in this chapter (Tables 22.1, 22.2 and 22.3).

Carbon Disulfide

Along with neurological and ophthalmological, cardiovascular effects appear to be the most sensitive target of toxicity following exposure to carbon disulfide. It has been shown that cardiovascular system is highly susceptible to carbon disulfide toxicity in both humans and animals. The majority of human studies on carbon disulfide toxicity originate from the viscose rayon industry. According to a systematic review, moderate evidence in humans and high evidence in laboratory animals suggest that inhalation of carbon disulfide targets the cardiovascular system [10].

Cardiovascular disease (e.g. coronary heart disease, ischemic heart disease, arteriosclerotic heart disease) has been identified as the most common risk of increased **mortality** among viscose rayon workers in several cohort studies from Finland, Czechoslovakia, the United Kingdom, the United States, Netherlands, and Poland [10]. Early proofs on cardiovascular-related death in viscose rayon workers, date back to decades prior to 1980 from a 15-year follow-up Finnish cohort. Increased mortality due to coronary heart disease was associated with much higher

Table 22.2 Comparison of BTEX exposure: occupational vs environmental

Aspect	Occupational exposure	Environmental exposure
Exposure type	Low-level BTEX (below TWA/STEL)	Ambient BTEX in air (ppb range)
Population	Petroleum refinery workers, oil spill cleanup crews	General population (e.g., Hong Kong, Taiwan, US, Iran)
Biomarkers Measured	S-phenylmercapturic acid, hippuric acid, methyl hippuric acid in urine	Ambient BTEX levels, urinary VOC metabolites
Cardiovascular Effects	↑ Diastolic blood pressure, ↑ myocardial infarction and ischemic heart disease (in high exposure)	↑ Heart failure hospitalizations, ↑ mortality, endothelial dysfunction
Key studies	Bingxian et al. [10], Chen et al. [11]	Ran et al. [12], Li et al. [13], Riggs et al. [14]

BTEX benzene, toluene, ethylbenzene, xylene; *TWA* time-weighted average; *STEL* short-term exposure limit

Table 22.3 Proposed mechanisms of cardiovascular toxicity of selected VOCs

VOC/group	Proposed mechanism(s)	Evidence type	Key references
Carbon disulfide	– Endocrine-dependent: ↓ Thyroid hormones → dyslipidemia → atherosclerosis – Oxidative stress (ROS) → oxidized LDL – Direct interaction with LDL → altered cholesterol homeostasis	Systematic review, human and animal	Printemps et al. [15], Jhun et al. [16]
Benzene	– ↓ Circulating angiogenic cells → impaired endothelial repair – ↑ LDL:HDL ratio – Endothelial apoptosis via reactive metabolites (e.g., mucondialdehyde) – Endothelial activation and granulocyte recruitment	Animal, in vitro, human	Abplanalp et al. [17], Malovichko et al. [18], Zelko et al. [19]
Toluene	– Cardiac electrophysiological changes: ↑ QTc, ↑ P-wave → arrhythmia risk – Potential myocardial toxicity → dilated cardiomyopathy	Case report	Karaca et al. [20], Denault et al. [21], Vural and Ogel [22]
BTEX (general)	– Oxidative stress and systemic inflammation – ↓ Angiogenic progenitor cells – Endothelial dysfunction – Dyslipidemia and leukocytosis	Biomonitoring, animal	Riggs et al. [14], He et al. [23], Wang et al. [24]
Acrolein, 1,3-butadiene	– Vascular dysfunction via endothelial injury and oxidative stress – ↑ Blood pressure, ↓ Reactive Hyperemia Index (endothelial function)	Human biomonitoring (urinary metabolites)	McGraw et al. [25]

occupational exposure levels of carbon disulfide, being estimated at above 10 ppm, while it was not obtained for the period following the reduction in exposure levels. Later studies confirmed positive association of cardiovascular disease-related death in higher exposed groups of workers, such as spinners, or did not differentiate exposure levels, which historically ranged from 2.6 to 48 ppm, in relation to the followed outcome and resulted in the absence of positive exposure-effect association.

In addition to mortality, increased **prevalence of cardiovascular disease** has been reported in several occupational studies of workers exposed to carbon disulfide. Most prominent result came from meta-analysis of 11 individual cohort studies, demonstrating a positive association with a relative risk of 1.56 (95% confidence interval 1.12–2.1) for cardiovascular outcomes, such as mortality, disease incidence, or increased cardiovascular disease risk [11]. Additionally, prevalence odds ratios of coronary heart disease in 91 male viscose rayon workers from Bulgaria were significant in highly exposed (based on job history and exposure duration), but not in moderately exposed workers, compared to non-exposed controls [12]. However, numerous evaluations of **hypertension, elevated systolic/diastolic blood pressure, abnormalities in electrocardiograms (ECGs)** and **measures of heart rate variability** in relation to carbon disulfide resulted in non-consistent results, despite substantial overlap in occupational exposure scenarios [10]. Similarly, although studies associated carbon disulfide-exposure via the inhalation route to hypercholesterolemia or metabolic syndrome, as known **risk factors for cardiovascular disease**, evidences are inadequate in humans and moderate in animals [10]. Retired viscose rayon factory workers ($n = 170$) in Korea, with a history of "carbon disulfide poisoning", had an increased overall risk of metabolic syndrome (prevalence ratio 1.57, 95% CI: 1.25–1.98), characterized by abdominal obesity, reduced serum HDL-C levels, elevated serum triglycerides, elevated blood pressure, and elevated fasting blood glucose levels, compared to non-exposed controls [13]. Recently, increased risk of diabetes (odds ratio 1.282, 95% CI: 1.055,1.558) was found in a Chinese population-based, cross-sectional study ($n = 3,338$) related to urinary levels of carbon disulfide metabolite, 2-thiothiazolidine-4-carboxylic acid (TTCA) set up as a continuous variable [14]. However, stratifying exposure (urinary TTCA levels) into quartiles, did not show significant exposure-effect association. In addition, the same study has shown positive correlation of fasting plasma glucose levels to urinary TTCA levels (regression coefficient 0.080, 95% CI: 0.002,0.157), while quartile analysis did not reveal a clear relationship between exposure and effect.

Considering potential **mechanisms of** carbon disulfide **cardiotoxicity** a systematic review proposed involvement of complex endocrine-dependent (ED) and/or non-ED modes of action (MoA) [15]. Hypothyroidism as an initial event, followed by inflammation, oxidized LDL, reactive oxygen species generation, might finally lead to the development of atherosclerosis. However, excessive oxidative damage, unrelated to hypothyroidism, which have been found could be a non-endocrine-dependent MOA contributing to altered cholesterol homeostasis and atherosclerosis. Given the findings on direct interactions between carbon disulfide and LDL cholesterol, there is stronger support for the non-endocrine-dependent MOA.

Benzene

A cross-sectional study in general adult population (n = 210, middle-aged 51 ± 10 years) associated benzene exposure, represented by urine levels of its metabolite, *trans, trans*-muconic acid (t,t-MA) with a risk of cardiovascular disease (Framingham Risk Score) [16]. Significantly higher levels of urinary t,t-MA were measured in subgroup of high cardiovascular disease risk compared to low risk category and this difference was more obvious when only non-smokers subgroup was evaluated. More specifically, in total cohort (adjusted for ethnicity, hyperlipidemia, and cotinine) for each 0.1 mg/g creatinine increase in t,t-MA, a significant decrease (~9%) was obtained in circulating angiogenic cells, being important for vascular regeneration and repair, as well as predictive for cardiovascular events and mortality. However, given that urinary t,t-MA is also a metabolite of preservative sorbic acid or sorbates found in beverages and food, these results should be interpreted with caution [7]. Nevertheless, consistent significant decrease of circulating angiogenic cells (3.4% lower levels of Flk-1+/Sca-1+ cells per 1.0 mg t,t-MA/g creatinine) was obtained in accompanying experimental study in male C57BL/6 mice exposed to benzene (50 ppm for 6 h/day, 5 days/week during 6 weeks), along with 23% increase in the plasma LDL:HDL ratio. Another experimental study, has proposed blood levels of endothelial cells, platelet-derived microparticles and platelet-leukocyte aggregates as a potential subclinical markers of cardiovascular toxicity of benzene inhalation, based on their significant elevation in mice exposed to concentration of 50 ppm, 6 h/day, 5 days/week during 6 weeks, compared to HEPA-filtered air exposed control group [17]. In addition, when human aortic endothelial cells were exposed to benzene metabolites *in vitro* (10 μM, 18 h), *trans,trans*-mucondialdehyde was shown to be the most potent inducer of endothelial cell apoptosis and microparticles formation, suggesting that endothelial injury is likely to be mediated by reactive benzene metabolites. Further, prolonged benzene exposure, using the same scenario as in prevous study (50 ppm, 6 h/day, 5 days/week, 6 weeks), has been found to worsen cardiac dysfunction induced in mice by transverse aortic constriction (TAC, commonly used model for pressure overload-induced cardiac hypertrophy and heart failure) [18]. Significant endothelial activation and enhanced recruitment of granulocytes and differential regulation of genes involved in neutrophil degranulation, inflammatory signaling, and stress response were found in cardiac tissue TAC/benzene-exposed mice hearts representing possible molecular mechanisms underlying cardiac dysfunction. In addition, *in vitro* finding on induction of the P-selectin expression in cardiac microvascular endothelial cells and increase the adhesion of neutrophils to these endothelial cells by benzene metabolites, hydroquinone and catechol (5 μM, 24 h), that was attenuated by anti-P-selectin antibody, suggested at least partially, involvement of active benzene metabolites in facilitation of neutrophil infiltration in the TAC hearts. However, prolonged benzene exposure in this study did not induce cardiac remodeling in placebo surgery-group mice (Sham).

Toluene

In a case-control study, including 32 toluene inhalant addicts for at least 1 year, significant changes were found in electrocardiographic (ECG) analysis, along with no differences in echocardiographic measurements compared to control, indicating that volatile solvent addiction was more of an electrical issue than a structural one [19]. Higher values of corrected (QTc), uncorrected QT intervals, and widest P-wave values were obtained in abusers group comapred to the control. Given that subjects had no history of systemic diseases nor evidence of cardiomyopathy, heart valve disease, arrhythmia or previous cardiac surgery, prolonged QTc and increased P-wave found in addicts group were suggested as useful predictors for higher risk of sudden cardiac death, as well as atrial and ventricular dysrhythmias. Dilated cardiomyopathy has been associated to toluene-occupational exposure in a 56-year-old woman of South American origin, presented to emergency department with a 9-month history of progressive dyspnea, orthopnea, and paroxysmal nocturnal dyspnea [20]. A transthoracic echocardiogram showed a left ventricular ejection fraction of 13%, with a severely dilated and globally hypokinetic left ventricle and moderate-to-severe mitral regurgitation, mild tricuspid regurgitation and normal pulmonary artery pressure. History showed that she was working in a plastic cases factory for the past 34 years without personal protective equipment, being exposed to numerous industrial solvents, including glue and glue vapors containing toluene. Exposure was subsequently confirmed in workers from the same factory by elevated urinary ortho-cresol levels exceeding the biological exposure index of 0.3 mmol/mmol of creatinine, as recommended by the American Conference of Governmental Industrial Hygienists (ACGIH), despite the absence of VOCs exceeding occupational air limits in the workplace. Treatment included intravenous diuretics and introduction of a beta-blocker, an angiotensin-converting enzyme inhibitor, and a mineralocorticoid receptor antagonist. No improvement in the left ventricle ejection fraction has been observed in a control transthoracic echocardiogram performed at 9 and 11 months. Despite genetic etiology could not be completely excluded in this case, toulene exposure might constitute a risk factor for heart disease [21, 22].

Benzene, Toluene, Ethylbenzene, Xylene (BTEX)—Occupational Exposure

Increased blood pressure among 649 petroleum refininig workers in a cross-sectional study in China, significantly correlated with low-concentration BTX exposure, along with years of work longer than 5 and obesity compared to non-exposed control, indicating importance of regular blood pressure monitoring and enhanced health interventions in this occupational group [23]. Low-BTX exposure in this study corresponded to ambient monitoring concentrations of BTX below the national limits by either time-weighted average (TWA) or short-term exposure limit

(STEL). However, urinary levels of solvent metabolites — S-phenylmercapturic acid (S-PMA), hippuric acid (HA), and methyl hippuric acid (MHA)— were significantly higher in exposed workers compared to the non-exposed control group. Moreover, low-concentration BTX exposure was found to be one of the main influencing factor for elevated diastolic blood pressure, resulting in a 1.337 mmHg increase compared to the control group. Myocardial infarction or death attributed to ischemic heart disease was linked to individual or combined exposure to benzene, toluene, ethylbenzene, xylene, n-hexane (BTEX-H), during a Deepwater Horizon (DWH) Oil Spill disaster in the United States among response and cleanup workers in a 10 years follow-up prospective study [24]. Comperhensive exposure esimates—matching full-shift personal-air plus direct-reading area measurements with job/activity, location and time period in order to obtain worker-groups with similar exposure distributions—were found to be below occupational guidelines set by the American Conference of Governmental Industrial Hygienists. Although, 509 workers from 22,655 in total, with no previous myocardial infarction diagnoses, experienced cardiovascular disease event till the end of the observed period, hazard ratio (HR) analysis did not show significant increase in risks nor potential exposure-effect relationship. However, the highest and significant HRs were obtained in the top quintile exposure group for total hydrocarbons (HR = 1.42, 95% CI: 1.01, 2.01) and benzene (HR = 1.44, 95% CI: 1.02, 2.02) in relation to cardiovascular disease outcome.

Benzene, Toluene, Ethylbenzene, Xylene (BTEX)—Environmental Exposure

Ran and coauthors estimated 4.2% (95% CI: 1.18%–7.26%) and 2.7% (95% CI: 0.39%–5.04%) increases of heart failure (HF) emergency hospitalizations in Hong Kong popullation (n = 54,003), associated with an interquartile range (IQR) increment of atmosphere alkyne (1.17 ppb) and benzene (0.43 ppb), respectively. These findings suggested VOCs as environmental stressor for heart failure [25]. In addition, short term elevation of atmosphere BTEX was suggested as a trigger event for premature circulatory deaths in Hong Kong population, given that 5.8% (95%CI: 1.0% to 10.8%) and 3.5% (95%CI: 1.0% to 6.1%) increases of mortality were associated with ambient IQR increments of benzene (1.4 μg/m3) and combined TEX (7.9 $\mu g/m^3$), respectively [25]. Supporting data were found in Taiwan population (n = 154,165), showing 1.21% (95% CI: 0.36%–2.07%) increase in heart failure hospitalizations associated with a rise of different VOCs subgroups in air represented by total carbon (IQR 102.6 ppb) [26]. Although pretty weak, an excess 3% risk of cardiovascular disease (pooled RR =1.03, 95% CI =1.01–1.05, $I^2 = 25\%$, p = 0.27) has been associated with benzene indoor exposure (maximum concentration 3.8 $\mu g/m^3$) in a systematic review and meta-analysis of observational studies published between 1980 and 2017 [27]. Giving the stronger estimates of exposure compared to studies described above, several recent human biomonitoring studies have provide evidences of VOCs potential cardiovascular toxicity. Vascular disfunction was associated with urinary metabolite levels of acrolein and 1,3-butadiene, in nonsmoking cohort of a cross-sectional study in Kentucky [28]. More specifically, significant increase in systolic blood pressure was associated with, the level of of urinary N-Acetyl-S-(3-hydroxypropyl)-L-cysteine (3HPMA) (0.98 mmHg per IQR 3HPMA, 95%CI: 0.06, 1.91, n = 308), along with significant decrese in reactive hyperemia index as a measure of peripheral endothelial function (approximately 4% for IQR of 3HPMA or N-AcetylS-(3,4-dihydroxybutyl)-L-cysteine (DHBMA), n = 70). Additionally, significant depletion of circulating angiogenic cells (CACs) in periferal blood has been associated with urinary levels of VOCs metabolites in a cross-sectional study among 603 participants with low-to-high cardiovascular disease risk, indicating compromised endothelial repair and angiogenesis [29]. Metabolites of ethylbenzene/styrene and xylene, were negatively associated with CAC levels in both the total sample and non-smokers. The metabolite of acrylonitrile was negatively associated with CD45dim/CD146+/CD34+/AC133+ cells and CD45+/CD146+/AC133+, and the toluene metabolite with AC133+ cells. In analysis of non-smokers (n = 375), inverse association has been shown between blood CD45dim/CD146+/CD34+ cells and urinary metabolites of ethylbenzene/ styrene, benzene and xylene. Cumulative VOC risk score showed a strong negative association with CD45dim/CD146+/CD34+ cells, suggesting that total VOC exposure has a cumulative effect on pro-angiogenic cells. In a two cross-sectional studies among US adult population (n = 5,213–6,814) different types of regression analysis have identified positive associations between specific urinary VOCs metabolites (among 15–20 types measured) and incidence of cardiovasular diseases, such as congestive heart failure, coronary heart disease, angina, or heart attack [30, 31]. Further, increased mortality risks related to blood benzene, ethylbenzene, o-xylene, m−/p-xylene and combined BEX levels were obtained using Cox proportional hazards models in a prospective cohort study among 16,968 US participants with a median follow-up of 8 years [32]. Cause-specific mortality analysis revealed that blood benzene or combined BEX levels were associated with significant 39% or 34% increased risk of the mortality from heart disease, respectively, followed by consistent result in non-smokers subpopulation. Weighted geometric means of blood benzene concentrations in this cohort were 0.375 (0.353–0.397) ng/L. Moreover, mortality related to ischemic heart disease (IHD) has been associated also with levels of urinary VOC metabolites (20 measured, independent of tobacco smoking) using Cox regression models in general population of Iran (n = 50,045, age 40–75 years, 8.4 years follow-up) [33]. Finally, blood levels of individual and mixed BTEX were associated with increased cardiovascular disease risk (congestive heart failure, angina pectoris, and heart attack), and cardiovascular risk factors (triglycerides, cholesterol, low-density lipoprotein, and WBC) among 17,007 US participants [34].

Conclusion

Compared to neurotoxicity, hepatotoxicity and carcinogenicity, cardiotoxicity has generally not been considered a primary or sensitive toxic effect of volatile organic compounds in humans and available data have often been scarce or inconsistent. Among occupational exposures, the most extensive evidences were available for carbon disulfide, which has enabled a robust systematic review by the US Agency of toxic substances and disease registry (ATSDR) leading to reliable conclusions regarding its association to cardiovascular outcomes, supported by human, animal and mechanistic studies. In recent years,growing scientific interest has shifted toward environmental exposure to VOCs, particularly the BTEX group, with emerging evidence suggesting their contribution to cardiovascular dysfunction in general population. Studies reporting blood pressure changes, endothelial damage, and increased risk of heart failure linked to VOC exposure, are likely to stimulate further experimental research in order to reveal molecular mechanisms underlying cardiotoxic effect. These findings highlight the urgent need to reduce exposure to VOCs - both in occupational and environmental settings - as an important strategy to protect cardiovascular health in the broader population.

References

1. ILO (2021) Exposure to hazardous chemicals at work and resulting health impacts: a global review
2. Dick FD (2006) Solvent neurotoxicity. Occup Environ Med 63:221–226. https://doi.org/10.1136/oem.2005.022400
3. Ceballos DM, Fellows KM, Evans AE, Janulewicz PA, Lee EG, Whittaker SG (2021) Perchloroethylene and dry cleaning: it's time to move the industry to safer alternatives. Front Public Health 9:638082. https://doi.org/10.3389/fpubh.2021.638082
4. Winterton N (2021) The green solvent: a critical perspective. Clean Techn Environ Policy 23:2499–2522. https://doi.org/10.1007/s10098-021-02188-8
5. Hessel V, Tran A, Tran QD, Long E-G (2022) Sustainability of green solvents–review and perspective. Green Chem 24:410–437
6. Semple S (2004) Dermal exposure to chemicals in the workplace: just how important is skin absorption? Occup Environ Med 61:376–382. https://doi.org/10.1136/oem.2003.010645
7. IARC (2018) IARC monographs on the evaluation of carcinogenic risks to humans volume 120: benzene. IARC
8. IARC (2014) Trichloroethylene, Tetrachloroethylene and Some Other Chlorinated Agents. In: IARC Monographs on the Evaluation of Carcinogenic Risks to Humans, vol 106. IARC
9. IARC (2017) IARC monographs on the evaluation of carcinogenic risks to humans. In: Volume 110: dichloromethane, 1,2-dichloropropane, and 1,3-propane sultone. IARC
10. ATSDR (2024) Toxicological profile for carbon disulfide (Draft for Public Comment)
11. Tan X, Peng X, Wang F, Joyeux M, Hartemann P (2002) Cardiovascular effects of carbon disulfide: meta-analysis of cohort studies. Int J Hyg Environ Health 205:473–477. https://doi.org/10.1078/1438-4639-00174

12. Kotseva K, Braeckman L, De Bacquer D, Bulat P, Vanhoorne M (2001) Cardiovascular effects in viscose rayon workers exposed to carbon disulfide. Int J Occup Environ Health 7:7–13. https://doi.org/10.1179/107735201800339713
13. Jhun H-J, Lee S-Y, Yim S-H, Kim M-J, Park K-K, Cho S-I (2009) Metabolic syndrome in carbon disulfide-poisoned subjects in Korea: does chemical poisoning induce metabolic syndrome? Int Arch Occup Environ Health 82:827–832. https://doi.org/10.1007/s00420-008-0363-8
14. Xu T, Wang B, Wang X, Yang S, Cao L, Qiu W, Cheng M, Liu W, Yu L, Zhou M, Wang D, Ma J, Chen W (2021) Associations of urinary carbon disulfide metabolite with oxidative stress, plasma glucose and risk of diabetes among urban adults in China. Environ Pollut 272:115959. https://doi.org/10.1016/j.envpol.2020.115959
15. Printemps N, Le Magueresse-Battistoni B, Mhaouty-Kodja S, Viguié C, Michel C (2022) How to differentiate general toxicity-related endocrine effects from endocrine disruption: systematic review of carbon disulfide data. Int J Mol Sci 23:3153. https://doi.org/10.3390/ijms23063153
16. Abplanalp W, DeJarnett N, Riggs DW, Conklin DJ, McCracken JP, Srivastava S, Xie Z, Rai S, Bhatnagar A, O'Toole TE (2017) Benzene exposure is associated with cardiovascular disease risk. PLoS One 12:e0183602. https://doi.org/10.1371/journal.pone.0183602
17. Malovichko MV, Abplanalp WT, McFall SA, Taylor BS, Wickramasinghe NS, Sithu ID, Zelko IN, Uchida S, Hill BG, Sutaria SR, Nantz MH, Bhatnagar A, Conklin DJ, O'Toole TE, Srivastava S (2021) Subclinical markers of cardiovascular toxicity of benzene inhalation in mice. Toxicol Appl Pharmacol 431:115742. https://doi.org/10.1016/j.taap.2021.115742
18. Zelko IN, Dassanayaka S, Malovichko MV, Howard CM, Garrett LF, Uchida S, Brittian KR, Conklin DJ, Jones SP, Srivastava S (2022) Chronic benzene exposure aggravates pressure overload-induced cardiac dysfunction. Toxicol Sci 185:64–76. https://doi.org/10.1093/toxsci/kfab125
19. Karaca G, Hatipsoylu E, Ekmekci A, Yazıcı ÖK, Kimiaei A, Safaei S, Sözen AB (2024) Toluene inhalant addiction and cardiac functions in young adults: a comparison of electrocardiographic and echocardiographic parameters. Clin Cardiol 47:e70037. https://doi.org/10.1002/clc.70037
20. Denault G, Ouazani Chahdi H, Patry L, Costa J-P (2022) A case of dilated cardiomyopathy associated with chronic toluene exposure. CJC Open 4:1024–1026. https://doi.org/10.1016/j.cjco.2022.09.001
21. Vural M, Ogel K (2006) Dilated cardiomyopathy associated with toluene abuse. Cardiology 105:158–161. https://doi.org/10.1159/000091344
22. Yasar S, Yildirim E, Koklu M, Gursoy E, Celik M, Yuksel UC (2016) A case of reversible cardiomyopathy associated with acute toluene exposure. Turk J Emerg Med 16:123–125. https://doi.org/10.1016/j.tjem.2016.09.002
23. Bingxian Z, Qisheng WU, Shiheng FAN, Zhuna SU, Jianye P, Chunyun LU, Nengde Z, Lei JIN, De'e YU, Jing Z (2024) Effect of low-concentration benzene, toluene, and xylene exposure on blood pressure of workers in a petroleum refining enterprise. J Environ Occup Med 41:1012–1019. https://doi.org/10.11836/JEOM24024
24. Chen D, Sandler DP, Keil AP, Heiss G, Whitsel EA, Edwards JK, Stewart PA, Stenzel MR, Groth CP, Ramachandran G, Banerjee S, Huynh TB, Jackson WB, Blair A, Lawrence KG, Kwok RK, Engel LS (2023) Volatile hydrocarbon exposures and incident coronary heart disease events: up to ten years of follow-up among Deepwater horizon oil spill workers. Environ Health Perspect 131:057006. https://doi.org/10.1289/EHP11859
25. Ran J, Qiu H, Sun S, Tian L (2018) Short-term effects of ambient benzene and TEX (toluene, ethylbenzene, and xylene combined) on cardiorespiratory mortality in Hong Kong. Environ Int 117:91–98. https://doi.org/10.1016/j.envint.2018.04.049
26. Qiu H, Bai C-H, Chuang K-J, Fan Y-C, Chang T-P, Yim SH-L, Ho K-F (2021) Association of cardiorespiratory hospital admissions with ambient volatile organic compounds: evidence from a time-series study in Taipei, Taiwan. Chemosphere 276:130172. https://doi.org/10.1016/j.chemosphere.2021.130172

27. Liu N, Bu Z, Liu W, Kan H, Zhao Z, Deng F, Huang C, Zhao B, Zeng X, Sun Y, Qian H, Mo J, Sun C, Guo J, Zheng X, Weschler LB, Zhang Y (2022) Health effects of exposure to indoor volatile organic compounds from 1980 to 2017: a systematic review and meta-analysis. Indoor Air 32:e13038. https://doi.org/10.1111/ina.13038
28. McGraw KE, Riggs DW, Rai S, Navas-Acien A, Xie Z, Lorkiewicz P, Lynch J, Zafar N, Krishnasamy S, Taylor KC, Conklin DJ, DeFilippis AP, Srivastava S, Bhatnagar A (2021) Exposure to volatile organic compounds—acrolein, 1,3-butadiene, and crotonaldehyde—is associated with vascular dysfunction. Environ Res 196:110903. https://doi.org/10.1016/j.envres.2021.110903
29. Riggs DW, Malovichko MV, Gao H, McGraw KE, Taylor BS, Krivokhizhina T, Rai SN, Keith RJ, Bhatnagar A, Srivastava S (2022) Environmental exposure to volatile organic compounds is associated with endothelial injury. Toxicol Appl Pharmacol 437:115877. https://doi.org/10.1016/j.taap.2022.115877
30. Ma M, Zhu X, Li F, Guan G, Hui R, Zhu L, Pang H, Zhang Y (2024) Associations of urinary volatile organic compounds with cardiovascular disease among the general adult population. Int J Environ Health Res 34:3876–3890. https://doi.org/10.1080/09603123.2024.2331732
31. Wang X, Chen Z, Cheng D, Cao Y, Xie X, Zhou J, Wu Y, Li X, Yu J, Yang B (2023) Association between urinary metabolites of volatile organic compounds and cardiovascular disease in the general population from NHANES 2011–2018. Ecotoxicol Environ Saf 264:115412. https://doi.org/10.1016/j.ecoenv.2023.115412
32. Li W, Ruan W, Cui X, Lu Z, Wang D (2022) Blood volatile organic aromatic compounds concentrations across adulthood in relation to total and cause specific mortality: a prospective cohort study. Chemosphere 286:131590. https://doi.org/10.1016/j.chemosphere.2021.131590
33. Nalini M, Poustchi H, Bhandari D, Chang CM, Blount BC, Wang L, Feng J, Gross A, Khoshnia M, Pourshams A, Sotoudeh M, Gail MH, Graubard BI, Dawsey SM, Kamangar F, Boffetta P, Brennan P, Abnet CC, Malekzadeh R, Freedman ND, Etemadi A (2024) Volatile organic compounds and mortality from ischemic heart disease: a case-cohort study. Am J Prev Cardiol 19:100700. https://doi.org/10.1016/j.ajpc.2024.100700
34. He Y, Qiu H, Wang W, Lin Y, Ho KF (2024) Exposure to BTEX is associated with cardiovascular disease, dyslipidemia and leukocytosis in national US population. Sci Total Environ 919:170639. https://doi.org/10.1016/j.scitotenv.2024.170639

Chapter 23
Cardiovascular Toxicity of Metals

Zorica Bulat and Petar Bulat

Abstract Toxic metals affect all organs and organ systems, including the cardiovascular system. The aim of the chapter is to compile and present recent research on the impact of the single metals and metal mixtures on cardiovascular disease (CVD) in humans. Most frequently, studies have focused on the effects on blood pressure and the development of hypertension. Apart from a rise in blood pressure and hypertension following exposure to increasing levels of lead, the precise role that hazardous metals play in the development of CVD has not yet been determined. Data on the cardiovascular effects of exposure to cadmium, arsenic, and mercury are conflicting, and even in studies that show a correlation between exposure to these metals and CVD it is generally modest. Exposure to metal mixtures, with varying concentrations and compositions, makes the research of these effects more difficult. High-quality prospective studies examining the effects of metal mixture cardiotoxicity in populations exposed to a wider range of metal exposures are necessary. In light of all of above, a thorough plan to avoid morbidity as well as deaths from CVD should include efforts to reduce toxic metal exposure.

Keywords Arsenic · Blood pressure · Cadmium · Hypertension · Lead · Mercury · Metal mixture

Z. Bulat (✉)
Department of Toxicology "Akademik Danilo Soldatović", University of Belgrade—Faculty of Pharmacy, Belgrade, Serbia
e-mail: zorica.bulat@pharmacy.bg.ac.rs

P. Bulat
University of Belgrade—Faculty of Medicine, Belgrade, Serbia

Serbian Institute of Occupational Health, Belgrade, Serbia

D. K. Agrawal et al. (eds.), *Cardiovascular Toxicity*, Advances in Biochemistry in Health and Disease 33, https://doi.org/10.1007/978-3-031-90598-8_23

Introduction

Toxic metals are ubiquitous pollutants, present in air, water, food as well as in many consumer products, medical devices, and in trace levels even in supplements, drugs etc. Chronic exposure to toxic metals is of utmost importance, affecting all organs and organ systems. Some organs are well recognised as targets of metal toxicity, like the kidneys or nervous system. Although the cardiovascular system (CVS) is not considered one of them, actions of toxic metals on it may play important roles in mediating the pathophysiologic effects [1]. According to the American Heart Association scientific statement [2], toxic metals, lead (Pb), cadmium (Cd) and arsenic (As) pose cardiovascular risk, contribute to atherosclerosis and coronary artery stenosis, as well as the risk of ischemic heart disease (IHD) and stroke, heart failure, and peripheral artery disease. In spite of the fact that toxic metals have their specificity, they exert the effects by similar mechanisms of toxicity, because they can be considered through the mixture toxicity approach. This chapter is focused on new findings on the cardiovascular toxicity of metals Pb, Cd, As and mercury (Hg), as well as metal mixtures in humans.

Lead

The most extensively researched toxic metal with the most well detailed harmful consequences is Pb. A lot of human studies dealing with Pb effects contribute that it is the metal with the best clarified toxic effects in humans. Especially, Pb exposure is established as toxic for CVS, with most evaluated effect on blood pressure (BP) and hypertension, which arise in higher blood Pb levels (BLL), up to 50 μg/dL. As a result of banned Pb gasoline, BLL in the population decreased to <10 μg/dL, but still the estimation was that doubling of BLL, systolic blood pressure (SBP) and diastolic blood pressure (DBP) would increase by 1 and 0.6 mmHg, respectively [3]. However, the public and occupational health risks related to Pb exposure need to be updated because a significant portion of the available literature has become outdated due to the substantial decline in exposure levels over recent decades. Specifically, BLL decreased from 1.76 μg/dL during 1999–2004 to 0.93 μg/dL in 2017–2020, while the percentage of individuals with BLL $<$ 1 μg/dL increased from 19.2% to 63.0% [4].

A cross-sectional study using database from the National Health and Nutrition Examination Survey (NHANES, 1999–2018; 32,289 participants) demonstrated the association of BLL with SBP and DBP in non-Hispanic white and Black participants, but not with hypertension [5]. On the other side, Qu et al. [6] study included 11,037 adults older than 18 years from China National Human Biomonitoring (2017–2018) and found that higher BLLs in Chinese adults (median BLL 22.4 μg/L) were associated with higher odds of pre-hypertension, with more pronounced associations observed among people whose BP was in the range of 130–139/

80–89 mmHg and hypertension when compared with the lowest quartile. In a study performed in Koreans [7], the relationship between BLL (< 10 μg/dL) and pre-hypertension and between BLL and hypertension in 7341 subjects was examined. BLL ≥ 6.87 μg/dL was associated with hypertension. In support of these results, a Swedish cohort study used cross-sectional data (4172 middle-aged men and women) and confirmed the association between Pb exposure and the prevalence of atherosclerotic plaque in the carotid artery. BLL was 25 μg/L (range: 1.5–258), and 36% of the cohort had any atherosclerotic plaque, confirming association in men and postmenopausal women but not in premenopausal women and persons who never smoked [8].

The recent Study for Promotion of Health in Recycling Lead (SPHERL; NCT02243904), a longitudinal, prospective study of newly hired Pb workers without known previous occupational exposure [9], accounted for interindividual variability between participants with respect to their vulnerability to the toxic effects of Pb exposure by assessing their health status before and after occupational lead exposure. 2-year BP responses and incident hypertension were not associated with the BLL increase on first occupational exposure, but the small sample size, the short follow-up of the cohort, the mean age of 29.7 years of participants, healthy workers, without co-exposures, warrant a cautious interpretation of the findings. The limitations of this study in terms of short exposure and young participants are confirmed in Wang et al. [10] study who observed that exposure to heavy metal mixtures, particularly Pb, may elevate the likelihood of hypertension, SBP, and DBP abnormalities. They found that BLL showed a slow increase with age and were associated with age and biological aging, using PhenoAge and PhenoAgeAccel Assessments, and revealed that Pb significantly mediated the relationship between biological aging and hypertension and between biological aging and SBP abnormalities, accounting for 64% of the effects.

In addition to the effects on BP and hypertension, some studies point out Pb effects on heart rhythms. Jing et al. [11] found that when the log of BLL increased by one unit, the odds of an abnormal QRS-T angle increased by 34% in men and 4% in women. Based on these findings, support QRS-T angle deviation, among those who are exposed to Pb, to identify individuals at risk for adverse cardiovascular outcomes such as arrhythmias. In Pb-exposed workers, Pmax, P min, P Wave Dispersion, and QT Dispersion values, which predictors of cardiac arrhythmias, were significantly lower after chelation therapy compared to values before chelation therapy suggesting that they are under the risk of ventricular and atrial arrhythmias [12].

The importance of considering gender in environmental epigenetics and mechanistic toxicology studies was evaluated by Svodoba et al. [13], who found that a single gene, Galnt2, showed differential methylation in both sexes in mice hearts, and at different time points, at weaning in a cohort of Pb-exposed mice in adulthood. In a human cohort study examining the effects of prenatal lead exposure on the epigenome also revealed an inverse correlation between first trimester Pb concentrations and adolescent blood leukocyte DNA methylation at a locus in Galnt2, indicating that this gene may serve as a biomarker for Pb exposure across species.

Collectively, these data demonstrate that Pb exposure promotes sex-specific programming of the cardiac epigenome and offers potential mechanistic insights into how Pb causes cardiovascular disease (CVD).

Cadmium

Cd is a toxic metal with well-established target organs, kidneys and the skeletal system, it is classified as human carcinogen based on lung cancer development and nowadays recognised as endocrine disruptor. Considering that renal dysfunction is main Cd toxic effect, and that hypertension has a direct influence on the development of kidney disease, it could be hypothesised that occurrence of hypertension could be related to Cd-induced kidney damage. Nevertheless, there is currently insufficient evidence to support this link in case of Cd exposure, bearing in mind that kidney damage only occurs after long-term Cd exposure and that it manifests itself predominantly by damage to tubular reabsorption.

Inhalation exposure to Cd does not appear to have significant effects on the CVS, which is confirmed in occupational exposure studies where no Cd-related cardiovascular toxicity was observed. This was expanded with wide range of environmental exposure to Cd where BP, prevalence of hypertension or other CVD was not correlated with the urine or blood Cd levels. On the other side, based on the data from the NHANES associations between blood and urine Cd levels and cardiovascular effects suggests a relationship between elevated Cd exposure and increased risk of cardiovascular mortality in men, but not in women. This should be carefully analysed especially having in mind that women, characterized by higher Cd exposure due to increased gastrointestinal absorption given by their low iron status [14]. The results also summarised in Toxicological profile for Cadmium indicate the difference between Cd blood and Cd urine levels in relation to elevated BP levels, so a modest but positive association between blood Cd, but not of urine, could be related to recent exposure having a short-term effect on BP. In addition, it could also be a consequence of the fact that Cd accumulates in the kidney and that only after the capacity of the kidney to excrete Cd is exceeded does a significant appearance of Cd in the urine occur. In a cross-sectional study [15], the association between low serum Cd concentration and blood pressure or hypertension was investigated in a general population in the Iwaki area, Japan, in 1144 volunteers aged over 19 years old. Median serum Cd concentration was 0.06 ng/mL, and the group with the highest quintile of serum Cd concentration had a 4.9 mmHg higher SBP and a 2.4 mmHg higher DBP compared to the group with the lowest quintile. This study identified that higher serum Cd concentration was significantly, positively, associated with SBP and DBP and hypertension prevalence. In the study using urine for biomonitoring [16], no association was found between urinary Cd, BP and hypertension in the total population in a representative French sample (2015 subjects from the ESTEBAN survey, 2014–2015). On the other side, urinary Cd was linearly and positively associated with cardiac infarction/injury score (CIIS) and subclinical myocardial injury (SC-MI) in a large-scale cross-sectional study in the general population without CVD and chronic kidney disease [17].

Although effect on BP and hypertension are most often in focus, Mirónczuk et al. [18] study showed that the disturbance of the metal balance of Pb and Cd plays a crucial role in the pathogenesis of acute ischemic stroke (AIS); in particular, blood-Cd concentration is connected with the incidence of AIS. They suggest that assessment of changes in the molar ratios of essential and toxic metals could serve as a valuable indicator of the levels of oxidative stress in AIS patients, especially Cd/Zn molar ratio which has the potential to serve as a useful biomarker for atherosclerosis in AIS patients.

Recent meta-analyses [19], based on 18 studies published between 2006 and 2024, found epidemiologic evidence that Cd exposure is associated with an increased risk of hypertension, with a generally linear relation based on urine concentrations and a non-linear relation based on blood concentrations. However, the data were insufficient to assess the dose-response relationship within some strata of gender and smoking status. The systematic review by Martins et al. [20], which focused on studies published between January 2010 and March 2020, included 38 studies and found a positive association between Cd blood levels and BP and/or hypertension independent of smoking. However, in the case of urinary Cd, the association with BP and/or hypertension remains uncertain due to conflicting results, including inverse relationships with lack of strong mechanistic support. Both reviews point out that the inconsistency of the results is due to the differences in Cd blood and Cd urine as biomarkers of Cd exposure. The associations varied by the marker of Cd levels (mainly blood *vs.* urine), two biomarkers generally linked to recent and long-term exposure to this toxic metal. Urinary Cd concentrations are a valid biomarker of long-term exposure, especially at low levels and in non-occupationally exposed individuals. It is possible that the inconsistent results are due to genetic differences. So, Wu et al. [21] found that the associations between Cd exposure and hypertension risk differed by genetic polymorphisms in MMPs genes in the Chinese population (497 hypertension cases and 497 healthy controls). Elevated urinary Cd concentrations were positively associated with hypertension risk, and the associations were modified by the rs14070 and rs7201 polymorphism in MMP-2 gene.

Arsenic

As is a toxic metalloid presented as inorganic in valence states As (III), As (V), and organic form. The toxic effect of chronic exposure to As are exerts on central and peripheral nervous system, and skin, it is linked to the onset of noncommunicable diseases including diabetes and cardiovascular diseases, and furthermore As is carcinogenic to humans, inducing lung and skin cancer. Chronic exposure to inorganic As is connected with hypertension, despite inconsistent results from studies conducted in different populations, as well as with effects on blood vessels. The most specific effect is endemic Blackfoot disease in an area of Taiwan (As in drinking water 0.17–0.80 ppm, 0.014–0.065 mg As/kg/day), characterized by changes leading to necrosis and gangrene, while other vascular effects increases in the incidences of Raynaud's disease and of cyanosis of fingers and toes [22]. As-induced cardiovascular toxicity is result of numerous mechanisms: oxidative stress and reactive oxygen species generation, endoplasmic reticulum stress and impairment of calcium homeostasis, autophagy dysregulation, mitochondrial dysfunction, inflammatory pathway.

The actuality of the topic is also indicated by Han et al. [23] study, comprehensive overview of current research focused on As-related concerns regarding cardiovascular health, based on bibliometric analysis. Hypertension and atherosclerosis are the most extensively studied topics, with redox imbalance, apoptosis, and methylation being the primary mechanistic clues. Cardiovascular damage caused by As includes arrhythmia, cardiac remodelling, vascular leakage, and abnormal angiogenesis. In support of these findings, Qu et al. [6] found that blood As was linked to the development of pre-hypertension in the Chinese population in addition to the most notable Pb influence on hypertension. In Tang et al. [24] study found urinary As level was associated with increased DBP level among general non-Hispanic Asians aged 20 years and older, using the 2011–2018 NHANES data (n = 5175). Sánchez-Rodríguez et al. [25] study also support these statements by revealing connection between chronic endemic regional hydroarsenicism and hypertension in a region in north-central Mexico where As exceeds the WHO guideline of 10 μg/L. Citizens of that region were found to have a higher risk of hypertension and diabetes mellitus type 2, compared to controls, and men is more likely to be diagnosed than women. Furthermore, less efficient As metabolism (higher % inorganic As and % monomethyl arsenate in the urine) was associated with increased BP among never smokers with high rice consumption, suggesting that As exposure through rice may contribute to high BP in the Hispanic/Latino community (12,609 participants) [26]. The dose–response relationship between As exposure and hypertension with vascular NO levels as a potential mediator of As-related hypertension in individuals exposed to a broad range of As was investigated in a total of 828 participants recruited from low-As and high-As exposure in Bangladesh [27]. Increasing exposure to As (based on participants' drinking water, hair, and nail As) was associated with increasing odds of hypertension. The association was mediated through the reduction of vascular NO bioavailability, suggesting that impaired NO bioavailability was a plausible underlying mechanism of As-induced hypertension in the Bangladeshi population.

In the performed meta-analysis, the association of As exposure with hypertension or BP, based on cross-sectional, case-control and longitudinal studies among general adult population was performed [28]. The meta-analysis revealed the significant association of chronic As exposure with hypertension among cross-sectional studies, mainly associated with increasing SBP but not significantly related to DBP (the differences in SBP levels between the exposed and reference groups were 4.03 mmHg). A dose–response assessment demonstrated that the OR of hypertension increased with an increasement of As exposure, and also statistical evidence of a nonlinear dose–response association. Xu et al. [29] conducted a systematic review and a dose-response meta-analyses to estimate the pooled association between the relative risk of each CVD endpoint and low-level As concentration in drinking water, lower then WHO provisional guideline value (10 μg/L). Significant increased risks of coronary heart disease (CHD) mortality and CVD mortality as well as combined fatal and non-fatal CHD, CVD, carotid atherosclerosis disease and hypertension in those exposed to 10 μg/L As in drinking water compared to the referent population (1 μg/L) suggests further lowering of this guideline value.

Mercury

Mercury (Hg) has several chemical forms, but most responsible for toxicity are Hg (+2), and organic Hg, especially methyl-Hg. Hg can pass through various membranes and barriers in the body like the blood-brain barrier, primarily affecting the central nervous system and kidneys, but also other organs and systems. Evaluated cardiovascular outcomes include BP, cardiac function, or diagnosis of clinical hypertension or CVD, although epidemiologic study results are inconsistent for all chemical forms [30]. For BP, the most studied outcome, the evidence for effects is conflicting, and studies showing positive associations indicate that the magnitude of changes is small or without clear evidence of a dose-response relationship between exposure and elevated BP in the population. Data on effects on cardiac function are insufficient to draw conclusions or data are lacking (inorganic Hg salts), although some studies have reported an inverse association between methyl Hg and heart rate variability, which may lead to more serious data on cardiac effects. Overall, the results of recent epidemiologic studies do not provide conclusive evidence that the CVS is a highly sensitive target for Hg. Nevertheless, some studies suggest effects on the CVS. In Tang et al. [24] study, high expose to Hg (and its specific species methyl-Hg) was associated with increased BP levels and prevalence of hypertension among general non-Hispanic Asians. In the New Hampshire Birth Cohort Study [31], prenatal and childhood Hg exposure in 395 young children at multiple time points was investigated. Neither early prenatal nor mid-gestation Hg exposure, as measured by maternal toenails, were related to any changes to child BP. However, the simultaneous inclusion of both child urine Hg and child toenail Hg in models suggested a potentially stronger association between urinary Hg at age 5–6 with DBP and mean arterial pressure compared with toenail Hg at age 3. These results indicate that childhood Hg exposure is associated with alterations in BP, pointed that childhood may be an important window of opportunity to reduce the impacts of Hg exposure on children's BP and thus long-term health. Additionally, the association between the presence or replacement of dental amalgams and the risk of gestational hypertension (GH) was assess in the 2001 pregnant women recruited between 2008 and 2011 in 10 Canadian cities, in a prospective cohort MIREC Study (Maternal-Infant Research on Environmental Chemicals) [32]. Although weak correlations between dental amalgams and blood Hg were observed, replacement of dental amalgam before or during pregnancy was associated with decreased SBP, but the presence of dental amalgams (1–4 or ≥ 5) or their replacement in the first or third trimester was not linked to GH. In the Indonesian coastal pregnant mother population, Armatussolikha et al. [33] evaluate whether GSTT1 (glutathione S-transferase T1) deletion variants and maternal hair Hg concentration are associated with BP and birth outcomes. While the GSTT1 homozygous deletion genotype has no role in diastolic hypertension and birth outcomes, or in hair Hg level, maternal Hg exposure and history of CVD are independent risk factors for pregnant hypertension. According to Pamphlett et al. [34], because Hg is found frequently within the adult adrenal medulla and increases with age, it could contribute to elevated plasma noradrenaline in aging. These findings provide a potential link between Hg exposure, increased sympathetic activity, and the pathogenesis of essential hypertension and the metabolic syndrome, leading to the severe health consequences of chronically raised sympathetic tone.

In a dose–response meta-analysis, the findings of published studies that reporting Hg biomarkers and hypertension or BP were summarized, extracted, and pooled to examine the potential relationship [35]. The inclusion of 29 research revealed a strong positive correlation between Hg and both BP and hypertension. For both hypertension and SBP, a non-linear dose-response relationship was found, with an inflection point at 3 μg/g hair Hg. This is because exposure dose plays a significant role in determining the harmful effects of Hg on hypertension. In the case of increased risk of all-cause mortality and fatal/nonfatal IHD, which begins to increase consistently at a hair Hg concentration of 2 μg/g, the same discovery was made in another meta-analysis, which was conducted by the same group of authors [36].

Metals Mixture

In everyday exposure to toxic substances, and consequently in epidemiological studies, numerous co-factors affect cardiovascular function, including co-exposure to toxic substances, which could affect the CVS either directly or indirectly through effects on other systems (renal, neurological). The collective evidence currently suggests possible cardiotoxic effects of metal mixture, with identifiable differences between single metal and multi-metal exposure. As most individuals are exposed simultaneously to multiple metals, studying the cardiotoxicity of metal mixture exposures is critical. Since the National Institute of Environmental Health Sciences (NIEHS) declared conducting combined exposure research as a priority issue [37], literature on chemical mixtures has expanded significantly. For instance, the scoping review by Yim et al. [38] summarizes the existing epidemiological research on the cardiotoxicity that assessed the effect of metal combinations on CVD risk factors and outcomes among nonoccupationally exposed populations. Twentynine original research studies, mostly cross-sectional in design, published from 1998 to 2021, were identified for review, while a notable increase in relevant mixtures publications was observed starting in 2019. The sample sizes varied from 127 to 10,818 and the majority of the eligible research were carried out in China and the United States. BP and/or diagnosis of hypertension, risk of preeclampsia, dyslipidemia, and CVD outcomes, including stroke incidence or CHD were primary focus areas. Usually analysed in blood, the metals that were most commonly studied were Cd, Pb, As, and cobalt (Co).

In order to examine the relationship between metal mixtures and malignant neoplasms, hypertension, diabetes, and heart diseases (NHANES 2011–2016 studies, 7236 adults), Wang et al. [39] found patterns of mixed metal exposure. Blood metal levels revealed three different patterns of exposure: "low Cd-Pb-Hg-Selenium (Se)", "low Cupper (Cu) & high Hg-Se", and "high Cu-Cd-Pb". Of the health outcomes examined, all three patterns had the strongest correlation with hypertension. Furthermore, compared to "low Cu & high Hg-Se," "low Cd-Pb-Hg-Se" showed stronger associations with heart disease and hypertension, indicating a possible interaction between these metals. Chen et al. [40] also used the NHANES data (1999–2018; 38,281 participants) to confirm that there was a substantial dose-response connection between blood Pb and Cd concentrations and an increased prevalence of resistant hypertension (RHTN). The mixture-exposure analyses also showed a positive correlation between a mixture of blood Pb, Cd, and Hg and elevated probability of RHTN, primarily driven by blood Cd and Pb, highlighted potential adverse impacts of exposure to environmental metals on BP management among hypertensive patients, particularly for middle-aged, male, and afflicted with kidney dysfunction. The blood metals mixture (As, Pb, Cd, Hg) was linked to an elevated DBP level in the previously stated Tang et al. [24] study, which was also based on NHANES data (2011–2018; blood $n = 10{,}177$, and urine $n = 5175$). Similarly, exposure to metal mixture of 13 metals as assessed in blood was positively related with higher SBP and DBP in the Qu et al. [6] study (China National Human Biomonitoring, 2017–2018); nonetheless, as previously mentioned, Pb is identified as crucial in the joint effects of metal mixture on the development of hypertension. The impact of environmental exposure to metals mixture (13 metals; whole blood levels) on hypertension, with obesity serving as a significant independent risk factor for hypertension, was evaluated in a cross-sectional study involving 3063 adults from 11 districts or counties in Guangdong, China [41]. The combined effect of As, Cd, Pb and manganese (Mn) were associated with the prevalence of hypertension, while potential interaction effects of Cd, Pb and obesity on hypertension risk may exist. Based to the systematic review of Nucera et al. [42] correlations between exposure to nonessential metals and increased incidence of CVD are reported in seven of eight reviews, published in the last 5 years through March 2023. Although it is clear that exposure to metals, more especially, Cd, Hg, and As, is linked to CVD, the variety of research has unavoidably produced findings that are not always comparable. The use of different biomarkers including blood, urine, nails, and hair could led to conflicting results, as well lack of homogeneity across studies, such as for exposure levels to various metals, chemical form of metals, etc.

In pregnant women, the overall combined effects of blood Pb, Cd, As, Hg, and essential metal Mn on the risk of developing GH and preeclampsia were not found in a cohort of Canadian women [43]. However, it was confirmed that even low BLL is a risk factor for preeclampsia, as each doubling of third trimester Pb and first trimester As was independently linked to an increased risk of preeclampsia. Moreover, GH was more likely to occur in women with lower Mn levels in the early stages of pregnancy and greater blood As concentrations. Geospatial patterns of early-onset HDP in primiparous mothers were evaluated in the cross-sectional study [44] using geocoded residential data from Kentucky live (N = 210,804) and still (N = 1247) birth records (2008–2017) and exposure to industrial emissions of As, Cd, chrome (Cr), Pb, Hg, Se, and Zn (Risk Screening Environmental Indicators model). Metal exposures were divided into four subgroups: low or no exposures; As, Cd, and Pb; Se and Zn; and Pb and Cr. The prevalence of early-onset HDP was considerably higher in women with a high chance of exposure to both Pb and Cr than in those with less or no exposure. This finding adds to the growing body of research on the relationship between metal exposures and pregnancy outcomes. The effect of a complex mixture of metals on HDP was also examined in Ma et al. [45] population-based nested case-control study (2013–2016, Wuhan, China, with 146 cases and 292 controls). Plasma Hg concentrations were identified as the most critical factors related to HDP, exposure to Al, Ni, and V might increase the risk, while plasma Al and V might be had a synergistic effect on HDP. Li et al. [46] investigated the effect of heavy metal exposure on congenital heart defects (CHDs), performing the meta-analysis that found significant associations between As, Cd, Hg and Pb exposure during pregnancy and an increased risk of certain CHDs in the offspring. Despite biomarkers (the amount of metal in the mother's blood and hair) demonstrated a greater correlation between gestational toxic metal exposure and CHD outcomes than environmental source assessment, the measurement method's limitations still make it impossible to accurately estimate exposure during pregnancy.

The effect of metal mixture on CVS was also considered in occupational exposure. In the cross-sectional study [47], Chinese molybdenum miners and iron and steel foundry workers (n = 395) classified in the highest quartiles of molybdenum (Mo), As and Pb had a 2.58-fold, 4.30-fold and 4.85-fold increased probability of having hypertension, respectively. The relationship was nonlinear between urinary Mo, Co, Cd, As and Pb concentrations and the prevalence of hypertension, based on threshold effect analyses. In addition, Pb, Mo, As and Co may have joint effect, and a strong positive correlation with the prevalence of hypertension. In hypertensive and non-hypertensive workers Qian et al. [48] the incidence of hypertension increased with an increase in Cr or Mn exposure level, based on metal concentration in the air. The risk of hypertension was 1.85 times higher in the highest exposure than in the lowest, indicated that Cr and Mn exposure in the work environment might increase the risk of hypertension, while no effect of Fe and Ni on BP was found. A prospective cohort Gulf Long-Term Follow-up (GuLF) study (957 participants) focused on potential health effects following the 2010 Deepwater Horizon oil spill did not confirmed evidence indicating cross-sectional associations between blood levels of Cd, Pb, Hg, or Se and either prevalent hypertension or BP overall. Since the relatively low blood levels of Cd, Pb, and Hg, the results of this study raise the possibility of no associations between these metals and hypertension under a certain threshold [49].

The primary objective of Guo et al. [50] meta-analysis was to summarize the available prospective studies and assess the associations between heavy metals in general, and Hg, Cd, Pb, As (in urine or blood) and mortality from all causes, CVD, and cancer in the general population. Based on 42 original studies which were identified, this study provided compelling evidence that Cd and Pb exposure was significantly associated with mortality from all causes, CVD, and cancer, for As moderate evidence was suggested, while the evidence for Hg was insignificant or indeterminate, in the general population.

Because of limited data available regarding the connection between hypertension and heavy metal exposure, study of Xu and Sun [51] intend to establish an interpretable machine learning (ML) model that identifies this connection. The authors used data sets from the US National Health and Nutrition Examination Survey (NHANES, 2013–2020) and included 19,368 participants in total. A best-performing eXtreme Gradient Boosting (XGB) with Genetic Algorithm (GA) for hypertension identification by 16 heavy metals was selected. Study participants' hypertension associated with heavy metal exposure was identified and finally, barium, Cd, Pb, Antimony, Tin, Mn, Thallium, Tungsten in urine, and Pb, Hg, Se, Mn in blood positively contributed to hypertension, while Cd in blood negatively correlated with hypertension.

Conclusion

Previous research indicates that only high exposer to Pb have a clear connection to the appearance of hypertension. In the case of exposed Cd, As and Hg, the data are contradictory, and even in the studies that indicate the existence of a link between exposure to these metals and CVD, the link is mostly weak. In concordance with previous, studies on Pb effect on development of CVD are numerous and have moderate to large sample sizes, while Cd, As and Hg are studied in smaller or more variable cohorts, leading to mixed or limited findings and weaker evidence. This is even more pronounced in the case of metal mixtures, where different levels of exposure to heterogeneous mixtures are singled out as the main problem. Having in mind growing arise of exposure to mixture of metals, there is a need for high-quality prospective studies investigating the impacts of cardiotoxicity of metal mixtures in populations exposed to a wider range of metal exposures. Based on all of that, a comprehensive strategy to prevent morbidity as well as deaths from CVD should include efforts to reduce toxic metal exposure.

References

1. Prozialeck WC, Edwards JR, Nebert DW, Woods JM, Barchowsky A, Atchison WD (2008) The vascular system as a target of metal toxicity. Toxicol Sci 102(2):207–218. https://doi.org/10.1093/toxsci/kfm263
2. Anon (2023) Contaminant Metals as Cardiovascular Risk Factors: A Scientific Statement From the American Heart Association. J Am Heart Assoc 12:e029852. https://doi.org/10.1161/JAHA.123.029852
3. Agency for Toxic Substances and Disease Registry (ATSDR) (2020) Toxicological profile for Lead. U.S. Department of Health and Human Services, Public Health Service, Atlanta, GA. https://doi.org/10.15620/cdc:95222
4. An DW, Yu YL, Hara A, Martens DS, Yang WY, Cheng YB, Huang QF, Asayama K, Stolarz-Skrzypek K, Rajzer M, Verhamme P, Nawrot TS, Li Y, Staessen JA (2024) Lead-associated mortality in the US 1999-2020: a time-stratified analysis of a national cohort. J Hypertens 42(8): 1322–1330. https://doi.org/10.1097/HJH.0000000000003713
5. Huang Z (2022) Association between blood Lead level with high blood pressure in US (NHANES 1999–2018). Front Public Health 10:836357. https://doi.org/10.3389/fpubh.2022.836357
6. Qu Y, Lv Y, Ji S, Ding L, Zhao F, Zhu Y, Zhang W, Hu X, Lu Y, Li Y, Zhang X, Zhang M, Yang Y, Li C, Zhang M, Li Z, Chen C, Zheng L, Gu H, Zhu H, Sun Q, Cai J, Song S, Ying B, Lin S, Cao Z, Liang D, Ji JS, Ryan PB, Barr DB, Shi X (2022) Effect of exposures to mixtures of lead and various metals on hypertension, pre-hypertension, and blood pressure: a cross-sectional study from the China National Human Biomonitoring. Environ Pollut 299:118864. https://doi.org/10.1016/j.envpol.2022.118864
7. Kim MG, Kim YW, Ahn YS (2020) Does low lead exposure affect blood pressure and hypertension? J Occup Health 62(1):e12107. https://doi.org/10.1002/1348-9585.12107
8. Harari F, Barregard L, Östling G, Sallsten G, Hedblad B, Forsgard N, Borné Y, Fagerberg B, Engström G (2019) Blood Lead levels and risk of atherosclerosis in the carotid artery: results from a Swedish cohort. Environ Health Perspect 127(12):127002. https://doi.org/10.1289/EHP5057
9. Yu YL, Yang WY, Hara A, Asayama K, Roels HA, Nawrot TS, Staessen JA (2023) Public and occupational health risks related to lead exposure updated according to present-day blood lead levels. Hypertens Res 46(2):395–407. https://doi.org/10.1038/s41440-022-01069-x
10. Wang C, Su J, Li J, Wei W, Yuan Z, Chen R, Wei W, Huang Y, Ye L, Liang H, Jiang J (2024a) Blood Lead mediates the relationship between biological aging and hypertension: based on the NHANES database. Nutrients 16:2144. https://doi.org/10.3390/nu16132144
11. Jing J, Thapa S, Delhey L, Abouelenein S, Morad W, Delongchamp R, Faramawi MF (2019) The relation of blood lead and QRS-T angle in American adults. Arch Environ Occup Health 74(5):287–291. https://doi.org/10.1080/19338244.2018.1488674
12. Karanfil M, Gündüzöz M, Karakurt M, Aruğaslan E, Özbay MB, Ünal S, Akbuğa K, Akdi A, Erdöl MA, Ertem AG, Yayla Ç, Özeke Ö (2022) Effect of chelation therapy on arrhythmogenic and basal ECG parameters of lead exposed workers. Arch Environ Occup Health 77(5): 382–388. https://doi.org/10.1080/19338244.2021.1910116
13. Svoboda LK, Wang K, Goodrich JM, Jones TR, Colacino JA, Peterson KE, Tellez-Rojo MM, Sartor MA, Dolinoy DC (2023) Perinatal Lead exposure promotes sex-specific epigenetic programming of disease-relevant pathways in mouse heart. Toxics 11:85. https://doi.org/10.3390/toxics11010085
14. Agency for Toxic Substances and Disease Registry (ATSDR) (2012) Toxicological profile for Cadmium. U.S. Department of Health and Human Services, Public Health Service, Atlanta, GA
15. Kaneda M, Wai KM, Kanda A, Ando M, Murashita K, Nakaji S, Ihara K (2022) Low level of serum cadmium in relation to blood pressures among Japanese general population. Biol Trace Elem Res 200(1):67–75. https://doi.org/10.1007/s12011-021-02648-8

16. Vallée A, Gabet A, Grave C, Blacher J, Olié V (2020) Associations between urinary cadmium levels, blood pressure, and hypertension: the ESTEBAN survey. Environ Sci Pollut Res Int 27(10):10748–10756. https://doi.org/10.1007/s11356-019-07249-6
17. Qiang YW, Hao MH, Yang J (2024) Urinary cadmium was linearly and positively associated with cardiac infarction/injury score and subclinical myocardial injury in the general population without cardiovascular diseases and chronic kidney disease. Int Urol Nephrol 56(5):1677–1685. https://doi.org/10.1007/s11255-023-03853-1
18. Mirończuk A, Kapica-Topczewska K, Socha K, Soroczyńska J, Jamiołkowski J, Chorąży M, Czarnowska A, Mitrosz A, Kułakowska A, Kochanowicz J (2023) Disturbed ratios between essential and toxic trace elements as potential biomarkers of acute ischemic stroke. Nutrients 15(6):1434. https://doi.org/10.3390/nu15061434
19. Verzelloni P, Giuliano V, Wise LA, Urbano T, Baraldi C, Vinceti M, Filippini T (2024) Cadmium exposure and risk of hypertension: a systematic review and dose-response meta-analysis. Environ Res 263(Pt 1):120014. https://doi.org/10.1016/j.envres.2024.120014
20. Martins AC, Almeida Lopes ACB, Urbano MR, Carvalho MFH, Silva AMR, Tinkov AA, Aschner M, Mesas AE, Silbergeld EK, Paoliello MMB (2021) An updated systematic review on the association between Cd exposure, blood pressure and hypertension. Ecotoxicol Environ Saf 208:111636. https://doi.org/10.1016/j.ecoenv.2020.111636
21. Wu W, Liu D, Jiang S, Zhang K, Zhou H, Lu Q (2019) Polymorphisms in gene MMP-2 modify the association of cadmium exposure with hypertension risk. Environ Int 124:441–447. https://doi.org/10.1016/j.envint.2019.01.041
22. Agency for Toxic Substances and Disease Registry (ATSDR) (2007) Toxicological profile for Arsenic. U.S. Department of Health and Human Services, Public Health Service, Atlanta, GA. https://doi.org/10.15620/cdc:11481
23. Han Y, Gao T, Li X, Wāng Y (2024) Didactical approaches and insights into environmental processes and cardiovascular hazards of arsenic contaminants. Chemosphere 352:141381. https://doi.org/10.1016/j.chemosphere.2024.141381
24. Tang J, Zhu Q, Xu Y, Zhou Y, Zhu L, Jin L, Wang W, Gao L, Chen G, Zhao H (2022) Total arsenic, dimethylarsinic acid, lead, cadmium, total mercury, methylmercury and hypertension among Asian populations in the United States: NHANES 2011-2018. Ecotoxicol Environ Saf 241:113776. https://doi.org/10.1016/j.ecoenv.2022.113776
25. Sánchez-Rodríguez BL, Castillo-Maldonado I, Pedroza-Escobar D, Delgadillo-Guzmán D, Soto-Jiménez MF (2023) Association of obesity, diabetes, and hypertension with arsenic in drinking water in the Comarca Lagunera province (North-Central Mexico). Sci Rep 13(1):9244. https://doi.org/10.1038/s41598-023-36166-5
26. Scannell Bryan M, Sofer T, Mossavar-Rahmani Y, Thyagarajan B, Zeng D, Daviglus ML, Argos M (2019) Mendelian randomization of inorganic arsenic metabolism as a risk factor for hypertension-and diabetes-related traits among adults in the Hispanic community health study/study of Latinos (HCHS/SOL) cohort. Int J Epidemiol 48(3):876–886. https://doi.org/10.1093/ije/dyz046
27. Khatun M, Haque N, Siddique AE, Wahed AS, Islam MS, Khan S, Jubayar AM, Sadi J, Kabir E, Shila TT, Islam Z, Sarker MK, Banna HU, Hossain S, Sumi D, Saud ZA, Barchowsky A, Himeno S, Hossain K (2024) Arsenic exposure-related hypertension in Bangladesh and reduced circulating nitric oxide bioavailability. Environ Health Perspect 132(4):47003. https://doi.org/10.1289/EHP13018
28. Zhao J, Li A, Mei Y, Zhou Q, Li Y, Li K, Xu Q (2021) The association of arsenic exposure with hypertension and blood pressure: a systematic review and dose-response meta-analysis. Environ Pollut 289:117914. https://doi.org/10.1016/j.envpol.2021.117914
29. Xu L, Mondal D, Polya DA (2020) Positive Association of Cardiovascular Disease (CVD) with chronic exposure to drinking water arsenic (as) at concentrations below the WHO provisional guideline value: a systematic review and meta-analysis. Int J Environ Res Public Health 17(7):2536. https://doi.org/10.3390/ijerph17072536

30. Agency for Toxic Substances and Disease Registry (ATSDR) (2024) Toxicological profile for mercury. U.S. Department of Health and Human Services, Public Health Service, Atlanta, GA
31. Farzan SF, Howe CG, Chen Y, Gilbert-Diamond D, Korrick S, Jackson BP, Weinstein AR, Karagas MR (2021) Prenatal and postnatal mercury exposure and blood pressure in childhood. Environ Int 146:106201. https://doi.org/10.1016/j.envint.2020.106201
32. Louopou RC, Trottier H, Arbuckle TE, Fraser WD (2020) Dental amalgams and risk of gestational hypertension in the MIREC study. Pregnancy Hypertens 21:84–89. https://doi.org/10.1016/j.preghy.2020.04.015
33. Armatussolikha HR, Winarni TI, Maharani N, Dewantiningrum J, Muniroh M (2024) Impact of GSTT1 AND GSTM1 variants and hair mercury concentration in maternal blood pressure among coastal pregnant women in Central Java, Indonesia. Reprod Toxicol 125:108574. https://doi.org/10.1016/j.reprotox.2024.108574
34. Pamphlett R, Kum Jew S, Doble PA, Bishop DP (2021) Mercury in the human adrenal medulla could contribute to increased plasma noradrenaline in aging. Sci Rep 11(1):2961. https://doi.org/10.1038/s41598-021-82483-y
35. Hu XF, Singh K, Chan HM (2018) Mercury exposure, blood pressure, and hypertension: a systematic review and dose-response meta-analysis. Environ Health Perspect 126(7):076002. https://doi.org/10.1289/EHP2863
36. Hu XF, Lowe M, Chan HM (2021) Mercury exposure, cardiovascular disease, and mortality: a systematic review and dose-response meta-analysis. Environ Res 193:110538. https://doi.org/10.1016/j.envres.2020.110538
37. Carlin DJ, Rider CV, Woychik R, Birnbaum LS (2013) Unraveling the health effects of environmental mixtures: An NIEHS priority. Environ Health Perspect 121:A6–A8
38. Yim G, Wang Y, Howe CG, Romano ME (2022) Exposure to metal mixtures in association with cardiovascular risk factors and outcomes: a scoping review. Toxics 10(3):116. https://doi.org/10.3390/toxics10030116
39. Wang X, Han X, Guo S, Ma Y, Zhang Y (2024b) Associations between patterns of blood heavy metal exposure and health outcomes: insights from NHANES 2011-2016. BMC Public Health 24(1):558. https://doi.org/10.1186/s12889-024-17754-0
40. Chen H, Zou Y, Leng X, Huang F, Huang R, Wijayabahu A, Chen X, Xu Y (2023) Associations of blood lead, cadmium, and mercury with resistant hypertension among adults in NHANES, 1999-2018. Environ Health Prev Med 28:66. https://doi.org/10.1265/ehpm.23-00151
41. Wu S, Li L, Ji G, Xing X, Li J, Ma A, Wei Y, Zhao D, Huang H, Ma W, Wu B, Dong M, Liu T, Chen Q (2023) Association of multi-metals with the risk of hypertension and the interaction with obesity: a cross-sectional study in China. Front Public Health 11:1090935. https://doi.org/10.3389/fpubh.2023.1090935
42. Nucera S, Serra M, Caminiti R, Ruga S, Passacatini LC, Macrì R, Scarano F, Maiuolo J, Bulotta R, Mollace R, Bosco F, Guarnieri L, Oppedisano F, Ilari S, Muscoli C, Palma E, Mollace V (2024) Non-essential heavy metal effects in cardiovascular diseases: an overview of systematic reviews. Front Cardiovasc Med 11:1332339. https://doi.org/10.3389/fcvm.2024.1332339
43. Borghese MM, Fisher M, Ashley-Martin J, Fraser WD, Trottier H, Lanphear B, Johnson M, Helewa M, Foster W, Walker M, Arbuckle TE (2023) Individual, independent, and joint associations of toxic metals and manganese on hypertensive disorders of pregnancy: results from the MIREC Canadian pregnancy cohort. Environ Health Perspect 131(4):47014. https://doi.org/10.1289/EHP10825
44. Walker CJ, Christian WJ, Kucharska-Newton A, Browning SR (2022) A cross-sectional examination of the early-onset hypertensive disorders of pregnancy and industrial emissions of toxic metals using Kentucky birth records, 2008-2017. PLoS One 17(9):e0274250. https://doi.org/10.1371/journal.pone.0274250
45. Ma J, Zhang H, Zheng T, Zhang W, Yang C, Yu L, Sun X, Xia W, Xu S, Li Y (2022) Exposure to metal mixtures and hypertensive disorders of pregnancy: a nested case-control study in China. Environ Pollut 306:119439. https://doi.org/10.1016/j.envpol.2022.119439

46. Li S, Wang Q, Luo W, Jia S, Liu D, Ma W, Gu H, Wei X, He Y, Cao S, Yuan Z (2022) Relationship between maternal heavy metal exposure and congenital heart defects: a systematic review and meta-analysis. Environ Sci Pollut Res Int 29(37):55348–55366. https://doi.org/10.1007/s11356-022-21071-7
47. Shi P, Jing H, Xi S (2019) Urinary metal/metalloid levels in relation to hypertension among occupationally exposed workers. Chemosphere 234:640–647. https://doi.org/10.1016/j.chemosphere.2019.06.099
48. Qian H, Li G, Luo Y, Fu X, Wan S, Mao X, Yin W, Min Z, Jiang J, Yi G, Tan X (2022) Relationship between occupational metal exposure and hypertension risk based on conditional logistic regression analysis. Meta 12(12):1259. https://doi.org/10.3390/metabo12121259
49. Xu J, Engel LS, Rhoden J, Jackson WB 2nd, Kwok RK, Sandler DP (2021) The association between blood metals and hypertension in the GuLF study. Environ Res 202:111734. https://doi.org/10.1016/j.envres.2021.111734
50. Guo X, Su W, Li N, Song Q, Wang H, Liang Q, Li Y, Lowe S, Bentley R, Zhou Z, Song EJ, Cheng C, Zhou Q, Sun C (2022) Association of urinary or blood heavy metals and mortality from all causes, cardiovascular disease, and cancer in the general population: a systematic review and meta-analysis of cohort studies. Environ Sci Pollut Res Int 29(45):67483–67503. https://doi.org/10.1007/s11356-022-22353-w
51. Xu S, Sun M (2024) The interpretable machine learning model associated with metal mixtures to identify hypertension via EMR mining method. J Clin Hypertens (Greenwich) 26(2): 187–196. https://doi.org/10.1111/jch.14768

Chapter 24
Excess Labile Zinc Ion Produces Toxic Effects in Cardiac Cells Through Oxidation and/or Phosphorylation of Proteins and Kinases

Belma Turan

Abstract Zinc is a micronutrient for mammalians and is essential for the normal growth, immune system, and reproduction of all mammalians including humans, and serves the fundamental role in producing hormones. Zinc as cation (Zn^{2+}) is widely present in various tissues, with its concentration in cardiomyocytes measured at less than 1-nM. In contrast, it is approximately five-fold higher in the sarcoplasmic/endoplasmic reticulum (S(E)R) and lower than the cytosolic level in mitochondria. Zinc, although a redox-inert element, also helps to prevent the high generation of free radicals and reactive oxygen species (ROS), providing a positive effect on protein stability and the proper functioning of antioxidative enzymes. However, several studies also demonstrated that zinc also has some serious toxic effects on biological systems, which are induced through various sources, including constant exposure to polluted sites as well as consumption of overdose zinc supplements, and other natural phenomena. It can be emphasized that taking too much zinc into the body can affect health via altered functions of various organs including the cardiovascular system in mammals. Several animal and human studies have shown that zinc is associated with cellular damage and cardiac dysfunction. Given the numerous functions of free Zn^{2+} in various cellular signal transduction, both intracellular and extracellular free Zn^{2+} increase affect phosphorylation state and subcellular localization of proteins and kinases although its cellular concentration regulation is tightly modulated particularly by specific Zn^{2+} transporters. Overall, taking into consideration the molecular mechanisms of Zn^{2+}-mediated regulation of cardiac function, including Zn^{2+} transporters, cellular free Zn^{2+} levels and Zn^{2+}/Ca^{2+} signaling in the induction of cellular Zn^{2+}-toxicity, the present chapter aims to document and bring new insights associated with specific possible Zn^{2+} inhibitors of the related mechanisms to understanding the Zn^{2+}-associated molecular mechanisms of cardiovascular diseases as well as the potential drug candidates.

B. Turan (✉)
Department of Biophysics, Faculty of Medicine, Lokman Hekim University, Ankara, Turkey
e-mail: belma.turan@lokmanhekim.edu.tr; belma.turan@medicine.ankara.edu.tr

D. K. Agrawal et al. (eds.), *Cardiovascular Toxicity*, Advances in Biochemistry in Health and Disease 33, https://doi.org/10.1007/978-3-031-90598-8_24

Keywords Nutrition · Cellular redox state · Cardiac function · Zinc transporters · Protein phosphorylation/oxidation

Introduction

Zinc is one of the most abundant essential trace elements in natural environments and plays important roles in all living organisms with multidimensional effects [1–5]. It has a redox-inert action and is typically buffered and bound to metalloproteins, but it may also exist in a labile (free ion; Zn^{2+}) form in cells. Zinc, as a free ion, contributes to cellular viability through structural roles and can function as an important signaling molecule [2, 3, 6, 7].

Zinc is a micronutrient for mammalians and is essential for the normal growth, immune system, and reproduction of all mammalians including humans as well as serves the fundamental role in the production of hormones [3, 8–10]. It is vital for the functionality of more than 300 enzymes, the stabilization of DNA, and gene expression as well as cell metabolism [11]. Zinc, although a redox-inert element, also helps to prevent the high generation of free radicals and reactive oxygen species (ROS), providing a positive effect on protein stability and the proper functioning of antioxidative enzymes [12]. Indeed, the human body can not store zinc for the long term, it is needed to aid the above functions through dietary intake and the exchangeable zinc pool [13]. It is also demonstrated that zinc deficiency causes detrimental effects on human health [14–16]. The cumulative concentration of zinc in the human body amounts to about 2–3 g, and more than its 80% is found in the bone and muscles. The transportation of zinc within the body is performed via various enzymes like superoxide dismutase and carbonic anhydrase in the red blood cells in the bloodstream [17]. Furthermore, it has been demonstrated in various studies that zinc can be used as a therapeutic supplement for the management of several pathological conditions and diseases, at most, through its antioxidant and anti-inflammatory actions [1, 18, 19].

However, several studies also demonstrated that zinc also has some serious toxic effects on biological systems, which are induced through various sources, including constant exposure to polluted sites as well as consumption of overdose zinc supplements, and other natural phenomena [20, 21]. Though zinc toxicity in humans is a treatable and non-life-threatening condition, several symptoms cause distress to human activities and lifestyle. Generally, it is accepted that two main types of zinc toxicity are classified as acute and chronic. However, acute zinc toxicity is often caused by consuming large amounts of zinc-containing foods/supplements in short periods, associated with headache, nausea, vomiting, diarrhea, abdominal discomfort, and, rarely metabolic imbalances and severe neurological symptoms [22, 23], conversely, chronic zinc toxicity occurs gradually over time as a result of extended exposure to elevated zinc levels, and appears with symptoms such as impaired immune function, decreased levels of high-density lipoprotein, and increased levels of low-density lipoprotein [24, 25].

Overall, it can be summarized that taking too much zinc into the body can affect health via altered functions of various organs including the cardiovascular system in mammals. Several animal and human studies have shown that Zn^{2+} is associated with cellular damage and cardiac dysfunction [26]. By using epidemiological, biochemical, and cell biological studies, an early review article documented how zinc has the potential to counteract the development of cardiovascular disease [27]. In later studies, authors mentioned that zinc supplements, even at fairly modest concentrations, might have adverse actions such as toxicity [20, 28–30]. Given the numerous functions of free Zn^{2+} in various cellular signal transduction, both intracellular and extracellular free Zn^{2+} increase affect phosphorylation state and subcellular localization of proteins and kinases although its cellular concentration regulation is particularly tightly modulated by specific Zn^{2+} transporters [1, 31–34].

The present chapter aimed to summarize early and recent studies associated with the role of excess labile Zn^{2+} in its toxicity-related cardiac dysfunction through oxidation and/or phosphorylation of excitation-contraction coupling in cardiomyocytes. In addition, it will be also documented the roles of Zn^{2+}-transporters in the induction of cellular Zn^{2+}-toxicity.

Role of Zinc in the Mammalian Heart: Beneficial or Harmful?

Zinc is a micronutrient in the mammalian diet and plays important roles in the physiology and pathology of many organ functions, at most, due to its important roles in proteins and enzymes [1, 31, 35, 36]. Correspondingly, there is a relationship between decreases in the Zn^{2+} levels of the whole body and/or individual tissues and increases in the risk for induction of various pathological conditions in mammalians [14–16, 37–39]. For instance, there has been shown a correlation between lower Zn^{2+} levels and several diseases such as atherosclerosis, coronary artery disease, angina, and cardiac ischemia [30, 38, 39]. Another notable finding showed a correlation between a high incidence of cardiac failure and patients with zinc deficiency [40]. Overall insights on this topic define zinc as an element having multifunctional roles in human health. In other words, Zn^{2+} behaves as a co-factor in various physiological processes, such as cell growth and development, metabolism, reproductive, and immune system function [41]. It also has important roles in the development, prevention, and treatment of various cancer statuses as well as becoming a biomarker in their diagnosis [42–46]. The body cannot make zinc and it is naturally found in cells throughout the body while it is mainly stored in the muscles and bones of mammals. Although Zn^{2+} as a cation is important for many functions of mammalians, however, low zinc intake and/or its uncontrolled excretion can induce some symptoms in healthy humans including hair loss, diarrhea, abnormalities in sexual activities, and problems in the eye and skin [41, 47, 48].

Zinc is accepted as a nontoxic trace element and an important micronutrient, however, an important number of studies demonstrated its toxicity in organ functions and arrest-type effect in cell cycles depending on the solubility and the size of zinc compounds by using zinc including metallic nanoparticles [49–51]. In this regard, the authors demonstrated that Zn^{2+} contributes to the development of pathological biomineralization in cancer tissue by affecting morphogenesis and increasing the rate of degenerative/necrotic alterations in tissues with an ability to enter into covalent bonds [28, 52, 53]. Furthermore, another interesting study has pointed out the possible preventive and therapeutic roles of zinc compounds in various human cancers [54]. To sum up, there are serious conflicts between publications associated with the body's zinc level, zinc intake with nutrition, and its role in human health.

Zinc as an essential microelement has roles in cardiovascular system structure and affects multiple physiologic pathways [1, 55–59]. Regarding the association between heart function and Zn^{2+} status, there has been reported an association between Zn^{2+} deficiency and significantly impaired cardiac function in humans, while some studies reported a marked attenuation in cardiac remodeling following myocardial infarction with Zn^{2+} supplementation [55, 56, 58, 60, 61]. Epidemiological studies are mostly accumulated on the effects of zinc supplementation or zinc intake on the risk of cardiovascular diseases, including the contribution of zinc intake to a decreased risk of type 2 diabetes-associated cardiovascular diseases [62–65].

Overall, the molecular mechanisms of Zn^{2+}-mediated regulation of cardiac function, including zinc transporters, cellular free Zn^{2+} levels, and Zn^{2+}/Ca^{2+} signaling, have not been clarified exactly yet. Further, the specific inhibitors and/or inhibitors of these related mechanisms seem to be useful tools for understanding the zinc-associated molecular mechanisms of cardiovascular diseases as well as the potential drug candidates.

Roles of Free Zn^{2+} in Cardiomyocytes

Zinc as an ion can induce marked conformational changes and thiol oxidations in proteins, and therefore, plays a critical role in various signaling pathways in cells [66–68]. Therefore, the intracellular free Zn^{2+} level ($[Zn^{2+}]_i$) has an important impact on the functions of proteins in living cells, while the total zinc level in these cells is in a range of 100~300 μM, while the $[Zn^{2+}]_i$ is about less than 1 nM in most of the cells. Not only its deficiency but also high $[Zn^{2+}]_i$ over its physiological level in cells have adverse effects on the cell.

The $[Zn^{2+}]_i$ in biological organisms are regulated by metallothioneins (MTs), the Zn^{2+} transporters inside (ZIPs) and outside (ZnTs) of cells, playing a role in the mechanism of cellular Zn^{2+} homeostasis. The free Zn^{2+} binds to MTs and is involved in regulating oxidative stress (mainly regulation of cellular ROS levels), inflammatory responses, and other types of various cellular processes. The $[Zn^{2+}]_i$ over its physiological levels induces many deleterious effects on proteins, enzymes, and other types of structural changes, which are irreversible processes and lead to the dysfunction of living cells, including cardiomyocytes. Interestingly, a low level of $[Zn^{2+}]_i$ is also detrimental to cell function. In this regard, studies demonstrated that transcription factors were adversely affected due to deficiency/low levels of $[Zn^{2+}]_i$ while there are close relationships between low $[Zn^{2+}]_i$ and decreased activation of nuclear factor-κB (NF-κB) associated decrease in gene expression of interleukin-2 (IL-2) and IL-2 receptor α (IL-2Rα) [69, 70].

The $[Zn^{2+}]_i$ plays an important role in heart function through the modulation of excitation-contraction coupling and morphogenesis of ventricular cardiomyocytes [71–77]. Since various studies mentioned how either excess or deficiency in cellular $[Zn^{2+}]_i$ can be one of the leading causes of cardiac dysfunction (i.e. changes in the electrical activity and contractile force), the importance of well-controlled $[Zn^{2+}]_i$ homeostasis in cardiovascular functioning is highlighted, particularly under any pathological stimulus [78–81]. The $[Zn^{2+}]_i$ in a ventricular cardiomyocyte isolated from a rat heart has been calculated to be less than 1 nM [74, 82], however, it is increased significantly under pathological conditions such as hyperglycemia and/or hyperinsulinemia [73, 83–92].

Experimental studies performed with Zn^{2+} exposure pointed out its important contribution to the electrical properties of cells. One of the earliest notable works in this area comes from research in the 1980s and 1990s that focused on how Zn^{2+} interacts with NMDA (N-Methyl-D-Aspartate) receptors and voltage-gated ion channels at most with a modulator role. In these regards, authors establish the role of Zn^{2+} as an important modulator of neuronal ion channels, affecting both ionic currents and synaptic transmission. Today, the importance of free Zn^{2+} in physiological and pathophysiological processes in nerve cells (at most, through its localization in nerve terminals and synaptic vesicles [93, 94] including cardiomyocytes [95], is more attractive through its effects, particularly on some voltage-dependent Ca^{2+} channels (i.e. L-type and N-type Ca^{2+} channels) [96] and some K^+ channels (KCNQ channels) [97, 98] besides its early known effects. Moreover, the studies in the neural cells provided an interesting and important event on the high sensitivity of K^+-channels to free Zn^{2+} levels than the Na^+-channels [99]. In addition to the above effects, other ion transporter mechanisms such as Na^+, K^+, and Cl^- channels can be regulated by free Zn^{2+} in various types of cells [100, 101]. There are others associated with the effects of Zn^{2+} in ionic mechanisms, such as the Na^+/H^+ exchanger for the control of intracellular pH in colonocytes, keratinocytes, and neurons besides regulation of K^+/Cl^- cotransporters for regulation of cell survival, migration, the permeability barrier, and neuronal excitability [100–102]. It has been demonstrated a Zn^{2+}-induced inhibition on L-type Ca^{2+} channels as well as Zn^{2+} influx via this type of channel in freshly isolated cardiomyocytes from either rabbit or rat ventricles [74, 76, 95]. Furthermore, a modulatory role of the Ca^{2+} influx and regulation of β-adrenergic stimulation by high levels of extracellular and intracellular Zn^{2+} has been shown in rat ventricular cardiomyocytes [103]. In this regard, a detailed study performed by either high extracellular or intracellular Zn^{2+} on the L-type Ca^{2+} current in cardiomyocytes has shown that these channel currents could

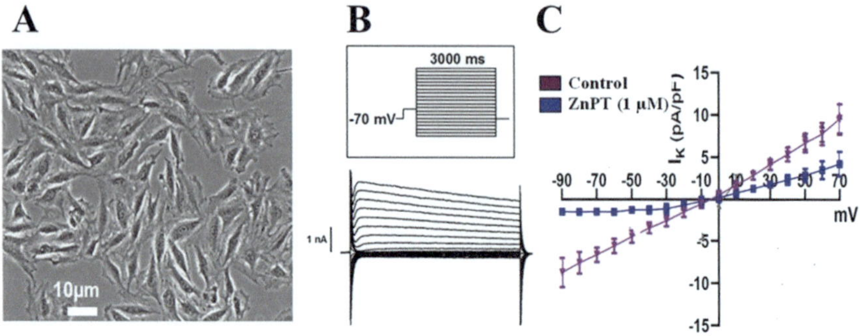

Fig. 24.1 The effect of acute intracellular Zn^{2+} increase on the current-voltage relationship of K^+-channel currents. The HL-1 mouse cardiomyocyte cell line was used and cells were grown in a Claycomb medium following its procedure. For electrophysiological examination measurements, HL-1 cells were first grown on coverslips and then kept in a conventional extracellular medium (**a**). Whole-cell voltage-dependent K^+-channel currents (I_K) in cells were recorded with a protocol given in (**b**). All recordings were performed at room temperature. The measured current values were divided by cell membrane capacitance (current density as pA/pF). A zinc-ionophore pyrithione (ZnPT; 1 µM) was added acutely in the recording medium

be inhibited with a marked leftward shift in their inactivation in the presence of external Ca^{2+}, as a manner of both concentration- and time-dependently [80]. In addition, free Zn^{2+} can affect the properties of other types of voltage-dependent ion channels such as K^+-channels. In Kv1.5 expressing HEK cells, Zn^{2+} induced a variable depolarizing shift and also reduced the current amplitude [104]. The effect of extracellular acute Zn^{2+} application as $ZnCl_2$ (1 µM) is also investigated in HL-1 cells (an immortalized mouse cardiac myocyte cell line) by determination of K^+-channel currents. As can be seen in Fig. 24.1, the $ZnCl_2$ induced significant inhibition of these currents at all depolarization potentials.

There are more detailed experimental data on the role of Zn^{2+} on the cellular crossroad providing information on the extracellular Zn^{2+} signaling [31]. In this study, authors proposed a signaling pathway to demonstrate how free Zn^{2+} released from a cell into the extracellular space of another cell can interact with proteins on the surface of another cell. Furthermore, a binding of Zn^{2+} to the proteins can initiate the activation of Ca^{2+} signaling via its release. This hypothesis fits the signaling between neurons when Zn^{2+} is released from vesicles in the presynaptic neuron into the synapse. Following this step, the interaction between Zn^{2+} and protein receptors is started on the postsynaptic neuron.

Experimental studies, at most in tissue and/or organ levels, demonstrated deleterious effects of dysregulation in $[Zn^{2+}]_i$ (at most its high level) on both electrical and mechanical functions in mammalian heart samples in a concentration-dependent manner. In this context, high $[Zn^{2+}]_i$ induced a marked prolongation in action potential duration, at most due to inhibited K^+-channel currents, which further leads to decreases in the time courses of contraction and relaxation in muscle strips from rat hearts [92]. Consequently, these *in vitro* studies pointed out the important role of $[Zn^{2+}]_i$ in excitation-contraction coupling in the heart. In this regard supporting studies were provided by Woodier and co-workers who [75] demonstrated how $[Zn^{2+}]_i$ can act as a high-affinity cardiac ryanodine receptors (RyR2) activator besides $[Ca^{2+}]_i$. Correspondingly, it has been demonstrated an ability of 1 nM $[Zn^{2+}]_i$ to directly activate RyR2 with significantly higher affinity than the $[Ca^{2+}]_i$ during excitation-contraction coupling [105]. Consideration of experimental data can emphasize an important relationship between $[Zn^{2+}]_i$ dyshomeostasis and certain types of cardiomyopathies, such as diabetic cardiomyopathy, in part, through Zn^{2+}-permeable ion channels affecting $[Zn^{2+}]_i$ dynamics [75, 84, 88, 89, 105].

Relationship Between Free Zn^{2+} Level and Oxidative Stress in Cardiomyocytes

The labile Zn^{2+} is one of the most important ions of mammalian cells including cardiomyocytes besides Ca^{2+}, mainly localized in the cytosol, vesicles, organelles, and nucleus of mammalian cells contributing to many signaling pathways [73, 82]. As discussed previously, although zinc as Zn^{2+}, a redox inactive element, contributes to the regulation of the redox state of the cells [17, 106–108]. An imbalance between oxidants and antioxidants can cause oxidative stress in cells, at most via an excessive production of ROS or depression in the antioxidant defense system. This metabolic disturbance further leads to serious oxidative damage in the cells and tissues and consequently to the development of several chronic diseases including diabetes [109]. The main mechanism between Zn^{2+} and oxidative stress is due to its role in the conversion of 2 superoxide radicals ($O_2^{\cdot -}$) to H_2O_2 and molecular oxygen (O_2) further reducing the toxicity of ROS in cells besides its role as a co-factor for some enzymes in the antioxidant defense system. In addition, zinc protects cells against oxidative damage and plays an important role in membrane stabilization [10].

There are various experimental data to demonstrate the Zn^{2+} toxicity in cardiomyocytes through intracellular stores-released free Zn^{2+} by an increase in not only intracellular ROS but also intracellular reactive nitrogen species (RNS) [74, 76, 84, 92, 110]. Consequently, a loss of signaling quiescence via a toxic effect of high in cardiomyocytes leads to apoptosis and death in cells [106, 108, 111, 112]. In addition, it has also been shown that an increase in cytosolic free Zn^{2+} level ($[Zn^{2+}]_{Cyt}$) contributes to arrhythmogenic action potentials in left ventricular cardiomyocytes through protein thiol oxidation and ATP depletion in cells [87]. Therefore, one can propose that $[Zn^{2+}]_i$ homeostasis in mammalian cells results from a coordinated regulation by different proteins involved in the uptake, excretion, and intracellular storage/trafficking of Zn^{2+} [113].

The level of $[Zn^{2+}]_i$ in isolated cardiomyocytes is about 100-fold less than that of $[Ca^{2+}]_i$ under physiological conditions, however, it can be increased about 30-fold with an about two-fold increase in $[Ca^{2+}]_i$ under oxidant exposure, which further leads to protein thiol oxidation [74, 114]. Supportingly, it has been demonstrated a release of toxic level ROS with increasing concentrations of Zn^{2+} applications [115]. These studies strongly pointed out the important contribution to oxidant-induced alterations of excitation-contraction coupling via increased $[Zn^{2+}]_i$. In *in vivo* studies on this topic, authors demonstrated an important level increase in cardiomyocyte $[Zn^{2+}]_i$ in diabetic [83, 88] or in aldosteronism conditions in animals [116]. In addition, a supporting study is provided a biphasic modulation of RyR2 in rat ventricular cardiomyocytes during oxidative stress by sulfhydryl oxidation through increased $[Zn^{2+}]_i$ besides increased $[Ca^{2+}]_i$ [117].

The studies in recent years emphasized the role of a close relationship between cardiomyocyte level of $[Zn^{2+}]_i$ and cellular oxidative stress status in heart function, particularly under pathological conditions. In these regards, authors examined first the $[Zn^{2+}]_i$ statuses in resting and stimulated states of cardiomyocytes by determination of Zn^{2+} sparks besides Zn^{2+} transients in FluoZin-3 loaded cells with a comparison of the Ca^{2+} sparks and Ca^{2+} transients in Fluo-3 loaded cells [92, 111]. In the same study, authors demonstrated an increase in $[Zn^{2+}]_i$ through Ca^{2+}-dependent Zn^{2+} release in freshly isolated rat ventricular cardiomyocytes under H_2O_2 application. However, an acidic pH status in the same cells inhibited the increase in $[Zn^{2+}]_i$. Although there are several unanswered questions associated with the role of intracellular $[Zn^{2+}]_i$ and it's status under pathological stimuli, it seems there is an important contribution of $[Zn^{2+}]_i$ in excitation-contraction coupling in cardiomyocytes, and thereby any alteration in $[Zn^{2+}]_i$ homeostasis can be associated in the induction of cardiac dysfunction.

Zinc is a redox-inert element, however, Zn^{2+} behaves as a pro-oxidant when its level is over the physiological level in cells, including cardiomyocytes [74, 92]. In this regard, one of the most affected organelles in cardiomyocytes is mitochondria while there are detectable levels of free Zn^{2+} together with free Ca^{2+} in mammalian cell mitochondria [86, 118, 119]. Since organelle-specific Zn^{2+}-fluorescence dyes exist, the measurements of organelle-free Zn^{2+} levels to be parallel to the cytosolic in cardiomyocytes have been performed. The highest free Zn^{2+} level in the sarcoplasmic/endoplasmic reticulum, S(E)R ($[Zn^{2+}]_{SER}$) was determined than those of its cytosolic ($[Zn^{2+}]_{Cyt}$) and mitochondrial ($[Zn^{2+}]_{Mit}$) levels [82]. Interestingly, the $[Zn^{2+}]_{Mit}$ was getting the highest increase under pathological conditions such as hyperglycemia although this value is lowest in cardiomyocytes under physiological conditions (Fig. 24.2a–c). Moreover, the structural analysis by electron microscopy supports a correlation between high $[Zn^{2+}]_{Mit}$ and functional/structural abnormalities in mitochondria. As can be seen in Fig. 24.2d, following image analysis by electron microscopy, the quantifications of the images are presented as the numbers of mitochondrion (left) and autophagosome level (right). The average mitochondrion number is significantly decreased in the HG group compared to the CON group (left), whereas the average number of autophagosomes in the HG group was significantly higher than that of the CON group (right). Supporting results from other studies have been presented in various pathological conditions in various heart tissue studies [82, 86, 120].

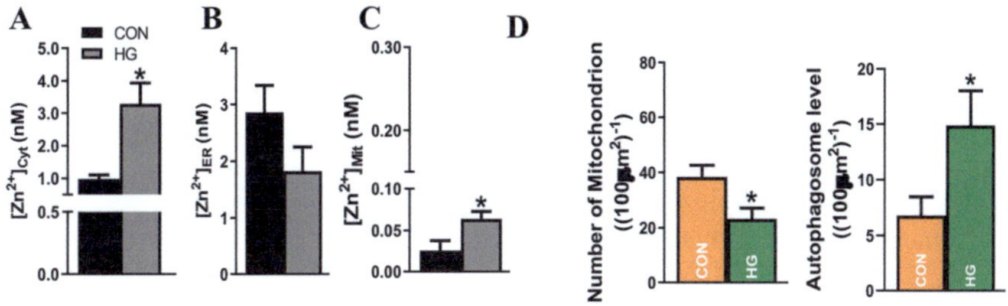

Fig. 24.2 Cellular distribution of free Zn^{2+} in cardiac cells under hyperglycemia and its effects on the ultrastructure of mitochondria. The free Zn^{2+} levels in (**a**) cytosol (Cyt), (**b**) S(E)R, and (**c**) mitochondria (Mit) were determined by using Zn^{2+} sensitive eCALWY FRET-sensors in embryonic rat heart–derived H9c2 cell lines (ventricular myoblasts). To induce hyperglycemia in cells, H9c2 cells were treated with 25 mM glucose for 24 h (HG-cells), while their controls were treated with 5.5 mM glucose for 24 h (CON-cells). (D) The quantifications of the electron microscopy analysis are presented as the numbers of mitochondrion (left) and autophagosome level (right). The average mitochondrion number in the HG group compared to the CON group (left), whereas the average number of autophagosomes in the HG group compared to the CON group (right). The total number of cells used per group was n = 15–20 cells. Significance level accepted at $^{*}p < 0.05$ *vs* CON-cells

It has been shown the important role of mitochondria in cellular regulation, particularly under pathological stimuli. For instance, authors have demonstrated an increase in $[Zn^{2+}]_{Mit}$ when cardiomyocytes are exposed directly to oxidants or hyperglycemia while this process, in turn, induced significant increases in intracellular ROS production levels parallel to depletion in cellular ATP levels [86, 87]. More importantly, as clinical sides of $[Zn^{2+}]_i$ in cells, studies pointed out the important contribution of high $[Zn^{2+}]_i$ in cardiovascular disorders, at most, through a relationship between impaired mitochondrial function and accumulation of cations in mitochondria. Correspondingly, one can suggest that mitochondrial-free Zn^{2+}-status may be a therapeutic target for heart disorders.

The other important consideration of free Zn^{2+} at the cellular level is its role in a cross-talk between mitochondria and S(E)R in cardiomyocytes through an exchange mechanism under physiological and pathological conditions. In this consideration, a focused study on this topic has demonstrated a close relationship between the regulation of $[Zn^{2+}]_i$ and the proper functioning of either S(E)R or mitochondria through a cross-talk between S(E)R and mitochondria in cardiomyocytes together with the important role of increased production of ROS into this cross-talk [119]. In addition supportingly, another *in vivo* experimental study examined the contributions of the increases in both $[Zn^{2+}]_i$ and $[Ca^{2+}]_i$ to diastolic dysfunction in diabetic rats and then a possible therapeutic role of an antioxidant (i.e. N-acetyl cysteine) treatment on cellular enhancements via regulation of redox state of the cardiomyocytes and preserving the antioxidant-defenses in the diabetic animals [84, 88]. In that study, authors demonstrated a significant recovery in both $[Zn^{2+}]_i$ and $[Ca^{2+}]_i$ via recoveries in RyR2 status besides other factors in the regulation of cell redox status. Furthermore, taking into consideration the role of extracellular free

Zn^{2+} for the regulation of intracellular free Ca^{2+} gradients among cytosol and organelles in cells including cardiomyocytes, the contents of intracellular free Zn^{2+} stores also affect Ca^{2+} mobilization from its stores [78, 111, 121]. This type of positive feedback process between two important cations in cardiomyocytes can support the important contribution of a well-controlled Zn homeostasis for proper cardiac function in mammals even if under physiological conditions.

Numerous proteins involved in cell redox signaling pathways, mitochondrial metabolism, and S(E)R functioning are common targets of $[Zn^{2+}]_i$ regulation. However, these regulatory actions of $[Zn^{2+}]_i$ are not limited to the function of the heart, but also extend to numerous other organ systems, at most, such as the central nervous system and immune system, as well.

Excess $[Zn^{2+}]_i$ Leads to Over-Oxidation and/or Phosphorylation of Proteins and Kinases in Cells

Zinc ion plays not only a structural and catalytic role in biological samples but also acts as an essential signaling mediator in various signaling pathways with fundamental biological functions, including gene and protein expression and lipid metabolism in mammalian cells under physiological conditions [122, 123]. However, free Zn^{2+} in cells with multiple functions including growth, differentiation, metabolism, and apoptosis, any disturbance in $[Zn^{2+}]_i$ can participate in the development of pathological conditions through distinct molecular mechanisms in different cell types [124]. In this regard, documents are pointing out the role of free Zn^{2+} as either an antioxidant or a prooxidant in association with the regulation of $[Zn^{2+}]_i$. Supporting studies on this statement emphasize positive feedback mechanisms between the prevalence of oxidants such as peroxynitrite and methylisothiazolinone and the released Zn^{2+} from metalloproteins or other Zn^{2+}-bound proteins provides Zn^{2+} as a prooxidant as a result of protein misfolding and the subsequent induction of oxidative stress [125–127]. Indeed, free Zn^{2+} is a powerful inhibitor of the redox-regulated pathways under an aberrant level of $[Zn^{2+}]_i$ with an effect on mitochondrial biogenesis. Protein phosphorylation is a posttranslational modification that regulates protein activity, degradation, and interaction with other molecules, while abnormal protein phosphorylation is a common pathologic event in various cellular dysfunctions. One of the responsible factors is high $[Zn^{2+}]_i$ in cells for the generation of protein phosphorylation by influencing the structures or activities of protein kinases and protein phosphatases [73, 84, 107, 123, 128–130]. Consequently, it can be summarized that $[Zn^{2+}]_i$ can play an important role in the phosphorylation and/or oxidation of proteins and kinases in cells including cardiomyocytes when its level increases over the physiological level. Furthermore, free Zn^{2+} has an important catalytic role in these proteins including being a structural component and/or a regulator or substrate of enzymatic activity [131–134]. The biological functions of proteins at about 10% in mammalian cells, at most, depend on its essential role in many biochemical pathways via its enzymatic roles in the redox signaling pathways, thereby triggering the release of Zn^{2+} from proteins/intracellular pools and causing tissue damage including myocardial tissue [1, 89, 135]. To sum up, Zn^{2+} has interactions with various proteins and enzymes with different degrees of affinity to bind them [17]. Among them, Zn^{2+} binds and permeates to ionotropic glutamate receptors, metabotropic glutamate receptors, and gamma-amino-n-butyric acid receptors [136–138]. Likewise, as mentioned in the previous section, free Zn^{2+} can modulate the responses of ligand-gated or voltage-gated ion channels via oxidation and/or phosphorylation of channel proteins [76, 100, 139, 140]. The free Zn^{2+} also regulates the activity of various ionic transporters and pumps in cells. Among them, there are effects of free Zn^{2+} on Na^+/K^+-ATPase, Ca^{2+}-ATPase, and the distribution and localization of Zn^{2+} transporters when they are exposed to either extracellularly or intracellularly [33, 141–144]. Among them, $[Zn^{2+}]_{Mit}$ overload is cytotoxic in cells, and then this finding can also explain why high $[Zn^{2+}]_i$ is associated with cardiotoxicity in eukaryotes. In addition, one important data among others also demonstrated that high $[Zn^{2+}]_i$ induced significant changes in cellular ATP level dependent in a reversible manner with a disulfide-reducing agent further leading to marked activation in ATP-sensitive K^+-channel currents [87]. There are many *in vitro* studies to demonstrate how and/or why increases in $[Zn^{2+}]_i$ affect the phosphorylation state, oxidation state, and/or subcellular localization of proteins in cells. In this context, it is known that the biological functions of proteins in cells are primarily controlled by their subcellular localization [145]. For instance, Slepchenko and co-workers showed that an $[Zn^{2+}]_i$ increase could induce important changes in the phosphorylation state and subcellular localization of PKCδ [32]. In this study, even during a short exposure of HeLa cells to exogenous Zn^{2+}, there was significant inhibition in the phosphorylation of PKCδ at Thr505, while that increase could affect its subcellular trafficking from the cytosol to Golgi. Molecular modeling simulations have also provided additional explanations for how a protein can use a zinc-binding pocket to respond to the changes in cellular $[Zn^{2+}]_i$ homeostasis via changes in $[Zn^{2+}]_i$. However, it should be noted that these types of studies can not be fully sufficient to explain a direct relationship between high cellular $[Zn^{2+}]_i$ and alterations in the localization, phosphorylation, and/or oxidation status of proteins. There are also different examples associated with modulations of various intracellular signaling pathways and high $[Zn^{2+}]_i$ at many levels. For instance, it has been demonstrated that there are important impacts of high $[Zn^{2+}]_i$ and the activity of proteins such as the mitogen-activated protein kinase, the phosphoinositide 3-kinase/protein kinase B, Ca^{2+}/calmodulin-activated protein kinase-2, cyclic nucleotide phosphodiesterases, and protein tyrosine phosphatases via promotion of phosphorylation reactions [121, 146–148]. Furthermore, Wilson and co-workers have shown modulation of signal transduction in endothelial cells by picomolar concentrations of free Zn^{2+} via affecting a receptor protein-tyrosine phosphatase β activity [33].

There are important numbers of studies documenting Zn^{2+}-mediated intracellular processes related to cardiovascular functions, at most related to oxidative stress-induced cellular dysfunctions [18, 80, 113, 141]. Multiple reports highlighted the involvement of high $[Zn^{2+}]_i$ and inflammation via high production of ROS in one of the molecular mechanisms underlying cardiovascular disorders. Among these mechanisms, there is activation of the NF-κB signaling pathway associated induction of inflammatory cytokines and inflammation-related enzymes (iNOS and COX2) [60]. On the other hand, another important side of high $[Zn^{2+}]_i$ in cardiomyocytes is the role of changes in the intracellular redox environment due to the changes in cellular $[Ca^{2+}]_i$, which can, in turn, affect many cellular processes, leading to cardiac pathologies, at most, associated with the central role of free Zn^{2+} in cellular redox homeostasis [74, 111, 149–152]. In these regards, experimental data obtained under both *in vivo* and *in vitro* conditions demonstrated that there could be a relationship between high and hyperphosphorylation of both RyR2 and calstabin2 in hyperglycemic cardiomyocytes, which could be reversed by an antioxidant treatment of diabetic rats [84]. These data were further confirmed with *in vitro* studies such as hyperphosphorylation in RyR2 as well as higher phosphorylation levels in both PKA and CaMKII as well as activation of NF-κB in ventricular cardiomyocytes when they were exposed to μM levels $ZnCl_2$ similar to data in hyperglycemic cells.

Therefore, experimental data strongly emphasized that there is a close relationship between high $[Zn^{2+}]_i$ and increases in ROS and RNS in the diseased heart. This is supported by data given in Fig. 24.2. Hyperglycemic cardiomyocytes with high $[Zn^{2+}]_{Cyt}$ and $[Zn^{2+}]_{Mit}$ evaluated by electron microscopy demonstrated marked dilatation of the ER cisterns and increases in autophagosome numbers with decreased numbers of mitochondrion along with signs indicating of fission-fusion defects in the mitochondria. Overall, it can be stated that any stimuli that trigger an increase in $[Zn^{2+}]_i$ are likely going to lead to deleterious changes in cardiomyocytes, further leading to heart dysfunction in mammals. In addition, our current knowledge reveals that the cellular $[Zn^{2+}]_i$ homeostasis including its subcellular distributions (i.e. $[Zn^{2+}]_{Cyt}$, $[Zn^{2+}]_{SER}$, $[Zn^{2+}]_{Mit}$) in mammalian cells is performed with different channels, carriers, and transporters proteins involved in the uptake, excretion, and intracellular storage/trafficking of Zn^{2+} while it is primarily regulated by opposing action of two families of metal transporters such as the ZIP (SLC39A) family that functions to increase $[Zn^{2+}]_{Cyt}$ and the ZnT (SLC30A) family that functions to reduce $[Zn^{2+}]_{Cyt}$. In addition, our current knowledge regarding the structures together with functions of ZIPs and ZnTs points to how protein phosphorylation seems to be important for the regulation of a Zn^{2+}-transporter activity, at most, depending on cellular $[Zn^{2+}]$ homeostasis [77, 80, 90, 113, 141, 153, 154].

The Role of Zn^{2+}-Transporters in Mediating the Control of Cellular $[Zn^{2+}]$ in Cardiomyocytes

Early and recent literature data display that both prokaryotes and eukaryotes have several Zn^{2+} transport systems to provide both influx and efflux of Zn^{2+} across biological membranes. In this context, cellular $[Zn^{2+}]_i$ homeostasis is primarily regulated by ZIPs and ZnTs which are membrane transport proteins of the solute carrier family, which control the membrane Zn^{2+} transport besides regulation of its cytosolic and subcellular concentrations (i.e. $[Zn^{2+}]_{SER}$, $[Zn^{2+}]_{Mit}$). Interesting review articles documented the emerging role of Zn^{2+} transporters in cellular homeostasis due to already knowns associated with the emergency of Zn^{2+} homeostasis considering Zn^{2+} as an important signaling molecule in the development and progression of pathological conditions [77, 80, 90, 113, 141, 153–158]. So, any breakdown in the Zn^{2+} transport systems plays a role in the development of diseases including cardiovascular diseases [60, 159–162].

Zn^{2+}-transporters have primary roles in the regulation and get under control of cellular-free Zn^{2+} in various cell types. In mammalian cells, there are several cellular components responsible for Zn^{2+} homeostasis providing a balance between extracellular and intracellular mediums as well as between intracellular and subcellular components with a well-controlled Zn^{2+} trafficking [121, 155–157, 163]. There are cell-dependent distributions of Zn^{2+}-transporters and the molecular functions of every Zn^{2+}-transporter can concert cellular responsiveness to pathophysiologic stimuli by Zn^{2+} trafficking. For instance, Zn^{2+}-binding to Zn^{2+}-transporters are coupled to its translocation across the membrane which can explain why Zn^{2+}-transporters are different from Zn^{2+}-metalloenzymes. In this regard, some studies demonstrated that ZnTs can act as Zn^{2+}/H^{+} antiporters, which is reasonable for ZnT2, ZnT3, ZnT4, and ZnT8. These transporters, localizing particularly to vesicles such as endosomes/lysosomes, synaptic vesicles, and insulin granules, have key functions in many cells [164–167]. Furthermore, the studies have shown promotion of Zn^{2+} influx into the cytoplasm either from the extracellular medium or from Zn^{2+}-enriched intracellular compartments by ZIPs in mammalian cells via using Zn^{2+} concentration gradients to move Zn^{2+} into the cytoplasm [168, 169].

Among Zn^{2+} transporters, the first cloned transporter is ZnT-1 which is found in all tissues of all organs while its expression is regulated by dietary Zn^{2+} intake [170]. Authors, in this study, demonstrated that ZnT-1 is localized to the basolateral membrane in enterocytes of the small intestine and renal tubular cells, while ZnT-2 exists in vesicles of the intestine, kidney, and testis. ZnT-3 is involved in the vesicles of neurons with a highly expressed ZnT-4 in the mammary gland and brain [171]. In addition, other studies also demonstrated the existence of other ZnTs, such as ZnT5, ZnT9, and ZnT10 which are localized in the nucleus, vesicles of pancreatic cells, and apical membrane of enterocytes, or plasma membranes of cells [172]. In a later study, Etzion and co-workers also showed a correlation between atrial ZnT-1 expression and atrial fibrillation in a human pilot study [173]. The ZnT-6 protein is normally localized to trans-Golgi network and vesicular compartments in cells [174, 175]. Later studies also discovered the existence of other ZnTs such as ZnT7 and ZnT8 which are expressed and localized into vertebrate cells and insulin secretory granules [176–180]. Moreover, there are documents in the literature about the expressions and involvements of some ZnTs such as ZnT5, ZnT6, and ZnT7 in the S(E)R of various cell types [176, 178].

Detailed studies on this topic mentioned that ZIP family members widely span from prokaryotic to eukaryotic organisms and can be divided into four subgroups based on their degree of sequence conservation [181]. ZIP1 and ZIP2 were the first transporters characterized as human ZIP Members into human erythroleukemia K562 cells [181]. The ZIP4 gene has been cloned as a Zn^{2+} transporter responsible for acrodermatitis and expressed in the small intestine (duodenum and jejunum), colon, stomach, and kidney enteropathica [182]. Although most ZIP proteins have been observed at the plasma membrane, later studies demonstrated organelle-localized ZIPs such as a ZIP7 localization at the Golgi apparatus [183]. In addition, studies pointed out an important event related to ZIP localizations in which some ZIPs may change their localizations according to Zn^{2+} availability or physiologic conditions. For instance, ZIP5 has a basolateral plasma membrane orientation in polarized cells during dietary Zn^{2+} sufficiency, but its regulation by Zn^{2+} is not defined [184, 185]. Another two members of Zn^{2+} transporters ZIP8 and ZIP14 are defined as broad-spectrum metal transporters mediating not only Zn^{2+} but also Fe^{2+}, $Mn^{2+,}$ and Cd^{2+} [186, 187]. Furthermore, the phosphorylation and proliferation roles of ZIP14 have been shown in liver regeneration in mice [188]. In animal studies with ZIP7 lacking intestinal epithelium, ER stress was triggered in proliferative progenitor cells with a significant increase in apoptotic cell death [189]. However, deleting ZIP7 in human KBM7 myeloid leukemia cells resulted in augmented ER stress via a change in its localization in the cell [190]. More importantly, other studies suggested a protective role of ZIP7 or ZIP14 against ER stress in different cells [191]. Consequently, these studies pointed out the protective roles of both ZIP7 and ZIP14 in promoting adaptive responses in mammalian cell types [192]. In these regards, it has been mentioned that ZIP7, similar to ZnT7, localizes to the S(E)R and is responsible for Zn^{2+} regulation between Zn^{2+} stores and cytosol of cells [183, 193]. It is abundantly expressed in pancreatic β cells and upregulated in response to glucose stimulation with a compensatory action with ZIP6 [194, 195]. Furthermore, a co-transporter activity of ZIP6 and ZIP7 in turn contributes to glucose-stimulated insulin secretion. In addition, it has been demonstrated a ZIP7-mediated Zn^{2+} release from a Zn^{2+} pool, ER into the cytosol via ZIP7 phosphorylation by casein kinase 2, CK2 [196]. Supporting this function of ZIP7, genetic ablation of ZIP7 in human osteosarcoma MG-63 cells caused a decrease in $[Zn^{2+}]_{Cyt}$ with a significant increase in $[Zn^{2+}]_{SER}$ [197]. Another finding highlights the critical role of ZIP7 as a manager of the stored $[Zn^{2+}]_i$ for the modulation of B-cell receptor signal intensity required for the development of B-cells [198], while the ZIP6–ZIP10 heterodimer might also contribute to the $[Zn^{2+}]_{Cyt}$ regulation with an association to cell division, and progression to mitosis [199]. To sum up, there is a concerted action of ZIP6–ZIP10 and ZIP7 in the regulation of cytosolic Zn^{2+} to control $[Zn^{2+}]_{Cyt}$ in cells, at least, through Zn^{2+}-dependent phosphorylation and oxidation processes [19]. A supporting recent study demonstrated that there ZIP family members ZIP1, ZIP4, and ZIP8 are significantly dysregulated in specific cancerous cells and there is a close association between their expression levels and their prognosis. Particularly, the mutation in the ZIP4 gene seems to be a potential immunomodulator in pancreatic adenocarcinoma via its strong correlation with immune cell infiltration [200]. Recent review articles documented and discussed cellular-free Zn^{2+} metabolism and Zn^{2+} signaling together with Zn^{2+} transporters and their functional integration in mammalian cells under physiological and pathological conditions as potential therapeutic targets [19, 156, 158, 201].

Following an understanding of the significance of free Zn^{2+} and Zn^{2+}-mediated signaling pathways is vital for mammalian cells via its targeting role in many molecular mechanisms, several basic and clinical studies have reported the relationship between $[Zn^{2+}]_i$ and the onset/pathology of cardiovascular diseases. In these regards, critical roles of ZIP family members such as ZIP2, ZIP8, ZIP12, and ZIP13 have been demonstrated *in vivo* and *in vitro* studies including ischemia/reperfusion injury and carotid artery disease in animal hearts and elderly humans [202, 203], in maintaining $[Zn^{2+}]_i$ of endothelial cells [71], Zn^{2+} uptake into the vascular wall [204], and positive effect on cardiovascular function via affecting the increased levels of ROS, $[Ca^{2+}]_{Mit}$, under ischemia/reperfusion injury and doxorubicin-induced cardiotoxicity [205–208].

Furthermore, there are even important numbers of early experimental data related to the role of ZIP7, at most being a co-transporter with ZnT7 in mammalian heart tissues [183, 209, 210]. Later studies further supported these early data related to the important contribution of alterations in the expression levels of ZIP7 and ZnT7 to experimentally-induced diabetic rat heart dysfunction [73]. This study for the first time demonstrated a subcellular re-distribution of free Zn^{2+} in diabetic cardiomyocytes (at most a co-localization of ZIP7 and ZnT7 in the S(E)R), being parallel to induction of ER stress, at most, via activated CK2-associated ZIP7 phosphorylation. This and other studies demonstrated the role these two Zn^{2+}-transporters in not only re-distribution of free Zn^{2+} among cytosol and organelles in cardiomyocytes being associated with the induction of heart dysfunction under diabetes, metabolic disorders or aging [73, 80, 85, 89, 92, 113, 118, 141]. Furthermore, the data in these studies pointed out an important contribution of ZIP7 and ZnT localizations to uncoupling between S(E)R and mitochondria which has an impact on the progression of cardiac dysfunction under pathological stimuli. The subcellular localizations of other transporters such as ZIP8, ZIP14, and ZnT8 in mammalian cardiomyocytes have been demonstrated emphasizing how sarcolemmal and S(E)R localizations of ZIP8 and ZIP14 shared by S(E)R and mitochondria, as well, with significantly high mitochondrial ZnT8 [90]. In addition, the expression levels of these three transporters were found to change in metabolic and elderly animal hearts as well as dilated and ischemic human hearts. In these studies, it has been shown that parallel to the high $[Zn^{2+}]_{Cyt}$, there was markedly upregulation in ER stress markers and PKCα indicating the correlation between high $[Zn^{2+}]_{Cyt}$ and increased oxidative stress [85, 90, 118].

The recent in vitro studies on this topic particularly focused on examining the roles of Zn^{2+}-transporters in mitochondria structure and function in cardiac cells. For instance, an important contribution of overexpression of ZnT7 to the regulation of not only $[Zn^{2+}]_{Cyt}$ but also both $[Zn^{2+}]_{Mit}$ and $[Ca^{2+}]_{Mit}$ has been demonstrated in cardiac ventricular cells while these effects further lead to change mitochondria function with an effect on histone modification [211]. Furthermore, another Zn^{2+} transporter ZnT6 overexpression can play an important role in the development of cardiomyocyte dysfunction by affecting mitochondria via activations in

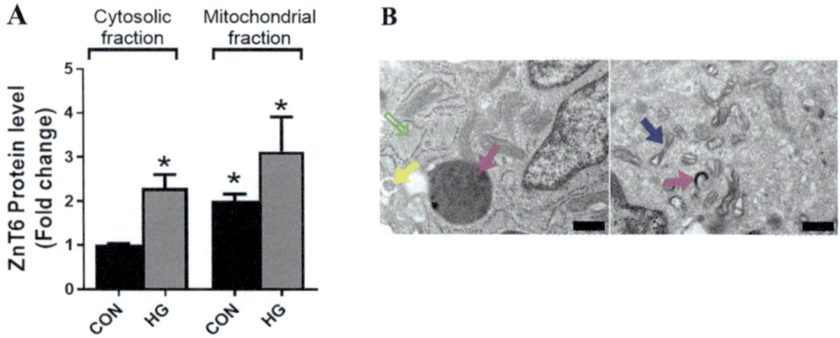

Fig. 24.3 The mRNA levels of ZnT6 in cytosolic and mitochondrial fractions of hyperglycemic cells. (**a**) The mRNA levels of ZnT6 are significantly higher in both cytosolic and mitochondrial parts of the HG cells compared to the CON cells. (**b**) Histological analysis of the HG cells (right) demonstrated marked alterations in the structure of mitochondria characterized by irregular and dilated cristae, high amount of autophagic vacuoles, numerous autophagosomes, late endosomes, multilamellar bodies, and lysosomes in the cytoplasm while normal appearance in the CON cells (left). Blue arrow: mitochondrion, green arrow: ER cisternae: pink arrow: autophagosomes, yellow arrow: endosome/multilamellar body, orange arrow: lysosome. Magnification: 20,000×, Scale bar; 500 nm

K-acetylation and epigenetic proteins [212]. Further analysis has demonstrated that the mRNA levels of ZnT6 are significantly high in both cytosolic and mitochondrial parts of hyperglycemic cardiac cells compared to those of controls (Fig. 24.3a). Histological analysis of these cells also demonstrated marked alterations in the structure of mitochondria characterized with an irregular and dilated cristae, high amount of autophagic vacuoles, numerous autophagosomes, late endosomes, multilamellar bodies, and lysosomes in the cytoplasm (Fig. 24.3b). In summary, all the above studies strongly support the importance of Zn^{2+} transporters in the regulation of $[Zn^{2+}]_i$ through its distribution between cytosol and organelles in cardiomyocytes under physiological conditions as well as their re-distributions under pathological conditions. Recent research area on this topic seems to turn their directions to understanding the molecular mechanisms of all above processes by using high-tech techniques including genomic analysis, cell transfection, and RNA interference in various cell and animal models.

Conclusions

Zinc is one of the essential trace elements required by humans, and its deficiency impairs zinc homeostasis in the human body, thereby affecting various organ functions. It also has important roles at cellular levels such as a cofactor for many transcription factors, proteins, and enzymes. Its cell levels in terms of its distribution

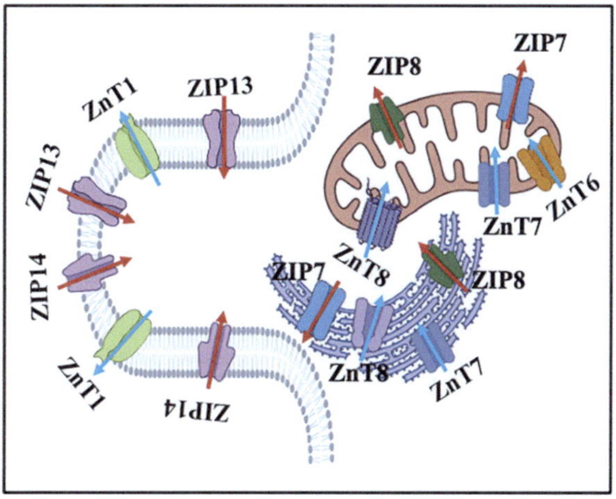

Fig. 24.4 A summary of Zn^{2+} transporters distribution in a mammalian cardiac cell. In mammalian cardiac cells, there are various cellular components responsible for Zn^{2+} storage, subcellular distribution, and its exchange between extracellular and intracellular mediums as well as intracellular and subcellular mediums with a very well-controlled intracellular Zn^{2+}-trafficking and Zn^{2+}-homeostasis systems, including Zn^{2+} transporters

among cytosol and organelles are precisely regulated mainly by Zn^{2+} transporters. The basal level of $[Zn^{2+}]_i$ in mammalian cardiac cells is less than one nanomolar under physiological conditions, however, it can abnormally increase under pathological conditions and even in aging. This increased level leads to deleterious effects in cardiomyocytes in many signaling pathways, at most, through its phosphorylation and oxidation actions. In mammalian cardiac cells, there are various cellular components responsible for Zn^{2+} storage, subcellular distribution, and its exchange between extracellular and intracellular mediums as well as intracellular and subcellular mediums with a very well-controlled intracellular Zn^{2+}-trafficking and Zn^{2+}-homeostasis systems, including Zn^{2+} transporters. Figure 24.4 briefly summarizes the types of Zn^{2+} transporters and their localizations in a cardiac cell.

Acknowledgments The author would like to present deep appreciation to Drs. Erkan Tuncay, Deniz Billur, and Sinan Degirmenci for their very kind contributions.

Declaration of Competing Interest The author has no conflicts of interest to declare.

Funding This study was supported by The Scientific and Technological Research Council of Turkiye (No. 2238N181).

References

1. Cuajungco MP, Ramirez MS, Tolmasky ME (2021) Zinc: multidimensional effects on living organisms. Biomedicines 9(2):208
2. Jarosz M et al (2017) Antioxidant and anti-inflammatory effects of zinc. Zinc-dependent NF-κB signaling. Inflammopharmacology 25:11–24
3. Roohani N et al (2013) Zinc and its importance for human health: an integrative review. J Res Med Sci 18(2):144

4. Frassinetti S et al (2006) The role of zinc in life: a review. J Environ Pathol Toxicol Oncol 25(3):597–610
5. Prasad AS et al (1963) Zinc, iron, and nitrogen content of sweat in normal and deficient subjects. J Lab Clin Med 62(1):84–89
6. Li X et al (2019) Acute toxicity and hazardous concentrations of zinc to native freshwater organisms under different pH values in China. Bull Environ Contam Toxicol 103:120–126
7. Yamasaki S et al (2007) Zinc is a novel intracellular second messenger. J Cell Biol 177(4): 637–645
8. Ozturk H et al (2023) The function of zinc in animal, plant, and human nutrition. J Res Appl Sci Biotechnol 2(2):35–43
9. Sangeetha V et al (2022) Zinc nutrition and human health: overview and implications. eFood 3(5):e17
10. Chasapis CT et al (2012) Zinc and human health: an update. Arch Toxicol 86:521–534
11. Rahman MT, Karim MM (2018) Metallothionein: a potential link in the regulation of zinc in nutritional immunity. Biol Trace Elem Res 182(1):1–13
12. Narayanan SE et al (2020) Molecular mechanism of zinc neurotoxicity in Alzheimer's disease. Environ Sci Pollut Res 27:43542–43552
13. Gibson RS, Anderson VP (2009) A review of interventions based on dietary diversification or modification strategies with the potential to enhance intakes of total and absorbable zinc. Food Nutr Bull 30(1_suppl1):S108–S143
14. Prasad AS (2013) Discovery of human zinc deficiency: its impact on human health and disease. Adv Nutr 4(2):176–190
15. Prasad AS (1991) Discovery of human zinc deficiency and studies in an experimental human model. Am J Clin Nutr 53(2):403–412
16. Prasad AS (1983) Clinical, biochemical and nutritional spectrum of zinc deficiency in human subjects: an update. Nutr Rev 41(7):197–208
17. Vallee BL, Falchuk KH (1993) The biochemical basis of zinc physiology. Physiol Rev 73(1): 79–118
18. Costa MI, Sarmento-Ribeiro AB, Gonçalves AC (2023) Zinc: from biological functions to therapeutic potential. Int J Mol Sci 24(5):4822
19. Chen B et al (2024) Cellular zinc metabolism and zinc signaling: from biological functions to diseases and therapeutic targets. Signal Transduct Target Ther 9(1):6
20. Plum LM, Rink L, Haase H (2010) The essential toxin: impact of zinc on human health. Int J Environ Res Public Health 7(4):1342–1365
21. Schoofs H, Schmit J, Rink L (2024) Zinc toxicity: understanding the limits. Molecules 29:13
22. Barceloux DG (1999) Zinc. J Toxicol Clin Toxicol 37(2):279–292
23. Kim YH et al (2010) Alveolar epithelial cell injury due to zinc oxide nanoparticle exposure. Am J Respir Crit Care Med 182(11):1398–1409
24. Fosmire GJ (1990) Zinc toxicity. Am J Clin Nutr 51(2):225–227
25. Doherty K, Connor M, Cruickshank R (2011) Zinc-containing denture adhesive: a potential source of excess zinc resulting in copper deficiency myelopathy. Br Dent J 210(11):523–525
26. Milton AH et al (2018) Prospective study of dietary zinc intake and risk of cardiovascular disease in women. Nutrients 10(1):38
27. Houtman JP (1996) Trace elements and cardiovascular diseases. J Cardiovasc Risk 3(1):18–25
28. Uski O et al (2017) In vitro toxicological effects of zinc containing nanoparticles with different physico-chemical properties. Toxicol In Vitro 42:105–113
29. Djoko KY et al (2015) The role of copper and zinc toxicity in innate immune defense against bacterial pathogens. J Biol Chem 290(31):18954–18961
30. Efeovbokhan N et al (2014) Zinc and the prooxidant heart failure phenotype. J Cardiovasc Pharmacol 64(4):393–400
31. Liang X, Dempski RE, Burdette SC (2016) Zn(2+) at a cellular crossroads. Curr Opin Chem Biol 31:120–125

32. Slepchenko KG, Holub JM, Li YV (2018) Intracellular zinc increase affects phosphorylation state and subcellular localization of protein kinase C delta (δ). Cell Signal 44:148–157
33. Wilson M, Hogstrand C, Maret W (2012) Picomolar concentrations of free zinc (II) ions regulate receptor protein-tyrosine phosphatase β activity. J Biol Chem 287(12):9322–9326
34. Markwell JP et al (1984) Use of zinc ions to study thylakoid protein phosphorylation and the state 1-state 2 transition in vitro. Plant Physiol 74(2):348–354
35. Livingstone C (2015) Zinc: physiology, deficiency, and parenteral nutrition. Nutr Clin Pract 30(3):371–382
36. Tsuji T et al (2017) Dissecting the process of activation of cancer-promoting zinc-requiring ectoenzymes by zinc metalation mediated by ZNT transporters. J Biol Chem 292(6): 2159–2173
37. Islamoglu Y et al (2011) The relationship between serum levels of Zn and Cu and severity of coronary atherosclerosis. Biol Trace Elem Res 144:436–444
38. Giannoglou GD et al (2010) Association of reduced zinc status with angiographically severe coronary atherosclerosis: a pilot study. Angiology 61(5):449–455
39. Shokrzadeh M et al (2009) Serum zinc and copper levels in ischemic cardiomyopathy. Biol Trace Elem Res 127:116–123
40. Eby GA, Halcomb WW (2006) High-dose zinc to terminate angina pectoris: a review and hypothesis for action by ICAM inhibition. Med Hypotheses 66(1):169–172
41. Kiouri DP et al (2023) Multifunctional role of zinc in human health: an update. EXCLI J 22: 809–827
42. Atakul T et al (2020) Serum copper and zinc levels in patients with endometrial cancer. Biol Trace Elem Res 195:46–54
43. Kazi Tani LS et al (2021) Copper isotopes and copper to zinc ratio as possible biomarkers for thyroid cancer. Front Med (Lausanne) 8:698167
44. Prasad N et al (2018) Immunity and inflammation in health and disease. Academic, Cambridge, MA
45. Zhang L et al (2022) Association between serum copper/zinc ratio and lung cancer: a systematic review with meta-analysis. J Trace Elem Med Biol 74:127061
46. Xie Y et al (2018) Higher serum zinc levels may reduce the risk of cervical cancer in Asian women: a meta-analysis. J Int Med Res 46(12):4898–4906
47. Zoroddu MA et al (2019) The essential metals for humans: a brief overview. J Inorg Biochem 195:120–129
48. Wong SH et al (2007) Zinc deficiency depresses p21 gene expression: inhibition of cell cycle progression is independent of the decrease in p21 protein level in HepG2 cells. Am J Physiol Cell Physiol 292(6):C2175–C2184
49. Truong-Tran AQ et al (2001) The role of zinc in caspase activation and apoptotic cell death. In: Zinc biochemistry, physiology, and homeostasis: recent insights and current trends. Springer, Dordrecht, pp 129–144
50. Maret W (2000) The function of zinc metallothionein: a link between cellular zinc and redox state. J Nutr 130(5S Suppl):1455s–1458s
51. Chasapis CT et al (2012) Zinc and human health: an update. Arch Toxicol 86(4):521–534
52. Romanjuk A et al (2016) The role of heavy metal salts in pathological biomineralization of breast cancer tissue. Adv Clin Exp Med 25(5):907–910
53. Brown MA et al (1964) Food poisoning involving zinc contamination. Arch Environ Health 8: 657–660
54. Hoang BX et al (2016) Zinc as a possible preventive and therapeutic agent in pancreatic, prostate, and breast cancer. Eur J Cancer Prev 25(5):457–461
55. Frustaci A et al (2012) Selenium- and zinc-deficient cardiomyopathy in human intestinal malabsorption: preliminary results of selenium/zinc infusion. Eur J Heart Fail 14(2):202–210
56. Gonçalves AF et al (2018) Zinc supplementation attenuates cardiac remodeling after experimental myocardial infarction. Cell Physiol Biochem 50(1):353–362

57. Darroudi S et al (2019) Association between hypertension in healthy participants and zinc and copper status: a population-based study. Biol Trace Elem Res 190(1):38–44
58. Rosenblum H et al (2020) Zinc deficiency and heart failure: a systematic review of the current literature. J Card Fail 26(2):180–189
59. Nakatani S et al (2021) Association of zinc deficiency with development of CVD events in patients with CKD. Nutrients 13(5):1680
60. Hara T et al (2023) Zinc in cardiovascular functions and diseases: epidemiology and molecular mechanisms for therapeutic development. Int J Mol Sci 24(8):7152
61. Huang JC et al (2021) Association between reduced serum zinc and diastolic dysfunction in maintenance hemodialysis patients. Nutrients 13(6):2077
62. Sun Q et al (2009) Prospective study of zinc intake and risk of type 2 diabetes in women. Diabetes Care 32(4):629–634
63. Song Y et al (2011) Multivitamins, individual vitamin and mineral supplements, and risk of diabetes among older US adults. Diabetes Care 34(1):108–114
64. de Oliveira Otto MC et al (2012) Dietary intakes of zinc and heme iron from red meat, but not from other sources, are associated with greater risk of metabolic syndrome and cardiovascular disease. J Nutr 142(3):526–533
65. Vashum KP et al (2013) Is dietary zinc protective for type 2 diabetes? Results from the Australian longitudinal study on women's health. BMC Endocr Disord 13:40
66. Lee SR et al (2015) The critical roles of zinc: beyond impact on myocardial signaling. Korean J Physiol Pharmacol 19(5):389–399
67. Maret W (2013) Zinc biochemistry: from a single zinc enzyme to a key element of life. Adv Nutr 4(1):82–91
68. Maret W et al (1999) Inhibitory sites in enzymes: zinc removal and reactivation by thionein. Proc Natl Acad Sci USA 96(5):1936–1940
69. Prasad AS et al (2001) Zinc activates NF-kappaB in HUT-78 cells. J Lab Clin Med 138(4): 250–256
70. Serfling E, Avots A, Neumann M (1995) The architecture of the interleukin-2 promoter: a reflection of T lymphocyte activation. Biochim Biophys Acta 1263(3):181–200
71. Lin W et al (2018) Zinc transporter Slc39a8 is essential for cardiac ventricular compaction. J Clin Invest 128(2):826–833
72. Reilly-O'Donnell B et al (2017) Dysregulated Zn2+ homeostasis impairs cardiac type-2 ryanodine receptor and mitsugumin 23 functions, leading to sarcoplasmic reticulum Ca2+ leakage. J Biol Chem 292(32):13361–13373
73. Tuncay E et al (2017) Hyperglycemia-induced changes in ZIP7 and ZnT7 expression cause Zn (2+) release from the sarco(endo)plasmic reticulum and mediate ER stress in the heart. Diabetes 66(5):1346–1358
74. Turan B, Fliss H, Désilets M (1997) Oxidants increase intracellular free Zn2+ concentration in rabbit ventricular myocytes. Am J Phys 272(5 Pt 2):H2095–H2106
75. Woodier J et al (2015) Intracellular zinc modulates cardiac ryanodine receptor-mediated calcium release. J Biol Chem 290(28):17599–17610
76. Atar D et al (1995) Excitation-transcription coupling mediated by zinc influx through voltage-dependent calcium channels. J Biol Chem 270(6):2473–2477
77. Hara T et al (2017) Physiological roles of zinc transporters: molecular and genetic importance in zinc homeostasis. J Physiol Sci 67(2):283–301
78. Pitt SJ, Stewart AJ (2015) Examining a new role for zinc in regulating calcium release in cardiac muscle. Biochem Soc Trans 43(3):359–363
79. Jackson EK et al (2017) SDF-1α (stromal cell-derived factor 1α) induces cardiac fibroblasts, renal microvascular smooth muscle cells, and glomerular mesangial cells to proliferate, cause hypertrophy, and produce collagen. J Am Heart Assoc 6(11):e007253
80. Turan B, Tuncay E (2017) Impact of labile zinc on heart function: from physiology to pathophysiology. Int J Mol Sci 18(11):2395

81. Dorward AM, Stewart AJ, Pitt SJ (2023) The role of Zn2+ in shaping intracellular Ca2+ dynamics in the heart. J Gen Physiol 155(7):e202213206
82. Chabosseau P et al (2014) Mitochondrial and ER-targeted eCALWY probes reveal high levels of free Zn2+. ACS Chem Biol 9(9):2111–2120
83. Ayaz M, Turan B (2006) Selenium prevents diabetes-induced alterations in [Zn2+]i and metallothionein level of rat heart via restoration of cell redox cycle. Am J Physiol Heart Circ Physiol 290(3):H1071–H1080
84. Tuncay E et al (2014) Enhancement of cellular antioxidant-defence preserves diastolic dysfunction via regulation of both diastolic Zn2+ and Ca2+ and prevention of RyR2-leak in hyperglycemic cardiomyocytes. Oxidative Med Cell Longev 2014:290381
85. Olgar Y, Turan B (2019) A sodium-glucose cotransporter 2 (SGLT2) inhibitor dapagliflozin comparison with insulin shows important effects on Zn(2+)-transporters in cardiomyocytes from insulin-resistant metabolic syndrome rats through inhibition of oxidative stress (1). Can J Physiol Pharmacol 97(6):528–535
86. Olgar Y, Tuncay E, Turan B (2019) Mitochondria-targeting antioxidant provides cardioprotection through regulation of cytosolic and mitochondrial Zn(2+) levels with re-distribution of Zn(2+)-transporters in aged rat cardiomyocytes. Int J Mol Sci 20(15):3783
87. Degirmenci S et al (2018) Cytosolic increased labile Zn2+ contributes to arrhythmogenic action potentials in left ventricular cardiomyocytes through protein thiol oxidation and cellular ATP depletion. J Trace Elem Med Biol 48:202–212
88. Tuncay E et al (2013) ß-blocker timolol prevents arrhythmogenic Ca^{2+} release and normalizes Ca^{2+} and Zn^{2+} dyshomeostasis in hyperglycemic rat heart. PLoS One 8(7):e71014
89. Billur D et al (2016) Interplay between cytosolic free Zn2+ and mitochondrion morphological changes in rat ventricular cardiomyocytes. Biol Trace Elem Res 174(1):177–188
90. Olgar Y et al (2018) Increased free Zn(2+) correlates induction of sarco(endo)plasmic reticulum stress via altered expression levels of Zn(2+) -transporters in heart failure. J Cell Mol Med 22(3):1944–1956
91. Tuncay E et al (2019) β3-adrenergic receptor activation plays an important role in the depressed myocardial contractility via both elevated levels of cellular free Zn2+ and reactive nitrogen species. J Cell Physiol 234(8):13370–13386
92. Tuncay E, Turan B (2016) Intracellular Zn(2+) increase in cardiomyocytes induces both electrical and mechanical dysfunction in heart via endogenous generation of reactive nitrogen species. Biol Trace Elem Res 169(2):294–302
93. Pérez-Clausell J, Danscher G (1985) Intravesicular localization of zinc in rat telencephalic boutons. A histochemical study. Brain Res 337(1):91–98
94. Slomianka L (1992) Neurons of origin of zinc-containing pathways and the distribution of zinc-containing boutons in the hippocampal region of the rat. Neuroscience 48(2):325–352
95. Turan B (2003) Zinc-induced changes in ionic currents of cardiomyocytes. Biol Trace Elem Res 94(1):49–60
96. Kerchner GA et al (2000) Zn2+ current is mediated by voltage-gated Ca2+ channels and enhanced by extracellular acidity in mouse cortical neurones. J Physiol 528(Pt 1):39–52
97. Gao H et al (2017) Intracellular zinc activates KCNQ channels by reducing their dependence on phosphatidylinositol 4,5-bisphosphate. Proc Natl Acad Sci USA 114(31):E6410–e6419
98. Aras MA, Saadi RA, Aizenman E (2009) Zn2+ regulates Kv2.1 voltage-dependent gating and localization following ischemia. Eur J Neurosci 30(12):2250–2257
99. Gilly W, Armstrong CM (1982) Slowing of sodium channel opening kinetics in squid axon by extracellular zinc. J Gen Physiol 79(6):935–964
100. Zhang C et al (2022) Neuronal signalling of zinc: from detection and modulation to function. Open Biol 12(9):220188
101. Laitakari A et al (2021) The zinc-sensing receptor GPR39 in physiology and as a pharmacological target. Int J Mol Sci 22(8):3872

102. Azriel-Tamir H et al (2004) Extracellular zinc triggers ERK-dependent activation of Na+/H+ exchange in colonocytes mediated by the zinc-sensing receptor. J Biol Chem 279(50): 51804–51816
103. Alvarez-Collazo J et al (2012) Zinc modulation of basal and β-adrenergically stimulated L-type Ca2+ current in rat ventricular cardiomyocytes: consequences in cardiac diseases. Pflugers Arch 464(5):459–470
104. Zhang S, Kehl SJ, Fedida D (2001) Modulation of Kv1.5 potassium channel gating by extracellular zinc. Biophys J 81(1):125–136
105. Stewart AJ, Pitt SJ (2015) Zinc controls RyR2 activity during excitation-contraction coupling. Channels (Austin) 9(5):227–229
106. Murakami M, Hirano T (2008) Intracellular zinc homeostasis and zinc signaling. Cancer Sci 99(8):1515–1522
107. Oteiza PI (2012) Zinc and the modulation of redox homeostasis. Free Radic Biol Med 53(9): 1748–1759
108. Maret W (2013) Zinc and human disease. Met Ions Life Sci 13:389–414
109. Butterfield DA, Di Domenico F, Barone E (2014) Elevated risk of type 2 diabetes for development of Alzheimer disease: a key role for oxidative stress in brain. Biochim Biophys Acta 1842(9):1693–1706
110. Malaiyandi LM, Dineley KE, Reynolds IJ (2004) Divergent consequences arise from metallothionein overexpression in astrocytes: zinc buffering and oxidant-induced zinc release. Glia 45(4):346–353
111. Tuncay E et al (2011) Intracellular free zinc during cardiac excitation–contraction cycle: calcium and redox dependencies. Cardiovasc Res 89(3):634–642
112. Cicek FA et al (2014) Beta-blocker timolol alleviates hyperglycemia-induced cardiac damage via inhibition of endoplasmic reticulum stress. J Bioenerg Biomembr 46(5):377–387
113. Turan B (2019) A brief overview from the physiological and detrimental roles of zinc homeostasis via zinc transporters in the heart. Biol Trace Elem Res 188(1):160–176
114. Turan B et al (1996) Oxidative effects of selenite on rat ventricular contractility and Ca movements. Cardiovasc Res 32(2):351–361
115. Maret W (2009) Molecular aspects of human cellular zinc homeostasis: redox control of zinc potentials and zinc signals. Biometals 22(1):149–157
116. Kamalov G et al (2010) Temporal responses to intrinsically coupled calcium and zinc dyshomeostasis in cardiac myocytes and mitochondria during aldosteronism. Am J Physiol Heart Circ Physiol 298(2):H385–H394
117. Xie H, Zhu PH (2006) Biphasic modulation of ryanodine receptors by sulfhydryl oxidation in rat ventricular myocytes. Biophys J 91(8):2882–2891
118. Olgar Y et al (2018) Aging related functional and structural changes in the heart and aorta: MitoTEMPO improves aged-cardiovascular performance. Exp Gerontol 110:172–181
119. Tuncay E et al (2019) Zn2+-transporters ZIP7 and ZnT7 play important role in progression of cardiac dysfunction via affecting sarco (endo) plasmic reticulum-mitochondria coupling in hyperglycemic cardiomyocytes. Mitochondrion 44:41–52
120. Olgar Y et al (2020) Ageing-associated increase in SGLT2 disrupts mitochondrial/sarcoplasmic reticulum Ca2+ homeostasis and promotes cardiac dysfunction. J Cell Mol Med 24(15): 8567–8578
121. Maret W (2017) Zinc in cellular regulation: the nature and significance of "zinc signals". Int J Mol Sci 18(11):2285
122. Fukada T, Kambe T (2011) Molecular and genetic features of zinc transporters in physiology and pathogenesis. Metallomics 3(7):662–674
123. Xu Z, Zhou J (2013) Zinc and myocardial ischemia/reperfusion injury. Biometals 26:863–878
124. Zhang H-L, Wang X-C, Liu R (2022) Zinc in regulating protein kinases and phosphatases in neurodegenerative diseases. Biomol Ther 12(6):785
125. Lee SR (2018) Critical role of zinc as either an antioxidant or a prooxidant in cellular systems. Oxidative Med Cell Longev 2018:9156285

126. Maret W (2008) Metallothionein redox biology in the cytoprotective and cytotoxic functions of zinc. Exp Gerontol 43(5):363–369
127. Land PW, Aizenman E (2005) Zinc accumulation after target loss: an early event in retrograde degeneration of thalamic neurons. Eur J Neurosci 21(3):647–657
128. Csermely P et al (1995) Autophosphorylation of grp94 (endoplasmin). J Biol Chem 270(11): 6381–6388
129. Csermely P et al (1988) Zinc can increase the activity of protein kinase C and contributes to its binding to plasma membranes in T lymphocytes. J Biol Chem 263(14):6487–6490
130. Krężel A, Maret W (2016) The biological inorganic chemistry of zinc ions. Arch Biochem Biophys 611:3–19
131. Banci L et al (2002) A new zinc-protein coordination site in intracellular metal trafficking: solution structure of the Apo and Zn(II) forms of ZntA(46-118). J Mol Biol 323(5):883–897
132. Debela M et al (2006) Crystal structures of human tissue kallikrein 4: activity modulation by a specific zinc binding site. J Mol Biol 362(5):1094–1107
133. Namuswe F, Berg JM (2012) Secondary interactions involving zinc-bound ligands: roles in structural stabilization and macromolecular interactions. J Inorg Biochem 111:146–149
134. Eriksson SE et al (2019) p53 as a hub in cellular redox regulation and therapeutic target in cancer. J Mol Cell Biol 11(4):330–341
135. Andreini C et al (2006) Counting the zinc-proteins encoded in the human genome. J Proteome Res 5(1):196–201
136. Christine CW, Choi DW (1990) Effect of zinc on NMDA receptor-mediated channel currents in cortical neurons. J Neurosci 10(1):108–116
137. Bresink I et al (1996) Zinc changes AMPA receptor properties: results of binding studies and patch clamp recordings. Neuropharmacology 35(4):503–509
138. Yin HZ, Weiss JH (1995) Zn(2+) permeates Ca(2+) permeable AMPA/kainate channels and triggers selective neural injury. Neuroreport 6(18):2553–2556
139. Cloues R (1995) Properties of ATP-gated channels recorded from rat sympathetic neurons: voltage dependence and regulation by Zn2+ ions. J Neurophysiol 73(1):312–319
140. Krall RF, Tzounopoulos T, Aizenman E (2021) The function and regulation of zinc in the brain. Neuroscience 457:235–258
141. Turan B, Tuncay E (2021) The role of labile Zn(2+) and Zn(2+)-transporters in the pathophysiology of mitochondria dysfunction in cardiomyocytes. Mol Cell Biochem 476(2): 971–989
142. Lees GJ, Leong W (1994) Brain lesions induced by specific and non-specific inhibitors of sodium-potassium ATPase. Brain Res 649(1–2):225–233
143. Kukic I et al (2013) Zinc-dependent lysosomal enlargement in TRPML1-deficient cells involves MTF-1 transcription factor and ZnT4 (Slc30a4) transporter. Biochem J 451(2): 155–163
144. Jacob C, Maret W, Vallee BL (1998) Control of zinc transfer between thionein, metallothionein, and zinc proteins. Proc Natl Acad Sci USA 95(7):3489–3494
145. Zhang YH et al (2022) Subcellular localization prediction of human proteins using multifeature selection methods. Biomed Res Int 2022:3288527
146. Hübner C, Haase H (2021) Interactions of zinc- and redox-signaling pathways. Redox Biol 41: 101916
147. Haase H, Maret W (2005) Protein tyrosine phosphatases as targets of the combined insulinomimetic effects of zinc and oxidants. Biometals 18(4):333–338
148. Kim B, Lee WW (2021) Regulatory role of zinc in immune cell signaling. Mol Cells 44(5): 335–341
149. Zima AV, Blatter LA (2006) Redox regulation of cardiac calcium channels and transporters. Cardiovasc Res 71(2):310–321
150. Dineley KE et al (2005) Zinc causes loss of membrane potential and elevates reactive oxygen species in rat brain mitochondria. Mitochondrion 5(1):55–65

151. Kuster GM et al (2010) Redox-mediated reciprocal regulation of SERCA and Na+-Ca2+ exchanger contributes to sarcoplasmic reticulum Ca2+ depletion in cardiac myocytes. Free Radic Biol Med 48(9):1182–1187
152. Bozym RA et al (2010) Free zinc ions outside a narrow concentration range are toxic to a variety of cells in vitro. Exp Biol Med (Maywood) 235(6):741–750
153. Aktay I et al (2024) An overexpression of SLC30A6 gene contributes to cardiomyocyte dysfunction via affecting mitochondria and inducing activations in K-acetylation and epigenetic proteins. Biochem Genet 62(4):3198–3214
154. Aksoy-Ozer ZB et al (2024) The role of zinc on liver fibrosis by modulating ZIP14 expression throughout epigenetic regulatory mechanisms. Biol Trace Elem Res 202(11):5094–5105
155. Bafaro E et al (2017) The emerging role of zinc transporters in cellular homeostasis and cancer. Signal Transduct Target Ther 2:17029
156. Kambe T, Taylor KM, Fu D (2021) Zinc transporters and their functional integration in mammalian cells. J Biol Chem 296:100320
157. Kambe T et al (2015) The physiological, biochemical, and molecular roles of zinc transporters in zinc homeostasis and metabolism. Physiol Rev 95(3):749–784
158. Bui HB, Inaba K (2024) Structures, mechanisms, and physiological functions of zinc transporters in different biological kingdoms. Int J Mol Sci 25(5):3045
159. Zhu B et al (2021) Increased expression of zinc transporter ZIP4, ZIP11, ZnT1, and ZnT6 predicts poor prognosis in pancreatic cancer. J Trace Elem Med Biol 65:126734
160. Sikora J, Ouagazzal AM (2021) Synaptic zinc: an emerging player in Parkinson's disease. Int J Mol Sci 22(9):4724
161. Lieberwirth JK et al (2021) Bi-allelic loss of function variants in SLC30A5 as cause of perinatal lethal cardiomyopathy. Eur J Hum Genet 29(5):808–815
162. Knez M, Glibetic M (2021) Zinc as a biomarker of cardiovascular health. Front Nutr 8:686078
163. Eide DJ (2006) Zinc transporters and the cellular trafficking of zinc. Biochim Biophys Acta 1763(7):711–722
164. Kawachi M et al (2008) Deletion of a histidine-rich loop of AtMTP1, a vacuolar Zn(2+)/H(+) antiporter of Arabidopsis thaliana, stimulates the transport activity. J Biol Chem 283(13): 8374–8383
165. Shusterman E et al (2014) ZnT-1 extrudes zinc from mammalian cells functioning as a Zn(2+)/H(+) exchanger. Metallomics 6(9):1656–1663
166. Girijashanker K et al (2008) Slc39a14 gene encodes ZIP14, a metal/bicarbonate symporter: similarities to the ZIP8 transporter. Mol Pharmacol 73(5):1413–1423
167. He L et al (2006) ZIP8, member of the solute-carrier-39 (SLC39) metal-transporter family: characterization of transporter properties. Mol Pharmacol 70(1):171–180
168. Eide DJ (2004) The SLC39 family of metal ion transporters. Pflugers Arch 447(5):796–800
169. Lin W et al (2010) Selective electrodiffusion of zinc ions in a Zrt-, Irt-like protein, ZIPB. J Biol Chem 285(50):39013–39020
170. McMahon RJ, Cousins RJ (1998) Mammalian zinc transporters. J Nutr 128(4):667–670
171. Cole TB et al (1999) Elimination of zinc from synaptic vesicles in the intact mouse brain by disruption of the ZnT3 gene. Proc Natl Acad Sci USA 96(4):1716–1721
172. Liuzzi JP et al (2004) Responsive transporter genes within the murine intestinal-pancreatic axis form a basis of zinc homeostasis. Proc Natl Acad Sci USA 101(40):14355–14360
173. Etzion Y et al (2008) Correlation between atrial ZnT-1 expression and atrial fibrillation in humans: a pilot study. J Cardiovasc Electrophysiol 19(2):157–164
174. Huang L, Kirschke CP, Gitschier J (2002) Functional characterization of a novel mammalian zinc transporter, ZnT6. J Biol Chem 277(29):26389–26395
175. Kambe T et al (2004) Overview of mammalian zinc transporters. Cell Mol Life Sci 61(1): 49–68
176. Suzuki T et al (2005) Two different zinc transport complexes of cation diffusion facilitator proteins localized in the secretory pathway operate to activate alkaline phosphatases in vertebrate cells. J Biol Chem 280(35):30956–30962

177. Ishihara K et al (2006) Zinc transport complexes contribute to the homeostatic maintenance of secretory pathway function in vertebrate cells. J Biol Chem 281(26):17743–17750
178. Ellis CD, Macdiarmid CW, Eide DJ (2005) Heteromeric protein complexes mediate zinc transport into the secretory pathway of eukaryotic cells. J Biol Chem 280(31):28811–28818
179. Chimienti F et al (2004) Identification and cloning of a beta-cell-specific zinc transporter, ZnT-8, localized into insulin secretory granules. Diabetes 53(9):2330–2337
180. Chimienti F et al (2006) In vivo expression and functional characterization of the zinc transporter ZnT8 in glucose-induced insulin secretion. J Cell Sci 119(Pt 20):4199–4206
181. Gaither LA, Eide DJ (2001) Eukaryotic zinc transporters and their regulation. Biometals 14(3–4):251–270
182. Wang K et al (2002) A novel member of a zinc transporter family is defective in acrodermatitis enteropathica. Am J Hum Genet 71(1):66–73
183. Huang L et al (2005) The ZIP7 gene (Slc39a7) encodes a zinc transporter involved in zinc homeostasis of the Golgi apparatus. J Biol Chem 280(15):15456–15463
184. Wang F et al (2004) Zinc-stimulated endocytosis controls activity of the mouse ZIP1 and ZIP3 zinc uptake transporters. J Biol Chem 279(23):24631–24639
185. Dufner-Beattie J et al (2004) The adaptive response to dietary zinc in mice involves the differential cellular localization and zinc regulation of the zinc transporters ZIP4 and ZIP5. J Biol Chem 279(47):49082–49090
186. Pinilla-Tenas JJ et al (2011) Zip14 is a complex broad-scope metal-ion transporter whose functional properties support roles in the cellular uptake of zinc and nontransferrin-bound iron. Am J Physiol Cell Physiol 301(4):C862–C871
187. Wang CY et al (2012) ZIP8 is an iron and zinc transporter whose cell-surface expression is up-regulated by cellular iron loading. J Biol Chem 287(41):34032–34043
188. Aydemir TB, Sitren HS, Cousins RJ (2012) The zinc transporter Zip14 influences c-Met phosphorylation and hepatocyte proliferation during liver regeneration in mice. Gastroenterology 142(7):1536–46.e5
189. Herbert TP, Laybutt DR (2016) A reevaluation of the role of the unfolded protein response in islet dysfunction: maladaptation or a failure to adapt? Diabetes 65(6):1472–1480
190. Fauster A et al (2019) Systematic genetic mapping of necroptosis identifies SLC39A7 as modulator of death receptor trafficking. Cell Death Differ 26(6):1138–1155
191. Kim MH et al (2017) Hepatic ZIP14-mediated zinc transport is required for adaptation to endoplasmic reticulum stress. Proc Natl Acad Sci USA 114(29):E5805–e5814
192. Merriman C, Fu D (2019) Down-regulation of the islet-specific zinc transporter-8 (ZnT8) protects human insulinoma cells against inflammatory stress. J Biol Chem 294(45): 16992–17006
193. Taylor KM et al (2004) Structure-function analysis of HKE4, a member of the new LIV-1 subfamily of zinc transporters. Biochem J 377(Pt 1):131–139
194. Bellomo EA, Meur G, Rutter GA (2011) Glucose regulates free cytosolic Zn^{2+} concentration, Slc39 (ZiP), and metallothionein gene expression in primary pancreatic islet β-cells. J Biol Chem 286(29):25778–25789
195. Liu Y et al (2015) Characterization of zinc influx transporters (ZIPs) in pancreatic β cells: roles in regulating cytosolic zinc homeostasis and insulin secretion. J Biol Chem 290(30): 18757–18769
196. Taylor KM et al (2008) ZIP7-mediated intracellular zinc transport contributes to aberrant growth factor signaling in antihormone-resistant breast cancer cells. Endocrinology 149(10): 4912–4920
197. Woodruff G et al (2018) The zinc transporter SLC39A7 (ZIP7) is essential for regulation of cytosolic zinc levels. Mol Pharmacol 94(3):1092–1100
198. Anzilotti C et al (2019) An essential role for the Zn(2+) transporter ZIP7 in B cell development. Nat Immunol 20(3):350–361
199. Falchuk KH, Fawcett DW, Vallee BL (1975) Role of zinc in cell division of Euglena gracilis. J Cell Sci 17(1):57–78

200. Liu Y et al (2024) Zinc transporters serve as prognostic predictors and their expression correlates with immune cell infiltration in specific cancer: a pan-cancer analysis. J Cancer 15(4):939–954
201. Hara T et al (2022) Zinc transporters as potential therapeutic targets: an updated review. J Pharmacol Sci 148(2):221–228
202. Du L et al (2019) The critical role of the zinc transporter Zip2 (SLC39A2) in ischemia/reperfusion injury in mouse hearts. J Mol Cell Cardiol 132:136–145
203. Giacconi R et al (2008) A novel Zip2 Gln/Arg/Leu codon 2 polymorphism is associated with carotid artery disease in aging. Rejuvenation Res 11(2):297–300
204. Zalewski PD et al (2019) Roles for endothelial zinc homeostasis in vascular physiology and coronary artery disease. Crit Rev Food Sci Nutr 59(21):3511–3525
205. Wang J et al (2021) Downregulation of the zinc transporter SLC39A13 (ZIP13) is responsible for the activation of CaMKII at reperfusion and leads to myocardial ischemia/reperfusion injury in mouse hearts. J Mol Cell Cardiol 152:69–79
206. Hirose T et al (2019) Morphometric analysis of thoracic aorta in Slc39a13/Zip13-KO mice. Cell Tissue Res 376(1):137–141
207. Traylor M et al (2021) Genetic basis of lacunar stroke: a pooled analysis of individual patient data and genome-wide association studies. Lancet Neurol 20(5):351–361
208. Hara T et al (2022) Role of Scl39a13/ZIP13 in cardiovascular homeostasis. PLoS One 17(10): e0276452
209. Kirschke CP, Huang L (2003) ZnT7, a novel mammalian zinc transporter, accumulates zinc in the Golgi apparatus. J Biol Chem 278(6):4096–4102
210. Yang J et al (2013) Gene profile identifies zinc transporters differentially expressed in normal human organs and human pancreatic cancer. Curr Mol Med 13(3):401–409
211. Tuncay E, Aktay I, Turan B (2023) Overexpression of Slc30a7/ZnT7 increases the mitochondrial matrix levels of labile Zn2+ and modifies histone modification in hyperinsulinemic cardiomyoblasts. J Trace Elem Med Biol 78:127198
212. Aktay I et al (2023) An overexpression of SLC30A6 gene contributes to cardiomyocyte dysfunction via affecting mitochondria and inducing activations in K-acetylation and epigenetic proteins. Biochem Genet:1–17

Chapter 25
Persistent Organic Pollutants and Cardiotoxicity

Aleksandra Buha Đorđević

Abstract Persistent Organic Pollutants (POPs) are a group of toxic chemicals that persist in the environment and accumulate in living organisms, posing significant risks to human health. Among these pollutants, polychlorinated biphenyls (PCBs), dioxins, polybrominated diphenyl ethers (PBDEs), and per- and polyfluoroalkyl substances (PFAS) have been identified as major contributors to cardiovascular diseases (CVD). These compounds, primarily entering the human body through food, water, and consumer products, cause adverse effects by inducing oxidative stress, inflammation, and endocrine disruption. Oxidative stress and inflammation are key mechanisms underlying cardiovascular damage, leading to conditions such as hypertension, atherosclerosis, and ischemic heart disease. Additionally, these pollutants interfere with lipid metabolism and alter gene expression, further exacerbating cardiovascular risk. Epidemiological studies have highlighted the link between POP exposure and CVD, although the long-term, cumulative effects remain poorly understood, especially regarding PFAS. The molecular mechanisms by which different POPs affect the cardiovascular system are still under investigation, with particular focus on how they disrupt lipid metabolism and hormone signaling. Despite significant progress in understanding their individual effects, the combined impact of multiple POPs, as well as their interaction with genetic and other environmental factors, is less explored. Future research should prioritize longitudinal studies to clarify causal relationships and explore potential synergistic effects of POPs mixtures. Improved regulatory measures and safer alternatives to POPs-containing products are essential to mitigate the health risks associated with these persistent pollutants.

Keywords Cardiotoxicity · Dioxins · Fluoroalkyl substances · Inflammation · Organic pollutants · Persistent organic pollutants · Polybrominated diphenyl ethers · Polychlorinated biphenyls · Perfluoroalkyl substances · Polyfluoroalkyl substances

---

A. B. Đorđević (✉)
Department of Toxicology "Akademik Danilo Soldatović", Faculty of Pharmacy, University of Belgrade, Belgrade, Republic of Serbia
e-mail: aleksandra.buha@pharmacy.bg.ac.rs

D. K. Agrawal et al. (eds.), *Cardiovascular Toxicity*, Advances in Biochemistry in Health and Disease 33, https://doi.org/10.1007/978-3-031-90598-8_25

Introduction

Persistent organic pollutants (POPs) are a class of synthetic organic chemicals that persist in the environment, bioaccumulate in the food chain, and pose significant risks to human health and ecosystems [1]. These chemicals, which include industrial compounds, pesticides, and unintentional by-products of combustion and industrial processes, exhibit high lipophilicity and low water solubility, contributing to their long-term environmental persistence [2]. The Stockholm Convention, an international treaty adopted in 2001, aims to eliminate or restrict the production and use of POPs due to their toxic effects and global distribution [3, 4].

Human exposure to POPs primarily occurs through dietary intake, inhalation of contaminated air, and dermal absorption. The most significant route of exposure is through the consumption of animal-derived foods, including fish, meat, and dairy products, where these compounds accumulate in lipids [1, 2]. Occupational and residential exposure, particularly in contaminated regions, further exacerbates the health risks associated with POPs [5]. Given their ability to undergo long-range atmospheric transport, POPs are detected in remote regions, including the Arctic, where they accumulate in indigenous populations relying on traditional diets [6, 7].

POPs are known for their broad spectrum of toxic effects, impacting multiple physiological systems. These compounds disrupt endocrine signaling, interfere with lipid metabolism, and exert immunotoxic, neurotoxic, and hepatotoxic effects [8–11].

The link between POP exposure and cardiovascular health has gained considerable attention in recent years. Several studies have reported associations between chronic exposure to POPs and increased risks of hypertension, atherosclerosis, and coronary artery disease [12, 13]. Mechanistically, POPs contribute to cardiovascular dysfunction through oxidative stress, inflammation, and endothelial dysfunction, which collectively promote vascular damage and increase the likelihood of ischemic events [14]. In particular, polychlorinated biphenyls (PCBs), dioxins, polybrominated diphenyl ethers (PBDEs), and per- and polyfluoroalkyl substances (PFAS) have emerged as key pollutants with strong evidence linking them to adverse cardiovascular outcomes and hence, will be discussed in more details in this chapter. Summary of key POPs is given in Table 25.1.

Polychlorinated Biphenyls (PCBs) and Cardiotoxicity

Polychlorinated biphenyls (PCBs) are synthetic organic chemicals that were widely used as dielectric fluids, coolants, and lubricants in electrical equipment due to their chemical stability and resistance to heat. Despite being banned in most countries since the late 1970s, PCBs persist in the environment and continue to pose significant health risks due to their bioaccumulative nature and long half-life in biological tissues [15, 16]. These compounds, composed of two benzene rings with varying

Table 25.1 Summary of exposure pathways, cardiovascular effects, toxic mechanisms, and knowledge gaps for key persistent organic pollutants (POPs)

Type of POPs	Exposure pathways	Cardiovascular effects	Mechanisms of toxicity	Knowledge gaps and future research
PCBs	Food, water, air, consumer products (e.g., non-stick cookware, textiles)	Hypertension, atherosclerosis, ischemic heart disease, arrhythmias	Oxidative stress, inflammation, endocrine disruption (PPAR signaling)	Lack of long-term cohort studies on the cardiovascular effects of PCBs
Dioxins	Food, air, soil, industrial processes	Hypertension, increased heart rate, atherosclerosis	Oxidative stress, altered thyroid hormone metabolism, inflammation	Limited understanding of synergistic effects with other environmental pollutants
PBDEs	Food, consumer products (e.g., electronics, textiles)	Altered lipid metabolism, increased blood pressure, arrhythmias	Disruption of lipid metabolism, oxidative stress, endocrine disruption	Role of PBDEs in chronic cardiovascular disease remains underexplored
PFAS	Food, drinking water, consumer products (e.g., food packaging, water-resistant textiles)	Hypertension, dyslipidemia, increased risk of cardiovascular diseases	Oxidative stress, inflammation, endocrine disruption (PPAR signaling), lipid metabolism disruption	Lack of longitudinal data on PFAS exposure and cardiovascular outcomes; need for studies on their synergistic effects

numbers of chlorine atoms, exist in 209 possible congeners, categorized as either dioxin-like or non-dioxin-like depending on their structural and toxicological properties.

Exposure to PCBs primarily occurs through the consumption of contaminated food, particularly fish, meat, and dairy products, where these lipophilic compounds accumulate in fat tissues. Additionally, exposure can result from inhalation of contaminated air, occupational contact, and dermal absorption in polluted environments. Given their hydrophobic nature, PCBs tend to bioaccumulate in adipose tissue, leading to long-term retention in humans and wildlife, and increasing the risk of chronic health effects [15, 16].

PCBs have been extensively studied for their toxicological effects, impacting multiple physiological systems. These compounds have been linked to endocrine disruption, neurotoxicity, hepatotoxicity, immunotoxicity, and metabolic disorders [10, 17]. Their disruption of thyroid hormone homeostasis and interference with lipid metabolism contribute to an array of health complications, including metabolic syndrome and obesity-related disorders [18].

Epidemiological studies have established a strong correlation between PCB exposure and cardiovascular disease. A systematic review and meta-analysis examining background PCB exposure and cardiovascular-specific mortality found a significant increase in the risk of mortality due to cardiovascular complications, particularly among populations with dietary exposure to PCBs [19]. The study by

Åkessonet al [20] utilized a prospective population-based cohort design involving 32,952 women and 36,546 men, free from cancer, cardiovascular disease, and diabetes at enrollment in 1997. Dietary exposure to PCBs was assessed using validated food frequency questionnaires. Incident heart failure cases were identified through linkage to national patient and death registers, tracking participants from 1998 to 2010. The study found a significant association between higher PCB exposure and a 42–48% increased risk of heart failure in both women and men. In contrast, higher intake of long-chain omega-3 fatty acids was linked to an 18–29% reduced risk of heart failure. Research conducted in communities exposed to PCBs via contaminated fish consumption has consistently shown elevated incidences of coronary artery disease, ischemic heart disease, and cardiac arrhythmias [13].

The cardiotoxic effects of PCBs are mediated through multiple interconnected mechanisms. These chemicals contribute to cardiovascular toxicity through multiple mechanisms, including chronic oxidative stress, which leads to cellular damage and inflammation, promoting cardiovascular diseases. PCB exposure also triggers inflammatory responses that damage vascular tissues and interfere with normal hormonal functions, affecting metabolism and fat deposition, further impacting cardiovascular health . Additionally, PCBs activate the aryl hydrocarbon receptor (AhR), alter gene expression, and modulate the renin-angiotensin system, increasing cardiovascular risk, underscoring the complexity of their toxic effects on heart health [21]. Furthermore, dioxin-like PCBs activate the aryl hydrocarbon receptor (AhR), a transcription factor involved in xenobiotic metabolism, which influences inflammatory gene expression, lipid metabolism, and endothelial function. Activation of AhR by PCBs has been implicated in vascular remodeling and the progression of cardiovascular disease [21, 22].

Another critical pathway involves the disruption of lipid metabolism. PCBs interfere with the homeostasis of cholesterol and triglycerides, leading to dyslipidemia, a major risk factor for cardiovascular disease [23, 24]. Studies have demonstrated that PCB exposure is associated with increased levels of low-density lipoprotein (LDL) cholesterol and triglycerides, while concurrently reducing high-density lipoprotein (HDL) cholesterol levels, further predisposing individuals to atherosclerotic cardiovascular disease [25].

Endothelial dysfunction is another major consequence of PCB exposure, as these compounds impair nitric oxide (NO) production, an essential molecule for vascular homeostasis [26]. Reduced NO bioavailability contributes to vascular constriction, increased blood pressure, and enhanced platelet aggregation, all of which heighten the risk of thrombotic events and ischemic damage. Moreover, PCBs alter calcium signaling in cardiac myocytes, potentially leading to arrhythmias and impaired contractility, further exacerbating cardiovascular risk [27].

In addition to these direct cardiovascular effects, PCBs contribute to systemic metabolic disturbances that indirectly exacerbate cardiometabolic risk. PCBs have been shown to promote insulin resistance and hyperglycemia, conditions that are tightly linked to both cardiovascular disease and metabolic syndrome [28]. Certain PCBs have been shown to exacerbate obesity-related cardiovascular complications by altering adipose tissue function and promoting pro-inflammatory cytokine release [29].

The cumulative evidence suggests that PCBs pose a substantial threat to cardiovascular health, necessitating ongoing monitoring and regulatory efforts to mitigate exposure. Despite global bans on their production, PCBs persist in the environment, continuing to enter the food chain and affecting human populations worldwide. Addressing this issue requires stringent policies for the remediation of contaminated sites and enhanced public health strategies to minimize dietary exposure, particularly in vulnerable populations.

Dioxins and Cardiotoxicity

Dioxins are a class of highly toxic POPs that include polychlorinated dibenzo-p-dioxins (PCDDs) and polychlorinated dibenzofurans (PCDFs). These compounds are primarily released as by-products of industrial combustion, waste incineration, and certain chemical manufacturing processes [30]. Their chemical stability and lipophilic nature enable them to persist in the environment for decades, accumulating in the fatty tissues of living organisms and biomagnifying along the food chain [30]. Among them, 2,3,7,8-tetrachlorodibenzo-p-dioxin (TCDD) is considered the most toxic and has been classified as a Group 1 carcinogen by the International Agency for Research on Cancer (IARC) [31]. Dioxins have been implicated in a variety of adverse health effects, including endocrine disruption, immune suppression, reproductive toxicity, and metabolic dysregulation, with increasing evidence supporting their role in cardiovascular diseases [30].

Human exposure to dioxins primarily occurs through dietary intake, mainly via consuming contaminated animal-derived foods such as dairy, meat, and fish. Given their lipophilic nature, dioxins accumulate in the adipose tissues of livestock and aquatic organisms, posing a significant exposure risk to human populations consuming these products. Additional sources of exposure include inhalation of contaminated air, occupational exposure in industries dealing with dioxin-generating processes, and unintentional environmental contamination [30–32]. Historical ecological disasters, such as the Seveso accident in Italy in 1976, highlighted the severe and long-term health effects of high-level dioxin exposure [33]. Epidemiological studies examining populations residing near industrial sites or hazardous waste facilities have consistently shown elevated dioxin levels in biological samples, correlating with increased risks of chronic diseases, particularly those affecting the cardiovascular system [34].

Epidemiological evidence has increasingly linked chronic dioxin exposure to an elevated risk of cardiovascular morbidity and mortality. The systematic review [35] indicates that dioxin exposure is associated with increased mortality from ischemic heart disease (IHD) and cardiovascular diseases, with a stronger link to IHD. The review incorporated 12 cohorts, mostly of occupationally exposed individuals, showing consistent, dose-related increases in IHD mortality, though many studies had limitations in adjusting for confounding factors. Animal studies further suggest that dioxins can negatively impact cardiovascular health, particularly affecting the fetal heart and adult animals after exposure, demonstrating their potential effects across life stages. The study, which included 914 participants aged 30–45 without pre-existing CVD, found that higher serum levels of PCDD/Fs were linked to an increased risk of CVD. After adjusting for confounding factors, the adjusted odds ratios showed significantly higher risks in individuals with higher serum PCDD/F levels. These findings suggest that elevated dioxin levels are associated with a higher risk of CVD, even before its clinical onset, supporting the idea that dioxins may contribute to the development of CVD [36]. Similarly, investigations of the Yusho and Yucheng poisoning incidents in Japan and Taiwan, where populations were exposed to polychlorinated biphenyls (PCBs) and dioxin-like compounds through contaminated rice oil, revealed persistent cardiovascular complications decades after the initial exposure [37]. These findings underscore the long-lasting and insidious effects of dioxin toxicity on human health.

Dioxins exert their toxic effects through the activation of the aryl hydrocarbon receptor (AhR), a ligand-activated transcription factor that regulates the expression of genes involved in xenobiotic metabolism, immune responses, and inflammatory pathways [30]. Upon binding to AhR, dioxins disrupt normal cellular signaling processes, contributing to systemic oxidative stress, chronic inflammation, and lipid metabolism alterations. These mechanisms are believed to be central to the cardiovascular toxicity of dioxins, as they promote endothelial dysfunction, arterial stiffness, and atherogenesis [22]. 2,3,7,8-Tetrachlorodibenzo-p-dioxin (TCDD) induces cardiotoxicity through several key mechanisms, particularly during embryonic development. TCDD exerts its toxic effects primarily through the activation of the AhR, which influences gene expression and disrupts cardiac development, though the specific downstream targets remain largely unclear. This activation leads to various developmental toxicities, including alterations in gene expression involved in cardiac function and growth. One significant effect is the generation of oxidative stress, which results in the overproduction of reactive oxygen species (ROS) that cause cellular damage, particularly in cardiac muscle cells, and compromise myocardial integrity [38]. Additionally, TCDD disrupts the expression of crucial genes such as vascular endothelial growth factor (VEGF), which is essential for proper coronary vascularization during heart development. This disruption leads to insufficient blood supply and a lack of growth factors necessary for developing cardiac tissues. Furthermore, TCDD inhibits cardiomyocyte proliferation while promoting apoptosis, contributing to inadequate heart tissue formation and conditions such as dilated cardiomyopathy observed in TCDD-exposed embryos. These toxic effects are most pronounced during specific developmental windows, where interference with normal signaling pathways can result in lasting cardiac malformations and dysfunction. Studies in various embryonic models, including zebrafish and chickens, suggest that these mechanisms of cardiotoxicity may be conserved across species, highlighting similar genetic pathways involved in cardiovascular development [39].

PBDEs and Cardiotoxicity

Polybrominated diphenyl ethers (PBDEs) are a class of brominated flame retardants widely used in consumer products such as textiles, plastics, and electronics to reduce flammability. Despite their phase-out in many countries, PBDEs persist in the environment due to their stability and bioaccumulation potential [40]. Human exposure occurs primarily through dietary intake, inhalation of contaminated dust, and dermal contact, with infants and young children being particularly vulnerable due to high exposure levels through breast milk and household dust [41]. PBDEs are detected in various biological matrices, including blood, adipose tissue, and breast milk, highlighting their widespread distribution and potential health risks [42].

PBDEs exhibit a range of toxic effects, including endocrine disruption, neurotoxicity, hepatotoxicity, and immunotoxicity [43, 44]. The study, which used data from the National Health and Nutrition Examination Survey (NHANES) conducted between 2003 and 2004, focused on adults aged 20 and older with available PBDE measurements and followed participants for about 16 years, with mortality data linked through 2019. The study found no significant association between PBDE exposure and cardiovascular mortality, despite adjusting for various confounders such as demographics, lifestyle, and health status. While higher PBDE levels were associated with increased cancer mortality, the lack of association with cardiovascular outcomes suggests that PBDEs may not specifically affect heart health. Although there are no data on epidemiological evidence of the connection of PBDEs and cardiotoxicity, emerging evidence suggests that PBDEs may contribute to cardiovascular diseases through mechanisms involving oxidative stress, inflammation, and lipid metabolism dysregulation [44].

PBDEs could exert their cardiotoxic effects through multiple pathways. The molecular mechanisms underlying PBDE-induced toxicity involve multiple pathways, including activation of the AhR, disruption of thyroid hormone signaling, and interference with lipid metabolism [40]. PBDEs structurally resemble thyroid hormones and can competitively bind to thyroid hormone receptors, disrupting normal endocrine function and metabolic homeostasis [41]. Given the persistence and widespread distribution of PBDEs in the environment and mechanisms of toxicity that could potentially lead to cardiotoxicity, their possible role in cardiovascular disease warrants further investigation.

PFAS and Cardiotoxicity

Per- and polyfluoroalkyl substances (PFAS) are a diverse group of synthetic chemicals extensively used in industrial and consumer products due to their hydrophobic and lipophobic properties. These substances have been incorporated into a wide range of applications, including non-stick cookware, water-repellent textiles, food packaging, and firefighting foams [45]. Their high resistance to degradation makes them persistent in the environment, earning them the nickname "forever chemicals". Due to their widespread use and environmental persistence, PFAS contamination is now detectable in drinking water, soil, air, and various food sources [46]. As a result, human exposure is nearly ubiquitous, with measurable concentrations of PFAS found in the blood of individuals worldwide [47]. Long-chain PFAS, such as perfluorooctanoic acid (PFOA) and perfluorooctane sulfonic acid (PFOS), have been of particular concern due to their extended biological half-lives in humans, ranging from several years to decades [48].

Recent reviews have shown a linkage of the PFAS exposure to an array of adverse health effects, including metabolic, endocrine, and cardiovascular dysfunction [49, 50]. A growing body of epidemiological data supports the association between PFAS exposure and adverse cardiovascular outcomes. The study, which examined first-level high school students in Northern Norway, with 1038 out of 1117 invited participants (93%) aged 15–19 years revealed that higher levels of PFAS, including PFOS, PFNA, and PFDA, were associated with dyslipidemia, as indicated by increased total cholesterol, LDL cholesterol, and apolipoprotein B [51]. Certain PFAS, such as PFHxS and PFOS, were also linked to a higher risk of hypertension, while PFHxS and PFHpS were associated with obesity in adolescents. The study suggests that PFAS exposure may have endocrine-disrupting effects, contributing to metabolic changes that increase cardiovascular risk early in life. Given its cross-sectional design, the study identifies associations rather than direct causation, highlighting the need for further longitudinal research.The other study that analyzed data from two large cohorts, the NEO Study (584 participants) and the Rhineland Study (1962 participants), examined the effects of PFAS exposure on lipid metabolism [52]. Blood samples revealed that higher concentrations of PFAS, particularly PFOS and PFHxS, were associated with increased total lipids, cholesterol, and phospholipid content across multiple lipoprotein subclasses, including HDL, LDL, IDL, and VLDL. The impact of PFAS on lipid profiles was found to be age-dependent, with younger individuals showing stronger associations between PFAS exposure and adverse lipid outcomes. PFAS exposure was also linked to elevated levels of apoB, fatty acids, phosphoglycerides, and lipoprotein lipid content, all of which are biomarkers for cardiovascular disease risk. These findings highlight the potential role of PFAS in disrupting lipid metabolism and increasing the risk of cardiovascular diseases, even at low exposure levels. In the study by Yang et al. [53] it was found that elevated serum levels of perfluorohexane sulfonic acid (PFHxS) and perfluorodecanoic acid (PFDeA) were associated with an increased risk of abdominal aortic calcification (AAC), with PFHxS showing the strongest correlation. A dose-dependent relationship was observed, suggesting that higher PFHxS concentrations significantly increased the likelihood of AAC, a known risk factor for CVDs. While the study did not find fatty acids or blood cell fractions to mediate this association, it suggests that PFAS may contribute to cardiovascular damage through hematological and inflammatory pathways. These findings highlight the need for further research to explore the biological mechanisms underlying PFAS-induced cardiovascular effects and their interactions with other environmental risk factors. The study also reinforces the importance of regulatory measures to limit

PFAS exposure, emphasizing the need for stricter monitoring of environmental and food contamination. Overall, these results contribute to growing evidence on the cardiovascular risks posed by PFAS, underscoring their significance as emerging environmental pollutants with long-term health implications. Additionally, the study, which examined the association between serum PFAS concentrations and hypertension in over 15,000 young adults (ages 20–39) from a highly exposed community in North-Eastern Italy revealed that higher PFAS levels were positively associated with increased systolic and diastolic blood pressure, with the prevalence of hypertension rising alongside PFAS exposure, particularly in men. Different PFAS compounds showed varying odds ratios for hypertension, with the strongest effect observed for PFNA in men. These findings suggest that even small increases in blood pressure due to PFAS exposure could contribute to long-term cardiovascular risks [54].

The cardiovascular effects of PFAS have been attributed to their ability to disrupt lipid homeostasis, promote oxidative stress, induce systemic inflammation, and interfere with key hormonal signaling pathways [55]. These mechanisms contribute to endothelial dysfunction, arterial stiffness, and increased risk of cardiovascular events, including myocardial infarction and stroke. Oxidative stress has been recognized as a critical pathway through which PFAS exert their cardiotoxic effects [56]. Studies have demonstrated that PFAS can impair mitochondrial function, leading to excessive production of reactive oxygen species (ROS) [55]. Elevated ROS levels contribute to lipid peroxidation, protein oxidation, and DNA damage, all of which compromise the structural and functional integrity of cardiovascular tissues [56]. Moreover, PFAS have been shown to disrupt calcium homeostasis in cardiomyocytes, potentially leading to arrhythmias and impaired cardiac contractility [57].

Inflammation is another key factor in PFAS-induced cardiotoxicity [58]. Prenatal exposure to PFAS has been associated with increased circulating levels of pro-inflammatory cytokines such as tumor necrosis factor-alpha (TNF-α), interleukin-6 (IL-6), and C-reactive protein (CRP) [59]. These inflammatory mediators play a central role in the pathogenesis of atherosclerosis by promoting endothelial cell activation, monocyte recruitment, and foam cell formation within arterial walls. Chronic inflammation driven by PFAS exposure not only accelerates plaque development but also increases plaque instability, raising the likelihood of acute cardiovascular events such as heart attacks and strokes [60].

Endocrine disruption is another crucial mechanism by which PFAS contribute to cardiovascular disease [61]. These chemicals have been shown to interfere with peroxisome proliferator-activated receptor (PPAR) signaling pathways, which are vital regulators of lipid metabolism, glucose homeostasis, and vascular function [62]. Activation of PPARα and PPARγ by PFAS has been linked to altered lipid profiles, including increased levels of low-density lipoprotein (LDL) cholesterol and triglycerides, along with reduced levels of high-density lipoprotein (HDL) cholesterol. Such dyslipidemia significantly enhances the risk of developing atherosclerosis and other cardiovascular complications [63]. Additionally, PFAS have been reported to disrupt thyroid hormone signaling, further exacerbating metabolic imbalances that contribute to cardiovascular pathology [64].

Future studies should focus on refining dose-response relationships, identifying potential biomarkers of PFAS-induced cardiotoxicity, and exploring therapeutic interventions to mitigate PFAS-associated cardiovascular risk. Given the growing recognition of PFAS as a public health concern, interdisciplinary efforts combining epidemiology, toxicology, and regulatory science will be essential to develop comprehensive strategies for reducing the burden of PFAS-related diseases.

Conclusions and Future Directions

Persistent Organic Pollutants, including PCBs, dioxins, PBDEs, and PFAS, represent a significant risk to cardiovascular health. As environmental contaminants with high bioaccumulation potential, POPs persist in the environment and continue to pose long-term health risks despite global efforts to limit their production and use. Epidemiological studies have consistently shown associations between POPs exposure and adverse cardiovascular outcomes, such as hypertension, atherosclerosis, and ischemic heart disease. These findings underscore the urgent need for comprehensive strategies to assess, reduce, and ultimately prevent the cardiovascular effects of POPs.

While the body of evidence linking POPs to cardiovascular dysfunction is substantial, significant knowledge gaps remain. One of the major limitations in current research is the lack of longitudinal studies that track the long-term effects of POP exposure on cardiovascular health. Most existing studies provide cross-sectional data, which, while valuable, cannot establish causal relationships or determine the cumulative impact of prolonged exposure. Future research must prioritize long-term cohort studies to better understand the delayed and chronic effects of POPs on heart health, particularly in vulnerable populations. Another key area of uncertainty is the molecular mechanisms through which different POPs contribute to cardiovascular toxicity. While oxidative stress and inflammation are recognized as primary pathways, the specific interactions of various POPs with cardiovascular cells and tissues remain underexplored. The role of individual congeners of PCBs and dioxins, for instance, may vary significantly, and more targeted research is needed to identify the distinct mechanisms of action for each compound. A deeper understanding of how these pollutants impact cardiac cells, endothelial function, and vascular remodeling will provide critical insights into their cardiotoxic effects.

Furthermore, the interaction between multiple POPs in real-world exposure contexts is not well understood. While individual POPs have been studied in isolation, people are generally exposed to mixtures of these chemicals, making it crucial to investigate the synergistic effects of multiple POPs on cardiovascular health. This research will be particularly important for understanding how combined exposures may amplify or mitigate the risks associated with these pollutants. Lastly, given the increasing recognition of endocrine disruption as a central mechanism of toxicity, future research should focus on elucidating the endocrine pathways affected by POPs and their long-term consequences for cardiovascular health. Investigating the role of hormones, lipids, and metabolic disorders in mediating POPs-induced cardiovascular toxicity could open new avenues for therapeutic interventions.

Future research should focus on understanding the long-term cardiovascular effects of POPs exposure. Further studies should aim to clarify dose-response relationships, identify potential biomarkers of POPs-induced cardiotoxicity, and explore therapeutic interventions to reduce cardiovascular risks associated with POPs exposure. Given the growing recognition of POPs as a public health concern, interdisciplinary efforts combining epidemiology, toxicology, and regulatory science will be crucial in developing comprehensive strategies to reduce the burden of diseases linked to POPs.

References

1. Negrete-Bolagay D, Zamora-Ledezma C, Chuya-Sumba C, De Sousa FB, Whitehead D, Alexis F, Guerrero VH (2021) Persistent organic pollutants: the trade-off between potential risks and sustainable remediation methods. J Environ Manag 300:113737. https://doi.org/10.1016/j.jenvman.2021.113737
2. Assey GE, Mogusu E (2024) Effects of persistent organic pollutants on environment, health and mountains: a review. J Environ Eng Sci 19(1):1–8. https://doi.org/10.1680/jenes.22.00089
3. Stockholm Convention on Persistent Organic Pollutants. United Nations Environment Programme. Accessed Jan 2025
4. United Nations Environment Programme (2017) Effectiveness evaluation of the Stockholm Convention. Accessed Jan 2025
5. Brender JD, Maantay JA, Chakraborty J (2011) Residential proximity to environmental hazards and adverse health outcomes. Am J Public Health 101(S1):S37–S52. https://doi.org/10.2105/AJPH.2011.300183
6. AMAP Assessment (2016) Chemicals of emerging Arctic concern. Arctic Monitoring and Assessment Programme (AMAP), Oslo
7. Wong F et al (2021) Time trends of persistent organic pollutants (POPs) and Chemicals of Emerging Arctic Concern (CEAC) in Arctic air from 25 years of monitoring. Sci Total Environ 775:145109. https://doi.org/10.1016/j.scitotenv.2021.145109
8. Lee YM et al (2016) Persistent organic pollutants in adipose tissue should be considered in obesity research. Obes Rev 18(2):129–139. https://doi.org/10.1111/obr.12481
9. Lind L, Lind PM (2012) Can persistent organic pollutants and plastic-associated chemicals cause cardiovascular disease? J Intern Med 271(6):537–553. https://doi.org/10.1111/j.1365-2796.2012.02536.x
10. Buha Djordjevic A, Antonijevic E, Curcic M, Milovanovic V, Antonijevic B (2020) Endocrine-disrupting mechanisms of polychlorinated biphenyls. Curr Opin Toxicol 19:42–49. https://doi.org/10.1016/j.cotox.2019.10.006
11. Rappazzo KM et al (2017) Exposure to perfluorinated alkyl substances and health outcomes in children: a systematic review. Environ Int 14(7):691. https://doi.org/10.3390/ijerph14070691
12. Grandjean P, Clapp R (2015) Perfluorinated alkyl substances: emerging insights into health risks. New Solut 25(2):147–163. https://doi.org/10.1177/1048291115590506
13. Blackowicz MJ, Persky VW, Sargis RM, Freels S, Anderson HA, Turyk ME (2024) Polychlorinated biphenyls and incident coronary heart disease-related outcomes in Great Lakes fish consumers. Environ Res 255:119071. https://doi.org/10.1016/j.envres.2024.119071

14. Chen JC, Baumert BO, Li Y, Li Y, Pan S, Robinson S, Rubbo B, Costello E, He J, Hampson H, Beglarian E, Rock S, Goodrich JA, Eckel SP, Aung MT, McConnell R, Conti DV, Chatzi L (2023) Associations of per- and polyfluoroalkyl substances, polychlorinated biphenyls, organochlorine pesticides, and polybrominated diphenyl ethers with oxidative stress markers: a systematic review and meta-analysis. Environ Res 239:117308. https://doi.org/10.1016/j.envres.2023.117308
15. Agency for Toxic Substances and Disease Registry (ATSDR) (2000) Toxicological profile for polychlorinated biphenyls (PCBs). U.S. Department of Health and Human Services, Public Health Service. https://www.atsdr.cdc.gov/toxprofiles/tp17.pdf
16. Zhang Y, Qiu X, Bian Y, Ma J, Feng XS, He Z, Tan Y (2024) Polychlorinated biphenyls in food: updates on occurrence, toxic effects, pretreatment, and analysis techniques. Food Biosci 59:104083. https://doi.org/10.1016/j.fbio.2024.104083
17. Raffetti E, Donato F, Speziani F, Scarcella C, Gai A, Magoni M, Magoni M (2018) Polychlorinated biphenyls (PCBs) exposure and cardiovascular, endocrine and metabolic diseases: a population-based cohort study in a North Italian highly polluted area. Environ Int 120:215–222. https://doi.org/10.1016/j.envint.2018.08.022
18. Gao X, Yan D, Li G, Wei Y, He H, Zhai J (2023) Polychlorinated biphenyls and risk of metabolic syndrome and comparison with the risk of diabetes: a systematic review and meta-analysis. Sci Total Environ 900:165773. https://doi.org/10.1016/j.scitotenv.2023.165773
19. Fiolet T, Mahamat-Saleh Y, Frenoy P, Kvaskoff M, Mancini FR (2021) Background exposure to polychlorinated biphenyls and all-cause, cancer-specific, and cardiovascular-specific mortality: a systematic review and meta-analysis. Environ Int 154:106663. https://doi.org/10.1016/j.envint.2021.106663
20. Åkesson A, Donat-Vargas C, Berglund M, Glynn A, Wolk A, Kippler M (2019) Dietary exposure to polychlorinated biphenyls and risk of heart failure—a population-based prospective cohort study. Environ Int 126:1–6. https://doi.org/10.1016/j.envint.2019.01.069
21. Perkins JT, Petriello MC, Newsome BJ, Hennig B (2016) Polychlorinated biphenyls and links to cardiovascular disease. Environ Sci Pollut Res 23(3):2160–2172. https://doi.org/10.1007/s11356-015-4479-6
22. Li K, Li K, He Y, Liang S, Shui X, Lei W (2023) Aryl hydrocarbon receptor: a bridge linking immuno-inflammation and metabolism in atherosclerosis. Biochem Pharmacol 216:115744. https://doi.org/10.1016/j.bcp.2023.115744
23. Aminian O, Moinfar Z, Eftekhari S, Esser A, Schettgen T, Felten M, Kaifie A, Kraus T (2020) Association of plasma levels of lipid and polychlorinated biphenyls in Iranian adults. Heliyon 6(12):e03775. https://doi.org/10.1016/j.heliyon.2020.e03775
24. Goncharov A, Haase RF, Santiago-Rivera A, Morsey G, McCaffrey RJ, Rej R, Carpenter DO (2008) High serum PCBs are associated with elevation of serum lipids and cardiovascular disease in a Native American population. Environ Res 106(2):226–239. https://doi.org/10.1016/j.envres.2007.10.006
25. Singh K, Chan HM (2018) Association of blood polychlorinated biphenyls and cholesterol levels among Canadian Inuit. Environ Res 160:298–305. https://doi.org/10.1016/j.envres.2017.10.010
26. Helyar SG, Patel B, Headington K, El Assal M, Chatterjee PK, Pacher P, Mabley JG (2009) PCB-induced endothelial cell dysfunction: role of poly(ADP-ribose) polymerase. Biochem Pharmacol 78(7):959–965. https://doi.org/10.1016/j.bcp.2009.06.019
27. Rebuzzini P, Zuccolo E, Civello C, Fassina L, Arechaga J, Izquierdo A, Faris P, Zuccotti M, Moccia F, Garagna S (2018) Polychlorinated biphenyls reduce the kinematics contractile properties of embryonic stem cells-derived cardiomyocytes by disrupting their intracellular Ca^{2+} dynamics. Sci Rep 8(1):17909. https://doi.org/10.1038/s41598-018-36333-z
28. Heindel JJ, Blumberg B, Cave M, Machtinger R, Mantovani A, Mendez MA, Nadal A, Palanza P, Panzica G, Sargis R, Vandenberg LN, vom Saal F (2017) Metabolism disrupting chemicals and metabolic disorders. Reprod Toxicol 68:3–33. https://doi.org/10.1016/j.reprotox.2016.10.001

29. Arsenescu V, Arsenescu RI, King V, Swanson H, Cassis LA (2008) PCB-77 induces adipocyte differentiation and proinflammatory adipokines and promotes obesity and atherosclerosis. Environ Health Perspect 116(6):761–768. https://doi.org/10.1289/ehp.10554
30. Mathew N, Somanathan A, Tirpude A, Pillai AM, Mondal P, Arfin T (2025) Dioxins and their impact: a review of toxicity, persistence, and novel remediation strategies. Anal Methods 17(4): 651–672. https://doi.org/10.1039/d4ay01767f
31. IARC (2012) Polychlorinated dibenzo-p-dioxins and polychlorinated dibenzofurans. In: IARC monographs on the evaluation of carcinogenic risks to humans, pp 135–145
32. Fiolet T, Nicolas G, Casagrande C, Horvath Z, Frenoy P, Weiderpass E, Gunter MJ, Manjer J, Sonestedt E, Palli D, Simeon V, Tumino R, Bueno-de-Mesquita B, Huerta JM, Rodriguez-Barranco M, Abilleira E, Sacerdote C, Schulze MB, Heath AK, Rylander C, Skeie G, Nøst TH, Tjønneland A, Olsen A, Pala V, Kvaskoff M, Huybrechts I, Mancini FR (2024) Dietary intakes of dioxins and polychlorobiphenyls (PCBs) and mortality: EPIC cohort study in 9 European countries. Int J Hyg Environ Health 255:114287. https://doi.org/10.1016/j.ijheh.2023.114287
33. Consonni D, Rognoni M, Cavalieri d'Oro L, Pesatori AC (2024) Mortality and cancer incidence in a population exposed to TCDD after the Seveso, Italy, accident (1976–2013). Occup Environ Med 81(5):349–358. https://doi.org/10.1136/oemed-2023-109167
34. Tait PW, Brew J, Che A, Costanzo A, Danyluk A, Davis M, Khalaf A, McMahon K, Watson A, Rowcliff K, Bowles D (2019) The health impacts of waste incineration: a systematic review. Aust N Z J Public Health 44(1):40–48. https://doi.org/10.1111/1753-6405.12939
35. Humblet O, Birnbaum L, Rimm E, Mittleman MA, Hauser R (2008) Dioxins and cardiovascular disease mortality. Environ Health Perspect 116(11):1443–1448. https://doi.org/10.1289/ehp.11579
36. Chang J-W, Chen H-L, Chang C-C, Su H-J, Liaw P-C, Lee C-C (2011) Predicting the risk of cardiovascular disease in people exposed to moderate to high levels of dioxin. J Hazard Mater 198:317–322. https://doi.org/10.1016/j.jhazmat.2011.10.046
37. Onozuka D, Nakamura Y, Tsuji G, Furue M (2020) Mortality in Yusho patients exposed to polychlorinated biphenyls and polychlorinated dibenzofurans: a 50-year retrospective cohort study. Environ Health 19:119. https://doi.org/10.1186/s12940-020-00680-0
38. Mohsenzadeh MS, Zanjani BR, Karimi G (2018) Mechanisms of 2,3,7,8-tetrachlorodibenzo-p-dioxin-induced cardiovascular toxicity: an overview. Chem Biol Interact 282:1–6. https://doi.org/10.1016/j.cbi.2018.01.002
39. Goldstone HMH, Stegeman JJ (2006) Molecular mechanisms of 2,3,7,8-tetrachlorodibenzo-p-dioxin cardiovascular embryotoxicity. Drug Metab Rev 38(1–2):261–289. https://doi.org/10.1080/03602530600570099
40. Xue J, Xiao Q, Zhang M et al (2003) Toxic effects and mechanisms of polybrominated diphenyl ethers. Int J Mol Sci 24:13487. https://doi.org/10.3390/ijms241713487
41. Dishaw L, Macaulay L, Roberts SC, Stapleton HM (2014) Exposures, mechanisms, and impacts of endocrine-active flame retardants. Curr Opin Pharmacol:125–133. https://doi.org/10.1016/j.coph.2014.09.018
42. Gao J, Xie Z, Wang Z et al (2004) Human toxicity of polybrominated diphenyl ethers (PBDEs) and their derivatives: a comprehensive review. Curr Res Food Sci 9:100918. https://doi.org/10.1016/j.crfs.2024.100918
43. Vuong AM, Yolton K, Dietrich KN, Braun JM, Lanphear BP, Chen A (2018) Exposure to polybrominated diphenyl ethers (PBDEs) and child behavior: current findings and future directions. Horm Behav 101:94–104. https://doi.org/10.1016/j.yhbeh.2017.11.008
44. Renzelli V, Gallo M, Morviducci L et al (2023) Polybrominated diphenyl ethers (PBDEs) and human health: effects on metabolism, diabetes and cancer. Cancers 15:4237. https://doi.org/10.3390/cancers15174237
45. Dhore R, Murthy GS (2021) Per/polyfluoroalkyl substances production, applications and environmental impacts. Bioresour Technol 341:125808. https://doi.org/10.1016/j.biortech.2021.125808

46. Kurwadkar S, Dane J, Kanel SR, Nadagouda MN, Cawdrey RW, Ambade B, Struckhoff GC, Wilkin R (2022) Per- and polyfluoroalkyl substances in water and wastewater: a critical review of their global occurrence and distribution. Sci Total Environ 809:151003. https://doi.org/10.1016/j.scitotenv.2021.151003
47. Göckener B, Weber T, Rüdel H, Bücking M, Kolossa-Gehring M (2020) Human biomonitoring of per- and polyfluoroalkyl substances in German blood plasma samples from 1982 to 2019. Environ Int 145:106123. https://doi.org/10.1016/j.envint.2020.106123
48. Agency for Toxic Substances and Disease Registry (ATSDR) (2022) PFAS exposure assessments final report: findings across ten exposure assessment sites. U.S. Department of Health and Human Services
49. Panieri E, Baralic K, Djukic-Cosic D, Buha Djordjevic A, Saso L (2022) PFAS molecules: a major concern for the human health and the environment. Toxics 10(2):44. https://doi.org/10.3390/toxics10020044
50. Panieri E, Buha Đorđević A, Saso L (2021) Endocrine disruption by PFAS: a major concern associated with legacy and replacement substances. Arhiv za farmaciju 71:429–454. https://doi.org/10.5937/arhfarm71-34197
51. Averina M, Brox J, Huber S, Furberg A-S (2021) Exposure to perfluoroalkyl substances (PFAS) and dyslipidemia, hypertension, and obesity in adolescents: the Fit Futures study. Environ Res 195:110740. https://doi.org/10.1016/j.envres.2021.110740
52. Faquih TO, Landstra EN, van Hylckama Vlieg A, Aziz NA, Li-Gao R, de Mutsert R, Rosendaal FR, Noordam R, van Heemst D, Mook-Kanamori DO, Willems van Dijk K, Breteler MMB (2024) Per- and polyfluoroalkyl substances concentrations are associated with an unfavorable cardio-metabolic risk profile: findings from two population-based cohort studies. Expo Health 16(4):1251–1262. https://doi.org/10.1007/s12403-023-00622-4
53. Yang J, Wang T, Li K, Wang Y (2024) Associations between per- and polyfluoroalkyl chemicals and abdominal aortic calcification in middle-aged and older adults. J Adv Res. https://doi.org/10.1016/j.jare.2024.04.022
54. Pitter G, Zare Jeddi M, Barbieri G, Gion M, Fabricio ASC, Daprà F, Russo F, Fletcher T, Canova C (2020) Perfluoroalkyl substances are associated with elevated blood pressure and hypertension in highly exposed young adults. Environ Health 19:102. https://doi.org/10.1186/s12940-020-00656-0
55. Fenton SE, Ducatman A, Boobis A, DeWitt JC, Lau C, Ng C, Smith JS, Roberts SM (2021) Per- and polyfluoroalkyl substance toxicity and human health review: current state of knowledge and strategies for informing future research. Environ Toxicol Chem 40(3):606–630. https://doi.org/10.1002/etc.4890
56. Tsutsui H, Kinugawa S, Matsushima S (2011) Oxidative stress and heart failure. Am J Physiol Heart Circ Physiol 301(6):H2181–H2190. https://doi.org/10.1152/ajpheart.00554.2011
57. Ko MY, Chon SH, Park H, Min E, Kim Y, Cha SW, Seo JW, Lee BS, Ka M, Hyun SA (2024) Perfluorooctanoic acid induces cardiac dysfunction in human induced pluripotent stem cell-derived cardiomyocytes. Ecotoxicol Environ Saf 273:116170. https://doi.org/10.1016/j.ecoenv.2024.116170
58. Zhang L, Louie A, Rigutto G, Guo H, Zhao Y, Ahn S, Dahlberg S, Sholinbeck M, Smith MT (2023) A systematic evidence map of chronic inflammation and immunosuppression related to per- and polyfluoroalkyl substance (PFAS) exposure. Environ Res 220:115188. https://doi.org/10.1016/j.envres.2022.115188
59. Siwakoti RC, Harris SM, Ferguson KK, Hao W, Cantonwine DE, Mukherjee B, McElrath TF, Meeker JD (2024) Prenatal exposure to per- and polyfluoroalkyl substances (PFAS) and their influence on inflammatory biomarkers in pregnancy: findings from the LIFECODES cohort. Environ Int 194:109145. https://doi.org/10.1016/j.envint.2024.109145
60. Meneguzzi A, Fava C, Castelli M, Minuz P (2021) Exposure to perfluoroalkyl chemicals and cardiovascular disease: experimental and epidemiological evidence. Front Endocrinol 12:706352. https://doi.org/10.3389/fendo.2021.706352

61. Mokra K (2021) Endocrine disruptor potential of short- and long-chain perfluoroalkyl substances (PFASs)—a synthesis of current knowledge with proposal of molecular mechanism. Int J Mol Sci 22(4):2148. https://doi.org/10.3390/ijms22042148
62. Szilagyi JT, Avula V, Fry RC (2020) Perfluoroalkyl substances (PFAS) and their effects on the placenta, pregnancy, and child development: a potential mechanistic role for placental peroxisome proliferator-activated receptors (PPARs). Curr Environ Health Rep 7(3):222–230. https://doi.org/10.1007/s40572-020-00279-0
63. Evans N, Conley JM, Cardona M, Hartig P, Medlock-Kakaley E, Gray LE Jr (2022) In vitro activity of a panel of per- and polyfluoroalkyl substances (PFAS), fatty acids, and pharmaceuticals in peroxisome proliferator-activated receptor (PPAR) alpha, PPAR gamma, and estrogen receptor assays. Toxicol Appl Pharmacol 449:116136. https://doi.org/10.1016/j.taap.2022.116136
64. Coperchini F, Croce L, Ricci G, Magri F, Rotondi M, Imbriani M, Chiovato L (2021) Thyroid disrupting effects of old and new generation PFAS. Front Endocrinol 11:612320. https://doi.org/10.3389/fendo.2020.612320

Chapter 26
Cardiovascular Toxicity of Natural Toxins

Danijela Đukić-Ćosić

Abstract Cardiovascular toxicity induced by natural toxins represents a critical area of research due to the potentially fatal consequences of toxin exposure on heart function. Natural toxins, derived from diverse sources such as animals, plants, bacteria, algae, fungi and mushrooms, exert significant effects on the cardiovascular system by targeting ion channels, disrupting cardiac cell membranes, and altering intracellular ion homeostasis. These interactions can lead to severe outcomes, including arrhythmias, myocardial ischemia, and heart failure. However, many natural toxins have been repurposed for therapeutic uses, serving as drugs, in vaccine production, and as tools for biomedical research. This chapter provides a comprehensive overview most of the natural toxins affecting heart function, their targets in the heart tissue, most important mechanisms of cardiovascular toxicity and clinical manifestations, as well as and treatment strategies for managing natural toxin poisoning.

Keywords Zootoxins · Phytotoxins · Bacterial and algal toxins · Mycotoxins · Mushroom toxins · Cardiac dysfunction · Therapeutic approaches.

Introduction

Natural toxins are toxic substances that are naturally synthesized by living organisms [1]. Typical sources of these toxins include animals, plants, microorganisms such as fungi, algae, and bacteria, as well as mushrooms [2, 3]. Natural toxins are bioactive compounds developed for defense, predation, or metabolic regulation [4]. These compounds are not harmful to the organisms that produce them but can pose significant risks to other species, including humans and animals. Some plant toxins are naturally synthesized as a defense against predators, insects, or microorganisms,

D. Đukić-Ćosić (✉)
Department of Toxicology "Akademik Danilo Soldatović", Centre for Toxicological Risk Assessment, Faculty of Pharmacy, University of Belgrade, Belgrade, Serbia
e-mail: danijela.djukic.cosic@pharmacy.bg.ac.rs

D. K. Agrawal et al. (eds.), *Cardiovascular Toxicity*, Advances in Biochemistry in Health and Disease 33, https://doi.org/10.1007/978-3-031-90598-8_26

while others arise from microbial infestations, such as mold. Similarly, microscopic algae and plankton in oceans—and occasionally in lakes—generate chemical compounds that are harmless to the fish or shellfish consuming them but toxic to humans. Illness can occur rapidly after consuming contaminated fish or shellfish [1]. This indicates that some natural toxins have species-specific effects. In addition to algal toxins, other natural toxin with this effect is spider venom, which is lethal to many animals but harmless to certain species like lizards [5]. Natural toxins exhibit a wide range of chemical structures and vary in their biological roles and levels of toxicity [6]. While these substances are beneficial in ecological contexts, poses a risk to human and health causing disease or death in certain cases [7]. Symptoms of natural toxin poisoning can vary widely depending on the specific toxin involved, the amount of exposure, and individual susceptibility. Certain toxins exhibit exceptional potency. Symptoms often develop rapidly, but some, such as organ damage or carcinogenic effects, may take longer to manifest, underscoring the need for prompt recognition and medical intervention [8]. Their impacts range from mild symptoms to severe and potentially fatal outcomes in humans and animals. Common symptoms may include gastrointestinal disturbances such as nausea, vomiting, diarrhea, and abdominal pain, which are frequently associated with plant, mushroom, or mycotoxin exposure [9, 10]. Neurological effects are also prevalent, including dizziness, confusion, headaches, seizures, and in severe cases, paralysis, which are commonly seen with zootoxins like snake venom or certain algal toxins [11, 12]. Respiratory difficulties, including shortness of breath or respiratory failure, may result from botulinum toxin, snake envenomation, or severe jellyfish stings [11, 13, 14]. Liver and kidney damage are hallmark effects of toxins such as amatoxins from mushrooms or aflatoxins from molds, leading to jaundice, decreased urine output, and potentially organ failure [15, 16]. Allergic reactions like skin rashes, itching, or swelling can occur, particularly with contact toxins from plants or marine organisms [17, 18]. Additionally, cardiovascular symptoms, such as irregular heart rhythms or hypotension, can occur with toxins like cardiac glycosides from plants or venoms that target the circulatory system. These effects can be severe and therefore cardiovascular toxicity is one of the most consequences of exposure to some natural toxins, involves toxic effects on heart structure and function [19]. Thus, the aim of this chapter is a comprehensive overview of the diverse sources of natural toxins, their molecular targets in the cardiovascular system, mechanisms of cardiovascular toxicity, and the clinical manifestations of toxin-induced cardiac dysfunction. Furthermore, it highlights current therapeutic approaches for managing of natural toxin-induced cardiovascular toxicity.

The Incidence of Intoxication with Natural Toxins

Natural toxin intoxications significantly impact global public health, with regional variations influenced by environmental exposure, dietary habits, and awareness. They are especially important in regions where contact with toxic plants, animals, fungi, or contaminated food and water is more common. Millions people are affected annually, with outcomes ranging from mild symptoms to severe illness and death [4].

Animal toxin intoxications pose a major global health challenge, especially in tropical and subtropical regions. Annually, 5.4 million snakebites lead to 1.8–2.7 million envenomation cases, causing 81,410–137,880 deaths and numerous disabilities, predominantly in Sub-Saharan Africa, South Asia, and Southeast Asia [11, 20]. In 2019, the World Health Organization (WHO) set an ambitious goal to reduce snakebite-related deaths and cases by 50% by 2030 [20]. Scorpion stings account for over 1.5 million cases and approximately 2600 deaths annually, primarily in areas with limited antivenom access like North Africa and Latin America [21]. Foodborne animal toxin poisonings affect nearly a million individuals annually, typically causing mild symptoms, though fatalities can occur without timely treatment [4].

Plant exposure accounts for approximately 5% of poisoning cases reported in North America and Europe, with over 50,000 annual cases in the U.S. alone [22, 23]. Most incidents involve young children due to accidental ingestion, though adolescents may consume plants recreationally or for self-harm. While the majority of cases are asymptomatic or result in mild symptoms, severe outcomes are rare but more likely in regions with limited healthcare access [22, 23].

Mycotoxins from fungal contamination in food are a major concern, especially in developing countries. These toxins are a threat to human health as they may cause severe and irrevocable damage to human health, such as cancer. Chronic exposure to aflatoxins contributes to an estimated 25,000–155,000 liver cancer cases annually, with significant mortality rates [24].

Fatal mushroom poisoning has increased globally, with regional differences. Western Europe reported 50–100 fatal cases annually in the early 2000s, while the U.S. averaged 7428 cases yearly from 1999 to 2016, with fewer than three fatalities per year [25].

Bacterial toxins also contribute significantly to global intoxication cases, with millions of cases reported annually. These toxins, produced by pathogenic bacteria, can cause severe illnesses with potentially fatal outcomes [26].

The global burden of natural toxin intoxication highlights the importance of knowledge of their toxic effects, prevention, early diagnosis, and treatment to mitigate its impact on public health.

Sources of Natural Toxins and Their Relevance to Cardiovascular Health

Natural toxins can be categorized based on their biological origins, highlighting the wide range of sources and mechanisms through which they impact living organisms. In the natural world, toxins primarily serve two purposes: to immobilize or capture prey, as seen in organisms like spiders, snakes, and scorpions, or to provide defense against predators, as exemplified by bee stings and frog poisons. Toxins that target cardiac physiology are particularly effective for both defense and predation. These compounds can disrupt the heart's electrical system, causing arrhythmias, or impair its mechanical function, compromising the pumping action and potentially leading to cardiac arrest [19].

Animal Toxins: Zootoxins

Zootoxins, produced by animals for hunting or defense, are categorized as venoms or poisons. Venoms, delivered through bites, stings, or specialized structures like snake fangs or scorpion stingers, contain complex mixtures of proteins, peptides, and enzymes targeting specific physiological systems. Poisons, in contrast, harm predators or humans through ingestion or contact, as seen in brightly colored frogs or pufferfish [27]. Potent zootoxins are also found in jellyfish, cone snails, and certain caterpillars, showcasing their biological diversity and significance [28, 29].

Zootoxins from animals such as snakes, scorpions, marine creatures, and insects often exhibit severe cardiotoxic effects. These venoms composed of polypeptides, enzymes, and other molecules, disrupt cardiac physiology by causing arrhythmias, myocardial damage, or cardiac arrest. Cardiotoxicity arises from ion channel interference, enzymatic damage, or systemic inflammation. Polypeptides primarily target ion channels and plasma membranes, while enzymatic components induce effects like membrane lysis and pore formation. Interestingly, venom components affecting the cardiovascular system have also advanced our understanding of their physiological mechanisms and certain medical applications [30].

Snake Venoms

Snake venoms, composed of peptides and proteins, target vital systems, particularly the cardiovascular system, with effects ranging from cardiotoxic to cardioprotective. Key cardiotoxic compounds include bradykinin-potentiating peptides, sarafotoxins, three-finger toxins (e.g., cobra cardiotoxins), and phospholipases A2. Venom proteins like vascular endothelial growth factors also promote angiogenesis and exhibit hypotensive and cardioprotective effects. These components offer potential for innovative cardiovascular drugs aimed at preventing or mitigating pathological processes in cardiovascular diseases, which remain the leading cause of mortality worldwide. For instance, the bradykinin-potentiating peptide from *Bothrops jararaca* venom led to the development of antihypertensive drugs like captopril and enalapril. Snake venoms impact heart muscle, vascular smooth muscle, and capillary networks [30].

Cytotoxins, or **cardiotoxins**, are three-finger toxins made up of about 60 amino acids, are ubiquitously present in cobra venoms [31]. These toxins interact directly with cell membranes, forming ionic pores that cause cell depolarization and death, particularly in cardiac tissue, hence their alternate name. Despite similar amino acid sequences, their biological effects vary significantly. Cardiomyocyte responses to cardiotoxins begin with increased contraction, followed by decreased contraction strength and heightened resting tension, effects observed even at concentrations below 1 μM. These actions are more severe in ventricular than atrial tissue and typically lead to irreversible contracture and cell death. The toxins trigger a rise in intracellular calcium levels from intra- and extracellular sources, activating peptidases and disrupting cardiomyocyte structure. This cascade of events results in necrotic cell death [19, 31].

Sarafotoxins are short peptide toxins found in the venom of *Atractaspis snakes*, known for their potent vasoconstrictive effects. These peptides, resembling endothelins, bind specifically to endothelin receptors. By activating endothelin receptors, sarafotoxins cause intense vasoconstriction, leading to left ventricular dysfunction, bronchospasm, and increased airway resistance [32]. When administered intravenously, sarafotoxins induce rapid cardiac arrest and death in mice. The vascular response is primarily driven by extracellular calcium influx through L-type calcium channels, with a minor contribution from intracellular calcium release [30].

Phospholipases A2 in snake venoms are the only enzymes with direct effects on the cardiovascular system. These secreted lipolytic enzymes cause cardiotoxicity primarily by damaging cell membranes, rather than through the products of lipolysis. Phospholipases A2 can lower blood pressure by producing arachidonic acid, which leads to the formation of prostaglandins and prostacyclins. High doses of phospholipases A2 can disrupt myocardial tissue, impair cardiac function, and cause bradycardia and atrioventricular block. The cardiotoxicity of phospholipases A2 varies among snake species, with some, like those from *Ophiophagus hannah* and *Naja nigricollis*, causing intracellular damage, while others, like phospholipases A2 from *Naja atra*, do not have cardiotoxic effects. Phospholipases A2 generally have a vasorelaxant effect on blood vessels, mediated by increased cGMP in smooth muscle cells [30].

Scorpion Venoms

Scorpion envenomation affects over two billion people globally, with more than one million stings reported annually. The most severe consequence is acute heart failure, which can lead to cardiogenic shock or pulmonary edema, contributing to a 0.27% fatality rate. Cardiovascular dysfunction follows two phases: the "vascular phase," characterized by intense catecholamine-induced vasoconstriction, increases left ventricular afterload, impairing ventricular emptying and raising filling pressure. The "myocardial phase" follows, marked by reduced LV contractility, hypotension, and reversible right ventricular dysfunction. This phase resolves either spontaneously or with inotropic treatment. Scorpion-induced cardiomyopathy is followed by an increase in catecholamines that causes myocardial ischemia via coronary vasomotor disturbances. Cardiovascular effects are linked to endothelial damage, arrhythmias, and myocardial injury. The mechanisms remain under research, with some suggesting a systemic inflammatory response or peripheral vascular and myocardial effects as primary causes. The venom's toxic components include polypeptides that disrupt sodium conductance, stimulating norepinephrine and acetylcholine release [33].

Marine Venoms

There is an enormous diversity and complexity of venoms and poisons in marine animals. Exposure to these toxins can occur through the consumption of contaminated seafood or direct contact with marine living organisms [8]. Intoxication has occurred from envenoming by sea snakes, jellyfish, and venomous fish such as stonefish, cone snails, and blue-ringed octopus. Although rare, fatalities that do occur usually do so within 5–20 min of the sting of marine animals, as consequences rapid manifestation of different toxic effects, among which is cardiotoxicity [34, 35].

Jellyfish, venomous cnidarians, use their potent venom for predation or self-defense. Human encounters with jellyfish stings during activities like swimming or fishing can lead to intense pain, itching, swelling, inflammation, shock, or even death. The venom contains various toxic components, including phospholipase A2, metalloproteinase, and protease inhibitors, with the lethal toxin being the most dangerous. Though the exact nature and mechanism of the lethal toxin remain unclear, toxicological studies suggest that cardiotoxic effects, such as myocardial infarction and severe reductions in heart rate and blood pressure, contribute to fatal outcomes. Research by Li et al. [36] confirmed that venom lethality is primarily linked to metalloproteinase activity, rather than phospholipase A2 or hemolytic activity [36, 37].

Palytoxin, a potent cardiotoxin initially isolated from the coral *Palythoa toxica*, increases intracellular calcium levels ([Ca^{2+}]) in embryonic chick ventricular cells, a process dependent on extracellular calcium. This effect is not linked to depolarization, activation of voltage-dependent calcium channels, sodium accumulation, or acidification. Palytoxin also enhances sodium uptake via the Na^{+}/Ca^{2+} antiporter [38]. Studies by Frelin et al. [39] suggest that palytoxins effects on cardiac cells differ from its actions in erythrocytes and fibroblasts, indicating distinct mechanisms. A primary effect of palytoxin is the opening of a calcium uptake pathway in the plasma membrane, triggering sodium uptake through the Na^{+}/Ca^{2+} antiporter [39].

Insect Venoms

The venom of the honey bee (*Apis mellifera*) induces cardiovascular dysfunction, primarily due to melittin, while other components like apamin, mastoparan, phospholipase A2, histamine, and dopamine show minimal effects. The cardiotoxicity of bee venom is fully neutralized by antimelittin antibodies, confirming melittin as the main agent. Okamoto et al. [40] found that melittin or whole venom exposure halted spontaneous cardiac beating, increased systolic intracellular calcium concentration $[Ca^{2+}]i$, and caused morphological damage such as balloon degeneration with further $[Ca^{2+}]i$ elevation. Melittin also depolarized maximum diastolic potentials, suppressed action potential generation, and raised intracellular sodium concentration $[Na^{+}]i$. These effects were attributed to extracellular calcium influx through the sarcolemmal sodium-calcium exchange system [40].

Plant Toxins: Phytotoxins

Plants produce a wide range of secondary metabolites with toxic properties called phytotoxins. These toxins include thousands of different compounds that have adverse effects on humans and animals [4]. Among phytotoxins, cardiac glycosides and certain alkaloids are particularly notable for their effects on the cardiovascular system.

Cardiac Glycosides

Cardiac glycosides are naturally occurring compounds found in plants such as *Digitalis purpurea* (foxglove) and *Nerium oleander* (oleander) as a defense mechanism against herbivory and predation. Interestingly, certain insect species exploit the toxicity of cardiac glycosides by absorbing these compounds, thereby gaining protection from predators. The primary mechanism of cardiac glycosides toxicity involves the inhibition of Na^+/K^+-ATPase Na^+/K^+-ATPase pump on cardiac myocytes, leading to increased intracellular sodium and subsequent calcium influx through the sodium-calcium exchanger. This results in enhanced myocardial contractility (positive inotropy). However, excessive inhibition of Na^+/K^+-ATPase disrupts ion gradients, leading to arrhythmias, conduction blocks, and even cardiac arrest. Clinical manifestations include nausea, vomiting, and bradycardia, progressing to ventricular tachycardia or fibrillation at toxic doses. Despite their toxicity to humans and some animals, cardiac glycosides have been utilized as therapeutic agents for cardiovascular diseases. Recent research has also highlighted their potential antiviral and anticancer properties [41].

Alkaloids

Plant-derived alkaloids, such as aconitines from *Aconitum* species and veratridine from *Veratrum* species, target voltage-gated sodium channels. By prolonging sodium channel activation, these alkaloids cause sustained depolarization, leading to increased excitability and abnormal cardiac rhythms [42, 43]. Aconitum and Veratrum have long been used as treatments for coronary heart disease and chronic heart failure, serving as cardiac agents and pain relievers. However, research has also demonstrated their considerable toxicity to the heart, nervous system, and liver [42, 43].

Aconitines, toxic alkaloids from aconite plants, act as neurotoxins by opening TTX-sensitive sodium (Na^+) channels in cardiac and other tissues. High-affinity aconitines bind to receptor site II on voltage-gated Na^+ channels, prolonging their open state and causing continuous Na^+ influx, sustained depolarization, and tachyarrhythmias. These compounds also enhance synaptosomal sodium and calcium activity. In contrast, low-affinity aconitines are less toxic, decrease intracellular calcium activity, and cause bradycardia. Aconitine toxicity typically manifests as hypotension, ventricular arrhythmias, and syncope, often leading to fatal outcomes if untreated [42, 43].

Veratrum species contain over 200 alkaloids, the primary toxic agents responsible for their effects. Their toxicity largely stems from the opening of voltage-gated sodium (Na^+) channels. In rat studies, intravenous **veratridine**, a *Veratrum* component, induced a Bezold-Jarisch-like effect, marked by transient hypotension and bradycardia. Persistent sodium currents, amplified during myocardial ischemia, contribute to ischemic arrhythmias. Veratridine-induced prolonged cardiac action potentials facilitate early after depolarizations, triggering tachyarrhythmias. Increased Na^+ influx activates the Na^+/Ca^{2+} exchanger, causing calcium overload in cardiomyocytes. This overload induces late depolarization after-potentials, which can trigger new action potentials when reaching threshold levels, often linked to cytosolic calcium oscillations and arrhythmias. Veratrum alkaloids are now under study for their promising pharmaceutical potential in oncology and cardiology [42].

Taxine alkaloids found in yew plants (*Taxus* species), inhibit calcium and sodium channel activity in cardiomyocytes, resulting in bradycardia, hypotension, fatal cardiac arrhythmias and acute cardiogenic shock. In severe cases, exposure leads to cardiac arrest due to asystole [42].

Toxins from Microorganisms

Microorganisms produce a wide variety of toxins that can affect humans, animals, and other organisms. These toxins serve different purposes, such as aiding in microbial survival, competing with other species, or defending against predators. Microorganism toxins include bacterial, algal and fungal toxins [45].

Bacterial Toxins

Certain bacteria produce toxins that adversely affect the cardiovascular system.

Vibrio parahaemolyticus is a gram-negative, salt-loving bacterium predominantly found in seafood such as fish, shrimp, and shellfish across the globe. It is the primary pathogen responsible for seafood-related diarrheal illnesses in humans worldwide [46]. *Vibrio parahemolyticus* toxin (haemolysin) is fatal to rats when injected intravenously at doses of 5 μg/kg or higher. It decreases conductivity in both intra-atrial and ventricular tissues, leading to atrioventricular block, and causes ventricular flutter before cardiac arrest occurs. The toxin is also harmful to cardiomyocytes in culture, where it increases the beating rhythm of the cells, which then abruptly stops [19].

Streptolysin O, a pore-forming toxin released by most group A streptococci and some group C and G streptococci, creates cation-permeable pores in cholesterol-rich membranes. When administered intravenously in high doses, it can cause sudden cardiac arrest, likely due to nonspecific binding to the lipid bilayers of cardiac cells [47, 48].

Saxitoxin, produced by certain marine dinoflagellates (*Alexandrium* sp., *Gymnodinium* sp.) and cyanobacteria species (*Anabaena* sp., some *Aphanizomenon* spp., *Cylindrospermopsis* sp.), is responsible for paralytic shellfish poisoning when ingested through contaminated shellfish. Saxitoxin primarily blocks sodium channels by binding to site 1 [49] and has also been found to inhibit L-type calcium currents in adult mouse ventricular myocytes [50].

Tetrodotoxin is a toxin of microbial origin, likely produced by certain marine bacteria, particularly those in the *Vibrio* genus [51]. However, it remains uncertain whether these bacteria are the definitive source of tetrodotoxin in animals. Tetrodotoxin has been found in various species, such as puffer fish, toads, and octopuses. Research has shown that rat cardiomyocytes contain both high and low affinity receptors for tetrodotoxin (sodium channels). However, only the low affinity binding site is functional on cardiac cells, with a dissociation constant for tetrodotoxin approximately three orders of magnitude higher than that for tetrodotoxin receptors in muscle and nerve. To block cardiac sodium channels, a very high concentration of tetrodotoxin is required. Since cardiac myocytes express fast voltage-gated sodium channels, tetrodotoxin can block action potential generation and electrical activity, leading to the loss of myocardial excitability and cardiac arrest, even though sodium channels are generally not impacted in cases of intoxication [19, 51, 52].

Algal Toxins

Algae are widespread microorganisms found in aquatic environments, some of which periodically produce harmful "blooms." These blooms are linked to the release of potent toxic compounds. Several algal toxins, including brevetoxin, yessotoxin, ciguatoxin, and maitotoxin, have notable effects on the cardiovascular system [19].

Brevetoxin, a polyether neurotoxin produced by marine dinoflagellates, targets voltage-gated sodium channels present in excitable membranes, including cardiac tissues. By opening these sodium ion channels, brevetoxin causes uncontrolled sodium influx into cells, leading to depolarization, cardiac excitotoxicity and arrhythmias [53]. Human exposure to brevetoxin can occur through ingestion of brevetoxin-contaminated shellfish. In doses exceeding 25 μg/kg, brevetoxin induces heart block, ventricular extrasystoles, and idioventricular rhythms in rats. Studies have concluded that brevetoxin disrupts the cardiac conduction system. In artificially ventilated cats injected with brevetoxin, systemic toxin accumulation results in cardiovascular collapse and death [19].

Yessotoxins are polycyclic ether compounds produced by dinoflagellate phytoplankton and can accumulate in shellfish, posing a risk to humans through seafood consumption. Yessotoxins, along with related toxins like homoyessotoxin and 45-hydroxy homoyessotoxin, are lethal when administered to mice. Their exact cardiotoxic mechanisms are not fully understood, but in vitro studies suggest that they affect intracellular calcium, cyclic AMP levels, cytoskeletal and adhesion molecules, activate caspases, and open mitochondrial permeability transition pores [54, 55]. These effects lead to ultrastructural changes in cardiac tissue, including edema and swelling of cardiomyocytes, particularly after high doses of yessotoxins (*i.p.* 300 μg/kg or oral 10 mg/kg) [56, 57].

Ciguatoxin is a potent toxin produced by dinoflagellates like *Gambierdiscus toxicus*, causing ciguatera poisoning (human poisoning caused by consuming contaminated seafood), and primarily in tropical regions. It is lipid-soluble and accumulates in the muscle tissue of marine finfish, passing up the food chain to humans. The condition is characterized by gastrointestinal, neurological, and cardiovascular symptoms, and severe cases can lead to paralysis, coma, or death. Ciguatoxin activate sodium ion (Na^+) channels, increasing membrane excitability and causing cardiac instability. Experimental studies show that ciguatoxin cardiovascular effects are dose-dependent: low doses cause bradycardia and atrioventricular block, sublethal doses lead to ventricular tachycardia, and high doses can result in heart failure [58].

Maitotoxin, a key contributor to ciguatera poisoning, is produced by the marine dinoflagellate Gambierdiscus toxicus. Maitotoxin exhibits dose-dependent effects on the heart, including positive inotropic effects on heart tissue preparations and irreversible contracture of isolated rat cardiomyocytes. This contracture can be prevented using verapamil, a specific blocker of voltage-dependent Ca^{2+} channels. Maitotoxin induces a dose-dependent increase in Ca^{2+} activity in freshly isolated cardiomyocytes, an effect that can be inhibited by lowering the extracellular Ca^{2+} concentration or using verapamil. These findings suggest that maitotoxin specifically activates voltage-dependent Ca^{2+} channels, with the resulting Ca^{2+} influx into cells playing a critical role in its cardiotoxic effects [59].

Microcystins are secondary metabolites produced by cyanobacteria and are considered the most toxic cyanotoxins found in aquatic organisms. Under typical daily exposure conditions, cyanotoxins do not pose a significant risk to cardiovascular health. The likelihood of developing diseases in other organs, especially the liver, is greater under similar exposure circumstances. However, indirect cardiovascular effects through systemic inflammation and oxidative stress may occur as a result of damage to other organs [60]. The cardiotoxic effects of microcystins are primarily driven by increased reactive oxygen species (ROS), mitochondrial and endoplasmic reticulum oxidative stress, inhibition of serine/threonine protein phosphatases 1 and 2A, and disruption of cytoskeletal structures. These factors collectively contribute to the development of cardiovascular diseases [61].

Fungal Toxins: Mycotoxins

Mycotoxins are secondary metabolites produced by filamentous fungi, including *Aspergillus, Fusarium*, and *Penicillium* species [62]. Humans can be exposed to these toxins either by consuming contaminated food (peanuts, corn, and grains) or by inhaling airborne fungal spores [63]. Prominent examples of mycotoxins include aflatoxins, ochratoxin A, fumonisins, trichothecenes, zearalenone, and patulin, all of which can have severe health effects on both humans and animals [62]. Among mycotoxins, aflatoxin one of the most prevalent fungal toxins produced by *Aspergillus flavus* can lead to liver damage and increases the risk of developing liver cancer. In addition to effects on the liver, mycotoxins can also cause damage to other organs, while trichothecenes and ochratoxin A are predominantly responsible for cardiotoxic effects.

T-2 toxin, a highly toxic trichothecene mycotoxin, is linked to various cardiovascular diseases, causing cardiomyocyte injury, necrosis, hemorrhage, and inflammatory infiltration in the heart. Prolonged exposure can lead to cardiac fibrosis and dysfunction. Its mechanisms include the generation of reactive oxygen species (ROS), mitochondrial dysfunction, activation of the PPAR-γ and TGF-β1/Smad2/3 pathways, ER stress, and inflammatory responses. Recent studies highlight ferroptosis as a key contributor to cardiovascular damage, with T-2 toxin increasing ROS and malondialdehyde levels, while depleting glutathione and ATP. Ferroptosis markers such as elevated iron levels and reduced glutathione peroxidase 4 and ferritin heavy chain 1 were also observed. Inhibition of ferroptosis with Liproxstatin-1 reversed cardiac injury, while T-2 toxin's downregulation of heme oxygenase-1 (HO-1) worsened oxidative damage and ferroptosis [64].

Mushroom Toxins

Cardiovascular toxicity in mushrooms poisoning, albeit rare, has been documented in literature [25]. A wide variety of cardiac manifestations were found after mushrooms poisoning, ranging from the simple elevation of cardiac enzymes to ventricular tachycardia, acute heart failure, and myocarditis. Several fungal genera were shown cardiovasculat toxicity, including *Amanita* but also rarer ones like *Russula* and *Tricholoma* [25]. A recent study by Lecot et al. in which an analysis of 204 cases of cyclopeptide mushroom poisoning showed that patients who died and developed severe consequences after poisoning had associated cardiovascular diseases compared to patients who recovered [66].

Ostreolysin, an acidic protein from the oyster mushroom (*Pleurotus ostreatus*), acts as a toxic, pore-forming cytolysin. Intravenous ostreolysin administration causes significant cardiovascular and respiratory effects, including electrocardiographic changes and arterial blood pressure fluctuations. It initially raises arterial blood pressure transiently, followed by a decline to mid-circulatory levels, accompanied by bradycardia, myocardial ischemia, and ventricular extrasystoles. Ostreolysin lyses rat erythrocytes in vitro and in vivo, leading to hyperkalemia, which significantly contributes to its cardiotoxic effects. Its pore-forming nature may also directly affect cardiomyocytes or cardiac circulation. Additionally, ostreolysin has a concentration-dependent contractile effect on elastic blood vessels like the aorta and coronary vessels, further contributing to its cardiotoxicity [19].

Summarise of Mechanisms and Clinical Manifestations of Cardiovascular Toxicity

Cardiovascular toxicity arises from interactions between natural toxins and cardiac physiology. Key mechanisms include ion channel modulation, where toxins like brevetoxins and aconitine prolong sodium channel activation, causing abnormal depolarization, and certain venoms increase calcium influx, leading to prolonged contractions and arrhythmias. Disruption of cellular homeostasis by cardiac glycosides elevates intracellular calcium, enhancing contractility but inducing arrhythmias. Oxidative stress and mitochondrial dysfunction, triggered by toxins like ochratoxin A, deplete energy and induce apoptosis. Additionally, inflammatory responses from toxins exacerbate myocardial inflammation and impair contractility via pro-inflammatory cytokine release. Clinically, cardiovascular toxicity manifests as arrhythmias, ranging from mild palpitations to severe ventricular tachycardia or fibrillation, with toxins like aconitine increasing the risk of sudden cardiac death. Myocardial ischemia and infarction can result from coronary vasospasm or direct myocardial injury induced by bacterial toxins and snake venoms. Chronic exposure to cardiotoxic agents such as mycotoxins can lead to progressive heart failure due to sustained oxidative damage and structural myocardial remodelling. The Fig. 26.1 shows the mechanisms of toxicity and clinical manifestations of the most significant cardiotoxic natural toxins.

Therapeutic Strategies

Therapeutic strategies for addressing cardiovascular toxicity from natural toxins focus on early detection, rapid intervention, and tailored treatments. These include supportive care, antidotes, toxin-specific therapies, pharmacological approaches, and emerging treatments, depending on the toxin type, exposure route, and severity. Supportive care is crucial in cases with cardiotoxic effects, aiming to stabilize vital functions, such as airway, breathing, and circulation. Antidotes and toxin-specific

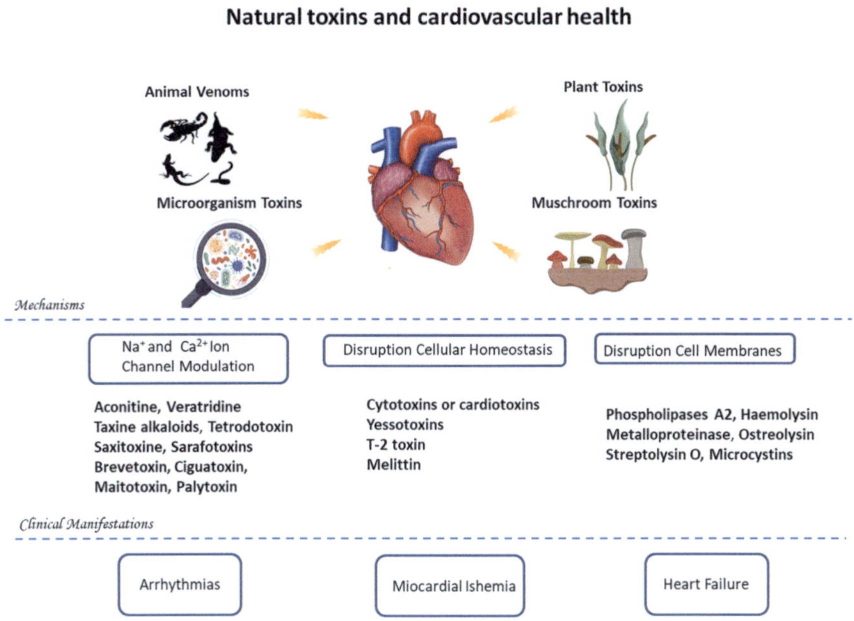

Fig. 26.1 Mechanisms of toxicity and clinical manifestations of the most significant cardiotoxic natural toxins

therapies help neutralize toxins, with antivenoms used for snake, scorpion, or spider envenomations, and toxin-specific therapies for ingested cardiac glycosides. Specific treatments often target disrupted biochemical pathways, such as drugs that modulate ion channels affected by cardiotoxins. Symptomatic management involves pain relief, anti-inflammatory agents, and medications like beta-blockers or calcium channel blockers to control arrhythmias or other cardiovascular issues. Emerging approaches, including monoclonal antibodies, small molecule inhibitors, and gene therapy, aim to counteract toxicity at the molecular level. Additionally, developing toxin-based drugs for cardiovascular or neurological conditions is being explored. Prevention through education, public awareness, proper food handling and protective measures in high-risk areas remains essential [60–62].

Conclusion

Natural toxins significantly impact cardiovascular health, with effects ranging from mild arrhythmias to severe and often fatal cardiac dysfunctions. These toxins, derived from diverse sources such as animals, plants, fungi, bacteria, algae and mushrooms operate through complex mechanisms, including ion channel modulation, disrupting cardiac cell membranes through oxidative stress, and disruption of cellular homeostasis. The clinical manifestations, including arrhythmias, myocardial infarction, and heart failure, underscore the critical importance of understanding these toxic effects. Effective management strategies depend on early detection and tailored therapeutic approaches, ranging from supportive care to innovative treatments like monoclonal antibodies and gene therapy. Additionally, advancements in toxin research not only improve therapeutic options but also inspire the development of toxin-derived drugs for treating cardiovascular and other systemic diseases. Despite their risks, natural toxins continue to hold biomedical potential, underscoring the importance of balancing their therapeutic benefits against their toxicological challenges.

Aknowledgement This research was funded by the Ministry of Science, Technological Development and Innovation, Republic of Serbia, through Grant Agreement with University of Belgrade-Faculty of Pharmacy (451-03-65/2024-03/200161).

References

1. World Health Organization (WHO) (2023) Natural toxins in food. https://www.who.int/news-room/fact-sheets/detail/natural-toxins-in-food
2. FDA Natural Toxins in Food 2024. https://www.fda.gov/food/chemical-contaminants-pesticides/natural-toxins-food
3. Fletcher MT, Netzel G (2020) Food safety and natural toxins. Toxins (Basel) 12(4):236. https://doi.org/10.3390/toxins12040236. PMID: 32276351; PMCID: PMC7232292
4. Nwaji AR, Arierib O, Anyangc AS, Nguediad K, Abiadee EB, Forcadosf GE, Oladipof OO, Makamaf S, Elishag IL, Ozelef N, Gotep JG (2022) Natural toxins and One Health: a review. Sci One Health 1:100013
5. Liu K, Wang M, Herzig V, Liu Z, Hu W, Zhou G, Duan Z (2016) Venom from the spider Araneus ventricosus is lethal to insects but inactive in vertebrates. Toxicon 115:63–69. https://doi.org/10.1016/j.toxicon.2016.03.010. Epub 2016 Mar 16. PMID: 26995209
6. Hansen HCB, Hilscherova K, Bucheli TD (2021) Natural toxins: environmental contaminants calling for attention. Environ Sci Eur 33:112. https://doi.org/10.1186/s12302-021-00543-6
7. Yang Y, Wei S, Zhang B, Li W (2021) Recent progress in environmental toxins-induced cardiotoxicity and protective potential of natural products. Front Pharmacol 12:699193. https://doi.org/10.3389/fphar.2021.699193. PMID: 34305607; PMCID: PMC8296636
8. Parak M, Asgari A, Hasani Nourian Y, Ghanei M (2024) A review of poisoning with various types of biotoxins and its common clinical symptoms. Toxicon 240:107629. https://doi.org/10.1016/j.toxicon.2024.107629. Epub 2024 Feb 8. PMID: 38336277
9. Akinniyi G, Akinboye AJ, Yang I, Lee JG (2024) Plant proteins, peptides, and non-protein amino acids: toxicity, sources, and analysis. Heliyon 10(15):e34890. https://doi.org/10.1016/j.heliyon.2024.e34890. PMID: 39145010; PMCID: PMC11320209
10. Still KR, Mohapatra A (2020, ISBN 9780128137246) Biotoxins. In: Wexler P (ed) Information resources in toxicology, 5th edn. Academic Press, pp 117–132
11. Afroz A, Siddiquea BN, Chowdhury HA, Jackson TN, Watt AD (2024) Snakebite envenoming: a systematic review and meta-analysis of global morbidity and mortality. PLoS Negl Trop Dis 18(4):e0012080. https://doi.org/10.1371/journal.pntd.0012080. PMID: 38574167; PMCID: PMC11020954

12. Dong H, Aziz MT, Richardson SD (2023) Transformation of algal toxins during the oxidation/disinfection processes of drinking water: from structure to toxicity. Environ Sci Technol 57(35):12944–12957. https://doi.org/10.1021/acs.est.3c01912. Epub 2023 Aug 21. PMID: 37603687
13. McGee RG, Webster AC, Lewis SR, Welsford M (2023) Interventions for the symptoms and signs resulting from jellyfish stings. Cochrane Database Syst Rev 6(6):CD009688. https://doi.org/10.1002/14651858.CD009688.pub3. PMID: 37272501; PMCID: PMC10240560
14. Rawson AM, Dempster AW, Humphreys CM, Minton NP (2023) Pathogenicity and virulence of Clostridium botulinum. Virulence 14(1):2205251. https://doi.org/10.1080/21505594.2023.2205251. PMID: 37157163; PMCID: PMC10171130
15. Horowitz BZ, Moss MJ (2024) Amatoxin mushroom toxicity. [Updated 2023 Jun 26]. In: StatPearls [Internet]. StatPearls Publishing, Treasure Island, FL. https://www.ncbi.nlm.nih.gov/books/NBK431052/
16. Tran HH, Juergens AL (2024) Mushroom toxicity. [Updated 2023 Aug 7]. In: StatPearls [Internet]. StatPearls Publishing, Treasure Island (FL). https://www.ncbi.nlm.nih.gov/books/NBK537111/
17. Peng X, Liu KT, Chen JB, Yan ZH, Danso B, Wang MK, Peng ZY, Xiao L (2024) Jellyfish stings: a review of skin symptoms, pathophysiology, and management. Med Sci Monit 30:e944265. https://doi.org/10.12659/MSM.944265. PMID: 39074073; PMCID: PMC11297359
18. Stegelmeier BL, Davis TZ, Clayton MJ (2020) Plant-induced photosensitivity and dermatitis in livestock. Vet Clin North Am Food Anim Pract 36(3):725–733. https://doi.org/10.1016/j.cvfa.2020.08.008. PMID: 33032702
19. Frangež R, Grandič M, Vrecl M (2012) Cardiovascular pathophysiology produced by natural toxins and their possible therapeutic implications. In: Cardiotoxicity of oncologic treatments. InTech. https://doi.org/10.5772/35079
20. World Health Organization (WHO) (2023) Snakebite envenoming. https://www.who.int/news-room/fact-sheets/detail/snakebite-envenoming
21. Feola A, Perrone MA, Piscopo A, Casella F, Della Pietra B, Di Mizio G (2020) Autopsy findings in case of fatal scorpion sting: a systematic review of the literature. Healthcare (Basel) 8(3):325. https://doi.org/10.3390/healthcare8030325. PMID: 32899951; PMCID: PMC7551928
22. Gummin DD, Mowry JB, Beuhler MC, Spyker DA, Rivers LJ, Feldman R, Brown K, Pham NPT, Bronstein AC, DesLauriers C (2024) 2023 Annual Report of the National Poison Data System® (NPDS) from America's Poison Centers®: 41st Annual Report. Clin Toxicol (Phila) 62(12):793–1027. https://doi.org/10.1080/15563650.2024.2412423. Epub 2024 Dec 17. PMID: 39688840
23. Nițescu GV, Grama A, Turcu T, Strătulă A, Dragomirescu A, Pană ES, Baciu A, Baconi DL, Crăciun MD, Ulmeanu CE (2024) Epidemiology and clinical characteristics of acute plant exposure in patients aged between 0 and 18 years—a six-year retrospective study. Children (Basel) 11(3):271. https://doi.org/10.3390/children11030271. PMID: 38539306; PMCID: PMC10969538
24. Benkerroum N (2020) Chronic and acute toxicities of aflatoxins: mechanisms of action. Int J Environ Res Public Health 17(2):423. https://doi.org/10.3390/ijerph17020423. PMID: 31936320; PMCID: PMC7013914
25. Balice G, Boksebeld M, Barrier Q, Boccalini S, Kassai-Koupai B, Paret N, Grenet G (2024) Mushroom poisoning-related cardiac toxicity: a case report and systematic review. Toxins (Basel) 16(6):265. https://doi.org/10.3390/toxins16060265. PMID: 38922159; PMCID: PMC11209194
26. GBD 2019 Antimicrobial Resistance Collaborators (2022) Global mortality associated with 33 bacterial pathogens in 2019: a systematic analysis for the Global Burden of Disease Study 2019. Lancet 400:2221–2248
27. Sharon M, Gwaltney-Brant. (2017) Zootoxins. In: Gupta RC (ed) Reproductive and developmental toxicology, 2nd edn. Academic Press, pp 963–972., ISBN 9780128042397. https://doi.org/10.1016/B978-0-12-804239-7.00049-4

28. Bjørn-Yoshimoto WE, Ramiro IBL, Yandell M, McIntosh JM, Olivera BM, Ellgaard L, Safavi-Hemami H (2020) Curses or cures: a review of the numerous benefits versus the biosecurity concerns of conotoxin research. Biomedicines 8(8):235. https://doi.org/10.3390/biomedicines8080235. PMID: 32708023; PMCID: PMC7460000
29. Nagy AL, Ardelean S, Chapuis RJJ, Bouillon J, Pivariu D, De Felice B, Bertazzo M, Fossati P, Spicer LJ, Dreanca AI, Caloni F (2024) Zootoxins and domestic animals: a european view. Toxins (Basel) 16(1):48. https://doi.org/10.3390/toxins16010048. PMID: 38251264; PMCID: PMC10818608
30. Averin AS, Utkin YN (2021) Cardiovascular effects of snake toxins: cardiotoxicity and cardioprotection. Acta Nat 13(3):4–14. https://doi.org/10.32607/actanaturae.11375. PMID: 34707893; PMCID: PMC8526186
31. Kalita B, Utkin YN, Mukherjee AK (2022) Current insights in the mechanisms of cobra venom cytotoxins and their complexes in inducing toxicity: implications in antivenom therapy. Toxins (Basel) 14(12):839. https://doi.org/10.3390/toxins14120839. PMID: 36548736; PMCID: PMC9780984
32. Neto A, Soares AM, Vale N, de C Gomes PA, Marangoni S, de A Calderon L, Da Silva SL (2017) Snake venom peptides and low mass proteins: molecular tools and therapeutic agents. Curr Med Chem 24(30):3254–3282. https://doi.org/10.2174/0929867323666161028155611. PMID: 27804880
33. Mazzei de Dàvila CA, Dàvila DF, Donis JH, de Bellabarba GA, Villarreal V, Barboza JS (2002) Sympathetic nervous system activation, antivenin administration and cardiovascular manifestations of scorpion envenomation. Toxicon 40(9):1339–1346. https://doi.org/10.1016/s0041-0101(02)00145-9. PMID: 12220720
34. Currie BJ (2003) Marine antivenoms. J Toxicol Clin Toxicol 41(3):301–308. https://doi.org/10.1081/clt-120021115. PMID: 12807313
35. Currie BJ, Jacups SP (2005) Prospective study of Chironex fleckeri and other box jellyfish stings in the "Top End" of Australia's Northern Territory. Med J Aust 183(11–12):631–636. https://doi.org/10.5694/j.1326-5377.2005.tb00062.x. PMID: 16336157
36. Li R, Yu H, Yue Y, Li P (2018) Combined proteome and toxicology approach reveals the lethality of venom toxins from jellyfish Cyanea nozakii. J Proteome Res 17(11):3904–3913. https://doi.org/10.1021/acs.jproteome.8b00568. Epub 2018 Oct 8. PMID: 30223649
37. D'Ambra I, Lauritano C (2020) A review of toxins from Cnidaria. Mar Drugs 18(10):507. https://doi.org/10.3390/md18100507. PMID: 33036158; PMCID: PMC7600780
38. Frelin C, Vigne P, Breittmayer JP (1990) Palytoxin acidifies chick cardiac cells and activates the Na^+/H^+ antiporter. FEBS Lett 264(1):63–66. https://doi.org/10.1016/0014-5793(90)80765-b. PMID: 1970964
39. Frelin C, Vigne P, Breittmayer JP (1990) Mechanism of the cardiotoxic action of palytoxin. Mol Pharmacol 38(6):904–909. PMID: 1979145
40. Okamoto T, Isoda H, Kubota N, Takahata K, Takahashi T, Kishi T, Nakamura TY, Muromachi Y, Matsui Y, Goshima K (1995) Melittin cardiotoxicity in cultured mouse cardiac myocytes and its correlation with calcium overload. Toxicol Appl Pharmacol 133(1):150–163. https://doi.org/10.1006/taap.1995.1136. PMID: 7597704
41. Bejček J, Jurášek M, Spiwok V, Rimpelová S (2021) Quo vadis cardiac glycoside research? Toxins (Basel) 13(5):344. https://doi.org/10.3390/toxins13050344. PMID: 34064873; PMCID: PMC8151307
42. Siegle J, Pietsch J (2024) Determination of Veratrum alkaloid contents in three Veratrum species by HPLC-MS/MS. Phytochem Anal 35(7):1577–1586. https://doi.org/10.1002/pca.3401. Epub 2024 Jun 11. PMID: 38863228
43. Mi L, Li Y-C, Sun M-R, Zhang P-L, Li Y, Yang H (2021) A systematic review of pharmacological activities, toxicological mechanisms and pharmacokinetic studies on aconitum alkaloids. Chinese Journal of Natural Medicines 19(7):505-520. ISSN 1875-5364. https://doi.org/10.1016/S1875- 5364(21)60050-X.

44. Schreiber N, Manninger M, Pätzold S, Reisinger AC, Hatzl S, Hackl G, Högenauer C, Eller P (2022) Cardiogenic shock due to yew poisoning rescued by VA-ECMO: case report and literature review. Channels (Austin) 16(1):167–172. https://doi.org/10.1080/19336950.2022.2104886. PMID: 35942511; PMCID: PMC9367666
45. Shier WT (2020) Handbook of toxinology (e-book.). CRC Press. https://doi.org/10.1201/9781003066538
46. Li L, Meng H, Gu D, Li Y, Jia M (2019) Molecular mechanisms of Vibrio parahaemolyticus pathogenesis. Microbiol Res 222:43–51. https://doi.org/10.1016/j.micres.2019.03.003. Epub 2019 Mar 8. PMID: 30928029
47. Harvey AL (1990) Cytolytic toxins. In: Shier TW, Mebs D (eds) Handbook of toxinology. Marcel Dekker Inc., New York, pp 1–66
48. Kobayashi M, Nishimura M, Kawamura M, Kamemura N, Nagamune H, Tabata A (2020) Change in membrane potential induced by streptolysin O, a pore-forming toxin: flow cytometric analysis using a voltage-sensitive fluorescent probe and rat thymic lymphocytes. Microbiol Immunol 64(1):10–22. https://doi.org/10.1111/1348-0421.12748. Epub 2019 Nov 7. PMID: 31584206
49. Mebs D, Hucho F (1990) Toxins acting on ion channels and synapses. In: Shier TW, Mebs D (eds) Handbook of toxinology. Marcel Dekker Inc., New York, pp 493–597
50. Su Z, Sheets M, Ishida H, Li F, Barry WH (2024) Saxitoxin blocks L-type ICa. J Pharmacol Exp Ther 308(1):324-329. https://doi.org/10.1124/jpet.103.056564. Epub 2003 Oct 17. PMID: 14566004.
51. Bane V, Lehane M, Dikshit M, O'Riordan A, Furey A (2014) Tetrodotoxin: chemistry, toxicity, source, distribution and detection. Toxins (Basel) 6(2):693–755. https://doi.org/10.3390/toxins6020693. PMID: 24566728; PMCID: PMC3942760
52. Lee CH, Ruben PC (2008) Interaction between voltage-gated sodium channels and the neurotoxin, tetrodotoxin. Channels (Austin) 2(6):407–412. https://doi.org/10.4161/chan.2.6.7429. Epub 2008 Nov 12. PMID: 19098433
53. Purkerson SL, Baden DG, Fieber LA (1999) Brevetoxin modulates neuronal sodium channels in two cell lines derived from rat brain. Neurotoxicology 20(6):909–920. PMID: 10693972
54. Dominguez HJ, Paz B, Daranas AH, Norte M, Franco JM, Fernández JJ (2010) Dinoflagellate polyether within the yessotoxin, pectenotoxin and okadaic acid toxin groups: characterization, analysis and human health implications. Toxicon 56(2):191–217. https://doi.org/10.1016/j.toxicon.2009.11.005. Epub 2009 Nov 17. PMID: 19925818
55. Tubaro A, Dell'ovo V, Sosa S, Florio C (2010) Yessotoxins: a toxicological overview. Toxicon 56(2):163–172. https://doi.org/10.1016/j.toxicon.2009.07.038. Epub 2009 Aug 4. PMID: 19660487
56. Aune T, Sørby R, Yasumoto T, Ramstad H, Landsverk T (2002) Comparison of oral and intraperitoneal toxicity of yessotoxin towards mice. Toxicon 40(1):77–82. https://doi.org/10.1016/s0041-0101(01)00192-1. PMID: 11602282
57. Terao K, Ito E, Oarada M, Murata M, Yasumoto T (1990) Histopathological studies on experimental marine toxin poisoning-5. The effects in mice of yessotoxin isolated from Patinopecten yessoensis and of a desulfated derivative. Toxicon 28(9):1095–1104. https://doi.org/10.1016/0041-0101(90)90148-z. PMID: 2260107
58. Lehane L, Lewis RJ (2000) Ciguatera: recent advances but the risk remains. Int J Food Microbiol 61(2–3):91–125. https://doi.org/10.1016/s0168-1605(00)00382-2. PMID: 11078162
59. Santostasi G, Kutty RK, Bartorelli AL, Yasumoto T, Krishna G (1990) Maitotoxin-induced myocardial cell injury: calcium accumulation followed by ATP depletion precedes cell death. Toxicol Appl Pharmacol 102(1):164–173. https://doi.org/10.1016/0041-008x(90)90093-a. PMID: 2296767
60. Svirčev Z, Chen L, Sántha K, Drobac Backović D, Šušak S, Vulin A, Palanački Malešević T, Codd GA, Meriluoto J (2022) A review and assessment of cyanobacterial toxins as cardiovascular health hazards. Arch Toxicol 96(11):2829–2863. https://doi.org/10.1007/s00204-022-03354-7. Epub 2022 Aug 23. PMID: 35997789; PMCID: PMC9395816

61. Cao L, Massey IY, Feng H, Yang F (2019) A review of cardiovascular toxicity of microcystins. Toxins (Basel) 11(9):507. https://doi.org/10.3390/toxins11090507. PMID: 31480273; PMCID: PMC6783932
62. Khan R, Anwar F, Ghazali FM (2024) A comprehensive review of mycotoxins: Toxicology, detection, and effective mitigation approaches. Helyion 10(8):e28361. https://doi.org/10.1016/j.heliyon.2024.e28361.
63. Brown R, Priest E, Naglik JR, Richardson JP (2021) Fungal toxins and host immune responses. Front Microbiol 12:643639. https://doi.org/10.3389/fmicb.2021.643639. PMID: 33927703; PMCID: PMC8076518.
64. Wang C, He J, Jin H, Xiao H, Peng S, Xie J, Zhang L, Guo J (2023) T-2 toxin induces cardiotoxicity by activating ferroptosis and inhibiting heme oxygenase-1. Chemosphere 341:140087. https://doi.org/10.1016/j.chemosphere.2023.140087. Epub 2023 Sep 5. PMID:37678596.
65. Lecot J, Cellier M, Courtois A, Vodovar D, Le Roux G, Landreau A, Labadie M, Bruneau C, Descatha A (2023) Cyclopeptide mushroom poisoning: a retrospective series of 204 patients. Basic Clin Pharmacol Toxicol 132(6):533–542. https://doi.org/10.1111/bcpt.13858. Epub 2023 Mar 20. PMID: 36908014
66. Bittenbinder MA, van Thiel J, Cardoso FC, Casewell NR, Gutiérrez JM, Kool J, Vonk FJ (2024) Tissue damaging toxins in snake venoms: mechanisms of action, pathophysiology and treatment strategies. Commun Biol 7(1):358. https://doi.org/10.1038/s42003-024-06019-6. PMID: 38519650; PMCID: PMC10960010
67. Găman MA, Egbuna C (2024.; ISBN 978-0-12-824472-2) Antidotes to toxins and drugs—from natural source to drug discovery in toxicology. Elsevier. https://doi.org/10.1016/C2020-0-02016-5
68. Shapira A, Benhar I (2010) Toxin-based therapeutic approaches. Toxins (Basel) 2(11): 2519–2583. https://doi.org/10.3390/toxins2112519. Epub 2010 Oct 28. PMID: 22069564; PMCID: PMC3153180

Part III
Genetics, Epigenetics, Environmental Pollution and Nutritional Effects in the Pathogenesis of Cardiovascular Toxicity

Chapter 27
Predisposing Role of Genetic and Epigenetic Factors in the Pathogenesis of Drug-Induced Cardiovascular Toxicity

Yssel Mendoza-Mari and Devendra K. Agrawal

Abstract The cardiovascular toxicity of drugs constitutes an important area of attention for clinicians, drug developers and general population. Numerous medications currently used to treat severe health issues like cancer, diabetes, AIDS and mental illness, have severe limitations due to mild to severe cardiac complications, directly attributable to the treatment. Some of these side effects disappear when the treatment is stopped. However, in other cases, cardiotoxic symptoms begin to appear much later. Susceptibility to cardiotoxic drugs is highly influenced by pre-existing conditions but during the last decades genetic and geographic studies have shown that genetic polymorphisms or ethnic background play significant predictive or prognostic roles in cardiovascular toxicity. The aim of this chapter is to summarize some examples that illustrate the predisposing role of genetic and epigenetic factors in increasing the pathogenesis of drug-induced cardiovascular toxicity.

Keywords Cardiotoxic drugs · Epigenetic factors · Gene polymorphism · Genetic factors

Introduction

Cardiotoxicity as a general term that refers to the damage of the structure and functioning of the heart caused by a toxic substance, that could be a pharmacological drug to treat a medical condition, recreational or illicit drugs, alcohol, among others. Each group of drug elicits its own mechanism of action in terms of toxicity (Fig. 27.1), but the final effect is similar, cardiac failure characterized by contractile dysfunction or ventricular arrhythmias, both of which lead to heart failure and sudden death. Sometimes, the consequences are not visible in the short term, but rather a cumulative effect occurs that can lead to cardiac dysfunction after a long

Y. Mendoza-Mari · D. K. Agrawal (✉)
Department of Translational Research, College of Osteopathic Medicine of the Pacific, Western University of Health Sciences, Pomona, CA, USA
e-mail: dagrawal@westernu.edu

D. K. Agrawal et al. (eds.), *Cardiovascular Toxicity*, Advances in Biochemistry in Health and Disease 33, https://doi.org/10.1007/978-3-031-90598-8_27

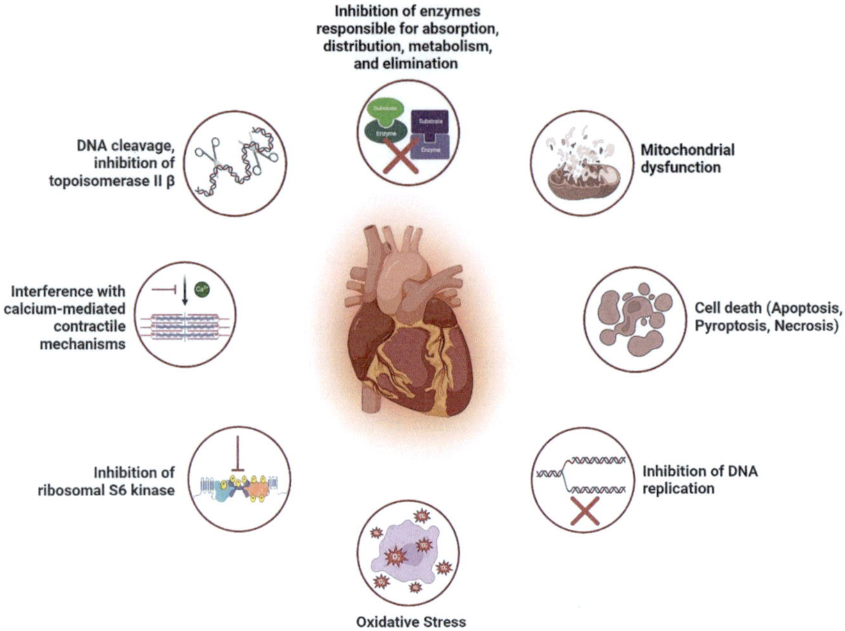

Fig. 27.1 Molecular mechanisms elicited by cardiotoxic drugs contributing to the development of cardiovascular complications. (Created in BioRender.com)

period of treatment, even long after it has stopped. This unpredicted drug-induced cardiotoxicity represents a significant social burden in terms of hospitalizations and deaths, as well as a substantial economic loss due to the possible withdrawal of drugs from the market after the occurrence of adverse cardiovascular events. That is why it is mandatory to carry out the study of possible toxic cardiovascular effects of any drug under study during the preclinical and clinical research of the product.

Known cardiotoxic drugs such as cytostatics or common use pills as non-steroidal anti-inflammatory drugs can cause damage to the heart. And not only pre-existing medical conditions, like diabetes or hypertension, are responsible for the increased susceptibility to cardiotoxic drugs in each patient. Also pre-existing genetic mutations and epigenetic modifications are directly implied in this mechanism. Single nucleotide polymorphisms (SNPs) are common, variants occurring either in coding or in non-coding DNA regions. SNPs represent variations in the genomic sequence that can contribute to the functional diversity of the proteins among human population. Genetic and geographic studies showed that gen polymorphisms or ethnic differences play significant predictive or prognostic roles in cardiovascular toxicity. Genetic variants may increase or reduce the potential toxicity of a given drug, determining how successful or detrimental the treatment will be. MicroRNAs (miRNAs) also play an important role in heart pathologies, their diagnosis and prevention [1] and, similarly, they have been found associated with cardiotoxicity [2]. miRNAs specific for cardiac tissue modify their expression under the toxic

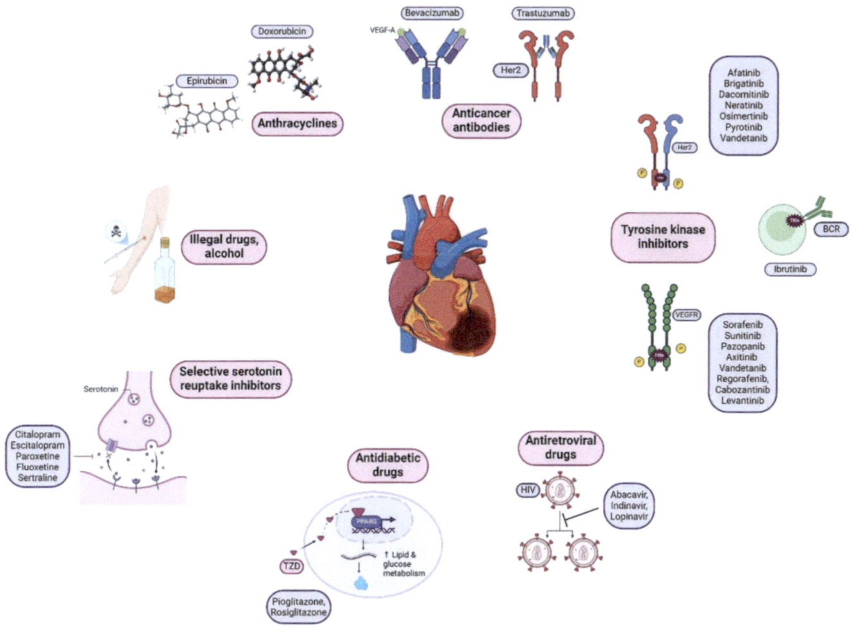

Fig. 27.2 Cardiotoxic drugs contributing to the development of cardiovascular complications. (Created in BioRender.com)

effects of drugs and this can be used for *in vitro* testing the toxicity of potential drug candidates. Besides, some of these drug-induced changes could be directly responsible for heart damage, considering their putative regulatory effect on the expression of physiological or cardioprotective genes. The present chapter addresses some examples of the genetic and epigenetic conditions associated to an increased susceptibility of different cardiotoxic substances like regularly used anticancer, antidiabetic and antiretroviral drugs, recreational and illegal drugs and alcohol (Fig. 27.2).

Anticancer Drugs

Cardiotoxicity associated with anticancer drugs constitutes an important source for non-cancer related mortality among these patients [3]. The incidence rate varies with dosage and type of drug and depends on pre-existing conditions, that is why specialists should carry out a meticulous assessment of the risk factors of the patient to significantly reduce cardiovascular complications. In the last years, it has been demonstrated that not only previous cardiovascular disease and cardiotoxic therapies, lifestyle risk factors, hypertension, and diabetes contribute to cardiotoxicity during an anticancer treatment [4]. Additionally, individual pre-existing genetic mutations in genes involved in the absorption, distribution, metabolism, and elimination (ADME) of drugs may contribute to affect drug response, in terms of pharmacokinetic and pharmacodynamic [5]. Epigenetic modifications also contribute to reinforcing the effects of anticancer drug-induced toxicity through DNA methylation, histone methylation, histone acetylation, RNA methylation, and non-coding RNA (ncRNA) [6]. Chemotherapy-associated cardiotoxicity is divided into type I and type II. Type I is caused by drugs like anthracyclines, is irreversible and typically associated with significant changes observed during biopsy analysis. Type II cardiotoxicity exerted by drugs like trastuzumab, is not dose-dependent, reversible upon discontinuation of the drug, and occurs without or with minimal ultrastructural changes [7].

Anthracycline

Anthracyclines induce cardiotoxicity by generating reactive oxygen species (ROS), damaging mitochondrial function and inhibiting topoisomerase II-beta (Top2β) [8]. Regarding drug transport genes, adenosine triphosphate-binding cassette (ABC) transporter proteins play active roles in transporting multiple drugs, including anthracyclines, across cellular membranes [9]. Within the myocardium, ABC transporters facilitate the export of various chemotherapeutic agents from cardiac cells. It has been demonstrated that at least eight different variants in five ABC genes have been related to defects in drug export, resulting in the accumulation of anthracycline within the cardiomyocytes and increasing the risk of cardiac dysfunction [10]. Gly671Val variants of doxorubicin (DOX) efflux transporter multidrug-resistant protein 1 and functionally similar multidrug resistant protein 2 are associated with acute anthracyclines induced cardiotoxicity [11]. From a metabolic point of view, it is well stablished that UDP-glucuronosyltransferases (UGTs) mediate the glucuronidation of endogenous and exogenous compounds, increasing their water solubility to facilitate elimination. Specifically, UGT1A6 plays a crucial role in the clearance of anthracene derivatives as it converts these molecules into water-soluble and excretable metabolites. In this sense, the *UGT1A6* rs17863783 variant has been associated with anthracyclines induced cardiotoxicity [12]. *GSTM1*, one of the glutathione S-transferases (GST), encodes the M1 enzyme and catalyzes the detoxification of several carcinogens and drugs, including anthracene derivatives and scavenges the free radicals generated by these compounds [13]. It has been shown that deletions on this gene in cancer patients (*GSTM1* null genotype) are associated with anthracycline-related cardiomyopathy [14]. Polymorphisms of the carbonyl reductase 3 coding gene (*CBR3*), which catalyzes the reduction of anthracyclines to cardiotoxic alcohol metabolites, can influence the synthesis of this metabolite. V244M polymorphism generates two protein isoforms which promote doxorubicinol formation at distinct catalytic rates making, accordingly, patients with the *CBR3*:GG genotype present a significantly increased risk of cardiotoxicity compared with patients with the CBR3:GA/AA genotype [15]. In another study, it was revealed that the single nucleotide polymorphism (SNP) rs1056892 is associated with a decreased left ventricular ejection fraction (LVEF) induced by anthracyclines [16].

There are some other enzymes that play key roles in general mechanisms in the cell like antioxidation, DNA repair, and sarcomere dysfunction, with genetic variants linked to an increased risk of cardiotoxicity after anthracycline treatment. For example, hyaluronic acid synthase (HAS), that protects myocardial cells from ROS damage, preventing the activation of death receptors, which ensures cardiomyocyte survival and function [17]. Wang et al. identified that the rs2232228 SNP in *HAS3* is closely associated with anthracycline dose-dependent cardiac injury [18]. On the other hand, the variant rs2229774 in the coding region of the retinoic acid receptor gene *RARG* induces the expression of *Top2b* gen, which encodes for the Top2β enzyme, resulting in a 4.7-fold increased risk of anthracycline related cardiotoxicity [19]. Splicing regulatory factors belonging to CELF family influence the cardiac structure by modulating tissue-specific splicing events. CELF activity promotes the appearance of distinct cardiac troponin T variants, leading to the dysregulated contraction of myocardial sarcomeres, and diminished myocardial contractility. A genome-wide association study in pediatric patients confirmed that the SNP rs1786814 located in the *CELF4* gene is associated with a higher cardiotoxic risk, due to its correlation with the presence of more than one alternatively spliced isoforms of the cardiac troponin T encoding gene *TNNT2* [20]. Besides, it has been demonstrated that truncated variants of titin encoding gene *TTN*, a sarcomeric scaffold protein which regulates cardiac contraction, are one of the most important causes of anthracycline induced cardiotoxicity in both pediatric and adult cancer patients [21].

Anticancer drugs also promote a change in the activity of epigenetic modifying enzymes, which is closely related to an increased susceptibility of a given patient to develop cardiotoxic symptoms. It has been shown that methylation of genes involved in cardiac dysfunction are downregulated in response to epirubicin [22] and DOX [23] anticancer drugs. This modification is driven by DNA methyltransferases, which catalyze the addition of a methyl group mostly in CpG islands located at the promoter regions. Also, histone methyltransferases and histone demethylases are involved in cardiotoxicity. Histone methylation typically occurs at arginine and lysine residues of H3 and H4 and this modification can activate or inhibit gene transcription. After a DOX treatment, the level of histone methylation modifying enzymes is changed, further contributing to pathological mechanisms like mitochondrial dysfunction and apoptosis in cardiomyocytes [24, 25]. Another set of enzymes, histone deacetylases (HDACs) and histone acetyl transferases mediate histone acetylation, which in turn regulates cardiac normal physiological functions and cardiovascular diseases [26]. Anti-cancer drugs like DOX modulate the activity of these enzymes, increasing cardiac inflammation, oxidative stress, fibrosis, apoptosis, and DNA damage in cardiomyocytes [27, 28]. A very detailed information regarding the impact of genetic modifications on anthracycline metabolism related genes on cardiovascular toxicity can be consulted at [29]. Other important epigenetic elements directly influenced by anti-cancer drugs are non-coding RNAs like micro

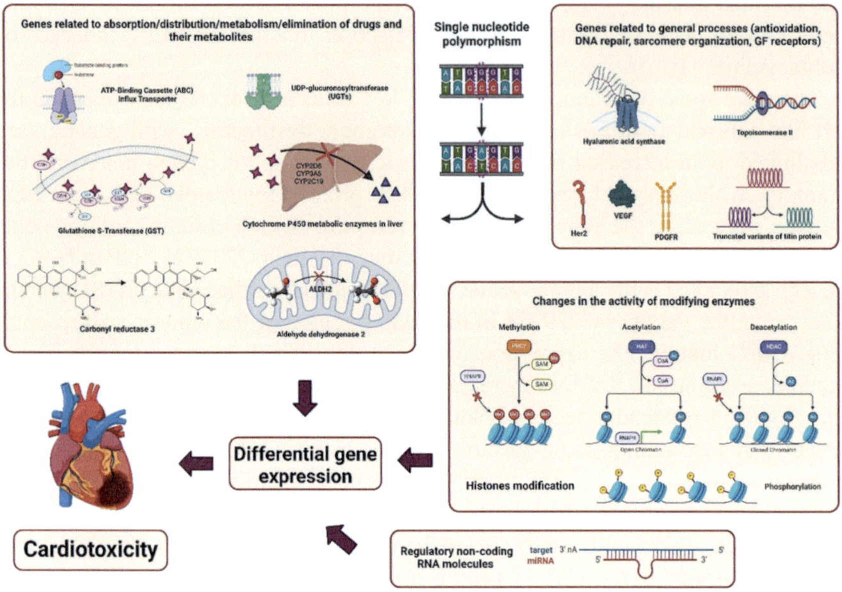

Fig. 27.3 Molecular mechanisms and target molecules responsible for the increased susceptibility to different cardiotoxic drugs. Single nucleotide polymorphisms appear in genes related to absorption, distribution, metabolism and elimination of drugs and their metabolites, as well as genes related to antioxidation, DNA repair, sarcomere organizations, growth factors and their receptors. The variation in the activity of the histone modifying enzymes and the levels of regulatory non-coding RNA molecules also inhibits or promote the expression of genes, contributing to cardiotoxicity. (Created in BioRender.com)

RNAs, circular RNAs, and long non-coding RNAs. DOX-induced cardiotoxicity is closely related to an increase in cardiotoxic ncRNAs levels, such Arhgap12, which promotes apoptosis, oxidative stress, and microvascular injury [30]. A wide group of miRNAs has been identified in animal studies in rats and mice in relation to DOX-induced cardiotoxicity (e.g., miR-1, miR-34a, miR-34c, miR-133a, miR-146a, and miR-208), and some others have been corroborated in human samples. A detailed overview of them and their clinical implications can be found in [31]. Some of the mechanisms and target molecules implied in the enhanced susceptibility to cardiotoxic drugs and that will be described throughout this chapter are depicted in Fig. 27.3.

Anticancer Antibodies

Among current therapies to treat lung cancer, the monoclonal antibody bevacizumab, that targets vascular endothelial growth factor (VEGF), constitutes a first-line treatment for non-small cell lung cancer (NSCLC) patients [32]. However, bevacizumab chemotherapy may result in hypertension, arrhythmia, decline in LVEF, syncope and heart failure [33, 34]. Different meta-analyses have evidenced that there are some ethnic differences related to susceptibility to cardiac complications after treatment with bevacizumab [35, 36]. Genetic polymorphisms like SV2C (71), and VEGF-A c.*237C > T are associated with bevacizumab-induced hypertension [37], and VEGF-634 CC to hypertension and thromboembolism. Although the VEGF-2578AA genotype correlated with superior survival, it was also linked to a higher incidence of hypertension. Genetic markers, such as *EGLN3*, *EGF* and *WNK1*, are closely associated with bevacizumab-induced hypertension [38]. On the other hand, there are some other VEGF polymorphisms that are related to less high-grade bevacizumab-induced hypertension, such as VEGF-1498 TT, VEGF-2578 C/C and VEGF-936 C/C [39].

From an epigenetic point of view, it has been observed that levels of miR1254 and miRNA579 are elevated in patients of colorectal cancer with bevacizumab-induced cardiotoxicity. These markers were selected among others from a small comparative study carried out between patients in bevacizumab treatment, with and without confirmed bevacizumab-induced cardiotoxicity. Then the results were validated in two subsequent studies, one in controls *vs* patients with bevacizumab-induced cardiotoxicity, and the second in patients with bevacizumab-induced cardiotoxicity *vs* patients with acute myocardial infarction. The high specificity of these miRNAs indicates they can be used in the diagnosis and prognosis of bevacizumab-induced cardiotoxicity. miR1254 along with miR-574-5p are also differentially expressed in lung cancer [40]. It was also shown that serum miR-30c levels increased during bevacizumab chemotherapy and this effect correlated to cardiotoxicity in NSCLC, evidenced by a reduction in the ejection fraction [41]. MiR-30c regulates proliferation, apoptosis and differentiation and plays a key role in cardiomyocyte hypertrophy [42] and control of structural changes in the extracellular matrix of the myocardium [43].

Another drug that has been described with cardiotoxic effects is trastuzumab, a humanized monoclonal antibody that selectively binds to the extracellular domain of the human epidermal growth factor receptor 2 (HER2) [44], particularly when is used in combination with anthracycline-based chemotherapy [45]. The HER2 Ile655Val polymorphism involves a nucleotide change in the coding region of the transmembrane domain at codon 655, where an adenine (A) is replaced by a guanine (G). The A variant may affect the dimerization capability of active HER2 proteins with other HER family members, potentially leading to reduced signal transduction. The presence of the G allele renders cardiomyocytes more dependent on HER2 signaling and sensitive to Trastuzumab [46, 47]. A recent study confirmed that HER2 655 A > G polymorphism is significantly linked to an increased risk of trastuzumab-induced cardiotoxicity but does not correlate with variations in disease-free survival or overall survival rates [48]. However, other studies did not find association between the HER2 Ile655Val polymorphism and trastuzumab-induced cardiotoxicity [49]. The Her2 SNP Pro 1170 Ala is also associated with the risk of trastuzumab cardiotoxicity. In this case, a significant connection was found between cardiotoxicity and codon 1170 when the genotype was Pro/Pro, compared to Pro/Ala or Ala/Ala cases [50]. Staropoli et al. identified 9 polymorphic variants in 8 ADME genes (UGT1A1, UGT1A6, UGT1A7, UGT2B15, SLC22A1, CYP3A5, ABCC4, CYP2D6) potentially associated with Trastuzumab-induced cardiotoxicity. The *UGT1A1*, *UGT1A6* and *UGT1A7* encode uridine 5′-diphospho-glucuronosyltransferase enzymes and are involved in inactivation of toxic substrates. The polymorphism found in *UGT2B15* has been associated with an increased risk of prostate cancer and with a decreased response to acetaminophen (paracetamol). The *CYP3A5* was associated with cardiotoxicity in several studies, and it has been suggested that changes in CYP3A5 activity could increase the accumulation of chemotherapeutics, such as DOX and cyclophosphamide, and the production of ROS in cardiomyocytes. The *ABCC4* missense variant has been associated with the metabolism of antiretroviral drug tenofovir. Finally, the cytochrome P450 2D6 (CYP2D6) is involved in the metabolism of up to 25% of the drugs that are commonly used [51]. The N9831 GWAS study identified genetic associations of the maximum decline in LVEF at loci *LDB2, BRINP1, chr6 intergenic, RAB22A, TRPC6,* and *LINC01060*, in the subset of patients who were treated with chemotherapy plus trastuzumab [52]. All these examples highlight the need of keep on studying the implication of genetic background in the responsiveness and safety of different drugs.

Tyrosine Kinase Inhibitors (TKIs)

TKI comprises a family of molecules that prevent tyrosine kinases activation by competitively binding to and inhibiting their ATP binding pocket. There are 52 TKIs that have been approved by the Food and Drug Administration, however, 73% of the approved drugs were associated with treatment-related cardiotoxicity [53]. Two classes of TKIs are currently available: monoclonal antibodies targeting receptor tyrosine kinases or their ligands (for example bevacizumab and trastuzumab, previously discussed) and small molecule inhibitors targeting both, receptor and non-receptor tyrosine kinases. Based on a wide background of experimental and clinical evidence it is evident to assure that polymorphisms in TKIs' targets are likely to play a role in determining the individual risk of cardiotoxicity.

There are several multi-targeting TKIs with an anti-VEGF effect (e.g. sorafenib, sunitinib, pazopanib, axitinib, vandetanib, regorafenib, cabozantinib, levantinib, etc.) approved for the treatment of different malignancies, but possess cardiovascular side effects including hypertension, ventricular dysfunction and cardiac heart failure [54]. Several genetic variations have been linked to an increased toxicity of these drugs. For example, VEGF SNP-634 G/G, VEGF ACG haplotype-2578A > C, VEGF ACG haplotype-460C > T and VEGF ACG haplotype-405C > G are related to an increased sunitinib-induced hypertension [55]. Ibrutinib is another TKI, specifically for Bruton's tyrosine kinase, was approved for B-cell malignancies like chronic lymphocytic leukemia. Besides a favorable side-effect profile has been obtained in clinical trials, studies in open population have revealed higher discontinuation rates due to drug-related toxicities, particularly cardiovascular side effects [56]. It has been found that ibrutinib-related cardiotoxicity is significantly associated with SNPs in *KCNQ1* and *GATA4* genes (*GATA4* rs804280, *KCNQ1* rs163182, and *KCNQ1* rs2237895) [57]. Nilotinib is a BCR-ABL kinase inhibitor approved for the treatment of Philadelphia chromosome positive chronic myelogenous leukemia. The UGT1A1 polymorphisms UGT1A1*28/*28 and *6/*28 have been linked to an increased risk of QT interval prolongation (grade 3), besides other signs of toxicity related to hepatic and lipid metabolism, in patients who receive nilotinib [58]. A wide analysis regarding genes or gene products affected by TKIs associated with heart failure showed that among 345 SNPs of selected genes significantly associated with cardiotoxicity below the genome-wide significant threshold ($P < 5 \times 10^{-8}$). Nine of them possessed strong regulatory potential, and particularly rs191188930, on platelet derived growth factor receptor, was associated with numerous TKI drugs (sunitinib, pazopanib, sorafenib, dasatinib and nilotinib) [59].

Antiretroviral Drugs

Antiretroviral drugs (ARVs) are essential for the survival of patients with human immunodeficiency virus (HIV), due to their efficacy in inhibiting viral replication. Unfortunately, patients living with HIV (PLWH) have a higher risk of developing cardiovascular complications such as ischemic stroke, heart failure, pulmonary hypertension, myocardial infarction, and venous thrombosis [60] and have increased cardiovascular diseases (CVD)-related mortality rates than those without HIV [61]. Factors like inflammation and immune activation in response to HIV infection and viremia, adverse effects of ARVs, and lifestyle risk factors (e.g., smoking, alcohol, and illicit drug use) do not fully explain the increased risk of CVD in PLWH. Some studies have demonstrated the contribution of genetic variation to cardiometabolic risk in PLWH [62, 63], as well as the increase in risk of myocardial infarction in patients recently exposed to abacavir, indinavir and lopinavir, protease inhibitors that decrease the amount of HIV in the blood [64]. *In vitro* experiments in cardiomyocytes have documented that ARVs induced pathological gene expression related to apoptosis, hypertrophy and fibrosis. In particular, homeodomain only-protein homeobox (HOPX), a non-DNA binding co-transcription factor, was significantly upregulated and caused a significant reduction of acetylation at H3K9 and H3K27 [65]. HOPX interacts with histone deacetylase protein 2 and regulates the acetylation level of histone 3 at the promoter region of cardiac-specific transcriptional factor GATA4, inhibiting its ability to transactivate cell cycle genes and induce cardiac myocyte proliferation [66]. These results suggest that ARVs might induce epigenetic modification through histone modification, which could contribute to the suppression of gene transcription and an increased susceptibility of HIV patients to cardiovascular complications.

Antidiabetic Drugs

The use of thiazolidinediones (TZDs) (e.g. pioglitazone and rosiglitazone) has been associated with increased heart failure risk [67]. A toxicoproteomics analysis carried out in human-derived cardiomyocytes showed that TZDs treated cells exhibited an oxidative phosphorylation impairment driven by mitochondrial dysfunction. Cytotoxicity assays showed a significant loss of mitochondrial adenosine triphosphate production upon exposure to both TZD agents. Each drug elicited its own mechanism of action, but all in one way or another contributed to the cardiotoxic effect [68]. It has been documented that rosiglitazone promotes cardiac hypertrophy by peroxisome proliferator-activated receptor gamma (PPARG)-independent mechanisms, increasing the risk of myocardial infarction and mortality [69]. Further studies documented that rosiglitazone treatment increased phosphorylation of histone H3 globally and at regulatory regions of fetal cardiac genes in primary neonatal rat cardiomyocytes, which constitutes a new mechanism of action of this anti-diabetic drug at the epigenetic level and has significant implications for its cardiotoxic effect [70]. Another transcriptome profiling study was carried out in human induced pluripotent stem cells-cardiomyocytes (hiPSC-CMs) from seven individuals lacking known cardiovascular disease-associated mutations, and in three isogenic human heart tissue and hiPSC-CMs patients. In this case, cells were treated with rosiglitazone as cardiotoxic drug. The results showed that, in low risk hiPSC-CMs, rosiglitazone doses did not induce cell death, changes in CM morphology and size, or sarcomere integrity and did induce a decrease in ROS production after treatment for 7 days. Contrarily, in high risk hiPSC-CMs, there was a significant increase in ROS levels and a lack of mitochondrial biogenesis, consistent with impaired PPARG/nuclear factor erythroid 2-like 2 (NRF2) signaling after rosiglitazone treatment. The results obtained suggested that patient-specific drug effects may arise from genomic variations that exert effects on multiple cell types. Comprehensive toxicology analysis based on gene expression profiles showed that patient-specific hiPSC-CM susceptibility to drug-induced cardiotoxicity can be computationally predicted and functionally validated [71].

Antidepressants and Antipsychotics

Citalopram and escitalopram, selective serotonin reuptake inhibitors, are prescribed for several psychiatric disorders including depression and anxiety. Although they are safer than previous similar drugs, they have been associated to prolongation of the QTc interval, cardiotoxicity, and cardiac related deaths [72]. Their interaction with ion channels like Kv11.1, Nav1.5, and Cav1.2 increase the risk of development of fatal arrhythmia like Torsades de Pointes. Citalopram and escitalopram are metabolized by the hepatic enzyme CYP2C19. Circulating levels of drug byproducts are low and do not contribute to the pharmacologic activity of the drug. However, patients with CYP2C19-mutations have higher serum ratio between the metabolites and the original compound [73]. Ji et al. identified genome-wide significant associations between escitalopram concentration and SNPs located in or near the CYP2C19 gene on chromosome 10 (rs1074145). Also, SNPs located in proximity to the CYP2D6 locus on chromosome 22 (rs1065852S) has a significant association with S-didesmethylcitalopram [74]. Venlafaxine is a serotonin-norepinephrine-dopamine reuptake inhibitor that also causes mild cardiotoxicity when ingested in overdose. It was found a direct relationship between the a prolonged metabolism an high levels of the drug and its metabolite e O -desmethylvenlafaxine, the development of acute heart failure and the presence of the specific polymorphisms CYP2D6 * 1/ * 5 (* 5 allele corresponding to a complete allele deletion) and CYP2C19 * 1/ * 17 [75]. Another drug that is highly sensitive to CYP2D6 polymorphism is risperidone, as this enzyme is involved in the metabolism of risperidone to 9-hydroxy (OH)-risperidone. A study carried out in European psychiatric patients evidences that CYP2D6, but not CYP2C9, was related to QTc lengthening [76].

Alcohol

Alcohol can induce cardiomyopathy, contractile dysfunction and arrhythmias [77]. Alcohol-induced cardiac toxicity is driven by oxidative stress mediated mitochondria dysfunction, sarcoplasmic reticulum Ca^{2+}-leak and disordered excitation-contraction coupling, changes in calcium sensitivity of myofilaments, deregulation of protein synthesis, disproportion between the different types of myofibrils, changes in the regulation of myosin ATPase, up-regulation of the L-type calcium channels, among others [78]. The individual level of susceptibility to alcohol-induced cardiotoxicity resides in genetic variants of different genes associated with alcohol metabolism, like those that encode alcohol dehydrogenase, aldehyde dehydrogenase 2 and cytochrome P450 (ADH1B, ALDH2, and CYP2E1 respectively). ALDH2 is a mitochondrial enzyme involved in the detoxification of alcohol-derived acetaldehyde and endogenous aldehydes. Specifically, the *ALDH2* rs671 polymorphism causes a change from glutamic acid to lysine at position 504, which results in a weaker NAD^+ binding site and a substantial reduction in dehydrogenase activity. This alteration is associated with an increased risk of coronary artery disease and alcohol-induced cardiac dysfunction [79]. Contrarily, polymorphisms in genes coding for components of the renin–angiotensin–aldosterone system are not associated with alcohol-induced cardiomyopathy [80].

Illicit Drugs

Cocaine is a strongly addictive psychostimulant and the second most used recreational drug in the world. Cocaine exerts a potent anesthetic effect by blocking sodium channels, along with a potent sympathomimetic effect which is achieved by the blockage of catecholamine reuptake at neuronal presynaptic terminals and constitutes the base for its cardiotoxic effects [81]. Cocaine increases the myocardial oxygen demand and, at the same time, decreases the myocardial oxygen flow. Consumption of this drug can produce irreversible structural damage to the heart, accelerate cardiovascular disease progression, and precipitate sudden cardiac death [82]. It has been proposed that the susceptibility to cocaine-induced cardiotoxicity is genetically regulated. More specifically, mutations or polymorphisms of genes related to ATP-sensitive potassium (K_{ATP}) channels have been implicated in the predisposition to human cardiac disease [83]. *KCNJ11* gene, coding the pore-forming Kir6.2 subunit, has been identified as a pro-survival molecule in response to the hyperadrenergic state imposed by acute cocaine overdose. Targeted deletion of this gen resulted in an amplified vulnerability to the toxic effects of chronic cocaine abuse [84].

MDMA (3,4-methylenedioxymethamphetamine) or ecstasy, besides its known effects on the brain, also provokes acute cardiotoxicity, including tachycardia and ventricular arrhythmia, which are associated pathogenetically with cardiomyopathy. A study carried out in mice revealed that MDMA reduced survival by almost 50% after 33 days of daily treatment and induced cardiac epigenetic changes in DNA methylation where hypermethylation predominated. Gene expression changes observed in the heart were associated with the circadian rhythm pathway and with an increase in cardiac heart weight, an indicator of cardiac hypertrophy, although this change was reversible following cessation of MDMA [85].

Methamphetamine (MA)-related cardiomyopathy includes oxidative stress and apoptosis [86]. Vulnerability to the development of methamphetamine-associated cardiomyopathy appears to be genetically influenced by an elevated risk in the subpopulation of cytochrome P45O CYP2D6 extensive metabolizers [87], although, as for the MDMD, some reversibility is possible after abstinence. QT prolongation leading to fatal arrhythmia has been implicated in sudden death during MA use [88]. Specifically, mutations in *KCNQ1* and *KCNH2*, long QT syndrome-associated genes which encode potassium channel proteins, have been involved in sudden death associated with antipsychotic drug [89] and new psychoactive substances [90]. An animal model of MA consumption revealed changes in the lncRNA expression profile and a differential expression of 347 genes compared with the control treatment after 2 weeks, while prolonged exposure time of 4 weeks increased this number to 854 genes; 147 genes were specifically related to progressive damage to the heart [91].

Conclusions

Cardiotoxicity is a very common phenomenon, inherent to daily use medical compounds, as well as recreational and illegal drugs. The pathological mechanisms involved are oxidative stress, mitochondrial dysfunction, lipid peroxidation, impaired DNA synthesis, interference with ion channels and intracellular calcium homeostasis or calcium-mediated contractile mechanisms, among many others. It is well known that individual response to these substances depends on several factors like age, gender, pre-existing medical conditions or comorbidities, co-administered drugs, life habits like diet, exercise, smoking, exposure to environmental stressors, etc. But internal specific factors like genetic background and epigenetic modifications also play an important and defining role in how a person will be able to respond to a certain medication. Knowing the mechanisms of action and the metabolism of these drugs allows us to identify the genes involved and to search for possible polymorphisms involved in increased susceptibility or detrimental effects such as cardiotoxicity. In this sense, clinical studies and analyses in open populations are crucial to address the identification of inter-individual variability in responses to current and future drugs.

Funding The research work of DKA is supported by R25AI179582 and R01 HL147662 grants from the National Institutes of Health, USA. The contents of this chapter are solely the responsibility of the authors and do not necessarily represent the official views of the National Institutes of Health.

Competing Interests Both authors have read the chapter and declare no conflict of interest. No writing assistance was utilized in the production of this manuscript.

Consent for Publication Both authors have read the chapter and consented for publication.

References

1. Viereck J, Thum T (2017) Circulating noncoding RNAs as biomarkers of cardiovascular disease and injury. Circ Res 120(2):381–399
2. Koturbash I et al (2015) microRNAs as pharmacogenomic biomarkers for drug efficacy and drug safety assessment. Biomark Med 9(11):1153–1176
3. Gilchrist SC et al (2019) Cardio-oncology rehabilitation to manage cardiovascular outcomes in cancer patients and survivors: a scientific statement from the American Heart Association. Circulation 139(21):e997–e1012
4. Lyon AR et al (2022) 2022 ESC guidelines on cardio-oncology developed in collaboration with the European Hematology Association (EHA), the European Society for Therapeutic Radiology and Oncology (ESTRO) and the International Cardio-Oncology Society (IC-OS). Eur Heart J 43(41):4229–4361
5. Arbitrio M et al (2022) Pharmacogenetics/pharmacogenomics of drug-metabolizing enzymes and transporters. In: Kenakin T (ed) Comprehensive pharmacology. Elsevier, Oxford, pp 657–697

6. Li D et al (2021) Role of acetylation in doxorubicin-induced cardiotoxicity. Redox Biol 46: 102089
7. Mauro C et al (2023) Exploring the cardiotoxicity spectrum of anti-cancer treatments: definition, classification, and diagnostic pathways. J Clin Med 12(4)
8. Wallace KB, Sardao VA, Oliveira PJ (2020) Mitochondrial determinants of doxorubicin-induced cardiomyopathy. Circ Res 126(7):926–941
9. Ni MM et al (2024) Association between genetic variants of transmembrane transporters and susceptibility to anthracycline-induced cardiotoxicity: current understanding and existing evidence. Clin Genet 105(2):115–129
10. Muckiene G et al (2023) The impact of polymorphisms in ATP-binding cassette transporter genes on anthracycline-induced early cardiotoxicity in patients with breast cancer. J Cardiovasc Dev Dis 10(6)
11. Wojnowski L et al (2005) NAD(P)H oxidase and multidrug resistance protein genetic polymorphisms are associated with doxorubicin-induced cardiotoxicity. Circulation 112(24): 3754–3762
12. Visscher H et al (2013) Validation of variants in SLC28A3 and UGT1A6 as genetic markers predictive of anthracycline-induced cardiotoxicity in children. Pediatr Blood Cancer 60(8): 1375–1381
13. Ferreira MJ et al (2023) Glutathione and peroxisome redox homeostasis. Redox Biol 67:102917
14. Singh P et al (2020) Association of GSTM1 null variant with anthracycline-related cardiomyopathy after childhood cancer—a Children's Oncology Group ALTE03N1 report. Cancer 126(17):4051–4058
15. Blanco JG et al (2012) Anthracycline-related cardiomyopathy after childhood cancer: role of polymorphisms in carbonyl reductase genes—a report from the Children's Oncology Group. J Clin Oncol 30(13):1415–1421
16. Serie DJ et al (2017) Genome-wide association study of cardiotoxicity in the NCCTG N9831 (Alliance) adjuvant trastuzumab trial. Pharmacogenet Genomics 27(10):378–385
17. Jeong SY et al (2024) Hyaluronic acid stimulation of stem cells for cardiac repair: a cell-free strategy for myocardial infarct. J Nanobiotechnol 22(1):149
18. Wang X et al (2014) Hyaluronan synthase 3 variant and anthracycline-related cardiomyopathy: a report from the Children's Oncology Group. J Clin Oncol 32(7):647–653
19. Aminkeng F et al (2015) A coding variant in RARG confers susceptibility to anthracycline-induced cardiotoxicity in childhood cancer. Nat Genet 47(9):1079–1084
20. Wang X et al (2016) CELF4 variant and anthracycline-related cardiomyopathy: a Children's Oncology Group Genome-Wide Association Study. J Clin Oncol 34(8):863–870
21. Hinson JT, Campbell SG (2024) TTN truncation variants produce sarcomere-integrating proteins of uncertain functional significance. J Clin Invest 134(2)
22. Nguyen N et al (2022) Epirubicin alters DNA methylation profiles related to cardiotoxicity. Front Biosci (Landmark Ed) 27(6):173
23. Robinson EL et al (2023) Differential expression of epigenetic modifiers in early and late cardiotoxic heart failure reveals DNA methylation as a key regulator of cardiotoxicity. Front Cardiovasc Med 10:884174
24. Hanf A et al (2019) The anti-cancer drug doxorubicin induces substantial epigenetic changes in cultured cardiomyocytes. Chem Biol Interact 313:108834
25. Wang P et al (2020) Histone demethylase JMJD3 mediated doxorubicin-induced cardiomyopathy by suppressing SESN2 expression. Front Cell Dev Biol 8:548605
26. Jang J et al (2022) Epicardial HDAC3 promotes myocardial growth through a novel microRNA pathway. Circ Res 131(2):151–164
27. Kuno A et al (2023) SIRT1 in the cardiomyocyte counteracts doxorubicin-induced cardiotoxicity via regulating histone H2AX. Cardiovasc Res 118(17):3360–3373
28. Li H et al (2022) Daidzein alleviates doxorubicin-induced heart failure via the SIRT3/FOXO3a signaling pathway. Food Funct 13(18):9576–9588

29. Ding Y et al (2022) Genetic susceptibility and mechanisms underlying the pathogenesis of anthracycline-associated cardiotoxicity. Oxidative Med Cell Longev 2022:5818612
30. Wang X et al (2021) Circular RNA Arhgap12 modulates doxorubicin-induced cardiotoxicity by sponging miR-135a-5p. Life Sci 265:118788
31. Skala M et al (2019) MicroRNAs in the diagnosis and prevention of drug-induced cardiotoxicity. Arch Toxicol 93(1):1–9
32. Lauro S et al (2014) The use of bevacizumab in non-small cell lung cancer: an update. Anticancer Res 34(4):1537–1545
33. Sandhu H, Maddock H (2014) Molecular basis of cancer-therapy-induced cardiotoxicity: introducing microRNA biomarkers for early assessment of subclinical myocardial injury. Clin Sci (Lond) 126(6):377–400
34. Chino H et al (2016) Cardiogenic syncope possibly related to bevacizumab-containing combination chemotherapy for advanced non-small cell lung cancer. J Thorac Dis 8(9):2646–2650
35. Polite BN et al (2012) Exploring racial differences in outcome and treatment for metastatic colorectal cancer: results from a large prospective observational cohort study (BRiTE). Cancer 118(4):1083–1090
36. Chen Z et al (2015) Specific safety profile of bevacizumab in Asian patients with advanced NSCLC: a meta-analysis. Medicine (Baltimore) 94(24):e975
37. Sibertin-Blanc C et al (2015) Vascular Endothelial Growth Factor A c.*237C>T polymorphism is associated with bevacizumab efficacy and related hypertension in metastatic colorectal cancer. Dig Liver Dis 47(4):331–337
38. Lambrechts D et al (2014) Genetic markers of bevacizumab-induced hypertension. Angiogenesis 17(3):685–694
39. Morita S et al (2013) Association between bevacizumab-related hypertension and vascular endothelial growth factor (VEGF) gene polymorphisms in Japanese patients with metastatic colorectal cancer. Cancer Chemother Pharmacol 71(2):405–411
40. Patz EF Jr et al (2007) Panel of serum biomarkers for the diagnosis of lung cancer. J Clin Oncol 25(35):5578–5583
41. Zhou F, Lu X, Zhang X (2018) Serum miR-30c level predicted cardiotoxicity in non-small cell lung cancer patients treated with bevacizumab. Cardiovasc Toxicol 18(3):284–289
42. Irani S, Hussain MM (2015) Role of microRNA-30c in lipid metabolism, adipogenesis, cardiac remodeling and cancer. Curr Opin Lipidol 26(2):139–146
43. Duisters RF et al (2009) miR-133 and miR-30 regulate connective tissue growth factor: implications for a role of microRNAs in myocardial matrix remodeling. Circ Res 104(2): 170–178, 6p following 178
44. Di Cosimo S (2011) Heart to heart with trastuzumab: a review on cardiac toxicity. Target Oncol 6(4):189–195
45. Vuger AT et al (2022) Anthracyclines in the treatment of early breast cancer friend or foe? Breast 65:67–76
46. Gomez Pena C et al (2015) Influence of the HER2 Ile655Val polymorphism on trastuzumab-induced cardiotoxicity in HER2-positive breast cancer patients: a meta-analysis. Pharmacogenet Genomics 25(8):388–393
47. Beauclair S et al (2007) Role of the HER2 [Ile655Val] genetic polymorphism in tumorogenesis and in the risk of trastuzumab-related cardiotoxicity. Ann Oncol 18(8):1335–1341
48. Blancas I et al (2024) HER2/neu 655 polymorphism, trastuzumab-induced cardiotoxicity, and survival in HER2-positive breast cancer patients. Clin Transl Oncol 26(10):2531–2540
49. Vazdar L et al (2021) Influence of Ile655Val polymorphism on trastuzumab-induced cardiotoxicity in early-stage HER2 positive breast cancer. Sci Rep 11(1):14395
50. Stanton SE et al (2015) Pro1170 Ala polymorphism in HER2-neu is associated with risk of trastuzumab cardiotoxicity. BMC Cancer 15:267
51. Staropoli N et al (2024) Identification of ADME genes polymorphic variants linked to trastuzumab-induced cardiotoxicity in breast cancer patients: case series of mono-institutional experience. Biomed Pharmacother 174:116478

52. Norton N, Weil RM, Advani PP (2021) Inter-individual variation and cardioprotection in anthracycline-induced heart failure. J Clin Med 10(18)
53. Jin Y et al (2020) A comprehensive review of clinical cardiotoxicity incidence of FDA-approved small-molecule kinase inhibitors. Front Pharmacol 11:891
54. Chen YC et al (2018) Genetic and ethnic modulation of cardiovascular toxicity of vascular endothelial growth factor inhibitors. Ann Med 50(1):46–56
55. Eechoute K et al (2012) Polymorphisms in endothelial nitric oxide synthase (eNOS) and vascular endothelial growth factor (VEGF) predict sunitinib-induced hypertension. Clin Pharmacol Ther 92(4):503–510
56. Mato AR et al (2018) Toxicities and outcomes of 616 ibrutinib-treated patients in the United States: a real-world analysis. Haematologica 103(5):874–879
57. Hamadeh IS et al (2023) Genetic predictors of ibrutinib-related cardiovascular side effects in patients with chronic lymphocytic leukemia. Clin Cancer Res 29(23):4941–4948
58. Shibata T et al (2014) Association between severe toxicity of nilotinib and UGT1A1 polymorphisms in Japanese patients with chronic myelogenous leukemia. Int J Clin Oncol 19(2): 391–396
59. Li Y et al (2021) Genome-wide prioritization reveals novel gene signatures associated with cardiotoxic effects of tyrosine kinase inhibitors. Oncol Lett 21(2):94
60. Islam FM et al (2012) Relative risk of cardiovascular disease among people living with HIV: a systematic review and meta-analysis. HIV Med 13(8):453–468
61. Croxford S et al (2017) Mortality and causes of death in people diagnosed with HIV in the era of highly active antiretroviral therapy compared with the general population: an analysis of a national observational cohort. Lancet Public Health 2(1):e35–e46
62. Rotger M et al (2013) Contribution of genetic background, traditional risk factors, and HIV-related factors to coronary artery disease events in HIV-positive persons. Clin Infect Dis 57(1):112–121
63. Cheng H et al (2020) Genetic architecture of cardiometabolic risks in people living with HIV. BMC Med 18(1):288
64. Bavinger C et al (2013) Risk of cardiovascular disease from antiretroviral therapy for HIV: a systematic review. PLoS One 8(3):e59551
65. Kashyap S et al (2021) HOPX plays a critical role in antiretroviral drugs induced epigenetic modification and cardiac hypertrophy. Cells 10(12)
66. Trivedi CM et al (2010) Hopx and Hdac2 interact to modulate Gata4 acetylation and embryonic cardiac myocyte proliferation. Dev Cell 19(3):450–459
67. Wallach JD et al (2020) Updating insights into rosiglitazone and cardiovascular risk through shared data: individual patient and summary level meta-analyses. BMJ 368:l7078
68. Al Sultan A, Rattray Z, Rattray NJW (2024) Cytotoxicity and toxicoproteomics analysis of thiazolidinedione exposure in human-derived cardiomyocytes. J Appl Toxicol 44(8): 1214–1235
69. He H et al (2014) Rosiglitazone causes cardiotoxicity via peroxisome proliferator-activated receptor gamma-independent mitochondrial oxidative stress in mouse hearts. Toxicol Sci 138(2):468–481
70. Pharaon LF et al (2017) Rosiglitazone promotes cardiac hypertrophy and alters chromatin remodeling in isolated cardiomyocytes. Toxicol Lett 280:151–158
71. Matsa E et al (2016) Transcriptome profiling of patient-specific human iPSC-cardiomyocytes predicts individual drug safety and efficacy responses in vitro. Cell Stem Cell 19(3):311–325
72. Nakatani Y, Amano T (2021) Contributions of S- and R-citalopram to the citalopram-induced modulation of the function of Nav1.5 voltage-gated sodium channels. Eur J Pharmacol 908: 174316
73. Farhat H et al (2024) Citalopram & escitalopram: mechanisms of cardiotoxicity, toxicology predisposition and risks of use in geriatric & hemodialysis populations. Glob Cardiol Sci Pract 2024(4):e202434

74. Ji Y et al (2014) Citalopram and escitalopram plasma drug and metabolite concentrations: genome-wide associations. Br J Clin Pharmacol 78(2):373–383
75. Vinetti M et al (2011) Severe acute cardiomyopathy associated with venlafaxine overdose and possible role of CYP2D6 and CYP2C19 polymorphisms. Clin Toxicol (Phila) 49(9):865–869
76. Llerena A et al (2004) QTc interval, CYP2D6 and CYP2C9 genotypes and risperidone plasma concentrations. J Psychopharmacol 18(2):189–193
77. Whitman IR et al (2017) Alcohol abuse and cardiac disease. J Am Coll Cardiol 69(1):13–24
78. Laurent D, Edwards JG (2014) Alcoholic cardiomyopathy: multigenic changes underlie cardiovascular dysfunction. J Cardiol Clin Res 2(1)
79. Zhang Y, Ren J (2011) ALDH2 in alcoholic heart diseases: molecular mechanism and clinical implications. Pharmacol Ther 132(1):86–95
80. Husemoen LL et al (2008) The association of ADH and ALDH gene variants with alcohol drinking habits and cardiovascular disease risk factors. Alcohol Clin Exp Res 32(11): 1984–1991
81. Schwartz BG, Rezkalla S, Kloner RA (2010) Cardiovascular effects of cocaine. Circulation 122(24):2558–2569
82. Phillips K et al (2009) Cocaine cardiotoxicity: a review of the pathophysiology, pathology, and treatment options. Am J Cardiovasc Drugs 9(3):177–196
83. Reyes S et al (2008) K(ATP) channel polymorphism is associated with left ventricular size in hypertensive individuals: a large-scale community-based study. Hum Genet 123(6):665–667
84. Reyes S et al (2009) Targeted disruption of K(ATP) channels aggravates cardiac toxicity in cocaine abuse. Clin Transl Sci 2(5):361–365
85. Koczor CA et al (2015) Ecstasy (MDMA) alters cardiac gene expression and DNA methylation: implications for circadian rhythm dysfunction in the heart. Toxicol Sci 148(1):183–191
86. Jafari Giv M (2017) Exposure to amphetamines leads to development of amphetamine type stimulants associated cardiomyopathy (ATSAC). Cardiovasc Toxicol 17(1):13–24
87. Sutter ME et al (2013) Polymorphisms in CYP2D6 may predict methamphetamine related heart failure. Clin Toxicol (Phila) 51(7):540–544
88. Haning W, Goebert D (2007) Electrocardiographic abnormalities in methamphetamine abusers. Addiction 102(Suppl 1):70–75
89. Kamei S et al (2014) Molecular analysis of potassium ion channel genes in sudden death cases among patients administered psychotropic drug therapy: are polymorphisms in LQT genes a potential risk factor? J Hum Genet 59(2):95–99
90. Nagasawa S et al (2018) Relationship between KCNQ1 (LQT1) and KCNH2 (LQT2) gene mutations and sudden death during illegal drug use. Sci Rep 8(1):8443
91. Zhang C et al (2022) New insight into methamphetamine-associated heart failure revealed by transcriptomic analyses: circadian rhythm disorder. Toxicol Appl Pharmacol 451:116172

Chapter 28
Pharmacogenetics and Pharmacogenomics in the Prediction of Cardiovascular Toxicity

Ranko Škrbić, Lana Nežić, and Miloš P. Stojiljković

Abstract Drug-induced cardiovascular toxicity remains a significant cause of heart-related morbidity and mortality. This could be clinical manifestations as arrhythmia, prolongation of QT interval, myocardial ischemia, cardiomyopathy, heart failure, or cardiac arrest. Traditional cardiovascular risk factors, such as age, gender, pre-existing cardiovascular diseases, renal failure or drug-drug interactions, only partly explain the risk of drug-induced cardiovascular toxicity. In addition to these factors, increasing evidence has emphasised the role of genetic variability as a crucial factor for the development of cardiovascular toxicity. Pharmacogenetics as a discipline studies the role of genetic variations on drugs response and could provide valuable input on identifying patients with inherent risk of drug-induce toxicity. The use of specific pharmacogenetic markers will unable better tailoring and personalisation of pharmacotherapy based on the clinical and genetic characteristics of a patient. This chapter reviews drugs that cause cardiovascular toxicity, the clinical manifestations and laboratory tests, as well as the genetic variations associated with the risk of drug-induced cardiovascular toxicity.

Keywords Cardiotoxicity · Pharmacogenetics · Pharmacogenomics · Gene polymorphism · Adverse drug reactions · Drug interactions · Chemotherapeutics · QT interval · Cardiovascular drugs

---

R. Škrbić (✉) · L. Nežić
Department of Pharmacology, Toxicology and Clinical Pharmacology, Faculty of Medicine, University of Banja Luka, Banja Luka, The Republic of Srpska, Bosnia and Herzegovina
e-mail: ranko.skrbic@med.unibl.org; lana.nezic@med.unibl.org

M. P. Stojiljković
Department of Pharmacology, Toxicology and Clinical Pharmacology, Faculty of Medicine, University of Banja Luka, Banja Luka, The Republic of Srpska, Bosnia and Herzegovina

Institute for Cardiovascular Diseases "Dedinje", Belgrade, Serbia
e-mail: milos.stojiljkovic@med.unibl.org

D. K. Agrawal et al. (eds.), *Cardiovascular Toxicity*, Advances in Biochemistry in Health and Disease 33, https://doi.org/10.1007/978-3-031-90598-8_28

Introduction

Cardiovascular toxicity is a very common adverse reaction related to pharmacotherapy that has important consequences in clinical practice. Drug-induced cardiotoxicity is a serious adverse drug reaction (ADR) that can occur at any time during treatment and may include different clinical manifestations like cardiomyopathy, myocarditis, pericarditis, valve injury, myocardial ischemia, arrhythmia, prolongation of the QT interval with severe ventricular tachycardia and death [1–3]. In recent years, cardio-oncology has been developed as a new interdisciplinary field that focuses on the cardiovascular side effect of chemotherapy and radiotherapy. Many factors are involved in cardiovascular toxicity such as gender, age, renal failure, drug-drug interactions, and pre-existing cardiovascular diseases, but all these factors cannot fully explain its occurrence [4]. In addition to common cardiovascular risk factors, increasing evidence has emphasised the genetic variability as a key factor for cardiovascular toxicity development [5]. Many genetic biomarkers associated with the increased risk of cardiovascular toxicity have been developed and translated into routine clinical practice to optimise pharmacotherapy. A new diagnostic method focused on the role of genetic factors in drug-related cardiovascular toxicity has been developed in recent years with the advancement of genotyping technology [6, 7].

Older people are particularly susceptible to ADRs or therapeutic failure. Increasing prevalence of multiple medical conditions in the elderly is associated with polypharmacy which increases the possibility of drug-drug and drug-gene interactions [8, 9]. The screening of patients with increased risk for ADRs and the implementation of personalised treatment regimens is based on studies of individual genetic variations. It combines genetic and genomic data with environmental and clinical factors in order to tailor the appropriate pharmacotherapy.

Pharmacogenetics as the discipline studies the role of genetic variation in pharmacokinetics and pharmacodynamics of drugs, but the term pharmacogenomics usually encompasses a broader scope of all genes involved [10]. The distinction is arbitrary and both terms can be used interchangeably. The most common type of genetic variation belongs to single-nucleotide polymorphisms (SNPs) [4]. SNP is a common type of genetic variation that occurs when a single nucleotide (adenine, guanine, cytosine or thymine) at a specific position in the genome varies among individuals in a population. In simpler terms, it is a point mutation where one base pair in the DNA sequence is replaced by another base pair. SNPs are usually present in at least 1% of the population and are the most common type of genetic variation in humans. SNPs could be associated with the risk or predisposition for many complex diseases, act as biomarkers to predict the risk for some diseases or predict how the patient could respond to certain drugs. Over the last few decades, there has been an increasing interest in pharmacogenetics and its application in personalised medicine.

Drug-induced cardiotoxicity is a serious clinical problem, that more often occur after the introduction of a new drug on the market. The possible explanation for these ADRs relates to the fact that current safety investigations of possible cardiac ADR are based on experimental and clinical studies performed in healthy individuals. It means that toxic drug effects can be concealed when safety testing is only done in healthy individuals but may become evident in the diseased conditions or associated with existing comorbidities ('hidden toxicity'). The 'hidden cardiotoxicity' has been defined as a toxicity that manifests itself only in individuals with certain pathological conditions [11].

In many cases, ADRs are the consequence of drug-drug interactions that involve pharmacokinetic or pharmacodynamic mechanisms [12]. The identification of genetic factors that influence major pharmacokinetic functions like drug absorption, distribution, metabolism, excretion, or action at the drug target level are crucial for personalised pharmacotherapy. Gene variants related to drug metabolism, drug transport proteins, or pharmacodynamic actions have been translated into clinical practice [13].

The list of drugs that can cause a certain type of cardiovascular toxicity is considerable, like anticancer drugs, antiretrovirals, anti-infective drugs, antipsychotics, antihistamines, bronchodilating drugs, the lipid lowering drugs, some oral antidiabetics, non-steroidal anti-inflammatory drugs, and even some cardiovascular drugs. The cardiotoxicity of most drugs is cumulative (dose-dependent), or time-dependent, and may develop years after the initial drug was given, and more frequently in patients that already have cardiovascular comorbidities.

Chemotherapeutic Drug-Induced Cardiovascular Toxicity

Cancer treatment has improved significantly in recent years having a tremendous impact on survival rates of cancer patients. However, at the same time this therapy has also increased morbidity and mortality of the cancer patients due to serious side effects such as left ventricular dysfunction, heart failure, hypertension and arrhythmias which can manifest many years after the completion of chemotherapy [14]. For example, the incidence of cardiotoxicity caused by anthracyclines is 9–18% [15], and the incidence of myocarditis induced by checkpoint inhibitors can be almost 14% [16]. Patients who develop heart failure following anticancer therapy have very poor prognosis with the mortality rate of 60% within the first 2 years [17]. The patient's responsiveness to chemotherapeutics is unpredictable and susceptibility to the cardiotoxic side effects are partly related to clinical and demographic factors, but individual differences related to genetics could provide insights into the mechanism for toxicity induced by chemotherapy. The characterisation and identification of specific genetic biomarkers is crucial for the prediction of risk of developing cardiac dysfunction induced by chemotherapy. Additionally, the progress in genetic research will contribute to elucidating the molecular mechanisms of chemotherapy-induced cardiotoxicity. Different intracellular signalling pathways, controlled by a large number of genes, are involved in cardiotoxicity. Below is an overview of signalling pathways and proteins regulated by a number of genes associated with drug-induced cardiotoxicity.

Oxidative Stress and Reactive Oxygen Species Chemotherapy induces significant production of reactive oxygen species (ROS) like hydroxyl radical ($^{\cdot}OH$), superoxide radical ($O_2^{\cdot -}$), and hydrogen peroxide (H_2O_2) that cause myocardial injury [18]. ROS easily interact and destroy lipid membranes, proteins or DNA. Cellular antioxidant enzymes such as catalase (CAT), superoxide dismutase 2 (SOD2), glutathione peroxidases (GPx), glutathione reductase (GR), or glutathione-S-transferases (GST) play important roles in the ROS metabolism. The disbalance between the high metabolic production of ROS and weak antioxidant defence due to low levels of antioxidative enzymes that neutralise ROS generated by chemotherapeutics in cardiac tissue make cardiomyocytes especially sensitive to free radical damage [17]. Cardiac damage was evaluated in relation to antioxidative genes polymorphism and a significant correlation was found between CC homozygosity for *CAT* (rs10836235 (c.66 + 78C > T)) and cardiac damage after anthracycline exposure in paediatric patients with acute leukaemia [19].

The NADPH oxidase (NOX) is a multienzyme complex that controls the ROS production in the myocardium [20]. It has been shown that genotypic variations of alpha-1 class glutathione S-transferase (*GSTA1*, rs3957357) and *NOX* p22phox (CYBA, rs4673) may predict patient prognosis after the rituximab-CHOP-(R-CHOP-)21 treatment [21]. It is also found that a variant of the *NOX* subunit NCF4 (rs1883112) may prevent haematological and nonhematological toxicity [22]. Additionally, the chronic anthracycline-induced cardiotoxicity was found to be related to the NCF4 (rs1883112) variant, while the acute form was found to be related to the p22phox subunit (rs4673) and the RAC2 (rs13058338) subunit of NOX. This was confirmed in experimental studies in which mice with insufficient NOX activity were resistant to chronic doxorubicin therapy [23, 24].

Adenosine Triphosphate (ATP)-Binding Cassette (ABC) Transporters ABC transporters are transmembrane proteins responsible for the uptake or export of molecules across cell membranes. These transporters have a significant impact on pharmacokinetics of drugs and may also influence drug resistance, [25, 26]. ABC transporters can export numerous substances from cardiac cells, including anthracyclines [27]. There have been numerous mutations of the *ABC* genes associated with drug exporting defects responsible for cardiotoxicity in cancer patients due to the accumulation of anthracyclines in cardiomyocytes [28].

The ABCC1 transporter, or multidrug resistance-associated protein 1 (MRP1), is highly expressed in human myocardium and its reduced activity can lead to intracellular accumulation of anthracycline [29]. It was found that the *ABCC1* rs4148350 TG genotype is associated with higher rates of anthracycline-induced cardiotoxicity in patients with breast cancer [26]. The SNPs of other ABC transporters like *ABCC1* (rs246221, rs45511401), *ABCC2* (rs8187710, rs8187694, rs3740066), *ABCB4* (rs1149222 and rs4148808), and *ABCC5* rs7627754 have been also associated with reduced activity. These mutations are associated with intracellular accumulation of anthracyclines and increased risk of cardiotoxicity [30–32]. Differently, the overexpression of ABCC1 transporter is associated with reduced cardiotoxicity due to increased drug efflux from cardiomyocytes.

The ABCB1 transporter, also known as multidrug efflux pump or P-glycoprotein, has been found to be protective in several tissues including heart [33, 34]. P-glycoprotein is expressed in the cardiac endothelium and acting as a blood–heart barrier protecting cardiac tissues from xenobiotics. Data showed that the *ABCB1* SNPs 1236C > T (rs1128503), 2677G > T/A (rs2032582), and 3435C > T (rs1045642) are cardioprotective toward ABCB1 substrates. The variant allele of *ABCB1* 3435C > T (rs1045642) had an additive protective effect in breast cancer patients treated with anthracyclines [35]. Moreover, the polymorphism of *ABCB1* gene is proposed as a marker for QT prolongation following the administration of drugs substrate for the ABCB1 protein [36].

Cytochrome P450 Oxidoreductase Cytochrome P450 oxidoreductase (also known as POR or CYPOR) is a flavoprotein that donates electrons from NADPH to all P450 enzymes [37]. The POR is crucial for doxorubicin transformation into hemiquinone radical form which stabilises the drug's crosslinking to DNA and thus greatly enhance its cytotoxicity [38]. The *POR* gene SNP rs13240755 was the most significant driver daunorubicin-induced cardiotoxicity. The patients with acute myeloid leukaemia carrying the G allele have increased risk of cardiotoxicity [39]. Another study examined the relationship between CYP3A5 genetic polymorphism and the plasma concentration of daunorubicin in patients with acute myeloid leukaemia. It was shown that *CYP3A5∗3* gene polymorphism is closely related to the enzyme activity, the mRNA expression of CYP3A5, and the daunorubicin plasma concentration, being responsible for different adverse drug reactions [40].

Glutathione S Transferase The glutathione S transferase (GST) family, as a drug metabolising enzymes, have a considerable effect for characterising the response of patients to chemotherapeutics [41]. The GST genes, such as *GSTM1*, glutathione s-transferase Pi (*GSTP1*), and *GSTT1*, as members of phase II metabolism, encode the detoxifying proteins that are involved in the conjugation of substrates that are toxic to cancer cells [42].

UDP Glucuronosyltransferases UDP-glucuronosyltransferase (UGT) exists as a superfamily of 22 proteins, which are divided into 5 families and 6 subfamilies. Members of the UGT1A and UGT2B subfamilies play a key role in terminating the biological actions in the process known as glucuronidation, a major part of phase II metabolism, and enhancing the renal elimination of lipophilic drugs. The study in paediatric patients showed that the variants of *UGT1A6* (rs17863783, V209 V) are associated with anthracycline-induced cardiotoxicity [43]. The glucuronidation of doxorubicine and daunorubicine by UGT1A6 enzyme might lead to accumulation of toxic metabolites in patients carrying *UGT1A6∗4*, resulting in an increased risk of cardiotoxicity [44]. However, the transformation of *UGT2B7-161* SNP C to T allele (*UGT2B7-161* C > T) is negatively associated with the risk of cardiotoxicity. Patients with breast cancer carrying the *UGT2B7-161* (rs7668258) T allele had decreased incidence of myocardial injury when treated with trastuzumab [7, 45]. The results of the Spanish study revealed that patients carrying mutations *UGT1A448Val*, *UGT2B7268Tyr* or with wild type genotypes for *UGT2B17nodel* and *UGT2B15523Lys* had a good response to tamoxifen therapy by eliciting effective plasma concentrations of active tamoxifen metabolite [46].

Solute Carriers The solute carriers (SLC) are membrane proteins family that play key roles in the absorption and transportation of different endogenous molecules across cell membranes. Anthracyclines are well known substrates for SLC transporters, increased expression of *SLC* genes with encoded proteins could accelerate anthracyclines transport into the cell leading to increased toxicity, while the decreased expression and reduced function would be protective. The *SLC22A6* rs6591722 AA genotype was associated with significant risk for a decreased cardiac function in paediatric patients with acute lymphoblastic leukaemia that were receiving anthracyclines [47]. Similarly, the *SLC22A17* SNP variants rs4982753 and rs4149178 were also associated with cardiotoxicity. However, the *SLC28A3* rs7853758 and rs885004 variants were associated with resistance to anthracycline related cardiotoxicity [43].

Carbonyl Reductases Carbonyl reductases (CBRs) are NADPH-dependent, cytosolic enzymes with broad substrate specificity for many endogenous and xenobiotic carbonyl compounds. CBR1 and CBR3 reduce the anthracycline anticancer drugs, doxorubicin and daunorubicin to their C-13 hydroxy metabolites, doxorucinol and daunorubicinol, respectively [48]. These metabolites exert cardiotoxicity by inhibition of Ca^{+2}/Mg^{2+}-ATPase in the sarcoplasmic reticulum, as well as by inactivation of the cytoplasmic aconitase/iron regulatory protein-1 complex [49]. The SNPs in *CBR1* and *CBR3* genes are responsible for the activity of these enzymes and consequently the production of the cardiotoxic metabolites. The *CBR1 V88I* polymorphism encodes for CBR1 protein isoforms (CBR1 V88 and CBR1 I88) with distinctive catalytic properties. The *CBR1 V88* isoform synthesised higher levels of the cardiotoxic metabolite daunorubicinol than *CBR1 I88* isoform [50]. At least one variant, rs1056892, in the *CBR3* gene was identified to be associated with an increased risk of cardiotoxicity in paediatric cancer patients treated with anthracycline [51]. Patients with *CBR3 V244M* homozygous G genotype are at increased risk of cardiomyopathy compared with individuals with *CBR3*:GA/AA genotypes when exposed to low- to moderate-dose of anthracyclines [52]. The *CBR3 Val244Met* SNP was associated with trastuzumab induced cardiotoxicity in breast cancer patients [31].

Aldo-Keto Reductases Aldo-keto reductases (AKR) are members of NAD(P)H dependent oxidoreductases. These enzymes have the ability to reduce ketones and aldehydes to their corresponding alcohols. AKR1C3 (17β-hydroxysteroid dehydrogenase type V) is considered as the main doxorubicin reductase [53]. Several SNPs in *AKR* genes have been identified. The *AKR1C3* variants showed a significantly diminished ability to reduce doxorubicin [48].

Inflammation Inflammation plays an important role in doxorubicin-induced cardiotoxicity due to the activation of the innate immune system. Doxorubicin treatment upregulates the expressions of pro-inflammatory cytokines such as tumour nuclear factor (TNF)-α, interleukin (IL)-1β, and IL-6, via activation of the nuclear factor-κB (NF-κB), triggering the progression of cardiovascular diseases [54]. The study related to patient's sensitivity to doxorubicin highlighting the important role of human leukocyte antigen (HLA) complex polymorphism in inflammatory response and immune trafficking [55].

Toll-like receptor 4 (TLR4) is part of the innate immune system that responds to the endogenous and exogenous signals known as pathogen-associated molecular patterns (PAMPs) and damage-associated molecular patterns (DAMPs) and triggers the pathophysiological functions [56]. Doxorubicin induces the release of PAMP and DAMP, which consequently result in TLR4-mediated cardiac inflammation that contributes to cardiotoxicity. Studies in vivo confirmed that doxorubicin induced cardiotoxicity is associated with enhanced activation of the TLR4 signalling pathway which further includes cardiac inflammation, remodelling, oxidative stress, and apoptosis [56, 57]. Additionally, clinical data showed that both, TLR4 and TLR2 expression are higher in patients with diastolic dysfunction associated with doxorubicin treatment [58].

Hyaluronan Hyaluronan, an essential component of the extracellular matrix, is synthesised by enzyme hyaluronan synthase (HAS) of which vertebrates have three: HAS1, HAS2, and HAS3. HAS2 is overexpressed in breast cancer cell lines and is associated with lymph node metastasis [59], and HAS3 has been linked to increased inflammation and atheroprogression [60]. HA is also known to decrease heart damage caused by ROS in cardiovascular disease. The study on childhood cancer survivors confirmed that SNP rs2232228 in the *HAS3* gene increased the risk of anthracycline-induced cardiomyopathy among patients exposed to high-dose anthracyclines. The rs2232228 AA genotype was associated with an 8.9-fold increased risk for cardiomyopathy compared with the GG genotype [61].

MicroRNAs (miRNAs) MicroRNAs (miRNAs) are universally expressed small, non-coding RNA molecules with 19–25 nucleotides in length that play a fundamental role in the post-transcriptional regulation of gene expression [62]. miRNAs bind to complementary sequences on target messenger RNAs (mRNAs), and induced their degradation or inhibition of translation. This regulatory capacity puts miRNAs as crucial players in diverse biological processes like cell cycle control, differentiation, apoptosis, and stress response [63]. In the context of cardiovascular diseases, alterations in the miRNA expression can reflect pathological conditions such as myocardial infarction, fibrosis, hypertrophy, heart failure, and cardiomyopathy [64]. Certain miRNAs have been linked to cardiotoxicity. serving as a promising tool for early detection, prognosis, and therapeutic targeting. Based on a greater number of studies focused on cardiotoxicity, two miRNAs, miR-1 and miR-133, have been identified as muscle-specific miRNAs crucial for cardiac function. Elevated miR-1 levels have been associated to myocardial infarction, heart failure, and arrhythmias, and miR-133 is involved in the regulation of cardiac hypertrophy [65]. Additionally, miR-133 also regulates the expression of genes encoding ion channels, including KCNJ2 and CACNA1C, and dysregulation of ion channels can lead to arrhythmias and other forms of cardiotoxicity [66]. The results of the clinical study on patients with colorectal carcinoma confirmed a specific rise in the expression of two miRNAs, miR1254 and miR579, in patients with developed cardiotoxicity induced by bevacizumab. It was shown that the increase in miR1254 had the strongest correlation with bevacizumab-induced cardiotoxicity [67]. Several miRNAs like Let-7 family, miR-140, miR-143, miR-200 family, miR-208, miR-21, and miR-34a-5p are widely studied as cardio specific markers [66]. The identification of specific miRNAs as biomarkers for cardiotoxicity has significant clinical implications.

Iron Homeostasis Iron is an important player in various cellular processes, including oxygen transport, energy metabolism and DNA synthesis. However, disruption of iron homeostasis, especially iron overload, can lead to cardiomyopathy and heart failure [30]. The *HFE* (high iron) gene encodes the HFE protein that is responsible for maintaining iron homeostasis within cells. This protein is in close interaction with transmembrane transferring receptor and thereby regulating the intracellular iron levels, including cardiomyocytes. Mutations in the *HFE* gene can dysregulate iron homeostasis, leading to iron overload or iron deficiency. Excessive iron accumulation in cardiomyocytes can initiate oxidative stress and mitochondrial dysfunction, ultimately leading to cardiomyopathy and heart failure [1]. The *HFE* gene contains two main functional polymorphisms, *C282Y* (rs1800562) and *H63D* (rs1799945) that can affect the function of the HFE protein and consequently alter the iron [68]. It was found that *H63D* polymorphism (rs1799945) in paediatric patients with leukemia on anthracyclines therapy may be particularly associated with more prominent heart damage and cardiotoxicity, compared to the *C282Y* polymorphism [69].

Human Epidermal Growth Factor Receptor Type 2 Human epidermal growth factor receptor 2 (HER-2) is an important target in cancer therapy. It is estimated that 25-30% of breast cancer patients have overexpression of HER-2 receptors [1]. Chemotherapeutics like trastuzumab and pertuzumab are well known HER-2 inhibitors that significantly improved the outcomes for patients with HER-2 positive cancer cells [6]. It has been shown that some patients have specific SNPs associated with trastuzumab-related cardiovascular toxicity. Five SNPs, including rs11932853, rs28415722, rs9316695, rs8032978 and rs7406710, can be predictable for patients at higher risk of trastuzumab-induced cardiotoxicity [70]. *Pro1170Ala* is one of the common variants in the *HER-2* gene associated with trastuzumab-induced cardiomyopathy [71]. The *HER-2* gene variations together with neuregulin 1 (NRG-1) could inhibit the process of hetero-dimerization of the HER4/HER2 enzyme, which further affects downstream signalling pathways such as the PI3K-Akt pathway. If HER2 signalling is blocked, then the blockade of PI3K-Akt pathway will cause the accumulation of ROS in cardiomyocytes, which further triggers apoptosis and structural tissue damage [72].

Tyrosine Kinase Receptor Tyrosine kinase receptors are single-pass transmembrane receptors essential for signal transduction pathways and cell-to-cell communication. These receptors play the key roles in biological processes such as cellular growth, differentiation, motility and metabolism. Inhibitors of tyrosine kinase are anticancer drugs that inhibit enzyme's ATP binding pocket. It is known that tyrosine kinases inhibitors may induce changes in gene expression that contribute to heart failure. Investigations in different genome integrated analysis suggest that genetic variations in RNA binding protein-mediated regulation may influence the cardiotoxicity of multiple tyrosine kinase inhibitors including sunitinib, pazopanib, sorafenib, dasatinib and nilotinib [73]. Cardiotoxicity induced by tyrosine kinase inhibitors is a consequence of inhibition of other co-regulating factors like the platelet-derived growth factor receptor (PDGFR) and the vascular endothelial growth factor (VEGF), which further downregulate protein kinase B (AKT) and extracellular signal-regulated kinase (ERK) phosphorylation leading to coronary microvascular dysfunction [74]. Inhibition of AKT signalling pathway further exacerbate the ischemia-reperfusion injury, induce pressure overload, and oxidative stress. This dysregulation in signalling pathways can ultimately lead to apoptosis and death of cardiomyocytes [75]. A recent study evaluated the genetic predictors of ibrutinib-induced cardiotoxicity in patients with chronic lymphocytic leukaemia. Ibrutinib is the first Bruton's tyrosine kinase inhibitor and single nucleotide polymorphisms in *KCNQ1* and *GATA4* were significantly associated with cardiotoxic events [76].

G Protein-Coupled Receptor 35 The G protein-coupled receptor 35 (GPR35) is the member of G protein-coupled receptors that mediates a wide range of physiological processes including inflammation [77]. Using genome-wide association studies GPR35 was associated with diseases like inflammatory bowel disease, type 2 diabetes mellitus, and coronary artery disease [78]. The *in vitro* analysis of cardiomyocytes suggested that the overexpression of *GPR35* significantly decreased cell viability and promoted morphological changes [79]. A gene-based analysis identified a significant association between *GPR35* SNP and coronary artery disease [80]. In another similar analysis, using exome array data, significant association between *GPR35* gene mutation and chronic anthracycline-induced cardiotoxicity was found in paediatric cancer patients [81]. The analyses of *GPR35* gene expression in myocardial tissue from heart failure patients showed a significant upregulation compared with that in healthy controls. This gene was selected for a follow-up analysis as a potential therapeutic target [78].

Pharmacogenetics provides crucial information related to the development of cardiotoxicity associated with anticancer drugs. The identification of the functional genetic variants to these toxicities will provide better understanding of the molecular mechanisms and pathways responsible for drug action that would be crucial for the development of novel therapeutic approaches for the prevention or treatment of drug-induced cardiotoxicity. Nowadays, the specific focus of scientific community is to provide new and reliable genetic markers that could identify patients with the risk of cardiotoxicity, and who would benefit from careful monitoring. By means of these genetic tastings the clinicians can choose specific treatments for patients according to the genotype studied.

Drug-Induced QT Interval Prolongation

The long QT (LQT) interval, either congenital or acquired, is a well-known risk factor for the development of ventricular arrhythmias and sudden cardiac death. Congenital LQT syndrome (cLQTS) is related to the mutations of genes that code ion channels responsible for the repolarisation current and can be divided into phenotypes, clinical forms according to the most significant gene mutations [82]. Acquired LQT syndrome (aLQTS) can be associated with other disorders and diseases, primarily cardiovascular diseases, such as ischemic heart disease, chronic heart failure, accompanied or not by electrolyte imbalances [83]. However, according to the epidemiological data, it is most often caused by drugs (drug-induced LQTS; diLQTS) [84]. Although the drug is a "trigger" for LQTS, it has always been believed that a genetic predisposition is also necessary for the development of aLQTS. There are two explanations for this claim. First, the heritability of QT interval duration in the general population (excluding cLQTS) is estimated to be around 35% [85], and second, the first-degree relatives of patients with cLQTS have a higher risk of developing diLQTS than individuals who are not related [84].

Considering that the duration of the QT interval is genetically determined, it is logical to conclude that the risk of diLQTS is basically determined by the same mechanism. Many drugs have been examined from a pharmacogenetic perspective, and diLQTS is no exception. In more recent years, during the SARS-CoV-2 virus pandemic, a therapeutic approach using azithromycin and hydroxychloroquine was widely accepted in the "treatment" of COVID-19, although it was well documented that both drugs can prolong the QT interval [86, 87]. The occurrence of adverse reactions such as LQTS and arrhythmias highlighted the importance and potential of pharmacogenetics in identifying and reducing the risk of LQTS. Furthermore, the application of personalised medicine in LQTS has become increasingly necessary, primarily for predicting and preventing fatal arrhythmias. Today, pharmacogenetic tests are more accessible and affordable, enabling their wider use for assessing the risk of diLQTS and individualising pharmacotherapy.

Clinical trials of drugs first provide data on the association of gene mutations with pronounced or frequent diLQTS, thus contributing to the risk stratification of drug use in certain populations, depending on the genotype. Later, pharmacovigilance in post-marketing surveillance further identifies clinical cases of diLQTS and may indicate the need for additional testing of genes that were previously thought to have a minor impact on the occurrence of diLQTS [88]. Since drugs are the most common cause of aLQTS and ventricular arrhythmias, such as the *Torsades de Pointes* (TdP), which is often a reason for the withdrawal of the risky drug from the market, the preclinical phases require mandatory testing of the arrhythmogenic potential, whereas in clinical trials of the drug, monitoring regarding the risk of prolonging the QT interval and inducing arrhythmias in general is required [11]. In their drug application guidelines and safety reports, the regulatory bodies require timely updates of data on the risks of diLQTS, interactions with other drugs, diseases or the administration in patients who have an increased risk of LQTS [89].

On the other hand, diLQTS most often occurs only in people with certain gene mutations. Consequently, withdrawing drugs from the market would deprive their use for those who are not at risk of developing LQTS. For this reason, by understanding the genetic background of drug-induced QT intervals, pharmacogenetic testing could identify patients who are at risk, i.e., patients who would benefit therapeutically and thereby prevent any unnecessary withdrawal of the drug from the market [88].

The Influence of Gene Polymorphisms on the Pharmacokinetics of Drugs and the Risk of Developing LQT

To date, significant gene mutations affecting the pharmacokinetics of drugs with a proven risk of QT interval prolongation have been identified. According to the epidemiological data, a significant number of patients who have experienced drug-induced TdP arrhythmia are carriers of mutations in genes responsible for cLQTS [90]. These patients often have cLQTS, probably as a result of incomplete penetrance of this syndrome, which can remain asymptomatic for a lifetime, but regardless of that, they have a significant risk of diLQTS and TdP.

Cytochrome P450 Substrates The microsomal enzymes, primarily the cytochrome P450 (CYP) family, are responsible for phase I metabolism of most drugs. Many drugs known to prolong the QT interval are metabolised by CYP2D6, CYP3A4/5 and CYP2C9 subfamilies [91]. A number of CYP P450-related genes have been studied regarding their impact on the pharmacokinetics and pharmacodynamics of drugs that have demonstrated the potential risk for the prolongation of QT intervals. Particularly interesting are carriers of polymorphisms for *CYP2D6* and/or CYP3A4/5 genes, known as poor metabolisers (PM), which are believed to be at a higher risk of diLQTS. Regarding the assessment of the risk for developing diLQTS, the gene most extensively studied is *CYP2D6*. A clear correlation has been demonstrated between mutations in the *CYP2D6* gene and LQTS induced by antipsychotics, risperidone, and iloperidone [92, 93]. However, no polymorphism of the *CYP2D6* gene and correlation with LQTS was found in patients treated with haloperidol and thioridazine [83]. The mutations of the *CYP2C19* gene have been also studied since the activity of CYP2C19 is proven to be a predictable factor for the prolongation of QT interval in patients using numerous antimicrobial drugs, except for quinolones [94]. Contrary to the expectations, the activity of CYP2C19 did not affect the length of the QT interval during the administration of drugs known to pose a risk of LQTS and TdP, such as nelfinavir or thioridazine [95, 96]. The CYP2B6 poor metabolisers are associated with the occurrence of LQTS during the administration of methadone having a 4.5 times higher risk of developing diLQTS, compared to normal or extensive metabolisers [97]. In addition, numerous studies have shown that the

CYP3A4/5 and *CYP2C9* polymorphisms are associated with an increased risk of LQTS or TdP induced by antimicrobial drugs, with the exception of quinolones [94].

P-Glycoprotein ABCB1 transporter, or P-glycoprotein, is a transmembrane protein that regulates the transport of xenobiotics and functions as an "efflux pump". It affects the bioavailability of numerous drugs, known as P-glycoprotein substrates. It is encoded by the ATP-binding cassette B1 (ABCB1) gene. The frequently studied polymorphisms of this gene are *C1236T*, *G2677T* and *C3435T*, and these are most often associated with diLQTS caused by P-glycoprotein substrates [98]. In a study conducted in patients suffering from schizophrenia and treated with risperidone, the QT interval was significantly longer in patients carrying the T allele of the *C3435T* polymorphism compared to homozygous carriers of the C allele [99]. The preclinical study with romidepsin, a chemotherapeutic agent that has the potential to prolong the QT interval, has shown that rodents lacking P-glycoprotein have higher drug concentrations in cardiomyocytes. This study also showed that carriers of certain *ABCB1* gene variants have less QT prolongation during the administration of romidepsin. Accordingly, P-glycoprotein has the potential to modify the action of drugs that have the potential to prolong the QT interval [36].

The Influence of Gene Polymorphisms on the Pharmacodynamics of Drugs and the Risk of Developing LQT

Although the focus of this chapter is diLQTS, it is important to make a brief description of cLQTS and genes related to this entity. The duration of the action potential is controlled by voltage- and time-dependent ion channels. The action potential prolongation occurs as a result of a decrease in the outward repolarisation current in K^+ channels or an increase in the inward current in Na^+ and Ca^{++} channels of cardiomyocytes [100]. cLQTS are caused by genetic mutations that disrupt electrical activity and consequently influence the normal cardiac function. It is now known that cLQTS are associated with mutations in the genes encoding K^+ and Na^+ channels. These syndromes are categorised into 15 subtypes (LQT1-15), and the subtypes LQT1-3 are the most common ones [101]. It is known that 17 genes are linked to cLQTS, but the results of genetic studies have shown that only three genes (*KCNQ1*, *KCNH2*, and *SCN5A*) are responsible for typical cLQTS [102]. Mutations in these three genes are believed to account for approximately 95% of all cases of cLQTS [86]. Both subtypes, LQT1 and LQT2 are associated with genes encoding K^+voltage-gated channels involved in the repolarisation of cardiomyocytes, while the LQT3 subtype is caused by mutations in a Na^+ voltage-gated channel [101]. The LQT1 type is related to a gene named *KCNQ1* which encodes for Kv7.1. protein, and loss-of-function mutations of this gene decrease the slow delayed rectifier current. On the other hand, the LQT2 type originates from loss-of-function mutations of the gene *KCNH2*. This gene produces a protein called *human Ether-á-go-go-related gene* (hERG, subunit of Kv11.1, a voltage-gated K^+ channel) that participates in the rapid delayed rectifier current of the cardiac action potential [103]. The gain-of-function mutations in *SCN5A* gene are responsible for the LQT3 subtype that is related to the potentiated voltage-dependent Na^+ current [86].

hERG Channel The hERG channel has become a subject of more intense studies in the last decade, after it was demonstrated that a mutation in the *KCNH2* gene causes reduced activity of the hERG channel, resulting in prolonged repolarisation of the myocardium (delayed I_{Kr} current) [103]. More to this, hERG is a target site of binding and action of some drugs, including many that prolong the QT interval. The hERG channel can be blocked by drugs in two different ways: through direct blockade or indirectly, by reducing the number of membrane channels. The resulting blockade of the I_{Kr} current in turn leads to excessive prolongation of the action potential that increased the risk of early depolarisation and consequently ventricular arrhythmia and sudden cardiac death [104]. Almost all drugs associated with TdP or LQTS block I_{Kr}, but conversely, not all I_{Kr}-blocking drugs cause the prolongation of the QT interval. In a study conducted in patients treated with antipsychotics, a significant correlation was demonstrated between the presumed hERG blockade and the prolongation of the QT interval induced by five I_{Kr}-blocking drugs (haloperidol, olanzapine, risperidone, thioridazine, and ziprasidone) [105].

NOS1AP Gene The *NOS1AP* gene (nitric oxide synthase 1 adaptor protein gene) encodes the nitric oxide synthase activating protein 1 and regulates neuronal NO synthase (nNOS), which is responsible for intracellular Ca^{++} metabolism and the contraction of cardiomyocytes. This enzyme inhibits the L-type Ca^{++} channels and affects the repolarisation of the myocardium and the propagation of impulses [106]. The expression of the *NOS1AP* gene affects the duration of the QT interval [107], and mutations in this gene are associated with an increased risk of LQTS induced by amiodarone, sotalol, and diuretics that cause hypokalaemia [108]. Given that NOS1AP is associated with the inhibition of the L-type Ca^{++} channels, it can be expected to be involved in the pharmacodynamic response and length of the QT interval during the administration of verapamil, and, to a lesser extent, diltiazem [109]. The results of a study examining the role of NOS1AP in the prolongation of the QT interval induced by Ca^{++} channel blockers showed that the rs10494366 GG genotype was associated with a greater prolongation of the QTc interval upon the administration of verapamil, compared to the homozygous wild-type genotype. In addition, mutations in the *NOS1AP* gene, which encodes proteins involved in β-adrenergic and sex hormone signalling pathways, may also be associated with a higher risk of diLQTS. Moreover, NOS1AP may be associated with LQT and TdP induced by antipsychotics, as well as by granisetron and ondansetron, particularly in heterozygous and homozygous carriers of the main allele rs10494366 [110].

Drugs That Cause Acquired Long QT Syndrome

Due to serious adverse effects, including fatal outcomes, there has been a demanding interest to categorise the drugs according to the relative risk of causing the TdP or QT prolongation. These categories include the drugs that prolong the QT interval and are associated with a known risk of TdP, those with a possible risk of TdP such as drugs that cause QT prolongation but no evidence for a risk of TdP, those with a conditional risk of TdP, meaning that some drugs can cause TdP but only under certain conditions, and drugs to be avoided in cLQTS [111]. As previously described, the majority of drugs can induce QT prolongation through I_{Kr} blockade by Kv11.1 inhibition, or by inhibition of Kv11.1 channel trafficking [100]. However, some additional risk factors should not be neglected in order to adequately prevent aLQTS such as female sex, electrolyte disbalances, over 65 years of age, heart disease, bradycardia, renal failure, diabetes, as well as concomitant use of two or more QT prolonging drugs [112]. The list of drugs with significant potential to cause serious adverse reactions (diLQT, TdP or sudden cardiac death) is presented in Table 28.1.

Table 28.1 The list of drugs associated with cardiotoxicity based on possible arrhythmogenic actions

Drug class	Drug	Possible arrhythmogenic mechanism(s)
Antibiotics	Erythromycin, clarithromycin	hERG inhibition
	Roxithromycin	I_{Kr} block
	Levofloxacin	I_{Kr} block, hERG inhibition
	Grepafloxacine, sparfloxacine	hERG inhibition
Antiarrhytmics	Amiodarone	I_{Kr} block
	Disopyramide	I_{Kr} block, I_{Na}
	Procainamide	I_{Kr} block
	Flecainide	I_{Kr} block, inhibition of CYP3A4
	Sotalol	I_{Kr} block, $I_{Na\text{-}L}$ augmentation
Antidepressants	Imipramine	$I_{Na,}$ hERG inhibition
	Fluoxetine	$I_{Na,}$ $I_{Ca,\ L,}$ hERG current and trafficking block
	Citalopram	hERG current and trafficking inhibition
Antiemetics	Domperidon	I_{Kr} block, hERG inhibition
	Ondansetron	I_{Kr} block
Antiepileptics	Retigabine	hERG, I_{Na} inhibition
	Lacosamide	I_{Na} inhibition
Antifungals	Fluconazole	hERG current and trafficking inhibition
Antimalarials	Chloroquine	I_{Kr} block, hERG inhibition
Antihistamines	Astemizole	hERG inhibition

(continued)

Table 28.1 (continued)

Drug class	Drug	Possible arrhythmogenic mechanism(s)
	Terfenadine	$I_{Na,}$ hERG inhibition
Antimuscarinics	Terodiline	hERG inhibition
Anestetics	Propofol	I_{Kr} block, I_{to} block
	Sevoflurane	I_{Kr} block, Ito block
Antipsyhotics	Haloperidol	hERG inhibition
	Risperidone	hERG inhibition
	Sulpiride	I_{Kr} block hERG inhibition
	Clozapine	hERG inhibition
	Thioridazine	I_{Kr} block, $I_{Na\text{-}L}$ augmentation
β_2-agonists	Salbutamol	hERG inhibition
Cholinesteraze inhibitors	Donepezil	IKr block, hERG inhibition
$NSAID_S$	Diclofenac	$I_{Na,}$ hERG, I_{Ks} inhibition
	Celecoxib	$I_{Na,}$ hERG, I_{Ks} inhibition
Opioid analgesics	Methadone	$I_{Na,}$ hERG inhibition
PDE inhibitors	Milrinone (PDE3 inhibitor)	cAMP dependent SR Ca^{2+} relase, I_f activation
	Vardenafil (PDE5 inhibitor)	hERG inhibition
Prokinetics	Cisapride	hERG inhibition
Vasodilators	Bepridil	hERG, I_{Na} inhibition

hERG human ether-a-go-go-related gene potassium current, I_f hyperpolarization-activated cyclic nucleotide gated pacemaker 'funny' current, I_{Ks} slow component of the delayed rectifier potassium current, I_{Na} voltage-gated sodium current, I_{to} transient outward potassium current, *NSAIDs* non-steroidal anti-inflammatory drugs, *PDE* phosphodiesterase, *SR* sarcoplasmic reticulum. Please note that the information in this table is compiled from many studies cited in this chapter

Drug-Drug Interactions and Risk of QT Prolongation

Concomitant use of two or more drugs is a significant risk factor for drug-induced arrhythmias in some patients. This is particularly important for drugs with QT-prolonging properties in which the arrhythmogenic risk increases as a function of plasma drug concentration. Therefore, the factors that can increase the plasma drug concentration, like overdose, advanced liver disease, renal disfunction, or the use of specific CYP inhibitors, could be the risk factor for aLQTS [113]. Many strong CYP inhibitors can significantly influence the metabolism of drugs known to have QT prolongation and TdP risk. This is known as type X interaction in which the administration of two drugs interacts with each other in a clinically significant manner. The inhibition of the drug's corresponding CYP can consequently increase the plasma concentration of the drug which can dramatically increase the risk of QT prolongation, TdP and a fatal outcome. The best example is the cardiotoxic interaction between terfenadine and antimycotic drug ketoconazole in which ketoconazole, as a strong CYP3A4 inhibitor, increased serum concentration of unmetabolised terfenadine and thus enhanced the QT-prolongation leading to TdP [114]. Terfenadine is a potent I_{Kr}-blocking drug that usually undergoes complete metabolism via CYP3A4 to a metabolite fexofenadine, which retains the antihistamine activity without cardiotoxic activity. Similarly, astemizole, as a second-generation antihistaminic drug, was also associated with diLQTS due to drug overdose or interactions related to the CYP3A4 inhibition [115]. The prolongation of QT interval and increased risk of TdP was also noticed for cisapride, especially as a consequence of overdosing or drug interactions [116]. Due to the serious ADR related to CYP drug interactions, a number of highly prescribed drugs such as terfenadine, astemizole, cisapride, thioridazine, grepafloxacin were withdrawn from the market [84]. The interactions between drugs with a well-known potential to induce TdP and strong CYP inhibitors are presented in Table 28.2.

Table 28.2 The list of drugs with known TdP risk that can interact with CYP inhibitors

CYP metaboliser	Drug with risk of sQT prolongation and TdP	Drug with CYP-inhibitory potency
CYP3A4	Amiodarone	Voriconazol
		Indinavir
		Ritonavir
		Saquinavir
		Clarithromycin
	Disopyramide	Itraconazol
		Clarithoromycin
		Ketoconazol (systemic)
		Saquinavir
		Talithromycin
	Dofetilide	Clarithomycin
		Itraconazole
		Ketoconazole
	Dronedarone	Clarithromycin
		Ceritinib
		Indinavir
		Itraconazole
		Nefazodone
		Ritonavir
		Saquinavir
		Telithromycin
		Voriconazole
	Ibutilide	Clarithromycin
	Quinidine	Clarithromycin
		Itraconazol
		Ketoconazol
		Ritonavir

(continued)

Table 28.2 (continued)

CYP metaboliser	Drug with risk of sQT prolongation and TdP	Drug with CYP-inhibitory potency
		Saquinavir
		Voriconazole
	Azithromycin	Certinib
		Clarithromycin
		Saquinavir
		Voriconazole
	Clarithromycin	Saquinavir
	Erythromycin	Saquinavir
	Cloroquine	Certinib
		Clarithromycin
		Saquinavir
		Voriconazole
	Chlorpromazine	Certinib Saquinavir
		Voriconazol
		Clarithomycin
	Haloperidol	Saquinavir
	Levomepromazine	Ritonavir
		Saquinavir
	Escitalopram	–
	Ondansetron	–
	Cisapride	Certinib
		Clarithromycin
		Indinavir
		Itraconazole
		Ketoconazole
		Nefazodone
CYP2D6	Flecainide	
	Ibutilide	Quinidine
	Procainamide	Quinidine
	Chloroquine	–
	Hydroxychloroquine	–
	Chlorpromazine	–
	Haloperidol	–
CYP2C19	Citalopram	–
	Escitalopram	
	Chlopromazine	–

Please note that the information in this table is compiled from many studies cited in this chapter

In genetic epidemiology, genome-wide association studies (GWAS) have identified and enhanced the knowledge of a large number of genes associated with the duration of the QT interval. The results of GWAS have demonstrated numerous gene polymorphisms in cLQTS, genes associated with intracellular Ca^{++} transport, associated with the activity of enzymes and receptors in the metabolism and activity of drugs, as well as genes for which the impact on myocardial repolarisation was not previously proven [83, 117] Genetic factors certainly have a key role in the risk of developing diLQTS in each patient. As previously mentioned, mutations in genes that have been proven to cause cLQTS are also associated with an increased risk of diLQTS. Furthermore, GWAS and whole-exome sequencing have been used to search for the genes responsible for diLQTS, revealing several newer candidates, such as members of the SLC22A23 and SLCO3A1 families [118], the Ceramide Kinase-Like Gene (*CERKL*), the adaptor protein gene *NOS1AP*, the paladin gene (*PALLD*), the *Bruno-like* 4 gene (*BRUNOL4*), the neuregulin 3 gene (*NRG3*), the nucleotide-binding protein-like gene (*NUBPL*), the voltage-gated potassium channel regulatory subunit 1 (*KCNE1*) gene, and the succinate dehydrogenase assembly factor 3 (*SDHAF3*) gene [83, 86]. However, the genetic mutations identified so far explain only a certain impact of heredity, while a significant part of it is considered as a "missing heredity". By applying the results of GWAS, the association of polymorphisms significant for drug pharmacokinetics and pharmacodynamics can be examined.

The importance of pharmacogenomics lies in the identification and assessment of the risk of drug-gene interactions for potentially induced diLQTS. From a practical point of view, testing specific genes is more feasible in proving the mechanism of drug-induced LQT and explaining interindividual variability in drug response, which contributes to safer use of many drugs. Accordingly, future studies could be focused on identifying gene polymorphisms with the best predictive value for the occurrence of ADR of the LQTS type, as well as, identifying new drug-gene interactions, epigenetics, clinical outcomes, or using alternative markers to predict the risk and occurrence of drug-induced arrhythmia and personalised pharmacotherapy.

Pharmacogenomics in Cardiovascular Drug Therapy

Drugs used in cardiology have very proven effectiveness and safety, but it is not rare that these drugs can also cause significant adverse reactions. Many factors can influence the interindividual variability in drug response and toxicity, but the genetic variability could be the most relevant one [119]. Pharmacogenetics consortia have issued guidelines for oral anticoagulants, antiplatelet agents, statins, and some beta-blockers [5]. Since the majority of pharmacogenetic recommendations are based on the assessment of single drug-gene interactions, it is imperative to develop tools for the prediction of multiple drug-drug-gene interactions. Several Scientific Societies and working groups, including the Clinical Pharmacogenetics Implementation

Consortium (CPIC), the Dutch Pharmacogenetics Working Group (DPWG), the French Network of Pharmacogenetics (RNPGx), and the Canadian Pharmacogenomics Network for Drug Safety (CPNDS) provide consensus recommendations for specific drug-gene pairs with proven scientific evidence that could significantly improve the therapy according to patient genotype [5]. Recently, the European Society of Cardiology Working Group on Cardiovascular Pharmacotherapy issued a position statement on the role of Pharmacogenomics in contemporary cardiovascular therapy [89].

Anticoagulants

Coumarin derivatives, such as warfarin, acenocoumarol and phenprocoumon are frequently used to prevent or treat thromboembolism. These drugs have narrow therapeutic window and due to significant inter- and intra-individual variability in patients' response, their use may be associated with the risk of bleeding or resistance to treatment. Coumarins inhibit vitamin K epoxide reductase complex subunit 1 (VKORC1), leading to decrease of clotting factors II, VII, IX, and X [120]. The S-warfarin enantiomer is preferentially metabolised by CYP2C9 and is approximately 3–5 times more potent than R-warfarin [121]. Both acenocoumarol enantiomers (S- and R-) are metabolised by CYP2C9. S-acenocoumarol is more active and has a very short elimination half-life of less than 2 h, and most of the anticoagulant effect is achieved by R-acenocoumarol. However, phenprocoumon is metabolised by both CYP2C9 and CYP3A4 [89]. In the Caucasian population, there are two main genetic variants of CYP2C9 associated with deficient CYP2C9 activity: polymorphism rs1799853 (*CYP2C9**2 or c.430C > T) and rs1057910 (*CYP2C9*3* or c.1075A > C) [122]. *CYP2C9* gene polymorphism differ between racial/ethnic groups, and differently from the Caucasian population these alleles are less prevalent in Asian and African-American populations [123]. The *CYP2C9*2* and *CYP2C9*3* polymorphisms are associated with significant reduction of S-warfarin metabolism by 30–40% and 80–90%, respectively. Patients with these polymorphisms are at an increased risk of bleeding when treated with warfarin and require lower doses as compared to patients with two normally-functioning alleles [124]. Other types of *CYP2C9* polymorphisms (*CYP2C9*5, *6, *8* and **11*) present in the African population, are also associated with the decreased CYP2C9 enzyme function and require lower warfarin doses [125].

Warfarin is a competitive inhibitor of the vitamin K epoxide reductase complex, subunit 1 VKORC1, which is encoded by the *VKORC1* gene [126]. Patients with the rs9934438 AA genotype are more sensitive to the warfarin treatment and they must have their therapeutic doses reduced by 70% [127]. However, patients with the *VKORC1* (GA) polymorphism are less sensitive to warfarin than normal individuals and require average doses of warfarin [124, 128]. It has been found that the G allele of *VKORC1*–1639 and the C allele of *VKORC1*–1173 were present in higher frequencies in the Caucasian population [129, 130]. CYP4F2 acts as an important counterpart of VKORC1 to limit excessive vitamin K accumulation. Patients with *CYP4F2* rs2108622 CT/TT genotypes required significantly higher warfarin dose than the CC carriers [129]. This polymorphism is present with higher frequency in Indians and Caucasians than in the Chinese and African-American population [131]. The inclusion of *CYP4F2* variants in warfarin dosing schemes together with CYP2C9, VKORC1 significantly improved the accuracy of dose prediction. Acenocoumarol, phenprocoumon or fluindione, as cumarin derivatives, also exert their anticoagulant action by inhibiting VKORC1, and mutations in the VKORC1 gene that lead to a reduced production of the VKORC1 protein would also require lower doses of these drugs to achieve the therapeutic effect. However, the use of these derivatives is highly variable across different countries: warfarin is predominantly used in the UK and Italy, phenprocoumon in Germany, acenocoumarol in Spain, and fluindione in France, while in the USA warfarin is the only vitamin K antagonist approved by the Food and Drug Administration (FDA) [5, 132].

Dabigatran, as a thrombin inhibitor, is a direct oral anticoagulant widely used for deep venous thrombosis and atrial fibrillation instead of vitamin K antagonists. As a prodrug, dabigatran etexilate is sequentially metabolised by intestinal carboxylesterase 2 (CES2) and in the second step by hepatic carboxylesterase 1 (CES1) to form its active metabolite dabigatran. It is reported that the *CES1* SNPs rs2244613 and rs8192935 were associated with lower plasma concentrations of dabigatran [133]. However, it was found that neither rs2244613 nor rs8192935 were associated with the dabigatran activation in human livers, but the *CES1 G143E* is a loss-of-function variant and is responsible for drug accumulation [134]. They also found that the hepatic CES1 activity was significantly higher in female liver samples than in male.

Antiplatelets

CYP2C19 is the main enzyme involved in the conversion of clopidogrel as a prodrug into its active form. Clopidogrel is an irreversible inhibitor of the P2Y12 receptor and thus prevents platelet activation and aggregation [135]. Approximately 30% of individuals possess the reduced or absent enzyme activity due to genetic polymorphisms of CYP2C19. Individuals who have 2 loss-of-function alleles of the *CYP2C19* gene (*CYP2C19*2* and**3*) are classified as poor metabolisers (PM) and they have a significantly reduced enzyme activity. These individuals cannot activate clopidogrel via CYP2C19, meaning the drug will have a reduced antiplatelet effect. It is estimated that 2% of Caucasians, 4% of African Americans, 14% of Chinese, and 57% of Oceanians are with *CYP2C19*2* and**3* phenotype [130, 136]. Patients with any two CYP2C19 loss-of-function alleles [*2, *3 (rs4986893), *4 (rs28399504), or *5 (rs56337013)] had a higher risk of major adverse cardiovascular events than others [137]. There are also individuals with the *CYP2C19*17* allele that is associated with the increased enzyme function, known as "ultra-rapid metabolisers". The frequency of the *CYP2C19*17* allele is approximately 22% for

Europeans, 17% in the African populations, 8% for the Americas, 0.5–5.7% for the Asian populations [130, 136]. According to the international guidelines, patients who are intermediate and poor metabolisers are recommended to switch from the clopidogrel to an alternative platelet inhibitor that is not metabolised by CYP2C19 (e.g., prasugrel, ticagrelor) due to the risk of ineffectiveness and consequent thrombotic events.

Statins

Statins are a class of drugs that act as inhibitors of 3-hydroxy-3-methylglutaryl-coenzyme A (HMG-CoA) reductase, which are commonly used to reduce lipid levels in patients with cardiovascular disease. Simvastatin is one of the most prescribed statins but its use is associated with an increased risk of myopathy that occur in about 1 per 10,000 patients per year [138].

The risk of myopathy is associated with genetic variation in the Solute carrier organic anion transporter family member 1B1 (*SLCO1B1*) gene. This gene encodes organic anion transporter polypeptide 1B1 (OATP1B1), a transporter that is essential for hepatic uptake and subsequent clearance of statins. The pharmacokinetic parameters like AUC and Cmax of simvastatin were 120 and 221% higher in individuals with the *SLCO1B1* c.521CC genotype than in those with the c.521TC and c.521TT genotypes, respectively. The Cmax of simvastatin occurred earlier in participants with the c.521CC and c.521TC genotypes than in those with the c.521TT genotype [139]. The reduced transporter activity is associated with significantly lower LDL-C reductions and a higher risk of myopathy [140]. The minor C allele at rs4149056 is contained within *SLCO1B1*5*, as well as the *15 and *17 haplotypes and is associated with lower plasma clearance of simvastatin [141]. It is recommended that patients carrying at least one C allele of the *SLCO1B1* reduced function should be prescribed with decreased daily dose of simvastatin or to be switched to another statin. However, if the homozygous variant (CC) is present, the alternative therapy should be prescribed.

Some data has shown associations between the C allele variant and higher rates of atorvastatin intolerance and myopathy. However, the literature is currently unclear as to whether *SLCO1B1* c.521 T > C is significantly associated with the discontinuation of atorvastatin due to myopathy [142]. Additionally, it has been found that the *SLCO1B1 521C* mutant allele is associated with an increased risk of rosuvastatin-associated myotoxicity in the Chinese population [89].

Beta-Blockers

Beta-blockers, such as metoprolol, carvedilol, and propranolol are metabolised by the CYP2D6 enzyme, which is encoded by the highly polymorphic gene, *CYP2D6*.

Of these, metoprolol is highly dependent on CYP2D6 for its elimination, with 70–80% of an oral dose undergoing the CYP2D6-mediated biotransformation [143]. Other beta-blockers, like atenolol and bisoprolol, undergo either no or negligible CYP2D6 metabolism. It is well known that the reduction in metoprolol metabolism leads to increased concentrations that dan consequently lead to adverse effects. Four phenotype groups are associated to the *CYP2D6* polymorphism: ultra-rapid metabolisers (UM), normal metabolisers (NM), intermediate metabolisers (IM), and poor metabolisers (PM) [144]. The *CYP2D6 PMs* (*3, *4, *5 alleles) have no enzyme activity, and patients with this phenotype can exhibit a several-fold increase in metoprolol concentrations compared to NMs [145]. Due to the increased metoprolol concentrations, patients with *CYP2D6* IM and especially PM phenotypes experience lower heart rates and blood pressure. According to clinical guidelines, if symptomatic bradycardia occurs, it is recommended to reduce the dose of metoprolol to 25–50% of the standard dose [89]. In the case of UM, it is recommended to prescribe the maximum standard dosage of metoprolol, or if needed, to increase the dose up to 2.5 the standard dose. Alternatively, the patients could be advised to other betablockers like atenolol, bisoprolol or carvedilol [5]. The doctors should be cautious when prescribing concomitant treatment with CYP2D6 inhibitors (e.g., bupropion, fluoxetine, paroxetine, and quinidine) since it can compromise the current therapy [89].

Antiarrhythmics

Among the antiarrhythmic group of drugs, flecainide and propafenone are of particular interest since the pharmacokinetic characteristics of these drugs are very much influenced by gene polymorphism.

Flecainide belongs to class Ic antiarrhythmic drug that blocks the cardiac Na^+ channels. Flecainide is used to prevent and treat paroxysmal supraventricular tachycardia, Wolff-Parkinson-White syndrome, atrioventricular re-entrant tachycardia, and atrial fibrillation/atrial flutter. Flecainide is also an option in treating life-threatening ventricular arrhythmias. Flecainide acts on the fast-inward Na^+ ion channel, particularly having a high affinity to activated Na^+ channels. It prolongs depolarisation and increases refractoriness due to the slow release from its binding site. It also works by inhibiting IKr channels, delaying potassium rectifier current, thus resulting in prolongation of action potential duration in both ventricular and atrial muscle fibres. Flecainide is shown to block the ryanodine receptor opening, which reduces the calcium release from the sarcoplasmic reticulum resulting after depolarisation and triggered activity [146].

Flecainide is mainly metabolised by CYP2D6 and patients who are IM and PM (*CYP2D6*1, CYP2D6*4, CYP2D6*5, CYP2D6*10, CYP2D6*21, CYP2D6*36*), who are at risk of drug accumulation and consequently to adverse events, due to the drug's relatively narrow therapeutic range. Due to this increased risk of adverse events, it is recommended to reduce the flecainide dose by half in PM, and by quarter in IM, and monitore the effect via the ECG and plasma concentration, if possible [5]. Similarly, patients characterised as UM should be also monitored due to the increased clearance of flecainide.

Propafenone is a class 1C antiarrhythmic drug that primarily blocks the cardiac Na^+ channel. It is also a weak K^+ channel blocker and has some beta-adrenergic receptor blocking effects. It works by slowing conduction and prolonging the refractory time of the cardiac conduction tissue. Propafenone is indicated for treatment of ventricular and supraventricular arrhythmias, and atrial fibrillation.

The primary metabolism of propafenone is to 5-hydroxypropafenone and N-des alkyl-propafenone mainly by the CYP2D6 enzyme, but CYP3A4 and CYP1A2 enzymes have also been reported to contribute to the drug's metabolism. Individuals characterised as PM are at risk of propafenone accumulation due to a decreased clearance of the drug [147]. Similarly, drugs with higher affinity can inhibit CYP2D6, CYP3A4, and CYP1A2 may also increase the propafenone levels, which may lead to cardiac arrhythmia episodes. It is recommended that patients characterised as PM should have a 70% reduction of the standard dose compared to NM. For those who are UM, a specific dose adjustment is needed, or patients should be switched to the alternative antiarrhythmic drug that is not a CYP2D6 substrate [5].

Antihypertensives

Losartan is an angiotensin II type 1 receptor (AT1) antagonist used for the treatment of hypertension and other cardiovascular diseases like hypertension with left ventricular hypertrophy or nephropathy with the type 2 diabetes mellitus. Losartan is mainly metabolised by CYP2C9 to its active metabolite E-317418. However, the polymorphic variants like *CYP2C9 *2* and **3* are associated with a reduced metabolism of losartan and consequently with les formation of its active metabolite, having significant influence on the treatment outcomes [148].

Additionally, ABCB1 gene polymorphism have been associated to response to losartan. The ATP-binding cassette sub-family B member 1 (ABCB1, MDR1, P-glycoprotein) is a transporter protein that is responsible for the cellular efflux of many drugs. Multidrug resistance 1 (*MDR1*) gene is expressed in the cells of different tissues and organs, and many drugs such as antihypertensives, antiarrhythmics, antineoplastics, HIV protease inhibitors, and immunosuppressants are MDR1 substrates. It has been found that the hypotensive response to losartan was significantly affected by the *C3435*T genetic polymorphism of *MDR1* and hypertensive patients with *MDR1 3435T* allele may present a better response to the losartan treatment [149].

ACE Polymorphism

Angiotensin-converting enzyme (ACE) plays a major role in the regulation of blood pressure by cleaving angiotensin I into angiotensin II. This dicarboxypeptidase is responsible for degradation of other vasoactive peptides like bradykinin, kallidin, substance P [150]. The insertion (I)/deletion (D) polymorphism of ACE gene is responsible for functional polymorphism which regulates the circulating and tissue ACE activity and this gene has been studied extensively with renal and cardiovascular complications of diabetic nephropathy [151]. It was suggested that the ACE genotype may predict the patient's response to the antiproteinuric and renoprotective effect of ACE inhibitors (ACEIs). Current data show that the *ACE* I/D polymorphism directly influences the circulating levels of ACE. Additionally, the II genotype has a protective role against the development of diabetic nephropathy, while the DD genotype predicts poor renal response to ACEIs [151, 152]. This genotype is also associated with lower effectiveness of captopril, enalapril or spironolactone [5]. The DD genotype of the rs1799752 *ACE* variant was associated with an increased requirement for mechanical ventilation in patients with COVID-19, and the low serum ACE activity could be related to the patient's severe outcome [153].

Conclusion

Drug-induced cardiovascular toxicity is a type of ADRs with serious clinical problems that show differential manifestations depending on age, sex, race, genotype, comorbidity, drug category, route of drug administration, or drug–drug interactions. Pharmacogenetics serves as a key factor for the drug response variability, with a significant role in the incidence and severity of ADRs. Pharmacogenetics provided data of drugs used for cancer treatment, the treatment of cardiovascular diseases, or neurological and psychiatric disorders are of valuable clinical importance for optimizing therapeutic and reducing cardiovascular toxicity. Gene polymorphisms are directly related to individual differences in the drug effects and toxicities, and the identification of the functional genetic variants related to these effects will significantly improve our understanding of the potential mechanisms and pathways related to drug metabolism or drug action. This can bring an additional value to clinicians for better selection of specific treatments for patients according to the genotype. Moreover, by using pharmacogenetic data it will be easier to study and compare the differences in drug efficacy, the ADRs or toxicity among patients with different genotypes. Clinician attitudes toward pharmacogenetics are vital to the success of translational programs and full implementation of this discipline into clinical practice.

References

1. Li MY, Peng LM, Chen XP (2022) Pharmacogenomics in drug-induced cardiotoxicity: current status and the future. Front Cardiovasc Med 9(1). https://doi.org/10.3389/fcvm.2022.966261
2. Porta-Sánchez A et al (2017) Incidence, diagnosis, and management of QT prolongation induced by cancer therapies: a systematic review. J Am Heart Assoc 6(12):1–19. https://doi.org/10.1161/JAHA.117.007724
3. Rhea IB, Oliveira GH (2018) Cardiotoxicity of novel targeted chemotherapeutic agents. Curr Treat Options Cardiovasc Med 20(7). https://doi.org/10.1007/s11936-018-0649-4
4. Osanlou O, Pirmohamed M, Daly AK (2018) Pharmacogenetics of adverse drug reactions. Adv Pharmacol 83:155–190. https://doi.org/10.1016/bs.apha.2018.03.002
5. García-González X, Salvador-Martín S (2021) Pharmacogenetics to avoid adverse reactions in cardiology: ready for implementation? J Pers Med. https://doi.org/10.3390/jpm11111180
6. Leong SL et al (2019) Roles of pharmacogenomics in non-anthracycline antineoplastic-induced cardiovascular toxicities: a systematic review and meta-analysis of genotypes effect. Int J Cardiol 280:190–197. https://doi.org/10.1016/j.ijcard.2018.12.049
7. Li J et al (2022) Value of UGT2B7-161 single nucleotide polymorphism in predicting the risk of cardiotoxicity in HER-2 positive breast cancer patients who underwent pertuzumab combined with trastuzumab therapy by PSL. Pharmacogenomics Pers Med 15:215–225. https://doi.org/10.2147/PGPM.S351718
8. Božina N et al (2020) Use of pharmacogenomics in elderly patients treated for cardiovascular diseases. Croat Med J 61:147–158. https://doi.org/10.3325/cmj.2020.61.147
9. Marković-Peković V et al (2016) Polypharmacy among the elderly in the Republic of Srpska: extent and implications for the future. Expert Rev Pharmacoecon Outcomes Res 16(5): 609–618. https://doi.org/10.1586/14737167.2016.1115347
10. Pirmohamed (2001) Pharmacogenetics and pharmacogenomics. http://snp.cshl.org
11. Ferdinandy P et al (2019) Definition of hidden drug cardiotoxicity: paradigm change in cardiac safety testing and its clinical implications. Eur Heart J 40:1771–1777. https://doi.org/10.1093/eurheartj/ehy365
12. Cacabelos R et al (2021) Genophenotypic factors and pharmacogenomics in adverse drug reactions. Int J Mol Sci 22(24). https://doi.org/10.3390/ijms222413302
13. Roden DM et al (2019) Pharmacogenomics. Lancet 394:521–532. https://doi.org/10.1016/S0140-6736(19)31276-0
14. Ferlay J et al (2013) Cancer incidence and mortality patterns in Europe: estimates for 40 countries in 2012. Eur J Cancer 49(6):1374–1403. https://doi.org/10.1016/j.ejca.2012.12.027
15. Cardinale D et al (2015) Early detection of anthracycline cardiotoxicity and improvement with heart failure therapy. Circulation 131(22):1981–1988. https://doi.org/10.1161/CIRCULATIONAHA.114.013777
16. Bajwa R et al (2019) Adverse effects of immune checkpoint inhibitors (programmed death-1 inhibitors and cytotoxic T-lymphocyte-associated protein-4 inhibitors): results of a retrospective study. J Clin Med Res 11:225–236. https://doi.org/10.14740/jocmr3750
17. Yang X, Li G et al (2020) Possible susceptibility genes for intervention against chemotherapy-induced cardiotoxicity. Oxid Med Cell Longev. https://doi.org/10.1155/2020/4894625
18. Yang X, An N et al (2020) Enhanced cardiomyocyte reactive oxygen species signaling promotes ibrutinib-induced atrial fibrillation. Redox Biol 30. https://doi.org/10.1016/j.redox.2020.101432
19. Rajić V et al (2009) Influence of the polymorphism in candidate genes on late cardiac damage in patients treated due to acute leukemia in childhood. Leuk Lymphoma 50(10):1693–1698. https://doi.org/10.1080/10428190903177212
20. Heymes C et al (2003) Increased myocardial NADPH oxidase activity in human heart failure. J Am Coll Cardiol 41(12):2164–2171. https://doi.org/10.1016/S0735-1097(03)00471-6

21. Rossi D et al (2009) Analysis of the host pharmacogenetic background for prediction of outcome and toxicity in diffuse large B-cell lymphoma treated with R-CHOP21. Leukemia 23(6):1118–1126. https://doi.org/10.1038/leu.2008.398
22. Wyche KE et al (2004) C242T CYBA polymorphism of the NADPH oxidase is associated with reduced respiratory burst in human neutrophils. Hypertension 43(6):1246–1251. https://doi.org/10.1161/01.HYP.0000126579.50711.62
23. Wojnowski L et al (2005) NAD(P)H oxidase and multidrug resistance protein genetic polymorphisms are associated with doxorubicin-induced cardiotoxicity. Circulation 112(24):3754–3762. https://doi.org/10.1161/CIRCULATIONAHA.105.576850
24. Shimo-Nakanishi Y et al (2004) Functional effects of NAD(P)H oxidase p22phox C242T mutation in human leukocytes and association with thrombotic cerebral infarction. Atherosclerosis 175(1):109–115. https://doi.org/10.1016/j.atherosclerosis.2004.01.043
25. Mandić D et al (2023) Overexpression of MRP1/ABCC1, survivin and BCRP/ABCC2 predicts the resistance of diffuse large B-cell lymphoma to R-CHOP treatment. Cancers 15(16). https://doi.org/10.3390/cancers15164106
26. Muckiene G et al (2023) The impact of polymorphisms in ATP-binding cassette transporter genes on anthracycline-induced early cardiotoxicity in patients with breast cancer. J Cardiovasc Dev Dis 10(6). https://doi.org/10.3390/jcdd10060232
27. Cascorbi I (2006) Role of pharmacogenetics of ATP-binding cassette transporters in the pharmacokinetics of drugs. Pharmacol Ther 112:457–473. https://doi.org/10.1016/j.pharmthera.2006.04.009
28. Ding Y et al (2022) Genetic susceptibility and mechanisms underlying the pathogenesis of anthracycline-associated cardiotoxicity. Oxid Med Cell Longev. https://doi.org/10.1155/2022/5818612
29. Jungsuwadee P et al (2009) Mrp1 localization and function in cardiac mitochondria after doxorubicin. Mol Pharmacol 75(5):1117–1126. https://doi.org/10.1124/mol.108.052209
30. Armenian SH et al (2013) Genetic susceptibility to anthracycline-related congestive heart failure in survivors of haematopoietic cell transplantation. Br J Haematol 163(2):205–213. https://doi.org/10.1111/bjh.12516
31. Kim Y, Seidman JG, Seidman CE (2022) Genetics of cancer therapy-associated cardiotoxicity. J Mol Cell Cardiol 167:85–91. https://doi.org/10.1016/j.yjmcc.2022.03.010
32. Sun S et al (n.d.) Pharmacogenomics in drug-induced cardiotoxicity: current status and the future
33. Fromm MF (2004) Importance of P-glycoprotein at blood-tissue barriers. Trends Pharmacol Sci 25:423–429. https://doi.org/10.1016/j.tips.2004.06.002
34. Meissner K et al (2002) Expression and localization of P-glycoprotein in human heart: effects of cardiomyopathy, the. J Histochem Cytochem. http://www.jhc.org
35. Hertz DL et al (2016) Evidence for association of SNPs in ABCB1 and CBR3, but not RAC2, NCF4, SLC28A3 or TOP2B, with chronic cardiotoxicity in a cohort of breast cancer patients treated with anthracyclines. Pharmacogenomics 17(3):231–240. https://doi.org/10.2217/pgs.15.162
36. Sissung TM et al (2011) Impact of ABCB1 allelic variants on QTc interval prolongation. Clin Cancer Res 17(4):937–946. https://doi.org/10.1158/1078-0432.CCR-10-0925
37. Bachur NR et al (1979) NADPH cytochrome P-450 reductase activation of quinone anticancer agents to free radicals. Med Sci
38. Kostrzewa-Nowak D et al (2005) The role of bioreductive activation of doxorubicin in cytotoxic activity against leukaemia HL60-sensitive cell line and its multidrug-resistant sublines. Br J Cancer 93(1):89–97. https://doi.org/10.1038/sj.bjc.6602639
39. Lubieniecka JM et al (2013) A discovery study of daunorubicin induced cardiotoxicity in a sample of acute myeloid leukemia patients prioritizes P450 oxidoreductase polymorphisms as a potential risk factor. Front Genet 4. https://doi.org/10.3389/fgene.2013.00231

40. Huang Z et al (2017) Effects of cytochrome P450 family 3 subfamily a member 5 gene polymorphisms on daunorubicin metabolism and adverse reactions in patients with acute leukemia. Mol Med Rep 15(6):3493–3498. https://doi.org/10.3892/mmr.2017.6470
41. Coles BF, Kadlubar FF (2003) Detoxification of electrophilic compounds by glutathione S-transferase catalysis: determinants of individual response to chemical carcinogens and chemotherapeutic drugs? Biofactors 17:115–130. https://doi.org/10.1002/biof.5520170112
42. McIlwain CC, Townsend DM, Tew KD (2006) Glutathione S-transferase polymorphisms: cancer incidence and therapy. Oncogene 25:1639–1648. https://doi.org/10.1038/sj.onc.1209373
43. Visscher H et al (2013) Validation of variants in SLC28A3 and UGT1A6 as genetic markers predictive of anthracycline-induced cardiotoxicity in children. Pediatr Blood Cancer 60(8):1375–1381. https://doi.org/10.1002/pbc.24505
44. Al-Otaibi TK et al (2022) Genetics of anthracycline-associated cardiotoxicity. Front Cardiovasc Med. https://doi.org/10.3389/fcvm.2022.867873
45. Lunardi M et al (2022) Genetic and RNA-related molecular markers of trastuzumab-chemotherapy-associated cardiotoxicity in HER2 positive breast cancer: a systematic review. BMC Cancer 22(1). https://doi.org/10.1186/s12885-022-09437-z
46. Romero-Lorca A et al (2015) Impacts of the glucuronidase genotypes UGT1A4, UGT2B7, UGT2B15 and UGT2B17 on tamoxifen metabolism in breast cancer patients. PLoS One 10(7). https://doi.org/10.1371/journal.pone.0132269
47. Sági JC et al (2018) Possible roles of genetic variations in chemotherapy related cardiotoxicity in pediatric acute lymphoblastic leukemia and osteosarcoma. BMC Cancer 18(1). https://doi.org/10.1186/s12885-018-4629-6
48. Piska K et al (2017) 'Metabolic carbonyl reduction of anthracyclines—role in cardiotoxicity and cancer resistance. Reducing enzymes as putative targets for novel cardioprotective and chemosensitizing agents. Invest New Drugs 35:375–385. https://doi.org/10.1007/s10637-017-0443-2
49. Charlier HA et al (2005) Anthraquinone substrates of carbonyl reductase reduction of 13-deoxydoxorubicin and daunorubicinol anthraquinones by human carbonyl reductase. Cardiovasc Toxicol. http://www.cardiotox.com
50. Gonzalez-Covarrubias V et al (2008) A functional genetic polymorphism on human carbonyl reductase 1 (CBR1 V88I) impacts on catalytic activity and NADPH binding affinity NIH public access. http://dmd.aspetjournals.org
51. Gándara-Mireles JA et al (2021) Association of genetic polymorphisms NCF4 rs1883112, CBR3 rs1056892, and ABCC1 rs3743527 with the cardiotoxic effects of doxorubicin in children with acute lymphoblastic leukemia. Pharmacogenet Genomics 31(5):108–115. https://doi.org/10.1097/FPC.0000000000000428
52. Blanco JG et al (2012) Anthracycline-related cardiomyopathy after childhood cancer: role of polymorphisms in carbonyl reductase genes—a report from the Children's Oncology Group. J Clin Oncol 30(13):1415–1421. https://doi.org/10.1200/JCO.2011.34.8987
53. Bains OS et al (2010) Naturally occurring variants of human aldo-keto reductases with reduced in vitro metabolism of daunorubicin and doxorubicin. J Pharmacol Exp Ther 335(3):533–545. https://doi.org/10.1124/jpet.110.173179
54. Amin MN et al (2020) Inflammatory cytokines in the pathogenesis of cardiovascular disease and cancer. SAGE Open Med. https://doi.org/10.1177/2050312120965752
55. Todorova VK et al (2017) Polymorphic variations associated with doxorubicin-induced cardiotoxicity in breast cancer patients. Oncol Res 25(8):1223–1229. https://doi.org/10.3727/096504017x14876245096439
56. Sumneang N, Tanajak P, Oo TT (2023) Toll-like receptor 4 inflammatory perspective on doxorubicin-induced cardiotoxicity. Molecules. https://doi.org/10.3390/molecules28114294
57. Lan Y et al (2020) Heat shock protein 22 attenuates doxorubicin-induced cardiotoxicity via regulating inflammation and apoptosis. Front Pharmacol 11. https://doi.org/10.3389/fphar.2020.00257

58. Pop-Moldovan AL et al (2017) Customized laboratory TLR4 and TLR2 detection method from peripheral human blood for early detection of doxorubicin-induced cardiotoxicity. Cancer Gene Ther 24(5):203–207. https://doi.org/10.1038/cgt.2017.4
59. Li P et al (2015) Hyaluronan synthase 2 overexpression is correlated with the tumorigenesis and metastasis of human breast cancer. Int J Clin Exp Pathol 8(10):12101–12114. http://www.ncbi.nlm.nih.gov/pubmed/26722395
60. Homann S et al (2018) Hyaluronan synthase 3 promotes plaque inflammation and atheroprogression. Matrix Biol 66:67–80. https://doi.org/10.1016/j.matbio.2017.09.005
61. Wang X et al (2014) Hyaluronan synthase 3 variant and anthracycline-related cardiomyopathy: a report from the children's oncology group. J Clin Oncol 32(7):647–653. https://doi.org/10.1200/JCO.2013.50.3557
62. Podyacheva E et al (2024) The role of microRNAs in the pathogenesis of doxorubicin-induced vascular remodeling. Int J Mol Sci. https://doi.org/10.3390/ijms252413335
63. O'Brien J et al (2018) Overview of microRNA biogenesis, mechanisms of actions, and circulation. Front Endocrinol. https://doi.org/10.3389/fendo.2018.00402
64. Sessa F et al (2023) miRNA dysregulation in cardiovascular diseases: current opinion and future perspectives. Int J Mol Sci. https://doi.org/10.3390/ijms24065192
65. Mitchelson KR (2015) Roles of the canonical myomiRs miR-1, -133 and -206 in cell development and disease. World J Biol Chem 6(3):162. https://doi.org/10.4331/wjbc.v6.i3.162
66. Moscoso I et al (2024) Transforming cardiotoxicity detection in cancer therapies: the promise of microRNAs as precision biomarkers. Int J Mol Sci. https://doi.org/10.3390/ijms252211910
67. Zhao Z et al (2014) Dysregulated miR1254 and miR579 for cardiotoxicity in patients treated with bevacizumab in colorectal cancer. Tumour Biol 35(6):5227–5235. https://doi.org/10.1007/s13277-014-1679-5
68. Vaitiekus D et al (2021) HFE gene variants' impact on anthracycline-based chemotherapy-induced subclinical cardiotoxicity. Cardiovasc Toxicol 21(1):59–66. https://doi.org/10.1007/s12012-020-09595-1
69. Hannuksela J et al (2005) Hereditary hemochromatosis gene (HFE) mutations C282Y, H63D and S65C in patients with idiopathic dilated cardiomyopathy. Eur J Heart Fail 7(1):103–108. https://doi.org/10.1016/j.ejheart.2004.03.007
70. Nakano MH et al (2045) A genome-wide association study identifies five novel genetic markers for trastuzumab-induced cardiotoxicity in Japanese population. Biol Pharm Bull. http://locuszoom.sph.umich.edu/
71. Stanton SE et al (2015) Pro1170 Ala polymorphism in HER2-neu is associated with risk of trastuzumab cardiotoxicity. BMC Cancer 15(1). https://doi.org/10.1186/s12885-015-1298-6
72. Lin M et al (2021) The research progress of trastuzumab-induced cardiotoxicity in HER-2-positive breast cancer treatment. Front Cardiovasc Med. https://doi.org/10.3389/fcvm.2021.821663
73. Li Y et al (2021) Genome-wide prioritization reveals novel gene signatures associated with cardiotoxic effects of tyrosine kinase inhibitors. Oncol Lett 21(2). https://doi.org/10.3892/ol.2020.12355
74. Yang Y, Bu P (2016) Progress on the cardiotoxicity of sunitinib: prognostic significance, mechanism and protective therapies. Chem Biol Interact 257:125–131. https://doi.org/10.1016/j.cbi.2016.08.006
75. Singh AP et al (2020) Cardiotoxicity of the BCR-ABL1 tyrosine kinase inhibitors: emphasis on ponatinib. Int J Cardiol 316:214–221. https://doi.org/10.1016/j.ijcard.2020.05.077
76. Patel JN, Singh J, Ghosh N (2024) Bruton's tyrosine kinase inhibitor-related cardiotoxicity: the quest for predictive biomarkers and improved risk stratification. Oncotarget 15:355–359. https://doi.org/10.18632/oncotarget.28589
77. Wu Y et al (2023) GPR35 acts a dual role and therapeutic target in inflammation. Front Immunol. https://doi.org/10.3389/fimmu.2023.1254446

78. Divorty N et al (2015) G protein-coupled receptor 35: an emerging target in inflammatory and cardiovascular disease. Front Pharmacol. https://doi.org/10.3389/fphar.2015.00041
79. Ronkainen VP et al (2014) Hypoxia-inducible factor 1-induced G protein-coupled receptor 35 expression is an early marker of progressive cardiac remodelling. Cardiovasc Res 101(1): 69–77. https://doi.org/10.1093/cvr/cvt226
80. Sun YV et al (2008) Application of machine learning algorithms to predict coronary artery calcification with a sibship-based design. Genet Epidemiol 32(4):350–360. https://doi.org/10.1002/gepi.20309
81. Ruiz-Pinto S et al (2017) Exome array analysis identifies GPR35 as a novel susceptibility gene for anthracycline-induced cardiotoxicity in childhood cancer. Pharmacogenet Genomics 27(12):445–453. https://doi.org/10.1097/FPC.0000000000000309
82. Barsheshet A, Dotsenko O, Goldenberg I (2013) Genotype-specific risk stratification and management of patients with long QT syndrome. Ann Noninvasive Electrocardiol 18(6): 499–509. https://doi.org/10.1111/anec.12117
83. Niemeijer MN et al (2015) Pharmacogenetics of drug-induced QT interval prolongation: an update. Drug Saf 38:855–867. https://doi.org/10.1007/s40264-015-0316-6
84. Kannankeril P, Roden DM, Darbar D (2010) Drug-induced long QT syndrome. Pharmacol Rev 62:760–781. https://doi.org/10.1124/pr.110.003723
85. Hong Y et al (2001) Familial aggregation of QT-interval variability in a general population: results from the NHLBI Family Heart Study. Clin Genet 59(3):171–177. https://doi.org/10.1034/j.1399-0004.2001.590305.x
86. Baracaldo-Santamaría D et al (2021) Genetic and molecular aspects of drug-induced QT interval prolongation. Int J Mol Sci. https://doi.org/10.3390/ijms22158090
87. Hodge C et al (2020) Drug interactions: a review of the unseen danger of experimental COVID-19 therapies. J Antimicrob Chemother 75:3417–3424. https://doi.org/10.1093/jac/dkaa340
88. Behr ER, Roden D (2013) Drug-induced arrhythmia: pharmacogenomic prescribing? Eur Heart J 34:89–95. https://doi.org/10.1093/eurheartj/ehs351
89. Magavern EF et al (2022) The role of pharmacogenomics in contemporary cardiovascular therapy: a position statement from the European Society of Cardiology Working Group on Cardiovascular Pharmacotherapy. Eur Heart J Cardiovasc Pharmacother 8:85–99. https://doi.org/10.1093/ehjcvp/pvab018
90. Sauer AJ, Newton-Cheh C (2012) Clinical and genetic determinants of torsade de pointes risk. Circulation 125:1684–1694. https://doi.org/10.1161/CIRCULATIONAHA.111.080887
91. Aerssens J, Paulussen ADC (2005) Pharmacogenomics and acquired long QT syndrome. Pharmacogenomics 6:259–270. https://doi.org/10.1517/14622416.6.3.259
92. Llerena A et al (2004) QTc interval, CYP2D6 and CYP2C9 genotypes and risperidone plasma concentrations. J Psychopharmacol 18(2):189–193. https://doi.org/10.1177/0269881104042618
93. Potkin SG et al (2013) A thorough QTc study of 3 doses of iloperidone including metabolic inhibition via CYP2D6 and/or CYP3A4 and a comparison to quetiapine and ziprasidone. J Clin Psychopharmacol 33(1):3–10. https://doi.org/10.1097/JCP.0b013e31827c0314
94. Owens RC (2004) QT prolongation with antimicrobial agents understanding the significance, drugs
95. Damle B et al (2009) Effects of standard and supratherapeutic doses of nelfinavir on cardiac repolarization: a thorough QT study. J Clin Pharmacol 49(3):291–300. https://doi.org/10.1177/0091270008329551
96. Thanacoody RHK et al (2007) Factors affecting drug concentrations and QT interval during thioridazine therapy. Clin Pharmacol Ther 82(5):555–565. https://doi.org/10.1038/sj.clpt.6100195
97. Eap CB et al (2007) Stereoselective block of hERG channel by (S)-methadone and QT interval prolongation in CYP2B6 slow metabolizers. Clin Pharmacol Ther 81(5):719–728. https://doi.org/10.1038/sj.clpt.6100120

98. Kurzawski M et al (2007) Impact of ABCB1 (MDR1) gene polymorphism and P-glycoprotein inhibitors on digoxin serum concentration in congestive heart failure patients
99. Suzuki Y et al (2014) Effect of risperidone metabolism and P-glycoprotein gene polymorphism on QT interval in patients with schizophrenia. Pharmacogenomics J 14(5):452–456. https://doi.org/10.1038/tpj.2014.6
100. Cubeddu LX (2016) Send orders for reprints to reprints@benthamscience.ae. Drug-induced inhibition and trafficking disruption of ion channels: pathogenesis of QT abnormalities and drug-induced fatal arrhythmias. Curr Cardiol Rev. www.crediblemeds.org
101. Bohnen MS et al (2017) Molecular pathophysiology of congenital long QT syndrome. Physiol Rev 97:89–134. https://doi.org/10.1152/physrev.00008.2016.-Ion
102. Adler A et al (2020) An international, multicentered, evidence-based reappraisal of genes reported to cause congenital long QT syndrome. Circulation 141(6):418–428. https://doi.org/10.1161/CIRCULATIONAHA.119.043132
103. Vandenberg JI et al (2012) hERG K channels: structure, function, and clinical significance. Physiol Rev 92:1393–1478. https://doi.org/10.1152/physrev.00036.2011.-The
104. Finlayson K et al (2004) Acquired QT interval prolongation and HERG: implications for drug discovery and development. Eur J Pharmacol 500:129–142. https://doi.org/10.1016/j.ejphar.2004.07.019
105. Crumb WJ et al (2006) Effects of antipsychotic drugs on Ito, INa, I sus, IK1, and hERG: QT prolongation, structure activity relationship, and network analysis. Pharm Res 23(6):1133–1143. https://doi.org/10.1007/s11095-006-0070-7
106. Oceandy D et al (2007) Neuronal nitric oxide synthase signaling in the heart is regulated by the sarcolemmal calcium pump 4b. Circulation 115(4):483–492. https://doi.org/10.1161/CIRCULATIONAHA.106.643791
107. Eijgelsheim M et al (2009) Identification of a common variant at the NOS1AP locus strongly associated to QT-interval duration. Hum Mol Genet 18(2):347–357. https://doi.org/10.1093/hmg/ddn341
108. Jamshidi Y et al (2012) Common variation in the NOS1AP gene is associated with drug-induced QT prolongation and ventricular arrhythmia. J Am Coll Cardiol 60(9):841–850. https://doi.org/10.1016/j.jacc.2012.03.031
109. Van Noord C et al (2009) Calcium channel blockers, NOS1AP, and heart-rate-corrected QT prolongation. Pharmacogenet Genomics 19(4):260–266. https://doi.org/10.1097/FPC.0b013e328324e556
110. Quraishi SA, Schuler GH, Janicki PK (2011) 5 HT3-receptor antagonists and cardiac repolarization time in patients expressing a novel genetic target associated with baseline QTc interval abnormalities. J Clin Anesth 23(4):297–302. https://doi.org/10.1016/j.jclinane.2010.11.003
111. Woosley RL et al (2017) Adverse drug event causality analysis (ADECA): a process for evaluating evidence and assigning drugs to risk categories for sudden death. Drug Saf 40(6):465–474. https://doi.org/10.1007/s40264-017-0519-0
112. Kallergis EM et al (2012) Mechanisms, risk factors, and management of acquired long qt syndrome: a comprehensive review. ScientificWorldJournal. https://doi.org/10.1100/2012/212178
113. Drew BJ et al (2010) Prevention of torsade de pointes in hospital settings. A scientific statement from the American Heart Association and the American College of Cardiology Foundation endorsed by the American Association of Critical-Care Nurses and the International Society for Computerized Electrocardiology. J Am Coll Cardiol 55:934–947. https://doi.org/10.1016/j.jacc.2010.01.001
114. Honig PK et al. (1993) Terfenadine-ketoconazole interaction. Pharmacokinetic and electrocardiographic consequences. JAMA 269(12):1513–8
115. Zhou Z et al (1999) Block of HERG potassium channels by the antihistamine astemizole and its metabolites desmethylastemizole and norastemizole. J Cardiovasc Electrophysiol 10(6):836–843. https://doi.org/10.1111/j.1540-8167.1999.tb00264.x

116. Piquette RK (1999) Torsade de pointes induced by cisapride/clarithromycin interaction. Ann Pharmacother 33(1):22–6. https://doi.org/10.1345/aph.18107
117. Pfeufer A et al (2009) Common variants at ten loci modulate the QT interval duration in the QTSCD study. Nat Genet 41(4):407–414. https://doi.org/10.1038/ng.362
118. Åberg K et al (2012) Genome-wide association study of antipsychotic-induced QTc interval prolongation. Pharmacogenomics J 12(2):165–172. https://doi.org/10.1038/tpj.2010.76
119. Yip VLM, Hawcutt DB, Pirmohamed M (2015) Pharmacogenetic markers of drug efficacy and toxicity. Clin Pharmacol Ther 98(1):61–70. https://doi.org/10.1002/cpt.135
120. Lutsey PL et al (2019) Direct oral anticoagulants and warfarin for venous thromboembolism treatment: trends from 2012 to 2017. Res Pract Thromb Haemost 3(4):668–673. https://doi.org/10.1002/rth2.12222
121. Verhoef TI et al (2014) Pharmacogenetic-guided dosing of coumarin anticoagulants: algorithms for warfarin, acenocoumarol and phenprocoumon. Br J Clin Pharmacol 77(4):626–641. https://doi.org/10.1111/bcp.12220
122. Al-Eitan LN, Almasri AY, Khasawneh RH (2018) Impact of CYP2C9 and VKORC1 polymorphisms on warfarin sensitivity and responsiveness in Jordanian cardiovascular patients during the initiation therapy. Genes 9(12). https://doi.org/10.3390/genes9120578
123. Takeuchi F et al (2009) A genome-wide association study confirms VKORC1, CYP2C9, and CYP4F2 as principal genetic determinants of warfarin dose. PLoS Genet 5(3). https://doi.org/10.1371/journal.pgen.1000433
124. Santos PCJL et al (2013) CYP2C9 and VKORC1 polymorphisms influence warfarin dose variability in patients on long-term anticoagulation. Eur J Clin Pharmacol 69(4):789–797. https://doi.org/10.1007/s00228-012-1404-5
125. Cavallari LH et al (2010) Genetic and clinical predictors of warfarin dose requirements in African Americans. Clin Pharmacol Ther 87(4):459–464. https://doi.org/10.1038/clpt.2009.223
126. Wadelius M et al (2005) Common VKORC1 and GGCX polymorphisms associated with warfarin dose. Pharmacogenomics J 5(4):262–270. https://doi.org/10.1038/sj.tpj.6500313
127. Eriksson N, Wadelius M (2012) Prediction of warfarin dose: why, when and how? Pharmacogenomics 13:429–440. https://doi.org/10.2217/pgs.11.184
128. Dean L (2012) Warfarin therapy and VKORC1 and CYP genotype. https://www.ncbi.nlm.nih.gov/books/
129. Li J et al (2018) Impact of VKORC1, CYP4F2 and NQO1 gene variants on warfarin dose requirement in Han Chinese patients with catheter ablation for atrial fibrillation. BMC Cardiovasc Disord 18(1). https://doi.org/10.1186/s12872-018-0837-x
130. Vidovic S et al (2021) Prevalence of five pharmacologically most important CYP2C9 and CYP2C19 allelic variants in the population from the Republic of Srpska in Bosnia and Herzegovina Prevalencija pet farmakoloski najznacajnijih CYP2C9 i CYP2C19 alelenih varijanti u populaciji Republike Srpske u Bosni i Hercegovini. Arh Hig Rada Toksikol 72(3):129–134. https://doi.org/10.2478/aiht-2021-72-3499
131. Singh O et al (2011) Influence of CYP4F2 rs2108622 (V433M) on warfarin dose requirement in asian patients. Drug Metab Pharmacokinet 26(2):130–136. https://doi.org/10.2133/dmpk.DMPK-10-RG-080
132. Le Heuzey JY et al (2014) Differences among western European countries in anticoagulation management of atrial fibrillation: data from the PREFER IN AF Registry. Thromb Haemost 111(5):833–841. https://doi.org/10.1160/TH13-12-1007
133. Paré G et al (2013) Genetic determinants of dabigatran plasma levels and their relation to bleeding. Circulation 127(13):1404–1412. https://doi.org/10.1161/CIRCULATIONAHA.112.001233
134. Shi J et al (2016) Dabigatran etexilate activation is affected by the CES1 genetic polymorphism G143E (rs71647871) and gender. Biochem Pharmacol 119:76–84. https://doi.org/10.1016/j.bcp.2016.09.003

135. Sangkuhl K, Klein TE, Altman RB (2010) Clopidogrel pathway. Pharmacogenet Genomics 20:463–465. https://doi.org/10.1097/FPC.0b013e3283385420
136. Dean L, Kane M (2012) Clopidogrel therapy and CYP2C19 genotype. In: Pratt V, Scott S, Pirmohamed M (eds) Medical genetics summaries. National Center for Biotechnology Information, Bethesda. https://www.ncbi.nlm.nih.gov/books/
137. Simon T et al (2009) Genetic determinants of response to clopidogrel and cardiovascular events. N Engl J Med 360(4):363–375. https://doi.org/10.1056/nejmoa0808227
138. SLCO1B1 Variants and Statin-Induced Myopathy—A Genomewide study (2008) N Engl J Med 359(8):789–799. https://doi.org/10.1056/nejmoa0801936
139. Pasanen MK et al (n.d.) SLCO1B1 polymorphism markedly affects the pharmacokinetics of simvastatin acid
140. Kitzmiller JP et al (2016) Pharmacogenomics of statins: understanding susceptibility to adverse effects. Pharmacogenomics Pers Med 9:97–106. https://doi.org/10.2147/PGPM.S86013
141. Lee HH, Ho RH (2017) Interindividual and interethnic variability in drug disposition: polymorphisms in organic anion transporting polypeptide 1B1 (OATP1B1; SLCO1B1). Br J Clin Pharmacol 83:1176–1184. https://doi.org/10.1111/bcp.13207
142. Linskey DW et al (2020) Association of SLCO1B1 c.521T>C (rs4149056) with discontinuation of atorvastatin due to statin-associated muscle symptoms. Pharmacogenet Genomics 30: 208–211. https://doi.org/10.1097/FPC.0000000000000412
143. Zisaki A, Miskovic L, Hatzimanikatis V (2015) Send orders for reprints to reprints@benthamscience.ae. Antihypertensive drugs metabolism: an update to pharmacokinetic profiles and computational approaches. Curr Pharm Des
144. Blake CM et al (2013) A meta-analysis of CYP2D6 metabolizer phenotype and metoprolol pharmacokinetics. Clin Pharmacol Ther 94(3):394–399. https://doi.org/10.1038/clpt.2013.96
145. Collett S et al (2023) Metoprolol and CYP2D6: a retrospective cohort study evaluating genotype-based outcomes. J Pers Med 13(3). https://doi.org/10.3390/jpm13030416
146. Bannister ML, MacLeod KT, George CH (2022) Moving in the right direction: elucidating the mechanisms of interaction between flecainide and the cardiac ryanodine receptor. Br J Pharmacol 179:2558–2563. https://doi.org/10.1111/bph.15718
147. Sunthankar SD et al (2022) Influence of CYP2D6 genetic variation on adverse events with propafenone in the pediatric and young adult population. Clin Transl Sci 15(7):1787–1795. https://doi.org/10.1111/cts.13296
148. Joy MS et al (2009) CYP2C9 genotype and pharmacodynamic responses to losartan in patients with primary and secondary kidney diseases. Eur J Clin Pharmacol 65(9):947–953. https://doi.org/10.1007/s00228-009-0707-7
149. Göktaş MT et al (n.d.) Relationship between genetic polymorphisms of drug efflux transporter MDR1 (ABCB1) and response to losartan in hypertension patients
150. Skrbic R, Igic R (2009) Seven decades of angiotensin (1939-2009). Peptides 30:1945–1950. https://doi.org/10.1016/j.peptides.2009.07.003
151. Ha SK (2014) ACE insertion/deletion polymorphism and diabetic nephropathy: clinical implications of genetic information. J Diabet Res. https://doi.org/10.1155/2014/846068
152. Parving H-H et al (n.d.) Effect of deletion polymorphism of angiotensin converting enzyme gene on progression of diabetic nephropathy during inhibition of angiotensin converting enzyme: observational follow up study Hans-Henrik Parving, chief physician
153. Fricke-Galindo I et al (2023) The ACE rs1799752 variant is associated with COVID-19 severity but is independent of serum ACE activity in hospitalized and recovered patients. Int J Mol Sci 24(8). https://doi.org/10.3390/ijms24087678

Chapter 29
Environmental Noise and Outdoor Light Pollution in the Pathogenesis of Cardiovascular Toxicity

Zubair Ahmed, Fihr Chaudhary, and Devendra K. Agrawal

Abstract Cardiovascular disease (CVD) remains the leading cause of death worldwide, despite 90% of risk factors deemed to be preventable for CVD. Novel therapeutics have proven effective in combating CVD, yet risk factors for the disease remain a concern with poor outcomes evident across the globe. This review article summarizes relevant risk factors pertaining to cardiovascular disease, of which include air pollution, outdoor light pollution, socioeconomic status, and environmental noise. Summarized are various studies pertaining to each risk factor and its relationship to CVD, along with possible pathophysiological mechanisms analyzing the associations. Public health policy for each risk factor is also made a concern in this article as a focus on preventative measures is made to help mitigate cardiovascular risk in the public.

Keywords Air pollution · Socioeconomic status · Environmental noise · Outdoor light pollution · Public health policy · Endothelial damage · Myocardial infarction · Ischemic heart disease · Oxidative stress · Mitochondrial dysfunction · Atherosclerosis · Inflammation · Hypertension · Thrombosis

Introduction

The cardiovascular system is a term used to refer to the heart and all the blood vessels in the human body [1]. A variety of complications can arise within this system, including but not limited to problems with conduction, rheumatic disease, endocarditis, or others [2]. The Cardiovascular diseases (CVD) can generally be divided up into the four entities listed below [3]: (i) coronary artery disease (CAD) resulting from decreased myocardial perfusion which causes angina, heart failure, or

Z. Ahmed · F. Chaudhary · D. K. Agrawal (✉)
Department of Translational Research, College of Osteopathic Medicine of the Pacific, Western University of Health Sciences, Pomona, CA, USA
e-mail: Zubair.ahmed@westernu.edu; Fihr.chaudhary@westernu.edu; DAgrawal@WesternU.edu

D. K. Agrawal et al. (eds.), *Cardiovascular Toxicity*, Advances in Biochemistry in Health and Disease 33, https://doi.org/10.1007/978-3-031-90598-8_29

myocardial infarction (MI). This typically accounts for anywhere between 33%–50% of all cases of CVD, (ii) cerebrovascular disease (CVD) including any cases of stroke or transient ischemic attacks (TIAs), (iii) peripheral artery disease (PAD) involving arteries in the periphery particularly limbs that may become claudicated, and (iv) aortic atherosclerosis including cases of thoracic and abdominal aneurysms that may develop.

Atherosclerosis stands as the leading cause of cardiovascular disease (CVD) and related fatalities [4]. The initial visible event in atherosclerosis is the buildup of plaque within major and medium-sized arteries. This plaque has the potential to grow and induce ischemia, either due to insufficient blood flow when oxygen demand is high or through rupturing, forming a thrombus that obstructs the arterial lumen [5]. Thrombi typically form only in high-risk or vulnerable plaques, which exhibit specific characteristics such as thin fibrous caps, limited smooth muscle cells, abundant macrophages (inflammatory cells), and a sizable lipid core [5]. The origin of plaque formation or atherogenesis lies in the endothelium of the artery wall. Normally, the endothelium promotes blood vessel dilation, inhibits excessive smooth muscle cell growth, and prevents an inflammatory response [6]. However, in atherosclerosis, endothelial dysfunction precedes the development of a more serious lesion called atheroma or plaque. This dysfunction leads to reduced production of nitric oxide, a crucial vasodilator, causing the blood vessel to constrict. Moreover, it becomes more permeable, allowing macrophages to uptake LDL cholesterol, leading to the accumulation of foam cells and eventually the formation of an early lesion known as a fatty streak.

By the mid twentieth century, cardiovascular disease had become the leading cause of mortality and morbidity among all Western countries (GBD). This further became a particular issue when towards the end of the twentieth century, CVD had become the leading cause of premature mortality and morbidity worldwide [7]. Chronic diseases have become increasingly significant in terms of their impact on global mortality. By the year 2005, the worldwide death toll from cardiovascular diseases (CVD), including coronary heart disease, stroke, and rheumatic heart disease, had risen to 17.5 million from 14.4 million in 1990. The industrialization of the economy that rapidly occurred starting in the twentieth century was a primary contributing factor to this phenomenon, as a shift from physically demanding labor to sedentary jobs in a largely technology dependent economy resulted in many lifestyle changes for the common folk. These included longer commutes, less leisure times, and longer working hours that left individuals to develop unhealthy eating as well as exercise habits impacting their everyday living conditions [8]. It is interesting to note that cardiovascular disease is considered the costliest disease known today, with estimated indirect costs totaling $237 billion dollars per year and a projected increase to $368 billion by the year 2035 [9].

Particularly, according to the INTERHEART study done to determine common risk factors prevalent in the development of cardiovascular disease in countries from around the word, the study found that there were nine modifiable risk factors that accounted for 90% of the risk of an individual experiencing their first MI: smoking, dyslipidemia, hypertension, diabetes, abdominal density, psychosocial factors, consumption of fruits and vegetables, regular alcohol consumption, and physical inactivity [7]. Physical inactivity is an important recent finding found in many studies [10] that has significant contributions to the risk for CVD as mentioned in ASCVD (atherosclerotic cardiovascular disease) guidelines [11]. By far, of all these factors, the study found that smoking had the largest impact as 36% of the risk associated with experiencing a first MI was contributed to smoking. Statistics on the probability of a first MI in the general population remains high, as there is a calculated 50% risk of an MI by the age of 45 for the public [12]. Onwards, the prevalence of CVD is known to increase significantly with each decade of life [13].

Treating cardiovascular disease has changed over the years, and plentiful evidence is now known that shows that drug treatment of known conventional risk factors is effective in reducing the occurrence of cardiovascular events [14]. There have been many trials done involving HMG CoA reductase inhibitors (statins) that have evidenced that the lowering of LDL cholesterol levels with these agents greatly reduces the occurrence of cardiovascular events [15]. In addition to these known benefits of statins and other anti-hypertensive drugs, increased attention has been put on herbal and dietary factors that may contribute to lowering the prevalence of cardiovascular disease. These supplements are nicely put together below in a chart showing doses and names of each respective to their dosages in known clinical studies.

This chapter will highlight the key information from published reports that deal with environmental factors in the pathogenesis of cardiovascular diseases. The components of the cardiovascular system are naturally susceptible to a wide range of environmental agents. These factors, just like other risk factors, are proven to advance mortality and risk of CVD through the initiation of pathophysiological processes that will be explained in detail in their respective sections [16]. Nonetheless, for these reasons, it is important to implement public guidelines reinforced by evidence-based studies that will aim to reduce these environmental exposures to better the health of the public by reducing the risk for CVD.

As mentioned above, environmental factors are not one of the nine modifiable risk factors that account for 90% of risk for a first myocardial infarction per the INTERHEART study cited [7]. However, they are integral components that contribute to cardiovascular disease and cannot be ignored. We will critically investigate the following environmental factors in detail, particularly in their relation to the impact as a risk factor for CVD: (i) socioeconomic status, (ii) air pollution, (iii) environmental noise, and (iv) outdoor light pollution.

Socio-Economic Status

Background

The socioeconomic status (SES) to which an individual belongs depends highly on the culture and geographic location in which they reside. This makes it difficult to evaluate the relationship between SES and CVD as conflicting studies become a possibility due to the nature of how SES is defined differently in accords to the region of which it describes. However, there are four markers for which a determined association between SES and CVD have been shown, especially concerning high income countries (HICs). These include income level, educational attainment, employment status, and environmental factors [17]. Focusing then on low- and middle-income countries (LMICs), they are known to carry approximately 80% of the global burden of CVD [18]. It is important to note the distinction between HICs and LMICs. The findings of a study done in an HIC country cannot be extrapolated to a LMIC country due to the differences in the way socioeconomic status is defined and how it would affect CVD in that area.

Epidemiology

Income Level

The risk of cardiovascular disease is inversely related to income level. A particular study based in Finland and the United States examined the incidence of non-fatal myocardial infarction and sudden cardiac death in the low-income individuals [19]. This study found an increased risk of non-fatal MI and sudden cardiac death among this low-income cohort examined. A separate study [20] found that every $10,000 increase in median income within a neighborhood corresponded to a decreased mortality risk by up to 10%. Another study conducted in the Netherlands found that 28-day and 1-year mortality rates were significantly higher for individuals admitted for MI that were of lower quartile income levels [21].

Educational Attainment

There is an inverse relationship between cardiovascular disease and educational status that has been well-studied for many decades now. Studies have proven that higher education may be the best predicator of good health [22]. More specifically, Winkleby's study included about 90,000 participants in the New Zealand and Australia regions, finding that individuals with tertiary levels of education had less incidence of cardiovascular disease and mortality in general. A separate study done involving a set of participants in South Korea showed that only low education (in the study, low education was defined as 6 or less years of schooling) was directly associated with increased mortality or heightened risk of cardiac incidence [23].

Employment Status

One of the primary ways that an individual's SES status is determined is through assigning employment status to an individual (individual can be either employed or unemployed). A particular study done amongst French population that were screened for any preexisting heart conditions found that there was a 20% increase in cardiovascular events in members of the population who were not employed [24]. A separate study was conducted within the United States and found that there was a 35% increase in risk status for AMI within the first year of being unemployed, after which this rate of incidence disappeared. The study went on to also show that the amount of time an individual was left unemployed for was also responsible for an increased risk of AMI [25].

Environmental Factors

Characteristics of neighborhoods relating to their appearance and geography have proven to be linked to individual levels of SES. A controlled population analysis was done of atherosclerosis risk in a community setting and evidenced that living in disadvantaged areas is linked to higher rates of cardiac incidence [26]. Other longitudinal studies have been done that show that neighborhoods with ample amounts of physical activity improved the prevalence of cardiovascular risk factors [27].

Analysis

It is important to discuss why these four factors of SES impact cardiovascular health, and the risk associated with it with such magnanimity. Firstly, income level had an inverse relationship with CVD risk: i.e. the higher the level of income, the lower the risk. This could be due to a variety of reasons, but one reason this may hold true is because low-income individuals may be resistant to surgical intervention to treat their cardiovascular health problems. Whether due to lack of insurance, distance away from hospitals, or lack of finances to pay for procedures, low-income individuals may elect to not have procedures done such as left heart catheterizations and other percutaneous coronary interventions. Compared to individuals with the financial means to perform such operations, it would be reasonable to see why this could be a strong possibility for this finding. Another possibility for why there are discrepancies in CVD risk between income levels could be differing standards of care for individuals of differing levels of income. An individual with a higher income level could reasonably be assumed to have access to better hospitals, and thus better physicians, while low-income individuals do not have this luxury. This could also contribute to the phenomenon seen, and thus differences in mortality rates are also supported by the studies cited in the literature.

Next, explaining the link of educational attainment to CVD risk can be explained by the health literacy component that an education provides. When an individual is highly educated, this person has been exposed to curriculum that includes health education, this informing this person of what healthy measures are and the benefits of maintaining a healthy lifestyle. Being exposed to and taught these benefits from a young age allows individuals to engage in behaviors that mitigate their CVD risk, unlike individuals who did not have this information in their disposal due to a lack of educational opportunities on these topics. Logically then, this is an ample explanation for this discrepancy as well.

Furthermore, employment status was found to have direct correlations to the risk of CVD. Although this could be because many poor people are also unemployed, and thus of a lower income background (a point already mentioned), another contending possibility is that the stress caused by the fact that an individual is unemployed could lead them to engage in behaviors that increase risk of CVD. Stress-induced behaviors such as smoking, binge eating, physical inactivity, and excessive alcohol consumption are all avenues associated with cardiovascular risk that an individual may consider due to being unemployed and under stress.

Lastly, environmental factors associated particularly with poor neighborhood socioeconomic characteristics are linked with increased rates of CVD. This makes sense, as neighborhoods with ample avenues for physical activity would have individuals in better health conditions than those that do not. In addition, it is a well-known fact that healthy foods tend to be more expensive than "junk" foods. Thus, the likelihood that healthy foods are found more prevalently in neighborhoods with higher SES than those with individuals having lower levels of SES is a strong possibility (this is a phenomenon known as "food deserts" and has been well-studied) [28].

Public Health Policy

Public policy has proposed a variety of interventions to help limit the incidence of CVD along with a fight to decrease disparities associated with low SES. As mentioned in this case report, there are environmental, behavioral, psychosocial, and geographic factors that result in individuals identifying with low SES having poorer health outcomes than their counterparts. Behavioral counseling, for one, has been recommended by the U.S Preventative Task Force to reduce CVD risk factors associated with hypertension and hyperlipidemia [29]. Other forms of counseling, such as smoking cessation programs, have been shown to be effective measures in reducing CVD risk among individuals with low SES. Smoking has become a policy issue, as flavored vaping products have been banned in many countries such as Massachusetts and New Jersey. There have been measures put in to increase health education in low SES areas as well, with programs and organizations catering towards achieving certain health outcomes within the demographic with which they are working. Whether it be goals tailored towards physical activity goals, smoking cessation, limiting alcohol consumption, or otherwise, these programs are all catered towards making behavioral changes in areas where they are needed most. Public policy must be combined with individual efforts to change lifestyles to create meaningful change particularly in these low SES settings [30].

Environmental Noise

Background

Environmental noise is a lesser known but important factor in taking in considerations for risk factors that may pertain to cardiovascular disease. Traffic noise, especially at night, is known to be associated with shortened time of sleep, elevation of stress hormone levels, and increased oxidative stress levels in the brain as well as vasculature of the human body [31]. According to the WHO, it is responsible for over one million lives lost per year in Europe. In addition to oxidative stress and vascular dysfunction, endothelial dysfunction has also been found to be a particular result of noise pollution [32] (Fig. 29.1).

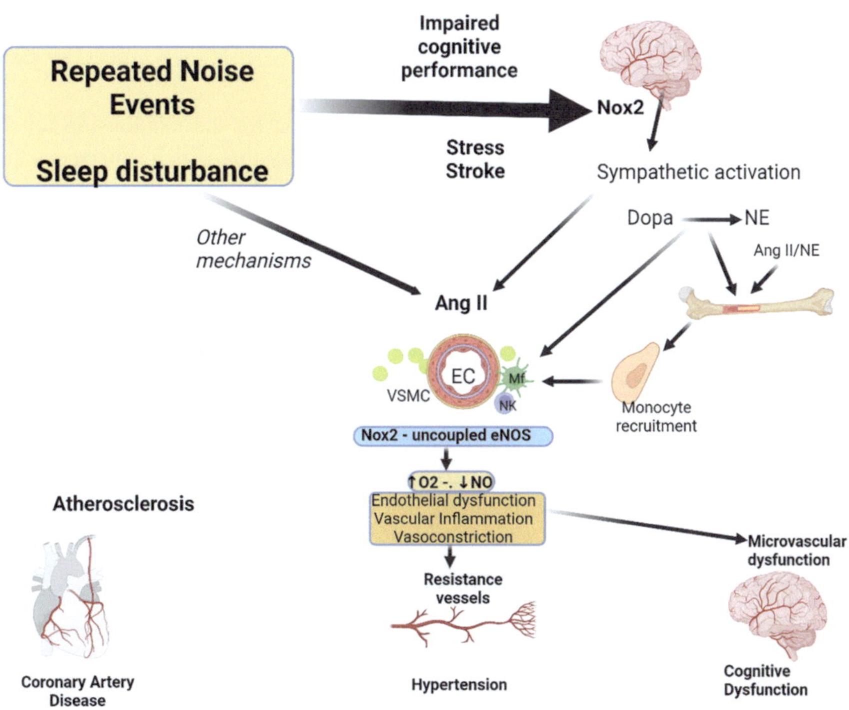

Fig. 29.1 Potential mechanisms linking noise to cardiovascular disease pathogenesis

Epidemiology

A particular case study was done in Denmark involving over 2.5 million participants and tracking them over a mean time of 9 years, looking specifically at the relationship between exposure to noise levels and CVD risk. The study found that increased road traffic and railway noise (Lden) at the most exposed façade in Denmark is linked to a higher risk of various heart conditions, including ischemic heart disease (IHD), myocardial infarction (MI), angina pectoris, and heart failure [33]. The hazard ratios (HRs) (95% confidence intervals (CI)) for these associations were as follows: 1.052 (1.044–1.059) for IHD, 1.041 (1.032–1.051) for MI, 1.095 (1.071–1.119) for angina pectoris, and 1.039 (1.033–1.045) for heart failure per 10 dB increase in 10-year mean exposure to road traffic noise [33]. Notably, these relationships followed a near-linear exposure-response pattern and remained robust even after accounting for air pollution with PM2.5 [33]. A different study occurred in Denmark and Sweden involving over 130,000 participants that looked at results of long-term exposure to noise pollution through using hazard ratios adjusted for SES and lifestyle factors. This study found that exposure to railway, road, and aircraft noise over the course of the 5 years the study took place was associated with an increased risk for ischemic heart disease (IHD) [34].

Pathophysiology

The noise-reaction model provides a likely explanation for the relationship found between noise pollution and CVD risk. Noise, especially that which supersedes the 55 dB average, is a psychosocial stressor that activates the endocrine and sympathetic systems [35]. Via synaptic interactions in the reticular activating system, catecholamines such as epinephrine, cortisol, and norepinephrine are released. This, taken with the emotional and cognitive perception of the sound taken in by the limbic system, triggers hypertensive effects in the body. A basic cause-effect chain can be summarized as the following: Noise creates sound, which can be perceived as annoying → hormones are released by the human body via physiological arousal →over time, these hormones compound and can result in biological risk factors (hypertension, hyperlipidemia, obesity, etc.) → these risk factors can manifest themselves in the form of a variety of cardiovascular diseases [35]. This pathway is best summarized in the figure below (Fig. 29.2).

Public Health Policy

Public policy to protect against noise pollution is severely limited at this time and must be enhanced. For example, in the United States, the Noise Control Act of 1972

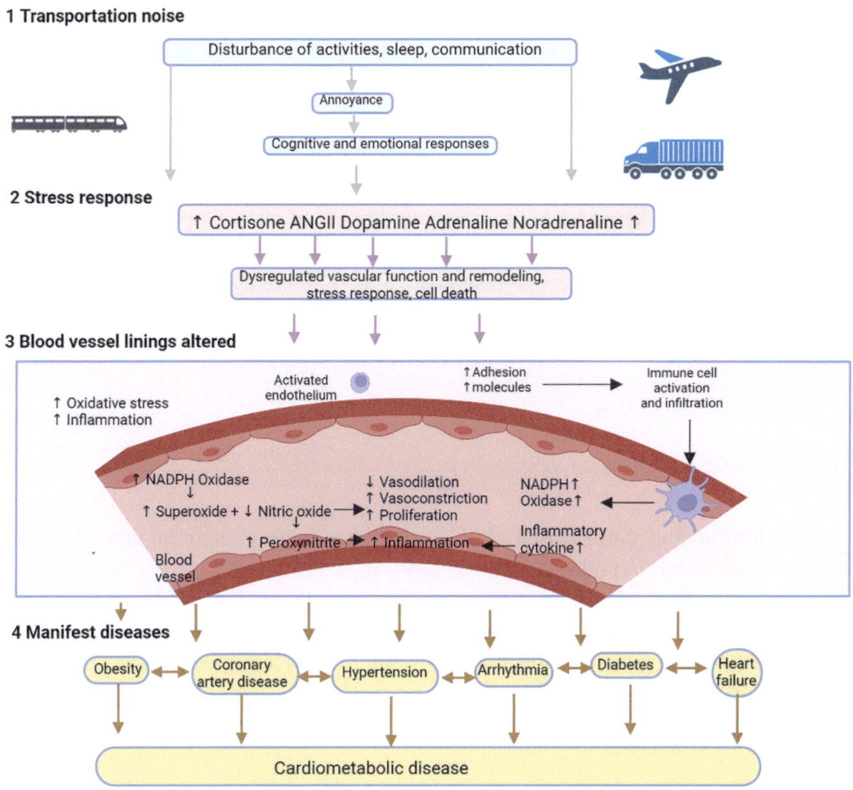

Fig. 29.2 The presence of traffic noise can elicit feelings of annoyance and tension, leading to the secretion of stress hormones such as cortisol, adrenaline, and angiotensin II (ANGII). The stress response, particularly in its chronic form, leads to changes in blood chemistry and interferes with the functioning of blood vessels. An increase in the quantity of deleterious oxidants and proteins known as adhesion molecules has the potential to induce remodeling, or 'activation,' of the endothelial lining of blood vessels. This exercise, along with the resulting inflammation, has the potential to contribute to or exacerbate several cardiovascular issues

was passed to prohibit any noise that would damage an American's health or welfare. Hearing loss remains the third most common chronic physical condition in the United States, and public policy has done little to nothing to act on any of such measures. According to the APHA, approximately 145 million Americans are at risk of noise-related hypertension annually. As more research shows direct links of noise pollution to CVD risk and other poor health outcomes, it will be critical to pass policies which reflect caution against such dangers to better protect the health of individuals across the globe.

Outdoor Light Pollution

Background

Light remains a constant factor in most people's lives. Whether it be natural in the form of sunlight or artificial in the form of light bulbs around commercial and living areas, it is found in some form in nearly every part of a developed country. Light pollution is well-understood now to cause several health issues, some of which include sleep deprivation, stress responses, and overall dysfunction of circadian rhythms. However, a lesser-known health effect that light pollution has is that it increases the incidence rate of cardiovascular disease.

Epidemiology

Light pollution and cardiovascular disease is not well-studied in literature, and several contending hypotheses are present. In a study done analyzing outdoor exposure of light particularly at night and incidental cardiovascular disease in a population study in Hong Kong involving over 58,000 participants, it was found that light exposure is associated with increased risk of CVD outcomes [36]. The study documented a total of 3772 hospitalizations and 1695 deaths related to cardiovascular disease over the course of the 11 years it followed.

Pathophysiology

Light exposure activates the sympathetic nervous system, thereby activating a hypertensive mechanism through the release of hormones that play important roles in cardiovascular risk stratification. A component of the medulla oblongata, namely the rostral ventrolateral medulla (RVLM), is responsible for maintaining a basal metabolic rate in the human body along with contributions towards a sympathetic tone [37]. The RVLM region is proposed to be responsible for causing sympatho-excitation, thus activating the nervous system and directly influencing the cardiovascular system as a result. Further research supports the notion that norepinephrine levels are also altered by differing light conditions via hormone alterations and the superior sympathetic ganglia, all part of the larger nervous system in the human body [38] (Fig. 29.3).

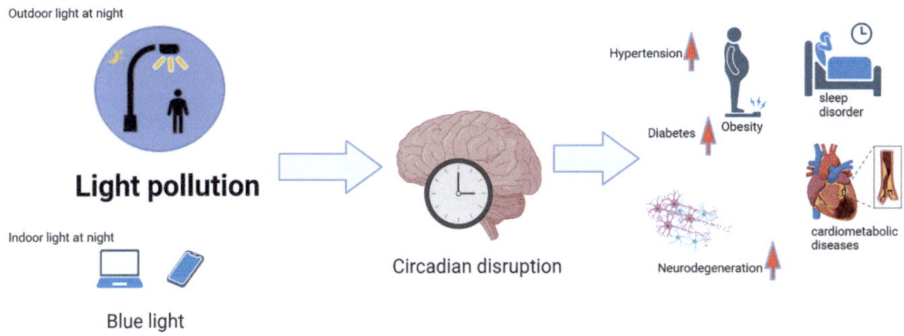

Fig. 29.3 Exposure to bedroom light pollution and cardiometabolic risk

Public Health Policy

Light pollution is typically defined as excessive or unwanted artificial light. At least 19 states in the United States have laws against light pollution in order to promote public safety, energy conservation, and other governmental research expositions. Several organizations within the United States, such as The Natural Sounds and Night Skies Division, are working to protect sky environments in Par systems and otherwise across the country. Still, better, and more relevant research is needed to better study direct linkages of light pollution to CVD risk so that public health policies can be implemented to prevent light pollution in cities.

Cardiotoxicity of Anticancer Drugs

Prognosis of cancer patients have greatly improved because to tailored medicines and chemotherapy. On the other hand, these cancer treatments come with the risk of cardiotoxicity, which can cause heart problems either right once or after some time has passed.

Factors unique to each patient, as well as the type, dosage, method, and length of time anticancer medications are administered, can influence when cardiotoxicity first manifests. The irreversible kind, marked by damage and cell death in the heart muscle, is distinct from the reversible kind, in which the effects on the heart are recovered when the treatment is stopped. In addition to causing subclinical cardiac dysfunction, anticancer treatments can cause congestive heart failure symptoms to ultimately manifest.

Cardiotoxicity Mechanisms Associated with Common Classes of Chemotherapeutic Drugs and Targeted Therapy

Anthracyclines

The anthracyclines are a group of anticancer medications derived from the Streptomyces bacteria. Some examples of these drugs are doxorubicin, daunorubicin, and epirubicin. Malignancies of the blood (leukemia, lymphomas, etc.), that affect both children and adults, and malignancies of the solid organs (breast, ovarian, bladder, lung, etc.) are all treated with these substances.

Importantly, there is a substantial amount of information that shows that in the presence of risk factors, such as hypertension, arrhythmias, coronary disease, combination with other anticancer treatments, and genetic predisposition to cardiotoxicity, cardiomyopathy can develop at lower doses of anthracyclines [39].

Research on cardiac biopsy samples taken a few hours following a single dosage of anthracycline has shown that cardiac damage happens with every dose [40]. Most cases of anthracycline cardiotoxicity are discovered within the first year after treatment ends [41], although it can also present months to years after chemotherapy ends [41]. Pathophysiological evidence suggests that anthracyclines cause cardiotoxicity via pathways involving various types of cardiac cells, including cardiomyocytes. Because cardiomyocytes die by a variety of biological processes, such as apoptosis, autophagy, necrosis, necroptosis, proptosis, and ferroptosis, anthracycline-induced cardiac damage may be irreversible [42].

Fluoropyrimidines

The second most common kind of cardiotoxicity caused by chemotherapy is exerted by fluoropyrimidines [43]. Incorporating 5-fluorouracil (5-FU) and its prodrug capecitabine into DNA or RNA makes this antimetabolite drug class effective as a cytostatic agent in the treatment of colorectal, breast, gastric, pancreatic, prostate, and bladder cancers [44]. There are a few potential cardiovascular side effects of fluoropyrimidines. One is myocardial ischemia, which is usually reversible. Another is cardiomyocyte death and loss, which can happen because of coronary artery thrombosis and myocardial infarction [45], or because of mechanisms that are specific to cardiomyocytes [46].

Taxanes

Antimitotic taxanes like paclitaxel stabilize microtubules in the mitotic spindle, preventing cell cycle progression. Tumors of the breast, lungs, and ovaries are among the many types of tumors treated with these chemotherapy medicines. The negative effects on heart function caused by taxanes include bradycardia, atrial fibrillation, and lengthening of the QT interval [47]. Among taxanes, paclitaxel is known to worsen toxicity caused by anthracyclines. Cardiomyopathy (HF) occurrences and histological changes (including severe necrosis) were both worsened by paclitaxel and doxorubicin administered together [40].

Cardiovascular Toxicity of Drugs

Unexpected cardiotoxicities occur in nearly every therapeutic medication class and are not exclusive to anticancer medicines. Because toxicity may not become apparent until the medication or its metabolites accumulate over an extended period, cardiotoxicity caused by persistently administered pharmaceuticals, such as neurologic/psychiatric treatments, is a significant concern [48].

Despite extensive attempts to identify cardiotoxicity during the preclinical stage of pharmaceutical product development, the lack of adequate understanding of the mechanisms of cardiotoxicity remains the primary cause of safety concerns, and nearly 10% of medications recalled from clinical markets globally in the past 40 years have been attributed to cardiovascular risks [48].

Some popular classical antineoplastic drugs have serious side effects, including drug-induced cardiotoxicity, which manifests as heart muscle malfunction and can lead to heart failure. Some of the cardiac side effects of anticancer drugs include heart failure, arrhythmias, pericarditis, myocarditis, hypertension, thromboembolism, and myocardial ischemia. One of the most common characteristics of cancer survivors is asymptomatic diastolic dysfunction, which is often the first cardiac defect that might be noticed [48].

Mechanisms of Drug-Induced Cardiotoxicity

Cardiotoxicity caused by antipsychotics, anthracyclines, or other drugs probably operates through a complex mechanism [49]. Declining mitochondrial activity causes cardiomyocyte and endothelial cell death, which in turn causes cardiovascular dysfunction [50]. Mitochondria play an important role in maintaining the homeostasis of cardiac tissues. Some antiretroviral nucleoside reverse transcriptase inhibitors, including zidovudine, can inhibit DNA polymerase-gamma and induce mutations in mitochondrial DNA, which can lead to cardiomyopathy [48]. Cardiac adverse effects can be broadly classified into two types: functional damages and structural damages. Impressive, significantly changed function may be entirely independent of structural impact, particularly in the early stages [48]. Anthracyclines not only impair cardiac function but also harm titin, a myofilament-forming protein that controls cardiac function and causes diastolic and systolic dysfunction [51].

Importantly, biological drugs that target proteins that regulate cancer cell proliferation can cause reversible (type 2) cardiotoxicity; this toxicity can go away after treatment ends or even while it's still going on; and when using multiple drugs that could be cardiotoxic, there is a chance that different types of toxicity will occur, either overlapping or adding to each other [52].

Environmental Factors Enhancing Drug-Induced Cardiovascular Toxicity

Air Pollution and Cardiac Arrhythmias

Heart arrhythmias, which include irregular heartbeats and abnormal heart rhythms in general, are among the leading causes of death and disability associated with cardiovascular disorders. Cardiovascular arrhythmias are common and can manifest in a variety of ways. Some examples include problems with the sinus node, atrial fibrillation (AF), conduction issues, ventricular tachycardia (VT), and ventricular fibrillation (VF). Medical research has confirmed that air pollution is linked to certain forms of arrhythmias [53].

The main theories proposed to explain arrhythmias caused by air pollution include systemic inflammatory response, malfunction of the cardiac autonomic nervous system, modification of the cardiac structure, and the transfer of particles into the cardiovascular systems [54] (Fig. 29.4).

Through systemic inflammatory responses, particulate particles can indirectly cause arrhythmias [55]. It is well-established that systemic inflammation has a role in the development of atherosclerosis [56], leading to both acute and chronic ischemia. Substrates for cardiac arrhythmias are provided by ischemic heart disease, which causes ion channel remodeling and increases the spatial dispersion of depolarization [57]. Second, myocardial dysfunction, brought on by elevated cytokine levels, may have a role in the development of arrhythmias in some severe inflammatory conditions including septic shock and sepsis. In addition, arrhythmogenesis may be indirectly influenced by the coagulation reaction associated with inflammation [58].

Particulate matter has been linked to a reduced HRV, which is a sign of cardiac autonomic dysfunction and a potential risk factor for deadly ventricular arrhythmias [49]. While particulate matter (PM) can have an indirect impact on cardiac electrical activity due to systemic inflammation, very small PM, like ultrafine particles, can penetrate the pulmonary epithelium and enter the bloodstream, where they can directly affect the cardiovascular system [59]. Finally, PM2.5 can promote structural remodeling of the heart, which can lead to pro-arrhythmias. Researchers found that people who were exposed to PM2.5 had a much higher risk of cardiac ventricular dilatation, which manifested as greater end-diastolic volumes in the left ventricle, end-systolic volumes in the right ventricle, and an overall increased risk of heart

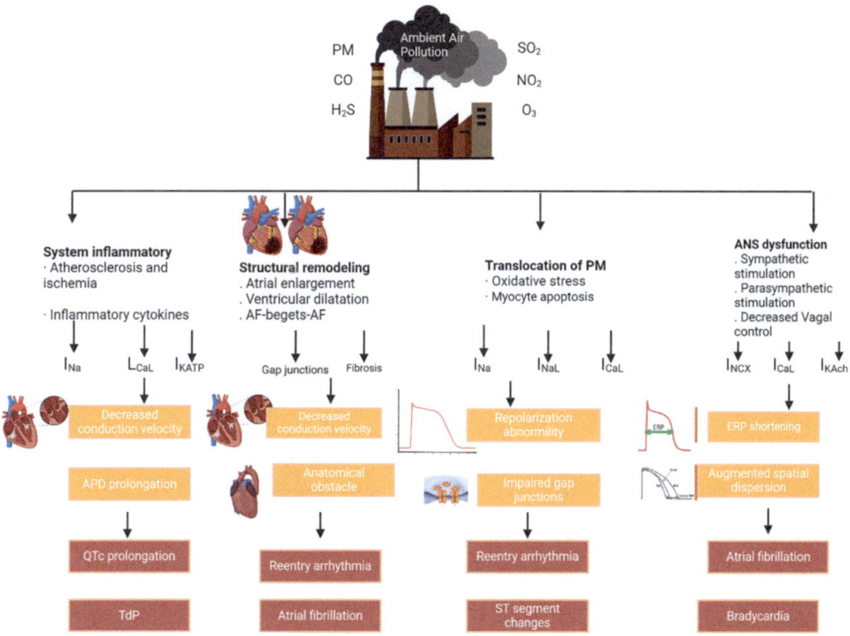

Fig. 29.4 Air pollution-induced arrhythmias have indirect causes, such as systemic inflammatory response, modification of the heart's structural components, particles' translocation into the circulatory systems, and disruption of the autonomic nervous system (ANS). There is a lot of overlap or closeness between the individual paths, and these are the main and most typical ones for each activity

disease [60]. Experimental investigations also validated the remodeling effect of PM2.5 [61].

Mechanism Underlying Carbon Monoxide-Induced Arrhythmia

Several studies have linked ambient carbon monoxide (CO) emissions, such as those from vehicles on the road, to worse cardiovascular health outcomes [62]. According to a meta-analysis that included nine million Americans, there is a link between CO_2 in the air and heart failure; specifically, for every 1 ppm increase in CO_2, the risk of heart failure-related hospitalizations or deaths increases by 3.52% [63]. An increase of 10 μg/m3 in CO was associated with a 21.1% rise in ischemic heart disease mortality, according to a recent study [64]. Arrhythmias and alterations to electrocardiograms (ECGs) can result from accidental exposure to elevated concentrations of carbon monoxide. One patient with ventricular arrhythmia and four with atrial

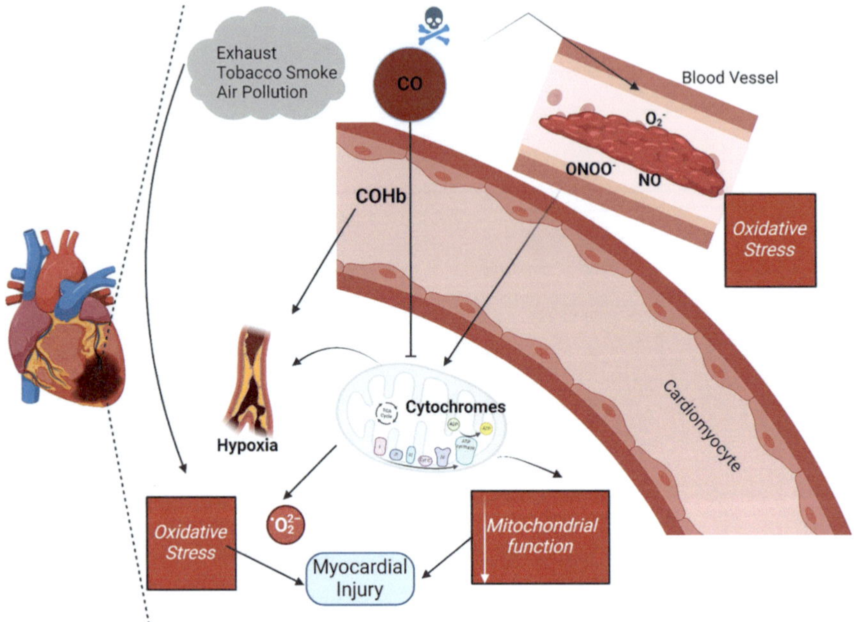

Fig. 29.5 Cardiovascular pathophysiology of carbon monoxide intoxication. Inhaled contaminants, which include carbon monoxide (CO), can quickly enter the circulation by diffusion. Aside from its binding to hemoglobin, carbon monoxide also influences platelet activity to enhance nitric oxide (NO) generation. Free radicals in the presence of oxygen react with NO

arrhythmias were among five patients with arrhythmias in a case-control study of 30 patients with acute CO poisoning [65].

The mitochondrial toxicity of CO has been demonstrated. In the myocardium, high CO binding affinity to intracellular myoglobin would reduce cardiac contractile force and lead to an energy crisis because less oxygen would reach the mitochondria [66].

Aside from ST-T alterations caused by myocardial damage, CO also causes QT alterations, which may be accomplished via separate processes than ischemia-induced ion channel remodeling. Due to the rapidity of depolarization, most of the QT duration on an electrocardiogram (ECG) is occupied by the repolarization phase, which is translated into the heart's QT interval. When the QT interval on the electrocardiogram (ECG) is long, it means that the ventricular repolarization is taking longer than it should, or that AP repolarization is taking longer from a cellular standpoint. Thus, the QT prolongation seen in CO poisoning patients is likely due to a diminished repolarization reserve caused by various ion channel abnormalities (Fig. 29.5).

Noise and Air Pollution as Risk Factors for Hypertension

A higher incidence of cardiovascular events has been associated with exposure to environmental stressors such as traffic noise and air pollution.

Environmental exposures may raise this risk through a variety of pathophysiological processes; one of them is hypertension, which can lead to both additive and synergistic harm [67]. Enhanced oxidative stress, systemic inflammation, and activation of central processes, such as sympathetic activation, which mediates exposures to a quick rise in resistance vessel tone, are all part of these common pathways [68]. New information from animal and human studies on controlled exposures to noise and air pollution as well as mitigation efforts has established irrefutable proof that exposure directly causes elevated blood pressure.

Mechanisms of Noise-Induced Arterial Hypertension and Cardiovascular Disease

When we hear noise, our brains respond by activating certain areas and releasing hormones that cause inflammation in our blood vessels and brain tissue, as well as oxidative damage. The development of cardiovascular disease (CVD) risk factors including diabetes, high cholesterol, hypercoagulable states, and hypertension may follow in due course [69]. Recent research has linked noise to arrhythmias like atrial fibrillation [70] and neurocognitive illnesses [69], in addition to increasing the risk of cardiovascular disease (CVD) due to stress reactions that disrupt sleep and communication. There are several pathophysiologic pathways that noise can affect, but two important components of noise pathophysiology are inflammation and oxidative stress caused by persistent activation of stress pathways and disruption of circadian rhythms [69].

Scientists have just recently discovered a connection between loud noise, inflammation in the blood vessels, and serious heart problems [71]. Indeed, a single night of exposure to aircraft and railway noise during the night can impair sleep quality, raise blood pressure and stress hormone levels, and cause vascular endothelial dysfunction, a subclinical indicator of atherosclerosis, according to field studies [72]. Patients who already had coronary artery disease were more likely to have endothelial dysfunction [69].

Mechanisms of Air Pollution-Induced Arterial Hypertension

Reversible peripheral conduit channel constriction, or fast vascular dysfunction characterized by microvascular endothelial dysfunction, can occur after acute exposure to particulate matter (PM)2.5 and diluted diesel exhaust (ultrafine particles)

[73]. Some of these investigations found that a delayed effect of concentrated PM2.5 exposure on endothelial-dependent vasodilation in conduit arteries was shown after 24 h, rather than immediately [74].

High concentrations of ultrafine, fine, and coarse PM2.5 and PM10 have been linked to elevated blood pressure in humans after brief exposure, with evidence from heart rate variability measurements suggesting sympathetic activation [68]. Based on postmortem research, it appears that particles can permeate the human body, including the central nervous system. Postmortem investigations may be impacted by pathological artifacts, and tracking the particles in vivo is difficult, therefore these remain tricky to do [75].

Air Pollution Enhances Incidence of Acute Myocardial Infarction

Acute myocardial infarction (AMI) is an epidemiologically linked event that is primarily explained by the pathophysiological mechanism of a rise in mean resting arterial blood pressure because of an increase in sympathetic tone and regulation of potential systemic vascular tone. The risk of intravascular thrombosis rises, which promotes the development of atherosclerosis, in response to a transitory rise in plasma viscosity and an amount of endothelial function degradation [76].

Several biological impacts of environmental particles can cause cardiac problems. As a first step, particles lodged in the alveoli trigger the recruitment of inflammatory cells and the activation of cytokine synthesis by alveolar macrophages [77] and epithelial cells. Secondly, during periods of high particle air pollution, plasma viscosity and C-reactive protein levels were shown to increase in healthy adults who were randomly selected [77]. In the third place, studies in both a randomly selected population and older adults have shown that exposure to air pollution is associated with faster heart rates and less variability in heart rate. According to one study, heart rate variability began to decrease within a few hours of being exposed: [77]. Results from electrocardiogram (ECG) studies in dogs showed changes within an hour after beginning exposure to concentrated ambient particles, according to controlled-exposure studies. Also, patients with implanted cardioverter-defibrillators are more likely to experience ventricular fibrillation and require more therapeutic treatments when exposed to ambient PM2.5 concentrations [77] (Fig. 29.6).

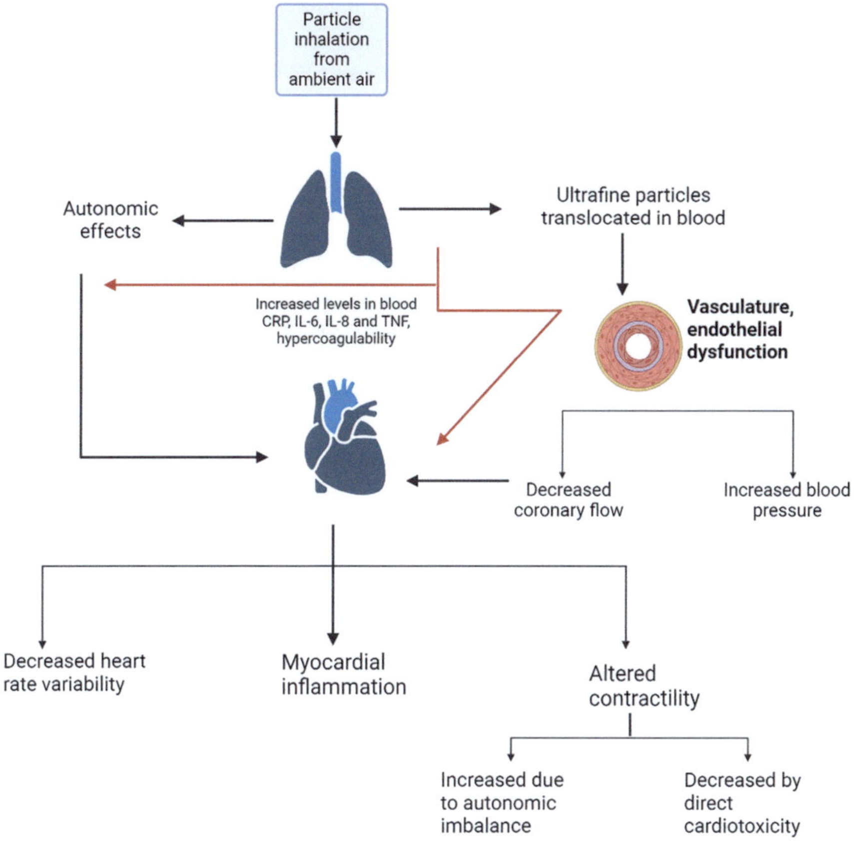

Fig. 29.6 Pathophysiological processes connecting ambient air pollution to the pathogenesis of acute myocardial infarction

Effect of Air Pollution on Ischemic Heart Disease

Ischemic heart disease (IHD) is primarily caused by coronary atherosclerosis, and there has been a lot of focus on identifying and treating the traditional CV risk factors for atherosclerosis—hypertension, smoking, dyslipidemia, and type 2 diabetes mellitus—early on.

Acute vascular changes, including arterial vasoconstriction and decreased vascular responsiveness, are linked to PM2.5 exposure and may serve as additional triggers for plaque destabilization in vulnerable people [78]. Chronic long-term exposure to PM2.5 (years) may increase the likelihood of acute PM2.5 exposure and, in turn, the development of atherosclerotic burden, vulnerable plaque features (such as inflammation, weakening of fibrous cap, and increased lipid content), and an aggravation of the harmful effects of other traditional CV risk factors (such as hypertension, dyslipidemia, and type 2 diabetes mellitus) [78]. Developing a vulnerable systemic condition with underlying vulnerable plaques can be accelerated by chronic PM2.5 exposure, which in turn increases the likelihood of acute ischemia events that are likely to be caused by fluctuations in PM2.5 exposure.

PM2.5 and Atherosclerosis: Pathogenetic Mechanisms

By disrupting many physiological processes, including as NADPH oxidases (NOXs), endothelial nitric oxide synthase (eNOS) uncoupling, and mitochondrial respiratory chain malfunction, PM2.5 can enhance the production of reactive oxygen species (ROS) and nitrogen species (RNS) [78]. Overproduction of superoxide radicals caused by NOX activation can damage macrophage mitochondria, leading to lipid buildup and the expansion of the necrotic core in atherosclerotic plaques, as well as foam cell apoptosis mediated by mitochondria [22, 23].

Exposure to PM2.5 has been linked to elevated levels of systemic inflammatory indicators such as C-reactive protein, plasma fibrinogen, and several pro-inflammatory cytokines. Additionally, PM2.5 could directly stimulate circulating leukocytes and the vascular endothelium, which in turn promotes cell adhesion and migration [75]. In addition to raising the risk of arterial thrombosis and plaque destabilization. Also, the pro-inflammatory effects of PM2.5 may hasten the onset of ACS [76].

Particulate Air Pollution and QT Interval Duration

Various studies have found that short-term exposure to PM2.5 increases the duration of the QT interval, a marker of ventricular repolarization and a risk factor for cardiac arrhythmia and sudden death. Arrhythmias and sudden cardiac deaths are associated with extension of the QT interval, which is the time it takes for the ventricles to depolarize and repolarize [78]. Prior research on the topic of particle air pollution and QT duration was limited. Researchers found that short-term particulate levels were associated with longer QT durations in both humans and animals [79]. However, a prior NAS analysis that looked at exposures up to 10 h before the study visit only found associations with black carbon, not PM2.5 overall [80]. Prolonged exposure to PM2.5 was positively associated with QT prolongation, according to a big cross-sectional investigation [81].

An increase in pulmonary oxidative stress caused by particulates is one possible mechanism that could explain the observed relationships [78]. This stress can then trigger the production of pro-inflammatory cytokines. Direct oxidative stress and inflammatory effects in the vasculature and myocardial substrate can be induced by particles as well [78]. Epidemiologic investigations, like the one conducted in the Normative Aging Study (NAS) cohort, have also connected exposure to PM to methylation alterations in the iNOS gene, which plays a role in regulating the inflammatory response [82]. Cardiac autonomic regulation can be altered by increased pro-inflammatory signals and oxidative stress, leading to a prolonged ventricular polarization phase through changes in sodium and calcium channels [78]. Inhibition of nitric oxide generation by oxidative stress may cause abnormalities in cardiac autonomic regulation and, in turn, cause QT prolongation [78].

Prevention and Treatment of Environmental Pollution-Induced Exacerbation of Cardiovascular Toxicity

Antioxidants

Many investigations have investigated the possibility that diets high in antioxidants can reduce the impact of air pollution on heart rate variability (HRV). There was an inverse relationship between PM2.5 exposure and some HRV measures. Those who took fish oil supplements saw a considerable reduction in these side effects (an "improvement" in HRV), while those who took soy oil supplements had only mild or no enhancement.

Additionally, researchers have examined the potential protective benefits of antioxidant supplements against pollution on HRV by conducting chamber studies, in which participants are exposed to controlled environments and pollutants for brief durations (typically 1–2 h) [83]. Supplementing with omega-3 fatty acids orally for 4 weeks mitigated the decreases in high-to-low frequency ratio measurements and increases in normalized low-frequency heart rate variability caused by cyclic alternative patterns (CAPs) during sleep, an effect that was not observed in the control group (olive oil supplements). Further, the lipid profiles of the participants were shown to be enhanced by the fish-oil supplementation, according to this study. Several animal studies have shown that antioxidants improve cardiac parameters after PM exposure. For instance, according to a study [84], the antioxidant N-acetylcysteine (NAC) reduced HRV alterations in rats because of pulmonary injection of urban PM. Also, beta-blockers prevented the PM-induced myocardial oxidative stress.

Researchers have investigated the connection between arrhythmia and oxidative stress in several lab research. Free radical scavengers (NAC or tiron) or inhibitors of cellular sources of free radicals (oxypurinol to inhibit xanthine oxidase; apocyanin to inhibit NAD(P)H oxidase) could partially mitigate the effect of direct addition of DEP to cultured cardiomyocytes on contractile function [84]. Diemethylthiourea and NAC both operate as hydroxyl radical and hydrogen peroxide scavengers, respectively, and can block the direct cytotoxicity and apoptotic effects of different PM on cardiomyocytes [79]. In rats, diesel exhaust particulate (DEP) inhaled into the lungs caused an accelerated heart rate and premature ventricular contractions [79]. A potential solution would be to administer NAC beforehand to forestall these side effects. Another study [85] utilized pulmonary instillation of DEP in rats to simulate myocardial ischemia caused by coronary artery ligation. Prolonged cardiac arrhythmias were commonly seen during ligation and were linked to a high rate of mortality. The effects were mitigated when a β1 receptor antagonist (metoprozol) was administered simultaneously or when the alveolar sensory receptors were blocked using a vanilloid receptor antagonist (transient receptor potential cation channel V1; TRPV1) was used.

By blocking alveolar TRPV1 receptors or co-administering a β1 receptor antagonist, the effects of myocardial infarction were mitigated. It was clear from these results that DEP can prime the heart to be more vulnerable to ischemic injury by increasing its sensitivity to sensory afferents and autonomic innervation. In a rat model of acute myocardial infarction, one group successfully prevented cardiac dysfunction caused by PM2.5 exposure by using a combination of β1-antagonist and β2-agonist treatment [86]. Urban PM10 lowered expression of endothelial nitric oxide synthase (eNOS) and antioxidant enzymes in cardiac tissue, while increasing expression of pro-inflammatory inducible nitric oxide synthase (iNOS) [87], as shown using a similar paradigm. With the use of the antioxidant vanillic acid, cardiac antioxidant levels were less affected by PM10, and mitochondrial abnormalities and cardiac performance were partially alleviated [88].

Beta-Blockers

Since HRV is mostly derived from the autonomic nervous system's regulation of the heart, beta-blockers can shield HRV from air pollution-related alterations. Outdoor PM2.5 (especially that associated with traffic) was associated with negative effects on HRV parameters, but only in patients who were not taking beta-blockers, according to a joint assessment of personal (measured on individuals using portable devices) and ambient (used of stationary monitoring networks) air pollution [89]. While exercising in Helsinki, Finland, the same group evaluated S-T segment depression in patients with stable coronary artery disease. This is an area of the electrocardiogram (ECG) that can suggest ischemia stress in coronary arteries… A higher incidence of S-T segment depression was linked to carbon monoxide, nitrogen dioxide, particulate matter 2.5 (a proxy for ultrafine particulate matter), and particle number, but not to the bigger coarse particulate matter. Subjects who did not use beta-blockers showed stronger connections, suggesting that these effects might be due to the heart's autonomic modulation [84].

Statins

Although pleotropic effects of statins (such as antioxidant and anti-inflammatory properties) are widely known, they are currently used almost universally to reduce blood cholesterol. Statins were observed to remove correlations between PM2.5 and

C-reactive protein (a measure of inflammation/acute phase response) levels in the blood of middle-aged women [90]. Those who are susceptible to the effects of oxidative stress due to a lack of the glutathione-S-transferase M1 (GSTM1) allele can take statins, which eliminate the effects of PM2.5 on high-frequency HRV measures [84]. Researchers hypothesize that favorable benefits were attributable, in part, to decreased oxidative stress.

Conclusion

Finally, environmental pollution enhances the cytotoxicity of anticancer medications and other pharmacological treatments, which in turn increases the dangers to cardiovascular health. Pollutants including particulate matter and environmental toxins can make the effects of these medications worse, leading to inflammation, oxidative stress, and damage to heart tissue. There are three possible approaches to address these side effects: beta-blockers to control cardiac function, statins to reduce inflammation and oxidative stress, and antioxidant-rich food and supplementation. Incorporating antioxidant-rich diets and supplements is likely the most effective of these strategies because they provide vital nutrients to fight oxidative damage and increase overall cardiovascular resilience against the negative effects of drugs and environmental pollutants.

Funding The research work of DKA is supported by R25AI179582 and R01 HL147662 grants from the National Institutes of Health, USA. The contents of this chapter are solely the responsibility of the authors and do not necessarily represent the official views of the National Institutes of Health.

Competing Interests All authors have read the chapter and declare no conflict of interest. No writing assistance was utilized in the production of this manuscript.

Consent for Publication All authors have read the chapter and consented for publication.

References

1. Farley A, McLafferty E, Hendry C (2012) The cardiovascular system. Nurs Stand 27(9):35–39
2. Olvera Lopez E, Ballard BD, Jan A (2023) Cardiovascular disease. [Updated 2022 Aug 8]. In: StatPearls [Internet]. StatPearls Publishing, Treasure Island (FL). https://www.ncbi.nlm.nih.gov/books/NBK535419/
3. Benjamin EJ, Virani SS, Callaway CW, Chamberlain AM, Chang AR, Cheng S, Chiuve SE, Cushman M, Delling FN, Deo R, de Ferranti SD, Ferguson JF, Fornage M, Gillespie C, Isasi CR, Jiménez MC, Jordan LC, Judd SE, Lackland D, Lichtman JH, Lisabeth L, Liu S, Longenecker CT, Lutsey PL, Mackey JS, Matchar DB, Matsushita K, Mussolino ME, Nasir K, O'Flaherty M, Palaniappan LP, Pandey A, Pandey DK, Reeves MJ, Ritchey MD, Rodriguez CJ, Roth GA, Rosamond WD, Sampson UKA, Satou GM, Shah SH, Spartano NL, Tirschwell DL, Tsao CW, Voeks JH, Willey JZ, Wilkins JT, Wu JH, Alger HM, Wong SS,

Muntner P, American Heart Association Council on Epidemiology and Prevention Statistics Committee and Stroke Statistics Subcommittee (2018) Heart disease and stroke statistics-2018 update: a report from the American Heart Association. Circulation 137(12):e67–e492
4. Amani R, Sharifi N (2012) Cardiovascular disease risk factors. https://doi.org/10.5772/34374
5. Rudd JHF et al (2005) Imaging of atherosclerosis—can we predict plaque rupture? Trends Cardiovasc Med 15:17–24
6. Davignon J, Ganz P (2004) Role of endothelial dysfunction in atherosclerosis. Circulation 109(23 suppl):III27
7. Yusuf S, Hawken S, Ounpuu S, Dans T, Avezum A, Lanas F, McQueen M, Budaj A, Pais P, Varigos J, Lisheng L, INTERHEART Study Investigators (2004) Effect of potentially modifiable risk factors associated with myocardial infarction in 52 countries (the INTERHEART study): case-control study. Lancet 364(9438):937–952
8. Vasan RS, Sullivan LM, Wilson PW, Sempos CT, Sundström J, Kannel WB, Levy D, D'Agostino RB (2005) Relative importance of borderline and elevated levels of coronary heart disease risk factors. Ann Intern Med 142(6):393–402
9. Dunbar SB, Khavjou OA, Bakas T, Hunt G, Kirch RA, Leib AR, Morrison RS, Poehler DC, Roger VL, Whitsel LP, American Heart Association (2018) Projected costs of informal caregiving for cardiovascular disease: 2015 to 2035: a policy statement from the American Heart Association. Circulation 137(19):e558–e577
10. Williams PT (2001) Physical fitness and activity as separate heart disease risk factors: a meta-analysis. Med Sci Sports Exerc 33(5):754–761. https://doi.org/10.1097/00005768-200105000-00012. PMID: 11323544; PMCID: PMC2821586
11. Goldsborough E 3rd, Osuji N, Blaha MJ (2022) Assessment of cardiovascular disease risk: a 2022 update. Endocrinol Metab Clin N Am 51(3):483–509. https://doi.org/10.1016/j.ecl.2022.02.005. Epub 2022 Jul 4. PMID: 35963625
12. Fox CS, Pencina MJ, Wilson PW, Paynter NP, Vasan RS, D'Agostino RB (2008) Lifetime risk of cardiovascular disease among individuals with and without diabetes stratified by obesity status in the Framingham heart study. Diabetes Care 31(8):1582–1584
13. Savji N, Rockman CB, Skolnick AH, Guo Y, Adelman MA, Riles T, Berger JS (2013) Association between advanced age and vascular disease in different arterial territories: a population database of over 3.6 million subjects. J Am Coll Cardiol 61(16):1736–1743
14. Walden R, Tomlinson B (2011) Cardiovascular disease. In: Benzie IFF, Wachtel-Galor S (eds) Herbal medicine: biomolecular and clinical aspects (Chapter 16), 2nd edn. CRC Press/Taylor & Francis, Boca Raton (FL). https://www.ncbi.nlm.nih.gov/books/NBK92767/
15. Baigent C, Keech A, Kearney PM et al (2005) Efficacy and safety of cholesterol-lowering treatment: prospective meta-analysis of data from 90,056 participants in 14 randomised trials of statins. Lancet 366:1267–1278
16. Cosselman KE, Navas-Acien A, Kaufman JD (2015) Environmental factors in cardiovascular disease. Nat Rev Cardiol 12(11):627–642. https://doi.org/10.1038/nrcardio.2015.152. Epub 2015 Oct 13. PMID: 26461967
17. Schultz WM, Kelli HM, Lisko JC, Varghese T, Shen J, Sandesara P, Quyyumi AA, Taylor HA, Gulati M, Harold JG, Mieres JH (2018) Socioeconomic status and cardiovascular outcomes. Circulation 137(20):2166–2178. https://doi.org/10.1161/CIRCULATIONAHA.117.029652. PMC 5958918. PMID 29760227
18. Murray CJL et al (2012) Disability-adjusted life years (DALYs) for 291 diseases and injuries in 21 regions, 1990–2010: a systematic analysis for the Global Burden of Disease Study 2010. Lancet 380(9859):2197–2223
19. Kucharska-Newton AM et al (2011) Socioeconomic indicators and the risk of acute coronary heart disease events: comparison of population-based data from the United States and Finland. Ann Epidemiol 21(8):572–579
20. Gerber Y et al (2008) Neighborhood income and individual education: effect on survival after myocardial infarction. Mayo Clin Proc 83(6)

21. Stirbu I et al (2012) Income inequalities in case death of ischaemic heart disease in the Netherlands: a national record-linked study. J Epidemiol Community Health 66(12):1159–1166
22. Winkleby MA et al (1992) Socioeconomic status and health: how education, income, and occupation contribute to risk factors for cardiovascular disease. Am J Public Health 82(6): 816–820
23. Kim JH et al (2014) The association of socioeconomic status with three-year clinical outcomes in patients with acute myocardial infarction who underwent percutaneous coronary intervention. J Korean Med Sci 29(4):536–543
24. Méjean C et al (2013) The contribution of diet and lifestyle to socioeconomic inequalities in cardiovascular morbidity and mortality. Int J Cardiol 168(6):5190–5195
25. Dupre ME et al (2012) The cumulative effect of unemployment on risks for acute myocardial infarction. Arch Intern Med 172(22):1731–1737
26. Roux AVD et al (2001) Neighborhood of residence and incidence of coronary heart disease. N Engl J Med 345(2):99–106
27. Müller-Riemenschneider F et al (2013) Neighborhood walkability and cardiometabolic risk factors in Australian adults: an observational study. BMC Public Health 13:1–9
28. Dubowitz T et al (2015) Healthy food access for urban food desert residents: examination of the food environment, food purchasing practices, diet and BMI. Public Health Nutr 18(12): 2220–2230
29. LeFevre ML, US Preventive Services Task Force* (2014) Behavioral counseling to promote a healthful diet and physical activity for cardiovascular disease prevention in adults with cardiovascular risk factors: US Preventive Services Task Force recommendation statement. Ann Intern Med 161(8):587–593
30. Pearson TA et al (2013) American Heart Association Guide for Improving Cardiovascular Health at the Community Level, 2013 update: a scientific statement for public health practitioners, healthcare providers, and health policy makers. Circulation 127(16):1730–1753
31. Münzel T, Sørensen M, Daiber A (2021) Transportation noise pollution and cardiovascular disease. Nat Rev Cardiol 18:619–636. https://doi.org/10.1038/s41569-021-00532-5
32. Schmidt FP et al (2013) Effect of nighttime aircraft noise exposure on endothelial function and stress hormone release in healthy adults. Eur Heart J 34(45):3508–3514. https://doi.org/10.1093/eurheartj/eht269
33. Thacher JD, Poulsen AH, Raaschou-Nielsen O, Hvidtfeldt UA, Brandt J, Christensen JH, Khan J, Levin G, Münzel T, Sørensen M (2022) Exposure to transportation noise and risk for cardiovascular disease in a nationwide cohort study from Denmark. Environ Res 211:113106. https://doi.org/10.1016/j.envres.2022.113106
34. Pyko A, Roswall N, Ögren M, Oudin A, Rosengren A, Eriksson C, Segersson D, Rizzuto D, Andersson EM, Aasvang GM, Engström G, Gudjonsdottir H, Jørgensen JT, Selander J, Christensen JH, Brandt J, Leander K, Overvad K, Eneroth K et al (2023) Long-term exposure to transportation noise and ischemic heart disease: a pooled analysis of nine scandinavian cohorts. Environ Health Perspect 131(1):017003. https://doi.org/10.1289/EHP10745
35. Babisch W (2002) The noise/stress concept, risk assessment and research needs. Noise Health 4: 1–11
36. Sun S, Cao W, Ge Y, Ran J, Sun F, Zeng Q, Guo M, Huang J, Lee RS-Y, Tian L, Wellenius GA (2021) Outdoor light at night and risk of coronary heart disease among older adults: a prospective cohort study. Eur Heart J 42:822–830
37. Guyenet PG (2006) The sympathetic control of blood pressure. Nat Rev Neurosci 7:335–346. https://doi.org/10.1038/nrn1902
38. Owman C, Edvinsson L, Lindvall M, Sjoberg NO (1982) Influence of external light conditions on norepinephrine levels in organs innervated by sympathetic nerves from different levels. Brain Res Bull 9:777–779. https://doi.org/10.1016/0361-9230(82)90185-x
39. Cardinale D, Colombo A, Bacchiani G, Tedeschi I, Meroni CA, Veglia F, Civelli M, Lamantia G, Colombo N, Curigliano G, Fiorentini C, Cipolla CM (2015) Early detection of

anthracycline cardiotoxicity and improvement with heart failure therapy. Circulation 131(22): 1981–1988. https://doi.org/10.1161/circulationaha.114.013777
40. Morelli MB, Bongiovanni C, Da Pra S, Miano C, Sacchi F, Lauriola M, D'Uva G (2022) Cardiotoxicity of anticancer drugs: molecular mechanisms and strategies for Cardioprotection. Front Cardiovasc Med 9. https://doi.org/10.3389/fcvm.2022.847012
41. Qiu S, Zhou T, Qiu B, Zhang Y, Zhou Y, Yu H, Zhang J, Liu L, Yuan L, Yang G, Duan Y, Xing C (2021) Risk factors for anthracycline-induced cardiotoxicity. Front Cardiovasc Med 8. https://doi.org/10.3389/fcvm.2021.736854
42. Renu K, Abilash VG, Tirupathi Pichiah PB, Arunachalam S (2017) Molecular mechanism of doxorubicin-induced cardiomyopathy—an update. Eur J Pharmacol 818:241–253. https://doi.org/10.1016/j.ejphar.2017.10.043
43. Jurczyk M, Król M, Midro A, Kurnik-Łucka M, Poniatowski A, Gil K (2021) Cardiotoxicity of fluoropyrimidines: epidemiology, mechanisms, diagnosis, and management. J Clin Med 10(19): 4426. https://doi.org/10.3390/jcm10194426
44. Alter P, Herzum M, Soufi M, Schaefer J, Maisch B (2006) Cardiotoxicity of 5-fluorouracil. Cardiovasc Hematol Agents Med Chem 4(1):1–5. https://doi.org/10.2174/187152506775268785
45. Madeddu C, Deidda M, Piras A, Cadeddu C, Demurtas L, Puzzoni M, Piscopo G, Scartozzi M, Mercuro G (2016) Pathophysiology of cardiotoxicity induced by nonanthracycline chemotherapy. J Cardiovasc Med 17:e12–e18. https://doi.org/10.2459/jcm.0000000000000376
46. Focaccetti C, Bruno A, Magnani E, Bartolini D, Principi E, Dallaglio K, Bucci EO, Finzi G, Sessa F, Noonan DM, Albini A (2015) Effects of 5-fluorouracil on morphology, cell cycle, proliferation, apoptosis, autophagy and ROS production in endothelial cells and cardiomyocytes. PLoS One 10(2):e0115686. https://doi.org/10.1371/journal.pone.0115686
47. Batra A, Patel B, Addison D, Baldassarre LA, Desai N, Weintraub N, Deswal A, Hussain Z, Brown S, Ganatra S, Agarwala V, Parikh PM, Fradley M, Ghosh A, Guha A (2021) Cardiovascular safety profile of taxanes and vinca alkaloids: 30 years FDA registry experience. Open Heart 8(2):e001849. https://doi.org/10.1136/openhrt-2021-001849
48. Kelleni MT, Abdelbasset M (2018) Drug induced cardiotoxicity: mechanism, prevention and management. InTech eBooks. https://doi.org/10.5772/intechopen.79611
49. Bellissima BL, Tingle MD, Cicović A, Alawami M, Kenedi C (2018) A systematic review of clozapine-induced myocarditis. Int J Cardiol 259:122–129. https://doi.org/10.1016/j.ijcard.2017.12.102
50. Varga ZV, Ferdinandy P, Liaudet L, Pacher P (2015) Drug-induced mitochondrial dysfunction and cardiotoxicity. AJP Heart Circ Physiol 309(9):H1453–H1467. https://doi.org/10.1152/ajpheart.00554.2015
51. Sadurska E (2015) Current views on anthracycline cardiotoxicity in childhood cancer survivors. Pediatr Cardiol 36(6):1112–1119. https://doi.org/10.1007/s00246-015-1176-7
52. Varricchi G, Ameri P, Cadeddu C, Ghigo A, Madonna R, Marone G, Mercurio V, Monte I, Novo G, Parrella P, Pirozzi F, Pecoraro A, Spallarossa P, Zito C, Mercuro G, Pagliaro P, Tocchetti CG (2018) Antineoplastic drug-induced cardiotoxicity: a redox perspective. Front Physiol 9. https://doi.org/10.3389/fphys.2018.00167
53. Yue C, Yang F, Li F, Chen Y (2020) Association between air pollutants and atrial fibrillation in general population: a systematic review and meta-analysis. Ecotoxicol Environ Saf 208: 111508. https://doi.org/10.1016/j.ecoenv.2020.111508
54. Fiordelisi A, Piscitelli P, Trimarco B, Coscioni E, Iaccarino G, Sorriento D (2017) The mechanisms of air pollution and particulate matter in cardiovascular diseases. Heart Fail Rev 22(3):337–347. https://doi.org/10.1007/s10741-017-9606-7
55. Liu Q, Gu X, Deng F, Mu L, Baccarelli AA, Guo X, Wu S (2019) Ambient particulate air pollution and circulating C-reactive protein level: a systematic review and meta-analysis. Int J Hyg Environ Health 222(5):756–764. https://doi.org/10.1016/j.ijheh.2019.05.005

56. Liang S, Zhang J, Ning R, Du Z, Liu J, Batibawa JW, Duan J, Sun Z (2020) The critical role of endothelial function in fine particulate matter-induced atherosclerosis. Part Fibre Toxicol 17(1): 61. https://doi.org/10.1186/s12989-020-00391-x
57. Zhang S, Zhang S, Fan X, Wang W, Li Z, Jia D, Wei Z, Zhang H (2019) Pro-arrhythmic effects of hydrogen sulfide in healthy and ischemic cardiac tissues: insight from a simulation study. Front Physiol 10. https://doi.org/10.3389/fphys.2019.01482
58. Yalta T, Yalta K (2017) Systemic inflammation and arrhythmogenesis: a review of mechanistic and clinical perspectives. Angiology 69(4):288–296. https://doi.org/10.1177/0003319717709380
59. Baranowska-Wójcik E, Szwajgier D, Oleszczuk P, Winiarska-Mieczan A (2019) Effects of titanium dioxide nanoparticles exposure on human health—a review. Biol Trace Elem Res 193(1):118–129. https://doi.org/10.1007/s12011-019-01706-6
60. Aung N, Sanghvi MM, Zemrak F, Lee AM, Cooper JA, Paiva JM, Thomson RJ, Fung K, Khanji MY, Lukaschuk E, Carapella V, Kim YJ, Munroe PB, Piechnik SK, Neubauer S, Petersen SE (2018) Association between ambient air pollution and cardiac morpho-functional phenotypes. Circulation 138(20):2175–2186. https://doi.org/10.1161/circulationaha.118.034856
61. Su X, Tian J, Li B, Zhou L, Kang H, Pei Z, Zhang M, Li C, Wu M, Wang Q, Han B, Chu C, Pang Y, Ning J, Zhang B, Niu Y, Zhang R (2020) Ambient PM2.5 caused cardiac dysfunction through FoxO1-targeted cardiac hypertrophy and macrophage-activated fibrosis in mice. Chemosphere 247:125881. https://doi.org/10.1016/j.chemosphere.2020.125881
62. Garg J, Krishnamoorthy P, Palaniswamy C, Khera S, Ahmad H, Jain D, Aronow WS, Frishman WH (2014) Cardiovascular abnormalities in carbon monoxide poisoning. Am J Ther 25(3): e339–e348. https://doi.org/10.1097/mjt.0000000000000016
63. Shah AS, Langrish JP, Nair H, McAllister DA, Hunter AL, Donaldson K, Newby DE, Mills NL (2013) Global association of air pollution and heart failure: a systematic review and meta-analysis. Lancet 382(9897):1039–1048. https://doi.org/10.1016/s0140-6736(13)60898-3
64. Dastoorpoor M, Riahi A, Yazdaninejhad H, Borsi SH, Khanjani N, Khodadadi N, Mohammadi MJ, Aghababaeian H (2020) Exposure to particulate matter and carbon monoxide and cause-specific cardiovascular-respiratory disease mortality in Ahvaz. Toxin Rev 40(4):1362–1372. https://doi.org/10.1080/15569543.2020.1716256
65. Hanci V, Ayoglu H, Yurtlu S, Yildirim N, Okyay D, Erdogan G, Abdusoglu M, Yilmaz M, Ozkocak I (2011) Effects of acute carbon monoxide poisoning on the P-wave and QT interval dispersions. Anadolu Kardiyol Dergisi/Anatolian J Cardiol. https://doi.org/10.5152/akd.2011.009
66. Hantson P (2018) Mechanisms of toxic cardiomyopathy. Clin Toxicol 57(1):1–9. https://doi.org/10.1080/15563650.2018.1497172
67. Kuntic M, Kuntic I, Krishnankutty R, Gericke A, Oelze M, Junglas T, Jimenez MTB, Stamm P, Nandudu M, Hahad O, Keppeler K, Daub S, Vujacic-Mirski K, Rajlic S, Strohm L, Ubbens H, Tang Q, Jiang S, Ruan Y et al (2022) Co-exposure to urban particulate matter and aircraft noise adversely impacts the cerebro-pulmonary-cardiovascular axis in mice. Redox Biol 59:102580. https://doi.org/10.1016/j.redox.2022.102580
68. Al-Kindi SG, Brook RD, Biswal S, Rajagopalan S (2020) Environmental determinants of cardiovascular disease: lessons learned from air pollution. Nat Rev Cardiol 17(10):656–672. https://doi.org/10.1038/s41569-020-0371-2
69. Münzel T, Schmidt FP, Steven S, Herzog J, Daiber A, Sørensen M (2018) Environmental noise and the cardiovascular system. J Am Coll Cardiol 71(6):688–697. https://doi.org/10.1016/j.jacc.2017.12.015
70. Hahad O, Beutel M, Gori T, Schulz A, Blettner M, Pfeiffer N, Rostock T, Lackner K, Sørensen M, Prochaska JH, Wild PS, Münzel T (2018) Annoyance to different noise sources is associated with atrial fibrillation in the Gutenberg Health Study. Int J Cardiol 264:79–84. https://doi.org/10.1016/j.ijcard.2018.03.126

71. Osborne MT, Radfar A, Hassan MZO, Abohashem S, Oberfeld B, Patrich T, Tung B, Wang Y, Ishai A, Scott JA, Shin LM, Fayad ZA, Koenen KC, Rajagopalan S, Pitman RK, Tawakol A (2019) A neurobiological mechanism linking transportation noise to cardiovascular disease in humans. Eur Heart J 41(6):772–782. https://doi.org/10.1093/eurheartj/ehz820
72. Herzog J, Schmidt FP, Hahad O, Mahmoudpour SH, Mangold AK, Andreo PG, Prochaska J, Koeck T, Wild PS, Sørensen M, Daiber A, Münzel T (2019) Acute exposure to nocturnal train noise induces endothelial dysfunction and pro-thromboinflammatory changes of the plasma proteome in healthy subjects. Basic Res Cardiol 114(6):46. https://doi.org/10.1007/s00395-019-0753-y
73. Münzel T, Gori T, Al-Kindi S, Deanfield J, Lelieveld J, Daiber A, Rajagopalan S (2018) Effects of gaseous and solid constituents of air pollution on endothelial function. Eur Heart J 39(38): 3543–3550. https://doi.org/10.1093/eurheartj/ehy481
74. Hahad O, Rajagopalan S, Lelieveld J, Sørensen M, Kuntic M, Daiber A, Basner M, Nieuwenhuijsen M, Brook RD, Münzel T (2023) Noise and air pollution as risk factors for hypertension: part II—pathophysiologic insight. Hypertension 80(7):1384–1392. https://doi.org/10.1161/hypertensionaha.123.20617
75. Miller MR, Raftis JB, Langrish JP, McLean SG, Samutrtai P, Connell SP, Wilson S, Vesey AT, Fokkens PHB, Boere AJF, Krystek P, Campbell CJ, Hadoke PWF, Donaldson K, Cassee FR, Newby DE, Duffin R, Mills NL (2017) Inhaled nanoparticles accumulate at sites of vascular disease. ACS Nano 11(5):4542–4552. https://doi.org/10.1021/acsnano.6b08551
76. Rao X, Zhong J, Brook RD, Rajagopalan S (2017) Effect of particulate matter air pollution on cardiovascular oxidative stress pathways. Antioxid Redox Signal 28(9):797–818. https://doi.org/10.1089/ars.2017.7394
77. Peters A, Dockery DW, Muller JE, Mittleman MA (2001) Increased particulate air pollution and the triggering of myocardial infarction. Circulation 103(23):2810–2815. https://doi.org/10.1161/01.cir.103.23.2810
78. Mordukhovich I, Kloog I, Coull B, Koutrakis P, Vokonas P, Schwartz J (2015) Association between particulate air pollution and QT interval duration in an elderly cohort. Epidemiology 1. https://doi.org/10.1097/ede.0000000000000424
79. Kim J, Kim C, Choi E, Park S, Park H, Pak H, Lee M, Shin DC, Hwang K, Joung B (2011) Particulate air pollution induces arrhythmia via oxidative stress and calcium calmodulin kinase II activation. Toxicol Appl Pharmacol 259(1):66–73. https://doi.org/10.1016/j.taap.2011.12.007
80. Baja ES, Schwartz JD, Wellenius GA, Coull BA, Zanobetti A, Vokonas PS, Suh HH (2010) Traffic-related air pollution and QT interval: modification by diabetes, obesity, and oxidative stress gene polymorphisms in the normative aging study. Environ Health Perspect 118(6): 840–846. https://doi.org/10.1289/ehp.0901396
81. Van Hee VC, Szpiro AA, Prineas R, Neyer J, Watson K, Siscovick D, Park SK, Kaufman JD (2011) Association of long-term air Pollution with ventricular conduction and repolarization abnormalities. Epidemiology 22(6):773–780. https://doi.org/10.1097/ede.0b013e31823061a9
82. Madrigano J, Baccarelli A, Mittleman MA, Sparrow D, Spiro A, Vokonas PS, Cantone L, Kubzansky L, Schwartz J (2012) Air pollution and DNA methylation: interaction by psychological factors in the VA normative aging study. Am J Epidemiol 176(3):224–232. https://doi.org/10.1093/aje/kwr523
83. Tong H, Rappold AG, Diaz-Sanchez D, Steck SE, Berntsen J, Cascio WE, Devlin RB, Samet JM (2012) Omega-3 fatty acid supplementation appears to attenuate particulate air pollution–induced cardiac effects and lipid changes in healthy middle-aged adults. Environ Health Perspect 120(7):952–957. https://doi.org/10.1289/ehp.1104472
84. Miller MR (2021) The cardiovascular effects of air pollution: prevention and reversal by pharmacological agents. Pharmacol Ther 232:107996. https://doi.org/10.1016/j.pharmthera.2021.107996
85. Robertson S, Thomson AL, Carter R, Stott HR, Shaw CA, Hadoke PWF, Newby DE, Miller MR, Gray GA (2014) Pulmonary diesel particulate increases susceptibility to myocardial

ischemia/reperfusion injury via activation of sensory TRPV1 and β1 adrenoreceptors. Part Fibre Toxicol 11(1). https://doi.org/10.1186/1743-8977-11-12

86. Gao Y, Lv J, Lin Y, Li X, Wang L, Yin Y, Liu Y (2014) Effects of β-adrenoceptor subtypes on cardiac function in myocardial infarction rats exposed to fine particulate matter (PM2.5). Biomed Res Int 2014:1–9. https://doi.org/10.1155/2014/308295
87. Dianat M, Radmanesh E, Badavi M, Mard SA, Goudarzi G (2015) Disturbance effects of PM10 on iNOS and eNOS mRNA expression levels and antioxidant activity induced by ischemia–reperfusion injury in isolated rat heart: protective role of vanillic acid. Environ Sci Pollut Res 23(6):5154–5165. https://doi.org/10.1007/s11356-015-5759-x
88. Zeng X, Liu J, Du X, Zhang J, Pan K, Shan W, Xie Y, Song W, Zhao J (2018) The protective effects of selenium supplementation on ambient PM2.5-induced cardiovascular injury in rats. Environ Sci Pollut Res 25(22):22153–22162. https://doi.org/10.1007/s11356-018-2292-8
89. De Hartog JJ, Lanki T, Timonen KL, Hoek G, Janssen NA, Ibald-Mulli A, Peters A, Heinrich J, Tarkiainen TH, Van Grieken R, Van Wijnen JH, Brunekreef B, Pekkanen J (2008) Associations between PM 2.5 and heart rate variability are modified by particle composition and beta-blocker use in patients with coronary heart disease. Environ Health Perspect 117(1):105–111. https://doi.org/10.1289/ehp.11062
90. Ostro B, Malig B, Broadwin R, Basu R, Gold EB, Bromberger JT, Derby C, Feinstein S, Greendale GA, Jackson EA, Kravitz HM, Matthews KA, Sternfeld B, Tomey K, Green RR, Green R (2014) Chronic PM2.5 exposure and inflammation: determining sensitive subgroups in mid-life women. Environ Res 132:168–175. https://doi.org/10.1016/j.envres.2014.03.042

Chapter 30
Nutrition, Gut Microbiota, and the Pathogenesis of Cardiovascular Toxicity

Mohamed M. Radwan

Abstract Cardiovascular diseases continue to be the leading cause of death worldwide. Research studies have emphasized the significant association between cardiovascular toxicity and dietary patterns as well as inflammatory processes; these factors critically influence cardiovascular health and contribute to major cardiovascular risk factors such as Type 2 diabetes Mellitus (T2DM) and obesity. Recent studies have elucidated a causal link between gut microbiota and cardiovascular toxicity, operating either directly or via its metabolic products. The intricate interplay between dietary intake, the gut microbiome, and the host's immune response presents a fascinating and complex puzzle. This complexity is further heightened because the underlying mechanisms driving inflammation and their contribution to cardiovascular pathologies remain incompletely defined. This discussion will delve into these interconnected components and the mechanisms by which they instigate cardiovascular toxicity and related diseases.

Keywords Cardiovascular toxicity · Gut microbiota · Immune responses · Western diet · Metabolites · Inflammation · Risk stratification

Introduction

The most common noncommunicable diseases of the present time are cardiovascular diseases, metabolic disorders, including type 2 diabetes mellitus, obesity and metabolic syndrome, and malignant diseases [1, 2]. These disorders share common themes, notably inflammation and unhealthy diet intake [3–9]. Recent research studies have shed light on the link between inflammation and gut microbiota (GM) [10], comprising 10–100 trillion microbes, the vast majority of which are bacterial cells belonging to several hundred species in adult humans [11–13]. Nutrition, a key

M. M. Radwan (✉)
Department of Translational Research, Western University of Health Sciences, Pomona, CA, USA
e-mail: mradwanahmed@westernu.edu

D. K. Agrawal et al. (eds.), *Cardiovascular Toxicity*, Advances in Biochemistry in Health and Disease 33, https://doi.org/10.1007/978-3-031-90598-8_30

player [7, 14], alters cellular metabolism and GM [15–17]; these complex interactions influence the immune system and may steer it toward an inflammatory phenotype that plays a central role in the pathogenesis of cardiovascular, metabolic [18], and malignant diseases [19], which are increasingly prevalent in today's society [2].

In addition to the host's metabolism [7], the gut microbiome metabolism plays a pivotal role in regulating the immune system [20]. Metabolic products of dysbiotic GM (imbalanced microbiota) can have detrimental effects not only on the intestinal tract [21] but also on distant organs such as the metabolically active liver and the cardiovascular system [20, 22, 23]. These distant effects underscore the systemic impact of gut health and the intricate relationship between nutrition, GM, and disease development [20].

Cardiovascular diseases are the most common cause of morbidity and mortality worldwide [2]; recently, a large body of research closely links the gut microbiome to cardiovascular health [24–29].

Nutrition and Cardiovascular Health

The Western diet, characterized by a high intake of processed foods, saturated fats, animal proteins, refined sugars, emulsifiers, and artificial additives [9], has been associated with a constellation of symptoms and signs known as metabolic syndrome [30, 31]. This syndrome is the precursor to cardiometabolic diseases [32], which encompass a combination of obesity, type 2 diabetes mellitus, and cardiovascular disease [33]. Cardiometabolic diseases serve as the critical link between diet and cardiovascular health [34], and their prevalence is significant on a global scale [35–37]. Research studies suggest that the Western diet's saturated fat, animal protein, sugars, and low fiber intake can lead to an imbalance in the GM [38], referred to as dysbiosis [39]. This dysbiosis is now recognized as a critical factor in developing cardiometabolic diseases and cardiovascular toxicity [40–44].

Role of Nutrients in GM Modification and Dysbiosis

Macronutrients

Saturated Fat

Saturated animal fat promotes the growth of inflammatory microbiota [45] such as Bilophila [46, 47] and inhibits beneficial microbiota such as Bifidobacterium, Bacteroidetes, Prevotella, and Lactobacillus [48], promoting insulin resistance and cardiometabolic disease [49]. In contrast, unsaturated plant fat containing omega 3 and 6 fatty acids supports the growth of beneficial bacteria such as Bifidobacterium and butyrate-producing bacteria, Roseburia and Faecilibacterium [50, 51]. Moreover, excessive dietary saturated fat intake increases the production of bile acids, which are transformed into secondary and unconjugated bile acids, such as deoxycholic acid (DCA), by the gut microbiota [52]. This further promotes the growth of the bile-resistant microbiome, which is found to be linked to the pathogenesis of obesity, metabolic syndrome [52, 53], and cardiovascular toxicity [54].

Animal Protein

Mouse model research found that the consumption of animal proteins increased the gut bacteria associated with gut diseases, including Streptococcus, Enterococcus, and Peptosterptoccocus [55], and inhibited the growth of the beneficial bacteria such as Bifidobacterium and hence inhibited the formation of short-chain fatty acids (SCFA), which are protective metabolites against cardiovascular diseases. It is noteworthy that plant proteins that are abundant in the Mediterranean diet are associated with increased bacteria belonging to the Bifidobacteriaceae, Desulfovibrionaceae, Lactobacillaceae, Desulfovibrionaceae, and Erysipleotrichaceae families with enhanced SCFA production [55].

Red meat intake is associated with the formation of trimethylamine N-oxide (TMAO) through the effect of gut microbiota [56]. Dietary animal protein is also found to be related to the formation of phenyl acetyl glutamine (PAG), a metabolite formed by gut microbiota [57, 58]. TMAO is highly linked to the incidence of cardiovascular toxicity [29, 59] and the formation of atheromatous plaques [42], and PGA is associated with thromboembolism [27], as mentioned below.

Carbohydrates

The carbohydrates found in the Westernized diet primarily consist of simple sugars (mono and disaccharides) and easily digestible starch [60]. These carbohydrates are quickly broken down and linked to a decrease in bacteria that produce short-chain fatty acids (SCFA) [61]. In contrast, the complex sugars and resistant starch found in plant fibers, which are abundant in vegetarian diets and less prevalent in Western diets, are not easily digestible. These carbohydrates, known as microbiota-accessible carbohydrates (MAC), can reach the distal colonic microbiota and undergo fermentation, producing cardiovascular-protective SCFAs [61, 62]. Dietary fibers promote health by increasing beneficial microbiota, such as Lactobacillus and Bifidobacterium [46].

Micronutrients

Although there is less clinical evidence than macronutrients, the research studies found that vitamins and trace elements affect the microbiota. Evidence shows that vitamin A, E, and D supplementation increases the beneficial microbiome, including

Bifidobacterium and Lactocabaccilus for vitamin A and E supplementation [46] and Bacteroids and Parbacteroids for vitamin D supplementation [63]. Interestingly, ingesting high iron formulas is associated with decreased beneficial Bifidobacterium and Lactobacillus [64].

Polyphenols

Research studies have shown that dietary polyphenols, such as flavonoids and phenolic acids, which are abundant in fruits (e.g., apples, berries) and tea, can promote the growth of beneficial microbiota in the gut [65–67], such as Bifidobacterium and lactobacillus. Simultaneously, these polyphenols have been found to inhibit the proliferation of harmful microbiota, including Clostridium perfringens, thus contributing to a healthier gut microbiome [67].

Gut Microbiota and Cardiovascular Toxicity

Over the last decade, GM has been primarily implicated in cardiovascular diseases, including atherosclerotic cardiovascular disease [68], hypertension [24, 69], myocardial fibrosis [70], and heart failure [25, 40, 44, 71]. Cardiovascular toxicity involves cardiovascular damage due to oxidative stress [72–74], inflammation [75], and toxin-induced irregularities in electrophysiology (arrhythmia) and cardiac muscle integrity (contractility) [76]. The link between dysbiosis and the development of these diseases is discussed below.

Pathogenesis of Cardiovascular Toxicity

The gut microbiota significantly influences the pathogenesis of cardiovascular damage [77–79], operating through both direct [80] and indirect mechanisms associated with its metabolites [81] (Fig. 30.1). Inflammation is a central feature of cardiovascular pathology [82], with systemic inflammation directly correlated to the development of myocardial injury and fibrosis, arrhythmias, and atherosclerosis [83–86], a condition characterized by plaque accumulation within the arterial lumen [87].

Direct Mechanisms

GM, particularly the dysbiotic microbiome, directly contributes to the generation of inflammatory responses primarily through the release of lipopolysaccharides (LPS) [88], complex molecules located in the outer membrane of Gram-negative bacteria

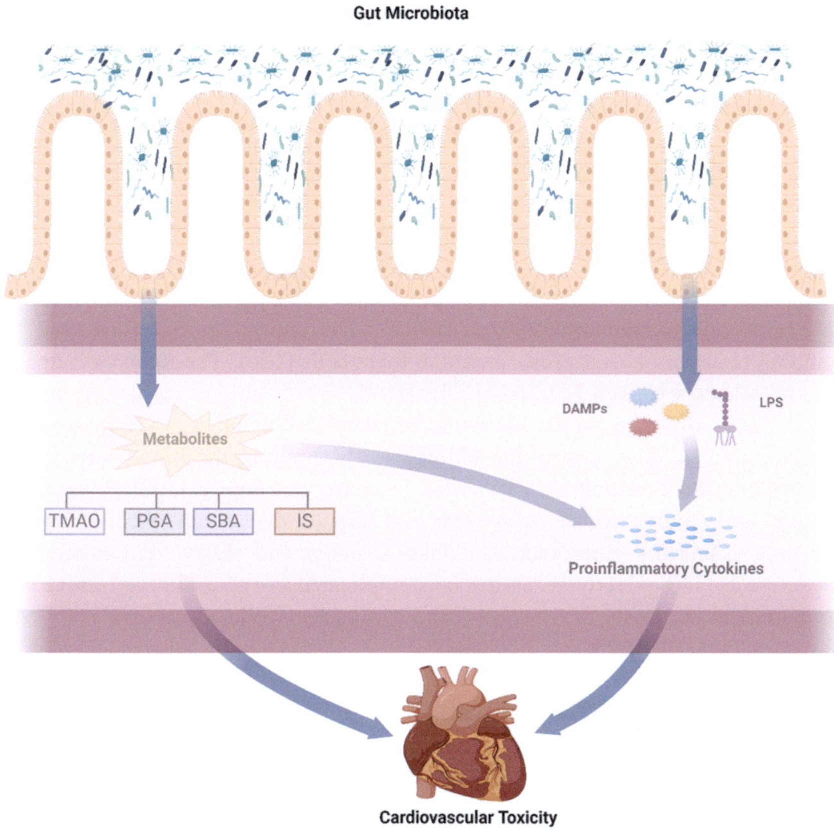

Fig. 30.1 Gut microbiota and cardiovascular toxicity pathogenesis. *GM* gut microbiota, *TMAO* trimethylamine N-oxide, *PAG* phenyl acetyl glutamine, *SBA* secondary bile acids, *IS* indoxyl sulfate

[89]. When the intestinal barrier becomes compromised, a condition commonly called "leaky gut," LPS enters the systemic circulation [90, 91]. LPS engage with Toll-like receptors (TLRs), specifically TLR4, located on the surface of various immune cells such as macrophages and dendritic cells [92, 93]. This interaction triggers a cascade of intracellular signaling pathways that activate NF-kB [94], upregulating the expression of proinflammatory cytokines, including IL1, IL6, IL27, and TNF-α [93, 95, 96]. LPS further compromises the gut epithelial barrier, releasing more LPS into the circulation [97]. Notably, ingesting saturated dietary fat and low dietary fibers (low SCFA levels), which characterize the Western diet, increase the release of LPS from the gut into the systemic circulation [98]. Furthermore, the gut microbiota bacterial and fungal toxins, viral RNA, extracellular matrix products resulting from damaged cells (Damage-Associated Molecular Patterns, or DAMPs), reactive oxygen species (ROS), cytokine receptors, and TLRs activate the NLRP3 (NOD-like receptor protein 3) inflammasome [99, 100], which is a critical component of the innate immune response. This activation further enhances the

inflammatory responses by upregulating the expression of additional proinflammatory cytokines, especially IL1β and IL 18 [101], which are known to affect the myocardium and produce myocardial injury and fibrosis [101]. High levels of LPS are found in the serum of heart failure patients presenting with acute exacerbation and recent onset of edema [102].

The proinflammatory cytokines also affect the endothelial cells lining the heart and the blood vessel wall, resulting in vascular endothelial dysfunction, a critical process in the pathogenesis of atherosclerosis [103]. The circulating low-intensity lipoprotein (LDL) cholesterol infiltrates the dysfunctional endothelium [104] and is internalized by macrophages, forming foam cells [104, 105]. LPS also increases cholesterol ester accumulation from native LDL and foam cell formation by inhibiting cholesterol ester efflux from the foam cells [106]. The foam cells aggregate and release oxidized LDL (ox-LDL) [107], significantly exacerbating inflammation and contributing to the formation of atheromatous plaques [108]. As these plaques progress, they narrow the blood vessels, limiting the blood flow to the myocardium and resulting in myocardial ischemia and injury [108]. Myocardial ischemia and cardiomyocyte injury can also happen due to embolization, where fragments of unstable atheromatous plaques dislodge and obstruct the small distal blood vessels [108]. Gut dysbiosis has been implicated as a contributing factor to the formation and instability of the atheromatous plaques [109].

Indirect Mechanisms Associated with The GM Metabolites

Among the various metabolites gut bacteria produce (Table 30.1), TMAO has garnered significant attention as a critical molecule influencing cardiovascular health [110–112]. TMAO is formed through the metabolic processes of GM when they

Table 30.1 GM metabolites and cardiovascular toxicity

Metabolite	Effects	Mechanism
TMAO	Impaired contractility and heart failure	Disruption of Ca^{2+} homeostasis and altered energy metabolism of the cardiomyocyte
		Cardiomyocyte T tubule injury
		Inflammation and myocardial fibrosis
	Atherosclerosis	Inflammation (NLRP3 inflammasome activation)
	Thrombosis, heart attacks and stroke	Platelet adhesion
PAG	Thrombosis, heart attacks and stroke	Platelet hyperreactivity through activation of the platelet adrenergic receptors
SBA	Increased cardiovascular disease risk in T2DM patients	Inflammation through activating FXR
IS	Heart failure & AFib in uremic patients	Myocardial fibrosis and pathological remodeling

TMAO trimethylamine N-oxide, *PAG* phenyl acetyl glutamine, *SBA* secondary bile acids, *FXR* farnesoid X receptor, *IS* indoxyl sulfate break down specific dietary components, particularly choline and carnitine. These compounds are prevalent in foods such as red meat, eggs, and certain dairy products [113, 114].

The process begins with specific strains of gut bacteria, often associated with dysbiosis, which convert choline and carnitine into trimethylamine (TMA) [113, 115]. This compound is then absorbed into the portal circulation and transported to the liver, where it undergoes oxidation. The enzyme flavin-containing monooxygenase 3 (FMO3) is crucial in transforming TMA into TMAO [113, 116].

TMAO has been associated with a range of adverse cardiovascular outcomes [117], mainly through its direct effects on cardiomyocytes. In their study, Savi et al. explored the influence of TMAO on murine cardiomyocytes, revealing that TMAO disrupts Ca2+ homeostasis and alters energy metabolism within the cardiac muscle [29]. Moreover, Jen et al. demonstrated the presence of cardiomyocyte T tubule injury in addition to the effect on Ca^{2+} hemostasis [59]. These effects lead to compromised contractility and oxidative injury to the cardiomyocyte [29, 118]. Studies also indicate that elevated blood levels of TMAO are associated with increased systemic inflammation, which leads to atherosclerosis [119], including atheromatous formations in the coronary arteries [120, 121]; high TMAO levels are also found to exacerbate the atheromatous plaque instability. Moreover, TMAO enhances platelet adhesion, promoting clot formation and raising the risk of thrombosis, resulting in further cardiovascular damage [119]. TMAO has also been correlated with the occurrence of hypertension [69]. Taken together, high TMAO levels increase the likelihood of major adverse cardiovascular events (MACE), such as heart attacks and strokes [122], as well as higher mortality rates in patients diagnosed with coronary artery disease [123]. TMAO-induced inflammation through the NLRP3 inflammasome pathway is linked to myocardial fibrosis [124]; studies have shown that heart failure patients tend to have significantly higher levels of TMAO [125], which has also been linked to the increased likelihood of requiring cardiac transplantation [126]. Notably, TMAO has emerged as a significant biomarker for cardiovascular diseases and can be used for cardiovascular risk stratification [112, 127–129].

Recently, another metabolite produced by microbiota from dietary protein, specifically phenylalanine amino acid called phenyl acetyl glutamine (PAG) [27, 130], was found to be associated with the risk of arterial thromboembolism leading to heart attack and stroke [27]; the gut microbiome was found to produce the precursor of phenyl acetyl glutamine, phenylacetic acid that is conjugated in the host liver with the amino acid glutamine to form the PAG [130] that acts through the blood platelets adrenergic receptors to induce platelet hyperreactivity that may result in thrombosis, that may contribute to heart attack, cerebrovascular accident, and death [27]. Higher blood levels of PAG were found in patients with a higher risk of MACE [27].

As mentioned above, increased levels of the secondary and unconjugated bile acid produced by GM, primarily DCA [131], are associated with high saturated fat intake [52]. Lu Q et al. found that high levels of secondary and unconjugated bile acids are associated with increased cardiac risk in newly diagnosed T2DM patients [28]. Bile acids may affect cardiovascular health through various receptors, including the farnesoid X receptor (FXR), the liver X receptor (LXR), and the G protein-coupled bile acid receptor (TGR5), modulating inflammation, lipid metabolism, and insulin resistance [132].

Indoxyl sulfate (IS), a dietary protein metabolite that arises from the GM metabolism of tryptophan, has been linked to the pathogenesis of cardiovascular diseases in uremic patients [133]. Elevated levels of this metabolite, often indicative of altered GM composition, contribute to the development of myocardial fibrosis [134]. This pathological remodeling has been linked to increased susceptibility to atrial fibrillation (Afib), a common arrhythmia, as well as heart failure [134–136]. Furthermore, high blood levels of indoxyl sulfate have been associated with higher mortality rates [137–139]. IS has been associated with endothelial dysfunction [140], NF-kB activation [141], and coronary atherosclerosis [142]; IS was found to be an independent risk factor for coronary artery disease incidence and severity [143].

Conclusion

The complex interactions between diet, GM, and inflammatory pathways—particularly those mediated by LPS, TLR activation, and the NLRP3 inflammasome—are the leading players in elucidating cardiovascular toxicity. Dietary choices can dramatically modify the GM, thereby influencing the synthesis of critical metabolites, such as TMAO, SCFAs, PAG, DCA, and IS; these metabolites are products of microbial metabolism and play well-documented roles in inflammatory responses and cardiovascular diseases.

The activation of TLRs occurs when LPS is translocated from the intestinal tract into the bloodstream, leading to the initiation of an inflammatory cascade. Concurrently, the NLRP3 inflammasome is critical for activating the immune system in response to cellular injury and stress signals. The interplay among these components forms a complex network that significantly contributes to cardiovascular toxicity and heightens the risk of cardiovascular diseases.

A comprehensive understanding of these mechanistic pathways is critical, as it can potentially inform innovative therapeutic strategies that alter GM composition and its metabolites or specifically target inflammatory pathways. Such interventions could reduce cardiovascular toxicity and lower the incidence of cardiovascular diseases, thereby advancing preventive and therapeutic efforts.

References

1. Putra IGNE et al (2024) Obesity, psychological well-being related measures, and risk of seven non-communicable diseases: evidence from longitudinal studies of UK and US older adults. Int J Obes 48(9):1283–1291

2. WHO (2023) Noncommunicable diseases. [cited 2024 1 November 2024]. https://www.who.int/news-room/fact-sheets/detail/noncommunicable-diseases
3. Li J et al (2020) Dietary inflammatory potential and risk of cardiovascular disease among men and women in the U.S. J Am Coll Cardiol 76(19):2181–2193
4. Ji W et al (2024) Association between neutrophil-percentage-to-albumin ratio (NPAR) and metabolic syndrome risk: insights from a large US population-based study. Sci Rep 14(1): 26646
5. Liu R et al (2015) Expression profile of innate immune receptors, NLRs and AIM2, in human colorectal cancer: correlation with cancer stages and inflammasome components. Oncotarget 6(32):33456–33469
6. Zitvogel L, Pietrocola F, Kroemer G (2017) Nutrition, inflammation and cancer. Nat Immunol 18(8):843–850
7. Kopp W (2019) How western diet and lifestyle drive the pandemic of obesity and civilization diseases. Diabetes Metab Syndr Obes 12:2221–2236
8. Furman D et al (2019) Chronic inflammation in the etiology of disease across the life span. Nat Med 25(12):1822–1832
9. Hitch TCA et al (2022) Microbiome-based interventions to modulate gut ecology and the immune system. Mucosal Immunol 15(6):1095–1113
10. Zhao M et al (2023) Immunological mechanisms of inflammatory diseases caused by gut microbiota dysbiosis: a review. Biomed Pharmacother 164:114985
11. Hugon P et al (2015) A comprehensive repertoire of prokaryotic species identified in human beings. Lancet Infect Dis 15(10):1211–1219
12. Almeida A et al (2019) A new genomic blueprint of the human gut microbiota. Nature 568(7753):499–504
13. Hou K et al (2022) Microbiota in health and diseases. Signal Transduct Target Ther 7(1):135
14. Sánchez-Tapia M, Tovar AR, Torres N (2019) Diet as regulator of gut microbiota and its role in health and disease. Arch Med Res 50(5):259–268
15. Arumugam M et al (2011) Enterotypes of the human gut microbiome. Nature 473(7346): 174–180
16. Deehan EC et al (2020) Precision microbiome modulation with discrete dietary fiber structures directs short-chain fatty acid production. Cell Host Microbe 27(3):389–404.e6
17. Zhang P (2022) Influence of foods and nutrition on the gut microbiome and implications for intestinal health. Int J Mol Sci 23(17)
18. Nesci A et al (2023) Gut microbiota and cardiovascular disease: evidence on the metabolic and inflammatory background of a complex relationship. Int J Mol Sci 24(10)
19. Rogers CJ, Prabhu KS, Vijay-Kumar M (2014) The microbiome and obesity—an established risk for certain types of cancer. Cancer J 20(3):176–180
20. Zheng D, Liwinski T, Elinav E (2020) Interaction between microbiota and immunity in health and disease. Cell Res 30(6):492–506
21. Toumi R et al (2021) Role of probiotics and their metabolites in inflammatory bowel diseases (IBDs). Gastroenterol Insights 12(1):56–66
22. Philips CA et al (2022) The role of gut microbiota in clinical complications, disease severity, and treatment response in severe alcoholic hepatitis. Indian J Gastroenterol 41(1):37–51
23. Zhang H et al (2022) The complex link and disease between the gut microbiome and the immune system in infants. Front Cell Infect Microbiol 12:924119
24. Hsu CN et al (2020) Perinatal resveratrol therapy prevents hypertension programmed by maternal chronic kidney disease in adult male offspring: implications of the gut microbiome and their metabolites. Biomedicines 8(12)
25. Zong X et al (2022) Trimethyllysine, a trimethylamine N-oxide precursor, predicts the presence, severity, and prognosis of heart failure. Front Cardiovasc Med 9:907997
26. Schuett K et al (2017) Trimethylamine-N-oxide and heart failure with reduced versus preserved ejection fraction. J Am Coll Cardiol 70(25):3202–3204

27. Nemet I et al (2020) A cardiovascular disease-linked gut microbial metabolite acts via adrenergic receptors. Cell 180(5):862–877.e22
28. Lu Q et al (2024) Gut microbiota-derived secondary bile acids, bile acids receptor polymorphisms, and risk of cardiovascular disease in individuals with newly diagnosed type 2 diabetes: a cohort study. Am J Clin Nutr 119(2):324–332
29. Savi M et al (2018) Trimethylamine-N-oxide (TMAO)-induced impairment of cardiomyocyte function and the protective role of urolithin B-glucuronide. Molecules 23(3)
30. Chassaing B et al (2015) Dietary emulsifiers impact the mouse gut microbiota promoting colitis and metabolic syndrome. Nature 519(7541):92–96
31. Kübeck R et al (2016) Dietary fat and gut microbiota interactions determine diet-induced obesity in mice. Mol Metab 5(12):1162–1174
32. Neeland IJ et al (2024) Metabolic syndrome. Nat Rev Dis Primers 10(1):77
33. Wang W et al (2023) Dietary patterns and cardiometabolic health: clinical evidence and mechanism. MedComm 4(1):e212
34. Mechanick JI et al (2020) Cardiometabolic-based chronic disease, adiposity and dysglycemia drivers: JACC state-of-the-art review. J Am Coll Cardiol 75(5):525–538
35. Abarca-Gómez L et al (2017) Worldwide trends in body-mass index, underweight, overweight, and obesity from 1975 to 2016: a pooled analysis of 2416 population-based measurement studies in 128· 9 million children, adolescents, and adults. Lancet 390(10113): 2627–2642
36. Federation ID (2022) IDF Diabetes Atlas. [cited 2024 Feb 8 2024]. 10th. https://diabetesatlas.org/
37. Roth GA et al (2020) Global burden of cardiovascular diseases and risk factors, 1990-2019: update from the GBD 2019 study. J Am Coll Cardiol 76(25):2982–3021
38. Armet AM et al (2022) Rethinking healthy eating in light of the gut microbiome. Cell Host Microbe 30(6):764–785
39. Cristofori F et al (2021) Anti-inflammatory and immunomodulatory effects of probiotics in gut inflammation: a door to the body. Front Immunol 12:578386
40. Jia Q et al (2019) Role and effective therapeutic target of gut microbiota in heart failure. Cardiovasc Ther 2019(1):5164298
41. Sun S et al (2019) Gut microbiota composition and blood pressure. Hypertension 73(5): 998–1006
42. Zhu B et al (2022) Trimethylamine N-oxide generated by the gut microbiota: potential atherosclerosis treatment strategies. Curr Pharm Des 28(35):2914–2919
43. Seldin MM et al (2016) Trimethylamine N-oxide promotes vascular inflammation through signaling of mitogen-activated protein kinase and nuclear factor-κB. J Am Heart Assoc 5(2)
44. Sata Y, Marques FZ, Kaye DM (2020) The emerging role of gut dysbiosis in cardio-metabolic risk factors for heart failure. Curr Hypertens Rep 22(5):38
45. Hamilton MK et al (2015) Changes in intestinal barrier function and gut microbiota in high-fat diet-fed rats are dynamic and region dependent. Am J Physiol Gastrointest Liver Physiol 308(10):G840–G851
46. Yang Q et al (2020) Role of dietary nutrients in the modulation of gut microbiota: a narrative review. Nutrients 12(2):381
47. Moszak M, Szulińska M, Bogdański P (2020) You are what you eat—the relationship between diet, microbiota, and metabolic disorders—a review. Nutrients 12(4):1096
48. Mujico JR et al (2013) Changes in gut microbiota due to supplemented fatty acids in diet-induced obese mice. Br J Nutr 110(4):711–720
49. Chen G et al (2018) Fuzhuan brick tea polysaccharides attenuate metabolic syndrome in high-fat diet induced mice in association with modulation in the gut microbiota. J Agric Food Chem 66(11):2783–2795
50. Muralidharan J et al (2019) Plant-based fat, dietary patterns rich in vegetable fat and gut microbiota modulation. Front Nutr 6:157

51. Abulizi N et al (2019) Gut mucosal proteins and bacteriome are shaped by the saturation index of dietary lipids. Nutrients 11(2):418
52. Wan Y et al (2020) Unconjugated and secondary bile acid profiles in response to higher-fat, lower-carbohydrate diet and associated with related gut microbiota: a 6-month randomized controlled-feeding trial. Clin Nutr 39(2):395–404
53. Wei M et al (2020) A dysregulated bile acid-gut microbiota axis contributes to obesity susceptibility. EBioMedicine 55
54. Zhang S et al (2023) The role of bile acids in cardiovascular diseases: from mechanisms to clinical implications. Aging Dis 14(2):261–282
55. Kostovcikova K et al (2019) Diet rich in animal protein promotes pro-inflammatory macrophage response and exacerbates colitis in mice. Front Immunol 10:446008
56. Papandreou C, Moré M, Bellamine A (2020) Trimethylamine N-oxide in relation to cardiometabolic health—cause or effect? Nutrients 12(5):1330
57. Dodd D et al (2017) A gut bacterial pathway metabolizes aromatic amino acids into nine circulating metabolites. Nature 551(7682):648–652
58. Saha PP et al (2024) Gut microbe-generated phenylacetylglutamine is an endogenous allosteric modulator of β2-adrenergic receptors. Nat Commun 15(1):6696
59. Jin B et al (2020) Destructive role of TMAO in T-tubule and excitation-contraction coupling in the adult cardiomyocytes. Int Heart J 61(2):355–363
60. Turnbaugh PJ et al (2008) Diet-induced obesity is linked to marked but reversible alterations in the mouse distal gut microbiome. Cell Host Microbe 3(4):213–223
61. Seo YS et al (2020) Dietary carbohydrate constituents related to gut dysbiosis and health. Microorganisms 8(3):427
62. Rowland I et al (2018) Gut microbiota functions: metabolism of nutrients and other food components. Eur J Nutr 57(1):1–24
63. Charoenngam N et al (2020) The effect of various doses of oral vitamin D3 supplementation on gut microbiota in healthy adults: a randomized, double-blinded, dose-response study. Anticancer Res 40(1):551–556
64. Sjödin KS et al (2019) Administration of ferrous sulfate drops has significant effects on the gut microbiota of iron-sufficient infants: a randomised controlled study. Gut 68(11):2095–2097
65. Moorthy M et al (2020) Prebiotic potential of polyphenols, its effect on gut microbiota and anthropometric/clinical markers: a systematic review of randomised controlled trials. Trends Food Sci Technol 99:634–649
66. Tomás-Barberán FA, Espín JC (2019) Effect of food structure and processing on (poly) phenol–gut microbiota interactions and the effects on human health. Annu Rev Food Sci Technol 10(1):221–238
67. Ma G, Chen Y (2020) Polyphenol supplementation benefits human health via gut microbiota: a systematic review via meta-analysis. J Funct Foods 66:103829
68. Qian B et al (2022) Update on gut microbiota in cardiovascular diseases. Front Cell Infect Microbiol 12:1059349
69. Ge X et al (2020) The gut microbial metabolite trimethylamine N-oxide and hypertension risk: a systematic review and dose-response meta-analysis. Adv Nutr 11(1):66–76
70. Nenna A et al (2024) Intestinal microbiota and derived metabolites in myocardial fibrosis and postoperative atrial fibrillation. Int J Mol Sci 25(11):6037
71. Shi M et al (2023) The role of the gut microbiota and bile acids in heart failure: a review. Medicine 102(45):e35795
72. Lefer DJ, Granger DN (2000) Oxidative stress and cardiac disease. Am J Med 109(4):315–323
73. Ceriello A, Motz E (2004) Is oxidative stress the pathogenic mechanism underlying insulin resistance, diabetes, and cardiovascular disease? The common soil hypothesis revisited. Arterioscler Thromb Vasc Biol 24(5):816–823
74. Kovacic P, Thurn LA (2005) Cardiovascular toxicity from the perspective of oxidative stress, electron transfer, and prevention by antioxidants. Curr Vasc Pharmacol 3(2):107–117

75. de Almeida AJPO et al (2020) Unveiling the role of inflammation and oxidative stress on age-related cardiovascular diseases. Oxidative Med Cell Longev 2020(1):1954398
76. Singh AK (2016) Chapter 7—Mechanisms of nanoparticle toxicity. In: Singh AK (ed) Engineered nanoparticles. Academic Press, Boston, pp 295–341
77. Koren O et al (2011) Human oral, gut, and plaque microbiota in patients with atherosclerosis. Proc Natl Acad Sci U S A 108 Suppl 1(Suppl 1):4592–4598
78. Karlsson FH et al (2012) Symptomatic atherosclerosis is associated with an altered gut metagenome. Nat Commun 3(1):1245
79. Yamashiro K et al (2017) Gut dysbiosis is associated with metabolism and systemic inflammation in patients with ischemic stroke. PLoS One 12(2):e0171521
80. Lepper PM et al (2011) Lipopolysaccharide-binding protein (LBP) is associated with total and cardiovascular mortality in individuals with or without stable coronary artery disease—results from the Ludwigshafen Risk and Cardiovascular Health Study (LURIC). Atherosclerosis 219(1):291–297
81. Tang WHW, Kitai T, Hazen SL (2017) Gut microbiota in cardiovascular health and disease. Circ Res 120(7):1183–1196
82. Lam YY et al (2012) Increased gut permeability and microbiota change associate with mesenteric fat inflammation and metabolic dysfunction in diet-induced obese mice. PLoS One 7(3):e34233
83. Paulus WJ, Zile MR (2021) From systemic inflammation to myocardial fibrosis. Circ Res 128(10):1451–1467
84. Taghdiri A (2024) Inflammation and arrhythmogenesis: a narrative review of the complex relationship. Int J Arrhythm 25(1):4
85. Moriya J (2019) Critical roles of inflammation in atherosclerosis. J Cardiol 73(1):22–27
86. Ridker PM et al (2017) Antiinflammatory therapy with canakinumab for atherosclerotic disease. N Engl J Med 377(12):1119–1131
87. Sakakura K et al (2013) Pathophysiology of atherosclerosis plaque progression. Heart Lung Circ 22(6):399–411
88. Mohr AE et al (2022) Lipopolysaccharide and the gut microbiota: considering structural variation. FEBS Lett 596(7):849–875
89. Caroff M, Karibian D (2003) Structure of bacterial lipopolysaccharides. Carbohydr Res 338(23):2431–2447
90. Cheng CK, Huang Y (2021) The gut-cardiovascular connection: new era for cardiovascular therapy. Med Rev 1(1):23–46
91. Polsinelli VB, Sinha A, Shah SJ (2017) Visceral congestion in heart failure: right ventricular dysfunction, splanchnic hemodynamics, and the intestinal microenvironment. Curr Heart Fail Rep 14:519–528
92. Creely SJ et al (2007) Lipopolysaccharide activates an innate immune system response in human adipose tissue in obesity and type 2 diabetes. Am J Physiol Endocrinol Metab 292(3): E740–E747
93. Kaisho T, Akira S (2002) Toll-like receptors as adjuvant receptors. Biochim Biophys Acta 1589(1):1–13
94. Arkan MC et al (2005) IKK-beta links inflammation to obesity-induced insulin resistance. Nat Med 11(2):191–198
95. Pålsson-McDermott EM, O'Neill LA (2004) Signal transduction by the lipopolysaccharide receptor, toll-like receptor-4. Immunology 113(2):153–162
96. Guzzo C et al (2012) IL-27 enhances LPS-induced proinflammatory cytokine production via upregulation of TLR4 expression and signaling in human monocytes. J Immunol 188(2): 864–873
97. Stephens M, von der Weid PY (2020) Lipopolysaccharides modulate intestinal epithelial permeability and inflammation in a species-specific manner. Gut Microbes 11(3):421–432
98. Tan JK, Macia L, Mackay CR (2023) Dietary fiber and SCFAs in the regulation of mucosal immunity. J Allergy Clin Immunol 151(2):361–370

99. Philpott DJ et al (2014) NOD proteins: regulators of inflammation in health and disease. Nat Rev Immunol 14(1):9–23
100. Jin Y, Fu J (2019) Novel insights into the NLRP3 inflammasome in atherosclerosis. J Am Heart Assoc 8(12):e012219
101. Abbate A et al (2020) Interleukin-1 and the inflammasome as therapeutic targets in cardiovascular disease. Circ Res 126(9):1260–1280
102. Niebauer J et al (1999) Endotoxin and immune activation in chronic heart failure: a prospective cohort study. Lancet 353(9167):1838–1842
103. Gorabi AM et al (2022) Implications for the role of lipopolysaccharide in the development of atherosclerosis. Trends Cardiovasc Med 32(8):525–533
104. Hansson GK, Robertson A-KL, Söderberg-Nauclér C (2006) Inflammation and atherosclerosis. Annu Rev Pathol Mech Dis 1(1):297–329
105. Canton J, Neculai D, Grinstein S (2013) Scavenger receptors in homeostasis and immunity. Nat Rev Immunol 13(9):621–634
106. Lakio L et al (2006) Pro-atherogenic properties of lipopolysaccharide from the periodontal pathogen Actinobacillus actinomycetemcomitans. J Endotoxin Res 12(1):57–64
107. Stocker R, Keaney JF Jr (2004) Role of oxidative modifications in atherosclerosis. Physiol Rev 84(4):1381–1478
108. Tabas I (2017) 2016 Russell Ross Memorial Lecture in vascular biology: molecular-cellular mechanisms in the progression of atherosclerosis. Arterioscler Thromb Vasc Biol 37(2): 183–189
109. Violi F et al (2023) Gut-derived low-grade endotoxaemia, atherothrombosis and cardiovascular disease. Nat Rev Cardiol 20(1):24–37
110. Gruppen EG et al (2017) TMAO is associated with mortality: impact of modestly impaired renal function. Sci Rep 7(1):13781
111. Roberts AB et al (2018) Development of a gut microbe–targeted nonlethal therapeutic to inhibit thrombosis potential. Nat Med 24(9):1407–1417
112. Zhen J et al (2023) The gut microbial metabolite trimethylamine N-oxide and cardiovascular diseases. Front Endocrinol 14
113. Canyelles M et al (2018) Trimethylamine N-oxide: a link among diet, gut microbiota, gene regulation of liver and intestine cholesterol homeostasis and HDL function. Int J Mol Sci 19(10):3228
114. Ramos S, Martín MÁ (2021) Impact of diet on gut microbiota. Curr Opin Food Sci 37:83–90
115. He S et al (2021) Trimethylamine/trimethylamine-N-oxide as a key between diet and cardiovascular diseases. Cardiovasc Toxicol 21(8):593–604
116. Ripp SL et al (1999) Species and sex differences in expression of flavin-containing monooxygenase form 3 in liver and kidney microsomes. Drug Metab Dispos 27(1):46–52
117. Park J et al (2019) Differential effect of short-term popular diets on TMAO and other cardio-metabolic risk markers. Nutr Metab Cardiovasc Dis 29(5):513–517
118. Lu X et al (2022) Microbial metabolites and heart failure: friends or enemies? Front Microbiol 13:956516
119. Zhang X, Gérard P (2022) Diet-gut microbiota interactions on cardiovascular disease. Comput Struct Biotechnol J 20:1528–1540
120. Almesned MA et al (2021) Temporal course of plasma trimethylamine N-oxide (TMAO) levels in ST-elevation myocardial infarction. J Clin Med 10(23)
121. Guo F et al (2020) The association between trimethylamine N-oxide and its predecessors choline, L-carnitine, and betaine with coronary artery disease and artery stenosis. Cardiol Res Pract 2020:5854919
122. Canyelles M et al (2023) Gut microbiota-derived TMAO: a causal factor promoting atherosclerotic cardiovascular disease? Int J Mol Sci 24(3)
123. Senthong V et al (2016) Intestinal microbiota-generated metabolite trimethylamine-N-oxide and 5-year mortality risk in stable coronary artery disease: the contributory role of intestinal microbiota in a COURAGE-like patient cohort. J Am Heart Assoc 5(6)

124. Li X et al (2019) Trimethylamine N-oxide exacerbates cardiac fibrosis via activating the NLRP3 inflammasome. Front Physiol 10:866
125. Dong Z et al (2021) Trimethylamine N-oxide is associated with heart failure risk in patients with preserved ejection fraction. Lab Med 52(4):346–351
126. Zhang Y et al (2021) TMAO: how gut microbiota contributes to heart failure. Transl Res 228: 109–125
127. Gabbianelli R, Bordoni L (2022) Trimethylamine N-oxide (TMAO) as a biomarker. In: Patel VB, Preedy VR (eds) Biomarkers in nutrition. Springer, Cham, pp 1–20
128. Guasti L et al (2021) TMAO as a biomarker of cardiovascular events: a systematic review and meta-analysis. Intern Emerg Med 16(1):201–207
129. Li XS et al (2018) Untargeted metabolomics identifies trimethyllysine, a TMAO-producing nutrient precursor, as a predictor of incident cardiovascular disease risk. JCI Insight 3(6)
130. Webster LT et al (1976) Identification of separate acyl-CoA:glycine and acyl-CoA:L-glutamine N-acyltransferase activities in mitochondrial fractions from liver of rhesus monkey and man. J Biol Chem 251(11):3352–3358
131. Haeusler RA et al (2013) Human insulin resistance is associated with increased plasma levels of 12α-hydroxylated bile acids. Diabetes 62(12):4184–4191
132. Zhang Z et al (2024) Role of the microbiota–gut–heart axis between bile acids and cardiovascular disease. Biomed Pharmacother 174:116567
133. Hung SC et al (2017) Indoxyl sulfate: a novel cardiovascular risk factor in chronic kidney disease. J Am Heart Assoc 6(2):e005022
134. Lekawanvijit S et al (2010) Does indoxyl sulfate, a uraemic toxin, have direct effects on cardiac fibroblasts and myocytes? Eur Heart J 31(14):1771–1779
135. Cao XS et al (2015) Association of indoxyl sulfate with heart failure among patients on hemodialysis. Clin J Am Soc Nephrol 10(1):111–119
136. Chen WT et al (2015) The uremic toxin indoxyl sulfate increases pulmonary vein and atrial arrhythmogenesis. J Cardiovasc Electrophysiol 26(2):203–210
137. Imazu M et al (2020) Plasma indoxyl sulfate levels predict cardiovascular events in patients with mild chronic heart failure. Sci Rep 10(1):16528
138. Lin XC et al (2022) The accumulation of gut microbiome-derived indoxyl sulfate and P-cresyl sulfate in patients with end-stage renal disease. J Ren Nutr 32(5):578–586
139. Aoki K et al (2015) Role of indoxyl sulfate as a predisposing factor for atrial fibrillation in renal dysfunction. J Am Heart Assoc 4(10):e002023
140. Faure V et al (2006) Elevation of circulating endothelial microparticles in patients with chronic renal failure. J Thromb Haemost 4(3):566–573
141. Ito S et al (2010) Indoxyl sulfate induces leukocyte-endothelial interactions through up-regulation of E-selectin. J Biol Chem 285(50):38869–38875
142. Gao H, Liu S (2017) Role of uremic toxin indoxyl sulfate in the progression of cardiovascular disease. Life Sci 185:23–29
143. Hsu CC et al (2013) Levels of indoxyl sulfate are associated with severity of coronary atherosclerosis. Clin Invest Med 36(1):E42–E49

Part IV
Prevention, Management and Treatment of Cardiovascular Toxicity

Chapter 31
Heart-Derived Stromal Cells and In Vitro 3D Models in Cardiotoxicity Assessment

Dongwei Sun and Finosh G. Thankam

Abstract Cardiovascular diseases, including heart failure and myocardial infarction, are leading causes of morbidity and mortality worldwide. Cardiotoxicity, a major concern in drug development, further emphasizes the need for accurate in vitro models for testing the potential drug candidates. The cardiac-specific cells are integral to the successful design of in vitro cardiotoxicity assessment models. Traditional two-dimensional (2D) systems and animal models fail to replicate the complex human cardiac environment, driving the development of three-dimensional (3D) models. These models, incorporating cardiomyocytes, fibroblasts, endothelial cells, and biomaterials including hydrogels, more accurately mimic human heart tissue. In this chapter, we explore the role of heart-derived stromal cells (HSCs) in cardiotoxicity assessment using in vitro 3D models, highlighting their advantages over traditional methods. The integration of in vitro 3D models into drug discovery/testing using HSCs promises the predictive accuracy of cardiotoxicity testing and improve therapeutic outcomes.

Keywords Heart-derived stromal cells · In vitro 3D models · Cardiotoxicity · Drug testing

Introduction

Cardiovascular diseases and heart failure remain leading causes of morbidity and mortality worldwide. Stromal cell therapy has emerged as a promising avenue for treating cardiovascular diseases, particularly in cases of myocardial infarction (MI) and ischemic heart disease. Various types of stem and progenitor cells, including

D. Sun
Department of Bioengineering, University of California, Riverside, CA, USA

F. G. Thankam (✉)
Department of Translational Research, College of Osteopathic Medicine of the Pacific, Western University of Health Sciences, Pomona, CA, USA
e-mail: FThankam@westernu.edu

D. K. Agrawal et al. (eds.), *Cardiovascular Toxicity*, Advances in Biochemistry in Health and Disease 33, https://doi.org/10.1007/978-3-031-90598-8_31

progenitor cells, mesenchymal stromal cells (MSCs), embryonic stem cell-derived cardiovascular progenitors, and induced pluripotent stem cell-derived cardiovascular cells (iPSCs), have been explored for their regenerative potential in heart repair [1–3]. These stem cells have shown potential to regenerate damaged heart tissue, either through direct replacement of lost cardiomyocytes or by supporting endogenous healing mechanisms. However, these cell types differ significantly in their mechanisms of action and the challenges they present.

In drug development, cardiotoxicity is a major concern as many therapeutic agents adversely affect cardiac function, leading to serious complications such as arrhythmias, myocardial injury, and heart failure. The isolation of human stem cells and subsequent establishment of methods for their differentiation into various lineages have provided a solid basis for toxicity evaluations. Human induced pluripotent stem cell-derived cardiomyocytes (iPSC-CMs) have been widely been used to predict cardiotoxicity [4]. The challenge of cardiotoxicity examinations includes the limitations of traditional animal models, which often fail to replicate the complex human cardiac physiology and genetic diversity. As a result, translating preclinical findings into successful human therapies has been a significant hurdle. The application of stromal cells with cardiac signatures in combination with 3D models promise addresses such unmet challenges.

To cite, iPSC-derived cardiomyocytes (iPSC-CMs), hold significant potential due to their pluripotency, allowing them to differentiate into functional cardiomyocytes [5–8]. Studies have shown that iPSC-CMs integrate into damaged heart tissue, restoring its contractile function. However, challenges such as arrhythmogenic risk, limited proliferative potential, and the possibility of tumorigenicity complicate their therapeutic use [8]. Despite their ability to form new myocardium, iPSC-CMs often suffer from low survival rates post-transplantation and face issues with immature cellular integration, establishing complex clinical application in heart repair [9–11]. Moreover, the risk of immune rejection and the need for extensive immunosuppressive treatment further limit their widespread adoption [12, 13].

In contrast, MSCs offer a more practical approach for cardiac repair due to their immunomodulatory properties and lower tumorigenic risk. MSCs are known for their ability to differentiate into osteoblasts, adipocytes, and chondrocytes, making them multipotent cells with a wide range of applications in tissue repair. Unlike iPSCs, MSCs support endogenous repair mechanisms through paracrine signaling. MSCs secrete growth factors, cytokines, and extracellular vesicles (EVs) that reduce inflammation, promote angiogenesis, and limit fibrosis [14]. Their low immunogenicity allows for easier transplantation, often without the need for significant immunosuppression. Given these advantages, MSCs have been more extensively studied in clinical trials for ischemic heart disease [15]. Within the MSC family, heart-derived stromal cells are of particular interest due to their natural adaptation to the cardiac microenvironment. These cells are naturally present in heart tissue and are therefore better adapted to the cardiac microenvironment than MSCs derived from other tissues [16]. Thus, they are better suited for the ischemic conditions present in the damaged heart tissue and have been shown to significantly improve outcomes by reducing infarct size, promoting vascularization, and enhancing overall cardiac function. Their proximity to the site of injury allows them to exert a more potent paracrine effect, further supporting tissue repair and regeneration [17–19].

This chapter summarizes the role of heart-derived stromal cells and *in vitro* 3D models in cardiotoxicity assessment. Also, explores the mechanisms by which these models are employed to evaluate drug-induced cardiac toxicity, challenges in their implementation, and potential future directions.

Approaches of Cardiotoxicity Testing

Traditionally, cardiotoxicity testing relies on *in vivo* animal models and two-dimensional (2D) cell culture systems [20, 21]. In 2D cultures, human-derived cardiomyocytes, often from induced pluripotent stem cells (iPSCs), are commonly used due to their ability to model specific cardiac functions. While these methods have been foundational for early-stage drug screening, they possess considerable limitations, especially in their ability to replicate the complex physiological environment of the human heart. These monolayer cultures often fail to replicate the three-dimensional (3D) architecture and biomechanical dynamics of the human heart (Fig. 31.1), limiting their predictive accuracy in cardiotoxicity testing [23]. Additionally, the immaturity of iPSC-derived cardiomyocytes remains a major challenge, as they fail to exhibit functional properties of adult human cardiomyocytes, such as their electrical and contractile behavior [24]. These limitations often result in false positives or negatives when evaluating drug-induced cardiotoxicity.

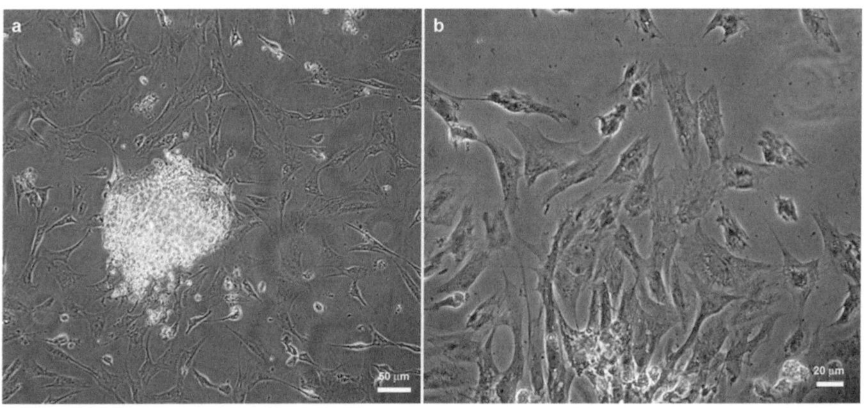

Fig. 31.1 3D-cardiac cells form round shaped-structures while 2D-cardiac cells are flattened. (Reprinted with permission from Carolina Pontes Soares et al. [22]). 3D-cardiac cell aggregates and 2D-cardiac cells were isolated from hearts of 11-day-old chick embryos by mechanical and enzymatic dissociation and plated onto collagen-coated substrates. Phase-contrast microscopy of live cells shows (**a**) 3D-organized cardiac aggregates and (**b**) 2D spread and flattened isolated cardiac cells

Traditionally, cardiotoxicity assessments have relied on *in vitro* methods, such as the human ether-a-go-go related gene (hERG) assay, and *in vivo* animal studies [21, 25–28]. For instance, hERG assays utilize single non-myocyte cells, which do not adequately represent the collective response of heart tissues. Animal models, while useful in studying systemic effects and organ-level responses, express limited translatability to human physiology due to species-specific differences [27–29]. For example, differences in ion channel expression, heart rate, and metabolic pathways in animals often lead to discrepancies in how cardiotoxicity manifests compared to humans. Moreover, ethical concerns, high costs, and time-consuming procedures further limit the practicality of animal models in high-throughput drug screening.

These challenges underscore the need for accurate and human-relevant systems for evaluating cardiotoxicity, which has led to the emergence of *in vitro* 3D models derived from human cells as a promising alternative. An overview of general differences between 2D and 3D culture are summarized in Table 31.1.

Table 31.1 Comparison of 2D and 3D cardiac cell culture in cardiotoxicity assessment

Feature	2D Culture	3D Culture
Cell–cell contacts	Limited contact at the periphery of cells or isolated cells	More tissue-like organization, with structured cell-cell contacts
Cell–ECM contacts	Only part of the cell body contacts ECM, rest exposed to liquid	Entire cell surface interacts with ECM, mimicking in vivo situation
Production of ECM	Limited production, adsorbed proteins used	Cells actively produce and deposit ECM, resembling in vivo conditions
Proliferative index	High proliferation, cells are often synchronized	Lower, and more variable depending on region, resembles natural tissue
Actin stress fibers	Extensive formation of stress fibers and focal adhesions	Few or no stress fiber-like structures
Differentiation	Rapid loss of specialized structures like T-tubules	Cells maintain differentiation, improved organotypic alignment
Cell morphology	Cells are flattened, display enhanced motility	Cells maintain a 3D shape, similar to natural tissues
drug diffusion	Direct exposure to drugs, rapid effect	Slower drug uptake, influenced by ECM and tissue structure
Exposure to drugs	Uniform exposure to drugs, effects are more pronounced	Gradients develop, simulating more realistic drug effects
Co-culture	Difficult due to uneven proliferation	Better support for co-cultures and tissue-like environments
Cardiotoxicity relevance	Overestimates drug effects due to direct exposure	More predictive of real-world responses, mimicking in vivo tissue
Contractility and beating	Rare, unsynchronized contractions	Regular, spontaneous contractions that mimic natural heart tissue

In Vitro 3d Models for Cardiotoxicity Assessment

In vitro 3D models of the heart, including spheroids, engineered heart tissues (EHTs), cardiac organoids, and organ-on-a-chip systems, provide a more physiologically relevant platform for cardiotoxicity assessment by mimicking the complex cell-cell and cell-matrix interactions of native cardiac tissue [30–32]. These 3D models simulate heart tissue architecture, such as contraction [33], electrical activity [34, 35], and drug responses [21, 36, 37]. Additionally, the use of biomaterials including hydrogels in constructing these models allows recreating the extracellular matrix (ECM) that supports the structural and functional integrity of the heart tissue [34, 38–40]. 3D models have revolutionized cardiotoxicity testing by providing accurate simulations of human cardiac physiology [34]. These models integrate multiple cell types, such as cardiomyocytes, fibroblasts, and endothelial cells, replicating the complex interactions found in the heart. This enables a more comprehensive evaluation of drug effects on various aspects of cardiac function, including electrical conductivity, contractility, and tissue structure; the factors often missed in 2D models.

Several types of 3D platforms have been developed for cardiotoxicity screening. Spheroids and organoids replicate the three-dimensional microenvironment of the heart, enhancing the realism of tissue architecture. Engineered heart tissues (EHTs) (Fig. 31.2) are hydrogel-based constructs that contain cells (typically cardiomyocytes) embedded in a matrix. These tissues are anchored between flexible supports, such as silicone posts, allowing them to contract and perform work, much like natural heart muscle [20, 21, 42–44]. EHTs are particularly valuable because they can be subjected to mechanical and electrical stimuli, helping to enhance the maturation of cardiomyocytes. This model enables the measurement of physiological parameters such as contractile force, calcium cycling, and electrical

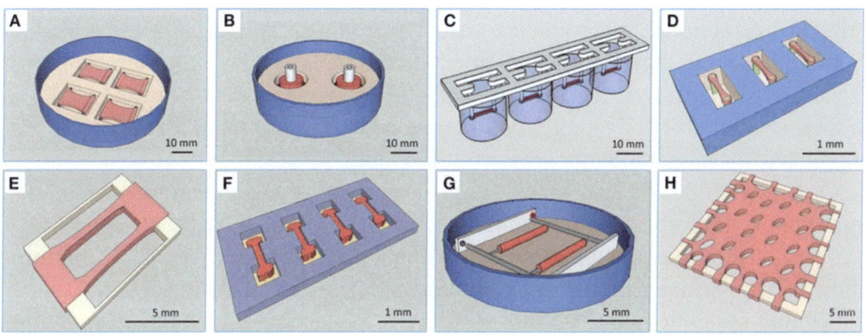

Fig. 31.2 Overview of different types of EHTs. (Reprinted with permission from Christian Zuppinger et al. [41]). (**a**) Plane EHT on Velcro-covered rods, (**b**) fabrication of ring-shaped EHTs, (**c**) fibrin-based mini-EHT on polydimethylsiloxane (PDMS) racks, (**d**) cardiac micro tissues (CMT) on fluorescent pillars, (**e**) cardiobundles on a PDMS frame, (**f**) micro heart muscle, (**g**) cardiac biowires, (**h**) cardiac patch [41]

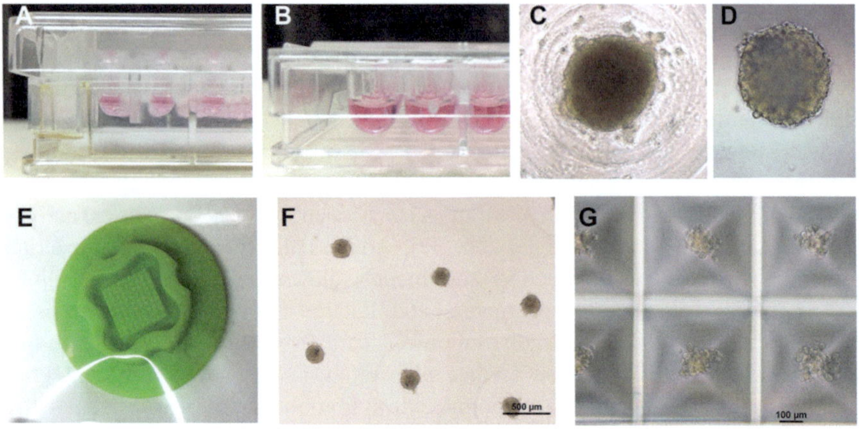

Fig. 31.3 Overview of different spheroid production methods. (Reprinted with permission from Christian Zuppinger et al. [41]). (**a**) Hanging drops (InSphero, GravityPlus), (**b**) U-shaped multi-well plate with non-adhesive coating (Greiner bio-one, Cellstar), (**c**) view inside a U-shaped well at day 2 of culture, (**d**) Spheroid made in hanging drop for 3 days was transferred to a non-adhesive multi-well plate for further culture (InSphero, GravityTrap), (**e**) Silicone micro-mold (3D Petri-dish, microtissues.com) for making an agarose cast for 81 spheroids, (**f**) Cardiac spheroids reside inside the 3D petri-dish agarose cast submerged in medium 4 days after seeding of the cell solution, (**g**) Small cardiac spheroids forming 3 days after seeding in a micro-patterned multi-well plate with 12 × 750 microwells (Kugelmeiers, Sphericalplate 5D)

conductivity, making EHTs ideal for disease modeling and drug testing. These systems are often used to simulate mechanical loads, which are crucial for studying the function of the heart under stress, such as in heart failure or hypertrophy [45, 46].

Spheroids (Fig. 31.3), on the other hand, are smaller self-assembled 3D aggregates of cells, formed in non-adhesive environments or hanging drops without the need for an external scaffold [47, 48]. They can include various cell types, such as cardiomyocytes, fibroblasts, and endothelial cells, enabling the modeling of complex cell-cell interactions. Spheroids are easier to produce in large quantities, making them suitable for high-throughput drug screening. They also replicate important aspects of cardiac tissue, such as spontaneous beating and calcium cycling [49]. Due to their smaller size and simpler production, spheroids are more practical for semi-automated analyses and large-scale drug screening [50, 51].

Organ-on-chip (OOC) technology combines human cardiac cells with microfluidic platforms to simulate the mechanical, electrical, and biochemical environment of the heart in a controlled and miniaturized format [52]. OOC platforms allow for the perfusion of nutrients and drugs through vascularized networks, closely mimicking systemic delivery. This system overcomes the limitations of traditional 2D models, such as the lack of nutrient and oxygen distribution, which often leads to tissue necrosis. By incorporating cardiac spheroids, which are microtissues formed from human iPSC-derived cardiomyocytes, endothelial cells, and fibroblasts, OOCs enable long-term studies of cardiac function, including contractility and tissue remodeling [53]. These systems allow real-time monitoring of drug responses and are highly valuable for high-throughput screening, upgrading as an essential tool in early drug development [48, 54].

Cellular Features of Human Cardiac Tissue for Toxicity Assessments

In native heart tissue, cardiomyocytes (CMs) are mechanically and electrically coupled, driving the contractile function. Although CMs account for 70–85% of the myocardial volume, they represent only 25–35% of total cardiac cells [55]. CMs exist in distinct subtypes, including sinoatrial and atrioventricular nodal cells, atrial cells, ventricular cells, and Purkinje fibers, each playing a specialized role in cardiac conduction [56]. The remaining 65–75% of cardiac cells are non-myocytes, such as endothelial cells, fibroblasts, and immune cells, which are crucial for supporting the structural and functional integrity of the heart.

The heart-derived stromal cells (HSCs) (Fig. 31.4) represent a heterogeneous population of multipotent progenitor cells that originate from various stromal compartments within the heart. These cells include cardiac fibroblasts, MSCs, and epicardial progenitor cells, all of which play critical roles in maintaining cardiac homeostasis and facilitating tissue repair and regeneration following injury [58]. HSCs are of particular interest due to their capacity to secrete bioactive factors through paracrine signaling, production of ECM, and ability to interact with other cardiac cells [59]. Hence, HSCs are the key regulators of both normal cardiac function and pathological remodeling processes. Also, HSCs exhibit several unique properties that contribute to their importance in cardiac research and regenerative therapies. Furthermore, HSCs are involved in fibrosis, angiogenesis, and immune

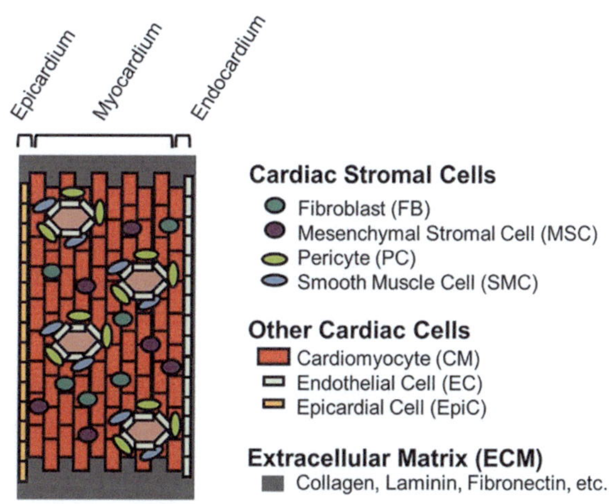

Fig. 31.4 Schematic of cardiac stromal cell composition and localization in cardiac wall. (Reprinted with permission from Martha E Floy et al. [57])

modulation. Thus, HSCs provide advantages in modeling cardiac pathology and studying the cardiotoxicity [59].

HSCs are critical mediators of the regenerative response following injury, such as myocardial infarction suggesting their promising applications in cardiotoxicity assessments. HSCs are activated upon injury and contribute to the complex process of tissue remodeling by producing ECM components and releasing factors that promote angiogenesis and recruit immune cells [60]. Furthermore, HSCs have the ability to differentiate into various cardiac cell types, including endothelial cells and smooth muscle cells, which are crucial for the formation of new blood vessels and the restoration of normal cardiac function [61]. Importantly, HSCs limit pathological fibrosis, preventing excessive scar formation, which is a key determinant in reducing adverse cardiac remodeling, and preserving long-term cardiac function [62]. Through these multifaceted roles, HSCs are recognized as essential players in both the maintenance of cardiac tissue integrity and the promotion of cardiac regeneration. Their regenerative potential and involvement in modulating immune responses make them prime candidates for therapeutic strategies aimed at repairing damaged cardiac tissue and mitigating the progression of heart failure.

The integration of heart-derived stromal cells (HSCs) into 3D cardiac models enhances the physiological relevance and functionality of these systems [63]. HSCs, with their ability to secrete ECM components, modulate immune responses, and support angiogenesis, play a vital role in replicating the native heart environment [64]. Additionally, HSCs contribute to the structural organization and dynamic remodeling of the tissue, helping to recreate key interactions between different cardiac cell types, such as cardiomyocytes and endothelial cells. This integration allows for the study of paracrine signaling, matrix deposition, and tissue repair processes, which are essential for understanding how the heart responds to toxic agents and injury. Hence, by incorporating HSCs, 3D models better mimic the cellular heterogeneity of the heart, facilitating more accurate assessments of cardiac function and drug responses.

With growing interest in creating physiologically relevant in vitro models, HSCs have become increasingly important in the development of three-dimensional (3D) cardiac models for drug screening and disease modeling. Inclision of HSCs in these models enhances the mimicry of in vivo heart tissue, allowing for a better understanding of cell-cell interactions, tissue dynamics, and responses to therapeutic agents, including their potential to cause cardiotoxicity.

Cardiac Adipose Tissue-Derived Stem Cells

Adipose tissue-derived stem cells (ADSCs) have garnered significant attention in recent years for their potential in myocardial regeneration [65–70]. ADSCs are attractive due to their easy accessibility from adipose tissue, typically obtained through liposuction, and their abundance, making them a promising source for regenerative medicine applications. ADSCs exhibit multipotent differentiation potential, allowing them to differentiate into various cell types, including cardiomyocyte-like cells, osteocytes, chondrocytes, and endothelial cells [13]. Unlike general ADSCs, cardiac ADSCs are particularly well-suited for cardiac repair due to their origin in pericardial or epicardial fat, which places them in close proximity to the heart. This proximity may offer a natural adaptation to the cardiac microenvironment, enhancing their therapeutic potential when used in treating myocardial infarction (MI) and other ischemic heart diseases [18]. Additionally, they have been utilized in the development of in vitro models for cytotoxicity assays [71, 72].

The immunophenotype profiling of cardiac ADSCs involved labeling cells with antibodies to assess the expression of surface markers typical of mesenchymal stem cells (MSC). These included positive markers: CD90, CD105, and CD73, which are associated with MSC identity, and negative markers: CD34, CD11b, CD45, CD19, and HLA-DR, indicating the absence of hematopoietic lineage markers. The labeled cells were analyzed using flow cytometry, confirming that ADSC retained their MSC characteristics, making them a suitable candidate for use in cytotoxicity assays [73]. In addition, ADSCs are induced to differentiate into adipocytes, and test substances are applied to evaluate their inhibitory effects on this process. Nile red staining is used to quantify lipid droplets and assess the degree of adipogenesis inhibition, allowing for the prediction of substance toxicity through dose-response curves [74].

For cardiotoxicity assessment, cardiac ADSCs have been integrated into 3D *in vitro* models to evaluate the effects of various drugs and toxins on cardiac function [75–78]. These 3D models provide a more physiologically relevant environment compared to traditional 2D cultures, enabling better simulation of the structural and functional properties of cardiac tissue. By using cardiac ADSCs in combination with cardiomyocytes and other relevant cell types in a 3D scaffold, researchers observed the interactions between cells, extracellular matrix components, and therapeutic agents, which is crucial for accurately assessing the cardiotoxic effects of new drugs.

Cardiac Fibroblasts

Cardiac fibroblasts play a crucial role in maintaining the structural integrity of the heart and regulating the ECM composition. In the context of in vitro 3D cardiotoxicity models, cardiac fibroblasts have emerged as essential components for mimicking the physiological environment of the heart. These cells are responsible for producing ECM proteins such as collagen, fibronectin, and elastin, which provide mechanical support to the myocardium and contribute to the stiffness and elasticity of the heart tissue [79, 80]. In 3D models, cardiac fibroblasts help create a more realistic microenvironment, as their interaction with other cardiac cells, particularly cardiomyocytes, is critical for understanding tissue-level responses to cardiotoxic drugs.

In cardiotoxicity assessment, 3D models incorporating cardiac fibroblasts allow for the investigation of how fibroblast activity changes in response to drug exposure. For instance, cardiotoxic drugs can induce fibrosis, a process in which fibroblasts become activated and excessively deposit ECM components, leading to stiffening of the heart tissue and impaired cardiac function [81]. Hence, cardiac fibroblasts in a 3D in vitro model better evaluate the pro-fibrotic effects of drugs, particularly those leading to adverse remodeling of the heart tissue. Additionally, these models enable fibroblast-cardiomyocyte interactions, which are critical in the context of drug-induced cardiotoxicity, as fibroblast activity influence cardiomyocyte survival, electrical conductivity, and contractile function [57].

Moreover, the use of 3D in vitro models with cardiac fibroblasts enhances the predictive power of cardiotoxicity assays by providing a more physiologically relevant system compared to traditional 2D cultures. Ability of fibroblasts in cardiac ECM remodeling and respond to mechanical stress in a 3D environment mirrors the conditions of the in vivo heart more accurately. This is particularly important for evaluating drugs that alter the biomechanical properties of the heart, leading to functional impairments such as heart failure or arrhythmias. Thus, incorporating cardiac fibroblasts into 3D cardiotoxicity models allows for a more comprehensive assessment of both direct and indirect drug effects on heart tissue, helping to identify potential risks before clinical application.

Endothelial Cells

Endothelial cells play a critical role in cardiovascular health by forming the inner lining of blood vessels, regulating vascular tone, permeability, and homeostasis, and responding to physiological and pathological stimuli [82]. In the context of in vitro 3D cardiotoxicity models, the inclusion of endothelial cells enables the assessment of drug-induced vascular toxicity, endothelial dysfunction, and their effects on the interaction between the vasculature and cardiac tissue. Incorporating endothelial cells into 3D cardiotoxicity models allows to study angiogenesis and maintain the balance between vasoconstriction and vasodilation, processes that are crucial for sustaining healthy cardiac function [83]. Endothelial cell damage or dysfunction, often a side effect of cardiotoxic drugs, lead to compromised blood flow, increased vascular permeability, and an inflammatory response, all of which contribute to the development of cardiovascular diseases such as hypertension, atherosclerosis, and myocardial infarction [84].

In 3D in vitro models, endothelial cells form vascular networks that interact with other cardiac cell types, such as cardiomyocytes and fibroblasts, to provide a more comprehensive system for evaluating drug effects [85]. For example, endothelial cells influence CM survival through paracrine signaling, and is critical for maintaining a functional and responsive vascular system within the heart. These responses are essential for understanding the overall impact of a drug on cardiac function, especially since cardiotoxicity is often associated with both direct effects on heart cells and indirect effects mediated by vascular impairment [86]. Furthermore, endothelial cells in 3D models simulate vascular toxicity in response to specific drugs, particularly those that target angiogenesis or induce endothelial cell apoptosis [87]. These models enable the study of how drugs affect endothelial barrier integrity, cause endothelial cell activation, or lead to vascular inflammation, which precipitate systemic effects such as blood pressure or thrombus formation [88]. Drugs known to cause vascular damage, such as certain chemotherapy agents, can be evaluated for their cardiotoxic effects more effectively when endothelial cells are included in the 3D system.

By using endothelial cells in 3D cardiotoxicity assays, the complex interaction between the heart and vasculature can be replicated, leading to a more accurate prediction of how drugs might impair cardiovascular function. These models offer insights not only into the direct cardiotoxic effects of drugs on the myocardium but also into how endothelial dysfunction exacerbate or mediate the overall toxicity, providing a more robust tool for preclinical safety testing.

Induced Pluripotent Stem Cells (iPSCs)

Induced pluripotent stem cells (iPSCs) offer a groundbreaking approach for assessing drug-induced cardiotoxicity, particularly in the context of anti-cancer therapies like anthracyclines [89]. iPSC-derived cardiomyocytes (iPSC-CMs) are highlighted for their ability to recapitulate key aspects of human cardiac physiology, which is crucial for studying drug safety [90]. iPSC-CMs successfully replicate several key mechanisms of cardiotoxicity, including oxidative stress, mitochondrial dysfunction, and disruptions in calcium homeostasis [32]. These iPSC-CMs, which can be generated in an unlimited supply, address the scarcity and ethical challenges of using primary human CMs, thus facilitating more robust drug screening and toxicology studies.

Since iPSCs can be derived from individual patients, this technology enables the creation of personalized cardiac models to predict specific responses to cardiotoxic drugs. Such personalized approach could drastically reduce the risks associated with cardiotoxic drugs and improve drug safety by identifying susceptible individuals to adverse cardiac effects. By offering a human-relevant, reproducible, and scalable platform, iPSCs have the potential to revolutionize drug-induced cardiotoxicity assessments, with implications for both preclinical drug development and personalized medicine.

While iPSC-CMs show great promise for assessing cardiotoxicity, several limitations remain [91]. iPSC lines, available from repositories such as the American Type Culture Collection (ATCC) and the European Bank for Induced Pluripotent Stem Cells, are expensive and require labor-intensive protocols for culture. Although advancements in differentiation protocols and commercially available differentiation kits have simplified the process, variability in differentiation efficiency between cell lines persists [92, 93]. This variability affects the physiological properties of the resulting CMs, with studies demonstrating a wide range of functional parameters, such as relaxation times and action potential durations (APDs). In addition, batch-to-batch differences in iPSC lines pose a challenge for standardization. However, despite these variations, canonical drug effects are still observed across different lines, and strategies comparing relative changes from baseline can help mitigate the impact of baseline variability [21].

A promising approach to overcoming the immaturity of iPSC-CMs is cardiac tissue engineering. This technique creates 3D cardiac tissue constructs, which better mimic the structure and function of the heart than traditional 2D cultures. Cardiac tissue engineering allows the inclusion of multiple cell types, such as endothelial and immune cells, in a physiologically relevant spatial arrangement [94]. Moreover, the contractility of the tissue, a key measure of heart function, can be assessed directly in these constructs. 3D tissue constructs provide more mature and functional CMs with improved sarcomere alignment and mitochondrial content [94]. Contrastingly, standard cardiac tissue constructs typically require a large number of cells (500,000 to 1 million), and the waiting time before experiments is between 5 and 14 days [89]. Despite these limitations, recent advancements in bioprinting techniques, such as Freeform Reversible Embedding of Suspended Hydrogels (FRESH), have enabled the creation of complex, functional heart structures like ventricle-shaped tissues and heart valves, further enhancing the potential of 3D tissue models in cardiotoxicity studies [95–97].

Cardiac Smooth Muscle Cells

Vascular smooth muscle cells (VSMCs) play a pivotal role in maintaining vascular tone and regulating blood pressure, making them critical in cardiovascular function and health. Chemotherapeutic drugs, such as tyrosine kinase inhibitors (TKIs) and other anti-cancer agents, significantly impact VSMC function, leading to changes in vascular stiffness and blood pressure [98]. Key biomarkers for assessing VSMC health in drug screening include alpha-smooth muscle actin (α-SMA), which is indicative of the contractile state of VSMCs, and proliferation markers such as Ki-67 or PCNA (proliferating cell nuclear antigen), which indicate pathological proliferation [99]. Pathological migration, another hallmark of disease states including atherosclerosis, can be evaluated using migration assays alongside markers such as matrix metalloproteinases (MMPs), which are involved in ECM remodeling and VSMC migration [100, 101].

In addition, VSMCs shift from a contractile to a synthetic phenotype show altered expression of osteopontin and calponin, which are with the hallmarks of synthetic state [102]. Synthetic VMSCs lose their normal contractility, and their pathological proliferation and migration lead to vascular complications such as hypertension and atherosclerosis [103]. hiPSC-derived VSMCs have been shown to model/replicate conditions such as pulmonary arterial hypertension, making them valuable tools for drug screening. Studies have demonstrated that several chemotherapy drugs alter blood pressure by affecting VSMCs through mechanisms such as increased vascular stiffness, further highlighting the importance of VSMCs in cardiotoxicity and vascular toxicity assessments [104]. Incorporating cardiac VSMCs into 3D models allows for a more comprehensive understanding of drug effects, particularly on the mechanical properties, contributing to the development of therapies targeting smooth muscle dysfunctions in heart disease.

Cardiomyocytes

CMs are the fundamental contractile cells of the heart, responsible for generating the force required for pumping blood throughout the body. CMs are highly specialized, with unique properties such as electrical excitability and contractile force generation, making them central to the heart function [105]. CMs have been shown to recapitulate clinical manifestations of doxorubicin-induced cardiotoxicity, including arrhythmias, elevated reactive oxygen species (ROS), apoptosis, and sarcomeric disarray, mirroring side effects observed in patients [106]. Additionally, studies have correlated the genetic backgrounds and the severity of doxorubicin-induced toxicity using CMs, demonstrating the potential of these models in personalized medicine. Also, gene expression analysis post-drug treatment provides valuable biomarkers for the early detection of cardiotoxic effects, aiding in the development of safer treatment strategies [30].

In 3D cardiac models, the use of CMs, either iPSCs or primary, enables the recreation of functional heart tissue to study drug-induced effects on contractility, electrical conduction, and overall cardiac performance [107, 108]. However, challenges remain in fully replicating the maturity and complex interactions of cardiomyocytes in vitro, particularly in ensuring long-term stability and responsiveness in 3D models. Nonetheless, cardiomyocytes remain a cornerstone of in vitro cardiac research, especially for evaluating the direct effects of pharmacological agents on heart function.

Future and Translational Significance

Despite the advancements in 3D in vitro models for cardiotoxicity assessment, challenges remain that hinder their routine use in drug development. Issues such as standardization, reproducibility, scalability, and economic viability are significant barriers to their broader application. These limitations need to be addressed for the models to fully realize their potential. However, these models still hold tremendous promise for improving the predictive accuracy of cardiotoxicity testing, ultimately contributing to safer drug development and reducing late-stage failures in clinical trials.

Advancements in bioprinting technologies drive innovation by enabling more precise control over the construction of 3D cardiac tissues allowing to create customized models that mimic specific regions of the heart or disease states with a high degree of accuracy. This is instrumental in understanding disease mechanisms and evaluating the cardiotoxic potential of new drug candidates. The integration of machine learning (ML) and artificial intelligence (AI) into cardiotoxicity models further enhance their potential. ML and AI analyze large datasets generated from 3D models to optimize experimental conditions and refine predictive algorithms, ultimately improving the accuracy of cardiotoxicity assessments. These tools could significantly accelerate drug discovery by providing real-time insights and reducing the likelihood of late-stage drug failures. Addressing the remaining challenges related to scalability, reproducibility, and cost-effectiveness is essential for the widespread adoption of 3D models in both research and pharmaceutical industries. Collaboration among bioengineers, pharmacologists, and regulatory agencies in establishing standardized protocols to ensure that 3D models meet the rigorous requirements for drug safety testing is warranted.

The translational significance of 3D in vitro cardiac models lies in their ability to bridge the gap between basic research and clinical application. By providing a more physiologically relevant platform for cardiotoxicity testing, these models significantly enhance the predictive accuracy of preclinical drug assessments. This shift has the potential to reduce the reliance on less predictive animal models and expedite the development of safer, more effective therapies. For drugs targeting the cardiovascular system, where cardiotoxicity has historically caused many promising candidate drugs to fail in clinical trials, 3D models offer a promising advantage. Notably, the ability to incorporate patient-specific cells paves the way for personalized medicine, where treatments can be tailored based on the genetic profiles of individual patients, optimizing therapeutic efficacy and minimizing adverse effects. As 3D models continue to evolve, they play an increasingly pivotal role in drug safety and efficacy testing, reducing the high costs associated with late-stage clinical trial failures. Additionally, these models advance our understanding of disease mechanisms in cardiovascular research, providing new insights into drugs interaction/mechanisms with human heart tissue. The continued development of 3D cardiac models promises to transform drug discovery and clinical practice, leading to the development of safer, more personalized therapies for patients with cardiovascular diseases.

Summary

Cardiovascular diseases, particularly heart failure, remain leading causes of mortality worldwide, driving the need for innovative therapeutic approaches. Stromal cell therapies, including those derived from mesenchymal stromal cells (MSCs), induced pluripotent stem cells (iPSCs), and heart-derived stromal cells (HSCs), have emerged as promising candidates for heart repair and regeneration, especially in the context of myocardial infarction and ischemic heart disease. This chapter summarizes the use of heart derived stromal cell types in designing 3D models for cardiovascular drug toxicity screening. This chapter highlights the integration of HSCs into 3D cardiac models and the translational significance of 3D models, which enhances the physiological relevance of these systems by supporting tissue remodeling and paracrine signaling. Overall, inclusion of HSCs, and other relevant cells in 3D platforms allows for a more comprehensive evaluation of cardiotoxicity, addressing the limitations of traditional models and paving the way for more personalized approaches to drug screening and offering new insights into disease mechanisms and therapeutic responses.

References

1. Meloni M et al (2015) Cardiac nerve growth factor overexpression induces bone marrow–derived progenitor cells mobilization and homing to the infarcted heart. Mol Ther 23(12): 1854–1866
2. Yan W et al (2017) C1q/tumor necrosis factor–related protein-9 regulates the fate of implanted mesenchymal stem cells and mobilizes their protective effects against ischemic heart injury via multiple novel signaling pathways. Circulation 136(22):2162–2177
3. Wang J et al (2019) Human embryonic stem cell-derived cardiovascular progenitors repair infarcted hearts through modulation of macrophages via activation of signal transducer and activator of transcription 6. Antioxid Redox Signal 31(5):369–386
4. Sharma A et al (2017) High-throughput screening of tyrosine kinase inhibitor cardiotoxicity with human induced pluripotent stem cells. Sci Transl Med 9(377):eaaf2584
5. Takahashi K, Yamanaka S (2006) Induction of pluripotent stem cells from mouse embryonic and adult fibroblast cultures by defined factors. Cell 126(4):663–676
6. Yu J et al (2007) Induced pluripotent stem cell lines derived from human somatic cells. Science 318(5858):1917–1920
7. Shiba Y et al (2016) Allogeneic transplantation of iPS cell-derived cardiomyocytes regenerates primate hearts. Nature 538(7625):388–391
8. Weinberger F et al (2016) Cardiac repair in guinea pigs with human engineered heart tissue from induced pluripotent stem cells. Sci Transl Med 8(363):363ra148
9. Gao L et al (2018) Large cardiac muscle patches engineered from human induced-pluripotent stem cell–derived cardiac cells improve recovery from myocardial infarction in swine. Circulation 137(16):1712–1730
10. Huo J et al (2017) Evaluation of batch variations in induced pluripotent stem cell-derived human cardiomyocytes from 2 major suppliers. Toxicol Sci 156(1):25–38
11. Takasuna K et al (2017) Comprehensive in vitro cardiac safety assessment using human stem cell technology: overview of CSAHi HEART initiative. J Pharmacol Toxicol Methods 83:42–54
12. Kawaguchi S et al (2021) Intramyocardial transplantation of human iPS cell–derived cardiac spheroids improves cardiac function in heart failure animals. Basic Transl Sci 6(3):239–254
13. Jiang Y, Lian XL (2020) Heart regeneration with human pluripotent stem cells: prospects and challenges. Bioact Mater 5(1):74–81
14. Zhang B et al (2020) Mesenchymal stem cell-derived extracellular vesicles in tissue regeneration. Cell Transplant 29:0963689720908500
15. Poomani MS et al (2022) Mesenchymal stem cell (MSCs) therapy for ischemic heart disease: a promising frontier. Glob Heart 17(1)
16. Bagno L et al (2018) Mesenchymal stem cell-based therapy for cardiovascular disease: progress and challenges. Mol Ther 26(7):1610–1623

17. Gao Q et al (2014) A cocktail method for promoting cardiomyocyte differentiation from bone marrow-derived mesenchymal stem cells. Stem Cells Int 2014(1):162024
18. Pittenger MF, Martin BJ (2004) Mesenchymal stem cells and their potential as cardiac therapeutics. Circ Res 95(1):9–20
19. Guo Y et al (2020) The therapeutic potential of mesenchymal stem cells for cardiovascular diseases. Cell Death Dis 11(5):349
20. Tadano K et al (2021) Cardiotoxicity assessment using 3D vascularized cardiac tissue consisting of human iPSC-derived cardiomyocytes and fibroblasts. Mol Ther Methods Clin Dev 22:338–349
21. Takeda M et al (2018) Development of in vitro drug-induced cardiotoxicity assay by using three-dimensional cardiac tissues derived from human induced pluripotent stem cells. Tissue Eng Part C Methods 24(1):56–67
22. Pontes Soares C et al (2012) 2D and 3D-organized cardiac cells shows differences in cellular morphology, adhesion junctions, presence of myofibrils and protein expression. PLoS One 7(5):e38147
23. Paz-Artigas L et al (2023) Current approaches for the recreation of cardiac ischaemic environment in vitro. Int J Pharm 632:122589
24. Sharma A et al (2014) Human induced pluripotent stem cell–derived cardiomyocytes as an in vitro model for coxsackievirus B3–induced myocarditis and antiviral drug screening platform. Circ Res 115(6):556–566
25. Sheng CC et al (2016) 21st century cardio-oncology: identifying cardiac safety signals in the era of personalized medicine. JACC: Basic Transl Sci 1(5):386–398
26. Fermini B et al (2016) A new perspective in the field of cardiac safety testing through the comprehensive in vitro proarrhythmia assay paradigm. J Biomol Screen 21(1):1–11
27. Enayetallah AE et al (2013) Assessing the translatability of in vivo cardiotoxicity mechanisms to in vitro models using causal reasoning. BMC Pharmacol Toxicol 14:1–12
28. Meng C et al (2022) Preparation and evaluation of animal models of cardiotoxicity in antineoplastic therapy. Oxidative Med Cell Longev 2022(1):3820591
29. Lamberti M et al (2014) Animal models in studies of cardiotoxicity side effects from antiblastic drugs in patients and occupational exposed workers. Biomed Res Int 2014(1): 240642
30. Schwach V, Slaats RH, Passier R (2020) Human pluripotent stem cell-derived cardiomyocytes for assessment of anticancer drug-induced cardiotoxicity. Front Cardiovasc Med 7:50
31. Ma Z et al (2014) Three-dimensional filamentous human diseased cardiac tissue model. Biomaterials 35(5):1367–1377
32. Kistamás K et al (2024) The current state of realistic heart models for disease modelling and cardiotoxicity. Int J Mol Sci 25(17):9186
33. Dou W et al (2022) Microengineered platforms for characterizing the contractile function of in vitro cardiac models. Microsyst Nanoeng 8(1):26
34. Camman M et al (2022) 3D models of dilated cardiomyopathy: shaping the chemical, physical and topographical properties of biomaterials to mimic the cardiac extracellular matrix. Bioact Mater 7:275–291
35. Tandon N et al (2009) Electrical stimulation systems for cardiac tissue engineering. Nat Protoc 4(2):155–173
36. Kofron CM et al (2021) A predictive in vitro risk assessment platform for pro-arrhythmic toxicity using human 3D cardiac microtissues. Sci Rep 11(1):10228
37. Tani H, Tohyama S (2022) Human engineered heart tissue models for disease modeling and drug discovery. Front Cell Dev Biol 10:855763
38. Keklikian A et al (2024) Chitosan–polyethylene glycol inspired polyelectrolyte complex hydrogel templates favoring NEO-tissue formation for cardiac tissue engineering. Gels 10(1):46
39. Heath DE (2019) A review of decellularized extracellular matrix biomaterials for regenerative engineering applications. Regen Eng Transl Med 5:155–166

40. Geckil H et al (2010) Engineering hydrogels as extracellular matrix mimics. Nanomedicine 5(3):469–484
41. Zuppinger C (2019) 3D cardiac cell culture: a critical review of current technologies and applications. Front Cardiovasc Med 6:87
42. Radhakrishnan J, Krishnan UM, Sethuraman S (2014) Hydrogel based injectable scaffolds for cardiac tissue regeneration. Biotechnol Adv 32(2):449–461
43. Li Z, Guan J (2011) Hydrogels for cardiac tissue engineering. Polymers 3(2):740–761
44. Alonzo M et al (2022) 3D biofabrication of a cardiac tissue construct for sustained longevity and function. ACS Appl Mater Interfaces 14(19):21800–21813
45. Thomas D, Shenoy S, Sayed N (2021) Building multi-dimensional induced pluripotent stem cells-based model platforms to assess cardiotoxicity in cancer therapies. Front Pharmacol 12: 607364
46. Eschenhagen T et al (1997) Three-dimensional reconstitution of embryonic cardiomyocytes in a collagen matrix: a new heart muscle model system. FASEB J 11(8):683–694
47. Polonchuk L et al (2017) Cardiac spheroids as promising in vitro models to study the human heart microenvironment. Sci Rep 7(1):7005
48. Di Cio S et al (2024) Vascularised cardiac spheroids-on-a-chip for testing the toxicity of therapeutics. Sci Rep 14(1):3370
49. Sharma P, Gentile C (2021) Cardiac spheroids as in vitro bioengineered heart tissues to study human heart pathophysiology. Vis Exp 167:e61962
50. Martin M et al (2023) Advances in 3D organoid models for stem cell-based cardiac regeneration. Int J Mol Sci 24(6):5188
51. Figtree GA et al (2017) Vascularized cardiac spheroids as novel 3D in vitro models to study cardiac fibrosis. Cells Tissues Organs 204(3–4):191–198
52. Weng K-C et al (2020) Human induced pluripotent stem-cardiac-endothelial-tumor-on-a-chip to assess anticancer efficacy and cardiotoxicity. Tissue Eng Part C Methods 26(1):44–55
53. Liu J, Wang Y (2023) Advances in organ-on-a-chip for the treatment of cardiovascular diseases. MedComm Biomater Appl 2(4):e63
54. Mozneb M et al (2024) Multi-lineage heart-chip models drug cardiotoxicity and enhances maturation of human stem cell-derived cardiovascular cells. Lab Chip 24(4):869–881
55. Pinto AR et al (2016) Revisiting cardiac cellular composition. Circ Res 118(3):400–409
56. Devalla HD, Passier R (2018) Cardiac differentiation of pluripotent stem cells and implications for modeling the heart in health and disease. Sci Transl Med 10(435):eaah5457
57. Floy ME et al (2020) Human pluripotent stem cell-derived cardiac stromal cells and their applications in regenerative medicine. Stem Cell Res 45:101831
58. Leong YY et al (2017) Cardiac stem cells for myocardial regeneration: they are not alone. Front Cardiovasc Med 4:47
59. Burchfield JS, Dimmeler S (2008) Role of paracrine factors in stem and progenitor cell mediated cardiac repair and tissue fibrosis. Fibrogenesis Tissue Repair 1:1–11
60. Kolesnikoff N, Chen C-H, Samuel MS (2022) Interrelationships between the extracellular matrix and the immune microenvironment that govern epithelial tumour progression. Clin Sci 136(5):361–377
61. Goichberg P et al (2014) Cardiac stem cells: biology and clinical applications. Antioxid Redox Signal 21(14):2002–2017
62. Ballard VL (2010) Stem cells for heart failure in the aging heart. Heart Fail Rev 15:447–456
63. Cui H et al (2018) 3D bioprinting for cardiovascular regeneration and pharmacology. Adv Drug Deliv Rev 132:252–269
64. Silva AC et al (2021) Bearing my heart: the role of extracellular matrix on cardiac development, homeostasis, and injury response. Front Cell Dev Biol 8:621644
65. Joo HJ, Kim JH, Hong SJ (2017) Adipose tissue-derived stem cells for myocardial regeneration. Korean Circ J 47(2):151–159

66. Kim J-H et al (2015) Sphere formation of adipose stem cell engineered by poly-2-hydroxyethyl methacrylate induces in vitro angiogenesis through fibroblast growth factor 2. Biochem Biophys Res Commun 468(1–2):372–379
67. D'andrea F et al (2010) Human CD34/CD90 ASCs are capable of growing as sphere clusters, producing high levels of VEGF and forming capillaries. Eur J Clin Investig 40:88
68. Gimble JM, Guilak F (2003) Adipose-derived adult stem cells: isolation, characterization, and differentiation potential. Cytotherapy 5(5):362–369
69. Zuk PA et al (2001) Multilineage cells from human adipose tissue: implications for cell-based therapies. Tissue Eng 7(2):211–228
70. Hong SJ, Traktuev DO, March KL (2010) Therapeutic potential of adipose-derived stem cells in vascular growth and tissue repair. Curr Opin Organ Transplant 15(1):86–91
71. Abud APR et al (2015) The use of human adipose-derived stem cells based cytotoxicity assay for acute toxicity test. Regul Toxicol Pharmacol 73(3):992–998
72. Corrêa NCR et al (2018) Human adipose-derived stem cells (ADSC) and human periodontal ligament stem cells (PDLSC) as cellular substrates of a toxicity prediction assay. Regul Toxicol Pharmacol 92:75–82
73. Dominici M et al (2006) Minimal criteria for defining multipotent mesenchymal stromal cells. The International Society for Cellular Therapy position statement. Cytotherapy 8(4):315–317
74. Abud APR et al (2021) Using inhibition of the adipogenesis of adipose-derived stem cells in vitro for toxicity prediction. MethodsX 8:101515
75. Doescher C et al (2022) Intelligent hydrogels in myocardial regeneration and engineering. Gels 8(9)
76. Chandrasekhar SK et al (2022) Myocardial tissue engineering: fundamentals and future. In: Tissue engineering, pp 33–51
77. de Barros NR et al (2024) Injectable nanoengineered adhesive hydrogel for treating enterocutaneous fistulas. Acta Biomater 173:231–246
78. Ghassemi K et al (2023) Engineered vesicles and hydrogel technologies for myocardial regeneration. Gels 9(10)
79. Pesce M et al (2023) Cardiac fibroblasts and mechanosensation in heart development, health and disease. Nat Rev Cardiol 20(5):309–324
80. LeBar K, Wang Z (2021) Extracellular matrix in cardiac tissue mechanics and physiology: role of collagen accumulation. In: Extracellular matrix-developments and therapeutics. IntechOpen
81. Osorio JM et al (2023) Senescent cardiac fibroblasts: a key role in cardiac fibrosis. Biochim Biophys Acta (BBA) Mol Basis Dis 1869(4):166642
82. Chatterjee S (2018) Endothelial mechanotransduction, redox signaling and the regulation of vascular inflammatory pathways. Front Physiol 9:524
83. He Z et al (2024) Research progress in the construction and application of in vitro vascular models. Appl Sci 14(15):6780
84. Hazra A, Swathi M, Kanthesh B (2024) Inflammatory responses in myocardial infarction and its therapeutic drugs. In: Drugs from nature: targets, assay systems and leads. Springer, pp 231–254
85. Wagner JUG et al (2020) Dissection of heterocellular cross-talk in vascularized cardiac tissue mimetics. J Mol Cell Cardiol 138:269–282
86. Zamani M, Karaca E, Huang NF (2018) Multicellular interactions in 3D engineered myocardial tissue. Front Cardiovasc Med 5:147
87. Ha DH et al (2020) Toxicological evaluation of exosomes derived from human adipose tissue-derived mesenchymal stem/stromal cells. Regul Toxicol Pharmacol 115:104686
88. Theofilis P et al (2021) Inflammatory mechanisms contributing to endothelial dysfunction. Biomedicines 9(7):781
89. Giacomelli E et al (2020) Human-iPSC-derived cardiac stromal cells enhance maturation in 3D cardiac microtissues and reveal non-cardiomyocyte contributions to heart disease. Cell Stem Cell 26(6):862–879.e11

90. Del Álamo JC et al (2016) High throughput physiological screening of iPSC-derived cardiomyocytes for drug development. Biochim Biophys Acta (BBA) Mol Cell Res 1863(7):1717–1727
91. Davis RP et al (2012) Cardiomyocytes derived from pluripotent stem cells recapitulate electrophysiological characteristics of an overlap syndrome of cardiac sodium channel disease. Circulation 125(25):3079–3091
92. Kattman SJ et al (2011) Stage-specific optimization of activin/nodal and BMP signaling promotes cardiac differentiation of mouse and human pluripotent stem cell lines. Cell Stem Cell 8(2):228–240
93. Moretti A et al (2013) Pluripotent stem cell models of human heart disease. Cold Spring Harb Perspect Med 3(11):a014027
94. Augustine R et al (2021) Stem cell-based approaches in cardiac tissue engineering: controlling the microenvironment for autologous cells. Biomed Pharmacother 138:111425
95. Stoter AMS et al (2020) Assessment of cardiotoxicity with stem cell-based strategies. Clin Ther 42(10):1892–1910
96. Criscione J et al (2023) Heart-on-a-chip platforms and biosensor integration for disease modeling and phenotypic drug screening. Biosens Bioelectron 220:114840
97. Williams NP (2020) Engineering 3D human cardiac ventricular models with controllable architecture. University of Washington
98. Fonoudi H, Burridge PW (2021) Cellular model systems to study cardiovascular injury from chemotherapy. J Thromb Thrombolysis 51:890–896
99. Tan M et al (2016) Thrombin stimulated platelet-derived exosomes inhibit platelet-derived growth factor receptor-beta expression in vascular smooth muscle cells. Cell Physiol Biochem 38(6):2348–2365
100. Hadzijusufovic E et al (2017) Nilotinib-induced vasculopathy: identification of vascular endothelial cells as a primary target site. Leukemia 31(11):2388–2397
101. Zhang J et al (2019) A human pluripotent stem cell-based screen for smooth muscle cell differentiation and maturation identifies inhibitors of intimal hyperplasia. Stem Cell Rep 12(6): 1269–1281
102. Riches-Suman K, Hussain A (2022) Identifying and targeting the molecular signature of smooth muscle cells undergoing early vascular ageing. Biochim Biophys Acta (BBA) Mol Basis Dis 1868(7):166403
103. Lacolley P et al (2017) Vascular smooth muscle cells and arterial stiffening: relevance in development, aging, and disease. Physiol Rev 97(4):1555–1617
104. Sa S et al (2017) Induced pluripotent stem cell model of pulmonary arterial hypertension reveals novel gene expression and patient specificity. Am J Respir Crit Care Med 195(7): 930–941
105. Burridge PW et al (2014) Chemically defined generation of human cardiomyocytes. Nat Methods 11(8):855–860
106. Knowles DA et al (2018) Determining the genetic basis of anthracycline-cardiotoxicity by molecular response QTL mapping in induced cardiomyocytes. elife 7:e33480
107. Burridge PW et al (2016) Human induced pluripotent stem cell–derived cardiomyocytes recapitulate the predilection of breast cancer patients to doxorubicin-induced cardiotoxicity. Nat Med 22(5):547–556
108. Maillet A et al (2016) Modeling doxorubicin-induced cardiotoxicity in human pluripotent stem cell derived-cardiomyocytes. Sci Rep 6(1):25333

Chapter 32
Circulating Biomarkers in the Detection of Cardiovascular Toxicity

Sanja Stankovic, Nina M. Radisavljevic, Zorislava Bajic, Marija Polovina, and Dragan M. Djuric

Abstract Cardiovascular (CV) toxicity is defined as the development of structural and/or functional abnormalities in the heart or vascular system resulting from the exposure to noxious substances, physical agents and/or medical treatments. It can occur due to the exposure to traditional CV risk factors (proatherogenic dyslypidaemia, homocysteine, tobacco smoke), pathogenic microorganisms and their toxins, medications, illicit drugs, radiation, etc. Particular interest has been recently dedicated to CV toxicity resulting from the exposure to environmental pollution (fine particulate matter, nitrogen dioxide, ozone), cancer therapy-related CV toxicity (antineoplastic medications and anticancer immunotherapy), and rare toxic effects of other medications and vaccines.

S. Stankovic
Center for Medical Biochemistry, University Clinical Center of Serbia, Belgrade, Serbia

Faculty of Medical Sciences, University of Kragujevac, Kragujevac, Serbia

N. M. Radisavljevic
Faculty of Medicine, Institute of Medical Physiology "Richard Burian", University of Belgrade, Belgrade, Serbia

Z. Bajic
Department of Physiology, Faculty of Medicine, University of Banja Luka, Banja Luka, The Republic of Srpska, Bosnia and Herzegovina

Centre for Biomedical Research, Faculty of Medicine, University of Banja Luka, Banja Luka, The Republic of Srpska, Bosnia and Herzegovina

M. Polovina
Department of Cardiology, University Clinical Center of Serbia, Belgrade, Serbia

Faculty of Medicine, University of Belgrade, Belgrade, Serbia

D. M. Djuric (✉)
Faculty of Medicine, Institute of Medical Physiology "Richard Burian", University of Belgrade, Belgrade, Serbia

Centre for Biomedical Research, Faculty of Medicine, University of Banja Luka, Banja Luka, The Republic of Srpska, Bosnia and Herzegovina
e-mail: dragan.djuric@med.bg.ac.rs

D. K. Agrawal et al. (eds.), *Cardiovascular Toxicity*, Advances in Biochemistry in Health and Disease 33, https://doi.org/10.1007/978-3-031-90598-8_32

This chapter discusses the definition, etiology, clinical manifestations, and diagnostic assessment of CV toxicity, with a particular focus placed on the role of circulating biomarkers in the diagnosis, risk stratification and prognosis of patients affected by CV toxicity. It provides a detailed overview of the role of traditional and emerging biomarkers of myocardial injury and remodeling, along with biomarkers of inflammation and oxidative stress and biomarkers of vascular endothelial dysfunction, damage and injury. The chapter adopts an integrative approach to evaluating CV toxicity by combining the findings from biomarkers, electrocardiographic alterations, multimodal imaging, and other diagnostic assessments. The detailed understanding of pathophysiological mechanisms and signalling pathways reflected by circulating biomarkers, alongside the requirements for their measurement and the clinical interpretation of their values, holds significant diagnostic and prognostic importance for researchers and clinicians alike.

Keywords Biomarker · Cancer · Cardiovascular toxicity · Homocysteine · Inflammation · Myocardial injury · Natriuretic peptides · Oxidative stress · Troponin

Introduction

The cardiovascular (CV) system is frequently affected by both natural diseases and damage due to the exposure to various endogenous and exogenous stressors, toxins, or infectious agents. Populational exposure to traditional CV risk factors, environmental pollution, pathogenic microorganisms, drugs, radiation, unhealthy dietary habits and many other factors contribute significantly to CV diseases (CVD) [1], which are the leading global cause of morbidity and mortality, responsible for 17.9 million deaths annually [2]. Over the recent years, with an improved control of traditional risk factors, particular interest has been dedicated to the cardiotoxic effects of environmental pollution [3], and cardiotoxicity associated with cancer treatment, vaccines, and other drugs [4, 5]. The underlying mechanisms include a complex interplay of increased production of inflammatory mediators, reactive oxygen species, disrupted cellular signalling, DNA damage, altered gene expression, posttranscriptional modifications and altered cellular energetics [6–9]. In addition, initiation of autoimmune and autoinflammatory mechanisms by certain drugs, vaccines and viruses also plays a role in CV toxicity [10]. The CV manifestations of the exposure to the toxic agents depend on the severity of the inflicted damage, and adaptability and resilience of CV system, whilst the pre-existing CV conditions and limited regenerative capacity increase individual vulnerability.

This chapter discusses the definition, etiology, clinical manifestations, and diagnostic assessment of CV toxicity, with a particular focus placed on the role of circulating biomarkers in the diagnosis, risk stratification and prognosis of patients affected by CV toxicity.

Definition

Despite the unequivocal clinical significance of CV toxicity, its universally accepted definition is currently lacking. As a consequence, various definitions and terminologies have been used across the documents, potentially leading to diagnostic uncertainties and underdiagnosis of CV toxicity. A frequently proposed definition stipulates that CV toxicity represents the occurrence of structural and/or functional abnormalities in the heart or vascular system caused by the exposure to certain substances, physical agents or medical treatments [11, 12].

Etiology

The list of etiologic agents is vast and includes various classes of medications, physical and biological agents, and environmental pollutants as summarised in Table 32.1.

Clinical Presentation

The clinical presentation of CV toxicity can vary depending on the duration, severity and the type of the inflicted CV damage as well as the patient's pre-existing risk factors and conditions. The most common clinical presentations include the following:

Cardiac Dysfunction and Heart Failure

According to the classification used by the cancer therapy-related CV toxicity [4], the severity of cardiac dysfunction is defined as follows:

(a) severe: new reduction in left ventricular ejection fraction (LVEF) $<40\%$
(b) moderate: new reduction in LVEF by ≥ 10 percentage points to an LVEF of 40–49%, or new LVEF reduction by <10 percentage points to an LVEF of 40–49% and either new relative decline in global longitudinal strain (GLS) by $>15\%$ from baseline or new rise in cardiac biomarkers.
(c) mild: LVEF $\geq 50\%$ and new relative decline in GLS by $>15\%$ from baseline and/or new rise in cardiac biomarkers.

The cardiac dysfunction can be either asymptomatic, or symptomatic, requiring institution/intensification of heart failure medications, hospitalisation for heart failure, or inotropic or mechanical circulatory support and consideration for durable ventricular assist devices or heart transplantation in the most severe cases.

Table 32.1 A list of common inciting agents and CV toxic effects[a]

Agent	Effect
Antineoplastic medications	
Anthracyclines (e.g., Doxorubicin, Daunorubicin)	Myocarditis; dose-dependent, irreversible cardiac damage, which can manifest as cardiomyopathy and heart failure
HER2-targeting monoclonal antibodies (trastuzumab and pertuzumab)	Cardiac dysfunction, especially when combined with anthracyclines
Tyrosine kinase inhibitors (e.g., lapatinib, dasatinib)	Cardiac dysfunction, arrhythmias and hypertension
Anti-vascular endothelial growth factor (VEGF) agents (e.g., bevacizumab)	Hypertension and arterial thrombotic events, due to the induction of vascular endothelial damage/dysfunction
Fluoropyrimidines (e.g., 5-fluorouracil)	Acute coronary syndromes due to endothelial injury and prothrombotic mechanisms
Cyclophosphamide	Cardiomyopathy, pulmonary hypertension through endothelial injury mechanisms
Alkylating drugs (e.g. cisplatin)	Acute coronary syndrome, cardiac dysfunction
Antipsychotic medications	
Haloperidol	QTc prolongation, arrhythmia
Clozapine	Myocarditis, hypertension
Chlorpromazine and thioridazine	Tachycardia, orthostatic hypotension
Risperidone, olanzapine	Bradycardia
Antibiotics	
Macrolides (erythromycin, azithromycin), fluoroquinolones (ciprofloxacin)	QTc prolongation
H1 antihistamines	
Terfenadine, astemizole	QTc prolongation
Antiviral medications	
Remdesivir, paxlovid, molnupiravir	Bradycardia, QTc prolongation, syncope
HIV protease inhibitors (lopinavir, ritonavir)	Cardiac dysfunction, AV conduction abnormalities
Antimalarial medications	
Hydroxychloroquine, chloroquine, and ivermectin	Bradycardia, QTc prolongation
Oral hypoglycaemic medications	
Pioglitazone, rosiglitazone	Cardiac dysfunction
Other medications	
Theophylline	Tachycardia, cardiac arrhythmia
Nonsteroidal anti-inflammatory drugs	Hypertension, cardiac dysfunction
Environmental toxins and pollutants	
Air pollution (fine particulate matter, nitrogen dioxide, ozone)	Systemic inflammation, oxidative stress → hypertension, accelerated atherosclerosis, acute coronary syndrome
Heavy metals (arsenic, lead, cadmium, mercury)	Endothelial dysfunction, hypertension, atherosclerosis

(continued)

Table 32.1 (continued)

Agent	Effect
Pesticides such as organophosphates	Endothelial dysfunction, hypertension, atherosclerosis, cardiac dysfunction
Substance abuse	
Alcohol	Alcoholic cardiomyopathy
Cocaine, methamphetamine	Tachycardia, cardiac arrhythmia, spontaneous coronary artery dissection, vasospasm, cardiac dysfunction
Biological agents	
Viruses	Myocarditis, inflammatory cardiomyopathy
Diphtheria toxin	Cardiac dysfunction
Vaccines	
Influenza, mRNA COVID-19, MMR, polio, etc. [13]	Myocarditis, pericarditis
Radiation	
Radiation therapy	Cardiac dysfunction, restrictive cardiomyopathy, pericardial disease, valvular heart disease
Tobacco smoking	
Carbon monoxide, polycyclic aromatic hydrocarbons, nicotine, and heavy metals	Endothelial dysfunction, vasospasm, accelerated atherosclerosis, hypertension

[a] The list is not exhaustive; Information is compiled from several published reports including references [14–19]

Myocarditis

Myocardial inflammation can occur as a result of autoimmune/autoinflammatory reaction triggered by a viral infection, drug administration (antipsychotic medications, antineoplastic cytotoxic medications, antineoplastic immunotherapies, salicylates) or vaccination, or by direct damage caused by the exposure to biologic toxins (snake venom, Diphtheria), organophosphates, heavy metals or radiation [20]. Myocarditisshould be suspected by a sudden occurrence of angina-like chest pain, symptoms of heart failure and/or cardiac rhythm abnormalities (ventricular arrhythmia, ventricular fibrillation, cardiac arrest, atrioventricular (AV) conduction abnormalities) associated with new-onset electrocardiographic abnormalities and elevated biomarkers in the absence of significant coronary artery disease (CAD). Additional diagnostic assessment with echocardiographic examination, cardiac magnetic resonance and/or endomyocardial biopsy followed by histological and immunohistological analysis is required for diagnostic confirmation and rule out of coronary artery disease either by invasive or non-invasive coronary angiography is necessary [21]. In most cases, patients with myocarditis are symptomatic but haemodynamically stable, without cardiac arrhythmias or evidence of severe impairment of myocardial function (uncomplicated myocarditis). Less frequently, myocarditis can take the complicated course characterised by significant LVEF decrease, development of acute heart failure, cardiac rhythm or conduction abnormalities, and haemodynamic instability requiring inotropic and/or mechanical circulatory support (fulminant myocarditis).

Vascular Toxicity

Exposure to various endogenous and exogenous agents can cause vascular toxicity manifesting as acute coronary syndrome, vasospastic angina, microvascular angina, Raynaud's phenomenon, vasculitis, stroke/transient ischemic attack, arterial and venous thrombosis, accelerated atherosclerosis and pulmonary arterial hypertension.

Arterial Hypertension

Hypertensive emergencies or arterial hypertension induced by exposure to toxic agents can also be the manifestations of CV toxicity. The most commonly implicated toxins include tobacco smoke, nicotine, cocaine, methamphetamine, alcohol, anti-neoplastic medications and nonsteroidal anti-inflammatory agents.

Cardiac Rhythm Disorders

Cardiac rhythm disorders including QTc prolongation (>500 ms), bradycardia, AV conduction abnormalities, supraventricular and ventricular arrhythmias, atrial fibrillation and cardiac arrest can occur as an isolated manifestation of CV toxicity or in conjunction with other clinical presentations, most frequently with concomitant cardiac dysfunction/heart failure, myocarditis and acute coronary syndrome.

Diagnostic Assessment

Clinical History and Physical Exam

A careful clinical history and physical examin are essential in the diagnostic assessment of patients with suspected CV toxicity. Pre-existing CV conditions (i.e. arterial hypertension, ischemic heart disease, cardiomyopathy, cardiac arrhythmias), traditional risk factors and the presence of comorbidities (i.e. chronic kidney disease, diabetes mellitus, chronic lung disorders, autoimmune/immune mediated disorders and cancer) increase the risk of CV toxicity. Patients should be asked about typical symptoms (e.g. chest pain, dyspnoea on exertion, orthopnoea, palpitations, syncope, peripheral oedema), whilst physical examination should document heart rate, blood pressure, signs of heart failure, heart murmurs, pericardial rub, alterations in arterial pulses, signs of vascular ischemia and vasculitis.

Electrocardiogram

A baseline 12-lead ECG is a readily available tool that can provide diagnostic "red flags" suggestive of an underlying CV disorder. ECG evidence of left or right ventricular hypertrophy, atrial enlargement, ischemic ST/T wave abnormalities, pathological Q waves, conduction abnormalities, supraventricular and/or ventricular arrhythmias, low voltages and QTc prolongation should be interpreted in the clinical context.

Multimodality Cardiovascular Imaging

CV imaging plays an important role in the diagnostic evaluation in patients with suspected CV toxicity. Transthoracic echocardiography (TTE) is the preferred technique for the initial and follow-up evaluation because of its widespread availability and non-invasive assessment. It can provide quantitative assessment of the left and right ventricular volumes and function, chamber dilation, left ventricular hypertrophy, wall motion abnormalities, diastolic function, presence of valvular heart disease, pulmonary arterial pressure and pericardial disease [22]. Accurate calculation of LVEF should be done with the best method available (i.e. 2D modified biplane Simpson method of disks) and, for the more accurate assessment, 3D echocardiography should be utilised [23]. In patients with inadequate TTE image quality, ultrasound-enhancing contrast agents should be used to improve the quantitative assessment of left ventricle (LV) volumes and function and for the detection of intracavitary masses [24]. It should be considered that LV volumes and LVEF can fluctuate due to factors unrelated to CV toxicity. Fox example, a transitory decline in LVEF can occur as a consequence of a decrease in LV end-diastolic volume due to decreased preload (e.g. dehydration, vomiting, diarrhoea, tachycardia etc.) without a change in LV end-systolic volume [25]. The findings of potential volume dependency of LVEF measurements are significant for both baseline and follow-up assessments, and in inconclusive cases it may be required to perform surveillance imaging when patients' intravascular volumes are less likely to be depleted. In addition to standard TTE measurements, assessment of left ventricular GLS is highly recommended [22]. Impairment in left ventricular GLS can precede the development of LV systolic dysfunction and it is therefore more sensitive in the detection of CV toxicity. Furthermore, it may have superior prognostic value to LVEF in the follow-up of patients exposed to CV toxicity [26].

Cardiac magnetic resonance imaging (CMR) can be considered as a reference technique in the assessment of left and right ventricular volumes and function, and therefore, it should be utilised when TTE is unavailable or non-diagnostic [27]. In addition, CMR offers the capability to assess tissue characteristics, enabling the detection of various myocardial abnormalities. Specifically, it can identify intracellular and interstitial edema, myocardial injury, and fibrosis. These features help in diagnosing and differentiating conditions like acute inflammation, chronic myocardial damage, or scarring [27]. CMR techniques employing T1- and T2-weighted imaging, as well as newer T2 and T1 mapping sequences can be utilised to detect myocardial injury/necrosis and expansion of extracellular volume (ECV) [27, 28]. Pre- and post-contrast weighted T1 imaging (late gadolinium enhancement) can be used to quantify ECV expansions due to oedema, cellular infiltration and/or fibrosis [28]. CMR is the preferred diagnostic technique for the evaluation of patients with new-onset progressive LV dysfunction, takotsubo syndrome, myocarditis or myocardial infarction with non-obstructive coronary arteries. Although the presence and extent of CMR-detected myocardial fibrosis (i.e. late gadolinium enhancement) have prognostic value in patients with ischemic and non-ischemic cardiomyopathy and myocarditis, its specific prognostic role after the exposure to CV toxins remains inconclusive.

Functional imaging techniques—such as stress echocardiographic exam, perfusion CMR, and nuclear myocardial perfusion imaging—are recommended for evaluating myocardial ischemia in patients experiencing symptoms like stable angina or debilitating shortness of breath. These tests are particularly useful when there is clinical suspicion of CAD, helping to confirm the diagnosis and assess the extent of ischemia. In addition, non-invasive or invasive coronary angiography can be performed to rule out or confirm CAD [28].

Circulating Biomarkers of Cardiovascular Toxicity

Biomarkers are essential tools in the research, assessment, and management of CV toxicity, offering in-depth insights into the mechanisms of toxicity of various agents, a more precise evaluation of damage, and tailored approaches to the prevention and treatment of cardiotoxic effects [29]. According to the National Institutes of Health (NIH) working group, a biomarker is defined as "a characteristic that is objectively measured and evaluated as an indicator of biological processes, pathogenic processes, or pharmacologic responses to a therapeutic intervention" [30]. Regarding CV toxicity, this definition is broad and not limited to circulating biomarkers. Nonetheless, circulating biomarkers are the focus of extensive research due to their potential for rapid and simple measurement, making them extremely valuable in clinical practice [31].

From a historic perspective, aspartate aminotransferase (AST) emerged as the pioneering biomarker for acute myocardial injury diagnosis in 1954, marking the beginning of laboratory-based cardiac diagnostics. By 1959, creatine kinase (CK) was introduced as a cardiac biomarker, followed by lactate dehydrogenase (LDH) in 1960. The cardiac-specific isoenzyme CK-MB was later introduced, and these markers were typically ordered as a panel, along with LDH cardiac-specific isoenzymes. While myoglobin was also recognized as a potential cardiac biomarker that showed elevation following acute myocardial infarction, it shared a common limitation with its predecessors: lack of cardiac specificity. These early biomarkers, though groundbreaking for their time, were hampered by high rates of false-positive results due to their presence in other tissues. This limited specificity ultimately led to their near-complete replacement by the more cardiac-specific biomarkers in modern clinical practice [32, 33].

The following sections will emphasize traditional circulating CV biomarkers as well as emerging methods for evaluating CV toxicity.

Markers of Myocardial Injury

Cardiac Troponins

Troponin represents a complex of three subunits. Troponin C binds calcium, troponin T binds tropomyosin and troponin I is an inhibitory component. Troponin C is not very specific for myocardial injury because it can be found in both skeletal and cardiac muscle. Troponin T and I are specific for cardiomyocyte injury because different isoforms of these troponins are present in skeletal and cardiac muscle. Troponin T is mainly bound to the contractile elements of cardiomyocytes but can also be found free in the cytoplasm. Troponin I is not isolated in skeletal muscle, so it is highly specific for cardiomyocytes. This is the reason why it is considered an ideal marker of myocardial injury. Troponins are released into circulation 6–8 h after the myocardial injury, they reach the peak after 12–24 h and after 7–10 days their level return to normal. Today, there are high-sensitivity tests that measure very low concentrations of cardiac troponin expressed in ng/mL (i.e. high sensitivity cardiac troponin - hs-cTn) [34, 35]. They allow earlier detection of troponin rise (within the first hour after the onset of acute myocardial injury, and remain elevated for 5–14 days [34]. Due to their high sensitivity, these test allow detection of even minimal myocardial injury caused by acute ischemia of other pathological processes (inflammation, cardiotoxicity).

Cardiac troponins, specifically troponin I and T, are considered the gold standard among circulating biomarkers for detecting cardiomyocyte injury and necrosis. They serve as reference parameters in assessing CV toxicity. The clinical significance of troponin monitoring in detecting CV toxicity has been extensively documented, particularly in studies involving oncology patients, with notable findings observed in those undergoing anthracycline-based therapeutic protocols [36]. The validity of using cardiac troponins to assess chemotherapy-induced CV toxicity was first confirmed in an experimental study on spontaneously hypertensive rats treated with doxorubicin. This study showed a positive correlation between elevated troponin levels, histopathological markers of cardiac injury, and the cumulative drug dose [37]. A meta-analysis examining the relationship between troponin levels and LV dysfunction in oncology patients treated with anthracyclines, found that patients with elevated troponin levels had a significantly higher risk of developing LV dysfunction. This association was most pronounced in patients treated with high-dose chemotherapy, highlighting troponins'essential role as predictors of LV dysfunction associated with cancer therapy [38]. A prospective cohort study monitoring breast cancer patients treated with anthracyclines and trastuzumab recorded a gradual increase in hs-cTnT levels throughout treatment, peaking 1 month after therapy completion. However, hs-cTnT levels did not predict cardiotoxicity development [39]. On the other hand, Sawaya and colleagues showed that in breast cancer patients receiving anthracyclines, taxanes, and trastuzumab, the combination of troponin measurements and GLS assessment after treatment completion demonstrated significant predictive value for cardiotoxicity development [40].

Regardless of the potentially predictive role of cardiac troponins in diagnosing clinical cardiac dysfunction, elevated levels of this marker indicate some degree of myocardial damage. In this context, some studies suggest the administration of angiotensin-converting enzyme (ACE) inhibitors and β-blockers in oncology patients with elevated troponin levels to prevent the development of LV dysfunction [41].

Certain studies have demonstrated a lack of association between elevated troponin levels and CV outcomes in chemotherapy-induced CV toxicity. The study by Onitilo and colleagues showed no correlation between troponin values and the risk of developing a clinically significant reduction in LV EF during trastuzumab therapy [42]. Similar results were reported by Goel and colleagues, confirming the absence of a significant association between elevated troponin levels and chemotherapy-induced CV damage [43]. In conclusion, while cardiac troponins are key biomarkers for evaluating myocardial damage, conflicting evidence regarding their role in assessing outcomes of chemotherapy-induced CV toxicity underscores the need for further research.

Beyond its significant role in evaluating chemotherapy-induced CV damage, cardiac troponin monitoring represents a reliable marker of CV toxicity of various other etiologies. Elevated cardiac troponin levels are characteristic of heart failure, which is attributed to pathophysiological mechanisms underlying chronic myocardial injury associated with heart failure. Additionally, in patients with acute decompensated heart failure, troponin elevation correlates with increased mortality and rehospitalization rates [44, 45]. Troponin elevation has been documented in acute coronary syndrome, myocarditis, and infiltrative cardiac diseases such as amyloidosis. Significant troponin elevation is also a hallmark of stress-induced cardiomyopathy (takotsubo syndrome), which develops due to the release of neurohumoral factors and catecholamines, leading to cardiomyocyte damage and transient LV dysfunction and dilatation with apycal balooning [46]. Cardiac troponin levels in takotsubo syndrome demonstrate a characteristic pattern of elevation with significant overlap with myocardial infarction. While troponin values are elevated in takotsubo syndrome, they typically peak at lower levels compared to acute myocardial infarction. However, there still persists a diagnostic challenge in distinguishing between these acute cardiac conditions based on the non-invasive assessment only and coronary angiography remains necessary to rule out significant CAD [47]. Although the potential clinical utility of natriuretic peptide to troponin ratio in differentiating takotsubo syndrome has been explored in multiple studies, results remain inconclusive. Based on a recent meta-analysis, combining natriuretic peptides with cardiac troponin does not provide substantial additional diagnostic benefit beyond troponin measurements alone [48]. These data confirm the reliable role of cardiac troponins as indicators of cardiac damage across various clinical entities.

Natriuretic Peptides

Natriuretic peptides (NPs) include a large group of peptides: atrial natriuretic peptides (ANP), B-type natriuretic peptides (BNP), and C-type natriuretic peptides (CNP). BNP and ANP are predominantly produced by the cardiac atria and ventricles, respectively, as a consequence of increased wall stress. They exert their actions systemically, whereas CNP is produced by the vascular endothelium, where it acts in a paracrine manner [49]. Increased levels of angiotensin II and endothelin can also promote NP secretion from the ventricles [50]. The structure of BNP is similar in different species, and the difference reflects in the length and amino acid composition of the "N"-terminal and "C"-terminal extensions. Under pathological conditions, a precursor of BNP with 134 amino acids, pre-proBNP, is synthesized. At the N-terminal part of pre-proBNP 26 amino acids separate and form proBNP with 108 amino acids. ProBNP convertases, corin and furin, cleave proBNP into an inactive form of 76 amino acids, NT-proBNP, and an active form of 32 amino acids, BNP [51]. NT-proBNP and BNP are produced in a ratio of 1:1 [52]. Because they are secreted in the same ratio, it is expected that their serum concentrations are the same, but the serum concentration of NT-proBNP is higher compared to BNP. The cause of this difference is the half-life of these peptides: NT-proBNP has a half-life of 120 min, and BNP has a half-life of only 20 min [53].

There are three types of membrane receptors for natriuretic peptides (natriuretic peptide receptors—NPR): NPR-A, NPR-B, and NPR-C. NPR-A is a receptor for BNP. When BNP binds to NPR-A, the level of cyclic guanylate monophosphate (cGMP) increases [51]. cGMP is a second messenger, which is formed from guanosine triphosphate (GTP) by the enzyme guanylyl cyclases (GCs). In mammals, GCs are found in two forms, soluble (sGC) and particulate (pGC). Natriuretic peptides stimulate pGC. There are at least 7 types of pGCs, marked as GC-A to GC-G. ANP and BNP are natural ligands for GC-A, and CNP is a ligand for GC-B. pGCs are present on the cell membrane. cGMP acts through cGMP-dependent protein kinases (PKGs). There are three isoforms of PKG, PKG type Iα (PKG Iα), PKG type Iβ (PKG Iβ), and PKG type II (PKG II) [54]. In the CV system, the major isoform is PKG I: PKG Iα is present in cardiomyocytes, and PKG Iβ in endothelial cells.

The BNP/cGMP/PKG signaling pathway leads to the initiation of biological activities against the renin-angiotensin-aldosterone system (RAAS) and the sympathetic nervous system, leading to the improvement of glomerular filtration and fractional filtration causing diuretic, natriuretic, and vasodilatory effects. NPR-C has an important role in the process of degradation and elimination of natriuretic peptides from the extracellular space [51].

The European Society of Cardiology (ESC) and American Heart Association (AHA) guidelines recognize BNP/NT-proBNP measurement as one of the key diagnostic criteria for heart failure. Elevated levels of these biomarkers serve as predictors of heart failure development and mortality, which is particularly significant for early detection and prevention of disease progression in high-risk populations [41]. Natriuretic peptides have demonstrated substantial prognostic value across various cardiac conditions, encompassing myocardial infarction, acute myocarditis, valvular heart disease, atrial fibrillation, and pulmonary embolism. Their capacity to detect myocardial stress and dysfunction has established their importance as biomarkers in the evaluation of specific conditions, including takostubo syndrome and cancer therapy-induced cardiotoxicity [55, 56].

Research has demonstrated a significant correlation between natriuretic peptide levels and cardiac dysfunction in anthracycline-treated patients. Early elevations in NT-proBNP, detected immediately following chemotherapy completion, may indicate compromised myocardial functional capacity preceding clinical or echocardiographic manifestation of the disease [57]. NT-proBNP has emerged as a reliable predictor of LV dysfunction, with persistently elevated levels at 1-year post-chemotherapy suggesting potentially irreversible chemotherapy-induced cardiotoxicity [58]. In their investigation of HER2-positive breast cancer patients, Bouwer and colleagues established a statistically significant correlation between NT-proBNP levels and trastuzumab-induced cardiotoxicity. Their findings indicated that a 10 pmol/L increase in NT-proBNP value corresponded to an approximate 4% reduction in LVEF. However, NT-proBNP failed to demonstrate reliability as an early cardiac dysfunction marker, as the observed changes were too subtle to be distinguished from physiological variations [59]. Furthermore, elevated levels of natriuretic peptides are observed in immune checkpoint inhibitor (ICI)-induced myocarditis [60]. Escudier and colleagues showed that NT-proBNP/BNP values were elevated in all patients diagnosed with ICI-related cardiotoxicity [61].

Despite high detection rates, BNP and NT-proBNP demonstrate limited specificity in identifying ICI-induced cardiotoxicity. Their diagnostic utility is constrained by the fact that these biomarkers may be elevated in various conditions of acute cardiac stress, while oncology patients frequently exhibit chronically elevated values due to inflammatory processes associated with malignancy [62].

However, findings from other studies have not established a consistent correlation between natriuretic peptides and chemotherapy-induced cardiotoxicity. The CARDIOTOX registry, encompassing 865 patients with moderate to high cardiotoxicity risk, demonstrated that baseline NT-proBNP values lacked predictive value in 27 patients who experienced either a decline in LVEF below 40% or developed symptomatic heart failure [60, 63]. Similar findings were reported in a study investigating the potential predictive role of NT-proBNP in early detection of cardiotoxicity associated with trastuzumab, lapatinib, and paclitaxel administration. The study's authors suggest that significant changes in circulating NT-proBNP levels are observed exclusively in patients treated with anthracyclines [64]. The conflicting results of these investigations can be attributed to population heterogeneity, diverse therapeutic protocols, varying sampling times, and methodological differences across studies. Moreover, the release kinetics of BNP/NT-proBNP in the context of chemotherapy-induced cardiotoxicity remains poorly understood [65].

Moreover, diagnostic and prognostic interpretation of natriuretic peptide levels should be considered in relation to other conditions that may affect their circulating levels. Advanced age, impaired renal function, atrial fibrillation and chronic "heart stress" are frequently associated with elevated levels of natriuretic peptides, even in the absence of overt heart failure [66–69]. By contrast, decreased natriuretic peptide secretion associated with chronic insulin resistance and hyperinsulinemia and increased BNP clearance in the adipose tissue lead to misleadingly low circulating levels of natriuretic peptides even in patients with established heart failure [70].

Heart Fatty Acid-Binding Protein

Heart fatty acid-binding protein (h-FABP), a molecule of low molecular weight, transports fatty acids from the plasma membrane to the site of β oxidation (mitochondria and peroxisomes) and to the site of lipid synthesis (endoplasmic reticulum). A larger amount of this protein is found in the myocardium, while a smaller amount is found in skeletal muscles, brain, and kidneys. In serum, h-FABP concentration increases within 30 min of myocardial injury, reaches its highest value after 6–8 h, and falls to normal within 24 h [32, 34].

Given its low molecular weight, water solubility, and high concentration in cardiomyocytes, h-FABP has been extensively studied as a marker of early myocardial injury [71, 72]. The high sensitivity of this biomarker makes it reliable for assessing subclinical myocardial ischemia and early detection of disease progression risk [73]. In an experimental study, h-FABP demonstrated high sensitivity for early detection of myocardial injury, showing significant elevation in serum levels before the development of maximal isoproterenol-induced lesions [74].

Furthermore, research has confirmed the prognostic significance of h-FABP in assessing chemotherapy-induced cardiac damage. In patients with non-Hodgkin lymphoma, elevated h-FABP levels significantly correlate with reduced LVEF following six cycles of chemotherapy [75]. In a prospective cohort study by Yuan and colleagues, changes in circulating biomarker levels were monitored in patients treated with ICI. Their findings revealed a significant increase in h-FABP levels 3 months after ICI administration, in the absence of a detectable reduction in LVEF. Conversely, changes in other circulating biomarkers were not observed, suggesting the high sensitivity of h-FABP in early detection of CV toxicity [76].

These results highlight the potential role of h-FABP in diagnosing chemotherapy-induced CV toxicity. While significant correlations with early cardiac injury detection have been demonstrated, further research is needed to elucidate the mechanisms and prognostic value of h-FABP across various treatment settings.

Homocysteine

Homocysteine represents a sulphur-containing, non-protein, semi-essential amino acid produced as an intermediary product in the metabolism of methionine and cysteine. Folic acid and B vitamins (B2, B6, B12) are essential for the proper metabolism of homocysteine, and their deficiency can result in accumulation of homocysteine. While it is an important metabolite in biological processes, elevated homocysteine concentrations can exert toxic effects on the CV system. Experimental and clinical evidence has demonstrated numerous harmful effects of homocysteine on CV function, including impaired endothelial relaxation mediated by nitric oxide (NO) and endothelium-derived hyperpolarizing factors (EDRF). Additionally, hyperhomocysteinemia induces smooth muscle cell proliferation, enhances the level and oxidation of low-density lipoproteins, and increases oxidative stress [77–80].

Beyond traditional CV risk factors, homocysteine levels serve as a significant biomarker in assessing predisposition to CVD development. Elevated homocysteine levels are associated with an increased likelihood of CV complications, highlighting the importance of monitoring its values for early detection and prevention [81]. In a recent study, Tan and colleagues showed a significant correlation between plasma total homocysteine levels and subclinical myocardial injury, as evidenced by troponin and NTproBNP levels. Notably, this association was observed independently of serum folate and vitamin B12 levels, suggesting that improving homocysteine metabolism may require strategies beyond supplementation of these vitamins alone [82, 83]. Moreover, meta-analysis revealed an informative relationship between homocysteine levels and risk of CAD. The study found that for every 5 μmol/L increase in homocysteine concentration, the risk of CAD related events increased by approximately 20% [84].

Homocysteine is an important biomarker of CV health, and its levels are influenced by various factors, including physical activity and CV system adaptability. Interestingly, exercise tolerance may be associated with homocysteine concentrations, suggesting that individuals with reduced exercise capacity could be at higher risk for CVD [85]. Recent study showed that adolescents with reduced exercise tolerance (poor performance) exhibited significantly higher levels of homocysteine compared to those with good or excellent performance. This observation implies that elevated homocysteine levels could serve as an indicator of impared CV adaptability and exercise capacity [86]. Given the well-established CV risks associated with hyperhomocysteinemia, understanding how homocysteine interacts with exercise tolerance could offer valuable insights for the early identification of individuals at higher risk for CVD.

In addition to its significant CV impact, hyperhomocysteinemia has been recognized as a potential risk factor for the development of malignancies, including breast cancer, colorectal cancer, and leukemia [87, 88]. In an experimental study by Fan et al., hyperhomocysteinemia was shown to accelerate the development of doxorubicin-induced cardiac dysfunction and cardiomyopathy. These findings suggest that elevated homocysteine levels may increase the risk of chemotherapy-induced cardiotoxicity, emphasizing the need for careful monitoring of blood homocysteine levels during and after chemotherapy [89].

Furthermore, homocysteine functions as a key biomarker of environmental health risk, linking factors such as inadequate nutrition, exposure to drugs, stress, and physical inactivity to CV pathogenesis through its central role in one-carbon metabolism and epigenetic regulation via DNA methylation. Elevated levels of homocysteine, especially in the form of highly toxic thiolactone metabolites, contribute to oxidative stress, endothelial dysfunction and inflammation by altering protein structure thorugh homocysteinylation and modifying critical epigenetic marks—specifically, reduced S-adenosylmethionine/S-adenosylhomocysteine (SAM/SAH) ratios and decreased methylation of DNA, RNA and histones [90].

Beyond that, recent research has identified epigenetic alterations as another important mechanism underlying CV dysfunction in heart failure. Oeing and colleagues demonstrated that DNA methylation patterns observed in ex vivo cardiomyocytes exposed to plasma from patients with dilated cardiomyopathy closely mirror the epigenetic landscape of the failing human myocardium. This study identified specific hypomethylation patterns of the ATG promoter that surpassed traditional biomarkers such as NT-proBNP in differentiating heart failure. These results suggest that DNA methylation patterns may serve as a biomarker of complex environmental changes that contribute to cardiac dysfunction [91]. Assessment of circulating epigenetic signatures in addition to homocysteine levels, may provide valuable insights into CV health and serve as a novel approach for the early detection of CV pathology.

Copeptin

Copeptin is a C-terminal part of provasopressin and it is secreted together with vasopressin. The pituitary gland releases copeptin at an early stage (within a few minutes) of acute myocardial injury. Therefore, it can be used both as a diagnostic and as a prognostic marker of myocardial injury [32, 92].

Research has demonstrated that copeptin serves multiple functions in CVD assessment. Its clinical utility extends across several domains, including differential diagnosis, stratification of risk levels, and prediction of outcomes. When combined with traditional cardiac biomarkers, copeptin demonstrates particular value in two key areas: the rapid exclusion of acute myocardial infarction and the assessment of heart failure prognosis. This dual capability makes it an increasingly important tool in CV medicine [93].

Experimental studies in rodent models have revealed an interesting pattern in copeptin dynamics following doxorubicin exposure. While endocardial and myocardial tissue showed reduced copeptin concentrations, circulating levels in the serum demonstrated an opposite trend, exhibiting marked elevation. These findings suggest that doxorubicin may interfere with the cellular production of copeptin in cardiac tissues. Although the elevated serum copeptin could potentially serve as an indicator for doxorubicin-associated cardiac damage, it's important to note that this phenomenon has yet to be confirmed in clinical studies [94, 95].

Despite the growing interest among healthcare professionals, additional research and supporting evidence are necessary before copeptin can be adopted as a standard clinical measurement in the future.

Inflammatory Biomarkers

C-Reactive Protein

C-reactive protein (CRP) is an acute phase protein. It is an inflammatory biomarker but also has proinflammatory effects (it causes the expression of adhesion molecules and inflammatory cells) [32]. Despite its limited specificity and sensitivity, CRP has a role as a biomarker a predictive of future CV events [96]. Furthermore, with the introduction of high-sensitivity tests which can detect low concentrations of CRP (hs-CRP), research has indicated that even minor increases in CRP, which might not be detected by traditional testing methods, can be predictive of CV events [97].

Elevated levels of CRP are frequently observed in patients who have undergone thoracic radiotherapy for cancer treatment, correlating with indicators of diminished cardiac function. These findings imply that alterations in CRP levels may be an early indicator of cardiotoxicity following radiotherapy and could serve as valuable biomarkers for predicting patient outcomes [98]. Similar results were reported in a recent study by Micheletti and colleagues, who documented increased CRP levels in patients receiving chemotherapy. Their findings indicate a significant correlation between elevated CRP and compromised CV health, reinforcing the notion that inflammatory markers may serve as critical indicators of cardiotoxicity in oncological settings [99].

Interleukin-6

Interleukin-6 (IL-6) plays a pivotal role in the inflammatory response during ischemia and reperfusion of the damaged myocardium. It activates inflammatory cells and stimulates the liver to secrete CRP. Like CRP, IL-6 is critical for diagnosis, risk assessment, and prognosis in acute myocardial injury. Elevated IL-6 levels are observed in acute coronary syndrome, serving as a prognostic indicator for adverse cardiac events [34].

In an experimental study, the administration of a single dose of IL-6 was found to significantly impact myocardial hemodynamics, resulting in a temporary decrease in both mean arterial pressure and LV end-systolic pressure. The effects observed were reversed upon the introduction of an inhibitor of endothelial nitric oxide (NO) synthase, suggesting that IL-6 plays a role in the regulation of NO synthesis, contributing to a decrease in cardiac contractility [100].

In addition, in individuals with metastatic melanoma receiving ICI therapy, higher initial levels of IL-6 have been correlated with reduced survival rates, diminished therapeutic responses to ICI, and an increased likelihood of immune-related adverse events [101]. Therapeutic strategies to reduce the CV toxicity of ICIs are urgently required, and emerging evidence indicates that inhibiting the IL-6 pathway may be a promising solution to this clinical challenge [101, 102].

Interleukin-1

Interleukin-1 (IL-1) family, which includes 11 cytokines and 10 receptors divided in three subgroups—IL-1α, IL-1β, IL-18, plays a central role in initiating and propagating inflammation within the injured myocardium. Through binding to the type 1 receptor, IL-1 initiates a cascade of proinflammatory cytokines and contributes to adverse cardiac remodelling, potentially leading to heart failure due to fibrosis and hypertrophy of cardiomyocytes [103–105].

Experimental research has revealed IL-1β as one of the key mediators in ICI associated cardiac toxicity. Notably, IL-1β inhibitors demonstrated dual benefits—reducing CV complications while enhancing the antitumor efficacy of immunotherapy treatments [106]. Another experimental study revealed that IL-1 signaling mechanisms are linked to doxorubicin-related cardiac damage, indicating their contributory role in the acute cardiotoxic effects of this chemotherapeutic agent. The proposed mechanisms through which IL-1 mediates doxorubicin-induced cardiotoxicity involve nuclear factor kappa B (NF-kB) pathway activation and subsequent cardiomyocyte apoptosis [107].

Tumor Necrosis Factor-α

Under physiological conditions, tumor necrosis factor-alpha (TNF-α) is not expressed in heart muscle cells. Its concentration increases in the myocardium in the conditions of ischemia, because of macrophage activation in the cardiac tissue, which causes an increase in TNF-α secretion. After an acute myocardial injury, it binds TNF-α receptor-1 (TNFR1) or TNF-α receptor-2 (TNFR2). When TNF-α binds TNFR1, it participates in inflammatory reactions and remodelling of the heart after injury, resulting in cardiomyocyte apoptosis and cardiotoxicity. On the other hand, when TNF-α binds TNFR2, inhibition of inflammation and heart remodelling occurs, thus protecting the heart from cardiomyocyte apoptosis [108].

Within cardiac tissue, TNF-α exerts its primary function by inducing programmed cell death, targeting both myocardial and endothelial cells [105]. Research has demonstrated significant correlation between circulating levels of TNF-α receptors and infarct size and LV function. Elevated TNF-α levels in the bloodstream serve as indicators of both cardiac dysfunction and vascular pathology [109].

Transforming Growth Factor-β

The role of transforming growth factor-beta (TGF-β) in CV toxicity extends beyond its traditional anti-inflammatory role, as this multifunctional cytokine also contributes to inflammatory processes and cellular deterioration in cardiac tissue [105]. The significance of TGF- β in CV pathology has gained increasing recognition, as mounting evidence reveals its fundamental role in promoting CV fibrosis and ventricular remodelling. This enhanced understanding of TGF-β signalling cascades in cardiac remodelling has identified novel therapeutic opportunities for targeting these pathways [110].

Growth Differentiation Factor-15

Growth differentiation factor-15 (GDF-15) is a member of a TGF-β superfamily. Under physiological conditions, GDF-15 maintains minimal expression throughout various tissues, including cardiac tissue. However, in response to tissue damage, GDF-15 production can increase markedly [111, 112].

Over the past decade, GDF-15 has gained attention as a significant prognostic biomarker in CVD, with elevated levels observed in both myocardial infarction and heart failure. GDF-15 levels correlate with adverse CV outcomes in both general population and patients without CVD [113]. Additionally, emerging evidence underscores the diagnostic and prognostic relevance of GDF-15 in takotsubo syndrome. Patients with takotsubo syndrome demonstrate significantly elevated GDF-15 levels upon hospital admission compared to those with ST-elevation myocardial infarction. Furthermore, increased GDF-15 concentrations correlate with increased risk of adverse clinical outcomes in patients with takotsubo syndrome. Notably, biventricular ballooning corresponds with elevated GDF-15 concentrations, emphasizing its potential as a biomarker for assessing disease severity and forecasting complications in takostrubo syndrome [114].

Furthermore, recent study has revealed higher GDF-15 levels in patients who developed cardiotoxicity during neoadjuvant dual anti-HER2 therapy compared to those without cardiac complications, suggesting GDF-15 value as a predictive biomarker for drugs induced CV toxicity [115]. Further investigation is warranted to establish GDF-15 optimal role in CV toxicity management, particularly its utility in guiding clinical decisions when used alone or combined with conventional biomarkers and predictors [33].

Myeloperoxidase

Myeloperoxidase (MPO) is a member of the heme peroxidase superfamily. Primarily found in polymorphonuclear neutrophils, myeloperoxidase is secreted upon leukocyte granulation [109].

Research has consistently shown a strong association between MPO and various CVDs, with elevated circulating levels of MPO correlating with a worse prognosis and an elevated risk of CV-related mortality. Consequently, MPO has been proposed as a biomarker for stratifying risk in CVD conditions such as acute coronary sindrome, atherosclerosis, heart failure, hypertension, and stroke [116].

Recent studies have highlighted an association between elevated MPO levels and CV toxicity in patients treated with anthracyclines and trastuzumab for breast cancer [117]. Moreover, another study highlighted that MPO may not only serve as a biomarker but also actively contribute to the pathophysiology of chemotherapy-induced CV toxicity [118]. These findings underscore MPO dual role in both detection and progression of chemotherapy-induced cardiac damage.

Soluble Urokinase Plasminogen Activator Receptor

Soluble urokinase plasminogen activator receptor (suPAR) represents a proinflammatory circulating biomarker derived from the membrane-bound urokinase-type plasminogen activator receptor (uPAR), predominantly expressed on endothelial and immune cells. Besides mediating plasminogen activation and fibrinolysis pathways, suPAR plays a crucial role throughout the atherosclerotic process, from early lesion formation to advanced plaque instability. Research has established that elevated plasma suPAR concentrations serve as a significant predictor of CVD progression, type 2 diabetes mellitus development, and cancer occurrence. Notably, suPAR demonstrates superior predictive value for vascular damage and adverse clinical outcomes compared to traditional inflammatory markers, including CRP. Additionally, suPAR levels have demonstrated strong prognostic value for overall mortality, establishing its importance as an integrative marker of disease burden and systemic inflammation [119–122].

Heat Shock Proteins

Heat shock proteins (HSPs) are highly conserved molecules present in prokaryotic and eukaryotic organisms. HSPs play a central role in maintaining proteostasis. They facilitate protein folding, prevent aggregation, and aid in the degradation of misfolded proteins. Research has demonstrated that cardiac tissue exhibits significant levels of HSPs expression. These molecular chaperones show remarkable compartmental diversity within cardiomyocytes, with their presence detected across multiple subcellular locations, including the mitochondrial compartment, cell membrane, cytoplasmic region, and extracellular space [123, 124]. Studies have revealed that cardiomyocytes in the LV exhibit reduced HSP60 expression under specific pathological conditions such as LV stress induced by hypertensive states, and in cases of diabetic metabolic alterations affecting myocardial tissue. These findings indicate a downregulation of HSP60 in response to both hemodynamic stress and metabolic cardiac stressors [125, 126]. Moreover, Demirel and colleagues reported significantly elevated levels of HSP70 in takotsubo syndrome patients compared to non-ST elevation myocardial infarction (NSTEMI) patients and healthy controls. The combination of HSP70 and NT-proBNP levels distinctly differentiated takotsubo syndrome from NSTEMI. While HSP70's predictive power was less pronounced than NT-proBNP, these findings align with previous research indicating elevated BNP/troponin ratios in takotsubo syndrome versus acute myocardial infarction. This suggests HSP70 could serve as a complementary biomarker for distinguishing takotsubo syndrome, particularly when used alongside NT-proBNP and hsTnT [127].

Cardiac Remodelling Biomarkers

Galectin-3

Galectin-3 (Gal-3) is a member of a lectin family, functioning as a galactoside-binding protein that plays a crucial role in a variety of biological processes. These include regulation of a cell-matrix interaction, cellular adhesion, proliferation apoptosis, immunity and inflammation [128]. In a healthy myocardium, Gal-3 expression is low. However, when cardiac injury occurs, Gal-3 expression increases significantly and rapidly, indicating its role in the response to myocardial stress and damage [129].

In CV pathology, Gal-3 serves both as a promoter of fibrotic changes and as a crucial biomarker. Its elevated levels signal ongoing cardiac remodeling and hypertrophy, enabling early detection of various cardiac conditions, including myocardial dysfunction, heart failure, acute coronary syndromes, and myocarditis [33]. Recent research has explored the role of Gal-3 as a potential diagnostic biomarker for cancer-therapy-induced cardiac dysfunction [53]. Shirzadi and colleagues identified markedly increased Gal-3 serum levels in patients who developed reduced LVEF following anthracycline therapy, suggesting its value as a potential biomarker for anthracycline cardiotoxicity detection [130]. Recent experimental studies have highlighted the protective role of Gal-3 in acute stages of doxorubicin-induced cardiac injury. Elevated intracellular levels of Gal-3 were observed in the myocardium of wild-type mice shortly after doxorubicin administration, which correlated with higher survival rates compared to Gal-3 knockout counterparts. Importantly, this early Gal-3 upregulation did not coincide with fibrosis or granulation tissue formation, suggesting that Gal-3's cardioprotective effects at this stage are primarily due to its intracellular activity, rather than its involvement in fibrotic processes [131].

Matrix Metalloproteinases

Matrix metalloproteinases (MMPs) are endopeptidases with ability to degrade collagen and elastin (composition of extracellular matrix). ProMMP-2 (gelatinase A) and proMMP-9 (gelatinase B) play a significant role in the development of acute myocardial injury [132]. MMP activity may participate in the pathogenesis of atherosclerosis by inducing migration of the vascular smooth muscle cells in the intima of the vessel. They proliferate there and participate in the formation of plaques [133]. MMP-2 and MMP-9 can also be used as potential biomarkers of cardiomyopathy progression [134].

MMP-2 has emerged as a reliable biomarker in heart failure, showing elevated plasma levels across various cardiac pathologies, including post-infarction states, dilated cardiomyopathy, and valvular disease. Its clinical significance is demonstrated by correlations with adverse outcomes, including higher New York Heart Association classifications and increased mortality risk, while its relatively stable temporal profile makes it particularly valuable among extracellular matrix proteases. Notably, MMP-2 also serves as an indicator of left ventricular remodeling in post-infarction heart failure patients, particularly those with extensive myocardial damage and compromised ejection fraction [135, 136].

ST2

The interleukin-1 receptor family member, suppression of tumorigenicity (ST2) is expressed in two distinct isoforms: the membrane-bound ST2 ligand (ST2L) and soluble ST2 (sST2). sST2 levels rise significantly in CVD, particularly in heart failure and acute myocardial infarction, correlating with injury severity. The critical role of sST2 in cardiac function is demonstrated by the adverse effects of its genetic deletion, including enhanced cardiac hypertrophy, fibrosis and ventricular dysfunction [33, 109, 137]. The Penn Heart Failure study established sST2 as a robust prognostic indicator. When combined with traditional circulating biomarkers, such as natriuretic peptides, sST2 enhanced risk stratification capabilities. Notably, elevated sST2 levels were linked to increased all-cause mortality and cardiac transplantation rates, with particularly pronounced effects observed in patients with non-ishemic heart failure [117, 138]. Moreover, sST2 serves as a valuable biomarker reflecting mechanical stress and neurohormonal activation in takotsubo syndrome. sST2 levels increase in response to regional akinesia, including apical, midventricular, or basal segments, observed during acute presentations of takotsubo syndrome Takotsubo cardiomyopathy [139]. Recent studies have demonstrated that sST2, particularly when measured in combination with solubile thrombomodulin (sTM), shows promising potential in differentiating takotsubo syndrome from acute anterior ST-elevation myocardial infarction, thereby enhancing diagnostic accuracy in acute cardiac conditions [140].

Recent research explored the potential role of sST2 in cancer therapies related to CV toxicity. Patients receiving anthracycline chemotherapy exhibit increased sST2 levels post-treatement, showing significant positive correlation with cTn levels. Furthermore, elevated sST2 concentrations at 6 months post chemotherapy were found to negatively correlate with LVEF, establishing sST2 as independent predictor of cardiac dysfunction [130, 141]. Further research is needed to fully elucidate the potential role of sST2 in CV toxicity. Specifically, additional studies should aim to clarify whether sST2 could serve as an early biomarker, offering insights sooner than traditional imaging methods such as echocardiography.

Biomarkers of Oxidative Stress

Oxidative stress represents a disruption in cellular redox balance, where reactive oxygen species (ROS) exceed antioxidant defence. While ROS production is a part of normal cellular metabolism, excessive levels can damage critical cellular components. In CV toxicity, oxidative stress emerges as a central mechanism, contributing to lipid peroxidation, endothelial dysfunction and chronic inflammation. Oxidative stress biomarkers include molecules modified by ROS interactions and components of the antioxidant defence system. Elevated ROS alters cellular DNA, lipids proteins and carbohydrates. These alterations are key indicators of oxidative damage and can significantly affect cellular function and tissue health [142].

Common biomarkers used to measure oxidative stress include lipid peroxidation products such as malondialdehyde and thiobarbituric acid reactive substances, while protein oxidation can be assessed through nitrotyrosine and carbonyl groups. For evaluating DNA oxidative damage, 8-hydroxy-2'deoxyguanosine serves as the primary indicator [143, 144].

Regarding antioxidant defense mechanisms, key enzymes such as catalase, superoxide dismutase (SOD), and glutathione peroxidase (GPx) play critical roles in mitigating oxidative stress. Catalase, reacts with hydrogen peroxide in the peroxisome, forming oxygen and water [145]. Transgenic expression of catalase in peroxisomes and mitochondria exhibits beneficial effects on myocardial structure and function in models of heart failure [146, 147]. SOD plays a pivotal role in neutralizing ROS, protecting tissues from oxidative damage. It catalyzes the conversion of superoxide radicals into hydrogen peroxide, which GPx further reduces to water. Increased SOD activity has been observed in myocardial injury models [148–153]. GPx, a selenium-dependent enzyme, removes hydrogen peroxide via oxidation of reduced glutathione (GSH), which is regenerated by glutathione reductase. Found in mitochondria and cytosol, GPx activity increases during acute myocardial injury [154–156]. GSH, a non-enzymatic antioxidant, mitigates oxidative stress during ischemia-reperfusion, although experimental results on its levels are inconsistent, ranging from decreased to normal values [150, 151, 157, 158].

Despite the availability of different oxidative stress biomarkers, none has emerged as superior in clinical practice. However, it remains crucial to evaluate oxidative stress parameters in correlation with established clinical risk assessment tools to determine their diagnostic validity and prognostic value [143].

Vascular Biomarkers

Endothelin-1

Endothelin-1 (ET-1), a potent vasoconstrictor, is synthesized predominantly in the endothelium and cardiomyocytes. ET-1 belongs to the endothelin family, which comprises three related peptides, ET-1, ET-2 and ET-3. Among them, ET-1 represents the primary CV isoform and has been the subject of intensive research in CV medicine [159]. ET-1 plays crucial role in vascular function by regulating blood pressure through its vasconstrictive properties and maintaining basal vascular tone [160].

Elevated ET-1 levels exacerbate endothelial dysfunction by enhancing vasoconstriction, inflammation and smooth muscle cell proliferation, key factors in the pathogenesis, atherosclerosis, heart failure and CAD. This relationship highlights the role of ET-1 not only as a biomarker but also as an active component to CVD progression and the compromised endothelial function with these conditions. Furthermore, ET-1 shows promise as a prognostic marker, particularly in predicting mortality following acute myocardial infarction and congestive heart failure [33, 161].

Beyond established CV role, ET-1 has been shown to potentiate tumor growth. ET-1 facilitates the epithelial-mesenchymal transition, promoting metastatic phenotype and enhances angiogenesis by stimulating VEGF production from tumors. Moreover, ET-1 contributes to resistance to chemotherapeutics [162]. Given its role in promoting tumor growth and angiogenesis, involvement of ET-1 in cancer therapy-induced CV toxicity has been investigated to understand potential adverse effects on the CV system during cancer treatment. In a recent study, Krishnarao and colleagues showed that ET-1 expression correlates with chemotherapy-induced CV toxicity, measured by reduced LVEF [163]. These findings underscore the need for further research to explore ET-1 as a potential predictive marker of CV toxicity, enhancing early risk assessment for patients undergoing cancer therapy.

P-Selectin

P-selectin, the largest transmembrane glycoprotein, is an adhesion molecule expressed by activated endothelial cells and platelets in response to different inflammatory stimuli [33, 162] P-selectin plays a crucial role in the initial recruitment of leukocytes to inflamed endothelial cells, marking the first step in leukocyte migration toward sites of infection [164].

Given its role in leukocyte rolling and attachment, P-selectin has been implicated in the pathogenesis of atherosclerosis, key risk factor for CVD [165, 166]. An increased levels of P-selectin are reported in corornary artery disease, acute myocardial infarction, stroke and peripheral artery diseases. Moreover, elevated levels of P-selectin correlate with an increased risk for major adverse CV events [167, 168].

Circulating Biomarkers of Cell Adhesion

The regulation of leukocyte adhesion to endothelial cells during acute and chronic inflammatory conditions is orchestrated by cell adhesion molecules (CAMs), members of the immunoglobulin superfamily. This diverse protein family includes key molecules such as intracellular adhesion molecules (ICAM-1, ICAM-2, and ICAM-3), vascular cell adhesion molecule-1 (VCAM-1), and platelet endothelial cell adhesion molecule-1 (PECAM-1), among others. These adhesion molecules, prominently expressed on endothelial cell surfaces, mediate critical processes, including firm leukocyte adhesion and subsequent transendothelial migration [169]. Their activity is integral to maintaining vascular homeostasis, yet excessive or deregulated expression contributes to endothelial dysfunction, a hallmark of vascular toxicity.

ICAM-1 and VCAM-1 are pivotal adhesion molecules whose overexpression signifies the onset of endothelial dysfunction, a precursor of atherosclerosis. Beyond their role as markers of endothelial inflammatory processes, these molecules hold potential as diagnostic indicators, with VCAM-1 demonstrating particular promise as a target for non-invasive detection of atherosclerotic changes [170]. Moreover, a prospective cohort study identified elevated levels of IL-6, ICAM-1, and VCAM-1 as significant predictors of post-acute myocardial infarction heart failure. Among these inflammatory markers, ICAM-1 demonstrated the strongest predictive value, while all three molecules exhibited high accuracy in identifying patients at risk for heart failure following an acute myocardial infarction [171]. Patients with coronary microvasculature dysfunction exhibit higher concentrations of ICAM-1, VCAM-1 and selectins compared to healthy individuals. The clinical significance of the elevation of these biomarkers is particularly relevant given that coronary microvasculature dysfunction is present across multiple cardiac conditions, including hypertrophic cardiomyopathy, aortic stenosis, takostubo syndrome, and heart failure with preserved ejection fraction [172]. Pospelova and colleagues demonstrated significantly elevated serum levels of ICAM-1 and PECAM-1 in 45 female breast cancer survivors compared to healthy controls, correlating with cognitive dysfunction, depression, vestibulocebellar ataxia. These findings suggest that endothelial dysfunction indicated by increased adhesion molecule levels, may contribute to white matter tract damage in post-breast cancer treatment patients. The identification of these adhesion molecules as a potential biomarkers could prove valuable for both early detection of central nervous system complications and monitoring therapeutic interventions in breast cancer survivors [173].

Von Willebrand Factor

Von Willebrand factor (vWF) is a large multimeric plasma glycoprotein. It connects platelets with the components of the subendothelial connective tissue. After synthesis, vWF is transported in megakaryocyte/platelet organelles (α granules) and endothelial cells (Weibel-Palade bodies) and stored there. The level of circulating vWF depends almost entirely on the amount of vWF from endothelial cells [174].

Endothelial cell disturbance triggers the release of vWF and its propeptide (vWFpp), with vWF serving as a particularly valuable biomarker due to its established role in endothelial dysfunction [175]. Elevated levels of vWF correlate with major CV risk factors, such as aging, smoking, high cholesterol, diabetes mellitus and hypertension [176]. Additionally, vWF serve as predictor of acute ischemic stroke [177].

In a prospective cohort study, Nuver and colleagues observed increased plasma levels of vWF in patients undergoing chemotherapy, indicating significant endothelial cell damage induced by cancer treatment [178]. Moreover, Panova-Noeva examined the long term effects of elevated vWF and fibrinogen levels in cancer survivors and found significant association between these biomarkers and an increased risk of vascular events and mortality. This association was independent of traditional CV risk factors [179]. Further research is needed to establish the role of vWF as a potential biomarker for monitoring vascular health in oncology patients.

Fibrinogen

Fibrinogen is a large, soluble glycoprotein consisted of three different polypeptides (Aα, Bβ, and γ chains) and it is mainly synthesized in the liver. It affects the viscosity of the blood by participating in the formation of blood clots and by stimulating the aggregation of erythrocytes. Its synthesis is stimulated by IL-6 and other proinflammatory cytokines, so its concentration is elevated in various inflammatory and thrombotic conditions, as well as in CVD [180].

Research demonstrates that elevated fibrinogen levels correlate with poor outcomes in CAD. Assessing fibrinogen concentrations may serve as a valuable tool for identifying high-risk CAD patients. Therefore, patients presenting with hyperfibrinogenemia warrant close monitoring and preventive interventions [181].

MicroRNAs as CV Biomarkers

MicroRNAs (miRNAs) are small, single-stranded, non-coding RNA molecules that regulate gene expression by influencing protein translation. As endogenous molecules, miRNAs impact nearly all biological processes and are known for their regulatory role in CV health. Through advances in precision medicine and next-generation sequencing, miRNAs have emerged as promising biomarkers, especially in CVD, where they influence critical aspects of heart function and vascular biology [182, 183].

Different miRNA profiles are increasingly associated with various CVD, with specific miRNAs identified as potential biomarkers for conditions such as heart failure, ishemic and dilated cardiomyopathy and aortic stenosis [184]. In experimental study, Ji and colleagues revealed that miR-208, miRNA specifically expressed in cardiac tissue, showed elevated plasma levels during myocardial injury induced by isoproterenol. The plasma concentrations of miR-208 correlated with serum troponin levels [185]. Furthermore, miRNAs can serve as predictive biomarkers for the development of complications following acute myocardial infarction. Matsumoto and colleagues reported that serum concentrations of miRNAs (specifically miR-155 and miR-380) were significantly higher in post-myocardial infarction patients who died from CV causes compared to those who did not experience adverse CV events [186]. These findings highlight the potential role of miRNAs as prognostic biomarkers for identifying patients with increased risk of CV mortality following myocardial infarction. Analysis of circulating miRNAs in takotsubo syndrome has revealed distinct expression patterns, with significant up-regulation of several key miRNAs, including miR-16, miR-26a, miR-1, and miR-133a, when compared to healthy controls [187]. Given the clinical challenge of differentiating takotsubo syndrome from acute coronary syndrome, analysis of circulating miRNAs represents a promising diagnostic approach [46]. Additionally, Hussein and colleagues concluded that miRNA-208b is a sensitive and specific biomarker for detecting drug induced cardiac injury in acute poisoning [188].

While cardiac troponins and natriuretic peptides currently serve as a gold standard biomarkers for assessing CV toxicity, emerging evidence suggests that miRNAs could provide more precise and time sensitive information about cardiac damage. miRNAs are promising candidates for potentially replacing traditional CV biomarkers in the future [109].

Conclusion

Biomarkers play a central role to the early detection, risk stratification, and management of cardiovascular injury across diverse clinical contexts, including but not limited to cardio-oncology. Validated markers like cardiac troponins and natriuretic peptides, along with emerging candidates such as homocysteine, ST2, galectin-3, and microRNAs, support more individualized care. Integration with imaging and multi-omics, supported by the use of artificial intelligence, offers new opportunities for precision medicine. To enhance clinical utility, future efforts must focus on robust study designs, harmonisation of measurement protocols, and multidisciplinary collaboration. This evolving field demands adaptable tools to meet the complexity of modern therapies and varied CV risk profiles.

References

1. Sagheer U, Al-Kindi S, Abohashem S et al (2024) Environmental pollution and CV disease. JACC Adv 3(2):100805. https://doi.org/10.1016/j.jacadv.2023.100805
2. Roth GA, Abate D, Abate KH et al (2017) Global, regional, and national age-sex-specific mortality for 282 causes of death in 195 countries and territories, 1980–2017: a systematic analysis for the Global Burden of Disease Study 2017. Lancet 392(10159):1736–1788. https://doi.org/10.1016/S0140-6736(18)32203-7
3. Ahmadi M, Lanphear B (2022) The impact of clinical and population strategies on coronary heart disease mortality: an assessment of Rose's big idea. BMC Public Health 22(1):14. https://doi.org/10.1186/s12889-021-12421-0
4. Lyon AR, López-Fernández T, Couch LS et al (2022) 2022 ESC guidelines on cardio-oncology developed in collaboration with the European Hematology Association (EHA), the European Society for Therapeutic Radiology and Oncology (ESTRO) and the International Cardio-Oncology Society (IC-OS): developed by the task force on cardio-oncology of the European Society of Cardiology (ESC). Eur Heart J 43(41):4229–4361. https://doi.org/10.1093/eurheartj/ehac244
5. Bitounis D, Jacquinet E, Rogers MA, Amiji MM (2024) Strategies to reduce the risks of mRNA drug and vaccine toxicity. Nat Rev Drug Discov 23(4):281–300. https://doi.org/10.1038/s41573-023-00859-3
6. Tousif S, Singh AP, Umbarkar P et al (2023) Ponatinib drives cardiotoxicity by S100A8/A9-NLRP3-IL-1β mediated inflammation. Circ Res 132(3):267–289. https://doi.org/10.1161/CIRCRESAHA.122.32150
7. Carrasco R, Castillo RL, Gormaz JG et al (2021) Role of oxidative stress in the mechanisms of anthracycline-induced cardiotoxicity: effects of preventive strategies. Oxid Med Cell Longev 2021:8863789. https://doi.org/10.1155/2021/8863789
8. Jomova K, Raptova R, Alomar SY et al (2023) Reactive oxygen species, toxicity, oxidative stress, and antioxidants: chronic diseases and aging. Arch Toxicol 97(10):2499–2574. https://doi.org/10.1007/s00204-023-03562-9
9. Babiker HM, McBride A, Newton M et al (2018) Cardiotoxic effects of chemotherapy: a review of both cytotoxic and molecular targeted oncology therapies and their effect on the CV system. Crit Rev Oncol Hematol 126:186–200. https://doi.org/10.1016/j.critrevonc.2018.03.014
10. König D, Läubli H (2020) Mechanisms of immune-related complications in cancer patients treated with immune checkpoint inhibitors. Pharmacology 106(3–4):123–136. https://doi.org/10.1159/000509081
11. Adão R, de Keulenaer G, Leite-Moreira A, Brás-Silva C (2013) Cardiotoxicity associated with cancer therapy: pathophysiology and prevention. Rev Port Cardiol (Engl Ed) 32(5):395–409
12. Taleb A, Ahmad KA, Ihsan AU, Qu J, Lin N, Hezam K et al (2018) Antioxidant effects and mechanism of silymarin in oxidative stress induced CV diseases. Biomed Pharmacother 102:689–698. https://doi.org/10.1016/j.biopha.2018.03.140
13. Kuntz J, Crane B, Weinmann S, Naleway AL (2018) Myocarditis and pericarditis are rare following live viral vaccinations in adults. Vaccine 36(12):1524–1527. https://doi.org/10.1016/j.vaccine.2018.02.030
14. Chen E, Xi L (2024) CV adverse effects of antiviral therapies for COVID-19: evidence and plausible mechanisms. Acta Pharmacol Sin. https://doi.org/10.1038/s41401-024-01382-w
15. Li XQ, Tang XR, Li LL (2021) Antipsychotics cardiotoxicity: what's known and what's next. World J Psychiatry 11(10):736–753. https://doi.org/10.5498/wjp.v11.i10.736
16. Gras J, Llenas J (1999) Effects of H1 antihistamines on animal models of QTc prolongation. Drug Saf 21(1):39–44; discussion 81–7. https://doi.org/10.2165/00002018-199921001-00006
17. Seferović PM, Polovina M, Rosano G et al (2023) State-of-the-art document on optimal contemporary management of cardiomyopathies. Eur J Heart Fail 25(11):1899–1922. https://doi.org/10.1002/ejhf.2979

18. Buoninfante A, Andeweg A, Genov G, Cavaleri M (2024) Myocarditis associated with COVID-19 vaccination. Npj Vaccines 9(1):122. https://doi.org/10.1038/s41541-024-00893-1
19. Wang WH, Wei KC, Huang YT et al (2023) The incidence of myocarditis following an influenza vaccination: a population-based observational study. Drugs Aging 40(2):145–151. https://doi.org/10.1007/s40266-022-00997-0
20. Nguyen LS, Cooper LT, Kerneis M et al (2022) Systematic analysis of drug-associated myocarditis reported in the World Health Organization pharmacovigilance database. Nat Commun 13(1):25. https://doi.org/10.1038/s41467-021-27631-8
21. Ammirati E, Frigerio M, Adler ED et al (2020) Management of acute myocarditis and chronic inflammatory cardiomyopathy. Circ Heart Fail 13(11):e007405. https://doi.org/10.1161/CIRCHEARTFAILURE.120.007405
22. Plana JC, Thavendiranathan P, Bucciarelli-Ducci C, Lancellotti P (2018) Multi-modality imaging in the assessment of CV toxicity in the cancer patient. JACC: CV Imaging 11(8):1173–1186. https://doi.org/10.1016/j.jcmg.2018.06.003
23. Hoffmann R, Barletta G, von Bardeleben S, Vanoverschelde JL, Kasprzak J, Greis C et al (2014) Analysis of left ventricular volumes and function: a multicenter comparison of cardiac magnetic resonance imaging, cine ventriculography, and unenhanced and contrast-enhanced two-dimensional and three-dimensional echocardiography. J Am Soc Echocardiogr 27(3):292–301. https://doi.org/10.1016/j.echo.2013.12.005
24. Porter TR, Mulvagh SL, Abdelmoneim SS, Becher H, Belcik JT, Bierig M et al (2018) Clinical applications of ultrasonic enhancing agents in echocardiography: 2018 American Society of Echocardiography guidelines update. J Am Soc Echocardiogr 31(3):241–274. https://doi.org/10.1016/j.echo.2017.11.013
25. Meléndez GC, Sukpraphrute B, D'Agostino RB Jr, Jordan JH, Klepin HD, Ellis L et al (2017) Frequency of left ventricular end-diastolic volume-mediated declines in ejection fraction in patients receiving potentially cardiotoxic cancer treatment. Am J Cardiol 119(10):1637–1642. https://doi.org/10.1016/j.amjcard.2017.02.008
26. Kalam K, Otahal P, Marwick TH (2014) Prognostic implications of global LV dysfunction: a systematic review and meta-analysis of global longitudinal strain and ejection fraction. Heart 100(21):1673–1680. https://doi.org/10.1136/heartjnl-2014-305538
27. Thavendiranathan P, Wintersperger BJ, Flamm SD, Marwick TH (2013) Cardiac MRI in the assessment of cardiac injury and toxicity from cancer chemotherapy: a systematic review. Circ Cardiovasc Imaging 6(6):1080–1091. https://doi.org/10.1161/CIRCIMAGING.113.000899
28. Vrints C, Andreotti F, Koskinas KC, Rossello X, Adamo M, Ainslie J et al (2024) 2024 ESC Guidelines for the management of chronic coronary syndromes: developed by the task force for the management of chronic coronary syndromes of the European Society of Cardiology (ESC) Endorsed by the European Association for Cardio-Thoracic Surgery (EACTS). Eur Heart J 45(36):3415–3537. https://doi.org/10.1093/eurheartj/ehae177
29. Kim SJ, Mesquita FCP, Hochman-Mendez C (2023) New biomarkers for CV disease. Tex Heart Inst J 50(5):e238178. https://doi.org/10.14503/THIJ-23-8178
30. Biomarkers Definitions Working Group (2001) Biomarkers and surrogate endpoints: preferred definitions and conceptual framework. Clin Pharmacol Ther 69(3):89–95. https://doi.org/10.1067/mcp.2001.113989
31. Gong FF, Cascino GJ, Murtagh G et al (2021) Circulating biomarkers for cardiotoxicity risk prediction. Curr Treat Options Oncol 22(6):46. https://doi.org/10.1007/s11864-021-00845-0
32. Mythili S, Malathi N (2015) Diagnostic markers of acute myocardial infarction. Biomed Rep 3(6):743–748. https://doi.org/10.3892/br.2015.500
33. Omran F, Kyrou I, Osman F et al (2022) CV biomarkers: lessons of the past and prospects for the future. Int J Mol Sci 23(10):5680. https://doi.org/10.3390/ijms23105680
34. Wang XY, Zhang F, Zhang C et al (2020) The biomarkers for acute myocardial infarction and heart failure. Biomed Res Int 2018035. https://doi.org/10.1155/2020/2018035
35. Bodor G (2016) Biochemical markers of myocardial damage. eJIFCC 27(2):95–111.

36. Ananthan K, Lyon AR (2020) The role of biomarkers in cardio-oncology. J Cardiovasc Transl Res 13(3):431–450. https://doi.org/10.1007/s12265-020-10042-3
37. Herman EH, Zhang J, Lipshultz SE et al (1999) Correlation between serum levels of cardiac troponin-T and the severity of the chronic cardiomyopathy induced by doxorubicin. J Clin Oncol 17(7):2237–2243. https://doi.org/10.1200/JCO.1999.17.7.2237
38. Michel L, Mincu RI, Mahabadi AA et al (2020) Troponins and brain natriuretic peptides for the prediction of cardiotoxicity in cancer patients: a meta-analysis. Eur J Heart Fail 22(2): 350–361. https://doi.org/10.1002/ejhf.1631
39. Díaz-Antón B, Madurga R, Zorita B et al (2022) Early detection of anthracycline- and trastuzumab-induced cardiotoxicity: value and optimal timing of serum biomarkers and echocardiographic parameters. ESC Heart Fail 9(2):1127–1137. https://doi.org/10.1002/ehf2.13782
40. Sawaya H, Sebag IA, Plana JC et al (2012) Assessment of echocardiography and biomarkers for the extended prediction of cardiotoxicity in patients treated with anthracyclines, taxanes, and trastuzumab. Circ Cardiovasc Imaging 5(5):596–603. https://doi.org/10.1161/CIRCIMAGING.112.973321
41. Michel L, Rassaf T, Totzeck M (2018) Biomarkers for the detection of apparent and subclinical cancer therapy-related cardiotoxicity. J Thorac Dis 10(35):S4282–S4295. https://doi.org/10.21037/jtd.2018.08.15
42. Onitilo AA, Engel JM, Stankowski RV et al (2012) High-sensitivity C-reactive protein (hs-CRP) as a biomarker for trastuzumab-induced cardiotoxicity in HER2-positive early-stage breast cancer: a pilot study. Breast Cancer Res Treat 134(1):291–298. https://doi.org/10.1007/s10549-012-2039-z
43. Goel S, Simes RJ, Beith JM (2011) Exploratory analysis of cardiac biomarkers in women with normal cardiac function receiving trastuzumab for breast cancer. Asia Pac J Clin Oncol 7(3): 276–280. https://doi.org/10.1111/j.1743-7563.2011.01422.x
44. Wettersten N, Maisel A (2015) Role of cardiac troponin levels in acute heart failure. Card Fail Rev 1(2):102–106. https://doi.org/10.15420/cfr.2015.1.2.102
45. D'Amato A, Severino P, Prosperi S et al (2024) The role of high-sensitivity troponin T regarding prognosis and CV outcome across heart failure spectrum. J Clin Med 13(12): 3533. https://doi.org/10.3390/jcm13123533
46. Osredkar J, Bajrić A, Možina H et al (2024) Cardiac troponins I and T as biomarkers of cardiomyocyte injury—advantages and disadvantages of each. Appl Sci 14(14):6007. https://doi.org/10.3390/app14146007
47. Airaksinen KEJ, Tuominen T, Paana T et al (2024) Novel troponin fragmentation assay to discriminate between Takotsubo syndrome and acute myocardial infarction. Eur Heart J Acute Cardiovasc Care. https://doi.org/10.1093/ehjacc/zuae115
48. Couch LS, Garrard JW, Henry JA et al (2024) Comparison of troponin and natriuretic peptides in Takotsubo syndrome and acute coronary syndrome: a meta-analysis. Open Heart 11(1): e002607. https://doi.org/10.1136/openhrt-2024-002607
49. Daniels LB, Maisel AS (2007) Natriuretic peptides. J Am Coll Cardiol 50(25):2357–2368. https://doi.org/10.1016/j.jacc.2007.09.021
50. Rademaker MT, Charles CJ, Espiner EA et al (2004) Combined inhibition of angiotensin II and endothelin suppresses the brain natriuretic peptide response to developing heart failure. Clin Sci (Lond) 106(6):569–576. https://doi.org/10.1042/CS20030366
51. Cao Z, Jia Y, Zhu B (2019) BNP and NT-proBNP as diagnostic biomarkers for cardiac dysfunction in both clinical and forensic medicine. Int J Mol Sci 20(8):18–20. https://doi.org/10.3390/ijms20081820
52. Fu S, Ping P, Zhu Q et al (2018) Brain natriuretic peptide and its biochemical, analytical, and clinical issues in heart failure: a narrative review. Front Physiol 9:692. https://doi.org/10.3389/fphys.2018.00692
53. Magnussen C, Blankenberg S (2018) Biomarkers for heart failure: small molecules with high clinical relevance. J Intern Med 283(6):530–543. https://doi.org/10.1111/joim.12756

54. Bork NI, Nikolaev VO (2018) cGMP signaling in the CV system—the role of compartmentation and its live cell imaging. Int J Mol Sci 19(3):21–24. https://doi.org/10.3390/ijms19030801
55. Attanasio U, Di Sarro E, Tricarico L et al (2024) CV biomarkers in cardio-oncology: antineoplastic drug cardiotoxicity and beyond. Biomolecules 14(2):199. https://doi.org/10.3390/biom14020199
56. Zhang X, Sun Y, Zhang Y et al (2022) Cardiac biomarkers for the detection and management of cancer therapy-related CV toxicity. J Cardiovasc Dev Dis 9(11):372. https://doi.org/10.3390/jcdd9110372
57. Sandri MT, Salvatici M, Cardinale D et al (2005) N-terminal pro-B-type natriuretic peptide after high-dose chemotherapy: a marker predictive of cardiac dysfunction? Clin Chem 51(8): 1405–1410. https://doi.org/10.1373/clinchem.2005.050153
58. De Iuliis F, Salerno G, Taglieri L et al (2016) Serum biomarkers evaluation to predict chemotherapy-induced cardiotoxicity in breast cancer patients. Tumour Biol 37(3): 3379–3387. https://doi.org/10.1007/s13277-015-4183-7
59. Bouwer NI, Liesting C, Kofflard MJM et al (2019) NT-proBNP correlates with LVEF decline in HER2-positive breast cancer patients treated with trastuzumab. Cardiooncology 5:4. https://doi.org/10.1186/s40959-019-0039-4
60. Murtagh G, Januzzi JL, Scherrer-Crosbie M et al (2023) Circulating CV biomarkers in cancer therapeutics-related cardiotoxicity: review of critical challenges, solutions, and future directions. J Am Heart Assoc 12(21):e029574. https://doi.org/10.1161/JAHA.123.029574
61. Escudier M, Cautela J, Malissen N et al (2017) Clinical features, management, and outcomes of immune checkpoint inhibitor-related cardiotoxicity. Circulation 136(21):2085–2087. https://doi.org/10.1161/CIRCULATIONAHA.117.030571
62. Chen DY, Huang WK, Chien-Chia Wu V et al (2020) CV toxicity of immune checkpoint inhibitors in cancer patients: a review when cardiology meets immuno-oncology. J Formos Med Assoc 119(10):1461–1475. https://doi.org/10.1016/j.jfma.2019.07.025
63. López-Sendón J, Álvarez-Ortega C, Zamora Auñon P et al (2020) Classification, prevalence, and outcomes of anticancer therapy-induced cardiotoxicity: the CARDIOTOX registry. Eur Heart J 41(18):1720–1729. https://doi.org/10.1093/eurheartj/ehaa006
64. Ponde N, Bradbury I, Lambertini M et al (2017) Cardiac biomarkers for early detection and prediction of trastuzumab and/or lapatinib-induced cardiotoxicity in patients with HER2-positive early-stage breast cancer: a NeoALTTO sub-study (BIG 1-06). Breast Cancer Res Treat 168(3):631–638. https://doi.org/10.1007/s10549-017-4628-3
65. Bracun V, Aboumsallem JP, van der Meer P et al (2020) Cardiac biomarkers in patients with cancer: considerations, clinical implications, and future avenues. Curr Oncol Rep 22(7):67. https://doi.org/10.1007/s11912-020-00930-x
66. Srisawasdi P, Vanavanan S, Charoenpanichkit C et al (2010) The effect of renal dysfunction on BNP, NT-proBNP, and their ratio. Am J Clin Pathol 133(1):14–23. https://doi.org/10.1309/AJCP60HTPGIGFCNK
67. Brady PF, Chua W, Nehaj F et al (2022) Interactions between atrial fibrillation and natriuretic peptide in predicting heart failure hospitalization or cardiovascular death. J Am Heart Assoc 11(4):e022833. https://doi.org/10.1161/JAHA.121.022833
68. Rørth R, Jhund PS, Yilmaz MB et al (2020) Comparison of BNP and NT-proBNP in patients with heart failure and reduced ejection fraction. Circ Heart Fail 13(2):e006541. https://doi.org/10.1161/CIRCHEARTFAILURE.119.006541
69. Bayes-Genis A, Docherty KF, Petrie MC et al (2023) Practical algorithms for early diagnosis of heart failure and heart stress using NT-proBNP: a clinical consensus statement from the Heart Failure Association of the ESC. Eur J Heart Fail 25(11):1891–1898. https://doi.org/10.1002/ejhf.3036
70. Madamanchi C, Alhosaini H, Sumida A et al (2014) Obesity and natriuretic peptides, BNP and NT-proBNP: mechanisms and diagnostic implications for heart failure. Int J Cardiol 176(3): 611–617. https://doi.org/10.1016/j.ijcard.2014.08.007

71. Rezar R, Jirak P, Gschwandtner M et al (2020) Heart-type fatty acid-binding protein (H-FABP) and its role as a biomarker in heart failure: what do we know so far? J Clin Med 9(1):164. https://doi.org/10.3390/jcm9010164
72. Das UN (2015) Heart-type fatty acid-binding protein (H-FABP) and coronary heart disease. Indian Heart J 68(1):16–18. https://doi.org/10.1016/j.ihj.2015.07.030
73. Ye XD, He Y, Wang S et al (2018) Heart-type fatty acid binding protein (H-FABP) as a biomarker for acute myocardial injury and long-term post-ischemic prognosis. Acta Pharmacol Sin 39(7):1155–1163. https://doi.org/10.1038/aps.2018.37
74. Hasić S, Jadrić R, Cosović E et al (2011) Heart-type fatty acid-binding protein and its relation with morphological changes in rat myocardial damage model induced by isoproterenol. Bosn J Basic Med Sci 11(4):240–244. https://doi.org/10.17305/bjbms.2011.2557
75. ElGhandour AH, Sorady ME, Azab S et al (2009) Human heart-type fatty acid-binding protein as an early diagnostic marker of doxorubicin cardiac toxicity. Hematol Rev 1(1):e6. https://doi.org/10.4081/hr.2009.e6
76. Yuan M, Zang L, Xu A et al (2021) Dynamic changes of serum heart type-fatty acid binding protein in cancer patients treated with immune checkpoint inhibitors. Front Pharmacol 12: 748677. https://doi.org/10.3389/fphar.2021.748677
77. Codoñer-Franch P, Alonso-Iglesias E (2016) Homocysteine as a biomarker in vascular disease. In: Patel V, Preedy V (eds) Biomarkers in CV disease. Biomarkers in disease: methods, discoveries and applications. Springer, Dordrecht. https://doi.org/10.1007/978-94-007-7678-4_11
78. Liu B, Ma S, Wang T et al (2016) A novel rat model of heart failure induced by high methionine diet showing evidence of association between hyperhomocysteinemia and activation of NF-kappaB. Am J Transl Res 8(1):117–124
79. Zivkovic V, Jakovljevic V, Djordjevic D et al (2012) The effects of homocysteine-related compounds on cardiac contractility, coronary flow, and oxidative stress markers in isolated rat heart. Moll Cell Biochem 370(1–2):59–67. https://doi.org/10.1007/s11010-012-1398-4
80. Djuric D, Jakovljevic V, Zivkovic V et al (2018) Homocysteine and homocysteine-related compounds: an overview of the roles in the pathology of the CV and nervous systems. Can J Physiol Pharmacol 96(10):991–1003. https://doi.org/10.1139/cjpp-2018-0112
81. Shih CC, Shih YL, Chen JY (2021) The association between homocysteine levels and CV disease risk among middle-aged and elderly adults in Taiwan. BMC Cardiovasc Disord 21(1): 191. https://doi.org/10.1186/s12872-021-02000-x
82. Tan X, Tang F, Tian W et al (2024) Homocysteine metabolism, subclinical myocardial injury, and CV mortality in the general population. JACC Asia 4(8):609–620. https://doi.org/10.1016/j.jacasi.2024.05.005
83. Spence JD (2024) Homocysteine and myocardial injury. JACC Asia 4(8):621–623. https://doi.org/10.1016/j.jacasi.2024.06.003
84. Wang B, Mo X, Wu Z et al (2022) Systematic review and meta-analysis of the correlation between plasma homocysteine levels and coronary heart disease. J Thorac Dis 14(3):646–653. https://doi.org/10.21037/jtd-22-78
85. Ruiz JR, Sola R, Gonzalez-Gross M et al (2007) CV fitness is negatively associated with homocysteine levels in female adolescents. Arch Pediatr Adolesc Med 161(2):166–171. https://doi.org/10.1001/archpedi.161.2.166
86. Shliakhova N, Rak L, Muzhanovskiy V (2024) Homocysteine as a biomarker of physical activity and exercise tolerance in adolescent children. Phys Rehabil Rec Health Technol 9(4). https://doi.org/10.15391/prrht.2024-9(4).07
87. Wu LL, Wu JT (2002) Hyperhomocysteinemia is a risk factor for cancer and a new potential tumor marker. Clin Chim Acta 322(1–2):21–28. https://doi.org/10.1016/s0009-8981(02)00174-2
88. Hasan T, Arora R, Bansal AK et al (2019) Disturbed homocysteine metabolism is associated with cancer. Exp Mol Med 51(2):1–13. https://doi.org/10.1038/s12276-019-0216-4

89. Fan R, Wang Y, Zhang J et al (2023) Hyperhomocysteinaemia promotes doxorubicin-induced cardiotoxicity in mice. Pharmaceuticals 16(9):1212. https://doi.org/10.3390/ph16091212
90. Djuric MD, Todorovic D, Bajic Z et al (2024) Is homocysteine a biomarker of environmental health risk and epigenetic-DNA methylation: links to CV pathogenesis and B vitamins. In: Djuric DM, Agrawal DK (eds) Environmental factors in the pathogenesis of CV diseases. Advances in biochemistry in health and disease, vol 30. Springer, Cham. https://doi.org/10.1007/978-3-031-62806-1_15
91. Oeing CU, Pepin ME, Saul KB et al (2023) Indirect epigenetic testing identifies a diagnostic signature of cardiomyocyte DNA methylation in heart failure. Basic Res Cardiol 118(1):9. https://doi.org/10.1007/s00395-022-00954-3
92. Keller T, Tzikas S, Zeller T et al (2010) Copeptin improves early diagnosis of acute myocardial infarction. J Am Coll Cardiol 55(19):2096–2106. https://doi.org/10.1016/j.jacc.2010.01.029
93. Mu D, Cheng J, Qiu L et al (2022) Copeptin as a diagnostic and prognostic biomarker in CV diseases. Front Cardiovasc Med 9:901990. https://doi.org/10.3389/fcvm.2022.901990
94. Aydin S, Eren MN, Kuloglu T et al (2015) Alteration of serum and cardiac tissue adropin, copeptin, irisin and TRPM2 expressions in DOX treated male rats. Biotech Histochem 90(3): 197–205. https://doi.org/10.3109/10520295.2014.977949
95. Tonry C, Russell-Hallinan A, McCune C et al (2023) Circulating biomarkers for management of cancer therapeutics-related cardiac dysfunction. Cardiovasc Res 119(3):710–728. https://doi.org/10.1093/cvr/cvac087
96. De Winter RJ, Bholasingh R, Lijmer JG et al (1999) Independent prognostic value of C-reactive protein and troponin I in patients with unstable angina or non-Q-wave myocardial infarction. Cardiovasc Res 42(1):240–245. https://doi.org/10.1016/s0008-6363(99)00018-8
97. Sellmayer A, Limmert T, Hoffmann U (2003) High sensitivity C-reactive protein in CV risk assessment. CRP mania or useful screening? Int Angiol 22(1):15–23
98. Canada JM, Thomas GK, Trankle CR et al (2020) Increased C-reactive protein is associated with the severity of thoracic radiotherapy-induced cardiomyopathy. Cardiooncology 6:2. https://doi.org/10.1186/s40959-020-0058-1
99. Micheletti PL, Carla-da-Silva J, Scandolara TB et al (2021) Proinflammatory circulating markers: new players for evaluating asymptomatic acute CV toxicity in breast cancer treatment. J Chemother 33(2):106–115. https://doi.org/10.1080/1120009X.2021.1873632
100. Comini L, Pasini E, Bachetti T et al (2005) Acute haemodynamic effects of IL-6 treatment in vivo: involvement of vagus nerve in NO-mediated negative inotropism. Cytokine 30(5): 236–242. https://doi.org/10.1016/j.cyto.2005.01.009
101. Delyon J, Lebbe C (2022) IL-6 blockade in cancer patients treated with immune checkpoint blockade: a win-win strategy. Cancer Cell 40(5):450–451. https://doi.org/10.1016/j.ccell.2022.04.010
102. Stroud CR, Hegde A, Cherry C et al (2019) Tocilizumab for the management of immune mediated adverse events secondary to PD-1 blockade. J Oncol Pharm Pract 25(3):551–557. https://doi.org/10.1177/1078155217745144
103. Frangogiannis N (2015) Interleukin-1 in cardiac injury, repair, and remodeling: pathophysiologic and translational concepts. Discoveries 3(1):e41. https://doi.org/10.15190/d.2015.33
104. Oprescu N, Micheu MM, Scafa-Udriste A et al (2021) Inflammatory markers in acute myocardial infarction and the correlation with the severity of coronary heart disease. Ann Med 53(1):1040–1046. https://doi.org/10.1080/07853890.2021.1916070
105. Zhang H, Dhalla NS (2024) The role of pro-inflammatory cytokines in the pathogenesis of CV disease. Int J Mol Sci 25(2):1082. https://doi.org/10.3390/ijms25021082
106. Jain R, Talele N, Kumra H et al (2022) 1241 IL-1 blockade prevents cardiac toxicity and improves immunotherapy efficacy in mouse models of pancreatic cancer. J Immunother Cancer 10(2):1284–1284. https://doi.org/10.1136/jitc-2022-SITC2022.1241

107. Zhu J, Zhang J, Zhang L et al (2011) Interleukin-1 signaling mediates acute doxorubicin-induced cardiotoxicity. Biomed Pharmacother 65(7):481–485. https://doi.org/10.1016/j.biopha.2011.06.005
108. Tian M, Yuan Y-C, Li J-Y et al (2015) Tumor necrosis factor-α and its role as a mediator in myocardial infarction: a brief review. Chronic Dis Transl Med 1:18–26. https://doi.org/10.1016/j.cdtm.2015.02.002
109. Zoltani CK (2019) CV toxicity biomarkers. In: Gupta RC (ed) Biomarkers in toxicology, 2nd edn. Academic Press/Elsevier, London, pp 209–228
110. Ma Y, Zou H, Zhu XX et al (2017) Transforming growth factor β: a potential biomarker and therapeutic target of ventricular remodeling. Oncotarget 8(32):53780–53790. https://doi.org/10.18632/oncotarget.17255
111. Kempf T, Eden M, Strelau J et al (2006) The transforming growth factor-beta superfamily member growth-differentiation factor-15 protects the heart from ischemia/reperfusion injury. Circ Res 98(3):351–360. https://doi.org/10.1161/01.RES.0000202805.73038.48
112. Kempf T, Zarbock A, Widera C et al (2011) GDF-15 is an inhibitor of leukocyte integrin activation required for survival after myocardial infarction in mice. Nat Med 17:581–588. https://doi.org/10.1038/nm.2354
113. Li M, Duan L, Cai YL et al (2020) Growth differentiation factor-15 is associated with CV outcomes in patients with coronary artery disease. Cardiovasc Diabetol 19(1):120. https://doi.org/10.1186/s12933-020-01092-7
114. Stiermaier T, Adams V, Just M et al (2014) Growth differentiation factor-15 in Takotsubo cardiomyopathy: diagnostic and prognostic value. Int J Cardiol 173(3):424–429. https://doi.org/10.1016/j.ijcard.2014.03.014
115. Tian C, Zhang H, Liu J et al (2024) GDF-15 is a potential candidate biomarker for an elevated risk of cardiotoxicity in breast cancer patients receiving neoadjuvant dual anti-HER2 therapy. Front Pharmacol 15:1396133. https://doi.org/10.3389/fphar.2024.1396133
116. Ramachandra CJA, Ja KPMM, Chua J et al (2020) Myeloperoxidase as a multifaceted target for CV protection. Antioxid Redox Signal 32(15):1135–1149. https://doi.org/10.1089/ars.2019.7971
117. Ky B, Putt M, Sawaya H et al (2014) Early increases in multiple biomarkers predict subsequent cardiotoxicity in patients with breast cancer treated with doxorubicin, taxanes, and trastuzumab. J Am Coll Cardiol 63(8):809–816. https://doi.org/10.1016/j.jacc.2013.10.061
118. Nettersheim FS, Schlüter JD, Kreuzberg W et al (2023) Myeloperoxidase is a critical mediator of anthracycline-induced cardiomyopathy. Basic Res Cardiol 118(1):36. https://doi.org/10.1007/s00395-023-01006-0
119. Sörensen NA, Nikorowitsch J, Neumann JT et al (2019) Predictive value of soluble urokinase-type plasminogen activator receptor for mortality in patients with suspected myocardial infarction. Clin Res Cardiol 108(12):1386–1393. https://doi.org/10.1007/s00392-019-01475-1
120. Goodchild TT, Li Z, Lefer DJ (2022) Soluble urokinase plasminogen activator receptor: from biomarker to active participant in atherosclerosis and CV disease. J Clin Invest 132(24):e165868. https://doi.org/10.1172/JCI165868
121. Hodges GW, Bang CN, Wachtell K et al (2015) suPAR: a new biomarker for CV disease? Can J Cardiol 31(10):1293–1302. https://doi.org/10.1016/j.cjca.2015.03.023
122. Torino C, Pizzini P, Cutrupi S et al (2018) Soluble urokinase plasminogen activator receptor (suPAR) and all-cause and CV mortality in diverse hemodialysis patients. Kidney Int Rep 3(5):1100–1109. https://doi.org/10.1016/j.ekir.2018.05.004
123. Duan Y, Tang H, Mitchell-Silbaugh K et al (2020) Heat shock protein 60 in CV physiology and diseases. Front Mol Biosci 30(7):73. https://doi.org/10.3389/fmolb.2020.00073
124. Zilaee M, Ferns GA, Ghayour-Mobarhan M (2014) Heat shock proteins and CV disease. Adv Clin Chem 64:73–115. https://doi.org/10.1016/b978-0-12-800263-6.00002-1

125. Sklifasovskaya AP, Blagonravov M, Ryabinina A et al (2023) The role of heat shock proteins in the pathogenesis of heart failure (Review). Int J Mol Med 52(5):106. https://doi.org/10.3892/ijmm.2023.5309
126. Blagonravov ML, Sklifasovskaya AP, Korshunova AY et al (2020) Heat shock protein HSP60 in left ventricular cardiomyocytes of hypertensive rats with and without insulin-dependent diabetes mellitus. Bull Exp Biol Med 170(1):10–14. https://doi.org/10.1007/s10517-020-04994-4
127. Demirel O, Paar V, Tolnai P et al (2024) Heat shock protein 70 constitutes a promising novel biomarker in differential diagnosis between Takotsubo syndrome and non-ST-segment elevation myocardial infarction. J Clin Med 13(14):4152. https://doi.org/10.3390/jcm13144152
128. Blanda V, Bracale UM, Di Taranto MD et al (2020) Galectin-3 in CV diseases. Int J Mol Sci 21(23):9232. https://doi.org/10.3390/ijms21239232
129. Mitic VT, Stojanovic DR, Deljanin Ilic MZ et al (2020) Cardiac remodeling biomarkers as potential circulating markers of left ventricular hypertrophy in heart failure with preserved ejection fraction. Tohoku J Exp Med. 250:233–242. https://doi.org/10.1620/tjem.250.233
130. Shirzadi S, Borazjani R, Attar A (2021) Efficacy of novel cardiac biomarkers in detecting anthracycline-induced cardiac toxicity. Int Cardiovasc Res J 15(2):e116347
131. Al-Salam S, Kandhan K, Sudhadevi M et al (2022) Early doxorubicin myocardial injury: inflammatory, oxidative stress, and apoptotic role of galectin-3. Int J Mol Sci 23(20):12479. https://doi.org/10.3390/ijms232012479
132. Matsumura SI, Iwanaga S, Mochizuki S et al (2005) Targeted deletion or pharmacological inhibition of MMP-2 prevents cardiac rupture after myocardial infarction in mice. J Clin Invest 115(3):599–609. https://doi.org/10.1172/JCI22304
133. Gopcevic K, Rovcanin B, Kekic D et al (2011) Matrix metalloproteinases and membrane damage markers in sera of patients with acute myocardial infarction. Mol Cell Biochem 350(1–2):163–168. https://doi.org/10.1007/s11010-010-0694-0
134. Caballero EP, Santamaría MH, Corral RS (2018) Endogenous osteopontin induces myocardial CCL5 and MMP-2 activation that contributes to inflammation and cardiac remodeling in a mouse model of chronic Chagas heart disease. Biochim Biophys Acta Mol Basis Dis 1864(1): 11–23. https://doi.org/10.1016/j.bbadis.2017.10.006
135. Gonçalves PR, Nascimento LD, Gerlach RF et al (2022) Matrix metalloproteinase 2 as a pharmacological target in heart failure. Pharmaceuticals (Basel) 15(8):920. https://doi.org/10.3390/ph15080920
136. Cogni AL, Farah E, Minicucci MF et al (2013) Metalloproteinases-2 and -9 predict left ventricular remodeling after myocardial infarction. Arq Bras Cardiol 100(4):315–321. https://doi.org/10.5935/abc.20130049
137. Ip C, Luk KS, Yuen VLC et al (2021) International Health Informatics Study (IHIS) Network. Soluble suppression of tumorigenicity 2 (sST2) for predicting disease severity or mortality outcomes in CV diseases: a systematic review and meta-analysis. Int J Cardiol Heart Vasc 37: 100887. https://doi.org/10.1016/j.ijcha.2021.100887
138. Sciatti E, Merlo A, Scangiuzzi C et al (2023) Prognostic value of sST2 in heart failure. J Clin Med 12(12):3970. https://doi.org/10.3390/jcm12123970
139. Topf A, Mirna M, Bacher N et al (2021) Analysis of selected CV biomarkers in Takotsubo cardiomyopathy and the most frequent cardiomyopathies. Front Cardiovasc Med 8:700169. https://doi.org/10.3389/fcvm.2021.700169
140. Rawish E, Stiermaier T, Santoro F et al (2021) Current knowledge and future challenges in Takotsubo syndrome: part 1-pathophysiology and diagnosis. J Clin Med 10(3):479. https://doi.org/10.3390/jcm10030479
141. Cartas-Espinel I, Telechea-Fernández M, Manterola Delgado C et al (2022) Novel molecular biomarkers of cancer therapy-induced cardiotoxicity in adult population: a scoping review. ESC Heart Fail 9(3):1651–1665. https://doi.org/10.1002/ehf2.13735

142. Ho E, Karimi Galougahi K, Liu CC et al (2013) Biological markers of oxidative stress: applications to CV research and practice. Redox Biol 1(1):483–491. https://doi.org/10.1016/j.redox.2013.07.006
143. Panda P, Verma HK, Lakkakula S et al (2022) Biomarkers of oxidative stress tethered to CV diseases. Oxid Med Cell Longev 2022:9154295. https://doi.org/10.1155/2022/9154295
144. Ng ML, Ang X, Yap KY et al (2023) Novel oxidative stress biomarkers with risk prognosis values in heart failure. Biomedicines 11(3):917. https://doi.org/10.3390/biomedicines11030917
145. Noichri Y, Chalghoum A, Chkioua L et al (2013) Low erythrocyte catalase enzyme activity is correlated with high serum total homocysteine levels in tunisian patients with acute myocardial infarction. Diagn Pathol 8(1):1–7. https://doi.org/10.1186/1746-1596-8-68
146. Dai DF, Johnson SC, Villarin JJ et al (2011) Mitochondrial oxidative stress mediates angiotensin II-induced cardiac hypertrophy and gαq overexpression-induced heart failure. Circ Res 108(7):837–846. https://doi.org/10.1161/CIRCRESAHA.110.232306
147. Sverdlov AL, Elezaby A, Qin F et al (2016) Mitochondrial reactive oxygen species mediate cardiac structural, functional, and mitochondrial consequences of diet-induced metabolic heart disease. J Am Heart Assoc 5(1):1–13. https://doi.org/10.1161/JAHA.115.002555
148. Yin Y, Han W, Cao Y (2019) Association between activities of SOD, MDA and Na+-K+-ATPase in peripheral blood of patients with acute myocardial infarction and the complication of varying degrees of arrhythmia. Hell J Cardiol 60(6):366–371. https://doi.org/10.1016/j.hjc.2018.04.003
149. Dhalla NS, Elmoselhi AB, Hata T et al (2000) Status of myocardial antioxidants in ischemia-reperfusion injury. Cardiovasc Res 47(3):446–456. https://doi.org/10.1016/S0008-6363(00)00078-X
150. Uzelac JJ, Djukic T, Mutavdzin S et al (2020) The influence of subchronic co-application of vitamins B6 and folic acid on cardiac oxidative stress and biochemical markers in monocrotaline-induced heart failure in male Wistar albino rats. Can J Physiol Pharmacol 98(2):93–102. https://doi.org/10.1139/cjpp-2019-0305
151. Ilic A, Todorovic D, Mutavdzin S et al (2021) Translocator protein modulation by 4′--chlorodiazepam and no synthase inhibition affect cardiac oxidative stress, cardiometabolic and inflammatory markers in isoprenaline-induced rat myocardial infarction. Int J Mol Sci 22(6):1–16. https://doi.org/10.3390/ijms22062867
152. Bajic Z, Sobot T, Uletilovic S et al (2023) Cardioprotective effects of liraglutide pretreatment on isoprenaline-induced myocardial injury in rats. Can J Physiol Pharmacol 101(5):258–267. https://doi.org/10.1139/cjpp-2022-0534
153. Sobot T, Bajic Z, Skrbic R et al (2024) Effect of folic acid on isoprenaline-induced myocardial injury in rats. Physiol Int 111(1):80–96. https://doi.org/10.1556/2060.2023.00291
154. Ursini F, Maiorino M (2013) Glutathione peroxidases, 2nd edn. Elsevier Inc. https://doi.org/10.1016/B978-0-12-378630-2.00383-2
155. Nikolic-Heitzler V, Rabuzin F, Tatzber F et al (2006) Persistent oxidative stress after myocardial infarction treated by percutaneous coronary intervention. Tohoku J Exp Med 210(3):247–255. https://doi.org/10.1620/tjem.210.247
156. Weinbrenner T, Cladellas M, Isabel Covas M et al (2003) High oxidative stress in patients with stable coronary heart disease. Atherosclerosis 168(1):99–106. https://doi.org/10.1016/s0021-9150(03)00053-4
157. Sudha M, Rajkumar D, Felix JW (2013) Protective effect of glutathione against isoproterenol induced myocardial injury in rats. Indian J Physiol Pharmacol 57(2):132–137
158. Jain PG, Mahajan UB, Shinde SD et al (2018) Cardioprotective role of FA against isoproterenol induced cardiac toxicity. Mol Biol Rep 45(5):1357–1365. https://doi.org/10.1007/s11033-018-4297-2
159. Dmour BA, Costache AD, Dmour A et al (2023) Could endothelin-1 be a promising neurohormonal biomarker in acute heart failure? Diagnostics (Basel) 13(13):2277. https://doi.org/10.3390/diagnostics13132277

160. Banecki KMRM, Dora KA (2023) Endothelin-1 in health and disease. Int J Mol 24(14):11295. https://doi.org/10.3390/ijms241411295
161. Dhaun N, Webb DJ (2019) Endothelins in CV biology and therapeutics. Nat Rev Cardiol 16(8):491–502. https://doi.org/10.1038/s41569-019-0176-3
162. Harrison M, Zinovkin D, Pranjol MZI (2024) Endothelin-1 and its role in cancer and potential therapeutic opportunities. Biomedicines 12(3):511. https://doi.org/10.3390/biomedicines12030511
163. Krishnarao K, Bruno KA, Di Florio DN et al (2022) Upregulation of endothelin-1 may predict chemotherapy-induced cardiotoxicity in women with breast cancer. J Clin Med 11(12):3547. https://doi.org/10.3390/jcm11123547
164. Baidildinova G, Nagy M, Jurk K et al (2021) Soluble platelet release factors as biomarkers for CV disease. Front Cardiovasc Med 8:684920. https://doi.org/10.3389/fcvm.2021.684920
165. Golubeva MG (2022) Role of P-selectin in the development of hemostasis disorders in COVID-19. Biol Bull Rev 12(4):406–413. https://doi.org/10.1134/S207908642204003X
166. Ridker PM, Buring JE, Rifai N et al (2001) Soluble P-selectin and the risk of future CV events. Circulation 103(4):491–495. https://doi.org/10.1161/01.cir.103.4.491
167. Sommer P, Schreinlechner M, Noflatscher M et al (2023) Increasing soluble P-selectin levels predict higher peripheral atherosclerotic plaque progression. J Clin Med 12(20):6430. https://doi.org/10.3390/jcm12206430
168. Marschang P, Friedrich GJ, Ditlbacher H et al (2006) Reduction of soluble P-selectin by statins is inversely correlated with the progression of coronary artery disease. Int J Cardiol 106(2): 183–190. https://doi.org/10.1016/j.ijcard.2005.01.042
169. Bouwens E, van den Berg VJ, Akkerhuis KM et al (2020) Circulating biomarkers of cell adhesion predict clinical outcome in patients with chronic heart failure. J Clin Med 9(1):195. https://doi.org/10.3390/jcm9010195
170. Kaur R, Singh V, Kumari P et al (2022) Novel insights on the role of VCAM-1 and ICAM-1: potential biomarkers for CV diseases. Ann Med Surg (Lond) 84:104802. https://doi.org/10.1016/j.amsu.2022.104802
171. Lino DOC, Freitas IA, Meneses GC et al (2019) Interleukin-6 and adhesion molecules VCAM-1 and ICAM-1 as biomarkers of post-acute myocardial infarction heart failure. Braz J Med Biol Res 52(12):e8658. https://doi.org/10.1590/1414-431X20198658
172. Chakrala T, Prakash R, Valdes C et al (2023) Circulating biomarkers in coronary microvascular dysfunction. J Am Heart Assoc 12(12):e029341. https://doi.org/10.1161/JAHA.122.029341
173. Pospelova M, Krasnikova V, Fionik O et al (2022) Adhesion molecules ICAM-1 and PECAM-1 as potential biomarkers of central nervous system damage in women breast cancer survivors. Pathophysiology 29(1):52–65. https://doi.org/10.3390/pathophysiology29010006
174. Wang X, Zhao J, Zhang Y et al (2017) Kinetics of plasma von Willebrand factor in acute myocardial infarction patients: a meta-analysis. Oncotarget 8(52):90371–90379. https://doi.org/10.18632/oncotarget.20091
175. Brott D, Gould S, Jones H et al (2005) Biomarkers of drug-induced vascular injury. Toxicol Appl Pharmacol 207(2):441–445. https://doi.org/10.1016/j.taap.2005.04.028
176. Spiel AO, Gilbert JC, Jilma B (2008) von Willebrand factor in CV disease: focus on acute coronary syndromes. Circulation 117(11):1449–1459. https://doi.org/10.1161/CIRCULATIONAHA.107.722827
177. Sabbah SA, Elkattan MM, Labib DM et al (2024) Role of Von Willebrand factor level as a biomarker in acute ischemic stroke. Egypt J Neurol Psychiatry Neurosurg 60:63. https://doi.org/10.1186/s41983-024-00837-5
178. Nuver J, Smit AJ, van der Meer J et al (2005) Acute chemotherapy-induced CV changes in patients with testicular cancer. J Clin Oncol 23(36):9130–9137. https://doi.org/10.1200/JCO.2005.01.4092

179. Panova-Noeva M, Schulz A, Arnold N et al (2018) Coagulation and inflammation in long-term cancer survivors: results from the adult population. J Thromb Haemost 16(4):699–708. https://doi.org/10.1111/jth.13975
180. Mjelva OR, Svingen GFT, Pedersen EKR et al (2018) Fibrinogen and neopterin is associated with future myocardial infarction and total mortality in patients with stable coronary artery disease. Thromb Haemost 118(4):778–790. https://doi.org/10.1055/s-0038-1629912
181. Cui Z, Zhao G, Liu X (2022) Blood fibrinogen level as a biomarker of adverse outcomes in patients with coronary artery disease: a systematic review and meta-analysis. Medicine (Baltimore) 101(33):e30117. https://doi.org/10.1097/MD.0000000000030117
182. Zhou S, Jin J, Wang J et al (2018) miRNAS in CV diseases: potential biomarkers, therapeutic targets and challenges. Acta Pharmacol Sin 39(7):1073–1084. https://doi.org/10.1038/aps.2018.30
183. Çakmak HA, Demir M (2020) MicroRNA and CV diseases. Balkan Med J 37(2):60–71. https://doi.org/10.4274/balkanmedj.galenos.2020.2020.1.94
184. Wang J, Chen J, Sen S et al (2016) MicroRNA as biomarkers and diagnostics. J Cell Physiol 231(1):25–30. https://doi.org/10.1002/jcp.25056
185. Ji X, Takahashi R, Hiura Y, Hirokawa G, Fukushima Y, Iwai N (2009) Plasma miR-208 as a biomarker of myocardial injury. Clin Chem 55(11):1944–1949. https://doi.org/10.1373/clinchem.2009.125310
186. Matsumoto S, Sakata Y, Nakatani D et al (2012) A subset of circulating microRNAs are predictive for cardiac death after discharge for acute myocardial infarction. Biochem Biophys Res Commun 427(2):280–284. https://doi.org/10.1016/j.bbrc.2012.09.039
187. Jaguszewski M, Osipova J, Ghadri JR et al (2014) A signature of circulating microRNAs differentiates takotsubo cardiomyopathy from acute myocardial infarction. Eur Heart J 35(15):999–1006. https://doi.org/10.1093/eurheartj/eht392
188. Hussien R, Ashry S, Agwa S et al (2018) Plasma MiRNA-208b as a biomarker for detection of cardiotoxicity induced by acute CV drugs poisoning. Ain Shams J Forensic Med Clin Toxicol 31(2):15–22. https://doi.org/10.21608/AJFM.2018.15873

Chapter 33
Cardiotoxicity, Cardioprotection, and Prognosis in Survivors of Anticancer Treatment Undergoing Cardiac Surgery

Melchior Luiz Lima

Abstract In this chapter the cardiovascular complications arising from anticancer treatments and their implications for cardiac surgery are explored. Cancer therapies, including chemotherapy, radiation, and targeted treatments, have significantly improved survival rates but pose risks of cardiotoxicity, leading to conditions such as heart failure, ischemic heart disease, arrhythmias, and valvular dysfunction. The study highlights the epidemiology and mechanisms of cardiotoxicity, emphasizing the impact of anthracyclines, HER2-targeted therapies, and immune checkpoint inhibitors. These agents contribute to myocardial injury through oxidative stress, immune-mediated damage, and endothelial dysfunction. The manuscript also discusses risk stratification, early detection methods, and the role of advanced imaging and biomarkers in identifying subclinical cardiotoxicity before it progresses to irreversible damage. A critical focus is on cardioprotection strategies, including pharmacologic agents like beta-blockers, ACE inhibitors, and dexrazoxane, alongside lifestyle interventions and cardiac rehabilitation. The document also examines cardiac surgery indications in cancer survivors, outlining the challenges of managing patients with prior cardiotoxic exposure. Multidisciplinary collaboration among oncologists, cardiologists, and surgeons is emphasized to optimize surgical outcomes. The review underscores the importance of long-term survivorship care plans to mitigate late-onset cardiotoxicity and improve quality of life. Future perspectives include advances in personalized medicine, targeted therapies with reduced cardiotoxicity, and innovations in surgical techniques. The study advocates for integrated cardio-oncology pathways to enhance patient care, ensuring both effective cancer treatment and cardiovascular health preservation.

Keywords Cardiotoxicity · Cardioprotection · Cancer survivorship · Cardiac surgery · Onco-cardiology · Chemotherapy-induced heart failure · Radiation-induced cardiac damage · Multidisciplinary care

---

M. L. Lima (✉)
Centrocor—Heart Clinic Ltd. Alfeu Alves Pereira, 60 Enseada do Sua, 29050285, Vitória, ES, Brazil

D. K. Agrawal et al. (eds.), *Cardiovascular Toxicity*, Advances in Biochemistry in Health and Disease 33, https://doi.org/10.1007/978-3-031-90598-8_33

Introduction

Overview of Cardiotoxic Effects of Anticancer Therapies

Cancer is the second leading cause of death overall in the United States, and while anticancer treatment aims to improve quality of life and increase life expectancy, cardiovascular toxicity remains a possible complication of acute or chronic cancer therapies [1]. Advances in modern therapies have shifted the perception of cancer from a death sentence to a manageable condition, leading to decreased mortality rates across various cancer types [2]. However, therapies used in cancer treatment have demonstrated a wide spectrum of cardiotoxic effects, making it essential to understand the mechanisms linking oncological drugs to myocardial injury and heart failure [3, 4]. Among the most common causes of heart failure in cancer patients is ischemic heart disease, with oncological patients frequently presenting comorbidities such as advanced age, sedentary lifestyle, systemic arterial hypertension, diabetes mellitus, smoking, dyslipidemia, alcoholism, and coronary artery disease [2, 5, 6]. These comorbidities may worsen during anticancer therapy, as many antineoplastic drugs can damage myocardial tissue through mechanisms that lead to ischemic disease or irreversible cardiac dysfunction [4]. Manifestations of cardiotoxicity include endothelial dysfunction, systemic arterial hypertension, thromboembolism, coronary artery disease, myocardial ischemia, ventricular dysfunction, and arrhythmias, all of which necessitate strict medical supervision [2, 5–7]. The collaboration between cardiologists and oncologists is critical for early identification and management of cardiovascular side effects, enabling better patient outcomes and facilitating research in cardiotoxicity [3]. Chemotherapeutic agents, such as anthracyclines and HER2-targeted therapies, are well-documented for their potential to cause left ventricular dysfunction, heart failure, and other myocardial injuries, while radiation therapy targeting thoracic regions can result in coronary artery disease, valvular dysfunction, and pericardial complications [2, 4, 8]. Novel therapies like immunotherapies and tyrosine kinase inhibitors, though effective against malignancies, introduce additional cardiovascular risks, such as myocarditis, arrhythmias, and hypertension. These challenges are compounded by preexisting risk factors, highlighting the need for vigilant monitoring, early detection, and cardioprotective strategies to optimize outcomes and maintain the continuity of cancer treatment [2, 4, 6].

Epidemiology and Prevalence of Cardiotoxicity in Cancer Survivors

Cardiotoxicity represents a significant clinical concern among cancer survivors, with its prevalence varying depending on the type of cancer, treatment modalities, and patient-specific risk factors. As advancements in oncologic therapies have increased survival rates, a growing number of patients face long-term cardiovascular complications. Cardiotoxicity associated with fluoropyrimidines has a reported prevalence of approximately 1.2–8% among patients, with severe and life-threatening events being less common. Clinical manifestations typically include chest pain, dyspnea, and, in more severe cases, myocardial infarction. These symptoms often arise during or shortly after the infusion of 5-FU or the initiation of capecitabine therapy, highlighting the need for careful monitoring during treatment [2, 9]. The incidence of cardiotoxicity is particularly high among survivors treated with anthracyclines, HER2-targeted therapies, or thoracic radiation, with left ventricular dysfunction and heart failure being the most commonly reported conditions. Radiation therapy has been associated with late-onset cardiovascular events, including coronary artery disease, valvular dysfunction, and pericardial disease, often manifesting decades after treatment. Novel treatments, such as immune checkpoint inhibitors and tyrosine kinase inhibitors, have expanded the spectrum of cardiotoxic risks, introducing complications like myocarditis, arrhythmias, and hypertension [5, 6]. Epidemiological data indicate that cancer survivors with preexisting conditions such as diabetes, hypertension, or baseline cardiovascular disease are at an elevated risk of developing cardiotoxicity. Women and older adults are particularly vulnerable, reflecting the interplay between demographic factors and therapy-related cardiac effects. Given the increasing number of cancer survivors globally, understanding the epidemiology and prevalence of cardiotoxicity is critical to developing strategies for prevention, early detection, and management to improve long-term outcomes [2–4].

Anticancer Treatments, Cardiac Toxicity, and Cardioprotection as of Today

Anthracyclines are among the most effective and widely used chemotherapeutic agents in oncology but are also associated with significant cardiac toxicity, presenting a major challenge in cancer treatment today. These drugs are known to induce dose-dependent cardiotoxicity, primarily through mechanisms involving oxidative stress and iron-mediated free radical damage, which impair myocardial function and lead to left ventricular dysfunction or heart failure. Acute effects, such as arrhythmias, may occur during or shortly after administration, while chronic cardiotoxicity can manifest years later as progressive heart failure. Strategies for cardioprotection in patients receiving anthracyclines have gained critical importance. Pharmacologic agents like dexrazoxane, a cardioprotective iron-chelating agent, have been shown to mitigate myocardial injury without compromising anticancer efficacy. Additionally, beta-blockers and ACE inhibitors are often employed prophylactically to preserve cardiac function. Advances in diagnostic imaging techniques, such as global longitudinal strain (GLS) via echocardiography and the use of cardiac biomarkers like troponins and NT-proBNP, facilitate the early detection of subclinical cardiac dysfunction, allowing for timely intervention.

Multidisciplinary care, involving oncologists, cardiologists, and specialized cardio-oncology clinics, is essential to optimizing treatment outcomes while minimizing the cardiovascular risks associated with anthracyclines. Current research continues to focus on developing less cardiotoxic analogs and refining precision medicine approaches to ensure both effective cancer control and long-term cardiovascular health [4, 10].

Trastuzumab, a monoclonal antibody targeting the HER2 receptor, has revolutionized the treatment of HER2-positive breast cancer and other malignancies, significantly improving survival rates. However, its association with cardiac toxicity has raised substantial clinical concerns, particularly regarding its effects on left ventricular function. Unlike anthracyclines, trastuzumab-induced cardiotoxicity is not dose-dependent and is often reversible upon discontinuation of the drug. The primary mechanism involves inhibition of HER2 signaling in cardiac myocytes, a pathway critical for cellular repair and survival under stress, leading to impaired myocardial function and heart failure in some cases. Risk factors for trastuzumab-related cardiotoxicity include advanced age, prior exposure to anthracyclines, preexisting cardiovascular conditions, and concurrent use of other cardiotoxic agents. Cardioprotection strategies in trastuzumab-treated patients focus on baseline cardiovascular risk assessment and close monitoring throughout therapy. Echocardiography, particularly with global longitudinal strain (GLS) analysis, and biomarkers like troponins and NT-proBNP are pivotal for early detection of cardiac dysfunction. Prophylactic or early use of beta-blockers and ACE inhibitors has been shown to mitigate cardiac damage and improve outcomes. Additionally, adopting treatment schedules that minimize overlap with anthracyclines or employing alternate dosing regimens may reduce the risk of cardiotoxicity. Multidisciplinary management, involving cardiologists and oncologists, is critical in optimizing therapeutic strategies for these patients. Research into HER2-targeted therapies with reduced cardiac side effects and ongoing development of cardio-oncology programs underscores the growing emphasis on balancing efficacy with safety in cancer care [4, 11].

Immune checkpoint inhibitors (ICIs), including CTLA-4 blockers (e.g., ipilimumab) and PD1/PDL1 blockers (e.g., nivolumab, pembrolizumab), have emerged as transformative therapies in oncology, offering prolonged survival and even potential cures in various malignancies. These agents work by unleashing the immune system to attack cancer cells, but their immune-enhancing effects can also lead to immune-related adverse events (irAEs), including cardiac toxicity. Cardiotoxicity associated with ICIs, though rare, can be severe or even fatal, manifesting as myocarditis, pericarditis, arrhythmias, and vasculitis. Myocarditis is the most serious cardiac irAE, with an estimated incidence of 0.27% but a high mortality rate of up to 50%, especially when diagnosis and treatment are delayed [12]. The pathophysiology of ICI-associated cardiotoxicity involves aberrant immune activation, with T-cell infiltration and subsequent inflammation and damage of myocardial tissue. Risk factors include combination ICI therapy (e.g., CTLA-4 and PD1/PDL1 inhibitors), older age, and preexisting cardiovascular conditions. Early diagnosis is crucial, and it relies on a high index of suspicion alongside diagnostic tools such as cardiac biomarkers (elevated troponins and NT-proBNP), echocardiography, cardiac MRI, and in some cases, endomyocardial biopsy [12]. Management of ICI-induced cardiotoxicity involves immediate discontinuation of ICIs and initiation of high-dose corticosteroids to suppress the immune-mediated damage. For steroid-refractory cases, additional immunosuppressive agents like mycophenolate mofetil, infliximab, or antithymocyte globulin may be required. Supportive care, including heart failure management and antiarrhythmic therapy, is tailored to individual clinical needs. Preventive strategies, including baseline cardiovascular risk assessment and close monitoring during ICI therapy, are critical to mitigate risks. As the use of ICIs expands, research into biomarkers for risk stratification, prophylactic measures, and treatment algorithms will play a pivotal role in enhancing the safety profile of these powerful therapies. Multidisciplinary collaboration between oncologists, cardiologists, and immunologists is essential to optimize outcomes for patients receiving ICIs [4, 12].

Chimeric Antigen Receptor T-cell (CAR-T) therapy represents a revolutionary advancement in oncology, providing targeted and personalized treatment for hematologic malignancies such as B-cell lymphomas and acute lymphoblastic leukemia. Despite its efficacy, CAR-T therapy is associated with unique toxicities, including significant cardiovascular complications. The primary mechanisms of cardiac toxicity in CAR-T therapy are linked to cytokine release syndrome (CRS) and immune effector cell-associated neurotoxicity syndrome (ICANS). CRS, observed in a substantial proportion of patients, is characterized by an overwhelming release of pro-inflammatory cytokines, including interleukin-6 (IL-6), which can lead to systemic inflammation, hypotension, tachycardia, and, in severe cases, myocardial dysfunction and heart failure [13, 14]. The pathophysiological basis of cardiac toxicity involves direct myocardial injury due to cytokine storm-mediated systemic inflammation, endothelial dysfunction, and vascular permeability. Risk factors include preexisting cardiovascular disease, high tumor burden, and the severity of CRS. Early diagnosis and management are critical to mitigate these effects. Diagnostic tools such as echocardiography, cardiac biomarkers (troponins, NT-proBNP), and clinical monitoring of hemodynamic parameters are essential in identifying and managing cardiac complications [13, 14]. Management strategies for CAR-T-associated cardiotoxicity focus on controlling CRS severity. Tocilizumab, an IL-6 receptor antagonist, is the frontline treatment for CRS and has demonstrated efficacy in reversing systemic and cardiac inflammation. Corticosteroids are often used for severe or refractory cases. Supportive care, including vasopressors for hemodynamic instability and heart failure management protocols, may also be required. Preventive strategies involve comprehensive pre-treatment cardiovascular assessment to identify high-risk patients and close monitoring during therapy [13, 14]. Emerging approaches to enhance cardioprotection include the use of prophylactic anti-inflammatory agents, improved CAR-T cell engineering to modulate cytokine release, and personalized therapeutic protocols. Collaborative care involving oncologists, cardiologists, and intensive care specialists is essential to optimize outcomes in patients undergoing CAR-T therapy. Continued research is necessary to refine strategies for mitigating cardiotoxicity while maintaining the efficacy of this transformative cancer therapy [4, 13, 14].

Pathophysiological Mechanisms of Cardiac Damage

The pathophysiological mechanisms underlying cardiac damage caused by cancer therapies are multifaceted and vary depending on the specific agents used. Chemotherapeutic agents like anthracyclines induce cardiac injury primarily through oxidative stress, which leads to the production of reactive oxygen species (ROS), mitochondrial dysfunction, and apoptosis of cardiomyocytes. These agents also disrupt topoisomerase-IIβ, further amplifying DNA damage and cardiomyocyte death. HER2-targeted therapies, such as trastuzumab, interfere with the HER2/ErbB2 signaling pathway, which is essential for cardiomyocyte survival, resulting in reversible cardiac dysfunction in some patients. Radiation therapy exerts its cardiotoxic effects through endothelial damage, promoting inflammation, fibrosis, and atherosclerosis, which can lead to coronary artery disease, valvular dysfunction, and myocardial fibrosis. Immune checkpoint inhibitors have introduced immune-mediated mechanisms, causing myocarditis through T-cell activation and infiltration into myocardial tissue. Tyrosine kinase inhibitors can provoke hypertension and microvascular dysfunction by disrupting vascular endothelial growth factor (VEGF) signaling [5, 6]. Additionally, the cumulative impact of these therapies is often exacerbated by patient-specific factors such as preexisting cardiovascular disease, advanced age, and metabolic disorders. The interplay of these mechanisms highlights the need for targeted cardioprotective strategies and ongoing research to mitigate therapy-induced cardiac damage while maintaining oncologic efficacy [3, 15–17].

Radiotherapy

Radiotherapy involving the chest can accelerate coronary artery disease, increasing the risk of myocardial infarction. Radiotherapy is a cornerstone in cancer treatment, effectively targeting malignant cells [2, 18]. However, when the heart or surrounding vasculature is within the radiation field, there is a risk of cardiotoxicity, which can manifest in various forms:

Mechanisms of Radiation-Induced Cardiotoxicity

The mechanisms of radiation-induced cardiotoxicity include endothelial damage, where radiation injures the endothelial cells lining blood vessels, triggering inflammation, fibrosis, and atherosclerosis, which increases the risk of coronary artery disease and other cardiovascular complications [2, 18]. Additionally, direct exposure to radiation can cause myocardial fibrosis, reducing the elasticity of the heart muscle and leading to diastolic dysfunction, further impairing cardiac function [2, 18].

Clinical Manifestations

Clinical manifestations of radiation-induced cardiotoxicity include coronary artery disease (CAD), where endothelial damage accelerates atherosclerosis and increases the risk of myocardial infarction. Pericarditis, characterized by inflammation of the pericardium, may present as acute pericarditis or progress to chronic constrictive pericarditis. Valvular heart disease can occur due to radiation-induced fibrosis and calcification of heart valves, leading to stenosis or regurgitation. Additionally, damage to the heart's electrical conduction pathways may result in conduction system abnormalities, such as arrhythmias or heart block, further complicating cardiac function [2, 18].

Risk Factors

Risk factors for radiation-induced cardiotoxicity include the radiation dose and volume, with higher doses and larger volumes of irradiated cardiac tissue being associated with increased risk. Patient age is also significant, as younger individuals may experience a longer latency period before symptom onset, whereas older patients often have pre-existing cardiovascular risk factors that exacerbate the effects of radiation. Additionally, concurrent chemotherapy, particularly with agents such as anthracyclines, can amplify the cardiotoxic effects of radiotherapy, further increasing the likelihood of adverse cardiac outcomes [2, 18].

Prevention and Mitigation Strategies

Prevention and mitigation strategies for radiation-induced cardiotoxicity include the use of advanced radiotherapy techniques, such as intensity-modulated radiotherapy (IMRT) and proton therapy, which help minimize radiation exposure to the heart. The deep inspiration breath hold (DIBH) technique further reduces cardiac exposure by increasing the distance between the heart and the radiation field during treatment. Although not routinely used, cardioprotective agents like ACE inhibitors or beta-blockers have shown potential protective benefits during treatment, as suggested by some studies, offering an additional layer of cardiac protection [2, 18].

Monitoring and Follow-Up

Monitoring and follow-up for radiation-induced cardiotoxicity involve conducting a thorough baseline cardiovascular evaluation before initiating radiotherapy, especially for patients with existing risk factors. Long-term surveillance is essential, as cardiotoxic effects can manifest years after treatment. Ongoing monitoring of cardiac function enables the early detection and management of late-onset complications, ensuring better outcomes and long-term cardiac health [2, 18].

In summary, while radiotherapy remains a pivotal component of cancer treatment, awareness and proactive management of its potential cardiotoxic effects are essential to optimize patient outcomes and maintain cardiovascular health [18].

Hypertension

Vascular Endothelial Growth Factor (VEGF) Inhibitors

VEGF inhibitors, such as bevacizumab, are integral in cancer therapy due to their ability to impede tumor angiogenesis. However, a notable side effect associated with these agents is the development of hypertension, which can be severe if not managed appropriately [2, 6, 7]. The incidence of hypertension in patients receiving bevacizumab varies, with reports indicating that all-grade hypertension occurs in approximately 23.6% of patients, while high-grade hypertension (grade 3 or 4) is observed in about 7.9% [2, 6, 7]. The exact mechanism by which VEGF inhibitors induce hypertension is not fully understood. However, it is believed that inhibition of VEGF signaling leads to decreased production of nitric oxide and prostacyclin, both of which are vasodilators. This reduction results in vasoconstriction and increased peripheral resistance, thereby elevating blood pressure [2, 6, 7].

Management Strategies

Management strategies for patients undergoing VEGF inhibitor therapy begin with a baseline assessment to identify and control pre-existing hypertension, reducing the risk of severe hypertensive events during treatment. Regular monitoring of blood pressure is crucial, with guidelines recommending measurements every 2–3 weeks to detect any elevations promptly. If hypertension develops, antihypertensive therapy should be initiated, with common options including ACE inhibitors, ARBs, and calcium channel blockers, chosen based on patient-specific comorbidities and potential drug interactions. In cases of severe or uncontrolled hypertension, dose adjustment of the VEGF inhibitor or temporary discontinuation of therapy may be necessary until blood pressure is adequately controlled [2, 6, 7]. Unmanaged hypertension can lead to serious cardiovascular complications, including stroke, myocardial infarction, and heart failure. Therefore, proactive management of blood pressure in patients receiving VEGF inhibitors is essential to minimize these risks and ensure the safe continuation of cancer therapy [2, 6, 7].

In summary, while VEGF inhibitors like bevacizumab are effective in cancer treatment, they are associated with an increased risk of hypertension. Appropriate monitoring and management strategies are vital to mitigate this side effect and maintain optimal patient outcomes [2, 6].

Thromboembolic Events

Cancer therapies can significantly increase the risk of thromboembolic events, encompassing both venous thromboembolism (VTE) and arterial thrombosis. These complications may lead to serious conditions such as deep vein thrombosis (DVT), pulmonary embolism (PE), or stroke [2, 19].

Venous Thromboembolism (VTE)

Venous thromboembolism (VTE), comprising deep vein thrombosis (DVT) and pulmonary embolism (PE), is a common and serious complication among cancer patients, representing a significant cause of mortality within this population. The mechanisms underlying VTE involve activation of the coagulation system through various pathways, including the release of procoagulant factors by tumor cells and endothelial damage caused by chemotherapy or radiotherapy, which collectively enhance the risk of thrombotic events in these patients [2, 19].

Arterial Thrombosis

Arterial thrombotic events (ATEs) are increasingly recognized as significant complications of cancer and certain chemotherapies. Their development is influenced by the interplay between individual cardiovascular risk factors, specific malignancies, and selected therapies, which together contribute to the elevated risk of arterial thrombosis in this patient population [2, 19].

Risk Factors

Risk factors for thromboembolic events in cancer patients include the type and stage of cancer, with malignancies such as pancreatic and lung cancers being particularly associated with a higher risk. Treatment modalities, including surgery, chemotherapy, and radiotherapy, further contribute to the likelihood of venous thromboembolism (VTE) and arterial thrombotic events (ATE). Additionally, patient-related factors such as advanced age, immobility, and pre-existing comorbidities significantly elevate the risk, necessitating careful monitoring and preventive strategies [2].

Clinical Implications

The clinical implications of thromboembolic events in cancer patients include deep vein thrombosis (DVT), which involves the formation of clots in deep veins, typically in the legs, causing pain, swelling, and potentially life-threatening complications if the clot dislodges. Pulmonary embolism (PE) occurs when a clot travels to the lungs, leading to symptoms such as shortness of breath, chest pain, and, in severe cases, can be fatal. Additionally, arterial thrombosis in cerebral arteries can result in ischemic stroke, causing neurological deficits and significantly impacting the patient's quality of life [2, 19].

Management Strategies

Management strategies for thromboembolic events in cancer patients include prophylactic anticoagulation in selected high-risk individuals to prevent such events. Early detection through vigilant monitoring of signs and symptoms is essential for prompt diagnosis and intervention. Once a thromboembolic event is diagnosed, therapeutic anticoagulation should be initiated, with careful consideration of the balance between effectively treating the thrombosis and minimizing the risk of bleeding [2, 19].

Preventive Measures

Preventive measures for thromboembolic events in cancer patients include the use of risk assessment models, such as the Khorana score, to identify individuals at high risk for venous thromboembolism (VTE) and guide the implementation of prophylactic strategies. Lifestyle modifications, including promoting mobility, maintaining adequate hydration, and encouraging smoking cessation, further help reduce the risk of developing thromboembolic complications [2, 19].

In summary, thromboembolic events are significant complications in cancer patients undergoing treatment. A comprehensive approach, including risk assessment, preventive measures, and timely management, is essential to mitigate these risks and improve patient outcomes [2].

Pericardial Disease

Pericardial disease encompasses conditions affecting the pericardium, the protective sac surrounding the heart. In oncology, certain treatments can induce pericardial complications, notably pericarditis (inflammation of the pericardium) and pericardial effusion (accumulation of fluid within the pericardial sac). If unaddressed, these conditions may escalate to cardiac tamponade, a life-threatening scenario where pressure from the fluid impedes the heart's ability to pump effectively [2, 20].

Pericarditis

Pericarditis in cancer patients can result from treatments such as chemotherapy and radiotherapy. Chemotherapeutic agents may directly injure the pericardium, while radiation induces inflammation and fibrosis of pericardial tissues. Clinically, patients often present with sharp, pleuritic chest pain that worsens with deep breathing or lying down and improves when sitting up or leaning forward. Additional symptoms may include fever and a pericardial friction rub, which can be detected upon auscultation [2, 20].

Pericardial Effusion

Pericardial effusion in cancer patients can arise from malignant infiltration of the pericardium or as a side effect of cancer therapies, with fluid accumulation that may be gradual or rapid, potentially impairing cardiac function. Small effusions are often asymptomatic, whereas larger accumulations can lead to symptoms such as dyspnea, chest discomfort, and signs of hemodynamic compromise, requiring timely recognition and management [2, 20].

Progression to Cardiac Tamponade

Progression to cardiac tamponade occurs as the pericardial effusion volume increases, causing a rise in intrapericardial pressure that compresses cardiac chambers. This compression restricts ventricular filling, leading to reduced cardiac output and systemic hypotension, a life-threatening condition known as cardiac tamponade. Symptoms include severe hypotension, jugular venous distension, muffled heart sounds, and pulsus paradoxus—a marked drop in blood pressure during inspiration—requiring immediate medical intervention to prevent fatal outcomes [2, 20].

Management Strategies

Management strategies for pericardial complications include anti-inflammatory therapy for pericarditis, typically using NSAIDs and colchicine to reduce inflammation and alleviate symptoms, with corticosteroids reserved for refractory cases or when NSAIDs are contraindicated. For pericardial effusion and cardiac tamponade, pericardiocentesis is performed to aspirate pericardial fluid, providing symptomatic relief and preventing tamponade progression. In cases of recurrent effusions, a pericardial window may be surgically created to allow continuous drainage. Addressing the underlying causes, such as effective cancer treatment, can help reduce malignant pericardial effusions [2, 21].

Monitoring and Prevention

Monitoring and prevention of pericardial involvement in cancer patients emphasize the importance of regular cardiac evaluations, including baseline and periodic echocardiography, to enable early detection. Additionally, educating patients to promptly report symptoms such as chest pain, dyspnea, or unexplained fatigue is crucial for facilitating timely intervention and reducing complications [2, 15, 20].

In summary, pericardial diseases like pericarditis and pericardial effusion are potential complications of cancer treatments. Timely recognition and appropriate management are crucial to prevent progression to cardiac tamponade and to ensure optimal patient outcomes [2].

Valvular Heart Disease

Radiotherapy can lead to valvular fibrosis and calcification, resulting in valvular stenosis or regurgitation over time. Valvular Heart Disease (VHD) is a recognized long-term complication of thoracic radiotherapy, particularly in patients treated for cancers such as Hodgkin lymphoma and breast cancer. Radiation exposure can lead to valvular fibrosis and calcification, resulting in valvular stenosis or regurgitation over time [2, 22].

Mechanisms of Radiation-Induced Valvular Disease

Radiation-induced valvular disease arises through mechanisms such as fibrosis and calcification, where radiation damages valvular endothelial cells, triggering inflammation that progresses to fibrosis and calcification, predominantly affecting the aortic and mitral valves. Clinically significant valvular disease usually manifests after a latency period of 10–20 years post-radiation exposure, highlighting the necessity for long-term surveillance of affected individuals [2, 22].

Clinical Manifestations

Clinical manifestations of radiation-induced valvular disease include valvular stenosis and regurgitation. Valvular stenosis, caused by thickening and calcification of valve leaflets, impedes blood flow and can lead to symptoms such as chest pain, syncope, and heart failure, commonly seen in aortic stenosis. Valvular regurgitation occurs when fibrotic changes prevent proper valve closure, resulting in blood backflow, volume overload of heart chambers, and subsequent chamber dilation and dysfunction [2, 22].

Risk Factors

Risk factors for radiation-induced valvular disease include the radiation dose and field, with higher doses and larger treatment areas involving cardiac structures significantly increasing the risk of valvular damage. Additionally, the concurrent use of certain chemotherapeutic agents during radiotherapy can exacerbate cardiac toxicity, further elevating the risk of developing valvular dysfunction [2, 22].

Diagnosis and Monitoring

Diagnosis and monitoring of radiation-induced valvular disease involve regular echocardiographic screenings to detect early valvular abnormalities in patients with a history of thoracic radiation. Echocardiography is critical for evaluating valve morphology and function. In more complex cases, multimodality imaging, such as cardiac MRI or CT, may be utilized to provide detailed anatomical information, particularly when planning for potential surgical interventions [2, 22].

Management

Management of radiation-induced valvular disease includes medical therapy to alleviate symptoms, although it does not stop disease progression, emphasizing the importance of regular monitoring and timely intervention. In severe cases, surgical interventions such as valve repair or replacement may be required. However, prior radiation exposure presents challenges, including tissue fibrosis and increased surgical risks. A thorough evaluation by a multidisciplinary team is essential to determine the most appropriate therapeutic approach [2, 22].

Preventive Strategies

Preventive strategies for radiation-induced valvular damage include the use of advanced radiation techniques, such as intensity-modulated radiotherapy (IMRT), to minimize cardiac exposure and reduce the risk of valvular injury. Additionally, managing cardiovascular risk factors, such as hypertension and hyperlipidemia, is crucial to lowering the overall risk of cardiovascular complications in patients undergoing thoracic radiotherapy. These measures help improve long-term outcomes and preserve cardiac health [2, 22].

In summary, radiation-induced valvular heart disease is a significant concern for cancer survivors who have received thoracic radiotherapy. Awareness of this potential complication, along with regular monitoring and appropriate management, is essential to optimize long-term cardiovascular health in these individuals [22].

Myocarditis

Immune checkpoint inhibitors used in immunotherapy have been associated with immune-mediated myocarditis, a potentially life-threatening condition. Myocarditis is an inflammatory condition of the heart muscle that can compromise cardiac function. In oncology, the use of immune checkpoint inhibitors (ICIs)—such as those targeting PD-1, PD-L1, and CTLA-4—has been associated with immune-mediated myocarditis, a potentially life-threatening adverse event [2, 23].

Incidence and Onset

Although relatively rare, ICI-associated myocarditis carries a high mortality rate, with estimates suggesting it occurs in approximately 1.1% of patients receiving these therapies. The median time to onset is around 30 days after initiation of treatment [23].

Pathophysiology

ICIs function by inhibiting regulatory pathways that restrain the immune system, thereby enhancing anti-tumor immunity. However, this immune activation can become dysregulated, leading to off-target effects where the immune system attacks healthy tissues, including the myocardium. The exact mechanisms remain under investigation, but T-cell–mediated cytotoxicity is believed to play a central role [2, 23].

Clinical Presentation

Patients may present with a range of symptoms, from mild fatigue and chest discomfort to severe manifestations such as heart failure, arrhythmias, or cardiogenic shock. Due to the potential for rapid progression, a high index of suspicion is necessary, especially in patients recently started on ICI therapy [2, 23].

Diagnosis

The diagnosis of myocarditis involves multiple approaches. Biomarkers such as elevated cardiac troponins are commonly observed and indicate myocardial injury, though they lack specificity and must be interpreted within the clinical context [24]. Imaging techniques like cardiac MRI are valuable for detecting myocardial inflammation and fibrosis, while echocardiography may show ventricular dysfunction but can remain normal in early stages [25]. Endomyocardial biopsy, considered the gold standard for diagnosis, can confirm myocarditis through identification of inflammatory infiltrates in the myocardium; however, it carries the risk of sampling error, and a negative result does not definitively exclude the condition [2, 25].

Management

The management of myocarditis, particularly immune checkpoint inhibitor (ICI)-associated cases, involves several critical steps. Immediate discontinuation of ICI therapy is essential upon suspicion to prevent further immune-mediated myocardial damage. Immunosuppressive treatment with high-dose corticosteroids is the cornerstone of therapy to suppress inflammation. For refractory cases, additional immunosuppressive agents like mycophenolate mofetil or infliximab may be utilized. Supportive care is tailored to the patient's clinical presentation, addressing symptoms such as heart failure, arrhythmias, and other related complications to optimize outcomes [2, 26].

Prognosis

The prognosis of immune checkpoint inhibitor (ICI)-associated myocarditis remains challenging, with reported mortality rates reaching up to 50% in some studies. Early recognition and prompt initiation of appropriate treatment are vital to improving patient outcomes and reducing the risk of severe complications or death [2, 27].

Pulmonary Hypertension

Pulmonary Hypertension (PH) is a condition characterized by elevated blood pressure in the pulmonary arteries, which can progress to right ventricular failure if left untreated. Certain cancer therapies have been associated with inducing pulmonary arterial hypertension (PAH), emphasizing the need for careful monitoring and management in oncology practice to mitigate potential cardiovascular complications [2, 28].

Mechanisms of Drug-Induced PAH

Mechanisms of drug-induced PAH involve endothelial dysfunction, where some anticancer agents can damage the endothelial cells lining the pulmonary vasculature, leading to vasoconstriction and increased pulmonary vascular resistance. Another mechanism is vascular remodeling, as chemotherapeutic drugs may induce structural changes in the pulmonary arteries, including smooth muscle proliferation and fibrosis, which contribute to increased arterial pressure [2, 28].

Notable Agents Associated with PAH

Notable agents associated with PAH include tyrosine kinase inhibitors such as dasatinib, used in the treatment of chronic myeloid leukemia, which have been linked to cases of PAH. Although the incidence is relatively low, it is significant enough to warrant monitoring during therapy. Additionally, chemotherapeutic agents like bleomycin and carmustine have been associated with pulmonary toxicity, including the development of PAH [2].

Clinical Presentation

Clinical presentation includes symptoms such as dyspnea, fatigue, chest pain, and syncope. These symptoms are often nonspecific and can be mistaken for other cardiopulmonary conditions. If unrecognized and untreated, drug-induced PAH can progress to right ventricular hypertrophy and eventual right heart failure, characterized by peripheral edema, ascites, and jugular venous distension [2, 28].

Diagnostic Evaluation

Diagnostic evaluation involves echocardiography, which serves as a non-invasive initial assessment tool to estimate pulmonary artery pressures and evaluate right ventricular function. Right heart catheterization, considered the gold standard for diagnosing PAH, provides direct measurement of pulmonary hemodynamics [2, 28].

Management Strategies

Management strategies include discontinuation of the offending agent, as ceasing the causative cancer therapy can lead to improvement in pulmonary pressures and clinical status. In some cases, pulmonary hypertension may be reversible, and more than half of the patients can be weaned off vasodilator therapy. Pharmacotherapy involving the use of pulmonary vasodilators, such as endothelin receptor antagonists or phosphodiesterase-5 inhibitors, may be considered to manage symptoms and hemodynamics. Regular follow-up with echocardiography and clinical assessments is essential to monitor disease progression and response to therapy [2, 28].

Preventive Measures

Preventive measures include conducting a baseline assessment prior to initiating potentially causative therapies, involving a thorough cardiovascular evaluation to identify pre-existing conditions that may predispose patients to pulmonary arterial hypertension. Patient education is also crucial, involving informing patients about the potential risks and encouraging prompt reporting of cardiopulmonary symptoms to facilitate early detection and intervention [2, 28].

In summary, while cancer therapies are essential for treating malignancies, they can sometimes lead to complications such as pulmonary arterial hypertension. Early recognition and appropriate management of this condition are crucial to prevent progression to right ventricular failure and to optimize patient outcomes [2].

Peripheral Vascular Disease

Peripheral Vascular Disease (PVD), particularly Peripheral Artery Disease (PAD), involves the narrowing or blockage of peripheral arteries, most commonly in the legs, leading to reduced blood flow. This condition can manifest as claudication—pain or cramping in the legs during physical activity—or progress to critical limb ischemia (CLI), characterized by severe obstruction of blood flow, leading to pain at rest, non-healing wounds, or gangrene [2, 28].

Impact of Cancer Treatments on Peripheral Arteries

Certain cancer therapies have been associated with an increased risk of developing peripheral arterial disease. Radiation therapy can cause damage to blood vessels, leading to inflammation and fibrosis, which may result in arterial narrowing or occlusion over time. Additionally, some chemotherapeutic agents can induce vascular toxicity, contributing to endothelial damage and accelerating atherosclerosis, thereby increasing the risk of peripheral arterial disease [2, 28].

Clinical Manifestations

Patients with peripheral arterial disease may experience muscle pain or cramping in the legs or arms that is triggered by activity, such as walking, and relieved after a few minutes of rest. This condition, known as claudication, occurs due to insufficient blood flow to meet the increased demands during exercise. As the disease progresses, patients may develop critical limb ischemia, characterized by ischemic rest pain, non-healing ulcers, or gangrene, indicating severe reduction in blood flow. This serious condition requires immediate medical attention to prevent limb loss [2, 28].

Management Strategies

Management strategies for peripheral arterial disease include encouraging lifestyle modifications such as smoking cessation, regular physical activity, and a healthy diet to help manage symptoms and slow disease progression. Pharmacotherapy with medications like antiplatelet agents and statins is recommended for all patients with PAD to reduce cardiovascular risk. In severe cases of claudication or critical limb ischemia, revascularization procedures such as angioplasty or bypass surgery may be necessary to restore adequate blood flow to the affected limbs [2, 28].

Preventive Measures

Preventive measures for peripheral arterial disease include regular monitoring for patients undergoing cancer treatments known to affect vascular health, ensuring early detection of PAD. Additionally, managing risk factors such as hypertension, diabetes, and hyperlipidemia is crucial to reducing the risk, particularly in patients receiving potentially vasculotoxic cancer therapies [2, 28].

In summary, while cancer treatments are essential for managing malignancies, they may contribute to the development of peripheral artery disease. Early recognition and comprehensive management of PAD are vital to prevent progression to critical limb ischemia and to improve patient outcomes [28].

Cardioprotection Strategies During Anticancer Therapy

Pharmacologic Agents for Cardioprotection

Cardioprotection strategies during anticancer therapy aim to mitigate the adverse cardiovascular effects associated with oncologic treatments, ensuring that patients can safely complete their therapeutic regimens. Pharmacologic agents form the cornerstone of these strategies, targeting specific pathways implicated in cardiotoxicity. Beta-blockers, such as carvedilol and metoprolol, reduce myocardial oxygen demand, prevent sympathetic overactivation, and mitigate left ventricular dysfunction caused by agents like anthracyclines. Angiotensin-converting enzyme (ACE) inhibitors and angiotensin II receptor blockers (ARBs), including enalapril and losartan, are effective in preventing maladaptive cardiac remodeling and preserving myocardial function. Statins, with their anti-inflammatory and antioxidant properties, enhance endothelial health and reduce myocardial stress, particularly in patients at risk for anthracycline or HER2-targeted therapy-related cardiotoxicity. Dexrazoxane, a potent iron chelator, is FDA-approved for patients receiving high-dose anthracyclines, offering robust protection by limiting oxidative stress and free radical formation. Emerging agents, such as neprilysin inhibitors like sacubitril-valsartan, have shown promise in promoting vasodilation and reducing cardiac stress in patients undergoing cardiotoxic treatments. These pharmacologic interventions, combined with routine monitoring and tailored patient care, are pivotal in optimizing cardiovascular outcomes during cancer therapy [4, 29, 30].

Role of Lifestyle Interventions and Cardiac Rehabilitation

Cardioprotection strategies during anticancer therapy encompass both pharmacologic and non-pharmacologic interventions aimed at minimizing cardiovascular risks while allowing patients to continue potentially life-saving cancer treatments. Pharmacologic measures include the use of beta-blockers, ACE inhibitors, ARBs, statins, and dexrazoxane to mitigate myocardial stress, prevent cardiac remodeling, and reduce oxidative damage. Non-pharmacologic strategies emphasize the critical role of lifestyle interventions and cardiac rehabilitation in optimizing cardiovascular health. A heart-healthy diet rich in fruits, vegetables, whole grains, and omega-3 fatty acids supports endothelial function and reduces inflammation, while regular physical activity, including aerobic and resistance training, improves cardiac fitness and reduces fatigue. Smoking cessation and alcohol moderation further mitigate risks, while stress management techniques like mindfulness and cognitive behavioral therapy address psychological well-being. Cardiac rehabilitation programs, integrating supervised exercise regimens, dietary counseling, and multidisciplinary care, provide a structured approach to recovery and prevention, ensuring long-term cardiovascular health for cancer patients and survivors. By combining these strategies, healthcare teams can enhance the overall prognosis and quality of life for individuals undergoing anticancer therapies [4, 29, 30].

Emerging Cardioprotective Therapies

Cardioprotection strategies during anticancer therapy involve both established pharmacologic interventions and emerging therapies aimed at mitigating cardiovascular risks without compromising the efficacy of cancer treatments. Established strategies include the use of beta-blockers, ACE inhibitors, ARBs, and statins to prevent myocardial damage, reduce oxidative stress, and maintain cardiac function during chemotherapy. Dexrazoxane, a chelating agent, is particularly effective in reducing anthracycline-induced cardiotoxicity by mitigating free radical formation. Concurrently, non-pharmacologic interventions such as lifestyle modifications and cardiac rehabilitation focus on optimizing cardiovascular health through diet, physical activity, and risk factor management [4, 29, 30].

Emerging cardioprotective therapies offer promising advancements in the field. Sodium-glucose co-transporter 2 (SGLT2) inhibitors, originally developed for diabetes, have shown potential for reducing myocardial stress and preventing heart failure in cancer patients. Novel antioxidants targeting mitochondrial reactive oxygen species (ROS) are under investigation for their ability to mitigate oxidative damage associated with chemotherapy. Additionally, neprilysin inhibitors, which enhance the effects of natriuretic peptides, are being explored for their role in reducing cardiac stress and improving ventricular function. The integration of these innovative therapies into existing cardioprotective frameworks could significantly enhance the cardiovascular safety profile of anticancer treatments, ensuring better outcomes for cancer patients [4, 29, 30].

Clinical Presentation and Diagnosis of Cardiotoxicity

Symptoms and Clinical Manifestations

The clinical presentation and diagnosis of cardiotoxicity in patients undergoing anticancer therapy require vigilance due to the broad spectrum of symptoms and manifestations. Symptoms often vary depending on the severity and underlying mechanism of cardiac injury but can range from subtle signs of dysfunction to life-threatening complications. Patients may present with symptoms such as fatigue, dyspnea, palpitations, and chest pain. Advanced cardiotoxicity often manifests as heart failure, characterized by fluid retention, peripheral edema, and orthopnea, or as arrhythmias, with symptoms including dizziness, syncope, or palpitations. In some cases, ischemic symptoms such as angina may indicate myocardial injury, particularly in patients receiving therapies like fluoropyrimidines or tyrosine kinase inhibitors. Pericardial involvement may lead to pericarditis or pericardial effusion, presenting with pleuritic chest pain and hemodynamic compromise [31, 32].

The diagnosis of cardiotoxicity is often confirmed through a combination of clinical assessment and diagnostic tools. Echocardiography remains the cornerstone of cardiac imaging, offering insights into left ventricular ejection fraction (LVEF), global longitudinal strain (GLS), and diastolic function. Cardiac MRI provides additional detail for detecting fibrosis or inflammation, particularly in cases of myocarditis. Biomarkers such as cardiac troponins and NT-proBNP are essential for detecting myocardial injury or stress, often preceding functional declines. Electrocardiography (ECG) is useful for identifying arrhythmias and conduction abnormalities. For definitive diagnosis, right heart catheterization or endomyocardial biopsy may be required in specific cases. Early identification of symptoms and timely diagnostic evaluation are critical to optimizing outcomes and guiding cardioprotective interventions in cancer patients [31, 32].

Diagnostic Tools: Imaging Modalities and Biomarkers

The clinical presentation of cardiotoxicity in patients undergoing anticancer therapy often varies based on the specific type of cardiac injury and its severity. Common symptoms include fatigue, dyspnea, chest pain, palpitations, and in advanced cases, signs of heart failure such as peripheral edema, orthopnea, and weight gain due to fluid retention. Arrhythmias may manifest as dizziness, syncope, or irregular heartbeats, while ischemic symptoms, such as angina, may indicate myocardial injury. Pericardial complications, including pericarditis or pericardial effusion, present with pleuritic chest pain, friction rub, or hemodynamic instability in severe cases. Prompt recognition of these symptoms is crucial to mitigate further cardiac damage and optimize patient outcomes [33, 34].

The diagnosis of cardiotoxicity employs a combination of clinical evaluation and advanced diagnostic tools. Imaging modalities play a pivotal role in assessing cardiac function and detecting early signs of injury. Echocardiography, the cornerstone of cardiac imaging, measures left ventricular ejection fraction (LVEF) and assesses diastolic function, while techniques like global longitudinal strain (GLS) can identify subclinical myocardial changes before significant LVEF decline. Cardiac MRI offers detailed visualization of myocardial fibrosis, inflammation, or edema, particularly useful in cases of myocarditis or advanced structural abnormalities. Nuclear imaging evaluates myocardial perfusion and function, aiding in the detection of ischemic changes induced by therapy [33, 34].

Biomarkers provide valuable insights into myocardial injury and stress. Cardiac troponins are highly sensitive indicators of acute myocardial injury, often rising before the onset of symptoms or functional decline. NT-proBNP or BNP levels reflect increased ventricular wall stress, serving as markers for heart failure or early cardiac dysfunction. Combining these biomarkers with imaging findings enhances diagnostic accuracy and facilitates early intervention. Electrocardiography (ECG) aids in identifying arrhythmias, conduction disturbances, or QT prolongation, which are potential side effects of certain cancer therapies. Utilizing these diagnostic tools collectively ensures timely detection, enabling personalized management strategies to mitigate the adverse cardiovascular effects of cancer treatment [33, 34].

Risk Stratification Methods for Early Detection

The clinical presentation of cardiotoxicity in cancer patients undergoing treatment is diverse and may range from asymptomatic subclinical findings to severe manifestations such as heart failure, arrhythmias, or ischemic heart disease. Common symptoms include fatigue, dyspnea, chest pain, and palpitations. In severe cases, patients may experience signs of heart failure, such as peripheral edema, orthopnea, or pulmonary congestion. Arrhythmias can present as syncope or dizziness, while ischemic symptoms may indicate myocardial injury or infarction. Early recognition of these signs is critical to prevent progression and optimize patient outcomes [35, 36].

Risk stratification methods for early detection of cardiotoxicity play a vital role in tailoring patient management. Baseline cardiovascular assessments, including a detailed medical history, physical examination, and imaging studies like echocardiography with global longitudinal strain (GLS), are essential to identify at-risk individuals. Biomarkers, such as cardiac troponins and NT-proBNP, provide sensitive indicators of myocardial injury and stress, even before clinical symptoms arise. Risk stratification tools, including the Revised Cardiac Risk Index (RCRI) and algorithms incorporating clinical, imaging, and biomarker data, help classify patients into low, moderate, or high-risk categories. High-risk patients, such as those with pre-existing cardiovascular disease or receiving cardiotoxic agents like anthracyclines or HER2-targeted therapies, require close monitoring with periodic imaging and biomarker assessments. This proactive approach ensures timely detection and intervention, reducing the likelihood of irreversible cardiac damage and supporting the continuity of cancer therapy [35, 36].

Long-Term Cardiovascular Risks in Cancer Survivors

Late-Onset Cardiotoxicity: Pathophysiology and Clinical Implications

Long-term cardiovascular risks in cancer survivors, particularly those who have undergone cardiotoxic therapies, remain a critical concern in oncology and cardiology. Survivors are at an elevated risk of developing conditions such as heart failure, ischemic heart disease, arrhythmias, and valvular dysfunction. These risks are influenced by factors such as the type of cancer therapy, pre-existing comorbidities, and individual patient susceptibility. Late-onset cardiotoxicity, a hallmark concern among cancer survivors, often emerges years or even decades after completing treatment. The pathophysiology involves cumulative myocardial injury resulting from therapies such as anthracyclines, HER2-targeted agents, and thoracic radiation. These treatments can induce oxidative stress, endothelial dysfunction, and myocardial fibrosis, leading to progressive ventricular dysfunction and a heightened risk of coronary artery disease. Radiation therapy, in particular, is associated with delayed cardiovascular effects, including accelerated atherosclerosis, pericardial disease, and valvular calcification. Clinically, late-onset cardiotoxicity often presents as declining cardiac function or new-onset symptoms of heart failure, arrhythmias, or ischemia. Its implications are profound, as it necessitates ongoing surveillance and proactive management to detect and address cardiovascular complications in cancer survivors. Tailored cardioprotective strategies, long-term monitoring, and multidisciplinary collaboration are essential to mitigate these risks and optimize long-term health outcomes for this growing patient population [30, 35, 37].

Progression to Heart Failure and Other Cardiovascular Complications

Long-term cardiovascular risks in cancer survivors are a significant concern due to the progression of treatment-induced cardiac damage, which can lead to heart failure and other complications. Cancer therapies such as anthracyclines, HER2-targeted agents, and thoracic radiation can initiate structural and functional changes in the myocardium, including ventricular remodeling, fibrosis, and ischemia. These changes often progress silently over time, eventually resulting in symptomatic heart failure, reduced ejection fraction, and diastolic dysfunction. Additionally, survivors are at an increased risk of developing coronary artery disease, arrhythmias, and valvular abnormalities due to radiation-induced endothelial damage and calcification. The compounding effects of preexisting cardiovascular risk factors, such as hypertension and diabetes, exacerbate the likelihood of severe outcomes [5, 6]. Clinically, this progression highlights the need for early detection through advanced imaging modalities and biomarkers, proactive risk management, and long-term monitoring to mitigate the impact on survivors' cardiovascular health and quality of life [38, 39].

Importance of Survivorship Care Plans

Long-term cardiovascular risks in cancer survivors necessitate the development and implementation of comprehensive survivorship care plans to address the growing burden of treatment-related cardiac complications. Cancer therapies, including anthracyclines, HER2-targeted agents, and radiation, contribute to progressive cardiovascular dysfunction, often leading to heart failure, coronary artery disease, and arrhythmias. Survivorship care plans play a crucial role in mitigating these risks by integrating regular cardiovascular monitoring, risk stratification, and preventive strategies tailored to the individual. These plans emphasize lifestyle interventions such as smoking cessation, physical activity, and dietary modifications to promote heart health. Furthermore, they facilitate early detection and management of late-onset cardiotoxicity through the use of advanced imaging modalities and biomarkers. A multidisciplinary approach involving oncologists, cardiologists, and primary care providers ensures that survivors receive holistic care aimed at optimizing their long-term cardiovascular outcomes and overall quality of life [40–42].

Indications for Cardiac Surgery in Cancer Survivors

Common Cardiac Conditions Requiring Surgery Post-anticancer Therapy

Indications for cardiac surgery in cancer survivors are often driven by complications arising from prior anticancer therapies or the progression of underlying cardiac conditions exacerbated by treatment. Common cardiac conditions requiring surgery post-anticancer therapy include valvular heart disease, coronary artery disease, and pericardial disease. Valvular dysfunction, often linked to radiation therapy, may necessitate valve repair or replacement due to fibrosis and calcification. Coronary artery bypass grafting (CABG) is indicated in survivors with significant radiation-induced coronary artery disease or chemotherapy-associated ischemia. Pericardial complications, such as constrictive pericarditis or large effusions resistant to medical management, may require pericardiectomy or pericardiocentesis. In select cases, surgery addresses heart failure secondary to structural damage or refractory arrhythmias. Comprehensive preoperative evaluation and collaboration between oncologists and cardiologists are essential to optimize surgical outcomes, ensuring that cardiac interventions are timed and tailored to the patient's oncologic and cardiovascular needs [2, 4, 43–45].

Decision-Making Framework for Surgical Intervention

Indications for cardiac surgery in cancer survivors often stem from complications induced by anticancer therapies or preexisting cardiovascular conditions exacerbated by treatment. These include valvular heart disease caused by radiation fibrosis or calcification, coronary artery disease requiring revascularization, and pericardial diseases like constrictive pericarditis or large, refractory effusions. The decision-making framework for surgical intervention involves a multidisciplinary approach, integrating inputs from cardiologists, oncologists, and cardiothoracic surgeons. Key factors include the patient's oncologic status, life expectancy, and overall health. Risk stratification tools assess the likelihood of perioperative complications, while functional capacity and symptom burden inform the urgency and feasibility of surgery. Preoperative optimization, including addressing modifiable risk factors and stabilizing cardiac function, ensures better outcomes. Decisions must balance the benefits of surgery against potential impacts on cancer treatment, with a goal to enhance survival and quality of life [2, 4, 43–46].

Multidisciplinary Approach to Patient Evaluation

Indications for cardiac surgery in cancer survivors are driven by cardiovascular complications resulting from anticancer therapies, such as valvular heart disease, coronary artery disease, pericardial diseases, or heart failure due to myocardial dysfunction. A multidisciplinary approach to patient evaluation is crucial for optimizing surgical outcomes. This collaborative process involves cardiologists assessing cardiac function and risks, oncologists determining cancer prognosis and treatment status, and cardiothoracic surgeons evaluating surgical feasibility. Additional specialists, such as anesthesiologists and rehabilitation experts, may contribute to perioperative planning and recovery strategies. Multidisciplinary discussions ensure that decisions consider both oncologic and cardiac priorities, balancing the benefits of surgery with the need to maintain continuity in cancer care. This integrated approach enhances personalized care, improves risk stratification, and supports better long-term outcomes for cancer survivors requiring cardiac surgery [2, 4, 43–46].

Preoperative Assessment and Risk Stratification

Key Components of Preoperative Cardiovascular Evaluation

Preoperative assessment and risk stratification in cancer survivors undergoing cardiac surgery are critical to minimize perioperative and postoperative complications. A thorough preoperative cardiovascular evaluation involves several key components. Comprehensive medical history-taking is essential to identify preexisting cardiac conditions, prior exposure to cardiotoxic therapies, and lifestyle factors such as smoking or physical inactivity. A detailed physical examination should assess signs of cardiovascular dysfunction, including heart failure, arrhythmias, or valvular abnormalities. Diagnostic tools such as electrocardiography (ECG), echocardiography, cardiac biomarkers (troponins, NT-proBNP), and advanced imaging modalities like cardiac MRI or stress testing are used to evaluate myocardial function and stratify risk. The Revised Cardiac Risk Index (RCRI) or other validated scoring systems help estimate perioperative cardiac risk. Additionally, a multidisciplinary team approach, incorporating cardiologists, oncologists, anesthesiologists, and surgeons, ensures personalized care tailored to the patient's cardiac and oncologic status. This comprehensive evaluation allows for optimized perioperative management and improved long-term outcomes [2, 4, 44–47].

Optimization Strategies for Cardiac Function Before Surgery

Preoperative assessment and risk stratification are crucial for optimizing cardiac function in cancer survivors before undergoing surgery. Optimization strategies aim to stabilize cardiovascular health and reduce the risk of perioperative complications. Pharmacologic interventions are fundamental, including the initiation or adjustment of cardioprotective medications such as beta-blockers, ACE inhibitors, or ARBs to improve myocardial function and control blood pressure. Statins may be introduced for their lipid-lowering and anti-inflammatory effects, while anticoagulants can be considered for patients with elevated thromboembolic risk. Addressing modifiable risk factors is equally important; smoking cessation, weight management, and promoting a heart-healthy diet are essential lifestyle modifications. Mild to moderate physical activity, tailored to the patient's condition, can enhance functional capacity. Psychosocial support, including stress management and counseling, addresses emotional factors that may impact perioperative outcomes. Timing surgery in coordination with oncologic treatment cycles can minimize overlapping toxicities, and close collaboration among cardiologists, oncologists, and surgeons ensures a multidisciplinary approach to preoperative care. These strategies collectively prepare the patient for better surgical outcomes and long-term cardiovascular health [2, 4, 44–47].

Tools for Predicting Surgical Risks in Cancer Survivors

Preoperative assessment and risk stratification for cancer survivors undergoing surgery involve utilizing tools to predict surgical risks and guide clinical decision-making. These tools assess cardiovascular risks and perioperative outcomes, helping to tailor interventions for individual patients. The Revised Cardiac Risk Index (RCRI) is commonly used to evaluate the risk of major adverse cardiac events, incorporating factors such as ischemic heart disease, heart failure, and diabetes. The American College of Surgeons National Surgical Quality Improvement Program (ACS NSQIP) risk calculator estimates mortality and morbidity risks based on patient-specific factors and procedural details. Functional capacity evaluations, such as the Duke Activity Status Index (DASI) or stress tests, assess a patient's ability to tolerate physical exertion and estimate perioperative resilience. Biomarkers, including troponins and NT-proBNP, provide insights into myocardial injury or stress, while advanced imaging techniques like echocardiography and cardiac MRI offer detailed evaluations of cardiac structure and function. Incorporating these tools into a multidisciplinary approach ensures a comprehensive evaluation of surgical risks, enabling the development of personalized strategies to optimize preoperative preparation and improve postoperative outcomes [4, 40, 46, 47].

Intraoperative Considerations for Cardiac Surgery

Specific Challenges in Managing Cancer Survivors Undergoing Surgery

Intraoperative considerations for cardiac surgery in cancer survivors require meticulous planning and multidisciplinary coordination to address specific challenges posed by their medical history and comorbidities. These patients often present with unique vulnerabilities due to prior exposure to cardiotoxic treatments, such as chemotherapy or radiation, which may have compromised myocardial, vascular, or valvular function. Maintaining hemodynamic stability is critical, with a focus on optimizing preload, afterload, and contractility to support cardiac output. The use of advanced intraoperative monitoring, including transesophageal echocardiography, can provide real-time assessments of ventricular function and guide fluid and medication management. Anesthetic strategies should be tailored to minimize myocardial oxygen demand while maintaining perfusion. Patients may have altered coagulation profiles due to cancer or its treatment, necessitating careful anticoagulation and hemostasis management. Additionally, addressing challenges such as tissue fibrosis from prior radiation or surgical adhesions can prolong operative times and increase the risk of complications. Collaboration among surgeons, anesthesiologists, and cardiologists ensures comprehensive care, enabling the successful management of these complex cases and improving surgical outcomes [2, 4, 40, 44–47].

Cardiopulmonary Bypass Techniques in Previously Treated Hearts

Cardiopulmonary bypass techniques in previously treated hearts require a tailored approach to address the challenges posed by prior oncologic therapies, such as chemotherapy and radiation, which often leave residual damage to cardiac and surrounding tissues. These treatments can cause myocardial fibrosis, impaired coronary perfusion, and reduced ventricular compliance, complicating the establishment and maintenance of effective cardiopulmonary bypass. To mitigate these risks, careful preoperative planning and intraoperative monitoring are essential. Modified perfusion protocols may be used to maintain optimal hemodynamic conditions, with a focus on minimizing ischemia-reperfusion injury. Lower flow rates and reduced pressures during bypass can help protect fragile myocardial and vascular structures. Additionally, anticoagulation strategies must be carefully managed, particularly in patients with thrombocytopenia or altered coagulation profiles from cancer or its treatments. The use of advanced monitoring techniques, such as transesophageal echocardiography, is critical for assessing real-time cardiac function and ensuring precise management of ventricular filling and output. Addressing these challenges through a multidisciplinary approach helps to optimize outcomes and minimize complications in this high-risk patient population [2, 4, 40, 44–47].

Role of Advanced Anesthetic Techniques and Monitoring

The role of advanced anesthetic techniques and monitoring in cardiac surgery for cancer survivors is critical to ensuring optimal outcomes in this high-risk population. These patients often present unique challenges due to the potential cardiopulmonary complications stemming from prior chemotherapy, radiation, or immunotherapy. Advanced anesthetic techniques, including total intravenous anesthesia (TIVA) and goal-directed hemodynamic management, allow for precise control of cardiovascular parameters, reducing myocardial stress and improving tissue perfusion. Monitoring technologies, such as transesophageal echocardiography (TEE), provide real-time visualization of cardiac function, enabling early detection and management of complications like ventricular dysfunction or valvular abnormalities. Continuous cardiac output monitoring and advanced hemodynamic monitoring systems, such as pulse contour analysis, help tailor fluid and vasopressor therapy to individual patient needs, minimizing perioperative hemodynamic instability. Additionally, the integration of multimodal analgesia and neuroprotective strategies can reduce the risks of perioperative myocardial ischemia and neurocognitive decline. A multidisciplinary approach involving anesthesiologists, cardiologists, and surgeons is essential to implement these techniques effectively and to address the complexities of cardiac surgery in cancer survivors [2, 4, 44–47].

Postoperative Management and Rehabilitation

Immediate Postoperative Care and Complications

Postoperative management and rehabilitation play a critical role in ensuring optimal recovery and outcomes following cardiac surgery, particularly for cancer survivors who often face unique challenges. Immediate postoperative care focuses on vigilant monitoring and management of complications, such as bleeding, arrhythmias, and hemodynamic instability, which are common during the early recovery phase. Intensive care unit (ICU) protocols typically include continuous cardiac monitoring, invasive hemodynamic support, and frequent assessment of respiratory and renal functions to detect and address complications promptly. Potential complications, including myocardial infarction, stroke, infection, and thromboembolism, require multidisciplinary coordination for effective management. Adequate pain control, fluid balance, and nutritional support further facilitate recovery. Early mobilization and respiratory therapy are critical to preventing pulmonary complications such as atelectasis or pneumonia. In cancer survivors, additional considerations include addressing the potential interactions between prior oncologic treatments and surgical outcomes, ensuring close collaboration between oncologists, cardiologists, and surgeons. Rehabilitation programs tailored to the patient's functional status and medical history are essential to improving long-term prognosis, enhancing quality of life, and reducing the risk of recurrent cardiovascular events [2, 4, 44–47].

Rehabilitation Programs Tailored for Cancer Survivors

Postoperative management and rehabilitation for cancer survivors undergoing cardiac surgery require specialized programs that address the unique interplay between oncologic and cardiovascular health. Rehabilitation programs are tailored to the individual needs of patients, considering their cancer history, prior treatments, and overall physical condition. These programs typically involve a multidisciplinary approach, integrating cardiologists, oncologists, physical therapists, and dietitians to ensure holistic care. Structured exercise regimens, including aerobic and resistance training, are introduced gradually to improve cardiovascular endurance and muscle strength while minimizing the risk of overexertion. Nutritional counseling emphasizes a heart-healthy diet rich in anti-inflammatory and antioxidant properties to support recovery and long-term health. Psychosocial support, including counseling or group therapy, helps address anxiety, depression, or fear of cancer recurrence, which are common among survivors. Education on recognizing signs of potential complications and the importance of adherence to prescribed therapies empowers patients to take an active role in their recovery. Regular follow-up with healthcare providers ensures ongoing monitoring and timely adjustments to rehabilitation plans, promoting improved quality of life and reduced risks of long-term complications [4, 46, 47].

Monitoring and Managing Late Complications

Monitoring and managing late complications in postoperative care for cancer survivors undergoing cardiac surgery are essential to ensure long-term health and quality of life. These patients are at risk of developing late-onset complications due to the combined effects of prior cancer treatments and cardiac surgery. Regular follow-up assessments are crucial, including cardiac imaging such as echocardiography or cardiac MRI to detect changes in ventricular function, valvular integrity, or pericardial abnormalities. Biomarkers such as NT-proBNP and troponins may be used to monitor myocardial stress or injury. Surveillance for arrhythmias, commonly associated with both cardiotoxic cancer therapies and surgical interventions, is typically conducted using ECGs or ambulatory Holter monitoring. Management strategies focus on optimizing cardiovascular health, including adjustments in pharmacologic therapy with beta-blockers, ACE inhibitors, or anticoagulants, as needed, based on the patient's evolving condition. Addressing other complications such as lymphedema, chronic fatigue, or reduced exercise tolerance involves interdisciplinary collaboration between cardiologists, oncologists, and physical therapists. Lifestyle interventions, including regular physical activity, smoking cessation, and dietary modifications, play a vital role in minimizing long-term risks. Psychosocial support and patient education further empower survivors to recognize symptoms of potential complications and seek timely medical care, ensuring a proactive approach to their ongoing health [4, 46, 47].

Prognostic Factors in Survivors Undergoing Cardiac Surgery

Impact of Preexisting Cardiotoxicity on Surgical Outcomes

Prognostic factors in survivors undergoing cardiac surgery are heavily influenced by the presence of preexisting cardiotoxicity, which significantly impacts surgical outcomes. Patients who have experienced cardiotoxic effects from prior cancer therapies, such as anthracyclines or HER2-targeted treatments, often present with compromised cardiac reserve and preexisting ventricular dysfunction, making them more vulnerable to perioperative complications. Additionally, these individuals are at an increased risk of arrhythmias, ischemic events, and slower recovery due to the cumulative strain on their cardiovascular systems. The extent of preexisting cardiotoxicity directly correlates with outcomes, as those with advanced heart failure or significant myocardial damage typically face higher morbidity and mortality rates post-surgery. Careful preoperative assessment, including advanced imaging and biomarker evaluation, plays a critical role in stratifying risks and tailoring interventions to optimize patient stability before surgery. Collaborative management by a multidisciplinary team, incorporating cardiologists, oncologists, and surgeons, is essential to mitigate risks and improve prognostic outcomes for cancer survivors requiring cardiac surgery [4, 46, 47].

Long-Term Survival and Quality of Life After Surgery

Long-term survival and quality of life after cardiac surgery in cancer survivors are influenced by various prognostic factors, including the severity of preexisting cardiotoxicity, the type of surgical intervention, and the patient's overall health and comorbidities. Survivors who successfully undergo cardiac surgery often experience improved functional capacity and symptom relief, contributing to enhanced quality of life. However, the long-term prognosis may be tempered by residual cardiovascular issues, such as persistent left ventricular dysfunction or the progression of other treatment-induced complications. Additionally, ongoing cancer-related therapies and their potential for cumulative cardiotoxic effects necessitate vigilant monitoring and follow-up care. Tailored rehabilitation programs and adherence to cardiovascular risk reduction strategies, including lifestyle modifications and pharmacologic interventions, play a pivotal role in optimizing outcomes. Collaborative, multidisciplinary approaches that integrate oncology, cardiology, and rehabilitation teams are essential for achieving favorable long-term survival and maintaining quality of life in this high-risk patient population [4, 46, 47].

Predictive Models for Postoperative Prognosis

Predictive models for postoperative prognosis in cancer survivors undergoing cardiac surgery are essential tools for guiding clinical decision-making and optimizing outcomes. These models typically incorporate a range of patient-specific factors, including age, comorbidities, baseline cardiac function, and the presence of prior cardiotoxicity. Advanced algorithms, such as the Revised Cardiac Risk Index (RCRI) and Cancer and Aging Research Group (CARG) score, are often used to stratify patients based on their perioperative risk and long-term survival probabilities. Incorporating data from biomarkers like NT-proBNP and troponins, along with imaging modalities such as echocardiography and cardiac MRI, enhances the accuracy of these models by providing insights into myocardial health and function. Additionally, machine learning-based predictive tools are emerging, offering dynamic and individualized risk assessments by analyzing large datasets and identifying subtle patterns in clinical variables. By employing these predictive models, clinicians can tailor preoperative optimization strategies, intraoperative management, and postoperative care to mitigate risks and improve the prognosis for cancer survivors requiring cardiac surgery [4, 46, 47].

Multidisciplinary Approaches in Cardio-Oncology and Surgery

Collaboration Between Oncologists, Cardiologists, and Surgeons

A multidisciplinary approach in cardio-oncology and surgery is essential for optimizing patient outcomes, emphasizing collaboration among oncologists, cardiologists, and surgeons. This collaborative model ensures a comprehensive assessment of the patient's oncological and cardiovascular status, allowing for integrated care planning that balances the urgency of cancer treatment with the management of cardiac risks. Oncologists provide insights into the therapeutic regimen, potential cardiotoxic agents, and cancer prognosis, while cardiologists assess cardiovascular risk, implement cardioprotective strategies, and monitor cardiac function throughout the treatment continuum. Surgeons play a crucial role in evaluating surgical candidacy and tailoring intraoperative and postoperative strategies to minimize complications. Regular interdisciplinary meetings and the use of shared decision-making frameworks facilitate seamless communication and coordinated care. By leveraging the expertise of each specialty, the multidisciplinary team can achieve better survival rates, reduced morbidity, and an improved quality of life for patients requiring both cancer therapy and cardiac surgery [4, 8, 42, 46, 47].

Role of Cardio-Oncology Clinics in Comprehensive Care

Cardio-oncology clinics play a pivotal role in providing comprehensive care for patients undergoing cancer treatment who are at risk of or experiencing cardiovascular complications. These specialized clinics serve as a centralized platform for integrating oncological and cardiological expertise, ensuring that patients receive personalized and holistic care throughout their cancer journey. The multidisciplinary team in a cardio-oncology clinic typically includes oncologists, cardiologists, radiologists, and other allied healthcare professionals who collaborate to assess, monitor, and manage cardiotoxic risks associated with cancer therapies. By offering tailored risk assessments, early detection of subclinical cardiac dysfunction through advanced imaging and biomarker evaluation, and implementing preventive measures such as cardioprotective pharmacologic agents and lifestyle interventions, these clinics enhance patient outcomes. Furthermore, cardio-oncology clinics provide a structured framework for long-term surveillance of cancer survivors, addressing late-onset cardiac complications and promoting seamless communication between care providers. This approach not only mitigates treatment-related cardiac risks but also enables patients to continue their cancer therapies without unnecessary interruptions, ultimately improving both survival and quality of life [4, 8, 42, 46, 47].

Case Studies Highlighting Best Practices

Case studies in cardio-oncology demonstrate the importance of multidisciplinary approaches in optimizing patient outcomes, showcasing best practices for managing complex cases involving both cancer and cardiovascular disease. In one case, a breast cancer patient receiving HER2-targeted therapy was monitored by a cardio-oncology team after developing early signs of left ventricular dysfunction. Through close collaboration, oncologists adjusted the cancer treatment plan, while cardiologists initiated cardioprotective therapy with beta-blockers and ACE inhibitors, successfully preventing progression to heart failure without compromising cancer treatment efficacy. Another example involves a lymphoma patient treated with anthracyclines who experienced arrhythmias during chemotherapy. The team employed advanced imaging and biomarker analysis to guide treatment adjustments, including a reduction in anthracycline dosage and the addition of antiarrhythmic therapy, allowing the patient to complete treatment safely. These cases highlight the value of integrated care, personalized treatment strategies, and proactive risk management, emphasizing how multidisciplinary collaboration in cardio-oncology can address the dual challenges of cancer and cardiovascular health effectively [4, 8, 42, 46, 47].

Future Perspectives in Cardiotoxicity and Cardioprotection

Advances in Personalized Medicine for Cardio-Oncology Patients

Future perspectives in cardiotoxicity and cardioprotection are increasingly focused on advances in personalized medicine, aiming to tailor cardiovascular care for cardio-oncology patients based on individual risk profiles, genetic predispositions, and specific cancer treatments. Personalized medicine leverages genomic, proteomic, and biomarker analyses to identify patients at higher risk of cardiotoxicity, enabling early interventions and more precise treatment strategies. For example, genetic screening can detect polymorphisms linked to increased susceptibility to drug-induced cardiotoxicity, guiding the selection of alternative therapies or preventive measures. Advances in imaging techniques, such as strain echocardiography and cardiac MRI, allow for the early detection of subclinical cardiac dysfunction, facilitating timely implementation of cardioprotective therapies. Additionally, the integration of artificial intelligence and machine learning into clinical decision-making processes promises to refine risk stratification and optimize treatment plans by analyzing large datasets from diverse patient populations. As the field progresses, the development of targeted therapies that mitigate cardiotoxicity without compromising anticancer efficacy represents a critical area of research, ultimately improving long-term outcomes and quality of life for cancer survivors [4, 8, 40–42, 46, 47].

Development of Targeted Therapies with Reduced Cardiotoxic Effects

Future perspectives in cardiotoxicity and cardioprotection emphasize the development of targeted therapies designed to minimize cardiotoxic effects while maintaining anticancer efficacy. Researchers are focusing on creating novel agents and modifying existing chemotherapeutic drugs to selectively target cancer cells without compromising cardiac health. This includes advancements in nanoparticle-based drug delivery systems, which allow for precise delivery of chemotherapeutic agents directly to tumor tissues, reducing systemic exposure and subsequent cardiac damage. Additionally, the integration of cardio-oncology biomarkers into drug development pipelines enables the early identification of potential cardiotoxic effects during preclinical and clinical trials. Immunotherapies and next-generation targeted therapies are being re-engineered to mitigate risks of myocarditis, arrhythmias, and vascular complications by modulating immune responses more specifically. These innovations, coupled with real-time monitoring technologies and personalized treatment algorithms, represent a promising shift toward safer and more effective cancer treatment regimens, ensuring improved long-term cardiovascular and oncological outcomes for patients [4, 8, 40–42, 46, 47].

Innovations in Cardiac Surgery Techniques for Complex Cases

Innovations in cardiac surgery techniques for complex cases have emerged as a promising frontier in addressing cardiotoxicity and optimizing outcomes for cancer survivors. Advanced minimally invasive procedures, such as robotic-assisted surgery and percutaneous interventions, offer reduced operative risks and faster recovery times for patients with prior exposure to cardiotoxic therapies. Hybrid operating rooms, integrating catheter-based and open surgical techniques, facilitate precise and individualized treatment strategies for complex cardiac conditions, such as radiation-induced valvular disease or coronary artery stenosis. Developments in biocompatible prosthetic materials and tissue-engineered valves have improved the durability and compatibility of cardiac implants, particularly for patients with prior thoracic radiation or chemotherapy. Additionally, the use of intraoperative imaging modalities, such as 3D echocardiography and cardiac MRI, allows surgeons to navigate challenging anatomical changes caused by cancer therapies with greater accuracy. Enhanced cardiopulmonary bypass techniques, tailored to preserve myocardial function, further support safer surgical outcomes in this vulnerable population. These advancements collectively represent significant progress in cardiac care for cancer survivors, emphasizing precision, safety, and interdisciplinary collaboration [4, 8, 40–42, 46, 47].

Conclusion and Recommendations

Summary of Key Findings and Practices

The interplay between cancer therapies and cardiovascular health underscores the critical importance of a multidisciplinary approach to managing cardiotoxicity. Advances in oncology have significantly improved cancer survival rates, but they have also introduced new challenges in cardiovascular care. Cardiotoxicity manifests in various forms, including heart failure, ischemic heart disease, arrhythmias, and valvular dysfunction, necessitating vigilant monitoring and proactive management. Effective strategies, such as the use of cardioprotective agents, personalized medicine, lifestyle interventions, and rehabilitation programs, play a vital role in mitigating cardiovascular risks. Furthermore, innovations in diagnostic tools, surgical techniques, and collaboration among oncologists, cardiologists, and surgeons enhance patient outcomes and quality of life [4, 8, 40–42, 46, 47].

Summary of Key Findings and Practices

1. **Epidemiology and Mechanisms of Cardiotoxicity**: Anticancer therapies, including anthracyclines, HER2-targeted treatments, and radiation, are associated with significant cardiovascular risks. Understanding pathophysiological mechanisms is critical for prevention and early intervention.
2. **Diagnostic and Monitoring Strategies**: Advanced imaging modalities, biomarkers, and risk stratification tools enable early detection and guide treatment decisions.
3. **Cardioprotection and Rehabilitation**: Pharmacologic agents, such as beta-blockers and ACE inhibitors, alongside lifestyle modifications and structured rehabilitation, support cardiovascular health during and after cancer therapy.
4. **Surgical Considerations**: Tailored approaches, including preoperative optimization, advanced intraoperative techniques, and postoperative management, address the complexities of cardiac surgery in cancer survivors.
5. **Multidisciplinary Approaches**: Collaborative efforts among healthcare providers ensure comprehensive care, addressing both oncological and cardiovascular needs.
6. **Future Directions**: Emerging therapies, personalized medicine, and innovations in surgical techniques hold promise for reducing cardiotoxicity and improving long-term outcomes.

Recommendations Continued research, education, and the establishment of cardio-oncology clinics are essential to address the growing burden of cardiotoxicity in cancer care. By integrating preventive strategies, early detection, and innovative treatments, healthcare systems can optimize survivorship care and improve the overall prognosis for cancer survivors.

Guidelines for Improving Outcomes in Cancer Survivors

The guidelines for improving outcomes in cancer survivors emphasize a comprehensive, multidisciplinary approach to managing the interplay between oncologic treatments and cardiovascular health. Key recommendations include conducting thorough pre-treatment assessments to stratify cardiovascular risk using advanced imaging and biomarkers, enabling early detection of vulnerabilities. Incorporating cardioprotective strategies such as pharmacologic agents (e.g., beta-blockers, ACE inhibitors, dexrazoxane) and lifestyle interventions (e.g., healthy diet, regular exercise, smoking cessation) is essential to mitigate cardiotoxicity during therapy. Vigilant monitoring through echocardiography and biomarkers ensures early identification of subclinical cardiac dysfunction, allowing for timely treatment modifications. Post-treatment rehabilitation programs tailored to survivors' needs focus on improving physical and psychological recovery, while long-term follow-up addresses late cardiovascular complications. Advanced surgical techniques and intraoperative strategies are crucial for survivors requiring cardiac surgery, supported by robust preoperative optimization and postoperative care. Finally, the integration of cardio-oncology clinics and research into innovative, targeted therapies with reduced cardiotoxic effects are vital for advancing personalized medicine and ensuring the best possible outcomes for cancer survivors [4, 8, 40–42, 46, 47].

Call for Integrated Cardio-Oncology Care Pathways

An integrated cardio-oncology care pathway is essential to address the growing prevalence of cardiovascular complications among cancer survivors. These pathways should incorporate seamless collaboration between oncologists, cardiologists, and other specialists to create individualized care plans that address both cancer and cardiovascular risks. Key components include early risk stratification using advanced diagnostic tools, implementation of evidence-based cardioprotective strategies, and continuous monitoring during and after cancer treatment. The establishment of cardio-oncology clinics can centralize expertise, ensuring timely interventions and comprehensive management. Such integrated approaches not only improve clinical outcomes but also enhance patients' quality of life, allowing them to complete cancer therapies safely while minimizing long-term cardiovascular morbidity. Developing standardized protocols and fostering multidisciplinary research will be critical in optimizing care pathways and advancing the field of cardio-oncology [4, 8, 40–42, 46, 47].

References

1. Chen R, Aschmann HE, Chen YH, Glymour MM, Bibbins-Domingo K, Stokes AC et al (2022) Racial and ethnic disparities in estimated excess mortality from external causes in the US, March to December 2020. JAMA Intern Med 182(7):776–778
2. Lyon AR, Lopez-Fernandez T, Couch LS, Asteggiano R, Aznar MC, Bergler-Klein J et al (2022) 2022 ESC guidelines on cardio-oncology developed in collaboration with the European Hematology Association (EHA), the European Society for Therapeutic Radiology and Oncology (ESTRO) and the International Cardio-Oncology Society (IC-OS). Eur Heart J 43(41): 4229–4361

3. Montisci A, Palmieri V, Liu JE, Vietri MT, Cirri S, Donatelli F et al (2021) Severe cardiac toxicity induced by cancer therapies requiring intensive care unit admission. Front Cardiovasc Med 8:713694
4. Palmieri V, Vietri MT, Montalto A, Montisci A, Donatelli F, Coscioni E et al (2023) Cardiotoxicity, cardioprotection, and prognosis in survivors of anticancer treatment undergoing cardiac surgery: unmet needs. Cancers (Basel) 15(8)
5. Butel-Simoes LE, Haw TJ, Williams T, Sritharan S, Gadre P, Herrmann SM et al (2023) Established and emerging cancer therapies and cardiovascular system: focus on hypertension-mechanisms and mitigation. Hypertension 80(4):685–710
6. Cohen JB, Brown NJ, Brown SA, Dent S, van Dorst DCH, Herrmann SM et al (2023) Cancer therapy-related hypertension: a scientific statement from the American Heart Association. Hypertension 80(3):e46–e57
7. Franczyk B, Rysz J, Lawinski J, Cialkowska-Rysz A, Gluba-Brzozka A (2023) Cardiotoxicity of selected vascular endothelial growth factor receptor tyrosine kinase inhibitors in patients with renal cell carcinoma. Biomedicines 11(1)
8. Plonska-Gosciniak E, Piotrowski G, Wojakowski W, Gosciniak P, Olszowska M, Lesiak M et al (2023) Management of valvular heart disease in patients with cancer: multidisciplinary team, cancer-therapy related cardiotoxicity, diagnosis, transcatheter intervention, and cardiac surgery. Expert opinion of the Association on Valvular Heart Disease, Association of Cardiovascular Interventions, and Working Group on Cardiac Surgery of the Polish Cardiac Society. Kardiol Pol 81(1):82–101
9. Fabin N, Bergami M, Cenko E, Bugiardini R, Manfrini O (2022) The role of vasospasm and microcirculatory dysfunction in fluoropyrimidine-induced ischemic heart disease. J Clin Med 11(5)
10. Dempke WCM, Zielinski R, Winkler C, Silberman S, Reuther S, Priebe W (2023) Anthracycline-induced cardiotoxicity—are we about to clear this hurdle? Eur J Cancer 185: 94–104
11. Dempsey N, Rosenthal A, Dabas N, Kropotova Y, Lippman M, Bishopric NH (2021) Trastuzumab-induced cardiotoxicity: a review of clinical risk factors, pharmacologic prevention, and cardiotoxicity of other HER2-directed therapies. Breast Cancer Res Treat 188(1): 21–36
12. Cozma A, Sporis ND, Lazar AL, Buruiana A, Ganea AM, Malinescu TV et al (2022) Cardiac toxicity associated with immune checkpoint inhibitors: a systematic review. Int J Mol Sci 23(18)
13. Patel M, Hudson O, Han J, Kondapalli L, Arora G, Hawi R et al (2023) Update on immunotherapy cardiotoxicity: checkpoint inhibitors, CAR T, and beyond. Curr Treat Options in Oncol 24(11):1489–1503
14. Karthikeyan B, Sunder SS, Puzanov I, Olejniczak SH, Pokharel S, Sharma UC (2024) Cardiotoxic profiles of CAR-T therapy and bispecific T-cell engagers in hematological cancers. Commun Med (Lond) 4(1):116
15. Curigliano G, Lenihan D, Fradley M, Ganatra S, Barac A, Blaes A et al (2020) Management of cardiac disease in cancer patients throughout oncological treatment: ESMO consensus recommendations. Ann Oncol 31(2):171–190
16. de Wit S, Glen C, de Boer RA, Lang NN (2023) Mechanisms shared between cancer, heart failure, and targeted anti-cancer therapies. Cardiovasc Res 118(18):3451–3466
17. Linders AN, Dias IB, Lopez Fernandez T, Tocchetti CG, Bomer N, Van der Meer P (2024) A review of the pathophysiological mechanisms of doxorubicin-induced cardiotoxicity and aging. NPJ Aging 10(1):9
18. Koutroumpakis E, Palaskas NL, Lin SH, Abe JI, Liao Z, Banchs J et al (2020) Modern radiotherapy and risk of cardiotoxicity. Chemotherapy 65(3–4):65–76
19. Palaskas NL, Ali HJ, Koutroumpakis E, Ganatra S, Deswal A (2024) Cardiovascular toxicity of immune therapies for cancer. BMJ 385:e075859

20. Curigliano G, Cardinale D, Dent S, Criscitiello C, Aseyev O, Lenihan D et al (2016) Cardiotoxicity of anticancer treatments: epidemiology, detection, and management. CA Cancer J Clin 66(4):309–325
21. Burkhart T, Keith MCL, Lenneman CAG, Fernando RR (2018) Bortezomib-induced cardiac tamponade in a 49-year-old man. Tex Heart Inst J 45(4):260–263
22. Podlesnikar T, Berlot B, Dolenc J, Goricar K, Marinko T (2022) Radiotherapy-induced cardiotoxicity: the role of multimodality cardiovascular imaging. Front Cardiovasc Med. 9: 887705
23. Jagielska B, Ozdowska P, Gepner K, Kubala S, Siedlecki JA, Sarnowski TJ et al (2020) Cardiotoxicity danger in immunotherapy. IUBMB Life 72(6):1160–1167
24. Chung WB, Youn JC, Youn HJ (2020) Cardiovascular complications of novel anti-cancer immunotherapy: old problems from new agents? Korean Circ J 50(9):743–753
25. Balanescu DV, Donisan T, Palaskas N, Lopez-Mattei J, Kim PY, Buja LM et al (2020) Immunomodulatory treatment of immune checkpoint inhibitor-induced myocarditis: pathway toward precision-based therapy. Cardiovasc Pathol 47:107211
26. Liu L, Yao W, Wang M, Wang B, Kong F, Fan Z et al (2024) A systematic review of cardiovascular toxicities induced by cancer immune therapies: underlying mechanisms, clinical manifestations and therapeutic approaches. Semin Cancer Biol 106-107:179–191
27. Varricchi G, Galdiero MR, Marone G, Criscuolo G, Triassi M, Bonaduce D et al (2017) Cardiotoxicity of immune checkpoint inhibitors. ESMO Open 2(4):e000247
28. Totzeck M, Aide N, Bauersachs J, Bucerius J, Georgoulias P, Herrmann K et al (2023) Nuclear medicine in the assessment and prevention of cancer therapy-related cardiotoxicity: prospects and proposal of use by the European Association of Nuclear Medicine (EANM). Eur J Nucl Med Mol Imaging 50(3):792–812
29. Moreno-Arciniegas A, Cadiz L, Galan-Arriola C, Clemente-Moragon A, Ibanez B (2024) Cardioprotection strategies for anthracycline cardiotoxicity. Basic Res Cardiol 120:71–90
30. Kourek C, Touloupaki M, Rempakos A, Loritis K, Tsougkos E, Paraskevaidis I et al (2022) Cardioprotective strategies from cardiotoxicity in cancer patients: a comprehensive review. J Cardiovasc Dev Dis 9(8)
31. Balough E, Ariza A, Asnani A, Hoeger CW (2025) Cardiotoxicity of anthracyclines. Cardiol Clin 43(1):111–127
32. Xiao J, Li X, Wang X, Guan Y, Liu H, Liang J et al (2023) Clinical characteristics and management of immune checkpoint inhibitor-related cardiotoxicity: a single-center experience. Front Cardiovasc Med 10:1093383
33. Berliner D, Beutel G, Bauersachs J (2020) Echocardiography and biomarkers for the diagnosis of cardiotoxicity. Herz 45(7):637–644
34. Murtagh G, Januzzi JL, Scherrer-Crosbie M, Neilan TG, Dent S, Ho JE et al (2023) Circulating cardiovascular biomarkers in cancer therapeutics-related cardiotoxicity: review of critical challenges, solutions, and future directions. J Am Heart Assoc 12(21):e029574
35. Tsiouris KM, Mitsis A, Grigoriadis G, Karanasiou G, Lakkas L, Mauri D et al (2023) Risk stratification for cardiotoxicity in breast cancer patients: predicting early decline of LVEF after treatment. Annu Int Conf IEEE Eng Med Biol Soc 2023:1–4
36. Valzania C, Paccagnella A, Spadotto A, Ruotolo I, Bonfiglioli R, Fallani F et al (2023) Early detection of cancer therapy cardiotoxicity by radionuclide angiography: an update. J Nucl Cardiol 30(5):2104–2111
37. Alshahrani AA, Kontopantelis E, Morgan C, Ravindrarajah R, Martin GP, Mamas MA (2024) Cardiovascular diseases in patients with cancer: a comprehensive review of epidemiological trends, cardiac complications, and prognostic implications. Chin Med J 138:143–154
38. Lyon AR, Dent S, Stanway S, Earl H, Brezden-Masley C, Cohen-Solal A et al (2020) Baseline cardiovascular risk assessment in cancer patients scheduled to receive cardiotoxic cancer therapies: a position statement and new risk assessment tools from the Cardio-Oncology Study Group of the Heart Failure Association of the European Society of Cardiology in

collaboration with the International Cardio-Oncology Society. Eur J Heart Fail 22(11): 1945–1960
39. Tocchetti CG, Farmakis D, Koop Y, Andres MS, Couch LS, Formisano L et al (2024) Cardiovascular toxicities of immune therapies for cancer—a scientific statement of the Heart Failure Association (HFA) of the ESC and the ESC Council of Cardio-Oncology. Eur J Heart Fail 26(10):2055–2076
40. Adams SC, Herman J, Lega IC, Mitchell L, Hodgson D, Edelstein K et al (2021) Young adult cancer survivorship: recommendations for patient follow-up, exercise therapy, and research. JNCI Cancer Spectr 5(1)
41. Ferreira VV, Ghosh AK (2023) Essentials of cardio-oncology. Clin Med 23(1):52–55
42. Ohlsen TJD, Chen Y, Baldwin LM, Hudson MM, Nathan PC, Snyder C et al (2023) Primary care utilization and cardiovascular screening in adult survivors of childhood cancer. JAMA Netw Open 6(12):e2347449
43. Huang L, Shi Y, Zhao YJ, Wang L, Hu WG, Zhu ZG et al (2022) Long-term cardiac disease- and cancer-associated mortalities in patients with non-metastatic stomach adenocarcinoma receiving resection and chemotherapy: a large competing-risk population-based cohort study. World J Oncol 13(2):69–83
44. Virani SS, Newby LK, Arnold SV, Bittner V, Brewer LC, Demeter SH et al (2023) 2023 AHA/ACC/ACCP/ASPC/NLA/PCNA guideline for the management of patients with chronic coronary disease: a report of the American Heart Association/American College of Cardiology Joint Committee on clinical practice guidelines. Circulation 148(9):e9–e119
45. Vahanian A, Beyersdorf F, Praz F, Milojevic M, Baldus S, Bauersachs J et al (2022) 2021 ESC/EACTS guidelines for the management of valvular heart disease. EuroIntervention 17(14): e1126–e1e96
46. Lorusso R, Vizzardi E, Johnson DM, Mariscalco G, Sciatti E, Maessen J et al (2018) Cardiac surgery in adult patients with remitted or active malignancies: a review of preoperative screening, surgical management and short- and long-term postoperative results. Eur J Cardiothorac Surg 54(1):10–18
47. Mistiaen WP (2015) Heart transplantation in patients with previous malignancy. An overview. Acta Cardiol 70(2):123–130

Chapter 34
Modulation of Drug-Induced Oxidative Stress in Cardiovascular Toxicity

Spencer Collins, Marcel P. Fraix, and Devendra K. Agrawal

Abstract Not only does cancer pose a serious threat to longevity, quality of life, and negatively impact social health of individuals worldwide, but various cancer treatments exacerbate this burden through the progression of cardiovascular toxicity. In 2017, startling statistics estimated about 17.8 million deaths worldwide associated with cardiovascular diseases (CVDs) and continues to have an upward trend, suggesting the importance of understanding CVDs and their progression. In this comprehensive review, the prevalence of cardiovascular toxicity will be investigated, and its causes, specifically from cancer treatments. This is particularly important due to numerous cancer treatments like Doxorubicin (DOX) inducing oxidative damage in a non-specific manner targeting healthy cells like cardiomyocytes, leading to atherosclerosis, heart failure, hypertension, arrhythmias, and more debilitating diseases. This review aims to evaluate mechanisms that drive the development of CVDs through cellular death pathways. It also highlights the relevance of potential therapeutic interventions by targeting signals and modulators that contribute to oxidative stress without compromising anti-cancer drug efficacy. Researchers provide remarkable evidence through H9C2 cell samples and animal studies guiding future advancements in clinical pharmacology. For instance, interest has grown in the antioxidant mimetic effects and reduction of lipid peroxidation observed with WGX50 and GC4419.

Keywords Antioxidants · Apoptosis · Arrhythmia · Arsenic trioxide · Atherosclerosis · Autophagy · Cisplatin · Dexrazoxane · Doxorubicin · Ferroptosis ·

S. Collins · D. K. Agrawal (✉)
Department of Translational Research, College of Osteopathic Medicine of the Pacific, Western University of Health Sciences, Pomona, CA, USA
e-mail: DAgrawal@WesternU.edu

M. P. Fraix
Department of Translational Research, College of Osteopathic Medicine of the Pacific, Western University of Health Sciences, Pomona, CA, USA

Department of Physical Medicine and Rehabilitation, College of Osteopathic Medicine of the Pacific, Western University of Health Sciences, Pomona, CA, USA

D. K. Agrawal et al. (eds.), *Cardiovascular Toxicity*, Advances in Biochemistry in Health and Disease 33, https://doi.org/10.1007/978-3-031-90598-8_34

Flavonoids · 5-Fluorouracil · Heart failure · Hypertension · Immunotherapy · Lipid peroxidation · Melatonin · Mitochondria · Necrosis · Oxidative stress · Polyphenols · Radiotherapy · Reactive oxygen species · TRIM21 · Verteporfin

Introduction

Overview of Free Radicals in Healthy Cells

In normal cells, a highly regulated balance exists between reactive oxygen species (ROS), a derivative of free radicals, and non-radical oxygen species. The importance of highly regulatory mechanisms is evident through the complexity of ROS being a metaphorical "double-edged sword," which will be further discussed later. Cells and their organelles often produce free radicals and derivatives like ROS for many reasons, including differentiation, tumor promotion, and cellular death. Mitochondria are responsible for a significant amount of ROS production via oxidative phosphorylation. This production occurs primarily through the leakage of electrons in the electron transport chain, which can react with oxygen to create superoxide radicals (O_2^-). This superoxide can then be further converted into hydrogen peroxide (H_2O_2) and hydroxyl radicals (•OH), contributing to increased oxidative stress in cells. The electrons tend to leak through complexes I, II, and III of the electron transport chain, but most ROS pass through complex I [1]. In the past, it was expected that the ROS produced only affected cells negatively, but modern research has introduced the other necessary favorable roles ROS plays.

Although the mitochondria are responsible for most of the ROS production, the Endoplasmic Reticulum (ER) triggers redox reactions while undergoing protein synthesis, anabolism/catabolism of lipids, and storage of ions such as Calcium (Ca^{2+}) [2]. However, other contributors to natural ROS production include peroxisomes, cytoplasm, cell membranes, and chloroplasts in plant cells, as shown in Fig. 34.1. The balance of ROS is maintained by complex antioxidant systems, including enzymes such as superoxide dismutase (SOD), catalase (CAT), glutathione peroxidase (GPX), and thioredoxin (Trx), as well as antioxidants like vitamin C, vitamin E, flavonoids, melatonin, and dioscin. These compounds are essential in neutralizing ROS and slowing its production, which, in turn, mitigates the accumulation of oxidative stress and controls the downstream pathways [3].

Mild levels of ROS initiate both beneficial and detrimental pathways of proliferation, differentiation, necrosis, and even apoptosis [4]. Conversely, higher levels of ROS promote cell death through factors such as ferroptosis, apoptosis, necrosis, and autophagy [5]. Due to the role of ROS in high quantities in inducing cell death, an overload of oxidative stress (OS) is commonly used for cancer therapies aiming to target and eliminate tumor cells. However, this approach can lead to unintended oxidative stress and damage in other tissues, particularly toxicity in the cardiovascular system and progression of cardiovascular diseases (CVDs). This relationship underscores the importance of developing clinical strategies and synthesizing drugs

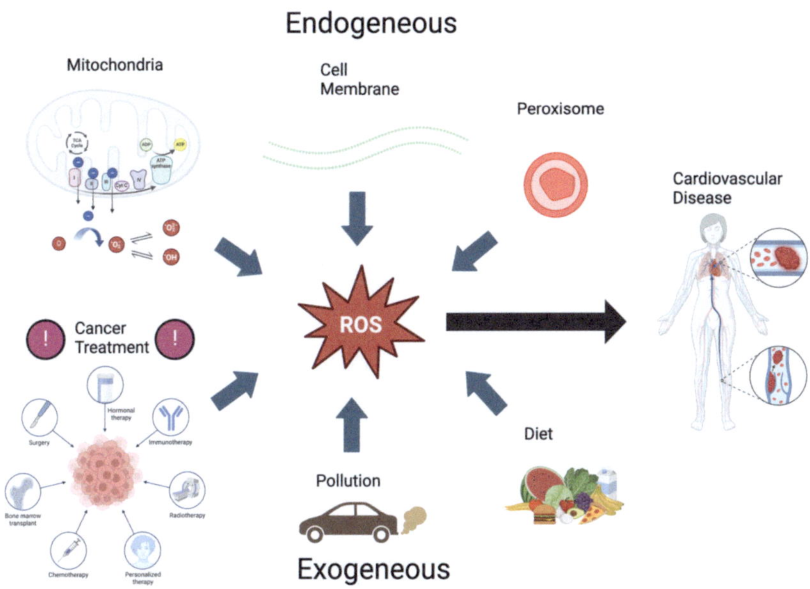

Fig. 34.1 The figure illustrates the complex interrelationship of reactive oxygen species (ROS) production via examples of endogenous (internal) and exogenous (external) sources. As highlighted by the pink exclamation marks on the bottom left quadrant, this comprehensive review extensively focuses on cancer treatments' specific role in ROS production, specifically immunotherapy, radiotherapy, and anticancer drugs that lead to oxidative stress of the cardiovascular system. Recognizing the multitude of sources contributing to ROS accumulation is vital because it demonstrates the bifunctional role of ROS in both intended therapeutic effects and unintended side effects on noncancerous cellular and biological systems. (Created with BioRender.com)

that precisely modulate the effects of cancer treatments and prevent the risk of progressive CVDs.

Toxicity Due to Reactive Oxygen Species

In instances in which ROS concentrations are raised, either by abnormal endogenous sources (such as mitochondria, ER, or cell membrane) or exogenous sources (such as cancer treatments, pollution, food, or cigarettes), cellular toxicity can occur through mechanisms such as DNA damage, voltage imbalance, lipid peroxidation, and changes of signaling pathways. It has been reported that ROS like O_2^- and H_2O_2 but mainly •OH cause DNA damage by oxidizing nucleosides, causing point mutations of purine to pyrimidine or vice versa, also known as transversion. A few examples of these mutations are mismatching base pairing of G-T or G-A. The correct base pairing for DNA is critical for the survival and development of cells; hence, the essence of exonuclease and incision for regulating pathways mismatched pairing are apparent. These various pathways, such as Base Excision Repair (BER), Mismatch Repair, and Single-Strand Annealing (SSA), are necessary for recognizing and repairing DNA damage and mutations. However, when damage co-occurs on opposing strands, mechanisms of BER attempt to repair by addressing single-strand breaks, inadvertently resulting in double-strand breaks that can cause chromosomal rearrangements or loss of genetic material [6]. Even though single-strand damage is generally less severe, it can accumulate when single-strand breaks are not properly repaired, especially when incorrect base pairs are involved. While single-strand damage may seem less significant, it still contributes to the overall toxicity associated with genomic mutations. There is a positive regulation of ROS that is activated leading to numerous results: high ROS → increase DNA damage → DNA damage response → Ataxia telangiectasia mutated (ATM) → Ataxia telangiectasia mutated and Rad3 related (ATR) → Chk2 (activated by ATM) and Chk1 (activated by ATR) → Cell Checkpoint, Immune Activation, DNA Repair, Apoptosis, Senescence. Common side effects of these genetic mutations include inflammation, mitochondrial dysfunction, and autophagy in cardiac tissue [7]. Furthermore, these lead to CVDs, such as myocardial hypertrophy, congestive heart failure (HF), cardiomyocyte oxidative stress, and ischemia-reperfusion (I/R) injuries.

Due to living organisms constantly maintaining equilibrium between oxidative and reductive agents, lipid peroxidation is a vital component of regulation, which is also a key element of downstream ferroptosis, a non-apoptotic method of cellular death. When there is an imbalance of available ROS, lipid peroxidation occurs, leading to several significant pathological degenerations and diseases, particularly compromising the cardiovascular system and promoting cancer progression through means of inhibiting tumor suppressors and enhancing oncogenes promoting proliferation. Lipid peroxidation occurs when ROS oxidizes lipid components necessary for membrane integrity and structure. For example, in polyunsaturated fatty acids (PUFAs), one hydrogen is lost from the methylene bridge (-CH_2-), creating an unstable lipid radical. The next step requires the structure to undergo resonance of the pi bond and react with oxygen to form lipid peroxide [8]. This final product of lipid peroxide is very detrimental to the membrane structure, permeability, fluidity, ion gradient, and the cells' overall health. In a study on the thermodynamics and kinetics of oxidized lipids effect on membrane and vascular integrity, scientists determined that common phospholipids like 1-palmitoyl-2-hydroxy-sn-glycero-3-phosphocholine (lysoPC) disrupt cell junctions. This disruption, in turn, increases the permeability of blood vessels, which harms arteries, veins, and capillaries through unnecessary inflammation, edema, and vascular leakage [9]. It is important to understand lipid peroxidation to discover methods of reducing the second messenger effects leading to ferroptosis, influencing inflammation, and potential cardiovascular dysfunction. The relevance of thoroughly investigating manifestation and damage to heart tissue due to lipid peroxidation, researchers are directed to explore methods to modulate unwanted oxygen species for patients undergoing cancer treatment.

Doxorubicin (DOX)

Among other anti-cancer drugs, DOX is particularly favored in clinical settings due to its effectiveness and strength. It is a part of the anthracycline antibiotic family, extracted from the Streptomyces bacterium. The indicators for DOX use for treating various cancers, including carcinomas, sarcomas, and even leukemias, make it highly effective for chemotherapeutic approaches. Despite the intended use of DOX, there are significant drawbacks due to its high reactivity and damage to cardiomyocytes, contributing to the toxicity of the cardiovascular system and leading to CVDs. The potency of DOX was evident in a study conducted by Guglin et al., showing that up to 30% of patients undergoing DOX treatment experienced acute toxicity. This was observed in ECGs within a 24-h period, indicating shocking results of atypical ST changes [10].

In general, like all anthracyclines, DOX has complex mechanisms that target and kill cancer cells, but many aspects of these processes are still not fully understood. Despite the need for future explorations, it was determined that DOX induces DNA damage via hydrogen bonding to adjacent Guanine nucleotides of complementary base pairs Cytosine-Guanine, creating DOX–DNA adducts, further destabilizing the formation of intricately twisted DNA and nucleosomes [11]. Furthermore, integration of DOX within DNA transcription and replication inhibits topoisomerases, inducing double and single strand breaks while also activating alternative downstream cell death. The most relevant cytotoxic technique is the ability of DOX to disrupt the balance of normal levels of ROS, leading to membrane and mitochondrial damage and cellular demise, contributing to the accumulation of cardiotoxicity. DOX directly interacts with the phospholipid cardiolipin on the inner mitochondrial membrane, stimulating excessive production of ROS. Also, DOX may contribute to the overproduction of waxy lipids known as ceramide, which highly promotes pro-apoptotic factors, caspases, and even the release of cytochrome C from the electron transport chain, resulting in further mitochondrial dysfunction [11]. To further demonstrate the effects of DOX, it is known to pose imbalances of iron by a reaction called the Fenton Reaction, producing anthracycline-iron complexes [12]. These imbalances contribute to lipid peroxidation and even ferroptosis. Ideally, the goal of DOX is to induce lethal effects on cancer cells to slow or stop malignancies. Unfortunately, the specificity of this damage encompasses substantial parts of the body, with cardiomyocytes being incredibly vulnerable to oxidative stress. Understanding how DOX facilitates an anticancer response to cancer cells but undesirable outcomes on healthy cells presents a dilemma for cancer treatment's benefits and consequences. Since the development of DOX was to diminish the negative impacts of cancer on one's life, there is a demand to relieve the unfavorable side effects suggested that patients face during treatment.

Verteporfin

Verteporfin, originally known for treating blood vessel disorders like macular degeneration in the eye, has shown promising clinical effectiveness against cancer cells through photodynamic therapy. This treatment increases ROS production via light-activated reactions, leading to targeted cell death. The efficacy of Verteporfin may be modulated by the highly conserved Hippo pathway, a kinase signaling cascade crucial for organ size, tissue growth, and life. In cancer cells, the Hippo pathway is inhibited, stimulating the downstream yes-associated protein (YAP), which regulates transcription factors involved in tumor proliferation, differentiation, and malignancies [13]. Interestingly, in human breast cancer, Verteporfin significantly reduced YAP and B-cell lymphoma 2 (Bcl-2) expression while increasing Bcl-2-associated X protein (Bax) expression, resulting in a 14-fold increase in apoptosis and a 50% reduction in cell viability through mitochondrial damage [13]. Side note, Bax is known to be a pro-apoptotic protein while Bcl-2 is an anti-apoptotic protein. However, despite these anti-cancer properties, the inhibition of YAP also contributes to significant oxidative stress in cardiomyocytes, potentially leading to the development of CVDs. By exploring the inner workings of Hippo pathway and Verteporfin's inhibition of YAP, clinical researchers can discover new therapeutic strategies to mitigate toxic effects while focusing on the anti-cancer characteristics.

Cisplatin

Cisplatin is the first platinum-based coordination complex, where the platinum (Pt) atom is coordinated to two ammonia (NH_3) ligands and two chloride (Cl^-) ions in a square planar configuration. The chloride leaving groups are positioned cis relative to each other, with the ammonia structures on the opposing trans plane, contributing to its stability. This basic structure of cisplatin holds significant potential that has yet to be fully uncovered. There is no evidence of its modified derivatives having stronger anticancer potential or even controlling oxidative stress, but the potential is there, waiting to be unlocked.

Analogous to DOX, Cisplatin and its derivatives are not selective to cancer cells only and have various modes in which they degrade normal cells. One of the main methods of inhibiting tumor growth is cisplatin's ability to target DNA by suppressing essential functions of DNA synthesis, repair, and modification. The presence of Pt atoms drives the prolonged period a cell stays in the G2 phase of the cell cycle, ultimately triggering apoptosis [8]. Different cisplatin analogs have proven to disrupt reactions of purine bases, break hydrogen bonds, disorganize DNA, and induce single-strand and double-strand breaks. Since these compounds target all forms of dividing cells, not just tumor cells, it leads to the accumulation of myocytes undergoing premature death and the oxidative stress of multiple body systems. Cisplatin interacts and activates signaling pathways, including MAPK, caspases, Bcl-2, p53, and p21, that are discussed in later sections and their progression of CVDs like heart failure, hypertension, and risk of stroke [14].

Arsenic Trioxide (ATO)

ATO was initially discovered when elemental arsenic is oxidized in the air, forming this complex compound and showing relatively minimal toxicity against leukemias. Evidence reveals that ATO inhibits tumor stem cells, promotes increased intracellular ROS production, and disrupts cell cycle progression. Researchers observed that even at a concentration as little as 5 μM, ATO significantly decreased the viability of rat C6 and 9L cell lines by disrupting the harmony of Bcl-2/Bax, leading to either autophagy or apoptosis [15]. Additionally, ATO impairs antioxidants such as heme oxygenase-1 (HO-1), glutathione, and Nrf2, contributing to the dangerous effects of free ROS [15]. ATO also facilitates the transportation of Fe^{2+} into the mitochondrial matrix via the Mitoferrin-2 channel protein, resulting in amplified Fe^{2+} influx and a subsequent elevation of ROS, leading to mitochondrial damage and oxidative stress. Despite its clinical success in depleting cancer cells, ATO also damages normal cells by obstructing cell cycle checkpoints, dismantling mitochondria, and directing cellular pathways toward autophagy and apoptosis.

5-Fluorouracil (5-FU)

5-FU, a lesser contributor to oxidative stress in cancer treatments, has its own specific mechanism for inducing apoptosis. 5-FU activates apoptotic pathways such as the tumor necrosis factor-related apoptosis-inducing ligand (TRAIL) by increasing p53 and Bax levels, elevating downstream cleaved caspase levels [16]. While 5-FU shows potential in treating aggressive cancers like lung cancer, it minimally contributes to oxidative stress. Its primary mechanism involves targeting DNA and programmed cell death through alternative pathways, separate from ROS-mediated implications. It is crucial to take note of these therapeutic effects because 5-FU induces far less oxidative stress compared to DOX, Verteporfin, Cisplatin, and ATO.

Radiotherapy and Immunotherapy

Most of the treatments mentioned previously are forms of chemotherapy; Verteporfin is particularly distinct due to its activation by light in photodynamic therapy. This section will focus on radiotherapy and immunotherapy, which involve a tumor's irradiation (IR) and enhancing immune response to tumors. Radiotherapy aims to use different modalities of electromagnetic radiation, such as X-rays and gamma rays, to target tumors and induce the production of ROS, specifically at the site of the abnormal tumor. This intervention causes DNA damage and ultimately leads to cellular death, but unfortunately, the heart is very susceptible to radiation [17]. Meanwhile, immunotherapy utilizes the body's natural ability to detect and present tumor-associated antigens and initiate specific cytotoxic T-cell responses [18]. These listed therapies exhibit promising outcomes and have practical potential for anticancer effects.

It is fascinating to notice that radiotherapy plays a significant role in enhancing the natural immune response to tumors through the application of IR. After exposure to IR, cells release damage-associated molecular patterns (DAMPs), leading to tumor recognition. Dysfunctional proteins and lipids are then identified and phagocytosed by antigen-presenting cells (APCs) [18]. These actions activate T-cell response through specific major histocompatibility complexes (MHC), allowing them to seek out and eliminate tumor cells. Furthermore, it has been revealed that innate immune responses are also activated, and various checkpoints are upregulated simultaneously. Complex pathways activate precursors of type 1 interferon, stimulating antigen-presenting cells like dendritic cells and the proliferation and differentiation of T-helper cells [18].

The relevance of these cancer treatments lies in their specificity to an area, but the disproportionate production of ROS causes oxidative stress in both cancer cells and nearby healthy ones. ROS is directly involved in autophagy, ferroptosis, and apoptosis, which will be explained deeply in the subsequent section. However, if cell death gets out of control and occurs in normal cells, it induces oxidative stress in systems, especially the cardiovascular system. The vast number of studies researching approaches to counteract the unwanted effects of cancer therapies while preserving anticancer effects demonstrates the demand and urgency to improve current cancer treatments.

Induction of Cellular Damage and Death

Cancer treatments like DOX aim to bombard tumor cells with ROS and induce cell death through several mechanisms, including ferroptosis, apoptosis, necrosis, and autophagy. Despite the intended goal of triggering cell death via ROS stimulation, in some cases, it may contribute to carcinogenesis via cell differentiation, angiogenesis of malignant cells, and tumor growth [19]. These further highlights ROS‘s unpredictability and emphasizes its duality, often metaphorically referred to as a "double-edged sword." Typically, elevated levels of ROS are very effective in promoting anti-tumorigenic pathways, but unfortunately, they are not specific to cancer cells only, initiating unnecessary oxidative stress through cellular damage and death (Fig. 34.2). Understanding the mechanisms behind uncontrolled and regulated apoptotic and non-apoptotic cell death can influence the development of ground-breaking drugs and strategies to mitigate these unwanted side effects while effectively treating cancer.

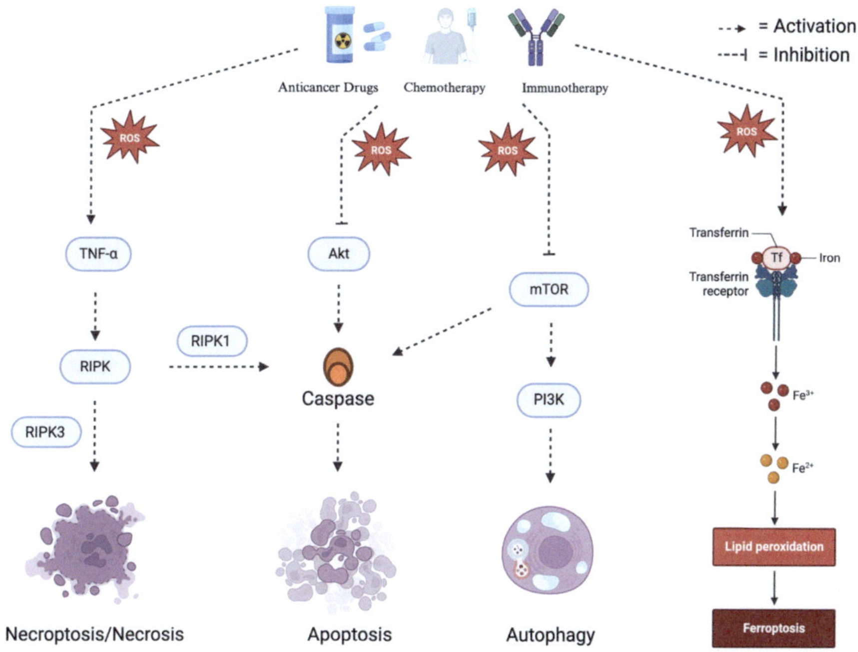

Fig. 34.2 This figure highlights the streamlined pathways and, most importantly, displays signaling molecules that are affected by ROS, leading to cell death. As illustrated by the legend in the top right corner, arrows represent activation, while lines denote inhibition. As revealed by RIPK1, there is crosstalk between necroptosis and apoptotic pathways. Additionally, the inhibition of mTOR plays a role in the activation of apoptotic pathways as well as stimulating PI3K. The right side of the figure illustrates the unique processes of ferroptosis, where ROS is generated because of Fe transported by transferrin, triggering lipid peroxidation and leading to additional damage and/or cell death. (Created with BioRender.com)

Apoptosis

DOX is a widely used chemotherapeutic treatment that is incredibly effective for cancer treatment due to its nature of inducing programmed cell death, also known as apoptosis. The stages of apoptosis involve the following key steps: shrinkage, DNA and nucleus condensation, formation of blebbing or bubbles in the cell membrane, division into apoptotic bodies, and engulfment via phagocytosis from other cell bodies and macrophages, marking the end of the cell's life. Apoptosis is crucial in quality control because it regulates damaged, dysfunctional, or potentially oncogenic (cancerous) cells. It is identified as a complex pathway that results in a cysteine protease family called caspases that cleave proteins in the cell and initiate cellular demise, beginning with shrinkage [20].

Additionally, the quantity of ROS generated by DOX leads to exaggerated apoptosis in cells and tissues that are essential for cardiovascular health, such as cardiomyocytes (Fig. 34.2). These cardiomyocytes are heart muscles responsible for persistent contraction and relaxation, ensuring efficient systemic circulation that delivers oxygen to the rest of the body and removal of waste products. Also, DOX exposure damages vital macromolecules within the cardiac muscles, disrupting integrity and function associated with developing adverse CVDs. Researchers studying the effects of FNDC5 on mitigating oxidative stress from DOX observed in a rodent model that DOX inhibits Protein kinase B (PKB/Akt), a serine/threonine kinase essential for cell survival and proliferation [21]. Furthermore, the importance of PKB/Akt activation is evident through its downstream mechanistic target rapamycin (mTOR), which stimulates P70 S6 kinase (P70) and 4E-binding protein 1 (4EBP1), promoting beneficial functions like protein synthesis and cell survival [21]. Likewise, inhibition of Glycogen Synthase Kinase 3 Beta (GSK3β) by PKB/Akt influences various cellular processes, including control of Nuclear factor erythroid 2-related factor 2 (Nrf2), a transcription factor involved in the cellular response to oxidative stress. The dysregulation provoked by ROS highlights its detrimental effects on cell death via apoptosis.

As indicated in vivo experiments conducted by this group of researchers, it was concluded that DOX's excessive bombardment of ROS inhibits PKB/Akt, which directly stops a signaling cascade relevant to cell survival and differentiation and causes apoptosis of imperative cells such as cardiomyocytes. The unfavorable outcome of inactivation of PKB/Akt from DOX is a very prevalent source of apoptosis leading to adverse CVDs, including atherosclerosis, hypertension, arrhythmias, and even hypertrophy. Remarkably, Zhang et al. observed via Immunohistochemistry staining that FNDC5 plays a very impactful therapeutic role in mitigating apoptosis of cardiomyocytes during cancer treatments and will be later discussed in greater detail [21].

Necroptosis/Necrosis

Often, apoptosis is the initial regulated form of programmed cell death, but in cases where it is inhibited or less effective, another process known as necroptosis commences.

Necroptosis is associated with inflammation and controlled cell death that involves stimuli like apoptosis. This process is mediated by cell death receptors such as FAS, Tumor necrosis factor receptor-1 (TNFR-1), and toll-like receptor-3 (TLR3). Researchers reported that a decrease in cell energy in the form of adenosine triphosphate (ATP) usually hinders apoptosis and favors necroptosis. However, if the cell death rate surpasses the ability to regulate itself, it may spiral out of control, transitioning into necrosis. The signaling cascade that modulates necroptosis begins with TNFα binding to its surface receptor TNFR1, which recruits several molecules like ubiquitinated RIPK1 to form "complex 1" and cellular inhibitors of apoptosis proteins (cIAP). Notably, cIAP is necessary for the activation of nuclear factor kappa-B (NF-κB) and mitogen-activated protein kinase (MAPK), playing pivotal roles in cell growth, differentiation, and proliferation and are generally traced back to poor cancer prognosis. In subsequent steps, the deubiquitination of RIPK1 by the enzyme CYLD causes "complex 1" to transition into "complex 2". This next step is the critical differentiation step, in which caspase-8 is either activated and promotes apoptosis or, at high levels of RIPK3 favors necroptosis. Necroptosis proceeds by RIPK1 interacting with RIPK3, in turn, phosphorylating mixed lineage kinase domain-like (MLKL) to complete necrosome assembly. Now that the process of necroptosis is in motion, the MLKL complex translocates to the cellular membrane, causing rupturing, increased permeability, and the release of DAMPs [22]. This simplified version of the multifaceted signaling pathway highlights the competitive relationship between apoptosis and necroptosis while stressing its crucial role in tumor suppression. Unfortunately, cancerous tumors tend to silence signals that weaken activation of necroptosis and sometimes allow evasion of the response, leading to the progression of cancers. By understanding the general mechanisms will aid in developing possible cancer treatments. For instance, Shikonin, a natural naphthoquinone, was reported early on as a drug that triggers necroptosis in osteosarcoma and pancreatic cancer due to its interference with the RIPK1/RIPK3 pathway [22].

While necroptosis can arise as a response to specific cancer treatments, it is not exclusive to cancer cells only. It can inadvertently affect healthy cardiovascular system cells, potentially leading to uncontrolled necrosis. It was concluded in a study that an antioxidant called.

N-Acetylcysteine significantly diminished cardiomyocyte deaths via necroptosis following aortic reconstruction and overload due to heart failure (HF), highlighting the role of ROS in driving necrosis [17]. Furthermore, a study carried out by Chi et al. established the dual relationship through a direct correlation between I/R-mediated RIP3 activation and ROS in leading to myocardial infarction (MI) and post-infarction HF. It was apparent that RIP3 is a modulator in ROS generators such as NOX. Regrettably, this creates an ongoing positive feedback loop where ROS increases RIP3 activation, which enhances NOX activity, further escalating the cell death of vital muscle cells in heart tissue. As demonstrated, RIP3 is a paramount player in downstream effects, including the opening of mPTP (mitochondrial permeability transition pore), causing membrane rupture, and dynamic processes like mitochondrial fission [23]. In summary, necroptosis serves as an alternative pathway of cellular death, with the interconnected role of ROS emphasizing the importance of understanding various cancer treatments and the unwanted collateral damage in the form of cardiovascular toxicity.

Autophagy

Autophagy, often referred to as "self-eating," involves degradation and recycling via lysosomic structures. However, when in excess, it leads cell death and tissue deterioration. It is the act of engulfing damaged and dysfunctional organelles or proteins, including pathogens. The process of autophagy plays an integral role in maintaining the effectiveness of the immune system and ensuring cellular quality control. Although apoptosis is an entirely different regulated cell death, autophagy contains numerous overlaps in the signaling checkpoint pathway. Hence, ROS can enhance tumor progression, but high levels of ROS can result in cellular and intracellular damage, activating autophagy (Fig. 34.2). The initiation begins with the inhibition of mTOR, specifically complex 1, or the activation of 5′-AMP-protein kinase (AMPK). These separate but vital signals are necessary for stimulating downstream signal vacuolar protein sorting 34 (VPS34) [22]. Additionally, through phosphorylation of phosphatidylinositol by VPS34, phosphatidylinositol 3-phosphate (PI3P) is formed, which is necessary for the formation of a complex structure of PI3P molecules that make up the foundations of a phagosome. Through complex lipidation, coupling, and docking, the complex becomes an autophagosome that fuses with local lysosomes, forming autolysosomes necessary for the degradation and recycling of essential material [22].

H_2O_2 is primarily known to increase autophagosome formations to manage destruction caused by oxidative stress. In addition to the normal H_2O_2 produced as a side product in ATP production, extrinsic ROS sources further upregulate the autophagic response. Researchers observed that H_2O_2 production leads to dysfunction in mitochondria, creating more ROS and promoting autophagy and cell death in human glioma cells [24]. These results show evidence that anti-tumorinergic approaches that bombard cells with ROS, like in DOX, may also be practical in controlling them. Additional studies determined that the exogenous H_2O_2 also stimulates cytokine, Tumor necrosis factor alpha (TNFα), leading to cellular demise by a continuous cycle of more ROS and more autophagy in response. Since cardiomyocytes are highly concentrated with mitochondria, the combination of ROS redox reactions and engulfment of affected mitochondria results in ineffective cellular function. The crosstalk between oxidative stress and autophagy is associated with the progression of CVDs, including atrial fibrillation, hypertrophy, and even cardiomyopathy (heart failure) [24]. As articulated, this complex relationship of activation of autophagy in protection against oxidative stress has unforeseen consequences, such as an accumulation of tissue degeneration.

Ferroptosis

Among the other more well-known mechanisms of cell death, ferroptosis has been recently discovered and is the most relevant when discussing oxidative stress caused by cancer treatments.

This complex non-apoptotic death is heavily dependent on iron homeostasis, lipid peroxidation, and, notably, the accumulation of ROS. Similarly to the other processes, ferroptosis has a two-fold effect on proliferation, differentiation, and senescence (aging of cells). Hence, much attention is drawn to ferroptosis in terms of the prevalence and development of CVDs [25]. The pathway initiating ferroptosis begins with iron accumulation due to transferrin upregulating iron uptake via its transferrin receptor (TFRC) and degradation of iron exporters via autophagic processes. Thus, the increase in intracellular metal iron ions leads to elevated ROS that stimulates iron-containing enzymes like arachidonic acid lipoxygenases (ALOXs). As mentioned in the introduction, this is the step in which PUFAs are incorporated into membrane phospholipids by the enzymes long-chain fatty acid–CoA ligase 4 (ACSL4) and lysophospholipid acyltransferase 5 (LPCAT3), making them susceptible to peroxidation and resulting in membrane leakiness and damage. This process leads to the formation of phospholipid hydroperoxides (PL-PUFA-OOH), which disrupt the balance of antioxidants and free ROS [22].

Additionally, there is a strong emphasis on maintaining equilibrium because a decrease in GPX effectiveness leads to uncontrolled oxidation of phospholipids. The specific morphological features are present when a cell undergoes ferroptosis, including swelling, shrinking of mitochondria, loss of mitochondrial crista, and mitochondrial membrane rupture [25]. Moreover, the common cancer treatment DOX can contribute to iron accumulation and increased concentration of ROS, which may lead to ferroptosis and downregulation of essential antioxidants, specifically GPX4. These changes, unfortunately, amplify lipid peroxidation and damage to essential mitochondria in cardiomyocytes [25]. Downregulation of GPX4 and the resulting increase in oxidative stress have been involved in various cardiovascular dysfunctions, including endothelial dysfunction and HF.

Interestingly, several studies with embryonic mouse fibroblast models have identified that knockout genes for the ACSL family, except ACSL4, do not cause ferroptosis [5]. These experiments indicate the deterministic effects of ACSL4 and introduce concepts of therapeutic interventions that may inhibit ferroptosis, ultimately reducing oxidative stress. Researchers, including Zhang et al., also mention that there is evidence of thiazolidinediones protecting mice with knockdown genes of ACSL4 that contribute to prolonged survival and reduction of oxidative stress [5]. Although this is only one example, after discussing the types of CVDs that arise, possible treatments will be compared and evaluated. As articulated, ferroptosis plays a key role in the balance between normal levels of ROS and excessive ROS leading to oxidative stress; the understanding of the steps and mechanisms is essential in therapeutic interventions for individuals undergoing cancer treatments to mitigate cardiotoxicity.

Types of Cardiovascular Diseases and Mortality from Oxidative Stress

Hypertension

The prevalence of hypertension among the population is alarming. As defined by the World Health Organization (WHO), hypertension is an unusually high blood pressure (typically 140/90 mmHg or higher) that disrupts arterial walls, blood vessels, and even organs. There is a considerable burden of untreated hypertension leading to the progression of CVDs, stroke, or coronary artery disease (CAD). Recent studies have found that underlying pathogenesis involves an imbalance of antioxidants and ROS, which affects signaling pathways and regulation of homeostasis. Notably, there is a link between the interconnectedness of available ROS and nitric oxide (NO) bioavailability.

In human and animal models, including case studies, researchers concluded that the integral antioxidant NO is depleted by ROS, causing hypertension and eventually leading to endothelial dysfunction [26]. NO is generated via redox reactions involving endothelial nitric oxide synthase (eNOS), but this enzyme is sensitive to uncoupling by highly reactive radicals [27]. NO production maintenance is crucial because this gas diffuses readily through cells, promotes vasodilation in smooth muscles, and is a cardioprotective agent. Unfortunately, when available NO levels decrease, vessels begin to undergo vasoconstriction, leading to increased resistance and hypertension. Key ROS generators within vascular walls include nicotinamide adenine dinucleotide phosphate oxidases (NOX), Cyclooxygenase (COX), and Xanthine oxidase (XO), which contribute to the reduction of NO bioavailability and the accumulation of oxidative stress. A study examining endothelial dysfunction reinforced this relationship; researchers found a significant link between oxidative stress and hypertension by analyzing COX presence in a hypertensive sample compared to a non-hypertensive control using small artery dissection and advanced Microrheological methods [28]. Conversely, there was evidence that physical activity like treadmill training or, alternatively, tai chi has some beneficial effects on oxidative stress. Still, future experiments need to be conducted to delve deeper into the intensity, duration, frequency, and type of exercise that would be most effective for alleviating symptoms and challenges of hypertension [26].

Atherosclerosis

Evidence from clinical studies indicates that atherosclerosis progresses and leads to several CVDs. This section will highlight ROS‘s complexity and roles in the context of atherosclerosis. It results from the accumulation of inflammatory cells and lipids in blood vessel walls that become unstable and rupture. Plaque rupture occurs when the accumulated contents are exposed to the bloodstream, inducing signals for clotting formation (thrombosis). The resulting thrombus can obstruct arteries or vessels, leading to inadequate blood flow to essential organs such as the brain and heart, resulting in detrimental conditions like MI or stroke [29]. Understanding ROS‘s effects on atherosclerosis can provide insight into preventing and slowing the progression of these debilitating diseases.

Immune cells play key roles in the attempt to mitigate the damage but, unfortunately, contribute to the increased disorganization and instability of the vessel walls. Exposure to low-density lipoprotein cholesterol (LDL) in the subendothelial space allows a redox reaction to begin [30]. ROS oxidizes LDL, yielding oxLDL, and drastically stimulates an inflammatory response. This response secretes cytokines and adhesion molecules, which recruit monocytes (leukocytes from the innate immune system) and additionally T-lymphocytes (leukocytes from the adaptive immune response). The monocytes move through the vessel via numerous mechanisms and differentiate into macrophages to engulf these oxLDL. This process of macrophages ingesting oxLDL produces even more cytokines and chemokines to recruit an abundance of immune response cells and create lipid-laden foam cells. The production of the foam cells results in fatty streaks and atherosclerotic plaque accumulation. With a persistent inflammatory response, the vascular smooth muscle cells form a fibrous cap over the plaque, but over time, this cap becomes unstable and ruptures, resulting in a massive thrombotic response and, hence, the obstruction of the vessels. Lastly, the foam cells also undergo apoptosis, while the accumulation of these dead cells contributes to furthering the instability of the plaque [29].

This cascade of events illustrates the detrimental effects of ROS oxidizing cholesterol. Furthermore, it shows that cancer treatments can contribute to the imbalance of the highly regulated homeostatic systems, such as the immune response. This oxidative stress has significant pro-atherogenic effects demonstrated through excessive inflammation, endothelial dysfunction, and alteration of lipid metabolism. Although therapeutic strategy research is limited, researchers like Batty et al. suggest increasing antioxidants via natural vitamins like E and D, but they mention that previous studies have shown minimal effectiveness. They suggest promising results with the integration of flavonoids and resveratrol. There is a lack of evidence due to limitations, challenges, and unknown short- and long-term side effects but this opens the door to future research [29].

Myocardial Infarction

Acute myocardial infarctions (Mis) are widespread CVDs that can give rise to severe outcomes such as heart failure and even death. Since MIs prevent blood flow to the myocardium, it is significantly involved in I/R injuries. This means that the tissue goes without oxygen for a period (ischemia), and then when the oxygen-rich blood supply returns (reperfusion), the tissue undergoes damage. MIs are detected via electrocardiograms (ECG) and found in the elevated ST segment that follows the QRS complex. To translate the findings of ECG to the actual physical condition of the heart, it must be understood that the ST segment represents the period between depolarization and repolarization. In normal heart muscles, there should be no change in electric potential; in other words, it is referred to as isoelectric. However, when the cells are damaged, the ECG reveals an abnormal deviation in membrane potential, indicating ischemia. This event is often referred to as a "STEMI" and tends to lead to irreversible damage to arterial walls and, in some cases, rupture [31]. As illustrated earlier, low levels of ROS have beneficial roles, in this circumstance, triggering pathways for the healing of myocardial tissue. Excessive production of ROS following traumatic events exacerbates tissue dysfunction, accelerates the progression of CVDs, and contributes to adverse ventricular remodeling. ROS act as a second messenger in activating the p38 mitogen-activated protein kinase (p38/MAPK) pathway, which stimulates tumor necrosis factor α (TNF-α), resulting in myocardial dysfunction [31].

Simultaneously, the antioxidant pathways, such as Nrf2, are upregulated to balance out the elevated ROS levels. Nevertheless, there are risks when the antioxidant response cannot neutralize the excessive ROS, including hypertrophy and myocardial cell death.

Moreover, platelets are involved in clotting and healing mechanisms following acute events such as a traumatic I/R injury. Platelets are activated through various mechanisms to enhance attachment and aggregation to stop a bleed following an injury. Although the upregulation is intended to prevent and conserve blood loss, it can lead to uncontrolled ROS production via enzymes like NOX. To combat the thrombotic formation caused by platelets, therapies for antiplatelets have been identified and put into effect to mitigate the risk of arterial clogging. In experimental clinical studies, researchers proposed that there is potentially a correlation between platelet downregulation and oxidative stress [31]. This underscores the importance of future research to delve deeper into the underlying mechanisms and potential therapeutic interventions, including drugs that can closely modulate the progression of CVDs that advance to MI. However, the integration of antiplatelet therapies in this study shows promising results.

Arrhythmias

Among the more prevalent CVDs, arrhythmias are characterized by varying degrees of irregularities and alterations in both atrial and ventricular rhythm. An imbalance of ROS causes chain reactions of lipid and protein oxidation, disrupting ion channel function and permeability and causing electrical instability in the heart. In ventricular arrhythmias, researchers have claimed that ROS-dependent disruptions of the physiological homeostasis of cations such as Ca^{2+}, K^{+}, and Na^{+} in cardiomyocytes contribute to defects of conductivity and electrophysiological responsiveness [32]. The sarcoplasmic reticulum is responsible for contraction due to its storage and release of Ca^{2+} ions during depolarization, resulting in contraction of the ventricles. For instance, ROS oxidize protein mediators responsible for Ca^{2+} channels, including the ryanodine receptor 2 (RyR2). Dysfunctional RyR2 contributes to Ca^{2+} leaking through and reduces the strength of the physiological opening of sarcoplasmic reticulum channels. This process increases the chances of developing ventricular arrhythmias such as catecholaminergic polymorphic ventricular tachycardia [32]. Researchers identified that ROS production in mitochondria is sustained through a positive feedback loop, where initial ROS generation stimulates further ROS production, thereby heightening cellular damage.

Pharmacological studies have investigated possible coping strategies, and researchers found that synthetic superoxide dismutase/catalase mimetic treatments mediated this paradigm, correlating to a lower incidence of arrhythmias [32]. Additionally, antagonists of a mitochondrial benzodiazepine receptor showed comparable effects, but the treatment of agonists of mitochondrial benzodiazepine receptors exacerbated the detrimental outcomes of arrhythmias [32].

Atrial fibrillation is a predominant form of arrhythmia in which rapid and chaotic electrical impulses create an irregular rhythm in the atrium. Some individuals describe this phenomenon as it almost looks like the heart is undergoing disorganized, 'quivering' facilitating a lack of oxygenated blood into ventricular chambers and diminishing oxygen delivery to the rest of the body, contributing to hypoxia. Oxidative stress remodels the heart's electrical system by reducing the refractory period, decreasing the velocity of conduction of action potentials, and reducing wavelengths [33]. Similarly to ventricular arrhythmias, RyR2, Na^+, and gap junctions are particularly impacted by accumulated oxidative stress. As Ca^{2+} ions leak from the sarcoplasmic reticulum, inward currents generation can be strong enough to propagate a depolarized action potential, leading to a premature contraction of the atrium, as shown in atrial fibrillation.

Next, gap junctions are formed by connexin proteins and play a vital role in the development of arrhythmias connexin proteins form them. Gap junctions are the intercellular channels between cardiomyocytes, promoting seamless transfer of action potentials throughout the heart and synchronizing contractions for coordinated heart function. Although researchers did not fully understand the complexity of mechanisms, their rat study concluded that hyperphosphorylation of connexin40 and connexin43 caused by oxidative stress plays a significant role in hypertrophy and enlargement of atria, leading to hypoxia [33].

Lastly, Na^+ is a crucial participant in the action potential formation due to its depolarization when entering the cell. Hence, a decrease of this inward current results in inflammatory cytokines like NF-κB, making the cell hypersensitive to oxidative stress. Exploring the role of ion imbalances and oxidative stress in the progression of atrial fibrillation could provide insights into potential preventive or therapeutic strategies. Researchers highlight the benefits of using antioxidants such as vitamin A and the pre-existing use of statins for post-atrial fibrillation management. Groundbreaking studies from Zhang et al. investigated the implementation of dipeptidyl peptidase inhibitors (DPP) because of their anti-inflammatory, antioxidant, and anti-fibrotic effects, as well as positive evidence in obese and diabetic rabbit models [34]. They found that DPP reverses electrical irregularities by restoring mitochondrial function and preventing the self-perpetuating cycle of ROS production. This highlights the necessity of ongoing research to reduce the detrimental outcomes caused by arrhythmia.

Heart Failure (HF)

As one of the leading causes of mortality worldwide, HF is identified as a chronic CVD that impairs the heart's normal ability to pump blood, thus resulting in insufficient oxygenation of the body's tissues. Most of the CVDs listed above contribute to the development of HF. There are several reasons and complex mechanisms for the occurrence of HF. Since mitochondria are essential for providing cardiomyocytes with the necessary ATP for optimal function, their dysfunction provoked by ROS lead to various forms of HF. The protective effects of scavenging ROS via peroxiredoxin (Prx) or Gpx systems were demonstrated in animal and human models by comparing antioxidant-deficient samples to the control treatment. These systems are reliant on the production of NADPH by enzymes isocitrate dehydrogenase 2 (IDH2) and nicotinamide nucleotide transhydrogenase (Nnt). Therefore, NADPH is essential for decreasing glutathione and thioredoxin oxidized species. Interestingly, observations from mammalian studies reveal that overexpression of NADPH-producing enzymes alters the production of ATP, which in turn contributes to hypertrophy, cardiomyopathy, and the progression of chronic HF [35].

Furthermore, there is evidence that ROS further contribute to the disruption of Ca^{2+} homeostasis, leading to ineffective contractions and maladaptive remodeling. These processes contribute to the progression of cellular death in cardiomyocytes and activate pathways, including MAPK, hypoxia-inducible factor signaling, and inflammatory responses. Researchers are interested in investigating therapeutic interventions to increase the NAD^+/NADH ratio because this could reduce contractile dysfunction, strengthen mitochondrial tolerance to oxidative stress, and even slow the progression of chronic HF [35]. Given the promising anti-inflammatory and antioxidant properties of SGLT-2 inhibitors, a recent meta-analysis illustrates that these drugs may have superior efficacy and favorable safety profile compared to other treatments, such as reducing hospitalizations for HF compared to other therapies like GLP-1 agonists and DPP-4 inhibitors [36]. This underlines the ongoing need to explore and refine therapeutic strategies that address the complex interplay between oxidative stress and HF advancement.

Drugs That Modulate ROS

Pharmacological and Drug-Based Interventions

WGX50

As aforementioned, anticancer treatments pose a significant risk of oxidative stress, leading to the overproduction of ROS, mitochondrial damage, iron accumulation, and various forms of cell death. These all are responsible for the development and progression of CVDs, which have a burden on patients undergoing cancer treatment. Tai et al. suggest that the only iron chelator currently approved by the U.S. Food and Drug Administration (FDA) is Dexrazoxane, which reduces the likelihood of cardiotoxicity induced by chemotherapies. Furthermore, the only FDA-approved antioxidant drug is Probucol, but it is not prevalent in clinical treatment anymore due to its severe side effects, such as hepatotoxicity and cardiotoxicity [3]. This indicates a need for safer and more efficient interventions to mitigate the damage. Hence, this research team investigates the efficacy of WGX50, a small molecule derived from a Chinese food flavoring and used in herbal medicine, Zanthoxylum bungeanum Maxim [2]. They underwent an in-depth analysis of male C57BL/6 mice with four treatment groups: control treatment, DOX only, DOX + Ferrostatin-1 (a potent ferroptosis inhibitor), and DOX + WGX50. Significant results show that the treatment of DOX + WGX50 had effects equivalent to those of DOX + Ferrostatin-1 when compared to DOX alone regarding cellular death via ferroptosis. Additionally, the outcomes of DOX + WGX50 exhibited antioxidant effects by protecting available GPX and reducing Fe^{2+} accumulation compared to the DOX-alone group [2]. The current drug, Dexrazoxane, has drawbacks, including the possible interference with the anticancer effects of DOX through the degradation and reduction of topoisomerase II. The importance of this study showed that WGX50 is a plausible potent alternative because it does not obstruct the effectiveness of DOX but demonstrates clinical potential by preserving mitochondrial function, reducing ROS by protecting GPX, and regulating ferroptosis and lipid peroxidation processes [2]. Furthermore, this study provided groundbreaking insights into WGX50 as a protective agent against the cardiotoxicity and cardiomyocyte damage that many people living with cancer face throughout their treatment.

GC4419/Ebselen

Given the principles of ROS generation, their central role in the differentiation and proliferation of all cells, and their detrimental effects by damaging cells like cardiomyocytes, ROS must be carefully regulated to maintain cellular health. Many researchers focus on redox-active drugs that balance the oxidation of chemotherapies while minimally disrupting treatment efficacy for cancers. Initial investigations were conducted on SOD mimetic drugs, specifically GC4419, which is still being developed to amplify the effects of radiation but also protect healthy cells, particularly during head and neck cancer treatments [12]. Furthermore, the drug Ebselen, a synthetic organoselenium compound that mimics GPX, shows potential in modulating the p53 pathways due to its GPX features. Additionally, it plays a dual role in eliminating cancer cells and protecting normal cells. Evidence suggests that Ebselen may have promoted cellular damage along with inducing apoptosis in cancer cells, both of which are essential steps for tumor elimination. Conversely, Ebselen demonstrated potential protection in normal cells against radiation-induced damage by reducing ROS levels and minimizing unnecessary inflammation [12]. This section highlights the potential of redox-active drugs that enhance anticancer treatment while also retaining antioxidant features critical in reducing OS, especially in sensitive tissues like the heart.

Tripartite Motif-Containing Protein 21 (TRIM21)

As demonstrated, DOX-induced toxicity and ferroptosis in cardiomyocytes impose a significant burden by leading to CVDs. Remarkable advancements have been made utilizing the RING finger domain-containing ubiquitin E3 ligase TRIM21 due to its negative regulation on Kelch-like ECH-associated protein-nuclear factor erythroid 2-related factor 2 (p62-Keap1-Nrf2) pathway homeostasis [37]. Using a TRIM21 knockout (TRIM21−/−) mice model with C57BL/6J background compared to control (TRIM21+/+) treatment. Notable results suggest that DOX treatment for TRIM21−/− mice led to longer survival, maintained body weight, and reduced obvious hair loss compared to their control counterparts. Furthermore, the TRIM21 +/+ mice exhibited irregularities in transthoracic echocardiography, indicating arrhythmias, decreased ventricular wall motion, enlarged heart, and dilated cardiomyopathy, all progressing to CVDs [37]. Another recent study revealed the mechanism by which TRIM21 ubiquitylates and inhibits p62 at the lysine residue (K7), thereby regulating autophagy. In the canonical p62-Keap1-Nrf2 pathway, p62 plays a pivotal influence by dimerizing, reducing the antioxidant response, and contributing to the surge of oxidative stress [38]. Conversely, researchers reported that overactivation of p62 and Nrf2 leads to tumorigenesis and proliferation, while TRIM21−/− markedly disrupts oncogenesis. TRIM21 deficiency plays a key role in the enhancement of antioxidant responses along with modulating pathways that determine cancer prognosis. Additionally, TRIM21 deficiency mice showed positive outcomes on the cardiovascular system compared to wild-type mice undergoing DOX treatments, likely due to their strengthened antioxidant defenses. Although a mouse study, it represented the potential use of interventions that can be transferred to humans undergoing potent anticancer treatments, such as DOX, to mitigate unwanted side effects of cardiovascular damage without compromising the desired anticancer properties.

Dipeptidyl Peptidase Inhibitors (DPP-4)

With a growing link between diabetes and the progression of CVDs via oxidative stress, researchers studied the effectiveness of DPP-4 inhibitors in treating arrhythmias, particularly atrial fibrillation. These agents are a class of drugs generally used for individuals diagnosed with diabetes mellitus because they inhibit the degradation of incretin hormones that enhance insulin production by pancreatic beta cells. Even though the desired effects of these enzyme inhibitors on blood glucose homeostasis, there is evidence that they also mediate pathways involved in antioxidant, antifibrotic, and anti-inflammatory responses [33]. For instance, Zhang et al. experimented with 90 rabbits, carried out by diagnosing and randomly assigning them to three conditions: a control group, a group induced with diabetes mellitus, and a group treated with Alogliptin (a DPP-4 inhibitor). Further results demonstrated significantly higher rates of atrial fibrillation in the diabetes mellitus-induced group compared to the control group. Moreover, the group treated with DPP-4 inhibitors showed a drastic reduction in atrial fibrillation rates compared to the untreated diabetic group. Additional evidence indicates that DPP-4 inhibitors mitigate myocardial damage by mitochondrial remodeling, maintenance of mitochondrial membrane permeability, and alleviation of inflammation [34]. The mechanism by which these agents scavenge ROS and reduce oxidative stress is still unknown, but experiments illustrate a relationship between Alogliptin and the enhancement of pathways like NRF2, PGC-1α, and Tfam, which were inhibited in the untreated diabetic condition [34]. This remarkable observation plays a crucial role in uncovering how DPP-4 reverses damage induced by oxidative stress, opening possibilities for the prevention or treatment of cardiovascular toxicity caused by cancer treatments.

Natural and Emerging Antioxidant Therapies

Melatonin

N-acetyl-5-methoxytryptamine, also commonly known as melatonin, is produced by the pineal gland and is known for regulating circadian rhythm. There is noteworthy evidence that melatonin may offer potential anticancer advantages along with the reduction of CVDs that are exacerbated by ROS production. In rat models with exposure to toxic agents, including anthracyclines DOX, Epirubicin, and Daunorubicin, as well as harmful elements like lead and dioxin, showed significant favorable outcomes when treated with various doses of melatonin. In the table outlined by Zare et al., rats had a decline in apoptosis activity, mitochondrial damage, lipid peroxidation, cardiac damage, and oxidative stress [39]. Furthermore, there is evidence that melatonin improves cardiac function and antioxidant activity through both direct and indirect mechanisms. Melatonin binds to intracellular proteins along with nuclear receptors, which stimulates the production of GPX and SOD [39]. On the molecular level, components like the acetyl group (-C-O-CH_3) contribute to lipophilicity, enabling it to permeate through cell membranes, while the stability of indoleamine allows it to donate electrons, neutralizing ROS [40]. The therapeutic advantages of melatonin are evident in mitigating stress caused by atherosclerosis, hypertension, and QT prolongation, highlighting the potential for drugs that stimulate increased production by the pineal gland or even supplementation of extra melatonin for patients undergoing cancer treatments. Moreover, not only does melatonin directly inhibit the production of new ROS, but it also scavenges existing ROS. Its indole ring allows it to delocalize electrons after neutralizing ROS and maintaining its stability. Experiments show that melatonin directly protects DNA from oxidative damage, specifically from H_2O_2, UV damage, and even metal oxidation [41]. As evidenced by several sources, melatonin has a versatile role in cardioprotection, which presents many opportunities for researchers to investigate the beneficial outcomes of weighing potential contraindications.

Dioscin

Dioscin is a steroid saponin often found in Chinese herbal medicine and can be extracted from various plants. With Dioscin's evidence of biological activity in reducing inflammation, cancers, diabetes, hyperlipidemia, and oxidative stress, researchers monitored its effectiveness on H9C2 cells derived from 50 rats and 50 mice. Zhao et al. are interested in Dioscin's effectiveness in alleviating DOX-induced oxidative stress that tends to lead to CVDs from DNA, mitochondrial, and overall myocardial tissue deterioration. Their research highlighted that microRNA-140-5p plays a pivotal role in modulating the Keap1-Nrf2 pathway, which regulates the expression of antioxidant enzymes like GPX and SOD, protecting cells from oxidative stress [42]. Despite the mechanism not being clearly understood, there was convincing evidence revealing that the concentration of Dioscin does not correlate to cellular toxicity but generally decreases the cardiotoxic effects of DOX. With the comparison of the combined effect of DOX + Dioscin treatment and ECG results to the control, there was a significant improvement in terms of ST intervals along with higher concentrations of free antioxidants, including glutathione and SODs [42]. In terms of microRNA-140-5p, there was drastic inhibition, in turn decreasing ROS and eventually diminishing oxidative damage. Dioscin demonstrated both prevention and treatment benefits for mitigating the progression of CVDs and highlights the saponins potential in clinical settings.

Herbal Monomers

Polyphenols have the potential to inhibit oxidative stress and are crucial for downstream activation of antioxidant pathways. Several polyphenols can be extracted from plants and have common features, such as aromatic rings interconnected by hydroxyl groups. The structures of polyphenols are particularly important, as their aromaticity ensures stability and the ability to delocalize electrons after donating the hydrogen atom from the OH group to neutralize ROS. It is worth mentioning that polyphenols can modulate a wide variety of signaling pathways as well as facilitate the inhibition of proteins, including downstream signals such as NF-κB. For example, curcumin, a lipophilic polyphenolic compound, has shown positive outcomes by reducing inflammation and atherosclerosis through means of inhibiting oxLDL and lipid peroxidation [3]. In addition to mitigating CVDs, experimenters observed that curcumin directly stimulates the PI3K-Akt pathway, thus promoting survival. It also inhibits p53 and activates Nrf2, promoting the production of antioxidants to scavenge ROS and reduce oxidative stress [3]. With researchers increasingly interested in curcumin's antioxidant, anti-cancer, and anti-inflammatory properties, they studied its positional isomerism to identify the isomer with the most prominent antioxidant activity. Although the placement of the hydroxyl groups in the aromatic rings at the para-substituted position exhibited significant antioxidant behavior, the ortho-orientation was much more stable, exemplifying the vital role of the methoxy group [40]. By demonstrating that curcumin generally has potential in therapeutic interventions for mitigating oxidative stress, future engineering of different isomers may result in more potent ways of reducing ROS-induced damage. With the vast range of polyphenols and possibilities, curcumin stands out for its clinical application, offering therapeutic benefits in combatting oxidative stress and specific diseases.

A subgroup of polyphenols that have meaningful potential are flavonoids. Their basic structure is notable due to its two benzene rings connected by a heterocyclic pyrane ring with diverse oxidative states depending on the flavonoid compound. Interestingly, since plants contain an abundance of polyphenolic compounds, humans naturally consume them as they contribute to the taste, color, protection against oxidation, and even production of vital vitamins and enzymes [43]. A wide variety of flavonoids, including flavanol, flavanone, and flavonol, exist in many consumable sources. For example, they are often consumed via teas, tomatoes, red peppers, fruit skin, onions, berries, and more. The prevalence of flavonoids in food sources and the lower incidence of disease in ancestors led researchers to wonder if it had to do with the health advantages of flavonoids. The prevalence of flavonoids in ancestral diets and the lower incidence of disease among early populations led researchers to question whether the health benefits could be attributed to flavonoids. Researchers observed that flavonoids play a pivotal role in pathways of antioxidation, antibacterial, anti-inflammation, and anti-viral responses. The antioxidation mechanisms are fascinating because the hydroxyl groups act as strong reducing agents, and their ability to stabilize the resulting radical through ring conjugation allows them to scavenge ROS. Notably, this process is favorable in a thermodynamic sense, as demonstrated by the lower reduction potential of flavonoids compared to ROS [44]. Beyond scavenging ROS, flavonoids participate in inhibiting superoxide producers like NOX, COX, and XO. The competitive and noncompetitive inhibition mechanisms underscore flavonoids'role in binding to NF-κB, disrupting Keap1 from binding to Nrf2 and preventing the Fenton reaction [44]. These mechanisms suggest a strong potential for therapeutic interventions in preventing oxidative stress during cancer treatments.

One of the most notorious examples of cancer treatment-induced oxidative stress is DOX, but there is evidence of flavonoids such as quercetin and apigenin having cardioprotective activity. Quercetin, commonly found in berries, apples, onions, pears, and spinach, has been shown to have a range of health benefits. In experimental trials involving DOX-induced toxicity in H9c2 cells, it was observed that quercetin added to DOX treatment drastically decreased apoptosis rates compared to the DOX-only control. Indeed, quercetin can neutralize a hydroxyl radical by donating hydrogen from its C4' hydroxyl group and thus scavenge ROS and prevent oxidative stress (Fig. 34.3). This protective activity is related to its downregulation of ERK and MAPK pathways in cardiac cell injuries [45]. Similarly, the flavone apigenin plays a role in a range of beneficial activities, including modulating ROS levels by upregulating the PI3K/Akt pathway and downregulating NF-κB [45]. Along with these features, apigenin plays a pivotal role in producing

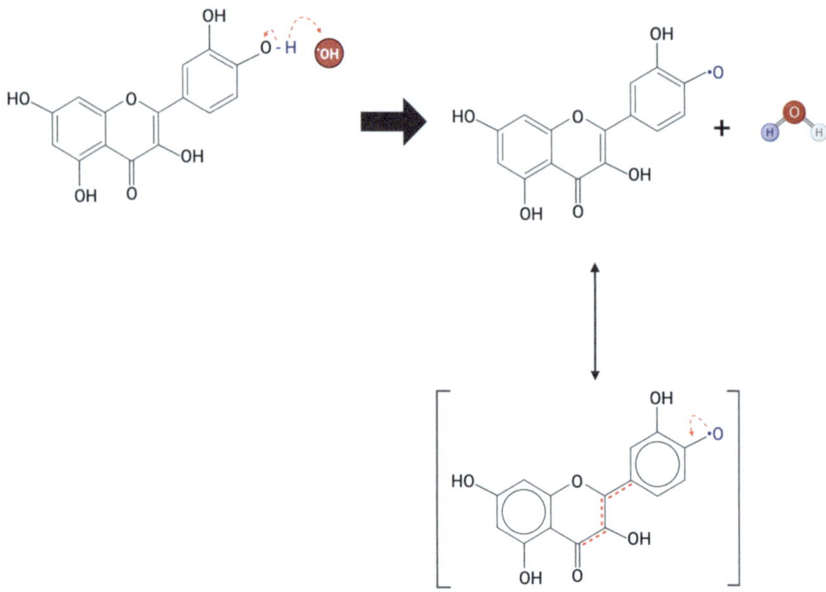

Fig. 34.3 This figure displays the process of quercetin neutralizing a hydroxyl radical by donating hydrogen from its C4′ hydroxyl group. This action extinguishes free ROS, which results in the production of water and formation of radicals in the B-ring intermediate structure of quercetin. The unstable electron is delocalized through resonance, as illustrated by the resonance hybrid in brackets, with the red dotted lines indicating the conjugation of pi bonds. The figure illustrates the antioxidant abilities that molecules can implement to scavenge ROS and prevent oxidative stress. (Created with BioRender.com)

antioxidants and has demonstrated significant cardioprotective activity when treated in conjunction with DOX.

Lastly, researchers identified Myricetin (3,5,7,3′,4′,5′-hexahydroxyflavone) as a central player in mitigating the stress caused by 5-FU-induced cardiotoxicity. Animal studies determined that Myricetin treatment significantly decreased caspase activity and Bax pathways, thereby reducing apoptosis. Additionally, direct measures of lipid peroxidation were significantly reduced [46]. Many of the results also show excessive ROS production in the control group of 5-FU alone. In comparison, in 5-FU treatments with Myricetin, there is substantial control and balance of ROS. These are just three prevalent flavonoids that highlight the versatility and benefits of a flavonoid-rich diet. Also, there is potential to extract and engineer these polyphenols with increased potency, thereby reducing the oxidative stress caused when undergoing cancer treatments.

Overall, natural and emerging antioxidant therapies can mitigate oxidative stress and thus inducing beneficial effects within an animal cell via several mechanisms, including DNA protection, prevention of lipid peroxidation, mitochondrial remodeling, neutralizing ROS, and converting O_2^- (Fig. 34.4).

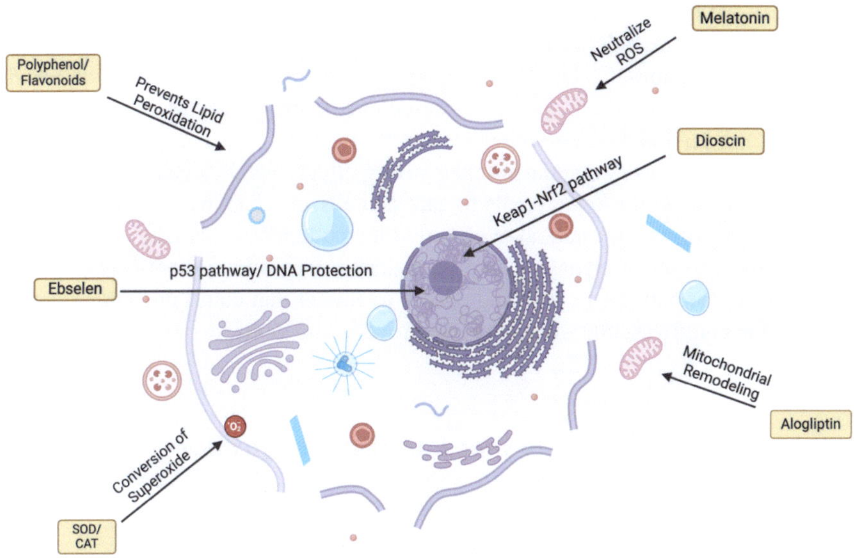

Fig. 34.4 Overview of drugs and natural antioxidants mitigating oxidative stress and inducing beneficial effects within an animal cell. This figure illustrates the site of action for each intervention and highlights their unique pathway to combat oxidative stress. The arrows indicate specific cellular targets for each compound, including DNA protection, prevention of lipid peroxidation, mitochondrial remodeling, neutralizing ROS, and converting O_2^-

Fibronectin Type III Domain-Containing Protein 5 (FNDC5)

The ongoing study of FNDC5 is particularly interesting to researchers as it is a protein that plays a key role in balancing metabolism and energy regulation while offering insight into CVDs. FNDC5 spans the membrane with its hydrophobic residues and two fibronectin domains, anchoring it in a transmembrane orientation. The structure illustrates its role in the essential transportation of molecules and signals. A pivotal hormonal signal that FNDC5 modulates is the cleavage and production of irisin, and researchers have identified that deficiencies in these processes correlate with the development of CVD, specifically atherosclerosis, hypertension, cardiac hypertrophy, and I/R injuries [21]. This relationship guided an interest in investigating the potential of preserving mitochondrial function and reducing apoptosis, which leads to oxidative stress. Zhang et al. explored the effects of FNDC5 on mice hearts and H9C2 cells in terms of reducing the oxidative stress induction of DOX by methods including Western blotting, immunohistochemistry, and echocardiograms [21]. Regarding heart functionality, a reduction of heart weight means poor prognosis through atrophy and dysfunction. Zhang et al. shows that heart weight/tibia length was approximately 6 mg/mm for saline samples, and DOX/FNDC5-treated samples also remained at about 6 mg/mm. Conversely, in the DOX alone sample, there is a significant decrease to about 5 mg/mm [21]. This evidence is critical to note because atrophy suggests toxicity and the potential failure of mitochondria to produce adequate energy for normal heart function, leading to the progression of serious CVDs. Furthermore, comparing trials of FNDC5-deficient samples to FNDC5-rich samples, an increased rate of apoptosis was indicated by the upsurge of Bax and C-Caspase3 concentration levels and a decrease of Bcl-2 [21]. These experiments demonstrate the strong effects FNDC5 can have on tissue samples, from musculoskeletal cells to cardiac cells. As displayed by the group of researchers, there was substantial evidence of its modulatory characteristics on both survival and cell death signals. There is evidence suggesting potential therapeutic interventions for patients undergoing cancer treatment that could prevent oxidative stress of the cardiovascular system or even reverse the damage.

Cardioprotective Strategies and Lifestyle Modifications

Diet

Beyond pharmaceutical interventions, lifestyle modifications, including dietary monitoring and exercise implementation, show significant cardioprotective effects. First, scientists have been intrigued by the red pigment commonly found in tomatoes and tomato-based food called lycopene. Lycopene sparked curiosity due to its natural ability to inhibit lipid peroxidation, oxidative stress, and oxLDL while activating SODs and GPXs. There are numerous meta-analyses on humans, along with experimental animal studies, demonstrating the positive impact of including lycopene in the diet. It was observed that lycopene has anti-atherosclerosis effects and decreases cases of hypertension through important signaling pathways. In human studies, researchers found that lycopene supplementation in CVD patients improved endothelium-dependent vasodilatation by 53%, lowering blood pressure [47]. In Male BALA/c mice trials, myocardial tissue was preserved after reducing the expression of cytokines like TNF-α and decreasing apoptotic signals like caspase [47]. Interestingly, natural compounds like lycopene could be enhanced through dietary behaviors by eating more tomato-based foods. There is strong evidence of the potential of cardioprotection via diet, but scientists do mention possible limitations, including bioavailability and current technology. In another experiment, the antioxidants CAT and SOD levels rose; CAT increased from 5.82 to 12.35 U/mg protein and SOD from 1.10 to 8.69 U/mg protein [48]. Lycopene has diverse natural applications, as it mitigates oxidative stress through mechanisms that directly impact pathways and improve levels of available antioxidants. The intake of foods containing lycopene has the potential to protect against cardiovascular toxicity but could also be used in conjunction with drugs. Although lycopene in tomatoes draws significant attention, foods including grapes, pomegranates, grapefruit, garlic, spinach, beetroot, wheat, saffron, sage, and ginger have beneficial antioxidant activities against DOX and other cancer treatments.

Exercise

Moreover, there are non-pharmacological interventions that prevent cardiotoxicity, and exercise has been recognized for its outstanding benefits and antioxidant properties. Exercising promotes the expression of ATP-binding cassette transporters that maintain homeostasis of anticancer drugs by pumping excess compounds out of the cell to deter accumulation and adverse effects from medications like DOX [49]. Additionally, researchers have identified that exercise defends mitochondria against amplification of permeability and swelling. Exercise interferes with the signaling cascades relevant to causing autophagy and apoptosis through Bax/Bcl-2 ratios [49]. Lastly, it is worth noting that exercise strongly regulates immune cell function, particularly with the adaptive immune system, B cells. There is evidence that B cells infiltrate cardiac tissue moderated by chemokines that are reliable for damage to tissue and other CVDs, as observed in myocardial infarction [50]. A study of mice showed that exercise contributes to cardiomyocyte apoptosis and fibroblast activation when compared to treatment of no exercise [50]. This study was groundbreaking as it introduced the interconnectedness of all body systems and how they relate to toxicity caused by cancer treatments. While finding an exact method to modulate cardiovascular toxicity is complex, the combined use of these subtle interventions can be highly effective. Exercise offers many benefits, including influencing signaling cascades, promoting anti-inflammation or anti-apoptotic signals, and reducing immune system damage, which exacerbates toxicity. This accentuates the critical role of physical activity in mitigating damage to normal cardiac cells, making it a necessary focus for future research.

Conclusion

The burden of cancer and intense treatments is enough to be overwhelming, but the additional strain of oxidative stress further exacerbates the challenge. On top of the normal production of ROS endogenously, cancer treatments such as DOX highly produce ROS, leading to premature cellular death via apoptosis, autophagy, necroptosis/necrosis, and ferroptosis. There are many intended effects of generating excessive ROS to eliminate cancer and deter the proliferation of cancer cells. In contrast to the intended effects, there are several negative activities, including damaging mitochondrial function, dysregulation of ions, and degradation of healthy cardiomyocytes that progress into CVDs, including hypertension, atherosclerosis, MI, arrhythmias, and even heart failure. It is important to identify antioxidant drugs and natural emerging therapies to mitigate damage in vulnerable patients fighting cancer and improve quality of life without reducing the effectiveness of anticancer treatments. There have been remarkable discoveries and strong evidence that drugs, including WGX50, GC4419/ Ebselen, TRIM21, and DPP-4, improve antioxidant behavior and mitigate oxidative stress caused by cancer treatments. Moreover, there have been studies involving natural therapeutic methods, including melatonin consumption, polyphenols, dioscin, and FNDC5, as well as the implementation of physical activity and nutrition strategies that reduce the detrimental effects of oxidative stress. As mentioned, there are many limitations to current rodent/animal models but there is evidence of effectiveness in humans, though it is difficult to experiment with individuals as there are ethical dilemmas.

Additionally, there are many unknowns when discussing mechanisms, crosstalk, and complex roles of compounds. Thus, there is an urgency to develop interventions that mitigate the damage of ROS with minimal side effects. Significant upcoming studies will aim to discover and establish suitable treatments that can be implemented in clinical practice to modulate oxidative stress for individuals undergoing cancer treatments. Future research will address questions concerning the long-term effects of antioxidant therapies, in-depth comparisons between synthetic and natural remedies, and investigate optimal timing and dosage. In conclusion, this review has covered significant information and hopefully paves the way for future insights for modulating ROS production and reducing cardiotoxicity exacerbated by various antitumor treatments.

Outstanding Questions and Challenges

While this review covered extensive insights ranging from mechanisms, prevention strategies, and treatment approaches, several questions remain unanswered. Previous studies were conducted in controlled laboratory environments, revealing challenges in translating findings into practical, patient-centered clinical applications. Furthermore, the studies involved samples of mice or other animal models, raising additional considerations about their application to human patients. This raises an essential question of how these conclusions can be transferred to real-world populations' diverse, dynamic conditions. The focus should be on addressing this challenge to bridge the gap between research and individualized patient outcomes in the future.

This scientific uncertainty underscores the daily ethical dilemmas healthcare professionals encounter, particularly in balancing treatment efficacy and potential long-term impacts on patient quality of life. How can clinicians justify using ROS-generating therapy when the treatment ultimately impacts a patient's longevity and quality of life? In contrast, there is an immediate threat if the team of healthcare providers does not address the cancer diagnosis. As articulated, when an ethical situation arises, there must be a clear indication that the benefits outweigh the unintended risks. With treatment ethics in mind, to what extent must patients be informed to properly determine if the cancer risk outweighs the potential for cardiovascular toxicity? Therefore, this is a complicated situation for patients, families, and even providers since all directions have the potential for adversity and cause a burden on lives.

Future research aspired to decrease CVD development during cancer treatments is promising, particularly in the development and implementation of drug alternatives that do not induce ROS generation. Moreover, this review highlighted the role of diet and exercise in scavenging ROS, though there are no definitive guidelines for duration or intensity. However, transferring these results to diverse populations is difficult since individuals differ in body type, environment, sex, and genetic background.

Key Points

- ROS exists in diverse forms, including O_2^-, H_2O_2, and •OH, all necessary for cell proliferation at physiological levels. However, excess ROS levels promote cellular death.
- Cell death occurs through specific processes known as ferroptosis, apoptosis, necrosis, and autophagy, which are influenced by various signaling molecules.
- As illustrated in Fig. 34.1, there are endogenous sources of ROS production, including mitochondria, cell membranes, and peroxisomes, while exogenous sources encompass factors such as cancer treatments (anticancer therapy, chemotherapy, immunotherapy), pollution, and diet.
- DOX is a well-known anthracycline chemotherapeutic agent, while verteporfin, cisplatin, ATO, and 5-FU are classified as non-anthracycline chemotherapeutic agents.
- ROS plays a significant role in the progression of CVDs, including hypertension, atherosclerosis, MI, and HF.
- WGX50, GC4419/ Ebselen, TRIM21, and DPP-4 are clinical pharmaceuticals that have antioxidant properties and demonstrate evidence of controlling oxidative stress.
- Melatonin, Dioscin, polyphenols, and FNDC5 are natural and herbal compounds with antioxidant effects demonstrating significant cardioprotective activity.
- Lifestyle changes, including exercise and diet, are crucial in slowing the progression of cardiovascular toxicity and can be easily implemented in individuals' lives.
- Future research should focus on exploring possible crosstalk, complex pathways, and the ethical considerations of translational research in clinical settings.

Funding The research work of DKA is supported by R25AI179582 and R01 HL147662 grants from the National Institutes of Health, USA. The contents of this chapter are solely the responsibility of the authors and do not necessarily represent the official views of the National Institutes of Health.

Competing Interests All authors have read the chapter and declare no conflict of interest. No writing assistance was utilized in the production of this manuscript.

Consent for Publication All authors have read the chapter and consented for publication.

References

1. Zhao RZ, Jiang S, Zhang L, Yu ZB (2019) Mitochondrial electron transport chain, ROS generation and uncoupling (Review). Int J Mol Med 44(1):3–15. https://doi.org/10.3892/ijmm.2019.4188. PMCID: PMC6559295, PMID: 31115493
2. Tai P, Chen X, Jia G, Chen G, Gong L, Cheng Y, Li Z, Wang H, Chen A, Zhang G, Zhu Y, Xiao M, Wang Z, Liu Y, Shan D, He D, Li M, Zhan T, Khan A, Li X, Zeng X, Li C et al (2023) WGX50 mitigates doxorubicin-induced cardiotoxicity through inhibition of mitochondrial ROS and ferroptosis. J Transl Med 21(1):823. https://doi.org/10.1186/s12967-023-04715-1. PMID: 37978379, PMCID: PMC10655295
3. Yan Q, Liu S, Sun Y, Chen C, Yang S, Lin M, Long J, Yao J, Lin Y, Yi F, Meng L, Tan Y, Ai Q, Chen N, Yang Y (2023) Targeting oxidative stress as a preventive and therapeutic approach for cardiovascular disease. J Transl Med 21(1):519. https://doi.org/10.1186/s12967-023-04361-7. Review. PMID: 37533007, PMCID: PMC10394930
4. Jomova K, Raptova R, Alomar SY, Alwasel SH, Nepovimova E, Kuca K, Valko M (2023) Reactive oxygen species, toxicity, oxidative stress, and antioxidants: chronic diseases and aging. Arch Toxicol 97(10):2499–2574. https://doi.org/10.1007/s00204-023-03562-9. Epub 2023 Aug 19. Review. PMID: 37597078, PMCID: PMC10475008
5. Zhang T, Deng W, Deng Y, Liu Y, Xiao S, Luo Y, Xiang W, He Q (2023) Mechanisms of ferroptosis regulating oxidative stress and energy metabolism in myocardial ischemia-reperfusion injury and a novel perspective of natural plant active ingredients for its treatment. Biomed Pharmacother 165:114706. https://doi.org/10.1016/j.biopha.2023.114706. Epub 2023 Jul 1. Review. PMID: 37400352
6. Srinivas US, Tan BWQ, Vellayappan BA, Jeyasekharan AD (2019) ROS and the DNA damage response in cancer. Redox Biol 25:101084. https://doi.org/10.1016/j.redox.2018.101084. Epub 2018 Dec 21. Review. PMID: 30612957, PMCID: PMC6859528
7. Quan Y, Xin Y, Tian G, Zhou J, Liu X (2020) Mitochondrial ROS-modulated mtDNA: a potential target for cardiac aging. Oxid Med Cell Longev 2020:9423593. https://doi.org/10.1155/2020/9423593. eCollection 2020. Review. PMID: 32308810, PMCID: PMC7139858
8. Tsvetkova D, Ivanova S (2022) Application of approved cisplatin derivatives in combination therapy against different cancer diseases. Molecules 27(8):2466. https://doi.org/10.3390/molecules27082466. PMCID: PMC9031877, PMID: 35458666
9. Heffern CT, Pocivavsek L, Birukova AA, Moldobaeva N, Bochkov VN, Lee KY, Birukov KG (2013) Thermodynamic and kinetic investigations of the release of oxidized phospholipids from lipid membranes and its effect on vascular integrity. Chem Phys Lipids 175–176:9–19. https://doi.org/10.1016/j.chemphyslip.2013.07.003. Epub 2013 Jul 30. PMID: 23911706, PMCID: PMC3869626
10. Songbo M, Lang H, Xinyong C, Bin X, Ping Z, Liang S (2019) Oxidative stress injury in doxorubicin-induced cardiotoxicity. Toxicol Lett 307:41–48. https://doi.org/10.1016/j.toxlet.2019.02.013. Epub 2019 Feb 25. Review. PMID: 30817977
11. Kciuk M, Gielecińska A, Mujwar S, Kołat D, Kałuzińska-Kołat Ż, Celik I, Kontek R (2023) Doxorubicin—an agent with multiple mechanisms of anticancer activity. Cells 12(4):659. https://doi.org/10.3390/cells12040659. PMID: 36831326; PMCID: PMC9954613
12. Jiang H, Zuo J, Li B, Chen R, Luo K, Xiang X, Lu S, Huang C, Liu L, Tang J, Gao F (2023) Drug-induced oxidative stress in cancer treatments: angel or devil? Redox Biol 63:102754. https://doi.org/10.1016/j.redox.2023.102754. Epub 2023 May 18. Review. PMID: 37224697, PMCID: PMC10220276
13. She G, Du JC, Wu W, Pu TT, Zhang Y, Bai RY, Zhang Y, Pang ZD, Wang HF, Ren YJ, Sadoshima J, Deng XL, Du XJ (2023) Hippo pathway activation mediates chemotherapy-induced anti-cancer effect and cardiomyopathy through causing mitochondrial damage and dysfunction. Theranostics 13(2):560–577. https://doi.org/10.7150/thno.79227. eCollection 2023. PMID: 36632235, PMCID: PMC9830444

14. Katanić Stanković JS, Selaković D, Rosić G (2023) Oxidative damage as a fundament of systemic toxicities induced by cisplatin-the crucial limitation or potential therapeutic target? Int J Mol Sci 24(19):14574. https://doi.org/10.3390/ijms241914574. PMID: 37834021; PMCID: PMC10572959
15. Fang Y, Zhang Z (2020) Arsenic trioxide as a novel anti-glioma drug: a review. Cell Mol Biol Lett 25:44. https://doi.org/10.1186/s11658-020-00236-7. PMCID: PMC7517624, PMID: 32983240
16. Nazim UM, Rasheduzzaman M, Lee YJ, Seol DW, Park SY (2017) Enhancement of TRAIL-induced apoptosis by 5-fluorouracil requires activating Bax and p53 pathways in TRAIL-resistant lung cancers. Oncotarget 8(11):18095–18105. https://doi.org/10.18632/oncotarget.14994. PMCID: PMC5392310, PMID: 28178647
17. Chi RF, Li L, Wang AL, Yang H, Xi J, Zhu ZF, Wang K, Li B, Yang LG, Qin FZ, Zhang C (2022) Enhanced oxidative stress mediates pathological autophagy and necroptosis in cardiac myocytes in pressure overload induced heart failure in rats. Clin Exp Pharmacol Physiol 49(1): 60–69. https://doi.org/10.1111/1440-1681.13583. Epub 2021 Sep 20. PMID: 34453856
18. Zheng Z, Su J, Bao X, Wang H, Bian C, Zhao Q, Jiang X (2023) Mechanisms and applications of radiation-induced oxidative stress in regulating cancer immunotherapy. Front Immunol 14: 1247268. https://doi.org/10.3389/fimmu.2023.1247268. eCollection 2023. Review. PMID: 37600785, PMCID: PMC10436604
19. Reuter S, Gupta SC, Chaturvedi MM, Aggarwal BB (2010) Oxidative stress, inflammation, and cancer: how are they linked? Free Radic Biol Med 49(11):1603–1616. https://doi.org/10.1016/j.freeradbiomed.2010.09.006. Epub 2010 Sep 16. Review. PMID: 20840865, PMCID: PMC2990475
20. Elmore S (2007) Apoptosis: a review of programmed cell death. Toxicol Pathol 35(4):495–516. https://doi.org/10.1080/01926230701320337. PMID: 17562483; PMCID: PMC2117903
21. Zhang X, Hu C, Kong CY, Song P, Wu HM, Xu SC, Yuan YP, Deng W, Ma ZG, Tang QZ (2020) FNDC5 alleviates oxidative stress and cardiomyocyte apoptosis in doxorubicin-induced cardiotoxicity via activating AKT. Cell Death Differ 27(2):540–555. https://doi.org/10.1038/s41418-019-0372-z. Epub 2019 Jun 17. PMID: 31209361, PMCID: PMC7206111
22. Gao W, Wang X, Zhou Y, Wang X, Yu Y (2022) Autophagy, ferroptosis, pyroptosis, and necroptosis in tumor immunotherapy. Signal Transduct Target Ther 7:196. https://doi.org/10.1038/s41392-022-01046-3. PMCID: PMC9208265, PMID: 35725836
23. Adameova A, Horvath C, Abdul-Ghani S, Varga ZV, Suleiman MS, Dhalla NS (2022) Interplay of oxidative stress and necrosis-like cell death in cardiac ischemia/reperfusion injury: a focus on necroptosis. Biomedicines 10(1):127. https://doi.org/10.3390/biomedicines10010127. PMID: 35052807; PMCID: PMC8773068
24. Yun HR, Jo YH, Kim J, Shin Y, Kim SS, Choi TG (2020) Roles of autophagy in oxidative stress. Int J Mol Sci 21(9):3289. https://doi.org/10.3390/ijms21093289. PMID: 32384691; PMCID: PMC7246723
25. Qin Y, Qiao Y, Wang D, Tang C, Yan G (2021) Ferritinophagy and ferroptosis in cardiovascular disease: mechanisms and potential applications. Biomed Pharmacother 141:111872. https://doi.org/10.1016/j.biopha.2021.111872. Epub 2021 Jul 7. Review. PMID: 34246187
26. Korsager Larsen M, Matchkov VV (2016) Hypertension and physical exercise: the role of oxidative stress. Medicina (Kaunas) 52(1):19–27. https://doi.org/10.1016/j.medici.2016.01.005. Epub 2016 Jan 29. Review. PMID: 26987496
27. Griendling KK, Camargo LL, Rios FJ, Alves-Lopes R, Montezano AC, Touyz RM (2021) Oxidative stress and hypertension. Circ Res 128(7):993–1020. https://doi.org/10.1161/CIRCRESAHA.121.318063. Epub 2021 Apr 1. Review. PMID: 33793335, PMCID: PMC8293920
28. Virdis A, Bacca A, Colucci R, Duranti E, Fornai M, Materazzi G, Ippolito C, Bernardini N, Blandizzi C, Bernini G, Taddei S (2013) Endothelial dysfunction in small arteries of essential hypertensive patients: role of cyclooxygenase-2 in oxidative stress generation. Hypertension 62(2):337–344. https://doi.org/10.1161/HYPERTENSIONAHA.111.00995. Epub 2013 Jun 3. PMID: 23734008

29. Batty M, Bennett MR, Yu E (2022) The Role of oxidative stress in atherosclerosis. Cells 11(23): 3843. https://doi.org/10.3390/cells11233843. PMID: 36497101; PMCID: PMC9735601
30. Kattoor AJ, Pothineni NVK, Palagiri D, Mehta JL (2017) Oxidative stress in atherosclerosis. Curr Atheroscler Rep 19(11):42. https://doi.org/10.1007/s11883-017-0678-6. Review. PMID: 28921056
31. Fuentes E, Moore-Carrasco R, de Andrade Paes AM, Trostchansky A (2019) Role of platelet activation and oxidative stress in the evolution of myocardial infarction. J Cardiovasc Pharmacol Ther 24(6):509–520. https://doi.org/10.1177/1074248419861437. Epub 2019 Jul 7. PMID: 31280622
32. Adameova A, Shah AK, Dhalla NS (2020) Role of oxidative stress in the genesis of ventricular arrhythmias. Int J Mol Sci 21(12). https://doi.org/10.3390/ijms21124200. Review. PMID: 32545595, PMCID: PMC7349053
33. Karam BS, Chavez-Moreno A, Koh W, Akar JG, Akar FG (2017) Oxidative stress and inflammation as central mediators of atrial fibrillation in obesity and diabetes. Cardiovasc Diabetol 16(1):120. https://doi.org/10.1186/s12933-017-0604-9. PMID: 28962617; PMCID: PMC5622555
34. Zhang X, Zhang Z, Zhao Y, Jiang N, Qiu J, Yang Y, Li J, Liang X, Wang X, Tse G, Li G, Liu T (2017) Alogliptin, a dipeptidyl peptidase-4 inhibitor, alleviates atrial remodeling and improves mitochondrial function and biogenesis in diabetic rabbits. J Am Heart Assoc 6(5):e005945. https://doi.org/10.1161/JAHA.117.005945. PMID: 28507060; PMCID: PMC5524117
35. Zhou B, Tian R (2018) Mitochondrial dysfunction in pathophysiology of heart failure. J Clin Invest 128(9):3716–3726. https://doi.org/10.1172/JCI120849. Epub 2018 Aug 20. Review. PMID: 30124471, PMCID: PMC6118589
36. Steven S, Frenis K, Oelze M, Kalinovic S, Kuntic M, Bayo Jimenez MT, Vujacic-Mirski K, Helmstädter J, Kröller-Schön S, Münzel T, Daiber A (2019) Vascular inflammation and oxidative stress: major triggers for cardiovascular disease. Oxid Med Cell Longev 2019: 7092151. https://doi.org/10.1155/2019/7092151. eCollection 2019. Review. PMID: 31341533, PMCID: PMC6612399
37. Hou K, Shen J, Yan J, Zhai C, Zhang J, Pan JA, Zhang Y, Jiang Y, Wang Y, Lin RZ, Cong H, Gao S, Zong WX (2021) Loss of TRIM21 alleviates cardiotoxicity by suppressing ferroptosis induced by the chemotherapeutic agent doxorubicin. EBioMedicine 69:103456. https://doi.org/10.1016/j.ebiom.2021.103456. Epub 2021 Jul 4. PMID: 34233258, PMCID: PMC8261003
38. Pan JA, Sun Y, Jiang YP, Bott AJ, Jaber N, Dou Z, Yang B, Chen JS, Catanzaro JM, Du C, Ding WX, Diaz-Meco MT, Moscat J, Ozato K, Lin RZ, Zong WX (2016) TRIM21 ubiquitylates SQSTM1/p62 and suppresses protein sequestration to regulate redox homeostasis. Mol Cell 61(5):720–733. https://doi.org/10.1016/j.molcel.2016.02.007. Erratum in: Mol Cell. 2016 Apr 7;62(1):149–51. https://doi.org/10.1016/j.molcel.2016.03.015. PMID: 26942676, PMCID: PMC4779181
39. Zare S, Heydari FS, Hayes AW, Reiter RJ, Zirak MR, Karimi G (2021) Melatonin attenuates chemical-induced cardiotoxicity. Hum Exp Toxicol 40(3):383–394. https://doi.org/10.1177/0960327120959417. Epub 2020 Sep 16. Review. PMID: 32935581
40. Charlton NC, Mastyugin M, Török B, Török M (2023) Structural features of small molecule antioxidants and strategic modifications to improve potential bioactivity. Molecules 28(3):1057. https://doi.org/10.3390/molecules28031057. PMID: 36770724; PMCID: PMC9920158
41. Galano A, Tan DX, Reiter RJ (2018) Melatonin: a versatile protector against oxidative DNA damage. Molecules 23(3):530. https://doi.org/10.3390/molecules23030530. PMID: 29495460; PMCID: PMC6017920
42. Zhao L, Tao X, Qi Y, Xu L, Yin L, Peng J (2018) Protective effect of dioscin against doxorubicin-induced cardiotoxicity via adjusting microRNA-140-5p-mediated myocardial oxidative stress. Redox Biol 16:189–198. https://doi.org/10.1016/j.redox.2018.02.026. Epub 2018 Mar 6. Erratum in: Redox Biol. 2019 Sep;26:101303. https://doi.org/10.1016/j.redox.2019.101303. PMID: 29524841, PMCID: PMC5953242

43. Kumar S, Pandey AK (2013) Chemistry and biological activities of flavonoids: an overview. ScientificWorldJournal 2013:162750. https://doi.org/10.1155/2013/162750. eCollection 2013. Review. PMID: 24470791, PMCID: PMC3891543
44. Slika H, Mansour H, Wehbe N, Nasser SA, Iratni R, Nasrallah G, Shaito A, Ghaddar T, Kobeissy F, Eid AH (2022) Therapeutic potential of flavonoids in cancer: ROS-mediated mechanisms. Biomed Pharmacother 146:112442. https://doi.org/10.1016/j.biopha.2021.112442. Epub 2022 Jan 4. PMID: 35062053
45. Arafah A, Rehman MU, Ahmad A, AlKharfy KM, Alqahtani S, Jan BL, Almatroudi NM (2022) Myricetin (3,3′,4′,5,5′,7-hexahydroxyflavone) prevents 5-fluorouracil-induced cardiotoxicity. ACS Omega 7(5):4514–4524. https://doi.org/10.1021/acsomega.1c06475. eCollection 2022 Feb 8. PMID: 35155943, PMCID: PMC8829927
46. Syahputra RA, Harahap U, Dalimunthe A, Nasution MP, Satria D (2022) The role of flavonoids as a cardioprotective strategy against doxorubicin-induced cardiotoxicity: a review. Molecules 27(4):1320. https://doi.org/10.3390/molecules27041320. PMID: 35209107; PMCID: PMC8878416
47. Przybylska S, Tokarczyk G (2022) Lycopene in the prevention of cardiovascular diseases. Int J Mol Sci 23(4). https://doi.org/10.3390/ijms23041957. Review. PMID: 35216071, PMCID: PMC8880080
48. Koss-Mikołajczyk I, Todorovic V, Sobajic S, Mahajna J, Gerić M, Tur JA, Bartoszek A (2021) Natural products counteracting cardiotoxicity during cancer chemotherapy: the special case of doxorubicin, a comprehensive review. Int J Mol Sci 22(18):10037. https://doi.org/10.3390/ijms221810037. PMID: 34576204; PMCID: PMC8467966
49. Gaytan SL, Lawan A, Chang J, Nurunnabi M, Bajpeyi S, Boyle JB, Han SM, Min K (2023) The beneficial role of exercise in preventing doxorubicin-induced cardiotoxicity. Front Physiol 14:1133423. https://doi.org/10.3389/fphys.2023.1133423. PMID: 36969584; PMCID: PMC10033603
50. Wang J, Liu S, Meng X, Zhao X, Wang T, Lei Z, Lehmann HI, Li G, Alcaide P, Bei Y, Xiao J (2024) Exercise inhibits doxorubicin-induced cardiotoxicity via regulating B cells. Circ Res 134(5):550–568. https://doi.org/10.1161/CIRCRESAHA.123.323346. Epub 2024 Feb 7. PMID: 38323433; PMCID: PMC11233173

Chapter 35
The Effect of Molecular Hydrogen on the Mitigation of Oxidative Stress in Cardiovascular Disorders

Ján Slezák, Tatiana Ravingerova, and Branislav Kura

Abstract The exploration of molecular hydrogen (H_2) as a therapeutic agent for reducing oxidative stress in cardiovascular disorders (CVD) has gained a substantial attention in recent decades. H_2 serves as an upstream signaling molecule for carbon monoxide (CO) and interacts with oxidized iron porphyrins, particularly hematin, to alleviate oxidative stress in different pathological situations including postischemic heart stunning, myocardial lethal injury and occurrence of malignant tachyarrhythmias. However, its role in the mechanisms of CVD, as well as in the protection against CV disorders remains not completely understood and requires further elucidation.

Keywords Molecular hydrogen · Oxidative stress · Cardiovascular diseases · Inflammation · Antioxidant

Introduction

Hydrogen (H_2) therapy has garnered significant attention in recent years for its diverse therapeutic effects across a range of health conditions, particularly in preclinical models. This innovative approach has demonstrated reducing oxidative stress, regulating blood pressure, facilitating injury recovery, potential in enhancing mitochondrial function and reprogramming metabolism [1–3]. Such findings underscore the necessity for further exploration of H_2's therapeutic capabilities in clinical environments.

Cardiovascular diseases (CVDs) are multifaceted conditions arising from various risk factors that inflict damage on cardiomyocytes and blood vessels, ultimately leading to functional impairment. In this context, H_2 therapy has emerged as a

J. Slezák (✉) · T. Ravingerova · B. Kura
Institute for Heart Research, Centre of Experimental Medicine, Slovak Academy of Sciences, Bratislava, Slovakia
e-mail: jan.slezak@savba.sk; tatiana.ravingerova@savba.sk

D. K. Agrawal et al. (eds.), *Cardiovascular Toxicity*, Advances in Biochemistry in Health and Disease 33, https://doi.org/10.1007/978-3-031-90598-8_35

promising intervention [3], especially concerning myocardial remodeling and fibrosis—two critical processes in the recovery following myocardial infarction.

Research indicates that H_2 can significantly mitigate fibrosis and apoptosis, both of which are closely associated with heart failure. This suggests that innovative therapeutic strategies targeting cardiac remodeling could enhance survival rates among ischemic patients. Moreover, reactive oxygen species (ROS), such as hydrogen peroxide (H_2O_2), play a pivotal role in the pathogenesis of vascular diseases by affecting vasomotor function [4]. Beyond conventional CVD risk factors like hypertension and diabetes, conditions such as rheumatoid arthritis independently contribute to atherosclerosis. The intricate interplay between inflammation and oxidative stress is particularly relevant in both rheumatoid arthritis and coronary artery disease (CAD), where free radicals and pro-inflammatory cytokines exacerbate their association [5]. Increased oxidative stress leads to the heightened expression of adhesion molecules like vascular cell adhesion protein 1 (VCAM-1) and intercellular adhesion molecule 1 (ICAM-1), which further promote vascular dysfunction [6].

Since the initial documentation of H_2's beneficial effects in 2007 [7], over 2000 studies have been published exploring its impact across various disease models. While early investigations primarily focused on H_2's radical scavenging properties, it is increasingly recognized that its therapeutic effects may arise from the modulation of multiple molecular pathways rather than merely neutralizing free radicals [8]. H_2 therapy thus presents a compelling avenue for addressing cardiovascular diseases by enhancing endothelial function and alleviating oxidative stress-related damage.

Molecular Hydrogen Characterization

H_2 is the simplest and the most abundant molecule in the universe consisting of two hydrogen atoms bonded together. Each hydrogen atom in H_2 has one proton and one electron, and they share their electrons through a covalent bond, resulting in a stable diatomic molecule. This molecule is colorless, odorless, and tasteless, making it imperceptible to human senses under normal conditions. H_2 gas is highly flammable and reacts explosively with oxygen to form water, releasing a significant amount of energy, which makes it a potential clean fuel source. It has a low density and is the lightest molecule, allowing it to escape Earth's atmosphere over geological timescales. Due to its small size, H_2 can diffuse rapidly through materials, which is a consideration in its storage and transport [9, 10].

In terms of chemical properties, H_2 is relatively inert at room temperature but becomes reactive at elevated temperatures or in the presence of catalysts. It serves as a reducing agent in various chemical reactions. H_2 plays a critical role in numerous scientific fields. In biochemistry, it is involved in metabolic processes in certain microorganisms [11].

Mechanism of Action of Molecular Hydrogen and Its Possible Targets

H_2 is engaged in catalytic reactions with hydroxyl radicals (•OH), facilitated by the oxidized form of porphyrins known as hematin. This interaction allows H_2 to repair various oxidized porphyrins through the following reaction:

$$\text{PrP-Fe(III)-OH} + H_2 \rightarrow \text{PrP-Fe(III)-H} + H_2O$$

This process highlights the role of H_2 as a biosensor and its capacity to convert harmful radicals into benign substances, thereby improving cellular function and protecting against oxidative damage [12].

Summarizing,

- H_2-targeted molecule is oxidized Fe-porphyrin
- Fe-porphyrin is a redox-related biosensor of H_2
- H_2 is an upstream signaling molecule of CO
- To stimulate the reaction of H_2 with •OH, a catalytic action of the oxidized form of a porphyrin (hematin) is required [12].
- H_2 is capable of repairing various types of porphyrins, from their oxidized form

Recent studies have highlighted the role of hematin as a crucial target for H_2, resolving discrepancies between reaction rates observed in homogeneous aqueous solutions and living cells. The oxidized form of Fe-porphyrins acts as a biosensor for H_2, facilitating rapid reactions that convert harmful radicals into harmless water [13].

H_2 exhibits numerous beneficial effects by reducing oxidative stress in various organs under different pathological situations. It also shows anti-inflammatory, anti-apoptotic, and energy metabolism-promoting properties. These diverse functions of H_2 may stem from its interactions with various types of porphyrins distributed throughout the body. Unlike other medical gases such as hydrogen sulfide (H_2S), which can be toxic at higher concentrations, H_2 is non-toxic even at elevated levels. In the study by Tomofuji et al. [14], it was observed no differences in morbidity or mortality were observed between the H_2 and control groups after drinking H_2-enriched water (0.19 mM H_2) for one year. In addition, the H_2 group exhibited reduced periodontal damage. Its safety has been confirmed by clinical trials, which demonstrated the efficacy of H_2 without adverse effects [15, 16]. Moreover, inhalation of H_2 gas has been approved as safe in Phase I clinical trial [17].

Effects of Molecular Hydrogen in Organism

H_2 differs from the conventional antioxidants and anti-inflammatory drugs due to its unique physical and chemical properties. Biomedical studies suggest that H_2 has a significant potential in alleviating the effects of excessive ROS production, such as postischemic induction of malignant arrhythmias and impaired functional recovery, due to reduced function of Na, K-ATPase and inflammation [18, 19]. As a small molecule, H_2 easily diffuses throughout the body without affecting signaling reactive species [20]. H_2 is stable and selectively scavenges cytotoxic hydroxyl and nitroxyl radicals, which are the most harmful molecules. It functions as a gas signaling modulator, effectively reducing oxidative stress and inflammation, and has shown therapeutic potential in over 170 disease models in animals and humans.

Animal studies indicate that H_2 enhances resistance and mitigates the negative effects of acute and chronic stress, such as inflammation, increased ROS, myocardial ischemia, anxiety, and depression [21]. Recent findings demonstrate pleiotropic therapeutic effects of H_2 in various disease models in animals and in numerous human diseases. It reduces the expression of several pro-inflammatory mediators and markers of oxidative stress and apoptosis, including tumor necrosis factor alpha (TNF-α), interleukins (IL-6, IL-1β, IL-10, IL-12), chemokine ligand 2 (CCL2), ICAM-1, nuclear factor kappa B (NF-κB), nuclear factor of activated T-cells (NFAT), high mobility group protein 1, prostaglandin E2, cyclooxygenase-2 (COX2), serum diamine oxidase, tissue malondialdehyde (MDA), carbonylated protein, TBARS, myeloperoxidase activity, cJun-N-terminal Kinase (JNK), and caspase-3 [1].

Antioxidant, anti-inflammatory, and anti-apoptotic effects of H_2 are crucial for providing cytoprotection against adverse effects of radiotherapy and chemotherapy in oncological patients. It has demonstrated radioprotective effects on cultured cells and in mice [2]. For example, in irradiated animals, lipid peroxidation marker MDA was significantly increased in the small intestine, but not in the H_2-enriched water group [22]. Similarly, increased oxidative stress (MDA) and inflammation (TNF-α) markers were observed in blood plasma after chest irradiation in rats, but H_2 application significantly reduced these markers [23]. The favorable chemical, physical, and biological properties of H_2 make it an excellent candidate for preventing and treating ischemia-reperfusion (I/R) injury [1, 2, 21–23].

Methods of Administration and Concentrations Used

H_2 can be delivered by different methods including inhalation, intravenous administration of H_2-rich saline, drinking H_2-enriched water, and topical applications such as eye drops [24]. In animal models, H_2 has shown efficacy in reducing myocardial injury during cold ischemia and reperfusion [25].

Administering H_2 for therapeutic purposes has garnered significant interest due to its potential antioxidant and anti-inflammatory properties. Some methods of administration and their typical concentrations are mentioned below:

(a) Inhalation is a direct method when H_2 gas is inhaled through a mask or nasal cannula. Concentrations typically range from 1 to 4% in air, as higher concentrations can pose flammability risks. Studies have shown that inhaling H_2 gas can have protective effects in conditions such as ischemia-reperfusion injury [7].

(b) H_2-rich water (HRW) involves dissolving metal magnesium powder in tablets-producing micro- or nanobubbles or directly H_2 gas into water, which is then consumed orally. The concentration of H_2 in water can vary, but it is often around 0.6–1.6 ppm (parts per million or more). This method is convenient and has been shown to improve outcomes in metabolic syndrome and other oxidative stress-related conditions [20].

(c) H_2 baths represent the method where H_2 gas is dissolved in water for bathing. The skin absorbs H_2, which can then exert systemic effects. Bathing in HRW has been reported to improve skin conditions and reduce fatigue [26].

(d) H_2 can also be dissolved in saline and administered intravenously. This method allows for controlled delivery of H_2 at therapeutic concentrations directly into the bloodstream. Research indicates potential benefits for acute medical conditions, such as strokes or myocardial infarction [27].

(e) H_2-infused gels or creams can be applied directly to the skin for localized effects. This method is explored for its anti-inflammatory and anti-aging effects on the skin [28].

(f) Oral capsules containing H_2-releasing compounds represent another method. These capsules release H_2 gas when ingested, providing a practical way to deliver H_2 internally.

It has been shown that in isolated rat heart exposed to I/R, a combination of H_2 present in the Krebs–Henseleit perfusion solution and hypoxic postconditioning substantially augmented anti-infarct cardioprotection afforded by hypoxic postconditioning alone (Fig. 35.1) [29, 30].

Each method varies in terms of bioavailability and therapeutic outcomes, and ongoing research continues to elucidate the optimal concentrations and delivery systems for different conditions. It's important to note that although H_2 therapy is promising, it should be considered as a complementary and not a replacement for conventional treatments.

Targeting Oxidative Stress

Recent studies have identified hematin as a key target for H_2, addressing inconsistencies between reaction kinetics observed in vitro and those occurring within living organisms. The presence of heme proteins throughout the body enables H_2 to exert its effects in various tissues, particularly under conditions characterized by oxidative stress [3, 7].

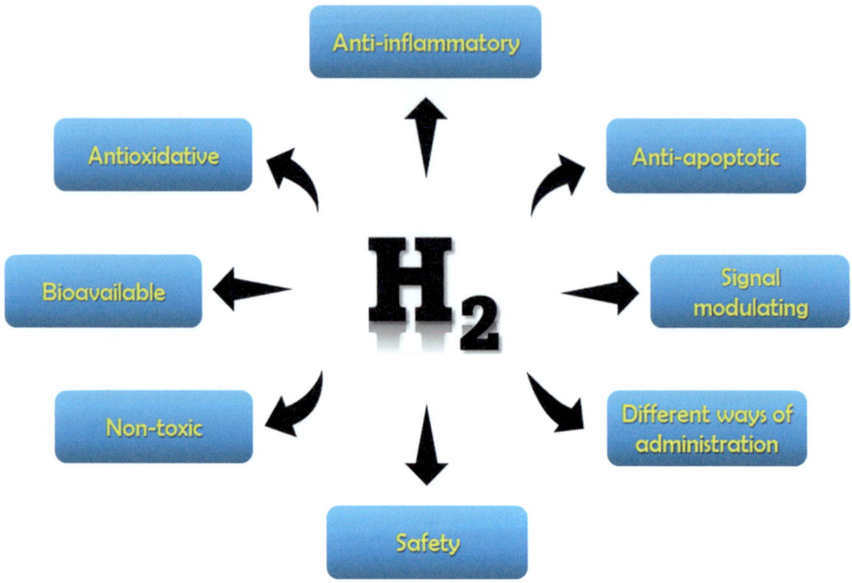

Fig. 35.1 Biological benefits of administering molecular hydrogen in the prevention and treatment of diseases

Oxidative stress occurs under conditions of an imbalance between ROS generation and the body's antioxidant defense mechanisms, leading to an excess of oxidants and reactive molecules. This imbalance is a contributing factor to numerous pathological conditions, as it impairs cellular and organ functions [31]. ROS are generated as byproducts of oxygen reduction during normal cellular metabolism. Key sources of ROS include mitochondrial respiration, NADH/NADPH oxidase, and xanthine oxidoreductase [32]. Reactive molecules encompass both ROS and reactive nitrogen species (RNS), with ROS comprising both radical and non-radical chemical entities. Free radicals, which contain unpaired electrons, include hydroxyl radicals (•OH), superoxide anion radicals ($\cdot O_2^-$), alkoxyl radicals (RO•), peroxyl radicals (R–OO•), carbon radicals (RC•), and various RNS- like nitric oxide (NO•). Non-radical reactive molecules include H_2O_2, singlet oxygen ($1O_2$), and RNS such as peroxynitrite ($ONOO^-$), nitrogen dioxide (NO_2), and dinitrogen trioxide (N_2O_3).

Most ROS are produced in the mitochondrial electron transport chain, particularly at the complex I, with other production sites including NADPH oxidase, nitric oxide synthase, and cytochrome p450, among others [33]. Pathological states can increase ROS production significantly, due to enhanced metabolism and activation of these systems (Fig. 35.2).

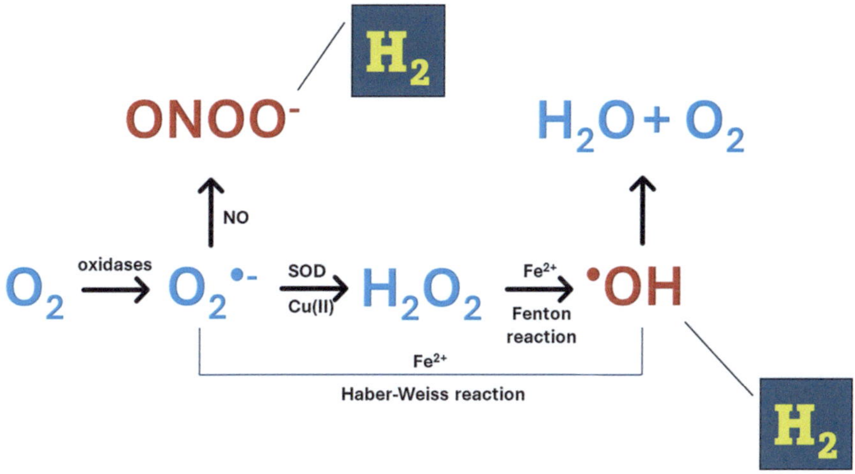

Fig. 35.2 Superoxide production and regulation. During immune responses and physical exertion, superoxide production increases due to elevated oxygen intake. This superoxide acts as a crucial signaling molecule, with its concentration tightly regulated. Superoxide is converted by superoxide dismutase into H_2O_2, which can further modulate gene expression and induce beneficial cellular changes before being transformed into water and oxygen by catalase. If superoxide or H_2O_2 levels exceed regulatory capacity, oxidative stress ensues. Elevated levels can also lead to the formation of toxic hydroxyl radicals (•OH) via the Haber–Weiss and Fenton reactions. $ONOO^-$ and •OH can oxidize cellular biomolecules, leading to diseases and damage. While antioxidants can reduce ROS concentration, H_2 is considered a selective scavenger of $ONOO^-$ and •OH due to its relative inertness. (Adapted from Slezák et al. [3])

Reactive Molecules and Their Dual Roles

Reactive molecules can have harmful or beneficial effects based on their origin, concentration, location, and duration. Under normal conditions, superoxide radicals are continuously generated in mitochondria, nuclear and plasma membranes, and the endoplasmic reticulum. Superoxide production rises during immune responses to eliminate pathogens and increase inflammatory cytokines [34]. Similarly, superoxide formation increases during physical exertion due to greater oxygen intake, acting as an important signaling molecule that activates transcription factors and enhances physical capacity. The concentration of superoxide is maintained within a narrow range through controlled production and clearance [35].

Superoxide dismutase converts superoxide into H_2O_2, which then regulates gene expression and induces positive cellular changes before being broken down by catalase [33]. Excessive superoxide or H_2O_2 leads to oxidative stress, potentially increasing toxic hydroxyl radical production through the Haber–Weiss and Fenton reactions. Iron and superoxide can regenerate, further promoting hydroxyl radical production. The nitric oxide radical (NO•) is crucial for endothelial function, gene expression, and cell proliferation but can form harmful peroxynitrite when in excess, particularly during hyperactivation of inducible nitric oxide synthase (iNOS). Peroxynitrite is a potent oxidant capable of oxidizing proteins, lipids, and DNA, and can generate toxic hydroxyl radicals independently of the Fenton reaction [36].

Excessive ROS production in pathological conditions or intense exertion can deplete the body's antioxidant defenses, leading to oxidative stress and its toxic consequences [34]. Various antioxidants can lower ROS levels, but H_2 is a selective scavenger of the aggressive radicals $ONOO^-$ and •OH, without disrupting ROS signaling functions [3]. Dysregulation between ROS and antioxidants is linked to various pathological conditions, including cardiovascular diseases, aging, and cognitive decline. The rapid production of free radicals can overwhelm the endogenous antioxidant system, damaging lipids, proteins, DNA, and RNA, and leading to cell degeneration. While antioxidants have not consistently yielded expected benefits in human medicine, H_2 offers advantages in mitigating oxidative stress due to its unique properties. It selectively reduces •OH radicals while preserving ROS necessary for normal cell signaling [3].

Clinical Applications

Research has increasingly focused on the effects of H_2 on diseases linked to free radical production, such as I/R injury, myocardial or brain infarction, heart transplantation, and radiation-induced heart disease. Application of H_2 in oxidative stress-related dysregulations was first reported over 16 years ago and has shown promising results in experimental and clinical studies [7]. H_2 selectively targets aggressive hydroxyl and nitrosyl radicals, suggesting its potential in preventing and treating cardiovascular diseases associated with oxidative stress.

Oxidative and inflammatory stress are key contributors to myocardial I/R injury. Cardiac myocytes require high ATP levels, necessitating numerous mitochondria. Mitochondria, filled with reactive intermediates and proapoptotic signals, play an important role in I/R injury. Stress-induced opening of the mitochondrial permeability transition pore (mPTP) leads to increased ROS production and subsequent damage. The accumulation of ROS, proteases, and growth factors during ischemia and reperfusion results in fibroblast proliferation, collagen accumulation, fibrosis, and tissue destruction, exacerbated by inflammation [37]. It has been shown that H_2 application can treat numerous diseases with oxidative stress background. A clinical trial revealed that patients inhaling 2% H_2 alongside oxygen exhibited significantly improved survival rates and neurological outcomes compared to controls in Post-Cardiac Arrest Syndrome [38].

The efficacy of H_2 therapy is exemplified by its impact on recovery after cardiac arrest. Clinical trials indicate that inhalation of H_2 significantly improves survival rates and neurological outcomes compared to standard therapies [38]. Long-term consumption of H_2 water has been associated with cognitive improvements in individuals with mild cognitive impairment, suggesting potential benefits for neurodegenerative disorders and Alzheimer's disease. Its ability to cross the blood–brain barrier enhances its therapeutic potential for neurodegenerative disorders [39, 40].

Preliminary investigations indicate that H_2 inhalation may enhance the efficacy of immunotherapy and improve the quality of life for patients with advanced cancer by modulating immune responses [24]. Research into H_2 therapy for advanced cancer suggests improvements in prognosis and quality of life while mitigating adverse effects from conventional treatments like chemotherapy and radiotherapy [41]. Ongoing investigations into metabolic syndrome focus on how H_2 can influence weight management and metabolic health, potentially offering new avenues for treatment in this growing public health concern [42].

H_2 represents a groundbreaking approach to addressing oxidative stress-related conditions across various medical fields. Its numerous roles highlight its potential as a safe and effective therapeutic option moving forward [43].

Molecular Hydrogen and Modulation of Signaling Pathways

H_2 modulates hormones and cytokines within specific tissues, influencing various signaling pathways that contribute to its beneficial effects. For instance, it can enhance the expression of ghrelin hormone, which plays a role in neuroprotection [44]. Recent studies have delineated pathways through which H_2 exerts multiple functions by modulating lipid peroxides (LPOs). By decreasing levels of LPOs, H_2 influences critical signaling pathways involved in inflammation and cell survival [45, 46].

H_2 not only affects gene expression but also alters protein levels without changing mRNA expression. For instance, H_2 reduces CD36 protein levels without affecting its mRNA, indicating a role in post-transcriptional regulation, such as enhancing protein degradation or inhibiting translation. H_2 influences protein levels of key biomolecules involved in signal transduction and phosphorylation cascades, including AMPK, ASK1, JNK, Lyn, FGF-21, FOXO1, HMGB1, mTOR, NLRP3, NF-κB, PGC-1α, NFATC1, mtUPR, STAT3, VEGF, SIRT, Nrf2, and miRNA. These molecules are indirectly affected by H_2, and the primary regulators behind these changes require further study. Nonetheless, they play crucial roles in mediating H_2's beneficial effects [47, 48].

A significant impact of H_2 is the activation of the Nrf2 pathway, which induces phase II enzymes through the Nrf2-antioxidant response element (ARE) signaling pathway. Nrf2 regulates over 200 genes related to antioxidation and detoxification, maintaining cellular integrity under oxidative stress [49]. Under normal conditions, Nrf2 is kept in the cytoplasm by Keap1, which promotes its degradation. Stress signals cause Nrf2 to dissociate from Keap1, allowing it to enter the nucleus and initiate transcription of antioxidant genes [50]. H_2 activates the Nrf2/EpRE pathway, preventing radiation-induced lipid peroxidation in the rat heart and increasing

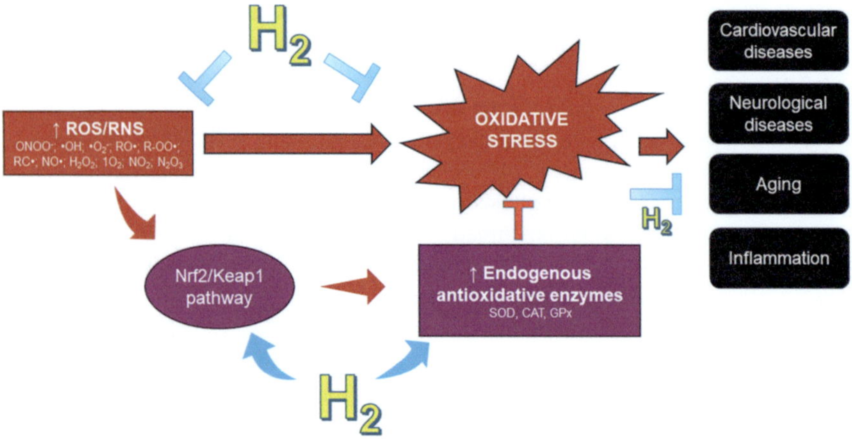

Fig. 35.3 Involvement of molecular hydrogen in Nrf2 signaling pathway. *H_2* molecular hydrogen, *ROS* reactive oxygen species, *RNS* reactive nitrogen species, *Nrf2* nuclear factor erythroid 2-related factor 2, *Keap1* Kelch domain of Kelch-like-ECH-associated protein 1, *SOD* superoxide dismutase, *CAT* catalase, *GPx* glutathione peroxidase

SOD-2 and Akt kinase phosphorylation. The activation of the Nrf2 pathway by H_2 may explain its therapeutic effects, including enhanced antioxidant production and reduced apoptosis and inflammation (Fig. 35.3) [51].

H_2 might also act as a hormetic effector, exhibiting pleiotropic effects through its hormetic mechanism of action. Hormesis involves adaptation to mild toxic stress, enhancing cellular protection [50]. Activation of the Nrf2 pathway by H_2, despite its reducing nature, may seem paradoxical, as Nrf2 is typically induced by electrophilic chemicals. However, oxidative stress activates the Nrf2 pathway, as seen with lipid peroxidation products like 4-hydroxy-2-nonenal. This mechanism may contribute to the protective effects of preconditioning, which involves mild oxidative stress and upregulation of proteins like cytokines, heat shock proteins, NF-κB, and Nrf2 [52, 53].

Microbiome, Aging, and the Microbiota–Gut–Brain Axis

The concept of the microbiome gained recognition in the late 1990s, highlighting its critical role in health and disease. The microbiome aids in digestion, immune regulation, protection against harmful bacteria, and vitamin production, including B vitamins and vitamin K. It is now regarded as an essential "organ" for proper bodily function. An imbalance in gut microbial flora, known as dysbiosis, is linked to various conditions such as inflammatory bowel disease, diabetes, obesity, cancer, cardiovascular issues, and central nervous system disorders, illustrating the microbiota–gut–brain axis. This bidirectional communication pathway, involving nervous, endocrine, immune, and humoral connections, is gaining attention for its role in psychiatric, neurodevelopmental, and age-related neurodegenerative disorders [54]. Gut microbes convert food into short-chain fatty acids (SCFAs), which interact with neurons that produce serotonin, a neurotransmitter crucial for mood regulation. Another neurotransmitter, gamma-aminobutyric acid (GABA), helps calm the nervous system and manage stress responses. Diet can, therefore, influence mental well-being by affecting bacterial function [54].

Conclusion

H_2 offers promising biological benefits for medical interventions. It has excellent permeability, allowing it to quickly reach subcellular compartments and protect critical cellular components like DNA, RNA, proteins, cell membranes, and mitochondria. Due to its relative inertness, H_2 selectively reacts with the most harmful radicals while preserving beneficial ROS signaling. It helps maintain redox homeostasis by modulating signals, such as downregulating the NADPH oxidase system, and activating the Nrf2 pathway, boosting the production of endogenous antioxidants like GSH, CAT, and GPx. H_2 also reduces pro-inflammatory mediators, such as cytokines and COX2. In some instances, H_2 may increase oxidant production, offering hormetic benefits. It enhances Sirt3 expression, maintains mitochondrial membrane potential, and boosts ATP production, acting as a mitohormetic effector by temporarily increasing mitochondrial superoxide production, followed by Nrf2 upregulation.

Ongoing research is essential to determine optimal H_2 administration methods and doses, as well as to elucidate its detailed molecular mechanisms. The diverse effects of H_2 on various proteins, molecules, and signaling pathways may partially explain its multifaceted preventive and therapeutic potential.

The mission of H_2 medicine in future is to study possible applications of H_2 across various fields for treatment and prevention with a potential to improve quality of life. Its potential extends beyond medicine into agriculture, promising safe and high-quality food production.

In summary, H_2 research represents a promising therapeutic avenue for addressing oxidative stress across various medical conditions. Its multifaceted roles in cellular signaling and protection against oxidative damage underscore its potential as a safe and effective treatment modality.

Further research will be focused on the elucidation of the mechanisms of H_2 underlying its beneficial effects and expanding of its clinical applications.

Funding Supported by the grants from the Slovak Research and Development Agency [APVV-19-0540 (T.R.), APVV-20-0242 (T.R.), APVV-19-0317 (B.K.)], grants from the Slovak Academy of Sciences [VEGA SR 2/0104/22 (T.R.), VEGA 2/0092/22 (B.K.), VEGA 2/0051/25 (J.S.)] and COST Action EU-METAHEART CA22169 (T.R.).

References

1. Ohta S (2011) Recent progress toward hydrogen medicine: potential of molecular hydrogen for preventive and therapeutic applications. Curr Pharm Des 17:2241–2252. https://doi.org/10.2174/138161211797052664
2. Barancik M, Kura B, LeBaron TW, Bolli R, Buday J, Slezak J (2020) Molecular and cellular mechanisms associated with effects of molecular hydrogen in cardiovascular and central nervous systems. Antioxidants 9:1281. https://doi.org/10.3390/antiox9121281
3. Slezak J, Kura B, LeBaron TW, Singal PK, Buday J, Barancik M (2021) Oxidative stress and pathways of molecular hydrogen effects in medicine. Curr Pharm Des 27:610–625. https://doi.org/10.2174/1381612826666200821114016
4. Chen Q, Wang Q, Zhu J, Xiao Q, Zhang L (2018) Reactive oxygen species: key regulators in vascular health and diseases. Br J Pharmacol 175:1279–1292. https://doi.org/10.1111/bph.13828
5. Medina-Leyte DJ, Zepeda-García O, Domínguez-Pérez M, González-Garrido A, Villarreal-Molina T, Jacobo-Albavera L (2021) Endothelial dysfunction, inflammation and coronary artery disease: potential biomarkers and promising therapeutical approaches. Int J Mol Sci 22:3850. https://doi.org/10.3390/ijms22083850
6. Scioli MG, Storti G, D'Amico F, Rodríguez Guzmán R, Centofanti F, Doldo E, Céspedes Miranda EM, Orlandi A (2020) Oxidative stress and new pathogenetic mechanisms in endothelial dysfunction: potential diagnostic biomarkers and therapeutic targets. J Clin Med 9:1995. https://doi.org/10.3390/jcm9061995
7. Ohsawa I, Ishikawa M, Takahashi K, Watanabe M, Nishimaki K, Yamagata K, Katsura KI, Katayama Y, Asoh S, Ohta S (2007) Hydrogen acts as a therapeutic antioxidant by selectively reducing cytotoxic oxygen radicals. Nat Med 13:688–694. https://doi.org/10.1038/nm1577
8. Tian Y, Zhang Y, Wang Y, Chen Y, Fan W, Zhou J, Qiao J, Wei Y (2021) Hydrogen, a novel therapeutic molecule, regulates oxidative stress, inflammation, and apoptosis. Front Physiol 12:789507. https://doi.org/10.3389/fphys.2021.789507
9. Cox KE, Williamson KD (2018) Hydrogen: its technology and implication. In: Utilization of hydrogen. CRC Press, Boca Raton
10. LeBaron TW, Sharpe R, Pyatakovich FA, Artamonov MY (2024) Hydrogen: from stars to fuel to medicine. In: Slezak J, Kura B (eds) Molecular hydrogen in health and disease. Springer Nature, Cham, pp 1–20
11. Yuan T, Zhao J-N, Bao N-R (2023) Hydrogen applications: advances in the field of medical therapy. Med Gas Res 13:99–107. https://doi.org/10.4103/2045-9912.344978
12. Jin Z, Zhao P, Gong W, Ding W, He Q (2023) Fe-porphyrin: a redox-related biosensor of hydrogen molecule. Nano Res 16:2020–2025. https://doi.org/10.1007/s12274-022-4860-y
13. Ohta S (2023) Molecular hydrogen may activate the transcription factor Nrf2 to alleviate oxidative stress through the hydrogen-targeted porphyrin. Aging Pathobiol Ther 5:25–32. https://doi.org/10.31491/APT.2023.03.104
14. Tomofuji T, Kawabata Y, Kasuyama K, Endo Y, Yoneda T, Yamane M, Azuma T, Ekuni D, Morita M (2014) Effects of hydrogen-rich water on aging periodontal tissues in rats. Sci Rep 4:5534. https://doi.org/10.1038/srep05534
15. Ono H, Nishijima Y, Ohta S, Sakamoto M, Kinone K, Horikosi T, Tamaki M, Takeshita H, Futatuki T, Ohishi W et al (2017) Hydrogen gas inhalation treatment in acute cerebral infarction: a randomized controlled clinical study on safety and neuroprotection. J Stroke Cerebrovasc Dis 26:2587–2594. https://doi.org/10.1016/j.jstrokecerebrovasdis.2017.06.012
16. Fontanari P, Badier M, Guillot C, Tomei C, Burnet H, Gardette B, Jammes Y (2000) Changes in maximal performance of inspiratory and skeletal muscles during and after the 7.1-MPa Hydra 10 record human dive. Eur J Appl Physiol 81:325–328. https://doi.org/10.1007/s004210050050
17. Cole AR, Sperotto F, DiNardo JA, Carlisle S, Rivkin MJ, Sleeper LA, Kheir JN (2021) Safety of prolonged inhalation of hydrogen gas in air in healthy adults. Crit Care Explor 3:e543. https://doi.org/10.1097/CCE.0000000000000543

18. Ravingerová T, Dzurba A, Vrbjar N, Tribulová N, Ziegelhöeffer A, Slezák J (1994) Free oxygen radicals-induced impairment of heart function: the role of Na, K-ATPase. Pharmacol (Life Sci Adv) 13:133–140
19. Ravingerová T, Slezák J, Tribulová N, Džurba A, Uhrík B, Ziegelhöffer A (1999) Free oxygen radicals contribute to high incidence of reperfusion-induced arrhythmias in isolated rat heart. Life Sci 65:1927–1930. https://doi.org/10.1016/S0024-3205(99)00449-X
20. Kajiyama S, Hasegawa G, Asano M, Hosoda H, Fukui M, Nakamura N, Kitawaki J, Imai S, Nakano K, Ohta M et al (2008) Supplementation of hydrogen-rich water improves lipid and glucose metabolism in patients with type 2 diabetes or impaired glucose tolerance. Nutr Res 28: 137–143. https://doi.org/10.1016/j.nutres.2008.01.008
21. Gao Q, Song H, Wang XT, Liang Y, Xi YJ, Gao Y, Guo QJ, Lebaron T, Luo YX, Li SC et al (2017) Molecular hydrogen increases resilience to stress in mice. Sci Rep 7:1–12. https://doi.org/10.1038/s41598-017-10362-6
22. Xiao H, Li Y, Luo D, Dong J, Zhou L, Zhao S, Zheng Q, Wang H, Cui M, Fan S (2018) Hydrogen-water ameliorates radiation-induced gastrointestinal toxicity via MyD88's effects on the gut microbiota. Exp Mol Med 50:e433. https://doi.org/10.1038/emm.2017.246
23. Gvozdjáková A, Kucharská J, Kura B, Vančová O, Rausová Z, Sumbalová Z, Uličná O, Slezák J (2020) A new insight into the molecular hydrogen effect on coenzyme Q and mitochondrial function of rats. Can J Physiol Pharmacol 98:29–34. https://doi.org/10.1139/cjpp-2019-0281
24. Johnsen HM, Hiorth M, Klaveness J (2023) Molecular hydrogen therapy—a review on clinical studies and outcomes. Molecules 28:7785. https://doi.org/10.3390/molecules28237785
25. Ichihara G, Katsumata Y, Moriyama H, Kitakata H, Hirai A, Momoi M, Ko S, Shinya Y, Kinouchi K, Kobayashi E et al (2021) Pharmacokinetics of hydrogen after ingesting a hydrogen-rich solution: a study in pigs. Heliyon 7:e08359. https://doi.org/10.1016/j.heliyon.2021.e08359
26. Kang K-M, Kang Y-N, Choi I-B, Gu Y, Kawamura T, Toyoda Y, Nakao A (2011) Effects of drinking hydrogen-rich water on the quality of life of patients treated with radiotherapy for liver tumors. Med Gas Res 1:11. https://doi.org/10.1186/2045-9912-1-11
27. Hayashida K, Sano M, Ohsawa I, Shinmura K, Tamaki K, Kimura K, Endo J, Katayama T, Kawamura A, Kohsaka S et al (2008) Inhalation of hydrogen gas reduces infarct size in the rat model of myocardial ischemia–reperfusion injury. Biochem Biophys Res Commun 373:30–35. https://doi.org/10.1016/j.bbrc.2008.05.165
28. Kato S, Saitoh Y, Iwai K, Miwa N (2012) Hydrogen-rich electrolyzed warm water represses wrinkle formation against UVA ray together with type-I collagen production and oxidative-stress diminishment in fibroblasts and cell-injury prevention in keratinocytes. J Photochem Photobiol B Biol 106:24–33. https://doi.org/10.1016/j.jphotobiol.2011.09.006
29. Noda M, Fujita K, Lee C-H, Yoshioka T (2011) The principle and the potential approach to ROS-dependent cytotoxicity by non-pharmaceutical therapies: optimal use of medical gases with antioxidant properties. Curr Pharm Des 17:2253–2263. https://doi.org/10.2174/138161211797052600
30. Zálešák M, Kura B, Graban J, Farkašová V, Slezák J, Ravingerová T (2017) Molecular hydrogen potentiates beneficial anti-infarct effect of hypoxic postconditioning in isolated rat hearts: a novel cardioprotective intervention. Can J Physiol Pharmacol 95:888–893. https://doi.org/10.1139/cjpp-2016-0693
31. Ďuračková Z (2010) Some current insights into oxidative stress. Physiol Res 59:459–469. https://doi.org/10.33549/physiolres.931844
32. Moris D, Spartalis M, Spartalis E, Karachaliou G-S, Karaolanis GI, Tsourouflis G, Tsilimigras DI, Tzatzaki E, Theocharis S (2017) The role of reactive oxygen species in the pathophysiology of cardiovascular diseases and the clinical significance of myocardial redox. Ann Transl Med 5: 326–326. https://doi.org/10.21037/atm.2017.06.27
33. Halliwell B, Gutteridge JMC (2015) Free radicals in biology and medicine, vol 73, 5th edn. Oxford University Press, New York

34. Finkel T (2011) Signal transduction by reactive oxygen species. J Cell Biol 194:7–15. https://doi.org/10.1083/jcb.201102095
35. Hong Y, Boiti A, Vallone D, Foulkes NS (2024) Reactive oxygen species signaling and oxidative stress: transcriptional regulation and evolution. Antioxidants 13:312. https://doi.org/10.3390/antiox13030312
36. Chaudhary P, Janmeda P, Docea AO, Yeskaliyeva B, Abdull Razis AF, Modu B, Calina D, Sharifi-Rad J (2023) Oxidative stress, free radicals and antioxidants: potential crosstalk in the pathophysiology of human diseases. Front Chem 11:1158198. https://doi.org/10.3389/fchem.2023.1158198
37. Kura B, Slezak J (2024) The protective role of molecular hydrogen in ischemia/reperfusion injury. Int J Mol Sci 25:7884. https://doi.org/10.3390/ijms25147884
38. Tamura T, Suzuki M, Homma K, Sano M, Iizuka R, Narimiya H, Tsuruta R, Kaneda K, Fujita M, Sasaki J et al (2023) Efficacy of inhaled hydrogen on neurological outcome following brain ischaemia during post-cardiac arrest care (HYBRID II): a multi-centre, randomised, double-blind, placebo-controlled trial. eClinicalMedicine 58:101907. https://doi.org/10.1016/j.eclinm.2023.101907
39. Iketani M, Ohsawa I (2017) Molecular hydrogen as a neuroprotective agent. Curr Neuropharmacol 15:324–331. https://doi.org/10.2174/1570159X14666160607205417
40. Nagata K, Nakashima-Kamimura N, Mikami T, Ohsawa I, Ohta S (2009) Consumption of molecular hydrogen prevents the stress-induced impairments in hippocampus-dependent learning tasks during chronic physical restraint in mice. Neuropsychopharmacology 34:501–508. https://doi.org/10.1038/npp.2008.95
41. Akagi J, Baba H (2018) Hydrogen gas restores exhausted CD8+ T cells in patients with advanced colorectal cancer to improve prognosis. Oncol Rep 41:301–311. https://doi.org/10.3892/or.2018.6841
42. Ohsawa I, Nishimaki K, Yamagata K, Ishikawa M, Ohta S (2008) Consumption of hydrogen water prevents atherosclerosis in apolipoprotein E knockout mice. Biochem Biophys Res Commun 377:1195–1198. https://doi.org/10.1016/j.bbrc.2008.10.156
43. Todorovic N, Fernández-Landa J, Santibañez A, Kura B, Stajer V, Korovljev D, Ostojic SM (2023) The effects of hydrogen-rich water on blood lipid profiles in clinical populations: a systematic review and meta-analysis. Pharmaceuticals 16:142. https://doi.org/10.3390/ph16020142
44. Kamimura N, Nishimaki K, Ohsawa I, Ohta S (2011) Molecular hydrogen improves obesity and diabetes by inducing hepatic FGF21 and stimulating energy metabolism in db/db mice. Obesity 19:1396–1403. https://doi.org/10.1038/oby.2011.6
45. Ohta S (2021) Direct targets and subsequent pathways for molecular hydrogen to exert multiple functions: focusing on interventions in radical reactions. Curr Pharm Des 27:595–609. https://doi.org/10.2174/1381612826666200806101137
46. Kamimura N, Ichimiya H, Iuchi K, Ohta S (2016) Molecular hydrogen stimulates the gene expression of transcriptional coactivator Pgc-1α to enhance fatty acid metabolism. npj Aging Mech Dis 2:1–7. https://doi.org/10.1038/npjamd.2016.8
47. Kura B, Bagchi AK, Singal PK, Barancik M, Lebaron W, Valachova K, Soltes L (2019) Molecular hydrogen (H_2): potential in mitigating oxidative stress-induced radiation injury. Can J Physiol Pharmacol 97:287–292. https://doi.org/10.1139/cjpp-2018-0604
48. Lebaron TW, Laher I, Kura B, Slezak J (2019) Hydrogen gas: from clinical medicine to an emerging ergogenic molecule for sports athletes. Can J Physiol Pharmacol 97:797–807. https://doi.org/10.1139/cjpp-2019-0067
49. Ngo V, Duennwald ML (2022) Nrf2 and oxidative stress: a general overview of mechanisms and implications in human disease. Antioxidants 11:2345. https://doi.org/10.3390/antiox11122345
50. LeBaron TW, Kura B, Kalocayova B, Tribulova N, Slezak J (2019) A new approach for the prevention and treatment of cardiovascular disorders. Molecular hydrogen significantly reduces the effects of oxidative stress. Molecules 24:2076. https://doi.org/10.3390/molecules24112076

51. Hammad M, Raftari M, Cesário R, Salma R, Godoy P, Emami SN, Haghdoost S (2023) Roles of oxidative stress and Nrf2 signaling in pathogenic and non-pathogenic cells: a possible general mechanism of resistance to therapy. Antioxidants 12:1371. https://doi.org/10.3390/antiox12071371
52. Baird L, Dinkova-Kostova AT (2011) The cytoprotective role of the Keap1-Nrf2 pathway. Arch Toxicol 85:241–272
53. López-Bernardo E, Anedda A, Sánchez-Pérez P, Acosta-Iborra B, Cadenas S (2015) 4-Hydroxynonenal induces Nrf2-mediated UCP3 upregulation in mouse cardiomyocytes. Free Radic Biol Med 88:427–438. https://doi.org/10.1016/j.freeradbiomed.2015.03.032
54. Rutsch A, Kantsjö JB, Ronchi F (2020) The gut–brain axis: how microbiota and host inflammasome influence brain physiology and pathology. Front Immunol 11:604179. https://doi.org/10.3389/fimmu.2020.604179

Chapter 36
SUR2A: A Potential Cornerstone for Cardioprotection

Aleksandar Jovanović

Abstract The study reviews the role of SUR2A in cardioprotection by regulating ATP-sensitive K+ (KATP) channels, which link cellular metabolism with membrane excitability. SUR2A is a regulatory subunit of the KATP channel and it is a rate-limiting factor in the channel formation and, consequently the levels of. SUR2A set up the number of assembled KATP channels in the sarcolemma. An increased number of fully assembled KATP channels protect cardiomyocytes from stress by optimizing KATP channel opening and catalyzing ATP production in the subsarcolemmal space as these channels include ATP-producing enzymes alongside SUR2A and Kir6.2 subunits. Increased fully-assembled channels enhance ATP levels, beneficial during ischemia. SUR2A expression is regulated by mechanisms such as AMPK, PI3K/Akt, ERK1/2, and intracellular NAD+/NADH and ATP levels. Various compounds can regulate SUR2A, with nicotinamide identified as a clinically viable and safe option for increasing SUR2A and providing cardioprotection. SUR2A-based cardioprotection is an effective and safe strategy for clinical application.

Keywords SUR2A · Cardioprotection · Ischemic heart disease

Introduction

The Sulfonylurea Receptor 2A (SUR2A) is a regulatory subunit of the ATP-sensitive potassium (KATP) channels, primarily found in cardiac and skeletal muscle tissues. It plays a crucial role in modulating the activity of the channel in response to metabolic changes. The structure of SUR2A can be described as follows:

A. Jovanović (✉)
Department of Basic and Clinical Sciences, University of Nicosia Medical School, Nicosia, Cyprus

Center for Neuroscience and Integrative Brain Research (CENIBRE), University of Nicosia Medical School, Nicosia, Cyprus
e-mail: jovanovic.a@unic.ac.cy

D. K. Agrawal et al. (eds.), *Cardiovascular Toxicity*, Advances in Biochemistry in Health and Disease 33, https://doi.org/10.1007/978-3-031-90598-8_36

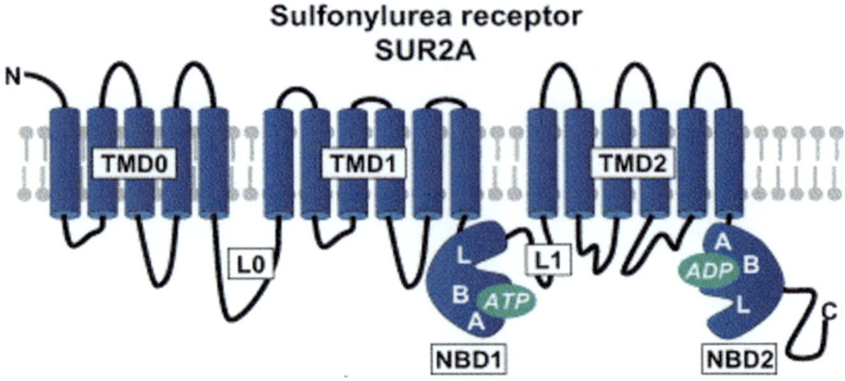

Fig. 36.1 Suggested structure of SUR2A (1549 amino acids), a regulatory subunit of sarcolemmal KATP channels. Sulfonylurea receptor (SUR) has three transmembrane domains (TMD0, 1 and 2), and two cytoplasmic nucleotide-binding domains (NBD1 and 2) including the Walker A, Walker B and Linker L consensus sequences. L0 and L1 refer to cytoplasmic loops. (This schematic structure of SUR2 is presented based on the well-established information on the structure of SUR2A in several published articles)

Transmembrane Domains: SUR2A contains multiple transmembrane domains that span the cell membrane.

1. TMD0: The initial transmembrane domain, which is unique to the SUR subunits and is involved in the interaction with the Kir6.x subunits.
2. TMD1: The first major transmembrane domain, consisting of six transmembrane helices.
3. TMD2: The second major transmembrane domain, also consisting of six transmembrane helices.

Nucleotide-Binding Domains (NBDs): SUR2A has two cytoplasmic nucleotide-binding domains that are critical for its regulatory function.

1. NBD1: Located between TMD1 and TMD2, this domain binds ATP and ADP, influencing the channel's activity.
2. NBD2: Located after TMD2, this domain also binds nucleotides and works in concert with NBD1 to regulate the channel.

Regulatory Sites: SUR2A contains several regulatory sites that interact with various pharmacological agents, such as sulfonylureas and potassium channel openers.

1. Sulfonylurea Binding Site: This site is responsible for the binding of sulfonylurea drugs, which inhibit the KATP channel.
2. Potassium Channel Opener Binding Site: This site binds potassium channel openers, which activate the KATP channel.

The overall structure of SUR2A (Fig. 36.1) allows it to sense changes in intracellular ATP and ADP levels, thereby modulating the activity of the KATP channel in response to the cell's metabolic state (reviewed by [1]).

Decades ago, the cardiac muscle was discovered to possess signaling pathways that, when activated, protect the myocardium against the damage induced by ischemia-reperfusion. This ability of the cardiac muscle to protect itself against injury has been termed 'cardioprotection'. One of the key players in this process is KATP channels. These channels are composed of two main types of subunits:

Kir6.x Subunits: These are the pore-forming subunits of the channel.

1. Kir6.1: One of the isoforms of the Kir6.x subunits.
2. Kir6.2: Another isoform of the Kir6.x subunits, more commonly found in cardiac and skeletal muscle tissues.

SUR (Sulfonylurea Receptor) Subunits: These are regulatory subunits that modulate the activity of the Kir6.x subunits.

1. SUR1: Typically found in pancreatic beta cells.
2. SUR2A: Predominantly expressed in cardiac and skeletal muscle tissues.
3. SUR2B: An isoform that can also be found in various tissues, including smooth muscle.

The functional KATP channel is a hetero-octamer, consisting of four Kir6.x subunits and four SUR subunits. The defining property of these channels is that they are closed by physiological levels of intracellular ATP. The specific combination of these subunits can vary depending on the tissue type, which influences the channel's pharmacological and physiological properties. SUR2A physically associates with Kir6.2 to form functional sarcolemmal KATP channels. These channels play a crucial role in the heart's response to metabolic stress by linking cellular energy metabolism to membrane excitability (reviewed by [2]). Creatine kinase, adenylate kinase, and glycolytic enzymes physically interact with SUR2A and Kir6.2 to form fully-assembled KATP channels in vivo [3–7]. This interaction ensures that the channels are properly assembled and functional, which is essential for their cardioprotective role [2]. It has been suggested that both the activation and increase of the number of functional KATP channels mediate cardioprotection [8, 9].

Cardioprotection Afforded by SUR2A

That the SUR2A could play cardioprotective was first hypothesized when it was found that in the hearts of female guinea pigs, there were higher levels of SUR2A than in their male counterparts. At the same time, there was a correlation between the number of KATP channels in sarcolemma and myocardial resistance to ischemia-reperfusion (Fig. 36.2). On the other hand, there were no sex-specific differences in levels of Kir6.2 and Kir6.1 subunits [14]. In strong support of the hypothesis that

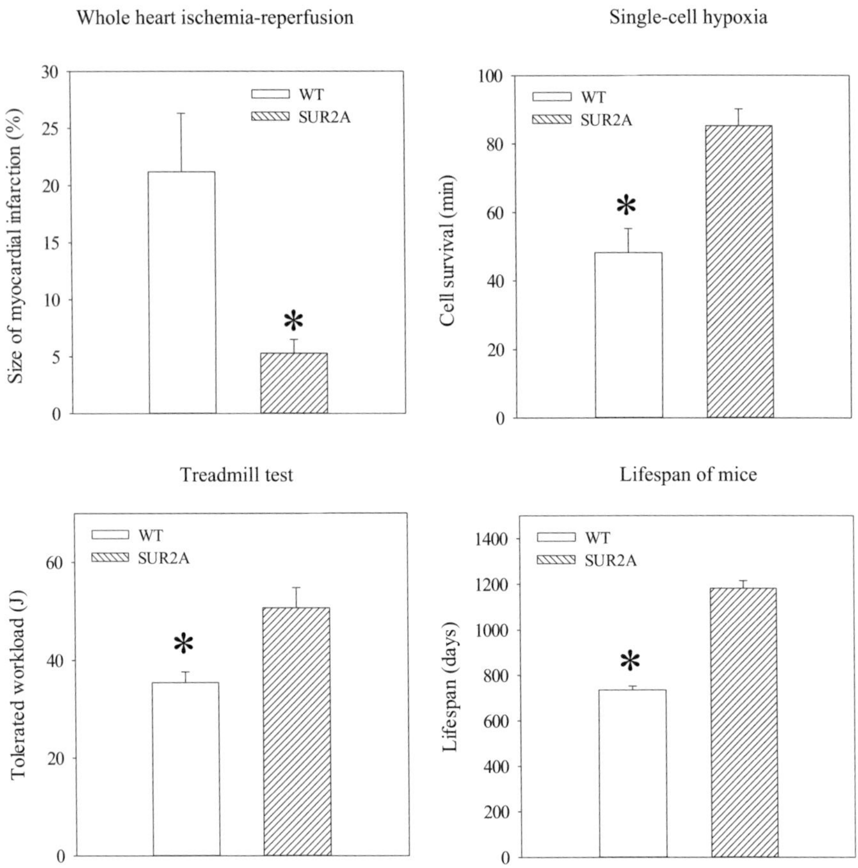

Fig. 36.2 Cardioprotection afforded by SUR2A. Graphs demonstrating the size of myocardial infarction following ischemia-reperfusion, cellular survival in hypoxia, tolerated workload in treadmill test, and lifespan of mice overexpressing SUR2A (SUR2A) and littermate controls (WT). Each bar represents mean ± SEM (n = 4–97). (Data in this figure are collected from several publications, including [10–13]), and cumulatively presented in this figure)

levels of SUR2A regulate cardioprotection, there were findings that the hearts of transgenic mice overexpressing SUR2A are more resistant to different types of metabolic stresses including ischemia-reperfusion than the hearts from their littermate wild types [10]. SUR2A-overexpressing phenotype was characterized by moderately increased levels of SUR2A, increased numbers of fully-assembled sarcolemmal KATP channels, increased myocardial resistance to different types of metabolic stresses, increased physical endurance, and prolonged lifespan [10–12] (Fig. 36.2). The mechanism of SUR2A-mediated cardioprotection seems to be associated with the SUR2A ability to regulate numbers of fully-assembled KATP channels [10]. It has been proposed that SUR2A is the rate-limiting protein in forming fully-assembled KATP channels and, consequently, the intracellular level of SUR2A determines the number of fully-assembled channels. Consequently, increased levels of SUR2A result in higher numbers of sarcolemmal KATP channels [10]. It has been demonstrated that an increased number of KATP channels results in their earlier activation during stress conditions, which, in turn, shortens the duration of action membrane potential, and decreases the influx of Ca^{2+}, preventing Ca^{2+} overload to confers cardioprotection [9, 10]. The way how increased number of channels leads to earlier channel activation is probably associated with the ability of lactate and 1,3-bisphosphoglycerate to open the channels despite the presence of millimolar intracellular ATP [4, 7, 15]. As enzymes producing these two KATP channel openers are physically associated with the channel subunits, more channel subunits mean the presence of a higher concentration of lactate and 1,3-bisphosphoglycerate in the microenvironment surrounding channels, which results in the channel activation [15, 16]. In addition to that, it has been shown that enzymes, such as creatine kinase, physically associated with sarcolemmal KATP channels produce ATP, whose presence in subsarcolemmal space is crucial for active transport and other important sarcolemma functions. Thus, the mechanism of cardioprotection by SUR2A is based on the regulation of the number of sarcolemmal KATP channels, which regulate the timing of channel opening during stress as well as levels of subsarcolemmal ATP. During metabolic stress, there is a decline in subsarcolemmal levels of ATP and increased levels of SUR2A counteract such decline by the catalytic activity of enzymes physically associated with the channels. It has been shown disrupting channel activity as well as the channel's ability to produce ATP results in the inhibition of cardioprotection [16–18]. Increased myocardial resistance to stress also has beneficial effects in physiological conditions. It has been demonstrated that mice moderately overexpressing SUR2A are characterized by a significant increase in physical endurance [12, 19] . It has been also shown that aging is associated with a decrease in cardiac SUR2A levels and this causes the aging-induced decline in physical fitness. In addition to that, it seems that increased SUR2A in the heart also significantly prolongs the lifespan in mice [11].

Regulation of SUR2A Level

So far, it has been reported that certain physiological conditions, estrogens, hypoxia, nicotinamide, and pyrazinamide increase SUR2A levels, enhancing cardiac resistance to stress [11, 14, 20–25]. It has been shown that the hearts of female experimental animals possess more SUR2A than their male counterparts and that correlates with myocardial resistance to ischemia-reperfusion [14]. This difference in KATP channel levels has been ascribed to estrogens that up-regulate SUR2A [24]. It has been also reported that mild hypoxia in vitro increases levels of SUR2A in a HIF1-apha-independent manner and that this process is mediated via phosphoinositide 3-kinase (PI3K), extracellular signal-regulated Kinases 1 and 2 (ERK1/2), and nicotinamide adenine dinucleotide oxidized form (NAD+) [20]. It has been shown that NAD+ itself increases SUR2A levels and that the NAD+/

Table 36.1 Conditions and factors regulating SUR2A levels in cardiomyocytes

Condition/factor	Effect on SUR2A levels	Mechanism/pathway	References
Estrogens	Increase	Not specified	[14, 24]
Hypoxia (mild, in vitro)	Increase	PI3K, ERK1/2, NAD+	[20]
NAD+	Increase	NAD+/NADH ratio	[20]
Hypoxia (in vivo)	Increase	Various signaling factors	[11, 21–23]
Insulin	Decrease	Not specified	[30]
Aging	Decrease	Not specified	[13]
ERK1/2	Increase	Mediates cardioprotection	[23, 26]
Akt	Increase	Mediates cardioprotection	[21]
LDH	Increase	Mediates cardioprotection	[21]
ATP	Increase	Not specified	[22]
AMPK	Increase	Mediates cardioprotection	[11]
c-jun	Increase	Transcriptional regulation	[20]

nicotinamide adenine dinucleotide reduced form (NADH) ratio is an important regulator of SUR2A levels in the heart [20]. Further research focused on in vivo hypoxia has shown that different degrees of hypoxia activate different signaling factors, but all of them have a single outcome and that is an increase in SUR2A [11, 21–23]. To this end, it has been shown that SUR2A levels are regulated by ERK1/2, protein kinase B (Akt), lactate dehydrogenase (LDH), and AMPK (AMP-activated protein kinase) [11, 21–23, 26]. The activation of all of these signaling factors is well-established to mediate cardioprotection and all of them regulate the expression of SUR2A, most likely via c-jun as the SUR2 promoter is controlled by c-jun, a transcription factor that regulates gene expression in response to stress and other stimuli [20, 27–29]. The involvement of c-jun in SUR2A regulation suggests that SUR2A expression is tightly controlled by transcriptional mechanisms in response to cellular stress [29]. On the other hand, it has been also demonstrated that basal levels of SUR2A could decrease and, under such circumstances, the resistance of cardiomyocytes to stress decreases. It has been suggested that aging as well as insulin downregulates SUR2A (Sudhir et al. 2011) [30], which could explain negative cardiovascular events associated with insulin therapy and aging-associated decreased myocardial resistance to metabolic stress [31–33]. Overall, it is fair to say that SUR2A is regulated by multiple intracellular factors (Table 36.1) and it looks like this protein serves as the end-effector of all/most cardioprotective signaling pathways.

The Clinical Potential of SUR2A-Based Therapeutic Strategy

Increasing myocardial resistance to metabolic stress is beneficial in many cardiac and non-cardiac conditions, including ischemia-reperfusion, heart failure, sarcopenia, and many others [34–38].

Ischemia-reperfusion, an injury underlying myocardial infarction, is characterized by impaired blood supply to the myocardium (ischemia), followed by the restitution of the blood flow (reperfusion). Reperfusion that occurs following ischemia exacerbates myocardial damage by inducing oxidative stress, Ca^{2+} overload, inflammation, and apoptosis [39]. So far, no generally accepted therapy is based on cardioprotective signaling. Despite extensive preclinical research, translating these findings into clinical practice has proven challenging [39]. The complexity of cardioprotective signaling pathways and the potential for adverse effects have hindered the development of effective therapies. Previous promising strategies, such as the activation of sarcolemmal KATP channels, were precluded due to adverse effects like arrhythmias and hypotension [40]. On the other hand, signaling factors such are protein kinases have multiple targets and their activation has multiple effects, many of which are undesirable [41]. The challenge lies in developing therapies that can provide cardioprotection without causing significant adverse effects. However, increased expression of SUR2A does not seem to be associated with significant adverse effects [11, 42]. On the contrary, studies in mice have shown that overexpression of SUR2A leads to increased physical endurance and prolonged lifespan without apparent negative consequences [11, 12]. The absence of significant adverse effects in animal studies suggests that SUR2A-based therapies could be safe for human use. Potential Therapeutic Approaches to manipulate myocardial SUR2A levels could be:

Gene Therapy: Adenoviral constructs of SUR2A increase cardiac levels of SUR2A and resistance to stress [16]. However, issues of gene delivery and potential adverse effects of viral vectors need to be addressed. Current advances in gene therapy techniques could make this a viable option [43, 44]. The development of safe and efficient gene delivery systems is crucial for the successful implementation of gene therapy for cardioprotection.

Hypoxia: While hypoxia increases SUR2A levels [11, 21–23], its long-term delivery to patients might not be practical [45]. Hypoxia as a therapeutic strategy requires further investigation to determine its feasibility and safety. The potential benefits of hypoxia-induced SUR2A upregulation need to be weighed against the challenges of implementing hypoxia therapy in clinical practice.

Nicotinamide: A form of vitamin B3, nicotinamide increases SUR2A levels and protects the heart against ischemia-reperfusion and increases physical endurance [19, 46] (Fig. 36.3). It is known to be safe and could be quickly introduced into clinical practice. Nicotinamide is the amide form of vitamin B3 (niacin) and is obtained through synthesis in the body or as a dietary source and supplement. Nicotinic acid is the other form of the water-soluble vitamin B3. Although also present in animal sources, the principal form of niacin in dietary plant sources is

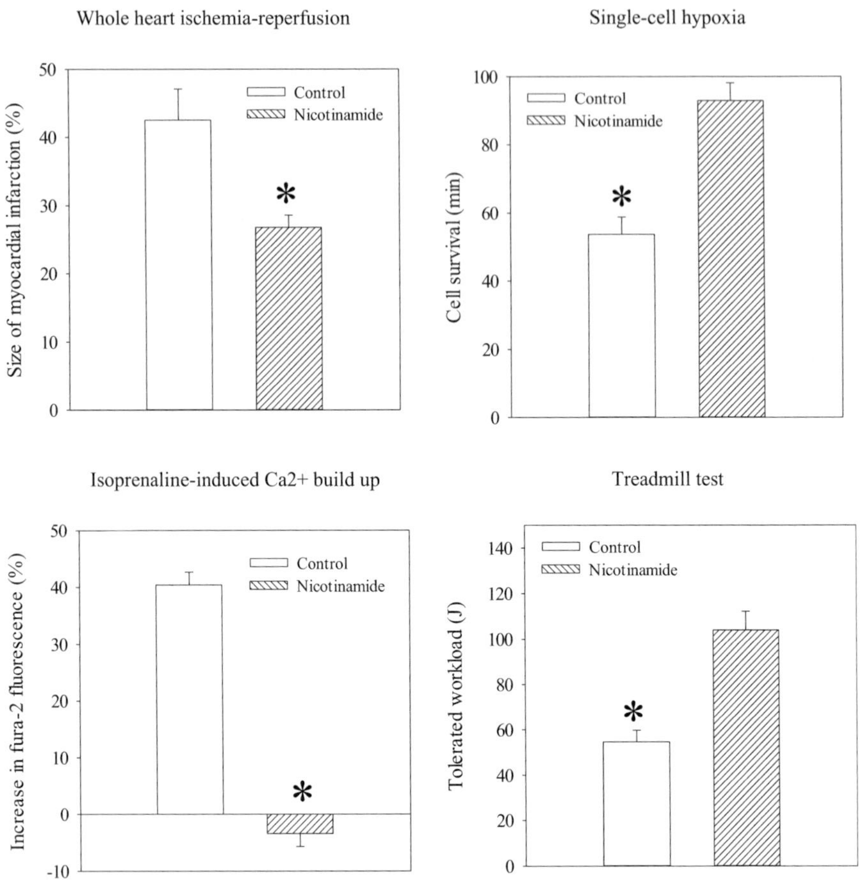

Fig. 36.3 Cardioprotection afforded by oral nicotinamide in mice. Graphs demonstrating the size of myocardial infarction following ischemia-reperfusion, cellular survival in hypoxia, intracellular Ca^{2+} levels when cardiomyocytes are challenged with isoprenaline (500 nM) as indicated by changes in Fura-2 fluorescence and tolerated workload in treadmill test in mice on regular (control) and nicotinamide-rich diet (nicotinamide). Each bar represents mean ± SEM (n = 6–12). (Data in this figure are collected from several publications, including [19, 46] and cumulatively presented in this figure)

nicotinic acid which is rapidly absorbed through the gastrointestinal epithelium. Nicotinamide is subsequently generated through the conversion of nicotinic acid in the liver or through the hydrolysis of NAD^+. Once nicotinamide is obtained in the body, it functions as the precursor for the coenzyme β-nicotinamide adenine dinucleotide (NAD^+) and also is essential for the synthesis of nicotinamide adenine dinucleotide phosphate ($NADP^+$). These cellular pathways are essential for energy metabolism and may directly impact normal physiology, as well as disease progression [47].

If an increase in NAD^+ up-regulates SUR2A and K_{ATP} channels in cardiac cells (as suggested in Crawford et al. [20]), then it was possible that a nicotinamide-rich diet would increase myocardial SUR2A/K_{ATP} channels and, consequently, heart resistance to metabolic stress. Therefore, we have tested this hypothesis by feeding mice for a week with 6 different doses of nicotinamide (ranging from 10 mg to 10 g of nicotinamide in 1 kg of diet; this is an estimated daily intake of nicotinamide ranging from 40 μg to 60 mg). We have measured the expression of SUR2A in the heart, the size of myocardial infarction in response to ischemia-reperfusion, and the survival of isolated cardiomyocytes in severe hypoxia. We have found that oral nicotinamide significantly decreases infarct size in response to ischemia-reperfusion and that increases cellular survival to severe hypoxia and β-adrenergic stress. The effect was dose-dependent and the maximal effect was observed when daily intake was 2 mg of nicotinamide [46]. This dose of nicotinamide increased the expression of SUR2A in the heart and the cardio-protective effect was largely inhibited by HMR1098, an inhibitor of sarcolemmal K_{ATP} channels [46] (it is also important to point out that nicotinamide has no K_{ATP}-channels opening properties), [19]. The activation of the cardioprotective signaling pathway protects the heart against a range of different stresses including ischemia without reperfusion as well as ischemia with reperfusion [28]. Clinically, there is evidence to suggest that oral antidiabetic drugs, blockers of K_{ATP} channels, worsen the outcome of chronic ischemic heart disease [48]. Conversely, in experimental animals, exercise was suggested to improve the outcome of ischemic heart disease IHD by increasing the number of sarcolemmal K_{ATP} channels [49]. Taken together, it is likely that nicotinamide would be a useful drug against ischemic heart disease. The maximal effective cardioprotective dose in mice is the dose that is equivalent to 500 mg/day of nicotinamide in humans (the human equivalent dose was calculated using mouse dose and dose translation according to Reagan-Shaw et al. [50]. Nicotinamide is known as one of the safest drugs in clinical practice. It should not be confused with nicotinic acid which is also a vitamin B3 form but exhibits more pharmacological and adverse effects than nicotinamide [51]. Current therapeutic use of nicotinamide is confined to the treatment of acne vulgaris, diabetes mellitus, and head and neck cancer, all of which are in doses above 500 mg/day [52, 53]. It has been suggested that chronic intake of nicotinamide below 3000 mg/day is safe and virtually devoid of adverse effects [51, 54]. Therefore, chronic treatment with 500 mg/day of nicotinamide is likely to be safe. At the same time, it is realistic to expect that nicotinamide taken orally at 500 mg/day would significantly increase myocardial resistance to ischemia. Taken all together, nicotinamide seems to be a good candidate to treat ischemic heart disease and such a drug could be chronically used for years without significant adverse effects.

However, an important issue is whether we can be confident that the effect we observed in mice is relevant for humans. It is well established that the pharmacokinetics of nicotinamide in humans and rodents including mice are similar [55]. In terms of pharmacodynamics, the therapeutic effect of a specific dose of nicotinamide for the treatment of head and neck cancer was first observed in mice [56] and then in humans [53]. In mice, 1000 mg/kg/day per mouse was used to treat head and neck cancer [56]. It was calculated from this dose human equivalent dose; the following formulae were used: HED (mg/kg) = mouse dose (mg/kg) × (mouse Km/Human Km) where Km = weight (kg)/body surface area (m^2); in humans, Km is 37 and in mice, Km is 3. Thus, the effective dose for humans for nicotinamide would be HED (mg/kg) = 1000 mg/kg × 3/37 = 1000 mg/kg × 0.081 = 81 mg/kg/day, which is what is used in clinical practice today in the therapy of head and neck cancer (80 mg/kg/day) [53]. The same applies to the use of nicotinamide in diabetes mellitus, i.e. use of nicotinamide to treat diabetes mellitus was also originally described in [57] and this was shown to have a bearing on patients suffering from this disease [52]. Thus, if nicotinamide is effective and safe in mice, it is likely that it would be safe and effective in humans. It is also encouraging that dose translation from mice to humans seems to be very reliable and that the human equivalent dose calculated based on the mouse dose would be as efficient/safe in humans as it is in mice. Taken together, nicotinamide deserves to be tested as a cardioprotective drug that could be useful in many conditions.

Pyrazinamide: An anti-tubercular agent that protects embryonic heart cells against metabolic stress [25]. Further studies are needed to establish its cardioprotective efficacy and pyrazinamide's potential as a cardioprotective agent warrants additional research to confirm its benefits and safety. The exploration of pyrazinamide as a cardioprotective agent could open new avenues for repurposing existing drugs for cardiac therapy.

Conclusion

SUR2A-based cardioprotective strategies hold significant potential for treating cardiac and non-cardiac conditions that benefit from increased myocardial resistance to stress. Nicotinamide, in particular, shows promise due to its established cardioprotective properties and safety profile. Further research and clinical trials are necessary to fully realize the potential of SUR2A-based therapies in clinical practice. The development of safe and effective methods to increase SUR2A levels could improve the treatment of ischemic heart disease and other related conditions, providing new hope for patients worldwide.

References

1. Nelson PT, Jicha GA, Wang WX, Ighodaro E, Artiushin S, Nichols CG, Fardo DW (2015) ABCC9/SUR2 in the brain: implications for hippocampal sclerosis of aging and a potential therapeutic target. Ageing Res Rev 24(Pt B):111–125. https://doi.org/10.1016/j.arr.2015.07.007
2. Foster MN, Coetzee WA (2016) KATP channels in the cardiovascular system. Physiol Rev 96(1):177–252. https://doi.org/10.1152/physrev.00003.2015

3. Crawford RM, Budas GR, Jovanović S, Ranki HJ, Wilson TJ, Davies AM, Jovanović A (2002a) M-LDH serves as a sarcolemmal K(ATP) channel subunit essential for cell protection against ischemia. EMBO J 21(15):3936–3948. https://doi.org/10.1093/emboj/cdf388
4. Crawford RM, Ranki HJ, Botting CH, Budas GR, Jovanovic A (2002b) Creatine kinase is physically associated with the cardiac ATP-sensitive K+ channel in vivo. FASEB J 16(1): 102–104. https://doi.org/10.1096/fj.01-0466fje
5. Elvir-Mairena JR, Jovanović A, Gomez LA, Alekseev AE, Terzic A (1996) Reversal of the ATP-liganded state of ATP-sensitive K+ channels by adenylate kinase activity. J Biol Chem 271(50):31903–31908. https://doi.org/10.1074/jbc.271.50.31903
6. Hong M, Kefaloyianni E, Bao L, Malester B, Delaroche D, Neubert TA, Coetzee WA (2011) Cardiac ATP-sensitive K+ channel associates with the glycolytic enzyme complex. FASEB J 25(7):2456–2467. https://doi.org/10.1096/fj.10-176669
7. Jovanović S, Du Q, Crawford RM, Budas GR, Stagljar I, Jovanović A (2005) Glyceraldehyde 3-phosphate dehydrogenase serves as an accessory protein of the cardiac sarcolemmal K(ATP) channel. EMBO Rep 6(9):848–852. https://doi.org/10.1038/sj.embor.7400489
8. Budas GR, Jovanović S, Crawford RM, Jovanović A (2004) Hypoxia-induced preconditioning in adult stimulated cardiomyocytes is mediated by the opening and trafficking of sarcolemmal KATP channels. FASEB J 18(9):1046–1048. https://doi.org/10.1096/fj.04-1602fje
9. Sukhodub A, Jovanović S, Du Q, Budas G, Clelland AK, Shen M, Sakamoto K, Tian R, Jovanović A (2007) AMP-activated protein kinase mediates preconditioning in cardiomyocytes by regulating activity and trafficking of sarcolemmal ATP-sensitive K(+) channels. J Cell Physiol 210(1):224–236. https://doi.org/10.1002/jcp.20862
10. Du Q, Jovanović S, Clelland A, Sukhodub A, Budas G, Phelan K, Murray-Tait V, Malone L, Jovanović A (2006) Overexpression of SUR2A generates a cardiac phenotype resistant to ischemia. FASEB J 20(8):1131–1141. https://doi.org/10.1096/fj.05-5483com
11. Mohammed Abdul KS, Jovanović S, Jovanović A (2017) Exposure to 15% oxygen in vivo up-regulates cardioprotective SUR2A without affecting ERK1/2 and AKT: a crucial role for AMPK. J Cell Mol Med 21(7):1342–1350. https://doi.org/10.1111/jcmm.13064
12. Sudhir R, Du Q, Sukhodub A, Jovanović S, Jovanović A (2020) Improved adaptation to physical stress in mice overexpressing SUR2A is associated with changes in the pattern of Q–T interval. Pflugers Arch 472(6):683–691. https://doi.org/10.1007/s00424-020-02401-5
13. Ranki HJ, Budas GR, Crawford RM, Jovanović A (2001) Gender-specific difference in cardiac ATP-sensitive K(+) channels. J Am Coll Cardiol 38(3):906–915. https://doi.org/10.1016/s0735-1097(01)01428-0
14. Jovanović S, Jovanović A (2005) High glucose regulates the activity of cardiac sarcolemmal ATP-sensitive K+ channels via 1,3-bisphosphoglycerate: a novel link between cardiac membrane excitability and glucose metabolism. Diabetes 54(2):383–393. https://doi.org/10.2337/diabetes.54.2.383
15. Du Q, Jovanović S, Sukhodub A, Jovanović A (2010) Infection with AV-SUR2A protects H9C2 cells against metabolic stress: a mechanism of SUR2A-mediated cytoprotection independent from the K(ATP) channel activity. Biochim Biophys Acta 1803(3):405–415. https://doi.org/10.1016/j.bbamcr.2010.01.018
16. Jovanović S, Du Q, Sukhodub A, Jovanović A (2009a) A dual mechanism of cytoprotection afforded by M-LDH in embryonic heart H9C2 cells. Biochim Biophys Acta 1793(8): 1379–1386. https://doi.org/10.1016/j.bbamcr.2009.04.007
17. Jovanović S, Du Q, Sukhodub A, Jovanović A (2009b) M-LDH physically associated with sarcolemmal KATP channels mediates cytoprotection in heart embryonic H9C2 cells. Int J Biochem Cell Biol 41(11):2295–2301. https://doi.org/10.1016/j.biocel.2009.05.012
18. Sukhodub A, Sudhir R, Du Q, Jovanović S, Reyes S, Jovanović A (2011) Nicotinamide-rich diet improves physical endurance by up-regulating SUR2A in the heart. J Cell Mol Med 15(8): 1703–1712. https://doi.org/10.1111/j.1582-4934.2010.01156.x
19. Crawford RM, Jovanović S, Budas GR, Davies AM, Lad H, Wenger RH, Robertson KA, Roy DJ, Ranki HJ, Jovanović A (2003) Chronic mild hypoxia protects heart-derived H9c2 cells

against acute hypoxia/reoxygenation by regulating expression of the SUR2A subunit of the ATP-sensitive K+ channel. J Biol Chem 278(33):31444–31455. https://doi.org/10.1074/jbc.M303051200
20. Mohammed Abdul KS, Jovanović S, Du Q, Sukhodub A, Jovanović A (2015a) Mild hypoxia in vivo regulates cardioprotective SUR2A: a role for Akt and LDH. Biochim Biophys Acta 1852(5):709–719. https://doi.org/10.1016/j.bbadis.2015.01.001
21. Mohammed Abdul KS, Jovanović S, Du Q, Sukhodub A, Jovanović A (2015b) A link between ATP and SUR2A: a novel mechanism explaining cardioprotection at high altitude. Int J Cardiol 189:73–76. https://doi.org/10.1016/j.ijcard.2015.04.069
22. Mohammed Abdul KS, Jovanović S, Sukhodub A, Du Q, Jovanović A (2014) Upregulation of cardioprotective SUR2A by sub-hypoxic drop in oxygen. Biochim Biophys Acta 1843(11): 2424–2431. https://doi.org/10.1016/j.bbamcr.2014.07.007
23. Ranki HJ, Budas GR, Crawford RM, Davies AM, Jovanović A (2002) 17Beta-estradiol regulates expression of K(ATP) channels in heart-derived H9c2 cells. J Am Coll Cardiol 40(2):367–374. https://doi.org/10.1016/s0735-1097(02)01947-2
24. Sinha S, Du Q, Jovanović S, Sukhodub A, Jovanović A (2019) Pyrazinamide may possess cardioprotective properties. J Antibiot (Tokyo) 72(9):714–717. https://doi.org/10.1038/s41429-019-0202-z
25. Mohammed Abdul KS, Rayadurgam J, Faiz N, Jovanović A, Tan W (2020) Cardioprotection by isosteviol derivate JC105: a unique drug property to activate ERK1/2 only when cells are exposed to hypoxia-reoxygenation. J Cell Mol Med 24(18):10924–10934. https://doi.org/10.1111/jcmm.15721
26. de Miranda DC, de Oliveira FG, Hermidorff MM, Dos Santos Silva FC, de Assis LVM, Isoldi MC (2021) Pre- and post-conditioning of the heart: an overview of cardioprotective signaling pathways. Curr Vasc Pharmacol 19(5):499–524. https://doi.org/10.2174/1570161119666201120160619
27. Jovanović A (2018) Cardioprotective signalling: past, present and future. Eur J Pharmacol 833: 314–319. https://doi.org/10.1016/j.ejphar.2018.06.029
28. Papavassiliou AG, Musti AM (2020) The multifaceted output of c-Jun biological activity: focus at the junction of CD8 T cell activation and exhaustion. Cells 9(11):2470. https://doi.org/10.3390/cells9112470
29. Du Q, Jovanović S, Sukhodub A, Ngoi YS, Lal A, Zheleva M, Jovanović A (2018) Insulin down-regulates cardioprotective SUR2A in the heart-derived H9c2 cells: a possible explanation for some adverse effects of insulin therapy. Biochem Biophys Rep 16:12–18. https://doi.org/10.1016/j.bbrep.2018.08.005
30. Díaz-Vesga MC, Zúñiga-Cuevas Ú, Ramírez-Reyes A, Herrera-Zelada N, Palomo I, Bravo-Sagua R, Riquelme JA (2021) Potential therapies to protect the aging heart against ischemia/reperfusion injury. Front Cardiovasc Med 8:770421. https://doi.org/10.3389/fcvm.2021.770421
31. Herman ME, O'Keefe JH, Bell DSH, Schwartz SS (2017) Insulin therapy increases cardiovascular risk in type 2 diabetes. Prog Cardiovasc Dis 60(3):422–434. https://doi.org/10.1016/j.pcad.2017.09.001
32. Liu J, Hu X (2022) Impact of insulin therapy on outcomes of diabetic patients with heart failure: a systematic review and meta-analysis. Diabetes Vasc Dis Res 19(3):14791641221093175. https://doi.org/10.1177/14791641221093175
33. Crimi G, Pica S, Raineri C, Bramucci E, De Ferrari GM, Klersy C, Ferlini M, Marinoni B, Repetto A, Romeo M, Rosti V, Massa M, Raisaro A, Leonardi S, Rubartelli P, Oltrona Visconti L, Ferrario M (2013) Remote ischemic post-conditioning of the lower limb during primary percutaneous coronary intervention safely reduces enzymatic infarct size in anterior myocardial infarction: a randomized controlled trial. JACC Cardiovasc Interv 6(10): 1055–1063. https://doi.org/10.1016/j.jcin.2013.05.011
34. Fath AR, Aglan M, Aglan A, Chilton RJ, Trakhtenbroit A, Al-Shammary OA, Oppong-Nkrumah O, Lenihan DJ, Dent SF, Otchere P (2024) Cardioprotective potential of

sodium-glucose cotransporter-2 inhibitors in patients with cancer treated with anthracyclines: an observational study. Am J Cardiol 222:175–182. https://doi.org/10.1016/j.amjcard.2024.04.032

35. Quagliariello V, Berretta M, Bisceglia I, Giacobbe I, Iovine M, Giordano V, Arianna R, Barbato M, Izzo F, Maurea C, Canale ML, Paccone A, Inno A, Scherillo M, Gabrielli D, Maurea N (2024) The sGCa Vericiguat exhibit cardioprotective and anti-sarcopenic effects through NLRP-3 pathways: potential benefits for anthracycline-treated cancer patients. Cancers (Basel) 16(8):1487. https://doi.org/10.3390/cancers16081487
36. Ravingerova T, Adameova A, Lonek L, Farkasova V, Ferko M, Andelova N, Kura B, Slezak J, Galatou E, Lazou A, Zohdi V, Dhalla NS (2023) Is intrinsic cardioprotection a laboratory phenomenon or a clinically relevant tool to salvage the failing heart? Int J Mol Sci 24(22): 16497. https://doi.org/10.3390/ijms242216497
37. Sabe SA, Harris DD, Broadwin M, Sellke FW (2024) Cardioprotection in cardiovascular surgery. Basic Res Cardiol 119(4):545–568. https://doi.org/10.1007/s00395-024-01062-0
38. Welt FGP, Batchelor W, Spears JR, Penna C, Pagliaro P, Ibanez B, Drakos SG, Dangas G, Kapur NK (2024) Reperfusion injury in patients with acute myocardial infarction: JACC Scientific Statement. J Am Coll Cardiol 83(22):2196–2213. https://doi.org/10.1016/j.jacc.2024.02.056
39. Jahangir A, Terzic A (2005) K(ATP) channel therapeutics at the bedside. J Mol Cell Cardiol 39(1):99–112. https://doi.org/10.1016/j.yjmcc.2005.04.006
40. Chen J, Li Y, Du C, Wei T, Shan T, Wang L (2022) Protein kinases in cardiovascular diseases. Chin Med J 135(5):557–570. https://doi.org/10.1097/CM9.0000000000001870
41. Sudhir R, Jaafar N, Du Q, Sukhodub A, Jovanović S, Kreouzi M, Jovanović A (2022) Increase in cardioprotective SUR2A does not alter heart rate and heart rate regulation by physical activity and diurnal rhythm. J Basic Clin Physiol Pharmacol 33(5):619–624. https://doi.org/10.1515/jbcpp-2021-0289
42. Bulcha JT, Wang Y, Ma H, Tai PWL, Gao G (2021) Viral vector platforms within the gene therapy landscape. Signal Transduct Target Ther 6(1):53. https://doi.org/10.1038/s41392-021-00487-6
43. Liu F, Su R, Jiang X, Wang S, Mu W, Chang L (2024) Advanced micro/nano-electroporation for gene therapy: recent advances and future outlook. Nanoscale 16(22):10500–10521. https://doi.org/10.1039/d4nr01408a
44. Burtscher J, Mallet RT, Pialoux V, Millet GP, Burtscher M (2022) Adaptive responses to hypoxia and/or hyperoxia in humans. Antioxid Redox Signal 37(13–15):887–912. https://doi.org/10.1089/ars.2021.0280
45. Sukhodub A, Du Q, Jovanović S, Jovanović A (2010) Nicotinamide-rich diet protects the heart against ischaemia-reperfusion in mice: a crucial role for cardiac SUR2A. Pharmacol Res 61(6): 564–570. https://doi.org/10.1016/j.phrs.2010.01.008
46. Liu Y, Clement J, Grant R, Sachdev P, Braidy N (2018) Quantitation of NAD+: why do we need to measure it? Biochim Biophys Acta Gen Subj 1862(12):2527–2532. https://doi.org/10.1016/j.bbagen.2018.07.023
47. Yin M, Westenbrink BD, Meissner M, van Gilst WH, de Boer RA (2013) Variable effects of anti-diabetic drugs in animal models of myocardial ischemia and remodeling: a translational perspective for the cardiologist. Int J Cardiol 169(6):385–393. https://doi.org/10.1016/j.ijcard.2013.10.005
48. Fitts RH, Wang X, Kwok WM, Camara AKS (2024) Cardiomyocyte adaptation to exercise: K+ channels, contractility and ischemic injury. Int J Sports Med 45(11):791–803. https://doi.org/10.1055/a-2296-7604. Epub ahead of print
49. Reagan-Shaw S, Nihal M, Ahmad N (2008) Dose translation from animal to human studies revisited. FASEB J 22(3):659–661. https://doi.org/10.1096/fj.07-9574LSF
50. Gille A, Bodor ET, Ahmed K, Offermanns S (2008) Nicotinic acid: pharmacological effects and mechanisms of action. Annu Rev Pharmacol Toxicol 48:79–106. https://doi.org/10.1146/annurev.pharmtox.48.113006.094746

51. Crinò A, Schiaffini R, Manfrini S, Mesturino C, Visalli N, Beretta Anguissola G, Suraci C, Pitocco D, Spera S, Corbi S, Matteoli MC, Patera IP, Manca Bitti ML, Bizzarri C, Pozzilli P, IMDIAB Group (2004) A randomized trial of nicotinamide and vitamin E in children with recent onset type 1 diabetes (IMDIAB IX). Eur J Endocrinol 150(5):719–724. https://doi.org/10.1530/eje.0.1500719
52. Kaanders JH, Bussink J, van der Kogel AJ (2002) ARCON: a novel biology-based approach in radiotherapy. Lancet Oncol 3(12):728–737. https://doi.org/10.1016/s1470-2045(02)00929-4
53. Knip M, Douek IF, Moore WP, Gillmor HA, McLean AE, Bingley PJ, Gale EA, European Nicotinamide Diabetes Intervention Trial Group (2000) Safety of high-dose nicotinamide: a review. Diabetologia 43(11):1337–1345. https://doi.org/10.1007/s001250051536
54. Horsman MR, Høyer M, Honess DJ, Dennis IF, Overgaard J (1993) Nicotinamide pharmacokinetics in humans and mice: a comparative assessment and the implications for radiotherapy. Radiother Oncol 27(2):131–139. https://doi.org/10.1016/0167-8140(93)90133-s
55. Wennerberg J, Kjellén E, Lybak S, Rydell R, Pero R (1993) Biochemical modulation of chemotherapy and radiotherapy in head and neck cancer. Anticancer Res 13(6B):2501–2506
56. Reddy S, Bibby NJ, Wu D, Swinney C, Barrow G, Elliott RB (1995) A combined casein-free-nicotinamide diet prevents diabetes in the NOD mouse with minimum insulitis. Diabetes Res Clin Pract 29(2):83–92. https://doi.org/10.1016/0168-8227(95)01109-9

Chapter 37
Repurposing of PARP Inhibitors for the Clinical Management of Acute Cardiotoxicity

Ferenc Gallyas Jr, Robert Halmosi, and Kalman Toth

Abstract Cardiotoxicity represents serious side effects of cancer therapy that affect quality of life and overall survival of patients significantly. Its most severe manifestation, heart failure, has a 5-year survival rate of 50%, worse than that of many types of cancer. However, medications aimed at ameliorating cardiotoxic side-effects of the antineoplastic treatment likely compromise efficacy of it. In this chapter, we propose repositioning inhibitors of the nuclear enzyme poly(ADP-ribose) polymerase, which are used for the treatment of cancers with defective homologous recombination DNA repair, for reducing cardiotoxicity of cancer therapy. We introduce the forms of cardiotoxicity, the involved mechanisms and the therapeutic agents, which cause them. We also present cellular effects of poly(ADP-ribose) polymerase and review the mechanisms involved in antineoplastic and cardioprotective effects of its inhibitors, respectively. Finally, we discuss the risks and benefits of the proposed repositioning.

Keywords Anthracyclines · Cardiotoxicity · Epidermal growth factor receptor · HER2 inhibitors · Human epidermal growth factor receptor · Fluorouracil · Mitochondrial membrane potential · PARP · Poly(ADP-ribose) polymerase · Proteasome inhibitors · Radiation therapy · Taxanes · Vascular endothelial growth factor · VEGF inhibitor

---

F. Gallyas Jr (✉)
Department of Biochemistry and Medicinal Chemistry, Medical School, University of Pecs, Pecs, Hungary
e-mail: ferenc.gallyas@aok.pte.hu

R. Halmosi · K. Toth
First Department of Internal Medicine, Clinical Centre, University of Pecs, Pecs, Hungary
e-mail: halmosi.robert@pte.hu; toth.kalman@pte.hu

D. K. Agrawal et al. (eds.), *Cardiovascular Toxicity*, Advances in Biochemistry in Health and Disease 33, https://doi.org/10.1007/978-3-031-90598-8_37

Introduction

Cardiotoxicity refers to damage to the myocardium that is induced by chemical agents or treatments, especially certain medications or toxins. The term is commonly associated with adverse side effects of cancer treatments, particularly chemotherapy, but can also result from a variety of drugs, environmental toxins, and other external stressors. Cardiotoxicity can occur through several mechanisms, depending on the agent involved. The underlying pathophysiological changes include direct damage to cardiac cells, disruption of metabolic processes, and imbalances in ion channels, leading to electrical disturbances. Key mechanisms include oxidative stress, mitochondrial disfunction, disruption of Ca^{2+} homeostasis, inhibition of DNA synthesis and repair, and malfunctioning of the immune system [1]. Among chemotherapeutic agents, anthracyclines are the most well-known for causing cardiotoxicity [2]. However, other antineoplastic drug families such as HER2 inhibitors [3], tyrosine kinase inhibitors [4], alkylating agents [5], 5-fluorouracil [6] and radiation therapy [7] were also demonstrated to have cardiotoxic side effects. Although pre-clinical toxicity studies on cultured cells have limited translational power, it is thought-provoking that 15 out of 35 different Food and Drug Administration (FDA)-approved anti-cancer drugs reduced viability of primary human cardiomyocytes significantly when applied at a concentration of 1 μM in the culturing medium for 72 h [8]. Nevertheless, the extent and severity of cardiovascular complications in cancer patients called for the creation of cardio-oncology, an emerging interdisciplinary field focusing on the prevention, diagnosis, and management of the said complications.

Ameliorating cardiotoxic side effects of antineoplastic drugs, however, is not straightforward. On the one hand, the aforementioned mechanisms underlying cardiotoxicity are involved in the primary anti-cancer effects of many cytostatic agents [9, 10]. Therefore, treatments mitigating the cardiotoxic side effects, likely reduce the efficacy of the cancer therapy [11]. On the other hand, despite of extensive preclinical research of the last 40 years and many different approaches claiming to limit cardiac damage in preclinical models, no notable clinical cardioprotective therapy is currently available to reduce the consequences of myocardial infarction [12]. Accordingly, novel therapeutic strategies are needed that target different mechanisms for cardioprotection and the anti-cancer effect.

Poly(ADP-ribose) polymerase1 (PARP1) is the first and principal member of the PARP family since it is responsible for the 85–90% of total PARP activity [13]. It, a zinc-finger DNA-binding enzyme, detects DNA strand breaks or nicks and using NAD^+ as substrate, PARylates substrate proteins including itself thereby initiating recruitment of DNA repair mechanisms [14]. Among other important cellular functions, it is a major regulator of cell death via maintenance of genomic stability, apoptosis and oxidative stress response [15]. Accordingly, PARP inhibitors were proposed for the therapy of many diseases including cancer, inflammation, autoimmunity, myocardial dysfunction and radiation/chemical exposure [16]. However, chronic inhibition of the enzyme leads to accumulation of DNA damage because of the compromised DNA repair culminating in apoptotic cell death [17].

Therefore, with the development of oral human PARP inhibitors, clinical studies aimed almost exclusively at using PARP inhibitors in different types of cancer as a monotherapy or in combination [18–23], and FDA-approved use of these drugs is restricted to cancer therapy [24]. Still, PARP activation is involved in many cellular functions besides DNA repair, and the well-established cytoprotective effects of the PARP inhibitors suggested their repurposing for non-oncological diseases [25].

In this chapter, we attempt to review the available data that supports repositioning of PARP inhibitors for the treatment of cancer therapy induced cardiotoxicity.

Cardiotoxic Side-Effects of Cancer Therapy

Cardiotoxicity of cancer therapy includes cardiac dysfunction, heart failure, myocarditis, hypertension, arrhythmia and vascular toxicity according to the 2022 European Society of Cardiology Guidelines for Cardiac Oncology [26]. In line with their respective molecular mechanism of action, the distinct types of antineoplastic drugs cause different treatment-related side effects.

Anthracyclines-Induced Cardiotoxicity

Among chemotherapeutic agents, anthracyclines such as Doxorubicin are the most well-known for causing cardiotoxicity. These drugs are used to treat a wide variety of cancers, including breast cancer, lymphomas, and leukemias [27]. They act by intercalating DNA strands and inhibiting topoisomerase II, which is crucial for DNA replication and repair [28]. However, compared to the general population, among Doxorubicin treated cancer patients, there is a 3.6-fold higher incidence of cardiovascular diseases [29]. Cardiotoxicity from anthracyclines can manifest as early, acute damage in the form of arrhythmia during or within weeks after treatment; early, chronic impairment in the form of heart failure or cardiomyopathy occurring within a year; or as late-onset insidious cardiac dysfunction, developing years after the treatment has ended [30]. Although cardiomyocyte apoptosis can result from the DNA intercalating effect of Doxorubicin, the crucial mechanisms in the cardiotoxic side-effects are tightly linked to mitochondria. The drug by inhibiting the mitochondrial electron-transport chain and by accumulating in association with the mitochondrial DNA reduces mitochondrial ATP production and generates large amounts of reactive oxygen species (ROS), which, in turn, cause oxidative DNA and protein damage as well as trigger lipid peroxidation. Malfunctioning of the electron-transport chain lowers the mitochondrial membrane potential (m$\Delta\Psi$) that shifts the balance of mitochondrial dynamics toward fission thereby promoting mitophagy and aggravating mitochondrial dysfunction. Because of the energy deficiency, intracellular Ca^{2+} homeostasis becomes maladjusted that causes impaired contractility, arrhythmias, and eventually heart failure. Additionally, increased intracellular Ca^{2+} may trigger m$\Delta\Psi$ depolarization and opening of the mitochondrial permeability transition pore (mPTP), via which water enters the matrix causing mitochondrial swelling and eventual rupture of the mitochondrial membranes. Prolonged opening of the mPTP results in necrotic cardiomyocyte death while destabilization of the outer mitochondrial membrane leads to the release of pro-apoptotic intermembrane proteins such as cytochrome c (Cyt C) triggering caspase-dependent cardiomyocyte apoptosis [31].

Tyrosine Kinase Inhibitor (TKI)-Induced Cardiotoxicity

Mainstay of treatment for chronic myeloid leukemia (CML) patients carrying breakpoint cluster region protein (BCR)-Abelson murine leukemia viral oncogene homolog (ABL)1 fusion gene sequence on their chromosome 22 is targeted TKI chemotherapy [32]. As with most antineoplastic therapies, treatment-resistance developed by the cancer cells represents a serious problem necessitating introduction of third-generation TKIs such as ponatinib and olverembatinib for handling first- and second-generation TKI-resistant CML [33]. Cardiovascular side-effects of TKI chemotherapy represent another considerable problem. The form and severity of TKI-induced cardiotoxicity differ considerably among TKIs, and involves congestive heart failure, pleural effusion, pulmonary hypertension, QT interval prolongation, hypertension and peripheral arterial disease [34]. The incidence of these side-effects among first-generation TKIs was about 45% for cardiovascular effects, 9% for athero-thrombotic adverse events, and 33% for hypertension according to a recent prospective study [35]. However, newer TKIs may have less cardiotoxic side effects, although we have less experience with them than with the first-generation ones, which have been on the market for more than 20 years. The first FDA-approved TKI, imatinib was reported to induce congestive heart failure in CML patients [36]. The mechanism involved inositol-requiring enzyme (IRE)1-mediated endoplasmic reticulum stress response activating c-Jun N-terminal kinase (JNK) and apoptosis signal-regulating kinase (ASK)1, which downregulate B-cell leukemia/lymphoma (Bcl)-2 protein leading to destabilization of the outer mitochondrial membrane, release of intermembrane cytochrome c resulting eventually in mitochondrial dysfunction and caspase-mediated cardiomyocyte apoptosis [37]. Cardiotoxicity of ponatinib, the most cardiotoxic FDA approved TKI, is produced by impeding growth signaling, inhibiting extracellular signal-regulated kinase (ERK) and Akt cytoprotective signaling pathways resulting in cardiomyocyte apoptosis [38].

Epidermal Growth Factor Receptor (HER)2 Inhibitors-Induced Cardiotoxicity

HER2 inhibitors are a class of targeted cancer therapies widely used to treat HER2-positive breast cancer, a subtype characterized by overexpression of HER2, a member of the growth hormone receptor family. For mediating proliferation promoting signals, HER2 needs to form heterodimers with other HER family proteins such as epidermal growth factor receptor HER1/HER2 and HER2/HER3 complexes, which are the most important for physiological functions and most relevant to tumorigenesis. The inhibitors used in cancer therapy are HER2 specific monoclonal antibodies that block these receptor complexes, TKI that prevent downstream signaling, or small molecule proliferation inhibitors coupled to HER2-specific monoclonal antibody [39]. Trastuzumab, a member of the first variety of inhibitors, was found to cause cardiac dysfunction in 3–7 and 28% of patients when used in monotherapy and in combination with anthracyclines, respectively [40, 41]. The cardiotoxic effect primarily manifests as a decline in left ventricular ejection fraction (LVEF), which can lead to symptomatic heart failure in severe cases. The mechanism is not fully understood but seems to involve inhibition of neuregulin 1 (NRG1) pathway that normally stimulates mature cardiomyocyte division and migration, induction of oxidative stress by downregulating the endogenous antioxidant defense system, and initiation of ferroptosis since its inhibitor, Ferrostatin-1 reversed the cardiotoxic processes [42]. HER2 itself seems to be essential for heart function since ventricular-restricted deletion of the receptor in mice led to the development of dilated cardiomyopathy [43]. Furthermore HER2 activation inhibits rapid endocytosis and degradation of HER1 thereby prolonging the phosphatase and tensin homolog (PTEN)/Akt/forkhead box O (FoxO)3a signaling that promotes growth and proliferation, which lead to tumorigenesis on the one hand but, on the other hand, is essential for cardiomyocyte survival and heart function [44].

Vascular Endothelial Growth Factor (VEGF) Inhibitors-Induced Cardiotoxicity

VEGFA, referred to as VEGF is the most studied member of a larger family of growth factors that have distinct expression pattern, receptor specificity and biological functions [45]. The protein has various splice variants that also differ in their receptor specificity and function [45]. Its role in cancer is not restricted to angiogenesis and vascular hyperpermeability since VEGF's autocrine signaling in tumor cells facilitates dedifferentiation and epithelial to mesenchymal transition and promotes cancer stem cell function. Additionally, acting as a chemoattractant, VEGF suppresses antitumor immune surveillance by attracting regulatory T cells to the tumor microenvironment [46]. Accordingly, targeting the VEGF signaling pathway by monoclonal antibody-based drugs that prevent binding of VEGF to its receptors and by TKI that block the downstream signaling was successfully applied in human therapy for the treatment of various solid tumors including metastatic colorectal cancer, non-small cell lung cancer, metastatic breast cancer, advanced gastric adenocarcinoma, renal and hepatocellular carcinoma, pancreatic neuroendocrine cancer, and gastrointestinal stromal tumor [47]. However, VEGF signaling pathway-targeting drugs have modest efficacy when used in monotherapy [48]. Accordingly, in the clinic, they are mostly used in combination with chemotherapeutics that results in a positive synergic effect [49]. Nevertheless, cardiotoxic complications in the form of hypertension, cardiac dysfunction and heart failure, arrhythmias and QT interval prolongation, cardiac ischemia, and thromboembolic disease can manifest upon treatment with VEGF inhibitors [50].

Hypertension is a comorbidity that exists in about 45% of cancer patients before treatment [51] and can exceed 90% during VEGF inhibitor therapy [52]. Severity depends on the presence of cardiovascular risk factors and can increase with the doses applied and with combination therapy [53]. The mechanisms underlaying VEGF inhibitors-induced hypertension are not fully elucidated. VEGF stimulates NO synthase activity. Accordingly, patients have reduced serum NO level during VEGF inhibitor therapy [54]. Furthermore, TKI elevate serum level of the vasoconstrictor endothelin-1, and capillary density is reduced by a yet unresolved mechanism increasing peripheral resistance that both contribute to the hypertensive effect of VEGEF inhibitors [54]. Additional mechanisms are oxidative stress and adverse regulation of Ca^{2+} channel activity and/or the renin–angiotensin–aldosterone system [52, 55]. Cardiac disfunction and heart failure could be the consequence of vascular side-effects of the VEGF inhibitors [56], however, may result from direct cardiotoxicity involving mitochondrial dysfunction and AMP-kinase (AMPK) pathway inhibition [57]. Although at an incidence of 1–2% only, vascular effects of VEGF inhibitors can lead to myocardial ischemia [26]. Arrhythmias and conduction defects are also frequent comorbidities present in 16–36% of cancer patients before starting chemotherapy [58]. VEGF inhibitors therapy mainly causes QT interval prolongation by activating cardiomyocyte potassium channels [59]. The risk of VEGF inhibitor therapy-related arterial thrombosis is very low [60], but for venous thrombosis and thromboembolism the incidence can reach up to 11% [61]. The latter requires special attention because of recurrence and mortality risk [61].

Immune Checkpoint Inhibitors (ICI)-Induced Cardiotoxicity

ICI are monoclonal antibody drugs that block inhibitory elements of cytotoxic T lymphocytes (CTL) such as CTL-associated protein (CTLA)4, programmed death protein (PD)-1 or PD ligand (PD-L)1. When these inhibitory elements interact with checkpoint proteins such as cluster of differentiation protein (CD)80/CD86 on antigen presenting cells (APC), they inhibit activation of the CTL thereby sparing cancer cells from elimination [62]. Although ICI are successfully improved overall survival rate and progression-free survival of patients in various tumors such as melanoma, hepatocellular carcinoma, lung cancer, gastric cancer, and intestinal cancer [63], they can cause cardiotoxic side-effects such as myocarditis, pericarditis, arrhythmias, heart failure, acute coronary syndromes, thromboembolism and vasculitis [64]. The incidence of these side-effects is low; according to a recent meta-analysis [65], only 32 out of 4751 patients receiving ICI therapy developed the most common form, myocarditis. However, 15 of these patients died that explains the clinical significance of ICI-induced cardiotoxicity. The underlaying mechanism is related to the activation of T-cells that involves multiple interactions between immune cells including binding of T-cell receptor and CD28 of CTL to antigen-major histocompatibility complex protein and CD80/86 of APC, respectively. When CTLA4 and/or PD-1 of CTL also bind to CD80/CD86 of APC, T-cell activation is suppressed that contributes to stabilization of the immune system under normal physiological conditions and prevention of autoimmune reactions [66]. However, by binding to CTLA4 or PD-1, ICI disrupts the binding of these inhibitory elements to CD80/CD86 thereby activating the immune response. Since the same said mechanism is responsible for both the antineoplastic effect and the cardiotoxic side effects of ICI, managing of the latter is not straightforward. It should be based on the type and severity of the toxic reaction and could mean termination of the ICI therapy in susceptible patients while alleviating the symptoms by combination therapy in less severe cases [26].

Cardiotoxicity Induced by Non-targeted Anticancer Drugs

Fluorouracil, a widely used chemotherapeutic agent, causes cardiovascular side-effects in the form of angina pectoris, ischemia-related electrocardiographic abnormalities, and myocardial infarction in about 10% of the patients [37]. The underlaying mechanism may involve protein kinase C-mediated vasospasm of coronary- and branch arteries, direct damage to vascular endothelial cells resulting in microthrombosis, and ferroptosis of cardiomyocytes [67, 68].

Taxanes induce arrhythmias primarily by increasing intracellular calcium ion concentration, reducing ATP synthesis, inducing mitochondrial ROS production and opening of the mPTP resulting in cardiomyocyte damage and apoptosis [69, 70].

Proteasome inhibitors prevent or diminish proteasomal degradation of excess proteins that accumulate, form aggregates that interfere with cellular functions eventually killing the cell. Cancer cells, which usually have elevated proteasome activity, are more susceptible to proteasome inhibitors than normal cells [71]. Proteasome inhibitors are mainstay of therapy for multiple myeloma; however, they can cause heart failure, atherosclerosis, myocardial infarction and cardiac arrest with a 6–7% incidence [72]. The underlaying mechanisms of these cardiotoxic side-effects are mostly associated with mitochondria; ultrastructural irregularities and diminished ATP production, which cause impaired cardiomyocyte contractility and dysregulated Ca^{2+} homeostasis [73].

Radiation Therapy is often a pivotal component in the therapy of many solid tumors including various lung cancers. Although advancements in radiotherapy planning and delivery have been made, the risk of harm to nearby thoracic tissues remains significant [74]. The heart is one of the most affected organs in thoracic radiation treatments, and cardiac complications, such as arrhythmia, heart failure, and ischemic events, affect 20% of lung cancer patients receiving radiotherapy [7]. The underlying mechanisms involve both acute and chronic damage to cardiac tissues. Radiation can damage the vasculature causing microvascular changes characterized by reduced capillary-to-myocyte ratio, endothelial dysfunction, damage to the epicardial vessels leading to a prothrombotic state, and induction of chronic inflammation by activating inflammatory proteins [75]. These processes promote atherosclerosis, where plaques occluding the blood vessels are resulting from increased infiltration of monocytes and macrophages to sites of active inflammation, the prothrombotic environment, and the formation of fatty deposits [76]. The adverse processes in the myocardial capillaries lead to myocardial fibrosis [77], further aggravated by the endothelial injury resulting eventually in decreased myocardial blood supply due to collagen deposition in the capillary walls and vascular stenosis [78]. Arrhythmic complications related to radiation therapy involve abnormal heart rate recovery times and elevated resting heart rates and are caused by radiation damages to components of the autonomous nervous system [79].

Cellular Effects of PARP Inhibitors Utilized for Cancer Therapy

Nuclear Effects of PARPs

PARP1 detects anomalous DNA structures, such as single- and double-strand breaks, crossovers, cruciforms, supercoils, and some specific double-stranded sequences, and binds to them that activates the enzyme [15]. It attaches PAR chains of heterogeneous length and extent of their branching on glutamate, aspartate, arginine, lysine, or serine residues of target proteins by using NAD^+ as a substrate [14]. However, PAR glycohydrolases and terminal ADP-ribose protein glycohydrolases catabolize the PAR chains very shortly after their synthesis to free ADP-ribose units [14, 80]. The PARP1 enzyme exists in the nucleus in a million-copy number and is accountable for the vast majority of PARylation occurring in mammalian cells [17]. In addition to the nuclear PARPs, a mitochondrial form was also reported to exist in many cell types; however, the functional role of mitochondrial PARP activity is yet to be resolved [81]. The primary target of PARP1's enzymatic activity is itself, but it PARylates many nuclear proteins including core histones, the linker histone H1, and a variety of transcription-related factors also [82]. Based on its nuclear localization, sensitivity toward DNA breaks, and enzymatic activity, PARP1 is regarded as a sensor and initiator of various DNA repair systems, thereby playing a crucial role in the maintenance of genome integrity [83]. During genotoxic stress, in response to DNA strand breaks, PARP1 PARylates itself and ataxia telangiectasia mutated kinase (ATM) to act as a scaffold for recruiting a small ubiquitin-like modifier (SUMO)-1 ligase, PIASy, and nuclear factor-kappa-B (NFκB) essential modulator (NEMO). Concertedly, PIASy SUMOylates and ATM phosphorylates NEMO that leads to inhibitor of NFκB kinase (IKK) activation resulting in NFκB activation, which, in turn, regulate resistance toward apoptosis [84]. PARP2 takes over PARP1's role when it is knocked out and vice versa. Nevertheless, the PARP1/PARP2 double knock out mutation leads to embryonic lethality [17]. Due to PARP1's prompt response to DNA breaks and its high copy number in the nucleus, genotoxic stress can trigger hyperactivation of the enzyme resulting in harmful consequences for the organism [85]. Substantial DNA damage induces excessive PARP1 activation, which is accompanied by a 40–70% decrease in cellular NAD^+ level within minutes [85]. Hence NAD^+-dependent metabolic pathways such as glycolysis, citric acid cycle and oxidative phosphorylation get impeded compromising energy homeostasis and causing cellular dysfunction. In an attempt to stabilize the metabolic situation, phosphoribosyl pyrophosphate synthetase and nicotinamide mononucleotide adenylyl transferase resynthesize NAD^+, which, being an energetically expensive process, lead to ATP depletion and predominantly necrotic cell death [86, 87]. The said mechanism can explain how DNA-alkylating agents such as high dose of N-methyl-N′-nitro-N-nitrosoguanidine (MNNG) kill cultured cells and provide an underlaying mechanism for the protective effects of PARP1 inhibition in animal models of oxidative stress associated diseases. However, some reports suggested that cytoprotective effects of PARP inhibition should include other mechanisms since MNNG induced similar NAD^+ depletion in $PARP1^{-/-}$ and $PARP1^{+/+}$ fibroblasts [88], and NAD^+ and ATP depletion was not prevented by PARP1 inhibition in carcinogen-treated hamster cells [89]. Indeed, it was found that hexokinase I binds to PAR and becomes inactivated [90] that significantly contributes to the collapse of energy homeostasis during genotoxic stress. Additionally, during parthanatos, following activation and self-PARylation, PARP1 is cleaved to 24- and 89-kDa fragments by caspase-3 [91]. The PARylated 89-kDa fragments translocate to the cytoplasm, bind apoptosis inducing factor (AIF) released from the mitochondrial intermembrane space and carry the PAR-bound AIF back to the nucleus, where it induces AIF mediated cell death [92].

Extranuclear Effects of PARP Activation

Akt pathway, one of the most powerful cytoprotective mechanisms, is inhibited by PARP activation as it was indicated in different experimental models [93–96]. In oxidative stress, this inhibitory effect of PARP counteracts the activation resulting from rapid oxidative deactivation of PTEN, the physiological regulator of Akt [97]. Missing elements of the underlying mechanism were elucidated by us

[98]. As it was described for NFκB-mediated apoptosis resistance in genotoxic stress [84], ATM phosphorylates itself and NEMO upon binding to the PAR scaffold produced by activated PARP [99]. However, when formation of the PAR scaffold is prevented by deactivation of PARP either by a pharmacological inhibitor or genetic manipulation, ATM forms a complex with NEMO, the complex translocates to the cytoplasm, binds to the outer mitochondrial membrane, creates an ATM–NEMO-Akt-mTOR signalosome, which activates Akt [98]. Many elements of this mechanistic model were described or confirmed by others, and the novel components were extensively proven by us (for the details see [100]). The significance of Akt activation among the mechanisms of PARP inhibitors' cytoprotective action is emphasized by the observation that transdominant expression of permanently active Akt rescued various ATM-deficient cancer cells from PARP1 inhibition induced death [98], although PARP1 inhibition is known to induce synthetic lethality in ATM-deficient cells [101].

In addition, the phosphatidylinositol-3 kinase (PI3K)-Akt pathway can regulate cell survival via directly affecting mitochondrial metabolism by phosphorylating Thr-473 of hexokinase II, the mitochondrially localized isoform of this glycolytic enzyme [102], or through Akt's downstream target, glycogen synthase kinase (GSK) 3β [103]. Phosphorylated hexokinase II binds to the voltage dependent anion channel (VDAC), an abundant outer mitochondrial membrane protein [104]. This couples glycolysis to oxidative phosphorylation facilitating the transfer of pyruvate in and ATP out of the mitochondria that enables the cells to cope with maintaining energy homeostasis and mΔΨ during hypoxic conditions [104]. Mitochondrial binding of hexokinase can be disrupted by GSK3β mediated phosphorylation of VDAC that results in collapse of mΔΨ, opening of the channel, destabilization of the outer mitochondrial membrane and the release of pro-apoptotic intermembrane proteins, which is prevented by Akt mediated inhibiting-phosphorylation of GSK3β [103]. Furthermore, PAR can also interrupt the binding between hexokinase and VDAC by forming a complex with the former [105]. Akt protects the mitochondria by phosphorylating Bcl-2 associated agonist of cell death (Bad), which binds to the cytoplasmic 14-3-3 protein thereby cannot form heterodimers with other pro-apoptotic members of the Bcl-2 family and destabilize the outer mitochondrial membrane allowing the release of pro-apoptotic intermembrane proteins [106]. Due to the powerful mitochondria- and cytoprotective effects of the PI3K-Akt pathway, the described activation of this pathway by PARP inhibition is beneficial for non-malignant diseases, but decreases the efficacy of PARP inhibitors in cancer therapy [107].

Mitogen-activated protein kinase (MAPK) pathways regulate various cellular processes such as proliferation, differentiation, apoptosis and stress responses. In most systems, extracellular signal-regulated kinase (ERK)1/2 are related to proliferation and differentiation, while c-Jun N-terminal kinases (JNK) and p38 MAPK pathways are associated with inflammation, stress responses and apoptosis [108]. MAPKs mediate the signaling of many trophic-, stress-, inflammatory- and environmental factors, however, ROS is a prominent activator for them in most systems [109]. Phosphatases that inactivate the MAPK pathways by dephosphorylating their respective upstream activating kinases, are inoperable in oxidative stress due to their oxidation-sensitivity [110]. Furthermore, the ROS can activate Ras by acting directly on growth receptors in the absence of their ligands [111] or even in a receptor-independent manner [110]. ROS-induced PARP activation was reported to cause m$\Delta\Psi$ collapse, opening of mPTP and eventual cell death [112–115], which effects are often mediated by the substantial contribution of JNK and p38 MAPK activation [114–116]. To support the latter concept, oxidative stress-induced damages were ameliorated by activation of MAPK phosphatase (MKP)1, which dephosphorylates thereby inactivates all classical MAPKs [117, 118]. Several reports suggested that the induction of PARP promotes MAPK activation, which mechanism significantly contributes to the oxidative stress situation-induced damages [114, 119–123]. Previously, we suggested a mechanism underlying this effect of PARP activation [124]. In oxidative stress, PARP1 becomes activated by the DNA strand breaks and PARylates activating transcription factor (ATF)4, the transcription factor responsible for regulating MKP1 expression [125]. PARylated ATF4 has lower affinity toward its responsive elements than the unPARylated one has; therefore, self-PARylated PARP1 replaces PARylated ATF4 on these sites resulting in decreased MKP1 expression [124]. Consequently, MKP1 mediated MAPK inhibition cannot counterbalance the MAPK-activating effect of oxidative stress resulting in MAPK activation, mitochondrial damage and eventual cell death [124]. Inhibition of PARP reverses all these processes leading to mitochondria- and cytoprotection that are beneficial for the patients with non-malignant diseases but hinder PARP inhibitor therapy in cancer [107].

Antineoplastic Effects of PARP Inhibitors

DNA Associated Antineoplastic Effects of PARP Inhibitors

Synthetic lethality is a biological concept in which the simultaneous disruption of two genes results in cell death, whereas the disruption of either gene alone does not significantly affect cell survival. Olaparib, the first FDA approved PARP1/2 inhibitor represents the first example for exploiting in the clinical practice a synthetic lethal approach to target ovary cancers with defective BRCA1 or BRCA2 tumor suppressor genes [126]. In cells with a BRCA1 or BRCA2 mutation, homologous recombination repair (HRR) of DNA damages is impaired. PARP inhibitors prevent PARylation of PARP1/2's target proteins thereby forestall DNA damage repair or trap PARP1/2 on DNA forming cytotoxic PARP1-DNA complexes leading to formation of double strand DNA breaks. Inhibiting PARP in BRCA-mutated cells results in catastrophic DNA damage and cell death, as both repair pathways are compromised. Normal cells, with functional BRCA genes, are less affected by PARP inhibition because they can rely on HRR (Fig. 37.1).

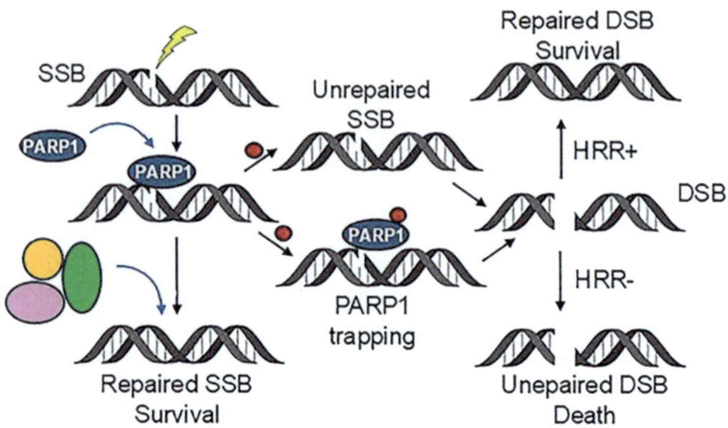

Fig. 37.1 Synthetic lethality of PARP1/2 inhibitors. PARP1/2 binds to single strand DNA breaks (SSB) induced by e.g. oxidative stress and recruits repair enzymes (colored shapes) that repair the damage resulting in prolonged survival of the cell. PARP1/2 inhibitors (red circles) prevent SSB repair (upper route) or trap PARP1 on DNA forming cytotoxic PARP1/2-DNA complexes (lower route) leading to formation of double strand DNA breaks (DSB). In cells with deficient homologous recombination repair (HRR−) such as BRCA 1/2-deficient cancers, DSB accumulate resulting eventually in death of the cells while HRR+ cells such as normal somatic cells survive by using BRCA1/2 for repairing DSB

The concept of using synthetic lethality for the treatment of cancer [127] much preceded the introduction of Olaparib to clinical practice and various tumor suppressor genes such as PTEN [128] or ATM [101] were suggested as targets for PARP inhibitor therapy. Recent FDA approved primary applications of PARP inhibitors include ovarian, primary peritoneum, fallopian tube, breast, pancreatic, and prostate cancers with BRCA1/2 or other HRR deficiency markers, and also advanced ovarian, primary peritoneum, and fallopian tube cancers that have previously shown a positive response to platinum-based treatments, regardless of the presence of genetic mutations in their HRR [129]. Olaparib can be used in combination therapy for the treatment of ovary and prostate cancer with intact HRR [129]. However, ongoing clinical trials aim to extend the use of PARP inhibitors for other cancer types such as triple negative breast cancer [130], glioma [131], non-small cell lung cancer [132] and gastric adenocarcinoma [133]. It is worth mentioning that PARP inhibitors' DNA repair-preventing activity is used in most combination therapy, where the DNA damage is induced by the other chemo- or radiotherapy component of the combination [134].

Non-DNA Targeted Antineoplastic Effects of PARP Inhibitors

Metabolism of cancer cells undergoes substantial reprogramming during carcinogenesis to provide the proliferating cells with both sufficient energy and supply of metabolic intermediates. Accordingly, most cancer cells use aerobic glycolysis rather than oxidative phosphorylation for ATP production since the former retains the atoms of glucose in the form of lactate even though its ATP yield is much lower than that of the latter [135]. However, the tumor cells of highly malignant, treatment resistant, metastasizing phenotype tend to shift their metabolism toward mitochondrial ATP synthesis [136]. This intricate balancing of their oxidative versus fermentative energy production makes them vulnerable to drugs that interfere with their energy metabolism [137]. PARP inhibitors are capable of doing it by several mechanisms. They can reduce NAD^+ consumption and PAR formation by inhibiting PARP's enzymatic activity. NAD^+ is an essential cofactor of glycolysis, accordingly, it is pivotal for both the oxidative and the fermentative ATP synthesis. Also, NAD^+/NADH balance is a major component of the cell's redox equilibrium, which can represent another therapeutic target for cancer therapy [138]. Additionally, NAD^+ is a substrate for sirtuins, which regulate various cellular functions ranging from genome stability and energy metabolism to stress response [139, 140]. Both the nuclear sirtuin-1 and the mitochondrial sirtuin-3 were implicated to mediate the positive effects of PARP inhibitors on mitochondrial metabolism [141, 142]. PAR, produced by PARP activation and cleaved off from the target proteins by glycohydrolases, can remove hexokinase from the VDAC, thereby destabilizing the outer mitochondrial membrane and disrupting the coupling between glycolysis and oxidative phosphorylation [105]. PARP inhibitors, by the mechanisms described above, activate the PI3K-Akt and inhibit the MAPK signaling pathways [98, 124] that stabilize all mitochondrial functions including energy production and reduce ROS generation. Unfortunately, together with the anti-apoptotic actions of these signaling pathways, all these mechanisms promote the survival of the cancer cells [143]. The same is true for increasing the binding of hexokinase to VDAC by Akt mediated phosphorylation of the enzyme and prevention of phosphorylation of VDAC by GSK3β [102, 103].

Inflammatory response involves activation of NFκB transcription factor in the early phase of the process [144]. DNA-strand break induced NFκB activation was described earlier when nuclear effects of PARP activation was presented. Accordingly, SUMOylated and phosphorylated NEMO, which is a modulator of IKK, translocates from the nucleus to the cytosol, forms a complex with IKKα and IKKβ that phosphorylates inhibitor of NFκB. It disrupts the complex of NFκB and its inhibitory protein, nuclear localization signal of NFκB becomes unmasked leading to nuclear translocation of NFκB, binding to its responsive element and expression of inflammatory response genes such as cytokines, chemokines, adhesion molecules, cell cycle regulators, anti-apoptotic factors, inducible NO synthase, and matrix metalloproteinases [84, 144]. As it naturally derives from the activation mechanism of NFkB, PARP inhibitors are effective anti-inflammatory agents

[25, 93]. However, the relationship between cancer and inflammation is not this straightforward. The malignant transformation and its response to therapeutic intervention is regulated by inflammation, which can hinder or facilitate the former while obviously having opposing effect on the latter depending on the type of cancer, the form, agent and stage of the therapy and many other patient-related factors such as age, co-morbidities etc. [145]. Intact immune surveillance and acute inflammatory reactions of the patients are indispensable elements of a successful chemo- or radiotherapy [146], while chronic inflammation often leads to cancer development [147]. The underlaying mechanisms are numerous, they involve many synergic and non-synergic cellular signaling pathways [148].

Oxidative stress and inflammation are closely related processes; ROS induced DNA breaks initiate NFκB activation, which, in turn, leads to ROS generation [144]. PARP inhibitors can normalize endogenous ROS production via their anti-inflammatory effect although they do not have any free-radical scavenging activity [123]. The relationship between oxidative stress and cancer development is more straightforward than in the case of inflammation. Generally, low to moderate levels of ROS are important for many cellular functions such as cell proliferation, migration, invasion, and angiogenesis, which processes are essential for the cancer cells [149]. Although, compared to normal cells, cancers usually have more elevated cellular ROS levels, they seem to cope with this shifted balance between their oxidants and antioxidants [150]. Regardless of the cell type, high ROS levels oxidize proteins, nucleic acids, and lipids, leading to membrane and organelle damages resulting eventually in cell death [151]. Antioxidant effects of PARP inhibitors mediated via their anti-inflammatory actions normalize elevated ROS levels thereby contributing to treatment resistance of the cancer cells [98, 124]. Recently, nuclear factor erythroid 2-related factor (Nrf)2, a transcription factor responsible for regulating the expression of the antioxidant defense system, thereby preserving cellular redox homeostasis has been suggested as therapeutic target in cancer [152]. In line with the detrimental effects of oxidative stress, elevated activation of Nrf2 promotes tumor formation and is involved in chemoresistance in many types of solid tumors [153]. Indirect connection between PARP1 and Nrf2 seem to exist via the Akt/AMPK/Nrf2 signaling pathway [154]. However, PARP2 was reported to PARylate Nrf2 thereby regulating its subcellular localization [155]. Nevertheless, the therapeutic significance of these findings is still elusive.

Repositioning of PARP Inhibitors for Treating Cardiotoxicity

Cardioprotective Effects of PARP Inhibitors

Most experimental evidence that indicates cardioprotective effects of PARP inhibitors derives from various in vitro, in vivo and ex vivo preclinical studies using cell culture, isolated heart and animal models including PARP1$^{-/-}$ mice [156]. All these models are well established for studying various features of the heart including its contractile function, energy metabolism, morphological and pathophysiological changes during and following myocardial infarct or ischemic insult, as well as the mechanisms underlaying these processes [157, 158]. Early inhibitors such as nicotinamide, 3-aminobenzamide, dihydroxyisoquinoline or hydroxyquinasoline derivatives had limited selectivity in inhibiting PARP's enzymatic activity and they had low affinity toward the enzyme; therefore, they had to be used at mM/sub mM concentrations. Still, the studies employing them provided compelling supporting data for the pivotal role of PARP activation in myocardial infarction and ischemia-reperfusion of the heart, and for demonstrating that inhibitors of the enzyme protect the heart in these conditions [156, 159]. It was demonstrated that PARP activation is continuously elevated up to, at least, 24 h after the start of reperfusion. The highest PARP activations were observed in the necrotic and peri-infarct zone in cardiac myocytes, endothelial cells, and circulatory mononuclear cells [160, 161]. During the ischemic phase, high-energy phosphate levels fall rapidly while inorganic phosphate concentration increases, intracellular pH decreases, and waste products accumulate. Persistence of this condition causes necrotic death for all types of cardiac cells leading to serious tissue damage and eventual death of the individual [162]. To prevent it, clinical or experimental intervention should re-establish the flow, but reperfusion itself causes tissue damage by many mechanisms, of which oxidative stress and inflammatory response are the most prominent ones [163]. Endogenous ROS formation during the reperfusion can originate from various sources including the mitochondrial electron transport chain, xanthine oxidases, NAD(P)H oxidases, cyclooxygenase, and NOS [164]. The ROS and peroxynitrite impair the heart tissue directly by damaging its macromolecules and indirectly via DNA strand break induced PARP activation (Fig. 37.2).

Hand in hand with the oxidative stress, early inflammatory response is induced by complement cascade activation and binding of necrotic debris to Toll-like receptors of local antigen presenting cells. It results in recruitment and activation of polymorphonuclear leukocytes, which escalate the primary inflammatory response and ROS production via activation of several downstream molecules, including NFκB, AP-1, p-eNOS and others, as shown in Fig. 37.2 [165–167]. PARP activation is involved in the process (Fig. 37.2) by regulating NFκB activation [144]. ROS and inflammation damage the mitochondria both directly and via perturbing intracellular signaling; most prominently, activating JNK and p38 MAPK while inhibiting PI3K-Akt pathways [120] (Fig. 37.2). Impairment of the mitochondrial functions results in reduced ATP and elevated ROS synthesis, decreased m$\Delta\Psi$, mitochondrial fragmentation, mitophagy, opening of mPTP, disruption of the mitochondrial membrane systems, release of pro-apoptotic intermembrane proteins such as AIF and cytochrome c, which can result in apoptotic cell death [168] and lead to cardiovascular complications.

Employing Langendorff-perfused isolated hearts of mice, rats, rabbits, dogs and pigs and ischemia-reperfusion setup, PARP inhibitors were reported to improve both systolic and diastolic functions, preserve cellular NAD^+ levels, facilitate faster and more complete recovery of the energy production, strengthen myocardial

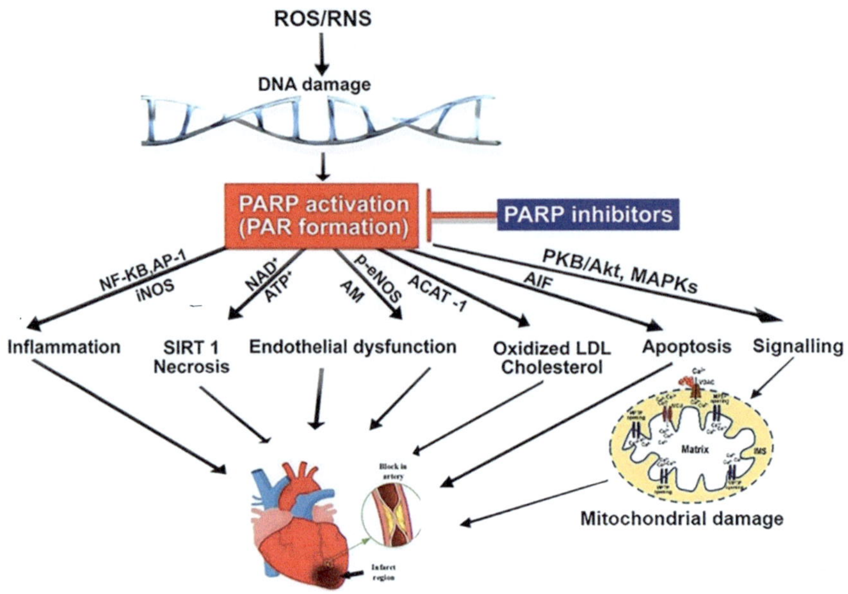

Fig. 37.2 Effects of PARP-1 activation in myocardial infarction. Oxidative stress activates PARP via DNA strand breaks formation that leads to tissue damage by several mechanisms including NFκB mediated inflammatory response, NAD^+ and ATP depletion resulting in necrotic death, peroxynitrite induced endothelial dysfunction, oxidized cholesterol causing atherosclerosis and inflammation, AIF mediated apoptosis, and mitochondrial damage caused mainly by oxidative damage to macromolecules, inhibition of Akt and activation of MAPKs. Red arrow with large flat head denotes inhibition by PARP inhibitors, which by preventing PARP's enzymatic activity turn the entire process around resulting in cardioprotection. *NFκB* nuclear factor κB, *ACAT-1* acetyl-CoA-cholesterol acetyltransferase-1, *AIF* apoptosis inducing factor, *AP-1* activator protein-1, as reported by Angel et al. [165]; *iNOS* inducible NO synthase, *LDL* low density lipoprotein, *MAPK* mitogen activated protein kinase, *p-eNOS* phosphorylated endothelial NO synthase, as reported by Corson et al. [166]; *PKB* protein kinase B

contractility, reduce infarct size, mitochondrial damage, and cardiomyocyte apoptosis by several groups [156, 159]. By using volume selective nuclear magnetic resonance (NMR) spectrometry for demonstrating real-time in situ time course of creatine phosphate and ATP levels during ischemia-reperfusion of isolated rat hearts [120, 169] and NMR imaging for visualizing inflammatory response in the hearts of doxorubicin or isoproterenol treated rats [120] we established the pivotal role of Akt activation and MAPK inhibition in the cardioprotective effect of PARP inhibitors [159]. Furthermore, we systematically explored the molecular events leading from the activation of the nuclear PARP to the regulation of the cytoplasmic kinase pathways and presented respective mechanisms for the activation of Akt and inhibition of JNK and p38 MAPK by PARP inhibition [98, 124]. We also established a positive correlation between Akt activation and PARP inhibitor mediated cardioprotection to emphasize the significance of the former in the latter by utilizing different PARP inhibitors, β-blockers and Ca^{2+}-channel blockers [170]. Early events of the myocardial infarct are followed by remodeling of the left ventricle that eventually progresses into heart failure. We demonstrated that Akt-, protein kinase C (PKC)-, ribosomal S6 kinases (RSK)-, or cAMP dependent protein kinase (PKA) mediated inhibitory phosphorylation of GSK3β play a significant role in the various processes of remodeling such as collagen deposition and cardiomyocyte hypertrophy [171]. All these observations were confirmed by other investigators [172]. Much less data is available about cardioprotective properties of potent, selective, third-generation PARP inhibitors since the research efforts were focused almost exclusively on cancer therapy. INO-1001 was used to protect against chronic heart failure induced by chronic coronary ligation in rats, doxorubicin treatment in mice, and a single dose of bacterial endotoxin in rats [173]. The PARP inhibitor decreased mortality, improved systolic and diastolic developed pressure, decreased cardiac decompensation and the consequent pulmonary edema [173]. In spontaneously hypertensive rats, a 46-week treatment with PARP inhibitor prevented the transition from hypertensive cardiopathy to heart failure; reduced the mortality, preserved systolic left ventricular function, moderated plasma-BNP levels and gravimetric parameters, and myocardial fibrosis without affecting the blood pressure [174]. According to the TIMI 37 trial, INO-1001 was well tolerated by 40 patients with ST-elevation myocardial infarction that provide a basis for further clinical studies employing PARP inhibitors for acute cardiac indications [175].

Expected Features of Agents Ameliorating Cardiotoxicity of Cancer Therapy

Efficacy in preventing cardiotoxic side-effects of antineoplastic therapy is one of the major features expected of a cardioprotective agent. Taking together, the preclinical evidence in favor of the PARP inhibitors is overwhelming. They proved to be efficient in a wide range of preclinical animal models from rodents to pigs, in wild type to genetically modified animals, in various experimental setups addressing acute and chronic aspects of the human cardiovascular disorders. The modern PARP inhibitors have a good therapeutic window [176] and their cardioprotective effects cover endpoints that are relevant for the human disorder such as infarct size, cardiac enzymes, contractility, inflammatory mediators, edema etc. Furthermore, PARP activation was indicated in many human diseases with cardiac involvement such as myocardial infarction [177], revascularization [178], cardiopulmonary bypass [179], septic shock [180], aortic valve stenosis [181], diabetic cardiomyopathy and chronic heart failure [182]. Considering the molecular mechanisms underlaying cardiotoxic side-effects of antineoplastic therapy, they highly overlap with those of cardiovascular diseases and the major patho-mechanisms, namely, oxidative stress, inflammatory response and adverse intracellular signaling are basically identical in both pathophysiological situations [183]. Accordingly, the data concerning cardioprotective effects of PARP inhibitors in preclinical models of cardiovascular diseases is likely highly translatable for the management of cardiotoxicity of cancer therapy [183].

No interference with the efficacy of cancer therapy is another obvious requirement. However, in this respect, the situation is not as straightforward as with cardioprotection. Targeted therapies such as TKI, HER2 inhibitors, VEGF inhibitors or ICI exert their antineoplastic effects with various molecular mechanisms, which do not necessarily overlap with those of their cardiotoxicity [184]. In these cases, a combination therapy with PARP inhibitors is not just feasible, but already recommended [185] or clinical trials are under way for getting FDA approval [176]. Nevertheless, when oxidative stress-, inflammatory response- and/or intracellular signaling-related mechanisms significantly contribute to the anti-cancer effects of targeted therapy, its efficacy can be reduced by the PARP inhibitor, although the cardioprotective effect of the latter may augment the former by increasing the patient's tolerance toward the therapy. In the case of non-targeted cancer therapies such as anthracyclines, taxanes, Pt-compounds or radiation therapy, the said mechanisms represent significant contribution to the primary antineoplastic effects of the therapy and the previous argument holds true to a more pronounced extent [183]. Even, toxicity issues limit the combinations of PARP inhibitors with genotoxic chemotherapy and radiotherapy [176]. Resistance of the cancer cells toward PARP inhibitor treatment may represent an additional complication. It involves reverting to $BRCA^{+/+}$ phenotype by $BRCA^{-/-}$ cancers, replication fork stabilization, increasing PARylation activity, and inhibitor removal by ATP-binding cassette transporters [186]. Another complication is created by the Akt activating effect of PARP inhibition. As we and others pointed out [143, 187], this effect reduces the efficacy of cancer therapy, which can be restored by the introduction of an Akt inhibitor into the therapy [143].

Side-effects are unwanted features of a cardioprotective agent. The most important side effect of the PARP inhibitors is the main hurdle in their non-oncological application; increased risk of malignant transformation because of compromised DNA repair. Accordingly, their repurpose was suggested to be restricted to severe diseases with limited therapeutic options and necessitating short PARP inhibitor administration [25]. However, this effect of PARP inhibitors per se does not represent a problem when alleviating cardiotoxic side-effects of cancer-therapy. In monotherapy, PARP inhibitors exert the usual side effects of chemotherapy including weakness, diarrhea, loss of appetite, headache, change of taste sense, cough, adverse cutaneous effects, pain in joints, anemia, and leukopenia, still, these effects seldom requiring termination of the treatment [188]. According to a recent systematic review and safety meta-analysis, PARP inhibitor therapy increases the risk of major adverse cardiac-, hypertension-, and thromboembolic events. However, because of the very low incidence and the absence of increased risk of high-grade events, the patients do not need any special monitoring [189]. Taking all data together, the side effects of PARP inhibitors should not represent a limiting factor in repositioning them for ameliorating cardiotoxic side-effects of antineoplastic agents.

Conclusions and Future Directions

FDA approval of Olaparib in 2014 and the positive experience with it in cancer patients carrying BRCA1/2 mutations opened a door for PARP inhibitors in the clinical practice. Nowadays, there are 34 ongoing phase III clinical trials, some of them started as early as 2013, utilizing 7 different PARP inhibitors, 14 in mono-, 20 in combination therapy for various oncological indications [176]. Additionally, extensive preclinical research is performed on non-oncological disease models ranging from third-degree burns to asthma and neurodegeneration by utilizing FDA approved PARP inhibitors [176]. All these studies should expand our knowledge about toxic interactions of PARP inhibitors with other anti-cancer drugs, provide additional therapeutic targets, solve therapy resistance issues and may lead to the development of novel, improved PARP inhibitors with more efficient therapeutic and less side-effects. That would solve the toxicity problems hindering combination therapy nowadays and would open the door for utilizing PARP inhibitors in these combinations for ameliorating cardiotoxicity of the therapy. Intracellular signaling is one of the fields in which our knowledge is insufficient to predict drug interactions. The human kinome contains 538 kinases and presently 76 kinase inhibitors are approved for cancer therapy by the FDA [190] that indicates the size of the challenge. Furthermore, kinase signaling regulates cellular metabolism and redox balance, which represent an emerging therapeutic target, especially for cancer cells of the most malignant phenotype [191]. Another aspect of the metabolism that can affect efficacy and side-effects of the antineoplastic therapy is the interaction of the therapeutic agent with the microbiome [192]. Some aspects, such as the roles of cancer–microbiome–immune axis in developing immunotherapy resistance are well characterized, however, we should achieve a comprehensive understanding of the field in the future [193]. In the field of cardiotoxicity, biological sex is an unavoidable factor since younger women are under-represented among patients with cardiovascular disease, which advantage is lost after menopause [194] suggesting a possibility of differential efficacy among the sexes of PARP inhibitor mediated cardioprotection in combination cancer therapy. Indeed, in many in vivo preclinical studies, PARP inhibitors were more effective in male than female animals [172]. However, future clinical studies should establish how these animal experiments translate to human cancer therapy and cardiotoxicity [195].

Repositioning a pharmaceutical agent requires a risk vs. benefit assessment. In this case, the risks are slight for a cancer patient. They mostly include side-effects of the PARP inhibitors and their limiting effects on therapy, which either occur or not depending on the nature of the other component of the combination. The benefits, on the other hand, are substantial starting with reduced cardiotoxic side-effects of the combination therapy. To emphasize the significance of this benefit, it is worth mentioning that 5-year survival rate of heart failure is close to 50% [196], worse than that of many types of cancers. Overwhelming preclinical results support the cardioprotective effects of the PARP inhibitors [172]. Human clinical studies have already established their safety, however, ongoing non-oncological clinical trials aim at non-cardiological indications [176]. Another benefit is a possible synergy between the PARP inhibitor and the other component of the combination antineoplastic therapy resulting in shorter chemotherapy and/or lower doses [172]. It is the very objective of the 20 ongoing phase III clinical trials employing PARP inhibitors in combination therapy [176]. Hopefully, future combinations will include PARP inhibitors with the goal of ameliorating cardiotoxicity of the anti-cancer therapy.

References

1. Vinas-Mendieta AE, Gallardo-Grajeda A, Lopez-Fernandez T (2024) Cardio-oncology: chances and challenges. Basic Res Cardiol 120(1):3–9
2. Camilli M, Cipolla CM, Dent S, Minotti G, Cardinale DM (2024) Anthracycline cardiotoxicity in adult cancer patients: JACC: CardioOncology State-of-the-Art Review. JACC CardioOncol 6(5):655–677
3. Zalaquett Z, Rita Hachem MC, Assi A, Mohanna R, Farhat M, Noujaim C et al (2024) Cardiac toxicity of HER-2 targeting antibody-drug conjugates: overview and clinical implications. Future Oncol 20(38):3151–3167
4. Ekram J, Rathore A, Avila C, Hussein R, Alomar M (2024) Unveiling the cardiotoxicity conundrum: navigating the seas of tyrosine kinase inhibitor therapies. Cancer Control 31: 10732748241285755
5. Brandao SR, Carvalho F, Amado F, Ferreira R, Costa VM (2022) Insights on the molecular targets of cardiotoxicity induced by anticancer drugs: a systematic review based on proteomic findings. Metabolism 134:155250
6. Rajaeinejad M, Parhizkar-Roudsari P, Khoshfetrat M, Kazemi-Galougahi MH, Mosaed R, Arjmand R et al (2024) Management of fluoropyrimidine-induced cardiac adverse outcomes following cancer treatment. Cardiovasc Toxicol 24(2):184–198
7. Walls GM, Bergom C, Mitchell JD, Rentschler SL, Hugo GD, Samson PP et al (2024) Cardiotoxicity following thoracic radiotherapy for lung cancer. Br J Cancer 132(4):311–325
8. Balachandran L, Haw TJ, Leong AJW, Croft AJ, Chen D, Kelly C et al (2024) Cancer therapies and cardiomyocyte viability: which drugs are directly cardiotoxic? Heart Lung Circ 33(5):747–752
9. Wu B, Zhang B, Li B, Wu H, Jiang M (2024) Cold and hot tumors: from molecular mechanisms to targeted therapy. Signal Transduct Target Ther 9(1):274
10. Liu B, Liu L, Liu Y (2024) Targeting cell death mechanisms: the potential of autophagy and ferroptosis in hepatocellular carcinoma therapy. Front Immunol 15:1450487
11. Gao F, Xu T, Zang F, Luo Y, Pan D (2024) Cardiotoxicity of anticancer drugs: molecular mechanisms, clinical management and innovative treatment. Drug Des Dev Ther 18:4089–4116
12. Bolli R, Tang XL (2022) New insights into cardioprotection, gained by adopting the CAESAR standards of rigor. Basic Res Cardiol 117(1):57
13. Ame JC, Spenlehauer C, de Murcia G (2004) The PARP superfamily. BioEssays 26(8): 882–893
14. D'Amours D, Desnoyers S, D'Silva I, Poirier GG (1999) Poly(ADP-ribosyl)ation reactions in the regulation of nuclear functions. Biochem J 342(Pt 2):249–268
15. Bouchard VJ, Rouleau M, Poirier GG (2003) PARP-1, a determinant of cell survival in response to DNA damage. Exp Hematol 31(6):446–454
16. Jagtap P, Szabo C (2005) Poly(ADP-ribose) polymerase and the therapeutic effects of its inhibitors. Nat Rev Drug Discov 4(5):421–440

17. Huber A, Bai P, de Murcia JM, de Murcia G (2004) PARP-1, PARP-2 and ATM in the DNA damage response: functional synergy in mouse development. DNA Repair (Amst) 3(8–9): 1103–1108
18. Luo L, Keyomarsi K (2022) PARP inhibitors as single agents and in combination therapy: the most promising treatment strategies in clinical trials for BRCA-mutant ovarian and triple-negative breast cancers. Expert Opin Investig Drugs 31(6):607–631
19. De Santis M, Breijo SM, Robinson P, Capone C, Pascoe K, Van Sanden S et al (2024) Feasibility of indirect treatment comparisons between niraparib plus abiraterone acetate and other first-line poly ADP-ribose polymerase inhibitor treatment regimens for patients with BRCA1/2 mutation-positive metastatic castration-resistant prostate cancer. Adv Ther 41(8): 3039–3058
20. Garofoli M, Maiorano BA, Bruno G, Giordano G, Di Tullio P, Maselli FM et al (2024) Androgen receptor, PARP signaling, and tumor microenvironment: the 'perfect triad' in prostate cancer? Ther Adv Med Oncol 16:17588359241258443
21. Morganti S, Marra A, De Angelis C, Toss A, Licata L, Giugliano F et al (2024) PARP inhibitors for breast cancer treatment: a review. JAMA Oncol 10(5):658–670
22. Purwar R, Ranjan R, Pal M, Upadhyay SK, Kumar T, Pandey M (2023) Role of PARP inhibitors beyond BRCA mutation and platinum sensitivity in epithelial ovarian cancer: a meta-analysis of hazard ratios from randomized clinical trials. World J Surg Oncol 21(1):157
23. Ragupathi A, Singh M, Perez AM, Zhang D (2023) Targeting the BRCA1/2 deficient cancer with PARP inhibitors: clinical outcomes and mechanistic insights. Front Cell Dev Biol 11: 1133472
24. Rana M, Thakur A, Kaur C, Pan CH, Lee SB, Liou JP et al (2023) Prudent tactics to sail the boat of PARP inhibitors as therapeutics for diverse malignancies. Expert Opin Drug Discov 18(10):1169–1193
25. Berger NA, Besson VC, Boulares AH, Burkle A, Chiarugi A, Clark RS et al (2018) Opportunities for the repurposing of PARP inhibitors for the therapy of non-oncological diseases. Br J Pharmacol 175(2):192–222
26. Lyon AR, Lopez-Fernandez T, Couch LS, Asteggiano R, Aznar MC, Bergler-Klein J et al (2022) 2022 ESC Guidelines on cardio-oncology developed in collaboration with the European Hematology Association (EHA), the European Society for Therapeutic Radiology and Oncology (ESTRO) and the International Cardio-Oncology Society (IC-OS). Eur Heart J 43(41):4229–4361
27. Song L, Qiu Q, Ju F, Zheng C (2024) Mechanisms of doxorubicin-induced cardiac inflammation and fibrosis; therapeutic targets and approaches. Arch Biochem Biophys 761:110140
28. Avagimyan A, Pogosova N, Kakturskiy L, Sheibani M, Challa A, Kogan E et al (2024) Doxorubicin-related cardiotoxicity: review of fundamental pathways of cardiovascular system injury. Cardiovasc Pathol 73:107683
29. Lopez-Sendon J, Alvarez-Ortega C, Zamora Aunon P, Buno Soto A, Lyon AR, Farmakis D et al (2020) Classification, prevalence, and outcomes of anticancer therapy-induced cardiotoxicity: the CARDIOTOX registry. Eur Heart J 41(18):1720–1729
30. Robert J (2007) Long-term and short-term models for studying anthracycline cardiotoxicity and protectors. Cardiovasc Toxicol 7(2):135–139
31. Elmorsy EA, Saber S, Hamad RS, Abdel-Reheim MA, El-Kott AF, AlShehri MA et al (2024) Mechanistic insights into carvedilol's potential protection against doxorubicin-induced cardiotoxicity. Eur J Pharm Sci 200:106849
32. Jabbour E, Kantarjian H (2022) Chronic myeloid leukemia: 2022 update on diagnosis, therapy, and monitoring. Am J Hematol 97(9):1236–1256
33. Frankfurt O, Licht JD (2013) Ponatinib—a step forward in overcoming resistance in chronic myeloid leukemia. Clin Cancer Res 19(21):5828–5834
34. Douxfils J, Haguet H, Mullier F, Chatelain C, Graux C, Dogne JM (2016) Association between BCR-ABL tyrosine kinase inhibitors for chronic myeloid leukemia and cardiovascular events,

major molecular response, and overall survival: a systematic review and meta-analysis. JAMA Oncol 2(5):625–632
35. Jain P, Kantarjian H, Boddu PC, Nogueras-Gonzalez GM, Verstovsek S, Garcia-Manero G et al (2019) Analysis of cardiovascular and arteriothrombotic adverse events in chronic-phase CML patients after frontline TKIs. Blood Adv 3(6):851–861
36. Kerkela R, Grazette L, Yacobi R, Iliescu C, Patten R, Beahm C et al (2006) Cardiotoxicity of the cancer therapeutic agent imatinib mesylate. Nat Med 12(8):908–916
37. Bellodi C, Lidonnici MR, Hamilton A, Helgason GV, Soliera AR, Ronchetti M et al (2009) Targeting autophagy potentiates tyrosine kinase inhibitor-induced cell death in Philadelphia chromosome-positive cells, including primary CML stem cells. J Clin Invest 119(5): 1109–1123
38. Singh AP, Glennon MS, Umbarkar P, Gupte M, Galindo CL, Zhang Q et al (2019) Ponatinib-induced cardiotoxicity: delineating the signalling mechanisms and potential rescue strategies. Cardiovasc Res 115(5):966–977
39. Swain SM, Shastry M, Hamilton E (2023) Targeting HER2-positive breast cancer: advances and future directions. Nat Rev Drug Discov 22(2):101–126
40. Seidman A, Hudis C, Pierri MK, Shak S, Paton V, Ashby M et al (2002) Cardiac dysfunction in the trastuzumab clinical trials experience. J Clin Oncol 20(5):1215–1221
41. Baselga J (2000) Current and planned clinical trials with trastuzumab (Herceptin). Semin Oncol 27(5 Suppl 9):27–32
42. Dugaucquier L, Feyen E, Mateiu L, Bruyns TAM, De Keulenaer GW, Segers VFM (2020) The role of endothelial autocrine NRG1/ERBB4 signaling in cardiac remodeling. Am J Physiol Heart Circ Physiol 319(2):H443–HH55
43. Crone SA, Zhao YY, Fan L, Gu Y, Minamisawa S, Liu Y et al (2002) ErbB2 is essential in the prevention of dilated cardiomyopathy. Nat Med 8(5):459–465
44. Fujio Y, Nguyen T, Wencker D, Kitsis RN, Walsh K (2000) Akt promotes survival of cardiomyocytes in vitro and protects against ischemia-reperfusion injury in mouse heart. Circulation 101(6):660–667
45. Koch S, Claesson-Welsh L (2012) Signal transduction by vascular endothelial growth factor receptors. Cold Spring Harb Perspect Med 2(7):a006502
46. Ellis LM, Hicklin DJ (2008) VEGF-targeted therapy: mechanisms of anti-tumour activity. Nat Rev Cancer 8(8):579–591
47. Filippelli A, Ciccone V, Donnini S, Ziche M, Morbidelli L (2021) Molecular mechanisms of resistance to anti-angiogenic drugs. Crit Rev Oncog 26(2):39–66
48. Cao Y, Arbiser J, D'Amato RJ, D'Amore PA, Ingber DE, Kerbel R et al (2011) Forty-year journey of angiogenesis translational research. Sci Transl Med 3(114):114rv3
49. Jain RK (2005) Normalization of tumor vasculature: an emerging concept in antiangiogenic therapy. Science 307(5706):58–62
50. Cohen JB, Brown NJ, Brown SA, Dent S, van Dorst DCH, Herrmann SM et al (2023) Cancer therapy-related hypertension: a scientific statement from the American Heart Association. Hypertension 80(3):e46–e57
51. Versmissen J, Mirabito Colafella KM, Koolen SLW, Danser AHJ (2019) Vascular cardio-oncology: vascular endothelial growth factor inhibitors and hypertension. Cardiovasc Res 115(5):904–914
52. Touyz RM, Lang NN, Herrmann J, van den Meiracker AH, Danser AHJ (2017) Recent advances in hypertension and cardiovascular toxicities with vascular endothelial growth factor inhibition. Hypertension 70(2):220–226
53. Di Lisi D, Madonna R, Zito C, Bronte E, Badalamenti G, Parrella P et al (2017) Anticancer therapy-induced vascular toxicity: VEGF inhibition and beyond. Int J Cardiol 227:11–17
54. Dobbin SJH, Cameron AC, Petrie MC, Jones RJ, Touyz RM, Lang NN (2018) Toxicity of cancer therapy: what the cardiologist needs to know about angiogenesis inhibitors. Heart 104(24):1995–2002

55. Cook KM, Figg WD (2010) Angiogenesis inhibitors: current strategies and future prospects. CA Cancer J Clin 60(4):222–243
56. Narayan HK, French B, Khan AM, Plappert T, Hyman D, Bajulaiye A et al (2016) Noninvasive measures of ventricular-arterial coupling and circumferential strain predict cancer therapeutics-related cardiac dysfunction. JACC Cardiovasc Imaging 9(10):1131–1141
57. Moslehi JJ, Deininger M (2015) Tyrosine kinase inhibitor-associated cardiovascular toxicity in chronic myeloid leukemia. J Clin Oncol 33(35):4210–4218
58. Zamorano JL, Lancellotti P, Rodriguez Munoz D, Aboyans V, Asteggiano R, Galderisi M et al (2016) 2016 ESC Position Paper on cancer treatments and cardiovascular toxicity developed under the auspices of the ESC Committee for Practice Guidelines: The Task Force for cancer treatments and cardiovascular toxicity of the European Society of Cardiology (ESC). Eur Heart J 37(36):2768–2801
59. Kloth JS, Pagani A, Verboom MC, Malovini A, Napolitano C, Kruit WH et al (2015) Incidence and relevance of QTc-interval prolongation caused by tyrosine kinase inhibitors. Br J Cancer 112(6):1011–1016
60. Pucci G, Milan A, Paini A, Salvetti M, Cerasari A, Vaudo G (2019) Acute blood pressure elevation associated with biological therapies for cancer: a focus on VEGF signaling pathway inhibitors. Expert Opin Biol Ther 19(5):433–442
61. Konstantinides SV, Meyer G, Becattini C, Bueno H, Geersing GJ, Harjola VP et al (2020) 2019 ESC Guidelines for the diagnosis and management of acute pulmonary embolism developed in collaboration with the European Respiratory Society (ERS). Eur Heart J 41(4): 543–603
62. Zhang A, Fan T, Liu Y, Yu G, Li C, Jiang Z (2024) Regulatory T cells in immune checkpoint blockade antitumor therapy. Mol Cancer 23(1):251
63. Carlino MS, Larkin J, Long GV (2021) Immune checkpoint inhibitors in melanoma. Lancet 398(10304):1002–1014
64. Laenens D, Yu Y, Santens B, Jacobs J, Beuselinck B, Bechter O et al (2022) Incidence of cardiovascular events in patients treated with immune checkpoint inhibitors. J Clin Oncol 40(29):3430–3438
65. Rubio-Infante N, Ramirez-Flores YA, Castillo EC, Lozano O, Garcia-Rivas G, Torre-Amione G (2021) Cardiotoxicity associated with immune checkpoint inhibitor therapy: a meta-analysis. Eur J Heart Fail 23(10):1739–1747
66. Smith-Garvin JE, Koretzky GA, Jordan MS (2009) T cell activation. Annu Rev Immunol 27: 591–619
67. Polk A, Vistisen K, Vaage-Nilsen M, Nielsen DL (2014) A systematic review of the pathophysiology of 5-fluorouracil-induced cardiotoxicity. BMC Pharmacol Toxicol 15:47
68. Li D, Song C, Zhang J, Zhao X (2022) ROS and iron homeostasis dependent ferroptosis play a vital role in 5-fluorouracil induced cardiotoxicity in vitro and in vivo. Toxicology 468:153113
69. Perez EA (1998) Paclitaxel and cardiotoxicity. J Clin Oncol 16(11):3481–3482
70. Varbiro G, Veres B, Gallyas F Jr, Sumegi B (2001) Direct effect of Taxol on free radical formation and mitochondrial permeability transition. Free Radic Biol Med 31(4):548–558
71. Mohty M, Bruno B, Gagelmann N, Mateos MV (2024) Multiple myeloma. In: Sureda A, Corbacioglu S, Greco R, Kroger N, Carreras E (eds) The EBMT handbook: hematopoietic cell transplantation and cellular therapies, 8th edn. Springer, Cham, pp 741–749
72. Georgiopoulos G, Makris N, Laina A, Theodorakakou F, Briasoulis A, Trougakos IP et al (2023) Cardiovascular toxicity of proteasome inhibitors: underlying mechanisms and management strategies: JACC: CardioOncology State-of-the-Art Review. JACC CardioOncol 5(1):1–21
73. Nowis D, Maczewski M, Mackiewicz U, Kujawa M, Ratajska A, Wieckowski MR et al (2010) Cardiotoxicity of the anticancer therapeutic agent bortezomib. Am J Pathol 176(6):2658–2668
74. Nygard L, Vogelius IR, Fischer BM, Kjaer A, Langer SW, Aznar MC et al (2018) A competing risk model of first failure site after definitive chemoradiation therapy for locally advanced non-small cell lung cancer. J Thorac Oncol 13(4):559–567

75. Yusuf SW, Venkatesulu BP, Mahadevan LS, Krishnan S (2017) Radiation-induced cardiovascular disease: a clinical perspective. Front Cardiovasc Med 4:66
76. Tapio S (2016) Pathology and biology of radiation-induced cardiac disease. J Radiat Res 57(5):439–448
77. Wang B, Wang H, Zhang M, Ji R, Wei J, Xin Y et al (2020) Radiation-induced myocardial fibrosis: mechanisms underlying its pathogenesis and therapeutic strategies. J Cell Mol Med 24(14):7717–7729
78. Liu LK, Ouyang W, Zhao X, Su SF, Yang Y, Ding WJ et al (2017) Pathogenesis and prevention of radiation-induced myocardial fibrosis. Asian Pac J Cancer Prev 18(3):583–587
79. Teng AE, Noor B, Ajijola OA, Yang EH (2021) Chemotherapy and radiation-associated cardiac autonomic dysfunction. Curr Oncol Rep 23(2):14
80. Steffen JD, Pascal JM (2013) New players to the field of ADP-ribosylation make the final cut. EMBO J 32(9):1205–1207
81. Brunyanszki A, Szczesny B, Virag L, Szabo C (2016) Mitochondrial poly(ADP-ribose) polymerase: the Wizard of Oz at work. Free Radic Biol Med 100:257–270
82. Kraus WL, Lis JT (2003) PARP goes transcription. Cell 113(6):677–683
83. Burkle A (2001) Poly(APD-ribosyl)ation, a DNA damage-driven protein modification and regulator of genomic instability. Cancer Lett 163(1):1–5
84. Stilmann M, Hinz M, Arslan SC, Zimmer A, Schreiber V, Scheidereit C (2009) A nuclear poly (ADP-ribose)-dependent signalosome confers DNA damage-induced IkappaB kinase activation. Mol Cell 36(3):365–378
85. Berger NA (1985) Poly(ADP-ribose) in the cellular response to DNA damage. Radiat Res 101(1):4–15
86. Virag L, Salzman AL, Szabo C (1998) Poly(ADP-ribose) synthetase activation mediates mitochondrial injury during oxidant-induced cell death. J Immunol 161(7):3753–3759
87. Ha HC, Snyder SH (1999) Poly(ADP-ribose) polymerase is a mediator of necrotic cell death by ATP depletion. Proc Natl Acad Sci USA 96(24):13978–13982
88. Shieh WM, Ame JC, Wilson MV, Wang ZQ, Koh DW, Jacobson MK et al (1998) Poly (ADP-ribose) polymerase null mouse cells synthesize ADP-ribose polymers. J Biol Chem 273(46):30069–30072
89. Kupper JH, Muller M, Wolf I (1999) NAD(+) consumption in carcinogen-treated hamster cells overexpressing a dominant negative mutant of poly(ADP-ribose) polymerase. Biochem Biophys Res Commun 265(2):525–529
90. Fouquerel E, Goellner EM, Yu Z, Gagne JP, Barbi de Moura M, Feinstein T et al (2014) ARTD1/PARP1 negatively regulates glycolysis by inhibiting hexokinase 1 independent of NAD+ depletion. Cell Rep 8(6):1819–1831
91. Perry DK, Smyth MJ, Stennicke HR, Salvesen GS, Duriez P, Poirier GG et al (1997) Zinc is a potent inhibitor of the apoptotic protease, caspase-3. A novel target for zinc in the inhibition of apoptosis. J Biol Chem 272(30):18530–18533
92. Mashimo M, Onishi M, Uno A, Tanimichi A, Nobeyama A, Mori M et al (2021) The 89-kDa PARP1 cleavage fragment serves as a cytoplasmic PAR carrier to induce AIF-mediated apoptosis. J Biol Chem 296:100046
93. Veres B, Gallyas F Jr, Varbiro G, Berente Z, Osz E, Szekeres G et al (2003) Decrease of the inflammatory response and induction of the Akt/protein kinase B pathway by poly-(ADP-ribose) polymerase 1 inhibitor in endotoxin-induced septic shock. Biochem Pharmacol 65(8):1373–1382
94. Tapodi A, Debreceni B, Hanto K, Bognar Z, Wittmann I, Gallyas F Jr et al (2005) Pivotal role of Akt activation in mitochondrial protection and cell survival by poly(ADP-ribose)polymerase-1 inhibition in oxidative stress. J Biol Chem 280(42):35767–35775
95. Veres B, Radnai B, Gallyas F Jr, Varbiro G, Berente Z, Osz E et al (2004) Regulation of kinase cascades and transcription factors by a poly(ADP-ribose) polymerase-1 inhibitor, 4-hydroxyquinazoline, in lipopolysaccharide-induced inflammation in mice. J Pharmacol Exp Ther 310(1):247–255

96. Koh SH, Park Y, Song CW, Kim JG, Kim K, Kim J et al (2004) The effect of PARP inhibitor on ischaemic cell death, its related inflammation and survival signals. Eur J Neurosci 20(6): 1461–1472
97. Lee SR, Yang KS, Kwon J, Lee C, Jeong W, Rhee SG (2002) Reversible inactivation of the tumor suppressor PTEN by H_2O_2. J Biol Chem 277(23):20336–20342
98. Tapodi A, Bognar Z, Szabo C, Gallyas F, Sumegi B, Hocsak E (2019) PARP inhibition induces Akt-mediated cytoprotective effects through the formation of a mitochondria-targeted phospho-ATM-NEMO-Akt-mTOR signalosome. Biochem Pharmacol 162:98–108
99. Kozlov SV, Graham ME, Jakob B, Tobias F, Kijas AW, Tanuji M et al (2011) Autophosphorylation and ATM activation: additional sites add to the complexity. J Biol Chem 286(11):9107–9119
100. Gallyas F Jr, Sumegi B, Szabo C (2020) Role of Akt activation in PARP inhibitor resistance in cancer. Cancers (Basel) 12(3):532
101. Williamson CT, Muzik H, Turhan AG, Zamo A, O'Connor MJ, Bebb DG et al (2010) ATM deficiency sensitizes mantle cell lymphoma cells to poly(ADP-ribose) polymerase-1 inhibitors. Mol Cancer Ther 9(2):347–357
102. Roberts DJ, Tan-Sah VP, Smith JM, Miyamoto S (2013) Akt phosphorylates HK-II at Thr-473 and increases mitochondrial HK-II association to protect cardiomyocytes. J Biol Chem 288(33):23798–23806
103. Pastorino JG, Hoek JB, Shulga N (2005) Activation of glycogen synthase kinase 3beta disrupts the binding of hexokinase II to mitochondria by phosphorylating voltage-dependent anion channel and potentiates chemotherapy-induced cytotoxicity. Cancer Res 65(22): 10545–10554
104. Azoulay-Zohar H, Israelson A, Abu-Hamad S, Shoshan-Barmatz V (2004) In self-defence: hexokinase promotes voltage-dependent anion channel closure and prevents mitochondria-mediated apoptotic cell death. Biochem J 377(Pt 2):347–355
105. Andrabi SA, Umanah GK, Chang C, Stevens DA, Karuppagounder SS, Gagne JP et al (2014) Poly(ADP-ribose) polymerase-dependent energy depletion occurs through inhibition of glycolysis. Proc Natl Acad Sci USA 111(28):10209–10214
106. Uchiyama T, Engelman RM, Maulik N, Das DK (2004) Role of Akt signaling in mitochondrial survival pathway triggered by hypoxic preconditioning. Circulation 109(24):3042–3049
107. Rajawat J, Shukla N, Mishra DP (2017) Therapeutic targeting of poly(ADP-Ribose) polymerase-1 (PARP1) in cancer: current developments, therapeutic strategies, and future opportunities. Med Res Rev 37(6):1461–1491
108. Zhang W, Liu HT (2002) MAPK signal pathways in the regulation of cell proliferation in mammalian cells. Cell Res 12(1):9–18
109. Papaconstantinou J (2019) The role of signaling pathways of inflammation and oxidative stress in development of senescence and aging phenotypes in cardiovascular disease. Cells 8(11): 1383
110. Assi M (2017) The differential role of reactive oxygen species in early and late stages of cancer. Am J Physiol Regul Integr Comp Physiol 313(6):R646–RR53
111. Lee M, Choy WC, Abid MR (2011) Direct sensing of endothelial oxidants by vascular endothelial growth factor receptor-2 and c-Src. PLoS One 6(12):e28454
112. Erdelyi K, Pacher P, Virag L, Szabo C (2013) Role of poly(ADP-ribosyl)ation in a 'two-hit' model of hypoxia and oxidative stress in human A549 epithelial cells in vitro. Int J Mol Med 32(2):339–346
113. Halmosi R, Berente Z, Osz E, Toth K, Literati-Nagy P, Sumegi B (2001) Effect of poly (ADP-ribose) polymerase inhibitors on the ischemia-reperfusion-induced oxidative cell damage and mitochondrial metabolism in Langendorff heart perfusion system. Mol Pharmacol 59(6):1497–1505
114. Mester L, Szabo A, Atlasz T, Szabadfi K, Reglodi D, Kiss P et al (2009) Protection against chronic hypoperfusion-induced retinal neurodegeneration by PARP inhibition via activation of PI-3-kinase Akt pathway and suppression of JNK and p38 MAP kinases. Neurotox Res 16(1):68–76

115. Xu Y, Huang S, Liu ZG, Han J (2006) Poly(ADP-ribose) polymerase-1 signaling to mitochondria in necrotic cell death requires RIP1/TRAF2-mediated JNK1 activation. J Biol Chem 281(13):8788–8795
116. Alano CC, Swanson RA (2006) Players in the PARP-1 cell-death pathway: JNK1 joins the cast. Trends Biochem Sci 31(6):309–311
117. Xu Q, Konta T, Nakayama K, Furusu A, Moreno-Manzano V, Lucio-Cazana J et al (2004) Cellular defense against H_2O_2-induced apoptosis via MAP kinase-MKP-1 pathway. Free Radic Biol Med 36(8):985–993
118. Racz B, Hanto K, Tapodi A, Solti I, Kalman N, Jakus P et al (2010) Regulation of MKP-1 expression and MAPK activation by PARP-1 in oxidative stress: a new mechanism for the cytoplasmic effect of PARP-1 activation. Free Radic Biol Med 49(12):1978–1988
119. Lin T, Mak NK, Yang MS (2008) MAPK regulate p53-dependent cell death induced by benzo[a]pyrene: involvement of p53 phosphorylation and acetylation. Toxicology 247(2–3):145–153
120. Palfi A, Toth A, Kulcsar G, Hanto K, Deres P, Bartha E et al (2005) The role of Akt and mitogen-activated protein kinase systems in the protective effect of poly(ADP-ribose) polymerase inhibition in Langendorff perfused and in isoproterenol-damaged rat hearts. J Pharmacol Exp Ther 315(1):273–282
121. Bartha E, Kiss GN, Kalman E, Kulcsar G, Kalai T, Hideg K et al (2008) Effect of L-2286, a poly(ADP-ribose)polymerase inhibitor and enalapril on myocardial remodeling and heart failure. J Cardiovasc Pharmacol 52(3):253–261
122. Meng F, Zhang H, Liu G, Kreike B, Chen W, Sethi S et al (2011) p38gamma mitogen-activated protein kinase contributes to oncogenic properties maintenance and resistance to poly(ADP-ribose)-polymerase-1 inhibition in breast cancer. Neoplasia 13(5):472–482
123. Radnai B, Antus C, Racz B, Engelmann P, Priber JK, Tucsek Z et al (2012) Protective effect of the poly(ADP-ribose) polymerase inhibitor PJ34 on mitochondrial depolarization-mediated cell death in hepatocellular carcinoma cells involves attenuation of c-Jun N-terminal kinase-2 and protein kinase B/Akt activation. Mol Cancer 11:34
124. Hocsak E, Szabo V, Kalman N, Antus C, Cseh A, Sumegi K et al (2017) PARP inhibition protects mitochondria and reduces ROS production via PARP-1–ATF4–MKP-1–MAPK retrograde pathway. Free Radic Biol Med 108:770–784
125. Turpaev K, Bouton C, Diet A, Glatigny A, Drapier JC (2005) Analysis of differentially expressed genes in nitric oxide-exposed human monocytic cells. Free Radic Biol Med 38(10):1392–1400
126. Banerjee S, Kaye SB, Ashworth A (2010) Making the best of PARP inhibitors in ovarian cancer. Nat Rev Clin Oncol 7(9):508–519
127. O'Connor MJ, Martin NM, Smith GC (2007) Targeted cancer therapies based on the inhibition of DNA strand break repair. Oncogene 26(56):7816–7824
128. Mendes-Pereira AM, Martin SA, Brough R, McCarthy A, Taylor JR, Kim JS et al (2009) Synthetic lethal targeting of PTEN mutant cells with PARP inhibitors. EMBO Mol Med 1(6–7):315–322
129. Kim D, Nam HJ (2022) PARP inhibitors: clinical limitations and recent attempts to overcome them. Int J Mol Sci 23(15):8412
130. Roskoski R Jr (2024) Targeted and cytotoxic inhibitors used in the treatment of breast cancer. Pharmacol Res 210:107534
131. Laemmerer A, Lehmann C, Mayr L, Bruckner K, Gabler L, Senfter D et al (2024) Alternative lengthening of telomere-based immortalization renders H3G34R-mutant diffuse hemispheric glioma hypersensitive to PARP inhibitor combination regimens. Neuro-Oncology 27(3):811–827
132. Dey P, Das R, Chatterjee S, Paul R, Ghosh U (2024) Combined effects of carbon ion radiation and PARP inhibitor on non-small cell lung carcinoma cells: Insights into DNA repair pathways and cell death mechanisms. DNA Repair (Amst) 144:103778

133. Khalid AB, Fountzilas C, Burney HN, Mamdani H, Schneider BP, Fausel C et al (2024) A phase II study evaluating safety and efficacy of niraparib in patients with previously treated homologous recombination defective metastatic esophageal/gastroesophageal junction/proximal gastric adenocarcinoma. Front Oncol 14:1435056
134. Drew Y, Zenke FT, Curtin NJ (2024) DNA damage response inhibitors in cancer therapy: lessons from the past, current status and future implications. Nat Rev Drug Discov 24(1): 19–39
135. Moindjie H, Rodrigues-Ferreira S, Nahmias C (2021) Mitochondrial metabolism in carcinogenesis and cancer therapy. Cancers (Basel) 13(13):3311
136. Hirpara J, Eu JQ, Tan JKM, Wong AL, Clement MV, Kong LR et al (2019) Metabolic reprogramming of oncogene-addicted cancer cells to OXPHOS as a mechanism of drug resistance. Redox Biol 25:101076
137. Weinberg SE, Chandel NS (2015) Targeting mitochondria metabolism for cancer therapy. Nat Chem Biol 11(1):9–15
138. Michels J, Obrist F, Castedo M, Vitale I, Kroemer G (2014) PARP and other prospective targets for poisoning cancer cell metabolism. Biochem Pharmacol 92(1):164–171
139. Hirschey MD (2011) Old enzymes, new tricks: sirtuins are NAD(+)-dependent de-acylases. Cell Metab 14(6):718–719
140. Vida A, Marton J, Miko E, Bai P (2017) Metabolic roles of poly(ADP-ribose) polymerases. Semin Cell Dev Biol 63:135–143
141. Bai P, Canto C, Oudart H, Brunyanszki A, Cen Y, Thomas C et al (2011) PARP-1 inhibition increases mitochondrial metabolism through SIRT1 activation. Cell Metab 13(4):461–468
142. Oppedisano F, Nesci S, Spagnoletta A (2024) Mitochondrial sirtuin 3 and role of natural compounds: the effect of post-translational modifications on cellular metabolism. Crit Rev Biochem Mol Biol 59(3–4):199–220
143. Juvekar A, Burga LN, Hu H, Lunsford EP, Ibrahim YH, Balmana J et al (2012) Combining a PI3K inhibitor with a PARP inhibitor provides an effective therapy for BRCA1-related breast cancer. Cancer Discov 2(11):1048–1063
144. Hassa PO, Hottiger MO (2002) The functional role of poly(ADP-ribose)polymerase 1 as novel coactivator of NF-kappaB in inflammatory disorders. Cell Mol Life Sci 59(9):1534–1553
145. Zhao H, Wu L, Yan G, Chen Y, Zhou M, Wu Y et al (2021) Inflammation and tumor progression: signaling pathways and targeted intervention. Signal Transduct Target Ther 6(1):263
146. Galluzzi L, Guilbaud E, Schmidt D, Kroemer G, Marincola FM (2024) Targeting immunogenic cell stress and death for cancer therapy. Nat Rev Drug Discov 23(6):445–460
147. Pan H, Liu P, Zhao L, Pan Y, Mao M, Kroemer G et al (2024) Immunogenic cell stress and death in the treatment of cancer. Semin Cell Dev Biol 156:11–21
148. Kroemer G, Galassi C, Zitvogel L, Galluzzi L (2022) Immunogenic cell stress and death. Nat Immunol 23(4):487–500
149. Pham L, Arroum T, Wan J, Pavelich L, Bell J, Morse PT et al (2024) Regulation of mitochondrial oxidative phosphorylation through tight control of cytochrome c oxidase in health and disease—implications for ischemia/reperfusion injury, inflammatory diseases, diabetes, and cancer. Redox Biol 78:103426
150. Li L, Xu T, Qi X (2024) Balanced regulation of ROS production and inflammasome activation in preventing early development of colorectal cancer. Immunol Rev 329(1):e13417
151. Di Carlo E, Sorrentino C (2024) Oxidative stress and age-related tumors. Antioxidants (Basel) 13(9):1109
152. Ge M, Papagiannakopoulos T, Bar-Peled L (2024) Reductive stress in cancer: coming out of the shadows. Trends Cancer 10(2):103–112
153. Mukherjee AG, Gopalakrishnan AV (2023) The mechanistic insights of the antioxidant Keap1-Nrf2 pathway in oncogenesis: a deadly scenario. Med Oncol 40(9):248

154. Zhang Q, Dang YY, Luo X, Fu JJ, Zou ZC, Jia XJ et al (2023) Kazinol B protects H9c2 cardiomyocytes from hypoxia/reoxygenation-induced cardiac injury by modulating the AKT/AMPK/Nrf2 signalling pathway. Pharm Biol 61(1):362–371
155. Janko L, Toth E, Laczik M, Rauch B, Janka E, Balint BL et al (2023) PARP2 poly (ADP-ribosyl)ates nuclear factor erythroid 2-related factor 2 (NRF2) affecting NRF2 subcellular localization. Sci Rep 13(1):7869
156. Pacher P, Szabo C (2007) Role of poly(ADP-ribose) polymerase 1 (PARP-1) in cardiovascular diseases: the therapeutic potential of PARP inhibitors. Cardiovasc Drug Rev 25(3):235–260
157. Vidavalur R, Swarnakar S, Thirunavukkarasu M, Samuel SM, Maulik N (2008) Ex vivo and in vivo approaches to study mechanisms of cardioprotection targeting ischemia/reperfusion (i/r) injury: useful techniques for cardiovascular drug discovery. Curr Drug Discov Technol 5(4):269–278
158. Bell RM, Mocanu MM, Yellon DM (2011) Retrograde heart perfusion: the Langendorff technique of isolated heart perfusion. J Mol Cell Cardiol 50(6):940–950
159. Halmosi R, Deres L, Gal R, Eros K, Sumegi B, Toth K (2016) PARP inhibition and postinfarction myocardial remodeling. Int J Cardiol 217(Suppl):S52–S59
160. Liaudet L, Szabo E, Timashpolsky L, Virag L, Cziraki A, Szabo C (2001) Suppression of poly (ADP-ribose) polymerase activation by 3-aminobenzamide in a rat model of myocardial infarction: long-term morphological and functional consequences. Br J Pharmacol 133(8): 1424–1430
161. Murthy KG, Xiao CY, Mabley JG, Chen M, Szabo C (2004) Activation of poly(ADP-ribose) polymerase in circulating leukocytes during myocardial infarction. Shock 21(3):230–234
162. Lejay A, Fang F, John R, Van JA, Barr M, Thaveau F et al (2016) Ischemia reperfusion injury, ischemic conditioning and diabetes mellitus. J Mol Cell Cardiol 91:11–22
163. Hausenloy DJ, Erik Botker H, Condorelli G, Ferdinandy P, Garcia-Dorado D, Heusch G et al (2013) Translating cardioprotection for patient benefit: position paper from the Working Group of Cellular Biology of the Heart of the European Society of Cardiology. Cardiovasc Res 98(1):7–27
164. Ungvari Z, Gupte SA, Recchia FA, Batkai S, Pacher P (2005) Role of oxidative-nitrosative stress and downstream pathways in various forms of cardiomyopathy and heart failure. Curr Vasc Pharmacol 3(3):221–229
165. Angel P, Imagawa M, Chiu R, Stein B, Imbra RJ, Rahmsdorf HJ et al (1987) Phorbol ester-inducible genes contain a common *cis* element recognized by a TPA-modulated *trans*-acting factor. Cell 49(6):729–739
166. Corson MA, James NL, Latta SE, Nerem RM, Berk BC, Harrison DG (1996) Phosphorylation of endothelial nitric oxide synthase in response to fluid shear stress. Circ Res 79(5):984–991
167. Ong SB, Hernandez-Resendiz S, Crespo-Avilan GE, Mukhametshina RT, Kwek XY, Cabrera-Fuentes HA et al (2018) Inflammation following acute myocardial infarction: multiple players, dynamic roles, and novel therapeutic opportunities. Pharmacol Ther 186:73–87
168. Tait SW, Green DR (2013) Mitochondrial regulation of cell death. Cold Spring Harb Perspect Biol 5(9):a008706
169. Kovacs K, Toth A, Deres P, Kalai T, Hideg K, Sumegi B (2004) Myocardial protection by selective poly(ADP-ribose) polymerase inhibitors. Exp Clin Cardiol 9(1):17–20
170. Kovacs K, Hanto K, Bognar Z, Tapodi A, Bognar E, Kiss GN et al (2009) Prevalent role of Akt and ERK activation in cardioprotective effect of Ca^{2+} channel- and beta-adrenergic receptor blockers. Mol Cell Biochem 321(1–2):155–164
171. Palfi A, Toth A, Hanto K, Deres P, Szabados E, Szereday Z et al (2006) PARP inhibition prevents postinfarction myocardial remodeling and heart failure via the protein kinase C/glycogen synthase kinase-3beta pathway. J Mol Cell Cardiol 41(1):149–159
172. Curtin NJ, Szabo C (2013) Therapeutic applications of PARP inhibitors: anticancer therapy and beyond. Mol Asp Med 34(6):1217–1256

173. Pacher P, Liaudet L, Mabley JG, Cziraki A, Hasko G, Szabo C (2006) Beneficial effects of a novel ultrapotent poly(ADP-ribose) polymerase inhibitor in murine models of heart failure. Int J Mol Med 17(2):369–375
174. Bartha E, Solti I, Kereskai L, Lantos J, Plozer E, Magyar K et al (2009) PARP inhibition delays transition of hypertensive cardiopathy to heart failure in spontaneously hypertensive rats. Cardiovasc Res 83(3):501–510
175. Morrow DA, Brickman CM, Murphy SA, Baran K, Krakover R, Dauerman H et al (2009) A randomized, placebo-controlled trial to evaluate the tolerability, safety, pharmacokinetics, and pharmacodynamics of a potent inhibitor of poly(ADP-ribose) polymerase (INO-1001) in patients with ST-elevation myocardial infarction undergoing primary percutaneous coronary intervention: results of the TIMI 37 trial. J Thromb Thrombolysis 27(4):359–364
176. Curtin NJ, Szabo C (2020) Poly(ADP-ribose) polymerase inhibition: past, present and future. Nat Rev Drug Discov 19(10):711–736
177. Toth-Zsamboki E, Horvath E, Vargova K, Pankotai E, Murthy K, Zsengeller Z et al (2006) Activation of poly(ADP-ribose) polymerase by myocardial ischemia and coronary reperfusion in human circulating leukocytes. Mol Med 12(9–10):221–228
178. Yao L, Huang K, Huang D, Wang J, Guo H, Liao Y (2008) Acute myocardial infarction induced increases in plasma tumor necrosis factor-alpha and interleukin-10 are associated with the activation of poly(ADP-ribose) polymerase of circulating mononuclear cell. Int J Cardiol 123(3):366–368
179. Ramlawi B, Feng J, Mieno S, Szabo C, Zsengeller Z, Clements R et al (2006) Indices of apoptosis activation after blood cardioplegia and cardiopulmonary bypass. Circulation 114(1 Suppl):I257–I263
180. Soriano FG, Nogueira AC, Caldini EG, Lins MH, Teixeira AC, Cappi SB et al (2006) Potential role of poly (adenosine 5′-diphosphate-ribose) polymerase activation in the pathogenesis of myocardial contractile dysfunction associated with human septic shock. Crit Care Med 34(4): 1073–1079
181. Nagy E, Caidahl K, Franco-Cereceda A, Back M (2012) Increased transcript level of poly (ADP-ribose) polymerase (PARP-1) in human tricuspid compared with bicuspid aortic valves correlates with the stenosis severity. Biochem Biophys Res Commun 420(3):671–675
182. Molnar A, Toth A, Bagi Z, Papp Z, Edes I, Vaszily M et al (2006) Activation of the poly (ADP-ribose) polymerase pathway in human heart failure. Mol Med 12(7–8):143–152
183. Karlstaedt A, Moslehi J, de Boer RA (2022) Cardio-onco-metabolism: metabolic remodelling in cardiovascular disease and cancer. Nat Rev Cardiol 19(6):414–425
184. de Boer RA, Hulot JS, Tocchetti CG, Aboumsallem JP, Ameri P, Anker SD et al (2020) Common mechanistic pathways in cancer and heart failure. A scientific roadmap on behalf of the Translational Research Committee of the Heart Failure Association (HFA) of the European Society of Cardiology (ESC). Eur J Heart Fail 22(12):2272–2289
185. Ye F, Dewanjee S, Li Y, Jha NK, Chen ZS, Kumar A et al (2023) Advancements in clinical aspects of targeted therapy and immunotherapy in breast cancer. Mol Cancer 22(1):105
186. Noordermeer SM, van Attikum H (2019) PARP inhibitor resistance: a Tug-of-War in BRCA-mutated cells. Trends Cell Biol 29(10):820–834
187. Szanto A, Bognar Z, Szigeti A, Szabo A, Farkas L, Gallyas F Jr (2009) Critical role of bad phosphorylation by Akt in cytostatic resistance of human bladder cancer cells. Anticancer Res 29(1):159–164
188. Fong PC, Boss DS, Yap TA, Tutt A, Wu P, Mergui-Roelvink M et al (2009) Inhibition of poly (ADP-ribose) polymerase in tumors from BRCA mutation carriers. N Engl J Med 361(2): 123–134
189. Palazzo A, Ciccarese C, Iacovelli R, Cannizzaro MC, Stefani A, Salvatore L et al (2023) Major adverse cardiac events and cardiovascular toxicity with PARP inhibitors-based therapy for solid tumors: a systematic review and safety meta-analysis. ESMO Open 8(2):101154
190. Roskoski R Jr (2022) Properties of FDA-approved small molecule protein kinase inhibitors: a 2022 update. Pharmacol Res 175:106037

191. Floor S, van Staveren WC, Larsimont D, Dumont JE, Maenhaut C (2011) Cancer cells in epithelial-to-mesenchymal transition and tumor-propagating-cancer stem cells: distinct, overlapping or same populations. Oncogene 30(46):4609–4621
192. Helmink BA, Khan MAW, Hermann A, Gopalakrishnan V, Wargo JA (2019) The microbiome, cancer, and cancer therapy. Nat Med 25(3):377–388
193. Jain T, Sharma P, Are AC, Vickers SM, Dudeja V (2021) New Insights into the cancer–microbiome–immune axis: decrypting a decade of discoveries. Front Immunol 12:622064
194. Lloyd-Jones D, Adams RJ, Brown TM, Carnethon M, Dai S, De Simone G et al (2010) Executive summary: heart disease and stroke statistics—2010 update: a report from the American Heart Association. Circulation 121(7):948–954
195. Blenck CL, Harvey PA, Reckelhoff JF, Leinwand LA (2016) The importance of biological sex and estrogen in rodent models of cardiovascular health and disease. Circ Res 118(8):1294–1312
196. Jones NR, Roalfe AK, Adoki I, Hobbs FDR, Taylor CJ (2019) Survival of patients with chronic heart failure in the community: a systematic review and meta-analysis. Eur J Heart Fail 21(11):1306–1325

Chapter 38
Potential Role of Nutraceuticals in the Prevention of Cardiovascular Toxicity

Resmi Rajalekshmi and Devendra K. Agrawal

Abstract The global burden of cardiovascular diseases (CVDs) is significant, often resulting from oxidative stress, inflammation, and endothelial dysfunction. Nutraceuticals, such as omega-3 fatty acids, polyphenols, phytosterols, Coenzyme Q10, and vitamins, have shown promise in preventing and managing cardiovascular toxicity (CVT). These bioactive compounds offer various protective effects, including improved lipid metabolism, reduced oxidative stress, modulation of inflammatory pathways, and enhanced endothelial function. Consumption of nutraceuticals through functional foods or supplements has been linked to a lower risk of hypertension, hyperlipidemia, atherosclerosis, and other CVD-related conditions. When used alongside lifestyle modifications and standard pharmacological therapies, nutraceuticals could serve as effective adjunct treatments to alleviate CVT. Despite promising findings, further research is needed to determine optimal dosages, long-term effects, and the precise mechanisms through which these compounds deliver cardiovascular benefits. This chapter delves into the current evidence on the role of nutraceuticals in preventing CVT and underscores their potential integration into clinical practice for managing CVDs.

Keywords Arrhythmia · Cardiovascular toxicity · Curcumin · Diabetes · Environmental toxicants · Flavonoids · Heart failure · Hyperglycemia · Hypertension · Inflammation · Metabolic disorder · Nutraceuticals · Nutrametabolomics · Oxidative stress · Phytosterols · Pollutants · Polyphenols · Radiation therapy · Weight management

R. Rajalekshmi · D. K. Agrawal (✉)
Department of Translational Research, Western University of Health Sciences, Pomona, CA, USA
e-mail: Rrajalekshmi@westernu.edu; DAgrawal@WesternU.edu

D. K. Agrawal et al. (eds.), *Cardiovascular Toxicity*, Advances in Biochemistry in Health and Disease 33, https://doi.org/10.1007/978-3-031-90598-8_38

Introduction

Cardiovascular toxicity (CVT) refers to a spectrum of adverse effects on the heart and blood vessels triggered by harmful substances, medications, or environmental factors. This condition can manifest in various forms, including hypertension, arrhythmias, cardiomyopathy, myocardial infarction, and heart failure. While cardiovascular toxicity is notably associated with cancer treatments like chemotherapy and radiation therapy, it can also result from chronic exposure to environmental toxins, poor dietary habits, and certain medications.

Understanding the mechanisms underlying CVT is crucial for developing effective prevention and management strategies. The cardiovascular system, comprising the heart and blood vessels, is essential for maintaining overall health by ensuring proper blood circulation, delivering oxygen and nutrients to tissues, and removing metabolic waste products. Disruption in this complex system can have extensive consequences, potentially leading to chronic diseases such as stroke, coronary artery disease, and diabetes. Therefore, safeguarding cardiovascular health is vital for preventing life-threatening conditions and promoting long-term wellness.

Given the rising prevalence of cardiovascular diseases globally, there is an urgent need to establish effective preventive and therapeutic measures. Nutraceuticals—products derived from food sources that offer health benefits beyond basic nutrition—represent a promising approach in this context. The term "nutraceutical" combines "nutrition" and "pharmaceutical," highlighting their role in health promotion and disease prevention. Nutraceuticals encompass dietary supplements, herbal products, functional foods, and fortified foods, which contain bioactive compounds like vitamins, minerals, antioxidants, polyphenols, and omega-3 fatty acids.

These bioactive compounds are known to support various physiological functions, including immune response, metabolic processes, and cardiovascular health. For instance, antioxidants combat oxidative stress, a key contributor to cardiovascular damage, while omega-3 fatty acids help modulate inflammatory responses and lipid profiles. Nutraceuticals have gained attention for their potential to complement traditional medicine, particularly in the prevention and management of chronic diseases such as cardiovascular disorders. Their generally favorable safety profile compared to pharmaceuticals makes them an appealing option for long-term health maintenance.

This book chapter aims to explore the potential role of nutraceuticals in mitigating cardiovascular toxicity. As cardiovascular diseases continue to rise and risk factors proliferate, there is increasing interest in alternative prevention and treatment approaches. Nutraceuticals, with their bioactive properties, offer a promising strategy to reduce the risk of cardiovascular toxicity, particularly in populations exposed to high-risk conditions such as chemotherapy, environmental pollutants, or poor dietary habits.

The chapter will provide a comprehensive review of the mechanisms through which nutraceuticals exert their protective effects, supported by evidence from preclinical and clinical studies. It will also address the challenges and opportunities associated with integrating nutraceuticals into cardiovascular health strategies. By examining current research and identifying future directions, this chapter aims to offer insights into the effective use of nutraceuticals for mitigating cardiovascular toxicity and enhancing overall cardiovascular health.

This exploration will highlight how nutraceuticals can be strategically utilized to complement conventional treatments, offering a broader approach to managing and preventing cardiovascular toxicity and associated diseases.

Cardiovascular Toxicity and Its Causes

CVT refers to damage to the heart and vascular system caused by exposure to toxic agents, resulting in both functional and structural changes. This toxicity can lead to various cardiovascular disorders, including myocardial injury, heart failure, arrhythmias, hypertension, and atherosclerosis [1]. It is a significant concern in clinical settings, especially among patients undergoing certain pharmacological treatments, exposed to environmental pollutants, or experiencing adverse lifestyle factors. Understanding the mechanisms and causes of CVT is crucial for developing effective prevention and treatment strategies as it has the potential to significantly impact morbidity and mortality.

Causes of Cardiovascular Toxicity

Cardiovascular toxicity arises from a variety of internal and external factors that disrupt normal heart and vascular function as shown in Fig. 38.1.

Pharmaceutical Agents

Various pharmaceutical agents, particularly chemotherapeutics and treatments for chronic conditions, are well-established contributors to cardiovascular toxicity. Agents such as anthracyclines, including doxorubicin [2], and targeted therapies like trastuzumab [3] are especially recognized for their cardiotoxic effects. These drugs induce cardiovascular damage primarily through mechanisms involving oxidative stress, mitochondrial dysfunction, and disruption of cellular signaling pathways, leading to compromised cardiac function and increased risk of long-term cardiovascular complications [4, 5]. Tyrosine kinase inhibitors (TKIs) including sunitinib and imatinib can lead to hypertension, heart failure, and arrhythmias [6]. Long-term use of nonsteroidal anti-inflammatory drugs (NSAIDs) increases the risk of cardiovascular events like myocardial infarction and stroke, primarily due to impaired prostaglandin synthesis and endothelial dysfunction

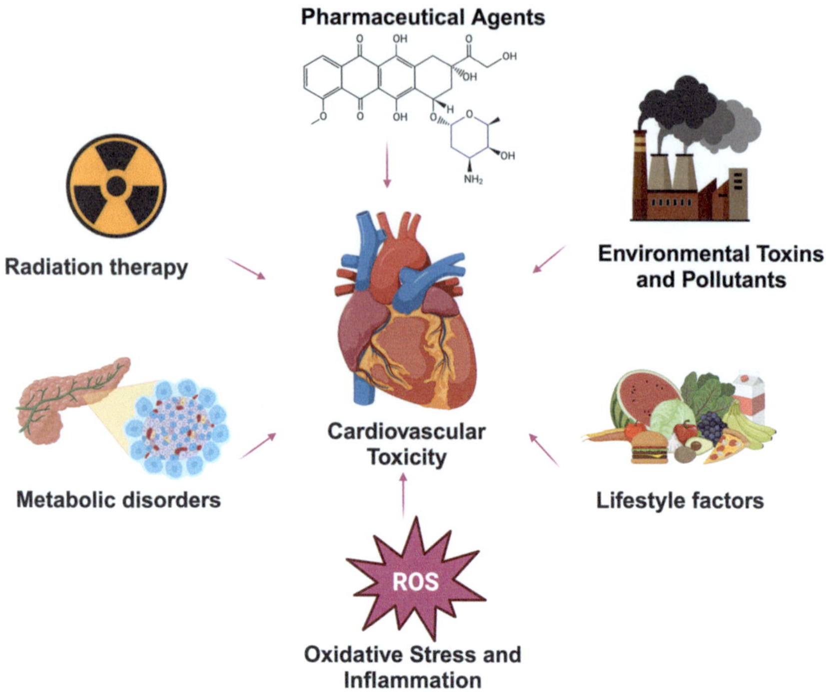

Fig. 38.1 Key factors and disorders that can induce cardiovascular toxicity. *ROS* reactive oxygen species

[7]. Furthermore, some cardiovascular drugs, including statins, may lead to side effects such as myopathy, exacerbating cardiovascular risk in certain populations [8].

Environmental Toxins and Pollutants

Environmental exposure to pollutants is a growing contributor to cardiovascular toxicity. Fine particulate matter (PM2.5) from air pollution induces oxidative stress and inflammation, leading to endothelial dysfunction and increased cardiovascular risk [9]. Heavy metals like lead, arsenic, and mercury also promote oxidative stress and lipid peroxidation, contributing to atherosclerosis and hypertension [10]. Tobacco smoke, both from active smoking and secondhand exposure, introduces numerous toxins that induce oxidative stress and vascular injury, significantly raising the risk of hypertension, atherosclerosis, and myocardial infarction [11].

Lifestyle Factors

Poor lifestyle choices further exacerbate cardiovascular toxicity. Diets high in processed foods, trans fats, and sugars contribute to dyslipidemia, hypertension, and systemic inflammation, promoting atherosclerosis [12]. Chronic excessive alcohol consumption is linked to cardiomyopathy, hypertension, and arrhythmias, with alcohol-induced oxidative stress and inflammation damaging cardiac tissue [13]. Physical inactivity is another critical risk factor, leading to obesity, hypertension, and metabolic disorders that increase the likelihood of cardiovascular dysfunction [14].

Oxidative Stress and Inflammation

Oxidative stress and inflammation play pivotal roles in driving cardiovascular toxicity and contributing to the progression of various cardiovascular diseases. Oxidative stress arises from an imbalance between reactive oxygen species (ROS) production and the body's antioxidant defenses, leading to cellular damage in the heart and blood vessels [15]. This damage impairs the functionality of lipids, proteins, and DNA, ultimately compromising myocardial function and vascular integrity. Chronic inflammation, often initiated by factors such as high-fat diets, environmental toxins, or metabolic disorders, further exacerbates endothelial dysfunction and promotes atherosclerosis by facilitating the formation of arterial plaques [16]. Collectively, oxidative stress and inflammation establish a detrimental cycle that accelerates cardiovascular injury and heightens the risk of conditions such as heart failure, hypertension, and atherosclerosis.

Metabolic Disorders

Metabolic disorders encompass a range of conditions, including hyperlipidemia, diabetes mellitus, and metabolic syndrome, all of which exert significant influence on cardiovascular health [17]. Hyperlipidemia, characterized by elevated levels of low-density lipoprotein (LDL) cholesterol, plays a pivotal role in the development of atherosclerosis. High LDL-C levels can lead to atherosclerosis, increasing the risk of heart attacks and strokes [18].

In diabetes mellitus, persistent hyperglycemia can lead to oxidative stress, inflammation, and endothelial dysfunction, which all contribute to the development of diabetic cardiomyopathy [19]. Additionally, metabolic syndrome, characterized by insulin resistance and dyslipidemia, further raises cardiovascular risk by promoting hypertension, inflammation, and oxidative damage. Elevated blood pressure in this context damages blood vessels, increasing the likelihood of atherosclerosis and elevating susceptibility to cardiovascular diseases [20].

Radiation Therapy

Radiation therapy-induced cardiotoxicity arises from multiple interconnected mechanisms involving oxidative stress, mitochondrial dysfunction, and inflammatory responses [21]. Its exposure enhances TGF-β1 and pro-inflammatory cytokine production, further exacerbating vascular injury and endothelial dysfunction [22]. Long-term consequences of radiation-induced cardiotoxicity include coronary artery disease, valvular heart disease, cardiomyopathies, and pericardial syndromes.

Introduction to Nutraceuticals

The historical use of food as a form of medicine dates back millennia, with ancient societies acknowledging the health-promoting properties of various foods and herbs. Traditional medicinal systems such as Ayurveda in India and Traditional Chinese Medicine (TCM) have long utilized natural products to both prevent and treat illnesses. For example, turmeric, which is rich in curcumin, was utilized in Ayurveda for its potent anti-inflammatory properties [23], while TCM employed ginseng for its immune-boosting and apoptogenic effects [24]. These ancient practices laid the groundwork for the present-day comprehension of the intricate relationship between diet and health, emphasizing the potential of natural compounds in influencing bodily functions and preventing disease.

The concept of "nutraceuticals" has a long history, spanning over 3000 years [25]. Hippocrates, the father of modern medicine, was the first to establish a significant link between food and its therapeutic benefits for health, laying the groundwork for further advancements in this field [26]. Coined in 1989 by Dr. Stephen DeFelice, the term combined "nutrition" and "pharmaceutical" to describe products derived from food sources that offer health benefits beyond basic nutritional value [27]. This concept arose from a growing understanding of the role that diet and specific nutrients play in health maintenance and disease prevention. In contrast to pharmaceuticals, which typically target specific symptoms or conditions, nutraceuticals are designed to promote overall health and well-being, making them an appealing option for preventive care.

The increasing prevalence of chronic diseases, coupled with the limitations and side effects of conventional pharmaceuticals, has further fueled interest in nutraceuticals. As lifestyle-related diseases such as obesity, diabetes, and heart disease continue to rise globally, there is a growing demand for safe and effective preventive strategies. Nutraceuticals offer a promising alternative, as they are generally associated with fewer side effects and can be integrated into daily diets [28]. They are now widely used as complementary therapies in various health conditions, including cardiovascular diseases, metabolic disorders, and neurodegenerative diseases [29, 30].

Classification of Nutraceuticals

The classification of nutraceuticals encompasses several categories based on their origin, composition, and intended purpose as shown in Fig. 38.2. One primary category is dietary supplements, comprising products such as vitamins, minerals, amino acids, and herbal extracts [31]. These supplements are designed to complement the diet, ensuring the intake of essential nutrients that may be deficient in regular food sources. Examples include vitamin D supplements for bone health, omega-3 fatty acids for cardiovascular health, and herbal supplements like echinacea for immune support [32]. Dietary supplements are commonly utilized to prevent or address nutrient deficiencies and promote overall well-being.

Another significant category is functional foods, which are foods enhanced with additional nutrients or bioactive compounds to provide health benefits beyond basic nutrition [33]. These foods are consumed as part of a regular diet but are specially formulated to improve specific health outcomes. Examples of functional foods include probiotic yogurt, which supports gut health, and fortified cereals with added vitamins and minerals [34]. Functional foods are designed to offer targeted

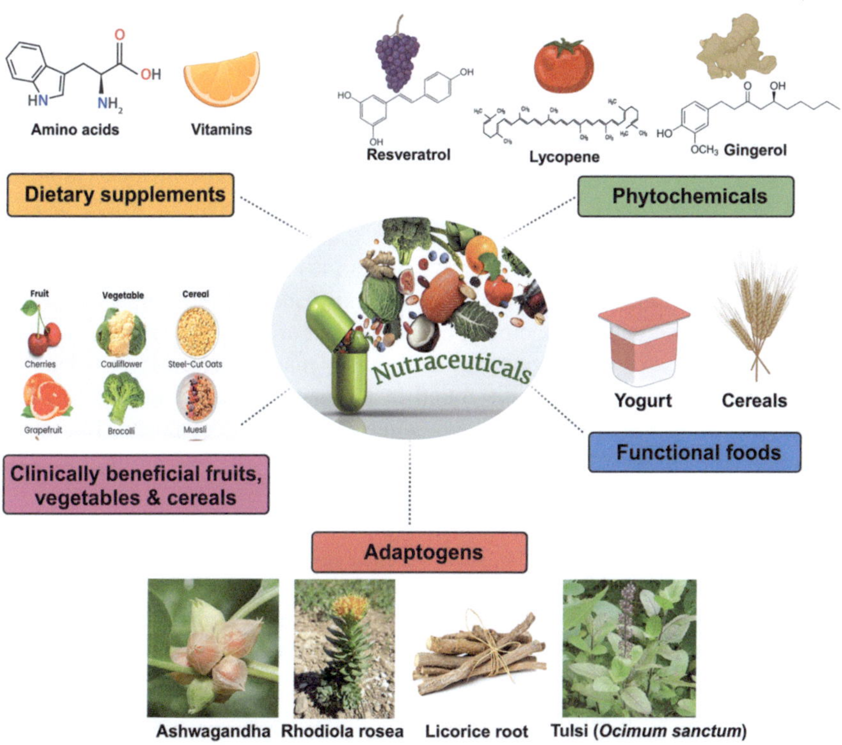

Fig. 38.2 List of potentially beneficial neutraceuticals with their classification in the prevention and/or treatment of cardiotoxicity

health benefits, such as improving digestive health, enhancing immune function, or reducing the risk of chronic diseases like heart disease [35, 36].

Medicinal or clinical foods, distinct from general functional foods, are specifically formulated to meet the unique nutritional needs of individuals diagnosed with certain medical conditions, such as metabolic disorders [37] or gastrointestinal diseases [38]. According to the FDA, these foods are designed to be administered internally under the supervision of a physician and are intended for the dietary management of diseases or conditions where specialized nutritional requirements are established based on recognized scientific principles [39]. Unlike other nutraceuticals, medicinal foods are regulated by the FDA [40] and often used under clinical guidance, such as high-protein shakes for patients with muscle-wasting conditions [41] or low-glycemic index foods for individuals with diabetes to help manage blood sugar levels [42].

Phytochemicals are bioactive compounds found in plants that offer health benefits [43]. They are divided into primary and secondary metabolites. Primary metabolites include simple sugars, carbohydrates, nucleic acids, lipids, and amino acids, and are essential for cellular processes. Secondary metabolites are primarily involved in stress responses and defense mechanisms [44]. Phytochemicals, when classified as nutraceuticals, are categorized based on their chemical structure and biological activity. Key categories include polyphenols like flavonoids (e.g., quercetin, resveratrol, catechins, and anthocyanins) found in tea, fruits, and vegetables, which have antioxidant, anti-inflammatory, and anticancer properties [45]. Other categories include phenolic acids like chlorogenic acid in coffee and fruits with antioxidant and anti-cancer effects [46], and tannins found in cranberries and pomegranates, which help prevent urinary tract infections [47]. Carotenoids include carotenes (alpha and beta-carotene) found in carrots and sweet potatoes, which can be converted into vitamin A and serve as potent antioxidants [48], and xanthophylls like lutein and zeaxanthin, crucial for eye health [49]. Saponins from legumes like soybeans lower cholesterol and fight cancer and diabetes [50]. Terpenes include terpineol in carrots with antimicrobial and anticancer effects [51], and limonoids in citrus fruits that have detoxifying and anticarcinogenic properties [52]. Alkaloids, such as caffeine and theobromine found in tea and cocoa, are known for their stimulant and antioxidant effects [53, 54], while organosulfur compounds like allicin in garlic and onions offer anti-inflammatory, antioxidant, and cardioprotective benefits [55].

Finally, other bioactive compounds like antioxidants and adaptogens also fall under the nutraceutical umbrella. Antioxidants, such as vitamin E and selenium, protect the body from oxidative stress and cellular damage, which are implicated in aging and chronic diseases [15]. Adaptogens, such as ashwagandha and *Rhodiola rosea*, are natural substances that help the body adapt to stress and maintain balance [56]. These compounds are increasingly recognized for their role in supporting mental and physical resilience, particularly in the context of modern lifestyle challenges.

Mechanisms of Action of Nutraceuticals in Preventing Cardiovascular Toxicity

Nutraceuticals have emerged as promising interventions for the prevention and management of CVDs, supported by substantial evidence of their efficacy. These dietary supplements are broadly classified based on their application in treating various cardiovascular conditions, including arrhythmias, congestive heart failure, angina, hypertension, and hyperlipidemias. Figure 38.3 illustrates the core mechanisms by which key nutraceuticals and dietary supplements contribute to the prevention and treatment of cerebral venous thrombosis (CVT).

Delineates the involvement of these nutraceuticals in modulating various cellular signaling pathways, thereby exerting therapeutic effects.

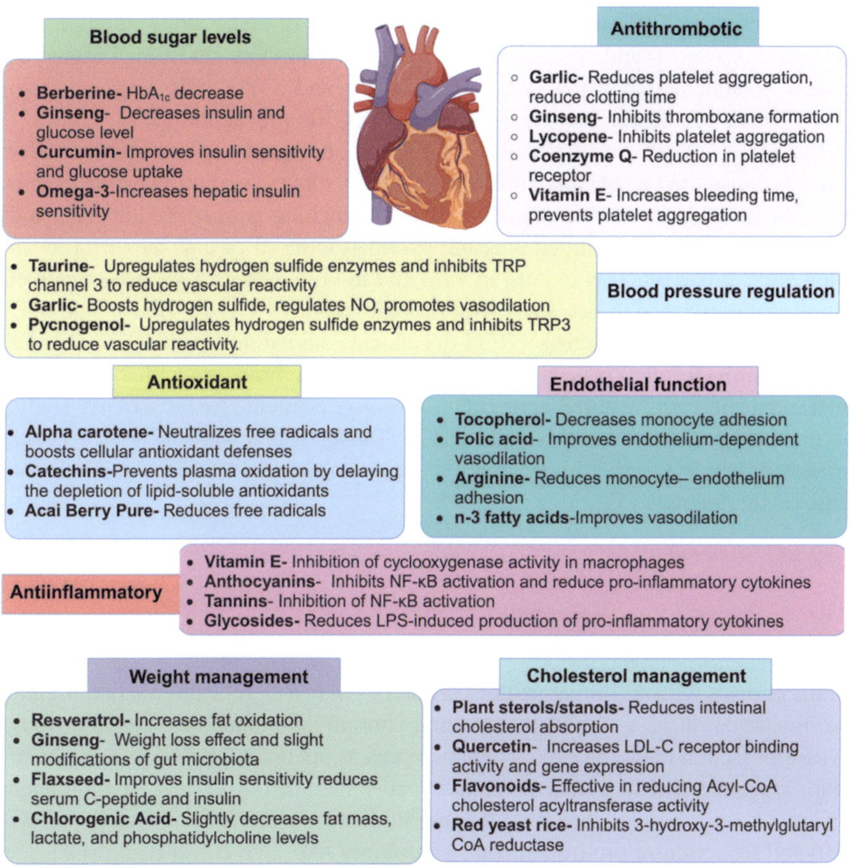

Fig. 38.3 Mechanisms of action of various neutraceuticals in the prevention of cardiovascular toxicity

Antioxidant Protection

Nutraceuticals with antioxidant properties help protect against cardiovascular toxicity (CVT) by neutralizing oxidative stress, a key contributor to cardiovascular diseases. Antioxidants like polyphenols, vitamins C and E, and flavonoids reduce reactive oxygen species (ROS), mitigating oxidative damage to lipids, proteins, and DNA. This preserves endothelial function, decreases inflammation, and prevents lipid peroxidation, which contributes to atherosclerosis and myocardial damage.

A recent study examined the impact of daily intake of vitamins A, E, C, and zinc on cardiovascular disease (CVD) risk in 5102 adults aged $\geq$30 years, with an average follow-up of 5.3 years. Dietary intake was assessed using a validated food frequency questionnaire, and CVD development was analyzed using multivariable Cox regression models. Participants included 2253 men and 2849 women, with average ages of 47.0 ± 11.6 and 45.6 ± 10.5 years, respectively. The main sources of these nutrients were fruits, vegetables, and legumes. The results revealed a significant reduction in CVD risk with increasing vitamin E intake while no significant associations were found for vitamins A, C, and zinc. The study highlighted the inverse relationship between vitamin E intake and CVD risk, suggesting the antioxidative benefits of fruits and vegetables in CVD prevention [57].

Lycopene, a potent natural antioxidant found abundantly in tomatoes and other fruits, is considered a valuable nutraceutical for its role in preventing cardiovascular diseases through its strong oxidative stress-quenching properties [58]. A study showed that lycopene inhibited pressure overload-induced cardiac hypertrophy and improved cardiac function in both in vitro and in vivo models. Lycopene treatment significantly reduced heart size, hypertrophic markers (ANP, BNP), and ROS production while inhibiting ROS-dependent signaling pathways (MAPK, Akt/GSK3β). It also preserved mitochondrial membrane potential, prevented mPTP opening, and restored antioxidant response element (ARE) activity, leading to the upregulation of antioxidant genes. These results indicate that lycopene protected against cardiac hypertrophy by reducing oxidative stress and boosting antioxidant defenses [59].

Anti-inflammatory Effects

Nutraceuticals with anti-inflammatory properties play a significant role in mitigating cardiovascular disease (CVD) by targeting chronic inflammation, a key contributor to cardiovascular pathologies. Compounds such as omega-3 fatty acids, polyphenols (e.g., resveratrol), and curcumin exhibit anti-inflammatory effects by modulating inflammatory pathways, including the inhibition of nuclear factor-kappa B (NF-κB) [60] and reduction of pro-inflammatory cytokines like TNF-α and IL-6 [61].

Curcumin, found in turmeric, has potent anti-inflammatory effects [62]. It's studied for preventing and managing the cardiovascular disease by inhibiting the

NF-κB signaling pathway and reducing expression of pro-inflammatory cytokines like TNF-α and IL-6, which lowers systemic inflammation and helps prevent CVD progression [63]. A randomized placebo-controlled study evaluated turmeric extract's effects on healthy volunteers during a 30-min exercise. Participants ($n = 22$) received either turmeric or placebo supplements. The turmeric group showed blunted BP response, reduced final BP and pulse wave velocity (PWV), increased urinary antioxidant power and polyphenol levels, and improved exercise performance. The placebo group showed no significant changes. The study concluded that turmeric concentrate reduced BP, enhanced antioxidant and anti-inflammatory status, improved arterial compliance, and boosted exercise performance [64].

Resveratrol, a polyphenolic compound found in grapes, berries, and red wine, has been widely studied as a nutraceutical due to its antioxidant, anti-inflammatory, and cardioprotective properties [65]. It exerted its effects by modulating key signaling pathways involved in oxidative stress, inflammation, and mitochondrial function. A study demonstrated that resveratrol (RES) protected against early doxorubicin (DOX)-induced cardiotoxicity and mitigated late-onset hypertension-induced cardiomyopathy in juvenile mice. Co-treatment with RES normalized blood pressure, restored heart weight, and reduced molecular markers of cardiac injury, such as the Myh7/Myh6 ratio. RES also reduced DOX-induced cardiac fibrosis and suppressed the NLRP3 inflammasome, a key regulator of inflammation, by downregulating inflammatory cytokines and markers like IL-1β, IL-18, and TNF-α. Additionally, RES decreased macrophage infiltration, cardiac inflammation, and systemic inflammatory cytokines, suggesting its protective effect was mediated by attenuating both cardiac and systemic inflammation [66].

Cholesterol Management

Managing cholesterol is crucial for preventing and treating cardiovascular disease. Nutraceuticals like plant sterols, stanols, omega-3 fatty acids, and soluble fibers, along with lifestyle changes and clinical treatments, can help lower LDL-C and improve lipid profiles, reducing the risk of cardiovascular disease.

Berberine, an alkaloid from *Coptis chinensis* and *Phellodendron chinense*, has traditionally treated diarrhea [67] and is now recognized for its cardiovascular benefits, including anti-hyperlipidemia, anti-inflammatory, and vasculoprotective effects. A study of 44 RCTs with 4606 patients found no significant differences in cholesterol levels or Crouse score between berberine alone and routine treatments or statins. Berberine alone significantly reduced the National Institute of Health Stroke Scale (NIHSS) score, hs-CRP, IL-6, TNF-α, and IMT compared to routine therapy. Combined with statins, berberine significantly lowered TC, TG, LDL-C, NIHSS, hs-CRP, TNF-α, IMT, Crouse score, and unstable plaques, but not HDL-C or IL-6. No significant differences in adverse reaction rates were observed. The study suggested berberine as a promising CVD treatment but called for higher-quality RCTs for more conclusive evidence [68].

Phytosterols (PS) are another class of natural compounds derived from plants that share a similar structure to cholesterol [69]. These compounds are renowned for their ability to lower cholesterol levels. Phytosterols effectively reduce low-density lipoprotein cholesterol (LDL-C) by competing with dietary cholesterol for absorption in the intestines, reducing cholesterol uptake by 30–50% [70]. This, in turn, helps to prevent the development of atherosclerosis and reduces the risk of cardiovascular disease (CVD). A recent meta-analysis, based on 125 studies, evaluated the impact of phytosterol (PS)-fortified foods on low-density lipoprotein cholesterol (LDL-C) reduction. The results showed that PS consumption lowered LDL-C by an average of 0.55 mmol/L, with higher PS doses leading to greater reductions. The food format influenced outcomes, with "butter, margarine, spreads" resulting in a larger LDL-C decrease compared to "bread, biscuits, cereals." No significant effects were observed for other factors such as treatment duration or concomitant statin use. Overall, PS-fortified foods were found to be effective in reducing LDL-C levels [71].

Blood Pressure Regulation

Nutraceuticals significantly impact blood pressure regulation and protect against cardiovascular toxicity. Omega-3 fatty acids, Coenzyme Q10, garlic extract, and polyphenols from green tea and dark chocolate support vascular health by enhancing endothelial function, reducing oxidative stress, and modulating inflammation (Fig. 38.4).

Sour tea a member of the Malvaceae family, is widely recognized in traditional medicine for its cardioprotective properties and is extensively utilized in the food industry as a beverage [72]. Abundant in polyphenols, anthocyanins, flavonoids, and other bioactive compounds, *Hibiscus sabdariffa* has been the subject of comprehensive research due to its anti-hyperlipidemic, hypotensive, anti-inflammatory, and antioxidant effects [73, 74]. A recent study analyzed the antihypertensive efficacy and toxicity of *H. sabdariffa L.* extract in animal models. The results indicated that a 250 mg/kg body weight dose notably reduced systolic and diastolic blood pressure. Ethyl acetate fractions were more effective than n-hexane and aqueous fractions. The extract's LD_{50} was 8.75 g/kg body weight for male rats and 7.5 g/kg body weight for female rats, classifying it as slightly toxic. Sub-chronic toxicity was observed, particularly in female rats. The study suggested therapeutic potential but advised cautious dosing due to its toxicity profile [75].

Another study aimed to determine the best method for extracting antihypertensive metabolites from *H. sabdariffa* and to confirm their in vivo efficacy. Dried calyces were extracted with cold and hot water, and the extracts were compared for in vitro renin and ACE inhibition activities, followed by testing in L-NAME–induced hypertensive rats. The results showed that both extracts significantly reduced angiotensin II, ACE, and aldosterone levels while improving nitric oxide regulation, with

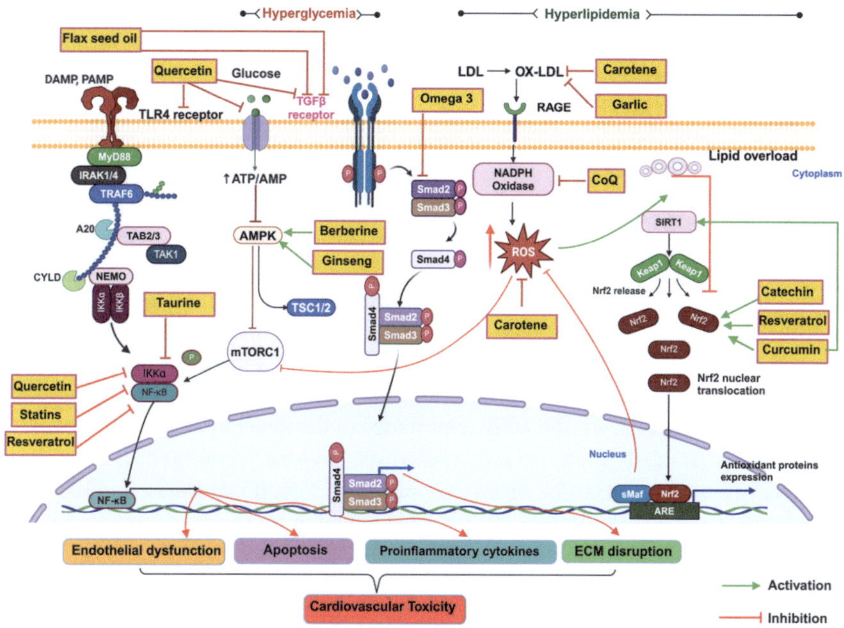

Fig. 38.4 Involvement of nutraceuticals in modulating signaling pathways associated with cardiovascular toxicity

the hot extract demonstrating superior antihypertensive and cardioprotective effects. Metabolomics analysis revealed that the hot extract contained higher levels of specific phenolics, likely responsible for its enhanced therapeutic properties [76].

Cocoa flavonoids, particularly those found in dark chocolate, have been extensively researched for their capacity to augment nitric oxide bioavailability, safeguard the vascular endothelium, and mitigate cardiovascular risk factors [77]. A recent Mendelian randomization study using genome-wide association data found that dark chocolate intake is causally linked to a reduced risk of essential hypertension (OR = 0.73; 95% CI 0.60–0.88; $p = 1.06 \times 10^{-3}$) and suggests a lower risk of venous thromboembolism (OR = 0.69; 95% CI 0.50–0.96; $p = 2.81 \times 10^{-2}$). No significant associations with other cardiovascular diseases were observed, highlighting dark chocolate's potential role in preventing essential hypertension [78].

Phytosterols and stanols, which are natural compounds found in plants and have a similar structure to cholesterol [70], have demonstrated promising potential in the management of high blood pressure and the reduction of the risk of cardiovascular disease (CVD). Recent studies indicated that phytosterols and stanols may regulate blood pressure by improving endothelial function and reducing oxidative stress. A systematic review and meta-analysis of 19 RCTs up to May 2019 found that phytosterol supplementation significantly reduced systolic (WMD: −1.55 mmHg) and diastolic blood pressure (WMD: −0.84 mmHg). The analysis revealed a nonlinear dose-response, with significant effects on systolic BP at doses

≥2000 mg and diastolic BP at doses <2000 mg, especially in interventions lasting less than 12 weeks. These results suggest that phytosterols could be a useful adjuvant therapy for hypertensive patients and a preventive option for prehypertensive and normotensive individuals, though further research is needed [79].

Improved Endothelial Function

Maintaining cardiovascular health relies heavily on the optimal function of the endothelium, which regulates blood flow, vascular tone, and overall cardiovascular balance [80]. Endothelial dysfunction, characterized by a decrease in the endothelium's ability to control vasodilation and an increase in pro-inflammatory and pro-thrombotic factors, significantly contributes to the development of cardiovascular diseases (CVDs) such as atherosclerosis, hypertension, and heart failure [81]. Nutraceuticals, including dietary supplements, functional foods, and bioactive compounds from natural sources, have gained attention for their potential to enhance endothelial function and mitigate the risk of CVDs.

Flavonoids, present in dark chocolate, green tea, and berries, are recognized for their ability to increase nitric oxide (NO) bioavailability [82], which is crucial for promoting vasodilation and improving endothelial function. A recent study explored the impact of acute cocoa flavanol intake on stress-induced vascular changes in 30 healthy men. Participants consumed a high-flavanol (150 mg) or low-flavanol (<4 mg) cocoa beverage before an 8-min mental stress task. The high-flavanol cocoa mitigated stress-induced impairments in endothelial function (FMD) and increased forearm blood flow (FBF) at rest and during stress. Despite similar cardiovascular and blood pressure responses in both groups, high-flavanol cocoa was more effective at counteracting stress-related vascular dysfunction, suggesting its potential for protecting vascular health under stress [83].

Plant-derived compounds like phytosterols (PS) and stanols improve endothelial function by reducing LDL cholesterol levels and thereby decreasing the risk of atherosclerosis. A study investigated the impact of plant sterols (PS) on atherosclerosis biomarkers independently of LDL-c changes. Thirty-eight volunteers with high LDL-c consumed either soy milk or soy milk plus PS (1.6 g/day) for 4 weeks. PS reduced endothelin-1 by 11% ($p = 0.02$) without affecting LDL-c levels. It also lowered total cholesterol (−5.5%), LDL-c (−6.4%), triglycerides (−8.3%), and apo B (−5.3%), while HDL-c remained unchanged. No significant changes were observed in fibrinogen, IL-6, hs-CRP, SAA, TNFα, or VCAM-1. The study suggests that PS can improve endothelial function and may aid in cardiovascular disease prevention independent of LDL-c reductions [84].

Furthermore, studies have linked vitamins and minerals such as vitamin D, vitamin C, and magnesium to endothelial health, suggesting that deficiencies in these nutrients may contribute to endothelial dysfunction. A study assessed whether oral vitamin C could reverse vascular dysfunction during acute inflammation in young (24 ± 4 years) and older adults (64 ± 5 years). Following 2 g of vitamin C intake, significant increases in interleukin-6 and C-reactive protein were observed 24 h after inflammation was induced by a typhoid vaccine. Vitamin C did not affect flow-mediated dilation (FMD) or pulse wave velocity (PWV) at rest, but restored FMD to baseline levels during inflammation ($\Delta + 1.0 \pm 1.8\%$, $p < 0.01$) without altering inflammatory markers or PWV. The study concluded that vitamin C can effectively restore endothelial function during acute inflammation [85].

Support for Heart Failure and Arrhythmias

Heart failure and arrhythmias are major cardiovascular issues affecting public health, with heart failure involving inadequate pumping and arrhythmias causing irregular rhythms that can lead to severe complications like stroke and sudden cardiac death. As traditional treatments continue to advance, there is growing interest in the potential of nutraceuticals—naturally occurring compounds that offer health benefits—to support heart health and manage these conditions.

Coenzyme Q10 (CoQ10) is an antioxidant essential for cellular energy production and has shown potential in improving heart failure symptoms and exercise capacity. A recent study investigated CoQ10 content and biosynthesis in two transgenic mouse models of cardiomyopathy, Mst1-TG and β2AR-TG. High-performance liquid chromatography and RNA sequencing revealed reduced cardiac CoQ10 levels and downregulated biosynthesis genes in both models. In Mst1-TG mice with a 70% reduction in cardiac CoQ10, oral CoQ10 treatment failed to increase levels, whereas injections restored CoQ10 in a time-dependent manner. However, this restoration did not improve mitochondrial dysfunction, oxidative stress, or cardiac function. Thus, while CoQ10 injections restored cardiac levels, they did not enhance functional outcomes in Mst1-TG mice [86].

In addition, the effectiveness of L-carnitine (L-C) in chronic heart failure (CHF) patients was assessed through a meta-analysis of randomized controlled trials (RCTs). A review of 17 RCTs involving 1625 CHF patients showed that L-C treatment significantly improved overall efficacy, left ventricular ejection fraction (LVEF), stroke volume (SV), cardiac output (CO), and E/A ratio. It also notably reduced serum levels of brain natriuretic peptide (BNP), N-terminal pro-brain natriuretic peptide (NT-proBNP), left ventricular end-systolic dimension (LVESD), left ventricular end-diastolic dimension (LVEDD), and left ventricular end-systolic volume (LVESV). However, L-C did not significantly impact all-cause mortality, 6-minute walk test results, or adverse events. In conclusion, L-C was effective in improving clinical symptoms and cardiac function in CHF patients, with good tolerance [87].

Blood Sugar Control

Maintaining optimal blood sugar levels is vital for reducing cardiovascular disease (CVD) risk, as elevated glucose from diabetes and prediabetes can lead to complications like coronary artery disease, heart failure, and stroke. Nutraceuticals offer potential benefits for managing blood sugar and reducing cardiovascular risk.

A meta-analysis of 41 randomized clinical trials assessed the impact of Alpha-Lipoic Acid (ALA) supplementation on glycemic and inflammatory parameters. The analysis included trials from multiple databases, with quality varying from good to poor. Results indicated that ALA significantly reduced fasting blood sugar (WMD $= -6.57$, $P = 0.016$), hemoglobin A1c (WMD $= -0.35$, $P = 0.004$), tumor necrosis factor alpha (WMD $= -1.57$, $P < 0.05$), interleukin 6 (WMD $= -1.15$, $P < 0.001$), and C-reactive protein (WMD $= -0.31$, $P < 0.001$). However, ALA showed no effect on insulin levels or the homeostatic model assessment of insulin resistance (HOMA-IR). The findings suggest that ALA is effective in improving certain glycemic and inflammatory biomarkers [88].

Another study explored hawthorn polyphenol extract (HPE) effects on insulin resistance, inflammation, and aortic injury in type 2 diabetes (T2D) rats, comparing it to metformin (MF). HPE, identified using HPLC-ESI-MS/MS, was administered at 300 mg/kg for 4 weeks. It significantly improved cholesterol, triglycerides, insulin, and lipopolysaccharide levels, and reduced lesions in muscle, liver, and aorta. HPE lowered inflammatory markers and insulin resistance, while also alleviating aortic injury. HPE's effects were similar to MF, suggesting its potential as an adjunct treatment for T2D [89].

Weight Management

Weight management plays a crucial role in reducing the risk of cardiovascular disease (CVD). Obesity, characterized by a body mass index (BMI) of 30 kg/m^2 or higher, is linked to an increased risk of cardiovascular disease compared to individuals with a normal BMI range of 18.8–24.9 kg/m^2 [90]. Integrating nutraceuticals into weight management approaches can further bolster cardiovascular health. Nutraceuticals, such as omega-3 fatty acids, polyphenols, and sterols, have demonstrated the ability to facilitate weight loss and enhance lipid profiles, reduce blood pressure, and alleviate inflammation.

Polyphenols are a diverse group of phytochemicals commonly found in plant-based foods. They have been widely studied in recent years due to their potential antioxidant and anti-inflammatory properties, as well as their ability to prevent metabolic obesity, type 2 diabetes, and cardiovascular disease [91]. In an 18-month DIRECT-PLUS trial, researchers evaluated the effects of a green-Mediterranean (green-MED) diet, rich in polyphenols and low in red/processed meat, on visceral adipose tissue (VAT) in 294 participants. Randomized into healthy dietary guidelines (HDG), Mediterranean diet (MED), or green-MED groups, all combined with exercise, the green-MED group consumed more polyphenols through green tea, walnuts, and *Wolffia globosa*. MRI results showed the green-MED group had the greatest VAT reduction (−14.1%) compared to MED (−6.0%) and HDG (−4.2%), linked to higher polyphenol intake [92].

Antithrombotic Effect

Antithrombotic effects are extremely important for preventing and managing cardiovascular disease (CVD) as they help to reduce the formation of blood clots that can block arteries and lead to heart attacks or strokes [93]. Nutraceuticals like omega-3 fatty acids, garlic, and polyphenols have been found to possess antithrombotic properties by inhibiting platelet aggregation, decreasing blood viscosity, and regulating the production of clotting factors.

Omega-3 fatty acids are beneficial in maintaining a healthy balance between clot formation and dissolution by increasing prostacyclin production and reducing thromboxane A2. A previous study on 15 atherosclerotic patients undergoing coronary bypass surgery found that preoperative supplementation with 3 g/day of eicosapentaenoic acid and 1.3 g/day of docosahexaenoic acid for 28 days resulted in reduced platelet aggregation, decreased thromboxane B2 levels, and a 40% increase in bleeding times. While serum triglycerides decreased by 50% without altering total cholesterol, prostacyclin (PGI2) production was significantly higher in the supplemented patients. Despite these changes, perioperative blood loss was not increased, indicating that omega-3 PUFAs exert antithrombotic effects through enhanced PGI2 production and platelet activity inhibition [94].

Garlic has also been observed to decrease platelet stickiness and improve fibrinolysis, further contributing to their antithrombotic activity [95]. The literature indicates that odorless garlic powder exhibits antithrombotic effects by promoting plasmin generation through tissue-type plasminogen activator (t-PA), without altering the secretion of t-PA or its inhibitor from endothelial cells. In a rat model, garlic notably decreased elevated thrombin–antithrombin III (TAT) complex levels, reflecting reduced coagulation. This reduction was associated with lower prolonged coagulation parameters compared to controls. These findings suggest that odorless garlic enhances fibrinolytic activity and suppresses coagulation, which may aid in preventing pathological thrombus formation in cardiovascular disorders [96].

Risks and Safety Considerations in Nutraceuticals for Cardiovascular Toxicity

Nutraceuticals can be beneficial for cardiovascular health but may pose risks of toxicity, particularly when consumed in excessive amounts. Overconsumption of certain nutraceuticals, such as fat-soluble vitamins like Vitamin E, can lead to adverse effects including organ damage and increased bleeding risk due to their anticoagulant properties. For example, a 31-year-old Indian man developed coagulopathy with oral bleeding, bruising, and gastroduodenal erosions after taking high-dose vitamin E and NSAIDs. Tests revealed vitamin K deficiency despite normal phylloquinone levels. Diagnosed with vitamin E toxicity-induced coagulopathy, he was treated with pantoprazole, vitamin K, and fresh frozen plasma, and his vitamin E supplementation was discontinued. This case underscores that even marginally elevated vitamin E levels can significantly increase bleeding risk, particularly when combined with other medications [97].

Similarly, high doses of calcium supplements have been associated with kidney stones and cardiovascular issues, such as arterial calcification, particularly when used beyond recommended daily allowances or without medical supervision. A study examining the effects of oral calcium supplementation on coronary artery calcification, using data from 9 randomized trials involving 447 patients receiving supplements and 4700 not receiving them, found no significant difference in atheroma volume changes (−0.02% vs. 0.02%, $p = 0.44$). However, calcium supplementation significantly increased the annualized calcium index (odds ratio: 1.15; $p = 0.004$), suggesting that it may raise calcium deposition in coronary vessels without affecting atheroma volume. This highlights the need for further research on the impact of calcium supplementation on plaque stability and cardiovascular outcomes [98].

Drug-nutrient interactions pose significant risks, as nutraceuticals can interfere with the absorption, metabolism, and efficacy of prescription medications, potentially diminishing therapeutic effectiveness or enhancing side effects. For instance, a study reviewed thirty-eight herbs for their interaction with warfarin, identifying four with highly probable interactions (Cranberry, Soya, St John's Wort, and Danshen) and three with probable interactions (coenzyme Q10, Chinese Angelica, and Ginger). Additionally, ten herbs were categorized as having possible interactions, while twenty-one were deemed doubtful [99]. The study strongly recommended avoiding concurrent use of herbs with highly probable and probable interactions with warfarin and suggested careful management and consultation with healthcare providers to mitigate adverse outcomes and ensure safe use of nutraceuticals [99].

Developing and following evidence-based guidelines for the use of nutraceuticals in cardiovascular health is crucial to maximize their potential benefits while minimizing risks. Nutraceuticals such as omega-3 fatty acids [100], epigallocatechin gallate (EGCG) [101] have been linked to cardiovascular health benefits, but their improper use can lead to adverse effects. Therefore, implementing standardized guidelines is vital for safe and effective usage.

Maintaining the quality of nutraceutical products varies significantly among manufacturers, making standardization a key factor in their safe use. It is preferable to use clinically tested and standardized formulations to ensure the active ingredients are present in the appropriate concentrations. High-quality, certified nutraceuticals reduce the risk of contaminants, impurities, or inaccurate dosages, which could otherwise compromise safety and efficacy. Without these standards, the variability in product quality may lead to unpredictable effects, especially when taken alongside cardiovascular medications.

The use of nutraceuticals should be tailored to individual patients, taking into account their medical history, current medications, and specific cardiovascular risk factors. Personalized nutraceutical regimens are essential, as individuals metabolize these products differently, and their interaction with prescribed cardiovascular medications can vary [102]. A personalized approach helps optimize treatment and avoid complications that might arise from generic recommendations.

Regular monitoring is critical for patients using nutraceuticals, particularly those with cardiovascular conditions. Healthcare providers should schedule periodic follow-ups to assess biomarkers such as cholesterol, lipid profiles, and other relevant cardiovascular indicators. Additionally, patients should be monitored for potential interactions between nutraceuticals and prescribed medications. This is particularly important for long-term therapy, as drug-nutrient interactions can evolve over time. Continuous monitoring allows for timely adjustments in therapy, ensuring the safe integration of nutraceuticals into cardiovascular care.

By following these evidence-based guidelines, nutraceuticals can be used safely and effectively to support cardiovascular health.

Conclusion

The inclusion of nutraceuticals and functional foods in the diet has been shown to provide beneficial effects not only on lipid metabolism but also on overall CVT. These dietary components can play a critical role in modulating cholesterol levels, improving endothelial function, and reducing inflammation—key factors in the development and progression of CVT. When used in conjunction with lifestyle modifications, such as increased physical activity and dietary changes, as well as standard pharmacological treatments, these interventions could enhance patient outcomes by addressing multiple pathways involved in CVT. Nonetheless, while current evidence highlights the potential of these approaches, further research is essential to establish definitive conclusions regarding their efficacy, optimal dosages, and long-term safety. This will help refine their role in clinical practice and support their integration into comprehensive cardiovascular care strategies.

Research Gaps

The lack of standardized dosages for nutraceuticals presents a challenge in establishing optimal therapeutic doses for cardiovascular protection, as studies employ varying amounts. While nutraceuticals are generally considered safe, further investigation is needed to assess their long-term effects and safety, particularly when used concurrently with conventional medications or over extended periods. Despite evidence supporting the cardiovascular benefits of nutraceuticals, the precise molecular mechanisms through which they exert their effects remain unclear. Moreover, existing research has predominantly focused on specific populations, such as cancer patients experiencing chemotherapy-induced cardiovascular toxicity, emphasizing the necessity for studies encompassing diverse populations with differing dietary habits, environmental exposures, and health conditions. The nutraceutical market encounters substantial regulatory and quality control challenges, manifested in inconsistent product quality, bioavailability, and efficacy. Addressing these issues is paramount to facilitate the clinical application of research findings.

Future Prospects

Advancements in nutrigenomics may soon enable personalized nutraceutical therapies tailored to an individual's genetic predisposition to cardiovascular diseases (CVDs) or specific metabolic and inflammatory profiles. This personalized approach could be further enhanced by integrating nutraceuticals with pharmacological treatments to explore potential synergistic effects in CVD prevention and management. However, large-scale, randomized clinical trials with standardized dosages and formulations are necessary to establish clear guidelines for the effective use of nutraceuticals in cardiovascular health. Additionally, the development of functional foods enriched with bioactive compounds targeting cardiovascular protection could make nutraceuticals more accessible in daily diets.

Future research should also prioritize mechanistic studies to better understand how nutraceuticals modulate cardiovascular toxicity at the molecular and cellular levels, which could lead to more targeted therapies. Public health strategies may benefit from incorporating nutraceuticals into dietary guidelines, with policymakers and healthcare providers playing a key role in promoting their use for CVD prevention. Moreover, investigating individual physiological responses to nutraceuticals remains a promising area of research. Although many nutraceuticals demonstrate cardioprotective properties, there is still a significant gap in understanding the molecular pathways they affect.

Metabolomics, particularly nutrametabolomics, offers a valuable approach to studying how phytochemical intake influences metabolic profiles and contributes to observed health benefits. Additionally, bioavailability and tissue distribution of these compounds after ingestion are critical areas for further study. However, data on the absorption and bioavailability of various nutraceuticals are limited, and clarifying their pharmacokinetics is essential to determine effective dosages for targeted therapeutic action.

Future research must also focus on clinical studies to address key questions regarding nutraceutical interventions, including identifying the appropriate target populations, optimal dosages, and timing for initiating or discontinuing treatment. A translational medicine approach is needed to investigate the underlying physiological mechanisms and therapeutic protocols for both single and combined nutraceutical therapies, as well as their integration with conventional medications, to maximize their potential in cardiovascular care. By addressing these research gaps and exploring future opportunities, the potential of nutraceuticals to mitigate cardiovascular toxicity and enhance cardiovascular health could be fully realized.

Key Points

- CVDs remain one of the leading causes of morbidity and mortality globally, often driven by oxidative stress, inflammation, and endothelial dysfunction. Addressing these mechanisms is crucial for reducing the burden of CVDs.
- Bioactive compounds like omega-3 fatty acids, polyphenols, phytosterols, Coenzyme Q10, and vitamins have demonstrated cardioprotective effects. These compounds help modulate oxidative stress, inflammation, lipid metabolism, and endothelial function.
- Nutraceuticals are believed to offer protective cardiovascular effects by improving lipid metabolism, reducing oxidative damage, modulating inflammatory pathways, and enhancing endothelial function. These mechanisms collectively lower the risk of CVTs.
- Research supports the efficacy of nutraceuticals in reducing CVT risk factors such as hypertension, hyperlipidemia, and atherosclerosis. Studies highlight their role in complementing pharmacological therapies and lifestyle modifications.
- Nutraceuticals hold potential as adjunct therapies alongside conventional pharmacological treatments, contributing to improved outcomes in managing cardiovascular toxicity and reducing the risk of chronic cardiovascular conditions.

Funding The research work of DKA is supported by R25AI179582 and R01 HL147662 grants from the National Institutes of Health, USA. The contents of this chapter are solely the responsibility of the authors and do not necessarily represent the official views of the National Institutes of Health.

Competing Interests Both authors have read the chapter and declare no conflict of interest. No writing assistance was utilized in the production of this manuscript.

Consent for Publication Both authors have read the chapter and consented for publication.

References

1. Mladěnka P, Applová L, Patočka J et al (2018) Comprehensive review of cardiovascular toxicity of drugs and related agents. Med Res Rev 38:1332–1403. https://doi.org/10.1002/med.21476
2. Phungphong S, Kijtawornrat A, Kampaengsri T et al (2020) Comparison of exercise training and estrogen supplementation on mast cell-mediated doxorubicin-induced cardiotoxicity. Am J Phys Regul Integr Comp Phys 318:R829–R842. https://doi.org/10.1152/ajpregu.00224.2019
3. Krikez I, Benmalek R, Bennouna G et al (2020) Reversibility of trastuzumab induced cardiotoxicity in breast cancer patients: a prospective observational study. Arch Cardiovasc Dis Suppl 12:42. https://doi.org/10.1016/j.acvdsp.2019.09.086
4. D'Oria R, Schipani R, Leonardini A et al (2020) The role of oxidative stress in cardiac disease: from physiological response to injury factor. Oxidative Med Cell Longev 2020:5732956. https://doi.org/10.1155/2020/5732956
5. Peoples JN, Saraf A, Ghazal N et al (2019) Mitochondrial dysfunction and oxidative stress in heart disease. Exp Mol Med 51:1–13. https://doi.org/10.1038/s12276-019-0355-7
6. Wang H, Wang Y, Li J et al (2023) Three tyrosine kinase inhibitors cause cardiotoxicity by inducing endoplasmic reticulum stress and inflammation in cardiomyocytes. BMC Med 21: 147. https://doi.org/10.1186/s12916-023-02838-2
7. Varga Z, Sabzwari SRA, Vargova V (2017) Cardiovascular risk of nonsteroidal anti-inflammatory drugs: an under-recognized public health issue. Cureus 9:e1144. https://doi.org/10.7759/cureus.1144
8. Vinci P, Panizon E, Tosoni LM et al (2021) Statin-associated myopathy: emphasis on mechanisms and targeted therapy. Int J Mol Sci 22:11687. https://doi.org/10.3390/ijms222111687
9. Bhatnagar A (2022) Cardiovascular effects of particulate air pollution. Annu Rev Med 73: 393–406. https://doi.org/10.1146/annurev-med-042220-011549
10. Yang A-M, Lo K, Zheng T-Z et al (2020) Environmental heavy metals and cardiovascular diseases: status and future direction. Chronic Dis Transl Med 6:251–259. https://doi.org/10.1016/j.cdtm.2020.02.005
11. Gallucci G, Tartarone A, Lerose R et al (2020) Cardiovascular risk of smoking and benefits of smoking cessation. J Thorac Dis 12:3866–3876. https://doi.org/10.21037/jtd.2020.02.47
12. Juul F, Vaidean G, Parekh N (2021) Ultra-processed foods and cardiovascular diseases: potential mechanisms of action. Adv Nutr 12:1673–1680. https://doi.org/10.1093/advances/nmab049
13. Domínguez F, Adler E, García-Pavía P (2024) Alcoholic cardiomyopathy: an update. Eur Heart J 45:2294–2305. https://doi.org/10.1093/eurheartj/ehae362
14. Liang Z, Zhang M, Wang C et al (2022) Association between sedentary behavior, physical activity, and cardiovascular disease-related outcomes in adults—a meta-analysis and systematic review. Front Public Health 10:1018460. https://doi.org/10.3389/fpubh.2022.1018460
15. Jomova K, Raptova R, Alomar SY et al (2023) Reactive oxygen species, toxicity, oxidative stress, and antioxidants: chronic diseases and aging. Arch Toxicol 97:2499–2574. https://doi.org/10.1007/s00204-023-03562-9
16. Kong P, Cui Z-Y, Huang X-F et al (2022) Inflammation and atherosclerosis: signaling pathways and therapeutic intervention. Signal Transduct Target Ther 7:1–24. https://doi.org/10.1038/s41392-022-00955-7
17. O'Neill S, O'Driscoll L (2015) Metabolic syndrome: a closer look at the growing epidemic and its associated pathologies. Obes Rev 16:1–12. https://doi.org/10.1111/obr.12229
18. Mortensen MB, Dzaye O, Bøtker HE et al (2023) Low-density lipoprotein cholesterol is predominantly associated with atherosclerotic cardiovascular disease events in patients with evidence of coronary atherosclerosis: the Western Denmark Heart Registry. Circulation 147: 1053–1063. https://doi.org/10.1161/CIRCULATIONAHA.122.061010

19. Funk SD, Yurdagul A, Orr AW (2012) Hyperglycemia and endothelial dysfunction in atherosclerosis: lessons from type 1 diabetes. Int J Vasc Med 2012:569654. https://doi.org/10.1155/2012/569654
20. Poznyak AV, Sadykhov NK, Kartuesov AG et al (2022) Hypertension as a risk factor for atherosclerosis: cardiovascular risk assessment. Front Cardiovasc Med 9:959285. https://doi.org/10.3389/fcvm.2022.959285
21. Siaravas KC, Katsouras CS, Sioka C (2023) Radiation treatment mechanisms of cardiotoxicity: a systematic review. Int J Mol Sci 24:6272. https://doi.org/10.3390/ijms24076272
22. Wei J, Xu H, Liu Y et al (2017) Effect of captopril on radiation-induced TGF-β1 secretion in EA.Hy926 human umbilical vein endothelial cells. Oncotarget 8:20842–20850. https://doi.org/10.18632/oncotarget.15356
23. Sharifi-Rad J, Rayess YE, Rizk AA et al (2020) Turmeric and its major compound curcumin on health: bioactive effects and safety profiles for food, pharmaceutical, biotechnological and medicinal applications. Front Pharmacol 11:01021. https://doi.org/10.3389/fphar.2020.01021
24. Jin L, Schmiech M, El Gaafary M et al (2020) A comparative study on root and bark extracts of *Eleutherococcus senticosus* and their effects on human macrophages. Phytomedicine 68:153181. https://doi.org/10.1016/j.phymed.2020.153181
25. Adeosun WB, Loots DT (2024) Medicinal plants against viral infections: a review of metabolomics evidence for the antiviral properties and potentials in plant sources. Viruses 16:218. https://doi.org/10.3390/v16020218
26. Witkamp RF, van Norren K (2018) Let thy food be thy medicine….when possible. Eur J Pharmacol 836:102–114. https://doi.org/10.1016/j.ejphar.2018.06.026
27. Puri V, Nagpal M, Singh I et al (2022) A comprehensive review on nutraceuticals: therapy support and formulation challenges. Nutrients 14:4637. https://doi.org/10.3390/nu14214637
28. Nasri H, Baradaran A, Shirzad H, Rafieian-Kopaei M (2014) New concepts in nutraceuticals as alternative for pharmaceuticals. Int J Prev Med 5:1487–1499
29. Dama A, Shpati K, Daliu P et al (2024) Targeting metabolic diseases: the role of nutraceuticals in modulating oxidative stress and inflammation. Nutrients 16:507. https://doi.org/10.3390/nu16040507
30. Şimşek H, Uçar A (2024) Polyunsaturated fatty acids as a nutraceutical for age-related neurodegenerative diseases: current knowledge and future directions. Clin Nutr Open Sci 56:65–73. https://doi.org/10.1016/j.nutos.2024.05.012
31. Djaoudene O, Romano A, Bradai YD et al (2023) A global overview of dietary supplements: regulation, market trends, usage during the COVID-19 pandemic, and health effects. Nutrients 15:3320. https://doi.org/10.3390/nu15153320
32. Bischoff-Ferrari HA, Vellas B, Rizzoli R et al (2020) Effect of vitamin D supplementation, omega-3 fatty acid supplementation, or a strength-training exercise program on clinical outcomes in older adults. JAMA 324:1855–1868. https://doi.org/10.1001/jama.2020.16909
33. Temple NJ (2022) A rational definition for functional foods: a perspective. Front Nutr 9:957516. https://doi.org/10.3389/fnut.2022.957516
34. Ballini A, Charitos IA, Cantore S et al (2023) About functional foods: the probiotics and prebiotics state of art. Antibiotics (Basel) 12:635. https://doi.org/10.3390/antibiotics12040635
35. Tur JA, Bibiloni MM (2016) Functional foods. In: Caballero B, Finglas PM, Toldrá F (eds) Encyclopedia of food and health. Academic, Oxford, pp 157–161
36. Cencic A, Chingwaru W (2010) The role of functional foods, nutraceuticals, and food supplements in intestinal health. Nutrients 2:611–625. https://doi.org/10.3390/nu2060611
37. Berry SA, Brown CS, Greene C et al (2020) Medical foods for inborn errors of metabolism: history, current status, and critical need. Pediatrics 145:e20192261. https://doi.org/10.1542/peds.2019-2261
38. Ciampa BP, Reyes Ramos E, Borum M, Doman DB (2017) The emerging therapeutic role of medical foods for gastrointestinal disorders. Gastroenterol Hepatol (N Y) 13:104–115

39. Chandra S, Saklani S, Kumar P et al (2022) Nutraceuticals: pharmacologically active potent dietary supplements. Biomed Res Int 2022:2051017. https://doi.org/10.1155/2022/2051017
40. Markowitz JS, Gurley PC, Gurley BJ (2020) Medical foods—a closer look at the menu: a brief review and commentary. Clin Ther 42:1416–1423. https://doi.org/10.1016/j.clinthera.2020.05.011
41. Kirk B, Iuliano S, Daly RM, Duque G (2020) Effects of protein supplementation on muscle wasting disorders: a brief update of the evidence. Australas J Ageing 39(Suppl 2):3–10. https://doi.org/10.1111/ajag.12853
42. Sabarathinam S (2023) A glycemic diet improves the understanding of glycemic control in diabetes patients during their follow-up. Future Sci OA 9:FSO843. https://doi.org/10.2144/fsoa-2022-0058
43. Kumar A, Nirmal P, Kumar M et al (2023) Major phytochemicals: recent advances in health benefits and extraction method. Molecules 28:887. https://doi.org/10.3390/molecules28020887
44. Salam U, Ullah S, Tang Z-H et al (2023) Plant metabolomics: an overview of the role of primary and secondary metabolites against different environmental stress factors. Life (Basel) 13:706. https://doi.org/10.3390/life13030706
45. Rudrapal M, Khairnar SJ, Khan J et al (2022) Dietary polyphenols and their role in oxidative stress-induced human diseases: insights into protective effects, antioxidant potentials and mechanism(s) of action. Front Pharmacol 13:806470. https://doi.org/10.3389/fphar.2022.806470
46. Murai T, Matsuda S (2023) The chemopreventive effects of chlorogenic acids, phenolic compounds in coffee, against inflammation, cancer, and neurological diseases. Molecules 28:2381. https://doi.org/10.3390/molecules28052381
47. Noce A, Di Daniele F, Campo M et al (2021) Effect of hydrolysable tannins and anthocyanins on recurrent urinary tract infections in nephropathic patients: preliminary data. Nutrients 13:591. https://doi.org/10.3390/nu13020591
48. Borges CV, Orsi RO, Maraschin M, Lima GPP (2023) Chapter 27-Oxidative stress in plants and the biochemical response mechanisms. In: Ghorbanpour M, Adnan Shahid M (eds) Plant stress mitigators. Academic, pp 455–468
49. Mrowicka M, Mrowicki J, Kucharska E, Majsterek I (2022) Lutein and zeaxanthin and their roles in age-related macular degeneration—neurodegenerative disease. Nutrients 14:827. https://doi.org/10.3390/nu14040827
50. Calderón Guzmán D, Juárez Olguín H, Veloz Corona Q et al (2020) Consumption of cooked common beans or saponins could reduce the risk of diabetic complications. Diabetes Metab Syndr Obes 13:3481–3486. https://doi.org/10.2147/DMSO.S270564
51. Masyita A, Mustika Sari R, Dwi Astuti A et al (2022) Terpenes and terpenoids as main bioactive compounds of essential oils, their roles in human health and potential application as natural food preservatives. Food Chem X 13:100217. https://doi.org/10.1016/j.fochx.2022.100217
52. Murthy KNC, Jayaprakasha GK, Safe S, Patil BS (2021) Citrus limonoids induce apoptosis and inhibit the proliferation of pancreatic cancer cells. Food Funct 12:1111–1120. https://doi.org/10.1039/D0FO02740E
53. Ősz B-E, Jîtcă G, Ştefănescu R-E et al (2022) Caffeine and its antioxidant properties—it is all about dose and source. Int J Mol Sci 23:13074. https://doi.org/10.3390/ijms232113074
54. Martínez-Pinilla E, Oñatibia-Astibia A, Franco R (2015) The relevance of theobromine for the beneficial effects of cocoa consumption. Front Pharmacol 6:30. https://doi.org/10.3389/fphar.2015.00030
55. El-Saadony MT, Saad AM, Korma SA et al (2024) Garlic bioactive substances and their therapeutic applications for improving human health: a comprehensive review. Front Immunol 15:1277074. https://doi.org/10.3389/fimmu.2024.1277074
56. Amir M, Vohra M, Raj RG et al (2023) Adaptogenic herbs: a natural way to improve athletic performance. Health Sci Rev 7:100092. https://doi.org/10.1016/j.hsr.2023.100092

57. Mirmiran P, Hosseini-Esfahani F, Esfandiar Z et al (2022) Associations between dietary antioxidant intakes and cardiovascular disease. Sci Rep 12:1504. https://doi.org/10.1038/s41598-022-05632-x
58. Shafe MO, Gumede NM, Nyakudya TT, Chivandi E (2024) Lycopene: a potent antioxidant with multiple health benefits. J Nutr Metab 2024:6252426. https://doi.org/10.1155/2024/6252426
59. Zeng J, Zhao J, Dong B et al (2019) Lycopene protects against pressure overload-induced cardiac hypertrophy by attenuating oxidative stress. J Nutr Biochem 66:70–78. https://doi.org/10.1016/j.jnutbio.2019.01.002
60. Saleh HA, Yousef MH, Abdelnaser A (2021) The anti-inflammatory properties of phytochemicals and their effects on epigenetic mechanisms involved in TLR4/NF-κB-mediated inflammation. Front Immunol 12:606069. https://doi.org/10.3389/fimmu.2021.606069
61. Liu W, Cui X, Zhong Y et al (2023) Phenolic metabolites as therapeutic in inflammation and neoplasms: molecular pathways explaining their efficacy. Pharmacol Res 193:106812. https://doi.org/10.1016/j.phrs.2023.106812
62. Peng Y, Ao M, Dong B et al (2021) Anti-inflammatory effects of curcumin in the inflammatory diseases: status, limitations and countermeasures. Drug Des Dev Ther 15:4503–4525. https://doi.org/10.2147/DDDT.S327378
63. Pourbagher-Shahri AM, Farkhondeh T, Ashrafizadeh M et al (2021) Curcumin and cardiovascular diseases: focus on cellular targets and cascades. Biomed Pharmacother 136:111214. https://doi.org/10.1016/j.biopha.2020.111214
64. Hajleh MNA, Al-Dujaili EAS (2023) Effects of turmeric concentrate on cardiovascular risk factors and exercise-induced oxidative stress in healthy volunteers; an exploratory study. Adv Pharm Bull 13:601–610. https://doi.org/10.34172/apb.2023.052
65. Farhan M, Rizvi A (2023) The pharmacological properties of red grape polyphenol resveratrol: clinical trials and obstacles in drug development. Nutrients 15:4486. https://doi.org/10.3390/nu15204486
66. Maayah ZH, Alam AS, Takahara S et al (2021) Resveratrol reduces cardiac NLRP3-inflammasome activation and systemic inflammation to lessen doxorubicin-induced cardiotoxicity in juvenile mice. FEBS Lett 595:1681–1695. https://doi.org/10.1002/1873-3468.14091
67. Germán-Acacio JM, Meza-Sánchez DE, Morales-Morales D (2020) Chapter 3-Therapeutically relevant natural products as AMPK activators in the treatment of diabetes. In: Atta-ur-Rahman (ed) Studies in natural products chemistry. Elsevier, pp 57–90
68. Yang L, Zhu W, Zhang X et al (2023) Efficacy and safety of berberine for several cardiovascular diseases: a systematic review and meta-analysis of randomized controlled trials. Phytomedicine 112:154716. https://doi.org/10.1016/j.phymed.2023.154716
69. Li X, Xin Y, Mo Y et al (2022) The bioavailability and biological activities of phytosterols as modulators of cholesterol metabolism. Molecules 27:523. https://doi.org/10.3390/molecules27020523
70. Cabral CE, Klein MRST (2017) Phytosterols in the treatment of hypercholesterolemia and prevention of cardiovascular diseases. Arq Bras Cardiol 109:475–482. https://doi.org/10.5935/abc.20170158
71. Fontané L, Pedro-Botet J, Garcia-Ribera S et al (2023) Use of phytosterol-fortified foods to improve LDL cholesterol levels: a systematic review and meta-analysis. Nutr Metab Cardiovasc Dis 33:1472–1480. https://doi.org/10.1016/j.numecd.2023.04.014
72. Kamyab R, Namdar H, Torbati M et al (2021) Pharmaceutical properties of *Hibiscus sabdariffa*, toward an ideal treatment for hypertension. Pharm Sci 28:506–516. https://doi.org/10.34172/PS.2021.68
73. Amer SA, Al-Khalaifah HS, Gouda A et al (2022) Potential effects of anthocyanin-rich roselle (*Hibiscus sabdariffa* L.) extract on the growth, intestinal histomorphology, blood biochemical parameters, and the immune status of broiler chickens. Antioxidants (Basel) 11:544. https://doi.org/10.3390/antiox11030544

74. Maciel LG, do Carmo MAV, Azevedo L et al (2018) *Hibiscus sabdariffa* anthocyanins-rich extract: chemical stability, in vitro antioxidant and antiproliferative activities. Food Chem Toxicol 113:187–197. https://doi.org/10.1016/j.fct.2018.01.053
75. Susilawati Y, Febriyanti RM, Febrina E et al (2023) A comprehensive in vivo study of the antihypertensive properties and toxicity of roselle (*Hibiscus sabdariffa* L.). Pharmacia 70: 1521–1530. https://doi.org/10.3897/pharmacia.70.e109119
76. Salem MA, Ezzat SM, Ahmed KA et al (2022) A comparative study of the antihypertensive and cardioprotective potentials of hot and cold aqueous extracts of *Hibiscus sabdariffa* L. in relation to their metabolic profiles. Front Pharmacol 13:840478. https://doi.org/10.3389/fphar.2022.840478
77. Katz DL, Doughty K, Ali A (2011) Cocoa and chocolate in human health and disease. Antioxid Redox Signal 15:2779–2811. https://doi.org/10.1089/ars.2010.3697
78. Yang J, Zhou J, Yang J et al (2024) Dark chocolate intake and cardiovascular diseases: a Mendelian randomization study. Sci Rep 14:968. https://doi.org/10.1038/s41598-023-50351-6
79. Ghaedi E, Foshati S, Ziaei R et al (2020) Effects of phytosterols supplementation on blood pressure: a systematic review and meta-analysis. Clin Nutr 39:2702–2710. https://doi.org/10.1016/j.clnu.2019.12.020
80. Félétou M (2011) Endothelium-dependent regulation of vascular tone. In: The endothelium: Part 1: multiple functions of the endothelial cells—focus on endothelium-derived vasoactive mediators. Morgan & Claypool Life Sciences
81. Medina-Leyte DJ, Zepeda-García O, Domínguez-Pérez M et al (2021) Endothelial dysfunction, inflammation and coronary artery disease: potential biomarkers and promising therapeutical approaches. Int J Mol Sci 22:3850. https://doi.org/10.3390/ijms22083850
82. Sadler DG, Draijer R, Stewart CE et al (2021) Cocoa-flavanols enhance moderate-intensity pulmonary kinetics but not exercise tolerance in sedentary middle-aged adults. Eur J Appl Physiol 121:2285–2294. https://doi.org/10.1007/s00421-021-04682-9
83. Baynham R, Veldhuijzen van Zanten JJCS, Johns PW et al (2021) Cocoa flavanols improve vascular responses to acute mental stress in young healthy adults. Nutrients 13:1103. https://doi.org/10.3390/nu13041103
84. Oliveira Godoy Ilha A, Sutti Nunes V, Silva Afonso M et al (2020) Phytosterols supplementation reduces endothelin-1 plasma concentration in moderately hypercholesterolemic individuals independently of their cholesterol-lowering properties. Nutrients 12:1507. https://doi.org/10.3390/nu12051507
85. Lefferts EC, Hibner BA, Lefferts WK et al (2021) Oral vitamin C restores endothelial function during acute inflammation in young and older adults. Physiol Rep 9:e15104. https://doi.org/10.14814/phy2.15104
86. Pu T-T, Wu W, Liang P-D et al (2023) Evaluation of Coenzyme Q10 (CoQ10) deficiency and therapy in mouse models of cardiomyopathy. J Cardiovasc Pharmacol 81:259–269. https://doi.org/10.1097/FJC.0000000000001401
87. Song X, Qu H, Yang Z et al (2017) Efficacy and safety of L-carnitine treatment for chronic heart failure: a meta-analysis of randomized controlled trials. Biomed Res Int 2017:6274854. https://doi.org/10.1155/2017/6274854
88. Rahimlou M, Asadi M, Banaei Jahromi N, Mansoori A (2019) Alpha-lipoic acid (ALA) supplementation effect on glycemic and inflammatory biomarkers: a systematic review and meta-analysis. Clin Nutr ESPEN 32:16–28. https://doi.org/10.1016/j.clnesp.2019.03.015
89. Liu S, Yu J, Fu M et al (2021) Regulatory effects of hawthorn polyphenols on hyperglycemic, inflammatory, insulin resistance responses, and alleviation of aortic injury in type 2 diabetic rats. Food Res Int 142:110239. https://doi.org/10.1016/j.foodres.2021.110239
90. Flegal KM, Graubard BI, Williamson DF, Gail MH (2005) Excess deaths associated with underweight, overweight, and obesity. JAMA 293:1861–1867. https://doi.org/10.1001/jama.293.15.1861

91. Zhang Z, Li X, Sang S et al (2022) Polyphenols as plant-based nutraceuticals: health effects, encapsulation, nano-delivery, and application. Food Secur 11:2189. https://doi.org/10.3390/foods11152189
92. Zelicha H, Kloting N, Kaplan A et al (2022) The effect of high-polyphenol Mediterranean diet on visceral adiposity: the DIRECT PLUS randomized controlled trial. BMC Med 20:327. https://doi.org/10.1186/s12916-022-02525-8
93. Barriuso I, Worner F, Vilahur G (2022) Novel antithrombotic agents in ischemic cardiovascular disease: progress in the search for the optimal treatment. J Cardiovasc Dev Dis 9:397. https://doi.org/10.3390/jcdd9110397
94. DeCaterina R, Giannessi D, Mazzone A et al (1990) Vascular prostacyclin is increased in patients ingesting omega-3 polyunsaturated fatty acids before coronary artery bypass graft surgery. Circulation 82:428–438. https://doi.org/10.1161/01.cir.82.2.428
95. El-Sabban F (2009) Garlic as an antithrombotic and antiplatelet aggregation agent. J Chin Clin Med 4:288–294
96. Fukao H, Yoshida H, Tazawa Y, Hada T (2007) Antithrombotic effects of odorless garlic powder both in vitro and in vivo. Biosci Biotechnol Biochem 71:84–90. https://doi.org/10.1271/bbb.60380
97. Abrol R, Kaushik R, Goel D et al (2023) Vitamin E-induced coagulopathy in a young patient: a case report. J Med Case Rep 17:107. https://doi.org/10.1186/s13256-023-03827-y
98. Bazarbashi N, Kapadia SR, Nicholls SJ et al (2021) Oral calcium supplements associate with serial coronary calcification. JACC Cardiovasc Imaging 14:259–268. https://doi.org/10.1016/j.jcmg.2020.06.030
99. Ge B, Zhang Z, Zuo Z (2014) Updates on the clinical evidenced herb–warfarin interactions. Evid Based Complement Alternat Med 2014:957362. https://doi.org/10.1155/2014/957362
100. Gross BW, Gillio M, Rinehart CD et al (2017) Omega-3 fatty acid supplementation and warfarin: a lethal combination in traumatic brain injury. J Trauma Nurs 24:15–18. https://doi.org/10.1097/JTN.0000000000000256
101. Rasheed NOA, Ahmed LA, Abdallah DM, El-Sayeh BM (2018) Paradoxical cardiotoxicity of intraperitoneally-injected epigallocatechin gallate preparation in diabetic mice. Sci Rep 8:7880. https://doi.org/10.1038/s41598-018-25901-y
102. de Hoogh IM, Winters BL, Nieman KM et al (2021) A novel personalized systems nutrition program improves dietary patterns, lifestyle behaviors and health-related outcomes: results from the Habit Study. Nutrients 13:1763. https://doi.org/10.3390/nu13061763

Chapter 39
Role of Tadalafil in Mitigating Cardiotoxicity Induced by Cancer Chemotherapy

Saisudha Koka, Inavolu Sriram Sandeep, and Rakesh C. Kukreja

Abstract Phosphodiesterase-5 (PDE5) inhibitors are drugs which improve the bioavailability of cyclic guanosine monophosphate (cGMP) by inhibiting its degradation by the PDE5 enzyme. Sildenafil, tadalafil and vardanefil are PDE5 inhibitors widely used for erectile dysfunction. In recent years, the clinical application of PDE5 inhibitors have expanded beyond their usage as erectile dysfunction drugs. Currently sildenafil and tadalafil are approved for the treatment of pulmonary hypertension. Mounting scientific evidence indicates beneficial use of PDE5 inhibitors in a plethora of diseases including ischemia/reperfusion injury, myocardial infarction, cardiac hypertrophy, cardiomyopathy, heart failure, benign prostatic hyperplasia, stroke and neurodegenerative diseases. We have demonstrated that PDE5 inhibitors, such as sildenafil and tadalafil, offer significant cardioprotection against ischemia-reperfusion injury, myocardial infarction, diabetic cardiomyopathy, metabolic syndrome and doxorubicin (DOX)-induced cardiomyopathy. Chemotherapy-induced cardiotoxicity is a significant concern in cancer treatment, often limiting the use of effective chemotherapeutic agents such as anthracycline antibiotics which are one of the most effective classes of chemotherapeutic agents. An anthracycline analogue, DOX is a widely used anticancer drug; however, its clinical use is limited owing to its cardiotoxicity. In this chapter, we review the critical role of long acting PDE5 inhibitor, tadalafil, in attenuating cardiotoxicity caused by chemotherapeutic agents. In particular, we will explore the key mechanisms underlying DOX-induced cardiomyopathy including oxidative stress, cardiomyocyte apoptosis and mitochondrial damage in the myocardium which ultimately contribute to heart failure. Additionally, we will highlight potential signaling pathways such as cGMP-PKG pathway and antioxidant mechanisms through which tadalafil may exert protective effects

S. Koka · I. S. Sandeep
Department of Pharmaceutical Sciences, Irma Lerma Rangel College of Pharmacy, Texas A&M University, Kingsville, TX, USA

R. C. Kukreja (✉)
Division of Cardiology, Department of Internal Medicine, Pauley Heart Center, Virginia Commonwealth University Medical Center, Richmond, VA, USA
e-mail: rakesh.kukreja@vcuhealth.org

D. K. Agrawal et al. (eds.), *Cardiovascular Toxicity*, Advances in Biochemistry in Health and Disease 33, https://doi.org/10.1007/978-3-031-90598-8_39

against DOX-induced cardiomyopathy and cardiotoxicity by other chemotherapeutic agents. The use of PDE5 inhibitors like tadalafil as a cardioprotective agent implicates a promising avenue for improving the cardiovascular outcomes of cancer patients undergoing chemotherapy. Future clinical trials will be essential for clarifying the role of tadalafil in cardioprotective strategies for cancer patients and for establishing its long-term safety and efficacy in this population.

Keywords Phosphodiesterase-5 inhibitors · Tadalafil · Doxorubicin · ROS · Protein kinase G · Mitochondrial dysfunction · Apoptosis

Introduction

Anthracycline antibiotics are among the most effective chemotherapeutic agents ever developed and have been widely used to treat a variety of cancers including breast, leukemia, lymphomas, ovarian, uterine, bladder and lung cancers. Some of the most clinically important anthracycline analogues include doxorubicin (DOX), daunorubicin, epirubicin, pirarubicin and idarubicin which are originally derived from *Streptomyces peucetius* bacterium [1, 2]. Despite their effectiveness against tumors, these drugs are known to induce cardiotoxicity, which varies based on dosage, administration method, patient age, cardiovascular history, previous cytotoxic and other associated treatments [3]. Doxorubicin (DOX), a prominent drug in this class sold under the brand name Adriamycin and has been used since the late 1960s to treat a range of cancers, including acute leukemias, lymphomas, cancers of the stomach, breast, and ovaries, as well as AIDS-related Kaposi's sarcoma and bone tumors [4, 5]. DOX is often used together with other chemotherapeutic agents. Adriamycin, bleomycin, vinblastine, dacarbazine (ABVD), taxotere, adriamycin, cyclophosphamide (TAC) and 5-fluorouracil, adriamycin, cyclophosphamide (FAC) are the most commonly used DOX-containing regimens utilized in the treatment of cancer patients. Despite its wide clinical utility cardiotoxicity remains the major side effect of DOX leading to cardiomyopathy and heart failure [6].

DOX-induced cardiomyopathy is dependent on the total cumulative dose [7] with incidence rates ranging from 4 to 36% as the doses exceed from 500 to 600 mg/m^2 [2, 6, 8]. DOX-induced cardiomyopathy typically results in irreversible dilated cardiomyopathy leading to both systolic and diastolic dysfunction which ultimately results in heart failure causing high mortality [8]. DOX-induced cardiotoxicity can manifest shortly after treatment or remain latent for years [9, 10]. In adults, cardiotoxic effects usually appear as congestive heart failure within weeks or months after treatment, while children are at higher risk of developing cardiomyopathy at any given cumulative dose [11]. Recent evidence shows that although lower doses can be effective, heart failure may remain undetected for up to 20 years after treatment [12] making long-term cardiac monitoring essential for patients who have received this potent chemotherapeutic agent [13, 14]. Understanding and mitigating the cardiotoxic effects of DOX remains a critical area of research to enhance the safety and effectiveness of this cancer therapy.

Over the past few decades, numerous efforts have been made to develop effective strategies to protect against DOX-induced cardiotoxicity. Antioxidants and free radical scavengers like probucol, amifostine, and dexrazoxane have shown some protective effects, underscoring the role of reactive oxygen species (ROS) in DOX-induced cardiotoxicity [15, 16]. However, these agents have significant clinical drawbacks, such as a marked decline in high-density lipoprotein levels, inability to prevent DOX-induced mortality, weight loss, and the potentiation of DOX-induced myelosuppression [17]. Other therapeutic strategies, including the use of β-blockers and renin–angiotensin system inhibitors, have been employed to reduce DOX-induced cardiotoxicity in its early stages. Nevertheless, these approaches are also limited by their notable side effects and disadvantages. The development of anthracycline analogues and alternative drug delivery methods, such as liposomal and nanosomal encapsulated DOX, are promising approaches aimed at improving antitumor efficacy while mitigating the toxic effects of DOX [18, 19]. Despite these advancements, cardiac transplantation remains the only definitive option for treating DOX-induced heart failure in its later stages. Therefore, there is an ongoing need to further investigate and develop efficient therapeutic agents to combat DOX-induced cardiac damage.

In this context, our research group has made enormous progress in understanding the basic biochemical and molecular mechanisms that underlie the cardiotoxicity associated with DOX and we have demonstrated the effective use of a novel class of drugs called phosphodiesterase 5 (PDE-5) inhibitors in attenuating DOX-induced cardiotoxicity [20, 21]. PDE5 inhibitors are drugs originally developed to treat erectile dysfunction (ED) in men. They work by inhibiting the PDE5 enzyme, which increases cyclic guanosine monophosphate (cGMP) levels, leading to smooth muscle relaxation in the penis [22]. Sildenafil (Viagra) and tadalafil (Cialis) are also approved for treating pulmonary hypertension due to their effects on the pulmonary vasculature. Landmark studies from our laboratory demonstrated that PDE5 inhibitors, including sildenafil, vardenafil (Levitra), and tadalafil provide cardioprotection against ischemia/reperfusion injury in rabbit and mouse hearts [23–25]. PDE5 inhibitors increase cGMP levels and activate protein kinase G (PKG), which play crucial roles in protecting cardiac cells from damage [26–28]. In this chapter, we will explore the novel role of PDE5 inhibitors as potential treatments for DOX-induced cardiotoxicity and present an overview of our recent findings and examine the potential mechanisms by which PDE5 inhibitor, tadalafil, attenuates cardiotoxicity induced by cancer chemotherapy.

Overview of the Mechanisms of Actions of Doxorubicin

DOX is a widely used chemotherapy drug with multiple mechanisms of action which contribute to its effectiveness against cancer [29]. Like other anthracycline analogues, DOX intercalates between DNA base pairs, disrupting the DNA double helix structure. This interference inhibits the replication and transcription processes, ultimately leading to cell death. DOX inhibits the enzyme topoisomerase II, which is essential for DNA replication and repair. It prevents the re-ligation of DNA strands by stabilizing the topoisomerase II-DNA complex, causing double-strand breaks and triggering apoptosis [30, 31]. DOX can intercalate into mitochondrial DNA (mtDNA), impairing mitochondrial function. This leads to a decrease in ATP production and an increase in ROS generation, contributing to cell death. DOX undergoes redox cycling, leading to the production of ROS such as superoxide hydrogen peroxide [32]. The ROS generated cause oxidative damage to cellular components, including lipids, proteins, and DNA. DOX can also trigger apoptosis through multiple pathways, including the intrinsic (mitochondrial) pathway and extrinsic, Fas/Fas L-mediated apoptosis [33, 34]. The presence of ROS increases intracellular calcium which induced calcium-dependent proteases to cleave caspases inducing apoptosis. Thus, DOX activates pro-apoptotic proteins, inhibits anti-apoptotic proteins, causing the release of cytochrome C from mitochondria and leading to the activation of caspases and subsequent apoptosis [35].

In addition, DOX can bind to the proteasome, a protein complex responsible for degrading unneeded or damaged proteins. This binding disrupts proteasome function, leading to the accumulation of damaged proteins and cell death. DOX can be metabolically activated by cellular oxidoreductases, such as NADPH cytochrome P450 reductase. This activation generates semiquinone radicals, which further contribute to ROS production and cytotoxicity [36]. Emerging evidence suggests that DOX can modulate the immune response. It can enhance the presence of tumor antigens and stimulate immune cells, contributing to its anticancer effects [37]. Overall, these multiple mechanisms of action of DOX collectively contribute to not only its potent antitumor activity but also to its side effects, such as cardiotoxicity and myelosuppression [29, 38]. Understanding these mechanisms can further help in developing novel strategies to mitigate the adverse effects while maintaining its therapeutic efficacy.

Pathogenesis of Doxorubicin-Induced Cardiomyopathy

DOX-induced cardiomyopathy and heart failure are complex conditions with multiple contributing mechanisms which include free radical formation, lipid peroxidation, inhibition of nucleic acid and protein synthesis, mitochondrial dysfunction, calcium overloading, and cardiomyocyte apoptosis [32, 39].

Among the above mentioned multiple mechanisms, the generation of ROS in cardiomyocytes and subsequent oxidative damage play a key role in the pathogenesis of DOX-induced cardiomyopathy and is clearly distinct from its anti-tumor activity, which primarily involves inhibiting of topoisomerase II enzyme [35, 40]. Studies using electron spin resonance spectroscopy have demonstrated direct ROS production by DOX, highlighting the role of oxidative stress in its cardiotoxic effects [41]. Additionally, increased levels of intracellular peroxynitrite, hydrogen peroxide, and malondialdehyde (a marker of lipid peroxidation) further support the increased production of oxidative stress by DOX [42, 43]. DOX inactivates the GSH-peroxidase-1 enzyme, reducing superoxide dismutase levels, and creating an antioxidant enzyme deficit that promotes hydroxyl radical formation. DOX's unique chemical structure makes it prone to redox cycling under aerobic conditions, leading to the formation of semiquinone free radicals that donate electrons to oxygen, forming superoxide radicals [44]. The heart tissue is particularly susceptible due to its low levels of antioxidants like catalase and glutathione peroxidase. This heightened oxidative stress causes cellular damage and various subcellular changes in the myocardium, such as the gradual loss of myofibrils and vacuolization of myocardial cells, which are characteristic of DOX-induced cardiomyopathy. Consequently, the increase in ROS and the decrease in antioxidant levels result in significant oxidative stress in the heart, leading to cardiomyopathy which eventually results in heart failure [43, 45].

Secondly, mitochondrial damage plays a crucial role in DOX-induced cardiotoxicity. Enhanced ROS production and oxidative stress impairs mitochondrial function and contributes to cardiomyocyte injury. DOX intercalates into mitochondrial DNA (mtDNA), disrupting its replication and transcription [46]. Damaged mtDNA compromises mitochondrial function. DOX affects the balance between mitochondrial fission and fusion, leading to an accumulation of dysfunctional mitochondria. This imbalance impairs mitochondrial quality control and exacerbates cardiac cell damage [44]. DOX inhibits mitophagy, the process by which damaged mitochondria are removed and recycled. This leads to the accumulation of defective mitochondria, contributing to cardiotoxicity. Mitochondrial dysfunction caused by DOX reduces ATP production, depriving cardiomyocytes of the energy needed for normal function and survival. Furthermore, DOX induces the opening of the Mitochondrial Permeability Transition Pore (MPTP) leading to loss of mitochondrial membrane potential, release of pro-apoptotic factors, and subsequent contributing to cardiomyocyte death [47]. The target organ of DOX-induced toxicity is the myocardium since it is highly enriched with mitochondria. Cardiomyocytes are particularly more susceptible to apoptosis than other tissues as they rely on oxidative phosphorylation and require continuous supply of high energy phosphates for appropriate myocardial function.

DOX is known to induce apoptosis both by extrinsic and intrinsic pathways. Hydrogen peroxide production remains the primary cause of DOX-induced apoptosis. Studies have demonstrated that in the intrinsic pathway, cytochrome c is released from mitochondria, leading to the activation of caspase-9 through the formation of apoptosome complexes composed of apoptosis-activating factor-1, cytochrome c, and pro-caspase-9 [48]. This process is facilitated by the upregulation of Bax or the depletion of Bcl-xl. Additionally, the activation of p53 further induces the activation of the Bax gene. These apoptotic responses, including the redox cycling of DOX, the induction of p53 by superoxide, hydrogen peroxide production and the transcriptional activation of the Bax gene, appear to be interlinked contributing to the opening of the MPTP by DOX [47]. Numerous studies showed strong evidence linking ROS production to cardiomyocyte apoptosis [49, 50]. Recent studies also suggest that cytokines such as TNFα may be involved in DOX-induced cardiotoxicity

[51, 52]. Elevated levels of TNFα levels have been reported in cancer patients treated with DOX [53]. TNFα can potentially drive innate signaling by pathogen activated molecular programs, as well as damage activated molecular programs [54]. Excessive TNFα can also regulate vital cellular processes that include cell survival and death [55]. Cardiomyocyte apoptosis contributes to cardiomyocyte loss, eventually leading to structural changes that are maladaptive to normal cardiac physiological demands which play a critical role in the development of DOX-induced cardiotoxicity.

Mechanism of Action of Tadalafil

Tadalafil is a long acting PDE5 inhibitor [56] approved by the US Food and Drug Administration (FDA) in 2003. It is mainly indicated for the treatment of erectile dysfunction (ED), pulmonary arterial hypertension (PAH) and benign prostatic hyperplasia (BPH) [57–59]. PDE5 enzyme is predominantly found in the smooth muscle cells of the corpus cavernosum in the penis, as well as in the pulmonary vasculature and the prostate gland. PDE5 plays a crucial role in the breakdown of cGMP, a molecule essential for vasodilation and smooth muscle relaxation. cGMP is synthesized in response to the release of nitric oxide (NO) during sexual stimulation. NO activates guanylate cyclase, an enzyme that converts guanosine triphosphate (GTP) into cGMP. Tadalafil increases the levels of cGMP by inhibiting PDE5 enzyme which hydrolyzes cGMP and leads to the activation of cGMP-dependent protein kinase (PKG), which initiates a cascade of events resulting in smooth muscle relaxation [60].

In the context of ED, tadalafil enhances the effects of sexual stimulation by increasing cGMP levels in the corpus cavernosum. During sexual arousal, NANC (non-adrenergic non-cholinergic) neurons release NO, which stimulates guanylate cyclase to produce cGMP. The accumulation of cGMP leads to the relaxation of smooth muscle cells in the corpus cavernosum, allowing blood to fill the penile tissue and produce an erection [57, 58]. By inhibiting PDE5, tadalafil prevents the degradation of cGMP, thereby prolonging its effects and supporting sustained erections. PAH is a condition marked by elevated blood pressure in the pulmonary arteries due to endothelial dysfunction which leads to decreased production of NO and prostacyclin, leading to reduced vasodilation, and increased production of endothelin-1 and thromboxane causing vasoconstriction. Inhibition of PDE5 by tadalafil in the pulmonary vasculature enhances NO-cGMP signaling, promoting vasodilation and reducing pulmonary arterial pressure [58].

The cardiac indications of tadalafil are relatively novel compared to its protective effects in the treatment of ED and PAH. Increasing the intracellular levels of cGMP by tadalafil via PDE5 inhibition activates PKG leading to various cardioprotective effects including myocardial contractility, vasodilation and inhibition of apoptosis [61]. Several studies from our laboratory and others have shown that PDE5 inhibitors including tadalafil induce powerful cardioprotective effects during ischemia/

reperfusion injury, heart failure, DOX-induced cardiomyopathy as well as diabetic cardiomyopathy [21, 24, 28, 62–64]. Our pioneering investigations demonstrated that PDE5 inhibitors, sildenafil and tadalafil increase eNOS/iNOS in the heart which leads to downstream protective mechanisms involving cGMP-dependent activation of PKG and opening of mitochondrial K_{ATP} channels [24–27]. We have demonstrated that in vivo treatment of mice with sildenafil before DOX administration conferred protective effects in the heart by significant attenuation of cardiomyocyte apoptosis, preservation of myofibrillar integrity and prevention of left ventricular dysfunction and ST prolongation consistent with chronic DOX-induced toxicity [20]. Administration of tadalafil (4 mg/kg, p.o. for 9 days) enhanced survival rates and decreased the ratio of heart weights to tibial length in the DOX treated mice. Tadalafil treated mice also showed improved left ventricular (LV) function (Fig. 39.1) by preserving the shortening (FS) and ejection fraction (EF) as compared to DOX alone group [21]. Antiapoptotic protein, Bcl-2 which is known to block mitochondrial pathway of apoptosis was downregulated in DOX group and co-treatment with tadalafil significantly preserved Bcl-2 level (Fig. 39.2a). Tadalafil decreased apoptosis of DOX as demonstrated by the decrease in TUNEL positive

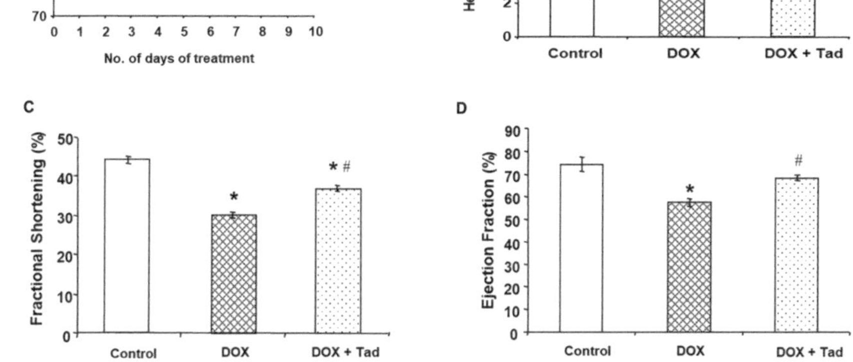

Fig. 39.1 Cardioprotective effects of tadalafil in DOX-induced cardiomyopathy. (**a**) Kaplan–Meier survival curve, (**b**) ratio of heart weight to tibia length (HW/TL). Percentages of surviving mice were plotted and compared by chi squares test (n = 15–24/group, $P < 0.05$). HW/TL were presented as means ± S.E. (n = 6–8/group, *$P < 0.05$ vs Control). The averaged data of fractional shortening (**c**) and ejection fraction (**d**) in the mice are presented as mean ± SE (n = 6 per group; *$P < 0.05$ vs Control, #$P < 0.05$ vs DOX). (Reprinted from Koka S et al. with permission)

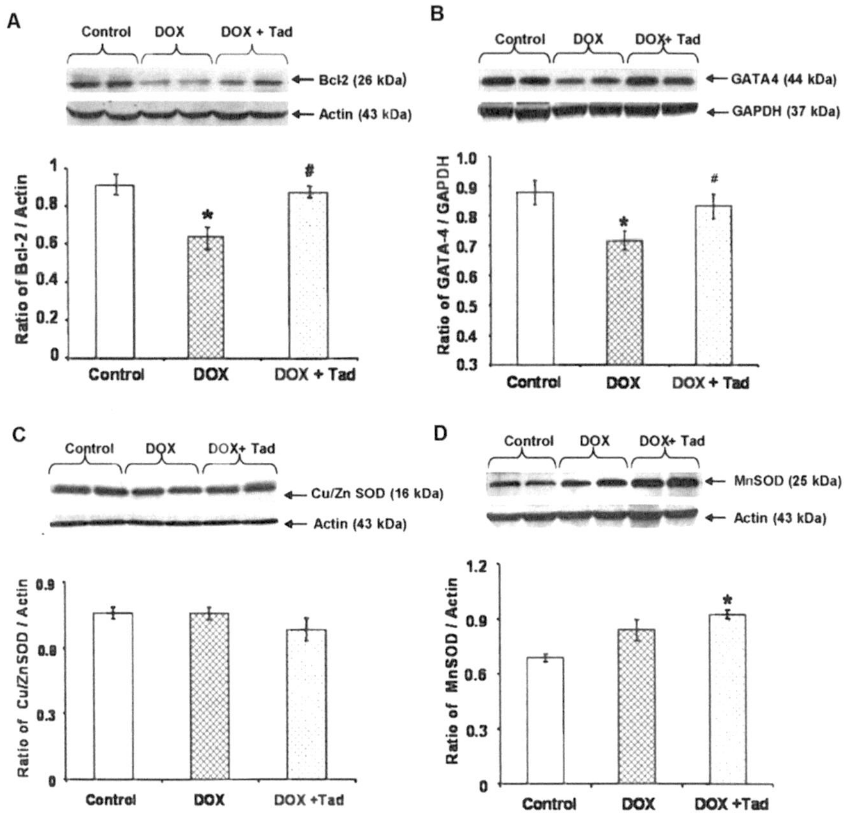

Fig. 39.2 Effect of tadalafil on anti-apoptotic proteins and superoxide dismutases (SOD) after DOX treatment. Representative western blots with specific bands of (**a**) Bcl-2 (**b**) GATA-4 (**c**) Cu/ZnSOD and (**d**) MnSOD are shown. Bar graphs show densiometric quantification from 4 individual hearts per group, which was normalized against actin or glyceraldehyde 3-phosphate dehydrogenase (GAPDH) level for each sample. Data were expressed as mean ± SE. *$p < 0.05$ vs Control, $^{\#}p < 0.05$ vs DOX. (Reprinted from Koka S et al. with permission)

nuclei without interfering with the anti-cancer efficacy [21]. Taken together, the cardioprotective effects by tadalafil conceptually support our previous work on sildenafil-induced cardioprotection in a chronic model of DOX-induced cardiomyopathy [20, 21].

The increased generation of ROS and consequent lipid peroxidation is a major pathogenic factor in DOX-induced cardiomyopathy [49, 65]. We demonstrated that tadalafil attenuated oxidative stress in DOX-induced cardiomyopathy. Antioxidant enzymes such as Cu/ZnSOD and MnSOD play a critical role in detoxification of ROS [66]. Tadalafil upregulated a key mitochondrial antioxidant enzyme MnSOD and reduced oxidative stress in chronic DOX-induced cardiomyopathic mouse model (Fig. 39.2c, d) [21]. Tadalafil treatment also attenuated the lipid peroxidation in DOX treated mice. The antioxidant properties of PDE5 inhibitors have not yet been well studied and understood in vivo models. Most recently, Serarslan et al. demonstrated that tadalafil reduced the spinal cord injury via increasing tissue/serum levels of NO and serum activity of SOD in a rat model of traumatic spinal cord injury [67] which is in line with our studies demonstrating antioxidative properties of tadalafil in a DOX-induced cardiomyopathic mouse model. GATA-4 is a member of the GATA family of zinc finger transcription factors, which plays important roles in transducing nuclear events that modulate cell lineage differentiation during development in the heart. GATA-4 depletion was shown to result in a progressive deterioration in the cardiac function and dilation in adulthood [68]. GATA-4 was reduced in DOX-treated mice as reported previously [69]. Our studies also showed significant downregulation of GATA-4 expression after DOX treatment which confirmed the previous reports [21]. Tadalafil significantly preserved GATA-4 expression in DOX treated mice suggesting that enhancement of GATA-4 transcription factor expression effectively contributed to tadalafil's cardioprotective effects in the setting of DOX-induced toxicity (Fig. 39.2b) [21, 70].

It is well established from ours as well other studies that PDE5 inhibitors increase cGMP levels and activate the cGMP/PKG dependent signaling pathway in the heart, which in turn plays a critical role in PDE5 inhibitors-induced cardioprotection [28, 61, 71–73]. We have earlier demonstrated the activation of NO-AMPK-Akt signaling pathway in diabetic mice in which tadalafil treatment restored AMPK activity and phosphorylation of Akt in the diabetic hearts [63]. Tadalafil treatment significantly increased PKG activity and cGMP levels in the DOX-induced cardiomyopathic mice which implicates the potential role of PKG/cGMP pathway in cardioprotection against DOX-induced cardiotoxicity [21, 70]. It has been reported that cGMP binding to PKG Iα attenuate oxidant-induced disulfide formation caused by DOX. Tadalafil was shown to reduce PKG Iα oxidation caused by DOX and protected against cardiac dysfunction and injury [74]. Activation of cGMP/PKG may counteract the detrimental effects of DOX such as enhanced ROS production and mitochondrial dysfunction. Excessive PKG Iα oxidation favors apoptosis and PKG activation by PDE5 inhibitors inhibit pro-apoptotic pathways and promotes cardiomyocyte survival. One of the targets of PKG Iα is RhoA which suppresses classical cGMP-PKG Iα pro-survival signaling, resulting in intensified apoptosis [75]. In fact, the DOX-induced apoptosis (Bcl2/Bax ratio) and attenuation of pro-survival kinase (phospho-ERK, phospho-AKT, and phospho-GSK3) signaling in mice hearts was shown to be blocked by co-treatment with tadalafil and PKG Iα knock in mice which were protected against DOX-induced cardiotoxicity (Fig. 39.3) [74]. Taken together, PKG/cGMP pathway plays a key role in tadalafil induced protection against DOX-induced cardiotoxicity.

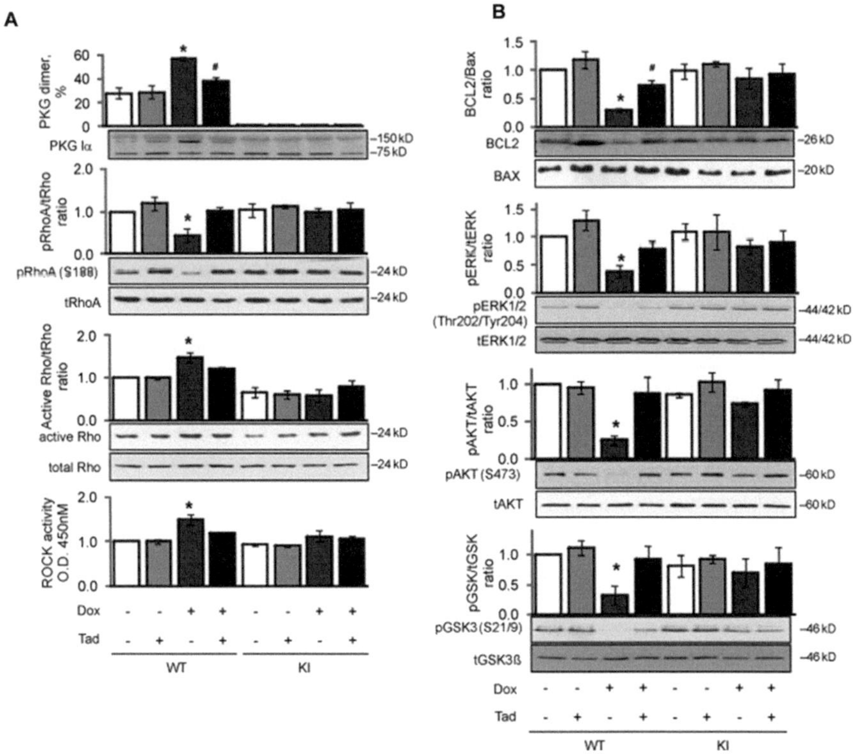

Fig. 39.3 Tadalafil limits DOX-induced PKG Iα oxidation and the associated apoptotic signaling in WT, whereas PKG Iα KI are basally protected. (**a**) PKG Iα disulfide dimerization, RhoA Ser-188 phosphorylation, RhoA, and ROCK activity in WT or KI hearts treated with doxorubicin (*Dox*) or vehicle, in the presence or absence of tadalafil (*Tad*), was indexed ($n = 4$). This demonstrated that doxorubicin triggered a loss of pro-survival phospho-RhoA (Ser-188) signaling in WT, and this was attenuated by PDE5 inhibition with tadalafil. (**b**) BCL2/Bax ratio, ERK, AKT, and GSK3β phosphorylation in hearts from WT or KI mice treated with doxorubicin or vehicle, in the presence or absence of tadalafil, was indexed. Consistent with the loss in anti-apoptotic phospho-RhoA induced by doxorubicin in WT hearts, there was a concomitant loss of pro-survival signaling in terms of BCL2/Bax ratio, ERK, AKT, and GSK3β. $^*p < 0.05$ between vehicle and DOX groups; $^\#p < 0.05$ between DOX alone and DOX + tadalafil groups. (Reprinted from Oleksandrs Prysyazhna et al. with permission)

Evidence of Tadalafil Protection Against Cardiotoxicity of Other Anticancer Agents

PDE5 inhibitors trigger anti-tumor signaling pathways by inducing NO signaling, which enhances cGMP formation and activates PKG. PKG can also induce β-catenin, leading to the attenuation of survival proteins and activation of pro-apoptotic proteins, resulting in apoptosis. These inhibitors also boost the effectiveness of chemotherapeutics by increasing drug accumulation inside cells and inhibiting drug efflux. We have earlier demonstrated that sildenafil increases chemotherapeutic efficacy of doxorubicin in prostate cancer and ameliorates cardiac dysfunction [76, 77]. Our group has earlier shown that tadalafil enhanced the lethality of a range of chemotherapeutic drugs in a number of gastrointestinal/genitourinary gastric cancer cell lines [78]. PDE5 inhibitors such as sildenafil and tadalafil interacted in potentiating manner with DOX/mitomycin C/gemcitabine/cisplatin/paclitaxel to cause cell death in bladder. Cisplatin is an antineoplastic agent used in the treatment of several cancers such as malignancy of testis, colon, ovarian, head, and neck [79]. In addition to nephrotoxicity, cisplatin causes vascular endothelial dysfunction leading to cardiovascular diseases which limits its usage [80]. Rasha Saleh et al. investigated the possible cardioprotective effect of tadalafil on cisplatin-induced cardiac and vascular damages in rats. Their study demonstrated that tadalafil was able to decrease blood pressure, heart rate, levels of serum cardiac troponin (cTn-I), malondialdehyde and increased levels of reduced glutathione (GSH) and nitric oxide (NO) in the heart homogenate sample from cisplatin treated rats [81]. In addition, histopathological examination showed that tadalafil reduced cardiac and aortic lesions indicating the cardioprotective role of tadalafil in cisplatin induced cardiac injury. Combretastatin A4 disodium phosphate (CA4DP) is a prodrug of combretastatin A4 (CA4), exhibits antitumor effects in vascular endothelial cells in tumors. However, administration of CA4DP increases systolic blood pressure, decreased cardiac function, lowered cGMP levels in the myocardium, and leading to cardiotoxicity. Co-administration of tadalafil prevented the decrease in cGMP levels in left ventricular wall and prevented CA4DP-induced cardiotoxicity without affecting the antitumor effects in CA4DP treated xenograft mice [82].

Elevated PDE5 expression has been reported in multiple human carcinomas including colon adenocarcinoma, bladder squamous carcinoma, metastatic breast, prostate, pancreatic, and lung cancers compared to healthy tissues and mounting evidence supports the usage of PDE5 inhibitors in the treatment of a variety of cancers and PDE5 inhibitors are being investigation in combination with other anticancer agents in a variety of cancers such as pancreatic cancers, head and neck squamous cell cancer, melanomas and gliomas [83]. PDE5 inhibitors enhance chemosensitization of existing cancer therapies and preclinical studies suggest that tadalafil and other PDE5 inhibitors may enhance the sensitivity of certain types of cancer to standard chemotherapeutic drugs, including DOX [77, 78, 84]. Currently there is an active ongoing phase II clinical trial investigating the combination of tadalafil with fluorouracil, leucovorin, oxaliplatin, and docetaxel (FLOT) chemotherapy in treating patients with stage I–III resectable gastric/gastroesophageal junction cancer (Table 39.1. NCT05709574). There is high possibility that PDE5 inhibitors can be repurposed for effective treatment of cancer patients. Future ongoing or planned clinical investigations are expected to reveal whether PDE-5 inhibitors have significant influence in improving the efficacy of existing cancer therapies.

Table 39.1 List of the studies on PDE5 inhibitors and cancer at www.clinicaltrials.gov[a]

Inhibitor	Indication	Status	ClinicalTrials.gov[a]
Sildenafil	Pancreatic cancer, cholangiocarcinoma	Phase 1 (recruiting)	NCT02106871
	Advanced solid tumors	Phase 1 (recruiting)	NCT02466802
	Bladder cancer	Phase 2 (recruiting)	NCT02422277
	Kidney cancer	Phase 2 (completed)	NCT01950923
	Breast, gastrointestinal, genitourinary, gynecological cancers	Phase 1 (active, not recruiting)	NCT01375699
	Non-small cell lung cancer	Phase 2, phase 3 (completed)	NCT00752115
	High-grade glioma	Phase 2 (recruiting)	NCT01817751
	Waldenstrom's macroglobulinemia	Phase 2 (completed)	NCT00165295
Tadalafil	Head and neck squamous cell carcinoma	Phase 1 (recruiting)	NCT02544880
	Head and neck cancer	Phase 2 (recruiting)	NCT02544880
	Head and neck cancer	Phase 3 (completed)	NCT00843635
	Head and neck squamous cell carcinoma	Phase 2 (completed)	NCT00894413
	Multiple myeloma	Phase 2 (recruiting)	NCT01858558
	Multiple myeloma	Phase 2 (terminated)	NCT01374217
	Prostate cancer	Phase 3 (completed)	NCT00931528
	Pancreatic cancer	Phase 1 (active, not recruiting)	NCT01342224
	Pancreatic cancer	Phase 1 (active, not recruiting)	NCT01903083
	Primary abdominal malignancy	Phase 1 (not yet recruiting)	NCT02998736
	Gastric/gastrooesophageal junction cancer	Phase 2 (recruiting)	NCT05709574
Vardenafil	Glioma, brain neoplasms, brain metastasis	Early phase 1 (recruiting)	NCT02279992
Udenafil	Sigmoid colon and rectal cancers	Phase 2 (completed)	NCT00607282

By courtesy from Ines Barone et al. [85].
[a]indicates the hyper link to clinical trials website.

Clinical Perspectives and Potential Applications

Ongoing clinical trials are evaluating tadalafil and other PDE5 inhibitors in patients undergoing chemotherapy in various types of cancers to assess their efficacy in potentiating chemotherapeutic effect and protection against cardiotoxicity [85] as shown in Table 39.1. In our studies on human osteosarcoma (OSA-1) cancer cell lines tadalafil did not impede the cell-killing efficacy of DOX in vitro or interfered with the in vivo anti-tumor effect of DOX in OSA-1 tumor xenograft model (Fig. 39.4). Tadalafil co-treatment did not change DOX-induced reduction in tumor weight, volume and body weight. Given that tadalafil does not interfere with DOX's anti-tumor effects, it holds promise as part of a combination therapy approach to minimize cardiac side effects while maintaining chemotherapy efficacy [21, 70]. Although both PDE5 inhibitors sildenafil and tadalafil, demonstrated protective effects against DOX induced cardiomyopathy in our studies, tadalafil

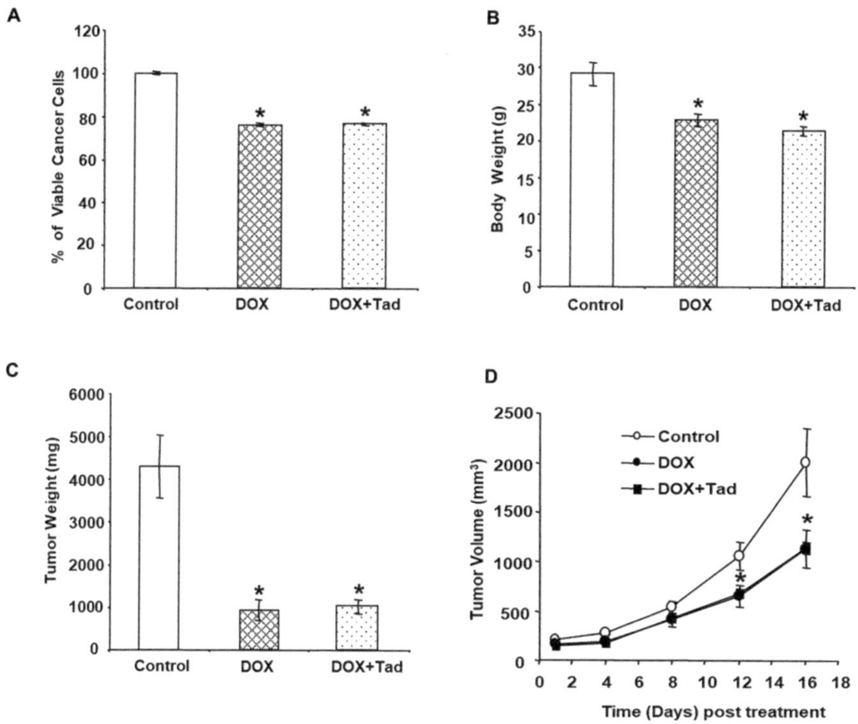

Fig. 39.4 Effect of tadalafil on the anti-tumor efficacy of DOX. (**a**) Cell viability in human osteosarcoma cancer cell lines (OSA-1) following 48 h of co-treatment with DOX and Tadalafil. (**b**) Effect on body weight, (**c**) tumor weight and (**d**) tumor volume in sarcoma (OSA-1) tumor xenograft model in BALB/cAnNCr-nu/nu mice following 16 days of tadalafil (4 mg/kg, p.o.) administration in DOX treated mice (n = 6/group). *P < 0.001 vs Control. (Reprinted from Koka S et al. with permission)

appears to be the preferred choice for long-term management of patients receiving DOX for malignant tumors owing to its favorable pharmacokinetic properties. Tadalafil has relatively slow metabolism than sildenafil and could be used at a relatively lower dosage regimen during chronic treatment. The rate and extent of its absorption is unaffected by food. It is well-absorbed after oral administration, with a bioavailability that varies among individuals. Tadalafil also has quick onset (16–17 min) and reaches peak plasma concentrations within 30 min to 6 h after ingestion. It has sustained action among this class of agents permitting in flexibility in dosing. It has a long half-life of approximately 17.5 h, allowing for prolonged therapeutic effects and the convenience of once-daily dosing. Tadalafil is >10,000-fold more potent for PDE5 than for other sub forms of PDE enzyme such as PDE1, PDE2, PDE3, PDE4 and PDE7 enzymes [86] and therefore exhibits fewer side effects compared to other PDE5 inhibitors. In clinical studies involving multicenter trials, at doses 10 mg and 20 mg tadalafil was reported to be safe, well tolerated and lower number of drug interactions [87]. All these pharmacokinetic factors associated

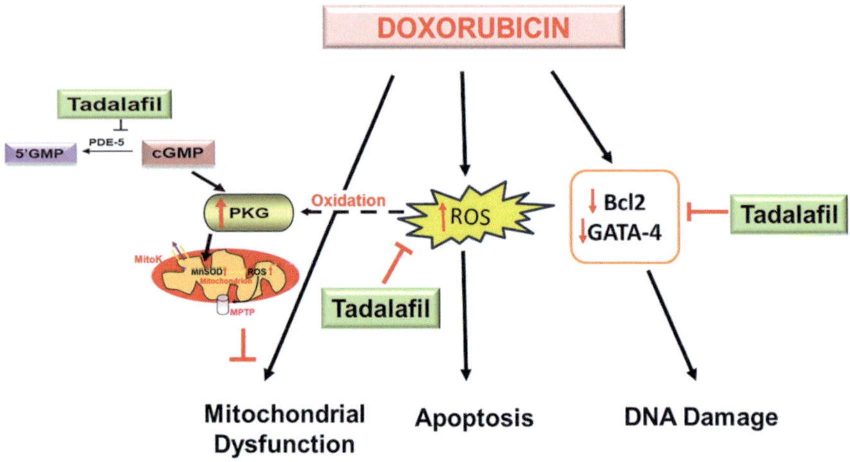

Fig. 39.5 Proposed pathway of inhibition of chemotherapy-induced cardiotoxicity by PDE5 inhibitors. PDE5 inhibitors activate downstream signaling pathways, mainly cGMP-PKG-mediated, inhibit ROS, mitochondrial damage and apoptosis of the myocardium. *ROS* reactive oxygen species, *PDE-5* phosphodiesterase 5, *cGMP* cyclic guanosine monophosphate, *PKG* protein kinase G, $mitoK_{ATP}$ mitochondrial ATP-sensitive potassium channels, *MPTP* mitochondrial permeability transition pore

with the tadalafil make it a clinical attractive agent in the long-term administration in DOX-indued cardiotoxicity patients.

Conclusion

PDE5 inhibitors can activate downstream signaling pathways, mainly cGMP-PKG-mediated, inhibit ROS, mitochondrial damage and apoptosis of the myocardium (Fig. 39.5), and thus could be beneficial in the inhibition of chemotherapy-induced cardiotoxicity. However, treatment of PDE5 inhibitors in cardiac and cancer patients has shown mixed results thus far. Nonetheless, there remains significant interest in exploring their novel clinical applications. This enthusiasm is largely driven by the established safety profile of PDE5 inhibitors and their widespread use in managing ED and PAH. While the outcomes of the ongoing clinical trials evaluating the role of PDE5 inhibitors in preventing cardiotoxicity associated with novel chemotherapeutic agents and their integration into the standard chemotherapy regimens are awaited, future research must be meticulously designed. Such trials should focus on carefully selecting patient populations based on clinical indications to ensure meaningful results. By adopting a strategic approach, the potential benefits of PDE5 inhibitors can be fully realized, paving the way for their clinical utility across diverse therapeutic contexts.

Acknowledgements This work was supported by grants from the National Institutes of Health RO1HL124366, RO1HL15895, RO1 CA221813, DK120866, HL118808 and Department of Defense Idea Award W81XWH-18-1-0308 (R.CK.) to Dr. Rakesh C. Kukreja.

Conflicts of Interest No potential conflicts of interest to disclose.

References

1. Fujiwara A, Hoshino T, Westley JW (1985) Anthracycline antibiotics. Crit Rev Biotechnol 3(2):133–157. https://doi.org/10.3109/07388558509150782
2. Minotti G, Menna P, Salvatorelli E, Cairo G, Gianni L (2004) Anthracyclines: molecular advances and pharmacologic developments in antitumor activity and cardiotoxicity. Pharmacol Rev 56(2):185–229. https://doi.org/10.1124/pr.56.2.6
3. Swain SM, Whaley FS, Ewer MS (2003) Congestive heart failure in patients treated with doxorubicin: a retrospective analysis of three trials. Cancer 97(11):2869–2879. https://doi.org/10.1002/cncr.11407
4. Hortobágyi GN (1997) Anthracyclines in the treatment of cancer. An overview. Drugs 54(Suppl 4):1–7. https://doi.org/10.2165/00003495-199700544-00003
5. Bristow MR, Billingham ME, Mason JW, Daniels JR (1978) Clinical spectrum of anthracycline antibiotic cardiotoxicity. Cancer Treat Rep 62(6):873–879
6. Lefrak EA, Pitha J, Rosenheim S, Gottlieb JA (1973) A clinicopathologic analysis of adriamycin cardiotoxicity. Cancer 32(2):302–314. https://doi.org/10.1002/1097-0142(197308)32:2<302::aid-cncr2820320205>3.0.co;2-2
7. Praga C, Beretta G, Vigo PL et al (1979) Adriamycin cardiotoxicity: a survey of 1273 patients. Cancer Treat Rep 63(5):827–834
8. Chatterjee K, Zhang J, Honbo N, Karliner JS (2010) Doxorubicin cardiomyopathy. Cardiology 115(2):155–162. https://doi.org/10.1159/000265166
9. Keefe DL (2001) Anthracycline-induced cardiomyopathy. Semin Oncol 28(4 Suppl 12):2–7
10. Simmons A, Vacek JL, Meyers D (2008) Anthracycline-induced cardiomyopathy. Postgrad Med 120(4):67–72. https://doi.org/10.3810/pgm.2008.11.1940
11. Grenier MA, Lipshultz SE (1998) Epidemiology of anthracycline cardiotoxicity in children and adults. Semin Oncol 25(4 Suppl 10):72–85
12. Steinherz LJ, Steinherz PG, Tan C (1995) Cardiac failure and dysrhythmias 6–19 years after anthracycline therapy: a series of 15 patients. Med Pediatr Oncol 24(6):352–361. https://doi.org/10.1002/mpo.2950240604
13. Fu LX, Waagstein F, Hjalmarson A (1990) A new insight into adriamycin-induced cardiotoxicity. Int J Cardiol 29(1):15–20. https://doi.org/10.1016/0167-5273(90)90267-9
14. Von Hoff DD, Layard MW, Basa P et al (1979) Risk factors for doxorubicin-induced congestive heart failure. Ann Intern Med 91(5):710–717
15. Nazeyrollas P, Prévost A, Baccard N, Manot L, Devillier P, Millart H (1999) Effects of amifostine on perfused isolated rat heart and on acute doxorubicin-induced cardiotoxicity. Cancer Chemother Pharmacol 43(3):227–232. https://doi.org/10.1007/s002800050888
16. Kumar D, Kirshenbaum LA, Li T, Danelisen I, Singal PK (2001) Apoptosis in adriamycin cardiomyopathy and its modulation by probucol. Antioxid Redox Signal 3(1):135–145. https://doi.org/10.1089/152308601750100641
17. Liu X, Chen Z, Chua CC et al (2002) Melatonin as an effective protector against doxorubicin-induced cardiotoxicity. Am J Physiol Heart Circ Physiol 283(1):H254–H263. https://doi.org/10.1152/ajpheart.01023.2001

18. Gabizon A, Shmeeda H, Barenholz Y (2003) Pharmacokinetics of pegylated liposomal doxorubicin: review of animal and human studies. Clin Pharmacokinet 42(5):419–436. https://doi.org/10.2165/00003088-200342050-00002
19. Zhu L, Lin M (2021) The synthesis of nano-doxorubicin and its anticancer effect. Anticancer Agents Med Chem 21(18):2466–2477. https://doi.org/10.2174/1871520621666201229115612
20. Fisher PW, Salloum F, Das A, Hyder H, Kukreja RC (2005) Phosphodiesterase-5 inhibition with sildenafil attenuates cardiomyocyte apoptosis and left ventricular dysfunction in a chronic model of doxorubicin cardiotoxicity. Circulation 111(13):1601–1610. https://doi.org/10.1161/01.Cir.0000160359.49478.C2
21. Koka S, Das A, Zhu SG, Durrant D, Xi L, Kukreja RC (2010) Long-acting phosphodiesterase-5 inhibitor tadalafil attenuates doxorubicin-induced cardiomyopathy without interfering with chemotherapeutic effect. J Pharmacol Exp Ther 334(3):1023–1030. https://doi.org/10.1124/jpet.110.170191
22. Goldstein I, Lue TF, Padma-Nathan H, Rosen RC, Steers WD, Wicker PA (1998) Oral sildenafil in the treatment of erectile dysfunction. Sildenafil Study Group. N Engl J Med 338(20): 1397–1404. https://doi.org/10.1056/nejm199805143382001
23. Kukreja RC, Ockaili R, Salloum F, Xi L (2003) Sildenafil-induced cardioprotection in rabbits. Cardiovasc Res 60(3):700–701. https://doi.org/10.1016/j.cardiores.2003.10.002. author reply 02–3
24. Kukreja RC, Ockaili R, Salloum F et al (2004) Cardioprotection with phosphodiesterase-5 inhibition—a novel preconditioning strategy. J Mol Cell Cardiol 36(2):165–173. https://doi.org/10.1016/j.yjmcc.2003.11.001
25. Kukreja RC, Salloum F, Das A et al (2005) Pharmacological preconditioning with sildenafil: basic mechanisms and clinical implications. Vasc Pharmacol 42(5–6):219–232. https://doi.org/10.1016/j.vph.2005.02.010
26. Salloum F, Yin C, Xi L, Kukreja RC (2003) Sildenafil induces delayed preconditioning through inducible nitric oxide synthase-dependent pathway in mouse heart. Circ Res 92(6):595–597. https://doi.org/10.1161/01.Res.0000066853.09821.98
27. Salloum FN, Abbate A, Das A et al (2008) Sildenafil (Viagra) attenuates ischemic cardiomyopathy and improves left ventricular function in mice. Am J Physiol Heart Circ Physiol 294(3): H1398–H1406. https://doi.org/10.1152/ajpheart.91438.2007
28. Salloum FN, Chau VQ, Hoke NN et al (2009) Phosphodiesterase-5 inhibitor, tadalafil, protects against myocardial ischemia/reperfusion through protein-kinase G-dependent generation of hydrogen sulfide. Circulation 120(11 suppl 1):S31–S36. https://doi.org/10.1161/CIRCULATIONAHA.108.843979
29. Singal PK, Li T, Kumar D, Danelisen I, Iliskovic N (2000) Adriamycin-induced heart failure: mechanism and modulation. Mol Cell Biochem 207(1–2):77–86. https://doi.org/10.1023/a:1007094214460
30. Zhang S, Liu X, Bawa-Khalfe T et al (2012) Identification of the molecular basis of doxorubicin-induced cardiotoxicity. Nat Med 18(11):1639–1642. https://doi.org/10.1038/nm.2919
31. Vejpongsa P, Yeh ET (2014) Topoisomerase 2β: a promising molecular target for primary prevention of anthracycline-induced cardiotoxicity. Clin Pharmacol Ther 95(1):45–52. https://doi.org/10.1038/clpt.2013.201
32. Singal PK, Iliskovic N (1998) Doxorubicin-induced cardiomyopathy. N Engl J Med 339(13): 900–905. https://doi.org/10.1056/nejm199809243391307
33. Ju BG, Lunyak VV, Perissi V et al (2006) A topoisomerase IIbeta-mediated dsDNA break required for regulated transcription. Science 312(5781):1798–1802. https://doi.org/10.1126/science.1127196
34. Kalivendi SV, Konorev EA, Cunningham S et al (2005) Doxorubicin activates nuclear factor of activated T-lymphocytes and Fas ligand transcription: role of mitochondrial reactive oxygen species and calcium. Biochem J 389(Pt 2):527–539. https://doi.org/10.1042/bj20050285

35. Wenningmann N, Knapp M, Ande A, Vaidya TR, Ait-Oudhia S (2019) Insights into doxorubicin-induced cardiotoxicity: molecular mechanisms, preventive strategies, and early monitoring. Mol Pharmacol 96(2):219–232. https://doi.org/10.1124/mol.119.115725
36. Minotti G (1989) Reactions of adriamycin with microsomal iron and lipids. Free Radic Res Commun 7(3–6):143–148. https://doi.org/10.3109/10715768909087936
37. Najibi AJ, Larkin K, Feng Z et al (2022) Chemotherapy dose shapes the expression of immune-interacting markers on cancer cells. Cell Mol Bioeng 15(6):535–551. https://doi.org/10.1007/s12195-022-00742-y
38. Mordente A, Meucci E, Silvestrini A, Martorana GE, Giardina B (2009) New developments in anthracycline-induced cardiotoxicity. Curr Med Chem 16(13):1656–1672. https://doi.org/10.2174/092986709788186228
39. Kciuk M, Gielecińska A, Mujwar S et al (2023) Doxorubicin-an agent with multiple mechanisms of anticancer activity. Cells 12(4):659. https://doi.org/10.3390/cells12040659
40. Myers C (1998) The role of iron in doxorubicin-induced cardiomyopathy. Semin Oncol 25(4 Suppl 10):10–14
41. Alegria AE, Samuni A, Mitchell JB, Riesz P, Russo A (1989) Free radicals induced by adriamycin-sensitive and adriamycin-resistant cells: a spin-trapping study. Biochemistry 28(21):8653–8658. https://doi.org/10.1021/bi00447a056
42. Vásquez-Vivar J, Martasek P, Hogg N, Masters BS, Pritchard KA Jr, Kalyanaraman B (1997) Endothelial nitric oxide synthase-dependent superoxide generation from adriamycin. Biochemistry 36(38):11293–11297. https://doi.org/10.1021/bi971475e
43. Singal PK, Pierce GN (1986) Adriamycin stimulates low-affinity Ca^{2+} binding and lipid peroxidation but depresses myocardial function. Am J Physiol Heart Circ Physiol 250(3):H419–HH25
44. Konorev EA, Kennedy MC, Kalyanaraman B (1999) Cell-permeable superoxide dismutase and glutathione peroxidase mimetics afford superior protection against doxorubicin-induced cardiotoxicity: the role of reactive oxygen and nitrogen intermediates. Arch Biochem Biophys 368(2):421–428. https://doi.org/10.1006/abbi.1999.1337
45. Octavia Y, Tocchetti CG, Gabrielson KL, Janssens S, Crijns HJ, Moens AL (2012) Doxorubicin-induced cardiomyopathy: from molecular mechanisms to therapeutic strategies. J Mol Cell Cardiol 52(6):1213–1225. https://doi.org/10.1016/j.yjmcc.2012.03.006
46. Abe K, Ikeda M, Ide T et al (2022) Doxorubicin causes ferroptosis and cardiotoxicity by intercalating into mitochondrial DNA and disrupting Alas1-dependent heme synthesis. Sci Signal 15(758):eabn8017. https://doi.org/10.1126/scisignal.abn8017
47. Green PS, Leeuwenburgh C (2002) Mitochondrial dysfunction is an early indicator of doxorubicin-induced apoptosis. Biochim Biophys Acta 1588(1):94–101. https://doi.org/10.1016/s0925-4439(02)00144-8
48. Kang PM, Izumo S (2000) Apoptosis and heart failure: a critical review of the literature. Circ Res 86(11):1107–1113. https://doi.org/10.1161/01.res.86.11.1107
49. Kim SY, Kim SJ, Kim BJ et al (2006) Doxorubicin-induced reactive oxygen species generation and intracellular Ca^{2+} increase are reciprocally modulated in rat cardiomyocytes. Exp Mol Med 38(5):535–545. https://doi.org/10.1038/emm.2006.63
50. Olson RD, Mushlin PS (1990) Doxorubicin cardiotoxicity: analysis of prevailing hypotheses. FASEB J 4(13):3076–3086
51. Bradley JR (2008) TNF-mediated inflammatory disease. J Pathol 214(2):149–160. https://doi.org/10.1002/path.2287
52. Gilliam LA, Ferreira LF, Bruton JD et al (2009) Doxorubicin acts through tumor necrosis factor receptor subtype 1 to cause dysfunction of murine skeletal muscle. J Appl Physiol (1985) 107(6):1935–1942. https://doi.org/10.1152/japplphysiol.00776.2009
53. Hayslip J, Dressler EV, Weiss H et al (2015) Plasma TNF-α and soluble TNF receptor levels after doxorubicin with or without co-administration of Mesna—a randomized, cross-over clinical study. PLoS One 10(4):e0124988. https://doi.org/10.1371/journal.pone.0124988

54. Kreckel J, Anany MA, Siegmund D, Wajant H (2019) TRAF2 controls death receptor-induced Caspase-8 processing and facilitates proinflammatory signaling. Front Immunol 10:2024. https://doi.org/10.3389/fimmu.2019.02024
55. Sun M, Dawood F, Wen WH et al (2004) Excessive tumor necrosis factor activation after infarction contributes to susceptibility of myocardial rupture and left ventricular dysfunction. Circulation 110(20):3221–3228. https://doi.org/10.1161/01.Cir.0000147233.10318.23
56. Daugan A, Grondin P, Ruault C et al (2003) The discovery of tadalafil: a novel and highly selective PDE5 inhibitor. 2: 2,3,6,7,12,12a-hexahydropyrazino[1′,2′:1,6]pyrido[3,4-b]indole-1,4-dione analogues. J Med Chem 46(21):4533–4542. https://doi.org/10.1021/jm0300577
57. Andersson KE (2018) PDE5 inhibitors—pharmacology and clinical applications 20 years after sildenafil discovery. Br J Pharmacol 175(13):2554–2565. https://doi.org/10.1111/bph.14205
58. Arif SA, Poon H (2011) Tadalafil: a long-acting phosphodiesterase-5 inhibitor for the treatment of pulmonary arterial hypertension. Clin Ther 33(8):993–1004. https://doi.org/10.1016/j.clinthera.2011.06.008
59. Mónica FZ, De Nucci G (2019) Tadalafil for the treatment of benign prostatic hyperplasia. Expert Opin Pharmacother 20(8):929–937. https://doi.org/10.1080/14656566.2019.1589452
60. Corbin JD, Francis SH (2002) Pharmacology of phosphodiesterase-5 inhibitors. Int J Clin Pract 56(6):453–459
61. Kukreja RC, Das A, Koka S, Samidurai A, Xi L (2023) Nitric oxide-cGMP-PKG signaling in the cardioprotective effects of phosphodiesterase 5 inhibitors. In: Ray A, Gulati K (eds) Nitric oxide: from research to therapeutics. Springer International Publishing, Cham, pp 111–126
62. Koka S, Xi L, Kukreja RC (2020) Chronic inhibition of phosphodiesterase 5 with tadalafil affords cardioprotection in a mouse model of metabolic syndrome: role of nitric oxide. Mol Cell Biochem 468(1):47–58. https://doi.org/10.1007/s11010-020-03710-0
63. Koka S, Aluri HS, Xi L, Lesnefsky EJ, Kukreja RC (2014) Chronic inhibition of phosphodiesterase 5 with tadalafil attenuates mitochondrial dysfunction in type 2 diabetic hearts: potential role of NO/SIRT1/PGC-1α signaling. Am J Physiol Heart Circ Physiol 306(11):H1558–H1568. https://doi.org/10.1152/ajpheart.00865.2013
64. Varma A, Das A, Hoke NN, Durrant DE, Salloum FN, Kukreja RC (2012) Anti-inflammatory and cardioprotective effects of tadalafil in diabetic mice. PLoS One 7(9):e45243. https://doi.org/10.1371/journal.pone.0045243
65. Šimůnek T, Štěrba M, Popelová O, Adamcová M, Hrdina R, Geršl V (2009) Anthracycline-induced cardiotoxicity: overview of studies examining the roles of oxidative stress and free cellular iron. Pharmacol Rep 61(1):154–171. https://doi.org/10.1016/S1734-1140(09)70018-0
66. Yen HC, Oberley TD, Vichitbandha S, Ho YS, St Clair DK (1996) The protective role of manganese superoxide dismutase against adriamycin-induced acute cardiac toxicity in transgenic mice. J Clin Invest 98(5):1253–1260. https://doi.org/10.1172/jci118909
67. Serarslan Y, Yönden Z, Ozgiray E et al (2010) Protective effects of tadalafil on experimental spinal cord injury in rats. J Clin Neurosci 17(3):349–352. https://doi.org/10.1016/j.jocn.2009.03.036
68. Oka T, Maillet M, Watt AJ et al (2006) Cardiac-specific deletion of Gata4 reveals its requirement for hypertrophy, compensation, and myocyte viability. Circ Res 98(6):837–845. https://doi.org/10.1161/01.RES.0000215985.18538.c4
69. Aries A, Paradis P, Lefebvre C, Schwartz RJ, Nemer M (2004) Essential role of GATA-4 in cell survival and drug-induced cardiotoxicity. Proc Natl Acad Sci U S A 101(18):6975–6980. https://doi.org/10.1073/pnas.0401833101
70. Koka S, Kukreja RC (2010) Attenuation of doxorubicin-induced cardiotoxicity by tadalafil: a long acting phosphodiesterase-5 inhibitor. Mol Cell Pharmacol 2(5):173–178
71. Takimoto E, Champion HC, Li M et al (2005) Chronic inhibition of cyclic GMP phosphodiesterase 5A prevents and reverses cardiac hypertrophy. Nat Med 11(2):214–222. https://doi.org/10.1038/nm1175

72. Das S, Maulik N, Das DK, Kadowitz PJ, Bivalacqua TJ (2002) Cardioprotection with sildenafil, a selective inhibitor of cyclic 3′,5′-monophosphate-specific phosphodiesterase 5. Drugs Exp Clin Res 28(6):213–219
73. Das A, Xi L, Kukreja RC (2008) Protein kinase G-dependent cardioprotective mechanism of phosphodiesterase-5 inhibition involves phosphorylation of ERK and GSK3β*. J Biol Chem 283(43):29572–29585. https://doi.org/10.1074/jbc.M801547200
74. Prysyazhna O, Burgoyne JR, Scotcher J, Grover S, Kass D, Eaton P (2016) Phosphodiesterase 5 inhibition limits doxorubicin-induced heart failure by attenuating protein kinase G Iα oxidation. J Biol Chem 291(33):17427–17436. https://doi.org/10.1074/jbc.M116.724070
75. Li F, Jiang Q, Shi KJ, Luo H, Yang Y, Xu CM (2013) RhoA modulates functional and physical interaction between ROCK1 and Erk1/2 in selenite-induced apoptosis of leukaemia cells. Cell Death Dis 4(7):e708. https://doi.org/10.1038/cddis.2013.243
76. Das A, Durrant D, Mitchell C et al (2010) Sildenafil increases chemotherapeutic efficacy of doxorubicin in prostate cancer and ameliorates cardiac dysfunction. Proc Natl Acad Sci U S A 107(42):18202–18207. https://doi.org/10.1073/pnas.1006965107
77. Das A, Durrant D, Mitchell C, Dent P, Batra SK, Kukreja RC (2016) Sildenafil (Viagra) sensitizes prostate cancer cells to doxorubicin-mediated apoptosis through CD95. Oncotarget 7(4):4399–4413. https://doi.org/10.18632/oncotarget.6749
78. Booth L, Roberts JL, Cruickshanks N et al (2014) Phosphodiesterase 5 inhibitors enhance chemotherapy killing in gastrointestinal/genitourinary cancer cells. Mol Pharmacol 85(3): 408–419. https://doi.org/10.1124/mol.113.090043
79. Kanotra SP, Kanotra S, Gupta A, Paul J (2011) Chemoradiation in advanced head and neck cancers: a comparison of two radiosensitizers, paclitaxel and cisplatin. Indian J Otolaryngol Head Neck Surg 63:229–236
80. Sekijima T, Tanabe A, Maruoka R et al (2011) Impact of platinum-based chemotherapy on the progression of atherosclerosis. Climacteric 14(1):31–40. https://doi.org/10.3109/13697137.2010.522278
81. Saleh RM, Awadin WF, El-Shafei RA, Elseady YY, Wehaish FE, Elshal MF (2015) Cardioprotective role of tadalafil against cisplatin-induced cardiovascular damage in rats. Eur J Pharmacol 765:574–581. https://doi.org/10.1016/j.ejphar.2015.09.015
82. Nagashima Y, Tochinai R, Sekizawa SI et al (2023) Pretreatment with tadalafil attenuates cardiotoxicity induced by combretastatin A4 disodium phosphate in rats. J Toxicol Pathol 36(3):151–158. https://doi.org/10.1293/tox.2022-0143
83. Pantziarka P, Sukhatme V, Crispino S, Bouche G, Meheus L, Sukhatme VP (2018) Repurposing drugs in oncology (ReDO)-selective PDE5 inhibitors as anti-cancer agents. Ecancermedicalscience 12:824. https://doi.org/10.3332/ecancer.2018.824
84. Jin Z, Zhang J, Zhi H et al (2013) The beneficial effects of tadalafil on left ventricular dysfunction in doxorubicin-induced cardiomyopathy. J Cardiol 62(2):110–116. https://doi.org/10.1016/j.jjcc.2013.03.018
85. Barone I, Giordano C, Bonofiglio D, Andò S, Catalano S (2017) Phosphodiesterase type 5 and cancers: progress and challenges. Oncotarget 8(58):99179–99202. https://doi.org/10.18632/oncotarget.21837
86. Kuan J, Brock G (2002) Selective phosphodiesterase type 5 inhibition using tadalafil for the treatment of erectile dysfunction. Expert Opin Investig Drugs 11(11):1605–1613. https://doi.org/10.1517/13543784.11.11.1605
87. Montorsi F, Verheyden B, Meuleman E et al (2004) Long-term safety and tolerability of tadalafil in the treatment of erectile dysfunction. Eur Urol 45(3):339–344. https://doi.org/10.1016/j.eururo.2003.11.010. discussion 44–5

Chapter 40
Physical Medicine and Rehabilitation Strategies in Cancer Treatment-Induced Cardiotoxicity

Manas Aavula, Sugeeth Kandikattu, Devendra K. Agrawal, and Marcel P. Fraix

Abstract Cardiotoxicity encompasses a plethora of different conditions that damage the heart muscle and are associated with diminished function of the heart. Common examples of diseases that this encompasses are arrhythmias, coronary artery disease (CAD), heart failure, and myocardial infarction. Cardiotoxicities are commonly caused by cancer treatments such as chemotherapy, radiation therapy, anthracyclines, etc. This is why it is very important to monitor cancer patients and cancer survivors and promote preventative/maintenance strategies to prevent the occurrence/reoccurrence of cardiotoxicity. A prominent therapy utilized is increasing physical activity. Exercise can provide an array of benefits that function to improve cardiovascular health and mitigate the effects of cardiotoxicity. The benefits of exercise can spread to other comorbidities of cancer patients and help improve the overall condition by helping alleviate those symptoms as well. These include comorbidities such as tobacco use side effects, psychosocial effects, lipid management, and diabetes management. Increased physical activity has been proven to show improvements in these comorbidities by increasing quality of life, preventing malignancy occurrence/reoccurrence, and improving a patient's overall well-being.

M. Aavula · S. Kandikattu · D. K. Agrawal
Department of Translational Research, College of Osteopathic Medicine of the Pacific, Western University of Health Sciences, Pomona, CA, USA
e-mail: Manas.aavula@westernu.edu; Sugeeth.kandikattu@westernu.edu; DAgrawal@WesternU.edu

M. P. Fraix (✉)
Department of Translational Research, College of Osteopathic Medicine of the Pacific, Western University of Health Sciences, Pomona, CA, USA

Department of Physical Medicine and Rehabilitation, College of Osteopathic Medicine of the Pacific, Western University of Health Sciences, Pomona, CA, USA

Section of Health Policy and Management, Department of Population Health Science, College of Osteopathic Medicine of the Pacific, Western University of Health Sciences, Pomona, CA, USA
e-mail: mfraix@westernu.edu

D. K. Agrawal et al. (eds.), *Cardiovascular Toxicity*, Advances in Biochemistry in Health and Disease 33, https://doi.org/10.1007/978-3-031-90598-8_40

Multiple facets of exercise have been shown to improve patient outcomes, but emphasis has been placed on personally tailored exercise regimens and resistance training which have shown better results. An issue that is hard to overcome is the physical limitations that patients experience when attempting to increase physical activity, but through personally tailored exercise regiments, those limitations can be considered and a proper routine that can be beneficial for the patient can be created without inducing extra psychosocial strain. Furthermore, if patients are not able to obtain this type of treatment, any increase in physical activity has been shown to improve patient outcomes. However, due to the lack of structure, patients are more likely to forgo exercise that can provide valuable health benefits. Due to these limitations, physicians do not emphasize exercise and its benefits. By prioritizing the benefits of exercise, physicians can push patients to actively engage in improving their own well-being, cardiovascular health/overall health, and reduce treatment for secondary comorbidities.

Keywords Anthracyclines · Cancer therapeutics · Cancer treatment · Chemotherapy · Exercise · Radiation therapy

Introduction

Cancer is an irregular growth of cells that can be benign or metastatic, which are cells that spread to other tissues or organs nearby. Cancers gain malignant potential through environmental and genetic mutations in important cellular mechanisms such as DNA damage and repair, cellular proliferation, and more. From 2016 to 2020 in the United States, the rate of occurrence of cancer was 3.75× greater in ages ≥65 years old in comparison to those <65 years old. There was a 15× greater mortality rate in ages ≥65 years old. The top 5 most common cancers were colorectal, lung, breast, prostate, and melanoma. There was a higher rate of cancer in men in comparison to females and in black individuals in comparison to white individuals, but certain racial groups may be at the forefront for incidence for certain types of cancer. For example, Native Americans and Hispanics have the highest rates of liver cancer. Incidence of cancer can increase in individuals due to exposure to various carcinogens such as cigarette smoking, ionizing radiation, UV radiation, viruses, bacteria, alcohol, genetics, and more. Of all the deaths associated with cancer in the United States, a staggering 48% are related cigarette smoking. Radiation exposure increases the chance of melanoma and nonmelanoma skin cancers. Infectious agents such as hepatitis B and hepatitis C increase the rate of liver cancer while *Helicobactor pylori* can increase the incidence of gastric cancer. Genetics are major risk factor for cancer growth and mutations in certain genes such as BRCA1 and BRCA2 increase the likelihood of breast and ovarian cancers. Cancers can be prevented in many ways with the main solution being limiting underlying risk factors. Decrease in tobacco use and obesity, vaccinations, limiting sun exposure, and more can decrease one's chance of having cancer [1].

If cancer does arise in individuals, there are many treatment strategies in society today to limit its spread or completely remove it from the body. If cancer is found in its earlier stages, surgery is on the best strategies while chemotherapy also has a high success rate but can lead to drug resistance and negatively impact healthy cells. With new advancements in technology, new forms of therapies are arising. Photodynamic therapy is an example of an immunotherapy that activates a patient's lymphocytes to fight cancer cells with the help of monoclonal antibodies that disrupt the function of immune checkpoints in cancer cells. Stem cell therapy involves the usage of undifferentiated cells from the bone marrow to regrow damaged tissues, but this method is still being studied. Pluripotent stem cells (iPSC) are involved in creating anti-tumor vaccines that stimulate the activation of effector T cells and natural killer cells. Gene therapy is based on the premise of removing a defective gene and replacing it with a healthy copy. Genes that are targeted in this method include tumor suppressor genes, creation of pro-apoptotic and chemo-sensitizing genes, and genes with the ability to initiate certain anti-tumor immune responses. A major problem with gene therapy is the immune response from the body that may tag viral vectors as foreign and attack them. Safer routes of introducing healthy genes into the body are being studied [2]. Another new therapy is the usage of cancer biomarkers to create personalized treatment based on each patient's molecular profile. With the molecular profile, doctors can figure out the exact cancer diagnosis and prognosis and create an effective treatment plan that attacks the specific type of tumor in a patient [3].

Cardiotoxicities

Cancer therapy has proven to be very toxic to the body and researching positive ways to mitigate the negative impact of these treatments on the body is of high importance to improve patient outcomes and standard of living. These therapies can be detrimental to many organ systems in the body such as the gastrointestinal, lymphatic, and dermatological systems, but this chapter is going to dive into the effects of cancer therapies on the cardiovascular system. Many of the typical cancer treatments can result in a plethora of damaging cardiotoxicities such as cardiomyopathy, arrhythmia, myocardial ischemia, hypertension, and ultimately heart failure [4].

Cardiomyopathy can be defined as a group of cardiac diseases that affect the heart muscles negatively, all of which make it more challenging for the heart to pump blood properly. Some examples of cardiomyopathies are dilated cardiomyopathy which is where the heart chambers become very enlarged and weak, hypertrophic cardiomyopathy where the heart muscles tend to fibrosis and thicken which impedes proper blood flow, and restrictive cardiomyopathy where the heart muscles become rigid, and resist expansion which limits the flow of blood [5]. While undergoing cancer therapies, drug interactions with cardiac tissue along with previous cardiac diagnoses increase the occurrence of cardiomyopathy and are even warned against for people with existing heart conditions due to its effects on cardiac tissue.

Arrhythmias are also a commonly noted side effect of cancer therapies and are defined as any variation in a heartbeat that disrupts the typical electrical impulses that are sent through the heart [6]. These can be deadly such as ventricular fibrillation which is known to cause sudden cardiac death or can be primarily asymptomatic such as premature heartbeats, also commonly known as "skipped beat" arrhythmias.

Myocardial ischemia during cancer therapy refers to a condition where blood flow to the heart muscle (myocardium) is decreased because of chemotherapy drugs, leading to a lack of sufficient oxygen for the proper function of the cardiac muscle [7]. This oxygen reduction can cause damage to the cardiac tissue, which resembles the effects of a heart attack, triggered by chemotherapy. Symptoms can include chest pain, shortness of breath, and other cardiac-related issues.

The most common pharmacological cancer therapies include anthracyclines, trastuzumab, and 5-fluorouracil [4]. Cardiomyopathies, arrhythmias, and myocardial ischemia are all induced by similar mechanisms when patients are given these chemotherapeutic drugs (Fig. 40.1). Anthracyclines such as doxorubicin are known to cause a reduction of NADH dehydrogenase in mitochondrial complex I leading to the formation of a semiquinone radical, which reacts with molecular oxygen to produce a superoxide radical [8]. This redox reaction results in the generation of hydrogen peroxide and the hydroxyl radical. Additionally, doxorubicin–iron complexes may facilitate a Fenton reaction which is an iron-driven conversion of hydrogen peroxide to hydroxyl radicals that contributes to the formation of reactive oxygen species [9]. Cardiomyocytes are more vulnerable to the oxidative stress induced by doxorubicin due to their dependence on oxidative metabolism and the high mitochondrial content compared to tumor cells are only affected at very high concentrations of doxorubicin [8]. As for Trastuzumab, analysis of its mechanism has shown that the interactions with the HER2 signaling pathway are directly responsible for its cardiotoxic effects. Trastuzumab interferes with HER signaling by phosphorylating specific sites on HER1 and HER2 at specific positions such as 845 and 1248, respectively. This phosphorylation activates the Erk/mTOR/Ulk 1 signaling pathway, which inhibits autophagy and increases the concentration of reactive oxygen species in cardiomyocytes, ultimately contributing to cardiotoxicity [10]. 5-Fluorouracil is the third leading option for cardiotherapies and its cardiotoxic effects are primarily hypothesized to be caused by vasospasm which triggers endothelial dysfunction. This occurs when the drug disrupts the function of the blood vessel lining, leading to an imbalance between vasoconstrictor and vasodilator substances, causing the vessels to constrict and trigger a vasospastic episode. This is believed to be due to the increased release of endothelin-1 (a vasoconstrictor) and reduced production of prostacyclin (a vasodilator) by the damaged endothelium [11].

Hypertension caused by cancer therapies refers to the onset of high blood pressure as a side effect of cancer treatment, typically triggered by chemotherapy drugs that impact blood vessel function, leading to elevated blood pressure [12]. This is more common with angiogenesis inhibitors, such as bevacizumab, which works

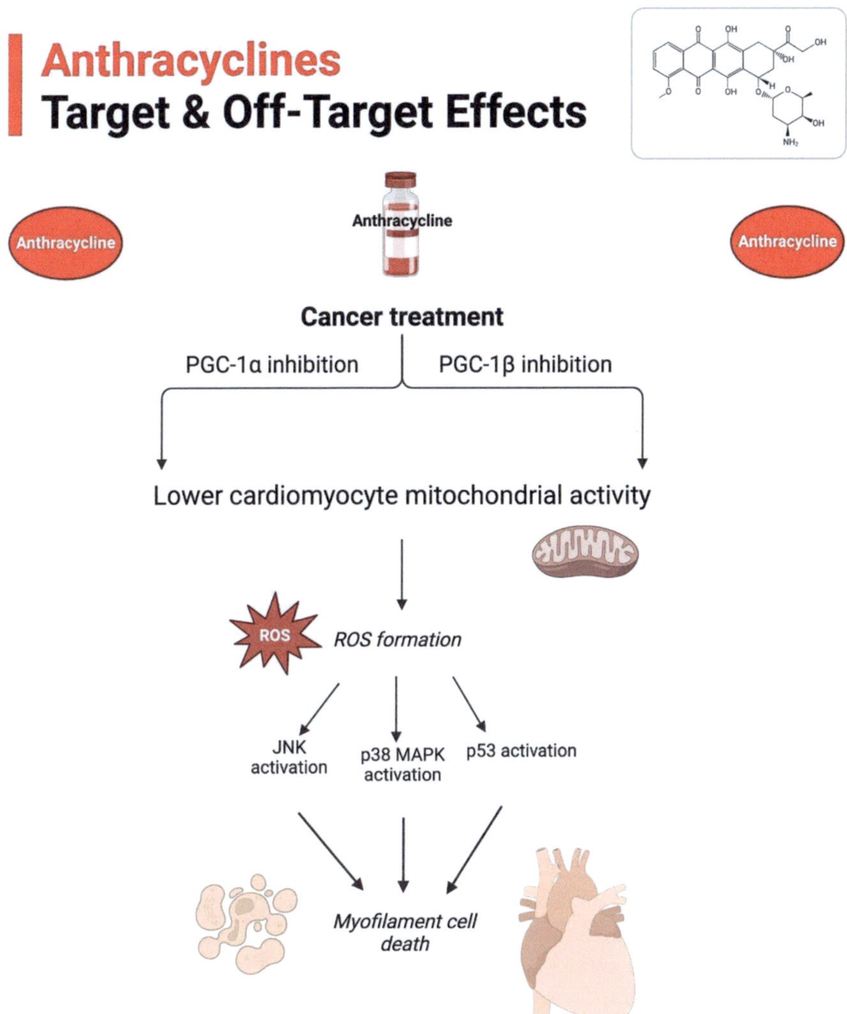

Fig. 40.1 Anthracycline directly inhibits certain genes, affecting mitochondrial activity which produces reactive oxygen species (ROS). This activates various proteins and signaling pathways, causing myofilament cell death

by blocking the formation of new blood vessels in tumors [13]. Angiogenesis inhibitors function by targeting Vascular Endothelial Growth Factor (VEGF) which can lead to hypertension by decreasing the production of nitric oxide (NO). Since NO functions as a potent vasodilator, the reduction in NO results in decreased vascular flexibility and increased peripheral vascular resistance which results in an increased prevalence of hypertension in patients on this medication [14]. As the other cardiotoxicities listed above are affected by pharmacological therapies through similar mechanisms, hypertension is slightly different. 5-Fluorouricil affects hypertension through the same mechanism of coronary vasospasm which eventually restricts blood flow and increases blood pressure [15]. Trastuzumab is shown to have little effect on hypertension and is not a major contributor to elevated blood pressure. Anthracyclines have a higher risk of causing hypertension by inducing oxidative stress, disrupting endothelial function, and interfering with the renin–angiotensin–aldosterone system (RAAS), which is an important hormone pathway that tightly regulates and could lead to elevated blood pressure if not functioning properly [16].

Physical Activity

Physical activity refers to an array of movements done by the skeletal muscles that yield energy expenditure. Categories of physical activity include sports, job-related, household, and others. Exercise is a category of physical activity that enhances and preserves one's physical fitness [17]. Physical activity has been shown to benefit physiological systems by preventing diseases or mitigating the impact of various health conditions. If a patient is diagnosed with a chronic illness, treatment outcome is better when exercise is part of the plan. Consistent exercise can lead to an improvement in one's well-being and longevity [18]. Physical activity can decrease fatigue and strengthen physical functioning. In patients with cancer, studies have shown that exercise can enhance the frequency of individuals who complete chemotherapy [19]. Exercise can range from low-intensity exercises, such as stretching and slow walks, to high-intensity exercises [20]. Combining endurance with weight training can help alleviate adverse cardiovascular effects from chemotherapy [21]. Creating a physical activity program that is tailored to each patient's limitations and motivations can motivate cancer patients to keep up with routine exercise. It can help limit the overall negative impact on the cardiovascular system for patients undergoing chemotherapy [22]. Although exercise has proven benefits, it is still not established as a nonpharmacological strategy in cardio-oncology care [23].

Anthracyclines are antineoplastic drugs that are used against a wide array of cancers such as solid tumors and hematological malignancies. Anthracyclines work to inhibit topoisomerase II, producing hydroxyl free radicals that affect myocardial tissue in the process. Dose-related effects of anthracyclines can lead to myelosuppression, a decrease in bone marrow activity which leads to a decrease in the production of red blood cells, and dose-related cardiotoxicity, meaning the more anthracyclines a cancer patient uses over time, the higher the risk for cardiotoxicity [24]. These drugs can decrease the efficacy of cancer chemotherapy due to possible cardiovascular dysfunctions. Overall myocardial damage and lower LV ejection fraction can occur through continued anthracycline damage [25]. Anthracyclines specifically repress two genes, PGC-1α and PGC-1β, lowering cardiomyocyte mitochondrial activity. This produces reactive oxygen species (ROS) that activates p53, a tumor suppressor gene, p38 mitogen-activated protein kinase (MAPK), and c-JUN N-terminal kinases (JNKs), the latter two of which are signaling pathways. These pathways initiate programmed cell death of myofilaments, modifying cardiac structure [26]. Exercise therapy activates various biochemical and physiological adaptations through the cardiovascular system which has proven to limit anthracycline-induced cardiotoxicities [27].

Newer studies are proposing the use of physical therapy, in addition to traditional therapies, to decrease the side effects of anti-neoplastic therapies. One study found that pretreatment of treadmill running led to an increase in HSP70, a cardioprotective marker, and superoxide dismutase (SOD) with a decrease in oxidants and cardiac damage, seen with a reduction in MDA, CK, CPK-MB, and CK/CPK-MB [28]. Other studies have found that exercise increases the amount of cardiac progenitor cells (CPC) in the heart. Aerobic training decreases the formation of ROS, and oxidative stress, and allows for myofilament protein synthesis [26]. To further prove this point, a study done by Bellissimo et al. used women diagnosed with breast cancer in stages I–III to be placed in two groups. The first group was the physical activity group; these women had better exercise capacity. They decreased cardiac dysfunction when studied over 3 months in comparison to patients with lower physical activity. The second group, who had lower physical activity, displayed decreased exercise capacity and a decline in cardiac function seen through left ventricle (LV) end-systolic volume, LV affection fraction, and LV circumferential strain [29].

Cardiorespiratory fitness (CRF) is an analysis of the cardiovascular function that decreases by 5–26% due to a wide array of treatments that may not revert to normal levels after treatment cessation. It was found that chemotherapy including anthracyclines rushed the cardiovascular aging process, as seen through the increase in expression of $p16^{INK4}\,\alpha$, a cellular senescence marker. The increase in this marker seen after 12 months was equivalent to an increase seen after 14.7 years of normal aging. One review analyzed various studies to determine the relationship between exercise and cardiovascular outcomes, during or after primary cancer therapy [30]. CRF is measured with VO_{2peak} which is maximal oxygen consumption in aerobic exercise. Poor CRF is related to treatment-associated toxicities, postoperative issues, and cancer mortality. CRF can be measured in three ways: cardiopulmonary exercise testing (CPET), incremental submaximal exercise testing, and the 6-minute walk test. The American Heart Association created an exercise training plan called CORE (Cardio-Oncology Rehabilitation) that suggests individualized programs that include aerobic and resistance components [23]. In a review that looked at high-intensity interval training (HIIT), which are short segments of high-intensity workouts, in breast cancer survivors, cardiovascular function was improved after chemotherapy [31]. The combined approach with aerobic and resistance exercise will increase one's muscle strength, CRF, physical function, and body structure [23]. One study that involved 280 cancer patients found that after 12 weeks of aerobic training with resistance training 2 days/week, patients displayed enhanced 6-minute walk duration, arm strength, and leg strength. Implementation of cardiac rehabilitation will improve CVD outcomes over a long timeframe, reduce potential risk factors in cancer patients, provide an approach that caters to each individual's needs, and provide a variety of other interventions [32]. In another study that studied breast cancer survivors who were treated with anthracyclines, an experimental group of women was required to perform aerobic and resistance exercises 3–4 days per week. The study looked to measure VO_{2peak}, functional disability, and cardiac reserve, which is the change in the heart's response to peak and resting exertion. Cardiac reserve was measured with exercise cardiac magnetic resonance (CMR) which determined the values of left and right ejection fraction, cardiac output, and stroke volume mean. Exercise saw improvements in VO_{2peak}, cardiac output, stroke volume, and left and right ventricular ejection fraction. Troponin levels rose less in exercise patients after chemotherapy [33].

Interestingly, another study that included patients >1 year and <5 years after their first chemotherapy looked to compare VO_{2peak} in patients with standard linear training (LET) and non-linear exercise training (NLET) to attention control (AC). The intensity of each exercise session was characteristic of each individual based on their pre-experimental performance on CPET. VO_{2peak} increased in the LET and NLET in comparison to AC but produced only moderate improvements in CRF. VO_{2peak} increased by about 20% in healthy middle-aged females but the same level of improvement was not seen in breast cancer patients, suggesting that the cardiotoxic effects of chemotherapy play a prominent role in affecting VO_{2peak} response [34]. In another case study that looked at two cancer survivors, one who had Hodgkin's lymphoma and one who had leukemia, both patients were put through a 16-week exercise regimen 3×/week for 30 min at an intensity that was 50% of the patient's heart rate reserve. Although a small sample size, both patients had better VO_{2peak} and quality of life, but only one patient's ejection fraction improved [35].

Trastuzumab is another chemotherapeutic agent that leads to cardiotoxicity, used in breast cancer patients who are ERBB2-positive. Trastuzumab used after anthracyclines increases the risk of cardiotoxicity and lead to a significantly lower VO_{2peak} and significant post-exercise decrease in LV ejection fraction. One group consisted of cardiotoxic patients and researchers found this group to have a higher resting heart rate, systolic blood pressure, and diastolic blood pressure in comparison to the non-cardiotoxic group that was also administered trastuzumab [36]. VO_{2peak} is associated with worsening quality of life because of increased fatigue and decreased exercise tolerance [36, 37]. Chemotherapy negatively impacts CRF due to worsening of oxygen delivery instead of improper metabolic muscle function [38]. As breast cancer outcomes continue to improve, patients will have a heightened risk for cardiotoxicity and impaired CRF, therefore, prevention measures are needed to mitigate these risks [36].

Doxorubicin is an anthracycline used for different types of cancers, such as lymphomas, sarcomas, and a broad spectrum of solid tumors, that produces reactive oxygen species as a side effect, leading to cardiotoxicity. Exposure to doxorubicin leads to a reduction in cardiac mass, leading to atrophy of cardiac muscle. Doxorubicin leads to myocardial fibrosis, dilated ventricles, and cardiac muscle degradation. Additionally, this drug activates caspase-3, a mediator of apoptosis that cleaves crucial cellular proteins, leading to the apoptosis of cardiac muscle cells [39]. Aerobic exercise has started to be used as a non-pharmacological approach that limits cardiotoxicities and helps increase systolic and diastolic function [40]. It prompts epigenetic modifications to reduce cardiovascular disease. This occurs through enzymes, such as histone deacetylases, signaling pathways, and metabolites. Certain proteins such as FGF21, osteocrin, and fibronectin type III domain-containing 5 (FNDC5) are associated with exercise and play a protective role in the heart by decreasing apoptosis and oxidative damage in the hearts of doxorubicin-treated patients. Additionally, exercise has been shown to have a protective impact on offspring during pregnancy from cardiotoxicity due to doxorubicin [41]. Exercise training protects the body from doxorubicin-induced cardiotoxicities through a multitude of pathways such as an increase in antioxidant machinery, proapoptotic signaling, cardiac progenitor cells, decrease in myocyte turnover, and the prevention of calcium overload [42]. Doxorubicin decreases mitochondrial biogenesis, complex I activity, and complex V while increasing acetylation and oxidative stress. A study that had patients perform physical exercise before and during the administration of doxorubicin has led to better mitochondrial bioenergetic fitness [43]. In another study that looked at mice treated with doxorubicin, they found that they had lower affection fraction and fractional shortening compared to a group treated with doxorubicin and exercise. Doxorubicin-induced cardiotoxicity included autophagy along with irregular mitochondria. A decline in the number of pericytes led to a change in the vascular architecture. After 12 weeks of running the study, 67% of control mice, 22% of mice treated with just doxorubicin, and 56% of mice that received doxorubicin and exercise survived. This proves that exercise helps limit blood vessel damage from doxorubicin, allowing normal cardiac blood flow [44]. Furthermore, glutathione, a non-enzymatic antioxidant, suppresses harmful free radical chain reactions, but doxorubicin increases the level of oxidized glutathione (GSSG) in cardiac tissue. Higher levels of GSSG suggest that there is a buildup of oxidative byproducts in the heart. However, mice put through an endurance swimming regimen saw an 18.1% decrease in oxidative protein damage compared to a control group who did not exercise. Endurance-trained mice had lower GSSG levels in comparison to the control group, proving the ability of the cardiac tissue to take in more GSH and limit damage to the heart from oxidative damage caused by doxorubicin [39].

Lipid Management

Poor lipid management can be defined as high cholesterol, high triglycerides, and low high-density lipoprotein levels. This is a highly prevalent issue in the world today as poor lipid management can lead to detrimental bodily functions such as increasing levels of cardiotoxicities which can impair the function of the heart. The heart primarily uses ATP as its source of energy which is oxidized from fatty acids for use in healthy myocardium. When poor lipid management in an individual occurs, this leads to an accumulation of fatty acids/lipids in the heart resulting in cardiac lipotoxicity [45]. The excess presence of lipids in the heart causes accumulation in areas that can impede the proper flow of blood and hinder the function of

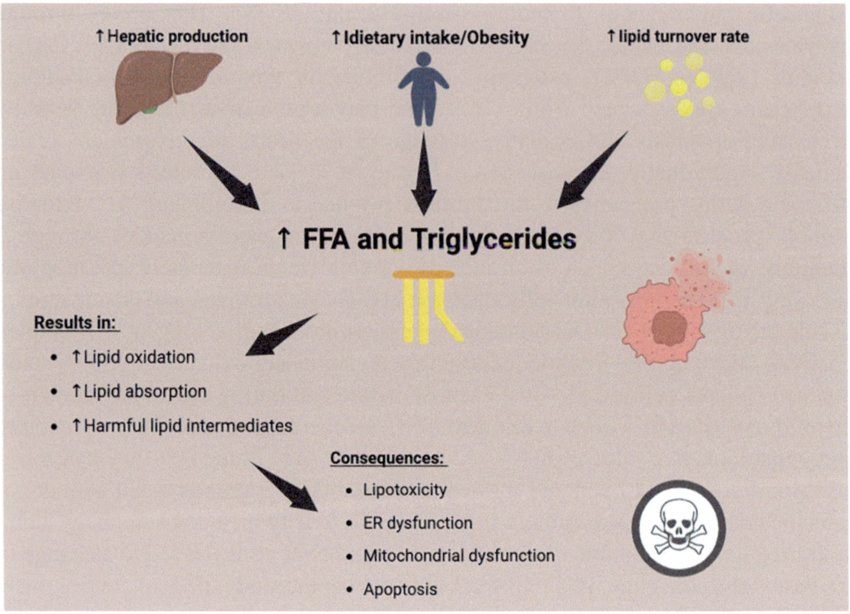

Fig. 40.2 Various effects of increased levels of free fatty acids and triglycerides causing lipotoxicities in the body

myocardial function causing cardiac fibrosis [46]. Another effect of lipotoxicity in the heart is mitochondrial dysfunction. The excess deposition of lipids in the mitochondria increases the release of ROS (reactive oxygen species), disrupts the mitochondrial structure which is directly associated with its function, and endoplasmic reticulum dysfunction. Additionally, it negatively affects RYR2 receptors, BCL2, and IP3R proteins which play key roles in the mitochondria to regulate the levels of Ca^{2+} which is the primary ion that is involved in cardiac contractility and regulating apoptosis of cells [47]. Cardiac lipotoxicity contributes to cardiotoxicity by causing an excessive buildup of lipids within heart muscle cells (cardiomyocytes). The build-up of lipids disrupts normal cellular functions, triggering stress pathways such as oxidative stress and endoplasmic reticulum stress. This results in continuous cardiomyocyte damage which goes on to impair heart muscle contractility and potentially lead to heart failure [48]. So as chemotherapeutic mechanisms disrupt the function of lipid metabolism and increase the deposition of lipids in the heart, it gives rise to more cardiotoxicity such as myocardial ischemia, arrhythmias, and cardiomyopathies [49]. These effects are primarily caused by poorly controlled diabetes, obesity, insulin resistance, and improper diet which can be corrected by lifestyle changes, pharmaceutical intervention, and novel treatments.

Increased levels of free fatty acids and triglycerides results in the development of lipotoxicities in the body (Fig. 40.2). The process of lipid deposition is exacerbated in certain types of cancer and by the treatments for cancer. Due to cancer cells relying on Fatty acids/lipids for energy, studies have shown that they undergo de novo lipogenesis to produce fatty acids and cholesterol to produce more ATP through the TCA cycle [50]. In certain cancers, despite adequate intake of fatty acids in a person's normal diet, cancer cells tend to synthesize up to 95% of the saturated and unsaturated fatty acid due to de novo lipogenesis (Fig. 40.2). Further studies have proven that cancer cells utilize both lipogenesis and lipolysis pathways to provide energy for cancer cell proliferation and survival [51]. A primary effector for these results has been the fatty acid synthase pathway (FASN). This is a metabolic process that involves the synthesis of long-chain saturated fatty acids, such as palmitate, from acetyl-CoA and malonyl-CoA. FASN affects not only fatty acid synthesis but also the glycolytic process. FASN overexpression has been shown in a plethora of cancers such as breast, colorectal, and ovarian cancers. This overexpression has been shown to increase the number of lipotoxic molecules in the body increase tumor burden, exacerbate later stages of disease, and have a poorer prognosis overall [52]. FASN also ties into other metabolic pathways and indirectly increases the required molecules to stimulate fatty acid production in cancer cells. In colorectal cancer cell lines, suppression of FASN expression has been proven to impair the glycolytic capacity and reserves of HT29 and HCT116 cells. Studies show that inhibition of the glycolysis pathway is linked to lipid synthesis-derived NADP+, which increases the cytosolic pool of NADP+ to maintain glycolysis activity and downregulates mitochondrial-bound hexokinase activity. In breast cancer cell lines, HER2 (protein receptor for cell growth and development) is activated by heregulin-β1 (HRG-β1-modulates cell growth and proliferation) to increase rates of glycolysis in high FASN and HER2-expressing SKBR-3 cells. HER2 overexpression in low FASN and HER2-expressing MCF-7 (human breast cell cancer line) cells, however, can increase FASN expression and glycolysis after the introduction of HRG-β1. Subsequently, the lack of FASN activity contradicts these observations, which shows the connection between FASN and glycolysis [53]. Another very prevalent protein that is highly associated with cancer is the sterol regulatory element-binding proteins the sterol regulatory element-binding proteins (SREBPs), which are a series of transcriptional factors that regulate cellular lipid homeostasis. Inappropriate activation of SREBPs has been proven to cause obesity, fatty liver disease, and insulin resistance, and could also be involved in cancer development [54]. SREBP-1 was revealed to be highly activated in cancers and has been proven to connect oncogenic signaling-regulated glucose and glutamine metabolism to de novo lipogenesis. SREBP-1 does this by activating and stabilizing the PI3K/atk pathway, a very important developmental pathway that controls growth, proliferation, and cell cycle progression [55]. Due to these pathways being heavily associated with an increased rate of cellular proliferation and production of FAs for energy, there is an increased risk of malignancy as well as the detrimental effects of increased lipids in the body stated above. Just as elevated lipid levels are elevated while cancers are rapidly dividing and using them for energy, chemotherapeutic drugs hindering the lipid breakdown pathways and can cause dyslipidemia. Doxorubicin has been linked to a decrease in mRNA transcription of ABCA1. Studies show that doxorubicin-treated myocytes showed elevated levels of cholesterol and cholesterol precursors. Additionally, studies show that after treatment with anthracyclines, there was a significant increase in apolipoprotein B, a known risk factor for atherosclerotic cardiovascular disease (ASCVD) [56]. Another example is the off-target effects of tyrosine kinase inhibitors (TKIs) which have been shown to disrupt lipid metabolism and contribute to cardiotoxic side effects. A popular TKI, Imatinib, targets BCR-ABL1, which reduces the cytoplasmic phosphorylation of LDL receptor-related protein, a crucial component in LDL signaling. This protein is involved in activating lysosomal enzymes, regulating glucose-induced insulin secretion, and controlling cholesterol metabolism [49]. Even though these prevalent side effects may cause some stress and limitations on the patients that are taking these therapies, but there are a variety if different ways that they can prevent and maintain their overall health.

Lifestyle changes are a modality that is universally promoted for the treatment of most morbidities and includes diet alterations, exercise, smoking cessation, and weight management. The effects of effective exercise have been proven to reduce lipid deposition and decrease the effects of cardiac lipotoxicity. For a plethora of patients who present with comorbidities such as diabetes and obesity, low or moderate-level exercise has been effective in reducing cholesterol levels and low-density lipoprotein levels [57]. Another benefit of exercise is increasing lipid turnover rates. Increased energy expenditures preferentially use lipids as the primary source of energy so engaging in any level of exercise can be beneficial. This can decrease the deposition of lipids in the heart and increase the breakdown/turnover rate of existing stores. This decreases the stress on the heart which can improve overall cardiovascular health and decrease the chance of cardiotoxicities from occurring [58]. Additional benefits are increased lipolysis, decreased fatty update, and increased mitochondrial activity which are all elevated by exercise levels and collectively also help decrease lipid accumulation in the body which prevents patients from experiencing the detrimental effects of cardiac lipotoxicity.

Diabetes Management

Sodium-glucose co-transporter-2 (SGLT-2) inhibitors are a type of medications that controls diabetes. Its function is to limit glucose reabsorption in the proximal convoluted tubule by inhibiting the SGLT-2 receptor. This receptor controls 90% of glucose reabsorption while SGLT-1 in the small intestine controls 10% of glucose reabsorption in the body. Empagliflozin, an SGLT-2 inhibitor, has limited serious cardiovascular events in diabetic patients who had previous cardiovascular problems. Empagliflozin led to lower heart failure hospitalizations while dapagliflozin exhibited a decreased heart failure risk and mortality from cardiovascular-related causes. Since SGLT-2 inhibitors work beyond just glycemic control, studies are being done to see if they lower anthracycline-related cardiotoxicity [59]. SGLT-2 inhibitors play a role in increasing the survival of cardiomyocytes, protect against degenerative abnormalities, promote autophagy, and lower cell death. They also reduce inflammation by limiting the effects of reactive oxygen species and the production of inflammatory cytokines. SGLT-2 inhibitors can maximize the effectiveness of chemotherapy by minimizing adverse reactions on the heart [60]. One study looked at patients who had cancer and diabetes mellitus treated with anthracyclines and SGLT-2 inhibitors in comparison to a control group with cancer and diabetes, but only treated with anthracyclines. The occurrence of cardiac events such as heart failure and cardiomyopathies was lower among patients treated with SGLT-2 inhibitors. A decrease in the rate of mortality was seen as well [61]. In another study, empagliflozin was given to mice who had doxorubicin-induced cardiomyopathies to determine its effect on left ventricular function. Results found that doxorubicin itself causes a significant impairment of systolic and diastolic left ventricular function. However, mice that were also given empagliflozin saw an improvement in ejection fraction and circumferential strain and a decrease in myocardial fibrosis [62]. In another study, cardiomyocytes exposed to doxorubicin had increased Ca^{2+} levels and pro-inflammatory markers related to cell death. Adding empagliflozin to the treatment lowered pro-inflammatory cytokines such as NLRP3, MyD88, and NF-kB. An increase in ejection fraction and fractional shortening was seen [63].

Tobacco Cessation

Tobacco use is a highly prevalent issue that has led to a plethora of dysfunctions in various cells and tissues in the body (Fig. 40.3). Approximately every 1 in 5 people are likely to use some form of tobacco. The detrimental effects of tobacco are well known through past research, and it has been shown to increase the risk of cancer, lead to increased cardiotoxicity, and decrease adherence to exercise regimens that could be beneficial in preventing worsening in symptoms. Tobacco causes cardiotoxicity by initiating a series of effects, including increased blood pressure, elevated heart rate, and artery narrowing due to vasoconstriction (Fig. 40.3). The vasoconstriction leads to ischemia throughout the cardiac muscles, increased rick if clot formation, and damage to the endothelial lining of blood vessels, all of which increase the chances of heart failure [64].

Tobacco use has been closely tied to causing many types of cancers such as the larynx, mouth, throat, lung, kidney, stomach, and liver, and increases the risk of acute myeloid leukemia [65]. This has been proven due to tobacco products containing carcinogens and nicotine. Though nicotine may not be a direct carcinogenic, it is addictive and is also associated with tumor cell proliferation. The primary cause of cancer from tobacco usage is carcinogens. These are chemicals that are known to cause many different types of cancer depending on the method of tobacco use (Fig. 40.3). The detrimental effects of carcinogens are that they cause DNA damage which inhibits its function, alter DNA repair due to their effects of hindering repair mechanism, and increase the speed of cell division which increases tumor cell production [66]. This is likely to increase proliferation rates of cancers while hindering repair pathways for the damaged endothelial cells that were caused by

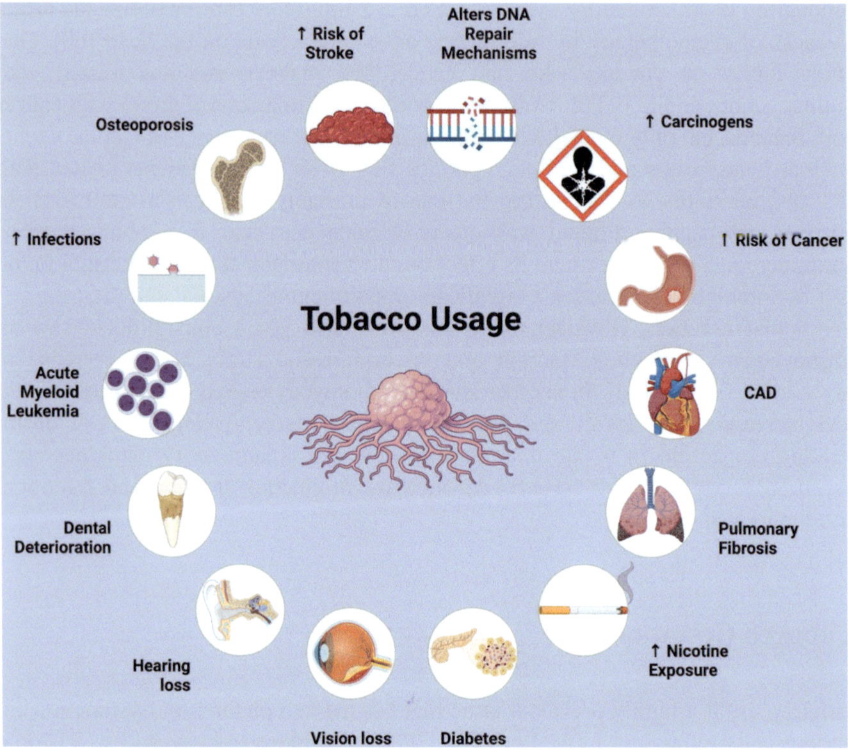

Fig. 40.3 This image shows the plethora of different effects of tobacco usage and delineates why it is in the patients' best interest to cease tobacco use

tobacco use. Some examples of the most powerful carcinogens are PAHs, *N*-nitrosamines, and aromatic amines, which are powerful enough to be present in lower amounts, and some of the weaker carcinogens that can occur in higher amounts, such as acetaldehyde and isoprene [67]. The presence of these strong and weak carcinogens contributes to the negative effects that are typically seen that eventually lead to cardiotoxicity that hinder the patient's quality of life and chances of recovery (Fig. 40.3). Due to the excessive carcinogen production using tobacco, the primary method of improving a patient's condition is to cease tobacco usage. There is a plethora of different therapies utilized by physicians to aid patients overcome this addictive tendency. Most popularly, pharmacological intervention is used where patients are provided a variety of pharmaceutical options designed to alleviate nicotine withdrawal symptoms, curb cravings, and enhance the prospects of successful smoking cessation, with common examples being nicotine replacement therapy (NRT), bupropion, and varenicline. Another therapy that is consistently used is behavioral therapy. The aim is to enhance the patient's self-control over their tobacco use behavior by altering their cognitive thought process and altering tobacco use behavior through various techniques, such as goal setting (e.g., establishing a quit day), self-monitoring, providing rewards, and offering skill training to help users resist their urges to engage in consistent use [68]. Lastly, a treatment that has had the most success is combination therapy which is a combination of different pharmacological and behavioral therapies that are personalized to the patients and have shown great improvement. Employing these various therapies not only decreases the deterioration of health in the patient but also decreases the rising risk of developing malignancies, and more importantly, always patients to pursue and engage in proactive treatments such as exercise. Exercise is an optimal choice of treatment for smoking cessation by relieving nicotine withdrawal symptoms and smoking cravings, while also being a highly accessible method. Consistent physical activity has been found to reduce nicotine withdrawal symptoms and smoking cravings, as well as increase beta-endorphin levels which are associated with a decrease in relapse [69]. Overall, it plays an important role in preventing cancers and cardiotoxicity, improving patient outcomes, and managing existing conditions as well.

Psychosocial Management

An overlooked aspect of cardiotoxicities and cancer is the psychosocial aspects and how those effects further deteriorate the patient's health. Prominent factors when considering analyzing the psychosocial aspect of these patients are depression, anxiety, and stress. Though depression itself cannot cause cancer, it does play a larger role in managing such conditions and more such as coronary heart disease (CAD). Depression has been shown to lead to a poorer quality of life (QOL) and higher rates of mortality in cancer. A meta-analysis revealed that minor or major depression increases mortality rates by up to 39% and even patients without depression but those who have displayed depressive symptoms have a 25% increased risk of mortality [70]. Anxiety also has a troubling effect on the patient's outcomes and maintenance of major conditions. Anxiety, along with the associated increase in physiological arousal, can heighten functions controlled by the autonomic nervous system, such as pain, nausea, and vomiting. This can create a cycle where anxiety intensifies further. Additionally, anxiety can cause behavioral paralysis, preventing individuals from utilizing effective coping strategies. As a result, anxiety may negatively impact a patient's ability to comply with treatment [71]. It has been shown that anxiety can lead to behavioral paralysis where patients become overwhelmed due to their condition/diagnosis, therefore they struggle to follow the physician's suggestions, avoid doctors' appointments, and even cease preventative treatments. This commonly leads to a poorer prognosis. Unlike the other two factors, chronic stress has been shown to increase the risk of cancer. Stress hormones have been shown to promote cancer and coronary diseases through mechanisms such as inducing DNA damage, accelerating the degradation of p53, triggering arrhythmias, and altering the tumor microenvironment [72]. Chronic stress has been shown to increase the inflammatory response, leading to crosslinking between inflammatory

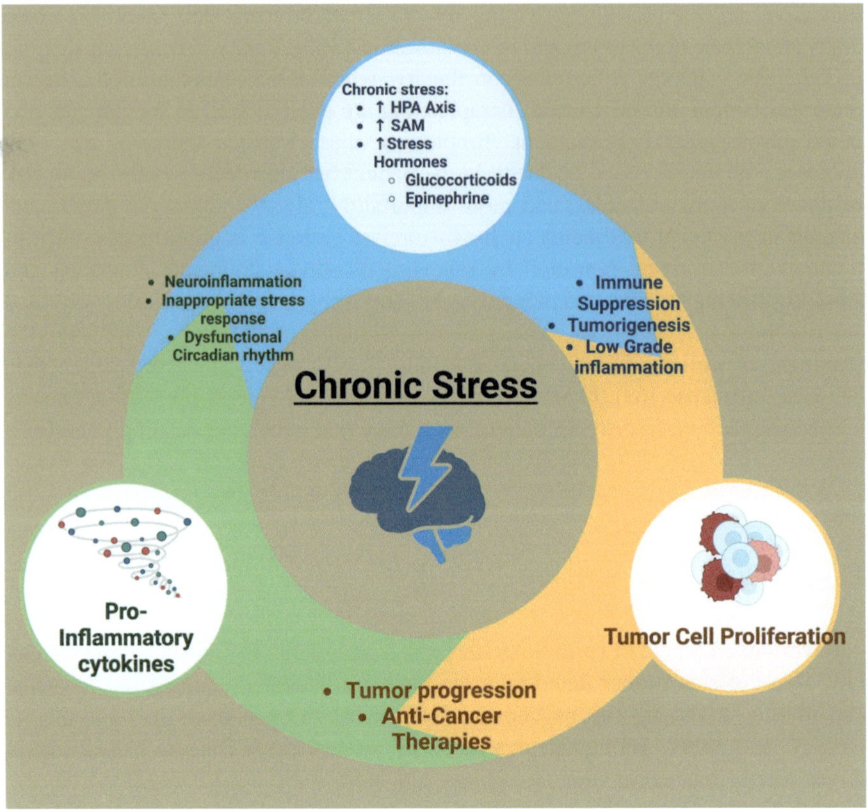

Fig. 40.4 Effect of chronic stress affecting multiple systems in a subject via increased pro-inflammatory cytokines, upregulation of hypothalamus–pituitary–adrenal (HPA) axis and sympathetic-adreno-medullary (SAM) axis resulting in high levels of various intracellular factors that could result in cellular dysfunction. Chronic stress may also suppress natural and adaptive immunity, induce dysfunctional circadian rhythm, and tumor cell proliferation

cells and cancer cells that create an inflammatory tumor microenvironment, which promotes tumor cell development (Fig. 40.4). Additionally, chronic stress can disrupt the cognitive abilities of the brain due to the continuous production of neuroinflammatory molecules [73].

Collectively, these psychosocial conditions have been shown to negatively affect the patient's adherence to provided treatments and even preventative/maintenance treatments such as increased physical activity. Patients who used physical activity as a preventative measure showed a smaller reduction in positive emotions during stressful situations, which shows evidence that physical activity can increase stress resilience and patients are less likely to be drawn away from their regiments due to it [74]. Anxiety is a condition that is commonly treated with pharmacological intervention but, exercise as a treatment option offers a similar level of advantage without the unruly side effects, such as weight loss and lower blood pressure. It is also more flexible, easy to incorporate into a patient's schedule which can eliminate unnecessary stress and is cost effective. In a series of studies that consisted of 675 individuals, it was proven that aerobic exercise was an effective treatment for patients with high anxiety compared to the control, where high-intensity exercise was more beneficial, but for patients who may not be in the physical condition to do so, intermediate physical activity still shows adequate benefits [75]. Finally, continuous physical activity is also beneficial in decreasing depressive episodes in the long term. A plethora of studies have shown how exercise programs involving three sessions per week over 12–24 weeks can cause a moderate to significant reduction in depression severity, as measured by symptom rating scales, and additionally, interventions are associated with a 22% higher likelihood of remission from depressive symptoms [76]. Overall, engaging in active, consistent physical exercise has been shown to reduce the negative effects of the comorbidities listed above, decrease the risk of cancer, and aid in the proper maintenance of existing symptoms. Moreover, it has also been shown that consistent physical activity increases overall health, health outcomes, and quality of life without subjecting patients to the detrimental side effects of preferred medications for these conditions.

Discussion

Cancer treatment-induced cardiotoxicities have affected patients' post-cancer treatment, leaving them with other adverse issues such as heart failure, cardiomyopathy, and more. In this review, we looked to explore how exercise, lipid management, diabetes management, tobacco cessation, and psychosocial management can play a protective role against cancer treatments.

Physical activity encompasses a variety of movements that help increase one's physical fitness, well-being, and longevity, and limit the effects of various chronic diseases. Tailoring exercise programs to each patient can help motivate them, but most importantly, reduce chemotherapy-induced cardiotoxicities. Anthracyclines are proven to be effective against an array of cancers but produce numerous cardiotoxicities such as oxidative stress and cardiac muscle apoptosis. Exercise therapy, which includes aerobic exercise and resistance exercises, limits this by increasing cardioprotective markers, reducing oxidative stress, and improving overall cardiac structure and function. CRF, which is measured through VO_{2peak}, decreases due to chemotherapy. As CRF gets worse, it is associated with poorer treatment outcomes and a higher mortality rate. Trastuzumab is another chemotherapeutic agent that increases cardiotoxicities, leading to worse left ventricular function and VO_{2peak}. Exercise therapy can improve oxygen delivery to the cardiac muscle. It also helps improve antioxidant defenses, reduce oxidative damage, and support mitochondrial biogenesis, helping to mitigate doxorubicin-induced cardiotoxicities. Despite the benefits of exercise therapy, it is still an underutilized method to limit the adverse effects of cancer treatments.

Poor lipid management is defined as having elevated cholesterol, triglycerides, and HDL levels. Some problems this leads to are cardiac lipotoxicity, mitochondrial dysfunction, and cardiac fibrosis. Once lipids accumulate to abnormal levels, they interrupt mitochondrial function, leading to an increase in ROS and reduced calcium regulation. It can also lead to endoplasmic reticulum dysfunction and cellular apoptosis. The fatty acid synthase pathway is overexpressed in a multitude of cancers and upregulation of SREBP increases malignancy risk. Cancer treatments further increase lipid deposition, causing an array of problems. Some of these problems can be limited through exercise which helps decrease cholesterol and LDL levels, cardiac lipotoxicity, and increases lipid turnover and breakdown. With a combination of effective lipid management and lifestyle changes, patients can limit the adverse effects of cardiac lipotoxicity, especially in patients who are at high risk of cancer.

SGLT-2 inhibitors are a class of drugs that inhibit the SGLT-2 inhibitor in the proximal convoluted tubule, and they have been shown to have numerous cardiovascular benefits. Empagliflozin reduced heart failure events while dapagliflozin decreased cardiovascular-related mortality. SGLT-2 inhibitors improved the survival of cardiomyocytes, increased autophagy, lowered inflammation by decreasing ROS and inflammatory cytokines, and limited chemotherapy-induced cardiotoxicities. Patients who were treated with both anthracyclines and SGLT-2 inhibitors had lower adverse cardiac events in comparison to patients treated with anthracyclines only. These drugs have shown promise in enhancing patient outcomes who have chemotherapy-induced cardiotoxicities.

Tobacco affects many individuals globally, leading to many societal and health dysfunctions such as cancer and increased cardiotoxicity. Carcinogens within tobacco can cause cancer in the larynx, mouth, throat, stomach, and more because they inhibit DNA repair and increase the production of tumor cells. Some of the most harmful carcinogens include PAHs, aromatic amines, and *N*-nitrosamines. Numerous types of therapies are available to help individuals stop smoking such as behavioral therapy, pharmacological intervention, or a mix between the two. Additionally, exercise can play a role in tobacco cessation because it can reduce withdrawal symptoms, cravings, and relapse rates. One's overall health will improve by reducing cardiotoxicities and the chance of having cancer.

Although many psychosocial factors may not directly cause cancer, they can play a role in worsening it. Depression and anxiety can lead to worse outcomes while chronic stress can promote cancer through p53 degradation and formation of a pro-inflammatory tumor microenvironment. Protective measures such as exercise can limit negative outcomes. Some of the benefits of physical exercise to prevent adverse psychosocial problems include decreasing the impact of negative emotions, more resistance to stress, better mood, and fewer depressive episodes.

Conclusion

Through the research of determining how various chemotherapy-induced cardiotoxicities can be mitigated, some questions still need to be answered. What is the best way to create exercise plans that fit for varying cancer types and treatment regimens? Will following an exercise plan provide long term cardioprotective effects? Can psychological interventions heighten treatment adherence and improve cardioprotective outcomes in cancer patients? After studying this topic, future research can be headed in a few different directions. Researchers can assess the viability of supervised exercise programs for cancer survivors who are at risk for various cardiotoxicities. Identify biochemical markers that will help understand which patients will gain the most from lipid and exercise management. Lastly, look at how the combination of exercise, lipid management, and diabetic medication can provide overall cardioprotective effect.

Key Points

- Physical therapy can assist cancer patients to finish their chemotherapy regiment, reduce cardiotoxicity, and improve cardiotoxic effects.
- Physical therapy has shown to combat anthracycline-induced cardiotoxicity which can diminish left ventricular ejection fraction, worsen myocardial tissue, and elevate oxidative stress.
- Exercise can improve lower oxidative stress, improve antioxidant defenses, and assist mitochondrial function.
- Doxorubicin can interrupt lipid metabolism and decrease cholesterol transport which can increase the risk of atherosclerosis.
- Physical exercise can limit lipid accumulation, increase lipid turnover, and limit cardiotoxic effects from lipid dysregulation.
- SGLT-2 inhibitors can increase the survival of cardiac cells limit chemotherapy-induced cardiotoxicities.
- Empagliflozin can improve the function of the left ventricle and decrease the inflammatory marker in doxorubicin-treated cardiomyocytes
- Pharmacological, behavioral, and physical exercise therapies help lower smoking rates and reduce cardiotoxic risks.

Funding The research work of DKA is supported by R25AI179582 and R01 HL147662 grants from the National Institutes of Health, USA. The contents of this chapter are solely the responsibility of the authors and do not necessarily represent the official views of the National Institutes of Health.

Competing Interests All authors have read the chapter and declare no conflict of interest. No writing assistance was utilized in the production of this manuscript.

Consent for Publication All authors have read the chapter and consented for publication.

References

1. Schwartz SM (2024) Epidemiology of cancer. Clin Chem 70(1):140–149. https://doi.org/10.1093/clinchem/hvad202
2. Kaur R, Bhardwaj A, Gupta S (2023) Cancer treatment therapies: traditional to modern approaches to combat cancers. Mol Biol Rep 50(11):9663–9676. https://doi.org/10.1007/s11033-023-08809-3
3. Passaro A, Al Bakir M, Hamilton EG, Diehn M, André F, Roy-Chowdhuri S, Mountzios G, Wistuba II, Swanton C, Peters S (2024) Cancer biomarkers: emerging trends and clinical implications for personalized treatment. Cell 187(7):1617–1635. https://doi.org/10.1016/j.cell.2024.02.041
4. Florescu M, Cinteza M, Vinereanu D (2013) Chemotherapy-induced cardiotoxicity. Maedica 8(1):59–67
5. Wexler R, Elton T, Pleister A, Feldman D (2009) Cardiomyopathy: an overview. Am Fam Physician 79(9):778–784
6. Desai DS, Hajouli S (2025) Arrhythmias. In: StatPearls. StatPearls Publishing. http://www.ncbi.nlm.nih.gov/books/NBK558923/
7. Badescu MC, Badulescu OV, Scripcariu DV, Butnariu LI, Bararu-Bojan I, Popescu D, Ciocoiu M, Gorduza EV, Costache II, Rezus E, Rezus C (2022) Myocardial ischemia related to common cancer therapy—prevention insights. Life 12(7):1034. https://doi.org/10.3390/life12071034
8. Volkova M, Raymond Russell III (2011) Anthracycline cardiotoxicity: prevalence, pathogenesis and treatment. Curr Cardiol Rev 7(4):214. https://doi.org/10.2174/157340311799960645
9. Thomas C, Mackey MM, Diaz AA, Cox DP (2009) Hydroxyl radical is produced via the Fenton reaction in submitochondrial particles under oxidative stress: implications for diseases associated with iron accumulation. Redox Rep 14(3):102–108. https://doi.org/10.1179/135100009X392566
10. Peshin S, Modi S, Namburu L, Rathod M (2024) Unpacking Trastuzumab-induced cardiomyopathy: a cardiac conundrum. Heart 5(2):2. https://doi.org/10.3390/hearts5020017
11. Sara JD, Kaur J, Khodadadi R, Rehman M, Lobo R, Chakrabarti S, Herrmann J, Lerman A, Grothey A (2018) 5-Fluorouracil and cardiotoxicity: a review. Ther Adv Med Oncol 10:1758835918780140. https://doi.org/10.1177/1758835918780140
12. Mouhayar E, Salahudeen A (2011) Hypertension in cancer patients. Tex Heart Inst J 38(3):263–265
13. de Souza VB, Silva EN, Ribeiro ML, de Martins WA (2015) Hypertension in patients with cancer. Arq Bras Cardiol 104(3):246–252. https://doi.org/10.5935/abc.20150011
14. Pandey AK, Singhi EK, Arroyo JP, Ikizler TA, Gould ER, Brown J, Beckman JA, Harrison DG, Moslehi J (2018) Mechanisms of VEGF-inhibitor associated hypertension and vascular disease. Hypertension (Dallas, Tex.: 1979) 71(2):e1–e8. https://doi.org/10.1161/HYPERTENSIONAHA.117.10271
15. Abou El Fadl M (2009) 5-Fluorouracil-induced cardiotoxicity during chemotherapy for adenocarcinoma of the small bowel. Gastrointest Cancer Res 3(4):167–170. https://pmc.ncbi.nlm.nih.gov/articles/PMC2739643/
16. Geisberg C, Sawyer DB (2010) Mechanisms of anthracycline cardiotoxicity and strategies to decrease cardiac damage. Curr Hypertens Rep 12(6):404–410. https://doi.org/10.1007/s11906-010-0146-y
17. Caspersen CJ, Powell KE, Christenson GM (1985) Physical activity, exercise, and physical fitness: definitions and distinctions for health-related research. Public Health Rep 100(2):126
18. Anderson E, Durstine JL (2019) Physical activity, exercise, and chronic diseases: a brief review. Sports Med Health Sci 1(1):3. https://doi.org/10.1016/j.smhs.2019.08.006
19. Courneya KS, Segal RJ, Mackey JR, Gelmon K, Reid RD, Friedenreich CM, Ladha AB, Proulx C, Vallance JKH, Lane K, Yasui Y, McKenzie DC (2007) Effects of aerobic and resistance exercise in breast cancer patients receiving adjuvant chemotherapy: a multicenter

randomized controlled trial. J Clin Oncol 25(28):4396–4404. https://doi.org/10.1200/JCO.2006.08.2024
20. Rock CL, Doyle C, Demark-Wahnefried W, Meyerhardt J, Courneya KS, Schwartz AL, Bandera EV, Hamilton KK, Grant B, McCullough M, Byers T, Gansler T (2012) Nutrition and physical activity guidelines for cancer survivors. CA Cancer J Clin 62(4):243–274. https://doi.org/10.3322/caac.21142
21. Baumann FT, Jensen W, Berling-Ernst A, Theurich S, Leitzmann M, Götte M (2024) Exercise therapy in oncology: the impact on quality of life and side effects. Dtsch Arztebl Int 121(10): 331. https://doi.org/10.3238/arztebl.m2024.0038
22. Ginzac A, Passildas J, Gadéa E, Abrial C, Molnar I, Trésorier R, Duclos M, Thivat E, Durando X (2019) Treatment-induced cardiotoxicity in breast cancer: a review of the interest of practicing a physical activity. Oncology 96(5):223–234. https://doi.org/10.1159/000499383
23. Wilson RL, Christopher CN, Yang EH, Barac A, Adams SC, Scott JM, Dieli-Conwright CM (2023) Incorporating exercise training into cardio-oncology care: current evidence and opportunities: JACC: CardioOncology State-of-the-Art Review. JACC CardioOncology 5(5): 553. https://doi.org/10.1016/j.jaccao.2023.08.008
24. Hortobágyi GN (1997) Anthracyclines in the treatment of cancer. An overview. Drugs 54(Suppl 4):1–7. https://doi.org/10.2165/00003495-199700544-00003
25. Shan K, Lincoff AM, Young JB (1996) Anthracycline-induced cardiotoxicity. Ann Intern Med 125(1):47–58. https://doi.org/10.7326/0003-4819-125-1-199607010-00008. PMID: 8644988
26. Tranchita E, Murri A, Grazioli E, Cerulli C, Emerenziani GP, Ceci R, Caporossi D, Dimauro I, Parisi A (2022) The beneficial role of physical exercise on anthracyclines induced cardiotoxicity in breast cancer patients. Cancers 14(9):2288. https://doi.org/10.3390/cancers14092288
27. Kang DW, Wilson RL, Christopher CN, Normann AJ, Barnes O, Lesansee JD, Choi G, Dieli-Conwright CM (2022) Exercise cardio-oncology: exercise as a potential therapeutic modality in the management of anthracycline-induced cardiotoxicity. Front Cardiovasc Med 8:805735. https://doi.org/10.3389/fcvm.2021.805735. PMID: 35097024; PMCID: PMC8796963
28. Shirinbayan V, Roshan VD (2012) Pretreatment effect of running exercise on HSP70 and DOX-induced cardiotoxicity. Asian Pac J Cancer Prev 13(11):5849–5855. https://doi.org/10.7314/apjcp.2012.13.11.5849. PMID: 23317268
29. Bellissimo MP, Canada JM, Jordan JH, Ladd AC, Reding KW, Moore TL, Ntim WO, Heiston EM, Brubaker P, Mihalko SL, Ralph D'Agostino J, O'Connell N, Ky B, Wagner LI, Hackney MH, Weaver KE, Lesser GJ, Avis NE, Sutton AL, Group, US et al (2023) Physical activity during breast cancer therapy associates with preserved exercise capacity and cardiac function (WF97415). JACC CardioOncology 5(5):641. https://doi.org/10.1016/j.jaccao.2022.12.011
30. Scott JM, Thomas SM, Peppercorn JM, Herndon JE 2nd, Douglas PS, Khouri MG, Dang CT, Yu AF, Catalina D, Ciolino C, Capaci C, Michalski MG, Eves ND, Jones LW (2020) Effects of exercise therapy dosing schedule on impaired cardiorespiratory fitness in patients with primary breast cancer: a randomized controlled trial. Circulation 141(7):560–570. https://doi.org/10.1161/CIRCULATIONAHA.119.043483. Epub 2020 Feb 17. PMID: 32065769; PMCID: PMC7032604
31. Tsuji K, Matsuoka YJ, Ochi E (2021) High-intensity interval training in breast cancer survivors: a systematic review. BMC Cancer 21:184. https://doi.org/10.1186/s12885-021-07804-w
32. Gilchrist SC, Barac A, Ades PA, Alfano CM, Franklin BA, Jones LW, La Gerche A, Ligibel JA, Lopez G, Madan K, Oeffinger KC, Salamone J, Scott JM, Squires RW, Thomas RJ, Treat-Jacobson DJ, Wright JS, American Heart Association Exercise, Cardiac Rehabilitation, and Secondary Prevention Committee of the Council on Clinical Cardiology; Council on Cardiovascular and Stroke Nursing; and Council on Peripheral Vascular Disease (2019) Cardio-oncology rehabilitation to manage cardiovascular outcomes in cancer patients and survivors: a scientific statement from the American Heart Association. Circulation 139(21):e997–e1012. https://doi.org/10.1161/CIR.0000000000000679. PMID: 30955352; PMCID: PMC7603804
33. Foulkes SJ, Howden EJ, Haykowsky MJ, Antill Y, Salim A, Nightingale SS, Loi S, Claus P, Janssens K, Mitchell AM, Wright L, Costello BT, Lindqvist A, Burnham L, Wallace I, Daly

RM, Fraser SF, La Gerche A (2023) Exercise for the prevention of anthracycline-induced functional disability and cardiac dysfunction: the BREXIT study. Circulation 147(7):532–545. https://doi.org/10.1161/CIRCULATIONAHA.122.062814. Epub 2022 Nov 7. PMID: 36342348

34. Scott JM, Nilsen TS, Gupta D, Jones LW (2018) Therapy and cardiovascular toxicity in cancer. Circulation 137(11):1176. https://doi.org/10.1161/CIRCULATIONAHA.117.024671
35. Hughes DC, Lenihan DJ, Harrison CA, Basen-Engquist KM (2011) Exercise intervention for cancer survivors with heart failure: two case reports. J Exerc Sci Fit 9(1):65–73. https://doi.org/10.1016/s1728-869x(11)60009-9. PMID: 21709755; PMCID: PMC3121107
36. Yu AF, Flynn JR, Moskowitz CS, Scott JM, Oeffinger KC, Dang CT, Liu JE, Jones LW, Steingart RM (2020) Long-term cardiopulmonary consequences of treatment-induced cardiotoxicity in survivors of ERBB2-positive breast cancer. JAMA Cardiol 5(3):309–317. https://doi.org/10.1001/jamacardio.2019.5586
37. Jones LW, Courneya KS, Mackey JR, Muss HB, Pituskin EN, Scott JM, Hornsby WE, Coan AD, Herndon JE 2nd, Douglas PS, Haykowsky M (2012) Cardiopulmonary function and age-related decline across the breast cancer survivorship continuum. J Clin Oncol 30(20): 2530–2537. https://doi.org/10.1200/JCO.2011.39.9014. Epub 2012 May 21. PMID: 22614980; PMCID: PMC3397786
38. Klassen O, Schmidt ME, Scharhag-Rosenberger F, Sorkin M, Ulrich CM, Schneeweiss A, Potthoff K, Steindorf K, Wiskemann J (2014) Cardiorespiratory fitness in breast cancer patients undergoing adjuvant therapy. Acta Oncol 53(10):1356–1365. https://doi.org/10.3109/0284186X.2014.899435. Epub 2014 May 16. PMID: 24837860
39. Gaytan SL, Lawan A, Chang J, Nurunnabi M, Bajpeyi S, Boyle JB, Han SM, Min K (2023) The beneficial role of exercise in preventing doxorubicin-induced cardiotoxicity. Front Physiol 14: 1133423. https://doi.org/10.3389/fphys.2023.1133423
40. Scott JM, Khakoo A, Mackey JR, Haykowsky MJ, Douglas PS, Jones LW (2011) Modulation of anthracycline-induced cardiotoxicity by aerobic exercise in breast cancer: current evidence and underlying mechanisms. Circulation 124(5):642–650. https://doi.org/10.1161/CIRCULATIONAHA.111.021774. PMID: 21810673; PMCID: PMC3154075
41. Li D, Yang Y, Wang S, He X, Liu M, Bai B, Tian C, Sun R, Yu T, Chu X (2021) Role of acetylation in doxorubicin-induced cardiotoxicity. Redox Biol 46:102089. https://doi.org/10.1016/j.redox.2021.102089
42. Kouzi SA, Uddin MN (2016) Aerobic exercise training as a potential cardioprotective strategy to attenuate doxorubicin-induced cardiotoxicity. J Pharm Pharm Sci 19(3):399–410. https://doi.org/10.18433/J3JS5R
43. Marques-Aleixo I, Santos-Alves E, Mariani D, Rizo-Roca D, Padrão AI, Rocha-Rodrigues S, Viscor G, Torrella JR, Ferreira R, Oliveira PJ, Magalhães J, Ascensão A (2015) Physical exercise prior and during treatment reduces sub-chronic doxorubicin-induced mitochondrial toxicity and oxidative stress. Mitochondrion 20:22–33. https://doi.org/10.1016/j.mito.2014.10.008
44. Wang F, Chandra J, Kleinerman ES (2021) Exercise intervention decreases acute and late doxorubicin-induced cardiotoxicity. Cancer Med 10(21):7572. https://doi.org/10.1002/cam4.4283
45. Schulze PC, Drosatos K, Goldberg IJ (2016) Lipid use and misuse by the heart. Circ Res 118(11):1736. https://doi.org/10.1161/CIRCRESAHA.116.306842
46. Watanabe S, Kumazaki S, Kusunoki K, Inoue T, Maeda Y, Usui S, Shinohata R, Ohtsuki T, Hirohata S, Kusachi S, Kitamori K, Mori M, Yamori Y, Oka H (2018) A high-fat and high-cholesterol diet induces cardiac fibrosis, vascular endothelial, and left ventricular diastolic dysfunction in SHRSP5/Dmcr rats. J Atheroscler Thromb 25(5):439. https://doi.org/10.5551/jat.40956
47. Kretzschmar T, Wu JMF, Schulze PC (2021) Mitochondrial homeostasis mediates lipotoxicity in the failing myocardium. Int J Mol Sci 22(3):1498. https://doi.org/10.3390/ijms22031498

48. Drosatos K, Schulze PC (2013) Cardiac lipotoxicity: molecular pathways and therapeutic implications. Curr Heart Fail Rep 10(2):109–121. https://doi.org/10.1007/s11897-013-0133-0
49. Bhatnagar R, Dixit NM, Yang EH, Sallam T (2022) Cancer therapy's impact on lipid metabolism: mechanisms and future avenues. Front Cardiovasc Med 9:925816. https://doi.org/10.3389/fcvm.2022.925816
50. Butler L, Perone Y, Dehairs J, Lupien LE, de Laat V, Talebi A, Loda M, Kinlaw WB, Swinnen JV (2020) Lipids and cancer: emerging roles in pathogenesis, diagnosis and therapeutic intervention. Adv Drug Deliv Rev 159:245. https://doi.org/10.1016/j.addr.2020.07.013
51. Zaidi N, Lupien L, Kuemmerle NB, Kinlaw WB, Swinnen JV, Smans K (2013) Lipogenesis and lipolysis: the pathways exploited by the cancer cells to acquire fatty acids. Prog Lipid Res 52(4):585. https://doi.org/10.1016/j.plipres.2013.08.005
52. Jones SF, Infante JR (2015) Molecular pathways: fatty acid synthase. Clin Cancer Res 21(24): 5434–5438. https://doi.org/10.1158/1078-0432.CCR-15-0126
53. Fhu CW, Ali A (2020) Fatty acid synthase: an emerging target in cancer. Molecules 25(17): 3935. https://doi.org/10.3390/molecules25173935
54. Baenke F, Peck B, Miess H, Schulze A (2013) Hooked on fat: the role of lipid synthesis in cancer metabolism and tumour development. Dis Model Mech 6(6):1353–1363. https://doi.org/10.1242/dmm.011338
55. Guo D, Bell EH, Mischel P, Chakravarti A (2014) Targeting SREBP-1-driven lipid metabolism to treat cancer. Curr Pharm Des 20(15):2619. https://doi.org/10.2174/13816128113199990486
56. Monzel JV, Budde T, Zu Schwabedissen HEM, Schwebe M, Bien-Möller S, Lütjohann D, Kroemer HK, Jedlitschky G, Grube M (2017) Doxorubicin enhances oxysterol levels resulting in a LXR-mediated upregulation of cardiac cholesterol transporters. Biochem Pharmacol 144: 108–119. https://doi.org/10.1016/j.bcp.2017.08.008
57. Albarrati AM, Alghamdi MSM, Nazer RI, Alkorashy MM, Alshowier N, Gale N (2018) Effectiveness of low to moderate physical exercise training on the level of low-density lipoproteins: a systematic review. Biomed Res Int 2018:5982980. https://doi.org/10.1155/2018/5982980
58. Arner P, Bernard S, Appelsved L, Fu K-Y, Andersson DP, Salehpour M, Thorell A, Rydén M, Spalding KL (2019) Adipose lipid turnover and long-term changes in body weight. Nat Med 25(9):1385–1389. https://doi.org/10.1038/s41591-019-0565-5
59. Basham, H. A., Keswani, S., Kumar, A., Rahol Rai, S. K.A Surkasha, FNU Kumari, A., Malik, J. (2024) Role of sodium-glucose co-transporter-2 inhibitor during anthracycline use: an updated review. Cardiol Rev https://doi.org/10.1097/CRD.0000000000000638
60. Vafa RG, Sabahizadeh A, Mofarrah R (2024) Guarding the heart: how SGLT-2 inhibitors protect against chemotherapy-induced cardiotoxicity: SGLT-2 inhibitors and chemotherapy-induced cardiotoxicity. Curr Probl Cardiol 49(3):102350. https://doi.org/10.1016/j.cpcardiol.2023.102350. Epub 2023 Dec 20
61. Gongora CA, Drobni ZD, Quinaglia Araujo Costa Silva T, Zafar A, Gong J, Zlotoff DA, Gilman HK, Hartmann SE, Sama S, Nikolaidou S, Suero-Abreu GA, Jacobsen E, Abramson JS, Hochberg E, Barnes J, Armand P, Thavendiranathan P, Nohria A, Neilan TG (2022) Sodium-glucose co-transporter-2 inhibitors and cardiac outcomes among patients treated with anthracyclines. JACC Heart Fail 10(8):559–567. https://doi.org/10.1016/j.jchf.2022.03.006. Epub 2022 Jun 8. PMID: 35902159; PMCID: PMC9638993
62. Sabatino J, De Rosa S, Tammè L, Iaconetti C, Sorrentino S, Polimeni A, Mignogna C, Amorosi A, Spaccarotella C, Yasuda M, Indolfi C (2020) Empagliflozin prevents doxorubicin-induced myocardial dysfunction. Cardiovasc Diabetol 19(1):66. https://doi.org/10.1186/s12933-020-01040-5. PMID: 32414364; PMCID: PMC7229599
63. Quagliariello V, De Laurentiis M, Rea D, Barbieri A, Monti MG, Carbone A, Paccone A, Altucci L, Conte M, Canale ML, Botti G, Maurea N (2021) The SGLT-2 inhibitor empagliflozin improves myocardial strain, reduces cardiac fibrosis and pro-inflammatory cytokines in non-diabetic mice treated with doxorubicin. Cardiovasc Diabetol 20(1):150. https://doi.org/10.1186/s12933-021-01346-y. PMID: 34301253; PMCID: PMC8305868

64. Benowitz NL, Burbank AD (2016) Cardiovascular toxicity of nicotine: implications for electronic cigarette use. Trends Cardiovasc Med 26(6):515–523. https://doi.org/10.1016/j.tcm.2016.03.001
65. Jacob L, Freyn M, Kalder M, Dinas K, Kostev K (2018) Impact of tobacco smoking on the risk of developing 25 different cancers in the UK: a retrospective study of 422,010 patients followed for up to 30 years. Oncotarget 9(25):17420. https://doi.org/10.18632/oncotarget.24724
66. Smith MT, Guyton KZ, Gibbons CF, Fritz JM, Portier CJ, Rusyn I, DeMarini DM, Caldwell JC, Kavlock RJ, Lambert PF, Hecht SS, Bucher JR, Stewart BW, Baan RA, Cogliano VJ, Straif K (2015) Key characteristics of carcinogens as a basis for organizing data on mechanisms of carcinogenesis. Environ Health Perspect 124(6):713. https://doi.org/10.1289/ehp.1509912
67. Xue J, Yang S, Seng S (2014) Mechanisms of cancer induction by tobacco-specific NNK and NNN. Cancers 6(2):1138. https://doi.org/10.3390/cancers6021138
68. Onwuzo CN, Olukorode J, Sange W, Orimoloye DA, Udojike C, Omoragbon L, Hassan AE, Falade DM, Omiko R, Odunaike OS, Adams-Momoh PA, Addeh E, Onwuzo S, Joseph-Erameh U (2024) A review of smoking cessation interventions: efficacy, strategies for implementation, and future directions. Cureus 16(1):e52102. https://doi.org/10.7759/cureus.52102
69. Klinsophon T, Thaveeratitham P, Sitthipornvorakul E, Janwantanakul P (2017) Effect of exercise type on smoking cessation: a meta-analysis of randomized controlled trials. BMC Res Notes 10(1):442. https://doi.org/10.1186/s13104-017-2762-y
70. Smith HR (2015) Depression in cancer patients: pathogenesis, implications and treatment (review). Oncol Lett 9(4):1509. https://doi.org/10.3892/ol.2015.2944
71. Goerling U, Hinz A, Koch-Gromus U, Hufeld JM, Esser P, Mehnert-Theuerkauf A (2023) Prevalence and severity of anxiety in cancer patients: results from a multi-center cohort study in Germany. J Cancer Res Clin Oncol 149(9):6371. https://doi.org/10.1007/s00432-023-04600-w
72. Dai S, Mo Y, Wang Y, Xiang B, Liao Q, Zhou M, Li X, Li Y, Xiong W, Li G, Guo C, Zeng Z (2020) Chronic stress promotes cancer development. Front Oncol 10:1492. https://doi.org/10.3389/fonc.2020.01492
73. Petrinović SV, Milošević MS, Marković D, Momčilović S (2023) Interplay between stress and cancer—a focus on inflammation. Front Physiol 14:1119095. https://doi.org/10.3389/fphys.2023.1119095
74. Childs E, de Wit H (2014) Regular exercise is associated with emotional resilience to acute stress in healthy adults. Front Physiol 5:161. https://doi.org/10.3389/fphys.2014.00161
75. Aylett E, Small N, Bower P (2018) Exercise in the treatment of clinical anxiety in general practice—a systematic review and meta-analysis. BMC Health Serv Res 18:559. https://doi.org/10.1186/s12913-018-3313-5
76. Murri MB, Ekkekakis P, Magagnoli M, Zampogna D, Cattedra S, Capobianco L, Serafini G, Calcagno P, Zanetidou S, Amore M (2019) Physical exercise in major depression: reducing the mortality gap while improving clinical outcomes. Front Psychiatry 9:762. https://doi.org/10.3389/fpsyt.2018.00762

Index

D. K. Agrawal et al. (eds.), *Cardiovascular Toxicity*, Advances in Biochemistry in Health and Disease 33, https://doi.org/10.1007/978-3-031-90598-8

D

N

O

P